In Focus – Special Book Series

Impact of Climate Change on Hydrology and Water Resources

Editors

Takeshi Takama, Muhammad Bilal, K. Srinivasa Raju

IWA PUBLISHING

Published by **IWA Publishing**
Unit 104–105, Export Building
1 Clove Crescent
London E14 2BA, UK
Telephone: +44 (0)20 7654 5500
Fax: +44 (0)20 7654 5555
Email: publications@iwap.co.uk
Web: www.iwapublishing.com

First published 2022

British Library Cataloguing in Publication Data
A CIP catalogue record for this book is available from the British Library

ISBN: 9781789063257

Contents

© 2022 The Authors Impact of Climate Change on Hydrology and Water Resources doi: 10.2166/wcc.2022.002

Editorial: Impact of climate change on hydrology and water resources

The impact of climate change on water resources is a research area that is gaining prominence over time. Related research topics in this direction are the prediction of rainfall, temperature, streamflow/discharge, and other meteorological variables.

A special issue on the theme *Impact of Climate Change on Hydrology and Water Resources* is initiated to help researchers and other relevant stakeholders working in this area. This issue deals with estimation of runoff, floods and droughts, intensity-duration frequency (IDF) curves and evapotranspiration (ET). In addition, some of the miscellaneous topics related to hydrology are also part of the special issue. Brief but relevant information on the papers included in this issue is as follows.

Maciel da Silva *et al.* (2022) analyzed the possible impacts of climate change (CC) on the projections of the naturalized streamflows and affluent natural energy for the Brazil hydropower sector. They discussed the trends of the same for different regions. Supharatid *et al.* (2022) estimated changes in rainfall and temperature over mainland Southeast Asia by 18 models with two shared socioeconomic pathways (SSPs). They employed multi-model ensembles. A robust increase in precipitation is expected. They have suggested appropriate adaptation measures. These two papers considered the CMIP6 framework.

Grover *et al.* (2022) examined the effect of changes in temperature and precipitation on the Chenab basin, western Himalaya. They employed the Hydrologiska Byråns Vattenbalansavdelning (HBV) model for discharge computation. An increase in discharge is expected. Abdulahi *et al.* (2022) discussed the response of CC impact on future streamflow availability in the Upper Awash River basin, Ethiopia, using HBV. Streamflow is expected to increase. Ayalew *et al.* (2022) evaluated the impacts of CC on the Ribb catchment, Lake Tana Basin, Ethiopia. They used the Soil & Water Assessment Tool (SWAT). Future streamflow has decreased due to temperature increase and decrease in rainfall. Haleem *et al.* (2022) studied the effects of land-use change and CC in the upper Indus basin, Pakistan, and found that they impacted runoff depth. Yang *et al.* (2022) employed SWAT in the upper reaches of the Minjiang River to study the impact of CC. Results show that the runoff has a unique response to CC. Song *et al.* (2022) analyzed the effects of CC on surface water resources in China's upper Ganjiang River Basin. Temperature and runoff are projected to increase. All these papers considered the CMIP5 framework.

Takele *et al.* (2022) assessed the impacts of CC in the Upper Blue Nile basin. They employed SWAT and CORDEX data. Runoff is impacted due to CC. Kwakye & Bárdossy (2022) quantified the hydrological consequences of CC in a Black Volta catchment using flow duration curves. They employed HBV and CORDEX data. It is concluded that most of the rivers in the catchment are running dry for most of the year.

Hernández *et al.* (2022) found that CC impacts water quality. Zango *et al.* (2022) studied the effects of urbanization and CC in the Carp River watershed, Ottawa, Ontario. They employed SWAT for CMIP5 outputs. It is concluded that discharge will increase. Qin *et al.* (2022) studied the effects of environmental factors on the greenhouse gas change in the middle section of the Three Gorges Reservoir. Wu *et al.* (2022) studied the water environmental capacity of the Yongzhou Section of Xiangjiang River Basin. It was based on the SWAT-EFDC model. It is found to be helpful for this type of problem. Abbasi *et al.* (2022a) studied the effects of land-use changes and CC on green water security in Kashafrood Basin in the CMIP5 framework. They employed SWAT. Decrease in groundwater scarcity and vulnerability are predicted. Hussain *et al.* (2022) modelled the effects of land use and CC on the performance of the Al-Ameer district Storm Water Sewer System, Iraq. They employed SWMM. The effect of CC on the stormwater drainage system was more adverse than that due to land-use change.

Zhou (2022a) studied the segmentation of flood season and scheme optimization in the Yellow River. He opined that segmentation plays a major role in flood prevention. Shu *et al.* (2022) assessed the flood risk of Ya'an, Sichuan, China, based on the energy theory. The study provided a scientific framework for disaster relief initiatives. Laddimath *et al.* (2022) assessed the impacts of CC in CMIP5 framework on drought-prone areas in the Bhima sub-basin, India. They used Standard Precipitation Index (SPI). SPI showed changes over various RCPs. Zhang *et al.* (2022) studied ecological drought and its state assessment of the Yellow River estuary. It is concluded that studies of this nature are beneficial to reduce loss due to drought. Waseem *et al.* (2022) studied the impact of meteorological drought on maize crop production in Punjab, Pakistan.

Tayşi & Özger (2022) generated future IDF curves with short-duration rainfalls in CMIP5. Increases in short-duration rainfalls are expected. Yamoat *et al.* (2022) studied the temporal change of extreme precipitation in Thailand.

Wang *et al.* (2022a) discussed wetland ET in China's Sanmenxia Reservoir area. Wang *et al.* (2022b) employed Landsat images and the METRIC model to study ET.

Zhou *et al.* (2022) studied extreme precipitation variations in the Guangdong-Hong Kong-Macao Greater Bay area from 1961 to 2018. Sun *et al.* (2022) analyzed future precipitation variations in China's Yellow River Basin using deterministic and probabilistic projections. Samantaray & Ghose (2022) developed a hybrid technique integrating phase-space reconstruction with a firefly algorithm and support vector machine and used it to predict runoff. Tiwari *et al.* (2022) modelled runoff in the Kolar river basin, India, using a wavelet with an artificial neural network. They also used performance indicators for comparison. Yeditha *et al.* (2022) investigated the spatiotemporal characterization of streamflow of six unregulated catchments in India. They used global climate indices for the same. Pervin & Khan (2022) analyzed the variability and trends of climate extreme indices in Chattogram City, Bangladesh.

Tumsa (2022) assessed the performance of six bias correction methods for the upper Awash basin, Oromia, Ethiopia. Gong *et al.* (2022) felt that strong relationships between elasticity and most factors are observed. Zhou (2022b) made an urban water dissipation calculation. The methodology is applied to Zhengzhou, the capital city of Henan Province, China, and concluded that water dissipation inside a building could not be neglected. Kui-Feng *et al.* (2022) studied Holocene environmental evolution history in the Yellow River Delta. They propose a reconstruction of the Holocene environment using multi-proxy indices.

ACKNOWLEDGMENTS

We are thankful to Prof D. Nagesh Kumar, Editor-in-Chief, Journal of Water and Climate Change, IWA Publishing and Ms. Emma Buckingham, Ms. Emma Gulseven, Ms. Natasha Healy of the editorial team, IWA Publishing for excellent support and co-ordination throughout this special issue.

Guest Editors
Takeshi Takama
University of Udayana, Indonesia

Muhammad Bilal
School of Marine Sciences, Nanjing University of Information Science and Technology, Nanjing 210044, China

K. Srinivasa Raju
Birla Institute of Technology and Science Pilani, Hyderabad Campus, India

REFERENCES

Abbasi, A., Amirabadizadeh, M., Afshar, A. A. & Yaghoobzadeh, M. 2022 Potential influence of climate and land-use changes on green water security in a semi-arid catchment. *Journal of Water and Climate Change* **13** (1), 287–303. doi: https://doi.org/10.2166/wcc.2021.055.

Abdulahi, S. D., Abate, B., Harka, A. E. & Husen, S. B. 2022 Response of climate change impact on streamflow: the case of the Upper Awash sub-basin, Ethiopia. *Journal of Water and Climate Change* **13** (2), 607–628. doi: https://doi.org/10.2166/wcc.2021.251.

Ayalew, D. W., Asefa, T., Moges, M. A. & Leyew, S. M. 2022 Evaluating the potential impact of climate change on the hydrology of Ribb catchment, Lake Tana Basin, Ethiopia. *Journal of Water and Climate Change* **13** (1), 190–205. doi: https://doi.org/10.2166/wcc.2021.049.

Gong, X., Xu, A., Du, S. & Zhou, Y. 2022 Spatiotemporal variations in the elasticity of runoff to climate change and catchment characteristics with multi-timescales across the contiguous United States. *Journal of Water and Climate Change* **13** (3), 1408–1424. doi: https://doi.org/10.2166/wcc.2022.242.

Grover, S., Tayal, S., Sharma, R. & Beldring, S. 2022 Effect of changes in climate variables on hydrological regime of Chenab basin, wes. Himalaya. *Journal of Water and Climate Change* **13** (1), 357–371. doi: https://doi.org/10.2166/wcc.2021.003.

Haleem, K., Khan, A. U., Ahmad, S., Khan, M., Khan, F. A., Khan, W. & Khan, J. 2022 Hydrological impacts of climate and land-use change on flow regime variations in upper Indus basin. *Journal of Water and Climate Change* **13** (2), 758–770. doi: https://doi.org/10.2166/wcc.2021.238.

Hernández, J. P. R., López, O. L. O., Pérez, P. T. G., Ortiz, F. G. G. & Gil, V. S. 2022 Perception of the inhabitants of the department of Caldas, Colombia on the effects of climate change on water quality. *Journal of Water and Climate Change* **13** (1), 43–55. doi: https://doi.org/10.2166/wcc.2021.200.

Hussain, S. N., Zwain, H. M. & Nile, B. K. 2022 Modeling the effects of land-use and climate change on the performance of stormwater sewer system using SWMM simulation: case study. *Journal of Water and Climate Change* **13** (1), 125–138. doi: https://doi.org/10.2166/wcc.2021.180.

Kui-Feng, W., Xue-Feng, Y., Jiang-Bao, X., Shu-Jian, X., Tai-Ping, Z., Yan, X. & Hong-Jun, Z. 2022 Holocene environmental evolution history based on sporopollenin and micropaleontological reconstruction of KY-01 in.the Yellow River Delta. *Journal of Water and Climate Change* **13** (1), 206–223. doi: https://doi.org/10.2166/wcc.2021.144.

Kwakye, S. O. & Bárdossy, A. 2022 Quantification of the hydrological consequences of climate change in a typical West African catchment using flow duration curves. *Journal of Water and Climate Change* **13** (1), 26–42. doi: https://doi.org/10.2166/wcc.2021.147.

Laddimath, R. S., Patil, N. S., Rao, P. S. & Nagendra. 2022 Assessing the impacts of climate change on drought-prone regions in Bhima sub-basin (India) using the Standard Precipitation Index. *Journal of Water and Climate Change* **13** (2), 817–838. doi: https://doi.org/10.2166/wcc.2021.195.

Maciel da Silva, M. V., da Silva Silveira, C., Cabral, S. L., Marcos, A. D., da Silva, G. K. & & Lima, C. E. S. 2002 Naturalized streamflows and Affluent Natural Energy projections for the Brazilian hydropower sector for the SSP2-4.5 and SSP5-8.5 scenarios of the CMIP6. *Journal of Water and Climate Change* **13** (1), 315–336. doi: https://doi.org/10.2166/wcc.2021.352.

Pervin, L. & Khan, M. S. K. 2022 Variability and trends of climate extremes indices from the observed and downscaled GCMs data over 1950–2020 period in Chattogram City, Bangladesh. *Journal of Water and Climate Change* **13** (2), 975–998. doi: https://doi.org/10.2166/wcc.2021.331.

Qin, Y., Gou, Y., Yu, Z. & Tan, W. 2021 Effects of environmental factors on the methane and carbon dioxide fluxes at the middle of Three Gorges Reservoir. *Journal of Water and Climate Change* **12** (8), 4007–4020. doi: https://doi.org/10.2166/wcc.2021.081.

Samantaray, S. & Ghose, D. K. 2022 Prediction of S12-MKII rainfall simulator experimental runoff data sets using hybrid PSR-SVM-FFA approaches. *Journal of Water and Climate Change* **13** (2), 707–734. doi: https://doi.org/10.2166/wcc.2021.221.

Shu, X., Ren, Y., Duan, Z., Liu, X., Hua, X. & Lei, H. 2022 Flood risk assessment in Ya'an, Sichuan, China based on the emergy theory. *Journal of Water and Climate Change* **13** (1), 247–259. doi: https://doi.org/10.2166/wcc.2021.133.

Song, P., Wang, C., Ding, G., Sun, J., Kong, L., Lu, M., Lei, X. & Wang, H. 2022 Evaluating the impact of climate change on surface water resources in the upper Ganjiang River Basin, China. *Journal of Water and Climate Change* **13** (3), 1462–1476. doi: https://doi.org/10.2166/wcc.2022.258.

Sun, Z., Liu, Y., Zhang, J., Chen, H., Shu, Z., Guan, T., Wang, G., Jin, J., Bao, Z. & Liu, C. 2022 Deterministic and probabilistic projections and their credibility in analyzing future precipitation variations in the Yellow River Basin, China. *Journal of Water and Climate Change* jwc2022359. doi: https://doi.org/10.2166/wcc.2022.359.

Supharatid, S., Nafung, J. & Aribarg, T. 2022 Projected changes in temperature and precipitation over mainland Southeast Asia by CMIP6 models. *Journal of Water and Climate Change* 1 January **13** (1), 337–356. doi: https://doi.org/10.2166/wcc.2021.015.

Takele, G. S., Gebrie, G. S., Gebremariam, A. G. & Engida, A. N. 2022 Future climate change and impacts on water resources in the Upper Blue Nile basin. *Journal of Water and Climate Change* **13** (2), 908–925. doi: https://doi.org/10.2166/wcc.2021.235.

Tayşi, H. & Özger, M. 2022 Disaggregation of future GCMs to generate IDF curves for the assessment of urban floods. *Journal of Water and Climate Change* **13** (2), 684–706. doi: https://doi.org/10.2166/wcc.2021.241.

Tiwari, D. K., Tiwari, H. L. & Raman Nateriya, R. 2022 Runoff modeling in Kolar river basin using hybrid approach of wavelet with artificial neural network. *Journal of Water and Climate Change* **13** (2), 963–974. doi: https://doi.org/10.2166/wcc.2021.246.

Tumsa, B. C. 2022 Performance assessment of six bias correction methods using observed and RCM data at upper Awash basin, Oromia, Ethiopia. *Journal of Water and Climate Change* **13** (2), 664–683. doi: https://doi.org/10.2166/wcc.2021.181.

Wang, J., Li, H. & Lu, H. 2022a An estimation of the evapotranspiration of typical steppe areas using Landsat images and the METRIC model. *Journal of Water and Climate Change* **13** (2), 926–942. doi: https://doi.org/10.2166/wcc.2021.432.

Wang, S., Liu, C., Tan, Y., Wang, J., Du, F., Han, Z., Jiang, Z. & Wang, L. 2022b Remote sensing inversion characteristic and driving factor analysis of wetland evapotranspiration in the Sanmenxia Reservoir area, China. *Journal of Water and Climate Change* **13** (3), 1599–1611. doi: https://doi.org/10.2166/wcc.2021.247.

Waseem, M., Khurshid, T., Abbas, A., Ahmad, I. & Javed, Z. 2022 Impact of meteorological drought on agriculture production at different scales in Punjab, Pakistan. *Journal of Water and Climate Change* **13** (1), 113–124. doi: https://doi.org/10.2166/wcc.2021.244.

Wu, L., Chen, Z., Ding, X., Liu, H.-y. & Wang, D.-q. 2022 Research on water environmental capacity accounting of the Yongzhou Section of Xiangjiang River Basin based on the SWAT-EFDC coupling model. *Journal of Water and Climate Change* **13** (2), 1106–1122. doi: https://doi.org/10.2166/wcc.2021.319.

Yamoat, N., Hanchoowong, R., Sriboonlue, S. & A. Kangrang, A. 2022 Temporal change of extreme precipitation intensity–duration–frequency relationships in Thailand. *Journal of Water and Climate Change* **13** (2), 839–853. doi: https://doi.org/10.2166/wcc.2021.348.

Yang, K., Chen, T., Ao, T., Zhang, X., Zhou, L. & Danyang Gao, D. 2022 Response of runoff in the upper reaches of the Minjiang River to climate change. *Journal of Water and Climate Change* **13** (1), 260–273. doi: https://doi.org/10.2166/wcc.2021.038.

Yeditha, P. K., Pant, T., Rathinasamy, M. & Agarwal, A. 2022 Multi-scale investigation on streamflow temporal variability and its connection to global climate indices for unregulated rivers in India. *Journal of Water and Climate Change* **13** (2), 735–757. doi: https://doi.org/10.2166/wcc.2021.189.

Zango, B.-S., Seidou, O., Sartaj, M., Nakhaei, N. & Stiles, K. 2022 Impacts of urbanization and climate change on water quantity and quality in the Carp River watershed. *Journal of Water and Climate Change* **13** (2), 786–816. doi: https://doi.org/10.2166/wcc.2021.158.

Zhang, J., Liu, B., Yang, L., He, L., Cao, X. & Shao, G. 2022 Ecological drought and its state assessment: a case study in the Yellow River estuary. *Journal of Water and Climate Change* **13** (1), 13–25. doi: https://doi.org/10.2166/wcc.2021.175.

Zhou, K. 2022a Flood season segmentation and scheme optimization in the Yellow River. *Journal of Water and Climate Change* **13** (1), 274–286. doi: https://doi.org/10.2166/wcc.2021.110.

Zhou, K. 2022b Urban water dissipation calculation based on the improved water balance models. *Journal of Water and Climate Change* **13** (1), 372–382. doi: https://doi.org/10.2166/wcc.2021.330.

Zhou, Y., Luo, Z., Li, S., Liu, Z., Shen, Y. & Zhuo, W. 2022 Temporal and spatial variations of extreme precipitation in the Guangdong-Hong Kong-Macao Greater Bay area from 1961 to 2018. *Journal of Water and Climate Change* **13** (1), 304–314. doi: https://doi.org/10.2166/wcc.2021.078.

© 2022 The Authors Impact of Climate Change on Hydrology and Water Resources doi: 10.2166/wcc.2021.352

Naturalized streamflows and Affluent Natural Energy projections for the Brazilian hydropower sector for the SSP2-4.5 and SSP5-8.5 scenarios of the CMIP6

Marx Vinicius Maciel da Silva, Cleiton da Silva Silveira, Samuellson Lopes Cabral, Antonio Duarte Marcos Junior, Greicy Kelly da Silva and Carlos Eduardo Sousa Lima

ABSTRACT

Climate change projects an increase in extreme weather events in the coming decades, which could significantly affect Brazil's water and energy security. Thus, this study sought to analyze possible impacts of climate change on the projections of naturalized streamflows and Affluent Natural Energy (ANE) for the Brazilian hydropower sector utilizing five models of the Coupled Model Intercomparison Projects version 6 (CMIP6), based on SSP-4.5 and SSP-8.5 scenarios for the 21st century. Naturalized streamflows for the 24 stations representing the National Interconnected System (NIS) were estimated through the concentrated hydrological model SMAP (Soil Moisture Accounting Procedure), while the streamflows for the other stations that comprise the NIS were obtained by linear regression. The streamflows, as well as the productivity of the reservoirs, were used to calculate the ANE. The results showed that most of the models project possible reductions in annual naturalized streamflows and ANE for the three periods analyzed and for the North, Northeast, and Southeast/Midwest sectors of Brazil. Meanwhile, the Northern and Southern sectors, for the period 2080–2099, most of the models indicated an increase of annual, precipitation, naturalized streamflows and ANE.

Key words | ANE, Brazilian hydropower sector, climate change, CMIP6, streamflow

Marx Vinicius Maciel da Silva (corresponding author)
Cleiton da Silva Silveira
Antonio Duarte Marcos Junior
Greicy Kelly da Silva
Carlos Eduardo Sousa Lima
Hydraulic and Environmental Engineering Department (DEHA),
Federal University of Ceará (UFC),
60440-593 Fortaleza, CE,
Brazil
E-mail: *marx.silva@alu.ufc.br*

Samuellson Lopes Cabral
National Center for Monitoring and Early Warning of Natural Disasters (CEMADEN),
São José dos Campos, SP,
Brazil

HIGHLIGHTS

- Most of the CMIP6 models showed that the NIS reservoirs are sensitive to increasing greenhouse gas emissions.
- The medians of the standardized anomalies of annual naturalized streamflows and ANE indicated an increase for the South and North sectors of the country.
- In the Northeast subsystem, most of the CMIP6 models indicated a reduction in naturalized streamflows and ANE.

doi: 10.2166/wcc.2021.352

INTRODUCTION

Climate change in the coming decades indicates significant impacts on the hydrological cycle with the intensification and higher frequency of extreme weather events that can be divided into three main categories: too little water (with longer periods of severe droughts), too much water (with torrential rainfalls and floods), and too dirty water (with pollution) (Babel *et al.* 2020). As a consequence of these impacts, studies point out that several essential sectors for society may be affected, and among the main ones is the electric power generation (Silveira *et al.* 2017; Hasan & Wyseure 2018; Qin *et al.* 2020; Silva *et al.* 2021).

In the modern world, the need for reliable and affordable electric power for society grows every year due to population growth and economic development. In the year 2018, the global electricity demand increased by 2.3%, which was the highest annual growth in the last decade, and for the first time in history, the population with no electricity access rate fell to less than one billion (IEA 2020). On the other hand, the same year saw a 1.7% increase in global CO_2 emissions, reaching an all-time high of 33.1 Gt (billion tons) (IEA 2020).

The Intergovernmental Panel on Climate Change (IPCC), through the fifth Assessment Report (AR5), stated that the climate system warming is evident and largely caused by the increasing concentration of CO_2 in the atmosphere, with anthropogenic factors being the main causes (IPCC 2014).

Therefore, there is a climate, water, food, and energy nexus where one may influence the other in the same proportion, i.e., energy generations that emit CO_2 into the atmosphere are positive feedback for climate change, which in turn intensifies extreme events, such as heat waves and severe droughts, causing changes in runoff and streamflow, and may significantly influence water availability, which directly impacts irrigation, urban water supply systems, and hydroelectric power production (Guimarães *et al.* 2016; Silveira *et al.* 2017; Silva *et al.* 2021). Thus, meeting the economic needs for electricity and protecting the environment from climate change impacts have become one of the main challenges of the modern world.

Renewable energy, including solar, wind, water, biofuels, and others, is part of the new, less carbon-intensive and more sustainable electric power system in transition. According to the IEA (2020), the net additions increase in renewable electricity capacity will be almost 4% higher in the year 2020 than in the year 2019. In quantitative totals, this represents a total of 198 GWh of installed renewable capacity in the year 2020, breaking another record and accounting for almost 90% of the total power capacity increase. Higher additions from wind (+8%) and hydropower (+43%) are expected in 2020, while photovoltaic solar growth remains stable (IEA 2020).

In Brazil, the water demand growth, especially for agriculture, threatens electricity generation, since the country's hydroelectric power plants are the main power generation source (Cunha *et al.* 2019; Silva *et al.* 2021). In 2019, for example, hydroelectric generation was responsible for 58% of the electricity generated by the sector, with the rest, approximately 17% coming from other renewable sources (except hydro, wind, and solar), 5% from fossil fuels, 8% from wind, 2% from nuclear, and 1% from solar sources. This is different from the reality found in the year 2000 when hydroelectric generation accounted for 87% of all electricity generated in the country (IEA 2020).

Therefore, the analysis of possible climate change impacts on the hydrological cycle, and consequently on the hydroelectric power generation in the next decades, is important information for the management of these resources. It brings subsidies for the elaboration of adaptation and mitigation strategies of these possible impacts in order to preserve the water and energy security, making the hydroelectric system more resilient.

The IPCC releases ARs on the possible climate change impacts, releasing projection scenarios based on the Greenhouse Gas emissions, using the Global Climate Models (GCM – General Circulation Model) (IPCC 2014). These models follow a pattern among various institutions which operate them, called Coupled Model Intercomparison Project (CMIP) (IPCC 2014). The last released AR by the IPCC, the AR5 in 2014, used the GCMs of the fifth phase of CMIP, called CMIP5.

Recently, updated data from the GCMs referring to the sixth phase of the CMIP, the CMIP6, were made available. CMIP6 climate projections differ from those of CMIP5's because they are not only produced with updates of the

GCMs versions, but also generated with new scenarios. The scenarios combine socioeconomic and technological developments, called Shared Socioeconomic Pathways (SSP), with future radiative forcings (RCP) scenarios based on updated data on 26 emission tendencies, in a matrix scenario architecture (Eyring *et al.* 2016; Gidden *et al.* 2019).

There are five different SSPs with quantifications of models that cover future sustainable or fossil fuel-powered growth potentials, SSP1 and SSP5, respectively; high inequality among or within countries, SSP3 and SSP4, respectively; and the SSP2 scenario, which is the 'halfway point' of the above-mentioned scenarios. For each SSP, a different RCP scenario can be achieved depending on the policies implemented, locally or globally, throughout the century (Eyring *et al.* 2016; Gidden *et al.* 2019).

According to Gidden *et al.* (2019), this new structure permits to standardize all socioeconomic premises (e.g., population, Gross Domestic Product, and poverty, among others) in the modeled representations of each scenario, and also permits a more subtle investigation of the variety of paths which climate results can follow.

However, Kundzewicz *et al.* (2018) point out several sources of uncertainty in the projections of climate change impacts on hydrological resources, including emissions and socioeconomic development scenarios, input data, and the hydrological models. The same studies show that in order to minimize these uncertainties, the authors indicate the use of multiple GCMs and the removal of biases, since GCMs are usually highly biased towards the current climate. Making use of multiple GCMs and applying bias removal techniques, that several studies show the possible impacts of climate change on water and energy resources.

Among these studies, the work of Ribeiro Neto *et al.* (2016) evaluated the impacts of climate change on water resources and hydrological processes throughout the Brazilian territory through the coupling of a hydrological model for large basins to a forced regional climate model with the HadGEM2-ES and MIROC5 models of CMIP5. The RCP4.5 and RCP8.5 scenarios were analyzed, and the results showed that water availability decreases almost entirely in the study area (except for the South) and the main hydroelectric power generation basins are affected.

The work of Silveira *et al.* (2017) analyzed the climate change impact on the estimated streamflow projections with the Soil Moisture Accounting Procedure (SMAP) hydrological model for 24 relevant hydrographic basins in the Brazilian hydropower sector from monthly precipitation projections of the IPCC-AR5 global models from 2010 to 2098 for the RCP4.5 and RCP8.5 scenarios. Their results showed streamflow reductions in the southeast/central-west regions with reduction magnitude ranging from 15 to 30% depending on the hydrographic basin in the RCP8.5 scenario. On the other hand, for the Southern sector, the results showed a streamflow increase over the extreme south of Brazil and a decrease at the intersection between the south and southeast regions.

On the other hand, works that analyze the entire National Interconnected System (NIS), with approximately 200 stations, have not yet been applied using the new CMIP6 scenarios. Therefore, this study aimed at evaluating the CMIP6 projections of the temperature, precipitation, naturalized streamflows, and Affluent Natural Energy (ANE) of the hydroelectric plants' reservoirs for the Brazilian electricity sector, based on the SSP2-4.5 and SSP5-8.5 scenarios. Results should help decision-makers prepare better for future scenarios and reduce climate change impacts on water and energy resources.

METHODOLOGY

In order to evaluate CMIP6 climate change projections for Brazil's hydroelectric plants usage, naturalized streamflows and ANE based on SSP-4.5 and SSP-8.5 scenarios were analyzed. The methodology consisted of choosing the basins representing the NIS, obtaining naturalized streamflows data, and ANE calculation. Such a process was divided in six steps, according to Figure 1.

In step 1, the choices of the hydrographic plants representing the NIS were made. For this, 24 stations of some hydroelectric plants were used to determine the 24 hydrographic basins (see Figure 2).

In step 2, mean, minimum, and maximum time series of precipitation and surface air temperatures (SAT) were extracted from the basins representing the NIS of the CMIP6 models related to the historical scenarios (1901–2000, 20th century), SSP-4.5 and SSP-8.5 (2015–2100, 21st century), and observation of precipitation of the Global

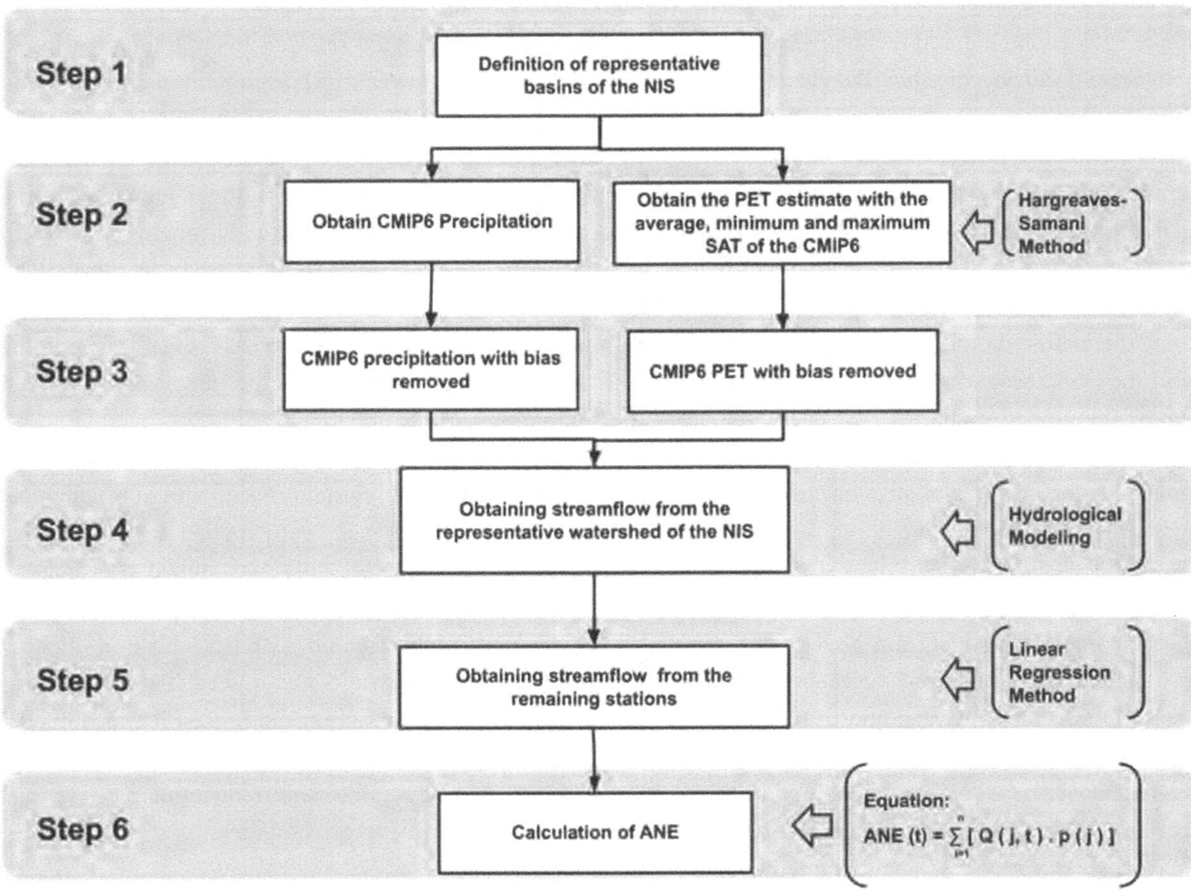

Figure 1 | Flowchart of the methodology steps.

Precipitation Climatology Centre (GPCC) and the mean, maximum, and minimum SAT of the Climate Research Unit (CRU), both in the period of 1931–2018. With the mean, minimum, and maximum SAT series of the CMIP6 models and observation of the CRU, time series of potential evapotranspiration (PET) were estimated using the Hargreaves–Samani method (Hargreaves & Samani 1985).

In step 3, statistical corrections for bias removal were performed using the Gamma distribution function of data precipitation and PET from CMIP6 scenarios based on the observational data.

In step 4, PET and unbiased precipitation data from CMIP6 were used as input by the SMAP hydrological model to generate naturalized streamflow data of the 24 basins representing the NIS. For that, the SMAP model had its parameters calibrated via an objective optimization procedure based on the Nash–Sutcliffe coefficient, which compares the streamflow series obtained by SMAP with the series provided by the National Power System Operator (ONS, in Portuguese). Further details are described in the subtopic 'Estimation of the naturalized streamflows of the CMIP6 models'.

Step 5 consisted of estimating streamflow data for the remaining stations, which do not have the calibrated SMAP model. The NIS, responsible for the Brazilian energy matrix, has 206 streamflow time series, divided into 185 stations of hydroelectric use, 161 natural stations, and 24 natural and artificial stations. Out of these 185 utilization sites, 16 started only in 2015 (ONS 2019). Therefore, it was decided to divide them into two groups: one has 24 stations and the other has 161, respectively. The 24 stations were used as predictors of the other station's streamflows, and thus, it was possible to obtain the regression parameters of the 185 stations by using stepwise. That would not be

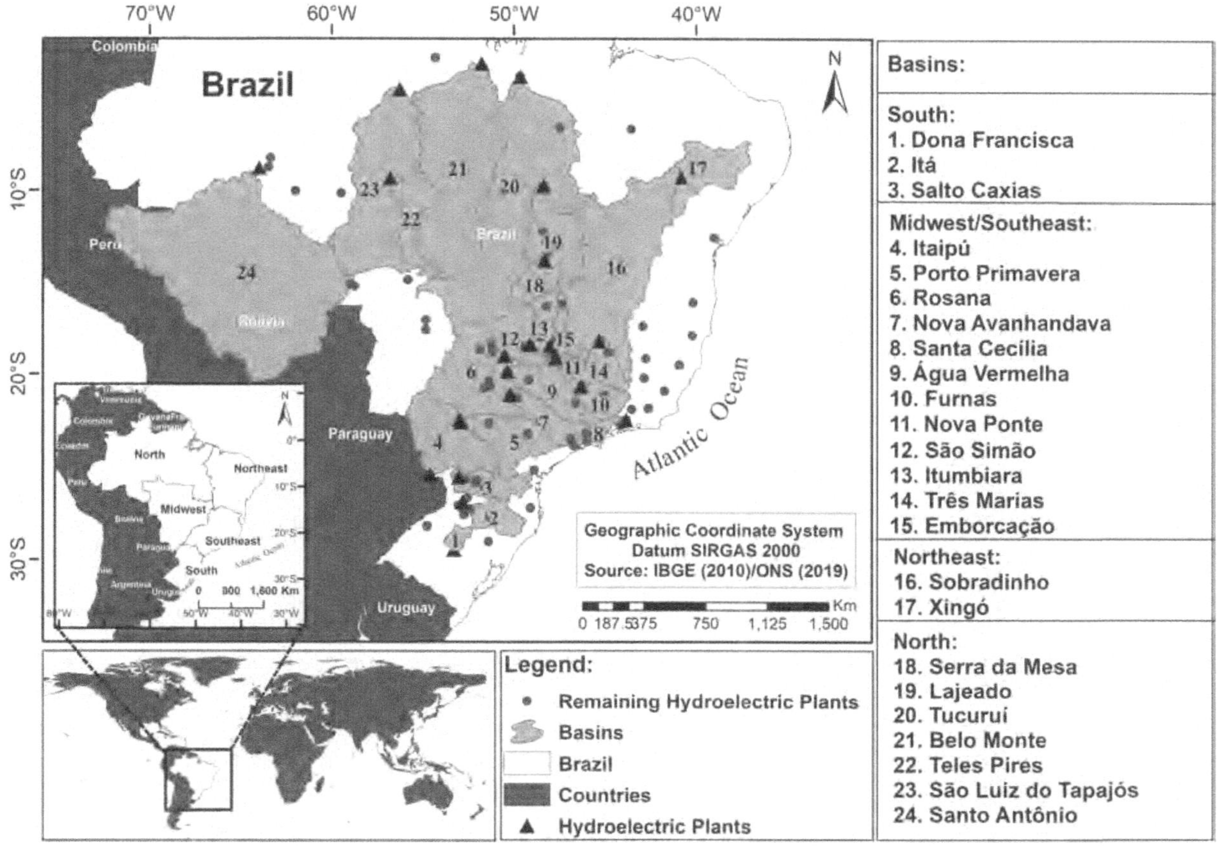

Figure 2 | Region of study with the hydrographic basins representing the NIS.

possible with the SMAP model alone, as there are missing data and many of the stations are nearby – with most of the 161 stations located within the 24 basins, and the downstream of the 24 predictor stations (see Figure 2).

In step 6, ANE was calculated through natural streamflows and productivity equivalent to 65% of the useful storage volume of hydroelectric plant reservoirs.

All steps, as well as the description of the study area, data used, and statistical analyses performed, will be discussed in the next topics.

Study area

The NIS is divided into four subsystems, namely Southwest/Midwest, South, Northeast, and North (see Figure 2), which are interconnected by an extensive transmission network, permitting surplus energy to be transferred, and favoring the optimization of the stocks stored of the hydroelectric plant's reservoirs. The Northern region produces 8% of the energy in GWh and demands 7%, the Northeastern region produces 11% and demands 18%, the Southern region produces 18% and demands 16%, the Southeastern/Midwestern region produces 47% and demands 61% (ONS 2019). From these reservoirs, 24 hydrographic basins of the NIS were generated, as illustrated in Figure 2. In addition to the stations of the reservoirs of the 24 hydrographic basins above mentioned, another 161 stations spread throughout the Brazilian territory were used. In total, 185 stations of the NIS were analyzed for streamflow and 120 for the ANE.

In this study, the 24 main reservoirs of the NIS were selected by the energy storage capacity of the basin/reservoir in relation to the subsystem, considering all reservoirs full, as shown in Table 1.

Electricity generation in Brazil went from approximately 200,000 Giga of kilowatts per hour (GWh) in 1990 to approximately 600,000 GWh in 2017, with an average growth rate of approximately 17% for every 5-year period, as shown in Figure 3. It is observed that hydraulic energy

Table 1 | Area, annual precipitation, annual streamflow and installed capacity for the basins representing the NIS

Basins	Area (km²)	Annual precipitation (mm)	Annual streamflow (m³/s)	Installed capacity (MW)
Emborcação	29,000	1,341	483	1,192
Nova Ponte	15,300	1,505	532	510
Itumbiara	51,011	1,432	1,548	2,280
São Simão	85,729	1,420	2,363	1,710
Furnas	50,464	1,754	929	1,312
Água Vermelha	89,436	1,357	2,089	1,396
Nova Avanhandava	62,300	1,300	747	347
Porto Primavera	190,760	1,300	7,130	1,540
Rosana	100,799	1,212	1,281	372
Itaipu	149,000	1,650	10,027	14,000
Santa Cecília	16,694	1,608	297	2
Salto Caxias	57,000	1,824	1,260	1,240
Itá	44,500	1,830	1,022	1,450
Dona Francisca	14,014	1,200	182	125
Três Marias	50,600	1,400	686	396
Sobradinho	447,825	1,000	2,706	1,050
Xingó	110,275	957	2,810	3,162
Serra da Mesa	50,975	1,500	784	1,275
Lajeado	134,543	1,500	2,484	902
Tucuruí	572,482	2,500	10,948	4,125
Belo Monte	480,000	1,971	8,045	11,233
Teles Pires	90,707	2,000	2,414	303
São Luiz do Tapajós	362,293	2,123	1,068	8,040
Santo Antônio	988,873	1,750	1,431	3,568

ONS (2019).

has always been predominant in the country, but it started to suffer a reduction in the participation of the energy power generation matrix especially in 2015, with 62%. This deficit in hydroelectric generation was replaced, mainly, by the wind and fossil energy with 3.7% and 23%, respectively (IEA 2020).

Observational data

For the calibration and validation of the SMAP and linear regression models and for the monthly Gamma statistical correction of the CMIP6, observed data in the period from 1901 to 2018 regularly spatialized on $0.5^\circ \times 0.5^\circ$ grid of precipitation from the GPCC (Schneider *et al.* 2017), and mean, minimum, and maximum SAT of the CRU were used. The choice of this data set was due to its high consistency and reliability when compared with data measured from meteorological stations (Limberger & Silva 2018; Eini *et al.* 2019). In addition, due to Brazil's extensive area, in some regions – such as the northern region – there are still low density of meteorological stations and short periods of time series (less than 30 years). Thus, reanalysis and satellite data, as in the case of CRU and GPCC, respectively, emerge as an excellent source of global data, and because they are regularly specialized, it is an excellent option for hydrological modeling of hydrographic basins, as reported by several studies (Jong *et al.* 2018; Eini *et al.* 2019). The affluent

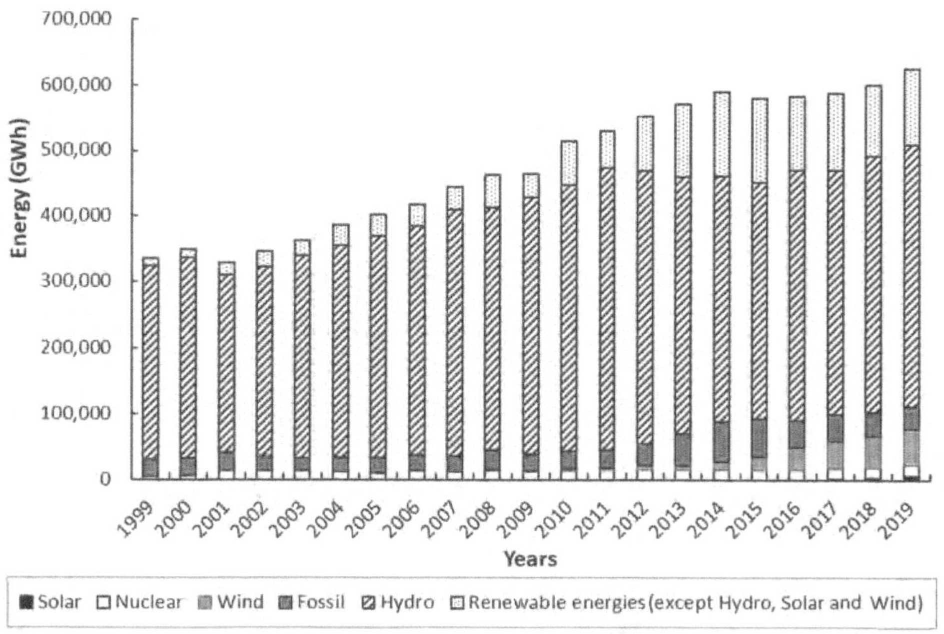

Figure 3 | Electricity generation in Brazil. Data source: IEA (2020).

natural streamflow data obtained for the period 1951–2008 from the National Power System Operator (ONS) were also used in the calibration and validation process of the SMAP and linear regression models.

CMIP6 data

CMIP6 prevenient data are results of simulations of five GCMs, conforming to Table 2.

This study used the precipitation variables, monthly mean, minimum, and maximum SAT for the historical scenario (1901–2000, 20th century), and projection scenarios SSP2-4.5 and SSP5-8.5 in the period from 2020 to 2100. In the coming decades, these results might serve as a parameter to measure in which scenario we find ourselves, if optimistic, with more satisfactory results than the SSP2-4.5; intermediate, with results close to those obtained in SSP2-4.5; or pessimistic, with results close to the SSP5-8.5 scenario, since the other scenarios, except for SSP1 and RCP2.6, are obtained from variations of the two chosen scenarios.

SSP2 is called the 'halfway' scenario, with moderate population growth and slower convergence of the income levels among countries (Gidden *et al.* 2019). Regarding the greenhouse gas emissions, it is expected that in the SSP2, electric power generation keeps on depending on fossil fuels close to the current rates, resulting in continuous growth of those gases (Gidden *et al.* 2019). Concerning pollution, according to Gidden *et al.* (2019), for SSP2, it is expected a decrease in

Table 2 | CMIP6 models with institutions, countries and citations where they talk about them

Models	Institute (country)	Grid
CanESM5	Canadian Earth System Model 5nd generation (Canada)	$2.81^\circ \times 2.81^\circ$
IPSL-CMSA-MR	Institut Pierre-Simon Laplace (France)	$2.5^\circ \times 2.5^\circ$
MIROC6	Atmosphere and Ocean Research Institute, National Institute for Environmental Studies, and Japan Agency for Marine-Earth Science and Technology (Japan)	$1.41^\circ \times 1.41^\circ$
BCC-CSM2-MR	Beijing Climate Center climate system model version 2 (China)	$1.13^\circ \times 1.13^\circ$
MRI-ESM2.0	Meteorological Research Institute Earth System Model version 2 (Japan)	$1.13^\circ \times 1.13^\circ$

polluting emissions – associated with efforts in continuing to diminish air pollution and with the developing economies reaching high-income countries. For SSP5, a world with strong economic growth with the use of fossil fuels is expected, with a substantial increase in income worldwide and a reduction in social inequality among countries. However, this growth occurs at the expense of potentially large impacts on climate change (Gidden *et al.* 2019).

PET estimation through the Hargreaves–Samani Method

PET was estimated using the Hargreaves–Samani Method with the observational data of the CRU and CMIP6 of the mean, maximum, and minimum SAT variables (in degrees Celsius) through Equation (1):

$$\mathrm{PET} = 0.0023(\mathrm{SAT_{max}} - \mathrm{SAT_{min}})^{0.5}(\mathrm{SAT_{med}} + 17.8)\mathrm{Ra} \quad (1)$$

where PET is given in mm/month, and Ra is the average external radiation, which was estimated from the latitude and month of the year, according to the work of Hargreaves & Samani (1985).

The reason why the Hargreaves–Samani method was chosen is that, among the PET estimation methods that use SAT as an input variable, the studies of Gurski *et al.* (2018) and Valle Júnior *et al.* (2020) pointed out the Hargreaves & Samani (1985) method as the one with the best performance for different Brazilian regions.

Statistical correction using the Gamma Cumulative Distribution Function

GCM's results have systematic errors when compared with the observational data. Hence, it is common to apply techniques which aim at correcting those errors, and among the most used ones is the statistical correction via Gamma Cumulative Distribution Function (CDF) (Silveira *et al.* 2017; Silva *et al.* 2021). The Gamma CDF was used in this study to correct the bias of the precipitation data and the PET of the CMIP6 models in the historical SSP2-4.5 and SSP5-8.5 scenarios. A CDF Gamma is expressed as:

$$\mathrm{f}_x(x) = \frac{1}{\beta^{\gamma}\Gamma(\gamma)}x^{\gamma-1}\mathrm{e}^{\frac{-x}{\beta}} \quad (2)$$

where x is the variable (precipitation or PET in this case); β is the scale parameter; γ is the shape parameter; and Γ is the Gamma function.

In this study, the process of bias correction of data can be mathematically expressed in the following five steps (Equations (3)–(7)):

$$(\gamma_s, \beta_s) = \mathrm{Gammafit}(x_s) \quad (3)$$

where γ_s and β_s are shape and scale parameters, respectively, for GCM's simulated historical data (20th century), and x_s is GCM-simulated historical data.

$$H_{SH} = F_{\gamma}(x_s, \gamma_s, \beta_s) \quad (4)$$

where F_{γ} is Gamma CDF.

$$(\gamma_o, \beta_o) = \mathrm{Gammafit}(x_o) \quad (5)$$

where γ_o and β_o are shape and scale parameter, respectively, for observed data (20th century), and x_o is GCM-observed data.

$$H_{OR} = F_{\gamma}(x_o, \gamma_o, \beta_o) \quad (6)$$

$$H_{CR} = F_{\gamma}^{-1}(H_{SH}, \gamma_s, \beta_s) \quad (7)$$

where F_{γ}^{-1} is inverse of Gamma CDF, and H_{CR} is historic bias-corrected data.

The following steps (Equations (8)–(10)) were then used to correct bias in future GCM's simulated data from SSP2-4.5 and SSP5-8.5 scenarios.

$$(\gamma_f, \beta_f) = \mathrm{Gammafit}(x_f) \quad (8)$$

where γ_f and β_f are shape and scale parameters, respectively, for GCM's simulated future data (21th century, SSP2-4.5 and SSP5-8.5 scenarios), and x_f is GCM-simulated future data.

$$F_{SR} = F_{\gamma}(x_f, \gamma_f, \beta_f), \quad (9)$$

$$F_{CR} = F_{\gamma}^{-1}\left(F_{SR}, \gamma_f \cdot \frac{\gamma_s}{\gamma_o}, \beta_f \cdot \frac{\beta_s}{\beta_o}\right), \quad (10)$$

F_{CR} is future bias-corrected data.

Estimation of the naturalized streamflows of the CMIP6 models

In this study, the naturalized streamflows for the 24 reservoirs representing the NIS were estimated through the SMAP hydrological model, while the naturalized streamflows for the 161 remaining stations were obtained by the linear regression model. These models have been widely used in several studies for the Brazilian electricity sector (Silveira *et al.* 2017; Gondim *et al.* 2018; Silva *et al.* 2021).

The rain streamflow-type SMAP conceptual hydrological model was developed by Lopes *et al.* (1982) in order to estimate streamflows in a hydrographic basin, having precipitation and PET as input variables. For this purpose, it seeks to represent water flows in the hydrographic basin via two fictional reservoirs, one superficial or soil R_{solo} and the other subterranean R_{sub}, as shown in Figure 4.

In the SMAP initial version, the hydrological processes encompassed daily time paces, that is, it had the daily information of precipitation about the basin as entry, the fictional reservoirs were updated, and the daily streamflows for the basin were obtained. In this version, SMAP presents a monthly time pace.

Transfers between reservoirs are governed by theoretical equations and calibrated deterministic parameters which represent the soil saturation (SAT) capacity in mm, the surface runoff through the exponent of the exponential equation (PES), the coefficient of recharge of the underground reservoir (C_{rec}) in %, and the recession equation of base runoff (K) in months. Besides, there are two initialization parameters of the initial soil moisture rate (T_u) which determines the initial level of the R_{solo}, and the initial surface runoff (EBin) defines the initial value of the R_{sub} which was manually adjusted based on calculated values for the Nash–Sutcliffe Efficiency (NSE) coefficient (NSE) on each simulation. NSE is expressed in Equation (11):

$$\mathrm{NSE} = 1 - \frac{\sum_{i=1}^{N} (o_i - m_i)^2}{\sum_{i=1}^{N} (o_i - \bar{o}_i)^2}, \tag{11}$$

in which N is the number of samples, o_i is the observed data, m_i is the modeled data, and $\bar{o}_i$ the average observed data (Nash & Sutcliffe 1970). The maximum value corresponds to 1, which indicates a perfect adjustment between the modeled and observed data.

Table 3 presents the SMAP model parameters calibrated for the NIS basins.

Both reservoirs are updated from month 1 to month $i+1$ through Equations (12) and (13):

$$R_{solo}(i+1) = R_{solo}(i) + P - E_r - E_s - R_{rec}, \tag{12}$$

$$R_{sub}(i+1) = R_{sub}(i) + R_{rec} - E_b, \tag{13}$$

in which E_s is the surface runoff in mm, P is the precipitation in mm, T_u is the initial moisture content, E_r is the actual evapotranspiration in mm, E_p is the PET in mm, P_{es} is the surface runoff parameter, C_{rec} is the recharge coefficient, E_b is the base runoff in mm, K is the recession constant in month^{-1}, and R_{ec} is the underground recharge.

The first parameter of the model is R_{solo}, which, by definition, receives as maximum value the same as the SATs. In turn, SAT can be expressed by Equation (14):

$$\mathrm{SAT} = \frac{R_{solo}}{T_u} \tag{14}$$

E_r is also calculated via T_u and E_p (which is one of the input variables) using Equation (15):

$$E_r = T_u \times E_p \tag{15}$$

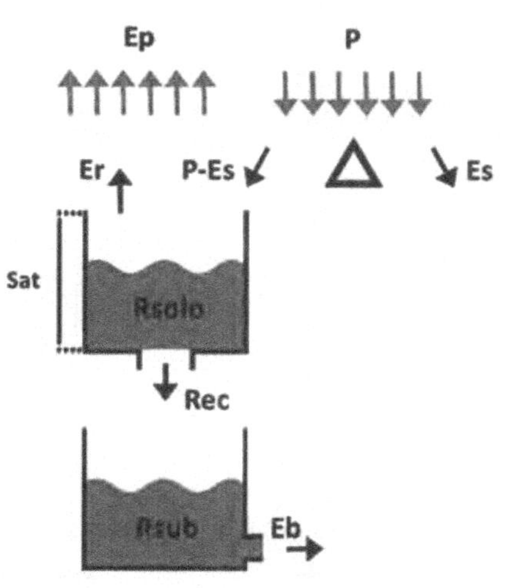

Figure 4 | Operation scheme of the monthly SMAP model. *Source*: Lopes *et al.* (1982).

Table 3 | Parameters of the SMAP model calibrated for the basins representing the NIS

Basins	Area (km²)	T_u	E_b (mm)	SAT (mm)	PES	C_{rec} (%)	K (month^{-1})
Emborcação	29,000	55	96	900	3.2	20	2
Nova Ponte	15,300	52	70	1,100	3	25	2
Itumbiara	51,011	52	170	2,000	3.1	9	2
São Simão	85,729	60	349	2,100	4.6	7	2
Furnas	50,464	48	221	1,400	2.2	26	2
Água Vermelha	89,436	55	449	1,200	3.3	20	2
Nova Avanhandava	62,300	67	352	1,200	4.6	10	3
Porto Primavera	190,760	54	1,071	1,400	3.3	20	3
Rosana	100,799	65	1,149	1,300	3.4	22	2
Itaipú	149,000	59	2,204	1,000	2.9	50	1
Santa Cecília	16,694	63	110	2,000	3.5	8	3
Salto Caxias	57,000	61	762	400	1.8	70	1
Itá	44,500	78	513	600	3.7	15	1
Dona Francisca	14,014	92	120	1,900	7	2	1
Três Marias	50,600	45	85	1,300	2.3	18	1
Sobradinho	447,825	31	541	1,800	2.5	50	6
Xingó	110,275	22	0	500	2.5	1	1
Serra da Mesa	50,975	63	107	1,800	3.7	3	2
Lajeado	134,543	59	163	1,350	3.9	4.5	1
Tucuruí	572,482	59	919	1,200	5	4	1
Belo monte	480,000	72	588	2,580	10	0.54	1
Teles pires	90,707	61	382	1,300	3.8	9.3	2
São Luiz do Tapajós	362,293	56	1,684	1,300	3.8	6	2
Santo Antônio	988,873	71	3,283	2,700	6.4	1.64	2

The R_{rec} recharge, which appears in Equations (12) and (13), is expressed by Equation (16):

$$R_{rec} = R_{solo} - C_{rec} \times T_u^4 \tag{16}$$

E_s is expressed by Equation (17):

$$E_s = P \times T_u^{PES} \tag{17}$$

E_b is expressed by Equation (18):

$$E_b = R_{sub} \times \left[1 - \left(\frac{1}{2}\right)^{\frac{1}{K}}\right] \tag{18}$$

E_s and E_b constitute together with the area and the affluent streamflow of the basin (Q) in m³/s, given by Equation (19):

$$Q = \frac{(E_b + E_s) \times \text{area}}{2,630} \tag{19}$$

With the estimated naturalized streamflows for the 24 basins representing the NIS, the linear regression model was used to estimate the naturalized streamflows for the other hydroelectric plant stations. To this end, the first stage consisted in the standardization of the naturalized streamflow monthly series of the ONS utilizing Equation (20):

$$Z_{i,j,k} = \frac{q_{i,j,k} - \overline{q_{i,j}}}{\sigma_{i,j}}, \tag{20}$$

in which Z is the normalized streamflow, the number of months (ranging from 1 to 12), k the number of years (ranging from 1931 to 2008), j the number of stations (in total 185), $q_{i,j,k}$

naturalized streamflow of station j in the month i in the year k, $\overline{q_{i,j}}$ is the matrix which represents the average of all months and stations, and $\sigma_{i,j}$ is the matrix which represents the standard deviation of the monthly series of all stations.

With the standardized time series, the linear regression of the naturalized streamflows observed in the ONS was used, and the parameters of each of the 161 stations that did not have the SMAP were obtained, considering the 24 basins representing the NIS as explanatory variables, as it is expressed in Equation (21):

$$Z_{i,pj,k} = \sum_{pk=1}^{pk=24} Z_{i,pk,k} \cdot \beta_{i,pk} \tag{21}$$

in which pk are the streamflows of the 24 basins obtained by the SMAP model (ranging from 1 to 24), pj are the stations that do not have the calibrated SMAP (ranging from 1 to 161), and $\beta_{i,pk}$ the regression coefficients.

The 24 stations were used as predictors of the other station streamflows, and thus, it was possible to obtain the regression parameters of the 185 stations by using stepwise. In this method, predictors are iteratively added and removed in the predictive model in order to find the subset of variables in the data set to have a better model performance, which is a method that reduces the prediction error.

The ANE calculation

With the naturalized streamflow data and the usage of 120 hydroelectric stations, it was possible to calculate the ANE by using Equation (22):

$$\text{ANE}_{\text{basins}}(t) = \sum_{i=1}^{m} [Q_{\text{nat}}(j, t) \cdot p(j)] \tag{22}$$

in which t is the time interval of the ANE calculation, i is the usage belonging to the usage system of the considered basin, Q_{nat} is the natural streamflow of the usage in the considered time interval, p is the average productivity of the turbine-generator set of the hydroelectric usage, related to the drop obtained by the difference between the level of amount, corresponding to a storage of 65% of the useful volume, and the average level of the escape channel; j is the usage belonging to the usage system of the considered subsystem; and m is the number of existing usages in the system (ONS 2019).

Statistical analysis of the projections

For the statistical analysis of projections, two methods were used, they are (i) the calculation of the annual anomalies and (ii) the non-parametric Mann–Kendall–Sen trend test.

The annual anomaly calculation of the SAT, precipitation, naturalized streamflow, and ANE of the CMIP6 model data for the projections related to the SSP2-4.5 and SSP5-8.5 scenarios from 2020 to the end of the 21st century was performed considering 20-year intervals, that is, from 2020 to 2039, 2040 to 2059, 2060 to 2079, and 2080 to 2099 in comparison to the historical scenario for the 20th century (period from 1901 to 2000), conforming Equation (23):

$$A_{\text{annual}} = \frac{(P^{\text{a}}_{\text{XXI}} - P^{\text{a}}_{\text{XX}})}{P^{\text{a}}_{\text{XX}}} \times 100 \tag{23}$$

where $P^{\text{a}}_{\text{XXI}}$ is the annual average of the projection variables for the 21st scenarios and P^{a}_{XX} is the annual average of the 20th century variables.

In this study, the non-parametric Mann–Kendall–Sen test was applied in the series of naturalized streamflows and ANE of the projections of the SSP2-4.5 and SSP5-8.5 scenarios for the 21st century to verify trends.

In an analogous way to the calculation of the anomaly, for the Mann–Kendall–Sen trend tests, the 21st century projections were compared with the 20th century characteristics. For this, they were standardized according to Equation (24):

$$Z = \frac{(X^{j}_{\text{XXI}} - \bar{X}_{\text{XX}})}{\sigma_{\text{XX}}} \tag{24}$$

where Z is the series of the 21st century scenarios standardized; X^{j}_{XXI} is the annual average series of the SSP2-4.5 and SSP5-8.5 scenarios for a year j; $\bar{X}_{\text{XX}}$ is the average of the annual average series of the 20th century; and the σ_{XX} is the standard deviation of the annual average series of the 20th century.

The standardized series of the naturalized streamflows and ANE were submitted to the Mann–Kendall test given by Equation (25):

$$S = \sum_{i=1}^{n=1} \sum_{j=i+1}^{n} \text{signal}(Z_j - Z_i) \tag{25}$$

where S is the value of the series in annual time intervals; i and j are time indexes, and n is the number of elements of the series (Moreira & Naghettini 2016). The signal term $(Z_j - Z_i)$ results from Equation (26):

$$\text{signal}(Z_j - Z_i) = \begin{cases} +1 \text{ se } (Z_j - Z_i) > 0 \\ 0 \text{ se } (Z_j - Z_i) = 0 \\ -1 \text{ se } (Z_j - Z_i) < 0 \end{cases} \tag{26}$$

The null hypothesis (H0) – referring to the absence of trend – is considered as accepted, when the test is less than a critical value named α, which for this study was $\alpha = 0.05$ (for a statistical significance of 95%), that is, for TAU $< \alpha$, the series has no positive trend. Otherwise, that is, for TAU $\geq \alpha$, the ST has a positive trend. In another way, the p-value of the S-statistic considers the H0 true, for p-value $> \alpha$ and false for p-value $\leq \alpha$ (Moreira & Naghettini 2016). The TAU variable is related with the classification and correlation coefficient and quantifies the monotonic association, being given by Equation (27):

$$\text{TAU} = \frac{S}{\frac{1}{2}n(n-1) - \sum_{i=1}^{g} t_i} \tag{27}$$

where n is the series size.

The estimator of sem provided the magnitude of the trends detected through the Q statistic, which is given by Equation (28):

$$Q_{ij} = \frac{X_j - X_i}{j - i} \text{com}\, i < j \tag{28}$$

where X_i and X_j are related to the values of the variable under study in times i and j (Moreira & Naghettini 2016). The positive or negative value for Q indicates increasing or decreasing trends, respectively.

RESULTS AND DISCUSSION

Temperature and precipitation

The median of the annual anomalies of the model projections for the 21st century in relation to the 20th century (1901–2000) indicates an increase in the SAT for the entire Brazilian territory, in all four periods and for both projection scenarios, as shown in Figure 5. The increase in SAT is more intense in the SSP5-8.5 scenario in contrast to the SSP2-4.5 scenario, with the 2060–2079 period showing anomaly magnitudes above 4% for part of the Northern and Southeast/Midwest sectors, and above 4% for the entire Brazilian territory in the 2080–2099 period. In addition, the median of the annual anomaly projections for the 21st century in relation to the 20th century from the models indicates a precipitation reduction for most sectors and for all periods and scenarios analyzed with anomaly magnitudes ranging from less than −50% to −10%, as per Figure 6.

Part of the coast of the Northeast sector and virtually the entire Southern sector are exceptions, for which the median of the models indicates a precipitation increase for most periods and scenarios analyzed. The precipitation increase indicated by the models for the Southern sector is greater than that shown in the northern Northeast sector, reaching a magnitude of 40% in the SSP5-8.5 scenario over the period 2080–2099, while, in the northern Northeast sector, the magnitude of the anomalies ranged from −5% to 30%.

These results coincide with several studies that have analyzed SAT projections and precipitation of previous versions of CMIP project models and regional models for various regions in Brazil (Guimarães *et al.* 2016; Jong *et al.* 2018; Silveira *et al.* 2019). Among them, the study by Jong *et al.* (2018), using the ensemble of three regional models, showed that for the period 2071–2100 in the RCP8.5 scenario, the projections for SAT in inland cities of the Northeast sector are projected to increase by about 4–5 °C, and the precipitations might decrease by about 25–50% in the semi-arid areas of Bahia and up to 80% in the coastal areas. Silveira *et al.* (2019), using the CMIP5 models, also showed a possible precipitation reduction for the periods 2015–2044, 2045–2074, and 2075–2098 in relation to the 1984–2003 period for most sectors of the NIS, except for

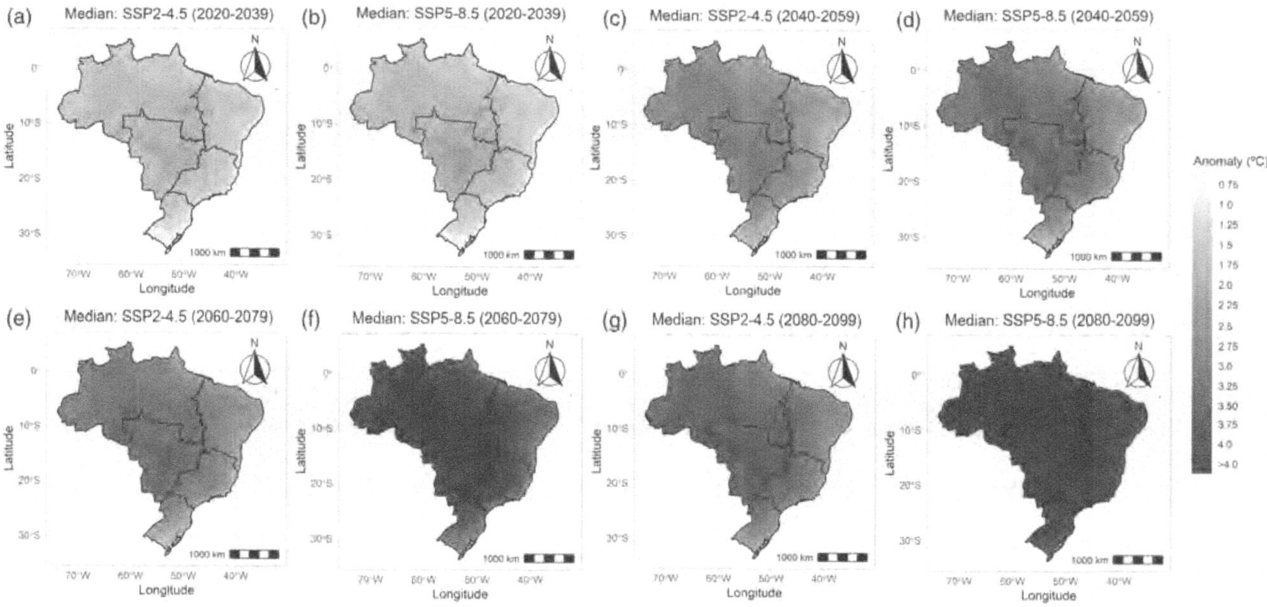

Figure 5 | Median of the average annual percentage anomalies of the SAT of the CMIP6 models in the SSP2-4.5 and SSP5-8.5 scenarios in relation to the 20th century.

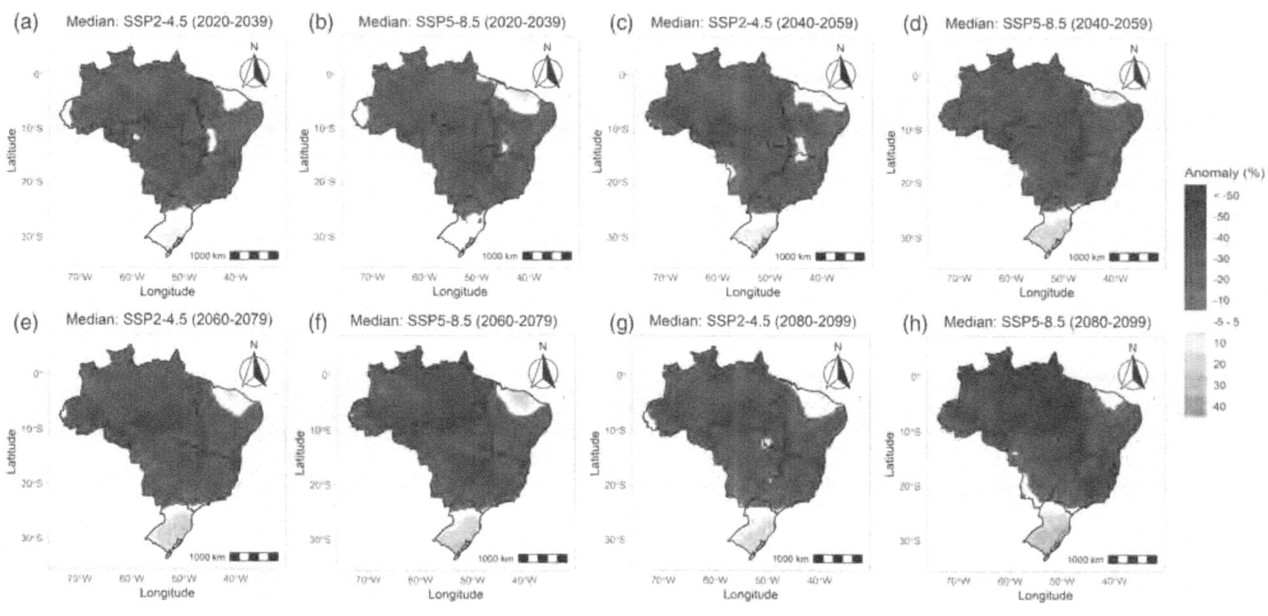

Figure 6 | Median of the average annual percentage anomalies of the precipitation of the CMIP6 models in the SSP2-4.5 and SSP5-8.5 scenarios in relation to the 20th century.

the Southern sector where the models indicated a possible increase in rainfall.

Figure 7 shows the model trends for the SSP2-4.5 and SSP5-8.5 scenarios for the SAT using the Mann–Kendall–Sen test.

The set of models indicates a significant and positive trend for all NIS subsystems. There is clear evidence that increased greenhouse gas emissions suggest a greater impact on SAT, as in all cases the module of the trend magnitude is always larger for the SSP5-8.5 scenario over the

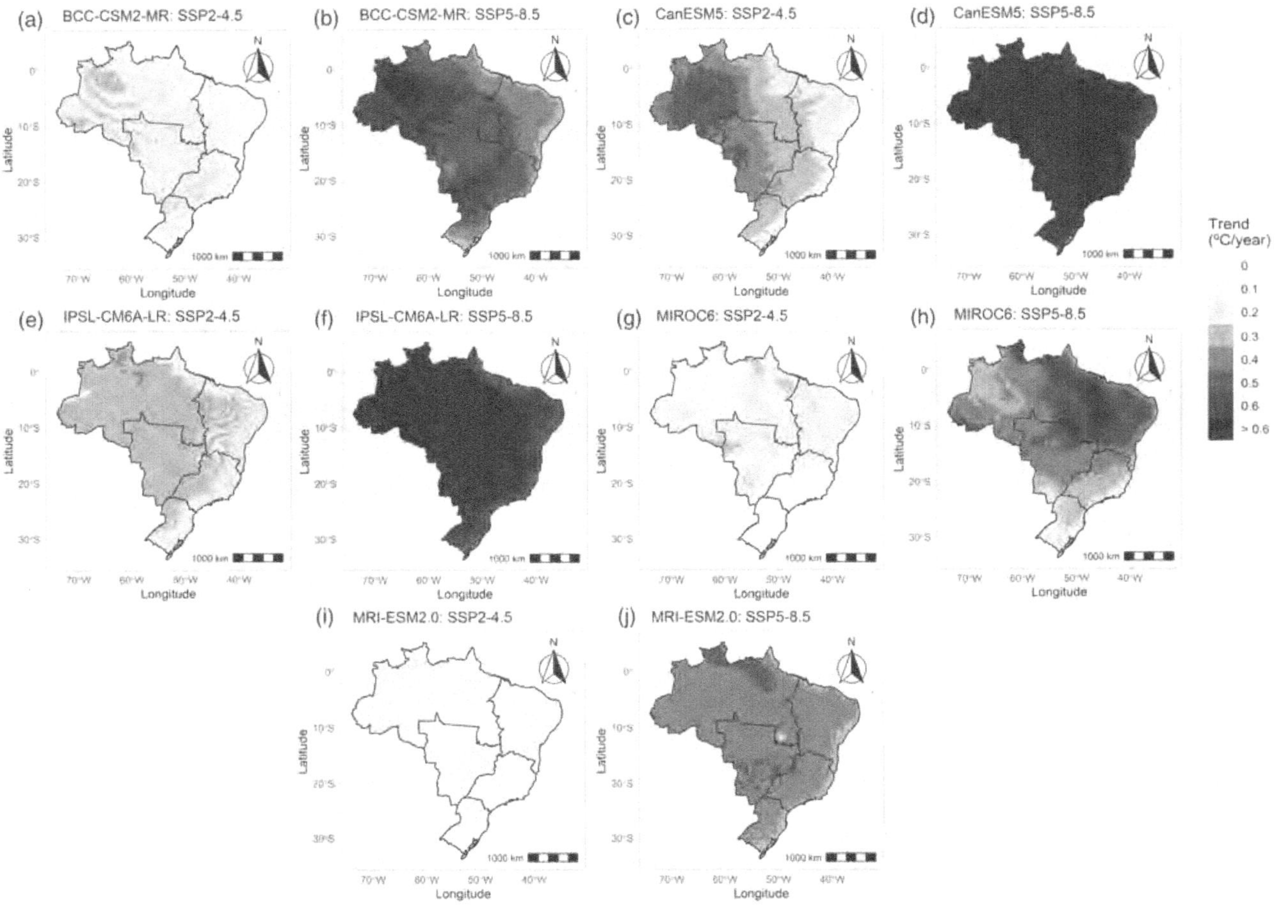

Figure 7 | Trends, according to the Mann–Kendall–Sen method, of the annual average temperature, referring to the SSP2-4.5 and SSP5-8.5 scenarios of the CMIP6 for the period from 2020 to 2100.

SSP2-4.5 scenario. The CanESM5 and IPSL-CM6A-LR models have demonstrated to be more pessimistic, showing a trend with an increase of more than 0.6 °C for the 21st century (2020–2100) for all subsystems in the SSP5-8.5 scenario.

Unlike the SAT, in which the models were unanimous in indicating a possible increase in the SAT for the NIS according to the Mann–Kendall–Sen test, for precipitation, the models presented divergent results, both spatially and in relation to the trend declivities for most subsystems (Figure 8). For the regions that presented significant trends, the MIROC6 and MRI-ESM2.0 models indicated a positive trend of precipitations above 6 mm/year for most subsystems, while the rest of the models indicated a negative trend of precipitation for most of the subsystems. However, even with the divergence among the models, all of them agreed with the possible increase of precipitation for the Southern subsystem for the SSP5-8.5 scenario, presenting a significant and positive trend with a value above 6 mm/year.

Calibration and validation of the SMAP and the linear regression model

Figures 9 and 10 present, respectively, the linear regression and SMAP model performance in the streamflow estimations of the 185 stations for the calibration and validation process through the NSE.

For this purpose, the streamflows' annual time series in the period from 1951 to 1990 were used for calibration and the period from 1991 to 2005 was used for validation. It is possible to observe that the stations calibrated with the SMAP model showed NSE values greater than 0.8 compared with the ONS naturalized flows, while the stations calibrated with the linear

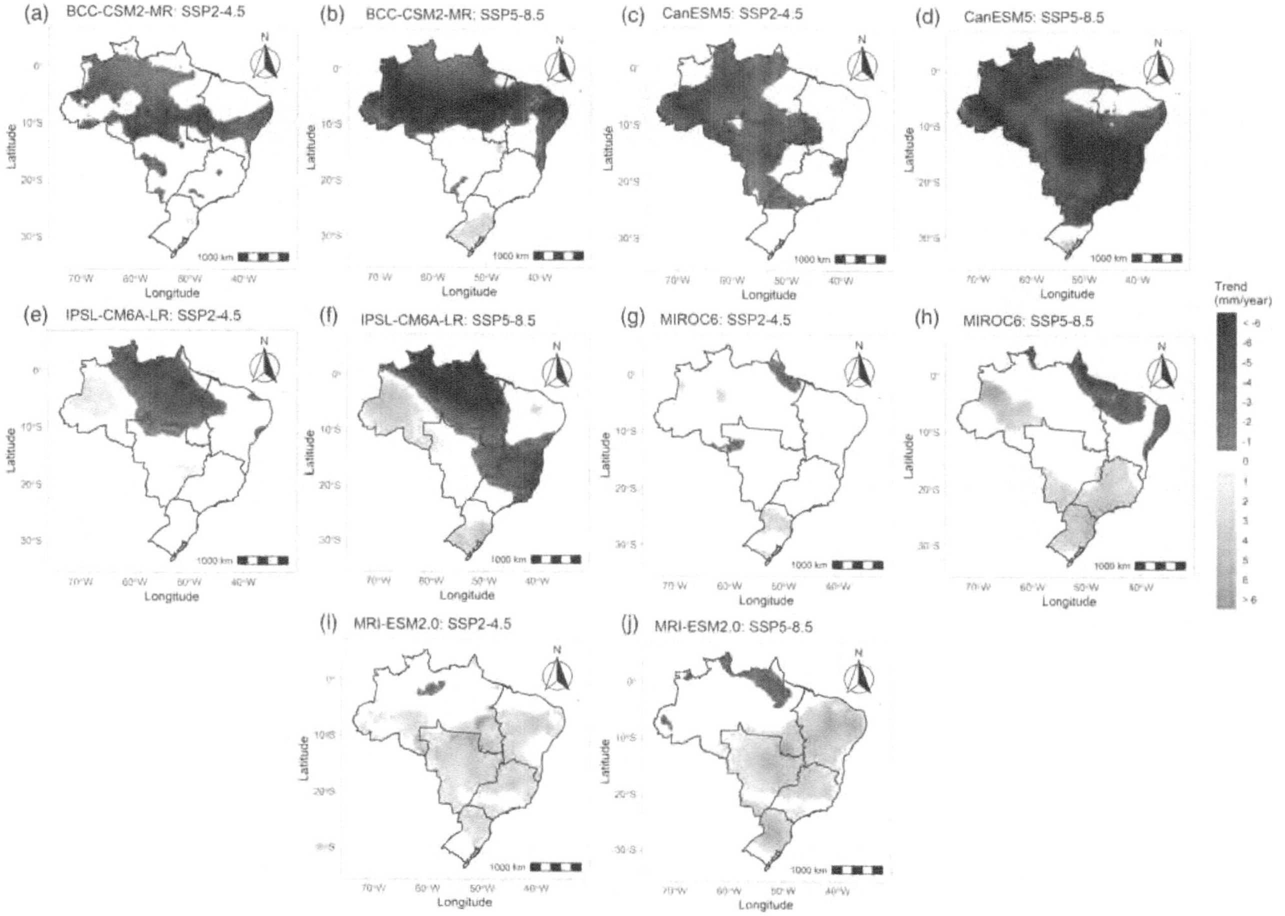

Figure 8 | Trends, according to the Mann–Kendall–Sen method, of the standardized annual average precipitation, referring to the SSP2-4.5 and SSP5-8.5 scenarios of the CMIP6 for the period from 2020 to 2100.

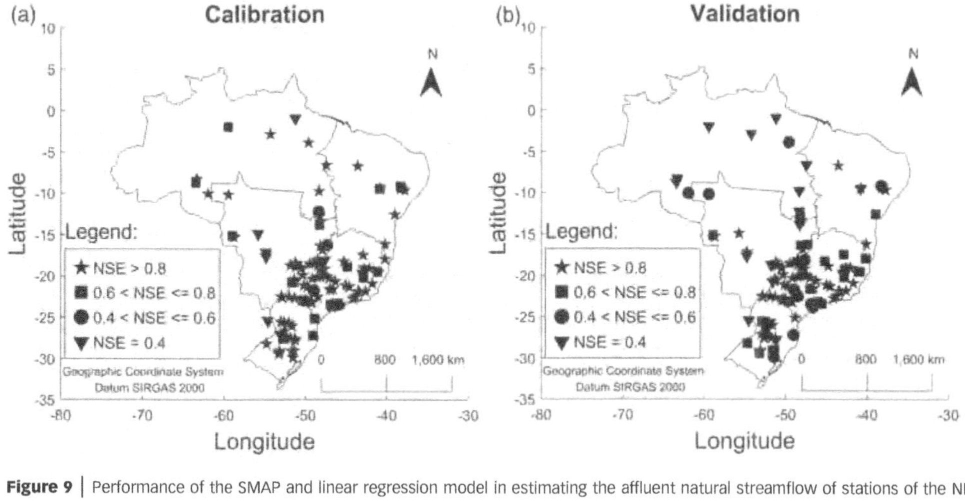

Figure 9 | Performance of the SMAP and linear regression model in estimating the affluent natural streamflow of stations of the NIS through the NSE.

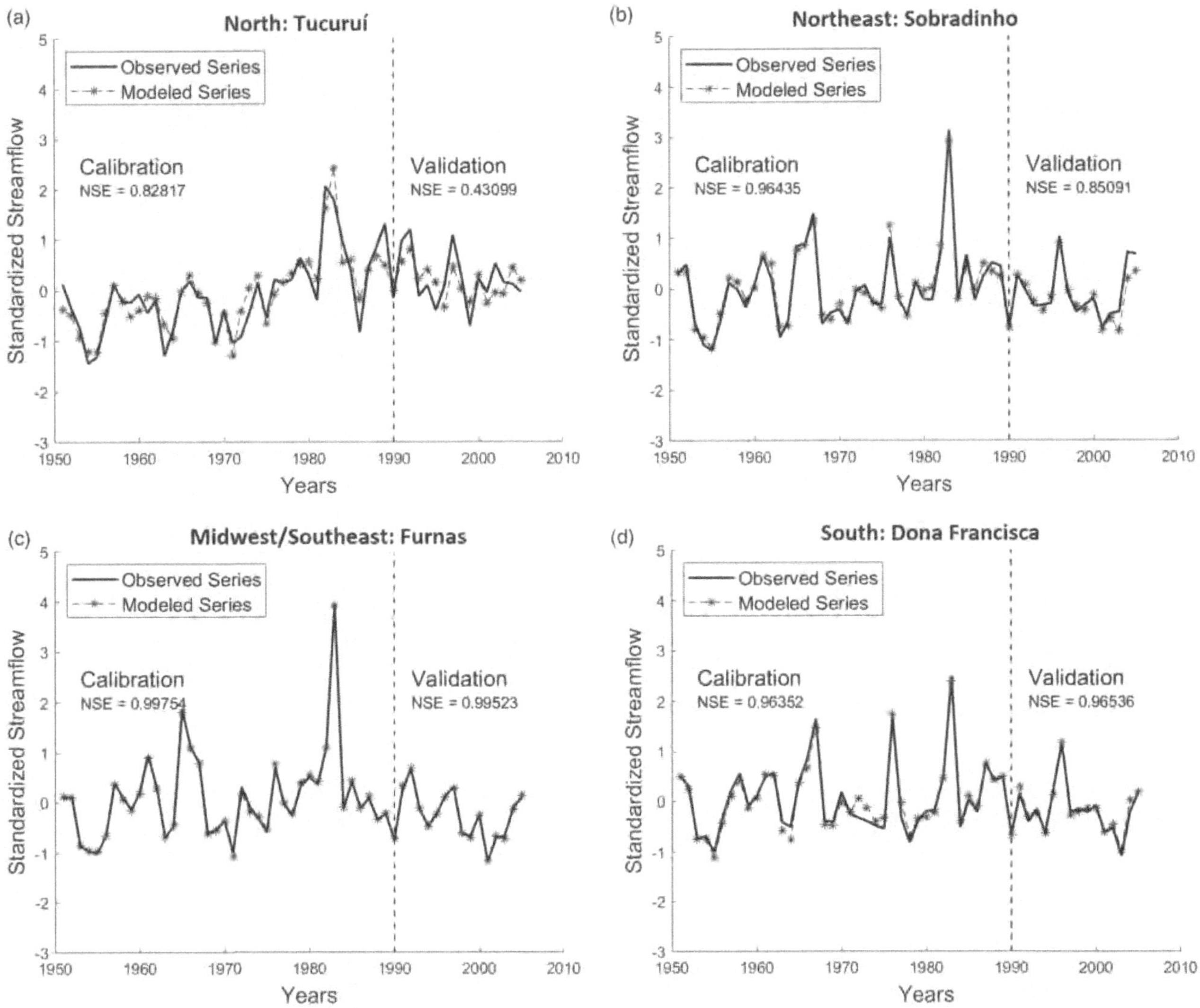

Figure 10 | SMAP model performance in estimating the affluent natural streamflow of the following stations: (a) Tucuruí, (b) Sobradinho, (c) Furnas and (d) Dona Francisca through the NSE.

regression model showed NSE values between 0.4 and 0.8. In the validation, most stations had a decrease in the NSE value, but continued to perform well with values equal to or above 0.4. Among the sectors, the Northern one presented the worst performance in the validation, with most stations presenting values equal to 0.4. The Tucuruí station presented the highest NSE value, even though it was much lower compared with the stations of the other sectors calibrated and validated with the SMAP model, with a value equal to 0.43099. This could be associated with the smaller number of rainfall measuring stations and SAT in this region, used by GPCC and CRU to generate their grids, respectively. Thus, for this region, one may be more uncertain regarding these data.

Even though the NSE indicates a good performance of the SMAP and linear regression models, due to the fact that it prioritizes maximum extreme values, this might cause the streamflows estimated by the models to have poor performance in representing minimum values.

Streamflow and ANE

The behavior of the mean percentage anomalies of the annual naturalized streamflows and ANE shown by the CMIP6 models for the periods 2020–2039, 2040–2059,

2060–2079, and 2080–2099, indicated projections with more intense modules in the SSP5-8.5 scenario than in the SSP2-4.5 scenario for the four analyzed periods and for most of the sectors, as shown in Figures 11 and 12. Such a fact suggests that most of the stations of the NIS reservoirs are sensitive to the increasing emissions of greenhouse gases.

In the North subsystem, for instance, most of the CMIP6 models indicated an increase in the naturalized streamflows and ANE for all the analyzed periods and scenarios. With emphasis on the magnitude of the median percentage anomalies in the naturalized flows of some reservoirs, they are: Santo Antônio, Tucuruí, and Belo Monte, with values above 5, 25, and 45%, respectively. This result is contrary to that obtained from SAT and precipitation, because with the SAT increase and the precipitation decrease for the northern sector, a streamflow and ANE decrease was

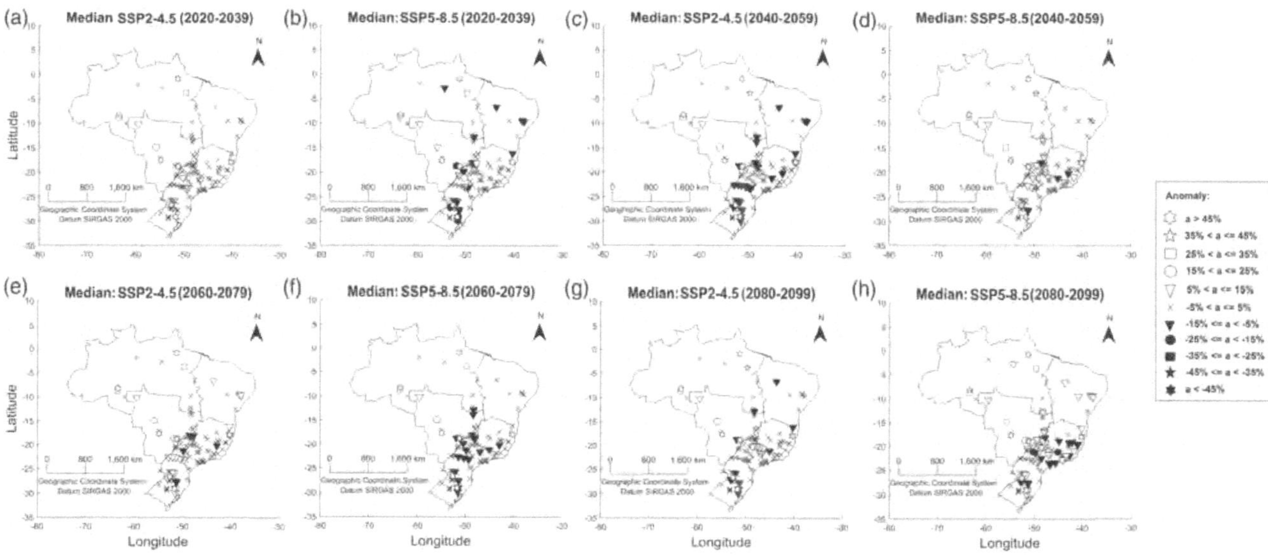

Figure 11 | Median of the average annual percentage anomalies of the naturalized streamflows of the CMIP6 models, referring to the SSP2-4.5 and SSP5-8.5 scenarios.

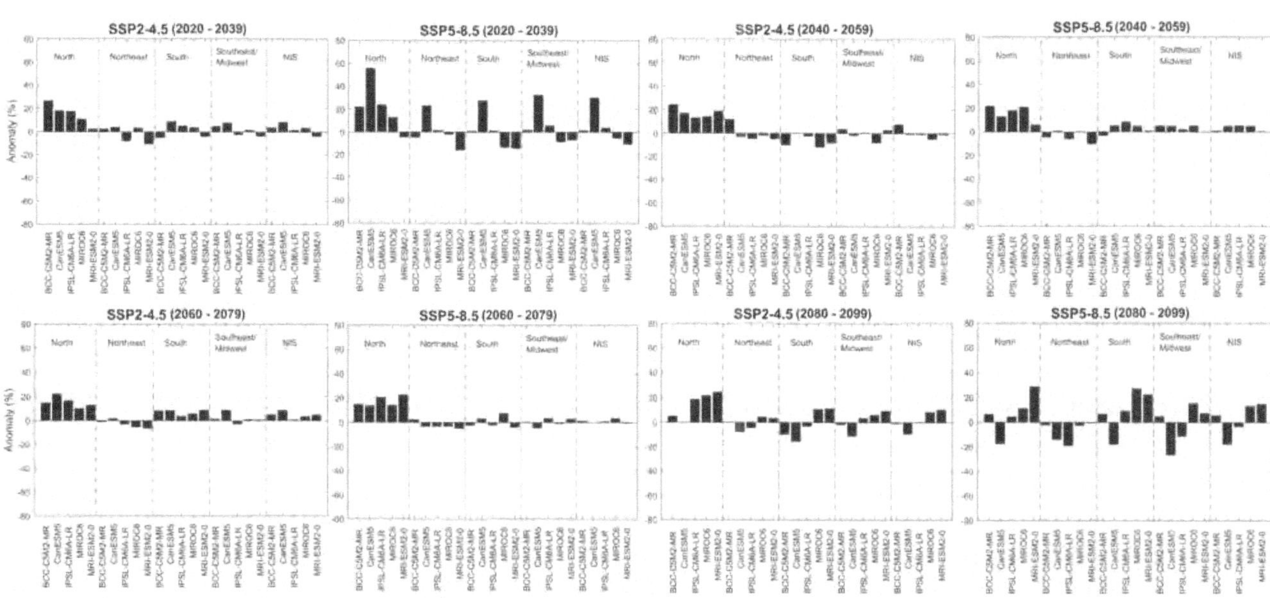

Figure 12 | Anomalies of the average annual ANE for the subsystems of the Brazilian electrical sector.

expected. Such fact may be associated with the poor performance of the linear regression and SMAP hydrological models during the validation of the stations of this sector, as demonstrated previously.

In the South subsystem, most of the CMIP6 models indicated a reduction in the naturalized streamflows and ANE for the first period analyzed in the SSP5-8.5 scenario. However, this projection changes in the last period analyzed, with most of the CMIP6 models suggesting an increase in the naturalized streamflows and ANE, with emphasis on the magnitudes of the average annual anomalies of the ANE, which presented values above 20% with the MIROC6 and MRI-ESM2-0 models. The projections of the possible increase in precipitation and naturalized streamflows had already been demonstrated in previous studies, using data from the CMIP3 and CMIP5 models (Ribeiro Neto *et al.* 2016; Silveira *et al.* 2017; Jong *et al.* 2018; Silva *et al.* 2021). Hence, similar results among the CMIP3, CMIP5, and CMIP6 databases show the cohesion in the methodologies adopted for the projection of climate variables. This way, recent studies like this one end up contributing to the confirmation of results obtained in previous studies, respecting the uncertainties associated with the process.

In the Northeast subsystem, most of the CMIP6 models indicated a reduction in naturalized streamflows and ANE. The CanESM5 and IPSL-CM6A-LR models, for example, indicated reductions in the average annual anomalies of the ANE of less than −10%. On the other hand, there was a variation in the magnitudes of the average annual anomalies of the naturalized streamflows of the Xingó and Sobradinho reservoirs, which showed magnitudes lower than −5% in the SSP5-8.5 scenario for the first period, non-significant in the periods 2040–2059 and 2060–2079, and higher than 5% in the period 2080– 2099.

This fact might be related to the greater increase in SAT projected by the SSP-8.5 scenario, resulting in a greater increase in PET projected for the 21st century, as demonstrated in this study and in previous studies with the use of previous versions of CMIP (Guimarães *et al.* 2016; Jong *et al.* 2018). The possible PET increase impacts directly on significant reductions in the naturalized streamflows for most of the stations from these sectors and consequently on the ANE.

In the Southeast/Midwest subsystem, there is a slight spatial dispersion between the response of reservoir streamflows and climate change. The reservoirs located further to the Midwest presented positive magnitudes of percentage anomalies during all the analyzed periods and scenarios, while there were variations in the values for the reservoirs located further to the Southeast. In those, during the periods 2020–2039 and 2040–2059, most reservoirs presented non-significant anomalies for both scenarios, in the period from 2060 to 2079 some reservoirs presented average percentage anomalies below −5% in both scenarios. In the period from 2080 to 2099, for the SSP5-8.5 scenario, the models suggest a greater possibility of reduction in streamflows or a slight increase, since the anomalies in the annual streamflow projections for this subsystem are between 5 and −25%.

These results, as in the Northern sector, also have contradictions with the SAT results of a possible increase and a possible decrease in precipitation for the sector. Thereby, streamflows and ANE reductions might also be expected, rather than values ranging from 5 to −25%. This may be associated with the calibration and validation process of the SMAP hydrological model using the NSE coefficient. Once it prioritizes maximum values, and with this, it may end up softening streamflows and, consequently, ANE results.

The naturalized streamflows and ANE trends, with the Mann–Kendall–Sen method, for the CMIP6 models in the SSP2-4.5 and SSP5-8.5 scenarios, reinforce that the climate change impact on the NIS is sensitive to the increase in the greenhouse gas emissions, since the magnitude of the trend declivities is more significant in the SSP5-8.5 scenario than in the SSP2-4.5 scenario, as shown in Figures 13 and 14.

Most of the CMIP6 models indicated a reduction in naturalized streamflows and ANE in the two analyzed scenarios for all the analyzed subsystems, except for the MIROC6 and MRI-ESM2-0 models, which indicated an increase for both scenarios.

Even with the uncertainties attributed to the GCMs, from the observed data used, the bias correction method, the calibration and validation of the SMAP, and linear regression models, it was possible to achieve consistent results with the reality shown in observational studies, which already show significant impacts of climate change in the last years in Brazil (Jong *et al.* 2018; IEA 2020), mainly over the Northeast and South regions of Brazil. In addition, new results were brought, covering not only the

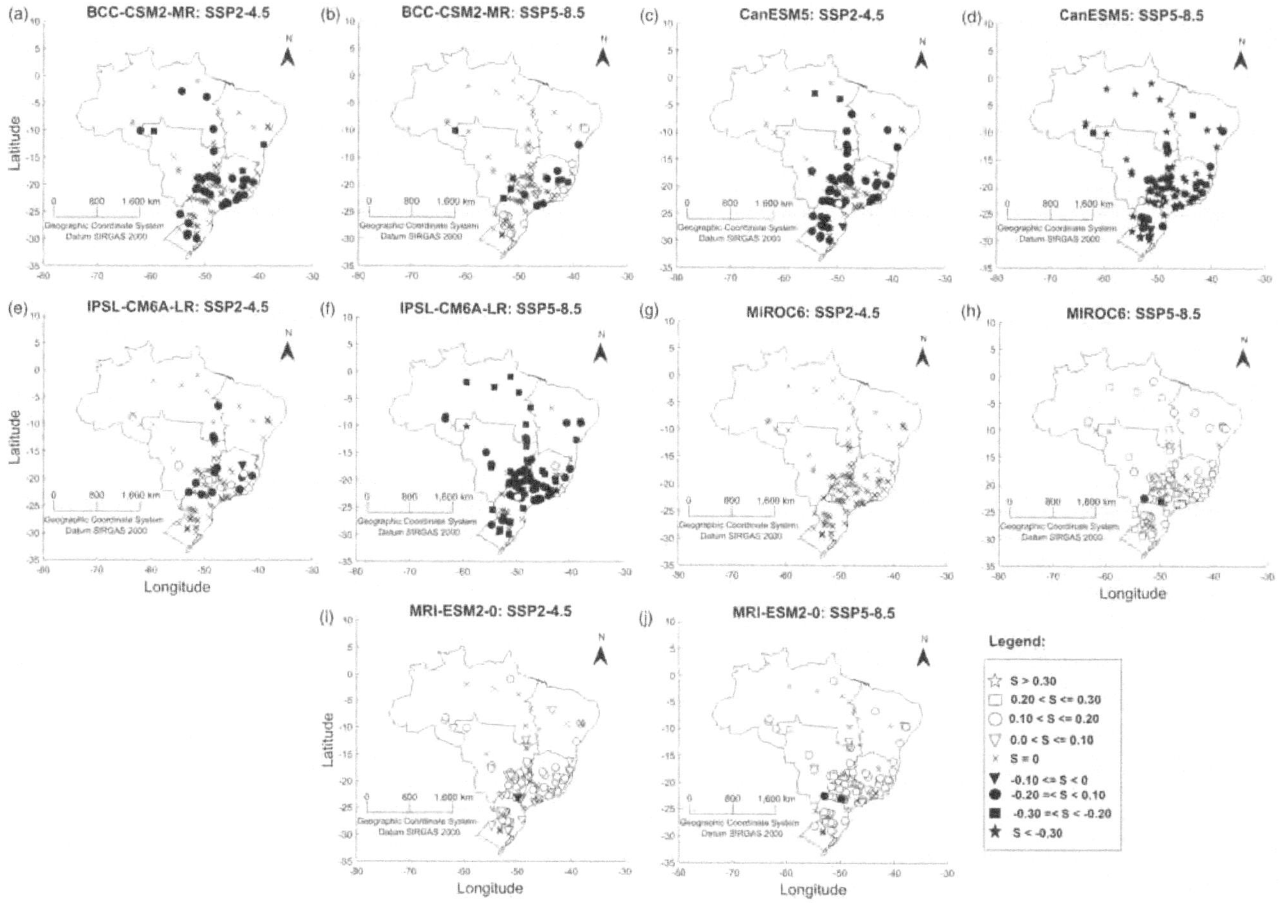

Figure 13 | Trends, according to the Mann–Kendall–Sen method, of the standardized annual average streamflow, referring to the SSP2-4.5 and SSP5-8.5 scenarios of the CMIP6 for the period from 2020 to 2100.

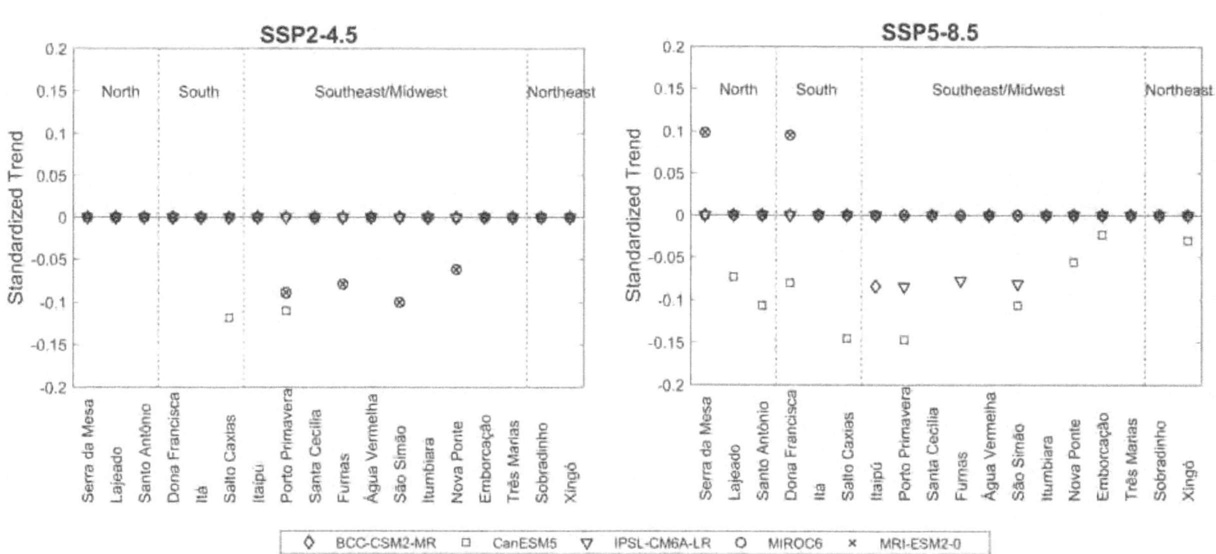

Figure 14 | Trends, according to the Mann–Kendall–Sen method, of the standardized annual average ANE, referring to the SSP2-4.5 and SSP5-8.5 scenarios of the CMIP6 for the period from 2020 to 2100.

main hydroelectric stations in the NIS, but with the use of SMAP and linear regression models, it was possible to estimate and project the streamflows and ANE for the almost 200 stations that constitute the NIS. These results are of great importance for government agencies and society, as they provide subsidies for the development of mitigation plans for the possible climate change impacts on the water and energy sectors.

CONCLUSION

The analyses proposed by this study aimed at identifying patterns of projections for SAT, precipitation, streamflows, and ANE for the NIS about the impacts associated with climate change of the CMIP6 models. Such information is of great importance to aid government agencies and society in making decisions concerning extreme hydrological events (floods and droughts), since climate has a strong influence on the development of society, interfering directly in the environment, agriculture, air quality, economy, energy sector, among others.

Most of the CMIP6 models showed that the NIS reservoirs are sensitive to increasing greenhouse gas emissions, as the climate change impacts were more intense in the SSP5-8.5 scenario than in the SSP2-4.5 scenario both in annual anomalies as in trend magnitudes.

The medians of the standardized anomalies of annual, precipitation, naturalized streamflows, and ANE indicated an increase for the South sector of the country for most of the models and periods analyzed.

The increase in the average annual precipitation, naturalized streamflows, and ANE in the South and North sectors of the country is not unanimously reflected in the trend of the Mann–Kendall–Sen test by the models. The CanESM5 and IPSL-CM6A-LR models, for example, signaled that most reservoirs might present negative trends for all the analyzed subsystems and scenarios, while the MIROC6 and MRI-ESM2-0 models signal positive trends. This dispersion in the results in the trend declivities might be associated with the greater occurrence of hydrological extremes in these regions.

In the Northeast subsystem, most of the CMIP6 models indicated an increase in SAT and a reduction in rainfall, naturalized streamflows, and ANE. The CanESM5 and IPSL-CM6A-LR models, for example, indicated reductions in average annual ANE anomalies of less than −10%. In the Southeast/Midwest sector, for the period from 2080 to 2099, in the SSP5 8.5 scenario, the models suggest a greater possibility of reducing flows or a slight increase, since anomalies in annual flow projections for this subsystem are between 5 and −25%.

The reduction in water availability in the Northern sector and possible reduction in the Southeastern/Midwestern sector, added to the increase in SAT shown in this work, reveal possible impacts due to conflicts among multiple uses, a possible slowdown of the economy due to water reduction for agriculture and industry, as well as shortage of cities. Thus, added to the possible growth in demand for electricity in Brazil in the next few years, it could lead to a crisis in the Brazilian electricity sector.

Therefore, investments in other forms of electricity generation will be necessary for the next decades. One of the forms of energy are nonrenewable ones that cause negative feedback for climate change and, consequently, increase its damage to the entire climate system. This compensation of hydroelectric energy is the one currently used, with the activations of thermoelectric plants – which make use of fossil fuels – and emit large quantities of CO_2 into the atmosphere.

Alternatively, and more beneficial, it could be the investment in renewable sources (wind and solar, for example), so that they can achieve a greater participation in the Brazilian electricity matrix. However, for the latter, a very complex policy is needed with investments in technology and labor qualification, so that in the long term the cost of this generation becomes cheaper for the country.

DATA AVAILABILITY STATEMENT

All relevant data are included in the paper or its Supplementary Information.

REFERENCES

Babel, M. S., Shinde, V. R., Sharma, D. & Dang, N. M. 2020 Measuring water security: a vital step for climate change adaptation. *Environmental Research* **185**, 109400. https://doi.org/10.1016/j.envres.2020.109400.

Cunha, A. P. M. A., Zeri, M., Leal, K. D., Costa, L., Cuartas, L. A., Marengo, J. A., Tomasella, J., Vieira, R. M., Barbosa, A. A., Cunningham, C., Cal Garcia, J. V., Broedel, E., Alvalá, R. & Ribeiro-Neto, G. 2019 Extreme drought events over Brazil from 2011 to 2019. *Atmosphere* **10** (11). https://doi.org/10.3390/atmos10110642.

Eini, M. R., Javadi, S., Delavar, M., Monteiro, J. A. F. & Darand, M. 2019 High accuracy of precipitation reanalyses resulted in good river discharge simulations in a semi-arid basin. *Ecological Engineering* **131** (March), 107–119. https://doi.org/10.1016/j.ecoleng.2019.03.005.

Eyring, V., Bony, S., Meehl, G. A., Senior, C. A., Stevens, B., Stouffer, R. J. & Taylor, K. E. 2016 Overview of the Coupled Model Intercomparison Project Phase 6 (CMIP6) experimental design and organization. *Geoscientific Model Development* **9** (5), 1937–1958. https://doi.org/10.5194/gmd-9-1937-2016.

Gidden, M. J., Riahi, K., Smith, S. J., Fujimori, S., Luderer, G., Kriegler, E., Van Vuuren, D. P., Van Den Berg, M., Feng, L., Klein, D., Calvin, K., Doelman, J. C., Frank, S., Fricko, O., Harmsen, M., Hasegawa, T., Havlik, P., Hilaire, J., Hoesly, R. & Takahashi, K. 2019 Global emissions pathways under different socioeconomic scenarios for use in CMIP6: a dataset of harmonized emissions trajectories through the end of the century. *Geoscientific Model Development* **12** (4), 1443–1475. https://doi.org/10.5194/gmd-12-1443-2019.

Gondim, R., Silveira, C., de Souza Filho, F., Vasconcelos, F. & Cid, D. 2018 Climate change impacts on water demand and availability using CMIP5 models in the Jaguaribe basin, semi-arid Brazil. *Environmental Earth Sciences* **77** (15), 1–14. https://doi.org/10.1007/s12665-018-7723-9.

Guimarães, S. O., Costa, A. A., Vasconcelos Júnior, F. d. C., da Silva, E. M., Sales, D. C., de Araújo Júnior, L. M. & de Souza, S. G. 2016 Projeções de mudanças climáticas sobre o nordeste brasileiro dos modelos do CMIP5 e do CORDEX. *Revista Brasileira de Meteorologia* **31** (3), 337–365. https://doi.org/10.1590/0102-778631320150150.

Gurski, B. C., Jerszurki, D. & de Souza, J. L. M. 2018 Alternative methods of reference evapotranspiration for Brazilian climate types. *Revista Brasileira de Meteorologia* **33** (3), 567–578. https://doi.org/10.1590/0102-7786333015.

Hargreaves, G. H. & Samani, Z. A. 1985 Reference crop evapotranspiration from temperature. *Applied Engineering in Agriculture* **1** (2), 96–99. https://doi.org/10.13031/2013.26773.

Hasan, M. M. & Wyseure, G. 2018 Impact of climate change on hydropower generation in Rio Jubones Basin, Ecuador. *Water Science and Engineering* **11** (2), 157–166. https://doi.org/10.1016/j.wse.2018.07.002.

IEA 2020 In *World energy balances 2020 edition* (issue January). Available from: https://ncdc.gov.ng/themes/common/docs/protocols/111_1579986179.pdf.

IPCC 2014 *Summary for Policymakers. Managing the Risks of Extreme Events and Disasters to Advance Climate Change Adaptation: Special Report of the Intergovernmental Panel on Climate Change, 9781107025*, pp. 3–22. https://doi.org/10.1017/CBO9781139177245.003.

Jong, P., Tanajura, C. A. S., Sánchez, A. S., Dargaville, R., Kiperstok, A. & Torres, E. A. 2018 Hydroelectric production from Brazil's São Francisco River could cease due to climate change and inter-annual variability. *Science of the Total Environment* **634**, 1540–1553. https://doi.org/10.1016/j.scitotenv.2018.03.256.

Kundzewicz, Z. W., Krysanova, V., Benestad, R. E., Hov, O., Piniewski, M. & Otto, I. M. 2018 Uncertainty in climate change impacts on water resources. *Environmental Science and Policy* **79** (October 2017), 1–8. https://doi.org/10.1016/j.envsci.2017.10.008.

Limberger, L. & Silva, M. E. S. 2018 Precipitação Observada Na Amazônia Brasileira: Redes Convencionais E Dados Da Reanálise I Do Ncep/Ncar, Cru E Gpcc. *Revista Brasileira de Climatologia* **22**, 20–37. https://doi.org/10.5380/abclima.v22i0.48841.

Lopes, J. E. G., Braga, B. P. F. & Conejo, J. G. L. 1982 SMAP – A simplified hydrological model, applied modelling in catchment hydrology. *Water Resourses Publications* **33** (2), 167–176.

Moreira, J. G. do V. & Naghettini, M. 2016 Detecção de Tendências Monotônicas Temporais e Relação com Erros dos Tipos I e II: Estudo de Caso em Séries de Precipitações Diárias Máximas Anuais do Estado do Acre. *Revista Brasileira de Meteorologia* **31** (4), 394–402. https://doi.org/10.1590/0102-778631231420140155.

Nash, J. E. & Sutcliffe, J. V. 1970 River flow forecasting through conceptual models part I – a discussion of principles. *Journal of Hydrology* **10** (3), 282–290. https://doi.org/10.1016/0022-1694(70)90255-6.

ONS 2019 *Plano de Operação Energética 2019–2023*. Available from: http://www.ons.org.br/AcervoDigitalDocumentosEPublicacoes/PEN_Executivo_2019-2023.pdf.

Qin, P., Xu, H., Liu, M., Du, L., Xiao, C., Liu, L. & Tarroja, B. 2020 Climate change impacts on Three Gorges Reservoir impoundment and hydropower generation. *Journal of Hydrology* **580**, 123922. https://doi.org/10.1016/j.jhydrol.2019.123922.

Ribeiro Neto, A., da Paz, A. R., Marengo, J. A. & Chou, S. C. 2016 Hydrological processes and climate change in hydrographic regions of Brazil. *Journal of Water Resource and Protection* **08** (12), 1103–1127. https://doi.org/10.4236/jwarp.2016.812087.

Schneider, U., Finger, P., Meyer-Christoffer, A., Rustemeier, E., Ziese, M. & Becker, A. 2017 Evaluating the hydrological cycle over land using the newly-corrected precipitation climatology from the Global Precipitation Climatology Centre (GPCC). *Atmosphere* **8** (3). https://doi.org/10.3390/atmos8030052.

Silva, M. V. M., Silveira, C. d. S., Costa, J. M. F. d., Martins, E. S. P. R. & Vasconcelos Júnior, F. d. C. 2021 Projection of climate change and consumptive demands projections impacts on hydropower generation in the São Francisco River Basin, Brazil. *Water* **13** (3), 332. https://doi.org/10.3390/w13030332.

Silveira, C. d. S., de Souza Filho, F. d. A. & Vasconcelos Júnior, F. d. C. 2017 Streamflow projections for the Brazilian hydropower sector from RCP scenarios. *Journal of Water and Climate Change* **8** (1), 114–126. https://doi.org/10.2166/wcc.2016.052.

Silveira, C. D. S., Vasconcelos Junior, F. D. C., Souza Filho, F. D. A. De, Guimarães, S. O., Marcos Junior, A. D., Dos Reis, G. N. L. & Porto, V. C. 2019 Performance evaluation of AR5-CMIP5 models for the representation of seasonal and multi-annual variability of precipitation in Brazilian hydropower sector basins under RCP8.5 scenario. *Hydrological Sciences Journal* **64** (11), 1279–1296. https://doi.org/10.1080/02626667.2019.1612521.

Valle Júnior, L. C. G., Ventura, T. M., Souza, R. S. R., Nogueira, J. d. S., Lobo, F. d. A., Vourlitis, G. L. & Rodrigues, T. R. 2020 Comparative assessment of modelled and empirical reference evapotranspiration methods for a Brazilian savanna. *Agricultural Water Management* **232**, 106040. https://doi.org/10.1016/j.agwat.2020.106040.

First received 24 December 2020; accepted in revised form 30 April 2021. Available online 18 May 2021

© 2022 The Authors

Impact of Climate Change on Hydrology and Water Resources doi: 10.2166/wcc.2021.015

Projected changes in temperature and precipitation over mainland Southeast Asia by CMIP6 models

S. Supharatid, J. Nafung and T. Aribarg

ABSTRACT

Five mainland SEA countries (Cambodia, Laos, Myanmar, Vietnam, and Thailand) are threatened by climate change. Here, the latest 18 Coupled Model Intercomparison Project Phase 6 (CMIP6) is employed to examine future climate change in this region under two SSP-RCP (shared socioeconomic pathway-representative concentration pathway) scenarios (SSP2-4.5 and SSP5-8.5). The bias-corrected multi-model ensemble (MME) projects a warming (wetting) over Cambodia, Laos, Myanmar, Vietnam, and Thailand by 1.88–3.89, 2.04–4.22, 1.88–4.09, 2.03–4.25, and 1.90–3.96 °C (8.76–20.47, 12.69–21.10, 9.54–21.10, 13.47–22.12, and 7.03–15.17%) in the 21st century with larger values found under SSP5-8.5 than SSP2-4.5. The MME model displays approximately triple the current rainfall during the boreal summer. Overall, there are robust increases in rainfall during the Southwest Monsoon (3.41–3.44, 8.44–9.53, and 10.89–17.59%) and the Northeast Monsoon (−2.58 to 0.78, −0.43 to 2.81, and 2.32 to 5.45%). The effectiveness of anticipated climate change mitigation and adaptation strategies under SSP2-4.5 results in slowing down the warming trends and decreasing precipitation trends after 2050. All these findings imply that member countries of mainland SEA need to prepare for appropriate adaptation measures in response to the changing climate.

Key words | climate change, CLMVT, CMIP6, precipitation, temperature

S. Supharatid (corresponding author)
J. Nafung
T. Aribarg
Climate Change and Disaster Center,
Rangsit University,
Pathumthani 12000,
Thailand
E-mail: *seree.s@rsu.ac.th*

J. Nafung
T. Aribarg
College of Digital Innovation and Information Technology,
Rangsit University,
Pathumthani,
Thailand

HIGHLIGHTS

- There is a robust increase in SWMR rainfall while the NEMR displays smaller increases.
- The temperature changes among CLMVT countries do not show any significant differences. However, there are significant precipitation changes.

INTRODUCTION

A large number of climate extremes (e.g., floods, droughts, heat waves, and extreme precipitation) have been recently observed across the globe in conjunction with global warming. Several studies reveal that continued greenhouse gas emissions will lead to long-term changes in the climate system and increasing the likelihood of widespread climate-extreme impacts on social and ecological systems (IPCC 2014; Easterling *et al.* 2016; Fischer & Knutti 2016; Giorgi *et al.* 2018; Oppenheimer *et al.* 2019; Cook *et al.* 2020). The near-surface air temperature has increased by approximately 0.78 °C (0.72–0.85 °C) on a global scale since 1900, with most of the increase occurring in recent decades (IPCC 2013). Impacts on natural and human systems from global warming have already been observed with high confidence (IPCC SR15).

In present-day climate simulations, general circulation models (GCMs) are the major tools used to project future climate based on well-established physical principles (Randall

doi: 10.2166/wcc.2021.015

et al. 2007). Both GCMs and observations on the global (Pall *et al.* 2007) and regional scales (Ghosh *et al.* 2012) have confirmed that rising extreme precipitation events are linked to climate warming, which leads to increased atmospheric moisture content and specific humidity (Willett *et al.* 2007). Extreme precipitation is projected to intensify in the future under a warming climate (Ali & Mishra 2017).

To better understand past, present, and future climate change, the Working Group on Coupled Modeling under the framework of the World Climate Research Programme (WCRP) established the Coupled Model Intercomparison Project (CMIP3, CMIP5, and now CMIP6). This serves as a fundamental basis for international climate research with a remarkable technical and scientific coordination effort among climate modeling centers (Eyring *et al.* 2015). Projections of potential changes in climate extremes are now being investigated by global climate models (GCMs) in many regions. Under IPCC AR4 (CMIP3), changes in temperature indices tend to agree for all seasons, but changes in precipitation are uncertain (Orlowsky & Seneviratne 2012). Projected changes in temperature and precipitation extremes are generally more pronounced in CMIP5 than in CMIP3 (Sillmann *et al.* 2013a, 2013b). Previous studies of climate model comparison revealed that CMIP5 models perform better than CMIP3 models, particularly over North and Central America including the Caribbean (Bukovsky *et al.* 2015; Hidalgo & Alfaro 2015; Koutroulis *et al.* 2016). Overall, comparisons between model results and observations indicated uncertainty in their projections (Alexander & Arblaster 2017). Most GCMs represent climatic variation at gross spatial resolutions (typically 100–300 km) which are not capable in impact assessments that require relatively fine spatial resolutions of just a few kilometers.

The latest CMIP6 models use the combination of shared socioeconomic pathways (SSPs) and representative concentration pathways (RCPs) that make more reasonable for future scenarios (Eyring *et al.* 2016). As CMIP6 has been improved in several aspects (e.g., higher horizontal resolution, better representation of synoptic processes, and better agreement with the estimation of global energy balance), more reasonable results can be obtained from climate-extreme studies (e.g., Di Luca *et al.* 2020; Kim *et al.* 2020; Laurie & Mathew 2020; Nie *et al.* 2020; Srivastava *et al.* 2020; Wild 2020).

Studies investigating the consequences of climate change over SEA are limited. Manton *et al.* (2001) analyzed extreme climate indices for the historical period over SEA and the South Pacific. The majority of SEA is influenced by the Asian–Australian monsoon and several regions within it are affected by extreme weather events, particularly tropical cyclones, droughts, and floods (Chang *et al.* 2005). The warm extremes increased while the cold extremes decreased over the Indo-Pacific region during 1971–2005 (Caesar *et al.* 2011). Most studies of the projected changes in SEA climate are embedded in global-scale domain carried out using GCMs (Chadwick *et al.* 2016). The projected changes in mean and extreme precipitation (using the NASA Earth Exchange Global Daily Downscaled Projection, NEX-GDDP dataset) over several parts of SEA show substantial increases in the 21st century (Mandapaka & Lo 2018). Suppari *et al.* (2019) found significant changes in consecutive dry day (CDD) and a decrease in total wet day precipitation (PRCPTOT) over most regions in SEA by using eight ensemble members of CORDEX-SEA simulations for RCP4.5 and RCP8.5 scenarios. A marked amplification in extreme precipitation over the Indochina Peninsula and the Maritime Continent were found under 1.5 and 2 °C global warming levels (Ge *et al.* 2019). Recently, Tangang *et al.* (2020) examined the projected rainfall changes in Southeast Asia in the 21st century based on seven regional climate models (RCMs) members of archived CORDEX-SEA simulations.

To our knowledge, few studies have analyzed the CMIP6 datasets to examine the future climate, especially in SEA. Grose *et al.* (2020) evaluated CMIP6 models and their future climate projects over Australia and compared the results to those obtained using CMIP5 models. Almazroui *et al.* (2020) examined the projected changes in temperature and precipitation over six South Asian countries during the 21st century using the latest CMIP6 dataset. Ukkola *et al.* (2020) revealed larger projected drought changes, which were more consistent in CMIP6 compared to CMIP5 models. Very recently, Almazroui *et al.* (2021) used CMIP6 models to examine the projected changes in temperature and precipitation over the United States, Central America, and the Caribbean. Su *et al.* (2020) analyzed the future drought characteristics over China using CMIP6 models. Mondal *et al.* (2021) examined the changing population exposure to drought across South Asia using 20 CMIP6 multi-model ensembles (MMEs).

In this study, we analyzes the changes in mean temperature and precipitation using the latest Couple Model Intercomparison Project Phase 6 (CMIP6) model simulation dataset over mainland SEA. It is still not well understood how the new CMIP6 models can effectively simulate the climate response to anthropogenic forcing in this region. The overall aim of this study is to reveal the ability of the CMIP6 model to simulate the climate response to anthropogenic forcing over five mainland Southeast Asian countries. The specific questions to address are: what are the long-term observed trends in temperature and precipitation over each country and how are they likely to change in the seasonal monsoon for the near-, mid-, and far-future periods? This is an initial step required for a decision on what appropriate level of adaptation measures to the impacts of projected climate-extreme events.

STUDY REGION AND DATA METHODOLOGY

Study region

Our region of interest consists of the five mainland SEA countries: Cambodia, Laos, Myanmar, Vietnam, and Thailand (CLMVT), as displayed in Figure 1. We first examined the historical climate over the full domain of SEA similar to CORDEX-SEA (14.8°S–27°N, 89.5–146.5°E). Then, we projected the change in climate and seasonal monsoon precipitation over each country. SEA experiences two distinct sub-monsoon seasons: wet and dry. The same weather system that delivers rain during India's monsoon season also affects Southeast Asia, but at different times (Kripalani *et al.* 2007). The southwest monsoon (SWM) causes the maximum rainfall during the boreal summer over the mainland SEA, while the northeast monsoon (NEM) causes maximum rainfall during the boreal winter over most areas in the Maritime Continent (Waliser & Gautier 1993; Chang *et al.* 2005). Furthermore, the complex terrain with several islands of different sizes causes significant regional variations in precipitation along the annual cycle (Chang *et al.* 2005).

Observation datasets

To estimate observational uncertainty, we examined several observation datasets (SA-OBS, APHRODITE, CPC-UNI, CRU, GPCP1DD, TRMM, ERA-Interim, JRA55, GPCC, CHIRPS, and CMORPH) over CLMVT. The observation datasets used in this study are the daily maximum and minimum temperature and daily precipitation. All observation datasets differ due to varying original observations, different

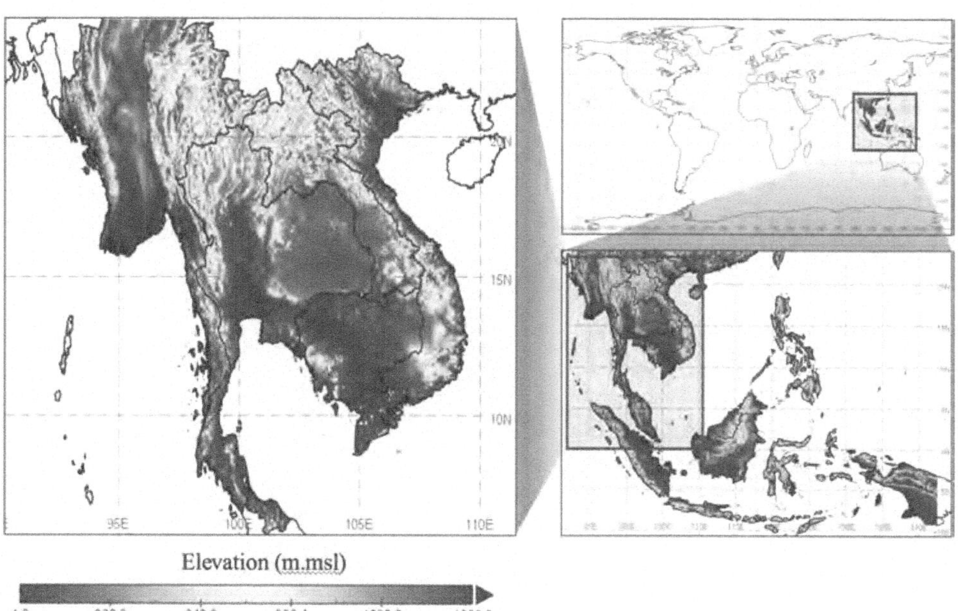

Figure 1 | Study area including five mainland SEA countries (CLMVT: Cambodia, Laos, Myanmar, Vietnam, and Thailand).

resolutions, and different methods employed. The common period of 1998–2014 was used for intercomparison due to data availability. Comparisons among these observation datasets (see Figure 2) revealed that SA-OBS is a station-based dataset that represents the median values and is suitable for use as a reference. Recently, Else *et al.* (2017) have concluded that SA-OBS is currently the best available daily gridded observational dataset for SEA. Owing to the different horizontal resolutions (0.25–1.0°) of the observations, we decided to regrid to $0.25^\circ \times 0.25^\circ$ common resolution using the bilinear interpolation technique.

Model data

We examined 18 CMIP6 models (available since February 2020 when we started this work) from the CMIP6 database website (https://esgf-node.llnl.gov/search/cmip6), as given in Table 1. The new generation of CMIP6 models differs from the CMIP5 in having a new set of specifications for concentration, emission, and land-use scenarios as well as a new start year (CMIP6: 2015 and CMIP5: 2006) for future scenarios. In this phase, SSPs are combined with the RCPs of CMIP5. The SSPs are based on five narratives that describe different levels of socioeconomic development (Riahi *et al.* 2017): sustainable development (SSP1), middle-of-the-road development (SSP2), regional rivalry (SSP3), inequality (SSP4), and fossil fuel-driven development (SSP5). The detailed descriptions of the SSPs are available in O'Neill *et al.* (2016).

To evaluate model performance and to set a baseline for assessing future changes, we used the historical runs of the same common period (1998–2014) of observations to determine the present-day climate. To assess future climate change, we analyzed the data for three periods (near-future: 2015–2039, mid-future: 2040–2069, and far-future: 2070–2099) derived from the climate projections under

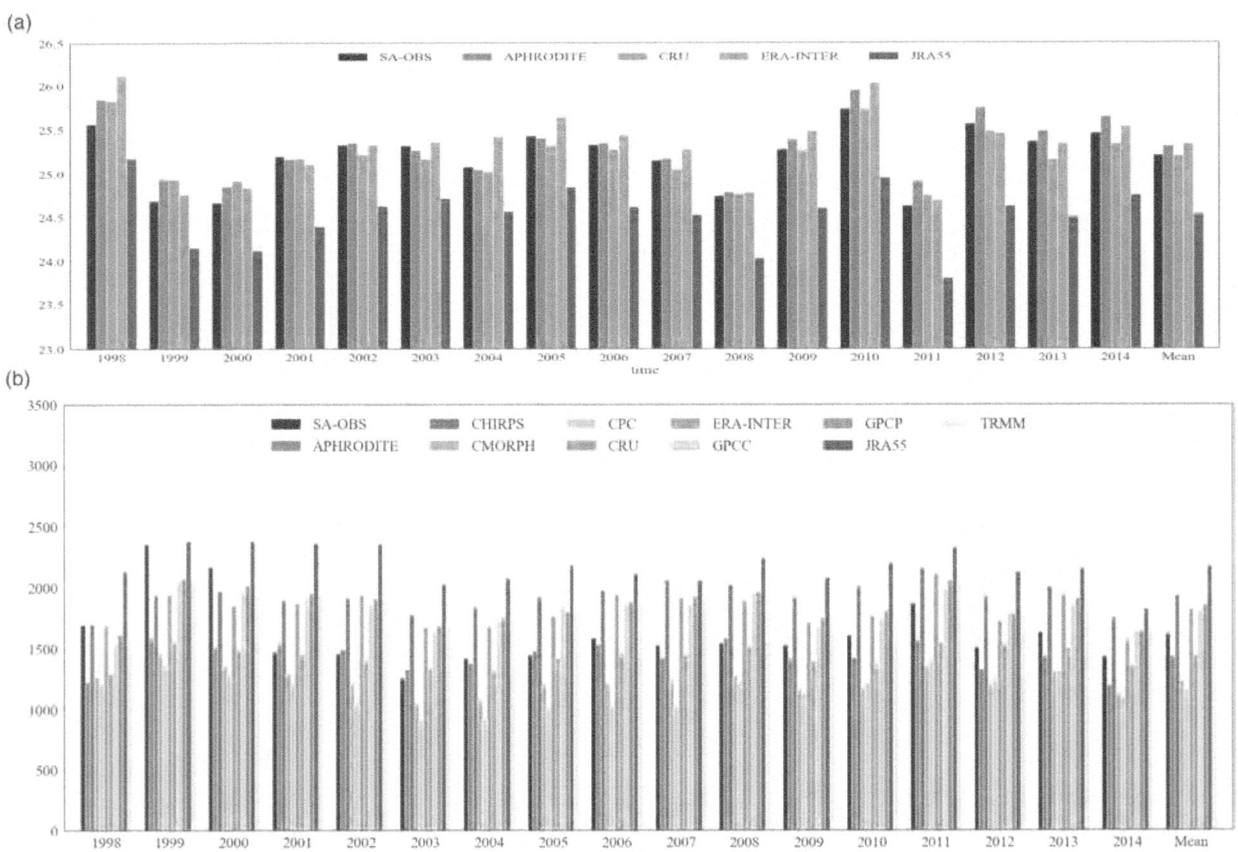

Figure 2 | Interannual variability of observations over CLMVT during 1998–2014. (a) Annual-mean temperature and (b) annual-mean precipitation.

Table 1 | List of CMIP6 models used in this study

GCM	Research Center	Resolution
ACCESS-CM2	Australian Community Climate and Earth System Simulator	1.88 × 1.25
ACCESS-ESM1-5	Australian Community Climate and Earth System Simulator	1.88 × 1.25
BCC-CSM2-MR	Beijing Climate Center, China Meteorological Administration, Beijing, China	1.12 × 1.11
CanESM5	Canadian Centre for Climate Modelling and Analysis, Environment and Climate Change Canada, Canada	2.81 × 2.77
CNRM-CM6-1	National Center for Meteorological Research, France	1.41 × 1.39
CNRM-ESM2-1	National Center for Meteorological Research, France	1.41 × 1.39
EC-Earth3	EC-Earth Consortium (EC-Earth)	0.70 × 0.70
FGOALS-g3	LASG, Institute of Atmospheric Physics, Chinese Academy of Sciences, Beijing 100029, China	0.70 × 0.70
GFDL-ESM4	NOAA Geophysical Fluid Dynamics Laboratory, USA	1.25 × 1.00
INM-CM4-8	Institute for Numerical Mathematics, Russia	2.00 × 1.50
INM-CM5-0	Institute for Numerical Mathematics, Russia	2.00 × 1.50
IPSL-CM6A-LR	The Institut Pierre Simon Laplace, France	2.50 × 1.27
MIROC6	JAMSTEC (Japan Agency for Marine-Earth Science and Technology, Japan), AORI (Atmosphere and Ocean Research Institute, The University of Tokyo), NIES (National Institute for Environmental Studies), and R-CCS (RIKEN Center for Computational Science), Japan	1.41 × 1.39
MIROC-ES2 L	JAMSTEC, AORI, NIES, and R-CCS, Japan	2.81 × 2.77
MPI-ESM1-2-LR	Max Planck Institute for Meteorology, Germany	1.88 × 1.85
MRI-ESM2-0	Meteorological Research Institute, Japan	1.12 × 1.11
NESM3	Nanjing University of Information Science and Technology, China	1.88 × 1.85
NorESM2-LM	NorESM Climate Modeling Consortium consisting of CICERO (Center for International Climate and Environmental Research), MET-Norway (Norwegian Meteorological Institute), NERSC (Nansen Environmental and Remote Sensing Center, Bergen), NILU (Norwegian Institute for Air Research), UiB (University of Bergen, Bergen), UiO (University of Oslo) and UNI (Uni Research), Norway	2.50 × 1.89

SSP-RCP scenarios, i.e., medium-emission (SSP2-4.5) and high-emission (SSP5-8.5). For a fair comparison, all models were regridded to $0.25^\circ \times 0.25^\circ$ resolution using a bilinear interpolation technique similar to the observation datasets.

Bias correction method

Bias correction is widely used in climate impact modeling. The aim is to adjust selected statistics (mean, variance, and/or quantile) in a climate model simulation to better match observed statistics during a reference period. Many bias correction methods have been employed in previous studies (Teutschbein & Seibert 2012; Supharatid 2016). In this study, we employed a 'variance scaling' method to correct the historical and projected temperature over SEA from CMIP6 models. This approach can guarantee that the adjusted model simulation in the reference period has the same mean and standard deviation as the observations.

The first step we used 'Delta change' approach to adjust the temperature at each grid point $(T_{i,j}(d))$ as given in Equation (1). Where $\bar{T}_{\text{obs},\,i,j}$ and $\bar{T}_{\text{model},\,i,j}$ are the climatological mean daily temperature of the observations and model in the reference period (1998–2014), respectively.

$$T_{i,j}(d) = T_{\text{model},\,i,j} + (\bar{T}_{\text{obs},\,i,j} - \bar{T}_{\text{model},\,i,j}) \tag{1}$$

The second step is to find the corresponding anomalies $(T'_{i,j})$ in the reference and projection periods by

$$T'_{\text{ref},\,i,j} = T_{\text{ref},\,i,j}(d) - \bar{T}_{\text{ref},\,i,j}(d) \tag{2}$$

$$T'_{\text{proj},\,i,j} = T_{\text{proj},\,i,j}(d) - \bar{T}_{\text{proj},\,i,j}(d) \tag{3}$$

Then, the anomalies from Equations (2) and (3) are scaled by the ratio of their observed ($\sigma_{obs,\,i,j}$) and reference ($\sigma_{ref,\,i,j}\,(d)$) standard deviations (SDs).

$$T^{//}_{ref,i,j} = T^{/}_{ref,i,j} \times \frac{\sigma(T_{obs,i,j})}{\sigma(T_{ref,i,j}\,(d))} \tag{4}$$

$$T^{//}_{proj,i,j} = T^{/}_{proj,i,j} \times \frac{\sigma(T_{obs,i,j})}{\sigma(T_{ref,i,j}\,(d))} \tag{5}$$

Finally, the corrected-adjust values during the reference and projection periods can be found:

$$T_{ref,i,j}(\text{cor},\,d) = T^{//}_{ref,i,j} + \bar{T}_{ref,\,i,j}\,(d) \tag{6}$$

$$T_{proj,i,j}(\text{cor},\,d) = T^{//}_{proj,i,j} + \bar{T}_{proj,\,i,j}\,(d) \tag{7}$$

In addition, we implement the 'Empirical quantile mapping (EQM)' method to remove the systematic precipitation biases in the GCMs simulation. The EQM, corrects the distribution shape of the monthly precipitation based on cumulative density functions (CDFs), is constructed for both the observed and the GCM simulation (1998–2014) for all months. For a given monthly precipitation, the cumulative density function of a control simulation is first matched with the CDF of the observations, generating a correction function depending on the quantile. Then, this correction function is used to unbias them from the climate simulation quantile by quantile. Finally, the monthly precipitation for the reference and future periods are obtained by Equations (8) and (9).

$$P_{ref,i,j}(\text{cor},\,d) = F^{-1}_{obs,i,j}\{F_{ref,i,j}(P_{ref,i,j})\} \tag{8}$$

$$P_{proj,i,j}(\text{cor},\,d) = F^{-1}_{obs,i,j}\{F_{proj,i,j}(P_{proj,i,j})\} \tag{9}$$

where F is the cumulative distribution function (CDFs) and F^{-1} is its inverse.

ANALYSIS OF THE PRESENT CLIMATE

Model simulation in the reference period

We first examined the CMIP6 models performance over CLMVT in the present climate (1998–2014) against SA-OBS. The spatial distributions of model bias (°C) in annual-mean daily temperature (T_{mean}) during 1998–2014 is displayed in Figure 3. The observed T_{mean} is computed from the mean of the maximum and minimum daily temperature. The CMIP6 models in qualitative behavior generally can capture climatological temperature distribution pattern over CLMVT. Most models commonly underestimate the temperature over Myanmar, Thailand (above lat 13°), and Cambodia by more than 3 °C except ACCESS-ESM1-5 and Miroc6. These cold biases are most pronounced in the simulation by CNRM-CM6-1, CNRM-ESM2-1, GFDL-ESM4, INM-CM4-8, INM-CM5-0, IPSL-CM6A-LR, MPI-ESM-2-LR, and NESM3. However, most models show warm biases over Laos and Vietnam, especially the simulation by ACCESS-ESM1-5. The MME shows better agreement (with bias within 2.5 °C) than any single model. A similar distribution pattern of bias is also evident in the MME. These may be linked to the uncertainties from low density of the observed stations (over northern CLMVT and central Vietnam) from SA-OBS (see Else *et al.* 2017).

The spatial distribution of bias (%) in annual-mean precipitation (P_{mean}) are displayed in Figure 4. In general, there are regional variations in P_{mean} among CMIP6 models and observations. Most CMIP6 models display wet biases along the coastline of Myanmar, Thailand, and southern Vietnam, especially BCC-CSM2-MR, CanESM5, INM-CM4-8, and INM-CM5-0. However, CNRM-CM6-1 and CNRM-ESM2-1 show significant dry biases over Thailand, Laos, Cambodia, and Vietnam. The dry biases are most pronounced over northern Laos and northern Vietnam except INM-CM4-8 and INM-CM5-0. The wet bias over northern CLMVT and central Vietnam may be linked to the uncertainties from low density of the observed rainfall stations. In general, most regions characterized by high topography give profound differences between the observation and the model.

Figure 5 displays the annual cycle and box plot of monthly mean temperature and precipitation of CMIP6 models during the reference period. The models of both temperature and

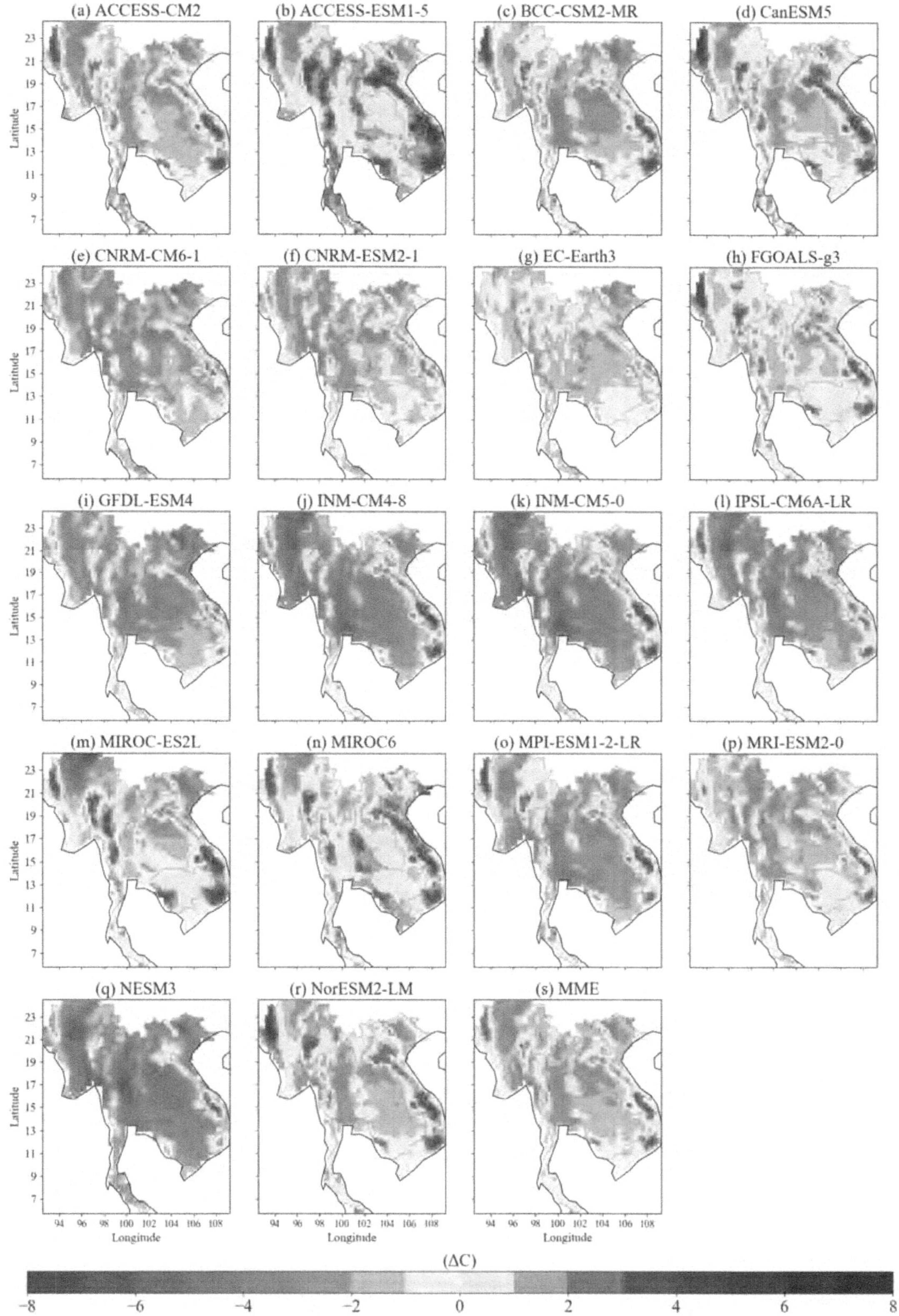

Figure 3 | Spatial distribution of model bias (°C) in T_{mean} during the reference period.

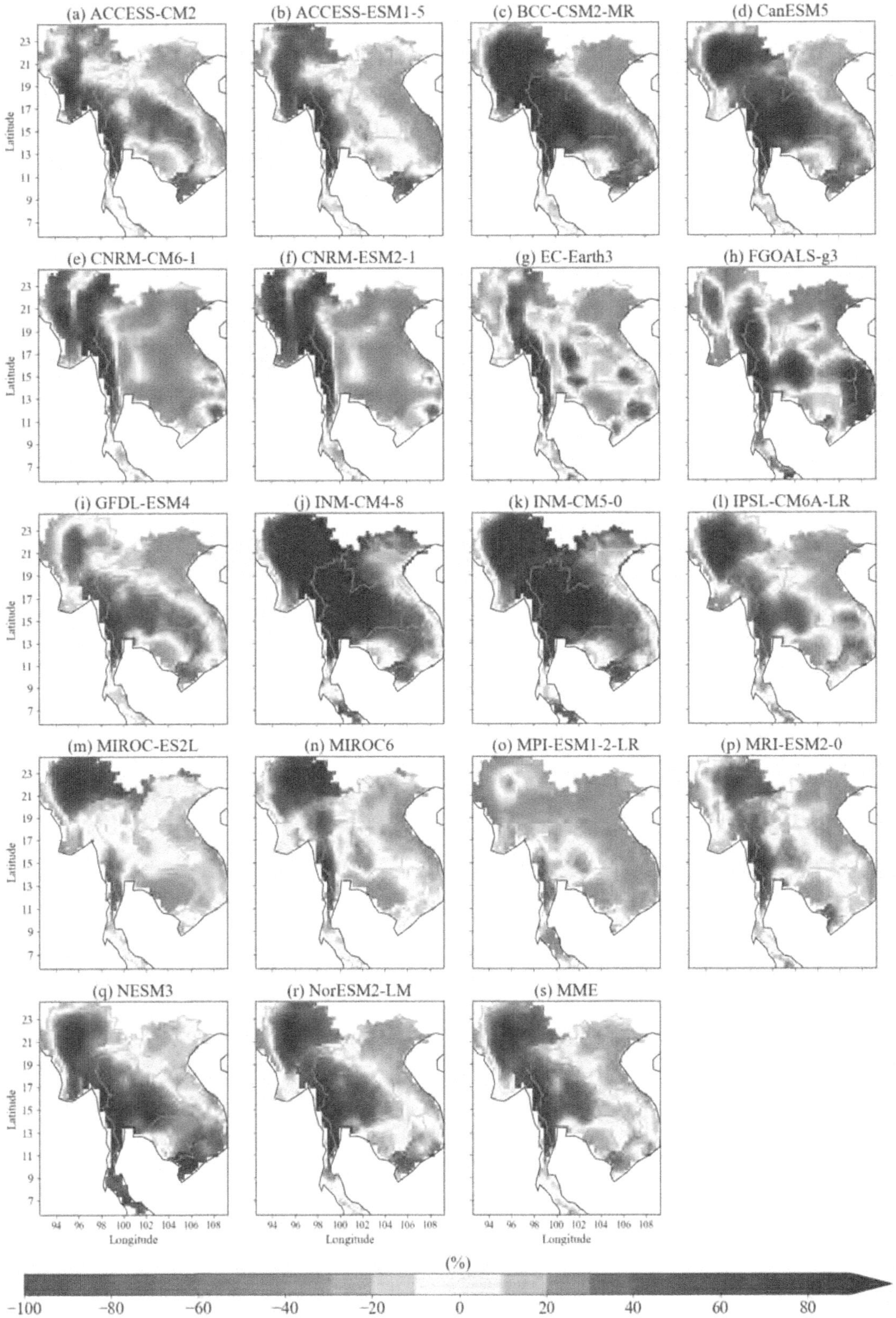

Figure 4 | Spatial distribution of model bias in P_{mean} (%) during the reference period.

precipitation show similarities on interannual variations with the observations. Most models show the highest mean temperature in April or May. ACCESS-ESM1-5 gives the largest warm bias (>30 °C) in May. Most models including the MME underestimate the mean temperature in months NDJF. For precipitation, most models display wet (Dry) biases in months JJAS (ONDJ). BCC-CSM2-MR gives the highest peak in June (>400 mm), while the observation peak is in August. The model median temperature is close to the observed values in months AMJ, but they underestimate in the rest season. The observed temperature is above the median and the 75th quantile of the box plots in months NDJF (Figure 5(a)). Larger spreads are found for the mean precipitation than the temperature during the peak season.

Evaluation of observation datasets and CMIP6 models in the present climate

From Figures 2–5, there are uncertainties in both T_{mean} and P_{mean} among the various observation datasets and CMIP6 models. Observation datasets give very different in precipitation, making it difficult to assess accurately, as previously reported by Herold *et al.* (2016). To assess the performance of individual CMIP6 models in reproducing the climatological spatial pattern, we performed bias correction by using 'Variance scaling' for T_{mean} and EQM for P_{mean} (Section 'Bias correction method'). We selected SA-OBS as a reference and performed a detailed intercomparison of all observation datasets and CMIP6 models (both T_{mean} and P_{mean}) in terms of correlation coefficient (R), center root-mean-square difference (RMSD), and SDs using the Taylor Diagram (Taylor 2001) during the reference period (see Figure 6). The center RMSD and the SD are normalized by their corresponding observations of SA-OBS. The observation datasets are represented by black symbols, and the CMIP6 models (both before and after bias corrections) are represented by other colors.

Most temperature observation datasets (Figure 6(a)) show high R (>0.9), with the lowest values for ERA-Interim and different magnitudes of RMSD and SD. The Era-Interim also displayed the highest RMSD. Most CMIP6 models before bias correction (red and blue symbols) give R in the

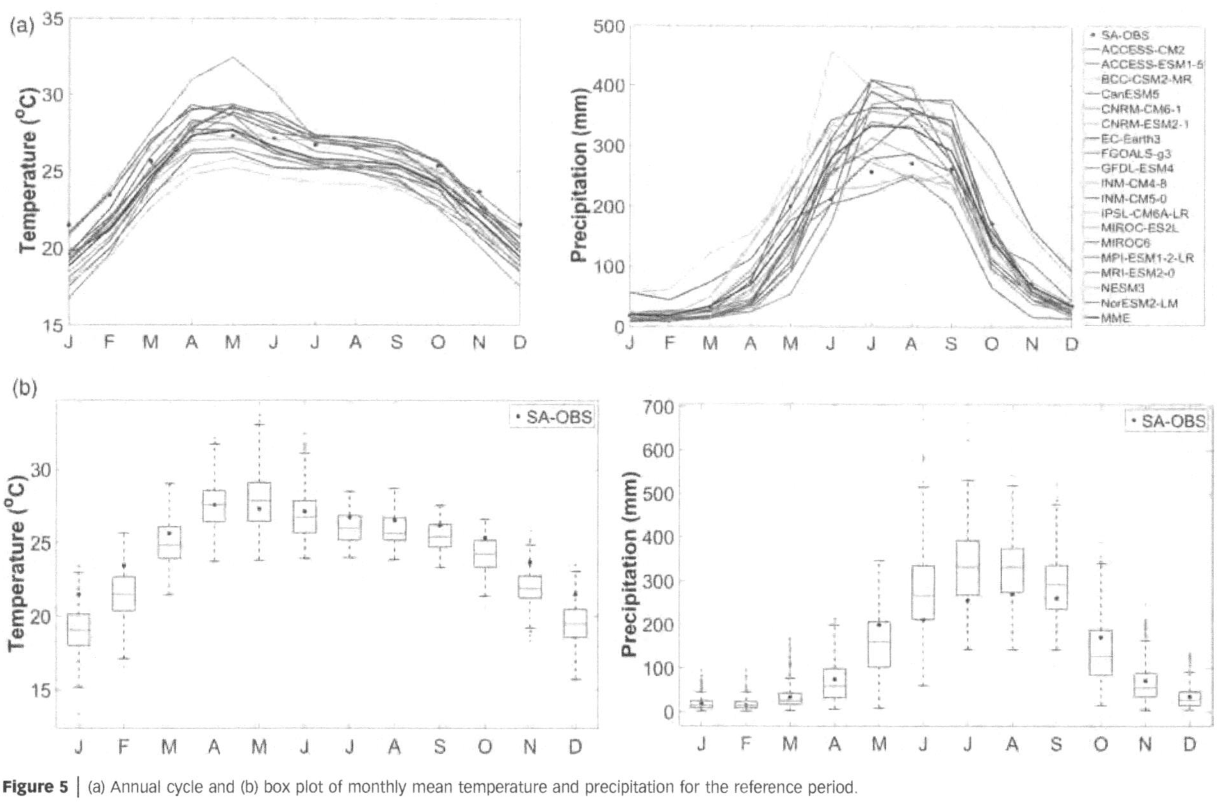

Figure 5 | (a) Annual cycle and (b) box plot of monthly mean temperature and precipitation for the reference period.

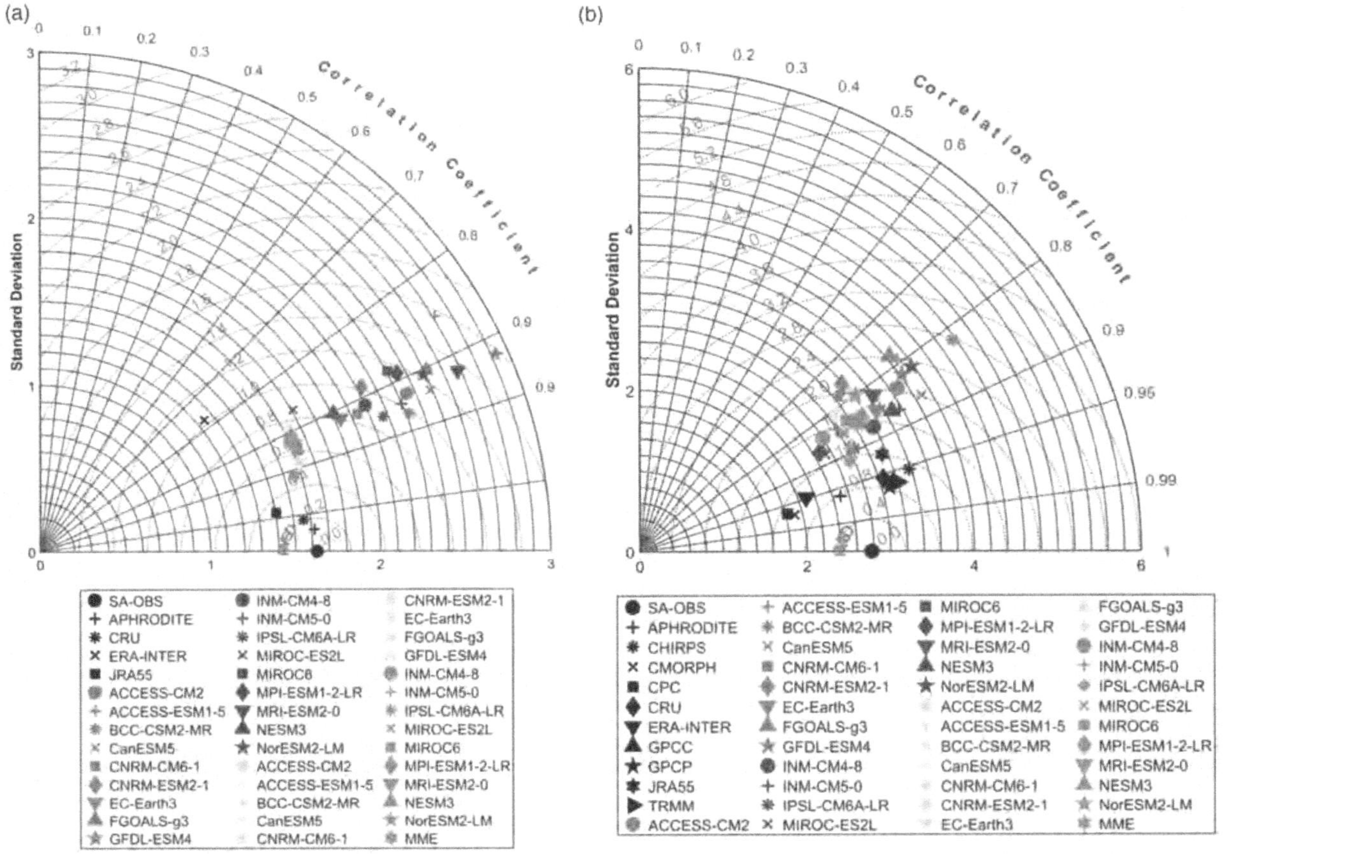

Figure 6 | Temporal changes in projected annual-mean precipitation. Taylor diagram of (a) T_{mean} and (b) P_{mean} of observations and CMIP6 models over CLMVT during the reference period. Please refer to the online version of this paper to see this figure in color: https://doi.org.10.2166/wcc.2021.015.

range 0.8–0.9, and show larger spread in SD than R and RMSD. After bias correction (green and pink symbols), R increases (>0.9) and RMSD and SD are significantly lower than their values before bias correction. Overall, the CMIP6 model has reliable capabilities in simulating the annual-mean temperature. For precipitation (Figure 6(b)), all observation datasets were found to give high R (>0.9) but lower than those of temperature. We observe distinctly different magnitudes of SD but similar RMSD values. The CMIP6 models before bias correction display more wide-spread in RMSD and SD with lower R compared with temperature. After bias correction, individual CMIP6 models do not show distinct improvement in R but significant improvement in RMSD and SD. The MME model yields the best results (highest R and lowest RMSD) among individual CMIP6 GCMs.

PROJECTED CHANGES IN ANNUAL-MEAN TEMPERATURE AND PRECIPITATION

The spatial distribution of the projected change in T_{mean} and P_{mean} under two different SSP-RCP scenarios is shown in Figure 7. The projected annual-mean temperature (Figure 7(a)) under two scenarios increases with time and shows little local difference in its pattern. However, it shows a larger increase over northern CLMVT, with the largest increase over northern Vietnam, Laos, and Myanmar. In the near-future, the annual-mean temperature averaged over CLMVT is projected to increase by 0.63 and 0.77 °C under SSP2-4.5 and SSP5-8.5, respectively. In the mid-future, it is projected to increase by 1.28 and 1.88 °C, respectively, and in the far-future, it is projected to increase by 1.80 and 3.36 °C, respectively.

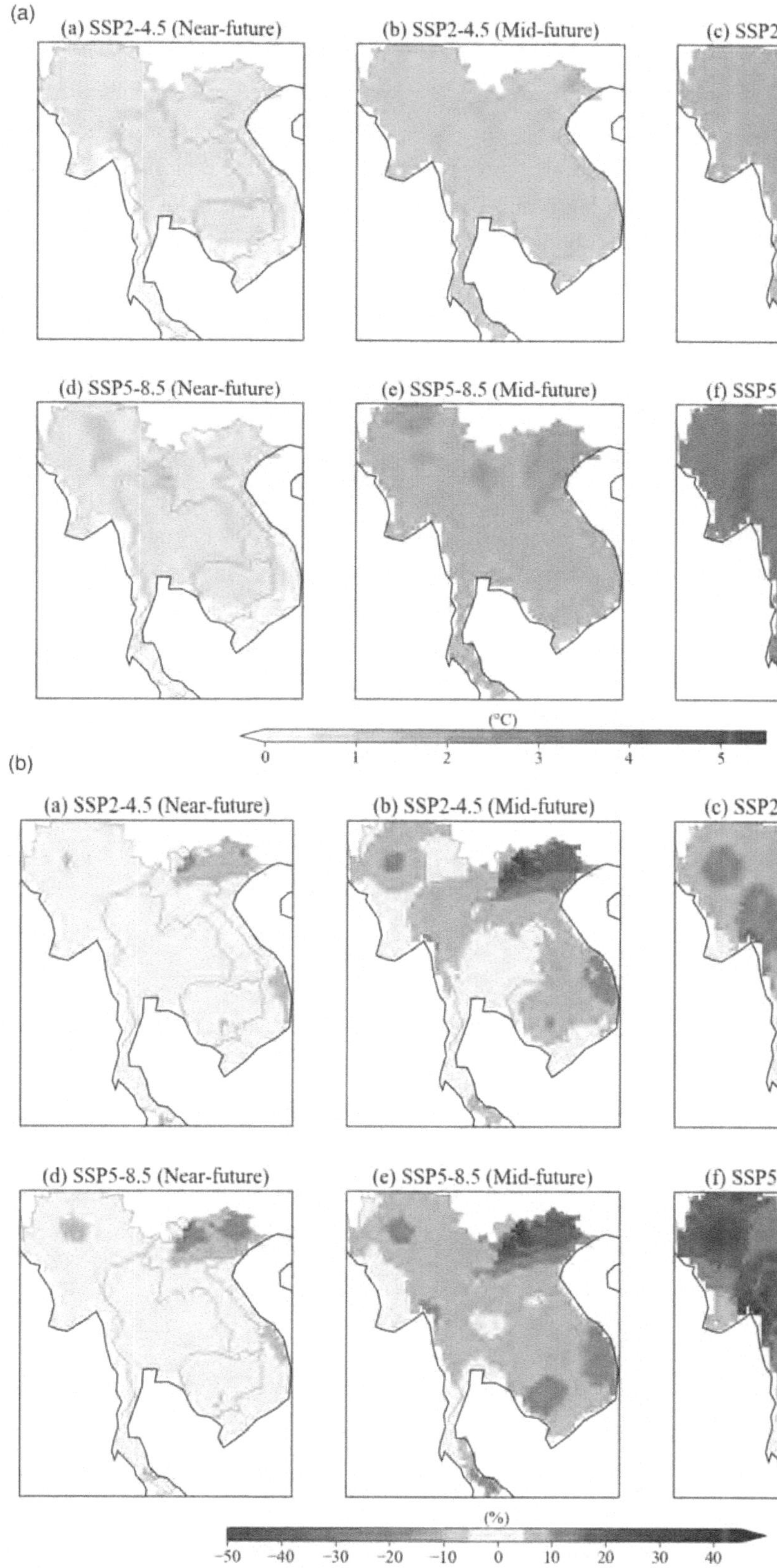

Figure 7 | Spatial distribution of future changes in T_{mean} and P_{mean}.

In contrast to T_{mean}, the changes in annual-mean precipitation show significant regional differences (Figure 7(b)). Overall, the projected annual-mean precipitation shows small reductions (<10%) over central Vietnam and Laos for the near-future and then increases in the far-future. In the near-future, the projected P_{mean} shows an increase of 2.36 and 2.88% under SSP2-4.5 and SSP5-8.5, respectively. For the mid-future period, it shows an increase of 6.76 and 8.09%, respectively, and in the far-future, it is projected to increase by 9.29 and 15.58%, respectively. Under the high-emission SSP5-8.5 scenario, most areas in CLMVT exhibit a significant increase in P_{mean} relative to the present climate. Overall, the temperature changes among CLMVT countries do not show any significant differences. However, there are significant precipitation changes with are minimum and maximum over Thailand and Vietnam, respectively. Mandapaka & Lo (2018) found similar results from the NEX-GDDP dataset under the RCP4.5 and RCP8.5 scenarios. Larger increases in precipitation are projected over northern and central Vietnam, northern Thailand, northern Myanmar, northern Laos, and Cambodia. These findings are generally consistent with those of Tangang *et al.* (2020), who used seven RCMs in the CORDEX-SEA domain and 11 driving CMIP5 GCMs and found robust increases in the 21st century over northern Vietnam, Cambodia, Laos, and northern Thailand.

Figures 8 and 9 display the temporal changes in projected T_{mean} and P_{mean} over CLMVT countries. The solid thick lines represent MME, and shaded regions denote the minimum and maximum changes from individual models. The change in T_{mean} was computed from the difference between the projected T_{mean} and the observed means during the reference period (1998–2014). The change in P_{mean} was computed using the normalized index between the projected P_{mean} and the mean observation during the reference period (1998–2014). Normalization was performed for each model at every grid to minimize the effect of large spatial variations on regional averages. The projected changes in T_{mean} over CLMVT countries show larger magnitudes and trends under SSP5-8.5 rather than SSP2-4.5 scenarios. The projected changes in 5-year running mean of temperature and precipitation over each country are summarized in Table 2. Under SSP2-4.5 (SSP5-8.5), the temperature over Cambodia, Laos, Myanmar, Vietnam, and Thailand is projected to increase by 1.21, 1.31, 1.19, 1.23, and 1.27 °C, respectively (1.58, 1.71, 1.70, 1.59, and 1.66 °C, respectively) in 2050 and 1.88, 2.04, 1.88, 2.03,

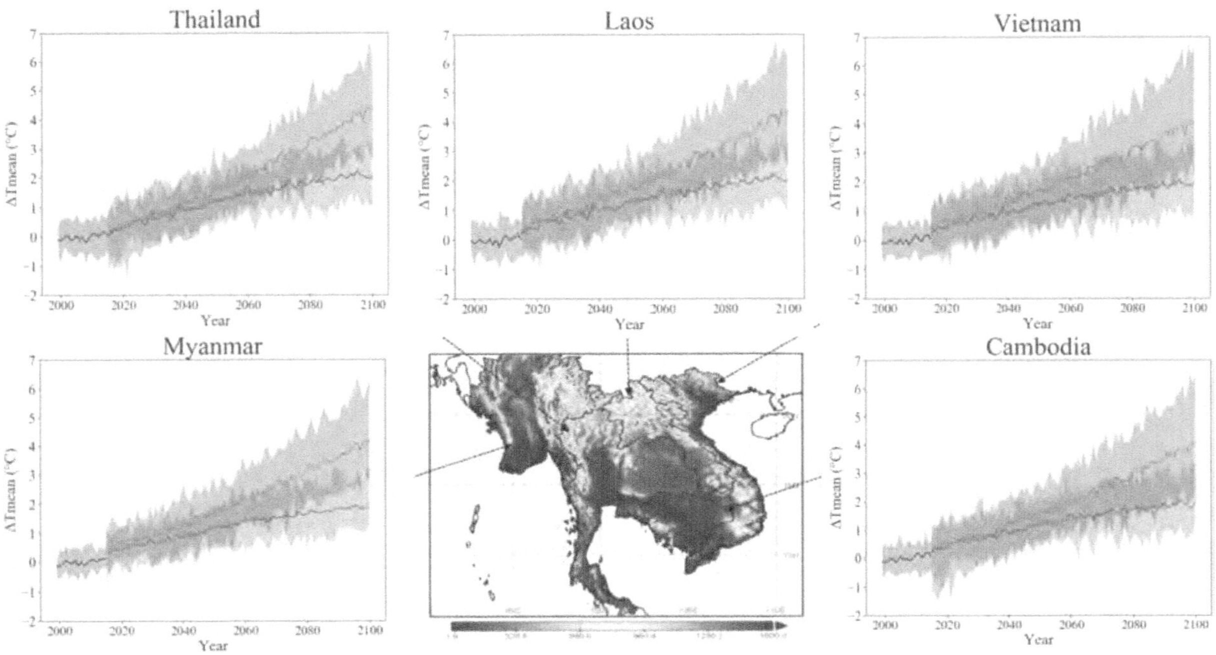

Figure 8 | Temporal changes in projected annual-mean temperature.

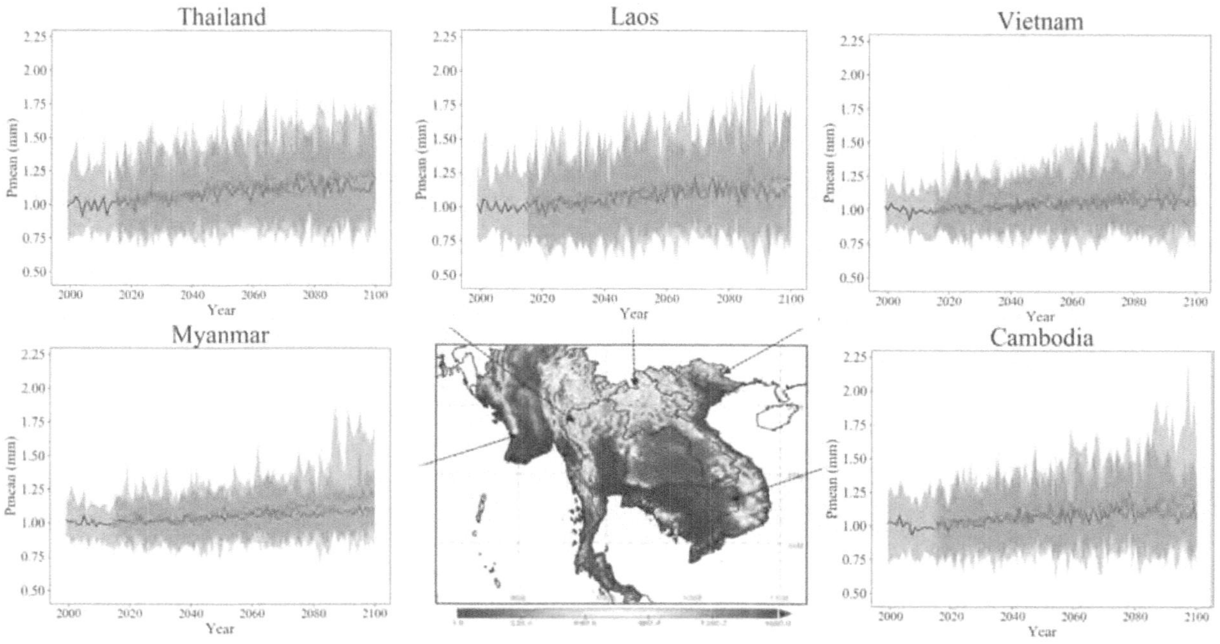

Figure 9 | Time series of projected changes in T_{mean}.

and 1.90 °C, respectively (3.89, 4.22, 4.09, 4.25, and 3.96 °C, respectively) in 2100. The warming trends start to slow down with decreasing trends over all countries after 2050 under SSP2-4.5. This reveals the effectiveness of anticipated climate mitigation and adaptation strategies under the combination of socioeconomic development and radiative forcing projection (SSP2-4.5: peak in 2040 and then decrease in magnitude).

The increasing trend in annual-mean precipitation is also seen for the SPS5-8.5 scenario, but the curves flatten under the SSP2-4.5 scenario as the climate stabilizes later in the century. Under SSP2-4.5 (SSP5-8.5), the precipitation over Cambodia, Laos, Myanmar, Vietnam, and Thailand is projected to increase by 8.12, 8.35, 5.03, 11.17, and 4.17%, respectively (10.52, 10.41, 7.03, 12.90, and 7.91%, respectively) in 2050 and 8.76, 12.69, 9.54, 13.47, and 7.03%, respectively (20.47, 21.10, 21.57, 22.12, and 15.17%, respectively) in 2100. Major differences are apparent between the lowest and the highest values of the individual models due to their projection uncertainties. Various sources of uncertainties (Almazroui *et al.* 2020) may arise from (1) internal variability of the climate system, (2) inter-model variability, and (3) variability among different emission scenarios. Uncertainties from (2) and (3) in this study are found to increase over time by the inter-modal spreads under both scenarios.

PROJECTED CHANGES IN SEASONAL MONSOON PRECIPITATION

Because CLMVT is located in the Asian-monsoon-dominated region, a large proportion of annual rainfall and its variability in the present and future climates comes from the summer monsoon, similar to South Asia (Kripalani *et al.* 2007). The monsoons and their variability critically influence people's livelihoods and socioeconomic status. Recent monsoon floods and droughts in CLMVT have resulted in the loss of lives, property damage, and decreased agricultural productivity. In this study, our focus is on the magnitude of change in precipitation that will affect the whole region. Therefore, we investigated the strength of the monsoon (only in terms of precipitation) by the Southwest and Northeast Monsoon rainfall (SWMR and NEMR).

Figure 10 displays the spatial-averaged mean seasonal monsoon precipitation over CLMVT for the MME model. The 1st, 2nd, 3rd, and 4th columns represent values for the historical, near-future, mid-future, and far-future periods,

Table 2 | Changes in temperature (°C) and precipitation (%) over CLMVT

Country	Variable	Scenario	2050	2100
Cambodia	Temperature	SSP2-4.5	1.21 (0.39–1.52)	1.88 (0.79–3.42)
		SSP5-8.5	1.58 (0.56–2.58)	3.89 (1.95–6.51)
	Precipitation	SSP2-4.5	8.12 (−14.77 to 34.24)	8.76 (−33.10 to 50.30)
		SSP5-8.5	10.52 (−22.52 to 46.90)	20.47 (−23.44 to 70.85)
Laos	Temperature	SSP2-4.5	1.31 (0.49–2.49)	2.04 (1.33–3.50)
		SSP5-8.5	1.71 (0.37–2.90)	4.22 (3.18–6.44)
	Precipitation	SSP2-4.5	8.35 (−12.48 to 50.36)	12.69 (−35.33 to 72.77)
		SSP5-8.5	10.41 (−28.87 to 56.97)	21.10 (−20.00 to 74.61)
Myanmar	Temperature	SSP2-4.5	1.19 (0.37–1.93)	1.88 (0.96–2.86)
		SSP5-8.5	1.70 (0.69–2.93)	4.09 (3.13–5.63)
	Precipitation	SSP2-4.5	5.03 (−13.25 to 27.89)	9.54 (−19.03 to 37.57)
		SSP5-8.5	7.03 (−23.46 to 28.94)	21.57 (−5.22 to 62.21)
Vietnam	Temperature	SSP2-4.5	1.23 (0.46–2.35)	2.03 (1.22–3.35)
		SSP5-8.5	1.59 (0.53–2.53)	4.25 (3.09–6.22)
	Precipitation	SSP2-4.5	4.17 (−15.50 to 25.57)	7.03 (−19.68 to 33.95)
		SSP5-8.5	7.91 (−16.38 to 43.26)	15.17 (−25.25 to 75.26)
Thailand	Temperature	SSP2-4.5	1.27 (0.35–1.96)	1.90 (0.90–3.62)
		SSP5-8.5	1.66 (0.15–2.95)	3.96 (2.22–6.45)
	Precipitation	SSP2-4.5	11.17 (−12.40 to 38.05)	13.47 (−27.43 to 75.18)
		SSP5-8.5	12.90 (−23.35 to 85.87)	22.12 (−24.94 to 72.87)

( ) denotes the minimum and maximum values of individual models.

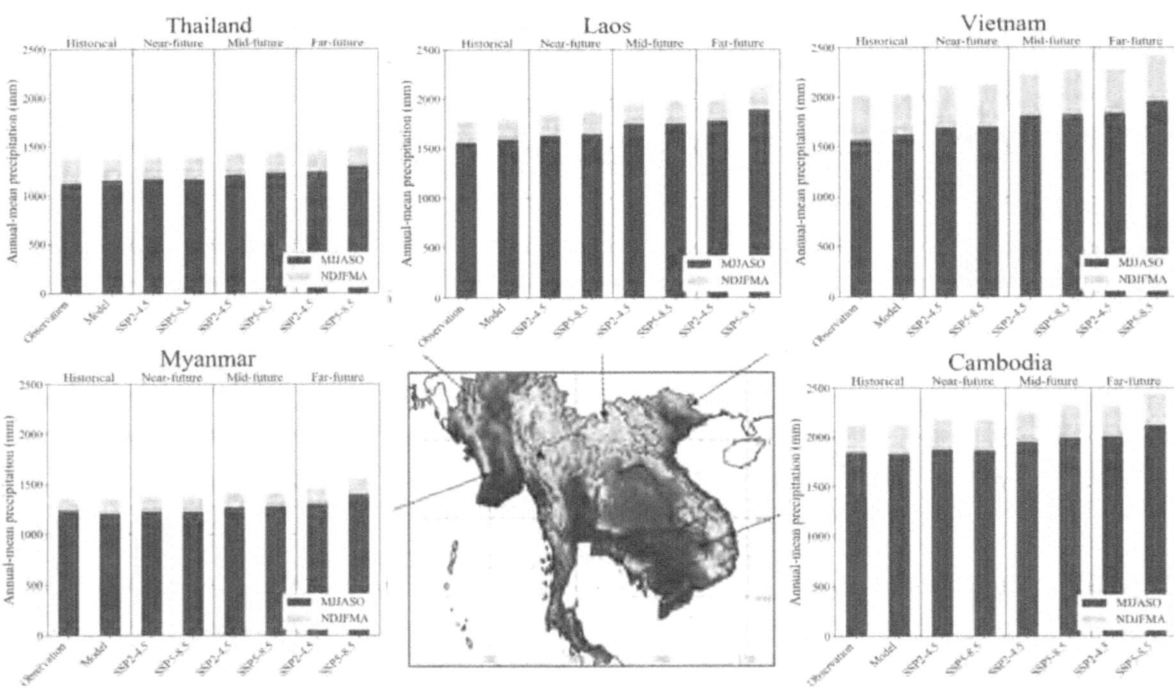

Figure 10 | Mean seasonal monsoon precipitation.

respectively. The projection in each column is displayed under both SSP2-4.5 and SSP5-8.5 scenarios. In general, CLMVT is significantly dominated by SWMR, similar to the findings of Chang *et al.* (2005). There is a robust increase in SWMR along the timeline of the 21st century, with larger increases found under SSP5-8.5 than SSP2-4.5. However,

NEMR does not display significant increases from the historical period. For SSP2-4.5, the SWMR (NEMR) increases by 3.41, 8.44, and 10.89% (−2.58, −0.43, and 2.32%) for the near-, mid-, and far-future, respectively. For SSP5-8.5, the SWMR (NEMR) increases by 3.44, 9.53, and 17.59% (0.78, 2.84, and 5.45%) for the near-, mid-, and far-future, respectively. Both SSP2-4.5 and SSP5-8.5 scenarios display larger increases in the boreal summer. Overall, these results correspond with Wang *et al.* (2020), who examined 15 CMIP6 models under the SSP2-4.5 scenario and revealed that during the boreal summer (JJAS), the projected monsoon precipitation significantly increases over mainland SEA.

Table 3 displays changes in mean seasonal monsoon precipitation for each country. The SWMR (NEMR) under SSP2-4.5 over Cambodia increases by 2.57, 6.79, and 9.63% (6.36, 5.82, and 8.44%) for the near-, mid-, and far-future, respectively. The SWMR (NEMR) under SSP2-4.5 over Laos increases by 2.42, 9.97, and 11.94% (−1.45, 0.98, and 5.12%) for the near-, mid-, and far-future, respectively. The SWMR (NEMR) for SSP2-4.5 under Myanmar increases by 2.06, 5.47, and 8.45% (−2.13, 2.03, and 3.84%) for the near-, mid-, and far-future, respectively. The SWMR (NEMR) under SSP2-4.5 over Vietnam increases by 4.43, 11.75, and 13.48% (1.49, 4.88, and 8.62%) for the near-, mid-, and far-future, respectively. The SWMR (NEMR) under SSP2-4.5 over Thailand increases by 1.15, 4.92, and 7.77% (0.59, 1.20, and 3.22%) for the near-, mid-, and far-future, respectively.

In addition, the SWMR (NEMR) under SSP5-8.5 over Cambodia increases significantly by 2.05, 9.09, and 16.10% (5.61, 12.89, and 14.33%) for the near-, mid-, and far-future, respectively. The SWMR (NEMR) under SSP5-8.5 over Laos increases significantly by 3.47, 10.14, and 18.82% (4.03, 4.85, and 7.06%) for the near-, mid-, and far-future, respectively. The SWMR (NEMR) under SSP5-8.5 over Myanmar increases significantly by 1.71, 6.31, and 15.90% (2.28, 1.97, and 9.46%) for the near-, mid-, and far-future, respectively. The SWMR (NEMR) under SSP5-8.5 over Vietnam increases significantly by 5.02, 12.52, and 20.95% (5.13, 8.33, and 11.38%) for the near-, mid-, and far-future, respectively. The SWMR (NEMR) under SSP5-8.5 over Thailand increases significantly by 1.10, 6.47, and 12.40% (4.24, 5.01, and 4.69%) for the near-, mid-, and far-future, respectively.

Table 3 | Mean seasonal monsoon precipitation over CLMVT

	Ref. period				Near-future				Mid-future				Far-future			
	Observation		Model		SSP2-4.5		SSP5-8.5		SSP2-4.5		SSP5-8.5		SSP2-4.5		SSP5-8.5	
Country	MJJASO	NDJFMA	MJJASO	NDJFMA	MJJASO	NDJFMA	MJJASO	NDJFMA	MJJASO	NDJFMA	MJJASO	NDJFMA	MJJASO	NDJFMA	MJJASO	NDJFMA
Cambodia	1,847	275	1,829	294	1,877	313	1,867	310	1,954	311	1,996	332	2,006	319	2,124	336
Laos	1,566	236	1,594	218	1,633	215	1,650	227	1,753	220	1,756	229	1,785	229	1,894	234
Myanmar	1,240	114	1,209	147	1,234	144	1,230	151	1,275	150	1,285	150	1,308	153	1,401	161
Vietnam	1,565	466	1,626	410	1,698	416	1,708	431	1,817	430	1,830	444	1,845	445	1,967	456
Thailand	1,128	243	1,154	217	1,171	219	1,166	227	1,210	220	1,228	228	1,243	224	1,297	228

We examine the future projection of extreme seasonal monsoon precipitation by the Gamma pdf plot in Figure 11. Projection of the NEMR does not fit well the Gamma pdf as compared to the southwest Monsoon rainfall (SWMR). The pdf curve during the wet season becomes flatter, with a peak reduction, increasing spread, and a mean value shift to the right relative to the historical curve. Its median does not show significant changes in magnitude and frequency for the near-future, mid-future, and far-future periods under SSP2-4.5 scenario. On the contrary, projection of the SWMR shows distinct increases (decreases) in magnitude (frequency). Therefore, we expect more (less) extreme rainfall in the wet season (dry season) from the near-future to the far-future periods.

CONCLUSIONS AND DISCUSSION

The present study analyzes the changes in mean temperature and precipitation using the latest Couple Model Intercomparison Project phase 6 (CMIP6) model simulation over Cambodia, Laos, Myanmar, Vietnam, and Thailand (CLMVT). The CMIP6 is based on community scenarios known as SSPs, which differ from CMIP3 and CMIP5 in a different start year of the future scenarios as well as a new set of specifications for emission and land-use scenarios. Here, 18 CMIP6 models were employed to assess future mean climate change for three periods (near-future: 2015–2039, mid-future: 2040–2069, and far-future: 2070–2099) derived from the climate projections under two SSP-RCP scenarios (SSP2-4.5 and SSP5-8.5).

Intercomparison among several observation datasets over southeast Asia (SA-OBS, APHRODITE, CPC-UNI, CRU, GPCP1DD, TRMM, ERA-Interim, JRA55, GPCC, CHIRPS, CMORPH, GsMap, and PERSIANN) reveal that SA-OBS is suitable for use as a reference due to its station-based dataset, which represents the median values (Figure 2). SA-OBS is the currently best available daily gridded observational dataset for Southeast Asia (Else *et al.* 2017). The model bias correction was employed by using the 'Variance scaling' method for the mean temperature and the 'EQM' method for the mean precipitation (Section 'Bias correction method'). The major results can be summarized below.

1. The CMIP6 models in qualitative behavior generally can capture climatological temperature distribution pattern over CLMVT (Figure 3). Most models commonly show cold biases (over Myanmar, Thailand (above lat 13°), and Cambodia) and warm biases (over Laos and Vietnam). Most CMIP6 models display wet biases (along the coastline of Myanmar, Thailand, and southern Vietnam) and dry biases (over northern Laos and northern Vietnam). The wet bias over northern CLMVT and central Vietnam may be linked to the uncertainties from low density of the observed rainfall stations. In general, most regions characterized by high topography give profound differences between the observation and the model. These may be linked to the uncertainties from low density of the observed stations (over northern

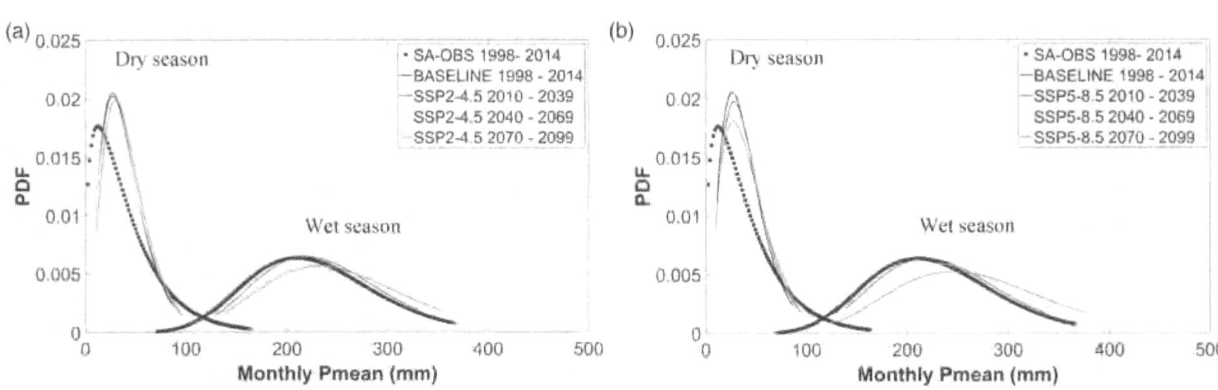

Figure 11 | Gamma Pdf plot of the seasonal monsoon precipitation. (a) SSP2-4.5 (b) SSP5-8.5.

CLMVT and central Vietnam) from SA-OBS (Else *et al.* 2017). The MME show better agreement than any single model.

2. The models of both temperature and precipitation show consistency monthly variability with the observations. Most models show the highest mean temperature in April or May and display wet (Dry) biases in months JJAS (ONDJ). Most CMIP6 models display the spatial pattern of climatological annual-mean temperature over CMLVT well (*R* typically >0.9 and RMSD and SD are significantly decreased). For the annual-mean precipitation, the CMIP6 models display more widespread in RMSD and SD with lower *R* compared to temperature. Individual CMIP6 model does not show distinct improvement in *R* but show significant improvements in RMSD and SD. The MME model yields the best results (highest *R* and lowest RMSD) among individual CMIP6 GCMs (Figure 6).
3. The projected changes in mean annual temperature over CLMVT countries show a continuous increase with larger magnitudes and trends under SSP5-8.5 rather than SSP2-4.5 scenarios. Overall, the temperature changes do not show any significant differences. However, there are significant precipitation changes with are minimum and maximum over Thailand and Vietnam, respectively (Figure 7). Major differences are apparent between the lowest and the highest values of the individual models due to their projection uncertainties. Uncertainties from inter-model variability and variability among different emission scenarios (Almazroui *et al.* 2020) are found to increase over time by the inter-modal spreads under both scenarios. The effectiveness of anticipated climate change mitigation and adaptation strategies under the combination of socioeconomic development and radiative forcing projection (SSP2-4.5: peak in 2040 and then decrease in magnitude) results in slowing down the warming trends overall countries after 2050. These findings are generally consistent with Mandapaka & Lo (2018) and Tangang *et al.* (2020).
4. CLMVT countries are significantly dominated by the SWMR. There is a robust increase in SWMR along the timeline of the 21st century, with larger increases are found under SSP5-8.5 than SSP2-4.5 (Figure 10). The NEMR displays smaller increases from the historical period compared to SWMR. In general, both SSP2-4.5 and SSP5-8.5 scenarios display larger increases during the boreal summer which generally agree with the work of Wang *et al.* (2020).
5. Different projected increase in the seasonal Monsoon rainfall are found over each country in CLMVT (Table 3). Overall, the largest (smallest) increases in SWMR and NEMR are found over Vietnam (Thailand) and Cambodia (Thailand), respectively, for the far-future period under SSP5-8.5 scenario. The projection of the NEMR does not fit well the Gamma pdf as compared to the SWMR (Figure 11). The pdf curve during the southwest Monsoon season becomes flatter, with a peak reduction, increasing spread, and a mean value shift to the right relative to the historical curve. Projection of the SWMR shows distinct increases (decreases) in magnitude (frequency). Therefore, we expect more (less) extreme rainfall in the wet season (dry season) from the near-future to the far-future periods.

Due to the temporarily limited number of available CMIP6 models which will be gradually released by the Scenario Model Intercomparison Project (Scenario MIP) (O'Neill *et al.* 2016), evaluation of more CMIP6 models still needs to be carried out in the future. However, based on the findings of the 18 CMIP6 models in this paper, significant climate projection in both precipitation and temperature have been improved and are benefitted to CLMVT countries. Therefore, the corresponding policymakers need to prepare for the appropriate level of adaptation measures in response to the projected changing climate.

ACKNOWLEDGEMENTS

We would like to thank the anonymous reviewers for their valuable comments. This research was conducted as part of the ASEAN-India Project Proposal on 'Climate Change Projection and Assessment of Impacts; Modelling and Capacity Building Programme' under the ASEAN-India Green Fund (Ref: ASCC/CSC/ENV/AWGCC_IISC/Vol.1/013). The authors also thank the Earth System Grid Federation (ESGF) for archiving and providing access to the CMIP6 dataset, including all observation dataset providers used in the research.

DATA AVAILABILITY STATEMENT

All relevant data are available from an online repository or repositories (https://esgf-node.llnl.gov/search/cmip6/).

REFERENCES

Alexander, L. V. & Arblaster, J. M. 2017 Historical and projected trends in temperature and precipitation extremes in Australia in observations and CMIP5. *Weather Clim. Extrem.* **15**, 34–56.

Ali, H. & Mishra, V. 2017 Contrasting response of rainfall extremes to increase in surface air and dewpoint temperatures at urban locations in India. *Sci. Rep.* 1–15. https://doi.org/10.1038/s41598-017-01306-1.

Almazroui, M., Saeed, S., Saeed, F., Islam, M. N. & Ismail, M. 2020 Projections of precipitation and temperature over the South Asian Countries in CMIP6. *Earth Syst. Environ.* https://doi.org/10.1007/s41748-020-00157-7.

Almazroui, M., Islam, M. N., Saeed, F., Ismail, M., Ehsan, M. A., Diallo, I., O'Brien, E., Ashfaq, M., Martínez-Castro, D., Cavazos, T., Cerezo-Mota, R., Tippett, M. K., Gutowski Jr, W. J., Alfaro, E. J., Hidalgo, H. G., Vichot-Llano, A., Campbell, J. D., Kamil, S., Rashid, I. U., Sylla, M. B., Stephenson, T., Taylor, M. & Barlow, M. 2021 Projected changes in temperature and precipitation over the United States, Central America, and the Caribbean in CMIP6 GCMs. *Earth Syst. Environ.* **5**, 1–24. https://doi.org/10.1007/s41748-021-00199-5.

Bukovsky, M. S., Carrillo, C. M., Gochis, D. J., Hammerling, D. M., McCrary, R. R. & Mearns, L. O. 2015 Toward assessing NARCCAP regional climate model credibility for the North American monsoon: future climate simulations. *J. Clim.* **28**, 6707–6728. https://doi.org/10.1175/JCLI-D-14-00695.1.

Caesar, J., Alexander, L. V., Trewin, B., Tse-ring, K., Sorany, L., Vuniyayawa, V., Keosavang, N., Shimana, A., Htay, M. M., Karmacharya, J., Jayasinghearachchi, D. A., Sakkamart, J., Soares, E., Hung, L. T., Thuong, L. T., Hue, C. T., Dung, N. T. T., Hung, P. V., Cuong, H. D., Cuong, N. M. & Sirabaha, S. 2011 Changes in temperature and precipitation extremes over the Indo-Pacific region from 1971 to 2005. *Int. J. Climatol.* **31**, 791–801. doi:10.1002/joc.2118.

Chadwick, R., Good, P., Martin, G. & Rowell, D. P. 2016 Large rainfall changes consistently projected over substantial areas of tropical land. *Nat. Clim. Change* **6**, 177–181. https://doi.org/10.1038/nclimate2805.

Chang, C.-P., Wang, Z., McBride, J. & Liu, C.-H. 2005 Annual cycle of Southeast Asia – maritime continent rainfall and the asymmetric monsoon transition. *J. Climate* **18**, 287–301. https://doi.org/10.1175/jcli-3257.1.

Cook, B. I., Mankin, J. S., Marvel, K., Williams, A. P., Smerdon, J. E. & Anchukaitis, K. J. 2020 Twenty-first century drought projections in the CMIP6 forcing scenarios. *Earth's Future* **8**, e2019EF001461. https://doi.org/10.1029/2019EF001461.

Di Luca, A., Pitman, A. J. & Elía, R. 2020 Decomposing temperature extremes errors in CMIP5 and CMIP6 models. *Geophys. Res. Lett.* **47**.

Easterling, D. R., Kunkel, K. E., Wehner, M. F. & Sun, L. 2016 Detection and attribution of climate extremes in the observed record. *Weather Clim. Extrem.* **11**, 17–27. https://doi.org/10.1016/j.wace.2016.01.001.

Eyring, V., Bony, S., Meehl, G. A., Senior, C., Stevens, B., Stouer, R. J. & Taylor, K. E. 2015 Overview of the coupled model intercomparison project phase 6 (CMIP6) experimental design and organization. *Geosci. Model Dev. Discuss.* https://doi.org/10.5194/gmdd-8-10539-2015.

Eyring, V., Bony, S., Meehl, G. A., Senior, C. A., Stevens, B., Stouffer, R. J. & Taylor, K. E. 2016 Overview of the coupled model intercomparison project phase 6 (CMIP6) experimental design and organization. *Geosci. Model Dev.* **9**, 1937–1958. https://doi.org/10.5194/gmd-9-1937-2016.

Fischer, E. M. & Knutti, R. 2016 Observed heavy precipitation increase confirms theory and early models. *Nat. Clim. Change* **6**, 986. https://doi.org/10.1038/nclimate3110.

Ge, F., Zhu, S., Peng, T., Zhao, Y., Sielmann, F., Fraedrich, K., Zhi, X., Liu, X., Tang, W. & Ji, L. 2019 Risks of precipitation extremes over Southeast Asia: does 1.5°C or 2°C global warming make a difference? *Environ. Res. Lett.* **14**, 044015. https://doi.org/10.1088/1748-9326/aaff7e.

Ghosh, S., Das, D., Kao, S. C. & Ganguly, A. R. 2012 Lack of uniform trends but increasing spatial variability in observed Indian rainfall extremes. *Nat. Clim. Change* **2**, 86–91. https://doi.org/10.1038/nclimate1327.

Giorgi, F., Coppola, E. & Raffaele, F. 2018 Threatening levels of cumulative stress due to hydroclimatic extremes in the 21st century. *npj Clim. Atmos. Sci.* **1**, 18. https://doi.org/10.1038/s4161 2-018-0028-6.

Grose, M. R., Narsey, S., Delage, F. P., Dowdy, A. J., Bador, M., Boschat, G., Chung, C., Kajtar, J. B., Rauniyar, S., Freund, M. B., Lye, K., Rashid, H., Zhang, X., Wales, S., Trenham, C., Holbrook, N. J., Cowan, T., Alexander, L., Arblaster, J. M. & Power, S. 2020 Insights from CMIP6 for Australia's future climate. *Earth's Future* **8**, e2019EF001469. https://doi.org/10.1029/2019EF001469.

Herold, N., Alexander, L. V., Donat, M. G., Contractor, S. & Becker, A. 2016 How much does it rain over land? *Geophys. Res. Lett.* **43** (1), 341–348. doi:10.1002/2015GL066615.

Hidalgo, H. & Alfaro, E. 2015 Skill of CMIP5 climate models in reproducing 20th century basic climate features in Central America. *Int. J. Climatol.* **35**, 3397–3421. https://doi.org/10.1002/joc.4216.

IPCC 2013 Climate change 2013: the physical science basis. In: *Contribution of Working Group I to the Fifth Assessment Report of the Intergovernmental Panel on Climate Change* (T. F. Stocker, D. Qin, G.-K. Plattner, M. Tignor, S. K. Allen, J. Boschung, A. Nauels, Y. Xia, V. Bex & P. M. Midgley, eds). Cambridge University Press, Cambridge.

IPCC 2014 Summary for policymakers. In: *Climate Change 2014: Impacts, Adaptation, and Vulnerability. Part A: Global and Sectoral Aspects. Contribution of Working Group II to the Fifth Assessment Report of the Intergovernmental Panel on Climate Change* (C. B. Field, V. R. Barros, D. J. Dokken, K. J. Mach, M. D. Mastrandrea, T. E. Bilir, M. Chatterjee, K. L. Ebi, Y. O. Estrada, R. C. Genova, B. Girma, E. S. Kissel, A. N. Levy, S. MacCracken, P. R. Mastrandrea & L. L. White, eds). Cambridge University Press, Cambridge and New York, NY, pp. 1–32.

IPCC SR15 2018 Summary for policymakers. In: *Global Warming of 1.5°C. An IPCC Special Report on the Impacts of Global Warming of 1.5°C Above pre-Industrial Levels and Related Global Greenhouse gas Emission Pathways, in the Context of Strengthening the Global Response to the Threat of Climate Change, Sustainable Development, and Efforts to Eradicate Poverty* (V. Masson-Delmotte, P. Zhai, H. O. Pörtner, D. Roberts, J. Skea, P. R. Shukla, A. Pirani, W. Moufouma-Okia, C. Péan, R. Pidcock, S. Connors, J. B. R. Matthews, Y. Chen, X. Zhou, M. I. Gomis, E. Lonnoy, T. Maycock, M. Tignor & T. Water eld, eds). World Meteorological Organization, Geneva, Switzerland, p. 32.

Kim, Y.-H., Min, S.-K., Zhang, X., Sillmann, J. & Sandstad, M. 2020 Evaluation of the CMIP6 multi-model ensemble for climate extreme indices. *Weather Clim. Extrem.* **29**. https://doi.org/10.1016/j.wace.2020.100269.

Koutroulis, A. G., Grillakis, M. G., Tsanis, I. K. & Papadimitriou, L. 2016 Evaluation of precipitation and temperature simulation performance of the CMIP3 and CMIP5 historical experiments. *Clim. Dyn.* **47**, 1881–1898. https://doi.org/10.1007/s00382-015-2938-x.

Kripalani, R. H., Oh, J. H., Kulkarni, A., Sabade, S. S. & Chaudhari, H. S. 2007 South Asian summer monsoon precipitation variability: coupled climate model simulations and projections under IPCC AR4. *Theor. Appl. Climatol.* https://doi.org/10.1007/s00704-006-0282-0.

Laurie, A. & Mathew, B. 2020 How well do CMIP6 historical runs match observed Northeast US precipitation and extreme precipitation-related circulation. *J. Clim.* **33**, 9835–9848.

Mandapaka, P. V. & Lo, E. Y. M. 2018 Assessment of future changes in Southeast Asian precipitation using the NASA Earth Exchange Global Daily Downscaled Projections dataset. *Int. J. Climatol.* **38**, 14. https://doi.org/10.1002/joc.5724.

Manton, M. J., Della-Marta, P. M., Haylock, M. R., Hennessy, K. J., Nicholls, N., Chambers, L. E., Collins, D. A., Daw, G., Finet, A., Gunawan, D., Inape, K., Isobe, H., Kestin, T. S., Lefale, P., Leyuk, C. H., Lwin, T., Maitrepierre, L., Ouprasitworng, N., Page, C. M., Pahalad, J., Plummer, N., Salinger, M. J., Suppiah, R., Tran, V. L., Trewin, B., Tibig, I. & Yee, D. 2001 Trends in extreme daily rainfall and temperature in Southeast Asia and the south pacific: 1961–1998. *Int. J. Climatol.* **21**, 269–284.

Mondal, S. K., Huang, J., Wang, Y., Su, B., Zhai, J., Tao, H., Wang, G., Fischer, T., Wen, S. & Jiang, T. 2021 Doubling of the population exposed to drought over South Asia: CMIP6 multi-model-based analysis. *Sci. Total Environ.* **771**, 145186. https://doi.org/10.1016/j.scitotenv.2021.145186.

Nie, S., Fu, S., Cao, W. & Jia, X. 2020 Comparison of monthly air and land surface temperature extremes simulated using CMIP5 and CMIP6 versions of the Beijing climate center climate model. *Theor. Appl. Climatol.* **140**, 487–502.

O'Neill, B. C., Kriegler, E., Ebi, K. L., Kemp-Benedict, E., Riahi, K., Rothman, D. S., van Ruijven, B. J., van Vuuren, D. P., Birkmann, J., Kok, K., Levy, M. & Solecki, W. 2016 The roads ahead: narratives for shared socioeconomic pathways describing world futures in the 21st century. *Global Environ. Change*. https://doi.org/10.1016/j.gloenvcha.2015.01.004.

Oppenheimer, M., Glavovic, B., Hinkel, J., Van de Wal, R., Magnan, A. K., Abd-Elgawad, A., Cai, R., Cifuentes-Jara, M., DeConto, R. M., Ghosh, T., Hay, J., Isla, F., Marzeion, B., Meyssignac, B. & Sebesvari, Z. 2019 Sea level rise and implications for low-lying islands, coasts and communities. In: *IPCC Special Report on the Ocean and Cryosphere in a Changing Climate* (H.-O. Pörtner, D. C. Roberts, V. Masson-Delmotte, P. Zhai, M. Tignor, E. Poloczanska, K. Mintenbeck, A. Alegría, M. Nicolai, A. Okem, J. Petzold, B. Rama & N. M. Weyer, eds). Cambridge University Press. https://www.ipcc.ch/site/assets/uploads/sites/3/2019/11/SROCC_Ch04-SM_FINAL.pdf.

Orlowsky, B. & Seneviratne, S. I. 2012 Global changes in extreme events: regional and seasonal dimension. *Clim. Change* **110** (3–4), 669–696.

Pall, P., Allen, M. R. & Stone, D. A. 2007 Testing the Clausius-Clapeyron constraint on changes in extreme precipitation under CO_2 warming. *Clim. Dyn.* **28**, 351–363. https://doi.org/10.1007/s00382-006-0180-2.

Randall, D. A., Wood, R. A., Bony, S., Colman, R., Fichefet, T., Fyfe, J., Kattsov, V., Pitman, A., Shukla, J., Srinivasan, J., Stouffer, R. J., Sumi, A. & Taylor, K. E. 2007 Climate models and their evaluation. In: *Climate Change 2007: The Physical Science Basis* (S. Solomon, D. Qin, M. Manning, Z. Chen, M. Marquis, K. B. Averyt, M. Tignor & H. L. Miller, eds). Cambridge University Press, Cambridge, pp. 589–662.

Riahi, K., Vuuren, D. P., Kriegler, E., Edmonds, J., O'Neill, B. C., Fujimori, S., Bauer, N., Calvin, K., Dellink, R., Fricko, O. & Lutz, W. 2017 The shared socioeconomic pathways and their energy, land use, and greenhouse gas emissions implications: an overview. *Global Environ. Change* **42**, 153–168. https://doi.org/10.1016/j.gloenvcha.2016.05.009.

Sillmann, J., Kharin, V. V., Zhang, X., Zwiers, F. W. & Bronaugh, D. 2013a Climate extremes indices in the CMIP5 multi-model ensemble: part 1. Model evaluation in the present climate.

J. Geophys. Res. Atmos. **118**, 1716–1733. http://dx.doi.org/10.1002/jgrd.50203.

Sillmann, J., Kharin, V. V., Zhang, X., Zwiers, F. W. & Bronaugh, D. 2013b Climate extremes indices in the CMIP5 multi-model ensemble: part 2. Future climate projections. *J. Geophys. Res. Atmos.* **118**, 2473–2493. http://dx.doi.org/10.1002/ jgrd.50188.

Srivastava, A., Grotjahn, R. & Ullrich, P. A. 2020 Evaluation of historical CMIP6 model simulations of extreme precipitation over contiguous US regions. *Weather Clim. Extrem.* **29**. https://doi.org/10.1016/j.wace2020.100268.

Su, B., Huang, J., Mondal, S. K., Zhai, J., Wang, Y., Wen, S., Gao, M., Lv, Y., Jiang, S., Jiang, T. & Li, A. 2020 Insight from CMIP6 SSP-RCP scenarios for future drought characteristics in China. *Atmos. Res.* 105375. https://doi.org/10.1016/j.atmosres.2020.105375.

Supharatid, S. 2016 Skill of precipitation projection in the Chao Phraya River Basin by multi-model ensemble CMIP3-CMIP5. *Weather Clim. Extrem.* http://dx.doi.org/10.1016/j.wace.2016.03.001i.

Supari, T. F., Juneng, L., Cruz, F., Chung, J. X., Ngai, S. T., Salimun, E., Mohd, M. S. F., Santisirisomboon, J., Singhruck, P., Tan, P. V., Ngo-Duc, T., Narisma, G., Aldrian, E., Gunawan, D. & Sopaheluwakan, A. 2019 Multi-model projections of precipitation extremes in Southeast Asia based on CORDEX-Southeast Asia simulations. *Environ. Res.* **184**. https://doi.org/10.1016/j.envres.2020.109350.

Tangang, F., Chung, J. X., Juneng, L., Supari, T. F., Salimun, E., Ngai, S. T., Jamaluddin, A. F., Mohd, M. S. F., Cruz, F., Narisma, G., Santisirisomboon, J., Ngo-Duc, T., Tan, P. V., Singhruck, P., Gunawan, D., Aldrian, E., Sopaheluwakan, A., Grigory, N., Remedio, A. R. C., Sein, D. V., Hein-Griggs, D., McGregor, J. L., Yang, H., Sasaki, H. & Kamar, P. 2020 Projected future changes in rainfall in Southeast Asia based on CORDEX-SEA multi-model simulations. *Clim. Dyn.* https://doi.org/10.1007/s00382-020-05322-2.

Taylor, K. E. 2001 Summarizing multiple aspects of model performance in a single diagram. *J. Geophys. Res.* **106**, 7183–7192. doi:10.1029/2000JD900719.

Teutschbein, C. & Seibert, J. 2012 Bias correction of regional climate model simulations for hydrological climate-change impact studies: review and evaluation of different methods. *J. Hydrol.* **456–457**, 12–29.

Ukkola, A., De Kauwe, M. G., Roderick, M. L., Abramowitz, G. & Pitman, A. J. 2020 Robust future changes in meteorological drought in CMIP6 projections despite uncertainty in precipitation. *Geophys. Res. Lett.* http://doi.org/10.1029/2020GL087820.

van den Besselaar, E. J. M., van der Schrier, G., Cornes, R., Iqbal, A. S. & Tank, A. M. G. K. 2017 SA-OBS: a daily gridded surface temperature and precipitation dataset for Southeast Asia. *J. Climate* **30**, 5151–5165. doi:10.1175/JCLI-D-16-0575.1.

Waliser, D. E. & Gautier, C. 1993 A satellite-derived climatology of the ITCZ. *J. Clim.* **6**, 2162–2174. https://doi.org/10.1175/1520-0442(1993)006 < 2162:Asdcot > 2.0.Co;2.

Wang, B., Jin, C. & Liu, J. 2020 Understanding future change of global monsoons projected by CMIP6 models. *Am. Meteorol. Soc.* doi:10.1175/JCLI-D-19-0993.1.

Wild, M. 2020 The global energy balance as represented in CMIP6 climate models. *Clim. Dyn.* **55**, 553–577.

Willett, K. M., Gillett, N. P., Jones, P. D. & Thorne, P. W. 2007 Attribution of observed surface humidity changes to human influence. *Nature* **449**, 710–712. https://doi.org/10.1038/nature06207.

First received 15 January 2021; accepted in revised form 30 April 2021. Available online 19 May 2021

© 2022 The Authors Impact of Climate Change on Hydrology and Water Resources doi: 10.2166/wcc.2021.251

Response of climate change impact on streamflow: the case of the Upper Awash sub-basin, Ethiopia

Salih Duri Abdulahi[a,*], Brook Abate[b], Arus Edo Harka [a] and Sead Burhan Husen[a]

[a] Department of Hydraulic and Water Recourses Engineering, Haramaya Institute of Technology (HIT), Haramaya University, P.O.Box: 138, Dire Dawa, Ethiopia
[b] College of Architecture and Civil Engineering, Addis Ababa Science and Technology University (AASTU), P.O.Box: 16417, Addis Ababa, Ethiopia
*Corresponding author. E-mail: salihduri22@gmail.com

AEH, 0000-0003-0117-219X

ABSTRACT

This paper discusses the response of climate change impact on future streamflow availability in Upper Awash River basin, Ethiopia. The change of climate was built using the CORDEX, RCM daily precipitation, maximum and minimum temperature under RCP4.5 and 8.5 scenarios. The climate model was examined in the historical period 1996–2015 for its ability to capture observed precipitation and temperature. Bias correction was performed on RCM temperature and precipitation to minimize the uncertainties that may occur from climate model projection. After the successful calibration and validation of the HBV hydrological model, streamflow was simulated for the periods of 2021–2040 and 2041–2060 and compared to streamflow of the baseline period 1996–2015 to investigate the changes. The results suggested that overall, following the precipitation increment, streamflow is expected to increase under both RCPs. The average monthly changes of streamflow are expected to increase by 40.1 and 29.4% under RCP4.5 and 16.9 and 18.5% under RCP8.5 scenarios for 2021–2040 and 2041–2060, respectively. Annual streamflow would increase by 28.5 and 23.95% under RCP4.5 and 8.5, respectively. The results of this work can help and inform the water resources planner and designer to frame an appropriate plan and management for the effective use of water in the future.

Key words: bias-correction, climate change, climate model, climate scenarios, streamflow

HIGHLIGHTS

- This study focus on stream-flow response to projected climate change in the Upper Awash sub-basin, Ethiopia.
- This paper coupled RCM with RCPs and HBV hydrological model.
- Bias correction was performed on RCM temperature and precipitation.
- Comparative analysis of the differences between projected and baseline stream-flow were discussed.
- Overall projected stream-flow would increase under both RCP scenarios.

GRAPHICAL ABSTRACT

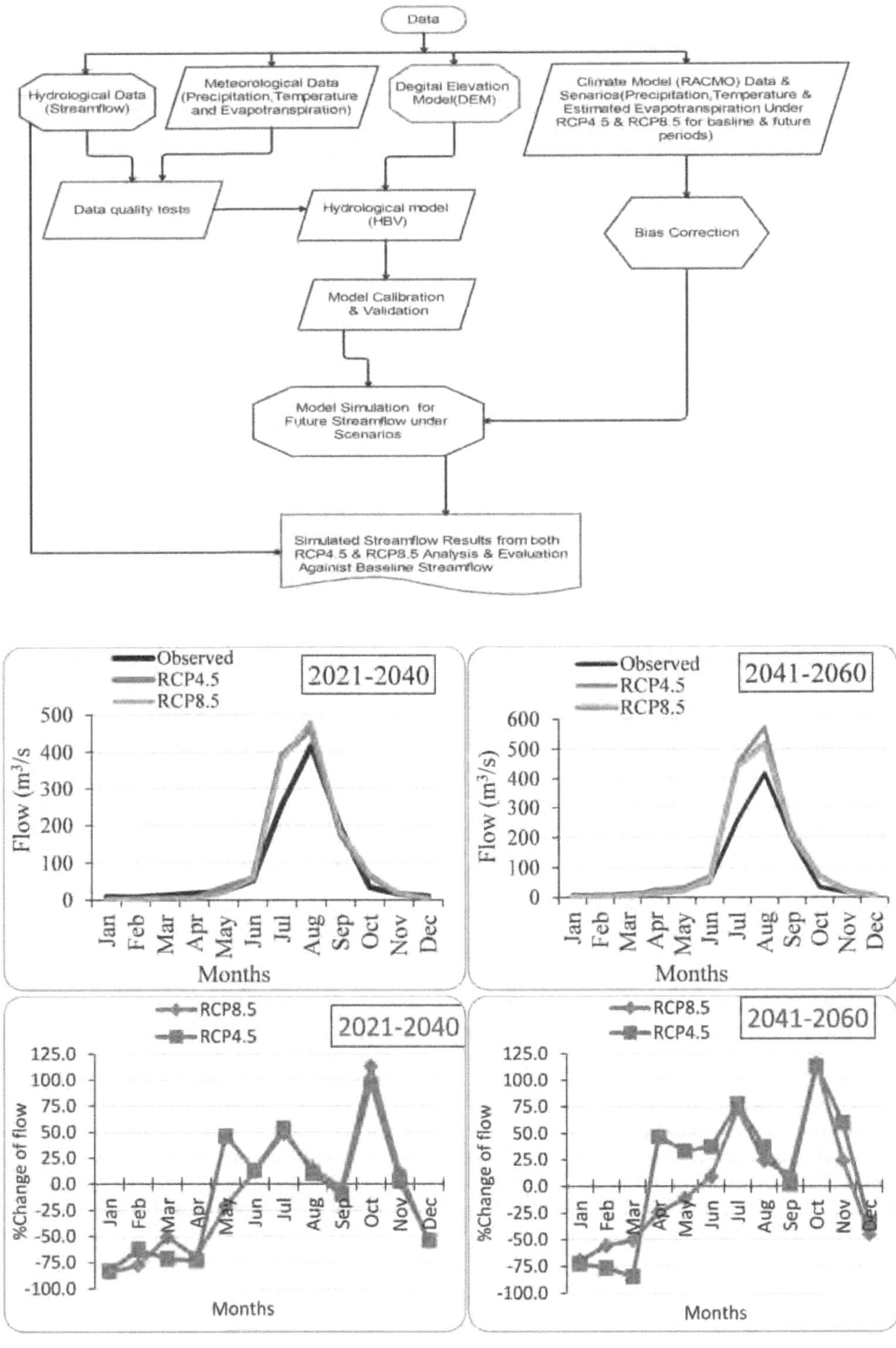

INTRODUCTION

Climate change refers to a change in the state of the climate that can be identified by changes in the mean and variability of its properties and that persists for an extended period, typically for a decade or more (IPCC 2007). Climate may be changed by internal processes such as changes in solar radiation and/or by external forcing within the climate system (IPCC 2007, 2014).

Currently, the effect of climate change is the major global concern as is evidenced by the increase of the greenhouse gases concentration in the atmosphere (Seneviratne *et al.* 2012; IPCC 2013). Based on different climate models and emission scenarios, the temperature is expected to increase continuously while precipitation shows mixed change (IPCC 2014; Negash *et al.* 2020). The IPCC (2013) also suggested that the global mean air temperature will likely increase, ranging from 1.4 to 4.8 °C, under all emission scenarios by the end of the 21st century. The high latitudes regions, the equatorial Pacific and the mid-latitudes of wet regions will probably experience an increase in precipitation under the RCP8.5 scenario while in the mid-latitudes and sub-tropical dry regions it is expected to decrease (Elena *et al.* 2019). Climate change will have a profound effect on streamflow availability and variability due to frequent changes in precipitation and temperature patterns which are now becoming the research areas that need attention (IPCC 2007; Babur *et al.* 2016; Tadese *et al.* 2019). Precipitation and temperature are highly affected by climate change among the components of the hydrological cycle that alter the temporal and spatial availability of streamflow magnitude, variability and timing of occurrence (Babur *et al.* 2016). Regional hydrology is highly sensitive to changing climate conditions that make climate change projections essential to assess the likely availability of water resources to meet future demands (Brekke *et al.* 2009; Claudia 2012; Negash *et al.* 2020). Understanding the climate change effect enhances the flexibility and has net benefits in water resources management (IPCC 2014). The uncertainty of the availability of streamflow in the future could affect agricultural production, socioeconomic systems and threaten environmental sustainability (Daba *et al.* 2017; Elena *et al.* 2019). The impact of climate change has a significunt effect on developing countries, especially those whose economy is dependent on agriculture (IPCC 2014). Ethiopia's economy is directly subjected to climate change effects as a large portion of the land is arid and semi-arid and is dependent on rainfall. Hence it is practical to understand the effect of climate change on future streamflow changes to adapt different mitigation strategies. Numerous studies have been conducted on the impact of climate change on potential availability of water resources using different climate secenarios and climate models, for instance Babel *et al.* (2014), Mulligan (2015), Wanders & van Lanen (2015), Babur *et al.* (2016), Daba *et al.* (2017) and Gizaw *et al.* 2017). Many of them used Special Report Emmission Scenarios (SRES) linked with GCMs outout in different regions, including the present study area. The outcomes of these studies suggested that the projected temperature shows continuously increasing trends under all emissions while the precipitation and respective streamflow show mixed patterns (increasing and decreasing). Currently, SRES are known to be old emission scenarios and may not possess the present climate conditions (Höök *et al.* 2010). To overcome the shortcomings of the SRES scenarios, new emission scenarios (representative concentration pathways, RCPs) have been developed by integrating technological enhancement, economic development, demography, policy and future institutional challenges and adaptation and mitigation linked to exclusive socioeconomic assumptions which shows the present and future climate conditions (IPCC 2007; Getahun 2015). The RCPs allow better resolution over SRES which helps in performing local and regional comprehensive studies (Höök *et al.* 2010). In spite of all these advantages, no study has so far been conducted in the Upper Awash by combining the new emission scenarios (RCPs) with a high resolution regional climate model (RCM). Hence, investigation of the response of climate change impact on future streamflow availability using the new emission scenarios (RCPs) incorporated in a regional climate model (RCM) is essential. The main scientific question to be addressed in this work is whether the streamflow in Upper Awash sub-basin could be affected by climate change and how much the change could be. A common approach to achieve this goal is to use hydrological models with scenarios provided by climate model outputs downscaled at catchment level (Daba *et al.* 2017; Negash *et al.* 2020). To minimize the uncertainties that may occur in climate projections, the application of bias correction is very important for better hydrological simulation (Claudia 2012; Fanget *et al.* 2015; Min *et al.* 2018). Thus, the intent of this study was to investigate the potential response of climate change on streamflow in the future. This finding will be useful to plan, design and apply suitable water management and mitigation strategies to minimize the impact of the change.

MATERIALS AND METHODS

Description of the study area

Awash River basin is one of the 12 Ethiopian River Basins which drains though the central and eastern highlands part of the country, covering a total catchment area of about 110,000 km^2. The river starts from Ginchi town, west of Addis Ababa, travels though Rift Valley, and ends in Lake Abe on the border between Ethiopia and Djibouti. The basin altitude ranges from 250 to 3,000 m above mean sea level. The total average annual rainfall ranges from 160 mm in the lowlands to 1,600 mm in the highlands. According to the Awash Basin Authority (2017) report, the basin's total annual water demand for irrigation, domestic water supply, livestock and industry is estimated to be around 3.4 BMC (billion metric cube). Based on the agricultural activities, socio-economic system, climatological, physical and water resources characteristics, Awash River basin is divided into upper valley, middle valley and lower valley (Edossa *et al.* 2010). The Upper Awash sub-basin (study area) is one of the most highly populated sub-basins in the western highland part of the Awash River basin, including the capital city of Ethiopia. Large mechanized and private irrigated agricultural farms and rapidly expanding industries are located in this sub-basin. The increasing rate of population together with other factors will intensify the water consumption rates in the future (Edossa *et al.* 2010; Tadese *et al.* 2019). Hence, investigation of streamflow changes in the future in response to climate change is economically and environmentally important for the development of the sub-basin as well as the whole country. The geographical location lies between latitude 8°10′57″–9°13′54″N and longitude 37°57′–39°11′E. The total area of the sub-basin is about 11,232 km^2 (Figure 1).

Data and method

The daily meteorological data, precipitation, and maximum and minimum temperatures from 1996 to 2015 were collected from the Ethiopian National Meteorological Service Agency (NMSA). The streamflow from 1996 to 2015 for the selected stations were collected from the Ethiopian Ministry of Water, Irrigation and Energy office (MoWIE). The meteorological and hydrological stations were selected based on the standards of the length of record years, completeness (or fewer missing records) and the extent of the catchment network. Hence six rainfall and temperature and three streamflow gauging stations were used for this work. Sometimes, the collected data from both meteorological hydrological gauging stations may have missing or inaccurate records (Daba *et al.* 2017). In those cases, to improve the quality of the collected data, we applied different data quality test mechanisms such as filling missing data, homogeneity tests and consistency. Predominantly, the rainfall pattern is unimodal with the main rain occurring from June to September and the other months are mainly dry except for

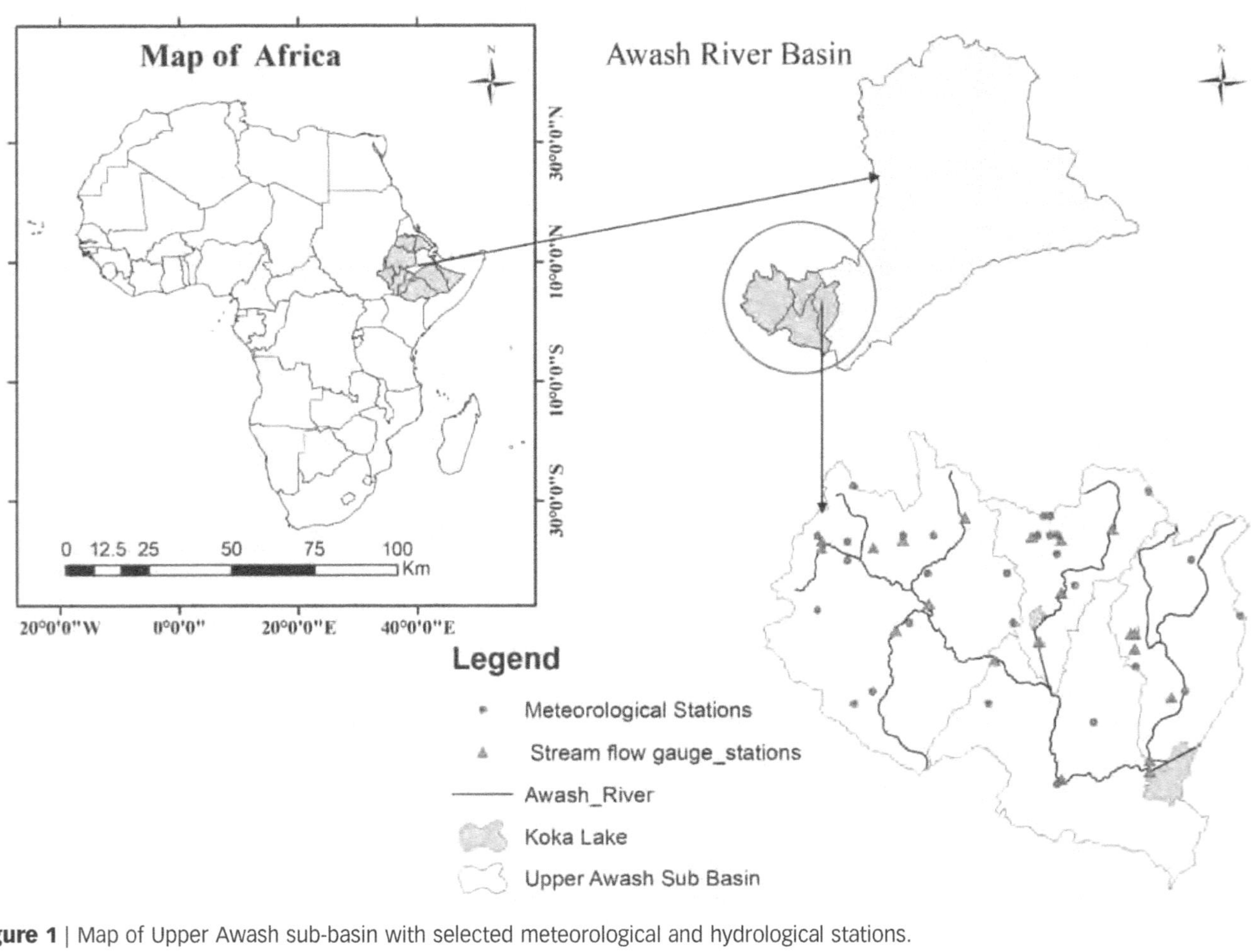

Figure 1 | Map of Upper Awash sub-basin with selected meteorological and hydrological stations.

small rains from March to May. The total average annual rainfall is about 1,120 mm. The annual average temperature varies from 4.8 to 31.4 °C (Figure 2). Koka dam streamflow gauging station, which is located at the outlet of the sub-basin, was used for calibration and validation of the HBV hydrological model (Figure 3). A digital elevation map (DEM) 30×30 m resolution was obtained from the Ethiopian Ministry of Water Irrigation and Energy Bureau Department of GIS and processed in ArcGIS to extract the drainage area and drainage network. Table 1 shows the meteorological and streamflow gauging stations used for this study.

Climate data were downloaded from the African CORDEX domain for present and future periods under RCP4.5 and 8.5 scenarios on a daily basis from http://cordexesg.dmi.dk/esgf-web-fe/ under a regional climate model, *RACMO* version 2.2, which was developed in the Koninklijk Meteorologisch Institute (KNMI) (The Netherlands) for climate projections. The selected climate model was based on vintage, resolution, validity and representativeness of the area (Gleckler *et al.* 2008; Räisänen *et al.* 2009). Also, the performance of the selected model to reproduce observed precipitation and temperature in the baseline period was evaluated by Yacouba *et al.* (2017). The downloaded data were minimum and maximum temperature and precipitation. The driving scenarios were RCP4.5 and 8.5 in W/m^2 which approximately represents the medium and high carbon emission concentrations scenarios, respectively. The data from 1996 to 2015 was considered as the baseline period and 2021–2040 and 2041–2060 data were used as near and far future periods, respectively, to impact studies. The data derived from the climate models may not exactly possess the statistical characteristics of the ground stations data of a particular catchment (Schaefli 2015). Hence, application of bias correction is very important to improve climate model prediction and to minimize the discrepancy between observed and simulated precipitation and temperature that may be in turn seen in the streamflow simulation (Graham *et al.* 2007; Fanget *et al.* 2015). This study employed linear scaling (LS) and a bias correction method in daily time steps. That is, the temperature was typically bias corrected by adding the monthly difference

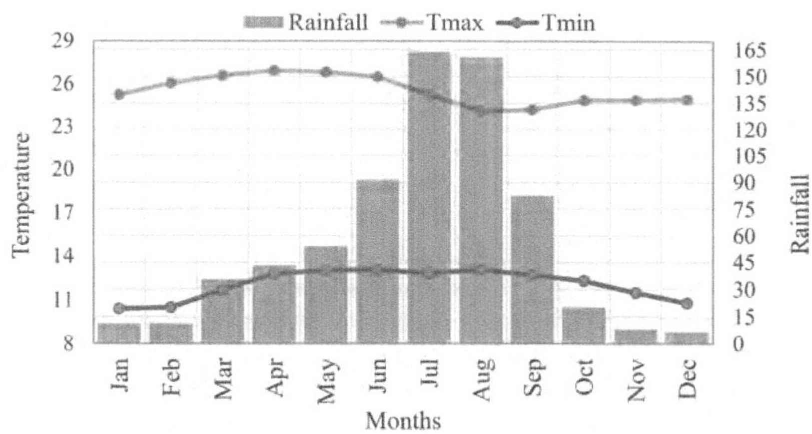

Figure 2 | Monthly average rainfall (mm) and temperature (°C) of study area (1996–2015).

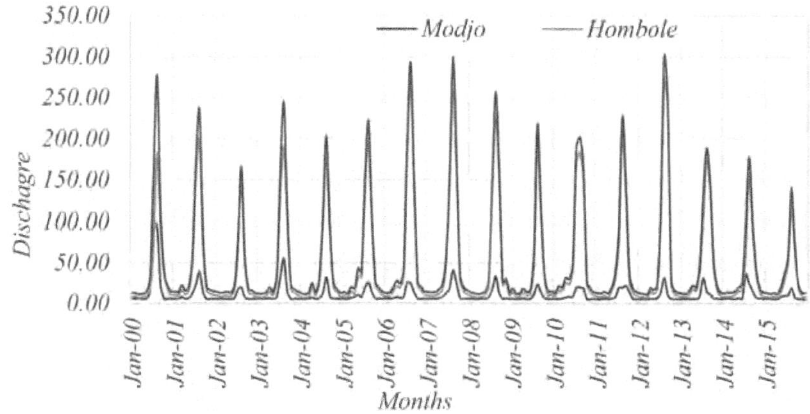

Figure 3 | Monthly streamflow (m^3/s) in the study area (1996–2015).

Table 1 | Selected meteorological and hydrological stations and their data periods (1996–2015)

No	Name of station	Latitude	Longitude	Elevation (m)
1	Addis Ababa	9.019	38.748	2,386
2	Akaki	8.870	38.786	2,057
3	Debre Zeiyt	8.733	38.950	1,900
4	Ginchi	9.017	38.133	2,132
5	Koka Dam	8.469	39.154	1,618
6	Tulu bolo	8.655	38.204	2,190
7[a]	@Modjo	8.400	39.023	2,180
8[a]	@Hombole	8.430	39.020	2,194
9[b]	@Kokadam	8.358	39.062	1,618

[a]Indicates streamflow gauging stations.
[b]Indicates the streamflow gauging station used for calibration and validation.

between the recorded and RCM in the baseline period to the daily RCM temperature in the future (Equation (1)) and precipitation was corrected by multiplying the ratio of monthly recorded and RCM precipitation in the baseline period by the RCM precipitation in the future (Equation (2)) (Min *et al.* 2018).

$$T_{daily,future} = T_{RCM,daily,future} + (mT_{obs,monthly} - mT_{RCM,obs,monthly}) \tag{1}$$

$$P_{daily.future} = P_{RCM,daily,future} * \left[\frac{mP_{obs,monthly}}{mP_{RCM,obs,monthly}}\right] \tag{2}$$

where $P_{daily,future}$ and $T_{daily,future}$ are bias corrected precipitation and temperature in future periods, respectively; $T_{RCM,daily,future}$ and $P_{RCM,daily,future}$ are bias raw temperature (°C) and precipitation (mm) in future periods, respectively; $T_{RCM,obs,monthly}$ and $T_{obs,monthly}$ are raw and observed temperature in baseline periods, respectively; $P_{RCM,obs,monthly}$ and $P_{obs,monthly}$ are raw and observed precipitation in baseline periods, respectively; and m denotes the mean value. Monthly differences are determined using observed and raw RCM in the baseline period. The performance of RCM/GCM(s) can be assessed by a number of statistical techniques. Not a single technique is taken as best, rather many statistical techniques are accompanied together to provide a comprehensive overview of model performance (Gordon & Shaykewich 2000). For this study, we applied percent bias (PBIAS), root mean square error (RSME) and correlation coefficient (R^2) to check the performance of selected climate models for climate prediction as illustrated by Equations (3)–(5), respectively. These statistical measures were recommended by the World Meteorological Organization (WMO) as reported in Gordon & Shaykewich (2000).

$$PBIAS = \left(\frac{\frac{1}{N} * \sum (RCM_i - OBS_i)}{\frac{1}{N} * \sum OBS_i}\right) * 100 \tag{3}$$

$$RMSE = \left(\frac{\sqrt{\frac{1}{N} * \sum (RCM_i - OBS_i)^2}}{\frac{1}{N} * \sum OBS}\right) * 100 \tag{4}$$

$$R^2 = \frac{(\sum (RCM_i - RCM_{avg}) * (OBS_i - OBS_{avg}))}{\sqrt{(\sum (RCM_i - RCM_{avg})2 * (OBS_i - OBS_{avg})^2}} \tag{5}$$

where RCM and OBS are model output and observed temperature and precipitation in the baseline period, respectively; i refers to the given time period and N is total number of data years in such pairs. P_{BIAS} measures whether the climate model overestimates or underestimates a particular climate variable in a given catchment. Positive values indicate

Table 2 | Summary of data used and sources

Data	Sources of data	Descriptions
Terrain	Ethiopian Ministry of Water, Irrigation and Energy (MoWIE) Department of GIS	DEM (digital elevation model) 30×30 m
Recorded climate data	Ethiopian National Meteorological Service Agency (ENMSA) from 1996 to 2015	Maximum and minimum temperature, rainfall
Climate model data (baseline and future)	http://cordexesg.dmi.dk/esgf-web-fe/ historical data (1996–2015) and future (2021–2060) for RACMO22T (RCM) model under RCP4.5 and 8.5	Maximum and minimum temperature and precipitation
Hydrological data	Ministry of Water, Irrigation and Energy (MoWIE)	Streamflow (m^3/s) (1996–2015)

overestimation whereas negative values indicate underestimation of the model output compared to that recorded in the reference period. The smaller the absolute value of *RMSE* the better climate model performance and vice versa. R^2 indicates the strength of the relationship between the climate model outputs (temperature and precipitation and the recorded station precipitation and temperature. R^2 value greater than 0.5 is taken as a good relationship. The performance of the model was checked for both precipitation and temperature. Table 2 shows a summary of the data and the sources used in this study.

HBV hydrological model

The HBV (Hydrologinska brans Vattenbalansavdelning) model is a semi-distributed conceptual hydrologic model developed at the Swedish Metrological and Hydrologic Institute (SHMI) in the 1970s. It is a tool that is used to forecast and simulate runoff in nearly 200 basins throughout Scandinavia and it has been applied in more than 30 countries, including Ethiopia. The advantage of this model over the others is the possibility of disaggregating the basin into a number of sub-basins, elevation and vegetation cover zones and it requires less data input for simulation. The HBV model consists of snow accumulation and snowmelt routine, soil moisture accounting routine, routines for runoff generation and a simple routing routine (SMHI 2008). The snow accumulation routine is an important process in the study of catchment where the snow is present and melts to contribute to the runoff generation. Another important process is the soil moisture accounting routine part which controls runoff formation. It is a process where rainfall goes to the root zone and to groundwater as recharge depending on the relation between field capacity (FC) and moisture content in the root zone (SM) (Equation (6), and actual evaporation is estimated depending on soil moisture availability using the relationship between SM and FC (Equation (7)). The impact of this routine in contribution to runoff from snowmelt and rain is high in wet conditions and small in dry soil. Likewise, the response routine is a function responsible for transforming excess water from soil moisture zones to generate runoff (Equation (8)). Finally, the routing routine (transformation routine) is the generated runoff from the response routine routed through a transformation routine in order to obtain the proper shape of the hydrograph at the outlet of the sub-basin. The transformation function is given in mm/day (Equations (9) and (10)), using a triangular weighting function defined by the parameter MAXBAS (Seibert 2005):

$$\frac{\text{Recharge}}{P(t)} = \left(\frac{SM_{(t)}}{FC}\right)^{\text{Beta}} \tag{6}$$

$$E_{\text{act}} = E_{\text{pot}}\,\text{Min}\left(\frac{SM_{(t)}}{FC,\, LP},\, 1\right) \tag{7}$$

$$Q_{GW(t)} = K_0 \max\,(SUZ - UZL,\, 0) + K_1 * SUZ + K_2 * SLZ \tag{8}$$

$$Q_{\text{sim}(t)} = \sum_{i=1}^{MAXBAS} C(i) * Q_{GW}(t - i + 1) \tag{9}$$

where

$$C_{(t)} = \int_{1}^{i} \frac{2}{MAXBAS} - abs\left(U - \frac{MAXBAS}{2}\right) * \frac{4}{MAXBAS^2} * du \tag{10}$$

where $P_{(t)}$ is the precipitation at time t, FC is the field capacity (maximum soil moisture storage in mm); β is a parameter that determines the relative contribution to runoff or to increase soil moisture storage from rain or snowmelt; E_{act} is the actual evapotranspiration; E_{pot} is the potential evapotranspiration; L_P is the soil moisture value above which E_{act} reaches E_{pot}; Q_{GW} is the groundwater recharge; Q_{sim} is the simulated runoff; and Ki is the recession constant.

The required input parameters for HBV model are; daily areal rainfall, daily average temperature (°C), daily estimated potential evapotranspiration and streamflow for calibration and validation of the model. For this study, potential evapotranspiration was estimated by the Hargreaves method. This method was adopted because of its simplicity and it requires less data compared to other methods (Hargreaves & Riley 1985). The Hargreaves method was compared to the Penman method to estimate evapotranspiration and the result indicated that the two methods showed almost similar output. The authors suggest that in the case of data limitation, it is possible to use Hargreaves's method to estimate evapotranspiration (Equation (11)):

$$\lambda Eo = 0.0023 * Ho * (T_{\max} - T_{m,n})^{-0.5} * (T_{avg} + 17.8) \tag{11}$$

where λ is the latent heat of vaporization (MJ kg^{-1}); Eo is the potential evapotranspiration (mm day^{-1}); Ho is the extraterrestrial radiation (MJ m^{-2} day^{-1}); and $T_{max,}$ T_{min} and T_{avg} are the maximum, minimum and average air temperature (°C) for a given day, respectively. The predictive ability, reliability and accuracy of the hydrological model to simulate the recorded streamflow has to be checked (Rientjies 2007). Normally 5–10 years of streamflow is sufficient to calibrate the HBV model (Bergström 1992). The total streamflow used to develop the model was 16 years (2000–2015). The first two years (2000–2001) were used for warming up periods to initialize the model before calibration starts. The calibration and validation were done manually by optimizing the parameters that could significantly affect the performance of the model for the years 2002–2010 and 2011–2015, respectively. The performance of the model was evaluated by three statistical parameters: correlation coefficient (R^2), Nash–Sutcliffe efficiency (NSE), and percent bias (P_{BIAS}), as suggested by Krause *et al.* (2005). Equations (12)–(14) describe the mathematical illustrations of the statistical parameters used. Nash–Sutcliffe efficiency shows how fine the graph of recorded flow plots fit to the simulated one. Both R^2 and NSE ranges from 0 to 1, with the high values indicating less error variance, and typically the value greater than 0.5 was accepted. Percentage bias measures the average tendency of the model simulated flow to be smaller or larger than their respective recorded flow. A positive value indicates overestimation while a negative value shows underestimation. Typically, the values between −15 and +15% are within acceptable ranges (Srinivasan *et al.* 2010):

$$R_{eff}(NSE) = 1 - \frac{\left(\sum (Q_{sim(t)} - Q_{obs(t)})\right)^2}{\sum (Q_{obs(t)} - Q_{obs(avg)})^2} \tag{12}$$

$$R^2 = \frac{\left(\sum (Q_{obs} - Q_{obs(avg)})\right) * \left(\sum (Q_{sim} - Q_{sim(avg)}\right)}{\sqrt{\sum (Q_{obs} - Q_{obs(avg)})2 * (Q_{sim} - Q_{sim(avg)})^2}} \tag{13}$$

$$P_{BIAS} = \left(\frac{\sum (Q_{obs(t)} - Q_{sim(t)})}{\sum Q_{obs}}\right) * 100 \tag{14}$$

where R_{eff} is coefficient of efficiency; $Q_{obs(t)}$ and $Q_{sim(t)}$ are recorded and simulated flow at a given time t; Q_{obs} is average recorded flow of respected time; and P_{BIAS} is percentage biases. The optimized parameters during the calibration and validation period were kept constant during future streamflow simulation. Then, the simulated streamflow changes were evaluated against streamflow in the reference period on the basis of monthly, seasonal and annual time steps. The general water balance of the HBV model is described in Equation (15) and Figure 4:

$$P - E - Q = \frac{d}{dt}(SP + SM + UZ + LZ + Lakes) \tag{15}$$

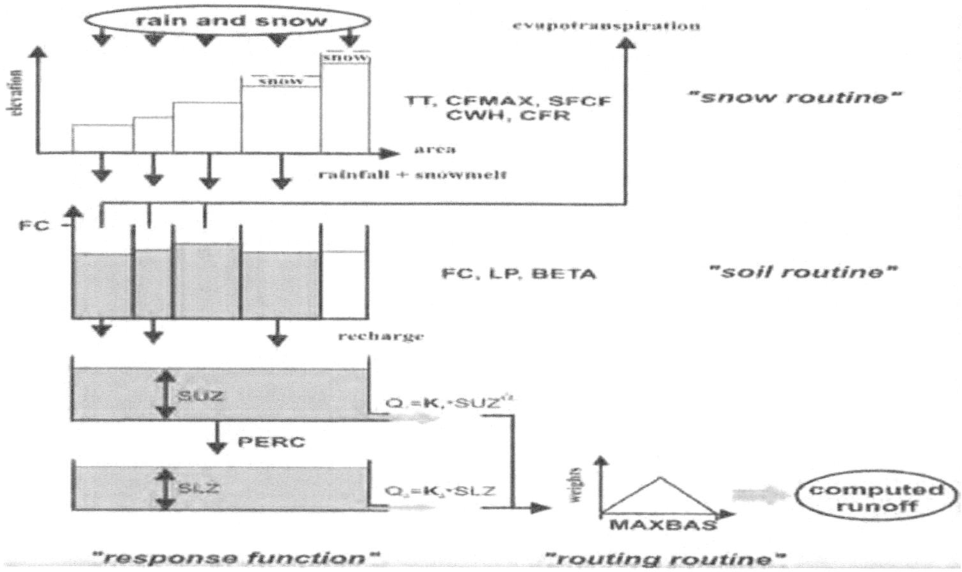

Figure 4 | General structure of the HBV model (Seibert 2005).

where P is precipitation; SM is soil moisture routine; E is evapotranspiration; UZ is upper ground water zone; Q=runoff; LZ is lower ground water zone; and SP=snowpack accumulation.

Finally, the trends of both rainfall and streamflow simulated from the RCM climate model and the HBV hydrological model, and recorded based on ground stations, were analyzed by non-parametric Mann–Kendall test. The Mann–Kendall trends test is used to detect whether the data series in hydro-climatic variables are consistently increasing or decreasing. The null hypothesis test shows there is no monotonic trend in the data series and the alternate hypothesis shows that the trend exists (Hess *et al.* 2001). The World Meteorological Organization (WMO) recommended a non-parametric Mann–Kendall trend test to assess the trends in meteorological and hydrological data as a result of its simplicity, allowing missing values and less sensitivity to outliers. There is no assumption of statistical distribution and it reduces the chance of errors during statistical analysis which makes it robust. However, the results of the test may have an error if autocorrelation exists in the data series. In such cases, a pre-whitening procedure is performed to remove autocorrelations in the time series (Equations (16) and (17)) (Hess *et al.* 2001; Tekleab *et al.* 2013). R represents the first autocorrelation (lag-1) coefficient of the time series X_t mean, X_t represents mean of the data and n is the number of data points in the time series:

$$R = \frac{\frac{1}{N-1}\sum_{t=1}^{n-1}(xt - xmean(t))(xt + 1 - xmean(t))}{\frac{1}{n}\sum_{t=1}^{n}(xt - xtmean)^2} \tag{16}$$

$$\frac{-1 - 1.645 * (N-2)^{0.5}}{N-1} \leq R \leq \frac{1 + 1.645 * (N-2)^{0.5}}{N-1} \tag{17}$$

If the lag-1 autocorrelation coefficient is found to be within the interval defined by Equation (17), it can be summarized that the time series does not exhibit a significant autocorrelation. On the other hand, if the estimated lag-1 autocorrelation coefficient is found to be outside of the interval, it can be said that the time series exhibits a significant autocorrelation at the 5% significance level. In this case, the Mann–Kendall test is applied on pre-whitening series obtained as (X_2–RX_1, X_3–RX_2…X_n–RX_{n-1}) where X_1, X_2, X_3… X_n are the data points of the series. Then, after autocorrelation was checked, the Mann–Kendall's statistical test was applied (Equations (18) and (19)):

$$S = \sum_{i=1}^{n} * \sum_{j=1+1}^{n} \mathrm{Sign}(x_j - x_i) \tag{18}$$

where S is the rating score (Mann–Kendall sum); X is the observation value; n is the number of observation values in the time series; X_i and X_j are the sequential hydro meteorological values in the years i and j ($j>i$), respectively. Positive and negative values of S dictate increasing and decreasing trends in the data series, respectively. The standard normal Z is calculated by:

$$Z = \begin{cases} \dfrac{1-S}{\sqrt{\mathrm{Var}(s)}}, & \text{if} \quad S > 0 \\ 0, & \text{if} \quad S = 0 \\ \dfrac{1+S}{\sqrt{\mathrm{Var}(s)}}, & \text{if} \quad S < 0 \end{cases} \tag{19}$$

The statistical value Z determines the significance of trend at a selected significance level. For this study the 5% significance level was selected. The slope of n pairs of data points was estimated using the Sen's slope estimator (Equation (20)):

$$T_i = \frac{X_j - X_k}{j-k}, \text{ for } i = 1, 2, 3 \ldots N \tag{20}$$

where X_j and X_k are data values at times j and i, respectively.

T_i is a Sen's estimator of slope which is the median of these N values. T indicates the increasing or decreasing values in trends. The whole methodology and procedures are summarized in the flowchart as shown in Figure 5.

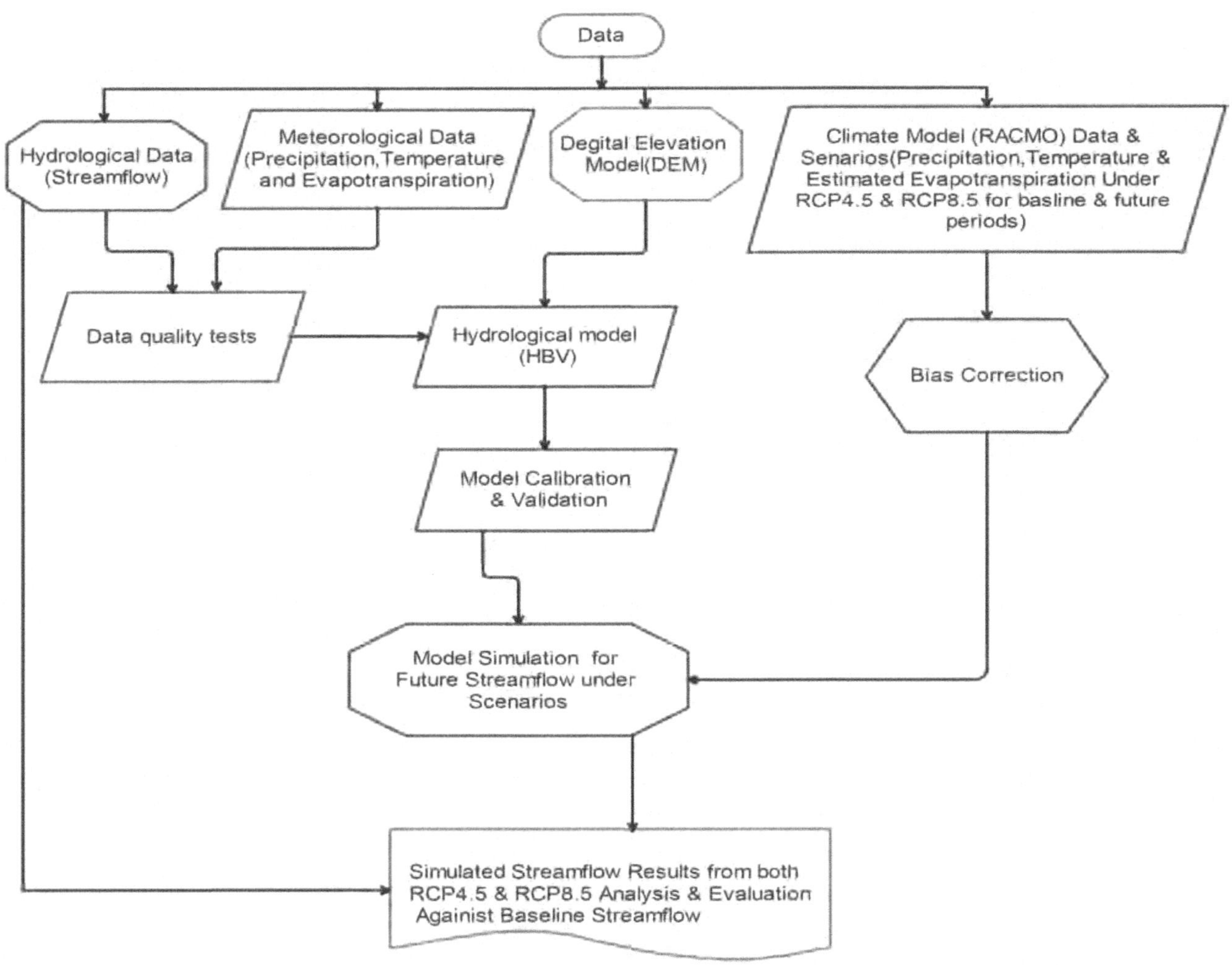

Figure 5 | Data used and framework of the methodology.

RESULTS

Performance of RCM in the baseline period (1996–2015)

One of the objectives of this study is to evaluate the quality of selected climate model to reproduce precipitation and temperature in the baseline period (1996–2015). Accordingly, Figure 6 shows the average monthly precipitation cycle and percentage change of RCM from that recorded in the baseline period. The result indicates that the precipitation cycle from RCM is overlapped with the recorded precipitation in all months except from the middle of June to September (Figure 6(a)). However, after bias correction the differences were significantly minimized. The difference before bias correction varied from –64.97 to +78.65%. However, this difference reduced to –6.45 to +8.72% after bias correction (Figure 6(b)). Similarly, Figure 7 depicts the average monthly RCM temperature cycle and its change compared to recorded temperature. The model totally underestimated both minimum and maximum temperatures. Nevertheless, after bias-correction, the shape of the simulated and observed graph overlapped well. The changes range from –46.6 to –14.1% before bias correction and from –3.2 to 3.5% after bias correction for maximum temperature. Bias correction significantly improved the quality of climate model prediction which may in turn minimize the uncertainties that may occur in hydrological simulation. Similar results were depicted by Fanget *et al.* (2015) and Ahmed *et al.* (2017). The correlation between observed rainfall and RCM rainfall, shown in Figure 8, indicates that strong correlation exists after an application of bias correction.

Table 3 provides a summary of statistical parameters to evaluate the performance of the climate model (RCM).

Projected precipitation and temperature results

Another specific objective of this study was to analyze the projected changes in precipitation and temperature based on the RCP4.5 and 8.5 scenarios for the period of 2021–2040 and 2041–2060 compared to the baseline period 1996–2015. Figure 9 depicts the general pattern of average monthly projected rainfall. The projected rainfall for both scenarios is expected to increase in the months of February, June, July, August and November, while in March, April, May, October, September, December and January it is expected to decline. Relatively, in summer (June–August) rainfall is projected to increase more than the other months (Figure 9(a) and 9(b)). Figure 10 shows the percentage change in precipitation between the reference period (1996–2015) and projected periods of 2021–2040 and 2041–2060. Overall, average annual rainfall is projected to increase by 11.97 and 20.67% under RCP4.5 and 8.5, respectively. For the period of 2021–2040, the projected rainfall is

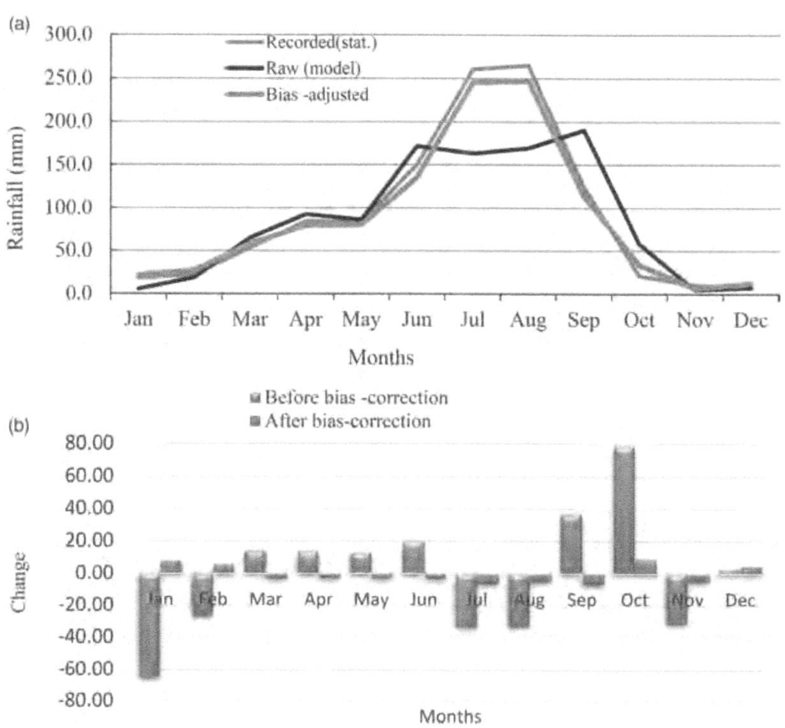

Figure 6 | Average monthly precipitation cycle (a) and percentage change (b) in the baseline period (1996–2015).

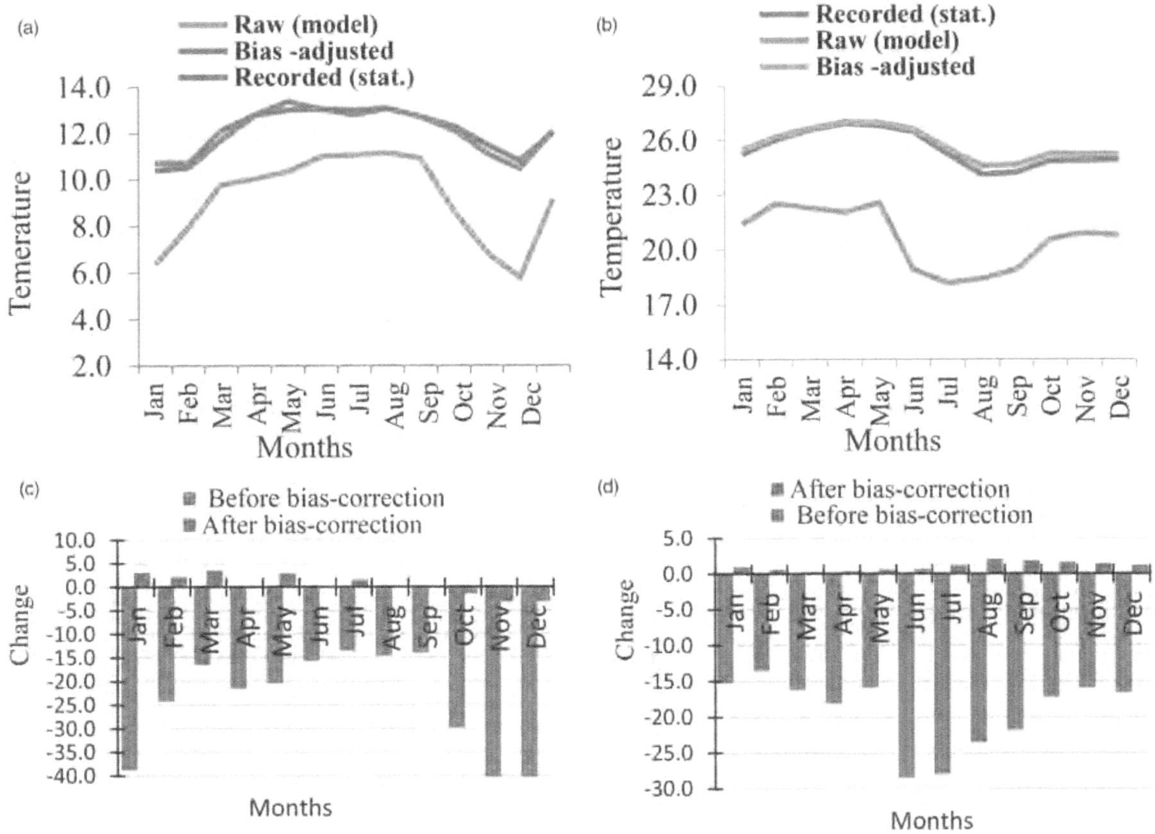

Figure 7 | Average monthly RCM versus recorded temperature (°C) (a) minimum, (b) maximum and percentage change (c) minimum and (d) maximum in the baseline period (1996–2015).

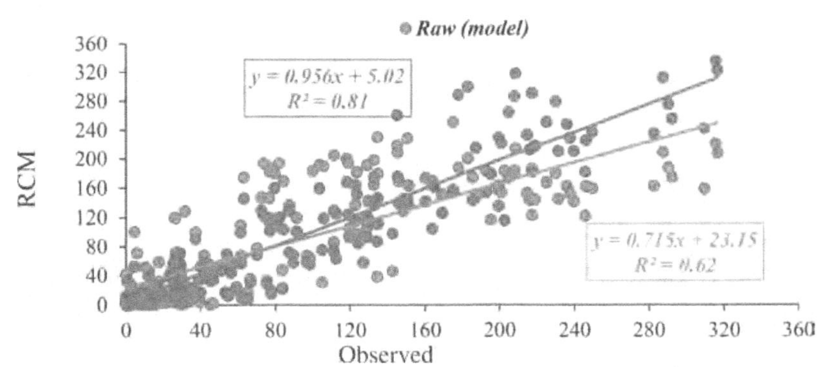

Figure 8 | Correlation plots of monthly recorded and RCM rainfall in mm (1996–2015).

Table 3 | Results of statistical parameters to evaluate RCM

	Bias corrected			Raw model uncorrected		
Statistical parameters	**Prep.**	T_{max}	T_{min}	**Prep.**	T_{max}	T_{min}
Correlation	0.86	0.78	0.72	0.71	0.59	0.67
PBIAS	+0.81	+1.02	+0.22	–18.4	–19.21	–24.43
RSME	18.7	6.3	8.75	69	20.68	26.62

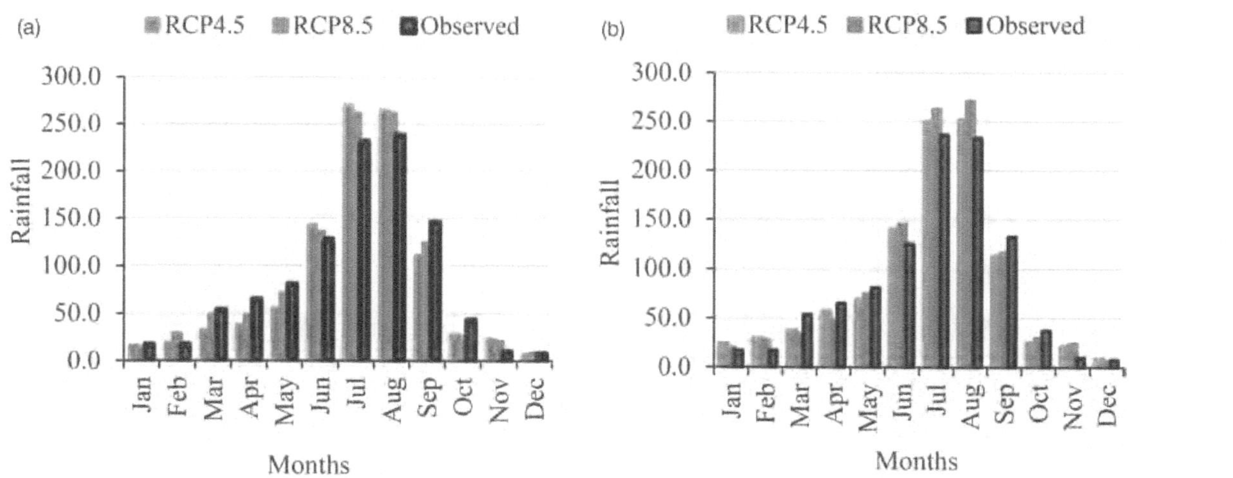

Figure 9 | Average monthly projected precipitations (mm) of the future periods, (a) 2021–2040 and (b) 2041–2060, compared to rainfall in the baseline period (1996–2015).

expected to increase to 3.22% under RCP4.5 and 16.21% under RCP8.5. The maximum increment will be expected in November (69.51%) under RCP4.5 and in February (66.6%) under RCP8.5. The maximum reduction will be expected in March under RCP4.5 and in April under RCP8.5 (Figure 10(b)). Likewise, for the period of 2041–2060, mean annual rainfall is projected to increase by 20.72 and 25.13% under RCP4.5 and 8.5, respectively. Relatively, in this time period, the rainfall increment will be high compared to the period 2021–2040. The maximum increment in this time period will be expected every November and the maximum reduction will be in March under both scenarios. Generally, the result clearly indicates that for both future periods and under both scenarios, the projected mean annual rainfall is expected to increase over the study area. However, the increment will vary among the scenarios and periods. The IPCC (2013) also suggested that, in the case of increasing temperature, the mean annual precipitation is expected to increase up to 20%.

Figure 11 shows monthly projected average temperature cycle and percentage change under both scenarios. As can be seen under both scenarios, both maximum and minimum temperature is expected to increase. On average, the projected minimum temperature is expected to increase by 9.6 and 11.7% under RCP4.5 and 8.5, respectively. Maximum temperature change could be +6.3% under RCP4.5 and +7.6% under RCP8.5. Moreover, the increment of minimum temperature is relatively high compared to the maximum. The results clearly show that the projected air temperature under RCP8.5 is higher than that of under RCP4.5 which confirms that RCP8.5 is a higher greenhouse carbon concentration emission scenario with a higher degree of global warming. The forecasted temperature in both the near and far future period is within the range forecast by IPCC (2007), i.e. average temperature increases from +1.4 to +20% at the end of the 21st century.

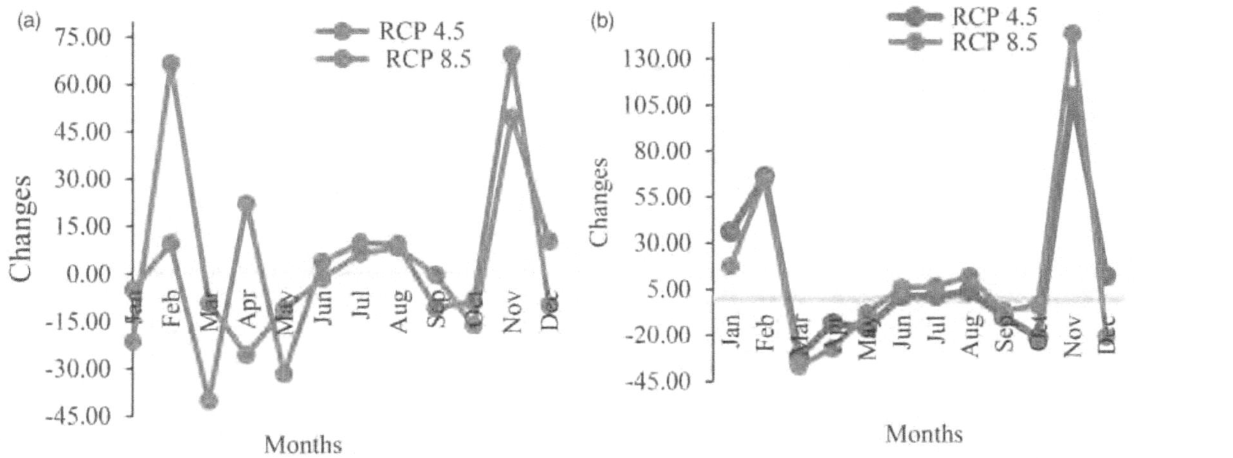

Figure 10 | Projected percentage changes in rainfall compared to baseline period for: (a) 2021–2040 and (b) 2041–2060.

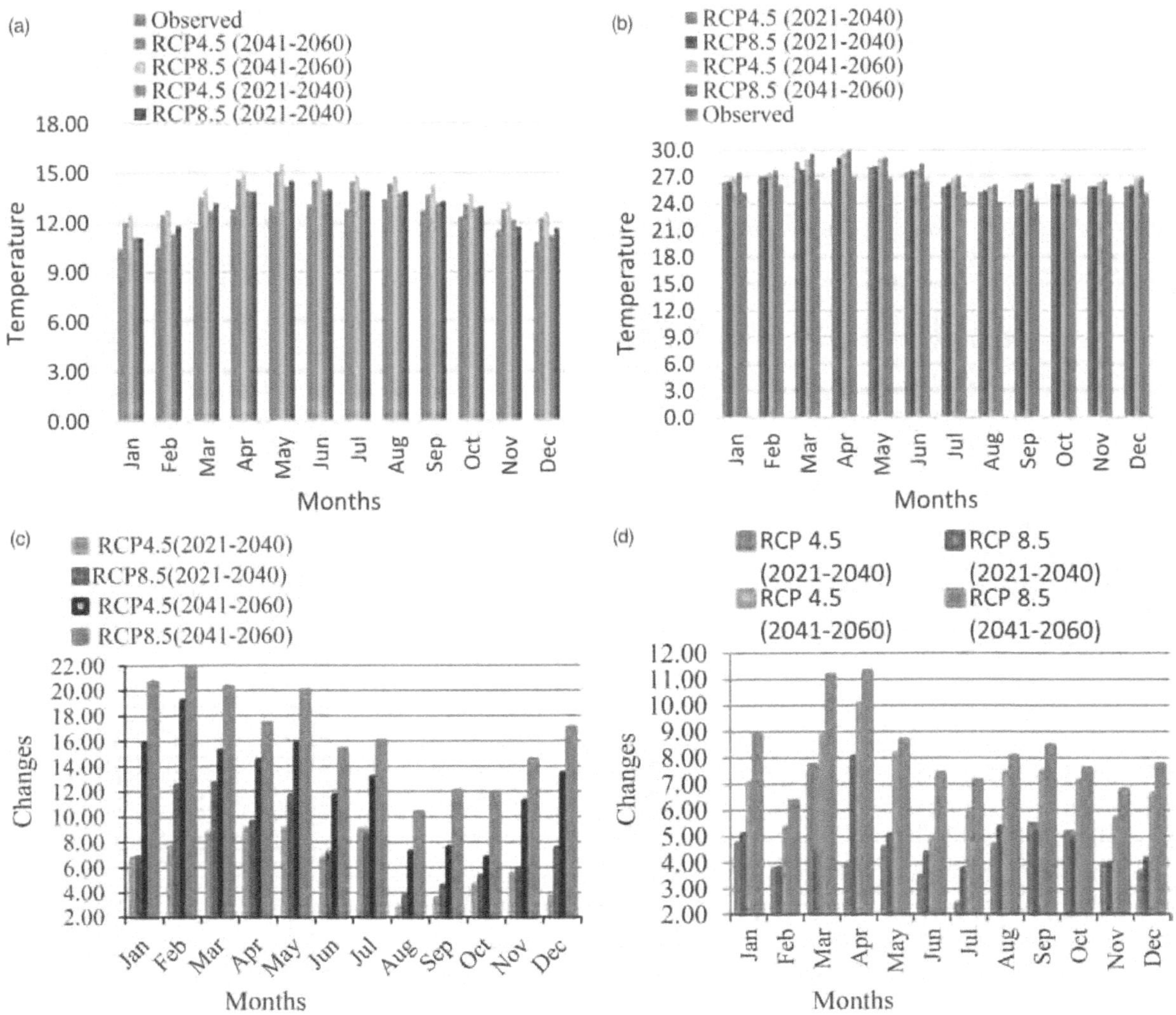

Figure 11 | Monthly projected average temperature: (a) minimum and (b) maximum, and percentage change (c) minimum and (d) maximum, compared to the baseline period 1996–2015.

Table 4 shows the monthly average change in percentage of precipitation, maximum and minimum temperature for the periods 2021–2040 and 2041–2060 compared to the baseline period (1996–2015). The result indicated that the temperature and precipitation increment is higher for the far future period (2041–2060) than the near future period (2020–2040) under both scenarios. The percentage increment of the maximum temperature would be less than that of minimum temperature under both scenarios. In general, both the projected temperature and precipitation would increase for the study area.

Hydrological model calibration and validation results

The HBV model calibration and validation was carried out manually by optimizing the model parameters in each subroutine that has a significant effect on the model performance. Based on this, several runs were performed to select the most optimum

Table 4 | Likely percentage changes for projected future precipitation and temperature

Scenarios	Time period	$T_{min.}$	$T_{max.}$	Prep.
RCP4.5	2020–2040	+6.34	+4.43	+3.21
	2041–2060	+12.58	+7.03	+20.31
RCP8.5	2020–2040	+7.94	+4.81	+16.20
	2041–2060	+16.39	+8.27	+25.01

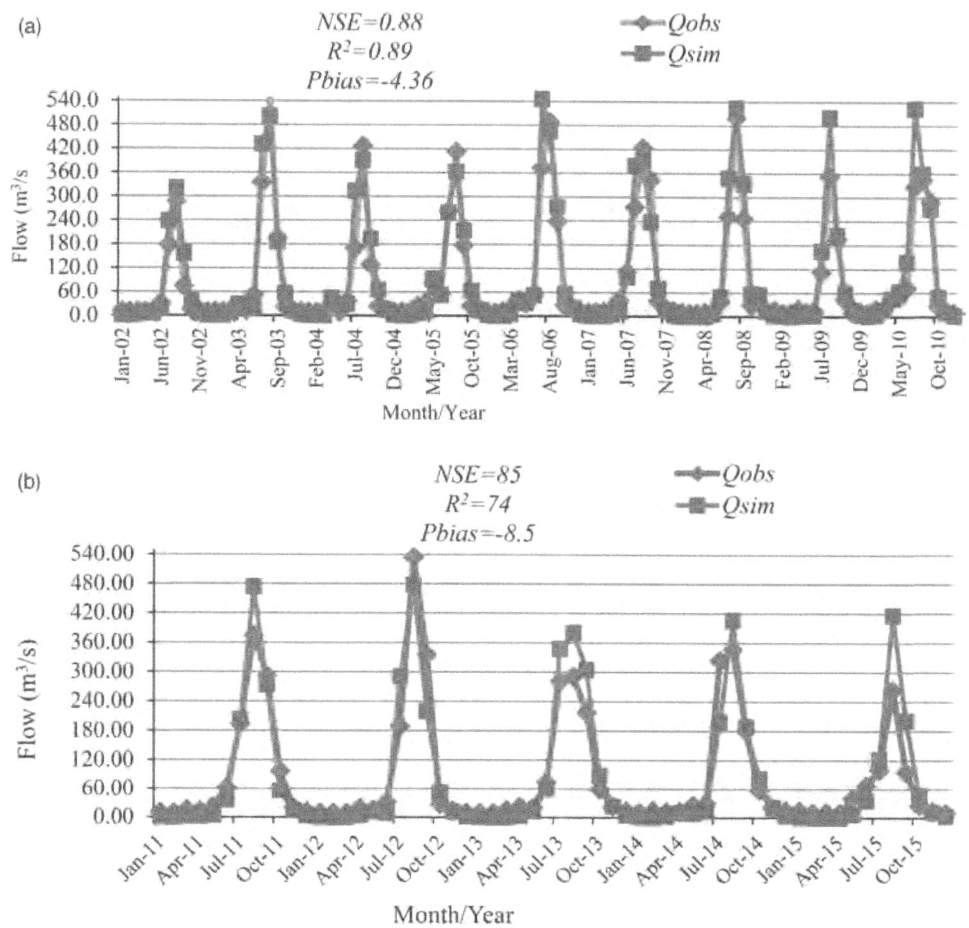

Figure 12 | Streamflow: (a) calibrated (2002–2010), (b) validated (2011–2015).

parameter sets in order to compare the observed to simulated hydrograph. The results of each run were evaluated based on: visual inspection by comparing the simulated and observed hydrograph, NSE, P_{BIAS} and R^2. The calibration and validation was conducted in monthly time series. Figure 12 shows monthly recorded and simulated runoff cycles for the calibration and validation period over the reference period. The shape of the simulated hydrograph is well captured by the shape of recorded flow using the optimized model parameters during the calibration period (2002–2010) and validation period (2011–2015). However, some of the peak and low flows were underestimated and overestimate recorded flow. All performance evaluation variables are within acceptable ranges for both calibration and validation periods, as displayed in Figure 12. The most sensitive parameters and their optimum values sets during the calibration and validation periods are shown in Table 5.

Table 5 | Most sensitive parameters and their optimum value sets during the calibration (2002–2010) and validation (2011–2015) periods

Parameters	Description of parameters	Optimum value during calibration period
FC	Maximum soil storage(mm)	200
PERC	Percolation (mm/day)	0.1
Beta (β)	It is a parameter that determines the relative contribution to run off from rain/snowmelt	1.4
LP	Limit for potential evaporation	0.8
K_1	Recession coefficient for upper response box	0.01
K_2	Recession coefficient for lower response box	0.02
MAXBAS	To transform rainfall to simulated runoff using a triangular weighting function	2.5

Projected streamflow result

One of the key objectives of this study was to evaluate simulated streamflow under RCP4.5 and 8.5 scenarios for two consecutive periods (2021–2040 and 2041–2060) compared to the baseline period (1996–2015). Thus, the time series of bias corrected precipitation, temperature and estimated evapotranspiration of the periods 2021–2040 and 2041–2060 under both RCPs were used as input for streamflow simulation. The optimized model parameters during calibration and validation were transferred with the intention to simulate future streamflow to investigate the changes from the baseline period. Projected streamflow in June, July and August is expected to increase more rapidly than the rest of the months under both scenarios (Figure 14(a) and 14(b)). Overall, streamflow is expected to increase under both scenarios and time periods (Figure 13(c) and 13(d)). On average, annual streamflow is expected to increase 28.5% under RCP4.5 and 23.9% under RCP8.5. In the near future period (i.e. 2021–2040), it is expected to increase up to 40.1% under RCP4.5 and 29.4% under RCP8.5. Likewise, for the far future period (2041–2060) it will increase to 18.5 and 16.9% under RCP4.5 and 8.5, respectively. However, for this period there will be a slight decrease in streamflow compared to the near future yet it will relatively increase compared to the reference period. Most of the change in streamflow is in line with the changes in rainfall and temperature (see also Figure 10). The maximum

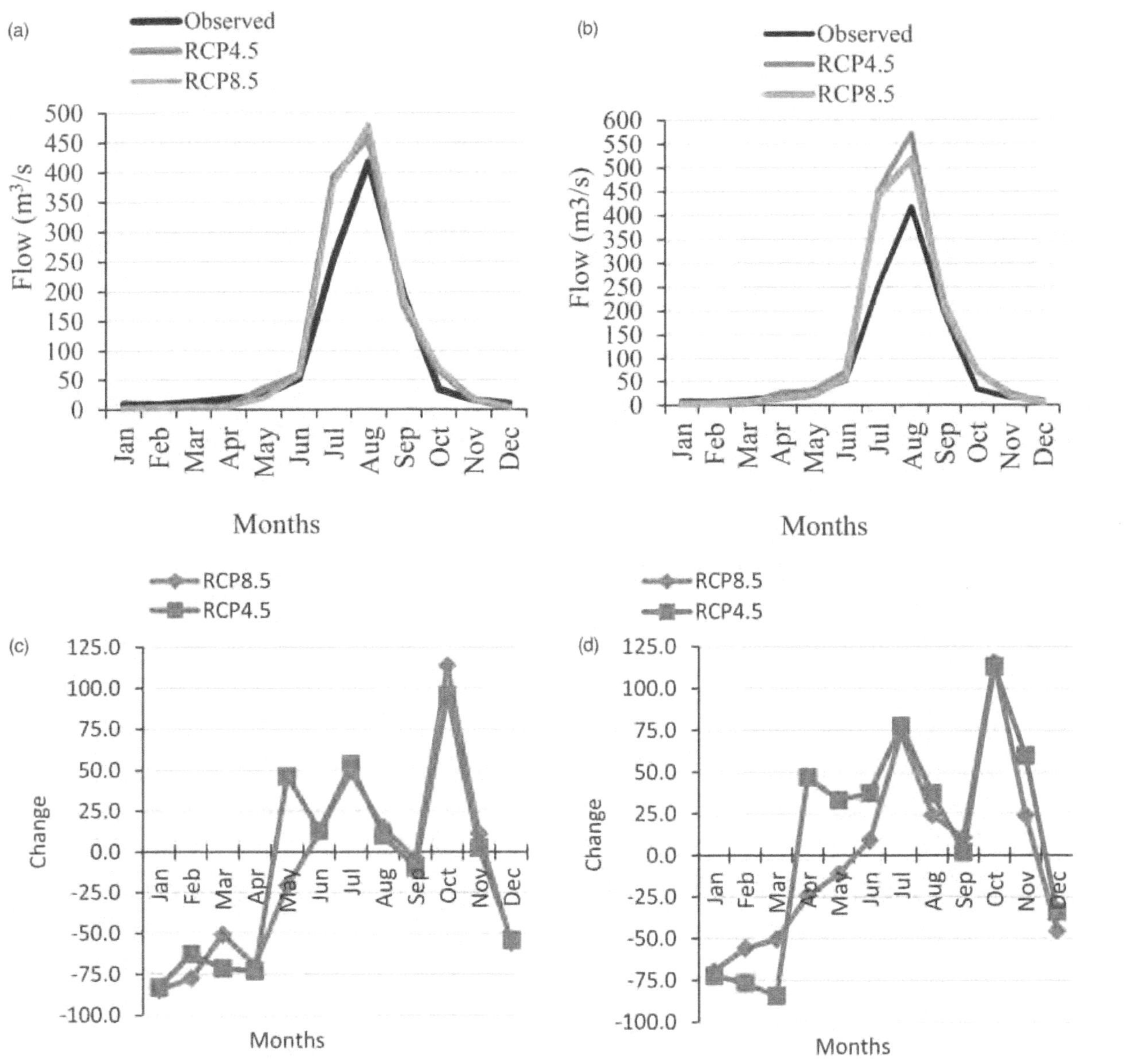

Figure 13 | Projected monthly average streamflow: (a) 2021–2040; (b) 2041–2060 and percentage change: (c) 2021–2040 and (d) 2041–2060, compared to the baseline period 1996–2015.

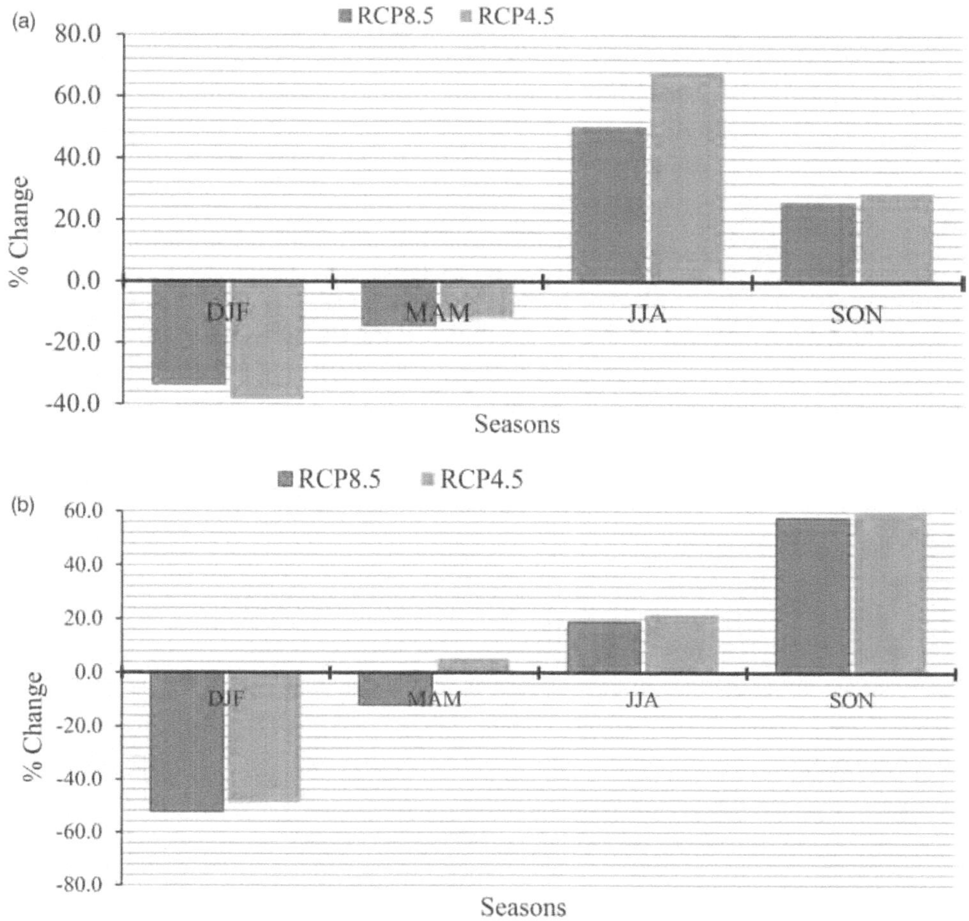

Figure 14 | Average seasonal change of streamflow: (a) 2021–2040 and (b) 2041–2060, compared to the baseline period (1996–2015).

increment will be expected in October under both scenarios while the maximum reduction will be expected in March under RCP4.5 and in January under RCP8.5. The increment suggested by the RCP4.5 scenario is more marked than under RCP8.5. This is due to the fact that increasing temperature under RCP8.5 increases evaporation which would lead to a reduction in streamflow. The projected seasonal change of streamflow from the baseline period was computed for winter, spring, summer and autumn seasons. In winter (DJF) and spring (MAM) the streamflow is expected to decrease except for the spring season under RCP4.5 for the period 2041–2060. Similarly, in summer (JJA) and autumn (SON) the projected seasonal streamflow is expected to increase under both climate scenarios and time periods (see Figure 14(a) and 14(b)). Generally, the average expected change in streamflow shows variations between months, seasons and annually. The average seasonal and annual change streamflow is relatively less compared to monthly change. In some months like October and July a higher magnitude of streamflow change may be expected which could lead to floods. On the contrary, in some months like January, February and March a significant reduction in streamflow may lead to water scarcity in the area. Table 6 shows the average change in streamflow for the periods 2021–2040 and 2041–2060 compared to the baseline period (1996–2015). The projected streamflow in all cases (monthly, seasonal and annual) is expected to increase. The increment of streamflow under RCP4.5 is higher than that under RCP8.5.

Table 6 | Average percentage changes of streamflow compared to base period

Scenarios	Monthly	Seasonal	Annual
RCP4.5	(+)34.75%	(+)18.51%	(+)28.5%
RCP8.5	(+)17.73%	(+)8.45%	(+)23.9%

Trend analysis results

This study was also aimed at examining trends. Accordingly, trends were tested in terms of annual time series for observed rainfall and streamflow and projected rainfall and corresponding streamflow using the non-parametric Mann–Kendall trend test at a significance level of 5%. The time series (trends) of observed annual streamflow and rainfall are presented in Figure 15 and projected rainfall and corresponding simulated streamflow for the period of 2021–2060 under RCP4.5 and 8.5 are presented in Figures 16 and 17. The observed mean annual flow and rainfall shows decreasing trends. The reduction was −7.73 mm/year for rainfall and −5.6 m^3/s/year for streamflow (Figure 15). However, the forecasted rainfall and streamflow under both RCP scenarios shows an increasing trend. The projected rainfall is expected to increase about 0.35 mm/year under RCP4.5 and 1.86 mm/year under RCP8.5. The forecasted streamflow under RCP4.5 is expected to increase by 6.34 and 7.63 m^3/s/year under RCP8.5. However, as presented in Table 7, the results of the calculated $|Z|$ values are less than the tabulated $|Z|$ at the significance level of 5%, which is 1.96. This shows that the trends for both recorded and simulated precipitation and streamflow are statistically insignificant under both RCPs.

DISCUSSION

The impact of climate change studies on streamflow by coupling climate models (RCM/GCM) with hydrological models involves a range of uncertainties (Graham *et al.* 2007; Fanget *et al.* 2015; Ahmed *et al.* 2017). Selection of a GCM/RCM model and plausible scenarios are the greatest sources of uncertainty on climate change analysis in addition to hydrological model simulation (Yacouba *et al.* 2017; Min *et al.* 2018). Despite these limitations, in this work every effort was made to minimize the uncertainties on climate prediction and hydrological simulation to understand the potential impact of climate

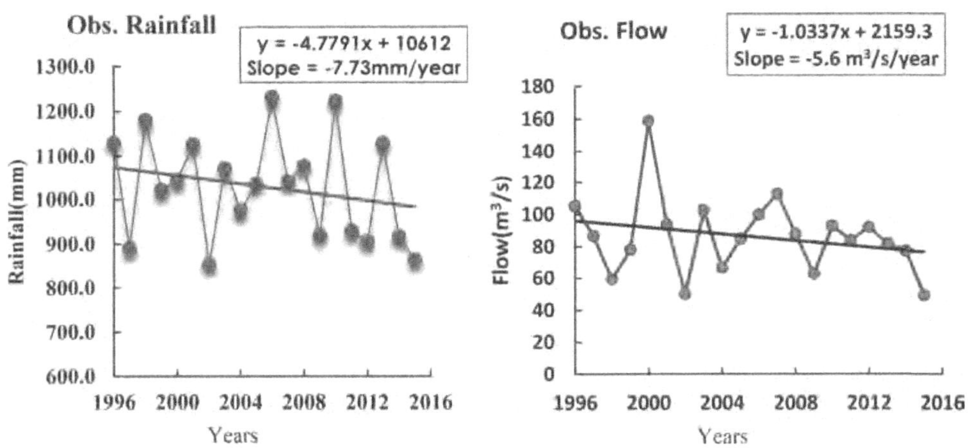

Figure 15 | Mean annaul trends of obseved rainfall and streamflow in the reference period (1996–2015).

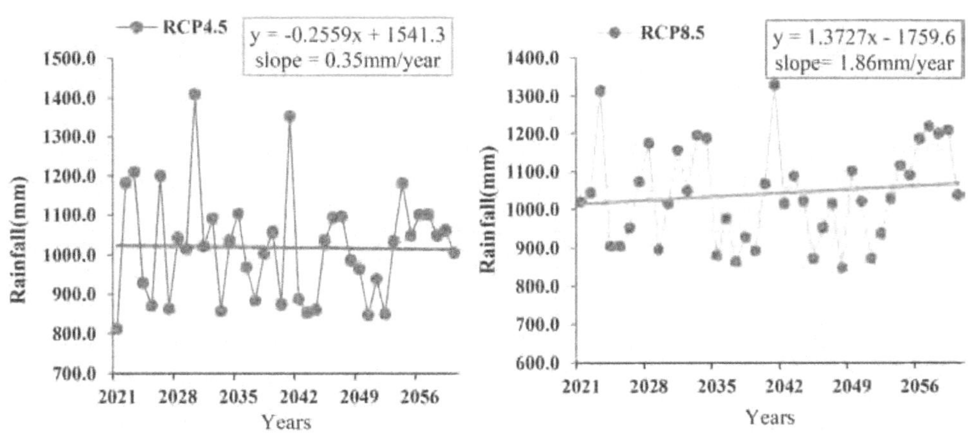

Figure 16 | Annual trends of projected rainfall from 2021 to 2060 under RCP4.5 and 8.5.

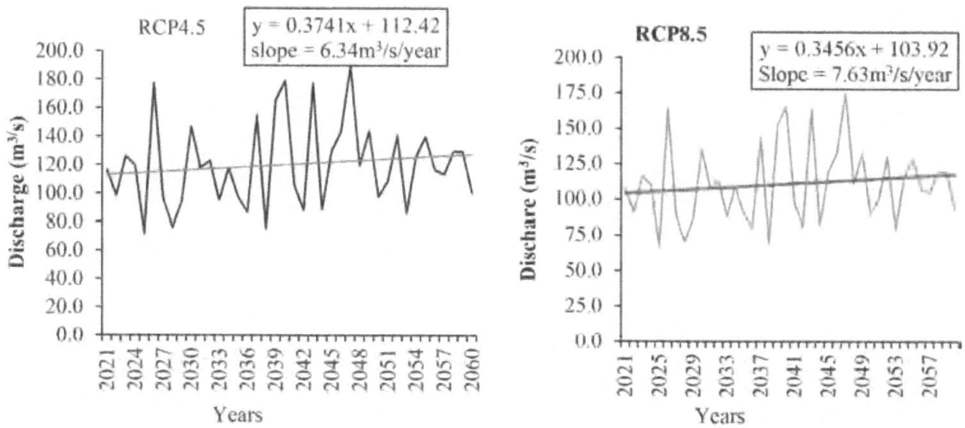

Figure 17 | Annual trends of simulated discharge from 2021 to 2060 under RCP4.5 and 8.5.

Table 7 | Statistical parameters for trend analysis and their results

Variables	Years	S	Z	Sens Slope	Trends
Rainfall (observed)	1996–2015	–34	–1.07	–7.73	Insignificant
Streamflow (observed)		–28	–0.88	–5.6	Insignificant
Rainfall (RCP4.5)	2020–2060	20	0.22	0.35	Insignificant
Rainfall (RCP8.5)		102	1.18	1.86	Insignificant
Streamflow (RCP4.5)	2020–2060	106	1.22	6.34	Insignificant
Streamflow (RCP8.5)		120	1.39	7.63	Insignificant

change on future streamflow changes, such as best climate model (RCM) selection, most plausible climate scenarios (RCP4.5 and 8.5), selection and bias correction climate model output. This study coupled a single climate model (RCM) under RCP4.5 and 8.5 emission scenarios with a semi-distributed hydrological model (HBV) to investigate future streamflow changes in the Upper Awash sub-basin. However, it is often recommended to apply different RCMs and ensemble models with all possible emission scenarios, including RCP2.6 and RCP6.0, to explore a wide range of climate changes. However, the results of this work are consistent with other similar studies (Getahun 2015; Mulligan 2015; Ahmed *et al.* 2017; Gizaw *et al.* 2017). Overall, projected temperature and precipitation is expected to increase in the study area. The increment in minimum temperature would be higher compared to maximum temperature and such changes are common globally. Similar findings were reported by IPCC (2013), Tekleab *et al.* (2013) and Negash *et al.* (2021). The temperature change is more severe under RCP8.5 than RCP4.5, confirming that RCP8.5 is the highest carbon emission scenario and this creates an active hydrological cycle which leads to heavy rainfall (IPCC 2013). The application of bias correction improved the projected temperature and precipitation accuracy. This finding is also in harmony with the results of Fanget *et al.* (2015), Ahmed *et al.* (2017) and Daba *et al.* (2017). On average, annual projected streamflow is expected to increase by 28.5% under RCP4.5 and 23.9% under RCP8.5. This result seems to be in line with similar studies that have been conducted by Mulligan (2015), Babur *et al.* (2016), Daba *et al.* (2017), Gizaw *et al.* (2017), Tadese *et al.* (2019) and Negash *et al.* (2021). However, it is in contradiction to the study by Kefeni *et al.* (2019). This may be due to differences in hydrological model, climate model resolution and bias-correction methods. Apart from this, further investigation is required to verify the results using a multi-climate model with different resolution and possible climate scenarios. The projected streamflow is expected to be high in July and October under both RCPs while a significant reduction is expected in January and April for the period 2021–2040 and in March for the period 2041–2060 under both RCPs emmision scenarios. A similar finding was depicted by Tadese *et al.* (2019). This study used an HBV hydrological model, which requires relatively less input data compared to the others to simulate streamflow. Thus, this method can be adapted and may be helpful especially to data limited regions to estimate likely streamflow changes in future (see also Endalkachew & Asfaw (2019)). Generally, on average, the expected change in streamflow varies among months, seasons and years. The expected streamflow variations in months and seasons are relatively less compared to annual changes.

CONCLUSION

Ethiopia is a country which is highly dependent on agricultural products for development. The availability of water resources in the future could be highly influenced by climate change. In this study, the potential response of climate change impact on streamflow availability in the Upper Awash sub-basin was assessed and quantified. A regional climate model (RCM) under RCP4.5 and 8.5 climate scenarios was used to obtain projected precipitation and maximum and minimum temperature in the future. Bias correction was employed to RCM temperature and precipitation to minimize uncertainties in climate projection before being fed into the HBV hydrological model to simulate streamflow. The application of bias correction improved the performance of climate model output (precipitation and temperature) in reproducing station-based recorded temperature and precipitation. Hence, it is practical to apply bias correction to the climate model output for better streamflow simulation for further study. The HBV hydrological model was calibrated and validated to simulate future streamflow. The streamflow response of two typical future periods (2021–2040 and 2041–2060) was predicted based on the bias corrected projected temperature and precipitation and estimated evapotranspiration was compared to the baseline period to assess the change. The predictive ability of the HBV model was evaluated by R^2 and *NSE* for both the calibration and validation periods. The results of R^2 and *NSE* were 0.89 and 0.88 for calibration and 0.74 and 0.85 for validation, respectively. Under both climate scenarios (RCP4.5 and 8.5) precipitation and temperature shows increasing trends and the temperature change is most severe under RCP8.5 confirming that RCP8.5 is the highest carbon emission scenario. The change of the maximum temperature is larger than the minimum temperature under both scenarios. Likewise, the change of precipitation is larger under RCP8.5 than the RCP4.5 change and the result agreed with the results of IPCC (2013). The average annual streamflow following an increase in precipitation is also expected to increase to 28.5 and 23.95% under RCP4.5 and 8.5, respectively. In the near future period (2021–2040), it could be increased to 40.1 and 29.4%, and 16.9% and 18.5% for the far future (2041–2060) under RCP4.5 and 8.5, respectively. A large increment of streamflow could be expected in October while a significant decline could be in March under both RCPs. There could be a slight decrease in streamflow in the far future (2041–2060) compared to the near future (2021–2040), yet will relatively increase compared to the reference period. Unlike the precipitation and temperature change, the projected streamflow suggested under the RCP4.5 scenario is increasingly more marked than that for RCP8.5. This may be due to the fact that increases in temperature under RCP8.5 resulted in high evaporation losses which could lead to a reduction in streamflow. The study conducted on the whole Awash River basin to evaluate climate change impact on future streamflow suggested that there will be slight decreases in streamflow. This may be due to the middle and lower part of the basin being historically subjected to water deficiency in all seasons and this could lead to an overall reduction in streamflow of the basin, but in the Upper basin (study area) projected streamflow is expected to increase. This study used the HBV hydrological model, which requires relatively less input data to predict future streamflow. Thus, adapting this kind of approach will be helpful for other basins, especially in data limited regions, to estimate likely future streamflow change. The results of this work can also help to inform the water resources planners and designers to frame appropriate planning and management strategies for the effective use of water in the future.

ACKNOWLEDGEMENTS

We express our sincere gratitude to the Ethiopian National Meteorological Services Agency (ENMSA) and Ministry of Water, Irrigation and Energy (MoWIE) for providing meteorological and hydrological data respectively. We would like to thank all colleagues for their encouragement throughout this study.

DATA AVAILABILITY STATEMENT

All relevant data are included in the paper or its Supplementary Information.

REFERENCES

Ahmed, M., Yves, T., Lahoucine, H., Denis, R. & Lionel, J. 2017 Climate change impacts on surface water resources in the Rheraya catchment (High Atlas, Morocco). *Hydrological Sciences Journal* **62**, 979–995. doi:10.1080/02626667.2017.1283042.

Awash Basin Autority 2017 *Awash River Basin Strategic Plan Main Report*. Awash Basin Authority, Addis Ababa, Ethiopia.

Babel, M. S., Bhusa, S. P., Wahid, S. M. & Agarwal, A. 2014 Climate change and water resources in the Bagmati River Basin, Nepal. *Journal of Theoretical and Applied Climatology* **115**, 639–654.

Babur, M., Mukand, B. S., Sangam, S., Akiyuki, K. & Tripathi, K. 2016 Assessment of climate change impact on reservoir inflows using multi climate-models under RCPs – the case of Mangla Dam in Pakistan. *Water* **8**, 389–409.

Bergström, S. 1992 *The HBV Model – Its Structure and Applications*. Swedish Metreological and Hydrological Institute (SMHI), Report Hydrology (RH) No 4. Norrköping, Sweden.

Brekke, L. D., Maurer, E. P., Anderson, J. D., Dettinger, M. D. & Townsley, S. E. 2009 Assessing reservoir operations risk under climate change. *Water Resource Journal* **45**. doi:10.1029/2008WR00694.

Claudia, A. 2012 Bias correction of regional climate model simulations for hydrological climate-change impact studies: review and evaluation of different methods. *Journal of Hydrology* **12**, 12–29.

Daba, M., Tadele, K. & Shimelis, A. 2017 Hydrological response to climate change impacts and adaptation strategies of Upper Awash Sub-Basin. In: *The International Conference on Climate Change Impacts for Scientists and Stakeholders*, 11–13 October 2017, Postdam, Germany. Available from: http://www.impactworld2017.org.

Edossa, D. C., Babel, M. S. & Gubta, A. D. 2010 Drought analysis in the Awash River Basin. *Water Resourse Management Journal* **24**, 1441–1460.

Elena, S., Chiara, B., Alexandre, S. G. & John, J. H. 2019 Mitigating climate change in the cultural built heritage sector. *Climate Journal* **7**, 90–108.

Endalkachew, A. & Asfaw, K. 2019 A comparation of hydrological models under climate change on the water resources of Megech River Catchment, Abbay Basin, Ethiopia. *Journal of Natural Sciences Research* **9**, 2224–3186. doi:10.7176/JNSR.

Fanget, G. H., Yang, J., Chen, Y. N. & Zammit, C. 2015 Comparing bias correction methods in downscaling meteorological variables for a hydrologic impact study in an arid area in China. *Hydrology Journal of Earth System* **19**, 2547–2559.

Getahun, Y. S. 2015 Impact of climate change on hydrology of the Upper Awash River Basin, Ethiopia. In: *Inter-Comparation of old SRES and new RCP Scenarios*. University of Wageningen, Wageningen. Available from: http://www.researchgate.net/puplication/316505332.

Gizaw, M. S., Biftu, G. F., Gan, T. Y., Moges, S. A. & Koivusalo, H. 2017 Potential impact of climate change on stream flow of major Ethiopian rivers. *Climate Change Journal* **143**, 371–383. doi:10.1007/s1584-017-2021-1.

Gleckler, P. J., Taylor, K. E. & Doutriaux, C. 2008 Performance metric for global climate models. *Journal of Geophysical Research* **113** (6). doi:10.1029/2007JD008972.

Gordon, N. & Shaykewich, J. 2000 *Guidelines on Performance Assessment of Public Weather Services. Report No. WMO/TD No.1023*. World Meteorological Organization, Geneva, Switzerland.

Graham, L. P., Haggemann, S., Jaun, S. & Beniston, M. 2007 On interpreting hydrological change from regional climate models. *Climate Change Journal* **81**, 97–122.

Hargreaves, G. & Riley, J. 1985 Agricultural benefits for Senegal River Basin. *Journal of Irrigation and Drainage Engineering* **111** (2), 113–124.

Hess, A., Iyer, H. & Malm, W. 2001 Linear trend analysis: a comparison of methods. *Atmospheric Environmental Journal* **35**, 5211–5222.

Höök, M., Sivertsson, A. & Aleklett, K. 2010 Validity of the fossil fuel production outlooks in the IPCC emission scenarios. *Natural Resources Research* **19**, 63–81.

IPCC 2007 *Climate Change: AR4 Synthesis Report*. Cambridge University Press, Cambridge. Available from: http://www.ipcc.ch.

IPCC 2013 Climate change: the physical science basis. In: *Contribution of Working Group I to the Fifth Assessment Report of the Intergovernmental Panel on Climate Change* (Stoker, T., Qin, D., Lattner, K., Tigner, M., Allen, K., Boschung, J., Nauels, A., Xia, Y. & Bex, V., eds). Cambridge University Press, Cambridge, UK.

IPCC 2014 Climate change 2014: synthesis report. In: *Contribution of Working Groups I, II and III to the Fifth Assessment Report of the Intergovernmental Panel on Climate Change* (Pachauri, R. K. & Meyer, L. A., eds). World Meteorological Organization, Geneva, Switzerland.

Kefeni, K., Dessalegn, T. & Roba, N. 2019 Evaluation of climate change impact on stream flow in Upper Awash River Basin, Ethiopia. *International Journal of Scientific & Engineering Research* **10** (8), 2229–2258.

Krause, P., Boyle, D. & Base, F. 2005 Comparation of different efficiency criteria for hydrological model assessment. *Advanced Geoscience Journal* **5**, 89–97.

Min, L., Tie, L., Fanhao, M., Yongchao, D., Amaury, F., Anming, B. & Philippe, M. D. 2018 Comparing bias correction methods used in downscaling precipitation and temperature from regional climate models: a case study from the Kaidu River Basin in Western China. *Water* **10**, 1046. doi:10.3390/w10081046.

Mulligan, M. 2015 Climate change and food-water supply from Africa's drylands: local impacts and teleconnections through global commodity flows. *International Journal of Water Resources* **31**, 450–460.

Negash, R. T., Dame, Y. G. & Asfaw, K. K. 2020 Modelling the effect of climate change on streamflow using climate and hydrological models: the case of the Kesem Sub-basin of the Awash River Basin, Ethiopia. *International Journal of River Basin Management* **19**, 6187–6207. doi:10.1080/15715124.2020.1755301.

Negash, R. T., Asfaw, K. K. & Dame, G. Y. 2021 Modeling climate change impacts on crop water demand, middle Awash River basin, case study of Berehet woreda. *Water Practice & Technology* **16**, 89–97. doi: 10.2166/wpt.2021.033.

Räisänen, J., Ruokolainen, L. & Yilhaisi, J. 2009 Weighting of model results for improving best estimate of climate change. *Climate Dynamics Journal* **35**, 407–422.

Rientjies, T. 2007 *Modeling Hydrology*. ITC Lecture Books, Uppsala, Sweden.

Schaefli, B. 2015 Projecting hydropower production under future climate: a guide for decision makers and modelers to interpret and design climate change impact assessment. *Wiley Interdisciplinary Reviews* **2**, 271–289.

Seibert, J. 2005 *HBV light. Version 2.* User's manual, Stockholm.

Seneviratne, S., Nicholls, N., Easterling, D., Goodess, C., Kanae, S., Kossin, J. & Zhang, X. 2012 *Change in Climate Extremes and Their Impacts on the Natural Physical Environment. A Special Report of Working Groups I and II of the Intergovermental Panel on Climate Change (IPCC)*. Cambridge University Press, Cambrige, UK. and New York, USA.

SMHI 2008 *Integrated Hydrological Modelling System: Manual Version 5.10 HBV-Model*. Swedish Metreological and Hydrological Institute (SMHI), Norrköping, Sweden.

Srinivasan, R., Zhang, X. & Arnold, J. 2010 Swat ungauged hydrological budget and crop yield predictions in the Upper Mississipi River Basin. *Transactions of the ASABE* **53**, 1533–1546.

Tadese, T. M., Kumar, L., Koech, R. & Zemadim, B. 2019 Hydro-climatic variability: a characterisation and trend study of the Awash River Basin, Ethiopia. *Hydrology Journal* **6**, 35–42. doi:10.3390/hydrology6020035.

Tekleab, S., Mohamed, Y. & Uhlenbrook, S. 2013 Hydro-climatic trends in the Abay/Upper Blue Nile Basin, Ethiopia. *Physical Chemistry Journal* **61**, 32–42.

Wanders, N. & van Lanen, H. A. J. 2015 Future discharge drought across climate regions around the world modelled with a synthetic hydrological modelling approach forced by three general circulation models. *Natural Hazards Earth System Journal* **15**, 487–504.

Yacouba, Y., Bernd, D., Gew, S. & Yaovi, B. 2017 Impact of climate on hydrological condition in tropical West African Catchment using an ensemble of climate simulation. *Hydrology and Earth System Sciences* **21**, 2143–2161.

First received 24 June 2021; accepted in revised form 11 October 2021. Available online 16 November 2021

© 2022 The Authors Impact of Climate Change on Hydrology and Water Resources doi: 10.2166/wcc.2021.049

Evaluating the potential impact of climate change on the hydrology of *Ribb* catchment, Lake Tana Basin, Ethiopia

Dessalegn Worku Ayalew [a,*], Tirusew Asefa[b], Mamaru Ayalew Moges [c,d] and Sileshie Mesfin Leyew [e]

[a] College of Agriculture, Woldia University, Woldia, Ethiopia
[b] Systems Decision Support, Tampa Bay Water, Clearwater, Florida, United States Minor Outlying Islands
[c] Bahir Dar Institute of Technology, Bahir Dar University, Bahir Dar, Ethiopia
[d] Amhara National Regional State Water Bureau, Bahir Dar, Ethiopia
[e] Amhara Design and Supervision Works Enterprise, Amhara, Ethiopia
*Corresponding author. E-mail: desuworku60@gmail.com, desuworku25@gmail.com

DWA, 0000-0003-3416-6394; MAM, 0000-0002-6471-741X; SML, 0000-0002-0989-5969

ABSTRACT

Scientific findings indicated there is climate change that affects given hydrology and, hence, water availability worldwide. To quantify its impact on a specific catchment scale, since spatial and temporal variability of climate change impact, this study was carried out at *Ribb* catchment, Lake Tana basin, Ethiopia. The catchment hydrology was represented by the Soil and Water Analysis Tool (SWAT) through using historical observed data. Regional Climate Model (RCM) projection data set for Nile Basin studies at Representative Concentration Pathway (RCPs) (RCP4.5 and RCP8.5) were used for future streamflow generation on three-time horizons; 2020s (2011–2040), 2050s (2041–2070), and 2080s (2071–2098). A baseline period (1976–2005) was used as a reference. SWAT was calibrated ($R^2 = 0.83$ and NSE = 0.74) and validated ($R^2 = 0.72$ and NSE = 0.71). The analysis was done based on the changes from the baseline period to the 2080s. Temperature showed an increasing trend but rainfall is decreasing. The mean annual streamflow could potentially reduce from 42.78 m^3/s to 40.24 m^3/s and from 42.78 m^3/s to 37.58 m^3/s based on RCP4.5 and RCP8.5 scenarios, respectively. On a monthly time scale, decreases in streamflow were found from March to August whereas there was a slight increase from September to February. Concerning individual months, June flows were found to have maximum impact in both scenarios (63.3% at RCP8.5 and 55.45% at RCP4.5 scenarios). The least impacted month was August based on the RCP8.5 scenario which is decreased by 6.64% and April based on the RCP4.5 scenario which is reduced by 1.21%. Looking at total volume, July showed a maximum decrease in both scenarios which is reduced by 21.08 m^3/s at the RCP4.5 scenario and 51.22 m^3/s at the RCP8.5 scenario. The maximum increase was found in October with 10.31 m^3/s and 11.26 m^3/s at RCP4.5 and RCP8.5 scenarios respectively. The future streamflow of *Ribb* River has decreased annually and monthly due to increasing temperature and reduction of rainfall.

Key words: climate change, RCP Scenarios, streamflow, SWAT

HIGHLIGHTS

- The research was done using the current RCP scenarios.
- The research work focuses on future climate change impacts of the streamflow.
- It is done on the different time horizons are predicted the climate change impact on the hydrology of a given catchment for a long period (up to 2100).
- It used direct observed data to the stations to calibrate the model.

ACRONYM AND ABBREVIATION

GCM: General Circulation Model; HRUS: Hydrological Response Units; IPCC: Intergovernmental Panel on Climate change; RCPs: Representative Concentration Pathways SWAT: Soil and Water Analysis Tool; SWAT-CUP: SWAT-Calibration and Uncertainty program; ALPHA_BNK: Baseflow alpha factor for bank storage; CANMX: Maximum canopy storage; CN2: Initial SCS CN II value; BIOMIX: Efficiency of soil biological mix; ESCO: Soil evaporation compensation factor; REVAPMN: Threshold water depth in the shallow aquifer for 'revap'; RCHRG_DP: Deep aquifer percolation fraction;

GW DELAY: Groundwater delay; GWQMN: Threshold water depth in the shallow aquifer for flow; GW_REVAP: Groundwater 'revap' coefficient; SOL_AWC: Available water capacity.

INTRODUCTION

Climate change refers to the change of climate variables, temperature, and precipitation. The average global temperature has risen by about 0.74 °C from 1906 to 2005 (IPCC 2007). This climate change happened in two phases, from the 1910 to 1940s and more strongly from the 1970s to 2005. An increase in temperature causes higher evapotranspiration and alters the global precipitation situation (Urrutia & Vuille 2009; Paparrizos *et al.* 2016). Increasing temperature also affects the occurrence, property, distribution, and movement of water resources and causes a higher frequency of hydrological events such as floods and droughts in the catchment level. A high concentration of greenhouse gases in the atmosphere causes increase in temperature and variability of rainfall distribution and amount (Houghton 2001). Rainfall and temperature (climate) change impacts surface hydrology and water resources which can change regional water balances and hydrological regimes.

The beginning of the 21st century is also the warmest (IPCC 2013). Global climate change impact will continue by increasing temperatures and variability of rainfall (Alexander *et al.* 2006; Zwiers & Wehner 2013). Climate change impact is the main issue in Africa, as the income of the continent depends on agriculture which is susceptible to climate change. The continent is one of the most impacted regions to climate change (IPCC 2007). Similarly, future climate change is shown to increase in Ethiopia. Conway and Schipper (2011) testified future climate change using medium to high Special Report on Emission Scenarios (SRES) of annual warming across Ethiopia could be 0.7–2.3 °C and 1.4–2.9 °C in the 2020 and 2050s respectively. Future climate change at the Lake Tana basin level showed variability of rainfall and an increase in temperature (Abdo *et al.* 2009; Gebremariame 2009; Adem *et al.* 2015).

IPCC reports showed the presence of climate change at the international level based on different scenarios data. The current report of IPCC, which is used for different impact assessment studies, is based on the Representative Concentration Pathways (RCPs) scenarios. It is the latest generation that provides input to climate models, which are time and space-dependent trajectories of concentrations of greenhouse gases and pollutants resulting from human activities. These scenarios provide a quantitative description of concentrations of climate change pollutants in the atmosphere over time, as well as their radioactive forcing in 2100. Four main scenarios have been defined with different targets of radioactive forcing. The RCP 8.5: It is considered increasing greenhouse gas emissions over time; will reach over 8.5 W/m^2 by 2100 and will continue to rise onwards after some time. RCP 6.0: total radioactive forcing is stabilized after 2100 without adopting any technologies and strategies to reduce greenhouse gas emissions. RCP 4.5: total radioactive forcing is stabilized before 2100 by applying a range of technologies and strategies to reduce greenhouse gas emissions. RCP 2.6: radioactive forcing peaks at approximately 3 W/m^2 before 2100 and then declines to approximately 2.6 W/m^2 in 2100.

Several empirical, physically based and conceptual models could be used for hydrological studies. Soil and Water Assessment Tool (SWAT) model, physically-based semi-distributed watershed model, uses modified Soil Conservation Service-Curve Number (SCS CN) method to estimate the runoff. This model is simple, user-friendly, used worldwide, and freely available. SWAT showed a good application performance in Ethiopia. It was calibrated and validated frequently in the Blue Nile basin and showed good performance (Kassa & Foerch 2007; Awulachew *et al.* 2008; Setegn *et al.* 2008; Shimelis 2008; Adamu 2013). Modeling of Gumara watershed in Lake Tana basin (Awulachew *et al.* 2008) indicated that streamflow and sediment yield simulated with SWAT was reasonably accurate.

The degree of climate change is spatially and temporally variable (Droogers & Aerts 2005). It is needed to study the impacts at a small catchment scale which has not been given much attention in developing countries like Ethiopia. Similarly, the previous studies in the Lake Tana sub-basin and also in Ethiopia mostly used Global Climate Models (GCMs) data of the third phase Coupled Model Intercomparison Project (CMIP3) downscaled to the local level based on A2 and B2 SRES. However, currently, available Representative Concentration Pathways (RCPs) are used for impact assessment and showed a better agreement to observed climate data (Fentaw *et al.* 2018). Estimation of future streamflow using projected changes of local-scale climate variables is necessary to properly plan for future climate change impact adoption and mitigation measures. Planning for sustainable utilization of water resources based on climate change impact assessment is critical for Ethiopia since the income of the country depends on agriculture which is highly vulnerable to climate change. Despite the significance of the Lake Tana basin for continuous settlement of the people, little has been done on future climate change impact assessment (Gebremariame 2009). *Ribb* River is among the main tributaries of Lake Tana, the origin of the Blue Nile. Blue Nile basin is the largest basin in Ethiopia with large population pressure, degraded land, and highly dependent on agricultural (Tsegaye 2006), and the Blue Nile (Abbay) River supplies water for many downstream African countries. The river is highly vulnerable to climate change which affects streamflow but the effects of climate change in water resource analysis and management at the Lake Tana sub-basin and the Blue Nile basin has not been adequately addressed (Setegn *et al.* 2011).

The main objective of this study was to estimate the potential impact of climate change on the water flow of Ribb River with the specific objectives: (1) Representing catchment hydrology through calibration and validation of Soil and Water Assessment Tool (SWAT); (2) Quantifying the potential impact of climate change on future streamflow relative to the present situation based on RCP4.5 and RCP8.5 scenarios data.

MATERIALS AND METHODS

Description of the study area

Ribb catchment is located in the Lake Tana basin, Ethiopia. The catchment stretches between 11°40′0″–12°10′0″N latitude and 37°30′0″–38°14′0″E (Figure 1). The total area of the catchment (gauged part) is about 131, 250 ha. Ribb River originates from the higher elevation of *Guna* Mountains, Southern Gondar. The catchment supports the total number of the population about 888,260 based on Statistical, (2012), and about 88.5% of the population mainly engaged in agriculture.

The elevation of a catchment developed from ArcGIS by using 30 m × 30 m resolution DEM; higher elevation is found at *Guna* Mountains, the southeastern part of the catchment, and also north and northeastern tips (Figure 2 left). The lower elevation is found at the western part of the catchment around *Fogera plain*. The catchment has also various slope features (Figure 2 right). The dominate slope is from 15–30% which covers 30% of the total area, about 22.5% of the area has a slope greater than 30% and only 8% has a slope of 0% to 3%.

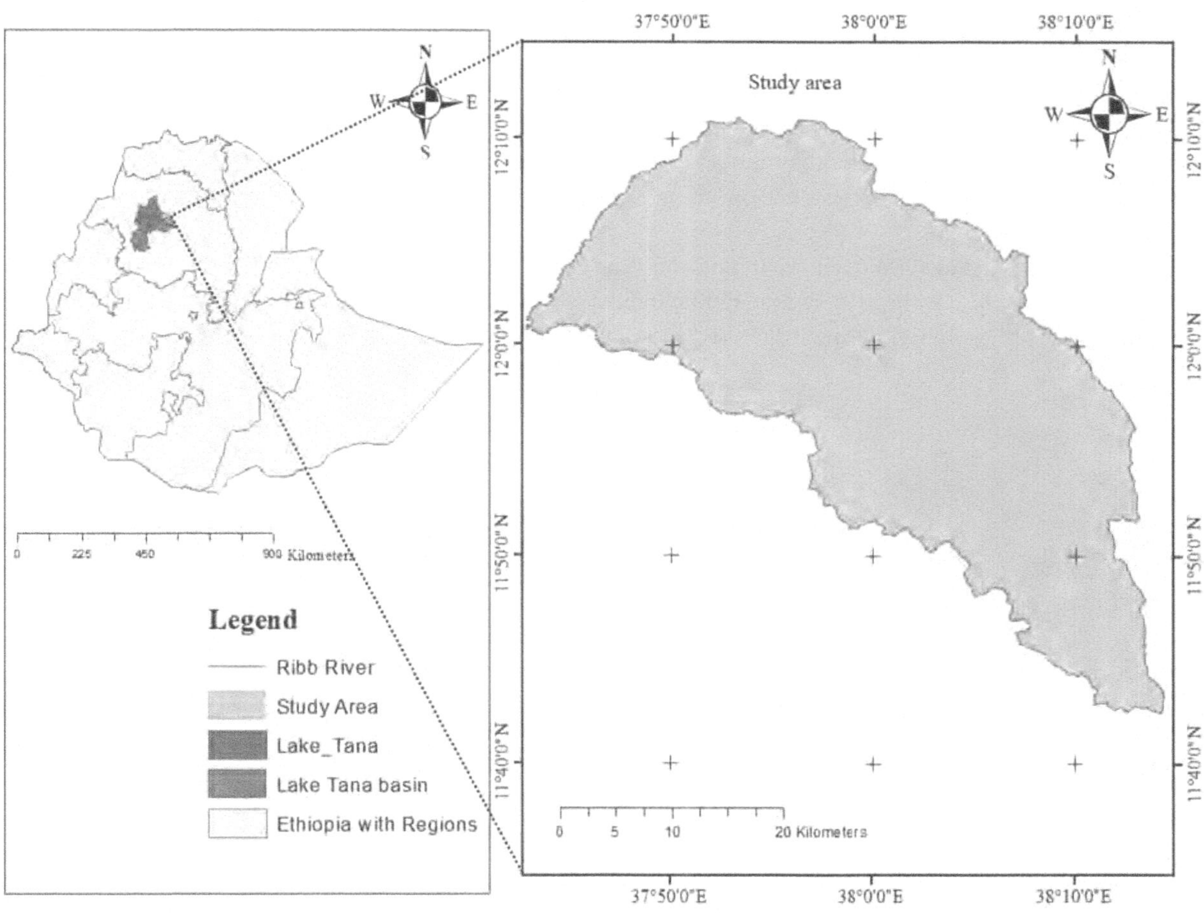

Figure 1 | Geographic locations of the Study Area.

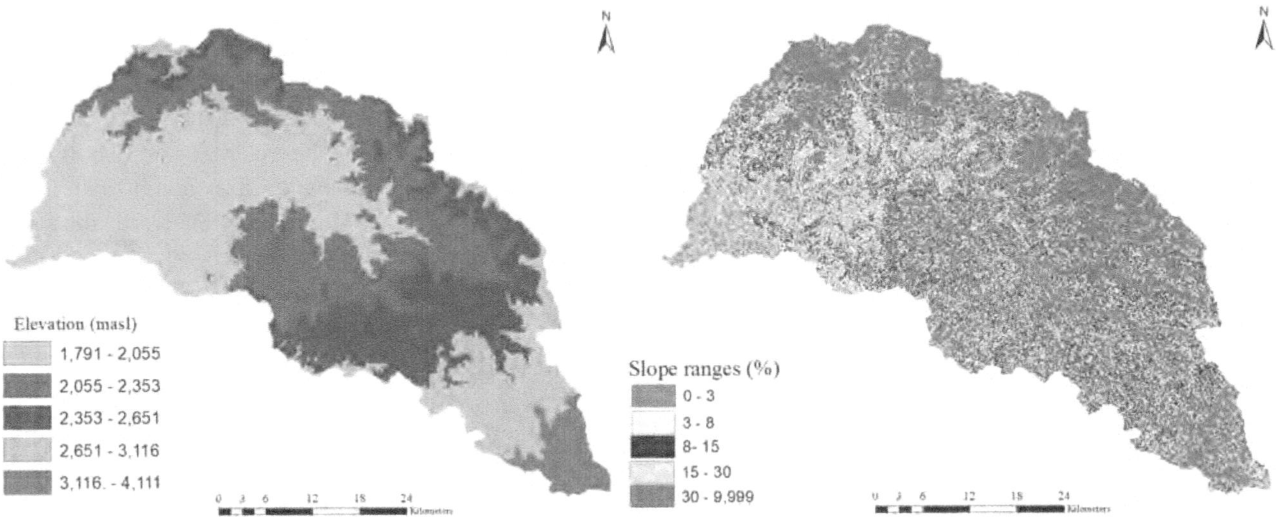

Figure 2 | Elevation and slope information.

Major soil types found in the catchment were Alisols, Luvisols, Nitosols, Vertisols, Regosols, Cambisols, Fluvisols, and Leptosols (Figure 3 left). Land use is changed to agricultural land due to rapid population growth. Since the catchment currently supports a large population, the dominant land use is for agriculture which covers about 95.5% (Figure 3 right).

The temperature of the catchment varies spatially and temporally. The mean annual maximum and minimum observed temperatures were about 24 °C and 8.77 °C respectively. The mean monthly maximum highest temperature record was found in March. April, February, October, November, and December also showed higher temperature records. The lowest maximum temperature was recorded in August and July. In monthly minimum temperature, the highest temperature was recorded in May and July and the lower record was found in December, November, and January.

Rainfall of the catchment also varies spatially, temporally, and seasonally. The mean annual rainfall of the catchment was about 1,350 mm per year. The mean monthly rainfall distribution showed maximum rainfall in July and August. The daily average discharge of the river was about 14.75 m^3/s, peak discharge of 160 m^3/s and minimum flow 0 m^3/s. The mean monthly flow is highest in August.

Study results obtained at Lake Tana sub-basin and the Blue Nile basins done by Conway (2000), Kebede *et al.* (2006), Sutcliffe & Parks (1999), and Tarekegn & Tadege (2005) indicated the hydrological year of the study areas are characterized by one main rainy season (summer) between June to September, in which 70% to 90% of the annual total rainfall occurs.

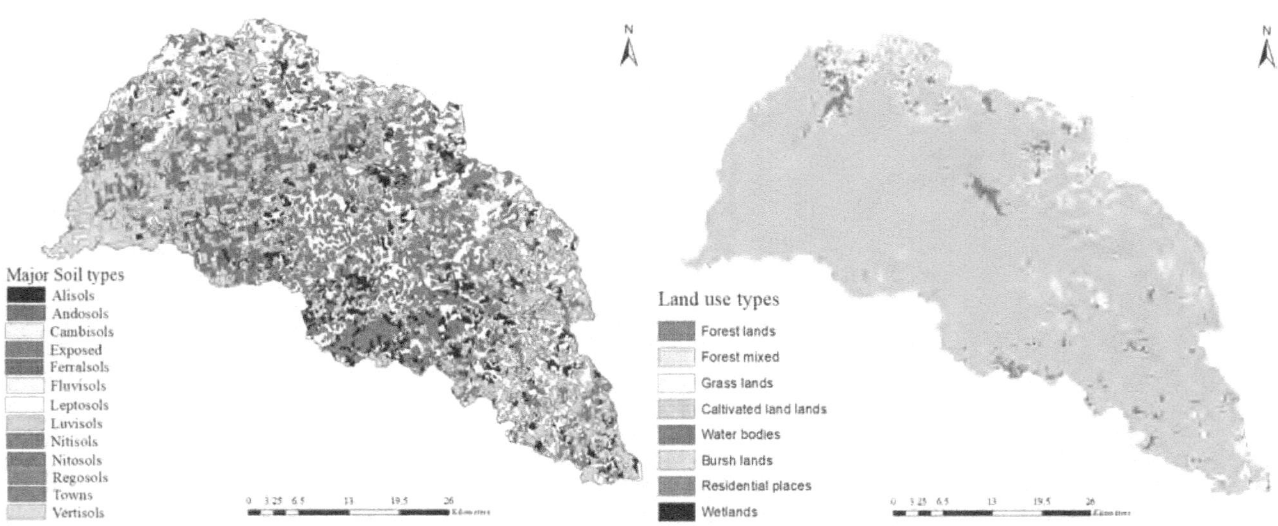

Figure 3 | Soil and land use/landcover.

Data sets

The observed meteorological data were rainfall, maximum and minimum temperatures, wind speed, relative humidity, sunshine hours, and solar radiation. These meteorological data were collected from the Ethiopian Meteorological Agency, Bahir Dar branches. However, only Debre Tabor meteorological station is found inside the catchment with the required quantity and quality of data. Taking only one meteorological station data may not rightly represent the entire area due to the diverse topographic feature of the study area. To alleviate this problem, additional meteorological data were collected from the surrounding stations: Addis Zemen, Amed Ber, Yifag, and Agere Genet meteorological stations (Table 1). Debre Tabor is the only principal station that has all the required data. Solar radiation was prepared from daytime sunshine hour data using the Angstrom method. The other model input data, the weather generator which is used to provide data for stations having missing data, was prepared manually by using PCPSTAT and dewPOINT2.0 version programs on 21-year rainfall, temperature, and wind data obtained from Debre Tabor meteorological station.

Digital Elevation Model (DEM), land use, and soil map data were spatial data used for this study. These data were obtained from Amhara Design and Supervision Works Enterprise (ADSWE) office. The observed meteorological and spatial data were used as input to SWAT to see its performance to represent the catchment hydrology. The observed discharge (flow) data of the river, collected from Amhara Water, Irrigation and Energy office, Ethiopia, was used for sensitivity analysis, calibration, and validation of the model.

The projected rainfall and temperature data at RCP4.5 and RCP8.5 scenarios were used to simulate the future discharge of the river. Nile Basin Initiative partnership prepared climate change projections data set for the Nile basin studies. It was established for the aims of the provision of well-vetted climate change projection data at an appropriate spatial and temporal resolution for key target end-users (Asefa 2019). Historical (1976–2005) and projected (2011–2098) temperature and rainfall data from SMHI-RCA4 Regional Climate Model (RCM) by 0.25-degree of resolution which derived from the MPI-M-MPI-ESM-LR Global Climate Model (GCM) were used for this study (Table 2).

Table 1 | Meteorological data availability

Stations	Rainfall	Temperature	Wind speed, RH & sunshine hours	Data available Year
Debre Tabor	x	x	x	1976–2015
Addis Zemen	x	x	–	1976–2015
Amed Ber	x	–	–	1976–2015
Yifag	x	–	–	1988–2015
Agere Genet	–	x	–	1988–2015

RH, Relative humidity.

Table 2 | Materials used for the study

No	Materials	Purpose
1	ArcGIS	Study area map preparation, Geo-referencing, rectification, and other various spatial analysis
2	ArcSWAT2012	Study are delineation, model development, run execution
3	XLSTAT2016	For meteorological and flow data quality testing
4	PCPSTAT	To prepare weather generator input data
5	DEWPOINT0.2	To prepare weather generator input data
6	SWAT_CUP2012	Used for model calibration and validation
7	Mintab 18	To stack the climate and flow data as SWAT input format
8	UTM Convertor	To convert the coordinates of the meteorological and gauge stations from latitude and longitude to UTM and vice versa.
9	Medley	Used to insert citations biography

Method

Data quality assessment, such as filling the missed data, data consistency checking, and data homogeneity and trend analysis, is needed to perform hydrological studies. The presence of missing data is a common and serious problem and producing bias in study results. No matter how much the data measuring system is well designed, the data recorder is professional and instruments are accurate; there are missing meteorological and hydrological data. Similarly, using inaccurate missing data filling methods also affects the accuracy of the study results. To perform a good analysis and simulation by using long time series continuous data, handling this problem by using the accepted missing data filling method is a pre-request. Many methods are employed to fill in missing data. Arithmetic mean, Normal ratio, Regression, and distance power methods are commonly used. Choosing an appropriate method depends on method simplicity to apply, length of missing data, and accuracy. The normal ration method was used based on the upper criteria.

The data homogeneity test was checked by Addinsoft's XLSTAT2016. It was evaluated based on *P* values estimated by the Monte Carlo method at a 95% confidence interval. As computed *p*-value is greater than alpha value, 0.05, it indicates data homogeneity. If a computed *P*-value is less than the Alpha value, it indicates the data is not homogeneous. Trend analyses of the observed data were evaluated by using the Mann-Kendall trend test in XLSTAT2016. The algorithm of this method is when computed *P*-value is greater than Alpha value; data has no significant trend (Arun & Sananda Kundu 2012) and vice versa. The data consistency was checked by the double mass curve method.

The meteorological and hydrological data were prepared based on the SWAT input data format by Mintab18. The meteorological data from 1988 to 2015 were used for SWAT modeling and from 1976 to 2005 for baseline period data to generate the streamflow which is used as a reference. The observed discharge data from 1990 to 2013 (1990–2005 for model calibration and 2006–2013 to model validation) was used with 2 years model warm-up period.

SWAT consists of sets of processes including project setup, watershed delineation, hydrologic response unit (HRU) definition, writing up the input table, editing SWAT input (if necessary), and SWAT simulation to get the final output. Watershed delineation was done by watershed delineator tool in ArcSWAT interface on ArcGIS by using Digital Elevation Model (DEM) as input. A watershed is delineated into many sub-basins, which are further divided into Hydrological Response Units (HRUs) that consist of homogeneous land use, slope, and soil characteristics. A detailed description of telling how SWAT is working is found in the user manual. SWAT was run by monthly data.

Many parameters affect the hydrologic system of a specific catchment but all parameters do not have equal effects. Therefore, sensitivity analysis was carried out to determine the number of optimized parameters to obtain a good fit between SWAT simulated and observed streamflow data.

Sensitivity analysis helps to know the relative ranking of which parameters mostly affect the output (Griensven *et al.* 2006), it reduces the uncertainty of model outputs and provides parameter estimation guidance for model calibration and validation. It was undertaken in SWAT-CUP by using the global sensitivity design method. Sensitivity analyses were conducted by using 23 flow parameters (Table 3) which were selected based on the reviewing different articles which were done at the Blue Nile and specifically Lake Tana sub-basin.

SWAT-CUP uses t-test and *p*-value to rank sensitive parameters that correspond to a greater change in output response. A t-stat provides a measure of sensitivity (larger in the absolute value of t-stat are more sensitive) and *p*-value determined significance of sensitivity, a value close to zero has more significance. Eleven most sensitive parameters were selected and model calibration and validation were conducted. SWAT_CUP using the SUFI_2 algorithm is commonly used in the Lake Tana sub-basin and Blue Nile system and showed the best-fitted result for *Gilgel* Abay, *Gumara*, *Ribb*, and *Megech* catchments (Setegn *et al.* 2009) for model calibration and validation. This algorithm was used for this study.

The level of goodness to fit of models has been evaluated by an objective function that measures the level of agreement between observed and model-simulated output. Model Performance was evaluated by using Nash-Sutcliffe efficiency (N_{SE}) and Coefficient of determination (R^2) (Table 4).

Future streamflow of the river was simulated by using the SWAT model after verifying the model performance by the observed data. The Nile Basin Initiative partnership projected rainfall and temperature data set for Nile basin studies were used as input to the model. This data set was developed based on the recently reported Representative Concentration Pathway scenarios. These projected data have a finer resolution but it was further downscaled to the nearest meteorological stations observed data. The downscaling was done by the linear scaling bias correction method. The future streamflow was simulated on three time horizons: 2020s (2011–2040), 2050s (2041–2070), 2080s (2071–2098), and baseline period (1976–2005) as a

Table 3 | Parameters used for sensitivity analysis

No	Parameters	No	Parameters
1	Base flow alpha factor (days)	13	Most soil albedo
2	The efficiency of soil biological mix (dimensionless)	14	Available water capacity (mm water/mm soil)
3	Maximum canopy storage (mm)	15	Saturated hydraulic conductivity (mm/hr)
4	Channel effective hydraulic conductivity (mm/hr.)	16	Soil depth (mm)
5	Manning's *n* value for the main channel	17	Surface runoff lag time (days)
6	Initial SCS CN II value	18	Temperature lapse rate (°C/km)
7	Plant uptake compensation factor	19	Channel cover factor
8	Soil evaporation compensation factor	20	Channel erodibility factor
9	Groundwater delay (days)	21	Baseflow alpha-factor [days]
10	Groundwater 'revap' coefficient (days)	22	Baseflow alpha factor for bank storage
11	Threshold water depth in the shallow aquifer for flow (mm	23	Deep aquifer percolation fraction
12	Threshold water depth in the shallow aquifer for 'revap' (mm)		

Table 4 | Performance ranking of R^2 and N_{SE}

Performance	N_{SE}	R^2
Very good	$0.75 < NSE \leq 1$	$0.75 < R^2 \leq 1$
Good	$0.6 < NSE \leq 0.75$	$0.6 < R^2 \leq 0.75$
Satisfactory	$0.36 < NSE \leq 0.6$	$0.36 < R^2 < 0.6$
Bad	$0 < NSE \leq 0.36$	$0.25 < R^2 \leq 0.5$

(*Continued.*)

reference. The model simulated discharge was used to analyze the effect of future climate change on the streamflow of the river on annual, monthly, seasonal, total volume, individual months, and on mean maximum and minimum discharge (Figure 4).

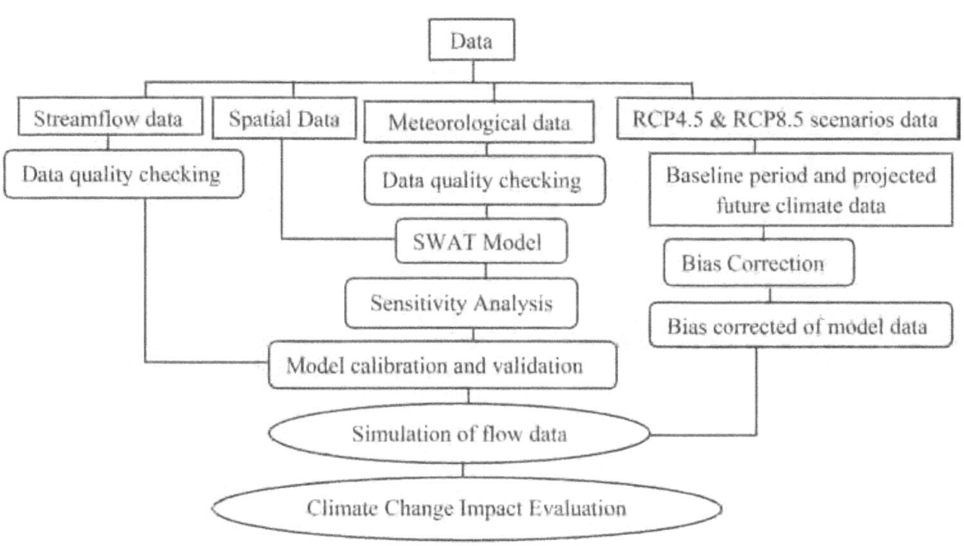

Figure 4 | Flow chart of the general methodology.

RESULTS AND DISCUSSION

The data quality checking was done and showed good results in all used parameters. SWAT sensitivity analysis on six iterations; 50, 300, 500, 800, 1,000, and 1,200 number of monthly simulations using 15 years (1990–2004) streamflow data were done and accepted N_{SE} which is 0.74 and R^2, and = 0.79, were obtained. Then eleven sensitive parameters, namely: GW_DELAY, CN2, GWQMN, GW_REVAP, RCHRG_DP, CANMX, ALPHA_BNK, REVAPMN, REVAPMN, SOL_AWC, ESCO, BIOMIX were selected based on t-test and *P*-value as shown in Figure 5.

Calibration helps to improve the result of model simulation and validation for result verification. Model calibration on four iterations; 100, 250, 400, 500, and 600 numbers of simulations provide acceptable results ($R^2 = 0.83$ and $N_{SE} = 0.72$). Model validation was also provided a very good result ($N_{SE} = 0.71$ and $R^2 = 0.72$) after four iterations with 100, 200, 300 and 400 simulations (Figure 6).

However, many research findings at Blue Nile Bain level obtained better N_{SE} and R^2 results during model calibration and validation. It is due to the *Ribb* River overflowing in the rainy season (SMEC 2007) and a traditional flow measuring instrument was used until seven years ago which may have inaccurate recorded flow data. However, the calibration and validation results showed SWAT can estimate the discharge of the river to see the climate change impact on the streamflow.

Monthly streamflow of the river was shown to decrease/fall from March to August and started to slightly rise/increase from September to February, from the baseline to 2080s, based on both RCP4.5 and RCP8. 5 scenarios data as shown in Figures 7 and 8.

The streamflow change is a function of temperature and rainfall changes. The above result showed decreasing streamflow from March to August may be due to temperature and rainfall change. The monthly temperature trend analysis of the catchment showed a high increase of maximum temperature from March to July but comparatively lower changes in the remaining months as is shown in Figure 9.

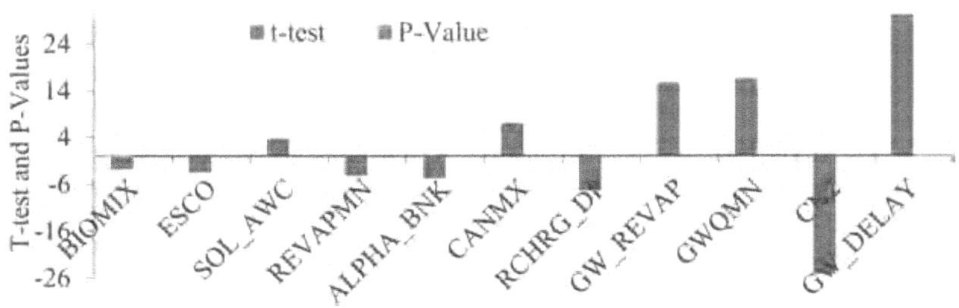

Figure 5 | Most sensitive parameters for model calibration and validation.

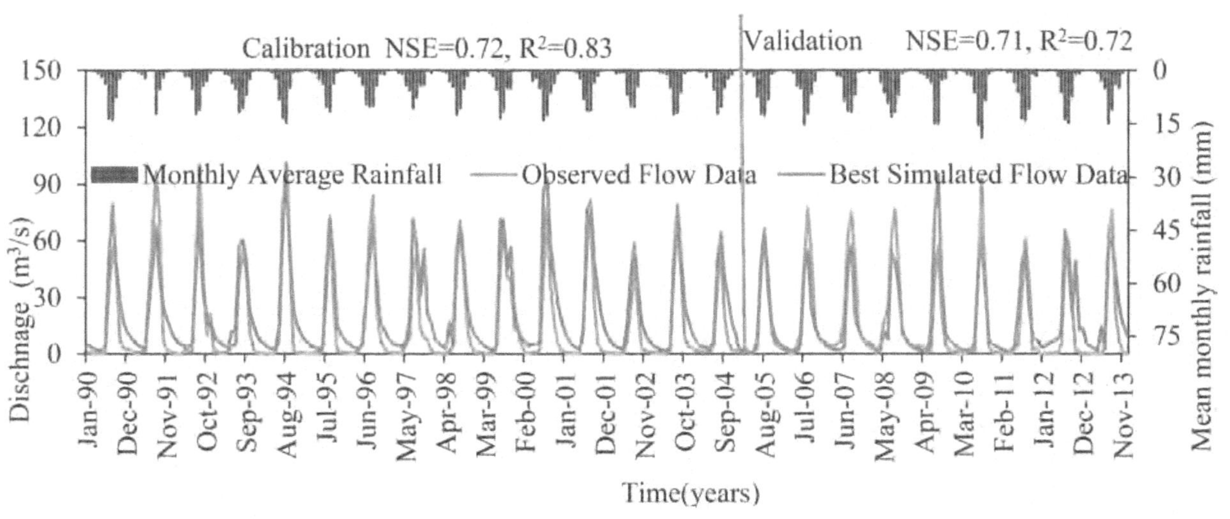

Figure 6 | Model calibration and validation results.

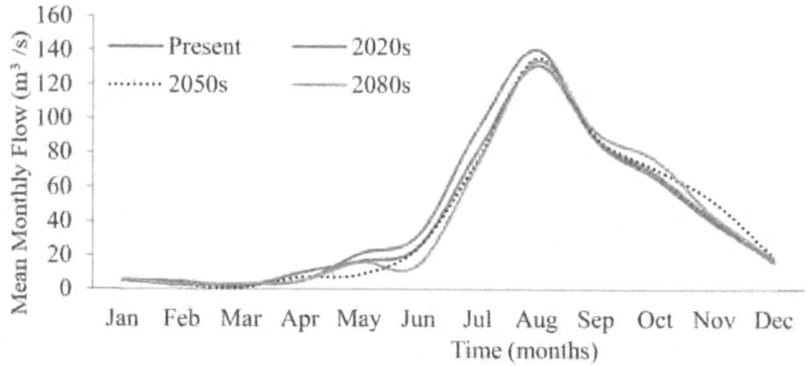

Figure 7 | Hydrograph of future streamflow based on the RCP4.5 scenario.

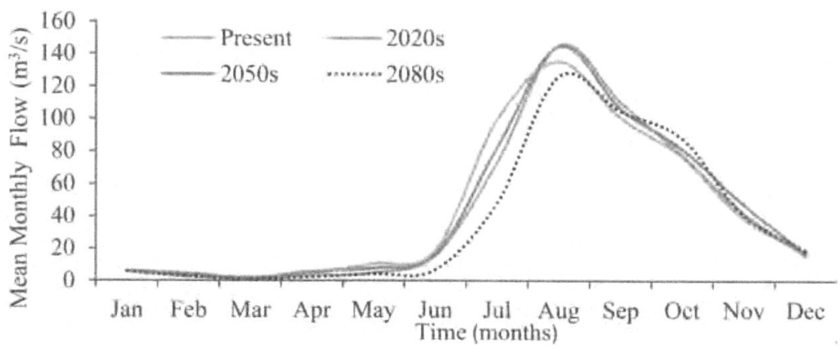

Figure 8 | Hydrograph of future streamflow based on the RCP8.5 scenario.

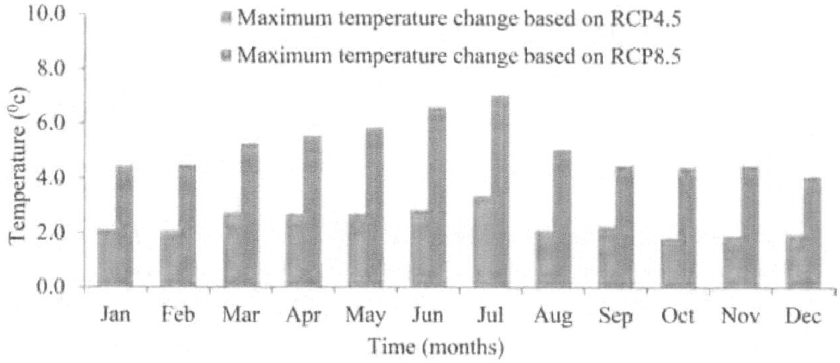

Figure 9 | Mean monthly maximum temperature change from the baseline period to 2080s.

However, the trend analysis of future monthly rainfall change showed a decreasing trend from March to July (except a slight rise in April based on RCP4.5 scenario at the three time horizons and RCP 8.5 scenario at 20250s) and showed more or less a rise from August onwards, based on RCP 4.5 and RCP 8.5 (Figure 10).

The study conducted by DeBoor (2007) at upper Nile basins showed a decrease of rainfall in May, June, and July whereas it was found to increase in September, October, and November which is more or less similar to this study result. The same result is obtained from the study conducted at Lake Tana (Abdo *et al.* 2009). This decrease of rainfall and increase of temperature leads to a decrease in the streamflow of the river because increasing temperature causes high evapotranspiration which leads to a decrease of the streamflow.

Concerning individual months streamflow change (Figure 11), June could show the highest magnitude of decrease (63.3% and 55.45% at RCP8.5 and RCP4.5 scenarios respectively) from baseline to the 2080s. A higher magnitude of decrease of

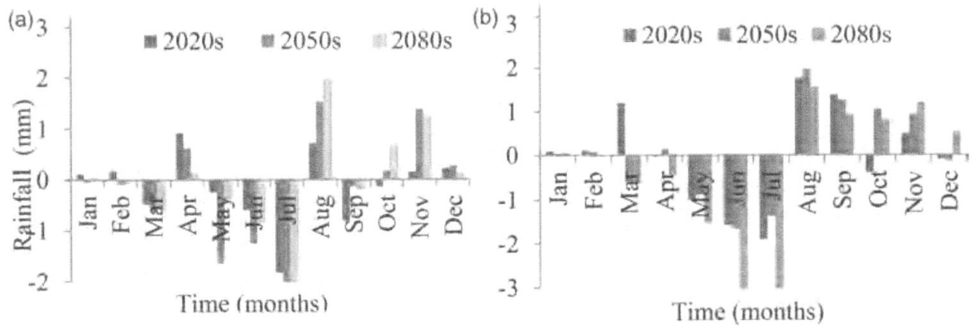

Figure 10 | Mean monthly rainfall change based on (a) RCP4.5. (b) RCP 8.5 scenarios.

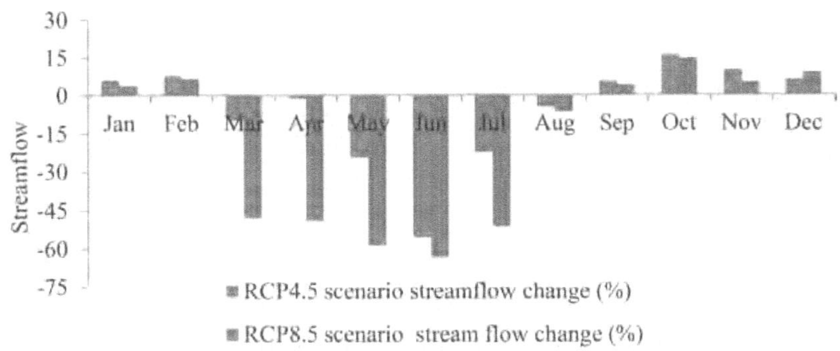

Figure 11 | Mean monthly streamflow change from the baseline to the 2080s.

streamflow would also observe in May (24.38% and 58.64% at RCP4.5 and RCP8.5 scenarios respectively) and July (22.31% and 51.33% at RCP4.5 and RCP8.5 scenarios respectively). The minimum magnitude of decrease would observe in August based on the RCP8.5 scenario which is found to be 6.64% and April (1.21%) based on the RCP4.5 scenario. Streamflow showed a comparatively maximum increase in October (15.87% and 14.66% at RCP 4.5 and RCP8.5 scenarios respectively). The minimum increase would observe in September (6.06% and 4.01% at RCP4.5 and RCP8.5 scenarios respectively).

Percentage change of mean monthly streamflow was obtained by multiplying the differences between the 2080s and baseline period by 100.

Based on the changes of maximum temperature and rainfall Figures 9 and 10, the highest magnitude of the decrease is expected to happen in July but it is in June. It may be due to the streamflow is affected by other factors such as the infiltration rate of the soil, land preparation, and the crop cover of the area. The soil infiltration rate increases at beginning of rain and gradually decreases when the rain continued. In the study area, rainfall mostly started in May and continued to rain up to the end of September/October. So, the result of the study may be due to the higher infiltration rate on the first rainy month cause for less runoff formation. Similarly, in the study area, land preparation for cultivation is mostly carried out in May and June. So, it may cause low runoff formation due to the high infiltration rate.

In total volume reduction (Figure 12), obtained by deducting baseline streamflow from streamflow obtained in the 2080s, July is the most impacted month in both scenarios. The total volume of water reduced in July is found to be 21.08 cubic meters per second and 51.22 cubic meters per second at RCP4.5 and RCP8.5 scenarios respectively. It is due to the maximum temperature increment and highest rainfall decrement in July. It may also be due to the total volume of the river in the baseline period is highest compared with other months. The maximum increase happen in October which is found to be 10.31 and 11.26 cubic meters per second at RCP4.5 and RCP8.5 scenarios respectively.

Concerning seasonal streamflow change of the river in the 2080s from the baseline period, it would decrease in summer (June, July, and August) and spring (March, April, and May) seasons. But, it would increase in the autumn (September, October, and November) and winter (December, January, and February) seasons of the country in both scenarios (Figure 13). The highest magnitude of the decrease is shown in spring based on the RCP8.5 scenario and summer

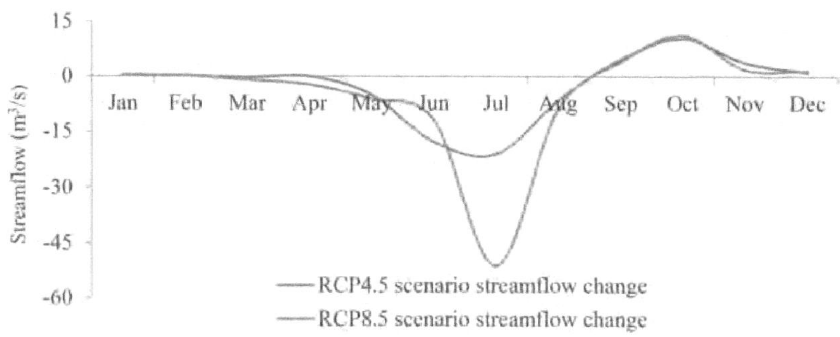

Figure 12 | Mean monthly total volume change from the baseline to the 2028s.

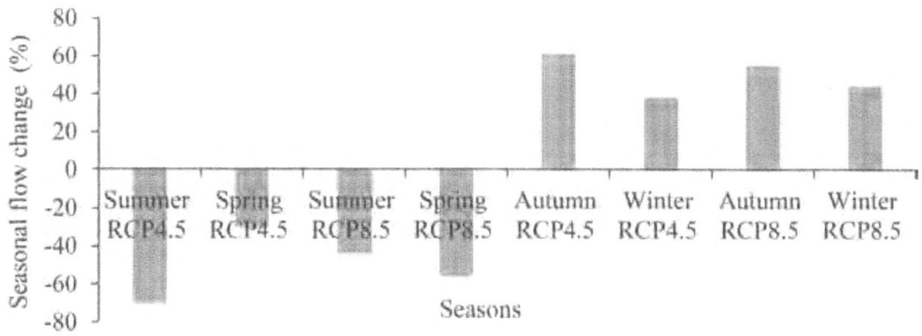

Figure 13 | Seasonal streamflow change from the baseline to 2080s. The percentage of seasonal streamflow change was calculated by multiplying the rate of change on the target season by 100 and divided by the sum of seasons that have shown the same increasing or decreasing trend from seasons.

based on the RCP4.5 scenario. The highest magnitude of increase should be seen in autumn rather than winter in both scenarios.

Streamflow would decrease by 69.87% and 30.13% in summer and spring seasons, respectively, based on the RCP4.5 scenario whereas 43.82% in summer and 56.18% in spring based on the RCP8.5 scenario in the 2080s from the current situation. Streamflow would increase by 61.29% and 38.71% in autumn and winter season respectively based on the RCP4.5 scenario whereas 55.19% in autumn and 44.81% in winter based on the RCP8.5 scenario in the 2080s from the baseline period.

Rivers in Ethiopia are mostly used for agricultural purposes in spring, winter, and autumn but rainfed agriculture is commonly adopted in summer. Future water resources planning and implementation for irrigation projects at *Ribb* River are advisable in autumn and winter since a slight increase of streamflow in the future allows production more than the current situations. In spring and summer, streamflow could decrease that shall construct reservoirs and other structures in the cause of different projects are planning to implement at the river based on current streamflow information.

Planning and implementing of projects at rivers shall be based on future maximum (peak) and minimum streamflow to increasing the life span, sustainability, and profitability of projects. Maximum streamflow may cause distraction of constructed structures if the constructions do not consider the peak discharge. It may also cause the construction of large structures which leads to unnecessary cost of construction. Knowing future minimum streamflow is also helpful to know the nature and type of projects to implement in the future.

Monthly mean maximum streamflow showed a decreasing trend from January to July and October whereas increasing in August, November, and December in both scenarios from baseline to 2080s. However, there is not the same result at RCP4.5 and RCP8.5 scenarios in September (Figure 14). The highest peak monthly mean streamflow is observed in August therefore structures constructed in the future should give consideration to the fact that peak flow could increase above that shown today. The same is true in November.

Besides, monthly mean minimum stream flow could change through time (Figure 15). The mean monthly minimum streamflow of the river could not show a significant change from January to May based on both scenarios. It showed a slightly increasing trend in September which could increase by 6.44 m^3/s and 2.65 m^3/s at RCP4.5 and RCP8.5 scenarios respectively. However, it would show a decreasing trend in July, August, October, November, and December at both scenarios. The

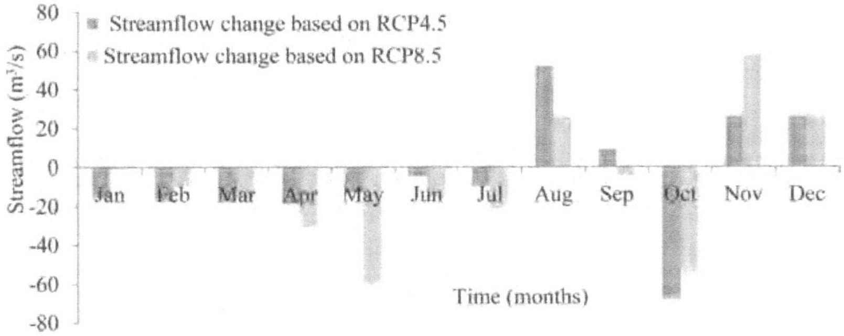

Figure 14 | Mean monthly change of maximum streamflow from baseline to 2080s.

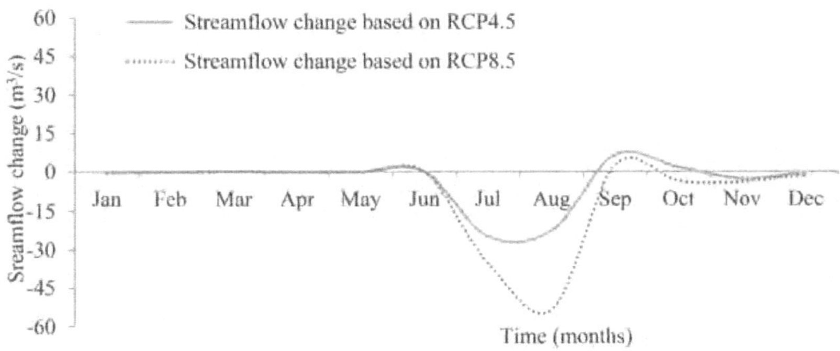

Figure 15 | Mean monthly change of minimum streamflow from the baseline to the 2080s.

maximum magnitude of the decrease is found in July and August which is 24.70 m^3/s at RCP4.5 and 35.19 m^3/s at RCP8.5 scenarios in July and by 22.45 m^3/s at RCP4.5 and 53.30 m^3/s at RC8.5 scenarios in August.

On the annual streamflow basis, the mean annual streamflow of the river was 42.78 m^3/s on the baseline period. It could be 40.75 m^3/s, 40.83 m^3/s, and 40.24 m^3/s at the RCP4.5 scenario and 41.73 m^3/s, 43.42 m^3/s, and 37.58 m^3/s at RCP8.5 scenario in the 2020s, 2050s, and 2080s respectively. This showed a decreasing trend of streamflow from baseline to the 2080s by 2.54 m^3/s and 5.2 m^3/s at RCP4.5 and 8.5 scenarios. The simulated streamflow in the 2080s is lower based on RCP8.5 scenario data than RCP4.5. It is due to RCP8.5 is increasing and RCP4.5 is stabilization scenarios.

The reduction of the streamflow from the baseline to the 2080s is due to increasing temperatures and decreasing rainfall (Figures 16a and b).

When looking at the changes between the consecutive time horizons (from the baseline to 2020s, the 2020–2050s, and the 2050–2080s), streamflow would not show a consistent trend. Streamflow would decrease from the baseline to the 2020s but increased from the 2020s to the 2050s then decreased again from the 2050s to the 2080s on both scenarios (Figure 17(a)).

The changes from baseline to each time horizon (the 2020s, 2050s, and 2080s) the streamflow becomes decreased in all scenarios but increased from baseline to 2050 based on the RCP8.5 scenario (Figure 17(b)). It seems out of logic to see the different results on RCP 4.5 and RCP 8.5. It may be due to climate projected data used which comes from a single model. So, it is recommended to see other models for future studies.

The annual average maximum streamflow of the river showed a decreasing trend from the baseline to the 2080s based on both scenarios simulation results (Figure 18). It was 294.60 m^3/s on the current situation and would be 234.8 m^3/s, 269.5 m^3/s, and 236.8 m^3/s at the RCP4.5 scenario and 288.1 m^3/s, 272.1 m^3/s, and 240.5 m^3/s at RCP8.5 scenario in the 2020s, 2050s, and 2080s. Therefore, the magnitude of annual peak streamflow showed a decreasing trend between the consecutive time horizons at both scenarios except showed a slightly increasing trend from the 2020 to 2050s at the RCP4.5 scenario.

Mean annual minimum streamflow also showed a decreasing trend from the baseline to the 2080s in both scenarios. It would also show a decreasing trend between consecutive time horizons except would show a slight increasing trend from the 2050s to the 2080s at the RCP4.5 scenario (Figure 19).

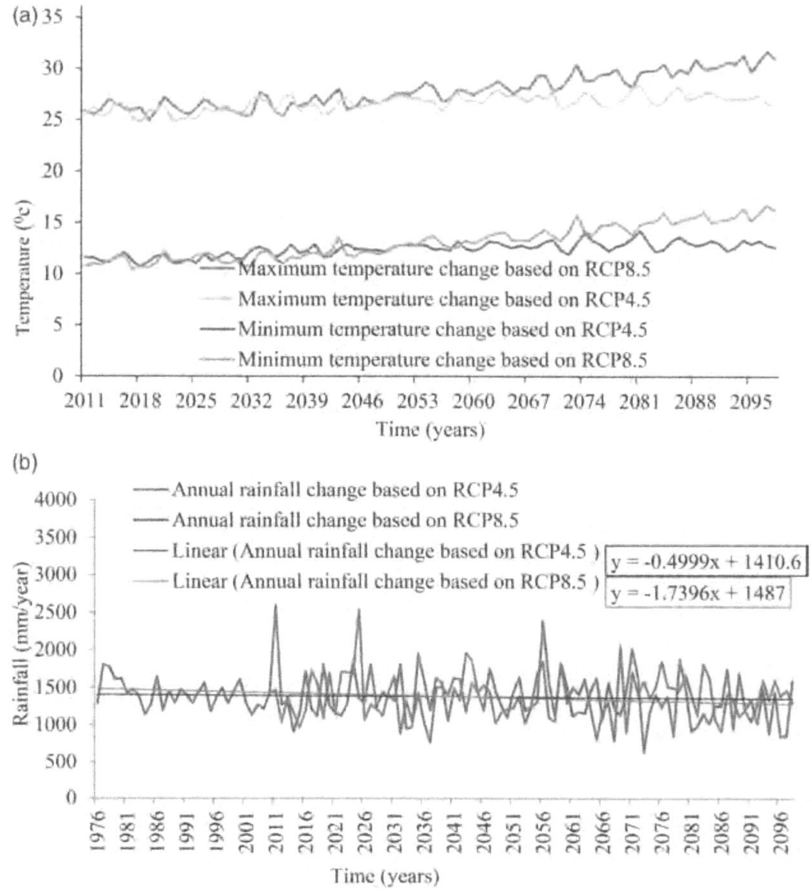

Figure 16 | Mean annual changes of (a) temperature, (b) rainfall.

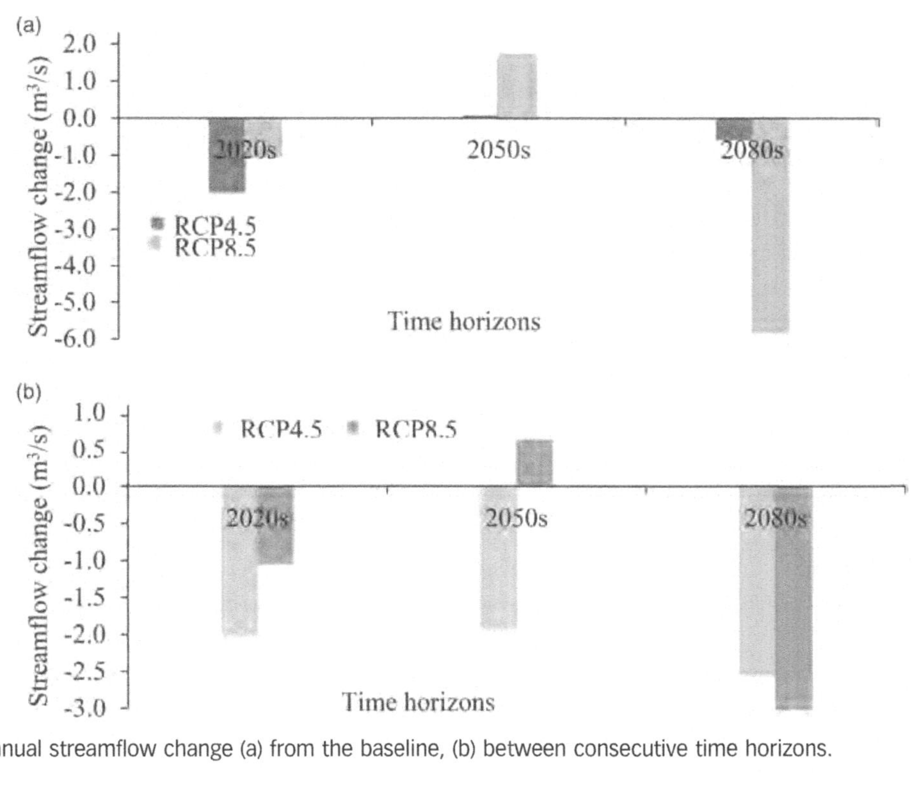

Figure 17 | Mean annual streamflow change (a) from the baseline, (b) between consecutive time horizons.

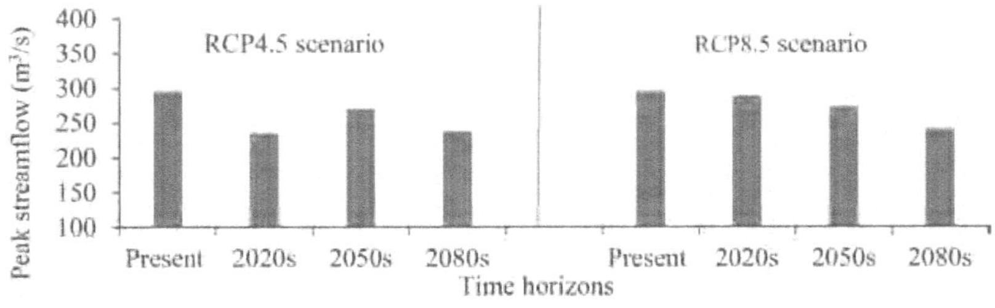

Figure 18 | Mean annual maximum streamflow change.

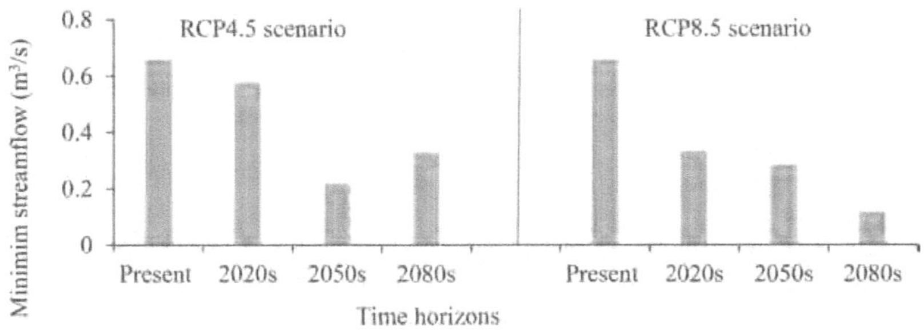

Figure 19 | Mean annual minimum streamflow change.

CONCLUSION AND RECOMMENDATIONS

This study evaluates the impacts of climate change on the Ribb River streamflow using the bias-corrected output of climate change projection data set for Nile Basin studies and the hydrological model simulation approach of the SWAT model. So the study reached the following conclusions.

SWAT calibration and validation were verified that the model can effectively estimate future streamflow of the river. The monthly streamflow of the river was used to calibrate and validate the model using performance measuring parameters of the coefficient of determination (R^2) and Nash Sutcliff (NSE). R^2 and NSE values were 0.72 and 0.83 for calibration and 0.71 and 0.72 for validation, respectively, which are deemed acceptable.

Climate change projections of RCM data set for Nile Basin studies were found to be biased and the dataset was bias corrected before it was used. A linear scaling method of bias correction was used for this study and was found to be effective in matching model data with observed data on a monthly time scale.

RCP4.5 and RCP 8.5 scenarios resulted in a somewhat similar projection of future streamflow conditions. The streamflow changes were looked at monthly, seasonal, and annual time scales. Annual streamflow showed a decreasing trend from the baseline to the 2080s. The simulated mean annual streamflow was 42.78 m^3/s for the baseline period whereas it became 40.75 m^3/s, 40.83 m^3/s, and 40.24 m^3/s at the RCP4.5 scenario and 41.73 m^3/s, 43.42 m^3/s, and 37.58 at the RCP8.5 scenario for the corresponding 2020s, 2050s, and 2080s time horizons, respectively. It showed a consistent decrease from the baseline to the 2080s but a slight rise and fall between successive time horizons.

Based on the monthly simulated flow increasing trend is shown from September until February and a decreasing trend from January until August. The maximum increase in magnitude is in October that showed an increase of 10.31 and 11.26 cubic meters per second in RCP4.5 and RCP8.5 scenarios, respectively. Whereas the maximum decrease was for July which is shown to be reduced by 21.08 and 51.22 cubic meters per second at RCP4.5 and RCP8.5 scenarios, respectively.

Based on the seasonal change in river flow, summer flows are shown to have a potential decrease by 69.87% and spring by 30.13% according to the RCP4.5 scenario whereas a decrease by 43.82% in summer and by 56.18% in spring is shown based on the RCP8.5 scenario from baseline to the 2080s. However, streamflow could increase by 61.29% and 38.71% in autumn and winter season respectively based on the RCP4.5 scenario and by 55.19% in autumn and 44.81% in winter based on the RCP8.5 scenario.

Monthly peak streamflow from the baseline to 2080s shows a decreasing trend from January to July and October and an increasing trend in August, November, and December in both scenarios. The minimum streamflow did not show a significant change from January to May in both scenarios. The slightly increasing trend of minimum streamflow in September was found

(6.44 and 2.65 m^3/s on RCP4.5 and RCP8.5 scenarios, respectively). However, it showed a decreasing trend from July to December in both scenarios. The maximum decrease in magnitude was in July and August. The annual maximum streamflow showed a decrease from baseline to 2080s in both scenarios. The mean annual minimum streamflow change also shows a decreasing trend in the future. Besides, both monthly maximum and minimum streamflow change have a decreasing trend in the future.

A result reported in this study is based on a single RCM and two emission scenarios. The generalization of these findings may have to be confirmed by looking at additional RCM scenarios. This will also allow an understanding of uncertainties associated with General Circulation Models for the study area.

SWAT is found to be a very important tool for simulating historical and future impacted hydrological processes at a watershed scale. However, in addition to climate change, future land-use change will also play a critical role in streamflow generation. The impact of potential future land-use change on streamflow for the study area was not within the scope of this study. Therefore, future studies should include both climate change impacts and land-use changes to have a complete picture of future stream flows.

Besides, there are also other factors affecting the streamflow generation such as soil type, evapotranspiration, solar radiation, etc., that may also be impacted by climate change and were not explicitly included. Hence, the results of this study should be taken with care and be considered as an indication for further studies than accurate predictions.

DATA AVAILABILITY STATEMENT

Data cannot be made publicly available; readers should contact the corresponding author for details.

REFERENCES

Abdo, K. S., Fiseha, B. M., Rientjes, T. H. M., Gieske, A. S. M. & Haile, A. T. 2009 Assessment of climate change impacts on the hydrology of Gilgel Abay catchment in Lake Tana basin, Ethiopia. *Hydrol Process* **23**, 3661–3669.

Adamu, A. 2013 *Assessing the Impacts of Land Use and Land Cover Change on Hydrology of Watershed: A Case Study on Gilgel–Abay Watershed, Lake Tana.*

Adem, A. A., Tilahun, S. A., Ayana, E. K., Worqlul, A. W., Assefa, T. T., Dessu, S. B. & Melesse, A. M. 2015 Climate change impact on stream flow in the Upper Gilgel Abay catchment. Blue Nile basin, Ethiopia. In: *Landscape Dynamics, Soils and Hydrological Processes in Varied Climates.* (A. M Melesse & W. Abtew, eds). Springer, London, pp. 645–673.

Alexander, L. V., Zhang, X., Peterson, T. C., Caesar, J., Gleason, B., Klein Tank, A. M. G., Haylock, M., Collins, D., Trewin, B., Rahimzadeh, F., Tagipour, A., Rupa Kumar, K., Revadekar, J., Griffiths, G., Vincent, L., Stephenson, D. B., Burn, J., Aguilar, E., Brunet, M., Taylor, M., New, M., Zhai, P., Rusticucci, M. & Vazquez-Aguirre, J. L. 2006 Global observed changes in daily climate extremes of temperature and precipitation. *Journal of Geophysical Research Atmospheres* **111**, 1–22. https://doi.org/10.1029/2005JD006290.

Arun, M. & Sananda Kundu, A. M. 2012 Rainfall trend analysis by Mann-Kendall test: a case study of north-eastern part of Cuttack District, Orissa. *International Journal of Geology, Earth and Environmental Sciences* **2** (1), 70–78.

Asefa, T. 2019 *Climate Change Projections Data Set for Impact Studies in Nile Basin.* Prepared for Nile Basin Initiative Secretariat (Nile-SEC), Entebbe, Uganda.

Awulachew, S. B., Tenaw, M., Steenhuis, T., Easton, Z., Ahmed, A., Bashar, K. E. & Hailesellassie, A. 2008 *Impact of Watershed Interventions on Runoff and Sedimentation in Gumera Watershed.* Arba Minch University, Ethiopia Research Service and Texas A M Backland Research Center.

Conway, D. 2000 A water balance model of the Upper Blue Nile in Ethiopia. *Hydrological Sciences Journal des Sciences Hydrologiques* **42** (2), 265–286. https://doi.org/10.1080/02626669709492024.

Conway, D. & Schipper, E. L. F. 2011 Adaptation to climate change in Africa : Challenges and opportunities identified from Ethiopia. *Global Environmental Change* **21** (1), 227–237. https://doi.org/10.1016/j.gloenvcha.2010.07.013.

DeBoor, B. 2007 *Climate Change and Impacts on the Extreme Rainfall Over the Blue Nile Region.* KNMI, Royal Netherlands Meteorological Institute, De Bilt, Netherlands.

Droogers, P. & Aerts, J. 2005 Adaptation strategies to climate change and climate variability: a comparative study between seven contrasting river basins. *Physics and Chemistry of the Earth, Parts A/B/C* **30**, 339–346. https://doi.org/10.1016/j.pce.2005.06.015.

Fentaw, F., Hailu, D., Nigussie, A. & Melesse, A. M. 2018 *Climate Change Impact on the Hydrology of Tekeze Basin, Ethiopia: Projection of Rainfall-Runoff for Future Water Resources Planning* (August). https://doi.org/10.1007/s41101-018-0057-3.

Gebremariame, Z. H. 2009 *Assessment of Climate Change Impact on the Net Basin Supply of Lake Tana Water Balance Assessment of Climate Change Impact on the Net Basin Supply of Lake Tana Water Balance.*

Griensven, A. V., Meixner, T., Grunwald, S. & Bishop, T. 2006 A global sensitivity analysis tool for the parameters of multi-variable catchment models. **324**, 10–23. https://doi.org/10.1016/j.jhydrol.2005.09.008.

Houghton, J. T. 2001 *Climate Changes the Scientific Basis: Contribution of Working Group I to the Third Assessment Report of the Intergovernmental Panel on Climate Change.* Cambridge University Press, Cambridge.

IPCC 2007 Climate change 2007: climate change impacts, adaptation, and vulnerability. Working Group II contribution to the Intergovernmental Panel on Climate Change Fourth Assessment Report. Summary for policymakers, 23.

IPCC 2013 Climate Change 2013: The Physical Science Basis -Summary for Policymakers. Contribution of Working Group I to the Fifth Assessment Report of the Intergovernmental Panel on Climate Change. – Intergovernmental Panel on Climate Change. IPCC Secretariat.

Kassa, T. & Foerch, G. 2007 Impacts of Land use/cover dynamics on streamflow: the case of Hare watershed, Ethiopia. In: *The Proceedings of the 4th International SWAT2005 Conference 2007.*

Kebede, S., Travi, Y., Alemayehu, T. & Ayenew, T. 2006 Water balance of Lake Tana and its sensitivity to fluctuations in rainfall, Blue Nile Basin, Ethiopia. *Journal of Hydrology* **316**, 233–247.

Liew, M. W. V., Arnold, J. G. & Garbrecht, J. D. 2003 H s a w: c b t m, 46(6), 1539–1551.

Moriasi, D. N., Arnold, J. G., Van Liew, M. W., Bingner, R. L., Harmel, R. D. & Veith, T. L. 2007 Model evaluation guidelines for systematic quantification of accuracy in watershed simulations. *Transactions of the ASABE* 50 (3), 885–900.

Paparrizos, S., Maris, F. & Matzarakis, A. 2016 Integrated analysis of present and future responses of precipitation over selected Greek areas with different climate conditions. *Atmospheric Research* **169** (Part A), 199–208.

Setegn, S., Srinivasan, R., Dargahi, B. & Melesse, A. 2008 Spatial delineation of soil erosion vulnerability in the Lake Tana Basin, Ethiopia. *Hydrological Processes* **3750**, 3738–3750. https://doi.org/10.1002/hyp.

Setegn, S. G., Srinivasan, R., Dargahi, B. & Melesse, A. M. 2009 Spatial delineation of soil erosion-prone areas: application of SWAT and MCE approaches in the Lake Tana Basin, Ethiopia. *Hydrol Process* **23** (26), 3738–3750. (Special issue: Nile Hydrology).

Setegn, S. G., Rayner, D., Melesse, A. M., Dargahi, B. & Srinivasan, R. 2011 Impact of climate change on the hydroclimatology of Lake Tana Basin, Ethiopia. *Water Resources Research* **47**, W04511.

Shimelis, G. 2008 Hydrological and sediment yield modeling in Lake Tana basin, Blue Nile Ethiopia. Royal Institute of Technology (KTH), SE-100 44 STOCKHOLM, Sweden TRITALWR.LIC 2042, ISSN 1650-8629, ISRN KTH/LWR/LIC 2042-SE, ISBN 978 - 91-7415 026-1.

SMEC, International Ptyltd 2007 *Hydrological Study of the Tana Beles sun-Basins Interim Report Snowy Mountain Engineering Corporation*. Australia, p. 77.

Sutcliffe, J. V. & Parks, Y. P. 1999 The Hydrology of the Nile, IAHS Special Publication no.5, pp. 127–141.

Tarekegn, D. & Tadege, A. 2005 *Assessing the Impact of Climate Change on the Water Resource of Lake Tana Sub-Basin Using the WATBAL Model, Ethiopia*.

Tsegaye, E. 2006 *Regionalization of Potential Evapotranspiration Prediction for the Blue Nile (Abbay) the River Basin, Ethiopia*. MSc Thesis, Arba-Minch University, Ethiopia.

Urrutia, R. & Vuille, M. 2009 Climate change projections for the tropical Andes using a regional climate model: temperature and precipitation simulations for the end of the 21st century. *Journal of Geophysical Research: Atmospheres* **114**, 1–15.

Zwiers, V. V. K. F. W. & Wehner, X. Z. M. 2013 Changes in temperature and precipitation extremes in the CMIP5 ensemble, 345–357. https://doi.org/10.1007/s10584-013-0705-8

First received 16 February 2021; accepted in revised form 18 July 2021. Available online 9 August 2021

Impact of Climate Change on Hydrology and Water Resources doi: 10.2166/wcc.2021.238

Hydrological impacts of climate and land-use change on flow regime variations in upper Indus basin

Kashif Haleem [a], Afed Ullah Khan [a,b,*], Sohail Ahmad[b], Mansoor Khan[b], Fayaz Ahmad Khan[a,b], Wisal Khan[b] and Jehanzeb Khan [c]

[a] National Institute of Urban Infrastructure Planning, University of Engineering and Technology, Peshawar, Peshawar 25000, Pakistan
[b] Department of Civil Engineering, University of Engineering and Technology Peshawar (Bannu Campus), 28100 Bannu, Pakistan
[c] Higher Education Department, Government Post Graduate College Kohat, KPK, Pakistan
*Corresponding author. E-mail: afedullah@uetpeshawar.edu.pk

KH, 0000-0002-3025-8593; AUK, 0000-0002-2742-0377; JK, 0000-0003-3517-2761

ABSTRACT

Investigating the effects of climate and land-use changes on surface runoff is critical for water resources management. The majority of studies focused on projected climate change effects on surface runoff, while neglecting future land-use change. Therefore, the main aim of this article is to discriminate the impacts of projected climate and land-use changes on surface runoff using the Soil and Water Assessment Tool (SWAT) through the lens of the Upper Indus Basin, Pakistan. Future scenarios of the land-use and climate changes are predicted using cellular automata artificial neural network and four bias-corrected general circulation models, respectively. The historical record (2000–2013) was divided into the calibration period (2000–2008) and the validation period (2009–2013). The simulated results demonstrated that the SWAT model performed well. The results obtained from 2000 to 2013 show that climate change (61.61%) has a higher influence on river runoff than land-use change (38.39%). Both climate and land-use changes are predicted to increase future runoff depth in this basin. The influence of climate change (12.76–25.92%) is greater than land-use change (0.37–1.1%). Global weather data has good applicability for simulating hydrological responses in the region where conventional gauges are unavailable. The study discusses that both climate and land-use changes impact runoff depth and concludes with some suggestions for water resources managers to bring water environment sustainability.

Key words: CA-ANN model, climate change, land-use change, runoff responses, SWAT model

HIGHLIGHTS

- The SWAT model was used to isolate the climate and land-use change impacts on streamflow.
- The general circulation model was used to assess climate change impacts on streamflow.
- The CA-ANN model was used to predict future land use.
- Multiple climate and land-use scenarios were set to isolate climate and land-use change impacts on streamflow.
- Climate change effects overweigh the land-use change effects on streamflow.

GRAPHICAL ABSTRACT

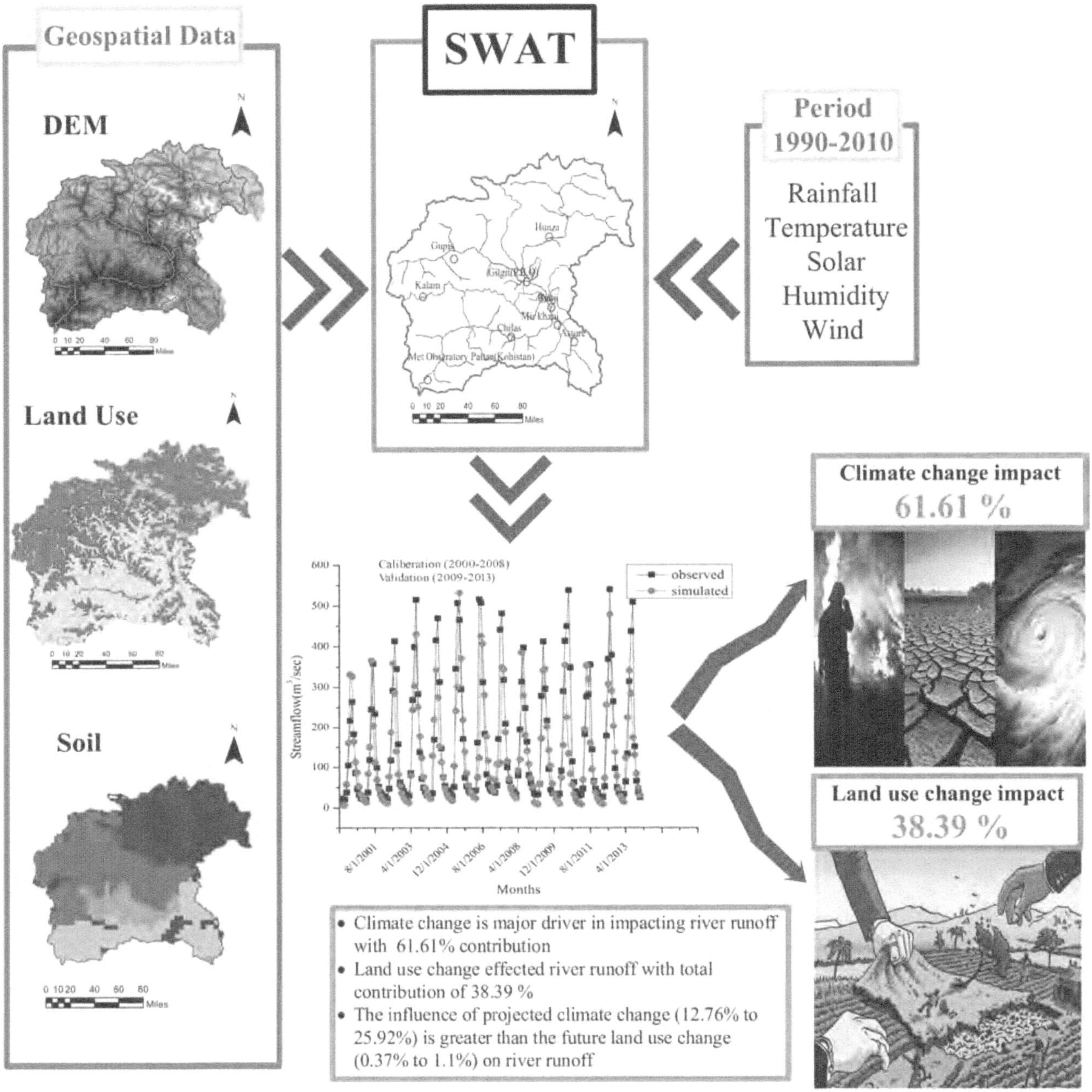

1. INTRODUCTION

Reduction in streamflow is the main concern in most of the river basins that supply water for agriculture, industry, and the growing population along the river (Zhang *et al.* 2017). The major concern of today's world is the global potential climate change impacts on water resources. It causes natural disasters, i.e. monsoonal flood hazards (Nanding *et al.* 2020), wildfire risk (Du *et al.* 2021), and severe drought (Pokhrel *et al.* 2021). This devastation is increased by changes in the watershed's land use (Stocker *et al.* 2001). Land-use and climate change have a major influence on water ecology, which are often reflected in long-term geographical and temporal changes in the water balance components, such as runoff, soil moisture, evapotranspiration, groundwater, and streamflow (Fang *et al.* 2013; Memarian *et al.* 2014; Deng *et al.* 2015). Runoff is the interface of climate and land-use change in a basin where climate change mostly affects the spatial distribution and temporal variability of runoff. Land-use change refers to the human intervention on natural processes which badly impact river flow (Li *et al.* 2009; Liu *et al.* 2010; Wang *et al.* 2014; Birkinshaw *et al.* 2017; Zhang *et al.* 2017). Therefore, it is critical to quantify the impacts of climate and land-use changes on river runoff in order to assist policymakers and watershed managers in achieving watershed sustainability (Liu *et al.* 2010; Shang *et al.* 2019).

Pakistan is one of the world's water-stressed countries, with an annual average rainfall of 240 mm. Rapid urbanization and population increase are the major sources of water quality impairment at the basin scale (Ghoraba 2015). Previously, the water supply systems were built considering climatic trends that cause variations in streamflow. Variations in precipitation pose a great challenge to water resources management. These negative consequences of anthropogenic activities – both qualitatively and quantitively on water resources – restrict the use of water to meet the growing demands of future populations (Mukheibir 2007; Obuobie *et al.* 2012).

In this study, the watershed hydrological model, Soil and Water Assessment Tool (SWAT), incorporated in Geographic Information System (GIS) was applied. The SWAT model is frequently used nowadays because it is a river basin scale model capable of simulating the impacts of land management practices in large, complex watersheds, including the impacts that occur in the smaller modeling units, called hydrologic response units (HRUs). The SWAT has been termed by many studies as a computationally highly efficient model for forecasting the effects of land use and climate change on streamflow (Neitsch *et al.* 2011). Numerous studies have evaluated the SWAT model's effectiveness while globally assessing hydrological processes in various areas (Khan *et al.* 2014; Shang *et al.* 2019).

The Intergovernmental Panel on Climate Change (IPCC) has developed the most advanced numerical-based general circulation models (GCMs) to understand the future global climatic systems, such as the atmosphere, the sea ice, and the ocean (Fowler *et al.* 2007; Mahmood & Babel 2013). These models are very efficient in forecasting climate changes across the globe; but, outputs obtained from these models are completely based on large grid sizes that range from 250 to 600 km, which restrict its use in the evaluation of hydrological impacts caused by climate change (Wilby *et al.* 2000; Gebremeskel *et al.* 2004). However, dynamical and statistical downscaling (SD) methods have been developed to include GCM data into hydrological models to predict climate change. SD is often used by various institutions working on climate change (Wilby *et al.* 2000; Fowler *et al.* 2007).

While many researchers have investigated water quality problems in Pakistan (Daud *et al.* 2017; Bashir *et al.* 2020), there is a deficiency in evaluating the impacts of both land use and climate changes on a large and complex watershed scale. The main purpose of this study is to understand how surface runoff responds to projected land use and climate change at a representative watershed scale to provide a solid background for long-term water resource prediction and planning in the upper region of the Indus River. Moreover, this study also separates the impacts of climate and land-use change on river runoff.

2. STUDY AREA DESCRIPTION, DATA COLLECTION, AND METHODS

2.1. Study area description

The study area comprises the Indus River which is shared by Pakistan, China, Afghanistan, and India and covers a total area of 1.1×10^6 km^2 (Lutz *et al.* 2014). Moreover, the largest part of the Indus River basin lies in Pakistan and the smallest part in Afghanistan. It is extended between 32.48–37.07°N and 67.33–81.83°E (Archer 2003). This research study is confined to the upper region of the Indus River of Pakistan at Astore (Gilgit Baltistan) as shown in Figure 1, whose basin area is 45,400 km^2 and elevation range is from 547 to 8,060 m.

2.2. Data collection

This research study is based on the following datasets whose details are given below.

2.2.1. Digital elevation model

Digital elevation model (DEM) data of the study area has been downloaded from the National Aeronautics and Space Administration (NASA) (https://earthdata.nasa.gov/). The resolution of the Advanced Spaceborne Thermal Emission and Reflection Radiometer (ASTER) GDEM (Global Digital Elevation Model) is 30 m $\times$ 30 m. DEM data is used for watershed delineation. The delineated watershed is demonstrated in Figure 1.

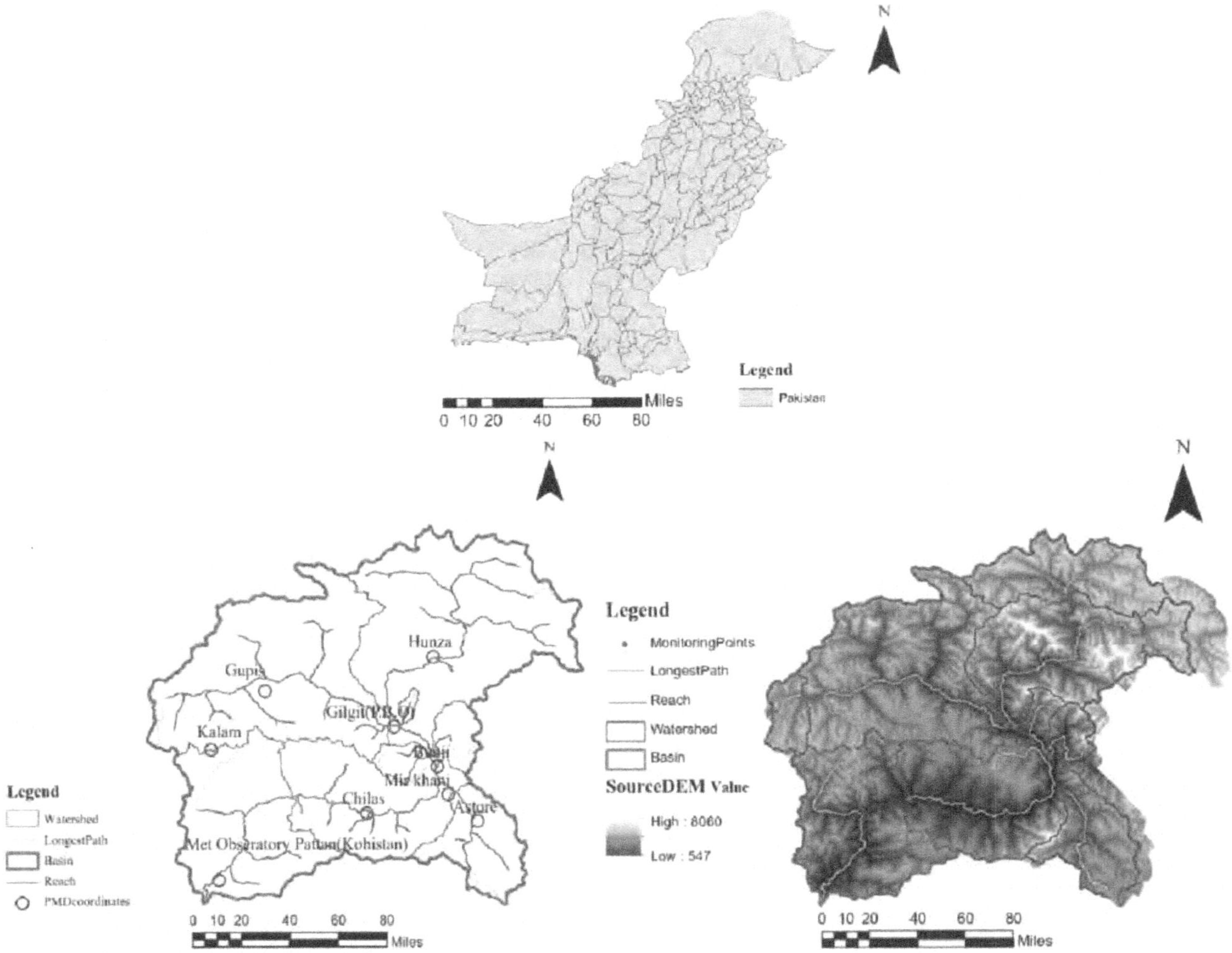

Figure 1 | Overview of study area showing the delineated watershed.

2.2.2. Land use

Land-use data of Moderate Resolution Imaging Spectroradiometer (MODIS) was downloaded from the United States Geological Survey (USGS) (https://earthexplorer.usgs.gov/). The details of land-use classes and their percentages composition for the delineated watershed are demonstrated in Figure 2.

2.2.3. Soil data

Soil data was downloaded from the Food and Agriculture Organization (FAO) (http://www.fao.org/). The details of soil classes and their percentage compositions for the delineated watershed are demonstrated in Figure 2.

2.2.4. Hydroclimatic data sets

Meteorological data, which includes precipitation, temperature (maximum and minimum), humidity, sunshine duration, evapotranspiration, and wind speed, were collected from Pakistan Meteorological Department (PMD) from 1990 to 2013. Similarly, hydrological data which includes daily and average flow were obtained from Water and Power Development Authority (WAPDA) for the period of 1990–2013. Our study area includes a total of nine PMD stations, in which only four stations' data are available. Because of the unavailability of meteorological data, Global Weather data was used for simulation purposes. The PMD coordinates and WAPDA monitoring stations are depicted in Figure 1.

Daily precipitation and temperature (min and max) of four GCMs, namely, GFDL-ESM2G, HadGEM2-ES, MIROC5, and CanESM2 were used for simulations in the SWAT model from 2005 to 2010, which were obtained from Coupled Model

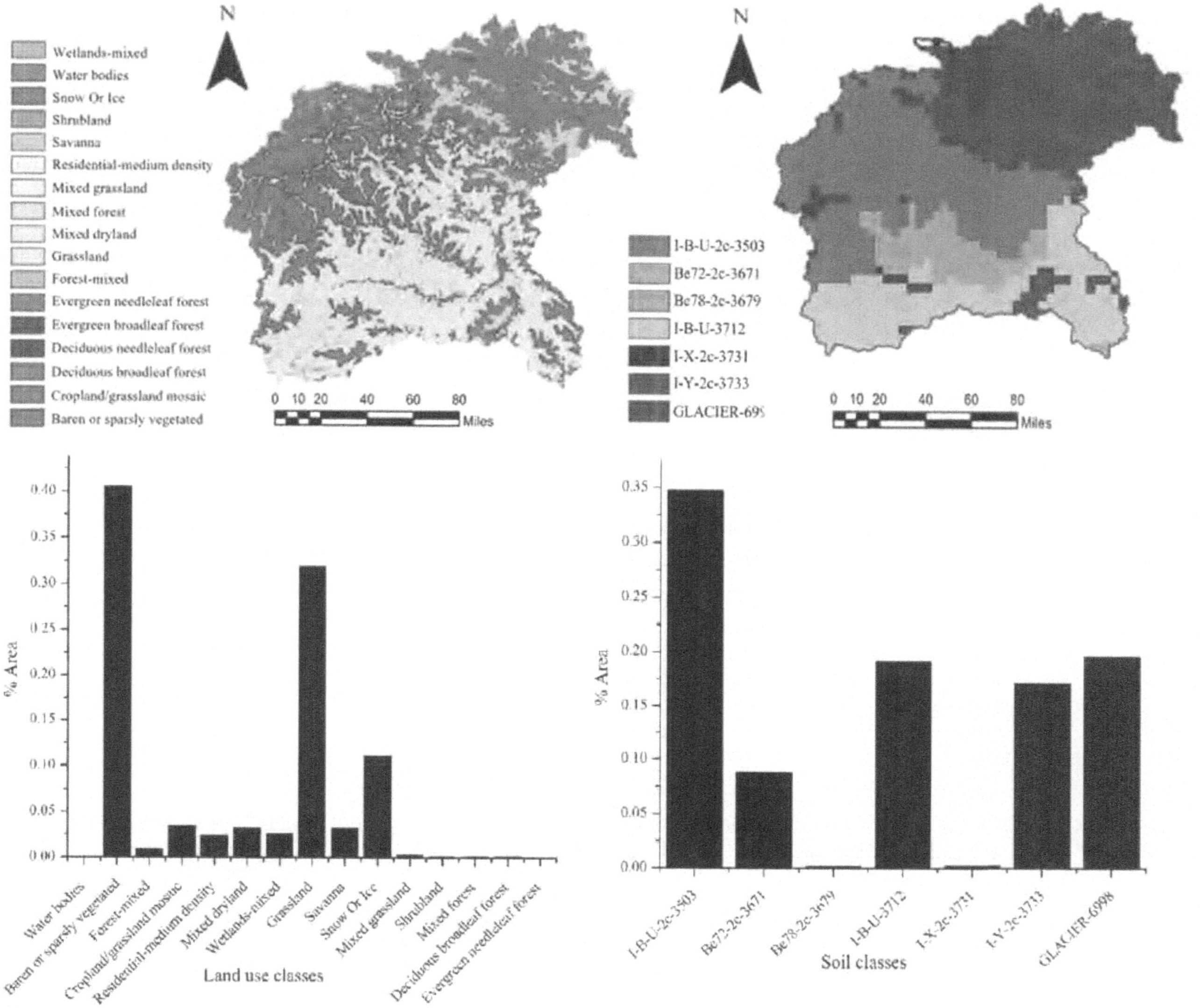

Figure 2 | Details of the land use and soil classes and their percentage compositions.

Intercom-parison Project Phase 5 (CMIP5) archive. These models are frequently used in the Asian region (Babar *et al.* 2015; Prasanna 2015; Mahmood & Jia 2016). The description of four GCMs is listed in Table 1.

2.3. Methods

2.3.1. Bias correction of GCMs

Bias corrections are often applied to climate models when they do not show any/little similitude to observed data. GCMs are corrected by bias corrections which use a transformation algorithm to relate outputs of GCMs. The main objective of the

Table 1 | Overview of the GCMs adopted for the current study

Modeling centre	Model	Institute ID	Resolution (km)
Geophysical Fluid Dynamics Laboratory, USA	GFDL-ESM2G	GFDL	90×144
Met Office Hadley Centre, UK	HadGEM2-ES	MOHC	145×192
Atmosphere and Ocean Research Institute, The University of Tokyo, Japan	MIROC5	MIROC	128×256
Canadian Centre for Climate Modelling and Analysis, Canada	CanESM2	CCCma	192×288

biases is to create a correlation between observed and simulated data, which will be the premise for correcting future projections of GCMs (Sun *et al.* 2011; Nguyen *et al.* 2020). The climate model data for hydrologic modeling (CMhyd) tool was used for biasing climate models (Rathjens *et al.* 2016b). There are plenty of biasing methods in the CMhyd tool, i.e. linear scaling (additive and multiplicative), delta change correction (additive and multiplicative), power transformation of precipitation, precipitation local intensity scaling, variance scaling of temperature, and distribution mapping of precipitation and temperature (Rathjens *et al.* 2016a). Utilizing GCM results at the regional scale is more straightforward and faster with linear scaling than with statistical or dynamic downscaling. These methods are used to correct biases in GCM or RCM findings. SD involves the development of empirical relationships between local-scale variables (e.g. temperature or precipitation) and large-scale variables (e.g. specific humidity, sea level pressure, and temperature of GCMs), and then simulating future data using these relationships. However, bias correction successfully corrects the simulated GCM results (e.g. temperature and precipitation) (Fang *et al.* 2015). We have used linear scaling (additive and multiplicative) bias correction for our study area (Worku *et al.* 2020).

2.3.2. Description of SWAT model

The SWAT model is a watershed hydrological model incorporated in the GIS developed by the United States Department of Agriculture in the 1990s. It was designed to predict the effects of weather, different land-use patterns, and soil conditions on river runoff (Yi & Sophocleous 2011; Yu *et al.* 2013; Wisal *et al.* 2020). The SWAT model requires topographic (DEM), land-use/land-cover, soil, and daily weather (precipitation and temperature) data as their minimum inputs (Ficklin & Barnhart, 2014). The SWAT model uses water balance equation for simulation as follows:

$$\mathrm{SW_t} = \mathrm{SW_o} + \sum_{i=o}^{t} (R_{\mathrm{day}} - Q_{\mathrm{surf}} - E_{\mathrm{a}} - W_{\mathrm{seep}} - Q_{\mathrm{gw}})_i \tag{1}$$

where $\mathrm{SW_t}$ is the final soil moisture content, mm; $\mathrm{SW_0}$ is the initial soil moisture content of the ith day, mm; t is the time, d; R_{day} is the precipitation of the ith day, mm; Q_{surf} is the surface runoff of day i, mm; E_{a} indicates the amount of evapotranspiration on day i, mm; W_{seep} indicates the amount of water entering the vale zone from the soil profile on day i, mm; and Q_{gw} indicates the return flow amount on day i, mm.

2.3.3. SWAT model calibration and validation

The Doyian station at Astore River was selected for model calibration to evaluate the dominant parameters of the streamflow using SWAT-CUP (SWAT Calibration and Uncertainty Procedures) automatic calibration with the Sequential Uncertainty Fitting program algorithm (SUFI-2) (Shang *et al.* 2019). The SWAT-CUP model was calibrated using monthly data from 2000 to 2008. Calibration is a process to assess the goodness of fit between the observed and simulated data to get the optimum values of representative functions such as Nash–Sutcliffe efficiency (NSE), coefficient of determination (R^2), root-mean-square error to the standard deviation ratio (RSR), and percent bias (PBIAS). For RSR and PBIAS, the most optimum value is zero, whereas for R^2 and NSE, it is 1. The R^2 value ranges from 0 to 1; values closer to 1 indicate the best results, while greater than 0.5 is in an acceptable range. Succeeding calibration, the model was validated for monthly data from 2009 to 2013. The model efficiency was tested in the validated period by evaluating the decision factors as shown in Table 2.

Table 2 | Statistical indicators for model performance evaluation

Performance rating	NSE	RSR	PBIAS (%)
Very good	$0.75 < \mathrm{NSE} \leq 1$	$0 \leq \mathrm{RSR} \leq 0.5$	$-10 < \mathrm{PBIAS} < 10$
Good	$0.65 < \mathrm{NSE} \leq 0.75$	$0.5 < \mathrm{RSR} \leq 0.6$	$\pm 10 \leq \mathrm{PBIAS} < \pm 15$
Satisfactory	$0.5 < \mathrm{NSE} \leq 0.65$	$0.6 < \mathrm{RSR} \leq 0.7$	$\pm 15 \leq \mathrm{PBIAS} < \pm 25$
Unsatisfactory	$\mathrm{NSE} \leq 0.5$	$\mathrm{RSR} > 0.7$	$\mathrm{PBIAS} \geq 25$

2.3.4. Climate and land-use change contribution rate assessment

The SWAT model was used to separate the percentage of climate change impacts from land use on river flow. The observed and simulated flow data were compared using preset climate and land-use scenarios. For instance, four scenarios were used in which scenario 1 and scenario 3 encountered the land-use change impacts on river runoff, while scenario 2 and scenario 4 encountered the climate change impacts on river runoff as given in Table 3. The contribution rate was computed by Equations (6) and (7) where Q_1, Q_2, Q_3, and Q_4 are the average river runoff obtained from the below-mentioned scenarios (Guo *et al.* 2016; Shang *et al.* 2019):

$$\Delta Q_C = Q_2 - Q_1 \tag{2}$$

$$\Delta Q_L = Q_3 - Q_1 \tag{3}$$

$$\Delta Q_M = Q_L + Q_C \tag{4}$$

$$\Delta Q = Q_4 - Q_1 \tag{5}$$

Hypothetically, ΔQ is equal to ΔQ_M, and thereafter, the impact of climate and land-use change on river runoff can be computed by η_C climate change and η_L land-use change:

$$\eta_C = \frac{\Delta Q_C}{\Delta Q_M} \times 100\% \tag{6}$$

$$\eta_L = \frac{\Delta Q_L}{\Delta Q_M} \times 100\% \tag{7}$$

2.3.5. Future land-use prediction

Land use was predicted for 2050 for the upper region of Indus River using the cellular automata artificial neural network (CA-ANN) model that uses QGIS plugin Modules for Land Use Change Simulations (MOLUSCE). Slope and aspects variables were used as land-use drivers. At the very outset, to check the model efficiency, the land use of 2001 and 2010 were used to predict the land use of 2015, and then that of 2050.

3. RESULTS AND DISCUSSION

3.1. Model calibration and validation

Calibration (2000–2008) and validation (2009–2013) were carried out on the observed flow data obtained from WAPDA at the Doyain outlet station at Astore River. A warm-up period of 2 years (1998–2000) was used to initialize the calibration process. The model calibration process was followed by sensitivity analysis to choose parameters that govern the observed river runoff. A total of 24 most effective parameters were selected for model calibration as demonstrated in Table 4. The model performance was evaluated to predict the watershed conditions by statistical indicators such as R^2, NSE, PBIAS, and RSR as given in Table 5, and simulated flow results are demonstrated in Figure 3.

3.2. Impacts of climate and land-use change on river runoff

The simulated annual streamflow results show the effects of climate and land-use change on river runoff over the preset scenarios as demonstrated in Table 6. The total variation observed in streamflow is 121.32 m^3/s. In the aforementioned total

Table 3 | Scenarios-based contribution rate analysis

Scenarios	Land use	Meteorological data
1	2001	2000–2008
2	2001	2008–2013
3	2013	2000–2008
4	2013	2008–2013

Table 4 | Parameters sensitive to river runoff, their fitted values, and initial ranges

Parameter	Description	Fitted value	Ranges
CN2.mgt	SCS runoff curve number	78.5	(35,98)
ALPHA_BF.gw	Baseflow alpha factor (days)	0.63	(0,1)
GW_DELAY.gw	Groundwater delay (days)	485.40	(0,500)
GWQMN.gw	Threshold depth of water in the shallow aquifer required for return flow to occur (mm)	534.95	(0, 5,000)
SURLAG.hru	Surface runoff lag time	12.52	(0.05,24)
SLSOIL.hru	Slope length for lateral subsurface flow	131.25	(0,150)
EPCO.hru	Plant uptake compensation factor	0.47	(0,1)
SOL_K(..).sol	Saturated hydraulic conductivity	283.33	(0, 2,000)
OL_AWC(..).sol	Available water capacity of the soil layer	0.52	(0,1)
CH_N2.rte	Manning's n value for the main channel	0.30	(−0.01,0.3)
SFTMP.bsn	Snowfall temperature	−4.56	(−20,20)
SMTMP.bsn	Snow melt base temperature	5.33	(−20,20)
SMFMX.bsn	Maximum melt rate for snow during year (occurs on summer solstice)	3.55	(0,20)
SMFMN.bsn	Minimum melt rate for snow during the year (occurs on winter solstice)	19.50	(0,20)
SNO50COV.bsn	Snow water equivalent that corresponds to 50% snow cover	0.25	(0,1)
CH_K2.rte	Effective hydraulic conductivity in main channel alluvium	481.32	(−0.01,500)
ESCO.hru	Soil evaporation compensation factor	0.60	(0,1)
OV_N.hru	Manning's n value for overland flow	17.45	(0.01,30)
LAT_TTIME.hru	Lateral flow travel time	49.07	(0,180)
CANMX.hru	Maximum canopy storage	97.50	(0,100)
CNOP{..}.mgt	SCS runoff curve number for moisture condition	29.13	(0,100)
RFINC(..).sub	Rainfall adjustment	63.25	(0,100)
TMPINC(..).sub	Temperature adjustment	64.98	(0,100)
SNO_SUB.sub	Initial snow water content	69.47	(0,150)

Table 5 | Statistical indicators for model calibration and validation which regulate the flow regime

Calibration (2000–2008)				Validation (2009–2013)			
R^2	NSE	PBIAS	RSR	R^2	NSE	PBIAS	RSR
0.7	0.69	9.7	0.56	0.66	0.61	21.3	0.62

variation in streamflow, climate and land-use changes contributed 74.75 and 46.58 m^3/s increase in the streamflow. Climate and land-use change caused 61.61 and 38.39% variations in the streamflow, respectively, in the upper region of the Indus River. It shows that climate change is the major driver in the understudy area which caused variation in the streamflow.

Climate and land-use changes significantly alter the hydrological cycle. Adaptation methods for water management should consider how climatic variability will impact the hydrological regime and water supplies (Krysanova & White 2015). Tan *et al.* (2015) used the SWAT model to investigate the effects of land-use change and climatic variability in the Johor River basin, Malaysia where temperature and precipitation showed a significant increasing trend. The results demonstrated that the anticipated climatic variability showed a greater impact on streamflow and evaporation, in comparison to the projected land-use changes.

3.3. Climatic effects on streamflow

The SWAT model was used to assess the influence of climate change, i.e. precipitation and temperature (min and max) on the streamflow using four different GCMs. The analysis was carried out for four different periods (2007–2020, 2021–2034, 2035–

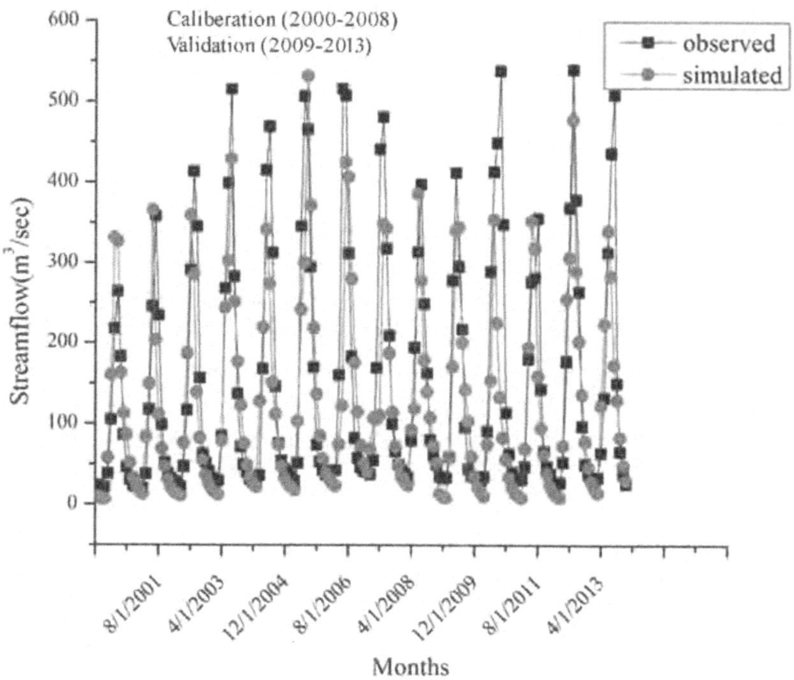

Figure 3 | SWAT model monthly streamflow calibration and validation results.

Table 6 | Climate and land use change impacts on streamflow under different scenarios

Scenarios	Period	Simulated annual runoff (m^3/s)	Variation (m^3/s)	% Climate contribution	% Land-use contribution
1	2000–2008	94.48	0		
2	2009–2013	151.23	56.75	61.61	38.39
3	2000–2008	129.84	35.36		
4	2009–2013	123.69	29.21		

2048, and 2086–2099). Table 7 shows the future mean annual streamflow of four GCMs at Astore River, Doyian station where a remarkable increase was observed in future streamflow in comparison to its base period (2007–2020).

Neupane *et al.* (2014) investigated the impacts of climate change on monsoon-driven streamflow in a central Himalayan watershed where predicted streamflow increased during the monsoon and post-monsoon seasons, while decreasing during the pre-monsoon season. Sood *et al.* (2013) used the SWAT model to assess the climate change impacts on the Volta River Basin in West Africa, where rainfall showed a decreasing trend while air temperature exhibited an increasing trend causing greater variability and reduced (40%) streamflow.

3.4. Land-use change effects on streamflow

Land use for the year 2050 was projected using QGIS plugin MOLUSCE. First, the land use of 2001 and 2010 was used to predict the land use of 2015, using CA-ANN. Later, the projected land-use map of 2015 was compared with NASA MODIS

Table 7 | Future prediction of streamflow (m^3/s) based on the four GCM models in the upper region of Indus River basin

Model	2007–2020	2021–2034	2035–2048	2086–2099
CanESM2	154.98	174.75 (+12.76%)	195.15 (+25.92%)	189.14 (+22.04%)
GFDL-ESM2G	150.12	178.25 (+18.74%)	183.36 (+22.14%)	173.47 (+15.55)
MIROC5	154.61	177.48 (+14.79%)	187.61 (+21.34%)	181.2 (+17.2%)
HadGEM2	158.69	184.45 (+16.23%)	193.85 (+22.16%)	186.77 (+17.7%)

land use for correctness and kappa value, which were found to be 80.26 and 0.72%, respectively. After validation, the land use of 2001 and 2018 was used to predict the land use of 2050. Increase and decrease are observed in different land-use classes as demonstrated. Figure 4 and Table 8 depict the area in percentages of variant land-use classes in different years.

Later, the projected land use of 2050 influence was evaluated on surface runoff using the SWAT model. The results showed that streamflow changes from 0.37 to 1.1% in response to the land use of 2018 and 2050.

The impacts of land-use/land-cover changes are evaluated using present trends and future scenarios. Getachew & Melesse (2012) studied the impacts of land-use change on the Angered watershed in Ethiopia. Decades of land-cover change affect runoff and reservoir inflow. Agricultural and urban areas increased, whereas forest and grassland areas shrank, affecting the mean monthly flow. Baker & Miller (2013) used the SWAT model to investigate how land development affects water resources in Kenya's River Njor. The results showed that dramatic alterations in land use have increased surface runoff and decreased groundwater recharge. Urbanization, agriculture, energy demand, and crop production are expected to grow in the Cedar River Basin of the Midwest, which will increase surface runoff and decrease baseflow (Wu *et al.* 2013). Schilling *et al.* (2014) examined the impacts of agricultural land use on flood risk reduction in Iowa's Raccoon River watershed. The study results unveiled that cropland conversion to permanent vegetation reduces the frequency of floods, but not their duration.

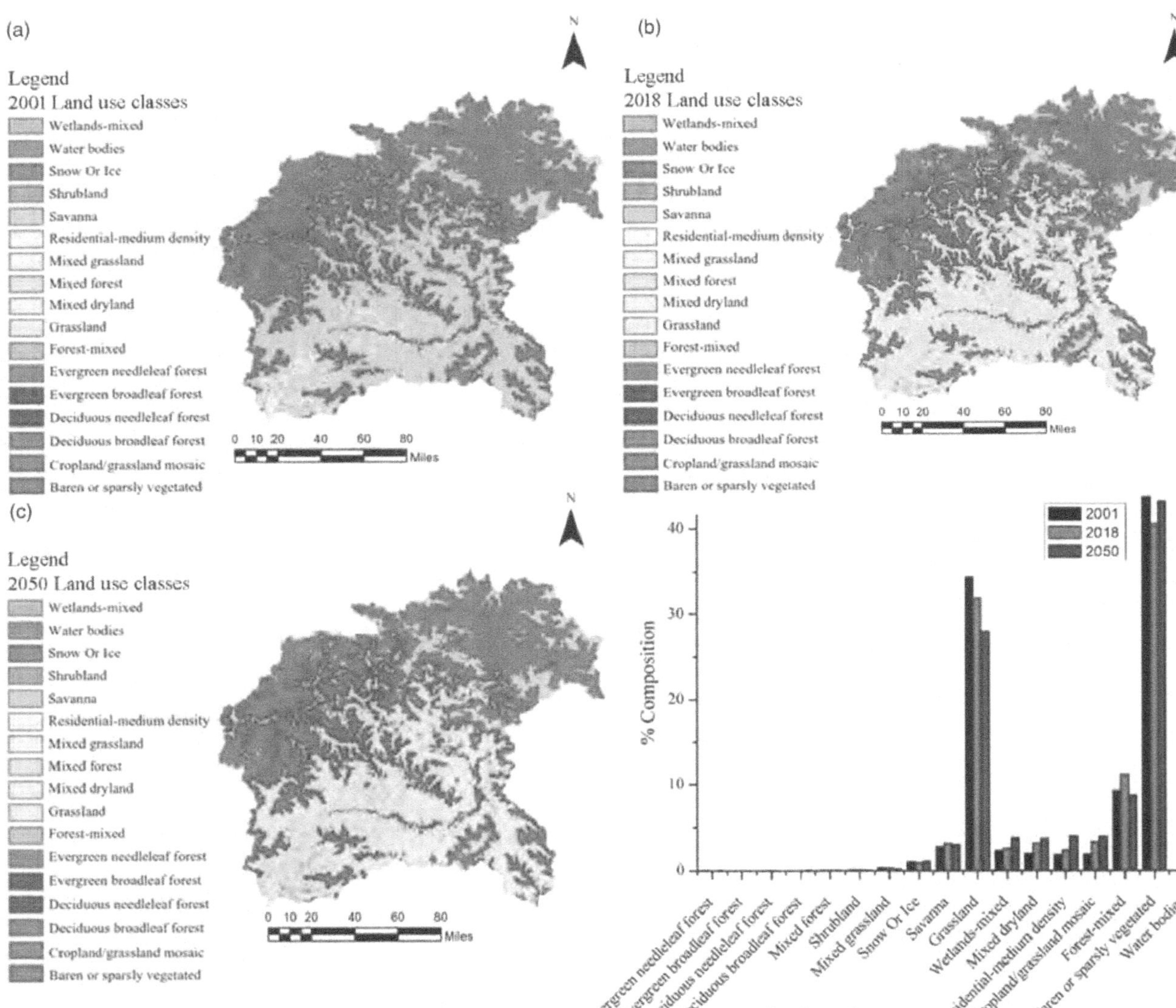

Figure 4 | Land-use maps of the study area: (a) 2001 land use, (b) 2018 land use, and (c) the predicted land use of 2050.

Table 8 | Area (%) change of different land-use classes

Land-use class	2001	2018	2050
Evergreen needleleaf forest	0.09%	0.05%	0.08%
Evergreen broadleaf forest	0.05%	0.03%	0.01%
Deciduous needleleaf forest	0.05%	0.08%	0.02%
Deciduous broadleaf forest	0.06%	0.04%	0.12%
Mixed forest	0.09%	0.09%	0.07%
Shrubland	0.14%	0.13%	0.12%
Mixed grassland	0.41%	0.35%	0.25%
Snow or ice	1.00%	0.90%	1.06%
Savanna	2.83%	3.19%	3.04%
Grassland	34.34%	31.90%	27.99%
Wetlands-mixed	2.31%	2.57%	3.77%
Mixed dryland	1.94%	3.20%	3.69%
Residential-medium density	1.86%	2.38%	3.96%
Cropland/grassland mosaic	1.88%	3.37%	3.89%
Forest-mixed	9.31%	11.13%	8.77%
Barren or sparsely vegetated	43.65%	40.55%	43.12%
Water bodies	0.02%	0.03%	0.03%

3.5. Watershed management implications

This study reported that climate change influence on the Indus flow regime overweighs land-use change. Climate change will increase water availability in the short term and decrease it in the long term. This dramatic change in flow regime is attributed to global warming. In the early 21st century, flow is expected to increase in summer followed by a decrease in the same in the late decades of the century. Summer and late spring discharges will increase in the short term because of excessive snowmelt, and in contrast, will decrease in the long term because of the disappearance of glaciers; but their effect will be lessened by rainfall. Some current aspects need re-assessment to improve waters condition in the basin. The sustainability of the current water system in the basin is questionable. Water management practices are suggested to store water which will help to lessen the influences of extreme conditions, i.e. flood and drought. The stored water will be used in the dry season for agriculture, hydropower, industrial, and domestic purposes. Furthermore, plantations are suggested in the study area to decline the increasing trend of air temperature as the sensitivity of streamflow is higher in the study area (Ahmad *et al.* 2020; Shah *et al.* 2021).

4. CONCLUSIONS

In this study, the SWAT model was used to evaluate the impacts of climate and land-use change on river runoff in the upper region of the Indus river basin. Future land-use and climate change scenarios are projected through the CA-ANN model and four bias-corrected GCM models, respectively. This study reported the percentage contribution of climate and land-use changes on runoff depth for the adopted as well as projected scenarios. The conclusions can be summarized as follows:

(1) The simulated results demonstrated that the SWAT model yields good statistical results during calibration (NSE $=0.69$, $R^2=0.7$, RSR $=0.56$, and PBIAS $=9.7$) and validation (NSE $=0.61$, $R^2=0.66$, RSR $=0.62$, and PBIAS $=21.3$).

(2) SWAT model was used to isolate the climate and land-use change impacts on river runoff in the upper region of the Indus river basin, Doyain station at Astore River, where the contribution rate of climate and land-use changes on surface runoff was 61.61 and 38.39%, respectively.

(3) GCMs (CanESM2, GFDL-ESM2G, MIROC5, and HadGEM2) based on CMIP5 have good applicability in the study area for the assessment of future climate change. Simulations with the calibrated model were conducted for 2007–2020, 2021–2034, 2035–2048, and 2086–2099 periods, where an increase was observed in the average annual river runoff.

(4) The predicted land use of 2050 by the CA-ANN model indicated an increasing trend of river runoff in the future by 1.1%.

DATA AVAILABILITY STATEMENT

All relevant data are included in the paper or its Supplementary Information.

REFERENCES

Ahmad, W., Khan, A. U., Khan, F. A., Farooq, M., Baig, A. A., Shah, L. A. & Khan, J. 2020 How vegetation spatially alters the response of precipitation and air temperature? Evidence from Pakistan. *Asian J. Atmos. Environ.* **14** (2), 133–145.

Archer, D. 2003 Contrasting hydrological regimes in the upper Indus Basin. *J. Hydrol.* **274**, 198–210.

Babar, Z. A., Zhi, X.-f. & Fei, G. 2015 Precipitation assessment of Indian summer monsoon based on CMIP5 climate simulations. *Arab. J. Geosci.* **8**, 4379–4392.

Baker, T. J. & Miller, S. N. 2013 Using the Soil and Water Assessment Tool (SWAT) to assess land use impact on water resources in an East African watershed. *J. Hydrol.* **486**, 100–111.

Bashir, N., Saeed, R., Afzaal, M., Ahmad, A., Muhammad, N., Iqbal, J., Khan, A., Maqbool, Y. & Hameed, S. 2020 Water quality assessment of lower Jhelum canal in Pakistan by using geographic information system (GIS). *Groundwater for Sustainable Development* **10**, 100357.

Birkinshaw, S. J., Guerreiro, S. B., Nicholson, A., Liang, Q., Quinn, P., Zhang, L., He, B., Yin, J. & Fowler, H. J. 2017 *Climate Change Impacts on Yangtze River Discharge at the Three Gorges Dam.*

Daud, M., Nafees, M., Ali, S., Rizwan, M., Bajwa, R. A., Shakoor, M. B., Arshad, M. U., Chatha, S. A. S., Deeba, F. & Murad, W. 2017 Drinking water quality status and contamination in Pakistan. *BioMed research international* **2017**, 7908183.

Deng, Z., Zhang, X., Li, D. & Pan, G. 2015 Simulation of land use/land cover change and its effects on the hydrological characteristics of the upper reaches of the Hanjiang Basin. *Environ. Earth Sci.* **73**, 1119–1132.

Du, J., Wang, K. & Cui, B. 2021 Attribution of the extreme drought-related risk of wildfires in spring 2019 over Southwest China. *Bulletin of the American Meteorological Society* **102** (1), S83–S90.

Fang, X., Ren, L., Li, Q., Zhu, Q., Shi, P. & Zhu, Y. 2013 Hydrologic response to land use and land cover changes within the context of catchment-scale spatial information. *J. Hydrol. Eng.* **18**, 1539–1548.

Fang, G., Yang, J., Chen, Y. & Zammit, C. 2015 Comparing bias correction methods in downscaling meteorological variables for a hydrologic impact study in an arid area in China. *Hydrology and Earth System Sciences* **19** (6), 2547–2559.

Ficklin, D. L. & Barnhart, B. L. 2014 SWAT hydrologic model parameter uncertainty and its implications for hydroclimatic projections in snowmelt-dependent watersheds. *Journal of hydrology* **519**, 2081–2090.

Fowler, H. J., Blenkinsop, S. & Tebaldi, C. 2007 Linking climate change modelling to impacts studies: recent advances in downscaling techniques for hydrological modelling. *Int. J. Climatol.* **27**, 1547–1578.

Gebremeskel, S., Liu, Y. B., De Smedt, F., Hoffmann, L. & Pfister, L. 2004 Analysing the effect of climate changes on streamflow using statistically downscaled GCM scenarios. *Int. J. River Basin Manag.* **2**, 271–280.

Getachew, H. E. & Melesse, A. M. 2012 The impact of land use change on the hydrology of the Angereb Watershed, Ethiopia. *Int. J. Water Sci.* **1** (6).

Ghoraba, S. M. 2015 Hydrological modeling of the Simly Dam watershed (Pakistan) using GIS and SWAT model. *Alex. Eng. J.* **54**, 583–594.

Guo, J., Su, X., Singh, V. P. & Jin, J. 2016 Impacts of climate and land use/cover change on streamflow using SWAT and a separation method for the Xiying River Basin in northwestern China. *Water* **8**, 192.

Khan, A., Ghoraba, S., Arnold, J. G. & Di Luzio, M. 2014 Hydrological modeling of upper Indus Basin and assessment of deltaic ecology. *Int. J. Mod. Eng. Res.* **4**, 73–85.

Krysanova, V. & White, M. 2015 Advances in water resources assessment with SWAT–an overview. *Hydrological Sciences Journal* **60** (5), 771–783.

Li, Z., Liu, W.-z., Zhang, X.-c. & Zheng, F.-l. 2009 Impacts of land use change and climate variability on hydrology in an agricultural catchment on the Loess Plateau of China. *J. Hydrol.* **377**, 35–42.

Liu, D., Chen, X., Lian, Y. & Lou, Z. 2010 Impacts of climate change and human activities on surface runoff in the Dongjiang River basin of China. *Hydrol. Process.* **24**, 1487–1495.

Lutz, A., Immerzeel, W., Shrestha, A. & Bierkens, M. 2014 Consistent increase in High Asia's runoff due to increasing glacier melt and precipitation. *Nat. Clim. Change* **4**, 587–592.

Mahmood, R. & Babel, M. S. 2013 Evaluation of SDSM developed by annual and monthly sub-models for downscaling temperature and precipitation in the Jhelum basin, Pakistan and India. *Theor. Appl. Climatol.* **113**, 27–44.

Mahmood, R. & Jia, S. 2016 Assessment of impacts of climate change on the water resources of the transboundary Jhelum River basin of Pakistan and India. *Water* **8**, 246.

Memarian, H., Balasundram, S. K., Abbaspour, K. C., Talib, J. B., Boon Sung, C. T. & Sood, A. M. 2014 SWAT-based hydrological modelling of tropical land-use scenarios. *Hydrol. Sci. J.* **59**, 1808–1829.

Mukheibir, P. 2007 *Access to Water – the Impact of Climate Change in Small Municipalities.*

Nanding, N., Chen, Y., Wu, H., Dong, B., Tian, F., Lott, F. C., Tett, S. F., Rico-Ramirez, M. A., Chen, Y. & Huang, Z. 2020 Anthropogenic influences on 2019 July precipitation extremes over the mid–lower reaches of the Yangtze River. *Frontiers in Environmental Science* **8**, 230.

Neitsch, S. L., Arnold, J. G., Kiniry, J. R. & Williams, J. R. 2011 *Soil and Water Assessment Tool Theoretical Documentation Version 2009.* Texas Water Resources Institute, College Station, TX.

Neupane, R. P., Yao, J. & White, J. D. 2014 Estimating the effects of climate change on the intensification of monsoonal-driven stream discharge in a Himalayan watershed. *Hydrol. Process.* **28** (26), 6236–6250.

Nguyen, H., Mehrotra, R. & Sharma, A. 2020 Assessment of climate change impacts on reservoir storage reliability, resilience, and vulnerability using a multivariate frequency bias correction approach. *Water Resour. Res.* **56**, e2019WR026022.

Obuobie, E., Kankam-Yeboah, K., Amisigo, B., Opoku-Ankomah, Y. & Ofori, D. 2012 Assessment of water stress in river basins in Ghana. *J. Water Clim. Change* **3**, 276–286.

Pokhrel, Y., Felfelani, F., Satoh, Y., Boulange, J., Burek, P., Gädeke, A., Gerten, D., Gosling, S. N., Grillakis, M. & Gudmundsson, L. 2021 Global terrestrial water storage and drought severity under climate change. *Nature Climate Change* **11** (3), 226–233.

Prasanna, V. 2015 Regional climate change scenarios over South Asia in the CMIP5 coupled climate model simulations. *Meteorol. Atmos. Phys.* **12**, 561–578.

Rathjens, H., Bieger, K., Srinivasan, R., Chaubey, I. & Arnold, J. 2016a *CMhyd User Manual.* Texas A&M University, College Station, TX.

Rathjens, H., Bieger, K., Srinivasan, R., Chaubey, I. & Arnold, J. 2016b *CMhyd User Manual: Documentation for Preparing Simulated Climate Change Data for Hydrologic Impact Studies.* Texas A&M University, College Station, TX.

Schilling, K. E., Gassman, P. W., Kling, C. L., Campbell, T., Jha, M. K., Wolter, C. F. & Arnold, J. G. 2014 The potential for agricultural land use change to reduce flood risk in a large watershed. *Hydrol. Process.* **28** (8), 3314–3325.

Shah, L. A., Khan, A. U., Khan, F. A., Khan, Z., Rauf, A. U., Rahman, S. U., Iqbal, M. J., Ahmad, I. & Abbas, A. 2021 Statistical significance assessment of streamflow elasticity of major rivers. *Civil Eng. J.* **7** (5), 893–905.

Shang, X., Jiang, X., Jia, R. & Wei, C. 2019 Land use and climate change effects on surface runoff variations in the Upper Heihe River Basin. *Water* **11**, 344.

Sood, A., Muthuwatta, L. & McCartney, M. 2013 A SWAT evaluation of the effect of climate change on the hydrology of the Volta River basin. *Water Int.* **38** (3), 297–311.

Stocker, T., Clarke, G., Le Treut, H., Lindzen, R., Meleshko, V., Mugara, R., Palmer, T., Pierrehumbert, R., Sellers, P. & Trenberth, K. 2001 *IPCC, 2001: Climate Change 2001: the Scientific Basis. Contribution of Working Group I to the Third Assessment Report of the IPCC.* Cambridge University Press, Cambridge, UK, pp. 417–470.

Sun, F., Roderick, M. L., Lim, W. H. & Farquhar, G. D. 2011 Hydroclimatic projections for the Murray-Darling Basin based on an ensemble derived from Intergovernmental Panel on Climate Change AR4 climate models. *Water Resour. Res.* **47**, 1–14.

Tan, M. L., Ibrahim, A. L., Yusop, Z., Duan, Z. & Ling, L. 2015 Impacts of land-use and climate variability on hydrological components in the Johor River basin, Malaysia. *Hydrological Sciences Journal* **60** (5), 873–889.

Wang, R., Kalin, L., Kuang, W. & Tian, H. 2014 Individual and combined effects of land use/cover and climate change on Wolf Bay watershed streamflow in southern Alabama. *Hydrol. Process.* **28**, 5530–5546.

Wilby, R. L., Hay, L. E., Gutowski Jr, W. J., Arritt, R. W., Takle, E. S., Pan, Z., Leavesley, G. H. & Clark, M. P. 2000 Hydrological responses to dynamically and statistically downscaled climate model output. *Geophys. Res. Lett.* **27**, 1199–1202.

Wisal, K., Asif, K., Ullah, K. A. & Mujahid, K. 2020 Evaluation of hydrological modeling using climatic station and gridded precipitation dataset. *MAUSAM* **71** (4), 717–728.

Worku, G., Teferi, E., Bantider, A. & Dile, Y. T. 2020 Statistical bias correction of regional climate model simulations for climate change projection in the Jemma sub-basin, upper Blue Nile Basin of Ethiopia. *Theor. Appl. Climatol.* **139**, 1569–1588.

Wu, Y., Liu, S., Sohl, T. L. & Young, C. J. 2013 Projecting the land cover change and its environmental impacts in the Cedar River Basin in the Midwestern United States. *Environ. Res. Lett.* **8** (2), 024025.

Yi, L. & Sophocleous, M. 2011 Two-way coupling of unsaturated-saturated flow by integrating the SWAT and MODFLOW models with application in an irrigation district in arid region of West China. *J. Arid. Land.* **3**, 164–173.

Yu, W., Zhao, Y., Nan, Z. & Li, S. 2013 Improvement of snowmelt implementation in the SWAT hydrologic model. *Acta Ecol. Sin.* **33**, 6992–7001.

Zhang, L., Karthikeyan, R., Bai, Z. & Srinivasan, R. 2017 Analysis of streamflow responses to climate variability and land use change in the Loess Plateau region of China. *Catena* **154**, 1–11.

First received 18 June 2021; accepted in revised form 11 October 2021. Available online 27 October 2021

Impact of Climate Change on Hydrology and Water Resources doi: 10.2166/wcc.2021.038

Response of runoff in the upper reaches of the Minjiang River to climate change

Kebi Yang[a,b], Ting Chen[a,c], Tianqi Ao[a,*], Xu Zhang[d,e], Li Zhou[a,d] and Danyang Gao[a]

[a] State Key Laboratory of Hydraulics and Mountain River Engineering, College of Water Resources and Hydropower, Sichuan University, Chengdu 610065, China
[b] Department of Ecological Environment of Sichuan Province, Government of China, Chengdu 610088, China
[c] Sichuan Meteorological Service Center, Heavy Rain and Drought-Flood Disasters in Plateau and Basin Key Laboratory of Sichuan Province, Chengdu 610072, China
[d] State Key Laboratory of Hydrology – Water Resources and Hydraulic Engineering, Nanjing Hydraulic Research Institute, Nanjing 210029, China
[e] Research Center for Climate Change, Ministry of Water Resources, Nanjing 210029, China
*Corresponding author. E-mail: atian21@tom.com

ABSTRACT

Climate change affects the water cycle in different regions. The response of annual runoff and seasonal distribution to climate change in the upper reaches of the Minjiang River during 2021–2050 was studied by coupling the Statistical Downscaling Model (SDSM) and the Soil and Water Assessment Tool (SWAT). This model was driven by the second-generation Canadian Earth System Model (CanESM2) under RCP2.6, RCP4.5, and RCP8.5 scenarios. The results show that the runoff in the upper reaches of the Minjiang River has a unique response to climate change. The maximum and minimum temperatures will increase with the increase in emissions, especially in December–January. The daily precipitation shows an upward trend, especially in July–August in the RCP4.5 scenario. The annual runoff shows an upward trend with the increase in emissions. Compared with the current increase of 13–26%, the most prominent period is November–April. Because the study area covers high mountains and gorge landforms, the altitude difference is great, and the influence of evapotranspiration and snow melting processes is more prominent, causing the monthly runoff to decrease in June–July with an increase in precipitation. From April to May, precipitation decreased while runoff increased.

Key words: change of runoff, Minjiang river, soil and water assessment tool, statistical downscaling model, typical concentration paths

HIGHLIGHTS

- The first study on annual runoff, seasonal distribution, and primary tributaries of the upper Yangtze River response to climate change in the Minjiang River upper reaches.
- Analysis of differences among the Minjiang upper reaches and other upper reaches of the Yangtze River.
- Runoff variation causes about seasons and three typical concentration paths.
- Coupling SDSM and SWAT analyzes application.

INTRODUCTION

The fifth Intergovernmental Panel on Climate Change (IPCC) assessment report pointed out that the average temperature has risen by 0.65–1.06 °C in the past 120 years (Qin & Thomas 2014). The global climate has been gradually warming. Climate change alters the original water cycle and changes the distribution and the total number of water resources, thereby increasing extreme hydrological events. The prediction of the water cycle and water resource fluctuations is one of the primary components in alleviating the impact of climate change on water resources (Githui *et al.* 2009; Moss *et al.* 2010; Taylor *et al.* 2012; Shrestha *et al.* 2017). The response of water resources to future climate change usually follows the future climate scenario settings, such as hydrological simulation of future events (Chen *et al.* 2014; Yang *et al.* 2015). Researchers worldwide have carried out several studies on this issue. Wang *et al.* (2015) showed that under the IPCC scenarios A2 and B2, the precipitation and runoff would be reduced in the Yangtze River. Su *et al.* (2017) analyzed the runoff changes in the Yangtze River using CMIP5 climate models. The results show that the runoff will increase due to increased precipitation. Birkinshaw *et al.* (2017) showed that under the RCP8.5 typical concentration paths, the precipitation in the upper reaches of the Three Gorges reservoir would increase by 4.1% between 2041 and 2070. However, due to the apparent increase in evapotranspiration caused, the inflow of the Three Gorges reservoir will decrease by 11.1%. Yang *et al.* (2019a, 2019b) showed that the runoff in the upper reaches of the Yangtze River would increase by 11.33–64.08% in winter and 8.64–40.97% in summer by using CMIP5 climate models. Chen *et al.* (2017) showed that the summer, spring, and annual runoff in the upper reaches of the Yangtze River would show an upward trend under the climate scenario, whereas the autumn and winter runoff will show a downward trend. The increase in runoff will increase the frequency and intensity of flood disasters, especially in summer. Yang *et al.* (2019a, 2019b) showed that runoff in the upper reaches of the Yangtze River shows a noticeable increase in the CanESM2 model. However, a downward trend is observed in the NorESM1-M model.

In general, annual and seasonal variations in the runoff in the same watershed are different. The uncertainty in research results is affected by the climate model, the hydrological model, the analysis period, and the station data. In addition, previous studies mainly focused on the total amount of water resources in the Yangtze River where the study area is located. The responses of different sub-basins in the upper reaches to regional climate change are different (Wang *et al.* 2015), but there are only a few studies on different sub-basins. Previous studies show that runoff changes are mainly affected by temperature and rainfall, and the changes in temperature and precipitation have strong spatial differences (Gu *et al.* 2015; Yang *et al.* 2019a, 2019b). For example, Luo *et al.* (2019) showed that precipitation and temperature in the western source of the Yangtze River were the most important climatic factors affecting the runoff during 1961–2016. In summer, the increases in precipitation, the annual average minimum temperature, and the amount of glacier and snowmelt affect the runoff. The apparent increase in spring runoff was related to the increase of the snow melting and the increase in the amount of snow melted due to temperature rise. In summary, the temporal and spatial differences in temperature and precipitation give the runoff unique characteristics. Previous studies mainly focused on analyzing annual runoff using multiple modes and multiple scenarios, and the studies that focus on the response of seasonal distribution of runoff to climate change are relatively scarce.

The Minjiang River Basin is the largest tributary of the upper reaches of the Yangtze River. The upper reaches of the Minjiang River are above Dujiangyan, which is the source of domestic and production water in the Chengdu Plain. The upper reaches of the Minjiang River have a large elevation difference, fragile geological conditions, and complex climate conditions. Climate change makes the difference between water demand and supply more obvious. Second, the Minjiang River is the ecological barrier of the Yangtze River. Further study of its future ecological environmental changes can provide a theoretical reference for constructing an ecological barrier. Furthermore, some changes in the Minjiang River Basin in recent years are prominent (Chen *et al.* 2019a, 2019b). Research on the change under the influence of climate change can provide useful data for hydropower energy development and flood control planning under climate change. In this study, the Statistical Downscaling Model (SDSM) is used to analyze the daily series of temperature and precipitation of the second-generation Canadian Earth System Model (CanESM2) in the future climate scenarios (RCP2.6, RCP4.5, and RCP8.5), and the possible changes under future climate scenarios in CMIP5 were analyzed. The Soil and Water Assessment Tool (SWAT) was used to simulate the runoff, and the sensitivity parameters were calibrated using SWAT-CUP. Finally, the SWAT derived by the temperature and precipitation series was used to analyze runoff changes in future climate scenarios. The possible trends of drought and flood events in the study area are analyzed by comparing the historical runoff data.

MATERIALS AND METHODOLOGY

Study area

The Minjiang River Basin is 790 km long, with a fall of 3,560 m and a drainage area of 135,881 km^2. The Minjiang River Basin originates from the Aba Prefecture of Sichuan Province and is rich in hydropower resources. According to the geographical characteristics of the mainstream, Dujiangyan City is upstream from the source of the basin, and hydropower generation is mostly concentrated upstream.

The upper reaches of the Minjiang River are located between 102°59′ E and 104°14′ E, and 26°33′ N and 33°16′ N, covering an area of 23,000 km^2 (as shown in Figure 1). The average annual precipitation in the upper reaches of the Minjiang River is 400–800 mm, and the average annual temperature is 11.2–12.9 °C. According to the data of the Zipingpu hydrological station in the Minjiang River, the average annual discharge in the upper reaches of the Minjiang River is 476 m^3/s. The ecological environment has been gradually degrading in recent decades, seriously threatening infrastructure security in the valley. The forest coverage is reduced, the ecological function is degraded, the soil and water loss is severe, and natural disasters occur frequently.

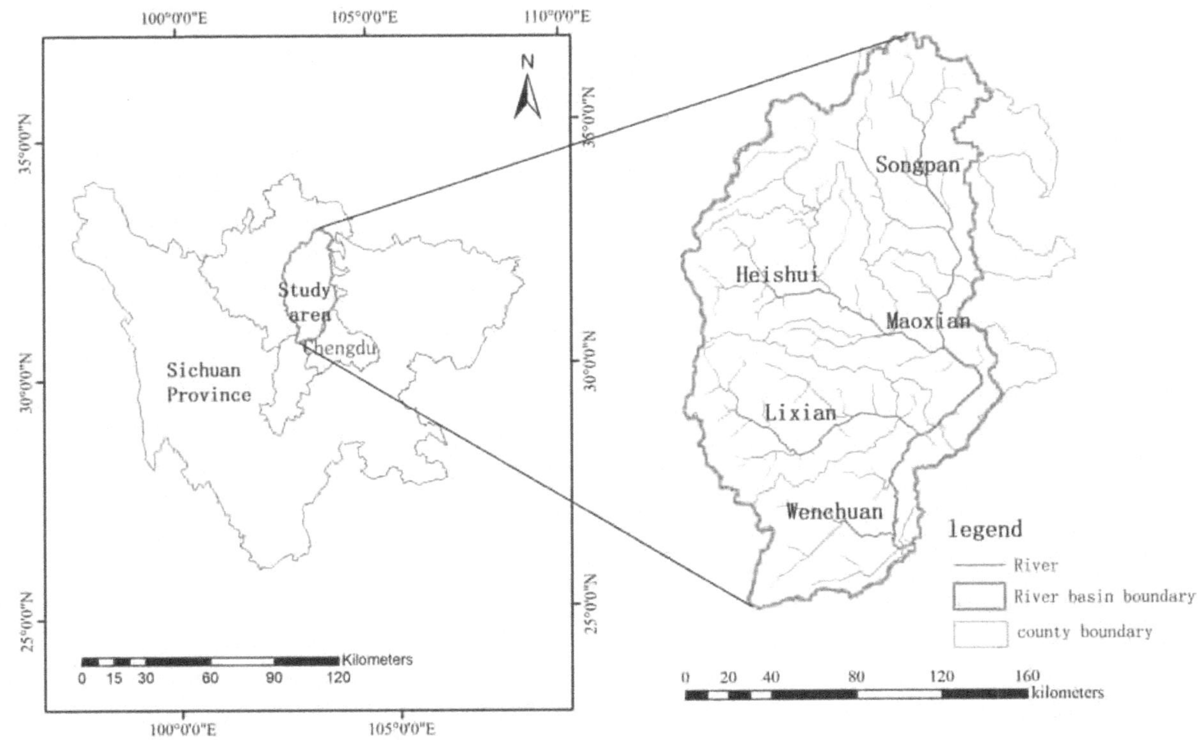

Figure 1 | Water system in the upper reaches of the Minjiang River.

Data sources

The daily temperature and precipitation data from different national weather stations were obtained from http://data.cma.cn/ (Table 1). The data period is 1961–2016, which has passed strict quality control. The daily average discharge of the Zipingpu hydrological station at the outlet section of the study area from 1961 to 2016 and the daily rainfall (Table 2) was selected for the study, and data were obtained from the hydrological yearbook.

Through literature research, it was understood that the CanESM2 model has a high spatial correlation coefficient in Southwest China, where the upper reaches of the Minjiang River are located ($r = 0.76$) (Sun *et al.* 2015; Zhan *et al.* 2017). Because the CanESM2 model can output more accurate large-scale variables, such as the upper atmospheric field and atmospheric circulation (Zhan *et al.* 2017), the spatial resolution of output products is $2.5° \times 2.5°$ and lacks regional climate information. We need to downscale the large-scale variables to obtain the information of the regional scale. The SDSM is a more accurate and widely used statistical downscaling model (Huang *et al.* 2008). NCEP reanalysis data were used in the downscaling analysis with SDSM. The NCEP data resolution is $2.5° \times 2.5°$, and each grid has 26 meteorological factors, including temperature, humidity, wind speed, and divergence. CanESM2 simulation results and NCEP reanalysis data were obtained from http://climate-scenarios.canada.ca/?page=pred-canesm2. Nine grids corresponding to the study area are selected, including the grid where the basin is located and eight adjacent grids.

Table 1 | Basic information regarding meteorological stations

Station	Longitude (°E)	Latitude (°N)	Altitude (m)
Songpan	103.57	32.65	2,850
Heishui	102.98	32.08	2,400
Maoxian	103.85	31.68	1,591
Wenchuan	103.58	31.47	1,326
Lixian	103.17	31.43	1,885

Table 2 | Basic information regarding hydrological stations and rainfall stations

	Station	Latitude (°N)	Longitude (°E)
Hydrological station	Zipingpu	31°02′	103°34′
Rainfall station	Songpan	32°38′	103°36′
	Maladun	32°23′	103°28′
	Shuzhuba	32°31′	103°37′
	Zhenjiangguan	32°18′	103°44′
	Jiaochangba	32°03′	103°40′
	Zhimulin	32°11′	103°11′
	Sandagu	32°15′	102°53′
	Heishui	32°04′	103°00′
	Cibushu	31°54′	103°24′
	Shaba	31°50′	103°40′
	Weimen	31°46′	103°50′
	Miyaluo	31°39′	102°49′
	Zhagunao	31°26′	103°08′
	Shangmeng	31°40′	103°08′
	Sangping	31°29′	103°35′
	Mianchi	31°22′	103°30′
	Huahongshu	31°02′	103°11′
	Yuzixi	32°04′	103°29′
	Sanjiangkou	30°55′	103°21′
	Shouxi	30°59′	103°29′
	Guanmenshi	31°08′	103°34′
	Zipingpu	31°02′	103°34′

The SWAT model's hydrological data calibrated and verified show the daily runoff at the Zipingpu hydrological station in the study area from 1967 to 1987. The sensitivity parameters were calibrated using SWAT-CUP (Abbaspour 2007), an official parameter calibration tool recommended by SWAT. Table 3 indicates the input data of the SWAT. It mainly includes spatial and attribute data. Spatial data mainly include the digital elevation model (DEM), land-use type distribution, and soil spatial distribution. Spatial data are uniformly projected with WGS coordinate_1984_UTM_Zone_49N. The DEM was developed from geospatial data. Land-use data were obtained from the International Geosphere-Biosphere Programme (IGBP) and correspond to land-use types in SWAT. Land-use types are divided into five categories. Soil data are provided by the Food and Agriculture Organization of the United Nations (FAO). Soil types are divided into 27 categories.

Methodology

In this study, SDSM (Wilby 2002; Wilby *et al.* 2003) was used to analyze the daily series of temperature and precipitation of the CanESM2 in the future scenario. SDSM is one of the more accurate and widely used statistical downscaling models, which combines multiple linear regression and random weather generator. Precipitation and temperature are the primary

Table 3 | Model input data list

Data name	Data description	Data source
DEM data	DEM 30 m dataset (elevation, slope)	www.gscloud.cn/search
Land-use data	1 km resolution (spatial distribution of land-use types) released by IGBP	http://www.igbp.net/
Soil data	1 km resolution (spatial distribution and soil properties)	http://www.crensed.ac.cn/portal/
Meteorological data	Daily meteorological data provided by CMA. The future meteorological data are from the highest, lowest temperature, and daily precipitation data from CanESM2 simulated by SDSM	http://data.cma.cn/
Hydrological data	Daily runoff data of the Zipingpu hydrological station from 1967 to 1987	Hydrological yearbook

influencing factors of flood season in the upper reaches of the Minjiang River, so SDSM is used to study the downscaling of precipitation and temperature in the upper reaches of the Minjiang River.

The relationship between SDSM generated predictor (R) and predictor (L) is as follows:

$$R = F(L) \tag{1}$$

where F is a deterministic or stochastic function. The prediction factor L is mainly selected according to the station data and research purposes.

SWAT was used to simulate the runoff, and the sensitivity parameters were calibrated using SWAT-CUP (Arnold *et al.* 1998; Olivera *et al.* 2006). The SWAT model is mainly composed of terrain sub-model, runoff sub-model, confluence sub-model, and water proton model. The model uses a grid to discretize the watershed, which can better reflect the uneven distribution of terrain, soil, land use, vegetation cover, and other time and space, and can reflect the response of water quantity and water quality brought by natural changes and human activities. The model parameters can reflect the physical characteristics of the underlying surface of the watershed. The SWAT model uses the hydrological response unit as the minimum calculation unit, which simplifies much calculation compared with other models, giving the SWAT model significant advantages in a long-term simulation of the large-scale watershed.

RESULTS

The prediction of future climate scenarios in the study area

The steps for constructing a downscaling model using SDSM in the study area are as follows. First, the optimal grid is selected using the correlation between NCEP and measured data. The optimal predictor is selected using the partial phase relation number of each prediction factor to the prediction variables, and the statistical relationship is established. Second, the independent observation data were used to test the statistical relationship, and the statistical relationship was used to predict the future daily meteorological information.

Prediction factor selection

The grid of meteorological stations in the study area and eight adjacent grids were selected. Based on the large-scale temperature, the grid with the highest correlation with temperature is selected. The grid with the highest correlation with precipitation is selected based on the surface absolute humidity. Referring to Wilby's research results (Wilby 2002; Wilby *et al.* 2003), the prediction factors of temperature and precipitation (Table 4), and rainfall station forecast factors are selected.

Table 4 | Predictor of temperature and precipitation at representative stations in the study area

	Maximum temperature		Minimum temperature		Precipitation	
Stations	**Grid**	**Predictor**	**Grid**	**Predictor**	**Grid**	**Predictor**
Maoxian	8	mslp;p1_u;p8_u;p850;s500;s850; temp	3	mslp;p1_u;p1_z;p5_z;p500; p850;s850;shum;temp	2	mslp;p1_v;p5zh;p1zh;p5_v;p5_z; p5zh;p8_v;prcp;s500;shum;temp
Songpan	5	mslp;p1_v;p5_u;p500;p8_v;p850; prcp;s850;shum;temp	2	mslp;p1_up1_v;p1zhp5_v; s500;s850;shum;temp;	1	mslp;p1_v;p1zh;p5_v;p500;p8_z; p850;p8zh;prcp; temp
Wenchuan	9	mslp;p1zh;p500;p8_z;s850;temp	3	mslp;p1_z;p500;p8zh;s850; shum;temp	1	p1_u;p1_v;p1zh;p5_u;p5_z;p500; p5zh;p8_u;p8zh;s500;shum
Lixian	8	mslp;p1_v;p850;s500;s850;temp	3	mslp;p1_z;p500;s850;shum; temp	1	mslp;p1_f;p1_u;p1zh;p5_z;p500; p5zh;p8_u;p8zh;prcp;shum;temp
Heishui	5	mslp;p1_v;p5_u;p5_z;p500;p5zh; prcp;s850;s500;shum;temp	1	mslp;p1zh;p5_v;p500;s850; shum;temp	2	p1_v; p1zh; p5_u; p5_z; p500; p8_ugl; p8zh; shum; temp

Note: p, p5, and p8 refer to near-surface, 500 and 850 hPa geopotential height; 500 and 850 refer to 500 and 850 hPa geopotential height. temp, f, z, u, v, zh, th, rhum, r, and p refer to average temperature, geostrophic wind speed, vorticity, zonal wind speed, meridional wind speed, divergence, wind speed, near-surface relative humidity, relative humidity, and potential height.

The effect of downscaling simulation

From the comparison of the monthly mean maximum and minimum temperatures observed and simulated in Figure 2, it was understood that the mean value and standard deviation of the observed and simulated values are similar. Figure 2 shows that the simulated monthly precipitation values are generally slightly higher than the measured values, which may be related to the complex physical mechanism of precipitation and the discontinuity of precipitation in time and space. However, the statistical relationship established by SDSM is relatively simple; thus, the precipitation at the simulated stations has a certain error (Zhao *et al.* 2008).

Generally, SDSM can simulate the monthly variation of temperature and precipitation in the study area. This variation may be related to the establishment of a downscaling model by month in SDSM so that the monthly variation characteristics can be reflected. The difference between the maximum and minimum temperature and precipitation prediction results may also be related to the strong spatial-temporal difference in precipitation. In contrast, temperature is the continuous distribution of space–time.

The average absolute error of variables in the validation period (Table 5) was calculated. According to Table 5, the average absolute errors of the highest and lowest temperatures have small seasonal variations, whereas the average absolute error of

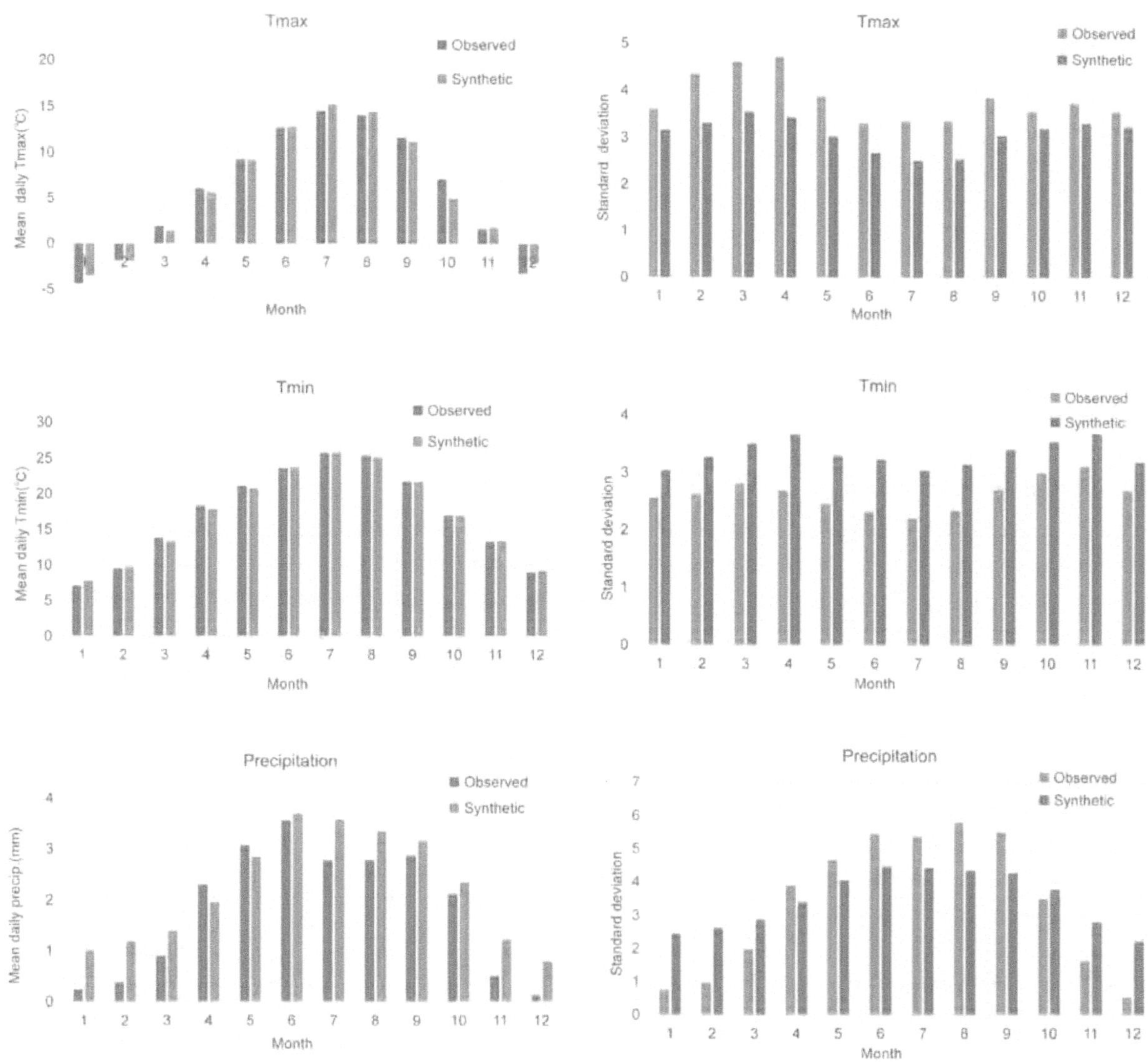

Figure 2 | Simulation results of temperature and precipitation, and precision of the SDSM in the study area during the validation period (1989–2005).

Table 5 | Average absolute error of variables during the validation period

Variable	Season			
	Spring	Summer	Autumn	Winter
Maximum temperature (°C)	1.4	1.4	1.4	1.4
Minimum temperature (°C)	1.4	1.3	1.4	1.5
Precipitation (mm)	1.3	2.2	1.3	0.4

Table 6 | Comparison of standard deviation and trends in temperature and precipitation between measured and simulated data

Variable	Annually average	
Maximum temperature	Measured standard deviation	7.40
	Simulated standard deviation	7.32
	Measured trend	0.11
	Simulation trend	0.06
Minimum temperature	Measured standard deviation	7.08
	Simulated standard deviation	7.43
	Measured trend	0.02
	Simulation trend	0.00
Precipitation	Measured standard deviation	4.12
	Simulated standard deviation	3.72
	Measured trend	−0.01
	Simulation trend	0.00

precipitation in summer is higher than that in other seasons. Generally speaking, it is feasible to use SDSM to analyze the variables in the study area.

Table 6 shows the difference in standard deviation and interannual trend of variables in the study area between the measured and simulated data. The results show that the standard deviation between the measured and simulated results is similar. The simulation effect of the maximum and minimum temperatures is better than that of the precipitation.

Prediction of future climate scenarios

Table 7 shows the interannual changes in variables in the future scenario. The maximum temperature increases by 12–14% compared with the historical period; the warming of the Songpan and Wenchuan stations shows the largest difference. The minimum temperature is 23–28% higher than the historical period, and the temperature at the Maoxian station shows the largest difference. The increase in temperature increases with an increase in emissions. Compared with the minimum and maximum temperature changes, the minimum temperature rise is more significant than the maximum temperature rise. In the future scenario, the annual precipitation will increase by 33–44%. The precipitation increment trends among the three

Table 7 | Interannual variation of temperature and precipitation in future scenarios (%)

Stations	Maximum temperature			Minimum temperature			Precipitation		
	RCP2.6	RCP4.5	RCP8.5	RCP2.6	RCP4.5	RCP8.5	RCP2.6	RCP4.5	RCP8.5
Totality	12.5	12.9	14.0	23.0	23.5	28.0	33.4	33.4	34.8
Maoxian	6.4	6.4	8.3	39.4	39.4	42.3	42.9	35.7	42.9
Songpan	18.3	19.0	19.7	3.8	3.8	11.5	42.1	42.1	42.1
Wenchuan	14.4	14.4	15.4	16.5	16.5	18.6	37.1	44.3	64.3
Lixian	9.8	10.3	11.5	16.2	16.2	20.6	29.4	29.4	29.4
Heishui	13.7	14.3	14.9	38.9	41.7	47.2	−4.8	−4.8	−4.8

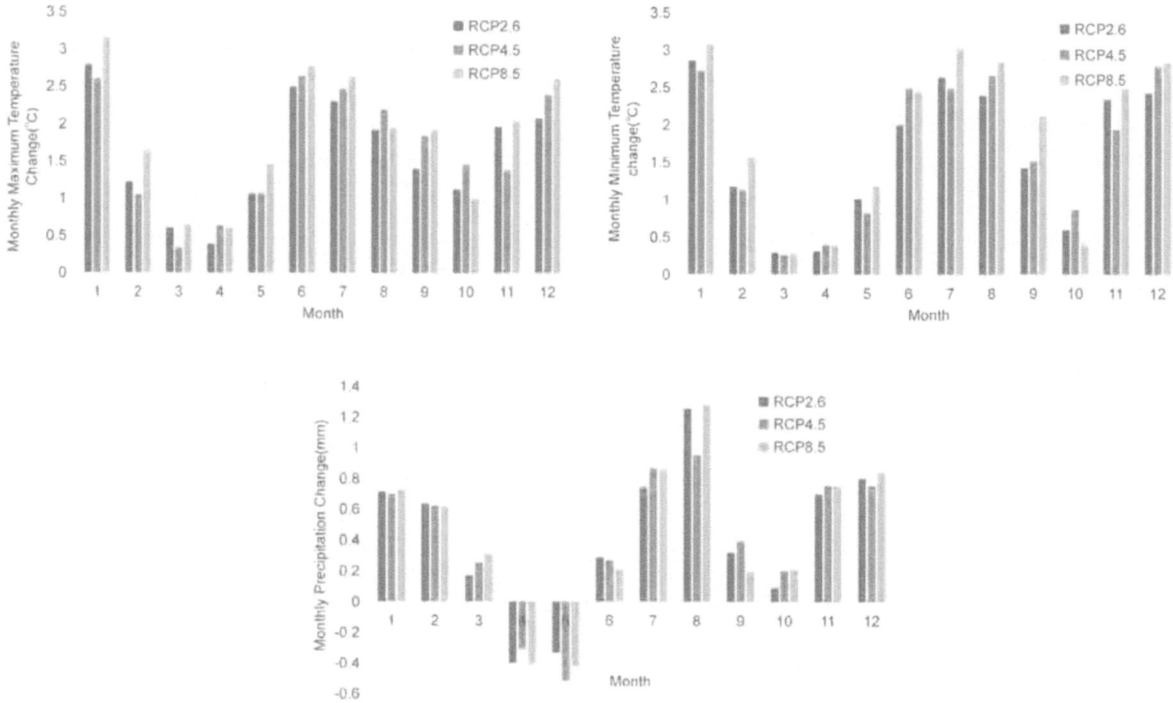

Figure 3 | Changes in monthly temperature and precipitation in future climate scenarios (2020–2050) compared with historical periods (1961–2016).

scenarios do not increase with the increase in carbon dioxide emissions, indicating that precipitation is more complicated due to climate change.

The calculation of the changes in monthly reveals that the maximum temperature is higher than the historical maximum temperature (Figure 3). The temperature rises most obviously from December to January in winter by 2.6–3.2 °C, followed by June–August (summer) by 2.5–2.7 °C. The least rise is 0.5–1.5 °C in March–May in spring.

Under future climate change, the following annual distribution characteristics of the minimum temperature are observed. The temperature rises most obviously from December to January in winter by 2.7–2.9 °C, followed by June–August in summer by 2–2.6 °C. The least obvious increase is in March–May in spring, ranging from 0.5 to 1.5 °C.

Under future climate change, the daily precipitation will increase compared with that of the historical period except for April–May in spring. The most obvious increase is from July to August in summer, with an increase of 0.8–1.2 mm/day.

Runoff simulation model in the study area

Model parameters

According to the DEM, land use, and soil distribution, the study area is divided into 422 hydrological response units. The model was constructed in the study area based on historical daily runoff data from 1967 to 1987. The warm-up period was from 1967 to 1968. The model was calibrated from 1969 to 1979 and verified from 1980 to 1987.

The sensitive parameters area is calibrated. The final values of the parameters are shown in Table 8. The result indicates that the runoff is most sensitive to the groundwater's base flow decline coefficient, followed by the base flow alpha factor and the minimum snow water equivalent under 100% snow cover. Under climate change, the exchange of surface water and groundwater changes significantly; the snow cover also changes, indicating that it is of great significance to study the response of the study area to climate change.

Calibration and validation of the SWAT model

The SWAT model was used to simulate the daily runoff in the study area. The simulation results showed that the determination coefficient of the regular rate was 0.87, the Naxi efficiency was 0.86, the determination coefficient in the validation period was 0.79, and the Naxi efficiency was 0.77. These results imply that the SWAT model is feasible for simulating the

Table 8 | The values of model parameters

Parameter name	Optimal value	*T*-stat	*P*-value
v__GW_DELAY.gw	13.06	34.63	0
v__ALPHA_BF.gw	0.63	−7.46	0
v__SNOCOVMX.bsn	358.99	−1.82	0.07
v__TLAPS.sub	18.31	1.53	0.13
r__SOL_K(1).sol_1	0.72	1.23	0.22
v__NPERCO.bsn	9.72	1.2	0.23
v__CH_N2.rte	−5.34	−1.16	0.25
v__SPEXP.bsn	1.45	−0.96	0.34
v__GW_REVAP.gw	0.02	−1.04	0.3
v__REVAPMN.gw_1	2.72	0.69	0.49
v__EPCO.hru	1.27	0.66	0.51
v__PHOSKD.bsn	147.57	0.44	0.66
r__SOL_ALB(1).sol	0.22	0.42	0.68
v__CH_K2.rte	−489.23	−0.42	0.68
v__ESCO.hru	1.09	−0.31	0.76
a__GWQMN.gw	910.29	−0.29	0.77
v__SMFMN.bsn	11.66	0.25	0.8
r__CN2.mgt	−0.58	−0.21	0.84
v__SMFMX.bsn	−0.88	0.13	0.9
r__SOL_AWC(1).sol	−3.88	−0.04	0.97
v__TIMP.bsn	0.76	0.03	0.98
v__SURLAG.bsn	719.8	0.01	0.99

r_ means an existing parameter value is multiplied by (1+ given value); v_ means the existing parameter value is to be replaced by a given value (Neitsch *et al.* 2011; Arnold *et al.* 2012).

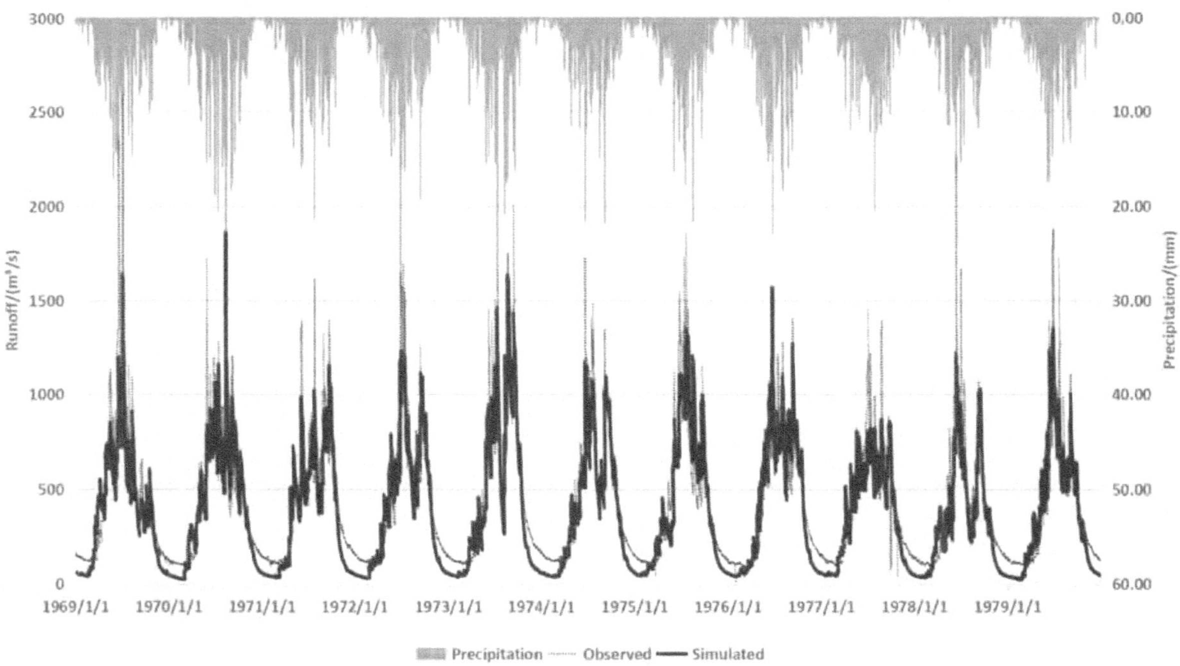

Figure 4 | Daily simulation results of runoff during the calibration period (1969–1979).

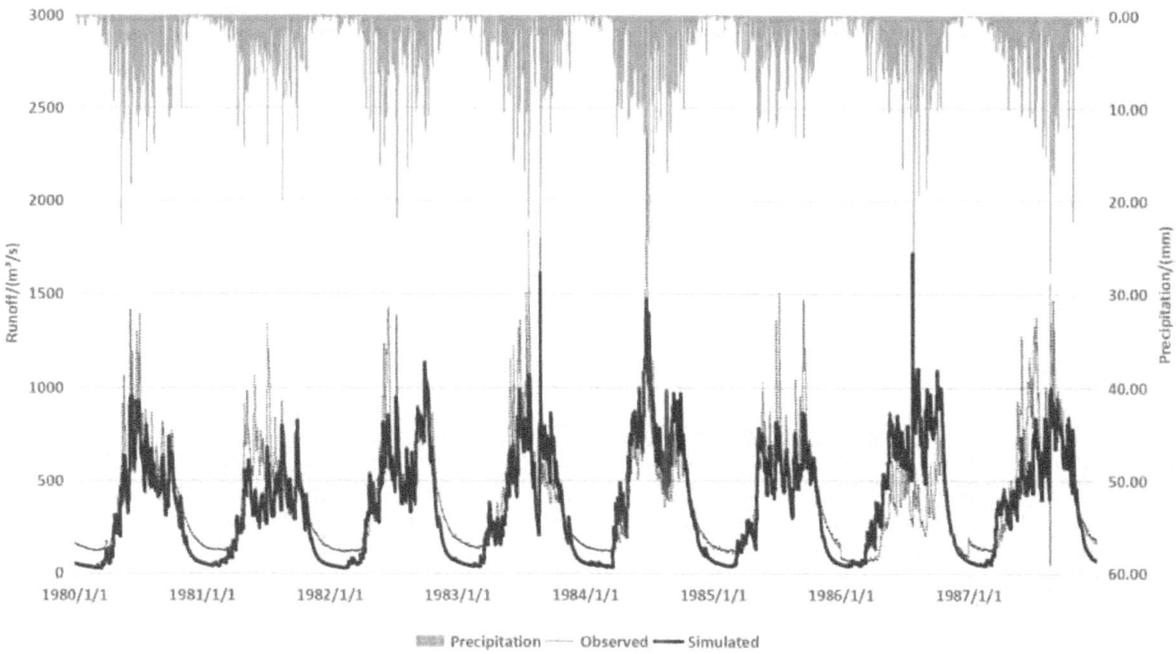

Figure 5 | Daily simulation results of runoff during the validation period (1980–1987).

daily runoff in the study area. The runoff hydrograph of the period and validation periods is shown in Figures 4 and 5. Compared with the actual runoff and the simulated runoff, it was found that the SWAT model after parameter calibration can simulate the seasonal distribution characteristics of runoff. SWAT is more accurate in simulating runoff during the flood season. The runoff simulated by SWAT is less than the measured runoff in the dry season. This lower runoff may be attributed to the large contribution of high flow in the flood season to the simulation error evaluation. Simultaneously, the contribution of low flow in the dry season to the simulation error evaluation is small. This phenomenon is also observed in similar studies (Chen *et al.* 2017; Mohammed *et al.* 2017).

Prediction of future runoff change in the study area

A comparison between the annual runoff in the study area in the historical period and in the future scenario (Figure 6) reveals that the annual runoff shows an upward trend in the future. With the increase in emissions, the upward trend of runoff is more prominent. Comparing the monthly runoff of the two periods (Figure 7), the increase in runoff is most evident from November to April of the following year. From June to July, under the trend of annual runoff increase in different scenarios, the monthly runoff decreases compared with the base period by more than 10%. From June to July in summer, the temperature and precipitation increase, and the runoff decreases, indicating that the precipitation rise cannot compensate for the influence of temperature rise on evapotranspiration. From April to May in spring, the temperature and runoff increases and the precipitation decreases, suggesting that the temperature rise on the snow melting process is more significant than that on evapotranspiration in spring. The increase in snow melting amount increases the runoff. The peak period of agricultural irrigation and water diversion from the Minjiang River downstream is from June to July. The decrease in runoff will affect the agricultural irrigation water, thereby reducing agricultural production. Simultaneously, it will affect the water distribution downstream and may worsen the improved ecology.

DISCUSSION

From 2020 to 2050, the temperature in the study area shows an upward trend under climate change. The upward trend is consistent with the increasing trend of emissions; this indicates that under the high emission scenario, there is more carbon dioxide emission, more greenhouse activity, and greater temperature rise (Qin et al. 2019). The temperature rise will enhance the snowmelt and evapotranspiration, of which the increase in snowmelt is more apparent in spring and evapotranspiration in summer.

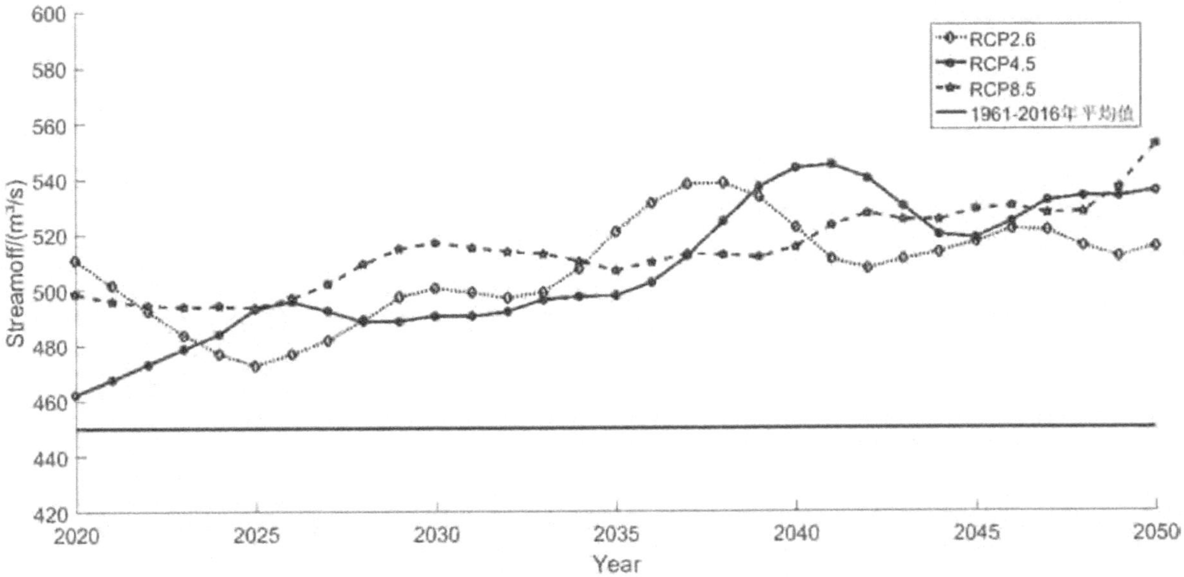

Figure 6 | Annual runoff in the historical period (1961–2016) and future scenario (2020–2050).

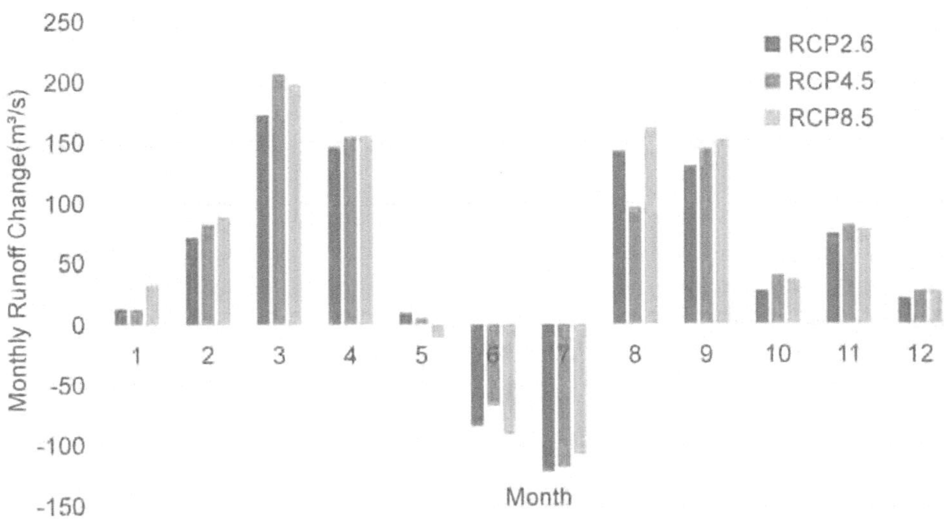

Figure 7 | Changes in monthly runoff over the base year in the future scenario (2020–2050).

From 2020 to 2050, the precipitation in the study area shows an increasing trend, which is consistent with the response results of most basins under climate change (Gu *et al.* 2015; Chen *et al.* 2017; Yang *et al.* 2019a, 2019b). Simultaneously, the study shows that the temporal and spatial variations show strong heterogeneity (Huang *et al.* 2010). The study area trend is different from the upper reaches of the Yangtze River, which has unique characteristics. The precipitation decreases from April to May in spring and increases in other months, especially in July and August. Such precipitation fluctuations reveal that the impact of climate change on precipitation is more complex, and there are large spatial differences in annual and seasonal changes.

The results show that the seasonal distribution of runoff change is consistent with the seasonal variation in precipitation, indicating that precipitation contributes greatly to the increase of the runoff (Gu *et al.* 2015; Yang *et al.* 2019a, 2019b). In autumn and winter, the rising trend of runoff in the study area is consistent with the rising trend of precipitation. The increase in runoff in winter is mainly due to the increase in precipitation by 20%, and the increase in temperature causes the precipitation to become runoff instead of snow (Gu *et al.* 2015). From June to July, the precipitation in the study area increased; however, the runoff decreased, suggesting that the increase in temperature increased evaporation, and the increase in precipitation could not compensate for the increase in evapotranspiration. The research results of the upper reaches of the Yangtze River show similar findings (Birkinshaw *et al.* 2017) but contrary to the summer runoff increase in the Minjiang River Basin (Chen *et al.* 2014; Yang *et al.* 2015). From April to May, the precipitation decreased, and the runoff increased in the study area, indicating that the increase in temperature increased the snowmelt and runoff. The seasonal variation of runoff in the study area is affected by the geographical location. The upper reaches of the Minjiang River are characterized by high mountain and gorge landforms, and the evapotranspiration and snow melting process are more vulnerable to climate change. From April to May, the precipitation in the study area decreased and the runoff increased, revealing that the increase in temperature increases snow melting. The temperature rise in the entire Yangtze River Basin will cause the precipitation to increase in the future; however, the uneven change in precipitation in time and space will cause uneven runoff change (Huang *et al.* 2010). At the same time, relevant research shows that the change in runoff is primarily affected by climate change, and changes in the annual distribution of runoff are mainly affected by human activities (Qin *et al.* 2019). The study area is rich in hydropower energy. To develop a local economy, utilize water resources, and avoid and mitigate floods, many water projects have been built in the mainstream and tributaries of the upper reaches of the Minjiang River, which play a key role in water storage and peak regulation. However, to a certain extent, especially in the dry season, the natural runoff pattern is affected by human activities.

In this study, SDSM is used to downscale the future climate scenarios predicted by CanESM2. However, there are some uncertainties in the prediction results and downscaling models of CanESM2. At the same time, SWAT is selected to simulate the runoff in the upper reaches of the Minjiang River, which also has some uncertainty. The uncertainties of future climate scenarios, downscaling models, and hydrological models make it uncertain to study the response of the upper Minjiang River to climate change. Based on the uncertainty of the study, the response of the upper reaches of the Minjiang River to climate change will be further studied using the results of multi-mode climate prediction.

CONCLUSIONS

From the simulation results of the study area, SDSM can simulate the monthly and annual variation trends of the temperature and precipitation at each station. The deterministic coefficient of runoff simulation by using SWAT is 0.79. The Naxi efficiency was 0.77 in the verification period, indicating that the SWAT model can accurately simulate the daily runoff and be used to study the response of water resources to climate change.

In the future scenario, the maximum temperature shows an upward trend. The largest temperature rise was from December to January by 2.6–3.2 °C, followed by June–August, with an increase of 2.5–2.7 °C. The increase in the minimum temperature is higher than the highest temperature. The temperature rise is the most obvious from December to January, up to 2.7–2.9 °C, followed by June to August, with an increase of 2–2.6 °C. Except for spring, the daily precipitation decreased from April to May and increased in other periods, especially from July to August, with an increase of 0.8–1.2 mm/day.

In the future scenario, the runoff in the study area will increase most significantly from November to April, whereas the runoff from June to July will decrease by more than 10% compared with the historical period. From June to July, the temperature increases, the precipitation increases, and the runoff decreases, indicating that the precipitation rise cannot compensate for the influence of temperature rise on evapotranspiration. From April to May, the temperature increases, the precipitation decreases, and the runoff increases, indicating that the influence of temperature rise on the snow melting process is greater than that on evapotranspiration during spring. Also, the increase in the amount of snow melting increases the runoff. June to July in summer is the peak period of agricultural irrigation in the Minjiang River; it is also the period of water diversion from the Minjiang River downstream. The decrease in runoff will affect the agricultural irrigation water, thereby reducing agricultural production; furthermore, it also affects the water distribution downstream and may worsen the improved ecology.

ACKNOWLEDGEMENTS

We gratefully acknowledge the Regional Innovation Cooperation Program from the Science & Technology Department of Sichuan Province (2020YFQ0013), and the China Scholarship Council (201806240035).

AUTHOR CONTRIBUTIONS

Kebi Yang (PhD candidate) conducted the writing of the original draft. Ting Chen (PhD) conducted the writing of review and editing. Tianqi Ao (PhD supervisor, professor) conducted the supervision of the draft. Xu Zhang (PhD candidate) conducted the part of the experiments. Li Zhou (PhD candidate) conducted the part of the experiments. Danyang Gao (Master candidate) conducted the data validation.

DATA AVAILABILITY STATEMENT

The daily temperature and precipitation data used in this paper are provided by the State Meteorological Administration of China. Relevant station data are available at https://data.cma.cn/ through registered scientific research users.CanESM2 simulation results and NCEP reanalysis data are obtained from http://climate-scenarios.canada.ca/?page=pred-canesm2.

REFERENCES

Abbaspour, K. C. (2007). User manual for SWAT-CUP, SWAT calibration and uncertainty analysis programs. *Eawag: Swiss Fed. Inst. of Aquat. Sci. and Technol.* Dübendorf, Switzerland, p. 93.

Arnold, J. G., Srinivasan, R., Muttiah, R. S. & Williams, J. R. 1998 Large area hydrologic modeling and assessment part i: model development. *JAWRA Journal of the American Water Resources Association* **34** (1), 73–89.

Arnold, J., Kiniry, J., Srinivasan, R., Williams, J., Haney, E. & Neitsch, S. 2012 *SWAT Input/Output Documentation Version 2012, TR-439.* Technical Report, Texas Water Resources Institute, College Station, TX, USA.

Birkinshaw, S. J., Guerreiro J. , S. B., Nicholson, A., Liang, Q., Quinn, P., Zhang, L., He, B., Yin, J. & Fowler, H. J. 2017 Climate change impacts on Yangtze River discharge at the Three Gorges Dam. *Hydrology and Earth System Sciences* **21** (4), 1911–1927.

Chen, L., Yu, B., Chen, Z., Li, B. & Wu, J. 2014 Investigating the temporal and spatial variability of total ozone column in the Yangtze River Delta using satellite data: 1978–2013. *Remote Sensing* **6** (12), 12527–12543.

Chen, J., Gao, C., Zeng, X., Xiong, M., Wang, Y., Jing, C., Krysanova, V., Huang, J., Zhao, N. & Su, B. 2017 Assessing changes of river discharge under global warming of 1.5 °C and 2 °C in the upper reaches of the Yangtze River Basin: approach by using multiple-GCMs and hydrological models. *Quaternary International* **453**, 63–73.

Chen, T., Ao, T. Q., Zhang, X., Li, X. D. & Yang, K. B. 2019a Climate change characteristics of extreme temperature in the Minjiang River Basin. *Advances in Meteorology* **2019**, 1935719.

Chen, T., Ao, T. Q., Zhang, X., Li, X. D., Zhou, L., Li, M. R. & Yang, K. B. 2019b Spatial and temporal variation and probability characteristics of extreme precipitation events in the Min River Basin from 1961 to 2016. *Applied Ecology and Environmental Research* **17** (5), 11375–11394.

Githui, F., Gitau, W., Mutua, F. & Bauwens, W. 2009 Climate change impact on SWAT simulated streamflow in western Kenya. *International Journal of Climatology* **29** (12), 1823–1834.

Gu, H., Yu, Z., Wang, G., Wang, J., Ju, Q., Yang, C. & Fan, C. 2015 Impact of climate change on hydrological extremes in the Yangtze River Basin, China. *Stochastic Environmental Research and Risk Assessment* **29** (3), 693–707.

Huang, J., Xu, Z., Liu, Z. & Zhao, F. 2008 Analysis of future climate change in the Taihu Basin using statistical downscaling. *Resoures Science* **30** (12), 1811–1817.

Huang, J., Zhang, J., Zhang, Z., Xu, C., Wang, B. & Yao, J. 2010 Estimation of future precipitation change in the Yangtze River basin by using statistical downscaling method. *Stochastic Environmental Research and Risk Assessment* **25** (6), 781–792.

Luo, Y., Qin, N., Pang, Y., Wang, C., Li, J. & Xisheng, L. 2019 Effect of climate factors on the runoff over Tuotuo River basin under the background of climate change. *Journal of Glaciology and Geocryology* **2019**, 952–964.

Mohammed, K., Islam, A. S., Islam, G. T., Alfieri, L., Bala, S. K. & Khan, M. J. U. 2017 Extreme flows and water availability of the Brahmaputra River under 1.5 and 2 °C global warming scenarios. *Climatic Change* **145** (1–2), 159–175.

Moss, R. H., Edmonds, J. A., Hibbard, K. A., Manning, M. R., Rose, S. K., van Vuuren, D. P., Carter, T. R., Emori, S., Kainuma, M., Kram, T., Meehl, G. A., Mitchell, J. F. B., Nakicenovic, N., Riahi, K., Smith, S. J., Stouffer, R. J., Thomson, A. M., Weyant, J. P. & Wilbanks, T. J. 2010 The next generation of scenarios for climate change research and assessment. *Nature* **463** (7282), 747–756.

Neitsch, S. L., Arnold, J. G., Kiniry, J. R. & Williams, J. R. 2011 *Soil and Water Assessment Tool Theoretical Documentation Version 2009, TR-406.* Technical Report, Texas Water Resources Institute, College Station, TX, USA.

Olivera, F., Valenzuela, M., Srinivasan, R., Choi, J., Cho, H. D., Koka, S. & Agrawal, A. 2006 ArcGIS-SWAT: A geodata model and GIS interface for SWAT. *Journal of the American Water Resources Association* **42** (2), 295–309.

Qin, D. & Thomas, S. 2014 Highlights of the IPCC Working Group I Fifth Assessment Report. *Climate Change Research* **10** (1), 1–6 (in Chinese).

Qin, P., Liu, M., Du, L., Xu, H., Liu, L. & Xiao, C. 2019 Climate change impacts on runoff in the upper Yangtze river basin. *Climate Change Research* **15** (04), 405–415.

Shrestha, N. K., Du, X. & Wang, J. 2017 Assessing climate change impacts on fresh water resources of the Athabasca River Basin, Canada. *Science of The Total Environment* **601–602**, 425–440.

Su, B., Huang, J., Zeng, X., Gao, C. & Jiang, T. 2017 Impacts of climate change on streamflow in the upper Yangtze river basin. *Climatic Change* **141** (3), 533–546.

Sun, Z., Jia, S., Lv, A., Zhu, W. & Gao, Y. 2015 Assessment on precision of temperature simulated by the IPCC AR5 GCMs in China, 1996–2005. *Progress in Geography* **34** (10), 1229–1240 (in Chinese).

Taylor, K., Stouffer, R. J. & Meehl, G. A. 2012 An overview of CMIP5 and the experiment design. *Bulletin of the American Meteorological Society* **93** (4), 485–498.

Wang, Y., Liao, W., Ding, Y., Wang, X., Jiang, Y., Song, X. & Lei, X. 2015 Water resource spatiotemporal pattern evaluation of the upstream Yangtze River corresponding to climate changes. *Quaternary International* **380–381**, 187–196.

Wilby, R. L. 2002 SDSM – a decision support tool for the assessment of regional climate change impacts. *Environmental Modeling & Software* **17**, 147–159.

Wilby, R. L., Tomlinson, O. J. & Dawson, C. W. 2003 Multi-site simulation of precipitation by conditional resampling. *Climate Research* **23**, 183–194.

Yang, S. L., Xu, K. H., Milliman, J. D., Yang, H. F. & Wu, C. S. 2015 Decline of Yangtze River water and sediment discharge, impact from natural and anthropogenic changes. *Scientific Reports – UK* **5**, 12581.

Yang, X., Yu, X. & Wang, Y. 2019a Estimating the response of hydrological regimes to future projections of precipitation and temperature over the upper Yangtze River. *Atmospheric Research* **230**, 104627.

Yang, X., Yu, X., Wang, Y., Liu, Y., Zhang, M., Ren, L., Yuan, F. & Jiang, S. 2019b Estimating the response of hydrological regimes to future projections of precipitation and temperature over the upper Yangtze River. *Atmospheric Research* **2019**, 104627.

Zhan, W., Wang, S. & Cen, S. 2017 Trends variability of annual runoff in the upper reaches of Yangtze River under future climate change scenarios. *Plateau and Mountain Meteorology Research* **37** (4), 34–39.

Zhao, C., Nan, Z., Cheng, G., Zou, S. & Zhang, Y. 2008 Prediction of the trend of the future climate change in northwestern China by statistical downscaling. *Journal of Lanzhou University (Natural Sciences)* **44** (5), 12–18 (in Chinese).

First received 9 February 2021; accepted in revised form 8 May 2021. Available online 12 June 2021

 Impact of Climate Change on Hydrology and Water Resources doi: 10.2166/wcc.2022.258

Evaluating the impact of climate change on surface water resources in the upper Ganjiang River Basin, China

Peibing Song[a], Chao Wang[b,*], Gongbo Ding[c], Jiahui Sun[d], Lingzhong Kong[e,f], Mengtian Lu[a], Xiaohui Lei[b] and Hao Wang[a,b]

[a] College of Civil Engineering and Architecture, Zhejiang University, Hangzhou 310058, China
[b] State Key Laboratory of Simulation and Regulation of Water Cycle in River Basin, China Institute of Water Resources and Hydropower Research, Beijing 100038, China
[c] State Key Laboratory of Hydraulics and Mountain River Engineering, Sichuan University, Chengdu 610065, China
[d] School of Civil Engineering, Shandong University, Jinan 250061, China
[e] College of Hydraulic Science and Engineering, Yangzhou University, Yangzhou 225009, China
[f] College of Water Conservancy and Hydropower Engineering, Hohai University, Nanjing 210098, China
*Corresponding author. E-mail: wangchao@iwhr.com

PS, 0000-0002-2857-0086; LK, 0000-0002-9213-7463

ABSTRACT

Projecting surface runoff is a meaningful task for rational allocation and optimal scheduling of water resources. Aimed at exploring the impact of climate change on surface water resources, the basin-scale water-balance model coupled with the Budyko-type equation was developed for the upper Ganjiang River Basin (UGRB). Studies demonstrated that the Budyko-CY function is the optimal water-balance model, with a Nash–Sutcliffe efficiency (*NSE*) of 0.843 in the calibration period and 0.653 in the validation period, respectively. The increase in temperature by 1 °C may bring about a 2.5–4.7% runoff decline, while the 10% increase in precipitation may lead to a 12.1–14.3% runoff ascent. The annual mean temperature is expected to grow by 0.69, 0.68 and 0.97 °C in the next 30 years relative to that during the reference period, respectively. Similarly, the surface runoff is estimated to increase by 8.4, 6.5 and 5.0% on a multi-year average scale, respectively. This study is beneficial to provide possibilities for climate scenarios that may occur in the future, and the results presented herein are capable of giving a reference for the planning and management of water resources in the UGRB.

Key words: Budyko hypothesis, climate change, hydrological response, upper Ganjiang River Basin

HIGHLIGHTS

- The water-balance model based on the Budyko-type equation is applicable in the study area.
- The annual temperature is expected to rise by 0.69, 0.68 and 0.97 °C in the next 30 years under three climate scenarios.
- Multi-year mean runoff increases slightly with the variability of 8.4% (RCP2.6), 6.5% (RCP4.5) and 5.0% (RCP8.5).

1. INTRODUCTION

Based on the IPCC AR5, global annual mean temperature exceeded 0.85 °C since 1880, and the annual mean temperature in China increased >1.0 °C in the past three decades, which was dramatically higher than the world's average (Warszawski *et al.* 2014; Fang *et al.* 2018). In the next 30 years, surface temperature is expected to rise ranging from 2.4 to 6.1 °C in China, while annual runoff may increase roughly by 3–10% (Liu *et al.* 2014). Influenced by the monsoon climate, China has become particularly sensitive to the global warming, which may bring about a descent in streamflow in some basins, such as the lower Yellow River Basin (Piao *et al.* 2010). Research demonstrated that climate change was detected to be a dominant driving factor yielding runoff reduction in major basins across China, and it was estimated to account for 53.5% of streamflow changes during the period 2001–2014 (Liu *et al.* 2017). Furthermore, a change in climate owing to ascending temperature and intensified evaporation has aggravated the situation of hydrological cycles, altered spatio-temporal distribution of water resources and further enlarged the evolution of climatic variables (e.g., precipitation and evaporation), and these, in turn, will inevitably produce an effect on the ecological system (John *et al.* 2020). Therefore, projecting climate change and its impacts on surface runoff are essential for allocation and scheduling of water resources, and researchers have conducted much significant work in recent years, basically following the pattern of future climate scenario design-hydrological model-impact assessment.

General circulation models (GCMs) have been used extensively to characterize future climate change, which can reflect the dynamic change of climate system through numerical simulation (Zinyengere *et al.* 2017). Employing the outputs provided by GCMs to drive a hydrological model is a widely adopted method in view of the fact that the hydrological model is a useful tool for simulating runoff processes (Chen *et al.* 2012). To compare the impact of two delta-change methods on different precipitation scenarios, Yuan *et al.* (2016) applied the Xinanjiang (XAJ) model to predict surface streamflow, and more accurate runoff estimates can be obtained in the watershed with stronger precipitation change signals in the Pearl River Basin. Zhu *et al.* (2019) used the Soil and Water Assessment Tool (SWAT) model to discuss the uncertainty of future runoff and concluded that monthly runoff increases are concentrated between May and October (July and August are not included) in the Biliu River Basin. Jin *et al.* (2020) employed the Variable Infiltration Capacity (VIC) model to depict the water cycle process, suggesting that annual precipitation may ascend while runoff may descend in the Yellow River source region. Yan *et al.* (2020) applied the Water and Energy Transfer processes in Large basins (WEP-L) model integrated with the global Bayesian model averaging (BMA) method to make the runoff projections become convincing and revealed that runoff in the Yellow River Basin during the period 2050–2070 would be extraordinarily lower than that in the baseline period. Although there are structural differences in different GCMs, the temperature changes they reflect are quite consistent, whereas precipitation features appear as large fluctuations across various models (Wen *et al.* 2016). Sun & Peng (2020) took the RCP4.5 scenario as an example to estimate the range of runoff variation and reported that future annual runoff varies between 268.9 and 544.2 mm from 2011 to 2100. In summary, the hydrological model makes it possible to obtain more potential changes and detailed information on various climatic characteristics of the catchment by predicting surface runoff and providing continuous simulations of daily and monthly runoffs. Moreover, this approach can enhance the simulation accuracy in surface runoff by making use of satellite-based data (Crow & Ryu 2009). Nevertheless, considering that the hydrological model is a data-intensive method, an estimation of future runoff is complicated and fraught with uncertainties. Also, this approach is always constrained by multiple data sets (e.g., vegetation, topography, soil) (Teng *et al.* 2012). Thus, it is evident that a variety of uncertainties are introduced through the hydrological model in the form of model parameters and structures, and the impact generated by the hydrological model may be unneglectable especially for the complex model (Thompson *et al.* 2013). As a consequence, the hydrological model may not be the most appropriate choice to estimate future runoff in some cases, and it may be inapplicable for the large-scale catchment with insufficient data or limited hydrological information.

In contrast with the hydrological model, the water-balance method is identified as an alternative technique owing to its fewer model parameters and simpler calculation principles, and this approach has received tremendous attention in speculating the hydrological process under changing environments. Chen *et al.* (2017) analyzed that >60% watersheds in China would face water deficiency and revealed the increasingly unbalanced distribution of water resources since the middle of the last century. Xing *et al.* (2018) established an elasticity method to estimate future runoff in 35 catchments across China, and the Wang–Tang equation was the most reasonable water-balance equation to strengthen the prediction reliability. Yuan *et al.* (2018b) constructed a basin-scale water-balance approach and investigated that natural runoff would increase by 14.4–16.8% under the RCP4.5 scenario over the Yellow River Basin. Zhang *et al.* (2018b) quantified the impact of leaf area index (LAI) changes on the streamflow, and it is estimated that the downtrend in runoff is likely to sustain. Li *et al.* (2020) applied the elasticity method coupled with the Blaney–Criddle equation into the upper Heihe River Basin and reported that the predicted runoff may decline by 5.6 and 6.7% during two periods 2021–2050 and 2051–2080, respectively. Guan *et al.* (2021) established the elasticity method to predict runoff and investigated that increasing runoff during the projection period may alleviate water shortages in northern China. Previous studies have illustrated the feasibility of the water-balance model coupled with GCMs in ascribing future streamflow patterns, and the estimated runoff calculated from the Budyko's framework is quite similar to that calculated from the hydrological model (Li *et al.* 2013). It is also worth pointing out that investigative assessments are always earlier than the hydrological modeling, where the Budyko hypothesis is widely accepted to address the trouble of flexibly distinguishing the influences of climate change on surface streamflow (Kazemi *et al.* 2019). Furthermore, the water-balance model based on the single-parameter Budyko-type equation is particularly suitable for ungauged catchments, because only estimates of long-term water availability are needed in these catchments (Chiew 2010).

The Ganjiang River Basin (GRB), located in southern China with abundant water resources, is the largest and longest river in Jiangxi province. Over the past decades, the GRB has undergone severe water shortages. The contradiction between water supply and demand is particularly remarkable, and the phenomenon of upstream and downstream competition for water resources is increasingly highlighted in the dry season (Zhang *et al.* 2018a). Multi-year mean runoff in the upper Ganjiang River basin (UGRB) accounts for about 22.5% of that in the Poyang Lake Basin (PLB), and it is the largest among all three-grade water resource partitions over the PLB (Yu *et al.* 2014). Projecting future surface runoff in the UGRB is crucial for minimizing these adverse impacts, as it would be beneficial to develop more effective adaptation measures and achieve sustainable water resource management. Aimed at probing into climate change and its effects on surface water resources in the UGRB, this study is organized as follows: (1) to calibrate a single parameter within the Budyko's framework and compare the feasibility of various water-balance models, and to seek for the optimal Budyko-type equation for capturing the runoff variation; (2) to build the statistical relation between temperature and potential evapotranspiration and to investigate the sensitivity of surface runoff to climatic variables; and (3) to further project the future changes of hydrometeorological factors during the period 2021–2050 under various GCMs, with taking multiple representative concentration pathways (RCPs) into account. The research framework of this paper is shown in Figure 1.

2. DATA AND METHODS

2.1. Study area

The GRB is situated in 113°34′–116°37′E and 24°31′–28°45′N, covering an area of >83,500 km^2. As the longest river of the PLB, the Ganjiang reaches a length of 766 km and lies mainly in the Jiangxi Province. The GRB has four distinctive seasons, that is, summers are hot and rainy while winters are dry and cold, with annual precipitation ranging from 1,400 to 1,800 mm and mean temperature of 17.8 °C (Song *et al.* 2020).

The Wan'an Reservoir is located 2 km upstream of Wan'an County, and it plays an important role in alleviating the water crisis and maintaining the availability of water resources. The Dongbei hydrological station is the outlet control station over the UGRB, which can well reflect the runoff changes of the study area. Annual rainfall amount, surface water resource amount, and groundwater resource amount in the UGRB are 633.2×10^8, 344.7×10^8 and $95.9 \times 10^8\,m^3$, equivalent to 1,574, 857 and 238 mm, respectively (Yu *et al.* 2014). The location of the UGRB is illustrated in Figure 2.

2.2. Data

Monthly runoff data from 1959 to 2017 were supplied by the hydrological yearbook issued by the local hydrological bureau. Observed daily meteorological data (1959–2017) of 13 meteorological stations (i.e., Ganxian, Changting, Guangchang, Ji'anxian, Xunwu, Longnan, Nanxiong, Ninghua, Ningdu, Nanfeng, Yongfeng, Suichuan and Jingangshan, labeled in Figure 2) were collected from the National Meteorological Information Centre (NMIC). The Penman–Monteith method is employed to calculate potential evapotranspiration for a single meteorological station. Daily precipitation, temperature and

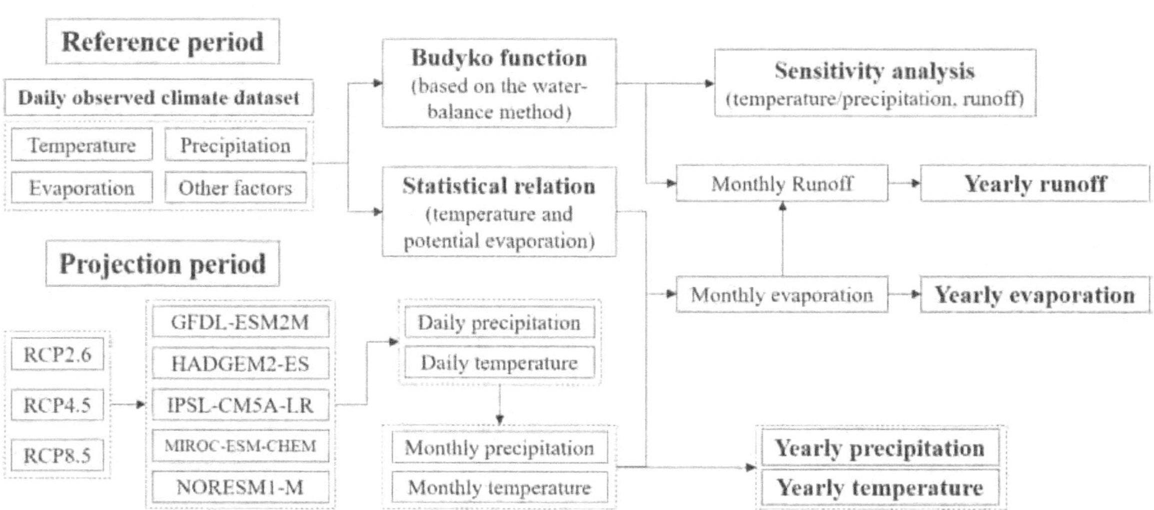

Figure 1 | Research framework of this paper.

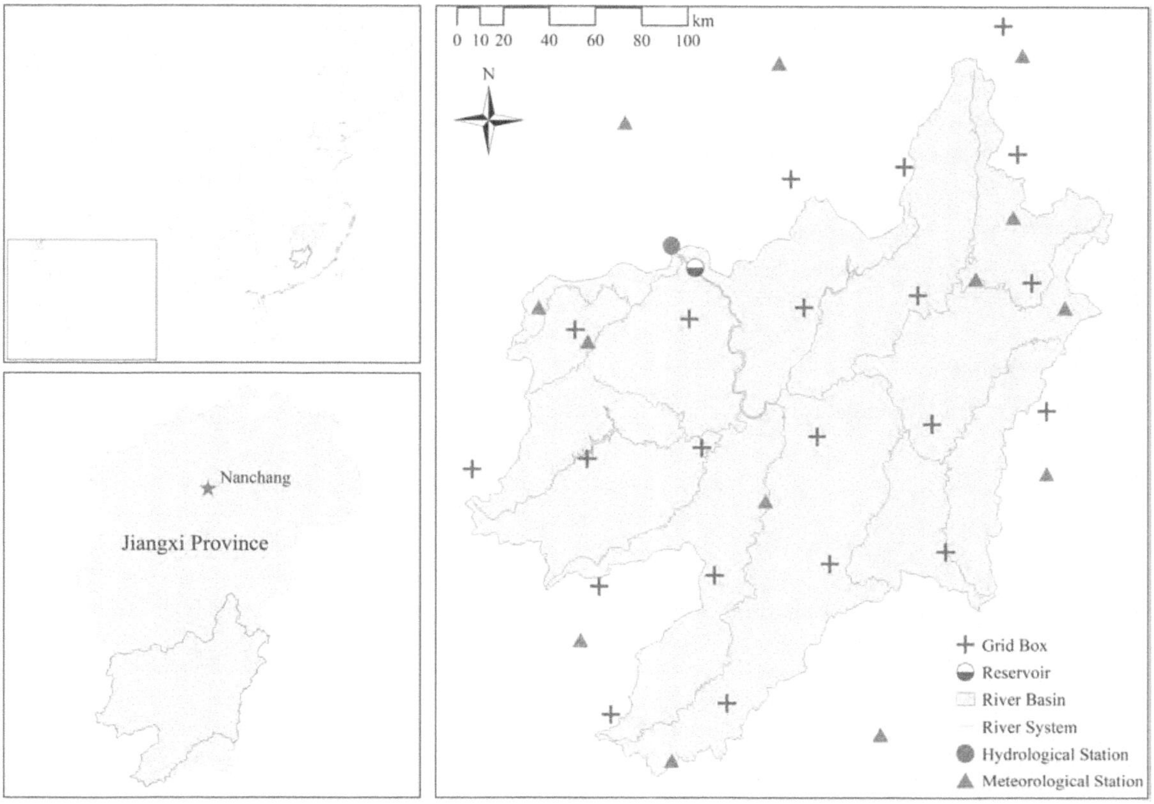

Figure 2 | Location of the UGRB.

evapotranspiration of the single station are reorganized into monthly precipitation, temperature and evapotranspiration, and then precipitation, temperature and evapotranspiration in the study area are calculated according to the Thiessen polygons method, respectively. It should be noticed that runoff depth is used to represent surface water resources. Precipitation, evapotranspiration and surface runoff during the reference period (1961–2010) are 1,560, 1,466 and 821 mm, respectively.

2.3. Climate scenarios

RCPs were defined in terms of a possible range of global radiation forcing value in 2100, and future mitigation emission scenarios are ranked as RCP2.6, RCP4.5, RCP6.0 and RCP8.5. Supposing that the forcing level will reach the highest point at ~3 W/m^2 before 2100, if this value drops to 2.6 W/m^2 by 2100, it stands for the RCP2.6 scenario; if this value stabilizes without overshoot pathways to 4.5 W/m^2 after 2100, it represents the RCP4.5 scenario; if this value stabilizes without overshoot pathways to 6.0 W/m^2 after 2100, it means the RCP6.0 scenario; if this value climbs rapidly to 8.5 W/m^2 by 2100, it symbolizes the RCP8.5 scenario (Yuan *et al.* 2018a).

Continuous daily-series meteorological data were derived from multiple GCMs. These climate models were chosen since they were widely accepted with persuasive feasibility in climate impact studies in China and can provide possible explanations concerning future climate which might make differences to some significant areas (Zhou *et al.* 2018). In addition, it should be noticed that Yuan *et al.* (2019) have demonstrated that these five climate models chosen in this paper are applicable in the similar catchment, the Fu River basin. Meteorological data contained the period from January 1951 to December 2050, with the geographical scope covering 70.25°–140.25°E and 15.25°–55.25°N (Piani *et al.* 2010). Precipitation and temperature data used in this study were bias-corrected and resampled, so as to ensure consistency between the observed data and the predicted data. In this paper, the statistical downscaling method and bilinear interpolation were used to uniformly interpolate the output data to a 0.5° × 0.5° grid, and these data were corrected based on the statistical deviation of probability distribution. A detailed description of five GCMs is listed in Table 1 (Taylor *et al.* 2012).

Three RCPs including RCP2.6 (low), RCP4.5 (medium) and RCP8.5 (high) scenarios were chosen to project the future precipitation, temperature and discharge variation, and 21 grid cells representing the future climatic variables in and around the UGRB are marked in Figure 2.

Table 1 | Five GCMs used in this study

Model name	Originating country	Resolution (atmosphere)	Resolution (ocean)
GFDL-ESM2M	United States, Geophysical Fluid Dynamics Laboratory (GFDL)	2.0° × 2.5°	0.9° × 1.0°
HadGEM2-ES	United Kingdom, Met Office Hadley Centre (MOHC)	1.3° × 1.9°	0.8° × 1.0°
IPSL-CM5A-LR	France, Institut Pierre Simon Laplace (IPSL)	1.9° × 3.8°	1.2° × 2.0°
MIROC-ESM-CHEM	Japan, Model for Interdisciplinary Research on Climate (MIROC)	2.8° × 2.8°	0.7° × 1.2°
NoRESM1-M	Norway, Norwegian Climate Centre (NCC)	1.9° × 2.5°	0.5° × 1.1°

2.4. Water-balance model

For a closed catchment, its hydrometeorological characteristics conform to the water-balance principle:

$$E = P - Q \tag{1}$$

where E, P and Q signify evapotranspiration (mm), precipitation (mm) and runoff (mm), respectively. E can be defined a function of aridity index (ϕ):

$$\frac{E}{P} = f(\phi) = f\left(\frac{E_0}{P}\right) \tag{2}$$

where E_0 denotes potential evapotranspiration, namely evaporation capacity, mm. Here, an empirical Budyko curve was attempted to quantitatively describe the relation among precipitation, evapotranspiration and evaporation capacity. In contrast with the non-parametric function, the single-parameter function has the advantage of enhancing the interpretation of climatic characteristics, as well as describing geographical characteristics (Wu *et al.* 2017). The Budyko-Fu function, with only a single parameter needing to be calibrated, is demonstrated to provide estimates of annual runoff similar to other models (Fu *et al.* 2007). Nevertheless, in view of the fact that existing studies always assign a single Budyko-type equation to quantify the response of runoff to climate change without giving convincing evidence, a variety of frequently used methods including Budyko-WT, Budyko-Zhang and Budyko-CY were presented to seek out the optimal equation in reproducing the runoff change (Wang & Tang 2014; Wu & Miao 2017). What these Budyko-type equations have in common is that they use fixed parameters instead of allowing them to vary with the catchment. As a result, data requirements can be reduced and this approach can be implemented automatically in other frameworks. Taking the Budyko-Fu function as an example, it describes the quantitative relation among precipitation, evapotranspiration and evaporation capacity as follows:

$$\frac{E}{E_0} = 1 + \frac{P}{E_0} - \left[1 + \left(\frac{P}{E_0}\right)^{\alpha}\right]^{1/\alpha} \tag{3}$$

Then, substituting Equation (1) into Equation (3), the above equation can be simplified into:

$$Q = E_0\left\{\left[1 + \left(\frac{P}{E_0}\right)^{\alpha}\right]^{1/\alpha} - 1\right\} \quad \text{(Budyko} - \text{Fu)} \tag{4}$$

Similarly, runoff can be described by other Budyko-type equations:

$$Q = \frac{P^2}{E_0 + \omega E_0^2/P + P} \quad \text{(Budyko – Zhang)} \tag{5}$$

$$Q = P - \frac{P + E_0 - \sqrt{P^2(1 + E_0/P)^{\partial} - 4\partial(2-\partial)E_0 P}}{2\partial(2-\partial)} \quad \text{(Budyko – WT)} \tag{6}$$

$$Q = P - \left[\left(\frac{1}{E_0}\right)^k + \left(\frac{1}{P}\right)^k\right]^{-k} \quad \text{(Budyko – CY)} \tag{7}$$

where α, ω, ∂ and k are dimensionless parameters and can be calibrated in a given time period.

Observed data during the period 1959–2017 were adopted to calibrate and verify the water-balance model, and the study period was divided into two parts: the calibration period (1959–1998) and the validation period (1999–2017). Then, a calibrated water-balance model was driven by climatic variables during the projection period (2021–2050) provided by the GCMs, and the relative variation characteristics of simulated runoff are used to characterize the trend of runoff.

2.5. Sensitivity analysis

The sensitivity of runoff to climate change is expressed as the proportional change of simulated runoff, and it can be obtained by comparing simulations driven by changes in precipitation or temperature, keeping other climatic variables unchanged simultaneously (Bao *et al.* 2012). In this case, the hypothetical scenario involving runoff to a climate change can be expressed as a function as follows:

$$\delta(\Delta T, \Delta P) = \frac{f(T + \Delta T, P + \Delta P) - f(T, P)}{f(T, P)} \times 100\% \tag{8}$$

where T denotes the mean temperature, °C; P denotes the precipitation, mm; ΔT denotes the hypothetical changes in mean temperature, °C; ΔP denotes the hypothetical changes in precipitation, mm; $\delta()$ represents the sensitivity function of runoff to climate change; $f()$ represents the relation function between runoff and the climate scenario.

2.6. Evaluation indices

Three indices, namely the relative error (*RE*), the Nash–Sutcliffe efficiency (*NSE*) and the deterministic coefficient (*DC*), were adopted to evaluate the performance of various Budyko-type equations:

$$RE = \left|\frac{Q_{sim},i - Q_{obs,i}}{Q_{obs},i}\right| \times 100\% \tag{9}$$

$$NSE = 1 - \frac{\sum_{i=1}^{n} (Q_{sim},i - Q_{obs,i})^2}{\sum_{i=1}^{n} (Q_{obs},i - \overline{Q_{obs}})^2} \tag{10}$$

$$DC = \frac{\left[\sum_{i=1}^{n} (Q_{sim},i - \overline{Q_{sim}})(Q_{obs},i - \overline{Q_{obs}})\right]^2}{\sum_{i=1}^{n} (Q_{sim},i - \overline{Q_{sim}})^2 \sum_{i=1}^{n} (Q_{obs},i - \overline{Q_{obs}})^2} \tag{11}$$

where $Q_{sim,i}$ is the ith simulated runoff, mm; $Q_{obs,i}$ is the ith observed runoff, mm; $\overline{Q_{sim}}$ is the average value of simulated runoff, mm; $\overline{Q_{obs}}$ is the average value of observed runoff, mm; n is the number of observed runoff series. The smaller *RE*, the larger *NSE* and the higher *DC* provides a better performance of the Budyko-type function. With reference to pre-existing studies, if $NSE > 0.5$, $DC > 0.6$ and $RE < 15\%$, it means that this approach is applicable in the catchment (Yuan *et al.* 2017).

3. RESULTS AND DISCUSSION

3.1. Runoff simulations by the water-balance model

For reasons of ensuring the water-balance model performs well in the study period, it is most important to calibrate the parameters in various Budyko-type equations during the period 1959–1998, followed by verifying these parameters during the period 1999–2017. After calibration, dimensionless parameters α, ω, ∂ and k are 1.70, 0.03, 0.02 and 0.99, respectively. The performance for different Budyko-type equations is shown in Table 2.

It can be concluded that all *RE* are <11%, all *NSE* are >0.84 and all *DC* are >0.84 in the calibration period, while all *RE* are <15%, all *NSE* are >0.64 and all *DC* are >0.72 in the validation period. Therefore, it is evident that four Budyko-type equations are highly consistent in runoff simulation for meeting the requirements of accuracy, and the water-balance model is applicable in the study area.

In contrast with other equations, the Budyko-CY equation has a better performance with respect to the *NSE* during two periods. As a consequence, the Budyko-CY function was screened as the optimal equation to project the runoff variation in the near future (2021–2050). Comparison between the simulated runoff and the observed runoff is shown in Figure 3.

3.2. Sensitivity analysis of runoff to climatic factors

The statistical relation between monthly mean temperature and monthly potential evapotranspiration was established via the MATLAB toolbox, taking the observed data during the period 1961–2010 into account. Figure 4 shows that monthly temperature and evapotranspiration conform to the non-linear relation composed of an exponential function and a linear function, and it was clear that this equation can fit the statistical relation well with the square of correlation coefficient of 0.894.

Table 2 | Performance for different Budyko-type equations

Time	Indices	Budyko-Fu	Budyko-Zhang	Budyko-WT	Budyko-CY
Calibration period (1959–1998)	*RE*	11.0%	10.9%	10.9%	10.9%
	NSE	0.843	0.840	0.840	0.843
	DC	0.843	0.843	0.843	0.843
Validation period (1999–2017)	*RE*	14.5%	14.9%	14.8%	14.5%
	NSE	0.651	0.645	0.650	0.653
	DC	0.720	0.720	0.720	0.721

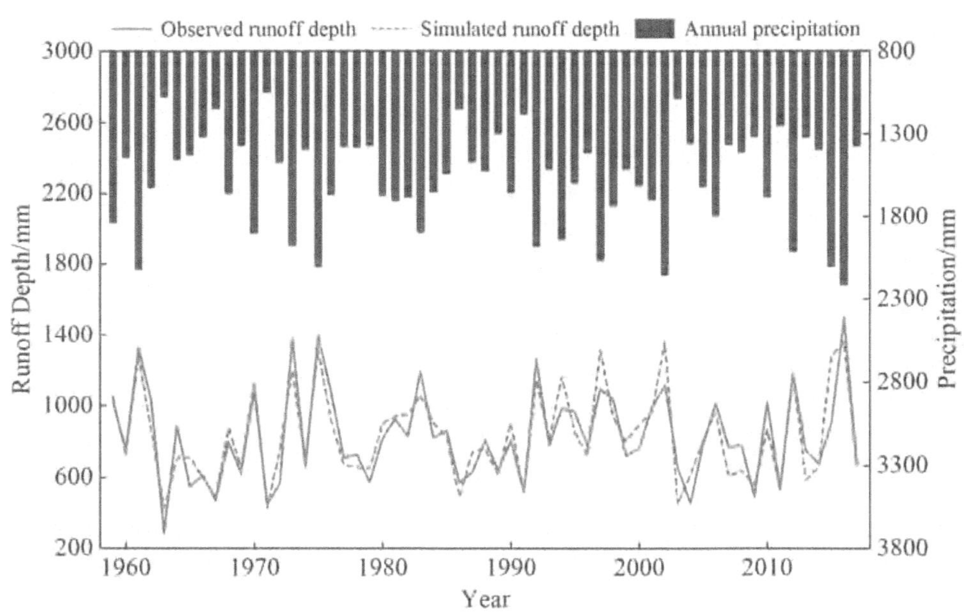

Figure 3 | Comparison between the simulated runoff and the observed runoff.

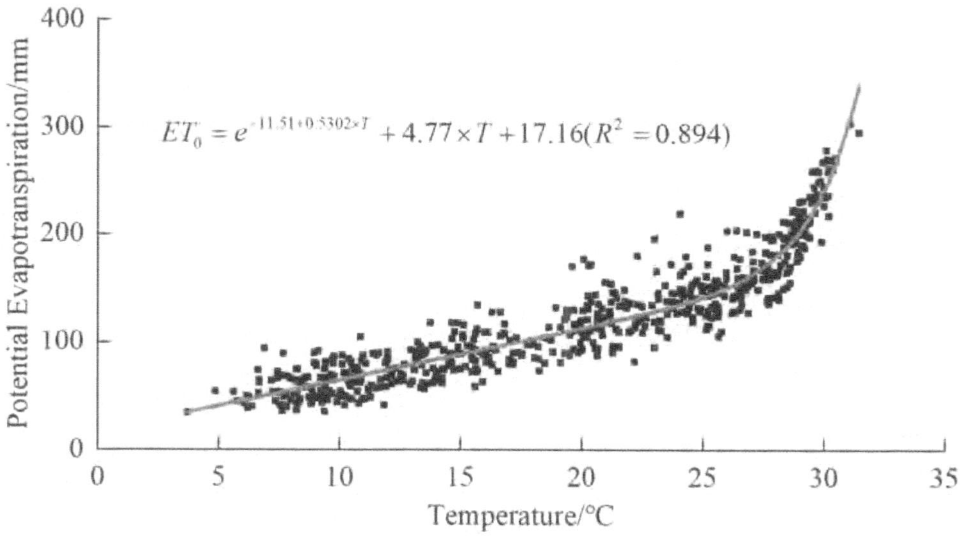

Figure 4 | Relations between monthly temperature and potential evapotranspiration.

Based on daily-series meteorological data of 21 grid cells in and around the UGRB generated by various GCMs, monthly precipitation and average temperature of 21 grid cells can be counted at first, and monthly precipitation and average temperature in the study area can be summarized. Then, monthly potential evapotranspiration in the study area can be calculated by means of the statistical relation proposed in Figure 4. Subsequently, surface runoff under multiple climate scenarios can be simulated with reference to the Budyko-CY function expressed in Equation (7). Ultimately, projected climate change and its possible influences on water resources can be estimated according to Equation (8), and the response of surface runoff to the variations in precipitation and temperature is shown in Figure 5.

In terms of the response of surface runoff to precipitation, on the condition of given precipitation, when annual temperature rises by 1 °C, the annual surface runoff is expected to decline by 2.5–4.7% on average, or 22.6–42.5 mm. In the case of a 30% ascent in precipitation, every 1 °C change in temperature may lead to the largest variation range in runoff, and corresponding runoff is expected to decrease by 7.8% when average temperature rises from 2 to 3 °C. Relatively speaking, in the case of a 30% reduction in precipitation, variation amplitude in runoff is the smallest for 1 °C change in temperature, and corresponding runoff is expected to decrease by 1.6% when average temperature changes from −3 to −2 °C. The foremost reason is that actual regional evaporation is relatively large under warm and humid conditions. Therefore, runoff is more sensitive to temperature rise under given precipitation.

In terms of the response of surface runoff to temperature, under the circumstance of given temperature, for every 10% increase in precipitation, annual surface runoff is expected to rise from 12.1 to 14.3% on average, that is, 110.7–130.6 mm. In the case of a 3 °C ascent in temperature, every 10% change in precipitation may cause the smallest variation range in runoff, and corresponding runoff is expected to increase by 11.0% when annual precipitation varies from −30 to −20%. Instead, in the case of a 3 °C reduction in temperature, variation amplitude in runoff is the largest for 10% variation in precipitation, and corresponding runoff is expected to increase by 15.0% when annual precipitation changes from 20 to 30%. Chiew (2010) revealed that the greater the runoff coefficient is in the watershed, the less sensitive it is to climate change. As far as the UGRB is concerned, its multi-year mean precipitation and streamflow during the period 1961–2010 are 1,560 and 821 mm, so the runoff coefficient is 0.53 over the UGRB. Therefore, the UGRB is not sensitive to climate change, and the analysis based on the water-balance model is consistent with it.

3.3. Future scenarios for hydrometeorological factors

3.3.1. Projected changes in monthly climatic factors

A comparison of three climatic factors is shown in Tables 3, where base represents the observed monthly average during the baseline period, while RCP2.6, RCP4.5 and RCP8.5 represent the projected ensemble mean value of five GCMs during the projection period (2021–2050).

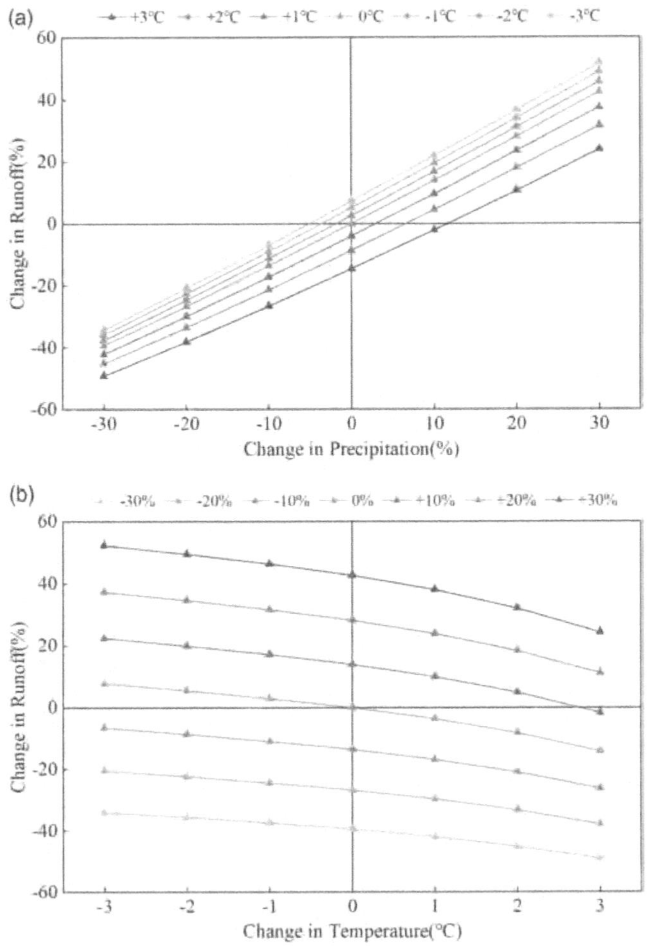

Figure 5 | (a) Response of surface runoff to the variation in precipitation and (b) the response of surface runoff to the variation in temperature.

In the next 30 years, the projected multi-year mean precipitation will vary as 10, −9 and −8 mm under three RCP scenarios, while the projected temperature will increase by 0.69, 0.68 and 0.97 °C, the projected potential evapotranspiration will increase by 162, 148 and 232 mm, respectively. Taking the temperature as an example, relative to the reference period, an average temperature will rise every month, especially in September and October, with warming by 1.1/1.1/1.6 °C and 1.0/0.9/1.2 °C under three RCPs. Similarly, the increased potential evaporation is mainly concentrated in July and August with monthly variation of 40/46/56 mm and 61/43/84 mm, respectively. The variation in monthly precipitation is specifically manifested as follows: rising in June, August and September, and falling in May, July and October.

3.3.2. Projected changes in annual precipitation

The projected annual precipitation in the UGRB is shown in Figure 6. The variation range and the ensemble mean for each RCP are shown concurrently, with the assumption that the ensemble mean is able to characterize the variation trend of annual precipitation, temperature and runoff, and five global climate models are equally plausible. It is worth noting that the solid lines in red, blue and green represent the multi-mode ensemble mean of annual runoff under three climate scenarios (i.e., RCP2.6, RCP4.5 and RCP8.5), respectively, whereas the dotted lines of corresponding colors represent the maximum and minimum annual runoffs in various GCMs during the projection period of 2021–2050. It is obvious that the annual precipitation under the RCP2.6 scenario presents the greater fluctuation of 558 mm due to its variation range, while that under both the RCP4.5 and RCP8.5 scenarios is 494 mm. The maximum and the minimum annual precipitation during the period of 2021–2050 are 2,465 mm (GFDL-ESM2M) and 900 mm (GFDL-ESM2M) under the RCP2.6 scenario, while they are 2,809 mm (GFDL-ESM2M) and 667 mm (MIROC-ESM-CHEM) under the RCP4.5 scenario and 2,560 mm (GFDL-ESM2M)

Table 3 | Comparison of (a) precipitation (mm), (b) temperature (°C) and (c) potential evapotranspiration (mm) between two periods

	Jan.	Feb.	Mar.	Apr.	May	Jun.	Jul.	Aug.	Sep.	Oct.	Nov.	Dec.	Sum
(a) Precipitation													
Base	64	100	171	208	241	244	131	141	94	69	54	43	1,560
RCP2.6	53	97	175	206	238	251	127	162	108	54	52	47	1,570
RCP4.5	61	98	164	211	231	244	129	151	101	60	58	43	1,551
RCP8.5	60	91	167	206	237	266	120	147	112	53	47	46	1,552
	Jan.	**Feb.**	**Mar.**	**Apr.**	**May**	**Jun.**	**Jul.**	**Aug.**	**Sep.**	**Oct.**	**Nov.**	**Dec.**	**Mean**
(b) Temperature													
Base	8.4	10.2	14.1	19.7	24	26.8	29.4	28.9	25.9	21.2	15.6	10.4	19.55
RCP2.6	8.8	10.6	14.7	20.1	24.6	27.6	29.9	29.7	27	22.2	16.4	11.3	20.24
RCP4.5	8.8	10.5	14.8	20.3	24.9	27.6	30	29.5	27	22.1	16.3	11	20.23
RCP8.5	9.1	10.7	14.9	20.7	24.8	27.9	30.3	30.1	27.5	22.4	16.6	11.2	20.52
	Jan.	**Feb.**	**Mar.**	**Apr.**	**May**	**Jun.**	**Jul.**	**Aug.**	**Sep.**	**Oct.**	**Nov.**	**Dec.**	**Sum**
(c) Potential evapotranspiration													
Base	60	60	79	108	139	151	226	195	155	131	91	71	1,466
RCP2.6	59	68	87	114	140	179	266	256	167	125	96	71	1,628
RCP4.5	59	67	88	114	142	176	272	238	169	124	95	70	1,614
RCP8.5	61	68	88	117	143	187	282	279	179	126	97	71	1,698

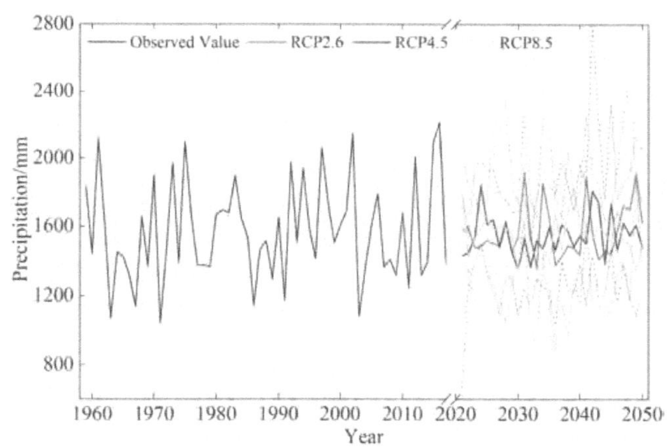

Figure 6 | Annual precipitation in the study area from 1959 to 2050.

and 693 mm (GFDL-ESM2M) under the RCP8.5 scenario, respectively. It is not difficult to find that the GFDL-ESM2M model is closely associated with annual rainfall extremes.

Figure 7 shows multi-year mean precipitation of various GCMs under three climate scenarios. It can be concluded that multi-year mean precipitation during the projection period changes by −8.5 to 8.7% (RCP2.6), −4.5 to 3.9% (RCP4.5) and −8.1 to 6.3% (RCP8.5), respectively. The average precipitation in the GFDL-ESM2M, HadGEM2-ES and NorESM1-M models will increase, with variation ranges of 1.6–7.2, 0.3–2.2 and 1.4–8.7%, while that in IPSL-CM5A-LR and MIROC-ESM-CHEM models will decrease, with variation ranges of −8.5 to −4.5% and −8.1 to −4.2%, respectively. In aggregate, the IPSL-CM5A-LR model has the minimum variation in future projected precipitation, with the average change of −6.9%, while the NorESM1-M model has the maximum variation with the average change of 5.4%.

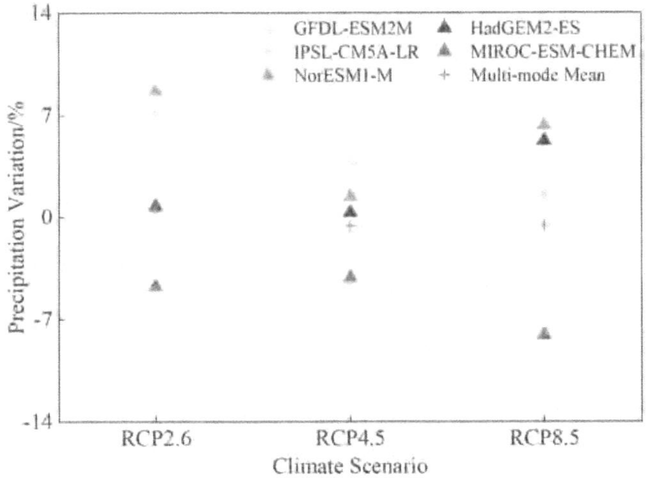

Figure 7 | Precipitation variation under various climate scenarios.

3.3.3. Projected changes in annual temperature

The projected annual temperature in the UGRB is shown in Figure 8. On a multi-year average scale, it is clear that temperature shows an upward trend in the projection period with the gradient of 0.29 °C/10a, 0.25 °C/10a and 0.61 °C/10a under three RCP scenarios, respectively. Especially in the last 10 years of the projection period, temperature is expected to increase by 1.03, 0.94 and 1.68 °C relative to that during the period of 1961–2010, suggesting that future temperature is expected to increase in the UGRB. The annual mean temperature under the RCP8.5 scenario presents the greater variation amplitude, with the maximum value of 21.74 °C in 2050 and the minimum value of 19.40 °C in 2040, while the variation amplitudes under RCP2.6 and RCP4.5 scenarios are 1.37 and 1.45 °C, revealing that the higher the temperature rises, the greater the variation range in temperature will be.

Figure 9 shows a multi-year mean temperature of various GCMs under three climate scenarios. From the perspective of RCP, the multi-year average temperature is expected to increase by 0.17–1.43 °C (RCP2.6), 0.22–1.18 °C (RCP4.5) and 0.53–1.72 °C (RCP8.5), respectively. When it comes to the GCM, all five models show a rise in temperature, with variation ranges of 0.22–0.53 °C (GFDL-ESM2M), 0.92–1.23 °C (HadGEM2-ES), 0.64–1.03 °C (IPSL-CM5A-LR), 1.18–1.72 °C (MIROC-ESM-CHEM) and 0.17–0.64 °C (NorESM1-M), respectively. The GFDL-ESM2M model has the minimum warming mean temperature of 0.33 °C, while the MIROC-ESM-CHEM model has the maximum warming mean temperature, reaching up to 1.44 °C under three RCPs. Different from the MIROC-ESM-CHEM and HadGEM2-ES models, the GFDL-ESM2M and NorESM1-M models both show a small increase in temperature under all RCP scenarios.

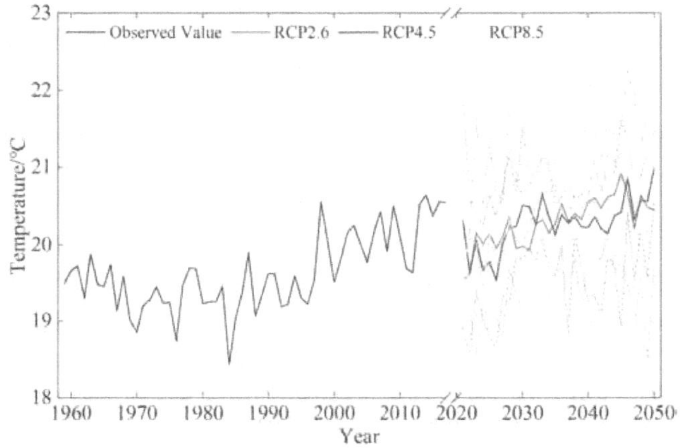

Figure 8 | Annual temperature in the study area from 1959 to 2050.

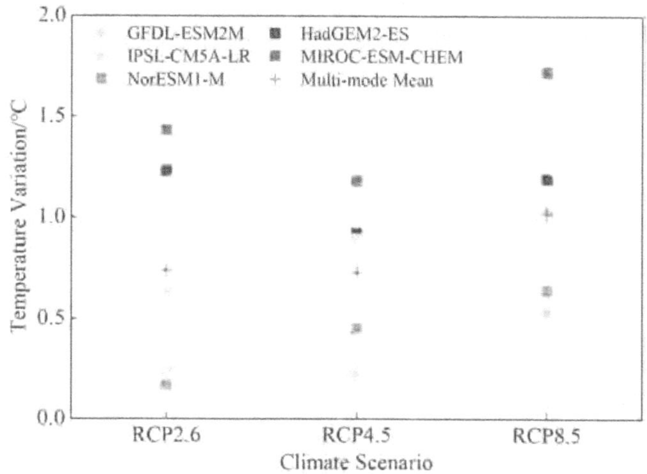

Figure 9 | Temperature variation under various climate scenarios.

3.3.4. Projected changes in annual runoff

The projected annual runoff in the UGRB is shown in Figure 10. The annual mean runoff is expected to increase by 8.4, 6.5 and 5.0% from 2021 to 2050 in contrast with that from 1961 to 2010, suggesting that surface water resources will increase in the next 30 years. The maximum runoff depth values are 1,190, 1,116 and 1,092 mm, while the minimum runoff depth values are 725, 692 and 662 mm under three RCPs, respectively. Studies demonstrated that a higher temperature may bring about a decrease in annual runoff in the UGRB. Nevertheless, different from RCP2.6 and RCP4.5 scenarios, the annual runoff under the RCP8.5 scenario shows a downward trend during the projection period, especially after 2040.

Figure 11 shows multi-year mean runoff depth of various GCMs under three climate scenarios. Surface water resources will increase in the future under various scenarios, except for the MIROC-ESM-CHEM model under three RCPs, as well as the IPSL-CM5A-LR model under RCP2.6 and RCP8.5 scenarios. The variation in projected runoff is similar between the GFDL-ESM2M model and the NorESM1-M model, because the former has the 21.2% ascent under the RCP2.6 scenario and 16.4% ascent under three RCPs, while those in the latter are 23.3% rise and 17.3% rise, respectively. On the contrary, the MIROC-ESM-CHEM model seems to be on a downward trajectory, with the variation amplitude of −12.5 to −2.1% and the mean variation value of −6.4%. The variation range under the RCP4.5 scenario is the smallest, varying from −2.05 to 16.02%, while variation ranges under RCP2.6 and RCP8.5 scenarios are −4.55 to 23.25% and −12.46 to 17.53%, respectively.

3.3.5. Discussions

The variations of the projected multi-year mean precipitation and temperature over the UGRB during the period 2021–2050 are relatively close to the conclusions proposed by Deng & Wang (2019), and both are precipitation declines and potential

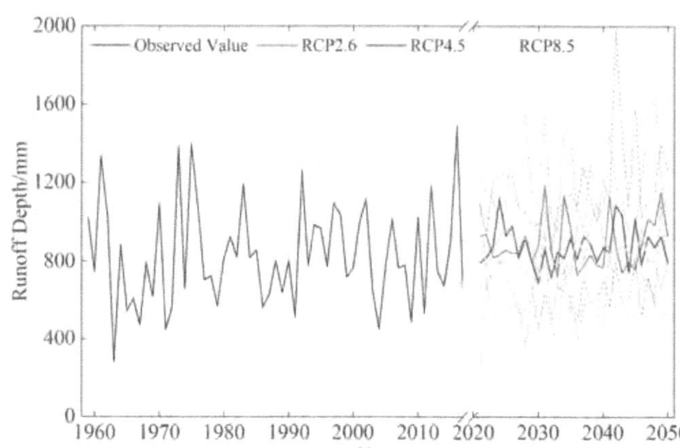

Figure 10 | Annual runoff depth in the study area from 1959 to 2050.

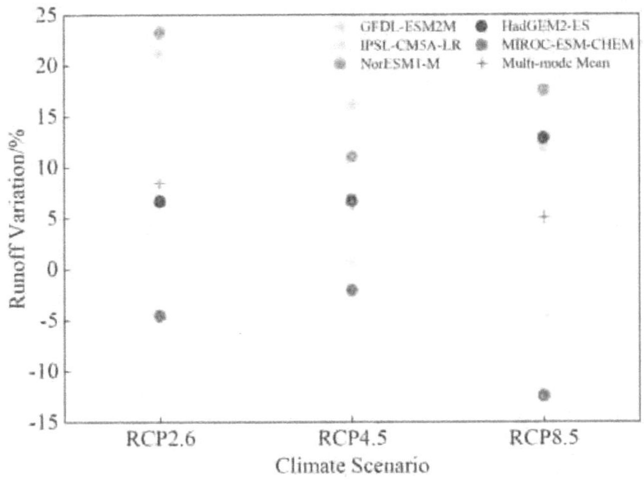

Figure 11 | Runoff variation under various climate scenarios.

evaporation increases; however, in terms of annual runoff, the opposite conclusions are drawn. In this paper, the annual mean runoff is expected to increase by 8.4, 6.5 and 5.0% under three RCPs in the next 30 years, while the latter reported a decreasing trend about −9.9, −19.5 and −11.4%. The reason for the opposite conclusions of projected runoff lies in the different climate models selected. Taking climate models as an example, multiple global GCMs from different countries were selected in this paper, while the latter just chose a climate model, namely BCC CSM1.1. Zhao *et al.* (2020) applied the HBV model and the same five climate models into the GRB to reveal that annual runoff showed an upward trend under the RCP4.5 scenario, while that may decrease after the middle of the 21st century under RCP2.6 and RCP8.5 scenarios. It is evident that the selected hydrological model, study area and study period all have a great influence on the projected runoff.

Since daily-series data of climatic factors are generated by multiple GCMs, the rationality of the results obtained is directly related to the selected climate models, which will create uncertainties inevitably. If different climate models are adopted, the different analyses and conclusions may be achieved in practice. Thus, it is still a challenge to quantitatively describe the variation trend and range for climatic variables.

4. CONCLUSIONS

In this study, precipitation, temperature and runoff in the UGRB were projected, and the impact of climate change on water resources was evaluated by employing the Budyko-type functions driven by multiple representative GCMs. The conclusions are summarized as follows:

- The constructed water-balance model performed well on simulating annual runoff, because *NSE* and *DC* are both >0.84 and >0.64 during two periods. The Budyko-CY function was the optimal water-balance equation screened from the Budyko-Fu, Budyko-Zhang, Budyko-WT and Budyko-CY functions, for its having the maximum *NSE*. Compared with the hydrological model, Budyko-type functions have the advantage of few parameters and simple calculations. Therefore, this approach is appropriate in predicting future streamflow in large-scale regions or ungauged catchments with limited hydrological information.
- When temperature rises by 1 °C, the surface runoff depth is expected to decline by 2.5–4.7% on average, that is 22.6–42.5 mm; when precipitation rises by 10%, surface runoff is expected to rise by 12.1–14.3% on average, that is, 110.7–130.6 mm. The rainfall elasticity of streamflow is strongly correlated to the runoff coefficient, annual rainfall and streamflow, where the runoff coefficient in the UGRB is 0.53, suggesting that this catchment is not sensitive to a climate change.
- The projected temperature over the UGRB presents an obvious escalating trend during the period 2021–2050 with a slope of 0.29 °C/10a, 0.25 °C/10a and 0.61 °C/10a under three RCP scenarios, respectively. The annual mean temperature is estimated to rise by 0.17–1.43, 0.22–1.18 and 0.53–1.72 °C, while the annual runoff is expected to increase slightly with the variability of 8.4, 6.5 and 5.0%, respectively. Five GCMs all show a rise in temperature, with variation ranges of 0.22–0.53 °C (GFDL-ESM2M), 0.92–1.23 °C (HadGEM2-ES), 0.64–1.03 °C (IPSL-CM5A-LR), 1.18–1.72 °C (MIROC-ESM-CHEM) and 0.17–0.64 °C (NorESM1-M), respectively.

As a whole, this research provides possibilities for climate scenarios that may occur in the future and results presented herein are capable of giving a reference to rational allocation and optimal scheduling of water resources in the UGRB. In addition, leveraging of the characteristics of a variety of climate models to prospect surface runoff will remain a tough task over the coming decades.

ACKNOWLEDGEMENTS

This research was financially supported by the Young Elite Scientists Sponsorship Program by CAST (2019QNRC001) and the National Key Research and Development Program of China (2018YFC0407405). The anonymous reviewers and the editor are thanked for providing insightful and detailed reviews that greatly improved the manuscript.

AUTHOR CONTRIBUTIONS

P.S., C.W. and G.D. conducted the research and wrote the paper. J.S., L.K. and M.L. helped to revise the paper. X.L. and H.W. gave the comments. All authors have read and agreed to the published version of the manuscript.

CONFLICTS OF INTEREST

The authors declare no conflict of interest.

DATA AVAILABILITY STATEMENT

All relevant data are included in the paper or its Supplementary Information.

REFERENCES

Bao, Z., Zhang, J., Liu, J., Wang, G., Yan, X., Wang, X. & Zhang, L. 2012 Sensitivity of hydrological variables to climate change in the Haihe River Basin, China. *Hydrological Processes* **26** (15), 2294–2306. doi:10.1002/hyp.8348.

Chen, H., Xu, C. & Guo, S. 2012 Comparison and evaluation of multiple GCMs, statistical downscaling and hydrological models in the study of climate change impacts on runoff. *Journal of Hydrology* **434–435**, 36–45. doi:10.1016/j.jhydrol.2012.02.040.

Chen, Z., Lei, H., Yang, H., Yang, D. & Cao, Y. 2017 Historical and future trends in wetting and drying in 291 catchments across China. *Hydrology and Earth System Sciences* **21** (4), 2233–2248. doi:10.5194/hess-21-2233-2017.

Chiew, F. H. S. 2010 Lumped conceptual rainfall-runoff models and simple water balance methods: overview and applications in ungauged and data limited regions. *Geography Compass* **4** (3), 206–225. doi:10.1111/j.1749-8198.2009.00318.x.

Crow, W. T. & Ryu, D. 2009 A new data assimilation approach for improving runoff prediction using remotely-sensed soil moisture retrievals. *Hydrology and Earth System Sciences* **13** (1), 1–16. doi:10.5194/hess-13-1-2009.

Deng, C. & Wang, W. 2019 Runoff predicting and variation analysis in upper Ganjiang Basin under projected climate changes. *Sustainability* **11** (21), 5885. doi:10.3390/su11215885.

Fang, J., Yu, G., Liu, L., Hu, S. & Chapin, F. S. 2018 Climate change, human impacts, and carbon sequestration in China. *Proceedings of the National Academy of Sciences* **115** (16), 4015–4020. doi:10.1073/pnas.1700304115.

Fu, G., Charles, S. P. & Chiew, F. H. S. 2007 A two-parameter climate elasticity of streamflow index to assess climate change effects on annual streamflow. *Water Resources Research* **43** (11), W11419. doi:10.1029/2007WR005890.

Guan, X., Zhang, J., Bao, Z., Liu, C., Jin, J. & Wang, G. 2021 Past variations and future projection of runoff in typical basins in 10 water zones, China. *Science of The Total Environment* **798**, 149277. doi:10.1016/j.scitotenv.2021.149277.

Jin, J., Wang, G., Zhang, J., Yang, Q., Liu, C., Liu, Y., Bao, Z. & He, R. 2020 Impacts of climate change on hydrology in the Yellow River source region, China. *Journal of Water and Climate Change* **11** (3), 916–930. doi:10.2166/wcc.2018.085.

John, A., Nathan, R., Horne, A., Stewardson, M. & Webb, J. A. 2020 How to incorporate climate change into modelling environmental water outcomes: a review. *Journal of Water and Climate Change* **11** (2), 327–340. doi:10.2166/wcc.2020.263.

Kazemi, H., Sarukkalige, R. & Badrzadeh, H. 2019 Evaluation of streamflow changes due to climate variation and human activities using the Budyko approach. *Environmental Earth Sciences* **78** (24), 1–17. doi:10.1007/s12665-019-8735-9.

Li, D., Pan, M., Cong, Z., Zhang, L. & Wood, E. 2013 Vegetation control on water and energy balance within the Budyko framework. *Water Resources Research* **49** (2), 969–976. doi:10.1002/wrcr.20107.

Li, Z., Li, Q., Wang, J., Feng, Y. & Shao, Q. 2020 Impacts of projected climate change on runoff in upper reach of Heihe River basin using climate elasticity method and GCMs. *Science of the Total Environment* **716**, 137072. doi:10.1016/j.scitotenv.2020.137072.

Liu, Y., Feng, J. & Ma, Z. 2014 An analysis of historical and future temperature fluctuations over China based on CMIP5 simulations. *Advances in Atmospheric Sciences* **31** (2), 457–467. doi:10.1007/s00376-013-3093-0.

Liu, J., Zhang, Q., Singh, V. P. & Shi, P. 2017 Contribution of multiple climatic variables and human activities to streamflow changes across China. *Journal of Hydrology* **545**, 145–162. doi:10.1016/j.jhydrol.2016.12.016.

Piani, C., Weedon, G. P., Best, M., Gomes, S. M., Viterbo, P., Hagemann, S. & Haerter, J. O. 2010 Statistical bias correction of global simulated daily precipitation and temperature for the application of hydrological models. *Journal of Hydrology* **395** (3), 199–215. doi:10.1016/j.jhydrol.2010.10.024.

Piao, S., Ciais, P., Huang, Y., Shen, Z., Peng, S., Li, J., Zhou, L., Liu, H., Ma, Y., Ding, Y., Friedlingstein, P., Liu, C., Tan, K., Yu, Y., Zhang, T. & Fang, J. 2010 The impacts of climate change on water resources and agriculture in China. *Nature* **467** (7311), 43–51. doi:10.1038/nature09364.

Song, P., Liu, W., Sun, J., Wang, C., Kong, L., Nong, Z., Lei, X. & Wang, H. 2020 Annual runoff forecasting based on multi-model information fusion and residual error correction in the Ganjiang River Basin. *Water* **12** (8), 2086. doi:10.3390/w12082086.

Sun, G. & Peng, F. 2020 Evaluation of future runoff variations in the north–south transect of eastern China: effects of CMIP5 models outputs uncertainty. *Journal of Water and Climate Change* **11** (4), 1355–1369. doi:10.2166/wcc.2019.261.

Taylor, K. E., Stouffer, R. J. & Meehl, G. A. 2012 An overview of CMIP5 and the experiment design. *Bulletin of the American Meteorological Society* **93** (4), 485–498. doi:10.1175/BAMS-D-11-00094.1.

Teng, J., Chiew, F. H. S., Vaze, J., Marvanek, S. & Kirono, D. G. C. 2012 Estimation of climate change impact on mean annual runoff across continental Australia using Budyko and Fu equations and hydrological models. *Journal of Hydrometeorology* **13** (3), 1094–1106. doi:10.1175/JHM-D-11-097.1.

Thompson, J. R., Green, A. J., Kingston, D. G. & Gosling, S. N. 2013 Assessment of uncertainty in river flow projections for the Mekong River using multiple GCMs and hydrological models. *Journal of Hydrology* **486**, 1–30. doi:10.1016/j.jhydrol.2013.01.029.

Wang, D. & Tang, Y. 2014 A one-parameter Budyko model for water balance captures emergent behavior in Darwinian hydrologic models. *Geophysical Research Letters* **41** (13), 4569–4577. doi:10.1002/2014GL060509.

Warszawski, L., Frieler, K., Huber, V., Piontek, F., Serdeczny, O. & Schewe, J. 2014 The inter-sectoral impact model intercomparison project (ISI–MIP): project framework. *Proceedings of the National Academy of Sciences* **111** (9), 3228–3232. doi:10.1073/pnas.1312330110.

Wen, X., Fang, G., Qi, H., Zhou, L. & Gao, Y. 2016 Changes of temperature and precipitation extremes in China: past and future. *Theoretical and Applied Climatology* **126** (1), 369–383. doi:10.1007/s00704-015-1584-x.

Wu, J. & Miao, C. 2017 Contribution analysis of the long-term changes in seasonal runoff on the Loess Plateau, China, using eight Budyko-based methods. *Journal of Hydrology* **545**, 263–275. doi:10.1016/j.jhydrol.2016.12.050.

Wu, J., Miao, C., Zhang, X., Yang, T. & Duan, Q. 2017 Detecting the quantitative hydrological response to changes in climate and human activities. *Science of the Total Environment* **586**, 328–337. doi:10.1016/j.scitotenv.2017.02.010.

Xing, W., Wang, W., Zou, S. & Deng, C. 2018 Projection of future runoff change using climate elasticity method derived from Budyko framework in major basins across China. *Global and Planetary Change* **162**, 120–135. doi:10.1016/j.gloplacha.2018.01.006.

Yan, Z., Zhou, Z., Liu, J., Han, Z., Gao, G. & Jiang, X. 2020 Ensemble projection of runoff in a large-scale basin: modeling with a global BMA approach. *Water Resources Research* **56** (7), e2019WR026134. doi:10.1029/2019WR026134.

Yu, Z., Tan, G. & Li, G. 2014 Characteristic analysis of inflow and water resources in the Poyang Lake Basin. *Journal of Water Resources Research* **3**, 465–472 (in Chinese).

Yuan, F., Tung, Y. & Ren, L. 2016 Projection of future streamflow changes of the Pearl River basin in China using two delta-change methods. *Hydrology Research* **47** (1), 217. doi:10.2166/nh.2015.159.

Yuan, Z., Yan, D., Yang, Z., Yin, J., Zhang, C. & Yuan, Y. 2017 Projection of surface water resources in the context of climate change in typical regions of China. *Hydrological Sciences Journal* **62** (2), 283–293. doi:10.1080/02626667.2016.1222531.

Yuan, Z., Xu, J. & Wang, Y. 2018a Projection of future extreme precipitation and flood changes of the Jinsha River Basin in China based on CMIP5 climate models. *International Journal of Environmental Research and Public Health* **15** (11), 2491. doi:10.3390/ijerph15112491.

Yuan, Z., Yan, D., Yang, Z., Xu, J., Huo, J., Zhou, Y. & Zhang, C. 2018b Attribution assessment and projection of natural runoff change in the Yellow River Basin of China. *Mitigation and Adaptation Strategies for Global Change* **23** (1), 27–49. doi:10.1007/s11027-016-9727-7.

Yuan, Z., Xu, J., Wang, Y., Hong, X., Zhou, Y. & Chen, J. 2019 Long-term projection of surface runoff in the Fu River basin based on water-energy balance. *Advanced Engineering Sciences* **51** (1), 60–67 (in Chinese).

Zhang, P., Lei, X., Wang, X. & Wang, C. 2018a Optimal water regulation in middle and lower reaches of Ganjiang River based on water demand control strategy. *Yangtze River* **49** (12), 43–48 + 86 (in Chinese).

Zhang, S., Yang, Y., McVicar, T. R. & Yang, D. 2018b An analytical solution for the impact of vegetation changes on hydrological partitioning within the Budyko framework. *Water Resources Research* **54** (1), 519–537. doi:10.1002/2017WR022028.

Zhao, M., Su, B., Wang, Y., Wang, A. & Jiang, T. 2020 Impacts of climate change on river runoff at the Ganjiang and Guanting River basins in the eastern monsoon region. *Climate Change Research* **16** (6), 679–689 (in Chinese).

Zhou, T., Wang, B., Yu, Y., Liu, Y., Zheng, W., Li, L., Wu, B., Lin, P., Guo, Z., Man, W., Bao, Q., Duan, A., Liu, H., Chen, X., He, B., Li, J., Zou, L., Wang, X., Zhang, L., Sun, Y. & Zhang, W. 2018 The FGOALS climate system model as a modeling tool for supporting climate sciences: an overview. *Earth and Planetary Physics* **2** (4), 276–291. doi:10.26464/epp2018026.

Zhu, X., Zhang, A., Wu, P., Qi, W., Fu, G., Yue, G. & Liu, X. 2019 Uncertainty impacts of climate change and downscaling methods on future runoff projections in the Biliu River Basin. *Water* **11** (10), 2130. doi:10.3390/w11102130.

Zinyengere, N., Theodory, T. F., Gebreyes, M. & Speranza, C. I. 2017 *Beyond Agricultural Impacts: Multiple Perspectives on Climate Change and Agriculture in Africa*. Academic Press, London, UK.

First received 2 July 2021; accepted in revised form 30 December 2021. Available online 24 January 2022

© 2022 The Authors Impact of Climate Change on Hydrology and Water Resources doi: 10.2166/wcc.2021.235

Future climate change and impacts on water resources in the Upper Blue Nile basin

Gebiyaw Sitotaw Takele [a,*], Geremew Sahilu Gebrie [b], Azage Gebreyohannes Gebremariam[a] and Agizew Nigussie Engida[b]

[a] Ethiopian Institute of Water Resources, Addis Ababa University, P.O. BOX. 150461, Addis Ababa, Ethiopia
[b] Addis Ababa Institute of Technology, Addis Ababa University, P.O. BOX. 150461, Addis Ababa, Ethiopia
*Corresponding author. E-mail: gebiyaw@gmail.com

GST, 0000-0001-6641-2559; GSG, 0000-0002-9548-6925

ABSTRACT

This study aims to assess the impact of climate change on the water resources of the Upper Blue Nile basin using an integrated climate and hydrological model. The impact of climate change on water resources is being assessed using the regional climate model (RCM) under the representative concentration pathway (RCP4.5 and RCP8.5) scenarios and the Soil and Water Assessment Tool (SWAT) hydrological model. Future climate scenarios have been developed for the 2030s (2021–2040) and the 2050s (2041–2060). The study found that the projected rainfall shows a decreasing trend and is not statistically significant, while the temperature shows an increasing trend and is statistically significant. Due to the sharp rise in temperature, the annual evapotranspiration increased by about 10.4%. This and the declining trend of rainfall will reduce streamflow up to 54%, surface runoff up to 31%, and water yield up to 31%. Climate change causes seasonal and annual fluctuations in the water balance components. However, the projected seasonal changes are much greater than the annual changes. Therefore, the results of this study will be useful to basin planners, policymakers, and water resources managers in developing adaptation strategies to offset the adverse effects of climate change in the Upper Blue Nile basin.

Key words: bias correction, climate change, regional climate model, SWAT, Upper Blue Nile basin, water resources

HIGHLIGHTS

- The baseline rainfall trend is increasing but insignificant, whereas the ensembled regional climate model (RCM)-simulated rainfall trend is decreasing.
- Despite some RCMs having a seasonal shift in rainfall, there is no discernible seasonal movement in surface runoff and water production.
- Ensembled rainfall, streamflow, surface runoff, and water output reveal a decrease in the 2030s and 2050s climate conditions.

GRAPHICAL ABSTRACT

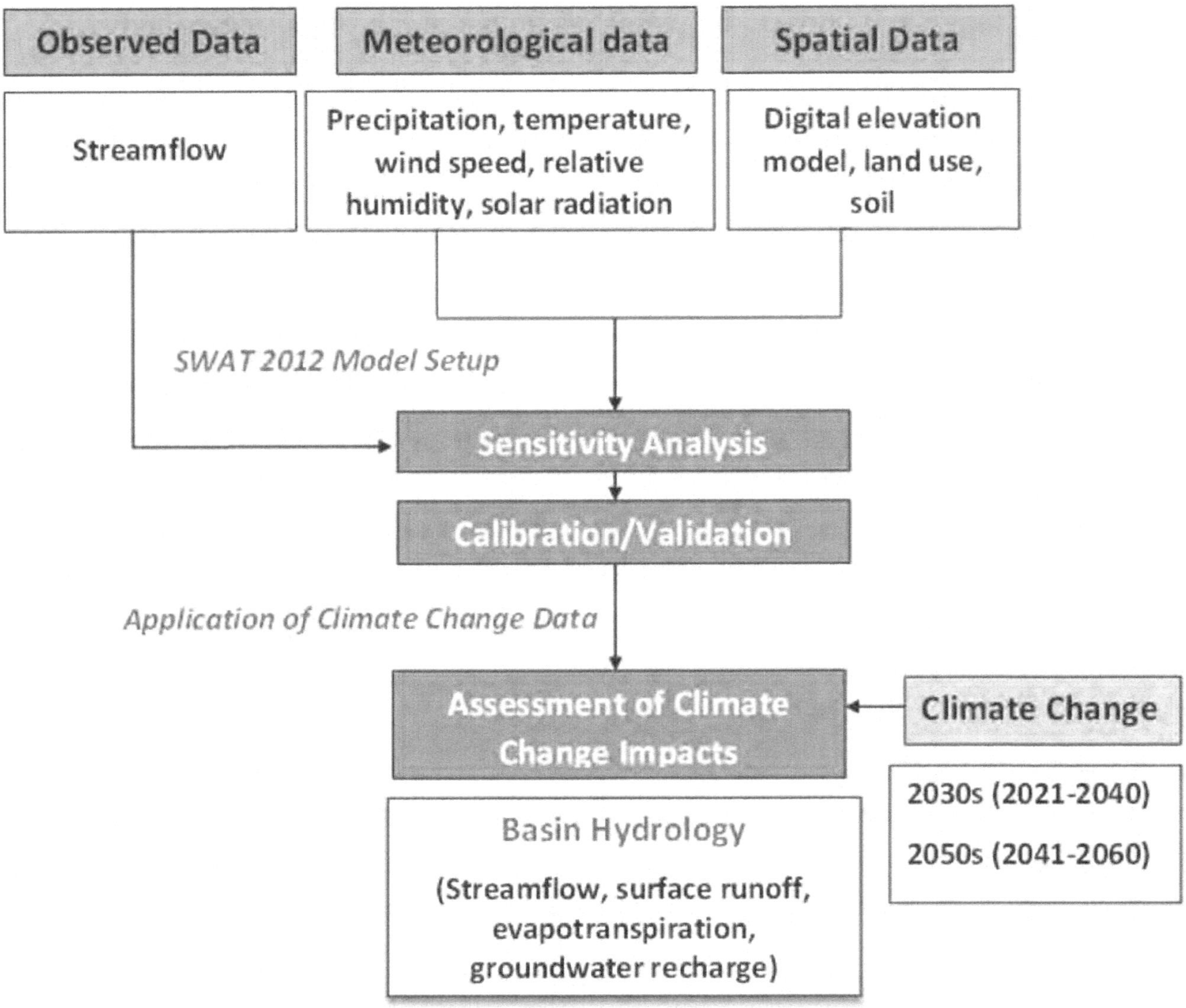

1. INTRODUCTION

Climate change has a significant impact on water resources, as it changes the components of the climate (Versini *et al.* 2016; Aduah *et al.* 2017). These changes are expected to have a significant impact on water resources, the water available in the region, and the temporal and spatial patterns of their distribution (UNESCO, UN-Water 2020). In addition, climate change can affect various characteristics of water resources, such as water quality and quantity to meet basic human needs, extreme hydrological events, event timing, and water temperature (Kim *et al.* 2008; Nan *et al.* 2011; Versini *et al.* 2016; Biao 2017). In addition, climate change not only leads to higher average temperatures, but also changes in annual and seasonal precipitation that affect the water cycle (Mohammed *et al.* 2016).

One of the rivers most affected by climate change is the Upper Blue Nile basin (Abdo *et al.* 2009; Taye *et al.* 2011). The basin covers 10% of the Nile basin and produces about 60% of the water that reaches Egypt and Sudan (Roth *et al.* 2018). Hundreds of millions of Ethiopians and those living downstream of the river depend on the availability of water in the Upper Blue Nile basin, which is strongly influenced by the sensitivity and variability of climatic conditions (Gelete *et al.* 2019; Mengistu *et al.* 2021). The basin is expected to experience year-to-year temperature and rainfall fluctuations in the future, which may change the patterns and trends of hydrology and water resources in the basin (Cherie 2013). It is very important to understand the possible effects of climate change on hydrological components to understand such changes in water resources (Singh & Saravanan 2020; Fatehifar *et al.* 2021).

Several studies, including studies by Taye *et al.* (2011), Dile *et al.* (2013), Gebre (2015), Fentaw (2018), Worqlul *et al.* (2018), and Boru *et al.* (2019), studied the effects of climate change on the water resources of the Upper Blue Nile sub-basin. Some studies such as Koch & Cherie (2013), Mellander *et al.* (2013), Haile *et al.* (2017), Roth *et al.* (2018), Mengistu *et al.* (2021), and Tariku *et al.* (2021) also studied the potential impact of climate change on the water resources of the entire basin. These studies generally investigated climate change and its consequent hydrological features. Some studies have used one or more emission scenarios, different downscaling methods (i.e., statistical or dynamical), and bias-corrected or raw data without bias correction. These methodological differences can really determine the outcome of an impact study. For example, some studies have simulated a decrease in runoff (Koch & Cherie 2013; Haile *et al.* 2017) and water availability (Koch & Cherie 2013; Mengistu *et al.* 2021) as a result of reduced water supply. Other studies (Roth *et al.* 2018) have simulated increases in discharge and water availability due to increased rainfall and water production.

This indicates that there is no general agreement on the impacts of climate change on water resources in the Upper Blue Nile basin. It challenges policymakers, water resource managers, and the entire community. So, continuous efforts are required to understand the hydrological processes and develop efficient water management strategies under changing environmental conditions (Zhang *et al.* 2018). To assess the impact of climate change on water resources, different emission scenarios and preferred bias correction techniques should be used. Hence, further research is needed to understand the impact of climate change on water resources and to expand knowledge of effective adaptation measures and other water-related development plans in the Upper Blue Nile basin.

The main objective of this study was to assess the impact of climate change on the water resources of the Upper Blue Nile basin. This study creates climate change scenarios for the 2030s (2021–2040) and 2050s (2041–2060), examines climate change trends, especially rainfall and temperature changes, and evaluates how changes in rainfall and temperature affect changes in streamflow and hydrological components using Soil and Water Assessment Tool (SWAT) model products. Since the livelihoods of the majority of the population in the Upper Blue Nile basin are based on rain-fed agriculture, there is an urgent need for detailed information on climate change and variability.

To this end, the bias-corrected regional climate models (RCMs) have been integrated into the SWAT model in various climatic scenarios. Therefore, the results of the study are important for the development of robust hydrological scenarios for water resource management, water security planning, and the design of optimal climate change strategies for the proposed water developments in the Upper Nile Blue basin. In addition, the methodology presented in this document may be a useful approach for studying the impact of climate change on the water resources of other river basins. The contents of the rest of this paper are as follows: Section 2 highlights the study area, data sets, and methods, while Section 3 presents the results and corresponding discussions. Finally, conclusions and recommendations are presented in Section 4.

2. METHODS AND MATERIALS

2.1. Description of the study area

The study was conducted in the Upper Blue Nile basin, the primary subbasin of the Nile River basin located in the highland region of Ethiopia. The absolute location of the basin covers a 7°44′32″–12°45′19″N latitude in the south-north direction and a 34°29′20″–39°48′17″E longitude in the west-east direction (Figure 1). The drainage area of the basin at the border station is about 174,166 km^2 (Takele *et al.* 2021). The basin covers 10% of the Nile basin and produces approximately 60% of the water that reaches Egypt and Sudan (Roth *et al.* 2018). Approximately 55.94% of the Upper Blue Nile basin is covered by agricultural land, 18.04% by closed forest, 14.5% by mixed forest, 6.12% by shrubs, and the remaining 5.4% by other land-use classes (Takele *et al.* 2021). Soil types in the basin include eutric nitosols (30%), eutric cambisols (24%), humic cambisols (16%), cambic arenosols (14%), dystric cambisols (5%), eutric regosols (3%), water bodies (2), eutric fluvisols (2), chromic vertisols (2%), pellic vertisols (1%), and orthic acrisols (1%).

The basin experiences a monomodal rainfall pattern from June to September and receives about 74% of the annual rainfall during the rainy season. The average annual rainfall in the basin ranges from 770 mm at the eastern end of the basin to 2,117 mm at the southern end. The seasons in this area are named concerning rainfall: from September to November is Tsedey (in Amharic) which means the post rainy of Autumn; October to February is Bega which means the dry winter

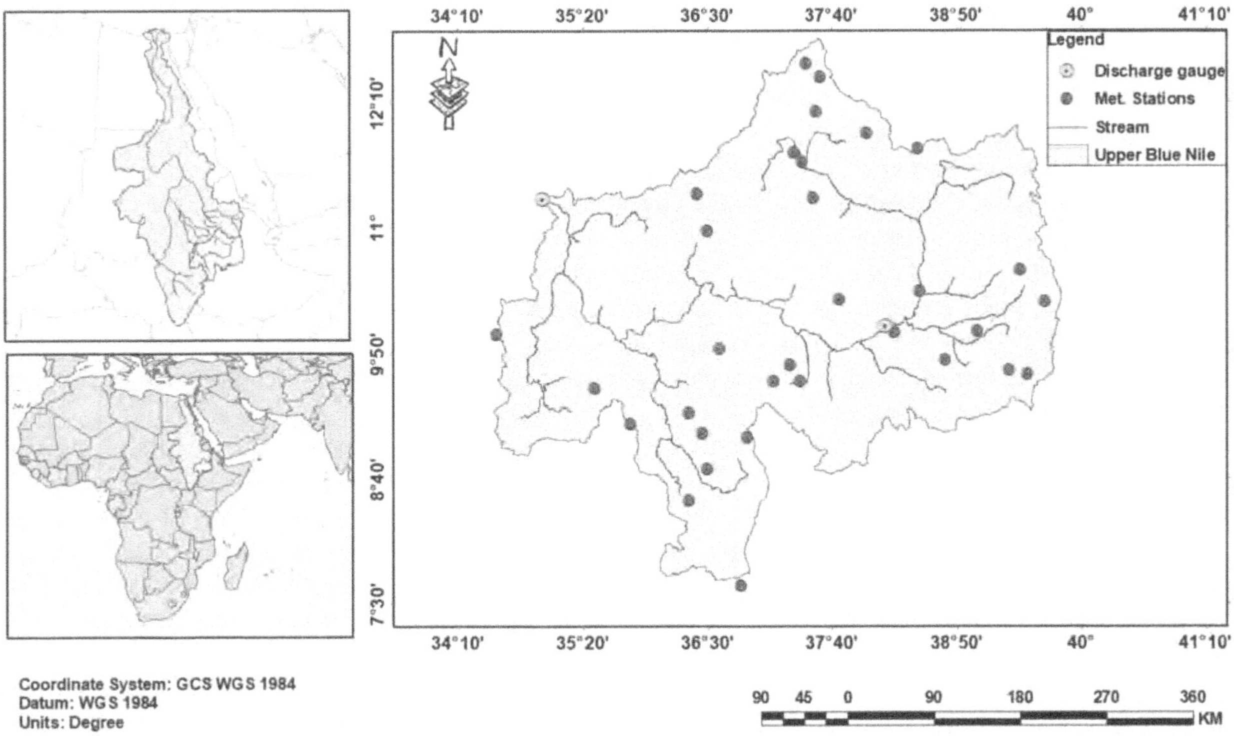

Figure 1 | Map of the Upper Blue Nile Basin.

season; from March to May is Belg which means the small rains of spring; June to September is Kiremt that means the wet summer season (Mellander *et al.* 2013).

2.2. Input data

SWAT hydrological simulations at the watershed or catchment level require spatial and temporal data. Temporal data, such as climatic and hydrological data, are used to build the model and hydrological processes, as well as spatial data such as the digital elevation model (DEM), the land-use map, and the soil map. The Alaska Satellite Facility (ASF) provided the DEM with a 12.5-m resolution, and the DEM was used to extract the drainage pattern of the catchment area of the basin. In addition, the DEM was used to derive the subarea parameters required for the SWAT model, such as the steepness of the slope, the length of the slope of the terrain, and the properties of the stream network.

The Copernicus Global Land Service (CGLS) provided a land-use map of the study basin with a resolution of 100 m. The extracted land-use map contains 10 land-use classes, which are agricultural land, bare land, closed forest, grassland, mixed forest, open forest, shrub, urban, water, and wetland (Takele *et al.* 2021). The physical and chemical properties of the soil map of the study basin were taken from the database of the World Food and Agriculture Organization of the United Nations (FAO).

The National Meteorological Services Agency (NMSA) of Ethiopia provided daily weather data for 32 stations from 1987 to 2016. A double mass curve procedure was used to verify the consistency of the data. Handling of missing data should be prioritized in data preparation procedures, especially in an event where all available resources, including partial information, must be used (Hamzah *et al.* 2021). The missing data were filled and homogeneity analysis was performed in R with the CLIMATOL v3.3 package (Takele *et al.* 2021). These data were used for bias correction for the RCM simulation, the impact assessment, and the SWAT model simulation. The Ministry of Water, Irrigation and Energy of Ethiopia provided daily streamflow data for the Kessie and Border stations from 2001 to 2014. These data were used to calibrate and validate the SWAT model and generate a water balance component for the reference period.

In this study, simulated rainfall and temperature data from the Coordinated Regional Climate Downscaling Experiment (CORDEX) Africa project were used. The latest RCM of the Rossby Centre regional atmospheric model (RCA) was used

Table 1 | Description of the selected RCMs

Institute	GCM name	Abbreviated name
Centre National de Recherches Météorologiques (France)	CNRM-CRAFACS-CNRM-CM5	CNRM-CM5
Consortium of European Research Institution and Researchers	ICHEC-EC-EARTH	EC-EARTH
National Institute for Environmental Studies, and Japan Agency for Marine-Earth Science and Technology (MIROC), Japan	MIROC-MIROC5	MIROC5
Max Planck Institute for Meteorology (Germany)	MPI-M-MPI-ESM-LR	MPI-ESM-LR

to generate these climate data sets, which were forced by four global climate models (GCMs): CNRMC-M5, EC-EARTH, MPI-ESM-LR, and MIROC5. Table 1 contains a description of the selected RCM. Two emission scenarios (RCP4.5 and RCP8.5) of the representative concentration pathway (RCP) were taken into account: RCP4.5 is a medium stabilization scenario with a carbon dioxide equivalent concentration range of 580–720 ppm in 2100, while RCP8.5 is a very high emission scenario with a carbon dioxide equivalent concentration of 1,000 ppm at 2100 (IPCC 2014).

2.3. Bias correction of the RCM simulation

GCMs are the most important source of information for the development of climate scenarios and form the basis for assessing the effects of climate change at all levels, from local to global. Although regional-scale climate models represent important mesoscale meteorological and climate characteristics much better than global models, the simulations are still subject to biases. RCM relies on higher-level GCM, so it inherits some, if not all, of the bias. Some may argue that the RCM is better suited for regional impact studies. However, because RCM is controlled by GCM as a boundary condition, it is affected by the robustness and accuracy of GCM for the particular area under investigation (Taye *et al.* 2018). And they are sensitive to the boundary conditions where the RCM is embedded (Tapiador *et al.* 2020).

A common drawback of dynamic RCM data is the overestimation of days with very low rainfall (Teutschbein & Seibert 2012). This issue is related to the size of the RCM grid cells combined with the convection of moist air. Humid air is completely saturated at certain altitudes as the temperature drops, resulting in massive precipitation within the RCM (Willkofer *et al.* 2018). In addition, RCM requires high computational work, and their effectiveness depends primarily on the input observation data (Willkofer *et al.* 2018).

Several statistical methods have been developed to correct these biases, including transformation algorithms that are statistically consistent with the observations (Willkofer *et al.* 2018). The bias correction methods, which require the correction algorithm and its parameterization for current weather conditions, also apply to future stationary weather conditions (Teutschbein & Seibert 2012). The purpose of using bias correction methods is to bring the climate model simulations closer to the actual observations during a specific reference period (Chen *et al.* 2020). Furthermore, a bias correction is ultimately performed to provide realistic climate change scenarios to assess the effects of climate change. To obtain acceptable hydrological simulations, these distortions must be eliminated.

In this study, the bias correction of the RCM simulations was carried out using the climate model for the hydrological modeling tool (CMhyd) at the station level. This tool provides simulated climate data representing meteorological gauges used in setting up a watershed model (Rathjens *et al.* 2016). The CMhyd is ideal for preparing simulated climate variables for SWAT model climate change impact studies (Zhang *et al.* 2018). The tool suggested several methods of bias correction, including linear scaling, nonlinear scaling, and distribution mapping. In this study, RCM-simulated rainfall and temperatures were extracted and corrected using the linear scaling method. It works with monthly corrections based on the difference between the observed data and the raw data (in this case the raw data simulated by RCM). Rainfall is usually corrected monthly by a multiplier, and temperature is corrected by an additive term (Fang *et al.* 2015). We chose this method based on simplicity, accuracy, parameter considerations, and reliability results from observational data (Tukimat 2018). In addition, the linear scaling method is suitable for water resources research in the context of climate change. Therefore, a simple method of bias correction is accessible to more users in the water resources research community (Shrestha *et al.* 2017).

2.4. Hydrological modeling

SWAT is a semi-distributed, physical based, continuous-time hydrological model (Arnold *et al.* 2012), and it was designed to study the impact of land management on water, sediment, and pesticide yields in ungauged catchments (Gassman *et al.* 2007; Devia *et al.* 2015). Physiological data used in the model include elevation, land and soil data, climate inputs such as daily rainfall, maximum, and minimum temperatures, humidity, solar radiation, and wind speed.

The model divides the study area into several subbasins, and the subbasins can be further divided into a series of hydrological response units useful for more detailed hydrological studies (Yu *et al.* 2018). The hydrological simulation of the SWAT model can be performed in two stages: the first controls the amount of water, sediments, and nutrients that flow from each subbasin into the main channel of the basin, and the second describes the movement of water, sediments, and nutrients through the canal from the watershed areas, which is determined by the surface topography (Neitsch *et al.* 2011; Arnold *et al.* 2012).

The hydrological effects of climate change in the Upper Blue Nile basin were evaluated using the SWAT model previously calibrated by Takele *et al.* (2021). The model ran from 1987 to 2016. The model was successfully calibrated and validated using the observed flow from 2001 to 2009 and 2010 to 2014 at the Kessie and Border stations. The SUFI2 optimization algorithm from SWAT CUP ver2012 was used to identify the sensitive parameters of the model and calibrate it. During the calibration and validation periods, the performance of the model is evaluated using the most used statistical evaluation indices such as the Nash–Sutcliffe efficiency (NSE) (Nash & Sutcliffe 1970), the coefficient of determination (R^2) (Krause *et al.* 2005), and the percent of bias (PBIAS) (Gupta *et al.* 1999). Further details on model parameterization and performance are available in Takele *et al.* (2021).

2.5. Estimating climate change impact on hydrology

The prediction of changes in climate variables was required to assess the impact of climate change on the hydrology of the study area. Therefore, this study used the bias-corrected RCM output (rainfall and temperature) under the RCP4.5 and RCP8.5 emission scenarios as inputs to the hydrological model. The major hydrological components such as surface runoff (SURQ), potential evapotranspiration (PET), water yield (WYLD), groundwater flow, and lateral flow were estimated. However, the first three key parameters were evaluated in detail using a SWAT model calibrated under baseline climate (1987–2016) and future climate scenarios (the 2030s (2021–2040) and the 2050s (2041–2060)). Furthermore, for streamflow, change detection is done annually, and for hydrological components, it is done monthly.

2.6. Trend analysis of climate variables

Trend analysis is considered one of the most critical topics in global climate studies (Yacoub & Tayfur 2018). Various statistical techniques have been used in studies to identify trends and other changes in climate and hydrological variables. These techniques are classified as parametric and non-parametric. Parametric methods assume a normal distribution of variables, non-parametric methods do not. Therefore, in this study, the non-parametric Mann–Kendall test was used to identify monotonic trends and statistical distributions of climatic variables (rainfall and temperature). The Mann–Kendall test was used for this study because it makes no assumptions about the distribution of the data, is resistant to outliers, and does not require the removal of outliers prior to trend detection.

The Mann–Kendall test is calculated as follows:

$$Z_s = \sum_{k=1}^{n-1} \sum_{j=k+1}^{n} \operatorname{sgn}(x_j - x_i) \tag{1}$$

$$\operatorname{sgn}(x) = \begin{cases} 1 & \text{if } x > 0 \\ 0 & \text{if } x = 0 \\ -1 & \text{if } x < 0 \end{cases} \tag{2}$$

where x_j and x_i are the time-series data, the annual values in year j and i correspondingly, sgn is the signum function that takes on the values 1, 0, or −1. A positive (negative) value of Z_s signifies an increasing (decreasing) trend in the time-series data.

In this study, the Theil–Sen slope estimator was used to compute the linear rate of change, or the magnitude of change in the variable at any given time, each year. As a result, the Theil–Sen test computes the median of the slopes (β) as follows:

$$\beta = \text{median}\left(\frac{x_j - x_i}{j - i}\right) \tag{3}$$

where i=1 to n−1, j=2 to n, and x_j and x_i are the data values at time j and i (j>i), respectively.

3. RESULTS AND DISCUSSION

3.1. SWAT model calibration and validation

The SUFI-2 algorithm was used to perform the sensitivity analysis for the entire catchment area. Twenty-six hydrological parameters were tested, of which 13 were found to be the most sensitive. SCS runoff curve number, deep aquifer percolation fraction, threshold depth of water in the shallow aquifer for 'revap' to occur, groundwater delay, soil evaporation compensation factor, effective hydraulic conductivity in main channel alluvium, baseflow α-factor, groundwater 'revap' coefficient, available water capacity of the soil layer, saturated hydraulic conductivity, threshold depth of water in the shallow aquifer required for return flow to occur, maximum canopy storage, and manning's n value for the main channel were the most hydrologically sensitive parameters. Takele *et al.* (2021) document the details of the sensitive flow parameters and their fitted values.

The selected parameters were calibrated and validated at the Kessie and Border gauging stations. There was a good agreement between the observed and simulated discharges, as indicated by the high values of the R^2 and NSE, and low PBIAS. Takele *et al.* (2021) summarized the performance of the model at the Kessie and Border stations during the calibration and validation periods. Takele *et al.* (2021) identified the components of the water balance of the basin. The results showed that evapotranspiration lost 50.17% of precipitation during the calibration period and 48.58% during the validation period. During the calibration and validation periods, SURQ contributes 44.95 and 45.26%, respectively, to the watershed's WYLD, while groundwater contributes 40.15 and 40.28%, respectively, to the WYLD.

3.2. Bias correction of the RCMs

In order to increase the predictability of climate models, it is often necessary to correct the deficiencies of the current GCM/RCM, which is useful as an input to assess the impacts of climate change. Figures 2 and 3 show the performance of the bias correction for the 30-year series of climatic averages for rainfall and temperature of the uncorrected (raw) values of the selected models and the bias-corrected values. The linear scaling method was used to adjust the discrepancy between the RCM-simulated temperature and rainfall data with the measured data.

The method has adjusted the extreme values of rainfall and has also adjusted the distribution of monthly rainfall, in particular the maximum values of rainfall. As a result, once the bias is corrected, the climate model simulations are closer to the values observed during a given reference period (1987–2016). In this case, the bias correction adjusts the RCMs simulation and the predictability of the average monthly rainfall (Figure 2). Other studies have examined how bias correction methods can be used to improve climate model simulations (Luo *et al.* 2018; Chen *et al.* 2020; Mendez *et al.* 2020). Similarly, the underestimation and overestimation of the RCM simulations for TMAX and the overestimation of TMIN were adjusted using the linear scaling bias correction method with the observed average monthly TMAX and TMIN (Figure 3).

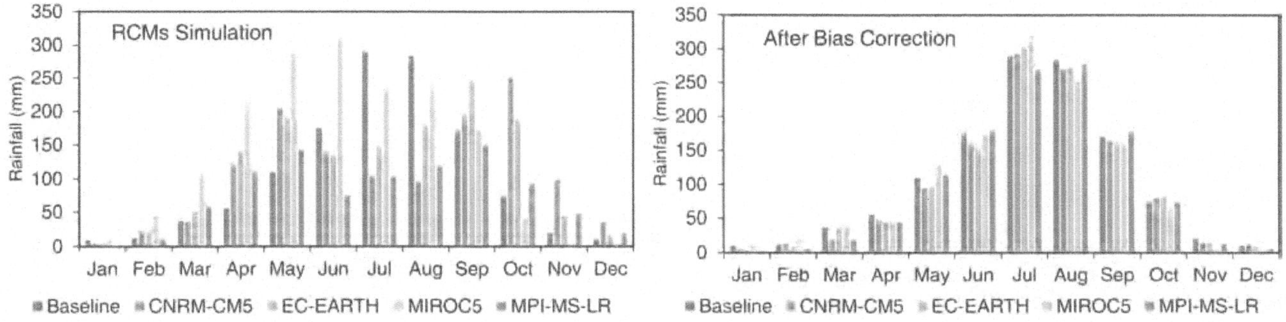

Figure 2 | Bias correction performance in adjusting the average monthly rainfall of the RCMs in the Upper Blue Nile basin.

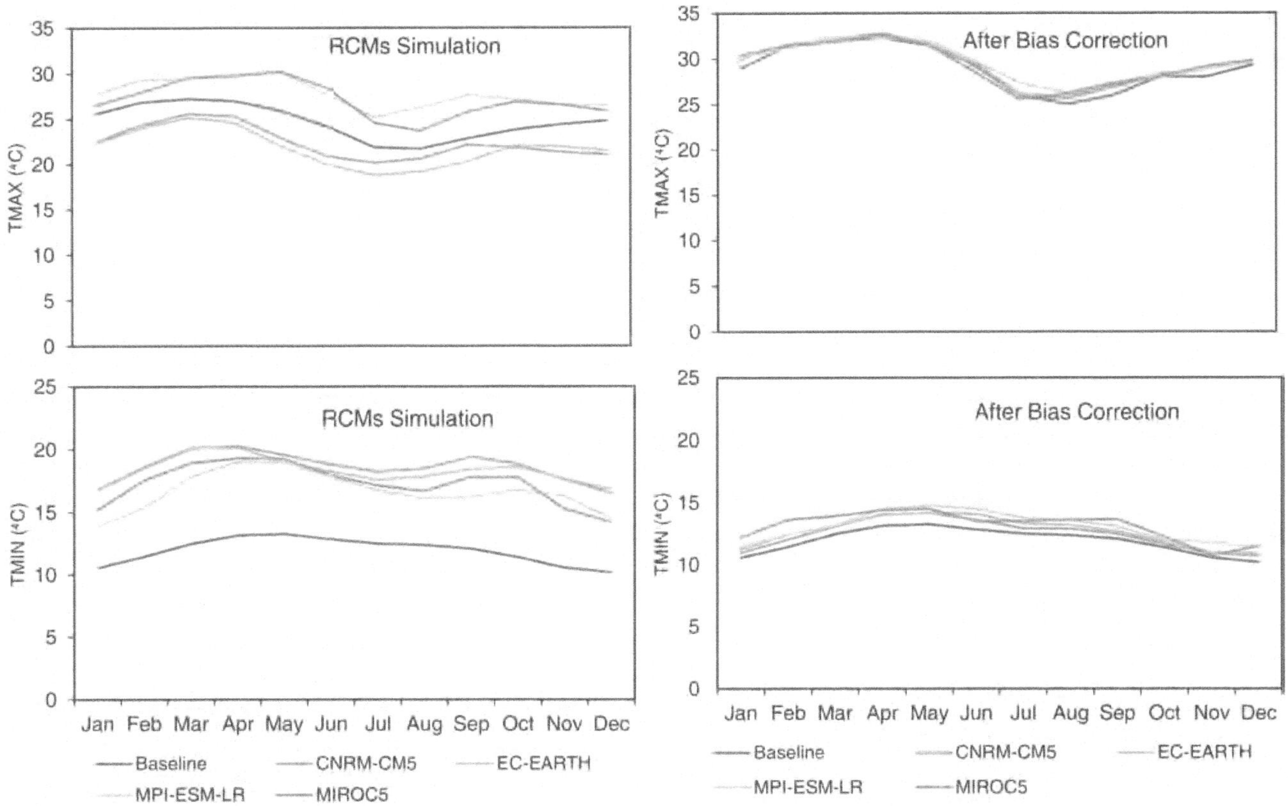

Figure 3 | Bias correction performance in adjusting the average monthly TMAX and TMIN of the RCMs in the Upper Blue Nile basin.

3.3. Future climate projection

Future climate change scenarios have been developed for the Upper Blue Nile basin, taking into account various RCM and emission scenarios. The average annual rainfall in RCMs shows a visible change of −13 to +6% in all scenarios and scenario periods (Table 2). The rainfall change signals show significant differences between the RCMs for the 2030s and 2050s under different emission scenarios. In the RCP8.5 emission scenario, changes range from −10 to +2% in the 2030s and from −13 to +6% in the 2050s. In this emission scenario, rainfall is expected to decline further in the 2050s (Table 2). In the RCP4.5 emission scenario, rainfall is expected to decline sharply (i.e., −8%) under the climatic conditions of the 2050s.

During the scenario periods, significant seasonal fluctuations in rainfall are observed. Table 3 shows the mean seasonal rainfall and the corresponding percentage changes predicted by all climate models in the 2030s and 2050s in both RCP emission scenarios, along with the reference period. In projection periods, rainfall is expected to decrease more in autumn (September–November), following spring (March–May). Rainfall forecasts for the main rainy season of the basin

Table 2 | Percentage change in the average annual rainfall of RCMs in the Upper Blue Nile basin under the RCP4.5 and RCP8.5 emission scenarios for the scenario periods (2030s and 2050s)

Change in Rainfall (%)

	2030s (2021–2040)		2050s (2041–2060)	
RCM runs	**RCP4.5**	**RCP8.5**	**RCP4.5**	**RCP8.5**
CNRM-CM5	−6	−3	−6	−6
EC-EARTH	−5	−10	−8	−13
MPI-ESM-LR	−3	−1	−7	−4
MIROC5	−1	2	−1	6
Average	−4	−3	−5	−4

Table 3 | Seasonal variation in projected rainfall under the RCP4.5 and RCP8.5 emission scenarios in the Upper Blue Nile basin for the baseline period

Season	Baseline (1987–2016)	2030s (2021–2040) RCP4.5	2050s (2041–2060) RCP4.5	2030s (2021–2040) RCP8.5	2050s (2041–2060) RCP8.5
Summer (June–August)	749	756 (1%)	754 (1%)	763 (2%)	749 (0)
Autumn (September– November)	264	222 (−16%)	199 (−25%)	218 (−17%)	212 (−20%)
Winter (December–February)	32	33 (3%)	32 (0%)	38(17%)	39(21%)
Spring (March–May)	202	187 (−7%)	192 (−5%)	190 (−6%)	188 (−7)

(June–August) show no significant changes. In the RCP8.5 emission scenario, winter precipitation is expected to increase by up to 17% in the 2030s, up to 21% in the 2050s, and 3% in the RCP4.5 emission scenario of the 2030s.

Individually, in the RCP8.5 emission scenario, almost all RCMs, with the exception of MIROC5 in the 2050s, show a decrease in average monthly rainfall in the 2030s and 2050s. In the RCP4.5 emission scenario, all RCMs show a decrease in average monthly rainfall over the scenario periods. Other studies have reported on the unpredictability of future rainfall in the basin. Elshamy *et al.* (2009), for example, reported expected changes in the basin's total annual rainfall ranging from −15 to +14% using 17 GCMs, but more models report reductions (10) than increase (7) by the end of the 21st century. According to Kim & Kaluarachchi (2009), the average annual rainfall over the basin will vary from −11 to +44% in the mid-century for the A2 emission scenarios. Tariku *et al.* (2021) estimate that the basin's precipitation will vary from −10.3 to +19.4% under the RCP4.5 and RCP8.5 emission scenarios in the 2050s and 2080s.

In future climate scenarios, each climate model predicts an increase in the average annual maximum and minimum temperatures in both emission scenarios (Table 4). The average annual maximum temperature of the basin rises from 0.96 to 2.04 °C, and the average annual minimum temperature rises from 0.95 to 2.1 °C. There are some differences in emission scenarios, and TMAX and TMIN are expected to constantly increase at RCP8.5 instead of RCP4.5. However, the rate of increase in TMAX and TMIN between RCMs is consistent across emission scenarios. TMAX and TMIN are expected to increase significantly in the climate of the 2050s compared to the climate of the 2030s. In both emission scenarios, all individual RCMs show an increase in TMAX and TMIN over the 2030s and 2050s. Previous research in the Upper Blue Nile basin shows an increasing trend in the average annual temperatures (maximum and minimum) (e.g., Kim *et al.* 2008; Beyene *et al.* 2009; Elshamy *et al.* 2009; Cherie 2013; Haile *et al.* 2017; Liersch *et al.* 2018; Roth *et al.* 2018; Mengistu *et al.* 2021; Tariku *et al.* 2021).

3.4. Trend of rainfall and temperature under baseline and future scenarios

Table 5 shows baseline and future rainfall, TMAX, and TMIN annual trend statistics (Mann–Kendall test and Sen's slope estimator) for various climate scenarios in the study area. Based on the *Z*-test and Sen's slope values, baseline rainfall (1987–2016) shows an increasing trend, but it is not statistically significant. Nevertheless, rainfall in the 2030s and 2050s is declining and not statistically significant (Figure 4). The rate of decrease in rainfall during the 2030s forecast period is 0.26 mm/year in the RCP4.5 emission scenario and 4 mm/year in the 2050s in the RCP8.5 emission scenario (Table 5).

Table 4 | Average annual maximum and minimum temperatures in the RCMs of the Upper Blue Nile basin under the RCP4.5 and RCP8.5 emission scenarios for the scenario period (2030s and 2050s)

	Change in maximum temperature (°C)				Change in minimum temperature (°C)			
	2021–2040		2041–2060		2021–2040		2041–2060	
RCM Runs	RCP4.5	RCP8.5	RCP4.5	RCP8.5	RCP4.5	RCP8.5	RCP4.5	RCP8.5
CNRM-CM5	0.8	0.9	1.5	1.8	0.5	0.7	1.1	1.4
EC-EARTH	0.9	1.0	1.5	2.0	0.7	0.9	1.3	1.8
MPI-ESM-LR	1.04	1.30	1.66	2.27	1.28	1.44	1.88	2.61
MIROC5	1.1	1.2	1.7	2.1	1.3	1.5	1.9	2.6
Average	0.96	1.10	1.59	2.04	0.95	1.14	1.55	2.10

The results of the Mann–Kendall test at 1 and 5% significance levels show an increasing trend for baseline and future TMAX and TMIN in both emission scenarios. The value of Sen's slope indicates that the rate of change of TMIN is higher than the future TMAX and baseline temperature. TMAX and TMIN trends in different climate scenarios suggest that climate change is responsible for the rise in temperature in the Upper Blue Nile basin, even small changes in temperature and climate (i.e., low greenhouse gas emissions scenarios) can have a significant impact on water availability, especially the extremes (UNESCO, UN-Water 2020). The study also examined the similarities between the results of the statistical tests of Mann–Kendall and Sen's slope. These results are similar to studies in the region that showed an upward trend in the mean annual temperature (Worku *et al.* 2021). In general, changes in temperature and rainfall patterns as a result of climate change are expected to affect the water resources of the basin. In addition, the maximum changes in rainfall and temperature rise are observed at the highest concentration of greenhouse gases, RCP8.5.

3.5. Impact of climate change on streamflow

Table 6 summarizes the SWAT model's estimate of the impact of climate change on streamflow in the Upper Blue Nile basin. In the event of future climate change, annual streamflow may be reduced compared to the reference period. In the RCP4.5 emission scenario, the change in average annual streamflow ranges from −48 to −10% in the 2030s and from −35 to −7% in the 2050s, whereas in the RCP8.5 emission scenario, the change in average annual streamflow ranges from −49 to −4% in the 2030s and from −54 to +15% in the 2050s.

In different scenarios, there are significant differences in streamflow changes between RCMs in the 2030s and 2050s. Under the RCP8.5 emission scenario, a significant reduction in streamflow is simulated in the 2050s climate condition. High streamflow reduction is projected in the 2030s climate condition under the RCP4.5 emission scenario. Individually, almost all RCMs except MIROC5 showed a decrease in annual streamflow in the 2030s and 2050s in the RCP8.5 emission scenario. In general, the direction of streamflow changes in response to changes in rainfall (Table 2). According to Elshamy *et al.* (2009), there is some evidence of the impact of reduced rainfall on reduced runoff and average annual streamflow. For example, in Tables 6 and 7, a 14.2% decrease in average annual rainfall by EC-EARTH leads to a 30.8% decrease in average annual runoff and a 54% decrease in mean annual streamflow. On the other hand, a 7.2% increase in average annual rainfall by MIROC5 leads to a 33.8% increase in average annual runoff and a 15% increase in annual streamflow.

3.6. Climate change impact on water balance components

This section describes the main components of the water cycle: SURQ, PET, and WYLD. It has been observed that future climate change in rainfall and temperature will cause a change in the components of the water balance in the Upper Blue Nile basin. Annual SURQ decreased in some climate scenarios and increased in others. The total annual average SURQ simulated by SWAT across the entire basin in the 2030s and 2050s is shown in Table 7. As can be seen from the table, changes in SURQ range from −18.7% (2030s) to +33.8% (2050s) in the RCP4.5 emission scenario and from −26.9% (2030s) to +18.3% (2030s) in the RCP8.5 emission scenario.

Independently, CNRM-CM5 and EC-EARTH RCM show a decrease in SURQ, while MPI-ESM-LR and MIROC5 RCM show an increase over the projection periods in both emission scenarios. This variation may be due to changes in rainfall in future climatic conditions. Figure 5 shows the monthly mean distribution of SURQ and rainfall in various emission

Table 5 | Average annual rainfall, TMAX and TMIN trends in the baseline and projected periods under the RCP4.5 and RCP8.5 emissions scenarios in the Upper Blue Nile basin

	Rainfall		TMAX		TMIN	
Period	Z_s	β	Z_s	β	Z_s	β
Baseline (1987–2016)	0.021	0.292	0.687**	0.057	0.453**	0.023
RCP4.5 (2021–2040)	−0.021	−0.256	0.347*	0.02	0.400*	0.027
RCP4.5 (2041–2060)	0.011	0.095	0.305*	0.018	0.432**	0.028
RCP8.5 (2021–2040)	−0.021	−0.508	0.463**	0.048	0.589**	0.04
RCP8.5 (2041–2060)	−0.263	−4.042	0.653**	0.062	0.716**	0.052

Z_S: Mann–Kendall test, β: Sen's slope estimator.
*Statistically significant trend at 5% significance level. **Statistically significant trend at 1% significance level.

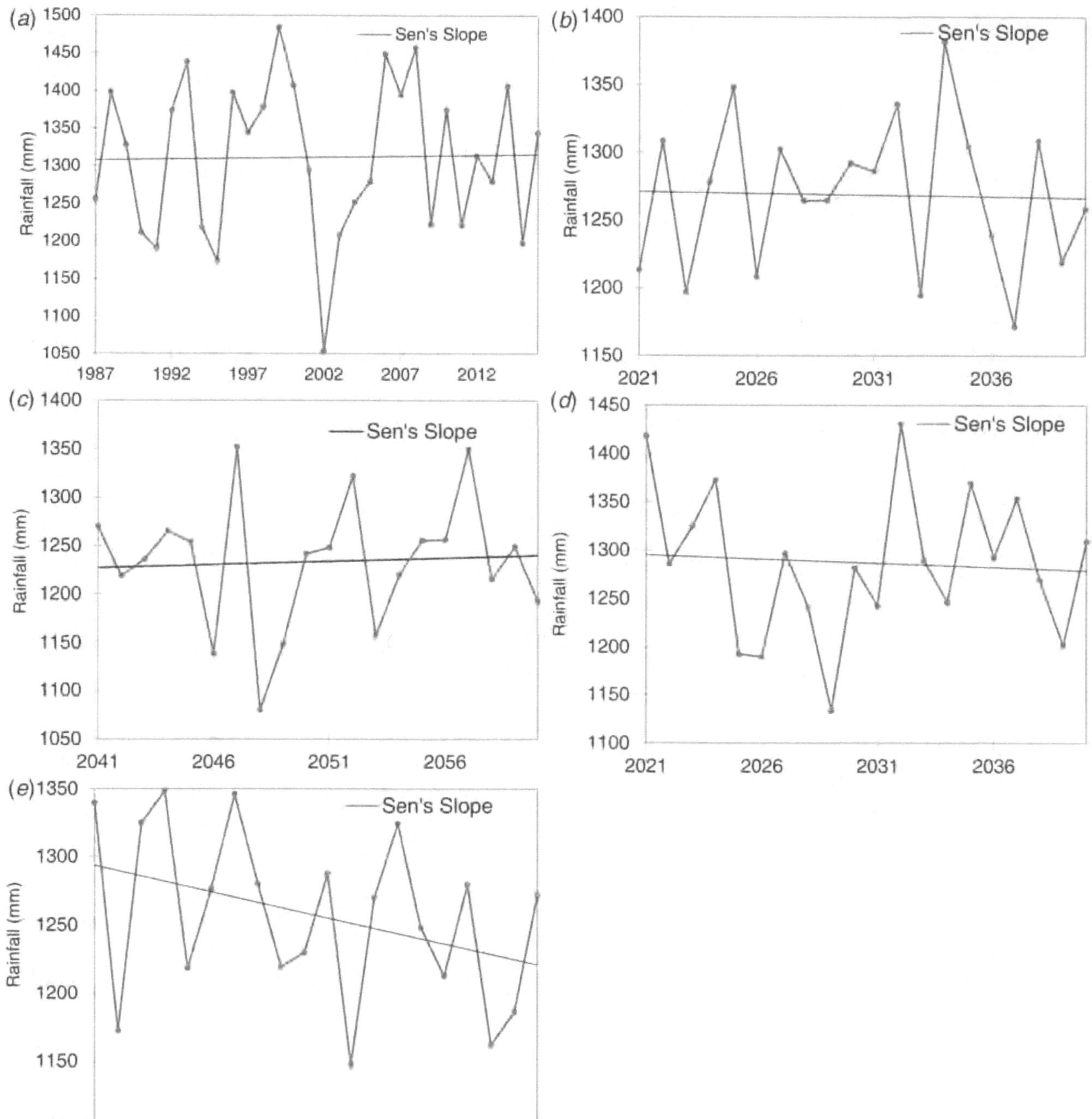

Figure 4 | Average annual precipitation trends in the base and future scenario periods under emission scenarios. a) Baseline period (1987-2016), (b) RCP4.5 (2030s), (c) RCP4.5 (2050s), (d) RCP8.5 (2030s), and (e) RCP8.5 (2050s).

scenarios, along with the reference period. As can be seen from the figure, in future climate scenarios, SURQ will decrease during the dry season (October–May) and increase during the rainy season (June–September). However, the MIROC5 run in the rainy season showed some changes in monthly SURQ. These results are consistent with other studies in the Upper Blue Nile basin, which predict an increase in SURQ during the rainy season (Mengistu *et al.* 2021). In general, the model result (Table 7) indicated an 8% change in rainfall, resulting in a 16% change in SURQ for the CNRM-CM5 and EC-EARTH RCMs, and a 3% increase in rainfall, resulting in a 22% increase in SURQ for the MPI-ESM-LR and MIROC5 RCMs.

Previous studies have linked changes in SURQ to changes in rainfall, temperature, and evapotranspiration (Kim *et al.* 2008; Elshamy *et al.* 2009; Mengistu & Sorteberg 2012; Cherie 2013). Mengistu & Sorteberg (2012) found that a 10% change in

Table 6 | Percentage change in the average annual streamflow of RCMs in the Upper Blue Nile basin under the RCP4.5 and RCP8.5 emission scenarios for the scenario periods (2030s and 2050s)

	Change in Streamflow (%)			
	2030s (2021–2040)		2050s (2041–2060)	
RCM Runs	RCP4.5	RCP8.5	RCP4.5	RCP8.5
CNRM-CM5	−14	−16	−16	−8
EC-EARTH	−48	−49	−35	−54
MPI-ESM-LR	−10	−8	−7	−2
MIROC5	−11	−4	−13	15
Average	−21	−19	−18	−12

Table 7 | Projected and percentage change in the water balance components of RCMs in the Upper Blue Nile basin under the RCP4.5 and RCP8.5 emission scenarios for the scenario periods (2030s and 2050s)

			Rainfall		Surf_Q		PET		Gw_Q		Lat_Q		WYLD	
		Time	mm	Δ%	mm	Δ%	mm	Δ%	mm	Δ%	mm	Δ%	mm	Δ%
RCMs	RCPs	Baseline	1,313	–	265	–	1,594	–	245	–	82	–	606	–
CNRM-CM5	RCP4.5	2030s	1,233	−6.1	229	−13.4	1,611	1.1	193	−21	71	−12.5	506	−16.5
	RCP8.5		1,281	−2.4	253	−4.5	1,595	0	202	−17.5	74	−9.3	539	−11.1
	RCP4.5	2050s	1,225	−6.7	236	−10.8	1,651	3.5	190	−22.3	72	−12.2	510	−15.9
	RCP8.5		1,227	−6.5	241	−9.1	1,667	4.6	183	−25.5	70	−13.8	509	−16
EC-EARTH	RCP4.5	2030s	1,239	−5.6	227	−14.2	1,608	0.9	200	−18.2	74	−9.4	513	−15.3
	RCP8.5		1,151	−12.3	193	−26.9	1,557	−2.4	178	−27.2	68	−16.9	450	−25.7
	RCP4.5	2050s	1,198	−8.7	215	−18.7	1,643	3	186	−24.1	71	−13.4	483	−20.3
	RCP8.5		1,127	−14.2	183	−30.8	1,664	4.4	158	−35.6	65	−20.1	416	−31.4
MPI-ESM-LR	RCP4.5	2030s	1,111	−15	283	6.9	1,687	5.8	190	−22.3	69	−15.0	554	−8.6
	RCP8.5		1,131	−14	306	15.4	1,698	6.5	204	−16.8	71	−13.4	592	−2.2
	RCP4.5	2050s	1,063	−19	280	5.7	1,728	8.4	190	−22.3	66	−24.1	527	−13.0
	RCP8.5		1,064	−19	289	9.0	1,760	10.4	167	−31.9	66	−19.6	531	−12.3
MIROC5	RCP4.5	2030s	1,323	0.7	313	18.3	1,604	0.6	265	8.3	86	5	679	12.1
	RCP8.5		1,356	3.3	320	20.9	1,605	0.7	269	9.8	87	6.1	691	14.1
	RCP4.5	2050s	1,314	0.1	310	17.1	1,632	2.3	263	7.4	86	4.8	674	11.2
	RCP8.5		1,407	7.2	354	33.8	1,638	2.7	276	12.8	90	10.1	737	21.6

rainfall resulted in a 19% change in SURQ in the Upper Blue Nile basin. Elshamy *et al.* (2009) found that a 10% change in PET resulted in a 14% change in SURQ. In addition, Elshamy *et al.* (2009) and Mengistu & Sorteberg (2012) assumed that a 1 °C increase in temperature would increase PET by 4% which leads to a decrease in SURQ. In addition, Krisnayanti *et al.* (2021) correlate changes in runoff depth with rainfall.

In addition to SURQ, other important factors in the water balance, such as PET and WYLD, are considered to investigate the effects of climate change in the Upper Blue Nile basin. PET increases in all climate scenarios due to rising temperature. Figure 6 shows the monthly mean distribution of PET and rainfall in various emission scenarios, along with the reference period. As a result of the driest and warmest conditions in the basin during the dry season (February–May), higher PET is estimated. In contrast, lower PET is estimated during the rainy season (June–September) in the basin as a result of increased cloud coverage and humidity associated with the rainy condition. According to Elshamy *et al.* (2009), there are signs that a significant decrease in rainfall may be associated with a significant increase in PET, probably as a result of a decrease in cloud cover. In Table 7, for example, a 14.2% decrease in mean annual rainfall by EC-EARTH results in a 4.4% increase in mean annual PET, while a 19% decrease in mean annual rainfall by MPI-ESM-LR results in a 10.4% increase in mean annual PET. A 7.2% increase in average annual rainfall with MIROC5 leads to a 2.7% increase in average annual PET.

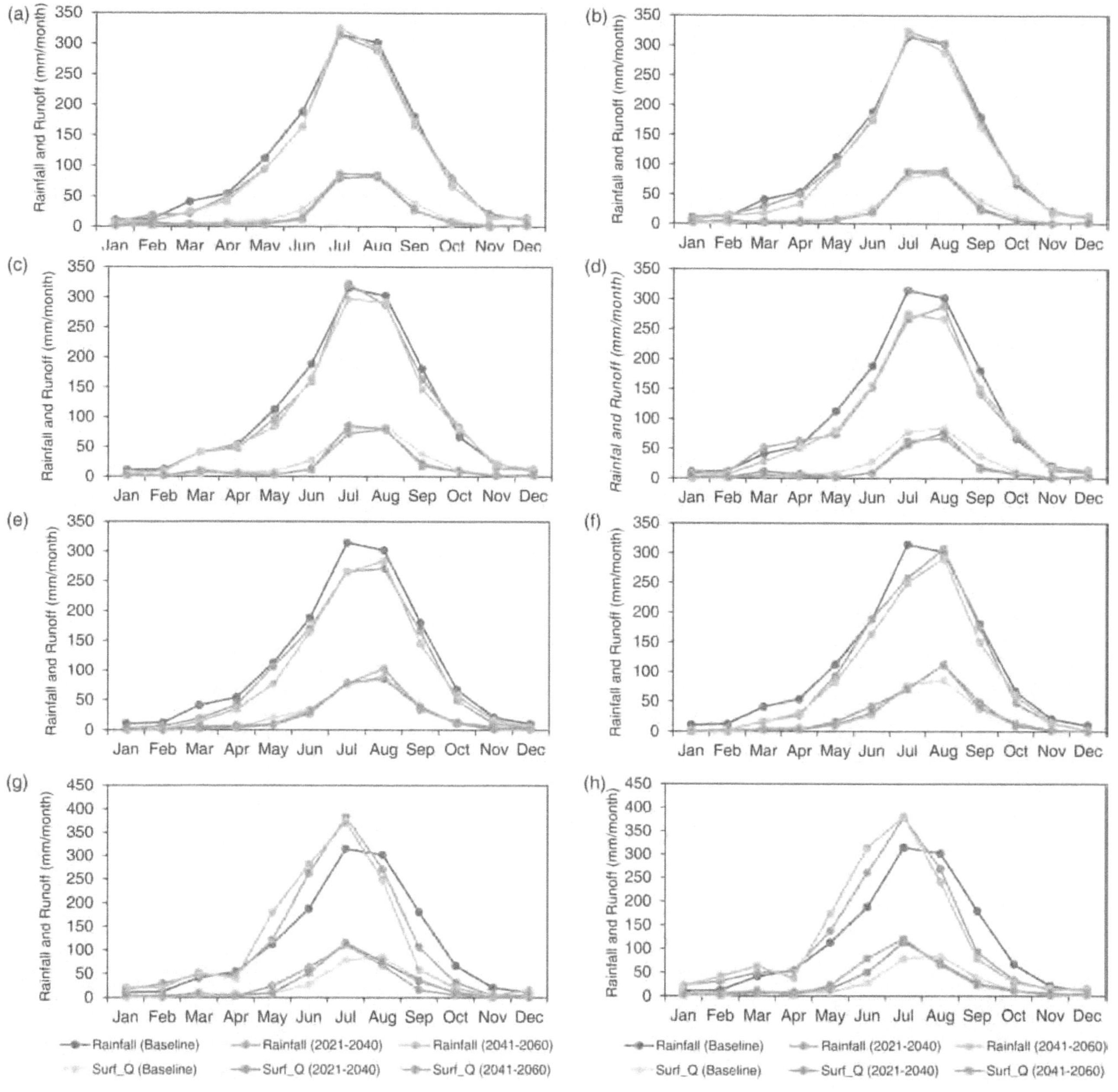

Figure 5 | Monthly distribution and changes in rainfall and surface runoff (Surf_Q) for (a) CNRM-CM5 under RCP4.5, (b) CNRM-CM5 under RCP8.5, (c) EC-EARTH under RCP4.5, (d) EC-EARTH under RCP8.5, (e) MPI-ESM-LR under RCP4.5, (f) MPI-ESM-LR under RCP8.5, (g) MIROC5 under RCP 4.5, and (h) MIROC5 under RCP 8.5.

WYLD was also taken into account in this study, as it can represent the availability of water resources in the basin. Due to climate change, the projected WYLD of the Upper Blue Nile basin may change in the 2030s and 2050s. With the exception of MIROC5 RCM, the decline in WYLDs under climatic conditions in the 2030s and 2050s was estimated under the RCP emission scenarios (Table 7). Predicted WYLDs in the 2030s are low in the RCP8.5 emission scenario and high in the RCP4.5 emission scenario. The decrease in WYLD is mainly due to the expected decrease in rainfall and the consequent SURQ.

Previous studies have linked a decrease in water from the basin to a decrease in rainfall, an increase in evaporation, and an increase in temperature. Mengistu *et al.* (2021) combine a decrease in WYLD and rainfall, an increase in evapotranspiration loss due to rising temperatures, followed by a decrease in groundwater recharge due to a decrease in lateral and

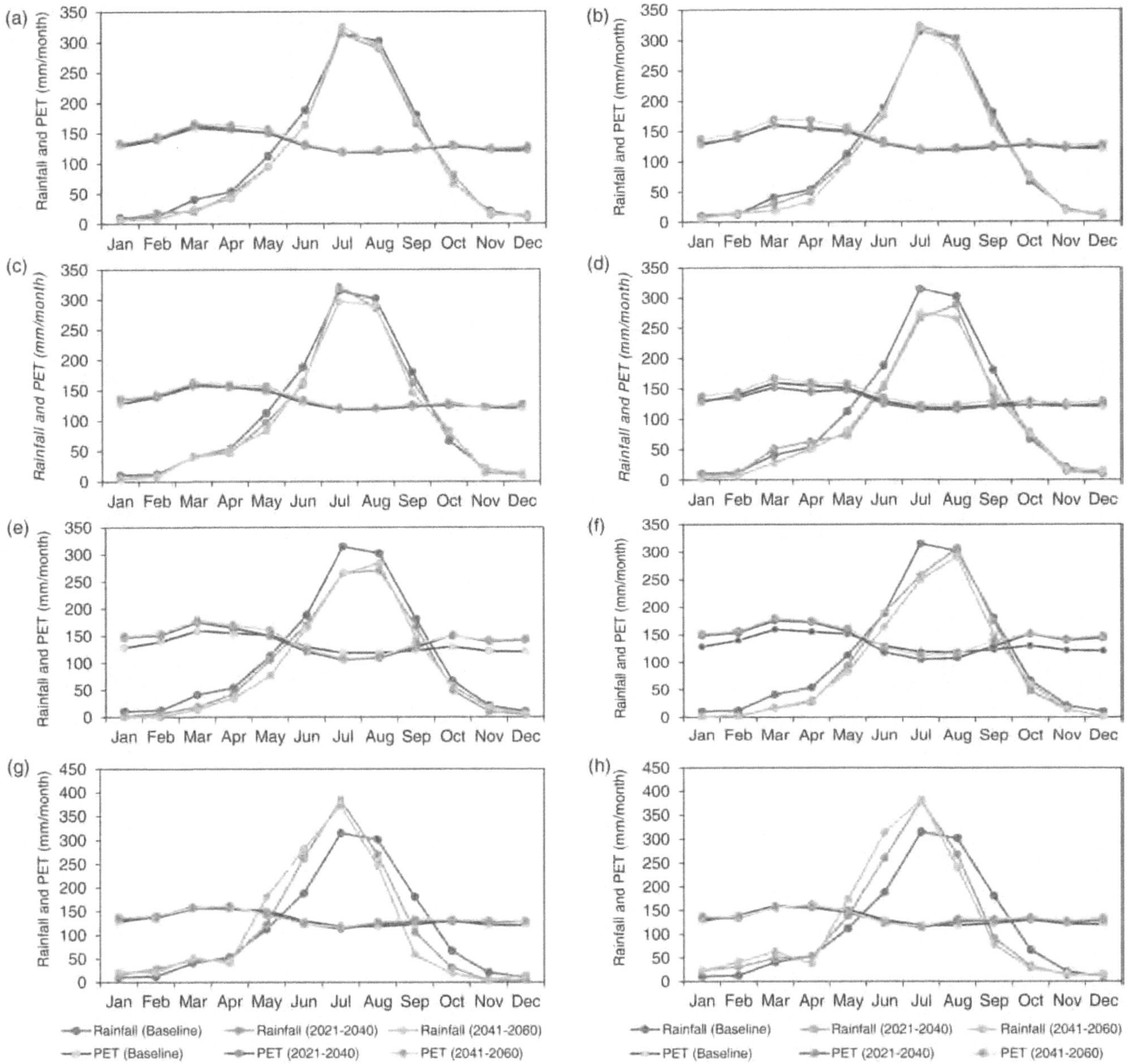

Figure 6 | Monthly distribution and changes in rainfall and potential evapotranspiration (PET) for (a) CNRM-CM5 under RCP4.5, (b) CNRM-CM5 under RCP8.5, (c) EC-EARTH under RCP4.5, (d) EC-EARTH under RCP8.5, (e) MPI-ESM-LR under RCP4.5, (f) MPI-ESM-LR RCP8.5, (g) MIROC5 under RCP 4.5, and (h) MIROC5 under RCP 8.5.

baseflow. MIROC5 is expected to have a higher WYLD under the RCP4.5 and RCP8.5 emission scenarios, owing to the higher projected rainfall and SURQ. According to RCM runs, the estimated monthly WYLDs in the basin are not constant throughout the year. Figure 7 shows the monthly mean distribution of WYLD and rainfall in various emission scenarios, along with the reference period. The result shows higher estimated WYLDs during the rainy season (June–September) due to lower PET, higher rainfall, and higher SURQ. Compared to the rainy season, the dry season (October–May) is projected to have less WYLD due to higher evapotranspiration, lower rainfall, and lower SURQ. Further according to Pulighe *et al.* (2021), a decrease in WYLD is associated with an increase in temperature and the consequent decrease in humidity.

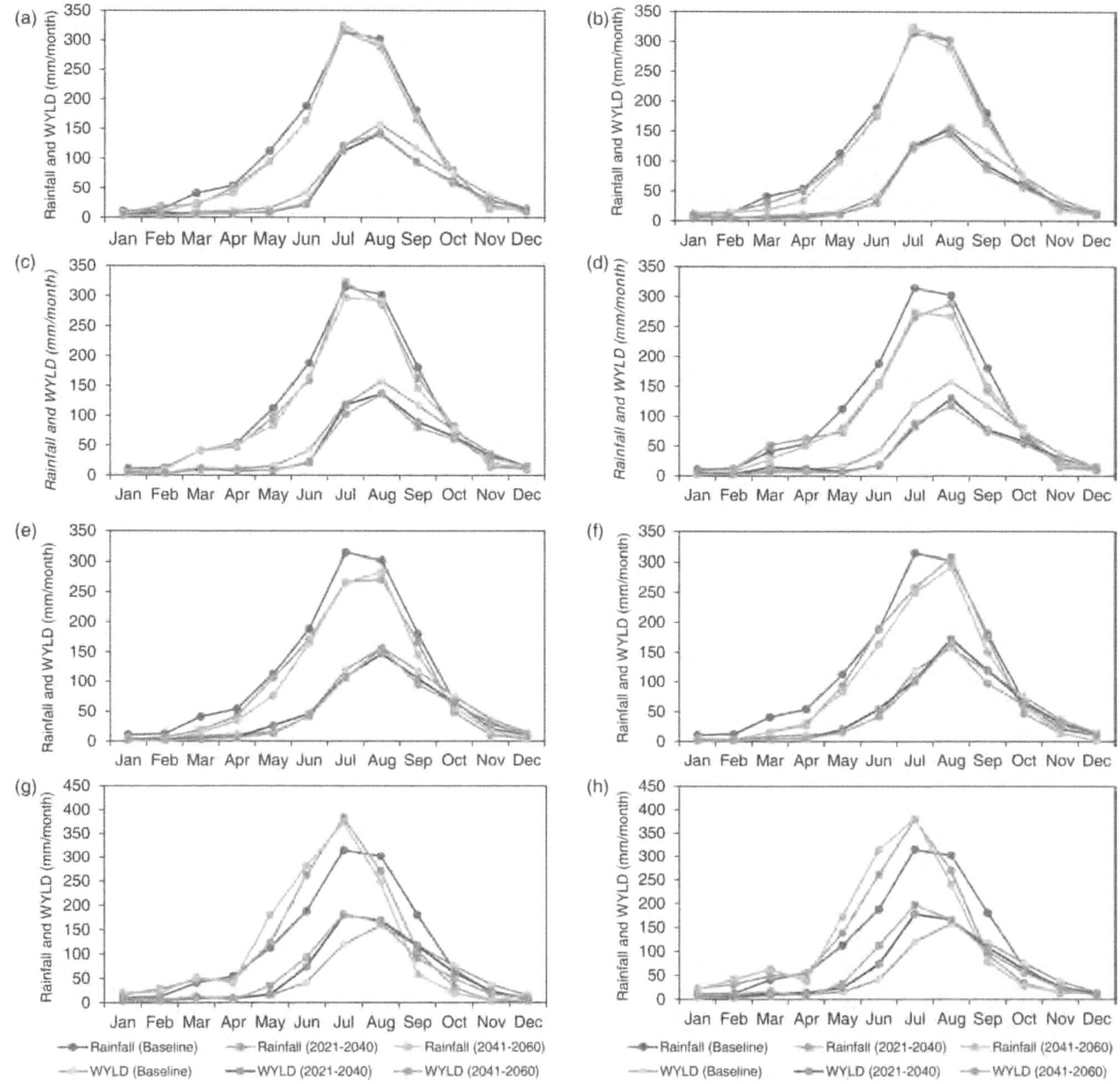

Figure 7 | Monthly distribution and changes in rainfall and water yield (WYLD) for (a) CNRM-CM5 under RCP4.5, (b) CNRM-CM5 under RCP8.5, (c) EC-EARTH under RCP4.5, (d) EC-EARTH under RCP8.5, (e) MPI-ESM-LR under RCP4.5, (f) MPI-ESM-LR under RCP8.5, (g) MIROC5 under RCP 4.5, and (h) MIROC5 under RCP 8.5.

4. CONCLUSION

During the rainy season, about 80% of the Nile water comes from the Upper Blue Nile basin. The water resources of the basin are used for irrigation, hydropower, and domestic purposes. Climate change can affect the water resources of the basin, as rainfall is the main source. The bias-corrected RCMs have been integrated into a semi-distributed SWAT hydrological model to assess the future climate and water resources of the Upper Blue Nile basin for the years 2021–2040 (2030s) and 2041–2060 (2050s).

During the scenario periods, the projected average climate shows a decrease in rainfall and an increase in temperature. The rainfall change signals show differences between the selected climate scenarios. Seasonally, predicted rainfall provides various signals of future climatic conditions. In all climate conditions, especially under the RCP8.5 scenario, the projected rainfall increases during the winter season. However, the projected rainfall does not change significantly during the rainy season (June–August). This results in a seasonal shift in rainfall in some RCM runs and emission scenarios.

Decreasing rainfall and rising temperatures can have a devastating impact on community life, as approximately 85% of the population of the basin depends on rain-fed agriculture. Decreasing rainfall and rising temperature change runoff regimes and thus change water availability. Hence, a strong correlation is found between rainfall, SURQ, and total WYLD. SURQ decreases sharply in the dry season and increases during the rainy season. This suggests that SURQs are particularly sensitive to changes in rainfall. Similarly, due to the dryness of the basin, high PET is expected in the dry season, and low PET is expected in the rainy season due to humid and rainy conditions.

In the rainy season, PET is low, rainfall is high, and SURQ is high, so the total amount of water is expected to increase. On the other hand, in the dry season, the possibility of evaporation in the basin is high, the amount of rainfall is small, and the SURQ is small, so the total amount of water is small. Overall, the declining trend of water resources in the study area suggests that water resources are inadequate to support the water development in the Upper Blue Nile basin. Therefore, strategies for managing the water resources of the basin are needed to prevent future impacts of climate change on water availability. In order to develop water resource management strategies, predicted climate change scenarios and simulated impacts of climate change need to be incorporated into basin adaptation decision analysis. Such strategies can provide optimal benefits in terms of maintaining the availability of water resources, mitigating and adapting to adverse climatic conditions, and thus maintaining or ensuring the development of water resources.

The limitation of this study is that it does not take into account future land-use/land-cover changes and the impact of socioeconomic conditions on water resources in river basins. Therefore, future studies need to consider the combined effects of climate change, land-use/land-cover, and the socioeconomic conditions of the Upper Blue Nile basin.

ACKNOWLEDGEMENTS

We would like to offer our special thanks to the Ethiopian Ministry of Water, Irrigation and Energy and the National Meteorological Service Agency (NMSA) of Ethiopia for providing the hydro-meteorological data of the study area.

AUTHORS' CONTRIBUTIONS

All authors contributed to the study conception and design. Material preparation, data collection, and analysis were performed by G.S.T., G.S.G., A.G.G., and A.N.E. The first draft of the manuscript was written by G.S.T., and all authors commented on previous versions of the manuscript. All authors read and approved the final manuscript.

FUNDING

This work was supported by Addis Ababa University, the office of the vice president for research and technology transfer, and Debre Birhan University.

CONFLICT OF INTEREST

There is no conflict of interest

DATA AVAILABILITY STATEMENT

Data cannot be made publicly available; readers should contact the corresponding author for details.

REFERENCES

Abdo, K. S., Fiseha, B. M., Rientjes, T. H. M., Gieske, A. S. M. & Haile, A. T. 2009 Assessment of climate change impacts on the hydrology of Gilgel Abay catchment in Lake Tana Basin, Ethiopia. *Hydrological Processes*. doi:10.1002/hyp.7363.

Aduah, M. S., Jewitt, G. P. W. & Toucher, M. L. W. 2017 Scenario-based impacts of land use and climate changes on the hydrology of a lowland rainforest catchment in Ghana, West Africa. *Hydrology and Earth System Sciences*. doi:10.5194/hess-2017-591.

Arnold, J. G., Moriasi, D. N., Gassman, P. W., Abbaspour, K. C., White, M. J. & Srinivasan, R., Santhi, C., Harmel, R. D., van Griensven, A., Van Liew, W. M., Kannan, N. & Ha, M. K. 2012 SWAT: model use, calibration, and validation. *Transactions of the ASABE* **55** (4), 1491–1508. doi:10.13031/2013.42256..

Beyene, T., Lettenmaier, D. P. & Kabat, P. 2009 Hydrologic impacts of climate change on the Nile River Basin: implications of the 2007 IPCC scenarios. *Climatic Change* **100** (3–4), 433–461. doi:10.1007/s10584-009-9693-0.

Biao, E. 2017 Assessing the impacts of climate change on river discharge dynamics in Oueme River Basin (Benin, West Africa). *Hydrology* **4** (4), 47. doi:10.3390/hydrology4040047.

Boru, G. F., Gonfa, Z. B. & Diga, G. M. 2019 Impacts of climate change on stream flow and water availability in Anger sub-basin, Nile Basin of Ethiopia. *Sustainable Water Resources Management* **5** (4), 1755–1764. doi:10.1007/s40899-019-00327-0.

Chen, J., Brissette, F. P. & Caya, D. 2020 Remaining error sources in bias-corrected climate model outputs. *Climatic Change* **162** (2), 563–582. doi:10.1007/s10584-020-02744-z.

Cherie, N. Z. 2013 *Downscaling and Modeling the Effects of Climate Change on Hydrology and Water Resources in the Upper Blue Nile River Basin, Ethiopia. Dissertation*, University of Kassel.

Devia, G. K., Ganasri, B. P. & Dwarakish, G. S. 2015 A review on hydrological models. *Aquatic Procedia* **4**, 1001–1007. doi:10.1016/j.aqpro.2015.02.126.

Dile, Y. T., Berndtsson, R. & Setegn, S. G. 2013 Hydrological response to climate change for Gilgel Abay River, in the Lake Tana Basin – Upper Blue Nile Basin of Ethiopia. *PLoS ONE* **8** (10), e79296. doi:10.1371/journal.pone.0079296.

Elshamy, M. E., Seierstad, I. A. & Sorteberg, A. 2009 Impacts of climate change on Blue Nile flows using bias-corrected GCM scenarios. *Hydrology and Earth System Sciences* **13** (5), 551–565. doi:10.5194/hess-13-551-2009.

Fang, G. H., Yang, J., Chen, Y. N. & Zammit, C. 2015 Comparing bias correction methods in downscaling meteorological variables for a hydrologic impact study in an arid area in China. *Hydrology and Earth System Sciences* **19** (6), 2547–2559. doi:10.5194/hess-19-2547-2015.

Fatehifar, A., Goodarzi, M. R., Montazeri Hedesh, S. S. & Siahvashi Dastjerdi, P. 2021 Assessing watershed hydrological response to climate change based on signature indices. *Journal of Water and Climate Change* **12** (6), 2579–2593. doi:10.2166/wcc.2021.293.

Fentaw, F. 2018 Impacts of climate change on the water resources of Guder Catchment, Upper Blue Nile, Ethiopia. *Waters* **1** (1), 16. doi:10.31058/j.water.2018.11002.

Gassman, P. W., Reyes, M. R., Green, C. H. & Arnold, J. G. 2007 The soil and water assessment tool: historical development, applications, and future research directions. *Transactions of the ASABE* **50** (4), 1211–1250. doi:10.13031/2013.23637.

Gebre, S. L. 2015 Potential impacts of climate change on the hydrology and water resources availability of Didessa Catchment, Blue Nile River Basin, Ethiopia. *Journal of Geology & Geosciences* **04** (01). doi:10.4172/2329-6755.1000193.

Gelete, G., Gokcekus, H. & Gichamo, T. 2019 Impact of climate change on the hydrology of Blue Nile basin, Ethiopia: a review. *Journal of Water and Climate Change* **11** (4), 1539–1550. doi:10.2166/wcc.2019.014.

Gupta, H. V., Sorooshian, S. & Yapo, P. O. 1999 Status of automatic calibration for hydrologic models: comparison with multilevel expert calibration. *Journal of Hydrologic Engineering* **4** (2), 135–143. doi:10.1061/(asce)1084-0699(1999)4:2(135).

Haile, A. T., Akawka, A. L., Berhanu, B. & Rientjes, T. 2017 Changes in water availability in the Upper Blue Nile basin under the representative concentration pathways scenario. *Hydrological Sciences Journal* **62** (13), 2139–2149. doi:10.1080/02626667.2017.1365149.

Hamzah, F. B., Mohd Hamzah, F., Mohd Razali, S. F. & Samad, H. 2021 A comparison of multiple imputation methods for recovering missing data in hydrological studies. *Civil Engineering Journal* **7** (9), 1608–1619. doi:10.28991/cej-2021-03091747.

Intergovernmental Panel on Climate Change 2014 *Climate Change 2014 Mitigation of Climate Change*. doi:10.1017/cbo9781107415416.

Kim, U. & Kaluarachchi, J. J. 2009 Climate change impacts on water resources in the Upper Blue Nile River Basin, Ethiopia. *JAWRA Journal of the American Water Resources Association* **45** (6), 1361–1378. doi:10.1111/j.1752-1688.2009.00369.x.

Kim, U., Kaluarachchi, J. J. & Smakhtin, V. U. 2008 Generation of monthly precipitation under climate change for the Upper Blue Nile River Basin, Ethiopia. *JAWRA Journal of the American Water Resources Association* **44** (5), 1231–1247. doi:10.1111/j.1752-1688.2008.00220.x.

Koch, M. & Cherie, N. 2013 SWAT modeling of the impact of future climate change on the hydrology and the water resources in the Upper Blue Nile River basin, Ethiopia. In: *Proceedings of the 6th International Conference on Water Resources and Environment Research, ICWRER*, Vol. 6, Issue 6, pp. 488–523.

Krause, P., Boyle, D. P. & Bäse, F. 2005 Comparison of different efficiency criteria for hydrological model assessment. *Advances in Geosciences* **5**, 89–97. doi:10.5194/adgeo-5-89-2005.

Krisnayanti, D. S., Bunganaen, W., Frans, J. H., Seran, Y. A. & Legono, D. 2021 Curve number estimation for ungauged watershed in semi-arid region. *Civil Engineering Journal* **7** (6), 1070–1083. doi:10.28991/cej-2021-03091711.

Liersch, S., Tecklenburg, J., Rust, H., Dobler, A., Fischer, M., Kruschke, T., Koch, H. & Hattermann, F. F. 2018 Are we using the right fuel to drive hydrological models? A climate impact study in the Upper Blue Nile. *Hydrology and Earth System Sciences* **22** (4), 2163–2185. doi:10.5194/hess-22-2163-2018.

Luo, M., Liu, T., Meng, F., Duan, Y., Frankl, A., Bao, A. & De Maeyer, P. 2018 Comparing bias correction methods used in downscaling precipitation and temperature from regional climate models: a case study from the Kaidu River Basin in Western China. *Water* **10** (8), 1046. doi:10.3390/w10081046.

Mellander, P.-E., Gebrehiwot, S. G., Gärdenäs, A. I., Bewket, W. & Bishop, K. 2013 Summer rains and dry seasons in the Upper Blue Nile basin: the predictability of half a century of past and future spatiotemporal patterns. *PLoS ONE* **8** (7), e68461. doi:10.1371/journal.pone.0068461.

Mendez, M., Maathuis, B., Hein-Griggs, D. & Alvarado-Gamboa, L.-F. 2020 Performance evaluation of bias correction methods for climate change monthly precipitation projections over Costa Rica. *Water* **12** (2), 482. doi:10.3390/w12020482.

Mengistu, D. T. & Sorteberg, A. 2012 Sensitivity of SWAT simulated streamflow to climatic changes within the Eastern Nile River basin. *Hydrology and Earth System Sciences* **16** (2), 391–407. doi:10.5194/hess-16-391-2012.

Mengistu, D., Bewket, W., Dosio, A. & Panitz, H.-J. 2021 Climate change impacts on water resources in the Upper Blue Nile (Abay) River Basin, Ethiopia. *Journal of Hydrology* **592**, 125614. doi:10.1016/j.jhydrol.2020.125614.

Mohammed, H., Jean, C. K. & Ahmad, W. A. 2016 Projections of precipitation, air temperature and potential evapotranspiration in Rwanda under changing climate conditions. *African Journal of Environmental Science and Technology* **10** (1), 18–33. doi:10.5897/ajest2015.1997.

Nan, Y., Bao-hui, M. & Chun-kun, L. 2011 Impact analysis of climate change on water resources. *Procedia Engineering* **24**, 643–648. doi:10.1016/j.proeng.2011.11.2710.

Nash, J. E. & Sutcliffe, J. V. 1970 River flow forecasting through conceptual models Part I – A discussion of principles. *Journal of Hydrology* **10** (3), 282–290. doi:10.1016/0022-1694(70)90255-6.

Neitsch, S. L., Arnold, J. G., Kiniry, J. R. & Williams, J. R. 2011 *Soil and Water Assessment Tool Theoretical Documentation Version 2009*. Texas Water Resources Institute, College Station, Texas.

Pulighe, G., Lupia, F., Chen, H. & Yin, H. 2021 Modeling climate change impacts on water balance of a Mediterranean watershed using SWAT+. *Hydrology* **8** (4), 157. doi:10.3390/hydrology8040157.

Rathjens, H., Bieger, K., Srinivasan, R., Chaubey, I. & Arnold, J. G. 2016 *CMhyd User Manual*.

Roth, V., Lemann, T., Zeleke, G., Subhatu, A. T., Nigussie, T. K. & Hurni, H. 2018 Effects of climate change on water resources in the Upper Blue Nile Basin of Ethiopia. *Heliyon* **4** (9), e00771. doi:10.1016/j.heliyon.2018.e00771.

Shrestha, M., Acharya, S. C. & Shrestha, P. K. 2017 Bias correction of climate models for hydrological modelling – are simple methods still useful? *Meteorological Applications* **24** (3), 531–539. doi:10.1002/met.1655.

Singh, L. & Saravanan, S. 2020 Simulation of monthly streamflow using the SWAT model of the Ib River watershed, India. *HydroResearch* **3**, 95–105. doi:10.1016/j.hydres.2020.09.001.

Takele, G. S., Gebre, G. S., Gebremariam, A. G. & Engida, A. N. 2021 Hydrological modeling in the Upper Blue Nile basin using soil and water analysis tool (SWAT). *Modeling Earth Systems and Environment*. doi:10.1007/s40808-021-01085-9.

Tapiador, F. J., Navarro, A., Moreno, R., Sánchez, J. L. & García-Ortega, E. 2020 Regional climate models: 30 years of dynamical downscaling. *Atmospheric Research* **235**, 104785. doi:10.1016/j.atmosres.2019.104785.

Tariku, T. B., Gan, T. Y., Li, J. & Qin, X. 2021 Impact of climate change on hydrology and hydrologic extremes of Upper Blue Nile River Basin. *Journal of Water Resources Planning and Management* **147** (2), 04020104. doi:10.1061/(asce)wr.1943-5452.0001321.

Taye, M. T., Ntegeka, V., Ogiramoi, N. P. & Willems, P. 2011 Assessment of climate change impact on hydrological extremes in two source regions of the Nile River Basin. *Hydrology and Earth System Sciences* **15** (1), 209–222. doi:10.5194/hess-15-209-2011.

Taye, M., Dyer, E., Hirpa, F. & Charles, K. 2018 Climate change impact on water resources in the Awash Basin, Ethiopia. *Water* **10** (11), 1560. doi:10.3390/w10111560.

Teutschbein, C. & Seibert, J. 2012 Bias correction of regional climate model simulations for hydrological climate-change impact studies: review and evaluation of different methods. *Journal of Hydrology* **456–457**, 12–29. doi:10.1016/j.jhydrol.2012.05.052.

Tukimat, N. N. A. 2018 Assessing the implementation of bias correction in the climate prediction. *IOP Conference Series: Materials Science and Engineering* **342**, 012004. doi:10.1088/1757-899x/342/1/012004.

UNESCO, UN-Water 2020 *The United Nations World Water Development Report 2020*. The United Nations World Water Development Report. doi:10.18356/e2014dcb-en.

Versini, P.-A., Pouget, L., McEnnis, S., Custodio, E. & Escaler, I. 2016 Climate change impact on water resources availability: case study of the Llobregat River basin (Spain). *Hydrological Sciences Journal* **61** (14), 2496–2508. doi:10.1080/02626667.2016.1154556.

Willkofer, F., Schmid, F.-J., Komischke, H., Korck, J., Braun, M. & Ludwig, R. 2018 The impact of bias correcting regional climate model results on hydrological indicators for Bavarian catchments. *Journal of Hydrology: Regional Studies* **19**, 25–41. doi:10.1016/j.ejrh.2018.06.010.

Worku, G., Teferi, E., Bantider, A. & Dile, Y. T. 2021 Modelling hydrological processes under climate change scenarios in the Jemma sub-basin of Upper Blue Nile Basin, Ethiopia. *Climate Risk Management* **31**, 100272. doi:10.1016/j.crm.2021.100272.

Worqlul, A. W., Ayana, E. K., Yen, H., Jeong, J., MacAlister, C., Taylor, R., Gerik, T. J. & Steenhuis, T. S. 2018 Evaluating hydrologic responses to soil characteristics using SWAT model in a paired-watersheds in the Upper Blue Nile Basin. *CATENA* **163**, 332–341. doi:10.1016/j.catena.2017.12.040.

Yacoub, E. & Tayfur, G. 2018 Trend analysis of temperature and precipitation in Trarza region of Mauritania. *Journal of Water and Climate Change* **10** (3), 484–493. doi:10.2166/wcc.2018.007.

Yu, Y., Liu, J., Yang, Z., Cao, Y., Chang, J. & Mei, C. 2018 Effect of climate change on water resources in the Yuanshui River Basin: a SWAT model assessment. *Arabian Journal of Geosciences* **11** (11). doi:10.1007/s12517-018-3619-y.

Zhang, B., Shrestha, N., Daggupati, P., Rudra, R., Shukla, R., Kaur, B. & Hou, J. 2018 Quantifying the impacts of climate change on streamflow dynamics of two major rivers of the Northern Lake Erie Basin in Canada. *Sustainability* **10** (8), 2897. doi:10.3390/su10082897.

First received 21 June 2021; accepted in revised form 12 December 2021. Available online 30 December 2021

Impact of Climate Change on Hydrology and Water Resources doi: 10.2166/wcc.2021.147

Quantification of the hydrological consequences of climate change in a typical West African catchment using flow duration curves

Stephen Oppong Kwakye [a,b,*] and András Bárdossy [b]

[a] Department of Environmental Science, Kwame Nkrumah University of Science and Technology, Kumasi, Ghana
[b] Institute for Modelling Hydraulic and Environmental Systems, University of Stuttgart, Stuttgart, Germany
*Corresponding author. E-mail: oppongsteven@yahoo.com

SOK, 0000-0002-5440-4524

ABSTRACT

The quantification of the consequences of climate change (CC) on the hydrology of the West Africa region was performed using a validated Hydrologiska Byrans Vattenbalansavdelning hydrological model and regional climate models which was driven by different general circulation models (GCMs) from the Coordinated Regional Downscaling Experiment (CORDEX) and the Regional Climate Division of the Institute of Meteorology and Climate Research at Karlsruhe Institute of Technology (IMK-IFU). The quantile mapping and linear-scaling bias adjustment methods were used to correct the inherent errors in the climate simulations. Flow duration curves (FDCs) and generic annual discharge cycles were used in determining the impacts of the change on hydrology (river flow) in the Black Volta catchment within the subregion. It was found out that, in the first segment of the FDCs representing high flows, there was a slight increase in the future flow characterizing a higher watershed water yield from high rainfall events in the future. The 10–40% exceedance probabilities of flow representing wet conditions; 40–60% relating to mid-range flows; 60–90% representing dry period conditions; and low flows (90–100%) all show a decrease in the future flows for four out of the five GCM driving models. Most worrying is the reduction in flows for the 90–100% exceedance probabilities in the future relating to the sustainability of streamflow in the long term. It was concluded that CC could negatively impact and decrease the hydrology of the subregion in the future with most of the rivers in the catchment running dry in most months of the annual discharge cycle.

Key words: climate change impacts, CORDEX, flow duration curves, GCM, hydrological modelling, West Africa

HIGHLIGHTS

- The study demonstrates the effectiveness of using simple and less data demanding tools in quantifying the hydrological consequences of climate change in the subregion.
- We were able to bias correct the climate simulations using measured (observed) rainfall and temperature data as the 'correct' data.
- We used two different correction methods in this study and their accompanying effects on the hydrology of the region.

1. INTRODUCTION

Precipitation and temperature variables are an integral part of the hydrological cycle; therefore, any changes that occur in the climate might also affect the hydrological regime, and there is a direct connection between the climate and the hydrology of an area. Several studies and reports (Jung & Kunstmann 2007; Solomon 2007; Wagner 2008; Speth *et al.* 2010; Patricola & Cook 2011; Bárdossy & Pegram 2012; Nikulin *et al.* 2012; Stocker *et al.* 2013; Laprise *et al.* 2013; Ibrahim *et al.* 2014) have projected that climate change (CC) has been happening and will continue to happen in the future. Changes in a river's eco-hydrological processes may have consequences for riparian countries whose livelihoods are derived from the hydro system (Kohler *et al.* 2015).

West Africa is a fast developing subregion of the African continent which depends largely on surface waters for its hydropower generation. Rain-fed agriculture is also the main source of occupation and livelihood in the region. The major water resource usage in the region includes hydropower, small-scale irrigation, and domestic water supply. Any slight changes in the hydrology of the basins in the subregion will significantly affect the citizens and the economy. The domestic water supply in West Africa has also been threatened because of changes in rainfall patterns and the drying-up of some of the major rivers.

Although population increases, pollution, and land-use changes could be factors for these problems, one cannot rule out the adverse effects of CC on hydrology and water resources.

Huge investments have been made in hydropower generation in this subregion over the past one to two decades but have yielded small results in terms of electricity production. An example is the Bui hydropower in Ghana which is working far below the designed capacity because of low levels of water in the reservoir of the dam. For the rapid development of the African Continent and its subregions, agriculture, water supply, and energy production are key to the realization of this dream. Quantifying the future hydrological regime in the subregion, taking into consideration CC, would be very instrumental in helping policymakers and water resource managers to make important decisions on investment, development, and management.

Studies by Kunstmann & Jung (2005), Jung (2006), Jung & Kunstmann (2007), and Jung *et al.* (2012) on the effects of CC on the hydrology of the Volta basin in Western Africa showed a very heterogeneous response of river runoff to changes in climate variables. Their work showed that there was no major change for the mean runoff regime of the future as compared with the current simulation.

Roudier *et al.* (2014) summarized the state of knowledge on potential future streamflow evolutions given global warming in West Africa by creating a quantitative database of 301 points of information on the impact of CC on runoff in the region. The results showed no clear general pattern of future runoff evolution in this area. The median of the distribution was 0% with a mean of +5.2%, after considering different time horizons, different rivers, scenarios, and models. The correlation analysis they performed revealed that runoff changes are tightly linked to changes in rainfall ($R = 0.49$). Kankam-Yeboah *et al.* (2013), however, projected a significant decrease in the projected mean annual streamflow for the years 2020 and 2050 for the White Volta and Pra Basins. Similar results of a downward trend of streamflow in the future for the West Africa subregion were obtained by Giertz *et al.* (2010) and McCartney *et al.* (2012). Götzinger (2007) reported similar results of a considerably reduced discharge, especially in the rainy season due to reduced precipitation and increased temperature for two different climate scenarios which were simulated for a basin in Benin (West Africa) in comparison to a simulated historical period. To improve the acceptance of these results by society and stakeholders, the continual update of the results with new models and scenarios is recommended (Götzinger 2007).

According to Vogel & Fennessey (1995), flow duration curves (FDCs) typify the old Chinese proverb 'one picture is worth a thousand words' through their ability to summarize a wealth of hydrologic information into a single graphic image and it could be used in summarizing the results of detailed and complex water resource studies. They are often used to summarize the impacts of potential CC scenarios on water resource systems (Schwarz 1977). The FDC shows the relationship between the magnitude and the frequency of daily streamflow for a river basin, and it provides an estimate of the percentage of time the streamflow was equaled or exceeded over a historical period. The majority of studies in FDCs are about the utilization domains of duration curves in water resources engineering (Cigizoglu & Bayazit 2000). They have been recommended for use in hydrological studies such as water supply, hydropower, waste-water treatment plant capacity, irrigation, and planning (Male & Ogawa 1984; Warnick 1984; Cigizoglu 1997). An approach based on the FDC for patching and extending observed time series of daily streamflow was developed by Hughes & Smakhtin (1996). Regional FDC procedures have been developed in different regions of the USA for ungauged sites by Fennessey & Vogel (1990) and Smakhtin *et al.* (1997). Although FDCs provide a convenient, simple, and powerful tool for studying the flow regime characteristics in a river basin under climate and land-use change (Pumo *et al.* 2013), the number of the studies related to CC impacts on hydrology (Arora & Boer 2001; Sellami *et al.* 2016; Langat *et al.* 2019) is limited.

Previous studies (Kunstmann & Jung 2005; Jung 2006; Jung & Kunstmann 2007; Solomon 2007; Wagner 2008; Biasutti *et al.* 2009) on the topic in the subregion had focused mainly on monthly to seasonal changes of temporal resolutions with shorter durations (i.e. 10 years), maybe because of the computational run-time demands of the models they used. In this study, we use a daily temporal resolution for the simulation and analysis and extended the simulation and climate periods to 45 years.

The objective of the study is to use simple, less data demanding, but robust tools such as the Hydrologiska Byrans Vattenbalansavdelning (HBV) hydrological model and FDCs are used to try and quantify the hydrological consequences of CC in the subregion after performing bias correction on the climate simulations. The bias correction of the various climate simulations used for the study was done using two different methods. The downscaled and corrected climate simulations were used in conjunction with a calibrated and validated HBV hydrological model to determine the consequences of CC on the hydrology of the Black Volta Basin. The study contributes to the knowledge of CC impacts and also provides a simple, rapid, and powerful tool for assisting water stakeholders in identifying the flow regime components and formulating mitigation and management strategies. The information gathered may be useful in the development of new sustainable water-related projects and river basin management systems.

2. DATA AND METHODOLOGY

2.1. Study area

In this study, general circulation model (GCM)/regional climate model (RCM) climate data from the Coordinated Regional Downscaling Experiment (CORDEX) and the Regional Climate Division of the Institute of Meteorology and Climate research at Karlsruhe Institute of Technology (IMK-IFU) were used to determine CC in the West Africa subregion. The study area is the Black Volta Basin in Western Africa with a catchment size of about 155,000 km^2 (Figure 1) and is shared by four other West African countries (Ghana, Mali, Burkina Faso, and Ivory Coast). The Black Volta River is the main river in the catchment which empties into Volta Lake.

2.2. GCM and RCM

The GCM driving models used for the study are (1) *CanESM2M*: the second edition of the climate model for the Canadian Center for Climate Modelling and Analysis (Chylek *et al.* 2011) and (2) *EC-EARTH*: an earth system model (ESM) developed and maintained by a consortium of European climate modelers, bringing together 27 research institutes from 10 European countries to collaborate on the development of an ESM (Hazeleger *et al.* 2012), (3) *ECHAM*: an atmospheric GCM developed by the Max Planck Institute (MPI) for Meteorology, forming the atmospheric component of the MPI–ESM (Stevens *et al.* 2013), (4) *GFDL*: climate and ESM developed by the Geophysical Fluid Dynamics Laboratory (GFDL) which is a laboratory

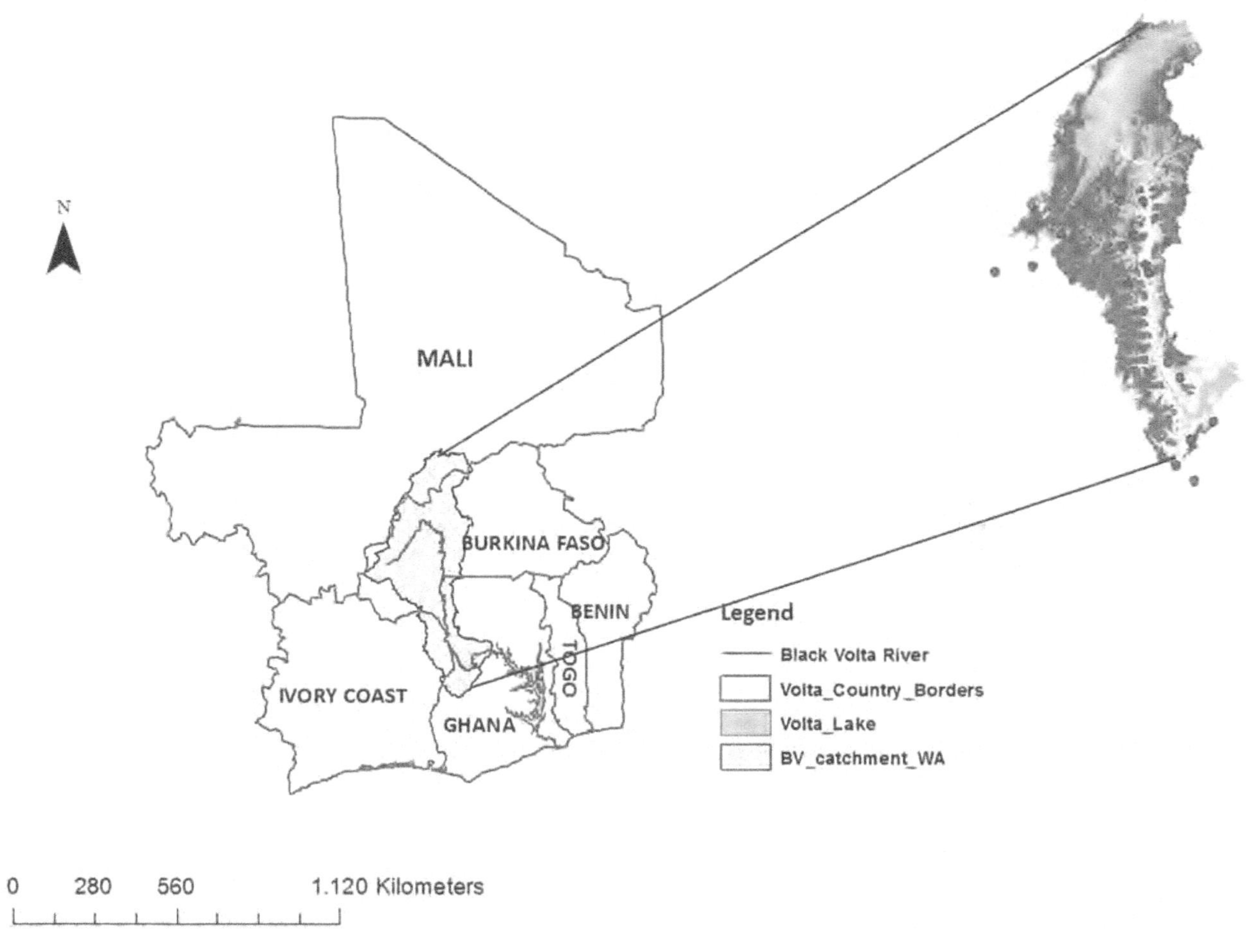

Figure 1 | Black Volta Basin with its riparian countries (insert: Black Volta Basin showing digital elevation map and the rainfall gauge stations).

in the National Oceanic and Atmospheric Administration (NOAA) Office of Oceanic and Atmospheric Research (OAR) (Dunne *et al.* 2012), and (5) *MIROC*: a climate model developed by the Center for Climate System Research in Japan (Watanabe *et al.* 2010).

The Rossby Centre Regional Atmospheric Climate Model (RCA4) (Kupiainen *et al.* 2011) and the Weather Research and Forecasting (WRF) (Skamarock *et al.* 2008) Model were used as the RCM for the data from CORDEX and IMK-IFU, respectively. The analysis was performed on a daily temporal resolution and a spatial resolution of 50 and 12 km, respectively, for the data from CORDEX and IMK-IFU. The GCM/RCM climate simulations often show the bias representation of observed climate (Varis *et al.* 2004; Christensen *et al.* 2008; Teutschbein & Seibert 2010, 2012).

2.3. Bias adjustment/correction of climate simulations

There are several bias adjustment methods which range from simple scaling methods to sophisticated approaches employing probability mapping or weather generators. The procedures employ a transformation algorithm for adjusting the RCM output.

In this study, the (multiple) linear-scaling (MC) method and the distribution mapping approach, which is also referred to as 'probability mapping' (Ines & Hansen 2006; Block *et al.* 2009), 'quantile–quantile mapping' (Boé *et al.* 2007; Bárdossy & Pegram 2011; Johnson & Sharma 2011; Sun *et al.* 2011; Pegram & Bárdossy 2013), 'statistical downscaling' (Piani *et al.* 2010), and 'histogram equalization' (Sennikovs & Bethers 2009; Rojas *et al.* 2011), were used to bias adjust the RCM outputs. The measured rainfall and temperature at the stations (Figure 1) were used as the observed or 'correct' climate variables and the bias correction is done for the closest grid points to stations. A precipitation threshold (for a day to be considered wet, precipitation should be greater than 0.1 mm) as used by Piani *et al.* (2010) was also used to adjust the wet-day frequencies of precipitation time series. 'MC' and 'QC' are used to represent the linear-scaling and quantile-quantile mapping methods of bias-correction respectively. The two methods are described in the subsequent subsections.

2.3.1. MC method

The MC method works with monthly adjustment values which are based on the differences between observed and historical simulated values (Teutschbein & Seibert 2012). The adjustment of precipitation is done with a factor based on the ratio of long-term monthly mean observed and historical (control) run data:

$$P^*_{\text{hist}}(d) = P_{\text{hist}}(d) \cdot \left[\frac{\mu_m(P_{\text{obs}}(d))}{\mu_m(P_{\text{hist}}(d))}\right] \tag{1}$$

$$P^*_{\text{rcp}}(d) = P_{\text{rcp}}(d) \cdot \left[\frac{\mu_m(P_{\text{obs}}(d))}{\mu_m(P_{\text{hist}}(d))}\right] \tag{2}$$

where P^*_{hist} is the adjusted precipitation values of the control or historical simulation; P_{hist} is the uncorrected (UC) control or historical simulation; P^*_{rcp} is the corrected/adjusted precipitation values of the future simulation; P_{rcp} is the UC future simulation; $\mu_m(P_{\text{obs}}(d))$ is the long-term monthly mean of observed data; $\mu_m(P_{\text{hist}}(d))$ is the long-term monthly mean of historical simulation.

Temperature is corrected/adjusted using an additive term which is based on the difference of the long-term monthly mean of observed and historical run data:

$$T^*_{\text{hist}}(d) = T_{\text{hist}}(d) + \mu_m(T_{\text{obs}}(d)) - \mu_m(T_{\text{hist}}(d)) \tag{3}$$

$$T^*_{\text{rcp}}(d) = T_{\text{rcp}}(d) + \mu_m(T_{\text{obs}}(d)) - \mu_m(T_{\text{hist}}(d)) \tag{4}$$

It should be noted that the applied correction/adjustment factors and terms are assumed to remain constant even for future conditions.

2.3.2. Quantile–quantile mapping method

Quantile–quantile mapping is to correct the distribution function of RCM-simulated climate values to agree with the observed distribution function and it is done by the creation of a transfer function to shift the occurrence distributions of precipitation and temperature (Sennikovs & Bethers 2009).

Various literature (Ines & Hansen 2006; Piani *et al.* 2010) reveal that the Gamma distribution (Thom 1958) with shape parameter α and scale parameter β is often assumed to be suitable and effective for distributions of precipitation events:

$$f_{\gamma}(x|\alpha, \beta) = x^{\alpha-1} \cdot \frac{1}{\beta^{\alpha} \cdot \Gamma(\alpha)} . e^{\frac{-x}{\beta}}; x \geq 0; \quad \alpha, \beta > 0 \tag{5}$$

The Gaussian distribution (Cramér 1945) with location parameter μ and scale parameter σ is usually to best fit temperature time series (Thom 1952; Schoenau & Kehrig 1990; Teutschbein & Seibert 2012):

$$f_N(x|\mu, \sigma^2) = x^{\mu-1} \cdot \frac{1}{\sigma \cdot \sqrt{2\pi}} \cdot e^{\frac{-(x-\mu)^2}{2\sigma^2}}; x \in < R \tag{6}$$

In this study, cumulative distribution functions (CDFs) were constructed for both the observed and RCM-simulated climate variables (1979–2005) for all days within a certain month (Supplementary material, Appendix A). The value of the RCM-simulated precipitation/temperature of day d within month m was then searched on the empirical CDF of the RCM simulations together with its corresponding cumulative probability (Supplementary material, Appendix A). The value of precipitation/temperature was thereafter located on the empirical CDF of observations. This value was finally used as the corrected value for the RCM control or historical run (1979–2005). For precipitation, this approach can be expressed mathematically in terms of the Gamma CDF (F_{γ}) and its inverse (F_{γ}^{-1}) as follows:

$$P^{*}_{\text{hist}}(d) = F_{\gamma}^{-1}(F_{\gamma}(P_{\text{hist}}(d)|\alpha_{\text{hist},m}, \beta_{\text{hist},m})|\alpha_{\text{obs},m}, \beta_{\text{obs},m}) \tag{7}$$

$$P^{*}_{\text{rcp}}(d) = F_{\gamma}^{-1}(F_{\gamma}(P_{\text{rcp}}(d)|\alpha_{\text{hist},m}, \beta_{\text{hist},m})|\alpha_{\text{obs},m}, \beta_{\text{obs},m}) \tag{8}$$

where P^{*}_{hist} is the corrected precipitation values of the control or historical simulation; P_{hist} is the UC control or historical simulation; P^{*}_{rcp} is the corrected precipitation values of the future simulation; P_{rcp} is the UC future simulation.

The same approach can be used in terms of the Gaussian (normal) CDF (F_N) and its inverse (F_N^{-1}) for the temperature time series as follows:

$$T^{*}_{\text{hist}}(d) = F_N^{-1}(F_N(T_{\text{hist}}(d)|\mu_{\text{hist},m}, \sigma^2_{\text{hist},m})|\mu_{\text{obs},m}, \sigma^2_{\text{obs},m}) \tag{9}$$

$$T^{*}_{\text{rcp}}(d) = F_N^{-1}(F_N(T_{\text{rcp}}(d)|\mu_{\text{hist},m}, \sigma^2_{\text{hist},m})|\mu_{\text{obs},m}, \sigma^2_{\text{obs},m}) \tag{10}$$

where T^{*}_{hist} is the corrected temperature values of the control or historical simulation; T_{hist} is the UC control or historical simulation; T^{*}_{rcp} is the corrected temperature values of the future simulation; and T_{rcp} is the UC future simulation.

The shape and scale parameters for each of the distributions were determined using the maximum likelihood estimation.

2.4. Hydrological modeling

The consequences of the changing climate on the hydrology of the region were performed using a validated HBV hydrological model and the UC and corrected climate simulations. The hydrological model performed well for the study area with an average Nash–Sutcliffe (NS) of 0.75 and 0.57 for calibration and validation, respectively. In all, results showed that the model was stable and reacted well to precipitation signals (Kwakye & Bárdossy 2020). Additional information on the application of the HBV hydrological model in the study area can be found in our earlier study (Kwakye & Bárdossy 2020). The observed hydrometeorological data used for the study were obtained from the Volta Basin Geoportal, Meteorological agencies of both Ghana and Burkina Faso, and from the Hydrological Services Department of Ghana. The duration of the time series is from 1961 to 2005 (45 years) with a daily temporal resolution. The areal precipitation, temperature, and potential evapotranspiration were used as inputs for the lumped HBV hydrological model to get the river flow (discharge) as the output for the historical (control) and future (rcp) time periods. FDCs and annual cycles were developed for the various driving models to quantify the hydrological consequences. A flowchart of the methodology is provided in Figure 2.

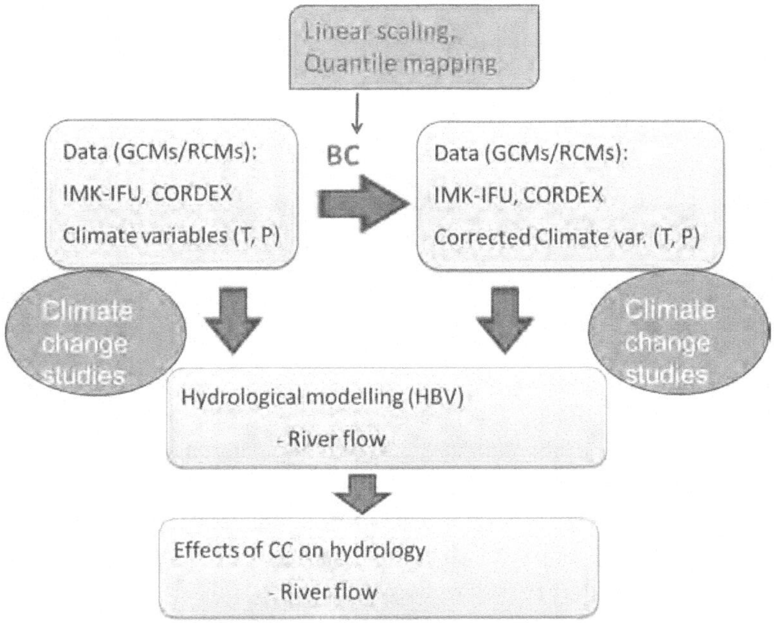

Figure 2 | Flowchart describing the methodology used in the study.

2.5. Flow duration curve

The FDC is formulated as $F = (R/(n+1))$ (100), where F is the frequency of occurrence expressed as percentage of the time a particular streamflow value is equaled or exceeded (exceedance probability), R is the rank corresponding to streamflow, Q and n is equal to the total number of days. To characterize the information in the FDC, it could be partitioned into different segments: (1) the first part representing high flows (0–10% exceedance probabilities of flow) characterizing watershed water yield from high rainfall events; (2) wet conditions (10–40%) described by flows from medium size rainfall events; and (3) mid-range flows (40–60%) represented by flows from moderate rainfall events. The other segments were dry period conditions (60–90%) related to the intermediate base flow relaxation response of the watershed water yield, and the low flows (90–100%) relating to the sustainability of streamflow in the long term (Langat *et al.* 2019).

3. RESULTS

3.1. CC studies

Supplementary Appendix B indicates the performance of the two different bias correction methods applied on the different GCM driving models used for this study. These figures show the annual cycle of areal rainfall for both methods with their UC data for the various driving models.

For precipitation change signal, the CanESM model does not show any clear change. The EC-EARTH, GFDL, and MIROC driving models predict a slight positive change signal in both the UC and bias-corrected GCM/RCM simulations. On the other hand, the ECHAM model indicates a slight decrease in the future precipitation in both the UC and bias-corrected simulations. The two correction methods gave similar precipitation change signals for all the GCM/RCM models used in this study. It could therefore be stated that the future precipitation change signal is difficult to predict for the subregion.

Both the linear-scaling MC and quantile mapping (QC) approaches tried to bring the historical simulations to the observed while maintaining the change signals between the present and the future in most of the models, but the MC method destroys part of the distribution in the UC data for some of the models in the process. The curve from the corrected historical data from the MC method was closer to the observed than the QC in the case of EC-EARTH and MIROC, but for the other driving models, the curve from the adjusted historical data from the QC was closer to the observed (Supplementary material, Appendix B).

3.2. Effects of CC on hydrology using the CanESM driving model

The FDC in Figure 3 shows that for the UC RCM simulations, the discharge (river flow) in the future will slightly increase, with magnitudes slightly higher for mean daily flow which is less than 100 m^3/s. For the flows which are higher than

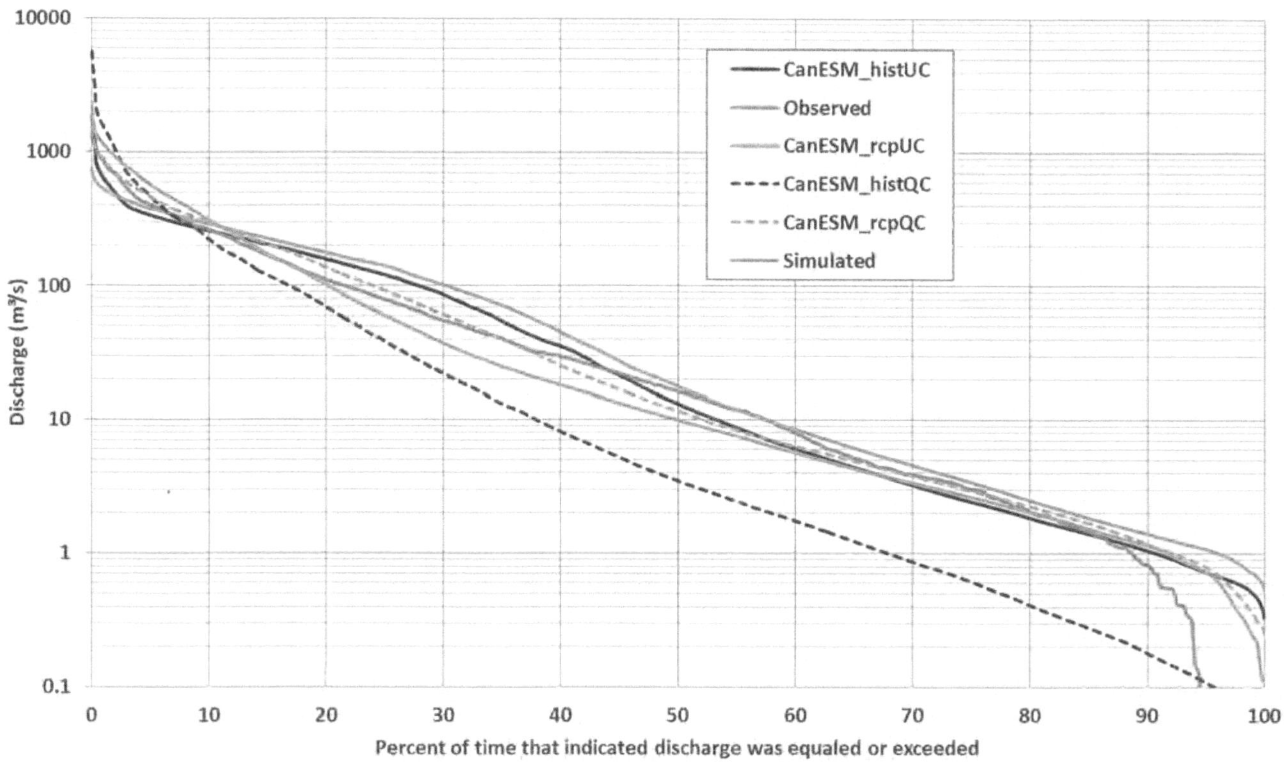

Figure 3 | FDCs of simulated discharges (historical (hist) (1980–2005) (*blue*); future (rcp) (2020–2045) (*green*); observed (1980–2005) (*red*); simulated (1980–2005) (*orange*); and bias-corrected RCM simulations using the QC method (*dashed blue and green*)) with CanESM as the driving model. Please refer to the online version of this paper to see this figure in colour: doi:10.2166/wcc.2021.147.

100 m^3/s, the increase is very minimal. However, the curve of the historical discharge simulations was quite different from the observed discharge curve and the reasons might be from the errors in observations and wrong parameterization of the climate and/or the hydrological models. It could also be seen from the curve that for 25% of the time, the daily discharge obtained from the RCM simulations equaled or exceeded 100 m^3/s. This number reduces to about 10–20 m^3/s for the 50% exceedance probability and reduces further to 2–3 m^3/s for the 75% exceedance probability. This shows that the rivers in this catchment virtually dry up about 25–30% of the time. The peak discharge is about 300 m^3/s and this is equaled or exceeded for about 10% of the time.

Figure 3 also shows the results from the bias-corrected simulations using the QC method. They are shown in dashed blue and dashed green colours for historical (1980–2005) and future (2020–2045) periods, respectively. The corrected data predict an increase in the future discharge (river flow) as compared with the historical with the positive signal more pronounced in the corrected data. For example, for 50% of the time, a discharge of 3 and 12 m^3/s is equaled or exceeded in the corrected historical and future, respectively. The change signal is reduced in higher discharges (i.e. flows greater than 100 m^3/s). In all three cases (i.e. the UC, corrected using QC, and corrected using MC), the change signal is maintained and reproduced. The simulated discharge curve (sim) is the simulation obtained by using the observed climate as input in the hydrological model to produce the runoff (discharge). This curve is close to the observed discharge but it underestimates the low flows and overestimates the high flows (Figure 3). A possible explanation of the under-/overestimation of the flows could be because models are imperfect representations of the real world and thus model uncertainties will always be present. Inherent complexities of natural mechanisms, as well as inappropriate assumptions within the entire modeling procedure, give rise to these uncertainties.

The discharge annual cycle from the RCM simulations using the CanESM driving model showed a bimodal discharge cycle for the UC and corrected data using QC with the first peak in May and the other in August. The observed annual cycle and the corrected data using the MC method showed a unimodal discharge cycle which peaks in August/September (Figure 4). From the RCM simulations, there is a decrease in the discharge for the first cycle and an increase during the second cycle for the future as compared with the historical simulation. It could be inferred that there will be a decreased discharge in the first/

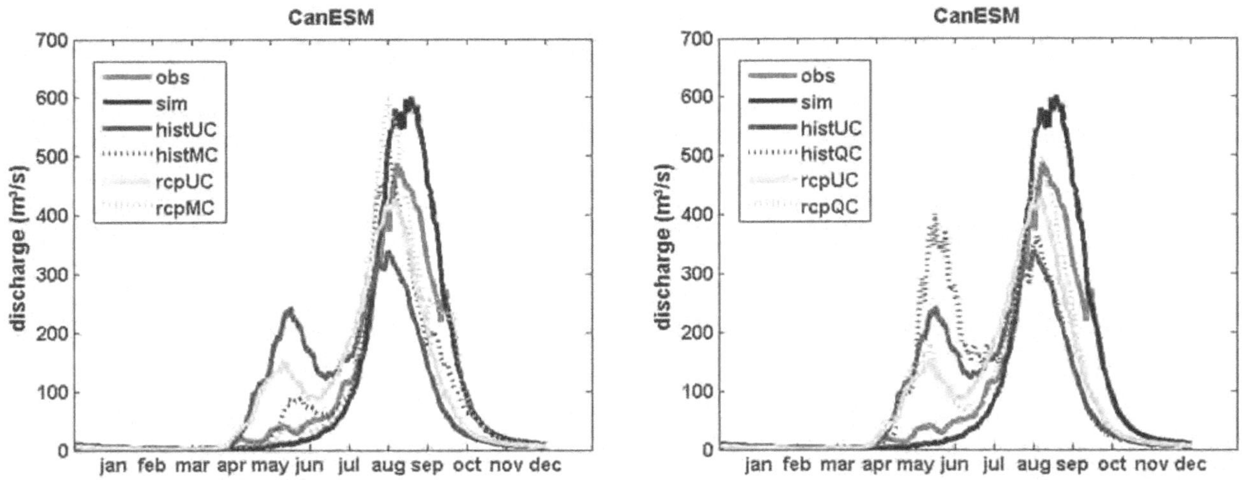

Figure 4 | Annual cycle of discharges (historical (hist) (1980–2005) (*blue*); future (rcp) (2020–2045) (*green*); bias-corrected RCM simulations using MC (left) and QC (right) methods (*dashed blue and green*) using CanESM as the driving model; observed (obs) (1980–2005) (*red*); and simulated discharges (sim) (1980–2005) (*black*)). Please refer to the online version of this paper to see this figure in colour: doi:10.2166/wcc.2021.147.

minor discharge cycle and an increased discharge during the second/major discharge cycle in the future. The rivers dry up for about 4 months of the year. Both the UC and corrected data show the discharge cycle reaching its highest amount in August as in the case of the observed discharge.

The discharge from the historical RCM simulations did not quite represent the observed discharge with the simulated discharge (sim) outperforming the discharge resulting from the historical RCM simulations. While the simulated discharge showed a unimodal annual cycle which is what happens in the observed data, the historical showed a bimodal cycle and there is a slight underestimation of the low flows and overestimation of the peaks in the simulated discharge (Figure 4).

3.3. Effects of CC on hydrology using the EC-EARTH driving model

The FDC from this driving model reveals that the discharge will decrease in the future as compared with the present which is in contrast to what the CanESM model predicted. In the first segment representing high flows (0–10% exceedance probabilities), the FDC shows a slight increase in the future flow characterizing a higher watershed water yield from high rainfall events in the future. The other segments of the curve (10–100% exceedance probabilities) for this driving model representing wet conditions (10–40%), mid-range flows (40–60%), dry period conditions (60–90%), and low flows (90–100%) all show a decrease in the future flows. Most worrying is the reduction of the low flows (90–100% exceedance probabilities) in the future relating to the sustainability of streamflow in the long term, which could lead to the drying-up of the watershed (Figure 5). For 50% of the time, a discharge of 18 and 28 m^3/s is equaled or exceeded for the future and historical periods, respectively (Figure 5).

The bias-corrected data also confirmed a negative signal for future discharge. At 25% of the time, a discharge of 70 m^3/s is equaled or exceeded, while 20 m^3/s is for 50% exceedance probability and 4 m^3/s for 75% in the future. The future flow for 50% exceedance probability is 20 and 8 m^3/s for QC and MC methods of bias correction, respectively (Figure 5).

The annual cycle of discharge (Figure 6) also confirms a slight decrease in the future flows and shows one discharge cycle and not two as seen in the case of the CanESM model. The flow begins to increase in April when the rainy season starts and peaks in August/September, then subsides in October–March with the high uncertainty in the first discharge months (April–June) evident in the historical, observed, and future flows. The change signal is maintained in the results from the bias-corrected data with both the MC and QC correction methods, showing a negative discharge signal in the future. The MC bias correction method predicted lesser discharges in the low flows and higher discharges in the peak flows as compared with the QC method.

3.4. Effects of CC on hydrology using the ECHAM driving model

The FDC (Figure 7) from this model shows a decrease in the future discharge as compared with the historical with a more pronounced negative change signal than what the EC-EARTH predicted. For the high flows (0–10% exceedance

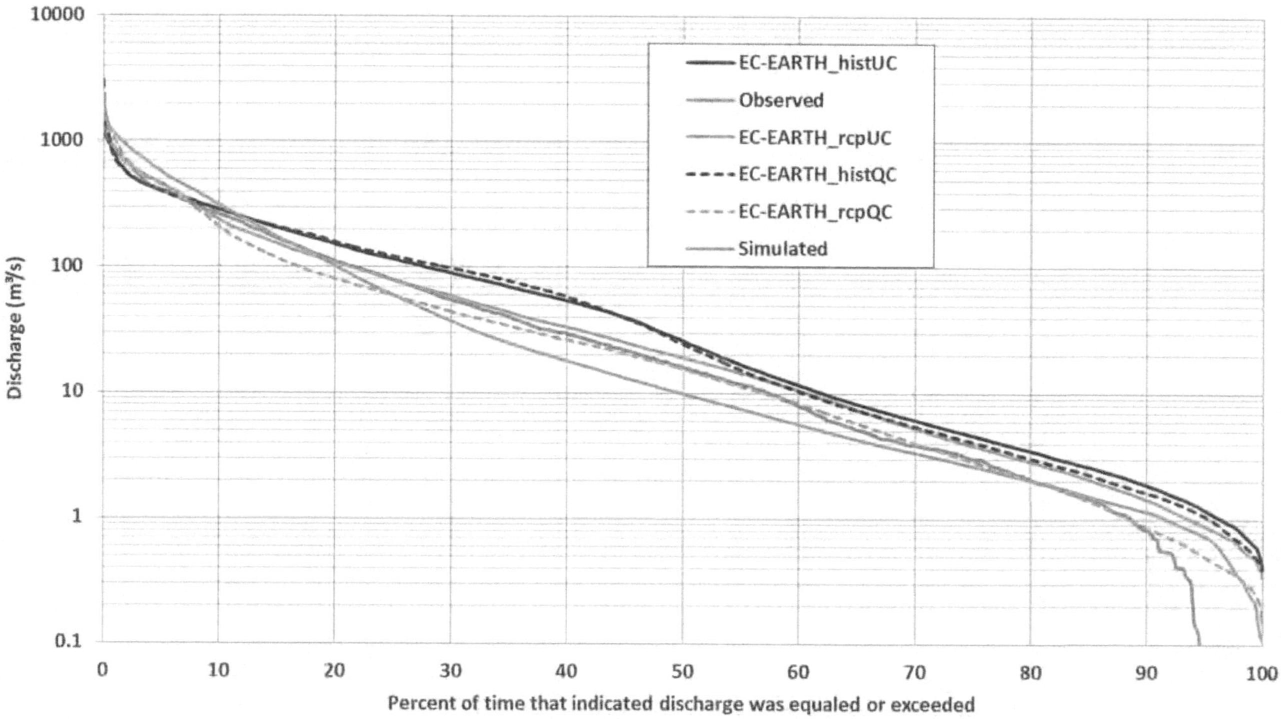

Figure 5 | FDCs of simulated discharges (historical (hist) (1980–2005) (*blue*); future (rcp) (2020–2045) (*green*); observed (1980–2005) (*red*); simulated (1980–2005) (*orange*); and bias-corrected RCM simulations using the QC method (*dashed blue and green*)) with EC-EARTH as the driving model. Please refer to the online version of this paper to see this figure in colour: doi:10.2166/wcc.2021.147.

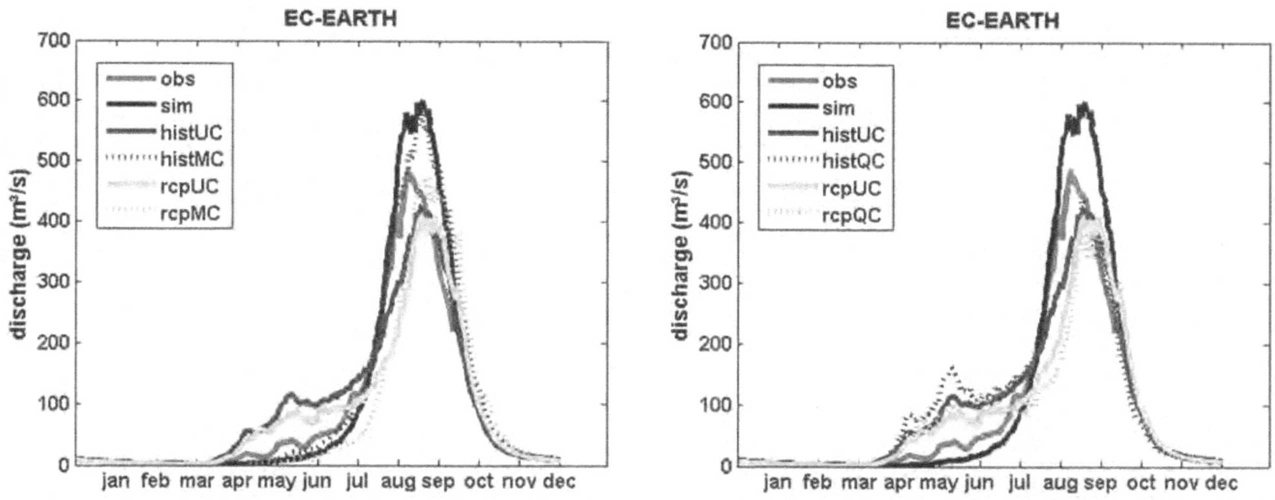

Figure 6 | Annual cycle of discharges (historical (hist) (1980–2005) (*blue*); future (rcp) (2020–2045) (*green*); bias-corrected RCM simulations using MC (left) and QC (right) methods (*dashed blue and green*) using EC-EARTH as the driving model; observed (obs) (1980–2005) (*red*); and simulated discharges (sim) (1980–2005) (*black*)). Please refer to the online version of this paper to see this figure in colour: doi:10.2166/wcc.2021.147.

probabilities), the UC climate simulations from this driving model predict a decline in the future flows and the corrected simulation shows a less significant change for this type of high flows relating to high rainfall events. The other segments of the FDC (10–100% exceedance probabilities) show a decrease and a negative change signal for the future flow with the UC data predicting a larger decrease in the future compared with the corrected climate simulations.

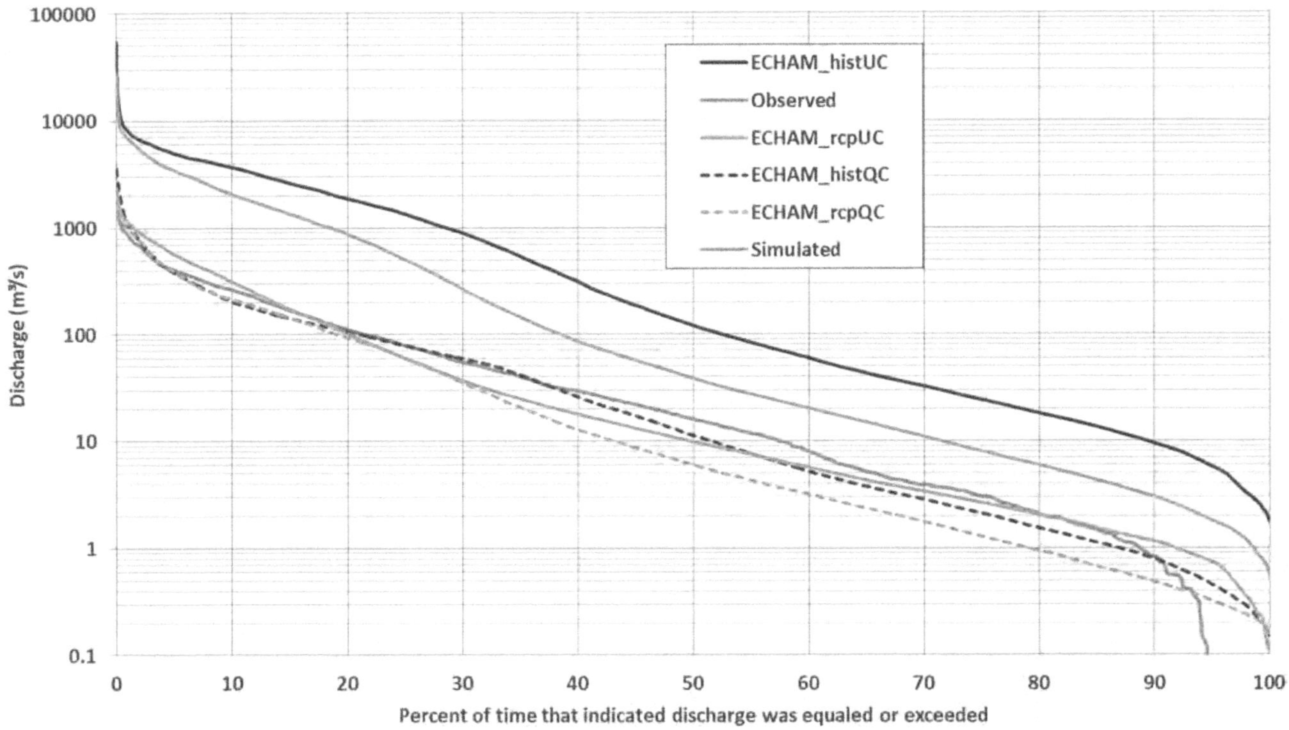

Figure 7 | FDCs of simulated discharges (historical (hist) (1980–2005) (*blue*); future (rcp) (2020–2045) (*green*); observed (1980–2005) (*red*); simulated (1980–2005) (*orange*); and bias-corrected RCM simulations using the QC method (*dashed blue and green*)) with ECHAM as the driving model. Please refer to the online version of this paper to see this figure in colour: doi:10.2166/wcc.2021.147.

Overall, this model yields higher discharges in comparison with the other models. For example, the 25, 50, and 75% exceedance probabilities give discharges of 400, 40, and 10 m^3/s, respectively. The discharge curve obtained from the historical RCM simulations as input is far away from the curve of the observed discharge, and the bias correction methods applied in this study tried to improve the RCM simulations. Both the bias correction methods maintained the negative change signal. The historical discharge as a result of correction using QC was better compared with the discharge from the MC correction method.

The annual cycle showed a decrease in the discharge for the future and a unimodal discharge cycle and emphasized that the corrected data using the QC method performed better than those using the MC method with the bias-corrected data performing better than the UC (Figure 8).

3.5. Effects of CC on hydrology using the GFDL driving model

In the FDC of this model (Figure 9), the UC climate data show both an increase and a decrease of flows for the future. The flows from both the UC and corrected climate simulations produced a positive change signal for the high flows (0–10% exceedance probabilities) characterizing high watershed water yield from high rainfall events in the future. However, for flows in the wet to low-flow conditions (10–100% exceedance probabilities), this driving model predicted a huge decrease in the future discharge for the corrected climate simulations, bringing into question the future sustainability of the streamflows in the subregion in the long term and the intermediate base flow relaxation response of the catchment water yield. For 50% of the time, the flows from the QC and MC method corrected data indicated that a discharge of 5 m^3/s will be equaled or exceeded in the future as compared with 15–20 m^3/s in the historical period (Figure 9).

The annual cycle shows a slight increase in the peak discharges for the future as compared with the historical (Figure 10). The discharge for the historical period has a similar flow pattern with the observed discharge but the peak discharge is almost doubled in September. The month of the peak discharge is shifted from August in the observed discharge cycle to September in both the historical and future discharge cycles.

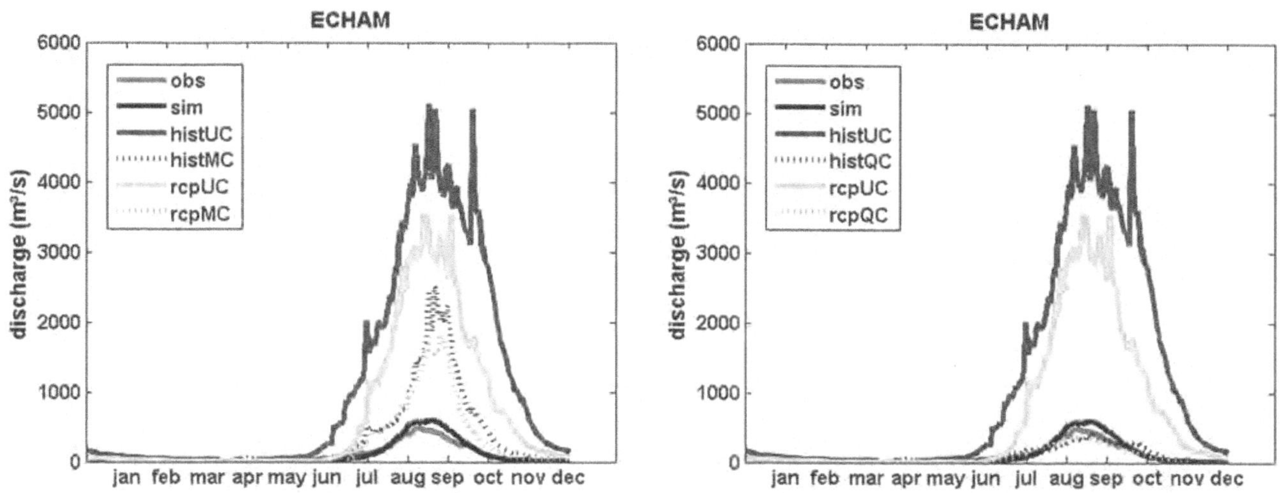

Figure 8 | Annual cycle of discharges (historical (hist) (1980–2005) (*blue*); future (rcp) (2020–2045) (*green*); bias-corrected RCM simulations using MC (left) and QC (right) methods (*dashed blue and green*) using ECHAM as the driving model; observed (obs) (1980–2005) (*red*); and simulated discharges (sim) (1980–2005) (*black*)). Please refer to the online version of this paper to see this figure in colour: doi:10.2166/wcc.2021.147.

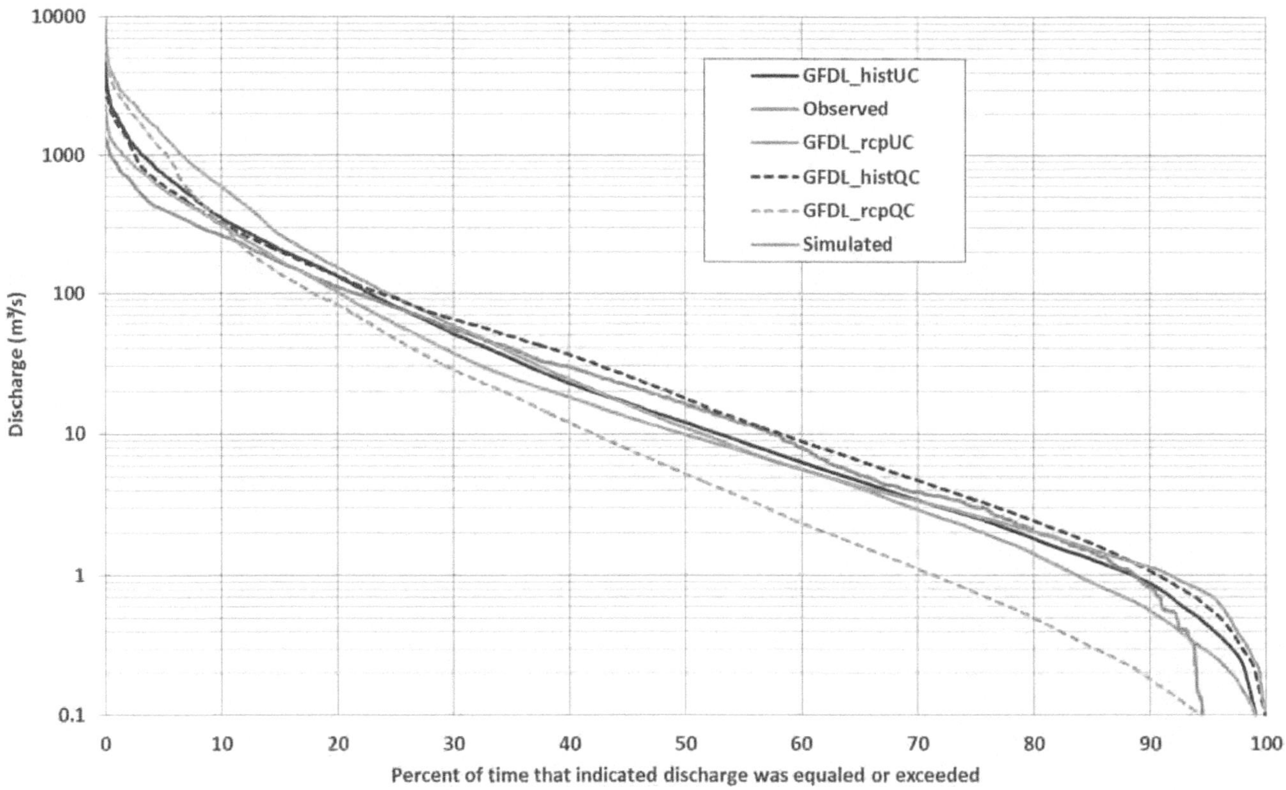

Figure 9 | FDCs of simulated discharges (historical (hist) (1980–2005) (*blue*); future (rcp) (2020–2045) (*green*); observed (1980–2005) (*red*); simulated (1980–2005) (*orange*); and bias-corrected RCM simulations using the QC method (*dashed blue and green*)) with GFDL as the driving model. Please refer to the online version of this paper to see this figure in colour: doi:10.2166/wcc.2021.147.

3.6. Effects of CC on hydrology using the MIROC driving model

Figure 11 shows the FDC obtained when the results of the MIROC driving model were used as the input to the hydrological model. For the high flows (0–10% exceedance probabilities of flow) as it occurred in the other driving models, the flows as a

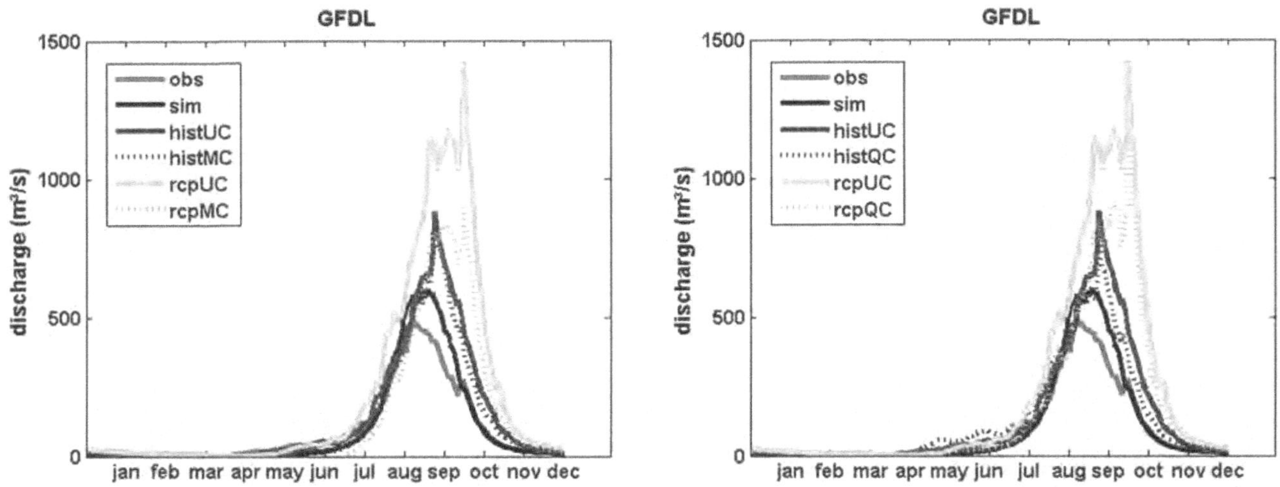

Figure 10 | Annual cycle of discharges (historical (hist) (1980–2005) (*blue*); future (rcp) (2020–2045) (*green*); bias-corrected RCM simulations using MC (left) and QC (right) methods (*dashed blue and green*) using GFDL as the driving model; observed (obs) (1980–2005) (*red*); and simulated discharges (sim) (1980–2005) (*black*)). Please refer to the online version of this paper to see this figure in colour: doi:10.2166/wcc.2021.147.

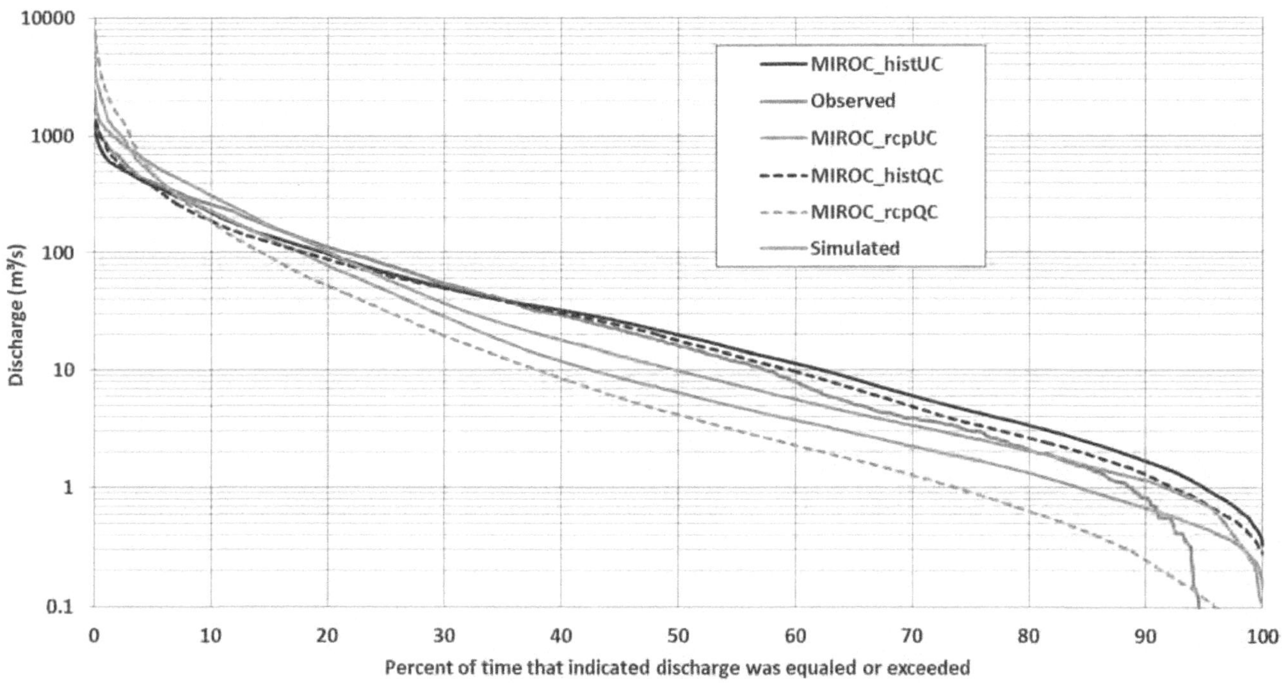

Figure 11 | FDCs of simulated discharges (historical (hist) (1980–2005) (*blue*); future (rcp) (2020–2045) (*green*); observed (1980–2005) (*red*); simulated (1980–2005) (*orange*); and bias-corrected RCM simulations using the QC method (*dashed blue and green*)) with MIROC as the driving model. Please refer to the online version of this paper to see this figure in colour: doi:10.2166/wcc.2021.147.

result of using the MIROC model climate simulations predict an increase in the future flows which characterizes higher water yield in the catchment from high rainfall events. The flows in the other segment (10–100% exceedance probabilities of flow) rather predict a negative change signal (decrease of flows) in the future in very huge amounts. The corrected data also predict a negative impact signal on the hydrology of the catchment in the future. For example, the 50% exceedance probability of flow gives a discharge of 8 m^3/s for the future and 20 m^3/s in the historical for the corrected climate simulations (Figure 11). The results from the correction using the QC method were better than the MC method for the historical corrected simulations.

The annual discharge cycle (Figure 12) for both the corrected and UC data showed the same decrease of flows for the future for flows below 150 m^3/s from January to the middle of August. Then for flows exceeding 300 m^3/s which occurs around August/September, there is an increased discharge for the future as compared with the present. It could therefore be inferred that in the future, the dry periods become drier and the wet periods become wetter with more extreme cases happening. The month for the peak discharge is shifted from August to September as seen in the observed discharge curve. The future from this GCM driving model shows that from November to April, there is very little discharge and the rivers in the catchment virtually dry up. The flow begins to increase from July which is very late as compared with the observed and historical cycle, indicating a shortened discharge regime from 7 months to just 4, which might have dire consequences for the agriculture, hydropower, and water supply of the countries within the subregion.

4. DISCUSSION

The study demonstrates the effectiveness of using simple and less data demanding tools such as the lumped version of the HBV hydrological model and FDCs in quantifying the hydrological consequences of CC in the subregion after performing bias correction on the climate simulations. In this study, downscaled and corrected climate simulations of five GCM driving models were used in conjunction with a calibrated and validated HBV model to determine the consequences of CC on the hydrology of the Black Volta Basin. The bias correction of the various climate simulations used for the study was done using QC and MC methods. Although the study was completed in the year 2016, we believe it is still relevant and it demonstrates some interesting and revealing results for the current study area.

The results of this study show that all except one of the GCM driving models project a decrease in discharge for the future. The bias-adjusted (corrected) data confirming a decrease (negative impact) for the future flows. In the high flows (0–10% exceedance probabilities of the FDCs), there is a slight increase in the future flow characterizing a higher watershed water yield from high rainfall events in the future. The 10–40% exceedance probabilities of flow, representing wet conditions; 40–60% relating to mid-range flows; 60–90% meaning dry period conditions; and low flows (90–100%) all show a decrease in the future flows for four out of the five GCM driving models. Most concerning is the negative change signal (decreased flows) for the low flows in the future relating to the sustainability of streamflow in the long term. This could lead to the drying-up of the watershed in the long term.

This study has several strengths. First, we were able to bias correct the climate simulations using measured (observed) rainfall and temperature data as the 'correct' data. Second, we used two different correction methods in this study and their accompanying effects on the hydrology of the subregion. However, a smaller subset of (i.e. five) GCM driving models, two

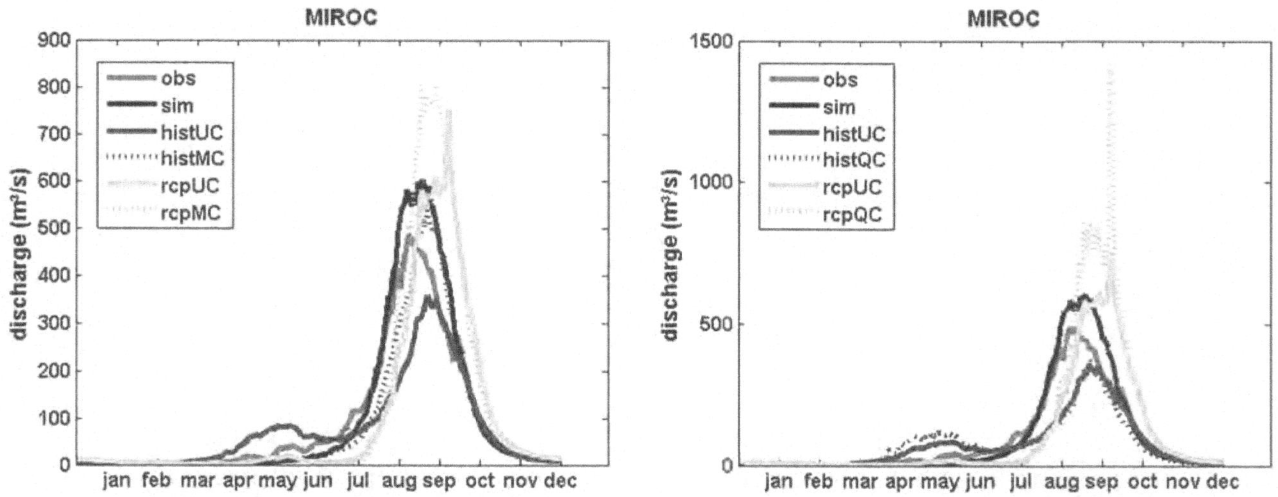

Figure 12 | Annual cycle of discharges (historical (hist) (1980–2005) (*blue*); future (rcp) (2020–2045) (*green*); bias-corrected RCM simulations using MC (left) and QC (right) methods (*dashed blue and green*) using MIROC as the driving model; observed (obs) (1980–2005) (*red*); and simulated discharges (sim) (1980–2005) (*black*)). Please refer to the online version of this paper to see this figure in colour: doi:10.2166/wcc.2021.147.

downscaling models, and one climate scenario (RCP 4.5) were used for the study, which might limit the overall results and conclusion of the study. Further research should be done with more GCM driving models and RCMs.

Our study slightly differs from the studies (Kunstmann & Jung 2005; Jung 2006; Jung & Kunstmann 2007; Jung *et al.* 2012; Roudier *et al.* 2014) in which the effects of CC on the hydrology of the Volta basin in Western Africa showed a very heterogeneous response of river runoff to changes in climate variables. Their work showed that there was no major change for the mean runoff regime of the future as compared with the current simulation. This discrepancy may partly result from differences in the climate models and the versions used for the various studies. However, our results are consistent with the results of the studies (Götzinger 2007; Giertz *et al.* 2010; McCartney *et al.* 2012; Kankam-Yeboah *et al.* 2013) that there is a significant decrease in the projected streamflow in the future for the West African subregion.

A particular strength of this study is the use of the daily temporal resolution of climate variables and streamflow datasets with a longer period (45 years) that provides a sufficient sample size for CC studies and impacts. Nevertheless, a major limitation of this study is the use of areal precipitation and areal climate variables which may be less precise to represent such a large catchment compared with previous studies that used distributed or semi-distributed climate datasets and hydrological models.

This study adds to the knowledge of CC impacts and also provides a simple, rapid, and powerful tool for assisting water stakeholders in identifying the flow regime components and formulating mitigation and management strategies. The information gathered may be useful in the development of new sustainable water-related projects and river basin management systems.

A number of questions remain unanswered, such as why some of the climate models give conflicting signal change results and as a result different impacts to hydrology and why most of the GCMs and RCMs are still not able to well capture the observed and historical climate. Therefore, additional research is needed to continually update the results with new models and scenarios to improve the acceptance of these results by society and stakeholders.

5. CONCLUSIONS

This study revealed that, except for the CanESM driving model which showed an increase in the future discharge, all the other GCMs (EC-EARTH, ECHAM, GFDL, and MIROC) projected a decrease in discharge for the future. The bias-adjusted data using QC and MC methods also confirmed a decrease (negative impact) for the future flows. The first segment of the FDCs, representing high flows (0–10% exceedance probabilities), shows a slight increase in the future flow characterizing a higher watershed water yield from high rainfall events in the future. The other segments of the curve, i.e. 10–40% exceedance probabilities of flow representing wet conditions; 40–60% relating to mid-range flows; 60–90% representing dry period conditions; and low flows (90–100%) all show a decrease in the future flows for four out of the five GCM driving models. Most worrying is the negative change signal (decreased flows) for the low flows (90–100% exceedance probabilities) in the future relating to the sustainability of streamflow in the long term. This could lead to the drying-up of the watershed in the long term.

It could therefore be concluded that CC will negatively affect and impact the hydrology of this subregion because there would be a decrease in the future discharge for the area. For flows occurring with 50% exceedance probabilities in the EC-EARTH driving model, a discharge of 18 and 28 m^3/s is obtained for the future and historical periods, respectively.

The annual discharge cycles from the driving models showed a mix of unimodal and bimodal cycles. The MIROC model showed a delayed discharge cycle from August to September and a shortened discharge regime from 6 to 4 months. The 25, 50, and 75% exceedance probabilities produced discharges of 30, 6, and 2 m^3/s, respectively, in the future for the MIROC model. The discharges using the ECHAM model were showing higher values in comparison with the other models.

In general, the decrease was more pronounced in the low flows (i.e. flows less than 100 m^3/s). The results from the bias-corrected data looked better than the UC with the corrected data using QC performing better than the MC method.

In ameliorating the effects of CC on the hydrology and water resources of the region, proper adaptation and water resources management measures need to be put in place. It is also recommended for the region to possibly look for alternative sources of electricity because there would not be enough water in the reservoirs for hydropower production in the future. The negative effects of CC on the water resources are also a regional security threat because it may lead to loss of jobs, occupation, and farming harvest. This may subsequently destabilize the already vulnerable subregion and increase food insecurity and poverty.

To mitigate food insecurity and poverty in the subregion, more irrigation dams need to be constructed to store the lower future water resource for farming activities.

ACKNOWLEDGEMENT

We wish to state that this paper is born out of the doctoral dissertation of the first author and as such there can be overlap between the dissertation, which can be accessed at elib.uni-stuttgart.de, and this journal article.

CONFLICT OF INTEREST

On behalf of all authors, the corresponding author states that there is no conflict of interest.

DATA AVAILABILITY STATEMENT

Data cannot be made publicly available; readers should contact the corresponding author for details.

REFERENCES

Arora, V. K. & Boer, G. J. 2001 Effects of simulated climate change on the hydrology of major river basins. *Journal of Geophysical Research: Atmospheres* **106** (D4), 3335–3348.

Bárdossy, A. & Pegram, G. 2011 Downscaling precipitation using regional climate models and circulation patterns toward hydrology. *Water Resources Research* **47** (4) doi : doi.org/10.1029/2010WR009689.

Bárdossy, A. & Pegram, G. 2012 Multiscale spatial recorrelation of RCM precipitation to produce unbiased climate change scenarios over large areas and small. *Water Resources Research* **48** (9) doi : doi.org/10.1029/2011WR011524.

Biasutti, M., Sobel, A. H. & Camargo, S. J. 2009 The role of the Sahara low in summertime Sahel rainfall variability and change in the CMIP3 models. *Journal of Climate* **22** (21), 5755–5771.

Block, P. J., Souza Filho, F. A., Sun, L. & Kwon, H. H. 2009 A streamflow forecasting framework using multiple climate and hydrological models. *JAWRA Journal of the American Water Resources Association* 45 (4), 828–843.

Boé, J., Terray, L., Habets, F. & Martin, E. 2007 Statistical and dynamical downscaling of the seine basin climate for hydro-meteorological studies. *International Journal of Climatology* **27** (12), 1643–1655.

Christensen, J. H., Boberg, F., Christensen, O. B. & Lucas-Picher, P. 2008 On the need for bias correction of regional climate change projections of temperature and precipitation. *Geophysical Research Letters* **35** (20) doi:10.1029/2008GL035694.

Chylek, P., Li, J., Dubey, M. K., Wang, M. & Lesins, G. 2011 Observed and model simulated 20th century Arctic temperature variability: Canadian earth system model CanESM2. *Atmospheric Chemistry and Physics Discussions* **11** (22), 893–822.

Cigizoglu, H. K. 1997 *Flow Duration Curve and Its Applications.*

Cigizoglu, H. K. & Bayazit, M. 2000 A generalized seasonal model for flow duration curve. *Hydrological Processes* **14** (6), 1053–1067.

Cramér, H. 1945 *Mathematical Methods of Statistics*, Vol. 9. Princeton University Press, Princeton, NJ.

Dunne, J. P., John, J. G., Adcroft, A. J., Griffies, S. M., Hallberg, R. W., Shevliakova, E., Stouffer, R. J., Cooke, W., Dunne, K. A., Harrison, M. J. & Krasting, J. P. 2012 GFDL's ESM2 global coupled climate–carbon earth system models. Part I: physical formulation and baseline simulation characteristics. *Journal of Climate* **25** (19), 6646–6665.

Fennessey, N. & Vogel, R. M. 1990 Regional flow-duration curves for ungauged sites in Massachusetts. *Journal of Water Resources Planning and Management* **116** (4), 530–549.

Giertz, S., Hiepe, C., Höllermann, B. & Diekkrüger, B. 2010 Impacts of global change on water resources and soil degradation in Benin. In: *Impacts of Global Change on the Hydrological Cycle in West and Northwest Africa.* (P. Speth, M. Christoph, B. Diekkrüger eds.) Springer, Berlin, pp. 484–511.

Götzinger, J. 2007 *Distributed Conceptual Hydrological Modelling-Simulation of Climate, Land use Change Impact and Uncertainty Analysis.*

Hazeleger, W., Wang, X., Severijns, C., Ştefănescu, S., Bintanja, R., Sterl, A., Wyser, K., Semmler, T., Yang, S., Van den Hurk, B. & Van Noije, T. 2012 EC-Earth V2.2: description and validation of a new seamless earth system prediction model. *Climate Dynamics* **39** (11), 2611–2629.

Hughes, D. A. & Smakhtin, V. 1996 Daily flow time series patching or extension: a spatial interpolation approach based on flow duration curves. *Hydrological Sciences Journal* **41** (6), 851–871.

Ibrahim, B., Karambiri, H., Polcher, J., Yacouba, H. & Ribstein, P. 2014 Changes in rainfall regime over Burkina Faso under the climate change conditions simulated by 5 regional climate models. *Climate Dynamics* **42** (5–6), 1363–1381.

Ines, A. V. & Hansen, J. W. 2006 Bias correction of daily GCM rainfall for crop simulation studies. *Agricultural and Forest Meteorology* **138** (1), 44–53.

Johnson, F. & Sharma, A. 2011 Accounting for interannual variability: a comparison of options for water resources climate change impact assessments. *Water Resources Research* **47** (4), 1–20.

Jung, G. 2006 *Regional Climate Change and the Impact on Hydrology in the Volta Basin of West Africa*. PhD Thesis, University of Augsburg, Germany.

Jung, G. & Kunstmann, H. 2007 High-resolution regional climate modeling for the Volta region of West Africa. *Journal of Geophysical Research: Atmospheres* **112** (D23), D23108.

Jung, G., Wagner, S. & Kunstmann, H. 2012 Joint climate–hydrology modeling: an impact study for the data-sparse environment of the Volta basin in West Africa. *Hydrology Research* **43** (3), 231–248.

Kankam-Yeboah, K., Obuobie, E., Amisigo, B. & Opoku-Ankomah, Y. 2013 Impact of climate change on streamflow in selected river basins in Ghana. *Hydrological Sciences Journal* **58** (4), 773–788.

Kohler, T. J., Stanish, L. F., Crisp, S. W., Koch, J. C., Liptzin, D., Baeseman, J. L. & McKnight, D. M. 2015 Life in the main channel: long-term hydrologic control of microbial mat abundance in McMurdo Dry Valley streams, Antarctica. *Ecosystems* **18** (2), 310–327.

Kunstmann, H. & Jung, G. 2005 Impact of regional climate change on water availability in the Volta basin of west Africa: regional hydrological impacts of climate variability and change. In: *Proceedings of Symposium S6 for the Seventh IAHS Scientific Assembly*, Foz de Iguacu, Brazil.

Kupiainen, M., Samuelsson, P., Jones, C., Jansson, C., Willén, U., Hansson, U., Ullerstig, A., Wang, S. & Döscher, R. 2011 Rossby Centre regional atmospheric model, RCA4. *Rossby Centre Newsletter*, June.

Kwakye, S. O. & Bárdossy, A. 2020 Hydrological modelling in data-scarce catchments: Black Volta basin in West Africa. *SN Applied Sciences* **2** (4), 1–19.

Langat, P. K., Kumar, L., Koech, R. & Ghosh, M. K. 2019 Hydro-morphological characteristics using flow duration curve, historical data and remote sensing: effects of land use and climate. *Water* **11** (2), 309.

Laprise, R., Hernández-Díaz, L., Tete, K., Sushama, L., Separović, L., Martynov, A., Winger, K. & Valin, M. 2013 Climate projections over CORDEX Africa domain using the fifth-generation Canadian regional climate model (CRCM5). *Climate Dynamics* **41** (11–12), 3219–3246.

Male, J. W. & Ogawa, H. 1984 Tradeoffs in water quality management. *Journal of Water Resources Planning and Management* **110** (4), 434–444.

McCartney, M., Forkuor, G., Sood, A., Amisigo, B., Hattermann, F. & Muthuwatta, L. 2012 *The Water Resource Implications of Changing Climate in the Volta River Basin*, Vol. 146. IWMI, Colombo, Sri Lanka, pp. 1–40.

Nikulin, G., Jones, C., Giorgi, F., Asrar, G., Büchner, M., Cerezo-Mota, R., Christensen, O. B., Déqué, M., Fernandez, J., Hänsler, A. & van Meijgaard, E. 2012 Precipitation climatology in an ensemble of CORDEX-Africa regional climate simulations. *Journal of Climate* **25** (18), 6057–6078.

Patricola, C. M. & Cook, K. H. 2011 Sub-Saharan Northern African climate at the end of the twenty first century: forcing factors and climate change processes. *Climate Dynamics* **37** (5–6), 1165–1188.

Pegram, G. & Bárdossy, A. 2013 Downscaling regional circulation model rainfall to gauge sites using recorrelation and circulation pattern dependent quantile–quantile transforms for quantifying climate change. *Journal of Hydrology* **504**, 142–159.

Piani, C., Haerter, J. & Coppola, E. 2010 Statistical bias correction for daily precipitation in regional climate models over Europe. *Theoretical and Applied Climatology* **99** (1–2), 187–192.

Pumo, D., Noto, L. V. & Viola, F. 2013 Ecohydrological modelling of flow duration curve in Mediterranean river basins. *Advances in Water Resources* **52**, 314–327.

Rojas, R., Feyen, L., Dosio, A. & Bavera, D. 2011 Improving pan-European hydrological simulation of extreme events through statistical bias correction of RCM-driven climate simulations. *Hydrology and Earth System Sciences* **15** (8), 2599.

Roudier, P., Ducharne, A. & Feyen, L. 2014 Climate change impacts on runoff in West Africa: a review. *Hydrology and Earth System Sciences* **18**, 2789–2801.

Schoenau, G. J. & Kehrig, R. A. 1990 Method for calculating degree-days to any base temperature. *Energy and Buildings* **14** (4), 299–302.

Schwarz, H. E. 1977 Climatic change and water supply: how sensitive is the Northeast? *Climate, Climatic Change, and Water Supply*, 111–120. The National Academies Press, Washington, DC.

Sellami, H., Benabdallah, S., La Jeunesse, I. & Vanclooster, M. 2016 Climate models and hydrological parameter uncertainties in climate change impacts on monthly runoff and daily flow duration curve of a Mediterranean catchment. *Hydrological Sciences Journal* **61** (8), 1415–1429.

Sennikovs, J. & Bethers, U. 2009 Statistical downscaling method of regional climate model results for hydrological modelling. In: *Proceedings of the 18th World IMACS/MODSIM Congress*, Cairns, Australia. Citeseer.

Skamarock, W. C., Klemp, J. B., Dudhia, J., Gill, D. O., Barker, D. M., Duda, M. G., Huang, X. Y., Wang, W. & Powers, J. G. 2008 *A Description of the Advanced Research WRF Version 3*. NCAR Tech. Note, No. NCAR/TN-475 + STR. National Center for Atmospheric Research: Boulder, CO, USA.

Smakhtin, V. Y., Hughes, D. A. & Creuse-Naudin, E. 1997 Regionalization of daily flow characteristics in part of the Eastern Cape, South Africa. *Hydrological Sciences Journal* **42** (6), 919–936.

Solomon, S. 2007 *Climate Change 2007 – the Physical Science Basis: Working Group 1 Contribution to the Fourth Assessment Report of the IPCC*, Vol. 4. Cambridge University Press, Cambridge, UK.

Speth, P., Christoph, M. & Diekkrüger, B. 2010 *Impacts of Global Change on the Hydrological Cycle in West and Northwest Africa*. Springer Science & Business Media, Berlin, Germany.

Stevens, B., Giorgetta, M., Esch, M., Mauritsen, T., Crueger, T., Rast, S., Salzmann, M., Schmidt, H., Bader, J., Block, K. & Brokopf, R. 2013 Atmospheric component of the MPI-M Earth system model: ECHAM6. *Journal of Advances in Modeling Earth Systems* **5** (2), 146–172.

Stocker, T. F., Qin, D., Plattner, G. K., Alexander, L. V., Allen, S. K., Bindoff, N. L., Bréon, F. M., Church, J. A., Cubasch, U., Emori, S. & Forster, P. 2013 *IPCC, 2013: Climate Change 2013: The Physical Science Basis: Contribution of Working Group I to the Fifth Assessment Report of the Intergovernmental Panel on Climate Change.*

Sun, F., Roderick, M. L., Lim, W. H. & Farquhar, G. D. 2011 Hydroclimatic projections for the Murray-darling basin based on an ensemble derived from intergovernmental panel on climate change ar4 climate models. *Water Resources Research* **47** (12), W00G02.

Teutschbein, C. & Seibert, J. 2010 Regional climate models for hydrological impact studies at the catchment scale: a review of recent modeling strategies. *Geography Compass* **4** (7), 834–860.

Teutschbein, C. & Seibert, J. 2012 Bias correction of regional climate model simulations for hydrological climate-change impact studies: review and evaluation of different methods. *Journal of Hydrology* **456**, 12–29.

Thom, H. 1952 Seasonal degree-day statistics for the United States 1. *Monthly Weather Review* **80** (9), 143–147.

Thom, H. C. 1958 A note on the gamma distribution. *Monthly Weather Review* **86** (4), 117–122.

Varis, O., Kajander, T. & Lemmelä, R. 2004 Climate and water: from climate models to water resources management and vice versa. *Climatic Change* **66** (3), 321–344.

Vogel, R. M. & Fennessey, N. M. 1995 Flow duration curves II: a review of applications in water resources planning 1. *JAWRA Journal of the American Water Resources Association* **31** (6), 1029–1039.

Wagner, S. 2008 *Water Balance in a Poorly Gauged Basin in West Africa Using Atmospheric Modelling and Remote Sensing Information.* PhD Thesis, University of Stuttgart, Germany.

Warnick, C. C. 1984 *Hydropower Engineering.*

Watanabe, M., Suzuki, T., O'ishi, R., Komuro, Y., Watanabe, S., Emori, S., Takemura, T., Chikira, M., Ogura, T., Sekiguchi, M. & Takata, K. 2010 Improved climate simulation by MIROC5: mean states, variability, and climate sensitivity. *Journal of Climate* **23** (23), 6312–6335.

First received 13 April 2021; accepted in revised form 24 August 2021. Available online 7 September 2021

Impact of Climate Change on Hydrology and Water Resources doi: 10.2166/wcc.2021.200

Perception of the inhabitants of the department of Caldas, Colombia on the effects of climate change on water quality

Jenny Paola Ríos Hernández [a,*], Olga Lucía Ocampo López [a], Paula Tatiana González Pérez [b], Fabián Guillermo Gaviria Ortiz[c] and Victoria Salazar Gil[d]

[a] Department of Mechanics and Production, Universidad Autónoma de Manizales, Old Railway Station, Manizales, Caldas, Columbia
[b] Department of Oral Health, Universidad Autónoma de Manizales, Old Railway Station, Manizales, Caldas, Columbia
[c] Regional Autonomous Corporation of Caldas – Corpocaldas, Cl. 21 # 23-22, Manizales, Caldas, Columbia
[d] Department of Territory and Peace, Universidad Autónoma de Manizales, Old Railway Station, Manizales, Caldas, Columbia
*Corresponding author. E-mail: jennyp.riosh@autonoma.edu.co

JPRH, 0000-0003-0750-7835; OLOL, 0000-0002-6394-977X; PTGP, 0000-0003-0445-9196

ABSTRACT

Climate change has direct effects on the availability and quality of water for human consumption. In order to propose actions aimed at reducing vulnerability caused by water shortages and risk management required due to extreme events, real knowledge of the community`s perception is vital. This study developed in the department of Caldas, in the Colombian Andean region, analysed the perception of the incidence of climate change particularly related to water resources. To achieve this, a survey was used with various actors based on the first National Survey of Public Perception of Climate Change. The results show that the respondents perceive that the availability and quality of water are indeed highly threatened by climate change. As actions for adaptation, they suggested the promotion of the protection of hydrographic basins and a greater control of dumping liquids into surface water sources. Finally, they requested increased opportunities to improve water governance and participation in decision-making bodies regarding climate change, which they see as a fundamental aspect to achieve a real climate empowerment that can lead to action and adaptation in the territories in emerging countries.

Key words: climate change, climate empowerment, governance, perception, water

HIGHLIGHTS

- Water quality is perceived to be affected by the effects of climate change.
- The inhabitants of the department of Caldas, Colombia observe that rainfall and temperature are increasingly variable due to climate change.
- The department of Caldas, Colombia has climatic conditions that make it vulnerable to climate change.

GRAPHICAL ABSTRACT

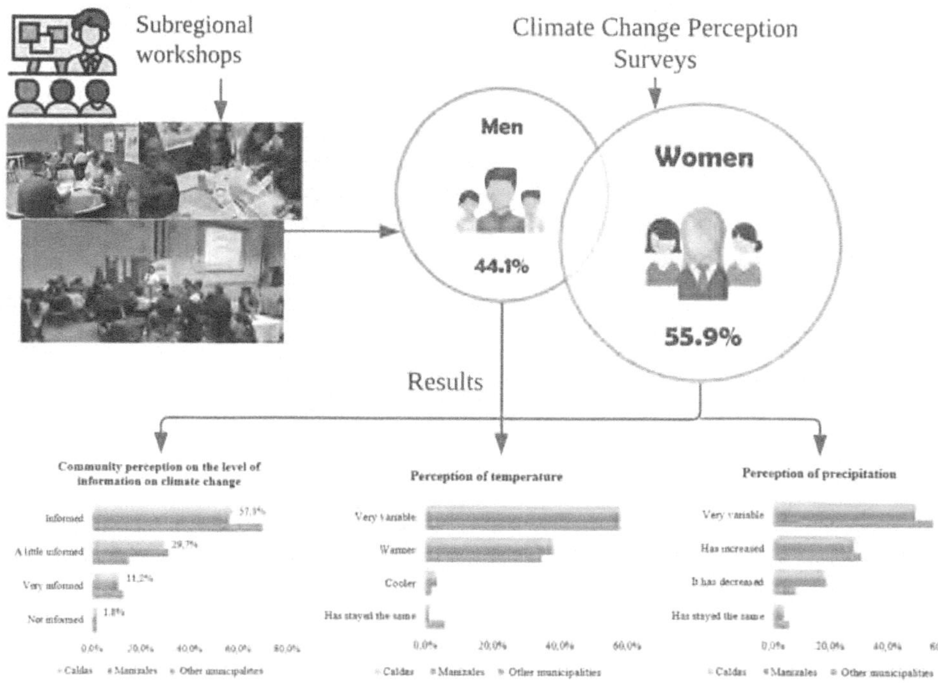

INTRODUCTION

One of the challenges for today's society is climate change, defined as climate change attributed directly or indirectly to human activity which alters the composition of the atmosphere due to greenhouse gases (GHGs) (CMNUCC 1992; Jol *et al.* 2009). Actions, such as industrialization, deforestation and large-scale agriculture, have contributed to the increase in GHGs in the atmosphere (IPCC 2014; ONU 2020).

The climate change crisis is strictly related to the vulnerability of water sources, especially their quality and availability (Ocampo López 2011; Leveque *et al.* 2020). The variability of the water cycle has increased due to climate change and has caused extreme meteorological phenomena, as well as reduced the possibility of forecasting the availability of water resources and also caused a marked decrease in water quality. This, therefore, constitutes a threat to sustainable development, biodiversity and the human right to safe drinking water and basic sanitation throughout the world (IPCC 2014; ONU 2019; OMM 2020). Some of the effects of climate change by sector are displayed in Table 1, based on Bárcena *et al.* (2020). The populations most affected by climate change are found in coastal regions and in tropical countries where fragile and highly susceptible ecosystems abound. In these regions, extreme climatic events and the spread of infectious diseases are evident (Becerra *et al.* 2020; Harper *et al.* 2020).

Geographical location increases the degree of vulnerability of tropical countries. This is why Colombia is so vulnerable to climate change and its effects (Ocampo & Vélez 2014; Sauchyn *et al.* 2016). Different studies (Pabón 2003; Poveda *et al.* 2007; Poveda 2009; Pabón Caicedo 2012; Ocampo-López *et al.* 2020) have analysed long-term trends in the precipitation and temperature variables in Colombia which had an effect on surface runoff. On average, the temperature increase was 0.13°C for the period between 1971 and 2000. Increases in the average air temperature of 1.4°C for 2040 have been estimated, as have their effects on the quantity and quality of surface water sources in the country (IDEAM 2020).

Colombia has low participation in GHG emissions; however, it does present a high vulnerability to the effects of climate change due to its environmental, social and economic conditions. Therefore, seeking adaptation measures which help to counteract the adverse effects of climate change is essential. For this purpose, the National Climate Change Policy has been defined, which incorporates management of emissions into public and private decisions to promote climate-resilient and low-carbon development, reduce risks and take advantage of opportunities (Minambiente 2017). At the regional level,

Table 1 | Potential impact and risks of climate change in Latin America

Threatened sector	Key risks	Climate drivers
Agricultural	Decrease in production and in the quality of food and income; price increases.	• Rising temperatures and extreme episodes. • Erratic precipitation and extreme episodes. • Fertilization due to increases in the concentration of CO_2.
Water resources	Less water availability in semi-arid regions and those dependent on the melting of glaciers; floods in rural and urban areas related to extreme rainfall.	• Tendency to increase in temperature. • Tendency to suffer droughts. • Covered with snow. • Increased rainfall.
Biodiversity and forests	Disappearance of forests, coral bleaching and loss of biodiversity and ecosystem services.	• Increase in deforestation. • Fertilization due to increases in the concentration of CO_2. • Tendency to increase temperature. • Acidification of the oceans.
Health	Spread of vector-borne diseases at higher altitudes and latitudes than in their original distribution.	• Temperature increase. • Increased rainfall.
Tourism	Loss of infrastructure, rise in sea level, appearance of invasive species and extreme phenomena in coastal areas.	• Sea level rises. • Extreme temperatures. • Extreme rainfall and flooding.
Socioeconomic factors	Decrease in the income of vulnerable populations, mainly agricultural, and increase in income inequality. Deterioration of the habitability conditions and quality of life of displaced populations or people living in poverty.	• Increase in temperature and extreme episodes. • Tendency to suffer droughts. • Erratic precipitation or precipitation outside the parameters of crop physiology.

Source: adapted from Bárcena *et al.* (2020).

this policy adopts the Comprehensive Territorial Climate Change Management Plans (PIGCCT – initials in Spanish) which are based on vulnerability analysis and GHG inventory performed by existing administrative units, which in the case of Colombia are the geographical departments. Based on this analysis, mitigation and adaptation measures and actions in the territory have been identified (Bárcena *et al.* 2020; Minambiente 2020).

In the department of Caldas, the PIGCCT was adopted to advance climate change management. The geographical location, the biophysical conditions and the socioeconomic dynamics make this territory vulnerable to climate change. Caldas is in the Colombian Andean region. Most of its towns and villages form part of the coffee-growing region declared by UNESCO to be a Cultural Heritage of Humanity (UNESCO 2009). Caldas is characterized by a marked altitudinal gradient which ranges from 5,286 meters above sea level (masl) in the Los Nevados National Natural Park down to 170 masl in the town of La Dorada located in the valley of the Rio Grande de La Magdalena, the main Colombian river. This area is made up of different strategic ecosystems, such as glaciers, wetlands, high Andean forest and tropical dry forest (Poveda *et al.* 2011). The region is extraordinarily biologically diverse. The avifauna is especially noteworthy having 50% of all the species reported at the national level and there are several different Important Areas for the Conservation of Birds and avitourism (Matallana Tobón 2017; Corpocaldas 2019). The Florence National Natural Park is in Caldas and has an average annual rainfall that reaches 8,000 mm and an average temperature of 19°C. The humid tropical forest with its climatic characteristics houses the greatest diversity of amphibians in the country with a high degree of endemism. It is a park of strategic importance, hydrologically speaking, because of the Magdalena river basins. All these characteristics make the department of Caldas a key hotspot for identifying signs of climate change (Pabón-Caicedo 2002).

In an area of 7,888 km^2 live 923,472 people, with 75.4% concentrated in urban areas. However, the geographical department is still agricultural and the majority of its territory is rural. The department has a high aging index (67.1%), one of the highest in the country, although the youth index is also significant at 23.74% (DANE 2019). This territory has 55,801 indigenous inhabitants, comprising 6% of the population, with the Emberá, Emberá Chamí and Cañamomo being the largest ethnic groups present, distributed in reservations, partialities and settlements. Also, 2% of the population are Afro-descendants and are distributed in 29 organizations.

This population diversity makes the community-based adaptation approach one of the guidelines of the National Climate Change Adaptation Plan which seeks to empower the most vulnerable communities to address the effects of climate change in the short, medium and long term (DNP 2016). Working with this adaptation approach, understanding the perception of the inhabitants regarding their vulnerability to climate change and its effects on their territories, ecosystems and resources is a necessity. This perception forms the first step in defining adaptation measures based on local priorities, needs, knowledge and capacities. Perception exercises have been prioritized at the national level as essential in reducing the vulnerability of communities (DNP 2016), because 'adaptation is specific to the place and context, and there is no single method to reduce risks that are appropriate for all situations' (IPCC 2015).

In Colombia, the Third National Communication on Climate Change (TCNCC – initials in Spanish) carried out the first national survey of public perception of climate change in 2016. The respondents perceived that climate change has had a direct influence on water sources and on the availability of water resources and reported the decrease in access to water as one of its main effects (IDEAM *et al.* 2016).

For Caldas, the community perception exercise was carried out in different workshops in the six sub-regions of the department. The National Study of Social Perception guidelines created by the TCNCC were used. This study presents the results of this perception exercise in relation to the knowledge of climate change, the change in climate variables, the threat and the impacts of climate change. The methodology is detailed below and the perception results are presented, with an emphasis on the effects of climate change on the availability and quality of water in the region.

MATERIALS AND METHODS

This study was carried out within the Department of Caldas' PIGCCT framework and also within the framework of the call for young researchers made by the Ministry of Sciences of Environmental Health (Minciencias en Salud Ambiental).

The research was both qualitative and quantitative, with a cross-sectional exploratory, descriptive and interpretative approach. The study was carried out to determine society's perception and especially the point of view of institutional actors, community leaders and experts on the effects of climate change on the department of Caldas' water resources.

Caldas is one of the 32 departments which make up the Republic of Colombia, it has an area of 7,888 km^2. Its physical geography is Andean, lying between the Western and Central mountain ranges (see Figure 1). It is divided into 27 towns which are distributed in the six subregions (DNP & PNUD 2009). The department has a population of 923,472 inhabitants, which represents 2.1% of the total national population. Of the total population, 48.5% are male and 51.5% are female. Manizales, the capital of the department, is home to 43.4% of the population, followed by La Dorada (8%), Villamaría (6%), Riosucio (6%) and Chinchiná (5%) (DANE 2019).

The six subregions of the department are Alto Occidente, Occidente Prospero, Centro Sur, Magdalena Caldense, Alto Oriente and Norte. All the thermal floors can be found in this department, from the hot valleys of the Magdalena River and the Cauca River to the perpetual snows of Nevado del Ruiz (Gobierno de Caldas 2017).

As reported by the Autonomous Regional Corporation of Caldas – Corpocaldas (2019), which is the environmental authority in the department, the communities are highly vulnerable and are exposed to hydrometeorological events and socio-natural threats, such as erosive processes and landslides; mass movements; flash floods, flooding and forest fires. The recurrence of extreme climatic events is of great concern due to the abundant rain precipitation which ranges from 1,200 to 8,000 mm.

A survey was designed to ascertain the perception of climate risk in the different communities, based on the first National Survey of Social Perception carried out by the TCNCC. The survey consisted of the following questions about the water resource dimension:

- What is your level of information on climate change?
    - o Very informed.

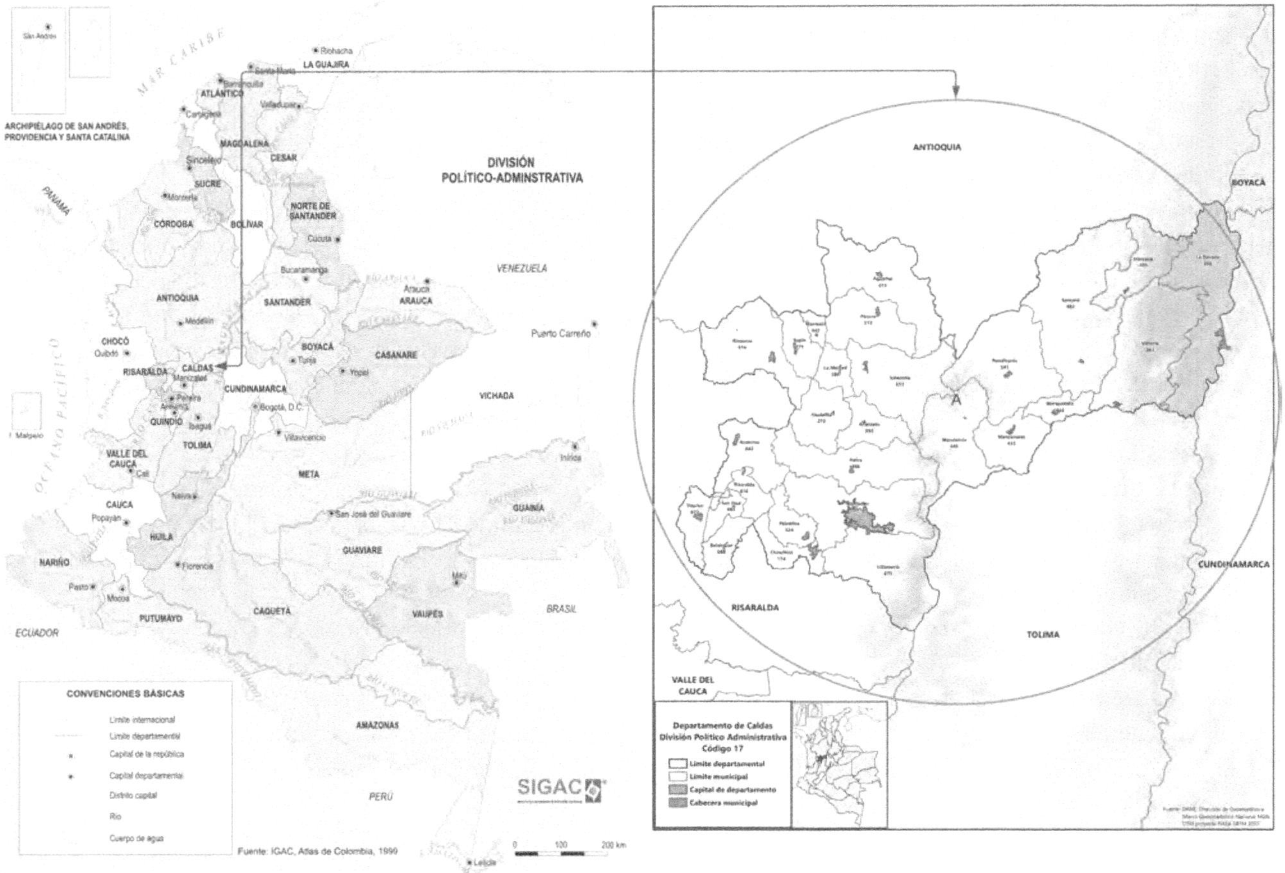

Figure 1 | Geographical location of the department of Caldas, source: IGAC. Atlas of Colombia (1999); DANE Geoportal. Map of Caldas (2018).

  - o Informed.
  - o Little Informed.
  - o Not Informed.
- Means by which you obtained information on climate change:
  - o Television.
  - o Radio.
  - o Newspapers.
  - o Internet.
  - o Specialized channels.
- Perception of the variation in temperature in your town or village in the last decade:
  - o Warmer.
  - o Cooler.
  - o Very variable.
  - o Has stayed the same.
- Perception of the amount of rain in your municipality in the last decade
  - o Has increased.
  - o Has decreased.
  - o Very variable.
  - o Has stayed the same.
- What is the threat of climate change to the availability of water? Consider your assessment on the following scale:
  - o Not threatened.
  - o A little threatened.

- o Threatened.
- o Highly threatened.
- What is your perception of the main impacts of climate change in Caldas? Mark the 5 (five) most important with an X
  - o Increase in food prices.
  - o Periods of extreme heat or cold.
  - o Unexpected temperature changes.
  - o Droughts.
  - o Increase in the cost of the domestic electricity service.
  - o Forest fires.
  - o Loss of fauna and flora.
  - o Water rationing.
  - o Severe rains or heavy storms.
  - o Increased incidence of tropical diseases (such as Zika, Dengue fever, Yellow Fever, Chikungunya).
  - o Soil erosion.
  - o Flooding.
  - o Landslides.
  - o Migratory movements of human beings.
  - o Increase in the number of earthquakes.
- What actions of mitigation and adaptation do you propose for your town or village?

Sampling: The survey was carried out by those attending the sub-regional workshops on Climate Change, which included the participation of community action boards, local action boards, territorial environmental health councils, municipal risk management councils, territorial planning councils, the Inter-institutional Committee for Environmental Education, citizen oversight bodies, municipal councils and/or committees for rural development, culture councils and/or committees, and councils and/or youth committees, indigenous communities and ethnic groups. In total, 583 people were surveyed, of which 257 were men (44.1%) and 326 were women (55.9%); 270 of these live in the capital city of Manizales (46.3%) and 313 in the towns and villages of the geographical department (53.7%).

RESULTS

According to the results obtained regarding the level of information on climate change, it is evident that in the department of Caldas, 57.3% of the community consider themselves informed, while 31.5% consider themselves either a little or not informed at all (see Figure 2). Highlighting that 68.5% of the inhabitants of the geographical department consider themselves either informed or very informed about climate change is important. This facilitates the correlation of the perception of the impacts of climate change in the region.

Figure 3 shows the most common means of communication by which people obtain information about climate change. The internet is the means by which the population in Caldas, Manizales and other towns obtain their information.

Over the last decade, 57.7% of the communities of the department consider that temperature has been highly variable, while 37.7% consider that it has been warmer (see Figure 4). On another point, the communities perceive highly variable rains or rainfall (50%), while 28.5% consider that rainfall has increased (see Figure 4).

Figure 5 illustrates the risk of climate change to the quality and availability of water. Forty-nine percent of the population feels this resource is very threatened and that it is essential for people's quality of life. Also, they perceive that the main impact of climate change is the rationing of water or non-purified water. In the city of Manizales, the capital of the department, 87% of those surveyed perceive the availability of water as threatened and highly threatened as an effect of climate change.

The main impacts of climate change for the department according to the perception study are landslides (11.3%); unexpected changes in temperature (11%); periods of extreme heat or cold (10.6%) and loss of fauna and flora (9.5%). In the subregions, residents chose impacts such as unexpected changes in temperature (10.6%); droughts (10.2%); landslides (10.2%) and periods of extreme heat or cold (10.2%) (see Figure 6). In contrast to these results, it can be seen that in other towns and villages of the geographical department, one of the greatest impacts of climate change chosen were droughts (10.2%); water rationing (7.5%) and severe rains or intense storms (7.3%).

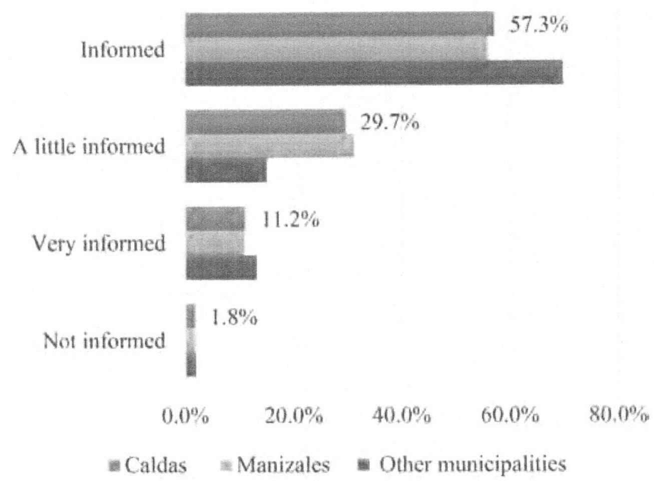

Figure 2 | Community perception on the level of information of the communities of Caldas, Colombia about climate change.

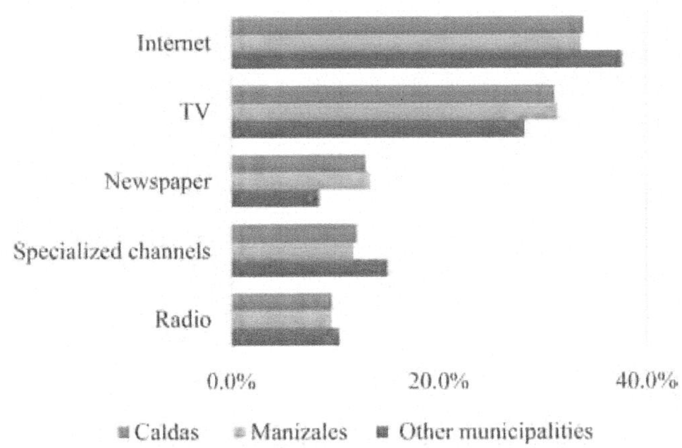

Figure 3 | Perception of community of information media about climate change.

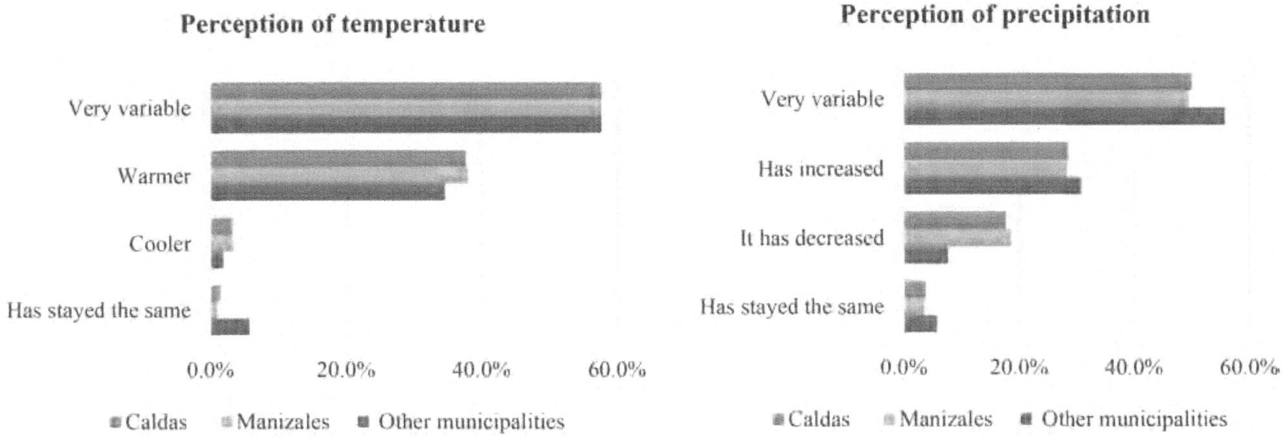

Figure 4 | Perception of the variation in temperature and rainfall in the last decade.

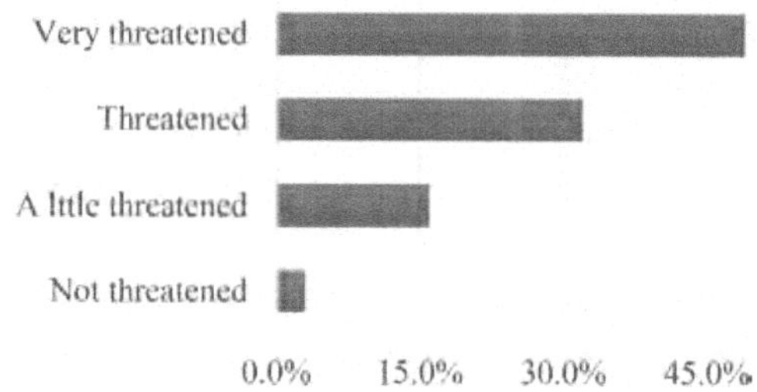

Figure 5 | Perception of the threat of change in water availability in Caldas, Colombia.

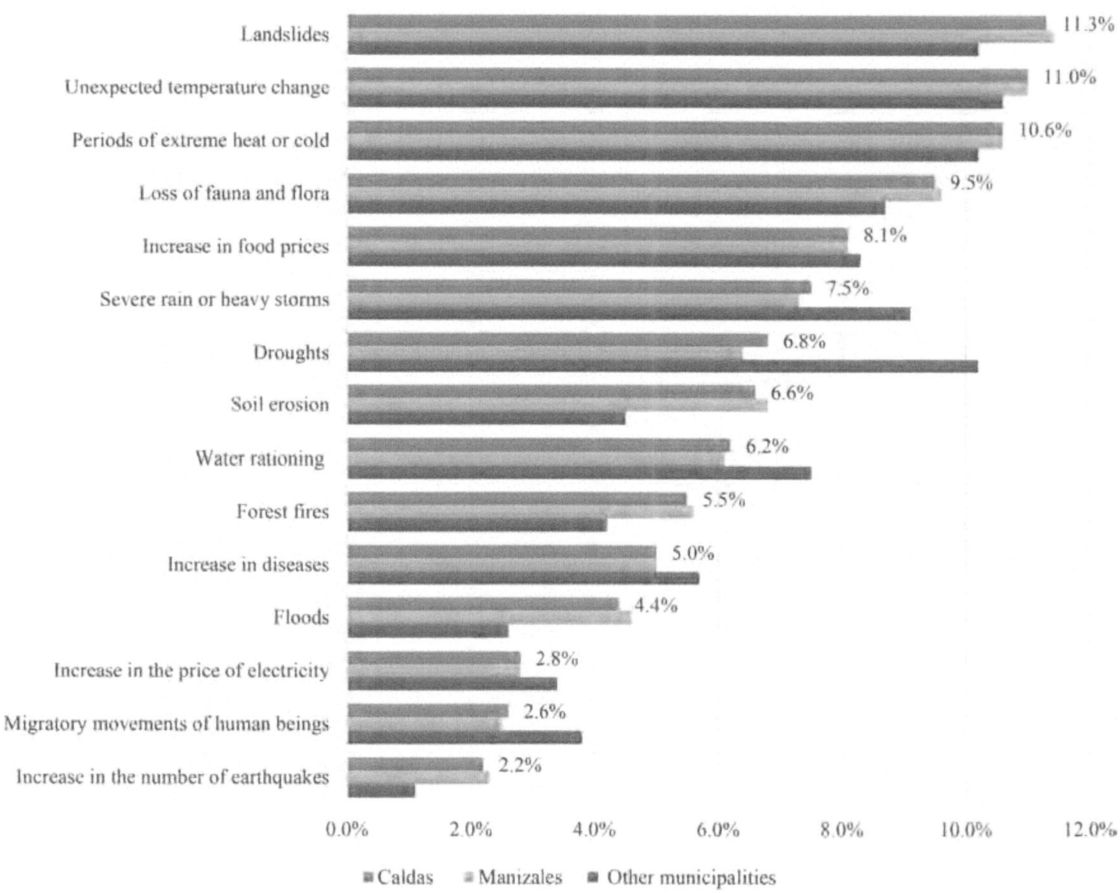

Figure 6 | Perception of the main impacts of climate change.

Some adaptation measures to climate change for the water resource dimension proposed by those surveyed are: defining watershed protection zones and areas supplying aqueducts, planting native trees, controlling logging and recovering water recharge areas, promoting protection and conservation watersheds, control and monitoring the quality of water sources, control of discharges from factories, promoting clean-up days for water sources, developing a specialized centre for work in training youth committees in water care, strengthening basin councils and early warning systems.

The adaptation and mitigation measures for the proposed sanitation dimension are: managing the construction of wastewater treatment plants; proper management of discharges; encouraging projects for the correct disposal of liquids caused by coffee processing; maintenance of sewerage systems; construction of septic tank systems and scupper cleaning, among others.

DISCUSSION

At the global level, various studies have been carried out concerning the perception of populations on climate change and how this global phenomenon has a direct influence on the availability and quality of water (Echavarren *et al.* 2019). The data obtained from these perception studies provide relevant information for designing adaptation strategies to respond to climate change. The main effects that communities perceive as being caused by climate change are droughts (Bishop 2013; Takakura *et al.* 2020; Thokchom 2020; Wang *et al.* 2021) and frequent floods (Yang *et al.* 2018; Thokchom 2020).

In this study, according to the perceptions provided by the inhabitants of the department of Caldas, in relation to the level of information, 57.3% of the surveyed population consider themselves informed regarding climate change. However, these results differ from those obtained in the first national survey of public perception called: *What do Colombians think about climate change?* where 75.1% state that they are a little or not at all informed and only 18.7% of the population consider themselves informed about the phenomenon. It should be noted that this survey was carried out in 57 towns throughout the country, where a total of 1,130 surveys were conducted (IDEAM *et al.* 2016). Caldas is one of the geographical departments, which has advanced the most in environmental education in the country. It has a well-defined policy, which integrates the subject of climate change in the School Environmental Projects and in the University Environmental Projects, as well as the Citizen Projects of Environmental Education. The theme of climate change is also incorporated in the ONDAS programme of the Ministry of Sciences as a line of research, which aims to promote in children and young people an interest in research and the culture of science, technology and innovation. Caldas has been a leader in this programme, which is functioning in all its 27 towns.

At the national level, approximately 90.9% of the population learned about the phenomenon from the television, followed by the internet (5.3%) (IDEAM *et al.* 2016). In Caldas, however, means of communication beyond the internet and television have been highlighted. The study shows that newspapers, radio and specialized channels are also sources of information regarding climate change. This diversity in communication channels is the product of the promotion of the Social Appropriation of Knowledge strategy led by the Ecosystem of Science, Technology and Innovation in the department of Caldas, which seeks to effectively reach the territories using all available means. In rural areas due to limitations in access to the internet, other communication media such as radio are essential.

Changes in the climatic variables are also perceived by the inhabitants of the department. Some respondents consider that the temperature has been highly variable (57.7%) and others consider that it has been warmer (37.7%) over the last decade. At the national level, the majority of respondents (56.7%) considered that the temperature was warmer and for others (36.6%) it was highly variable (IDEAM *et al.* 2016). Local studies of climate change signs agree with trends of increases in temperature, determined by the altitudinal gradient and changes in vegetation cover (McMartin *et al.* 2018) which alter the water resources. Likewise, climate variability is marked by the dynamics of the Southern Oscillation phenomenon (ENSO), where the inhabitants feel the increases in temperature during the El Niño phenomenon and the reductions in La Niña, influence the extreme events and the adaptation measures of productive systems (Quintero-Angel *et al.* 2012). At an international level, people perceive the hottest climate (Assan *et al.* 2020) and attribute that as a cause of periods of drought. The increase in temperature and prolonged periods of droughts are perceived as problems for the agricultural sector (Nnko *et al.* 2021). For the region, the respondents mentioned problems of drought and increases in food prices.

Caldas is vulnerable to increases in temperature, which have a negative impact on the moors, on endemic species, as well as on the dynamics of pests and invasive species. There is also evidence of a reduction in strategic ecosystems that affect surface water bodies (Corpocaldas 2019). Additionally, signs of climate change have been detected in hydrographic basins and in extreme events which affect the different economic activities and affect vulnerability and risk in the territory (Mussetta *et al.* 2017). Warmer air can retain more moisture; this is associated with stronger storms and weather-related disasters. More specifically, severe storms put pressure on water quality and increased suspended solids reduce the level of dissolved oxygen in the superficial waters (Fathy & Rosenbach 2020). In the department of Caldas, water rationing occurs in the rainy season due to the load of suspended solids and the level of turbidity of the water. Respondents report a higher incidence of water rationing in rural villages and areas.

Respondents perceive precipitation as highly variable (50%) and increasing (28.5%). At the national level, however, both decreases (54%) and high variability (31.2%) are perceived. In the Caldas region, the ENSO affects rainfall with increases during La Niña and decreases during El Niño. In 2010 and 2011, a La Niña event was registered with social, environmental and economic repercussions of great magnitude (BID – CEPAL 2012) and in the department of Caldas, there were events triggered by the heavy rain including mass movements, landslides and floods. In 2018, an El Niño event occurred which affected the entire country and in the department of Caldas, there were effects on the availability of water in several towns, according to the Regional Water Study. In another point, fires in vegetation cover were registered especially in the lower areas of the valleys of the Cauca and Magdalena rivers. Precipitation and storms have increased and these extreme weather events affect food security, biodiversity and the well-being of agricultural households (Assan *et al.* 2020). In the Caldas region, respondents report the presence of heavy rains and storms, landslides, erosion and floods.

Climate change is one of the main forces, which cause the loss of biodiversity (Ocampo López 2011) and essential resources for survival, such as water (McMartin *et al.* 2018). The inhabitants of Caldas report the loss of flora and fauna and problems with water resources. Also, climate change and climate variability affect the demand for water and intensify the pressures on this resource (Bárcena *et al.* 2020). In Caldas, local studies show these effects (Jiménez *et al.* 2017). Despite the availability of water resources, which are abundant for the area according to National Water Studies (IDEAM), 49% of the population considers this resource to be very threatened because of climate change. In the workshops, they expressed the importance of water for their quality of life and mentioned that the abrupt changes in some territories due to mismanagement of water have led to the drying streams and springs, polluted rivers, floods, landslides and avalanches. Consequently, they prioritize actions for risk reduction and adaptation to climate change with emphasis on the conservation of ecosystem services, and the contributions of nature, such as domestic water supply for people.

For indigenous communities, achieving harmony with Mother Earth and returning to ancestral practices such as agro-ecology is necessary. Local studies (Ulloa *et al.* 2008) of the perception of the indigenous peoples of Caldas on climate change show that this is not a situation which is directly attributed to nature but is seen more as a consequence of human actions. The territories of the indigenous communities of Caldas face environmental problems related to the use of wood, open permits for forest use, lack of protection of micro-watersheds, expansion of the agricultural frontier, use of pesticides and the mismanagement of non-reusable waste. Even though the Regional Autonomous Corporation of Caldas (Corpocaldas) has made progress in the planning and management of hydrographic basins, deforestation, illegal mining and the illegal commercialization of species remain latent problems in the territory requiring joint work and, the need for the strengthening territorial capacities and actions towards the governability and governance of water, especially.

Droughts, river erosion and floods are the main climatic risks which in turn have a significant effect on agriculture in the regions (Ahmed *et al.* 2021). The agricultural emphasis of Colombia and especially of the department of Caldas makes it imperative to look at adaptation in agriculture (Quintero-Angel *et al.* 2012). The residents' perception of the vulnerability and risks associated with landslides, droughts and floods prioritize risk management and adaptation to climate change actions. Although the environmental authorities have made progress in risk management, especially with the installation of the Caldas Environmental Monitoring System, for risk reduction processes with disaster mitigation and response work, a greater approach to knowledge of risk is still required. Early warning systems and the promotion of adaptation actions with an emphasis on community-based adaptation and ecosystem-based adaptation are also required. For this, involving residents, in consultative capacities, as well as in the decision-making exercises, is essential. This will allow for prioritizing actions based on local knowledge.

CONCLUSIONS

The people of the Caldas geographical department see climate change as a series of variations in climate that are clearly visible in changes in temperature and rainfall. The residents state that these effects on the climate have caused the loss of ecosystems and biodiversity and have affected hydrographic basins, especially the quality and availability of water. They see themselves as highly threatened by the consequences of climate change and mention landslides, loss of fauna and flora, increased food prices, droughts, soil erosion and water rationing among the main impacts on their communities.

The protection of the hydrographic basins, planting of trees, control of discharges, empowerment of young people and the strengthening of the hydrographic basin councils are all adaptation actions suggested by the communities. Knowledge of risk is essential in the territory to promote climate actions with community participation techniques, such as early warnings and adaptation strategies based on their ecosystems.

RECOMMENDATIONS

In order to achieve effective progress in the implementation of local adaptation and mitigation measures, continuing with environmental education strategies and the social appropriation of knowledge is necessary, as well as greater coordination between government, public–private companies, academia and communities. Joint work involves a complex and systemic approach which leads to a greater understanding of territorial dynamics and the contexts in which they develop. Understanding the perceptions of the inhabitants facilitates interventions and provides new information which must be incorporated so as to guarantee the effectiveness of the programmes.

DATA AVAILABILITY STATEMENT

All relevant data are included in the paper or its Supplementary Information.

ACKNOWLEDGEMENTS

We are grateful for the support of Minciencias and the research groups in Public Health, Sustainable Regional Development, Mechanical Design and Industrial Development; Corpocaldas; the departmental Government of Caldas and the Autonomous University of Manizales.

REFERENCES

Ahmed, Z., Guha, G. S., Shew, A. M. & Alam, G. M. M. 2021 Climate change risk perceptions and agricultural adaptation strategies in vulnerable riverine char islands of Bangladesh. *Land Use Policy* **103**, 105295. https://doi.org/10.1016/j.landusepol.2021.105295.

Assan, E., Suvedi, M., Olabisi, L. S. & Bansah, K. J. 2020 Climate change perceptions and challenges to adaptation among smallholder farmers in semi-arid Ghana: a gender analysis. *Journal of Arid Environments* **182**, 104247. https://doi.org/10.1016/j.jaridenv.2020.104247.

Bárcena, A., Cimoli, M., García-Buchaca, R., Samaniego, J. & Pérez, R. 2020 La emergencia del cambio climático en América Latina y el Caribe: ¿seguimos esperando la catástrofe o pasamos a la acción? In: *Libros de la Comisión Económica para América Latina y el Caribe (CEPAL), N° 160*. Available from: www.cepal.org/apps.

Becerra, M. J., Pimentel, M. A., De Souza, E. B. & Tovar, G. I. 2020 Geospatiality of climate change perceptions on coastal regions: a systematic bibliometric analysis. *Geography and Sustainability*. https://doi.org/10.1016/j.polar.2020.100556.

BID – CEPAL 2012 *Valoración Económica de Daños Y Pérdidas. Ola Invernal Colombia 2010–2011.*

Bishop, B. H. 2013 Drought and environmental opinion: a study of attitudes toward water policy. *Public Opinion Quarterly* **77** (3), 798–810.

CMNUCC 1992 *Convención Marco de las Naciones Unidas Para el Cambio Climatico. 50*. Available from: https://unfccc.int/files/essential_background/background_publications_htmlpdf/application/pdf/convsp.pdf.

Corpocaldas 2019 *Plan de Acción Institucional 2016–2019 |Actualización del Diagnóstico Ambiental de Caldas*. Available from: http://www.corpocaldas.gov.co/publicaciones/331/2016-2019/06-16/ActualizacionDiagnosticoPA-Web.pdf.

DANE 2018 *Geoportal DANE – Descarga de Mapas Departamentales Divipola*. Available from: https://geoportal.dane.gov.co/servicios/descarga-y-metadatos/descarga-divipola/.

DANE 2019 *Resultados Censo Nacional de Población Y Vivienda 2018 Manizales, Caldas*. Available from: https://www.dane.gov.co/files/censo2018/informacion-tecnica/presentaciones-territorio/191003-CNPV-presentacion-Huila.pdf.

DNP 2016 *Plan Nacional de Adaptación al Cambio Climático de Colombia*. Available from: https://www4.unfccc.int/sites/NAPC/Documents/Parties/Colombia NAP Spanish.pdf.

DNP & PNUD 2009 *El Departamento de Caldas Frente a los Objetivos de Desarrollo del Milenio*.

Echavarren, J. M., Balžekienė, A. & Telešienė, A. 2019 Multilevel analysis of climate change risk perception in Europe: natural hazards, political contexts and mediating individual effects. *Safety Science* **120**, 813–823. https://doi.org/10.1016/j.ssci.2019.08.024.

Fathy, R. & Rosenbach, M. 2020 Climate change and inpatient dermatology. *Current Dermatology Reports* **9** (4), 201–209. https://doi.org/10.1007/s13671-020-00310-5.

Gobierno de Caldas 2017 *Identificación del Departamento de Caldas*. Available from: https://caldas.gov.co/index.php/portfolio-2/informacion-general/identificacion-del-departamento.

Harper, S. L., Wright, C., Masina, S. & Coggins, S. 2020 Climate change, water, and human health research in the Arctic. *Water Security* **10**, 100062.

IDEAM 2020 *Escenarios De Cambio Climatico*. Available from: http://www.ideam.gov.co/web/tiempo-y-clima/escenarios-cambio-climatico.

IDEAM, PNUD, MADS, DNP, & CANCILLERÍA 2016 *¿Qué Piensan los Colombianos Sobre el Cambio Climático? Primera Encuesta Nacional de Percepción Pública del Cambio Climático en Colombia*.

IGAC 1999 *Mapas Nacionales | GEOPORTAL*. Available from: https://geoportal.igac.gov.co/contenido/mapas-nacionales.

IPCC2014 Cambio climático: Impactos, adaptación y vulnerabilidad. Resúmenes, preguntas frecuentes y recuadros multicapítulos. In: *Contribución del Grupo de Trabajo II al Quinto Informe de Evaluación del Grupo Intergubernamental de Expertos Sobre el Cambio Climático*. Available from: http://www.ipcc.ch/pdf/assessment-report/ar5/wg2/WGIIAR5-IntegrationBrochure_es.pdf.

IPCC2015 Cambio climático: Mitigación del cambio climático. Resumen para responsables de políticas y Resumen técnico. Contribución del Grupo de Trabajo III al Quinto Informe de Evaluación del Grupo Intergubernamental de Expertos sobre Cambio Climático. In: *Ipcc*.

Jiménez, D. F., Orrego, S. A., Cossio, D. M., Vásquez, F. A. & Ponce, R. D. 2017 Efecto de la variabilidad climática sobre la demanda de agua para uso residencial urbano: El caso de la ciudad de Manizales, Colombia. *Desarrollo y Sociedad* **2017** (79), 91–124. https://doi.org/10.13043/DYS.79.3.

Jol, A., Raes, F. & Menne, B. 2009 Impacts of Europe's changing climate – 2008 indicator based assessment. *IOP Conference Series: Earth and Environmental Science* **6** (29), 292042. https://doi.org/10.1088/1755-1307/6/9/292042.

Leveque, B., Burnet, J.-B., Dorner, S. & Bichai, F. 2020 Impact of climate change on the vulnerability of drinking water intakes in a northern region. *Sustainable Cities and Society* **66**, 102656.

Matallana Tobón, C. L. 2017 *Información Actualizada Sobre las AICAS Reconocidas Para el País, con la Identificación de la AICAS con Potencial Para el Aviturismo.* Available from: http://repository.humboldt.org.co/bitstream/handle/20.500.11761/34085/17-076PS-6.pdf?sequence=1&isAllowed=y.

McMartin, D. W., Hernani Merino, B. H., Bonsal, B., Hurlbert, M., Villalba, R., Ocampo, O. L., Upegui, J. J. V., Poveda, G. & Sauchyn, D. J. 2018 Limitations of water resources infrastructure for reducing community vulnerabilities to extremes and uncertainty of flood and drought. *Environmental Management* **62** (6), 1038–1047. https://doi.org/10.1007/s00267-018-1104-8.

Minambiente 2017 *Política Nacional de Cambio Climático.* https://doi.org/9789588901657.

Minambiente 2020 *Aproximación A Territorio: Planes Integrales de Gestión de Cambio Climático Territoriales – PIGCCT | Ministerio de Ambiente Y Desarrollo Sostenible.* Available from: https://www.minambiente.gov.co/index.php/gestion-territorial-de-cambio-climatico/aproximacion-a-territorio-planes-territoriales-de-cambio-climatico.

Mussetta, P., Barrientos, M. J., Acevedo, E., Turbay, S. & Ocampo, O. 2017 Vulnerabilidad al cambio climático: dificultades en el uso de indicadores en dos cuencas de Colombia y Argentina. *EMPIRIA – Revista de Metodología de Ciencias Sociales* **36**, 119–147. https://doi.org/empiria.36.2017.17862.

Nnko, H. J., Gwakisa, P. S., Ngonyoka, A. & Estes, A. 2021 Climate change and variability perceptions and adaptations of pastoralists' communities in the Maasai Steppe, Tanzania. *Journal of Arid Environments* **185** (February 2018), 104337. https://doi.org/10.1016/j.jaridenv.2020.104337.

Ocampo López, O. L. 2011 Climate change and its impact on the agriculture. *Revista de Ingeniería* **33**, 115–123.

Ocampo, O. L. & Vélez, J. J. 2014 Comparative analysis of continuous simulation hydrological models in high mountain watersheds: case of the Chinchiná River. *Revista Ingenierías Universidad de Medellín* **13** (24), 43–58.

Ocampo-López, O. L., Vélez-Upegui, J., Forero-Hernández, T. & Marín-Salazar, J. 2020 Análisis de tendencias climáticas con RClimdex en el departamento de Caldas, Colombia. *Scientia et Technica* **25** (04), 1–11. https://doi.org/10.22517/23447214.22771.

OMM 2020 *Preguntas Frecuentes – Agua | Organización Meteorológica Mundial.* Available from: https://public.wmo.int/es/preguntas-frecuentes-agua.

ONU 2019 *Informe de Políticas de ONU-AGUA Sobre el Cambio Climático Y el Agua. 28.* Available from: file:///C:/Users/Lorena Ramirez/Downloads/UN-Water_PolicyBrief_Water_Climate-Change_ES (2).pdf.

ONU 2020 *Cambio Climático | Naciones Unidas.* Available from: https://www.un.org/es/sections/issues-depth/climate-change/index.html.

Pabón, J. D. 2003 El cambio climático global y su manifestación en Colombia. *Cuadernos de Geografía: Revista Colombiana de Geografía* **12**, 111–119.

Pabón-Caicedo, J. D. 2002 Cambios en los patrones de temperatura media anual del aire y precipitación anual en los páramos de Colombia. In: *Transformación y Cambio en el Uso del Suelo en los Páramos de Colombia en las Ultimas Décadas* (J. Hincapié, C. Castillo & S. Argüello, eds). Páramos y Ecosistemas Alto Andinos de Colombia en Condición HotSpot & Global Climatic Tensor. MA, IDEAM, PNUD, Bogotá, Colombia, pp. 210–333.

Pabón Caicedo, J. D. 2012 Cambio climático en Colombia: tendencias en la segunda mitad del siglo XX y escenarios posibles para el siglo XXI. *Revista de la Academia Colombiana de Ciencias Exactas, Físicas Y Naturales* **36** (139), 261–278.

Poveda, G. 2009 Evidence of climate and environmental change on water resources and malaria in Colombia. *IOP Conference Series: Earth and Environmental Science* **6** (29), 292054. https://doi.org/10.1088/1755-1307/6/29/292054.

Poveda, G., Vélez, J. I., Mesa, O. J., Cuartas, A., Barco, J., Mantilla, R. I., Mejía, J. F., Hoyos, C. D., Ramírez, J. M., Ceballos, L. I., Zuluaga, M. D., Arias, P. A., Botero, B. A., Montoya, M. I., Giraldo, J. D. & Quevedo, I. D. 2007 Linking long-term water balances and statistical scaling to estimate river flows along the drainage network of Colombia. *Journal of Hydrologic Engineering* **12** (1), 4–13. https://doi.org/10.1061/(ASCE)1084-0699(2007)12:1.

Poveda, G., Álvarez, D. M. & Rueda, Ó. A. 2011 Hydro-climatic variability over the Andes of Colombia associated with ENSO: a review of climatic processes and their impact on one of the Earth's most important biodiversity hotspots. *Climate Dynamics* **36** (11–12), 2233–2249. https://doi.org/10.1007/s00382-010-0931-y.

Quintero-Angel, M., Carvajal-Escobar, Y. & Aldunce, P. 2012 Adaptación a la variabilidad y el cambio climático intersecciones con la gestión del riesgo. *Luna Azul* **34**, 16. Available from: https://www.redalyc.org/pdf/3217/321727348015.pdf.

Sauchyn, D., Velez Upegui, J. J., Masiokas, M., Ocampo, O., Cara, L. & Villalba, R. 2016 Exposure of rural communities to climate variability and change: case studies from Argentina, Colombia and Canada. *Implementing Climate Change Adaptation in Cities and Communities*, June. https://doi.org/10.1007/978-3-319-28591-7.

Takakura, H., Fujioka, Y., Ignatyeva, V., Tanaka, T., Vinokurova, N., Grigorev, S. & Boyakova, S. 2020 Differences in local perceptions about climate and environmental changes among residents in a small community in Eastern Siberia. *Polar Science* 100556. https://doi.org/10.1016/j.polar.2020.100556.

Thokchom, B. 2020 Chapter 3. Water-related problem with special reference to global climate change in India. En P. Singh, Y. Milshina, K. Tian, D. Gusain & J. P. Bassin (eds). *Water Conservation and Wastewater Treatment in BRICS Nations*. Elsevier, pp. 37–60. https://doi.org/10.1016/B978-0-12-818339-7.00003-5.

Ulloa, A., Escobar, E. M., Donato, L. M. & Escobar, P. 2008 *Mujeres Indígenas Y Cambio Climático. Perspectivas Latinoamericanas*. Available from: http://www.carbonoybosques.org/images/stories/esp/archivos/mujeres-indigenas-y-cambio-climatico-2008.pdf.

UNESCO 2009 *Paisaje Cultural Cafetero, Colombia | Diversidad de las Expresiones Culturales*. Available from: https://es.unesco.org/creativity/story/paisaje-cultural-cafetero-colombia.

Wang, K., Liu, J., Xia, J., Wang, Z., Meng, Y., Chen, H., Mao, G. & Ye, B. 2021 Understanding the impacts of climate change and socio-economic development through food-energy-water nexus: a case study of Mekong river delta. *Resources, Conservation and Recycling* **167**, 105390.

Yang, L. E., Chan, F. K. S. & Scheffran, J. 2018 Climate change, water management and stakeholder analysis in the Dongjiang River basin in South China. *International Journal of Water Resources Development* **34** (2), 166–191. https://doi.org/10.1080/07900627.2016.1264294.

First received 15 March 2021; accepted in revised form 21 August 2021. Available online 2 September 2021

© 2022 The Authors Impact of Climate Change on Hydrology and Water Resources doi: 10.2166/wcc.2021.158

Impacts of urbanization and climate change on water quantity and quality in the Carp River watershed

Baba-Serges Zango [a,*], Ousmane Seidou [a,b], Majid Sartaj [a], Nader Nakhaei [c] and Kelly Stiles [c]

[a] University of Ottawa, 75 Laurier Ave. E, Ottawa, ON K1N 6N5, Canada
[b] United Nations University Institute for Water, Environment, and Health, 204-175 Longwood Rd S, Hamilton, ON L8P 0A1, Canada
[c] Mississippi Valley Conservation Authority (MVCA), 10970 Hwy 7, Carleton Place, ON K7C 3P1, Canada
*Corresponding author. E-mail: szang068@uottawa.ca

B-SZ, 0000-0001-6978-9538

ABSTRACT

Pressure on water resources has reached unprecedented levels during the last decades because of climate change, industrialization, and population growth. As a result, vulnerability to inappropriate water availability and/or quality is increasing worldwide. In this paper, a Soil & Water Assessment Tool (SWAT) model of the Carp River watershed located in the city of Ottawa, Ontario was calibrated and validated. The model was then used to evaluate the individual and coupled impacts of urbanization and climate change on water quantity (discharge) and quality (nitrogen and phosphorus loads). While most of the watershed is currently rural, the headwaters will undergo rapid urbanization in the future, and there are concerns about possible negative impacts on water quantity and quality. Seven scenarios were developed to represent various watershed configurations in terms of land use and climate regime. Future climate time series were obtained by statistically downscaling the outputs of nine regional climate models, run under representative concentration pathways (RCP)4.5 and RCP8.5. The impacts were evaluated at the main outlet and at the outlet of an upstream sub-watershed that would be most affected by urbanization. Results show that climate change and urbanization's impacts vary greatly depending on the spatial scale and geographic location. Globally, the annual average discharge will increase between 6.75 and 9.34% by 2050, while changes in annual average nitrogen and phosphorus loads will vary between −1.20 and 24.84%, and 19.15 and 23.81%, respectively. Local impacts in sub-watersheds undergoing rapid urbanization would be often much larger than watershed-scale impacts.

Key words: Carp River, climate change, hydrological modeling, SWAT, urbanization

HIGHLIGHTS

- A global-local approach was used to assess the impacts.
- Local impacts can be an order of magnitude higher than watershed-scale impacts.
- Impacts at the local and watershed scale might have opposite directions of change.
- Climate change will likely drive the flow and nitrogen load in the Carp River watershed.
- Urbanization will control the phosphorus load.

1. INTRODUCTION

During the last decades, pressure on natural resources has been steadily increasing, mainly driven by industrialization and population growth. The United Nations estimated that the world population would increase by 2 billion in the next 30 years, from 7.7 billion in 2020 to 9.7 billion in 2050 (United Nations 2019). This demographic change will be associated with higher standards of living, increased demand for land and water, and increased pollution. This will inevitably affect water supply, possibly disrupting domestic and industrial water uses, as well as environmental sustainability. The consequence will be lower access to essential water-related services, decreased economic prosperity, and increased conflicts around water bodies. For instance, Munia *et al.* (2020) reported that the number of people exposed to water stress would increase by 50–100%, depending on the emission scenario and population growth rate.

The Mississippi Valley Conservation Authority (MVCA) is the responsible authority for the sustainable management of the Carp River watershed located in Ontario, Canada (Figure 1). The watershed is currently mainly in a rural state; however, the headwaters are expected to undergo rapid urbanization in the future. The MVCA anticipates significant urban developments in the upcoming decades. It is expected that both urbanization and climate change will induce substantial changes in the watershed's hydrology, which in turn will affect its ecosystem health through changes in water quantity and quality. The results of this study could support authorities in planning and managing water resources and providing information for sound decision-making.

The combined effects of climate change and land-use change on water quantity and/or quality have been examined by many authors (Tu 2009; Dunn *et al.* 2012; El-Khoury *et al.* 2015; Karlsson *et al.* 2016; Kundu *et al.* 2017; Čerkasova *et al.* 2018; Choukri *et al.* 2020; Guo *et al.* 2020; Wang *et al.* 2020). These studies differ in the location of the study area, the hydrological model, the type of assessed water quantity and quality variables, and the way climate change and land-use scenarios were generated. These studies involved numerical experiments where calibrated hydrological models incorporated climatic inputs representing one or several global warming scenarios and using current and/or projected land-use maps. Although these studies confirmed the impact of urban development and climate change on streamflow, the evaluation was only limited

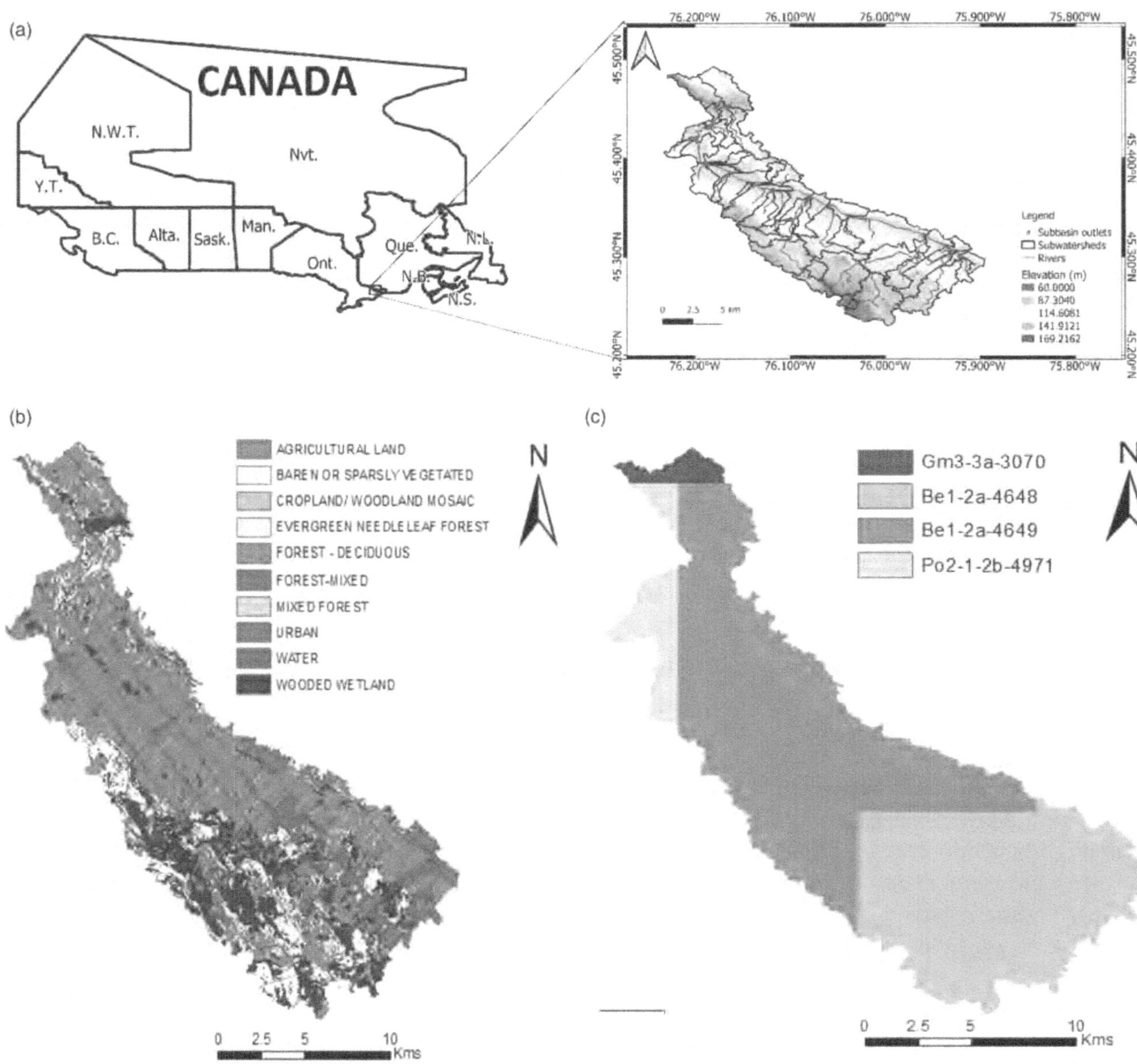

Figure 1 | Carp River watershed: (a) location and topography, (b) land-use, and (c) soil classes.

to the effects at the main outlet (global level). In this study, the impacts at the local scale were also considered by estimating the change at the outlet of the sub-watershed that was most affected by urbanization.

Since 2005, several studies related to the preparation of the Carp River model were conducted: City of Ottawa (2016), Greenland International Consulting Ltd (2009, 2010, 2011, 2014), Stantec Consulting Ltd (2006), and CH2M Hill (2005, 2006). All these works focused on flood level analysis and restoration plans, but not climate change or urbanization. A limited number of studies addressing the combined effects of climate change and urbanization in Canada can be found in the literature. Of particular interest is El-Khoury *et al.* (2015), who assessed the impact of climate change and deforestation on the South Nation watershed in opposition to urbanization.

Hydrological impacts are generally assessed by simulating hydrological processes and water exchange within an area. The impact of natural variability and anthropogenic factors on a hydrologic system can be assessed by changing the model's climatic inputs or watershed configuration (Yu 2015). The application of rainfall–runoff models varies from straightforward water balance models to more complex watershed models, such as the Annualized Agricultural Non-Point Source model (Young *et al.* 1989) and the Soil Water & Assessment Tool (SWAT) (Arnold *et al.* 1998), which can simulate loads and concentrations of various water quality variables. The selection of the appropriate model depends on several factors such as scale of the problem, computational cost, and robustness of the model. There is no single best model, as there are many plausible solutions (Ogden 2021). Therefore, a typical model selection is more guided by familiarity with the model than appropriateness (Addor & Melsen 2019). The SWAT model was selected in this research, as it has been widely used to assess the hydrology in small- and large-scale catchments: Huo & Li (2013), Khoiab & Thomb (2015), and Jajarmizadeh *et al.* (2017) in Asia; Arnold *et al.* (2012), El-Khoury *et al.* (2015), and Havrylenko *et al.* (2016) in America; Abbaspour *et al.* (2015) in Europe; and Mengistu *et al.* (2019), Choukri *et al.* (2020), and Muluneh (2020) in Africa. The SWAT is also used to investigate the relative impact of climate, land-use & land-cover (LULC) changes, and management practices on the available water resources (Arnold *et al.* 1998; Neitsch *et al.* 2009; Tirupathi & Shashidhar 2020). In addition, its structure and applicability to address different problems make it more flexible. The SWAT also allows for the simulation of water quantity and quality processes within the same model (e.g., flow, N, and P) without the contribution of other programs. For instance, the precedent model of the municipality developed by the municipality combines several hydrologic platforms and a separate hydraulic model.

Several approaches have been reported in the literature for generating climate change projections: Wang *et al.* (2020) utilized the Delta-change method, which consisted of adding an arbitrarily chosen (but realistic) variation to historical observations to represent global warming. Others used the outputs of global and regional climate models (RCMs) without any processing (Tu 2009; Čerkasova *et al.* 2018), or in combination with a statistical downscaling approach such as the Delta-change method (e.g., Dunn *et al.* 2012) and bias correction (El-Khoury *et al.* 2015; Karlsson *et al.* 2016; Kundu *et al.* 2017; Choukri *et al.* 2020; Guo *et al.* 2020). The Delta-change approach and the bias correction method have their advantages and weaknesses. They are straightforward in the generation of the climate output. However, data generated from raw global data outputs with the Delta method cannot capture changes in climate variability, and bias correction does not remove distortion because the approaches may be too simplistic to characterize climate trends (El-Khoury *et al.* 2015). In addition, the choice of the downscaling method is very determinant, as it significantly affects the outputs. For instance, Miralha *et al.* (2020) have demonstrated that the choice of bias correction method affects the direction of change and the magnitude of nutrient loads (N and P) and hydrological processes. Chen *et al.* (2012) have found that for the same global climate model (GCM), the simulated runoffs vary significantly when using rainfall provided by different statistical downscaling techniques as the input to the hydrological models. Therefore, reasonable care must be taken in the selection of downscaling method to represent future hydrological scenarios that are more realistic.

This paper aims to evaluate the individual and coupled impacts of urbanization and climate change on water quantity (discharge) and quality (nitrogen and phosphorus loads) in the carp watershed. The outputs of nine RCMs were downscaled using quantile mapping. The results are examined at two scales: the watershed scale (i.e., at the outlet of the watershed) and the local scale (i.e., at the outlet of the sub-watershed which undergoes the most drastic urbanization).

2. MATERIALS AND METHODS

2.1. Study area

The Carp River is an important natural ecosystem located west of the City of Ottawa, ON, Canada. It is 42 km long and drains an area of approximately 265 km^2. It has its headwaters in the Glen Cairn area of Kanata and flows north into the Ottawa

River at Fitzroy Harbour. The Carp River is one of the main sources of the City of Ottawa's drinking water. The annual average precipitation of the historical period of simulation (1990–2018) is 933 mm. The daily minimum, mean, and maximum of the observed streamflow at outlet 8 (Figure 2) during the same period are 0.032, 2.95, and 23.7 m^3/s, respectively. The highest elevation has an altitude of 203 m above sea level, while the outlet lies at 59 m. The average elevation measures 117 m above sea level. Based on the latest provincial land-use update in 2017 (Ministry of Natural Resources and Forestry 2019), the watershed is classified into 10 different land-use types with a dominance of agricultural land, covering more than half of the total area (52%). Urbanized area represents 10% (Figure 1). Mixed forests consisted of two types (FRST and FOMI) and only differ on biomass die-fraction. Tables 1 and 2 show the percentage of coverage of each land-use/-cover and soil type within the watershed.

2.2. Model setup

The SWAT (Arnold *et al.* 1998) is a physically-based, semi-distributed, and continuous model used to evaluate and predict the impacts of climate change and land-use management on water quantity and quality. The SWAT operates on a daily time step and can simulate continuously over a long period. The procedure used by the SWAT consists of dividing the watershed into several sub-watersheds, which are further subdivided into hydrologic response units (HRUs). An HRU is a unique combination of topographical, land-use management, and soil characteristics (Arnold *et al.* 2012).

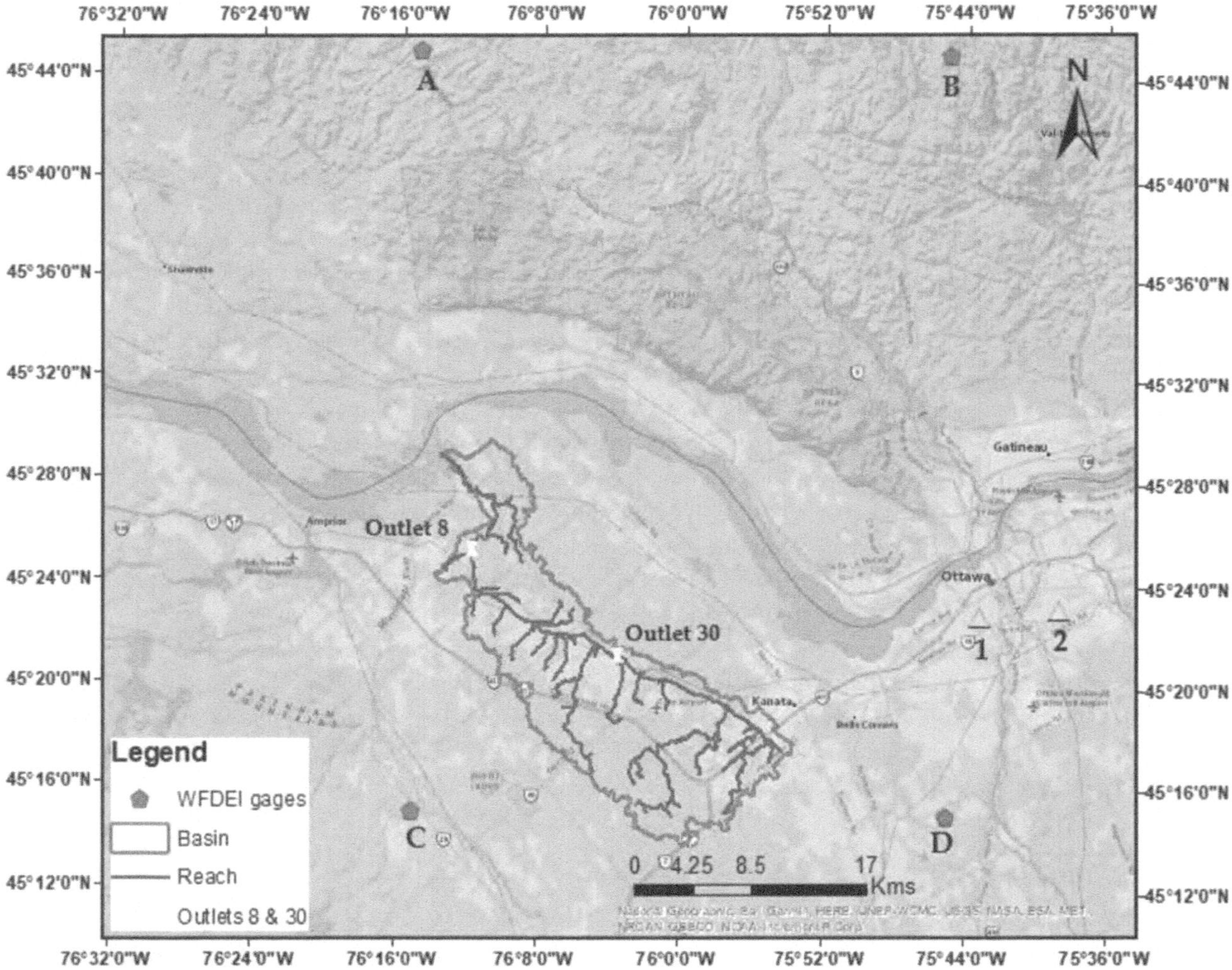

Figure 2 | Location of meteorological, hydrometric, and water quality stations.

Table 1 | Land-use characteristics of the Carp River watershed

Land use/cover type	SWAT code	Area (%)
1. Agricultural land	AGRC	52.34
2. Wooded wetland	WEWO	18.02
3. Evergreen needle leaf forest	FOEN	9.93
4. Urban	URBN	9.90
5. Forest – deciduous	FRSD	4.16
6. Forest-mixed[a]	FRST	3.25
7. Barren or sparsely vegetated	BSVG	0.82
8. Water	WATR	0.69
9. Mixed forest[a]	FOMI	0.55
10. Cropland/woodland mosaic	CRWO	0.34

[a]Two types of mixed forests defined in the SWAT with two different codes (FRST and FOMI).

Table 2 | Soil characteristics

Soil name	SWAT code	Area (%)
Loam^{*1}	Be1-2a-4649	51.29
Loam^{*2}	Be1-2a-4648	40.30
Sandy Loam	Po2-1-2b-4971	6.33
Clay	Gm3-3a-3070	2.08

Note: The two types of loam differ in their available water capacity: 0.08 mm/mm ($*^1$) and 0.175 mm/mm ($*^2$).

Hydrological simulations in the SWAT are based on the water balance equation, which is applied to the HRU scale at each time step:

$$\mathrm{SW}_t = \mathrm{SW}_0 + \sum_{i=1}^{t} (R_{day} - Q_{surf} - E_a - w_{seep} - Q_{gw}) \tag{1}$$

where SW is soil water content at the final time, t, and time 0, R_{day} is the amount of precipitation on day i, Q_{surf} is the amount of surface runoff on day i, E_a is the amount of evapotranspiration on day i, w_{seep} is the amount of water entering the vadose zone from the soil profile on day i, and Q_{gw} is the amount of return flow on day i, in mm. The calculated water, sediment, and nutrient fluxes are then routed through the stream network, where the SWAT tracks mass flow and the transformation of chemicals in the stream and streambed (Neitsch *et al.* 2009). Rainfall is partitioned into infiltration and surface runoff using the curve number (CN) method (Neitsch *et al.* 2009):

$$Q_{surf} = \frac{(P - 0.2)^2}{(P - 0.8S)} \tag{2}$$

where P is the rainfall (mm), Q_{surf} is the accumulated runoff (mm), and S is the retention parameter (mm) – representing the maximum amount of water that can be abstracted by watershed.

S is expressed in terms of a CN, a dimensionless watershed parameter ranging from 0 to 100. Groundwater is represented by two reservoirs (a shallow and a deep aquifer), which interact with surface water through infiltration, percolation, evaporation, lateral flow, and baseflow. The SWAT also simulates the complete nutrient cycle (transformation and movement) for nitrogen and phosphorus as well as the degradation of any pesticides applied in an HRU. Nutrient levels are computed on a mass basis, while it allows the input data as concentrations.

The SWAT model was set up with ArcSWAT, a GIS preprocessor of the SWAT model that combines the spatial and climatic inputs and converts them into a SWAT model. ArcSWAT requires a few subjective user inputs, including the location of sub-watershed outlets and the HRUs definition method (i.e., how soil, land use, and slope are classified and used to define HRUs). Options for HRU definition are:

(1) multiple HRUs per sub-watershed representing all combinations of the slope, land-use, and soil classes;
(2) a single HRU per sub-watershed corresponding to the dominant slope class, dominant land-use, and dominant soil classes; and
(3) a single HRU corresponding to the most extensive combinations of the slope, land-use, and soil classes.

In this paper, the first option was selected. Given the relatively flat topography of the area, a single slope class was considered.

Different land-use changes and climate change scenarios were developed by downscaling the outputs of nine RCMs; the projected climate time series were used to force the SWAT model, and changes in water quantity and water quality parameters were calculated.

2.2.1. Input data

The model setup requires three types of input data (described in Table 3). Observed water quantity and quality time series of the parameters of interest are required for calibration and validation. In this paper, time series of streamflow, total nitrogen, and total phosphorus were used. The initial land-use classes were reviewed and reclassified based on land-use types defined in the SWAT database. Observed daily precipitations and temperature were obtained from the Environment and Climate Change Canada (Environment and Natural Resources Canada 2020) at two locations in the area of Ottawa Airport. Other meteorological parameters (humidity, wind speed, and solar radiations) were retrieved from the Watch-Forcing Data ERA-Interim (Weedom *et al.* 2014). Streamflow, Total N, and Total P load data were available for different periods: measured streamflow at outlet 8 covers the 1990–2015 period, while records of total N and total P for outlets 8 and 30 were available for 2000–2018 and 2000–2007 (Figure 4). These observed water quantity and quality data were used for model calibration and validation. Table 4 presents the characteristics of the record stations used.

2.2.2. Model performance measures

The performance of the calibration/validation is generally evaluated using statistical and graphical methods. Recent works such as Jajarmizadeh *et al.* (2017) and Čerkasova *et al.* (2018) selected NS/R^2 and PBIAS, combined with time-series plots to evaluate the model performance. This study used Nash–Sutcliffe efficiency (NSE or NS, Nash & Sutcliffe 1970)

Table 3 | Data characteristics

Data	Description	Temporal scale	Spatial scale	Period	Source
Topography	Elevation – LiDAR	N/A	1 m	2015	City of Ottawa
Land use and cover	Categories of land occupation (wood, urban, and agricultural)	N/A	30 m	2017	Ministry of Natural Resources and Forestry of Canada (2019)
Soil	Soil types and physical properties	N/A	10 km	2007	Food and Agricultural Organization (2019)
Weather	Precipitation and temperature	Daily	Point	1990–2018	Environment and Climate Change Canada (ECCC)
	Solar radiation, relative humidity, and wind speed	Daily	0.5°	1990–2018	WFDEI
Water quantity	Streamflow	Daily (one measurement per month)	Point	1990–2015	HYDAT – Environment Canada
Water quality	N, P	Monthly		2000–2018 2000–2007	Ministry of Environment (Ontario)

N/A, not applicable.

Table 4 | Weather stations characteristics

	Station name (Climate ID)	Longitude	Latitude	Distance to the watershed boundary (km)
1	Ottawa CDA (6105976)	−75.72	45.38	16.5
2	Ottawa Macdonald-Cartier Int'l (6106000)	−75.67	45.32	22
A	–	−75.25	45.75	30
B	–	−75.75	45.75	42
C	–	−75.25	45.25	11
D	–	−75.75	45.25	11

and percentage bias (PBIAS, Gupta *et al.* 1999) as statistical measures. They are given by the following equations:

$$\mathrm{NS} = 1 - \frac{\sum_{i=1}^{n} (Q_m - Q_s)_i^2}{\sum_{i=1}^{n} (Q_{m,i} - \bar{Q}_m)^2} \tag{3}$$

$$\mathrm{PBIAS} = 100 * \frac{\sum_{i=1}^{n} (Q_m - Q_s)_i}{\sum_{i=1}^{n} Q_{m,i}} \tag{4}$$

where Q is the variable (e.g., discharge, N, and P quantity), the bar stands for average, m is the measured data, s is the simulated data, i is the time, and n is the total number of periods.

NS ranges from $-\infty$ to 1. The optimal value is 1, corresponding to the situation where the plot of observed data fits the simulation (Khoiab & Thomb 2015) perfectly, while values less than 0 indicate that the observed data mean is a more accurate predictor than the simulated output (Jajarmizadeh *et al.* 2017). The target interval for PBIAS is zero to $\pm 25\%$, with zero corresponding to the optimum. Positive values indicate the model is underestimating the observations, while negative values indicate an overestimation.

The statistical measures are compared to the criteria illustrated in Table 5 (Moriasi *et al.* 2007) to classify the model's performance.

2.2.3. Sensitivity analysis, calibration, and validation

Calibration consists of modifying model parameters to achieve a good fit between the simulated output and the observed data. Validation checks the model's performance using the calibrated parameters and an independent data set (not used in the calibration process). The simulation period is split into calibration and validation periods. Table 6 presents how the period of data was used for the calibration and the validation.

Given that complex models such as SWAT have hundreds of parameters, sensitivity analysis is used to select a small set of parameters to use for the calibration process. A parameter sensitivity analysis provides insights on which parameters contribute most to the output variance due to input variability (Holvoet *et al.* 2005). It is computed by altering each parameter, one by one, while all other parameters remain fixed. It is then evaluated using the values of t-stat and p-value determined with a

Table 5 | Model performance criteria from Moriasi *et al.* (2007)

	Satisfactory	Good	Very good
NS	0.5–0.7	0.7–0.8	0.8–1.0
PBIAS (%)	15–25	10–15	Less than 10%

Table 6 | Data repartition for calibration and validation

Variable		Period	
		Calibration	Validation
Discharge	Outlet 8	1990–2005	2007–2015
N	Outlet 8	2000–2012	2013–2018
	Outlet 30	2000–2003	2004–2006
P	Outlet 8	2000–2012	2013–2018
	Outlet 30	2000–2003	2004–2006

multiple regression analysis using the following equation (Abbaspour 2007):

$$g = \alpha + \sum_{i=1}^{m} \beta_i \, b_i \tag{5}$$

where g is the objective function (e.g., correlation coefficient, absolute or relative bias, and Nash–Sutcliffe model efficiency) value used to evaluate the model calibration effectiveness; b is the parameter; α is the regression constant; β corresponds to the technical coefficient attached to the variable b, and m is equal to the number of parameters.

The mean of the variations in the objective function estimates the sensitivity. The t-stat is the regression coefficient of a parameter divided by its standard error. The p-value for each parameter tests the null hypothesis that the regression coefficient is equal to zero. The higher the absolute value of t-stat and the smaller the value of p-value, the more sensitive the parameter is (Abbaspour 2007). A small p-value, typically lower than 0.05, means the sensitivity of the parameter is statistically significant (Verhagen *et al.* 2004), while a value of 0.05 indicates that there is a 95% probability that a parameter change will affect the dependent variable (Abbaspour *et al.* 2009).

Both the sensitivity analysis and the calibration were conducted using the SUFI2 (Sequential Uncertainty Fitting) algorithm implemented in the SWAT-CUP program (Abbaspour 2007). The parameters included in the sensitivity analysis are shown in Table 7.

2.3. Land-use projections

The projected land-use map was elaborated by the City of Ottawa in collaboration with the MVCA based on projections of the next 30 years (by 2050). The land-use scenario consisted of a change in the land-use area ratios and the type of urban area: from residential-medium density to commercial in the future. This conversion allows predicting impacts in severe conditions. The two land-use types differ in several characteristics, including the fraction of total impervious area, fraction directly connected to impervious, mass fraction of total N and total P in suspended solid load from the impervious area. No change in management operations in agricultural lands was completed between current and future projected conditions.

Table 8 presents the percentages of change between the current land uses and projected land uses within the watershed (map shown in Figure 3).

2.4. Climate change scenarios

Developing a climate change scenario involves considerable inherent uncertainties (Shrestha *et al.* 2020). As these scenarios will be used in planning and decision-making processes, it is crucial to estimate a range of plausible future climate conditions in studies of climate change impact. Therefore, using a single RCM or carbon emission scenario is not sufficient for assessment purposes, since it cannot provide an uncertainty range. Consequently, a combination of three RCMs with two emission scenarios were used for this study. Scenarios are defined based on IPCC AR5 representative concentration pathway (RCP)4.5 and RCP8.5 (Collins *et al.* 2013). The RCP4.5 represents medium emissions, while RCP8.5 provides future climate data with high emissions assuming high population and relatively slow income growth with modest rates of technological change and energy intensity improvements resulting in the long term to high-energy demand and greenhouse gas (GHG) emissions without any climate change policies (Wayne 2013). Various RCMs were used to generate the projected future climate data at smaller scale driven by GCMs from several institutes as listed in Table 9.

Table 7 | Parameters used for calibration and validation

Parameters	Description	Variation range
r__CN2.mgt	Initial SCS CNII value	−0.2 to 0.5
v__ESCO.hru	Soil evaporation compensation factor	0.01–1
v__GWQMN.gw	Threshold water depth in the shallow aquifer for flow	0–2,500
v__ALPHA_BF.gw	Baseflow alpha factor	0–1
a__REVAPMN.gw	Threshold depth of water in the shallow aquifer for 'revap' to occur (mm)	0–500
v__SOL_AWC.sol	Available water capacity	0–0.8
a__OV_N.hru	Manning's *n* value for overland flow	0.01–20
v__GW_REVAP.gw	Groundwater 'revap' coefficient	0.02–0.15
a__SLSUBBSN.hru	Average slope length	10–130
v__SOL_K.sol	Saturated hydraulic conductivity	0–1
a__RCN.bsn	Concentration of N in rainfall	0–15
a__ERORGN.hru	Organic N enrichment ratio	0–5
a__SHALLST_N.gw	Concentration of nitrate in groundwater contribution to streamflow from sub-basin (mg N/l)	0–1,000
a__N_UPDIS.bsn	N uptake distribution parameter	0–100
a__NPERCO.bsn	N percolation coefficient	0–1
a__ANION_EXCL.sol	Fraction of porosity (void space) from which anions are excluded	0.01–1
a__P_UPDIS.bsn	P uptake distribution parameter	0–100
a__PPERCO.bsn	P percolation coefficient	10–17.5
a__ERORGP.hru	Organic P enrichment ratio	0–5
a__GWSOLP.gw	Concentration of soluble P in groundwater contribution to streamflow from sub-basin (mg P/l)	0–1,000
a__PSP.bsn	P sorption coefficient	0.01–0.7
a__PHOSKD.bsn	P soil partitioning coefficient	100–200

Replace or *v_*: the existing parameter value is to be replaced by the given value.
Absolute or *a_*: the existing parameter value is added to a given value.
Relative or *r_*: (1+a given value) multiply the existing parameter value.

Table 8 | Projected land-use change in the Carp River watershed by 2050

Land-use type	Change (%)
1. Agricultural land	(*−2.63*)
2. Wooded wetland	(*−0.54*)
3. Evergreen needle leaf forest	(*−0.13*)
4. Urban	(+**3.64**)
5. Forest – deciduous	(*−0.17*)
6. Forest-mixed (FRST)	(*−0.17*)
7. Barren or sparsely vegetated	(+**0**)
8. Water	(+**0**)
9. Mixed forest (FOMI)	(*−0.02*)
10. Cropland/woodland mosaic	(+**0**)

bold: increase; italic: decrease

The 1985–2100 precipitation, temperature (maximum and minimum), solar radiation, relative humidity, and wind speed data sets were simulated using the RCMs under RCP4.5 and RCP8.5 scenarios. Data were downloaded from North America – CORDEX (Coordinated Downscaling Experiment: Mearns *et al.* 2017). The time series for the five variables were extracted for the period of 2021–2050 and downscaled using the quantile mapping, also known as quantile–quantile or quantile matching.

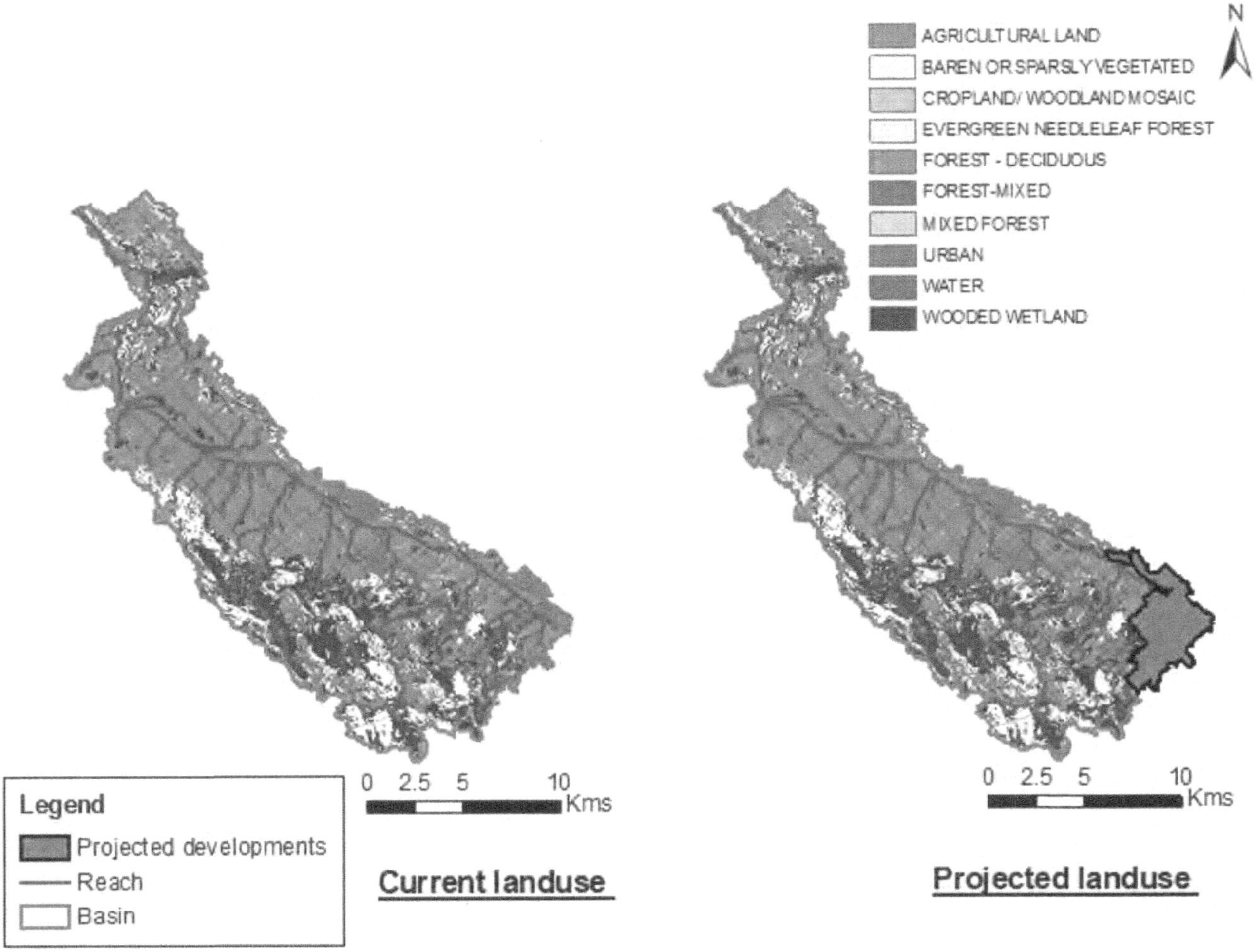

Figure 3 | Current and projected land uses within the Carp River watershed. The current land-use map corresponds to the land-use repartition in 2017; the projected map includes the anticipated extension of urban development in the watershed.

Table 9 | Characteristics of the GCM/RCM model combinations

No.	Institute	Institute acronym	Driving GCMs	RCMs
1	Canadian Centre for Climate Modeling and Analysis	CCCma	CanESM2	CANRCM4
2	Institut Catholique des Hautes Études Commerciales	ICHEC	EC-EARTH	RCA4, HIRHAM5
3	Met Office Hadley Centre	MOHC	HadGEM2-ES	WFR, RegCM4
4	Max Planck Institute for Meteorology	MPI-M	MPI-ESM-LR	WFR, RegCM4
5	National Oceanic and Atmospheric Administration – Geophysical Fluid Dynamics Laboratory	NOAA-GFDL	GFDL-ESM2M	WFR, RegCM4

2.5. Seven scenarios

Based on the climate change and land-use change projections, the seven following experiments, presented in Table 10, were developed: the first scenario (S0o) is the reference (baseline scenario) with observed climate, while S0 m is generated with simulated climate data of RCMs. S1 represents the impact of urbanization on the current model, while S0M45 and S0M85 represent climate change impacts under RCP4.5 and RCP8.5, respectively. The effects of combined urbanization

Table 10 | Description and characteristics of scenarios

No.	Scenarios	Climate data frame	Land-use map	Description
1.	**S0o**	Historical observations (1990–2018)	2017	Current model (Baseline)
2.	**S0m**	Historical model (1990–2018)	2017	Baseline scenario for climate change and combined effects scenarios
3.	**S1**	Historical observations (1990–2018)	2050	Land-use change
4.	**S0M45**	RCP4.5 (2021–2050)	2017	RCP4.5 climate change
5.	**S0M85**	RCP8.5 (2021–2050)	2017	RCP8.5 climate change
6.	**S1M45**	RCP4.5 (2021–2050)	2050	Combined land-use change and climate change under RCP4.5
7.	**S1M85**	RCP8.5 (2021–2050)	2050	Combined land-use change and climate change under RCP8.5

and climate change are assessed with S1M45 and S1M85 under RCP4.5 and RCP8.5, respectively. S1 outputs are compared to S0o, while the scenario involving climate change will be compared to S0 m. This is to ensure the comparison between the two scenarios is consistent with the climate data frame (observations or simulations).

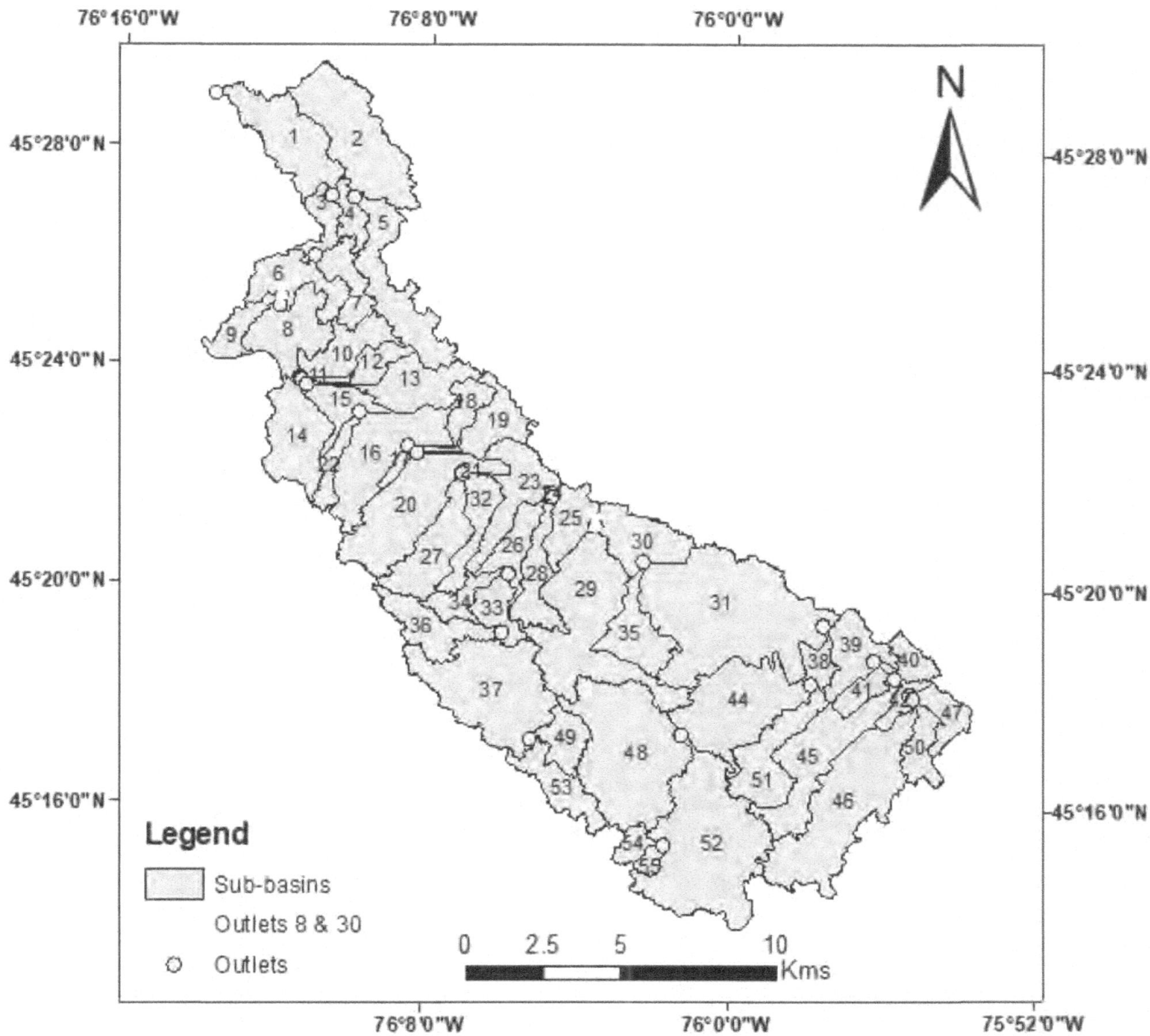

Figure 4 | Carp River watershed delineation. The numbers represent the assigned index to the sub-watersheds.

The simulated outputs were compared to the baseline model (current conditions) outputs at global and local levels using the percentage of change, estimated according to the following equation:

$$\text{Percent change} = \left(\frac{\text{Scenario}_{S1,\,S0M,\,S1M} - \text{Scenario}_{S0}}{S_{\text{Baseline}}}\right) \times 100 \tag{6}$$

A negative percent change corresponds to a reduction of the quantity from the current condition scenario to the corresponding future scenario, while a positive value is an indication that the quantity increases.

The local/watershed-scale approach consisted of evaluating the impact of each scenario at the main outlet (watershed scale) and the outlet of sub-watershed 46 (local scale). This upstream sub-watershed was selected because it is the most affected by urbanization.

Monthly averages are calculated by averaging the values for a specific month during the simulation period excluding the warmup period, as the model does not include it in the output time series. Annual maximum boxplots are created based on the maximum discharge observed at each year of simulation.

3. RESULTS AND DISCUSSION

3.1. Parameter sensitivity

A total of 55 sub-watersheds were delineated (Figure 4).

Table 11 | Sensitivity analysis results

Parameter name	*t*-stat	*p*-value	Rank
A__N_UPDIS.bsn	35.17	0.00	1 (**most sensitive**)
A__ERORGP.hru	−23.95	0.00	2
A__ERORGN.hru	19.98	0.00	3
R__CN2.mgt	−9.30	0.00	4
A__OV_N.hru	6.37	0.00	5
V__ESCO.hru	−2.04	0.04	6
A__RCN.bsn	−1.37	0.17	7
V__GW_REVAP.gw	−1.29	0.20	8
A__REVAPMN.gw	0.93	0.35	9
A__P_UPDIS.bsn	0.85	0.40	10
V__SOL_K(..).sol	−0.78	0.43	11
A__PPERCO.bsn	−0.77	0.44	12
A__SLSUBBSN.hru	0.65	0.52	13
R__GWSOLP.gw	0.64	0.52	14
V__SOL_AWC.sol	0.48	0.63	15
A__PHOSKD.bsn	0.47	0.64	16
A__PSP.bsn	0.47	0.64	17
A__ANION_EXCL.sol	−0.43	0.67	18
R__SHALLST_N.gw	−0.30	0.76	19
V__ALPHA_BF.gw	−0.26	0.79	20
A__NPERCO.bsn	0.15	0.88	21
A__GWQMN.gw	−0.03	0.98	22 (**least sensitive**)

3.1.1. Parameter selection and sensitivity analysis

Recommendations of similar studies in the literature (e.g., Abbaspour 2007; Arnold *et al.* 2012; Abbaspour *et al.* 2015; El-Khoury *et al.* 2015; Khoiab & Thomb 2015) led to the selection of 22 parameters to use in the calibration. First, a sensitivity analysis was performed to assess the influence of these parameters on the objective function, which in this case is the sum of the NS coefficient for streamflow, total nitrogen, and total phosphorus in their respective calibration periods. The sensitivity analysis provides information about the most important process drivers in the study region, depending on local characteristics (Abbaspour *et al.* 2018). The results are shown in Table 11, where it can be observed that the most sensitive parameter is N_UPDIS with a *p*-value and *t*-stat of 0.00 and 35.17, respectively. On the other hand, GWQMN.gw is the parameter with the least influence on the performance of the model. In this case, the uptake distribution of N seemed to affect the model performance the most.

3.1.2. Model performance

The calibration was performed using the 22 parameters identified in the previous section. The optimal range for each parameter as well as the fitted values are shown in Table 12. The outputs of the calibrated model are shown in Figure 5.

Overall, the model shows a good performance for the discharge at outlet 8 in calibration: NS is in the satisfactory range ($0.5<NS=0.67<0.7$) according to Moriasi *et al.* (2007), while the PBIAS is excellent (close to 0). In validation, a relatively satisfactory performance was reached with an NS of 0.48 and a PBIAS of +16.9% (Figure 5(a)). However, the graphical comparison shows the model tends to underestimate the peaks and overestimate low flows. This is a common problem in rainfall–runoff modeling, which is linked to the quality of climatic inputs and the spatial discretization of the study area (Ghosh & Hellweger 2012; Sharma & Regonda 2021). In addition, the spatial repartition of rain gauges is a factor (Lobligeois *et al.* 2014) as a sparse number of rain gauges (as in this paper) or very large sub-watershed may lead to some extreme precipitation

Table 12 | Calibration of parameters

Parameters	Optimal range	Fitted values
r__CN2.mgt	[0.11713, 0.12033]	0.11873
a__GWQMN.gw	[−1347.63, −1136.19]	−1241.9
v__GW_REVAP.gw	[0.097516, 0.100094]	0.098805
v__ESCO.hru	[0.997944, 1.010913]	1.0044
v__ALPHA_BF.gw	[0.482128, 0.54378]	0.51295
v__SOL_AWC().sol	[0.974373, 1.001545]	0.98796
a__REVAPMN.gw	[187.1984, 212.1717]	199.69
v__SOL_K().sol	[0.577949, 0.600627]	0.58929
a__OV_N.hru	[7.371342, 8.932384]	8.1519
a__SLSUBBSN.hru	[135.1207, 137.2007]	136.16
a__RCN.bsn	[14.5449, 16.35953]	15.452
a__N_UPDIS.bsn	[−20.5479, −16.0446]	−18.296
r__SHALLST_N.gw	[−135.764, −86.191]	−110.98
a__ERORGN.hru	[−0.13804, 0.148926]	0.005445
a__NPERCO.bsn	1.144399, 1.161047]	1.1527
a__ANION_EXCL.sol	[0.763784, 0.786918]	0.77535
a__PSP.bsn	[0.570577, 0.594221]	0.5824
a__PHOSKD.bsn	[116.2567, 120.4304]	118.34
a__P_UPDIS.bsn	[89.41322, 99.38607]	94.400
a__PPERCO.bsn	[12.34229, 12.44181]	12.392
a__ERORGP.hru	[0.011739, 0.019001]	0.01537
r__GWSOLP.gw	[297.5318, 319.4124	308.47

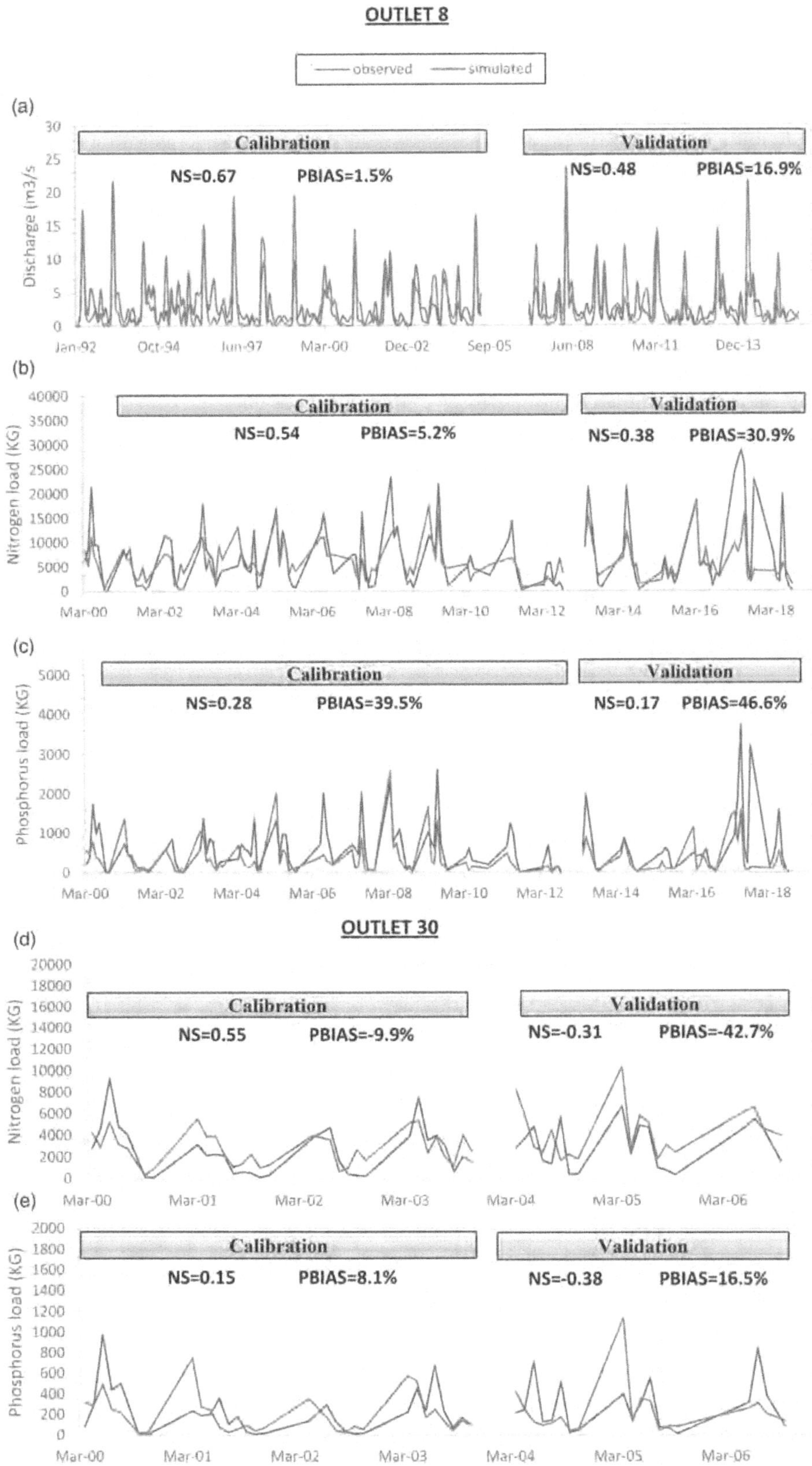

Figure 5 | Calibration and validation performance at the two outlets: outlet 8 – (a) discharge, (b) nitrogen, and (c) phosphorus; Outlet 30 – (d) nitrogen and (e) phosphorus.

events being overlooked or extreme occurring in one small area that happens to have a rain gauge assumed to occur over a larger area.

As expected, a lower performance was obtained for water quality parameters, especially for P. According to the NS (0.54 for outlet 8 and 0.55 for outlet 30), the simulation of the N at both outlets showed acceptable results in calibration. Meanwhile, both PBIAS were excellent: 5.2 and −9.9%. Unsurprisingly, discharge is better simulated than nutrients by the model. A few studies (El-Khoury *et al.* 2015; Zhou *et al.* 2015; Čerkasova *et al.* 2018) have also obtained better performance for streamflow calibration. The reason for the low performance in predicting water quality parameters is the limited available data for a large number of parameters, highlighted in Hoang *et al.* (2019) and Noori *et al.* (2020). There are also greater uncertainties in nutrient data associated with errors in streamflow measurements, sample collection, storage, and analysis (Harmel & Smith 2007). For instance, in this study, the length of validation period for

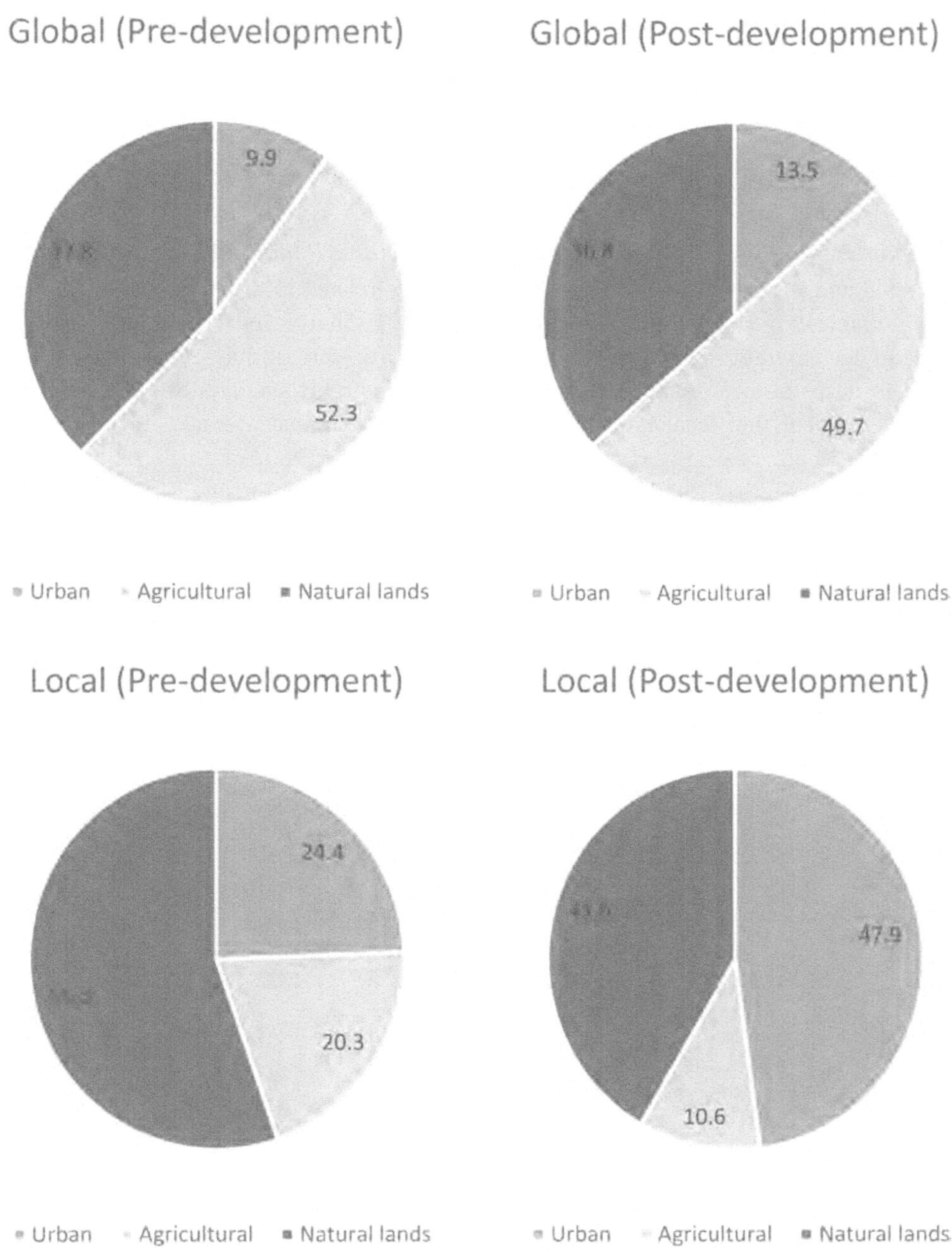

Figure 6 | Repartition of mainland uses/covers in the entire watershed (watershed scale) and the selected sub-watershed (local, sub46) under urbanization.

discharge was 11 years (2007–2015), while only 6 and 3 years of data were used for N and P loads at outlets 8 and 30, respectively.

All the simulations for water quality parameters showed unsatisfactory performance in the validation phase except for P at outlet 30, where PBIAS was 16.5%. The graphical comparison shows the model better simulates the average value of N at outlet 8 (Figure 5(b)), but not the peaks. It is important to note that these results are in the same range (or even better) as some published studies such as El-Khoury *et al.* (2015) with NS=0.24 and 0.28 in the calibration of nitrite-nitrogen and organic P, respectively; Shrestha *et al.* (2021) with NS=−0.71 and PBIAS=60.3% for P calibration.

3.1.3. Impacts of the scenarios on discharge, N load, and P load

This section presents the impacts of each scenario in terms of annual maximum discharge and monthly average discharge, N load, and P load at the main outlet (global) and at the outlet of sub-watershed 46 (local). This sub-watershed, located upstream (Figure 4), was selected because it was one of the most affected by the developments, with an important difference in the urbanization ratio compared to the global watershed. This allows an observable difference in the impacts at the two scales when comparing them. It also helps to investigate how different the effects between the global and the local scales would be. Figure 6 illustrates the repartition of the land uses before and after the developments in terms of watershed and local scales.

3.1.3.1. Impact of urbanization (S0o vs. S1) on discharge, N load, and P load. Figure 7 shows the annual maximum discharge, while Figures 8 and 9 present the average monthly outputs related to discharge, N, and P for the experiments S0o and S1. Globally, urbanization slightly increased the averages of discharge and P load by 1.9 and 23%, respectively, while the average N load decreased by 4.4%. The N load decreases particularly in July–August, where higher precipitation amounts are observed. In terms of local impacts, the variation of the monthly averages has the same direction of change for all the three variables except that local impact is more significant. Annual averages in local increased by 9.6 and 69.5% for discharge and P load, respectively, and decreased by 30.2 for N load (higher than the previous percent changes). The high contrast change in P load compared to the N load is not surprising, since the model performance in simulating P was low (Figure 5(c) and 5(e)).

Peak values for discharge and P are observed during the summer–spring period (April–August), while the same is observed in July–August for N. This is due to the increase of total precipitations (combination of snow melting and rainfall). In terms of

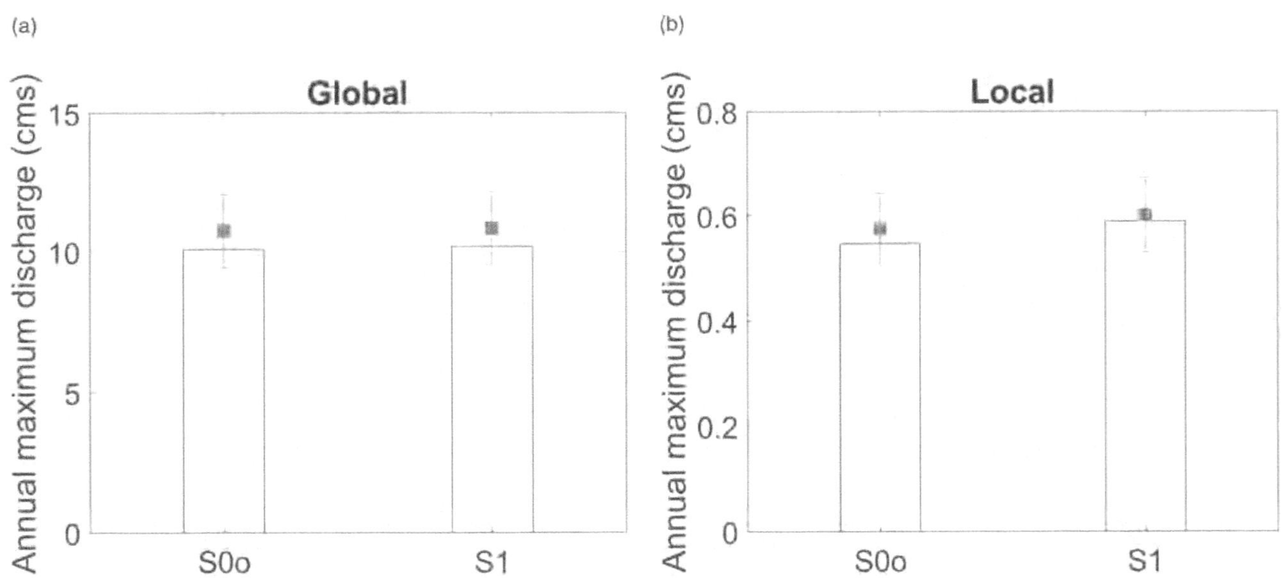

Figure 7 | (a) Global and (b) local (sub46) impacts of land-use change on annual maximum discharge. Column height corresponds to the median, top and bottom whiskers are the errors with 95% confidence, and red markers represent the mean discharge. Please refer to the online version of this paper to see this figure in color: http://dx.doi.org/10.2166/wcc.2021.158.

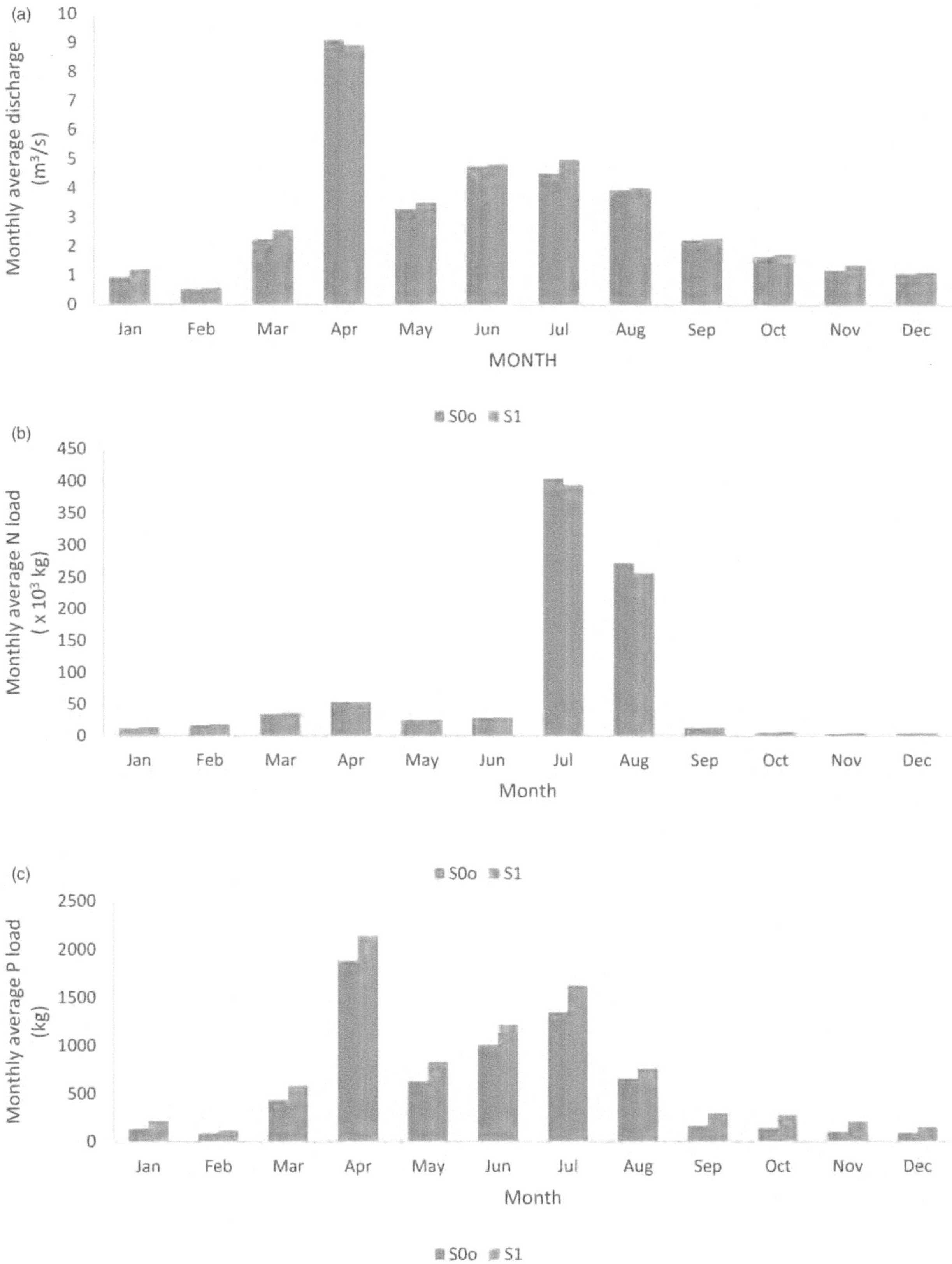

Figure 8 | Watershed-scale impact of land-use change on monthly average discharge (a), nitrogen load (b), and phosphorus load (c).

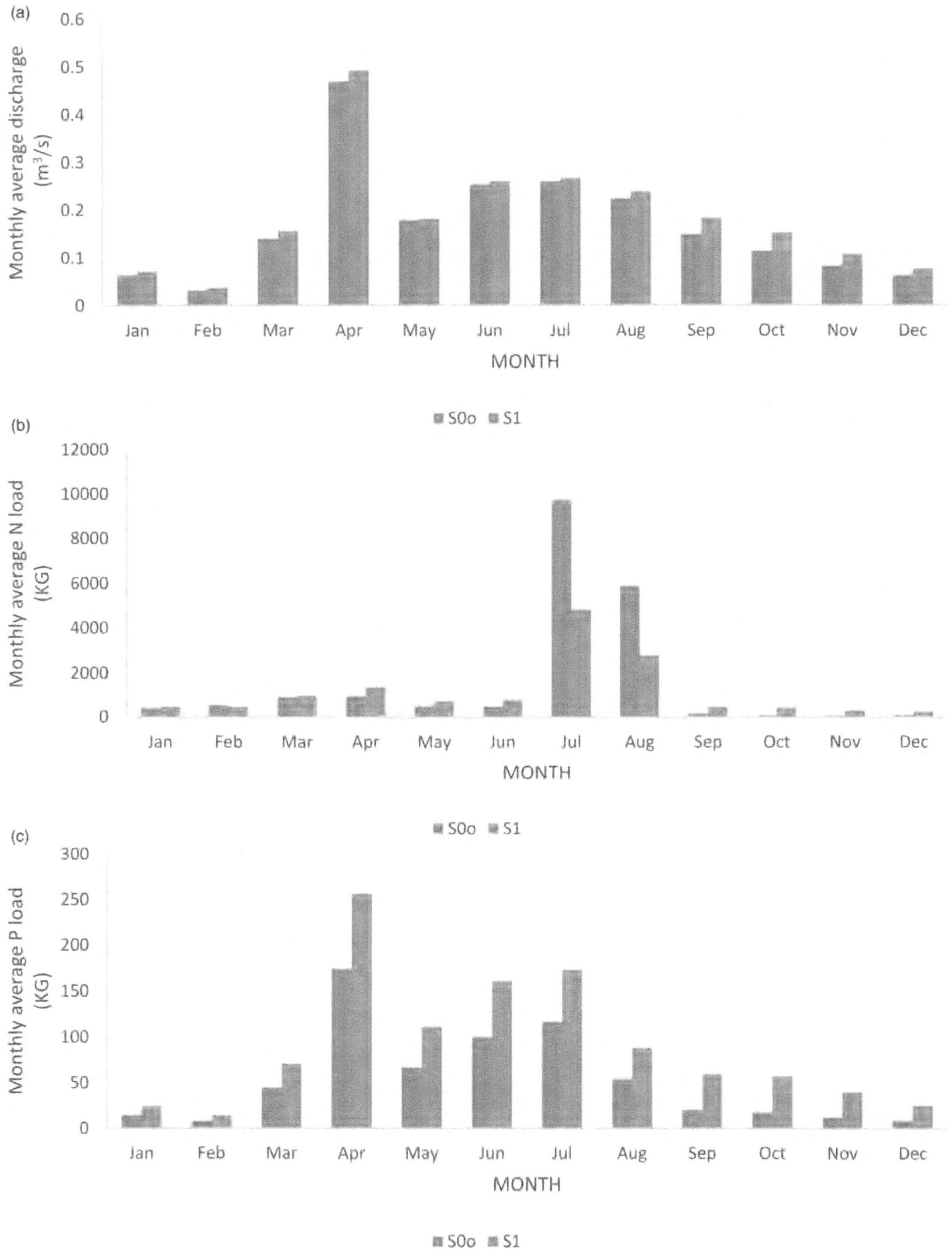

Figure 9 | Local impact (sub46) of land-use change on monthly average discharge (a), nitrogen load (b), and phosphorus load (c).

water quantity, extreme events are more likely to occur around April. They are expected to vary by +4.78% at the local scale, while a relatively small difference of +0.73% is estimated at the watershed scale (Figure 7).

3.1.3.2. Impacts of climate change (S0 m vs. S0M45/85) on discharge, N load, and P load. Unlike the land-use change scenario, the change in climate change scenario is calculated and compared to S0 m output instead of S0o, although the S0o results are also presented on graphs only as a reference.

In terms of annual maximum, the discharge has experienced a relatively low change under the two RCPs, as presented in Figure 10. On average, at the watershed scale, the maximum discharge reached 12.63 m^3/s in S0 m, and the percentages of change under RCP4.5 and RCP8.5 scenarios are +3.76 and +0.90%, respectively. However, an opposite change is observed at the local scale with −1.06 and −3.54% for S0M45 and S0M85, respectively. Annual average discharge presented a similar trend in S0M45 at the watershed scale and the local scale with an increase of 5.49 and 2.61%, respectively, while in S0M85, the change was 7.52% (watershed scale) and 1.13% (local) increase. In terms of water quality, overall the climate change seems to have a significant impact on N in S0M45 with an increase of 29.62% (+265 tons) globally and 28.24% (+5 tons) locally, while the change is less important in S0M85 with +2.03% (+18 tons) and −0.96% (−0.18 tons) globally and locally, respectively. Monthly results are illustrated in Figure 11 and Figure 12.

3.1.3.3. Combined impacts of land-use and climate changes (S0 m vs. S1M45/85) on discharge, N load, and P load. Separately assessing the impacts of land-use change and climate change provides a better understanding of the level of impact caused by each of the two scenarios on the water quantity and quality. However, it is also important to determine and quantify the impact when the two effects are coupled. Figure 13 shows that the maximum discharge is not significantly affected by the combined effects of climate and land-use changes. The difference between the means (cross mark inside the boxes) of the scenarios S0 m, S1M45, and S1M85 is relatively small, both in watershed and local scales, although the RCP4.5 has a higher impact compared to the RCP8.5.

In contrast, the annual discharge is more sensitive to the RCP8.5 than RCP4.5. The annual average changed from 3.15 m^3/s in the baseline conditions to 3.44 m^3/s (+9.34%) under RCP8.5, while RCP4.5 caused an increase of 0.21 m^3/s (+6.75%). This is also the same at the local scale, where an increase of 13.07 and 11.10% is observed under RCP8.5 and RCP4.5, respectively. The coupled effect caused a reduction of the annual N load except in S1M45, where a significant increase of 24.84% was observed at the watershed scale. The local impact under RCP8.5 is also likely to be important, as the load decreased by

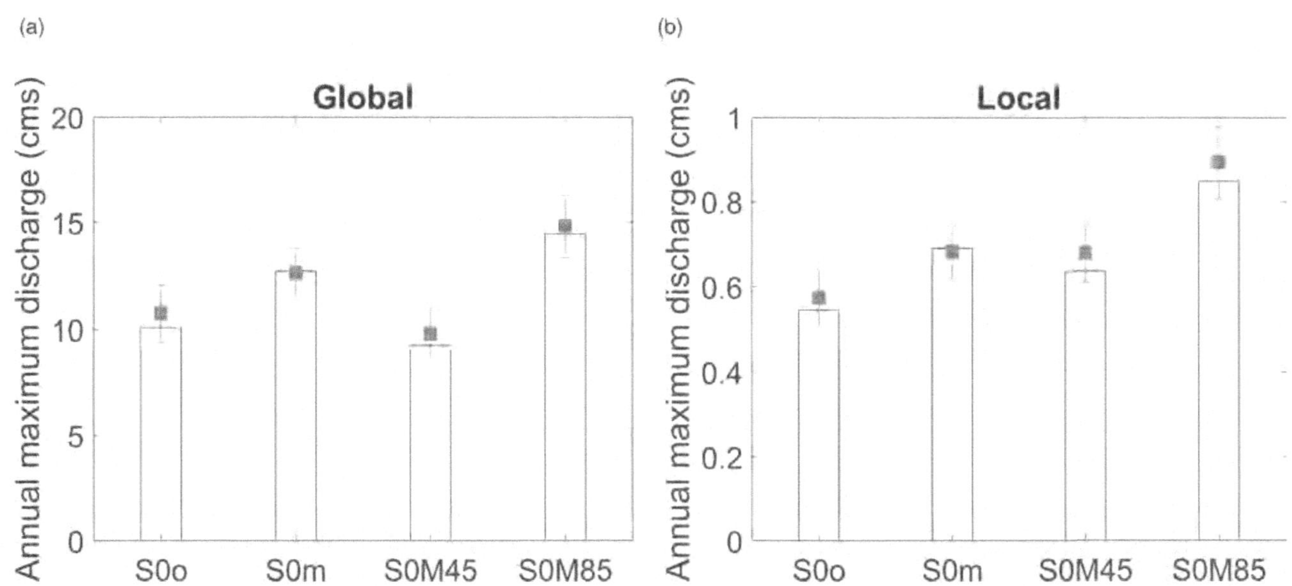

Figure 10 | (a) Global and (b) local (sub46) impacts of climate change on annual maximum discharge under RCP4.5 and RCP8.5. Column height corresponds to the median, top and bottom whiskers are the errors with 95% confidence, and red markers represent the mean discharge. Please refer to the online version of this paper to see this figure in color: http://dx.doi.org/10.2166/wcc.2021.158.

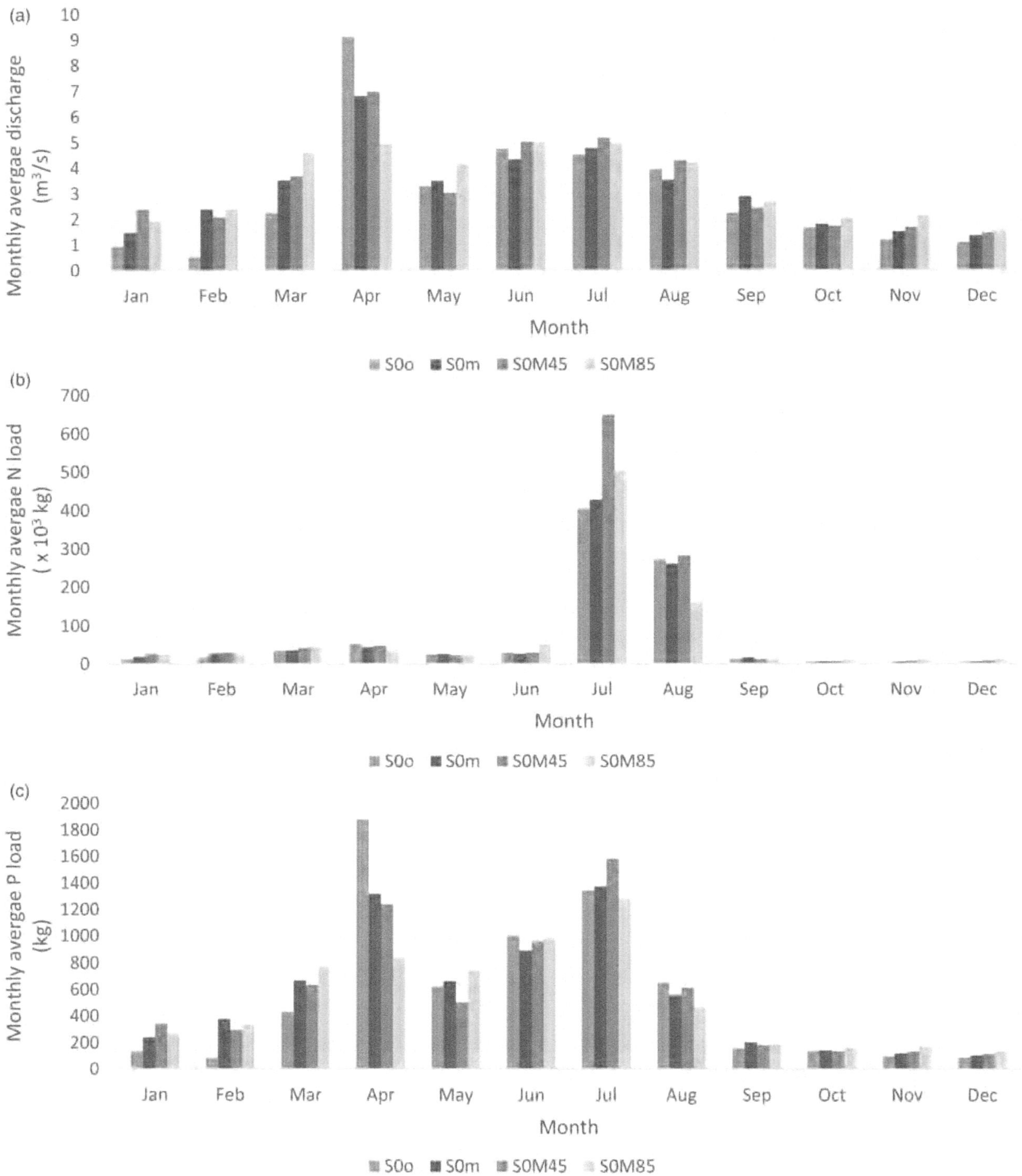

Figure 11 | Watershed-scale climate change impact on annual average discharge (a), nitrogen (b), and phosphorus (c) loads under RCP4.5 and RCP8.5.

24.80% compared to scenario S0 m. However, the results indicated the annual load at the main outlet under RCP8.5 was not significantly impacted, and only a 1.20% decrease (−10.8 tons) of the baseline condition load is expected. The quantity of P faced a watershed-scale significant increase and was even higher at the local outlet (in terms of percentage of change). The load increase at the main outlet was estimated at 1.6 tons (23.81%) in S1M45 and 1.3 tons in S1M85 (19.15%), while in the upstream sub-watershed, it was 0.411 tons (66.15%) and 0.360 tons (57.97%). Monthly results are illustrated in Figure 14 and Figure 15.

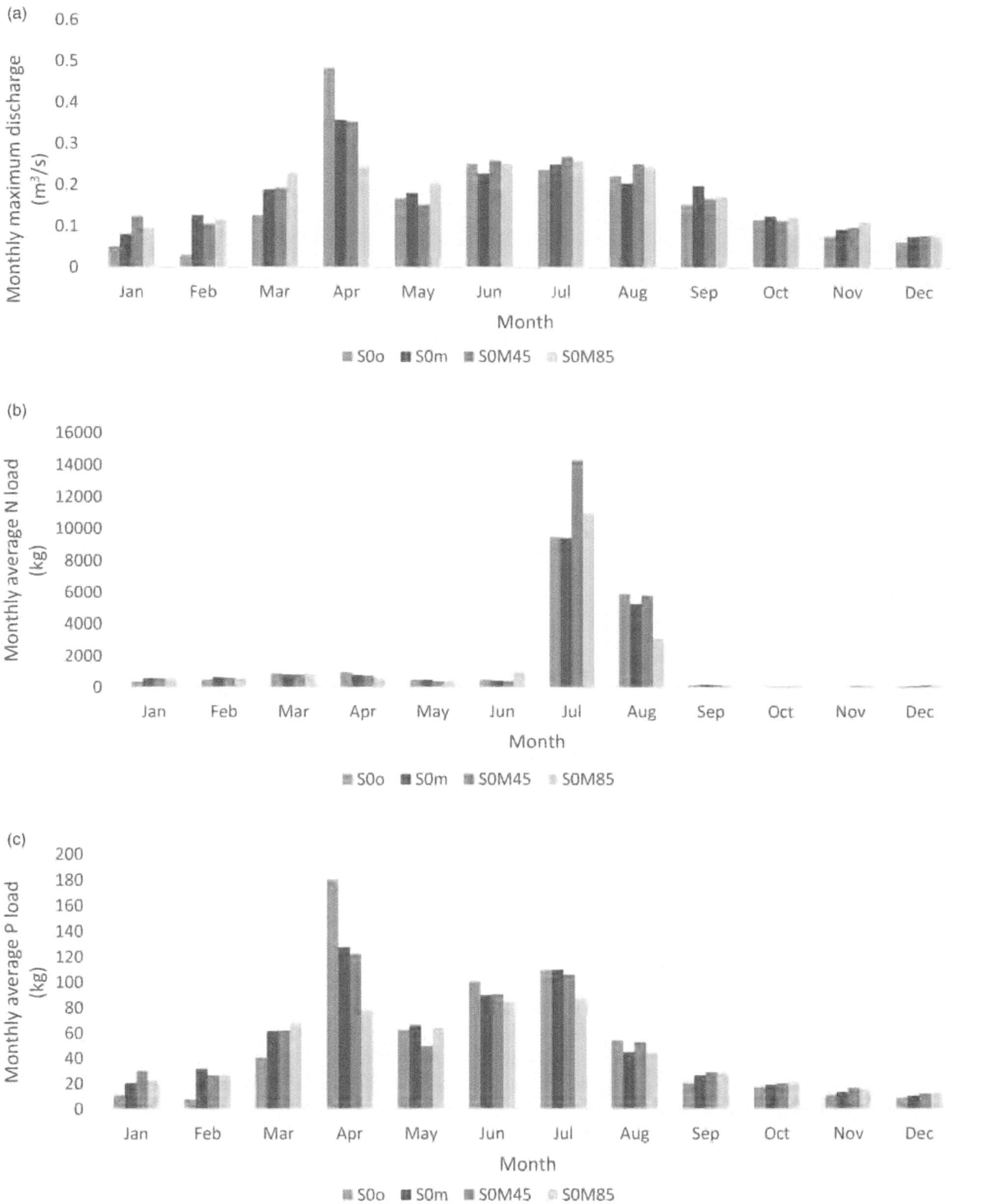

Figure 12 | Local (sub46) climate change impact on annual average discharge (a), nitrogen (b), and phosphorus (c) loads under RCP4.5 and RCP8.5.

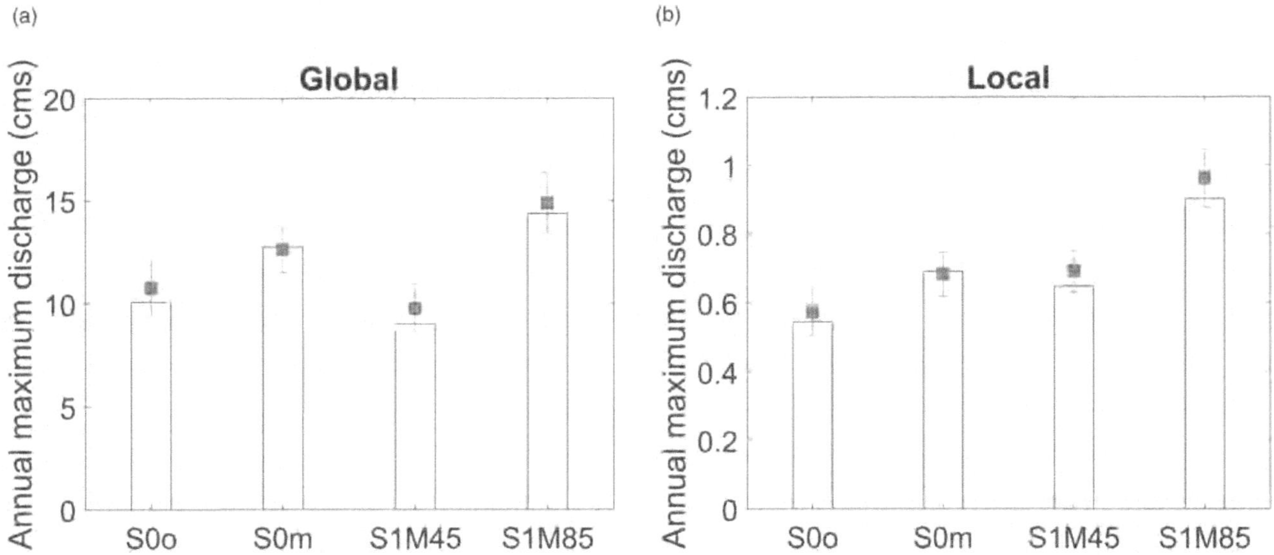

Figure 13 | (a) Global and (b) local (sub46) impacts of combined climate and land-use changes on annual maximum discharge under RCP4.5 and RCP8.5. Column height corresponds to the median, top and bottom whiskers are the errors with 95% confidence, and red markers represent the mean discharge. Please refer to the online version of this paper to see this figure in color: http://dx.doi.org/10.2166/wcc.2021.158.

3.1.3.4. Summary of global and local impacts. The summary of the results is presented in Figure 16 and Table 13.

3.2. Discussion

Results suggest that water quantity was more sensitive to climate change than urbanization in the Carp River watershed when the factors are taken individually (S1 and S0M45/85). When climate change alone is considered in the watershed, the increase in maximum discharge is expected to be within 0.90–3.76% at the main outlet. Urbanization will only lead to a 0.76% increase in the maximum discharge. Similarly, urbanization is expected to increase the average discharge by 1.57%, while this change is expected to be between 5.49 and 7.42% under climate change. This is not surprising as similar studies on areas located in Canada have come up with the same conclusion: Kaykhosravi *et al.* (2020) in Montreal and Toronto; El-Khoury *et al.* (2015) for the South Nation watershed in Ontario. For instance, the impact caused by the land-use and climate changes in the South Nation watershed was, respectively, +1.2 and +11.2%. Although the percentage of the urban area increased in the urbanization scenario, the agricultural lands are expected to remain dominant with 49.7% of the total area against only 13.5% for urban zones in 2050. This land-use repartition minimized the upstream developments' effect and could explain the small impact observed in the Carp River watershed. A different urbanization rate or a different location of the developments within the same watershed may lead to very different results.

Monthly output analysis indicated extreme events were likely to occur at the end of the winter (around April) and during the summer. This is very common in cold areas. This study has shown that snow contribution to the total precipitations is more than rainfall, as the peaks of April are higher than during the summer, meaning that floods were expected to be more severe in April. According to the model, this characteristic of the watershed is not expected to change by 2050 because the predicted monthly hydrograph in the future conditions (S1M45/85) follows the same variation as the historical period (1990–2018). The variation of the peak flow during April is expected to be within −27.1 to 2.38%. Therefore, no significant increase is expected, but on the contrary, peaks may decrease up to 27.1% on average over the next 30 years. Although the peaks of July (due to rainfall events) are less severe, unlike April peaks, they were predicted to increase on average by 8.60%.

In terms of water quality, the sources of pollution of N and P are generally related to land uses such as urban and agricultural. Based on the impacts observed in the urbanization scenario, N seemed more sensitive to agricultural lands during July–August when agricultural activities occur. As an important amount of fertilizers is used for these activities, it is unsurprising to observe a drastic increase in the monthly graphs for all the scenarios. The sensitivity is also demonstrated by a 1.88% decrease

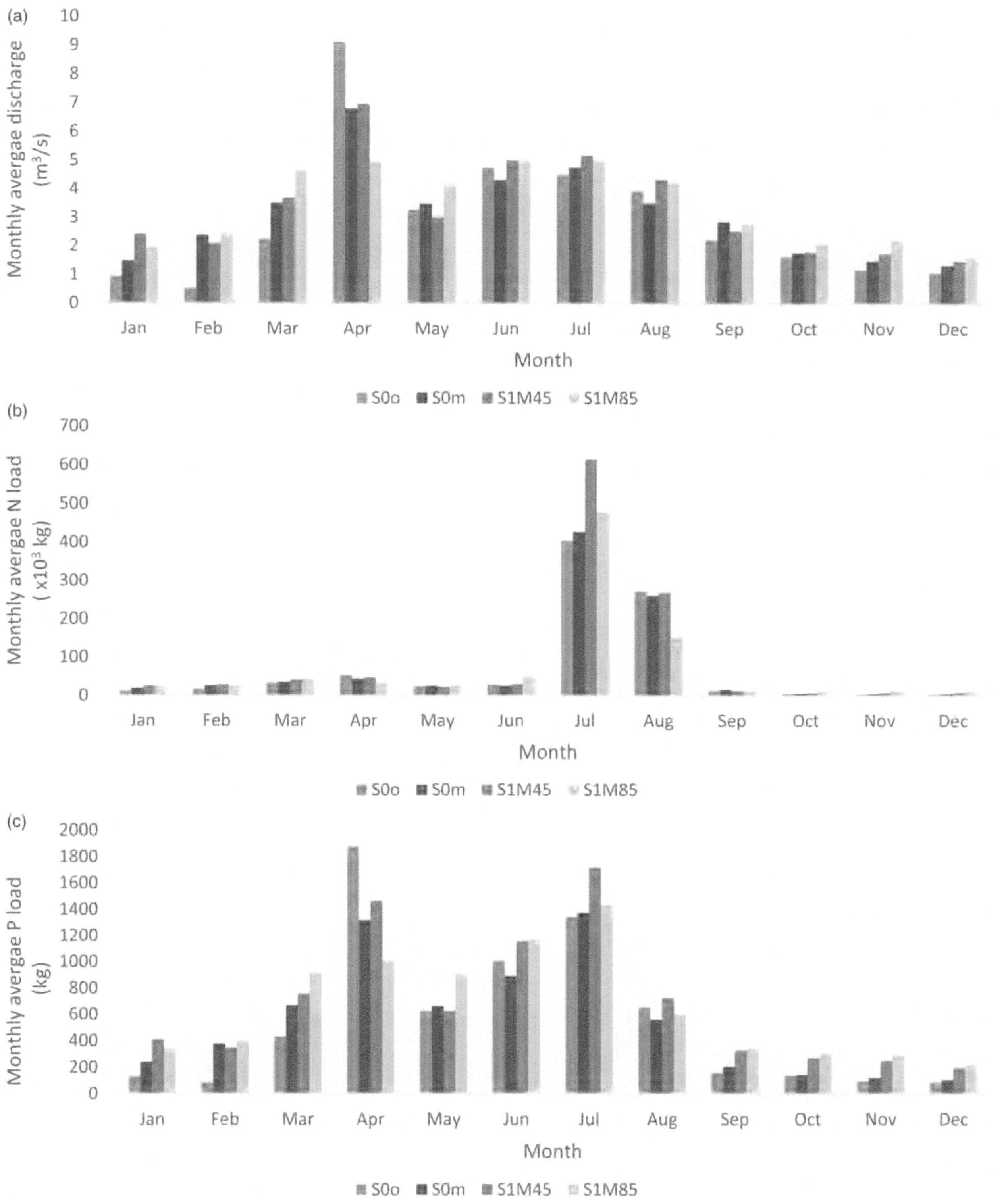

Figure 14 | Watershed-scale impacts of combined climate and land-use changes on annual average discharge (a), nitrogen (b), and phosphorus (c) loads under RCP4.5 and RCP8.5.

in the annual total load at the main outlet when the ratio of the agricultural land is reduced by 2.63% (see Table 8). Similarly, a more significant decrease was observed at the local scale when the agricultural land ratio dropped from 49.7 to 10.6%, resulting in a reduction of the impact from 1.88 to 29.00%.

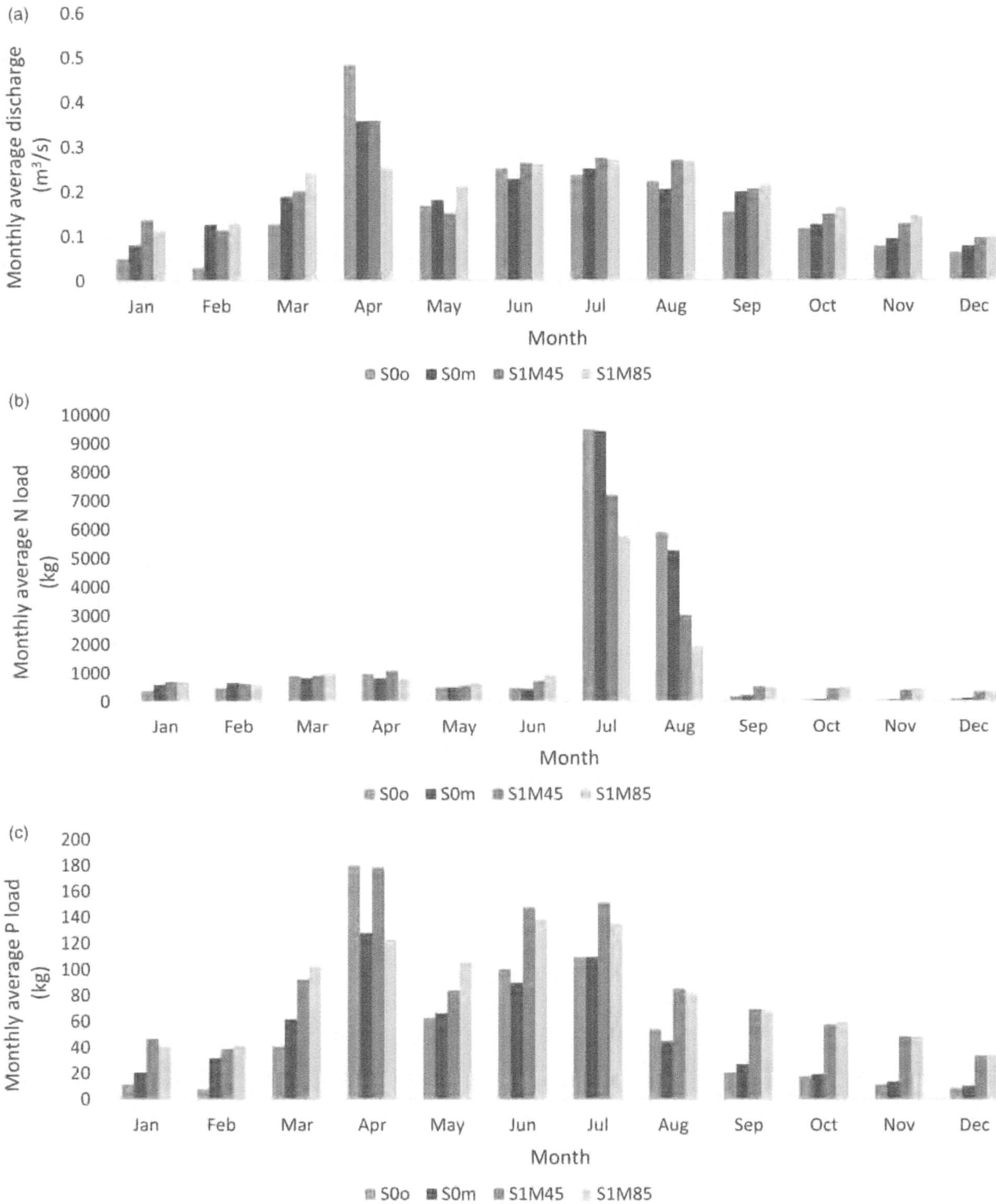

Figure 15 | Local impacts (sub46) of combined climate and land-use changes on annual average discharge (a), nitrogen (b), and phosphorus (c) loads under RCP4.5 and RCP8.5.

In the same perspective, as the increase of urban areas led to an increase of the annual P load (Figure 16(a)), the pollution of P seems to be dominated by the urban areas. This is consistent with what is observed at the local scale where a higher urban area ratio caused a larger increase (from 26.49% globally to 73.56% locally when urban ratio passed from 13.5 to 47.9%, Figure 6).

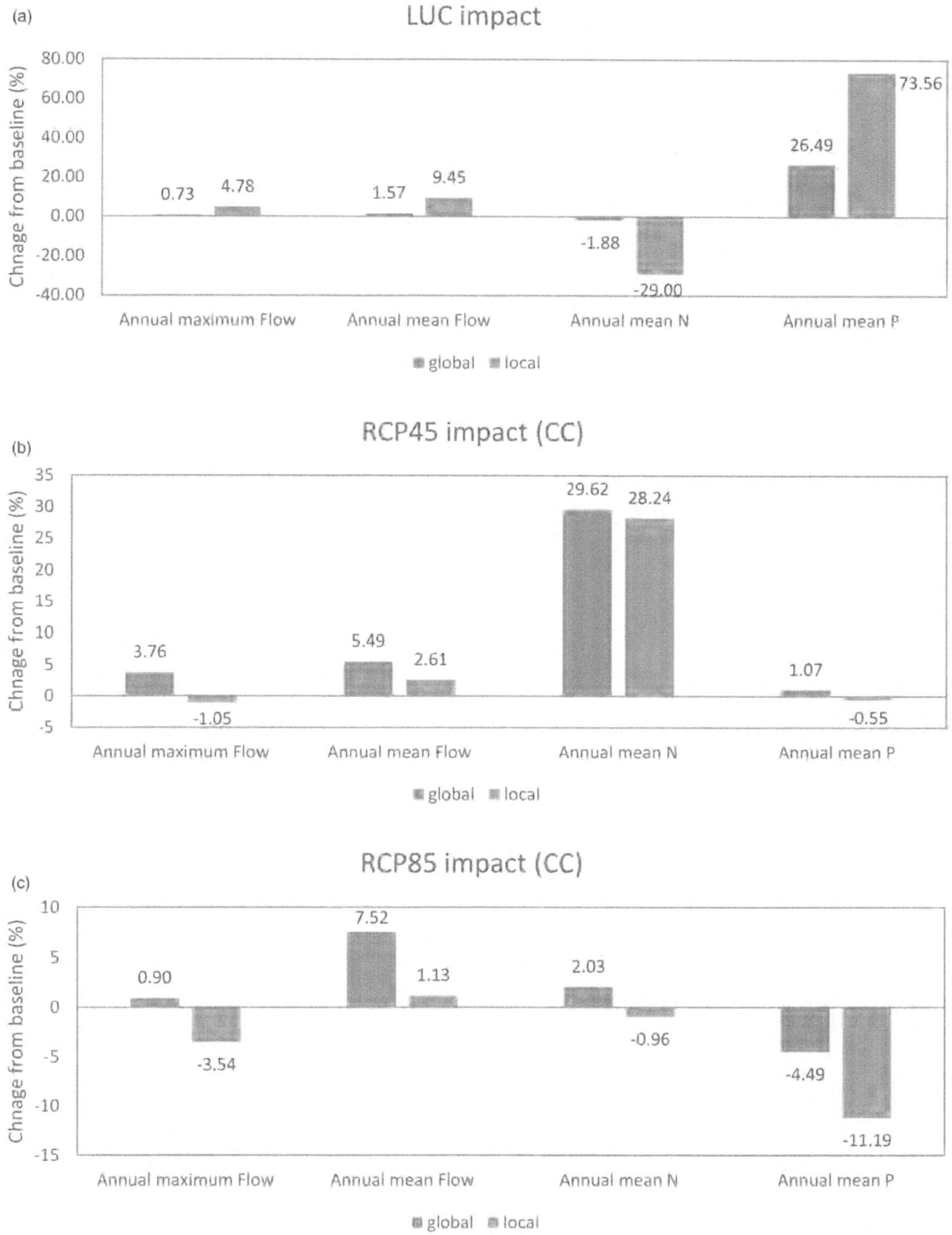

Figure 16 | Annual average variation of discharge, N and P. Variation due to land-use change (a), climate change (b,c), and combined effects (d,e) compared to the reference conditions. LUC, land-use change, CC, climate change. (*continued*).

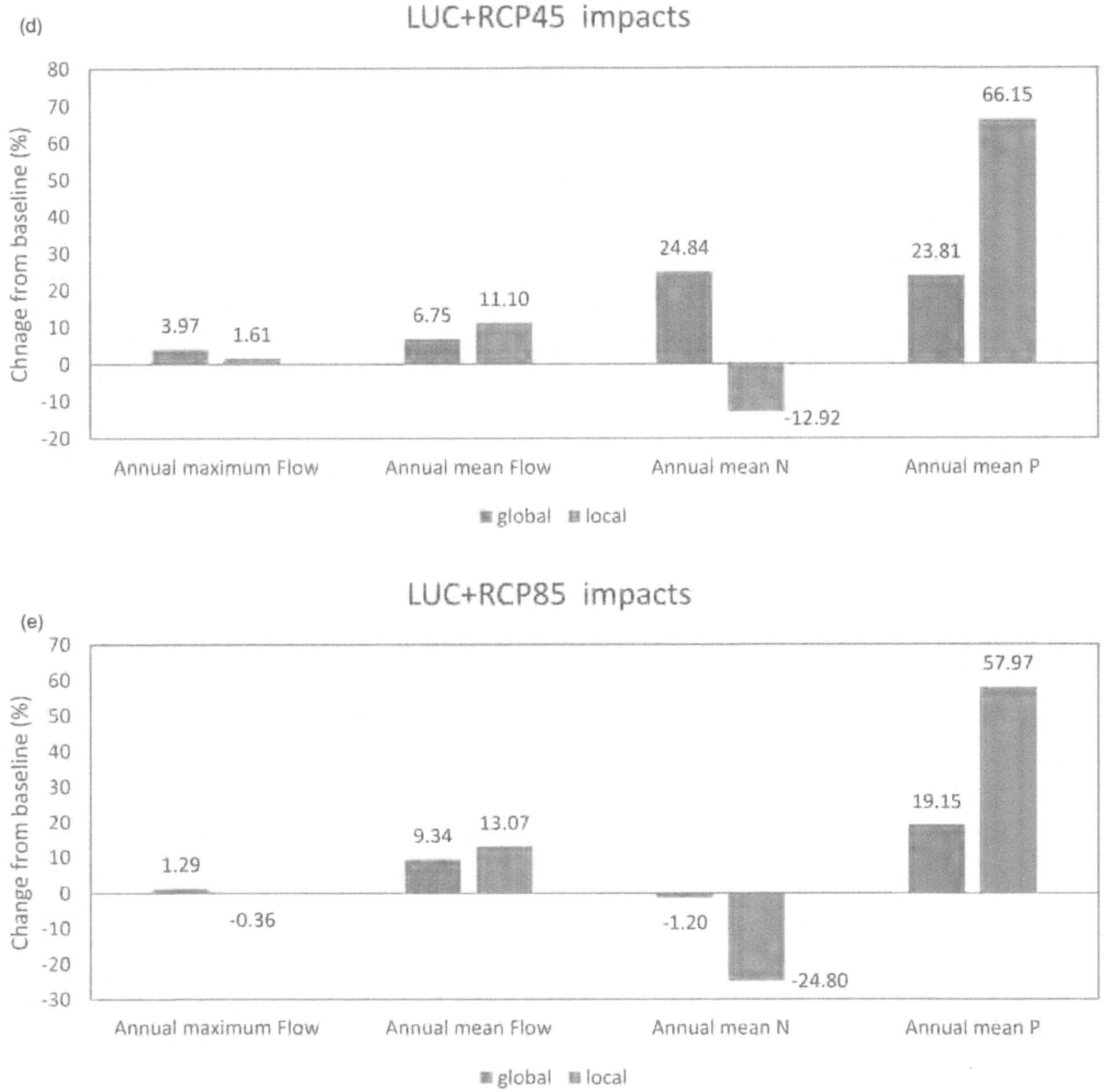

Figure 16 | Continued.

Considering the most severe conditions in the future (S1M85), a decrease in the annual average of N load (−1.20%) is expected in opposition to an increase of 24.84% in the scenario S1M45. Therefore, a contrasting impact on N is likely to occur under the two RCPs. As a justification, the increase of the temperatures in the future conditions, especially in RCP8.5, influences the evaporation and transpiration (indispensable for the growth of plants) from natural (forests and water bodies) and agricultural lands. Given that plant cells control the openings where water is released to the atmosphere, higher temperatures stimulate them to open, enhancing water release. Consequently, the rate of transpiration, defined as the water loss from living plant surfaces (Manashi 2016), goes up with the temperature (USGS 2020). Plant growth is influenced by the availability of N, a vital element in the photosynthesis process. The higher the plant growth rate, the higher the demand in N. Considering the expected increase in transpiration and N availability, the plant's demand for N is likely to increase in the future. Consequently, this will decrease the annual N load, as observed in the S1M85 scenario with a −1.20% reduction.

The above justification is also valid for P, as the increase of the load at the main outlet is smaller in the hottest scenario (S1M85) than that of the coolest (S1M45) with 19.15 and 23.81%, respectively. The difference is that P is less soluble

Table 13 | Summary of the results describing the impacts of land-use change, climate change, and combined effects at the watershed scale and local levels

		Annual maximum flow		Annual mean flow		Annual nitrogen load		Annual phosphorus load	
Scenarios	**Impact**	**Value (m^3/s)**	**Change (%)**	**Value (m^3/s)**	**Change (%)**	**Value ($\times 10^5$ Kg)**	**Change (%)**	**Value ($\times 10^3$ Kg)**	**Change (%)**
S0o	Watershed scale	10.75	–	35.40	–	8.71	–	6.62	–
	Local	0.57	–	1.97	–	0.19	–	0.62	–
S0 m	Watershed scale	12.63	–	3.15	–	8.96	–	6.65	–
	Local	0.68	–	0.18	–	0.19	–	0.62	–
S1	Watershed scale	10.83	0.73	3.09	1.57	8.55	– 1.88	8.38	26.49
	Local	0.60	4.78	0.18	9.45	0.14	– 29.00	1.08	73.56
S0M45	Watershed scale	13.11	3.76	3.32	5.49	11.61	29.62	6.72	1.07
	Local	0.67	– 1.05	0.18	2.61	0.24	28.24	0.62	– 0.55
S0M85	Watershed scale	12.75	0.90	3.38	7.52	9.14	2.03	6.35	– 4.49
	Local	0.66	– 3.54	0.18	1.13	0.19	– 0.96	0.55	– 11.19
S1M45	Watershed scale	13.13	3.97	3.36	6.75	11.19	24.84	8.24	23.81
	Local	0.69	1.61	0.19	11.10	0.16	– 12.92	1.03	66.15
S1M85	Watershed scale	12.79	1.29	3.44	9.34	8.85	– 1.20	7.93	19.15
	Local	0.68	– 0.36	0.20	13.07	0.14	– 24.80	0.98	57.97

than N, and consequently, it is less sensitive to temperature. N and P are both partially lost in plant uptake, but in N's case, the losses also include denitrification and volatilization. Given that higher temperatures enhance their decomposition and accumulation (Hong & Tang 2014; Geng *et al.* 2017), denitrification and volatilization are promoted, and consequently, N loss may be more significant, which is the case in the Carp River watershed. In addition, even though having similar sources of pollution, N and P seemed to have different transport pathways. Due to its solubility, N is more likely to be retained within the watershed by infiltration into the groundwater and/or in the atmosphere by volatilization, while the P will mostly go into the surface water, especially for urbanized areas. This was demonstrated for the Mississippi River watershed in the USA (Hobbie *et al.* 2017) where only 22% of P inputs was retained vs. 80% of the N inputs. Nieder *et al.* (2018) also stated that nitrate is the most common chemical contaminant in the world's groundwater aquifers, and surface water is particularly affected by P, confirming the findings in the Carp River watershed.

The contrast in N and P variation (Figure 16) observed in our study area differs from some published studies such as Mehdi & Lehner (2014), El-Khoury *et al.* (2015), Eum *et al.* (2016), and Luo *et al.* (2020) and with the different direction of change. For instance, El-Khoury *et al.* (2015) found out that P will increase regardless of the type of conditions (LUC, CC, or the two combined), which contradicts what is observed in the Carp River watershed. This divergence indicates the impact of climate change, land-use change, or both must be evaluated case by case. It cannot be generalized even for watersheds located in the same region because every study area has different characteristics in terms of topography, land-use occupation, and climate conditions. Therefore, water quality is considerably affected by the entire watershed characteristics (Liu *et al.* 2018).

4. CONCLUSION

This study examined the influence of the land-use change and climate change individually and their coupled impact on the watershed of the Carp River using a hydrological model SWAT. The model was applied using input data from several sources and different resolutions, and overall, the calibrated model showed a satisfactory performance. The impacts were evaluated by developing realistic scenarios based on the 2050 year land-use map projected by the local authority and future climate conditions under RCP4.5 and RCP8.5. Results showed that climate change and urbanization's impacts vary greatly depending on the spatial scale and geographic location. In addition to that, the following points list the main other findings that emerged from this study: (a) climate change is likely to be the most dominant factor affecting the flow (peaks and trend) and N, while urbanization will control the P quantity in the future; (b) global and local impacts may significantly differ and it is crucial to check the impacts at both levels; (c) the local impact can be more significant than the watershed-scale impact; (d) impacts are not additive; and (e) for impacts assessment study, evaluating one effect alone without considering the other is incorrect and leads to a wrong understanding of the variation. In terms of quantity and quality:

1. Expected global increase of the annual maximum flow from 1.29 to 3.97%.
2. Expected increase of the annual average flow (global) from 6.75 to 9.34%.
3. Expected global variation of the annual average N load from −1.20 to 24.84%.
4. Expected global variation of the annual average P load from 19.15 to 23.81%.

As a recommendation, P management should emphasize reducing watershed inputs and transport from urban zones. In contrast, N management should reduce watershed inputs and transport from cultivated zones when they are active.

ACKNOWLEDGEMENTS

The authors gratefully acknowledge the data and financial support of the MVCA and financial support from MITACs. The authors also expressed their gratitude to the City of Ottawa for providing us with high-resolution topographic data and for sharing with us the future urban development map of the municipality.

AUTHOR'S CONTRIBUTION

B.-S.Z. is involved in conceptualization, data curation, formal analysis, methodology, software, visualization, writing – original draft, and writing – review & editing. O.S. is involved in conceptualization, supervision, resources, visualization, and writing – review & editing. M.S. is involved in conceptualization, supervision, resources, visualization, writing – review & editing, and funding acquisition. N.N. and K.S. are involved in investigation, data curation, and project administration.

FUNDING

This work was supported by the MVCA through the Mitacs Accelerate (No. IT15546) program.

DATA AVAILABILITY STATEMENT

All relevant data are included in the paper or its Supplementary Information.

REFERENCES

Abbaspour, K. 2007 *User Manual for SWAT-CUP, SWAT Calibration and Uncertainty Analysis Programs*. Swiss Federal Institute of Aquatic Science and Technology, Eawag, Dübendorf, Switzerland.

Abbaspour, K., Faramarzi, M., Ghasemi, S. S. & Yang, H. 2009 Assessing the impact of climate change on water resources in Iran. *Water Resources Research* **45**, 1-16. doi.org/10.1029/2008WR007615.

Abbaspour, K., Rouholahnejad, E., Vaghefi, S., Srinivasan, R., Yang, H. & Kløve, B. 2015 A continental-scale hydrology and water quality model for Europe: calibration and uncertainty of a high-resolution large-scale SWAT model. *Journal of Hydrology* **524**, 733–752. https://doi.org/10.1016/j.jhydrol.2015.03.027.

Abbaspour, K., Vaghefi, S. A. & Srinivasan, R. 2018 A guideline for successful calibration and uncertainty analysis for soil and water assessment: A review of papers from the 2016 International SWAT Conference. *Water* **10** (1), 6. doi.org/10.3390/w10010006.

Addor, N. & Melsen, L. 2019 Legacy, rather than adequacy, drives the selection of hydrological models. *Water Resources Research* **55** (1), 378–390. doi:10.1029/2018WR022958.

Arnold, J., Srinivasan, P., Muttiah, R. & Williams, J. 1998 Large area hydrologic modeling and assessment, part I: model development. *Journal of the American Water Resources Association* **34** (1), 73–89. doi:10.1111/j.1752-1688.1998.tb05961.x.

Arnold, J. G., Moriasi, D. N., Gassman, P. W., Abbaspour, K., White, M. J., Srinivasan, R., Santhi, C., Harmel, R. D., Griensven, A., van Liew, M. W., Kannan, N. J. & Jha, M. K. 2012 SWAT: model use, calibration, and validation. *American Society of Agricultural and Biological Engineers* **55** (4), 1491–1508. doi: 10.13031/2013.42256.

Čerkasova, N., Umgiesser, G. & Ertürk, A. 2018 Development of a hydrology and water quality model for a large transboundary river watershed to investigate the impacts of climate change – a SWAT application. *Ecological Engineering* **124**, 99–115.

Chen, H., Xu, C.-Y. & Guo, S. 2012 Comparison and evaluation of multiple GCMs, statistical downscaling and hydrological models in the study of climate change impacts on runoff. *Journal of Hydrology* **434–435**, 36–45. https://doi.org/10.1016/j.jhydrol.2012.02.040.

CH2M Hill 2005 *Existing Conditions, Flow Characterization and Flood Level Analysis – Carp River, Feedmill Creek, and Poole Creek: Final Report*. CH2M Hill, Ottawa.

CH2M Hill 2006 *Post Development Flow Characterisation and Flood Level Analysis for Carp River*. CH2M Hill, Ottawa.

Choukri, F., Raclot, D., Naimi, M., Chikhaoui, M., Nunes, J. P., Huard, F., Hérivaux, C., Sabir, M. & Pépin, Y. 2020 Distinct and combined impacts of climate and land use scenarios on water availability and sediment loads for a water supply reservoir in northern Morocco. *International Soil and Water Conservation Research* **8** (2), 141–153.

City of Ottawa 2016 *Carp River PCSWMM Model Documentation*. City of Ottawa, Ottawa.

Collins, M., Knutti, R., Arblaster, J., Dufresne, J.-L., Fichefet, T., Friedlingstein, P., Gao, X., Gutowski, W. J., Johns, T., Krinner, G., Shongwe, M., Tebaldi, C., Weaver, A. J. & Wehner, M. 2013 Long-term climate change: projections, commitments and irreversibility. In: *Climate Change 2013: The Physical Science Basis. IPCC Working Group I Contribution to AR5*. (IPCC eds.), Ch. 12. Cambridge University Press, Cambridge, UK, pp. 1029–1136.

Dunn, S., Brown, I., Sample, J. & Post, H. 2012, April Relationships between climate, water resources, land use and diffuse pollution and the significance of uncertainty in climate change. *Journal of Hydrology* **434–435**, 19–35. https://doi.org/10.1016/j.jhydrol.2012.02.039.

El-Khoury, A., Seidou, O., Lapen, D. R., Que, Z., Mohammadian, M., Sunohara, M. & Bahram, D. 2015 Combined impacts of future climate and land use changes on discharge, nitrogen and phosphorus loads for a Canadian river basin. *Journal of Environmental Management* **151**, 76–86. https://doi.org/10.1016/j.jenvman.2014.12.012.

Environment and Natural Resources Canada 2020 *Historical Climate Data*. Environment and Natural Resources Canada: Available from: https://climate.weather.gc.ca/.

Eum, H.-I., Dibike, Y. & Prowse, T. 2016 Comparative evaluation of the effects of climate and land-cover changes on hydrologic responses of the Muskeg River, Alberta, Canada. *Journal of Hydrology: Regional Studies* **8**, 198–221.

Food and Agricultural Organization, UNESCO 2019 FAO/UNESCO Soil Map of the World, November. http://www.fao.org/soils-portal/soil-survey/soil-maps-and-databases/faounesco-soil-map-of-the-world/en/

Geng, Y., Baumann, F., Song, C., Zhang, M., Shi, Y., Kühn, P., Scholten, T. & He, J.-S. 2017 Increasing temperature reduces the coupling between available nitrogen and phosphorus in soils of Chinese grasslands. *Scientific Reports* **7**, 43524. doi:10.1038/srep43524.

Ghosh, I. & Hellweger, F. L. 2012 Effects of spatial resolution in urban hydrologic simulations. *Journal of Hydrologic Engineering* **17** (1). https://doi.org/10.1061/(ASCE)HE.1943-5584.0000405.

Greenland International Consulting Ltd 2009 *Third-Party Review – Carp River Restoration Plan*. City of Ottawa, Ottawa.

Greenland International Consulting Ltd 2010 *Carp River Restoration Plan, Widening Alternatives*. City of Ottawa, Ottawa.

Greenland International Consulting Ltd 2011 *Carp River Model Calibration Validation Exercise Final Report*. Ottawa.

Greenland International Consulting Ltd 2014 *Model Development Program – Carp River Restoration Plan Draft Report*. Ottawa.

Guo, Y., Fang, G., Xu, Y.-P., Tian, X. & Xie, J. 2020 Identifying how future climate and land use/cover changes impact streamflow in Xinanjiang Basin, East China. *Science of the Total Environment* **710**. https://doi.org/10.1016/j.scitotenv.2019.136275.

Gupta, H., Sorooshian, S. & Yapo, P. 1999 Status of automatic calibration for hydrologic models: comparison with multi-level expert calibration. *Journal of Hydrology* **4** (2), 135–143.

Harmel, R. D. & Smith, P. K. 2007 Consideration of measurement uncertainty in the evaluation of goodness-of-fit in hydrologic and water quality modeling. *Journal of Hydrology* **337** (3–4), 326–336. https://doi.org/10.1016/j.jhydrol.2007.01.043.

Havrylenko, S. B., Bodoque, J. M., Srinivasan, R., Zucarelli, G. V. & Mercuri, P. 2016 Assessment of the soil water content in the Pampas region using SWAT. *CATENA* **137**, 298–309. doi.org/10.1016/j.catena.2015.10.001.

Hoang, B. H., Hien, H. N., Dinh, N. T., Thao, N. A., Ha, P. T., Kandasamy, J. & Nguyen, T. V. 2019 Integration of SWAT and QUAL2 K for water quality modeling in a data scarce basin of Cau River basin in Vietnam. *Ecohydrology & Hydrobiology* **19** (2), 210–223. https://doi.org/10.1016/j.ecohyd.2019.03.005.

Hobbie, S. E., Finlay, J. C., Janke, B. D., Nidzgorski, D. A., Millet, D. B. & Baker, L. A. 2017 Contrasting nitrogen and phosphorus budgets in urban watersheds and implications for managing urban water pollution. *National Academy of Sciences* **114** (16), 4177–4182. doi.org/10.1073/pnas.1618536114.

Holvoet, K., Van Griensven, A., Seuntjens, P. & Vanrolleghem, P. 2005 Sensitivity analysis for hydrology and pesticide supply towards the river in SWAT. *Physics and Chemistry of the Earth, Parts A/B/C* **30** (8–10), 518–526. doi.org/10.1016/j.pce.2005.07.006

Hong, F. & Tang, J. 2014 *Influence of Temperature and Moisture on Nitrogen Cycling in Soils From Experimentally Heated and Control Plots at the Harvard Forest, MA*. Mount Holyoke College, Chicago, Marine Biological Laboratory. Available from: https://www.mbl.edu/ses/files/2015/04/Fangyuan-Hong.pdf (accessed December 2020).

Huo, A. & Li, H. 2013 Assessment of climate change impact on the stream-flow in a typical debris flow watershed of Jianzhuangcuan catchment in Shaanxi Province, China. *Environmental Earth Sciences* **69**, 1931–1938.

Jajarmizadeh, M., Sidek, L. M., Harun, S. & Salarpour, M. 2017 Optimal calibration and uncertainty analysis of SWAT. *Air, Soil and Water Research* **10**, 1–14. https://doi.org/10.1177%2F1178622117731792.

Karlsson, I. B., Sonnenborg, T. O., Refsgaard, J. C., Trolle, D., Børgesen, C. D., Olesen, J. E., Jeppesen, E. & Jensen, K. H. 2016 Combined effects of climate models, hydrological model structures and land use scenarios on hydrological impacts of climate change. *Journal of Hydrology* **535**, 301–317. https://doi.org/10.1016/j.jhydrol.2016.01.069.

Kaykhosravi, S., Khan, U. T. & Jadidi, M. A. 2020 The effect of climate change and urbanization on the demand for low impact development for three Canadian cities. *Water* **12** (5), 1280. https://doi.org/10.3390/w12051280.

Khoiab, D. N. & Thomb, V. T. 2015 Global ecology and conservation. *Parameter Uncertainty Analysis for Simulating Streamflow in a River Catchment of Vietnam* **4**, 538–548. https://doi.org/10.1016/j.gecco.2015.10.007.

Kundu, S., Khare, D. & Mondal, A. 2017 Individual and combined impacts of future climate and land use changes on the water balance. *Ecological Engineering* **105**, 42–57. https://doi.org/10.1016/j.ecoleng.2017.04.061.

Liu, L., Ma, C., Huo, S., Xi, B., He, Z., Zhang, H., Zhang, J. & Xia, X. 2018 Impacts of climate change and land use on the development of nutrient criteria. *Journal of Hydrology* **563**, 533–542. https://doi.org/10.1016/j.jhydrol.2018.06.039.

Lobligeois, F., Andréassian, V., Perrin, C., Tabary, P. & Loumagne, C. 2014 When does higher spatial resolution rainfall information improve streamflow simulation? An evaluation using 3620 flood events. *Hydrology and Earth System Sciences* **18** (2), 575–594. https://doi.org/10.5194/hess-18-575-2014.

Luo, C., Li, Z., Liu, H., Li, H., Wan, R., Pan, J. & Chen, X. 2020 Differences in the responses of flow and nutrient load to isolated and coupled future climate and land use changes. *Journal of Environmental Management* **256**, 109918.

Manashi, P. 2016 Impacts of Land Use and Climate Changes on Hydrological Processes in South Dakota Watersheds. South Dakota State University, Agricultural and Biosystems Engineering. ProQuest Dissertations Publishing. Available from: https://search.proquest.com/docview/1837118008?pq-origsite=summon.

Mearns, L., McGinnis, S., Korytina, D., Arritt, R., Biner, S., Bukovsky, M., Chang, H.-I., Christensen, O., Herzmann, D., Jiao, Yanjun, K., Slava Lazare, M., Nikulin, G., Qian, Minwei, Q., Scinocca, J. & Winger, J. 2017 The NA-CORDEX dataset, version 1.0. NCAR Climate Data Gateway, Boulder, CO. Retrieved November 20, 2020. https://doi.org/10.5065/D6SJ1JCH.

Mehdi, B. B. & Lehner, B. 2014 *Scenarios and Implications of Land Use and Climate Change on Water Quality in Mesoscale Agricultural Watersheds*. McGill University, Department of Geography, Montreal. doi:10.3390/w6103152.

Mengistu, A. G., Van Rensburg, L. D. & Woyessa, Y. E. 2019 Techniques for calibration and validation of SWAT model in data scarce arid and semi-arid catchments in South Africa. *Journal of Hydrology: Regional Studies* **25**, 100621.

Ministry of Natural Resources and Forestry 2019 *Ontario Flow Assessment Tool*. Retrieved from Ontario Flow Assessment Tool. Available from: http://www.gisapplication.lrc.gov.on.ca/OFAT/Index.

Miralha, L., Muenich, R. L., Scavia, D., Wells, K., Steiner, A. L., Kalcic, M., Apostel, A., Basile, S. & Kirchhoff, C. J. 2020 Bias correction of climate model outputs influences watershed model nutrient load predictions. *Science of the Total Environment*. https://doi.org/10.1016/j.scitotenv.2020.143039.

Moriasi, D., Arnold, J., Van Liew, M., Bingner, R., Harmel, R. & Veith, T. 2007 Model evaluation guidelines for systematic quantification of accuracy in watershed simulations. *Transactions of the ASABE* **50** (3), 885–900. doi:10.13031/2013.23153.

Munia, H. A., Guillaume, J. H., Wada, Y., Veldkamp, T., Virkki, V. & Kummu, M. (2020, May). Future Transboundary Water Stress and Its Drivers Under Climate Change: A Global Study. *Earth's Future* **8** (7), e2019EF001321. doi:10.1029/2019EF001321.

Muluneh, A. 2020 Impact of climate change on soil water balance, maize production, and potential adaptation measures in the Rift Valley drylands of Ethiopia. *Journal of Arid Environments* **179**, 104195. https://doi.org/10.1016/j.jaridenv.2020.104195.

Nash, J. & Sutcliffe, J. 1970 River flow forecasting through conceptual models part I – a discussion of principles. *Journal of Hydrology* **10** (3), 282–290. https://doi.org/10.1016/0022-1694(70)90255-6.

Neitsch, S. L., Arnold, J. G., Kiniry, J. R. & Williams, J. R. 2009 Soil and Water Assessment Tool Theoritical Documentation Version 2009. Grassland, Soil and Water Research Laboratory, Agricultural Research Service,Blackland Research Center, Texas AgriLife Research , US Department of Agriculture. Texas Water Research Institute, College Station, TX. https://swat.tamu.edu/media/99192/swat2009-theory.pdf.

Nieder, R., Benbi, D. K. & Reichl, F. X. 2018 Reactive water-soluble forms of nitrogen and phosphorus and their impacts on environment and human health. *Soil Components and Human Health* **2018**, 223–255.

Noori, N., Kalin, L. & Isik, S. 2020 Water quality prediction using SWAT-ANN coupled approach. *Journal of Hydrology* **590**. https://doi.org/10.1016/j.jhydrol.2020.125220.

Ogden, F. L. 2021 Geohydrology: hydrological modeling. In: *Encyclopedia of Geology*, 2nd edn. pp. 457–476. https://doi.org/10.1016/B978-0-08-102908-4.00115-6.

Sharma, V. C. & Regonda, S. K. 2021 Multi-spatial resolution rainfall-Runoff modelling – a case study of Sabari River Basin, India. *Water* **13** (9), 1224. https://doi.org/10.3390/w13091224.

Shrestha, S., Sattar, H., Khattak, M. S., Wang, G. & Babur, M. 2020 Evaluation of adaptation options for reducing soil erosion due to climate change in the Swat River Basin of Pakistan. *Ecological Engineering* **158**, 106017. https://doi.org/10.1016/j.ecoleng.2020.106017.

Shrestha, N. K., Rudra, R. P., Daggupati, P., Goel, P. K. & Shukla, R. 2021 A comparative evaluation of the continuous and event-based modelling approaches for identifying critical source areas for sediment and phosphorus losses. *Journal of Environmental Management* **277**. Article 111427. https://doi.org/10.1016/j.jenvman.2020.111427.

Stantec Consulting Ltd 2006 *Kanata West Master Servicing Study*. Stantec Consulting Ltd, Ottawa.

Tirupathi, C. & Shashidhar, T. 2020 Investigating the impact of climate and land-use land cover changes on hydrological predictions over the Krishna river basin under present and future scenarios. *Science of the Total Environment* **721**. https://doi.org/10.1016/j.scitotenv.2020.137736.

Tu, J. 2009 Combined impact of climate and land use changes on streamflow and water quality in eastern Massachusetts, USA. *Journal of Hydrology* **379** (3–4), 268-283. doi.org/10.1016/j.jhydrol.2009.10.009.

United Nations 2019 *World Population Prospects 2019 – Volume II: DemographicProfiles*. Economic and Social Affairs, PopulationDivision. United Nations. Available from: https://population.un.org/wpp/Publications/Files/WPP2019_Volume-II-Demographic-Profiles.pdf (accessed May 2020).

USGS 2020 *Water Science School – Evapotranspiration and the Water Cycle*. USGS – Science for a Changing World. Available from: https://www.usgs.gov/special-topic/water-science-school/science/evapotranspiration-and-water-cycle?qt-science_center_objects=0#qt-science_center_objects (accessed 12 May 2020).

Verhagen, A. P., Ostelo, R. W. & Rademaker, A. 2004 Is the p value really so significant?*. *Australian Journal of Physiotherapy* **50** (4), 261–262. doi.org/10.1016/S0004-9514(14)60122-7.

Wang, Q., Xu, Y., Wang, Y., Zhang, Y., Xiang, J., Xu, Y. & Wang, J. 2020 Individual and combined impacts of future land-use and climate conditions on extreme hydrological events in a representative basin of the Yangtze River Delta, China. *Atmospheric Research* **236**, 104805. https://doi.org/10.1016/j.atmosres.2019.104805.

Wayne, G. 2013 *The Beginner's Guide to Representative Concentration Pathways*. Skeptical Science. Available from: https://www.skepticalscience.com/docs/RCP_Guide.pdf.

Weedom, G., Balsamo, G., Bellouin, N., Gomes, S., Best, M. & Viterbo, P. 2014 The WFDEI meteorological forcing data set: WATCH forcing data methodology applied to ERA-Interim reanalysis data. *Water Resources Research* **50**. doi:10.1002/2014WR015638.

Young, R., Onstad, C. & Bosch, D. 1989 AGNPS: a nonpoint source pollution model for evaluating agricultural watersheds. *Journal of Water and Soil Conservation* **44** (2), 168–173.

Yu, Z. 2015 Hydrology, floods and droughts: modeling and prediction. In: *Encyclopedia of Atmospheric Sciences*, 2nd edn. pp. 217–223. doi:10.1016/B978-0-12-382225-3.00172-9.

Zhou, X. V., Clark, C. D., Nair, S. S., Hawkins, S. A. & Lambert, D. M. 2015 Environmental and economic analysis of using SWAT to simulate the effects of switchgrass production on water quality in an impaired watershed. *Agricultural Water Management* **160**, 1–13. https://doi.org/10.1016/j.agwat.2015.06.018.

First received 21 April 2021; accepted in revised form 21 September 2021. Available online 8 October 2021

Impact of Climate Change on Hydrology and Water Resources doi: 10.2166/wcc.2021.081

Effects of environmental factors on the methane and carbon dioxide fluxes at the middle of Three Gorges Reservoir

Yu Qin*, Yujia Gou, Zongtai Yu and Wei Tan
Key Laboratory of Hydraulic and Waterway Engineering of the Ministry of Education, Chongqing Jiaotong University, 400074 Chongqing, China
*Corresponding author. E-mail: qinyu54001@163.com

ABSTRACT

In recent years, scientists have paid special attention to the greenhouse gas emissions of the Three Gorges Reservoir. This study took Fuling to Wanzhou, located at the middle section of the Three Gorges Reservoir, as the research locations. From August 2017 to August 2018, the partial pressure of carbon dioxide (CO_2) and methane (CH_4) and the flux of CO_2 and CH_4 at the water–gas interface were studied. Spearman's correlation between the partial pressure and discharge flux of CO_2 and CH_4 in water and environmental variables was analyzed. The research results showed that the variation of partial pressure and flux was consistent. At different sample locations, there was no statistical difference in CO_2 and CH_4 fluxes, but under different operating periods, the CO_2 and CH_4 fluxes were significantly different. The highest values happened during the drainage period as well as the low-water period, respectively. Total organic carbon and total nitrogen were significantly positively correlated, and dissolved oxygen was extremely negatively correlated. The high value of CH_4 flux in the middle of the reservoir area was related to the spatial distribution of the sediment and the amount of sediment deposition.

Key words: GHG flux, influencing factors, middle section of reservoir area, sediment deposition, surface waters, Three Gorges Reservoir

HIGHLIGHTS

- This paper focuses on the greenhouse gas change in the middle section of the Three Gorges Reservoir area, which has received relatively little attention at present.
- The research time span of this paper is long, and the analysis of greenhouse gas (GHG) is comprehensive.

GRAPHICAL ABSTRACT

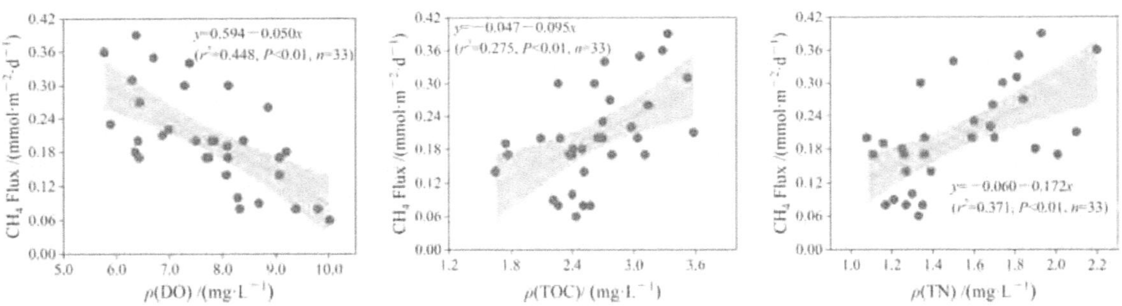

1. INTRODUCTION

The study of greenhouse gas (GHG) source and sink changes in reservoirs has been an interdisciplinary frontier and hotspot in the field of global river ecological environment research and water conservancy and hydropower engineering in the past 20 years (Abril *et al.* 2014; Sawakuchi *et al.* 2014). According to some studies, reservoirs emit about 70 million tons of methane (CH_4) and 100 million tons of carbon dioxide (CO_2) every year, and reservoirs have become an important source of GHG in the atmosphere; what is more, the amount of CH_4 released from reservoirs accounts for 20% of the total CH_4 emissions from human activities (Mccully 2002). The infrared absorption function of CH_4 molecule is stronger than that of CO_2, which makes the global warming efficiency of CH_4 20–32 times that of CO_2 (Rodhe 1990; Lemer & Roger 2001), and plays an important role in global warming. At present, the preliminary consensus is that the reservoir will have a high GHG flux in the initial 10–20 years and gradually stabilize after 20 years (Tremblay *et al.* 2004). Therefore, the main factors affecting the change of GHG source and sink of reservoirs are the age of reservoirs and the climate zone of reservoirs (Barros *et al.* 2011; Kumar *et al.* 2019). At the same time, the differences in rainfall, air temperature, wind speed and relative humidity among the crops on the two sides verified by the SAWT model also have indirect effects on the reservoir (Dash *et al.* 2021).

Dissolved oxygen (DO), chlorophyll concentration, temperature, surface water temperature, salinity and turbidity are known as water quality variables (Shamshirband *et al.* 2019). According to the existing studies, the release form of carbon is mainly CO_2 in the case of abundant oxygen in water, while the form is mainly CH_4 and CO_2 in the case of anoxic or anaerobic conditions. Therefore, an important indirect factor in the production, diffusion and release of GHG is the concentration of DO in water. In the tropical reservoirs, part of CH_4 generated in sediments rose to the water surface through diffusion (Rosa *et al.* 2003). However, due to the gradual increase in the concentration of DO in the rising process, most of the CH_4 was oxidized to CO_2 by CH_4-oxidizing bacteria on the anaerobic–aerobic critical surface. As a result, the CH_4 release from the water–gas interface was significantly reduced, while the CO_2 release was increased. Changes in concentrations of C and N in water also affect the primary productivity of aquatic plants and the metabolic process of plankton, and the concentration of pollutants changes along the river, thus indirectly affecting the concentrations of CO_2 and CH_4 in water and the release fluxes of CO_2 and CH_4 at the water–air interface (Yang *et al.* 2013; Kargar *et al.* 2020).

After the completion of the Three Gorges Reservoir, several kilometers to tens of kilometers of river water were inundated to form slow or still water reaches in the lower reaches of the mainstream and tributaries of the Yangtze River, and river water was significantly reduced in the tributaries. The original diversified and fast-flowing river system disappeared, and a complex lake–river system with slow or even stationary water flow appeared (Nilsson & Berggren 2000). It is worth noting that the Three Gorges Reservoir is a typical channel canyon reservoir. After the establishment of the reservoir, the water level of the natural channel increases, and the flow velocity gradually slows down. The changes of the natural river hydrodynamic environment brought about by the construction of the dam and the water impoundment process also lead to the accumulation of sediment in the reservoir with the increase of the operating years. While the self-purification capacity of water body is weakened, for the river–lake transitional water body, the carbon which is from biogeographic transformation and GHG source and sink in the reservoir also presents transition and high heterogeneity in time and space (Morales-Pineda *et al.* 2014). For example, in the river area at the tail of the reservoir, the reservoir mainly degrades carbon from different sources because the upstream river has more organic matter from different sources. However, in the lake area at the head of the reservoir, due to the decrease of flow rate, the enhancement of light transmittance and the growth of algae, the reservoir mainly presents the process of endogenous carbon synthesis. The differences described above will, to some extent, lead to gradient changes in the source and sink of GHG in reservoirs in time and space (Li 2018; Li *et al.* 2018b). In the past 10 years, many teams in China have been monitoring and analyzing GHG and reservoir carbon cycle in the Three Gorges Reservoir (Li *et al.* 2016a; Liu *et al.* 2017; Li 2018). The main views are as follows: (1) The GHG source and sink volume of the water body in the Three Gorges Reservoir does not have the characteristics of high release flux as previously reported. On the contrary, it is at a low level in the global case base (Li *et al.* 2018b). (2) The mainstream of the Three Gorges Reservoir mainly degrades substances of different sources, while the algae growth in the backwater area of tributaries may present a local carbon sink phenomenon (Qin *et al.* 2017a).

At present, there are few reports on the characteristics and influencing factors of GHG sources and sinks in the middle section of the Three Gorges Reservoir. The middle section of the Three Gorges Reservoir (Fuling–Wanzhou section), as a transition section from 'fluvial facies' to 'lacustrine facies', is different from rivers and lakes in terms of channel characteristics ('S' type) (Li *et al.* 2016b), habitat characteristics, ecological structure and function, and key source factors (Li *et al.* 2018b). In addition, the current literature reports mainly focus on the macro characteristics of reservoir GHG sources and sinks, and the relevant literature reports focus on the spatial heterogeneity of reservoir GHG and its influencing factors are still limited. Therefore, the study of the typical sections of the Three Gorges Reservoir can effectively improve the understanding of GHG emissions in the reservoir ecosystem and clarify its impact on the global carbon cycle, which is of self-evident importance. In the preliminary investigation, the gas concentration in the water body of the Three Gorges Reservoir was initially tracked and monitored. The study found that the highest concentration of GHG in the water body appeared in the middle section of the reservoir, and there was no significant change in the concentration along with the depth of the river. It is preliminarily judged that this characteristic is related to the spatial distribution of sediment deposition in the mainstream of the

Three Gorges Reservoir to some extent and may be affected by the operation of the reservoir 'storing clear water and discharging muddy water' (Maeck *et al.* 2013).

Therefore, this study selected three typical sections of Fuling (FL), Zhongxian (ZX) and Wanzhou (WZ), which are in the middle section of the Three Gorges Reservoir, as the objects to carry out a 1-year *in situ* monitoring experiment. Combined with environmental factors such as meteorological and climatic conditions, the study revealed the main factors affecting the change of GHG source and sink in the middle section of the Three Gorges Reservoir. This study provides theoretical and data support for further study of GHG effects of the reservoir and provides a basis for further revealing the biogeochemical process of GHG source and sink changes in the middle section of the Three Gorges Reservoir under the operation of the reservoir.

2. MATERIALS AND METHODS

2.1. Study site

The length of the Three Gorges Reservoir is about 663 km. The average water surface width of the mainstream reservoir section is 1,100 m, and the water area is about 1,084 km^2 (Li *et al.* 2013; Luo *et al.* 2018). According to the operational plan of the Three Gorges Project about 'storing clear and releasing muddy', the water level of the reservoir will be lowered to 145 m before the flood season begins at the end of June and gradually be increased to175 m after the flood season ends at the end of September. The range is between 145 and 175 m. The drainage period is from February to May, while the low-water-level operation period is from June to September, and from October of each year to January of next year is the period of impoundment and high-water-level operation (short for 'high-water-level period'). When the water level is at a low running stage, the backwater of the mainstream reaches the downstream of Chongqing Longevity area, 524 km away from the leading edge of the dam, which is called the 'perennial backwater area' (Li *et al.* 2015a).

In the preliminary investigation, our team tracked and monitored the gas concentration in the water body of the Three Gorges Reservoir. The study found that the highest concentration of GHG in the water body appeared in the middle section of the reservoir (Wanzhou–Fuling section), and there was no significant change in the concentration along the depth of the river (Figure 1). Based on the above analysis, in this study, we chose three sections, Fuling (FL), Zhongxian (ZX) and Wanzhou (WZ), as key research areas in the 'perennial backwater area' mentioned to carry out *in situ* observation for 1 year.

FL, which is located in the middle of Chongqing and located at the conjunctions of the Yangtze River and the Wujiang River, is 120 km from Chongqing above and 482 km from the Three Gorges Dam below. This reach, with a flow of 86 km and an average annual flow of about 11,200 m^3/s, is a typical mountainous river course in the upper reaches of the Yangtze River. The annual runoff distribution of this reach is extremely uneven, with an average flow of 1,050 m^3/s from June to October in the flood season, and its runoff accounts for 74.5% of the annual runoff.

ZX, 190 km above Chongqing and 386 km below the Three Gorges Dam on the Yangtze River, is located in the eastern and central part of Chongqing. The reach is flat and wide, with a uniform flow of 88 km and an average annual flow of about 12,400 m^3/s.

WZ, located in the northeast of Chongqing, is a central city in the upper reaches of the Yangtze River. It is 328 km from Chongqing and 283 km from the Three Gorges Dam. This reach is a stable section of large mountain rivers, with a relatively straight course of 84.3 km and an average annual flow of about 13,700 m^3/s.

The data obtained by the study will be divided according to the above three water-level running periods to analyze the changes of GHG flux source and sink in the Three Gorges Reservoir section. Total amount of water will influence the

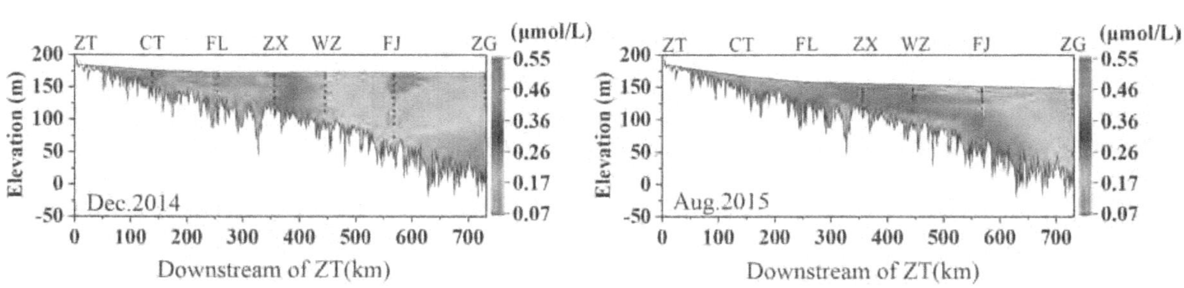

Figure 1 | Concentration distribution of CH_4 in water of the Three Gorges Reservoir (Li *et al.* 2020).

turbidity of the reservoir (Alizadeh *et al.* 2018). From August 2017 to August 2018, the water level changes during the study period of the Three Gorges Reservoir (Figure 2) (*source*: https://www.ctg.com.cn/sxjt/sqqk/index.html).

2.2. Sampling and analysis

From August 2017 to August 2018, this study conducted monthly monitoring and analysis on three sampling sites for a period of 12 months. The monitoring time of all sampling sites was controlled from 20th to 22nd of each month. Massive amounts of GHG gathering research will be at 9:00–11:00 (Wu 2016) as the representative of the flux of the day time, Wu (2012) to choose 9:00–11:00 in the morning in the Three Gorges Reservoir with monitoring, Zhao *et al.* (2011) monitored GHG emissions in the Xiangxi River Basin of the Three Gorges Reservoir from 9:00 to 14:00. Due to spatiotemporal differences in GHG fluxes, GHG fluxes at the same point and at different times will also change. To obtain more accurate data, the sampling time of this study is as much as possible controlled at 9:00–11.00 am. The sampling tool is a 5-L water sampler to collect the 0.5 m surface water of the above three sections, respectively (Gong *et al.* 2019; Liu *et al.* 2019), the water samples will be sent back to the laboratory immediately on the same day, all the test and analysis work will be completed within 48 h, and the relevant water environmental indicators and the characteristics of GHG flux at the water–air interface will be monitored simultaneously.

At present, there are two commonly used methods for monitoring the gas emission flux at the water–gas interface: one is static chamber method – gas chromatography; the other is the combination of headspace balance method – gas chromatography and thin boundary layer (TBL) model estimation method. Static box-gas chromatography is usually only able to obtain flux data at points, with high labor intensity and high cost of gas analysis, which makes it unsuitable for large area and long-term observation. Therefore, static box-gas chromatography is only suitable for the observation of static water bodies (Tremblay *et al.* 2005). However, the model estimation method is most commonly used in field monitoring due to its simplicity, flexibility and easy operation (Qin *et al.* 2017b). Considering the strong fluidity of the water body, the long distance between sampling points, the long duration of monitoring and the poor geographical and climatic conditions of the points, the feasibility of the static box method is weak. In addition, due to the disturbance of water outside the box, the flux data obtained by the static box method is relatively discrete. The comparative experiment shows that the flux results obtained by the TBL estimation method are generally more stable than the former method (Yao *et al.* 2015). Therefore, in this study, the GHG flux at the water–gas interface in the middle section of the Three Gorges Reservoir was obtained by using the headspace balance TBL model estimation method.

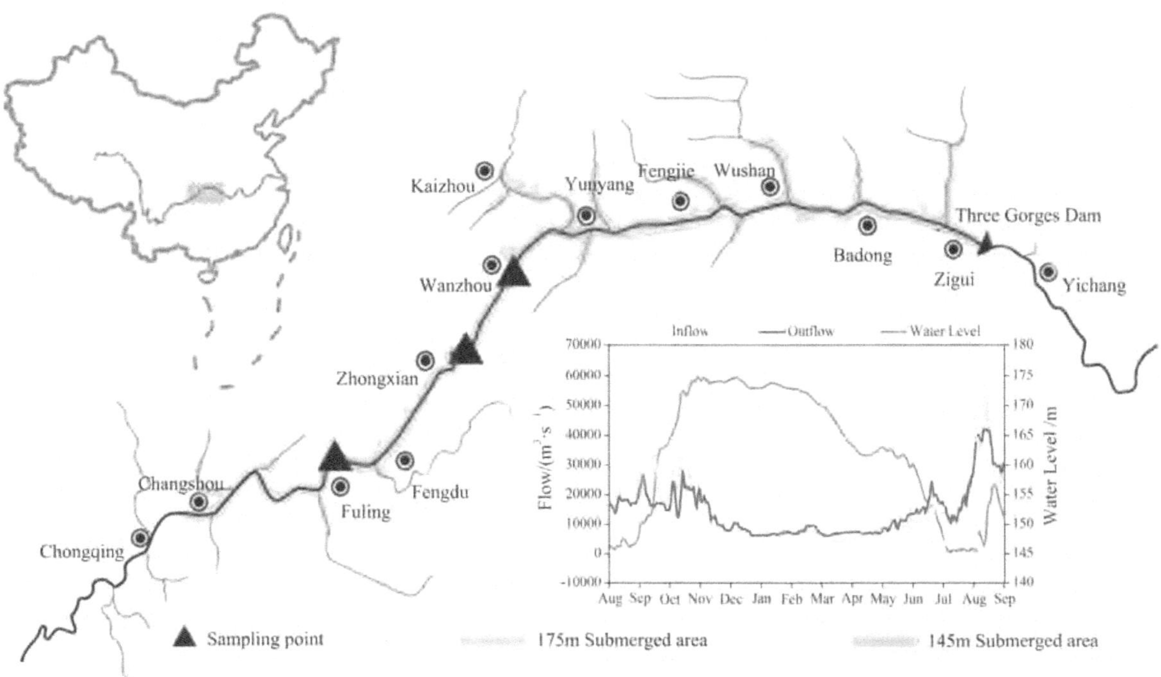

Figure 2 | Sampling point and flow water-level map of the Three Gorges Reservoir; sample list is showed in Table 1.

Table 1 | Basic information of the sampling sites

Sample sites	Location
FL	29°48′00″N, 107°27′00″E
ZX	30°24′57.63″N, 108°12′40.86″E
WZ	30°46′26.66″N, 108°24′46.74″E

At the sampling place, the headspace bottle is placed in a water collector filled with water and sealed in a completely submerged state. It is confirmed that there is no bubble in the headspace, put it upside down and send it back to the laboratory for analysis. The bottle is sealed in a completely submerged state. At the same time, it is confirmed that there is no bubble in the head empty bottle, it is placed upside down, and sent back to the laboratory for analysis. After the headspace bottle filled with water sample is taken out, two injection needles are inserted into the headspace bottle at the same time at room temperature, to the headspace bottle injection of 5 mL of high-purity nitrogen mechanical shock after 20 min, after waiting for two phase of water–gas balance to extract the upper air bottle, into an agilent7820A gas chromatograph, after TDX-01 chromatographic column separation in an Ni catalyst with a FID detector test, after obtaining the CO_2 and CH_4 concentration, Equation (1) is plugged in, to get water samples under test of CO_2 and CH_4 partial pressure (Kolb & Ettre 2006), the specific method to calculate the GHG see literature (Qin *et al.* 2017b):

$$P(\text{Gas}) = \frac{(P_{\text{final}} \times K_{\text{equailibrium}}) + \left((\text{HS}/S) \times \frac{(P_{\text{final}} - P_{\text{initial}})}{V_{\text{m}}}\right)}{K_{\text{sample}}} \tag{1}$$

where P(Gas) is the gas partial pressure in the water sample to be measured, μatm; P_{initial} and P_{final} are the partial pressure of the gas to be measured in the air above the bottle before and after equilibrium, μatm; HS/S is the volume ratio of the gas in the bottle to the water; K_{sample} and $K_{\text{equilibrium}}$ are the corresponding solubility of the gas to be measured at the time of sampling and water temperature in the bottle before sample analysis, respectively, $\text{mol·L}^{-1}\text{·atm}^{-1}$; V_{m} is the mole of the gas volume, mol·L^{-1}; the unit of partial pressure is uniformly converted into Pa,1 μatm = 0.101325 Pa.

According to Fick's law (Rantakari *et al.* 2015), for freshwater, the gas exchange flux at the water–air interface can be calculated by the following equation:

$$\text{Flux} = k_{\text{x}}(C_{\text{water}} - C_{\text{air}}) \tag{2}$$

where Flux is the GHG diffusion flux, $\text{mmol·m}^{-2}\text{·d}^{-1}$; k_{x} is the gas exchange coefficient, cm·h^{-1}; C_{water} is the GHG concentration in water, $\mu\text{mol·L}^{-1}$; C_{air} is the GHG concentration in the atmosphere, $\mu\text{mol·L}^{-1}$.

For the gas exchange coefficient k_x, the empirical model formula (Rik 1992) is adopted, the instantaneous wind speed (U_1) above the water surface measured in the field is used and the wind speed at 10 m above the water surface (U_{10}) is converted according to the following equation:

$$U_{10} = 1.22 \times U_1 \tag{3}$$

At the sampling place, the water temperature and DO were measured by YSI® Pro ODO (the accuracy of 0.1 °C, 0.01 mg·L^{-1}, respectively); the pH value was measured by the multi-parameter water quality analyzer HACH® MS5 (the accuracy of 0.01); the instantaneous wind speed above the water surface was measured by AR-826 anemometer (the accuracy was 0.01 m·s^{-1}); the temperature was measured by the thermometer on site (the accuracy was 0.1 °C); the atmospheric pressure is measured by a barometer on site (with an accuracy of 0.01 bpa); the field measuring instruments have passed the standard calibration procedures before sampling; and at the end of the sampling field test, the water samples from each section are taken back to the laboratory to complete the analysis and testing of other indicators, including total organic carbon (TOC) and total nitrogen (TN), within 48 h (Yang & Lv 2016). Inside the TOC was determined by Whatman GF/F glass fiber filter membrane, which was dried at 450 °C for 4 h, and the TN was determined by alkaline potassium persulfate oxidation ultraviolet spectrophotometry.

2.3. Data analysis method

In this study, all the experimental and calculated data were entered into SPSS® or Origin® for statistical analysis, and one-way ANOVA was used to explain the differences of the data in different periods and spaces. Spearman's correlation analysis was used to analyze the correlation between CO_2 and CH_4 concentration and various environmental factors (DO, TOC and TN), or Origin was used for linear regression analysis, indicating the linear correlation between data changes. There are also Pearson's correlation analysis methods for correlation analysis, but Spearman's analysis is more suitable and more accurate because GHGs belong to non-normal distribution. Pearson's correlation analysis is more suitable for data with normal distribution (Dash *et al.* 2019).

3. RESULTS AND DISCUSSION

3.1. Results

3.1.1. Changes of CO_2 partial pressure in water and flux at the water–air interface

As shown in Figure 3, the average annual partial pressure of CO_2 in the middle of the Three Gorges Reservoir is 155.8 ± 38.2 Pa. The average annual partial pressure of CO_2 at FL point was 148.8 ± 34.3 Pa, reaching its maximum value in May of the next year; the average annual partial pressure of CO_2 at ZX point was 157.5 ± 38.3 Pa, reaching its maximum value in April of the next year; and the average annual partial pressure of CO_2 at WZ point was 161.2 ± 40.7 Pa, reaching its maximum value in April of the next year. The monthly changes of FL, ZX and WZ were similar. There was no significant difference in CO_2 partial pressure at each point (ANOVA, $P > 0.05$). The mean partial pressures of CO_2 in the high-water

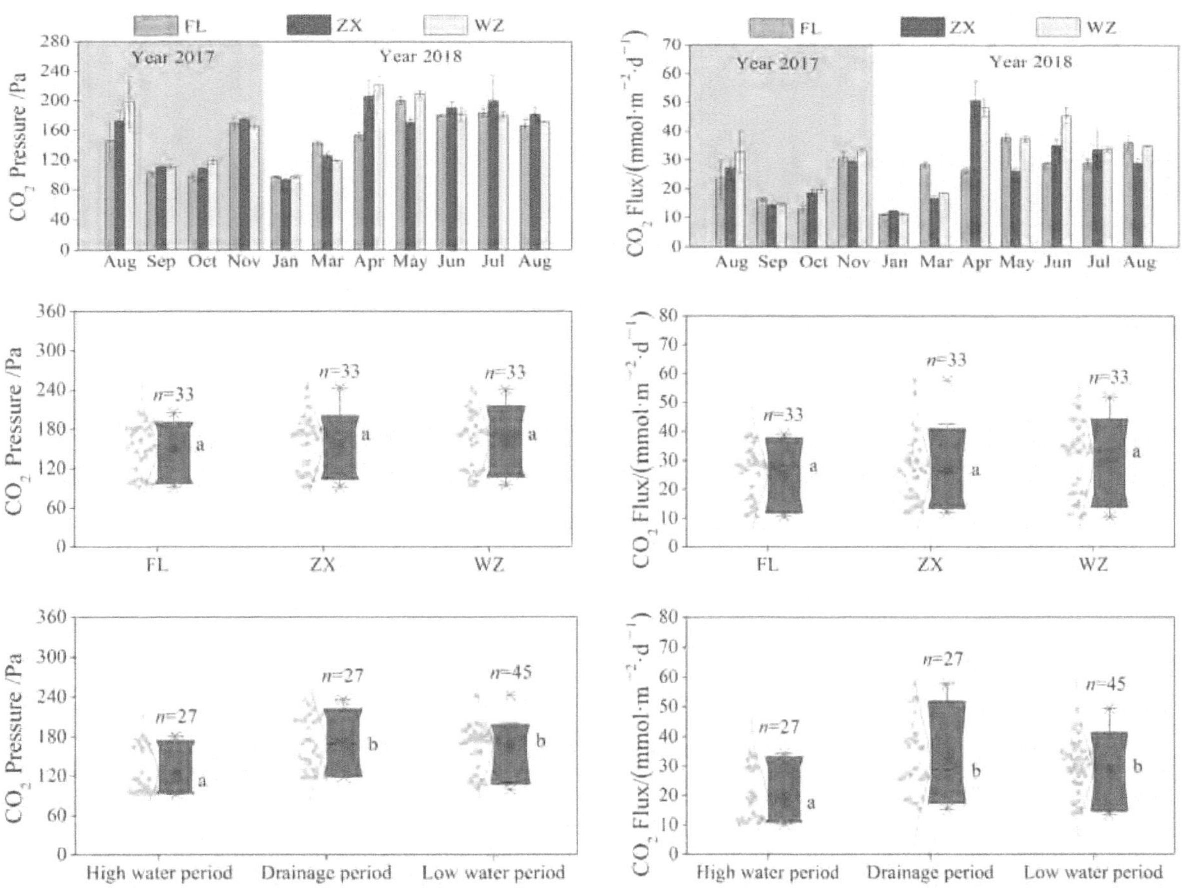

Figure 3 | CO_2 partial pressure of water and CO_2 flux at the water–air interface. (*Note*: a and b indicate differences between groups.)

stage, drainage stage and low-water stage were 124.6 ± 32.5, 171.8 ± 36.3 and 165.1 ± 31.1 Pa, respectively. Under different operating conditions, the partial pressure of CO_2 changed significantly (ANOVA, $P = 0.03 < 0.05$), and the drainage stage and the low-water stage were significantly higher than the high-water stage.

The average annual CO_2 flux in the middle of Three Gorges Reservoir is 27.4 ± 10.8 mmol·m^{-2}·d^{-1}. The average annual CO_2 flux at FL point was 25.6 ± 8.6 mmol·m^{-2}·d^{-1}, reaching its maximum value in May of the following year; the average annual CO_2 flux at ZX point was 26.6 ± 11.0 mmol·m^{-2}·d^{-1}, reaching its maximum value in April of the next year; and the average annual CO_2 flux at WZ point was 30.0 ± 11.9 mmol·m^{-2}·d^{-1}, reaching its maximum value in April of the next year. The annual CO_2 flux is positive, which is the 'source' of atmospheric CO_2. The monthly changes of FL, ZX and WZ were similar. There was no significant difference in CO_2 flux at each point (ANOVA, $P > 0.05$). The average CO_2 fluxes were 19.9 ± 8.5, 32.2 ± 11.6 and 28.9 ± 9.0 mmol·m^{-2}·d^{-1} in high-water stage, discharge period and low-water stage, respectively. Under the conditions of different operating periods, CO_2 flux changes significantly (ANOVA, $P = 0.03 < 0.05$), and the drainage stage and low-water stage are significantly higher than the high-water stage, which are similar laws to the partial pressure of CO_2.

3.1.2. Changes of CH_4 partial pressure in water and flux at the water–air interface

As shown in Figure 4, the average annual CH_4 partial pressure in the middle period of the Three Gorges Reservoir is 21.7 ± 8.4 Pa. The average annual CH_4 partial pressure at FL point was 18.6 ± 4.9 Pa, reaching its maximum value in May of the next year; the average annual CH_4 partial pressure at ZX point was 23.8 ± 9.3 Pa, reaching its maximum value in August; and the average annual CH_4 partial pressure at WZ point was 22.7 ± 9.4 Pa, reaching its maximum value in August of the next year. There was no statistical difference in CH_4 partial pressure at each point (ANOVA, $P > 0.05$). The mean CH_4 partial

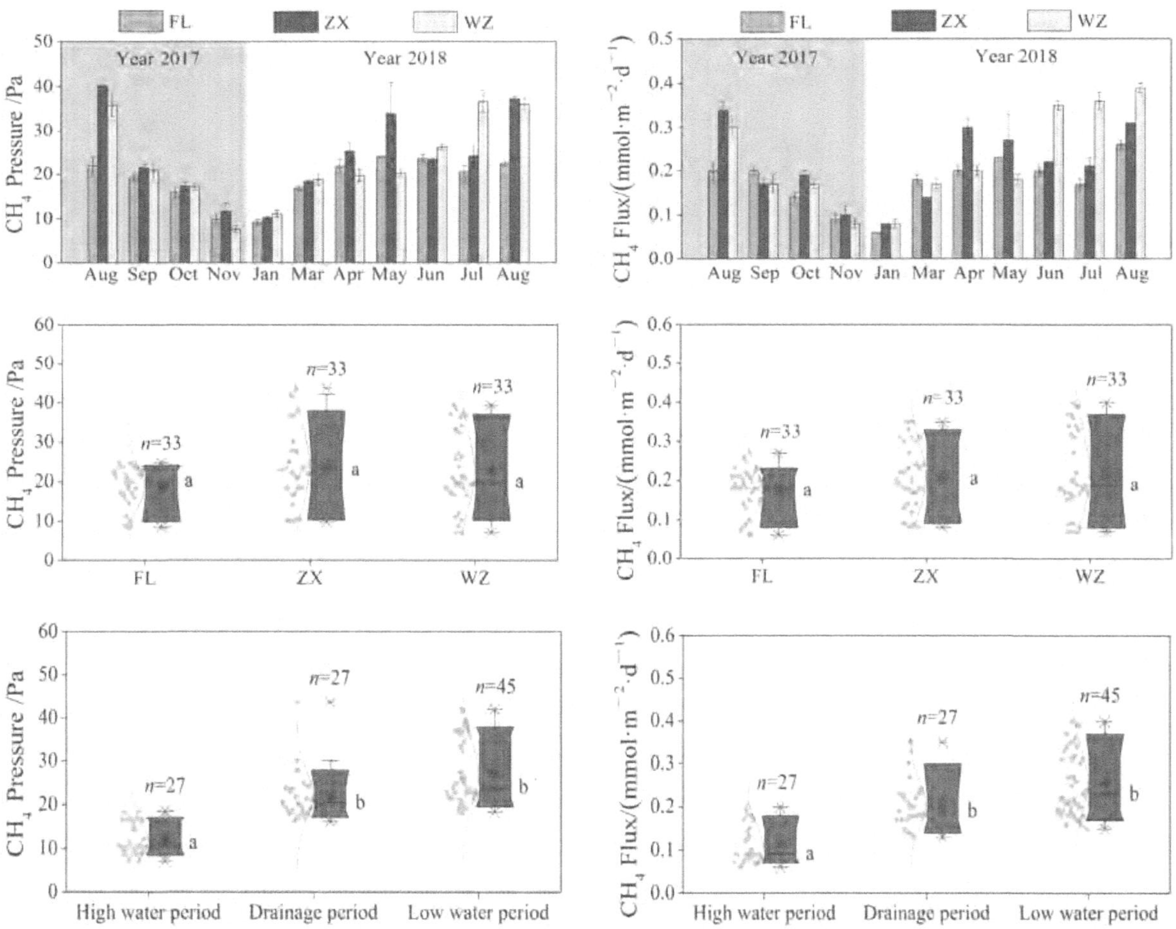

Figure 4 | CH_4 partial pressure of water and CH_4 flux at the water–air interface. (*Note*: a and b indicate differences between groups.)

pressure in high-water stage, drainage stage and low-water stage was 12.2 ± 3.4, 22.0 ± 4.8 and 27.2 ± 7.1 Pa, respectively. Under different operating conditions, there were significant differences in CH_4 partial pressure (ANOVA, $P = 0.000 < 0.05$), and the low-water stage and drainage stage were significantly higher than the high-water stage.

The average annual CH_4 flux in the middle of Three Gorges Reservoir is 0.20 ± 0.09 mmol·m^{-2}·d^{-1}. The average annual CH_4 flux at FL point was 0.17 ± 0.05 mmol·m^{-2}·d^{-1}, reaching its maximum value in August of the next year; the annual average CH_4 flux at ZX point was 0.21 ± 0.08 mmol·m^{-2}·d^{-1}, reaching its maximum value in August; and the annual average CH_4 flux at WZ point was 0.22 ± 0.10 mmol·m^{-2}·d^{-1}, reaching its maximum value in August of the next year. The annual CH_4 fluxes of FL, ZX and WZ are all positive and are the 'source' of atmospheric CH_4. There was no statistical difference in CH_4 flux at each point (ANOVA, $P > 0.05$). The average CH_4 flux in high-water stage, drainage stage and low-water stage is 0.11 ± 0.04, 0.20 ± 0.04 and 0.25 ± 0.07 mmol·m^{-2}·d^{-1}, respectively. In different operating periods, CH_4 fluxes were significantly different (ANOVA, $P = 0.000 < 0.05$), the low-water level and drainage stage were significantly higher than the high-water-level stage and the partial pressure of CH_4 fluxes and CH_4 presented similar laws.

3.1.3. Changes in physical and chemical indices of water quality

The change rules of DO, TOC and TN in the middle of the Three Gorges Reservoir at different points and different operating periods are shown in Figure 5. At different points, DO, TOC and TN showed similar variation patterns, in which DO reached a higher value the next January, and TOC and TN both reached a higher value the next July. There was no significant difference in annual DO, TOC and TN values at FL, ZX and WZ. In different operation periods, DO reaches the highest value in the high-water stage and the lowest value in the low-water stage. Both TOC and TN reached the lowest value in the high-water stage and the highest value in the low-water stage.

3.1.4. Main influencing factors of GHG flux at the water–gas interface

The correlation between the partial pressure of GHG and the GHG flux at the water–gas interface and environmental indicators in the middle section of the Three Gorges Reservoir is shown in Table 2. Correlation analysis shows that the partial pressures of CO_2, CH_4 and fluxes of CO_2 and CH_4 at the water–gas interface in the middle part of the Three Gorges Reservoir are significantly positively correlated with TOC and TN, and extremely significantly negatively correlated with DO (Table 2). Correlation analysis of DO, TOC, TN and CH_4 fluxes is shown in Figure 6.

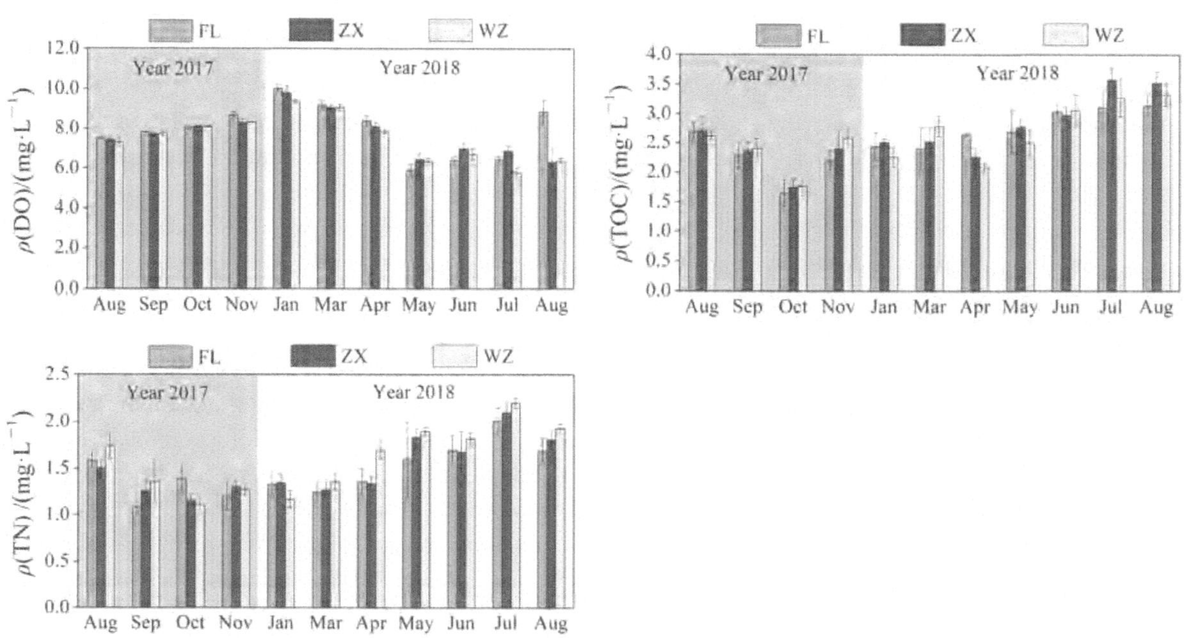

Figure 5 | Distribution of DO, TOC and TN in the water body of the middle part of the Three Gorges Reservoir.

Table 2 | Correlation between CO_2 and CH_4 partial pressure and flux and various physical and chemical indicators

	CO_2 partial pressure	CH_4 partial pressure	CO_2 flux	CH_4 flux
DO	−0.632**	−0.752**	−0.481**	−0.682**
TOC	0.425*	0.658**	0.380*	0.572**
TN	0.662**	0.699**	0.549**	0.621**

*indicates $P \leq 0.05$, indicating significant correlation.
**indicates $P \leq 0.01$, indicating extremely significant correlation.

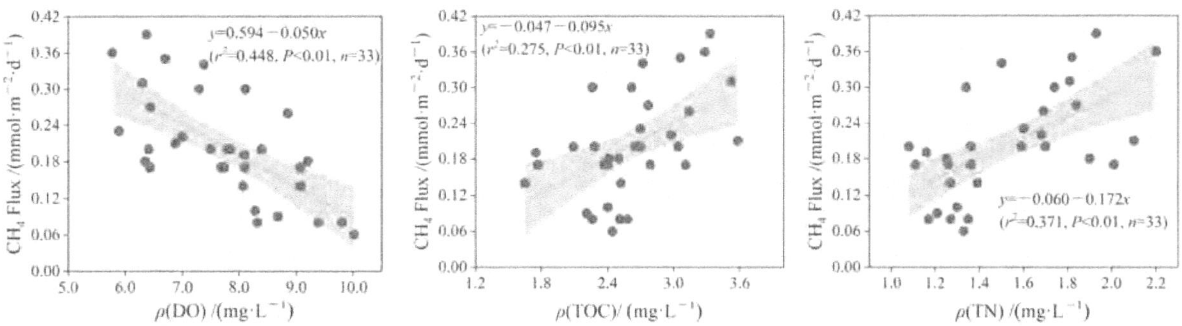

Figure 6 | Correlation between TOC, TN and CH_4 fluxes.

3.2. Discussion

3.2.1. Sediment deposition and its effects

More than 90% of suspended sediment in the Three Gorges Reservoir has been deposited in the perennial backwater area and in the 'point' form in the wide valley and curved reach in the past 10 years (Hu *et al.* 2013). Sediment deposition is related to river plane morphology, average flow rate, hydraulic residence time, sand-carrying capacity and relative sediment concentration (Li *et al.* 2016b). Sand-carrying capacity is greater than sediment concentration, resulting in scour, and conversely, sedimentation (Li *et al.* 2015b).

As shown in Figure 7, in the Zhutuo–Cuntan section of the tail river of the reservoir, the hydraulic retention time is short, the flow rate is large and the sand-carrying capacity is greater than the sediment content. Therefore, scour is dominant in this section. The Cuntan–Fuling section belongs to the fluctuating backwater area, and the sand-carrying capacity is weakened. In the Fuling–Wanzhou section, the perennial backwater area of the reservoir, the average flow velocity decreased, the hydraulic retention time increased and the sand-carrying capacity greatly weakened. In addition, the overall shape of the channel was 'S', resulting in a large number of sediment deposition, which was the most serious section in the reservoir area (Hu *et al.* 2013; Liu *et al.* 2014; Li *et al.* 2015b, 2016b). In the perennial backwater area of the first reservoir, affected by the impoundment of the Three Gorges Reservoir, the change of water level is basically synchronous with the water level in front of the Three Gorges Dam, the sand-carrying capacity is greatly reduced, the sand-carrying capacity is less than the sediment concentration and the sediment deposition is large.

According to the research (Yang *et al.* 2016), the average cumulative sediment deposition in the trunk stream of the Three Gorges Reservoir area was 3,430 ± 644.0 tons. In the Zhutuo–Cuntan section of the reservoir end river area, the sediment caused scour and the sediment flushing amount was 255.2 ± 734,000 tons. In the Cuntan–Fuling section of fluctuating backwater area, there was a small amount of silt deposition, which was 285 ± 2.588 million tons. In the perennial backwater area of the first part of the reservoir, the sediment state is deposition, the deposition amount is 960 ± 2.485 million tons, but the deposition is lower than that in the middle part of the reservoir. As shown in Figure 7, from 2014 to 2017, due to the comprehensive influence of water conservancy projects' sand blocking, temporal and spatial distribution changes of rainfall, sand mining in river channels and other factors, the sediment transport quantity decreased significantly and the deposition rate slowed down somewhat.

The high CH_4 flux in the middle section of the Three Gorges Reservoir (Fuling–Wanzhou section) is correlated to the spatial distribution and deposition of sediment (Maeck *et al.* 2013). Compared to published literature, as shown in Figure 8,

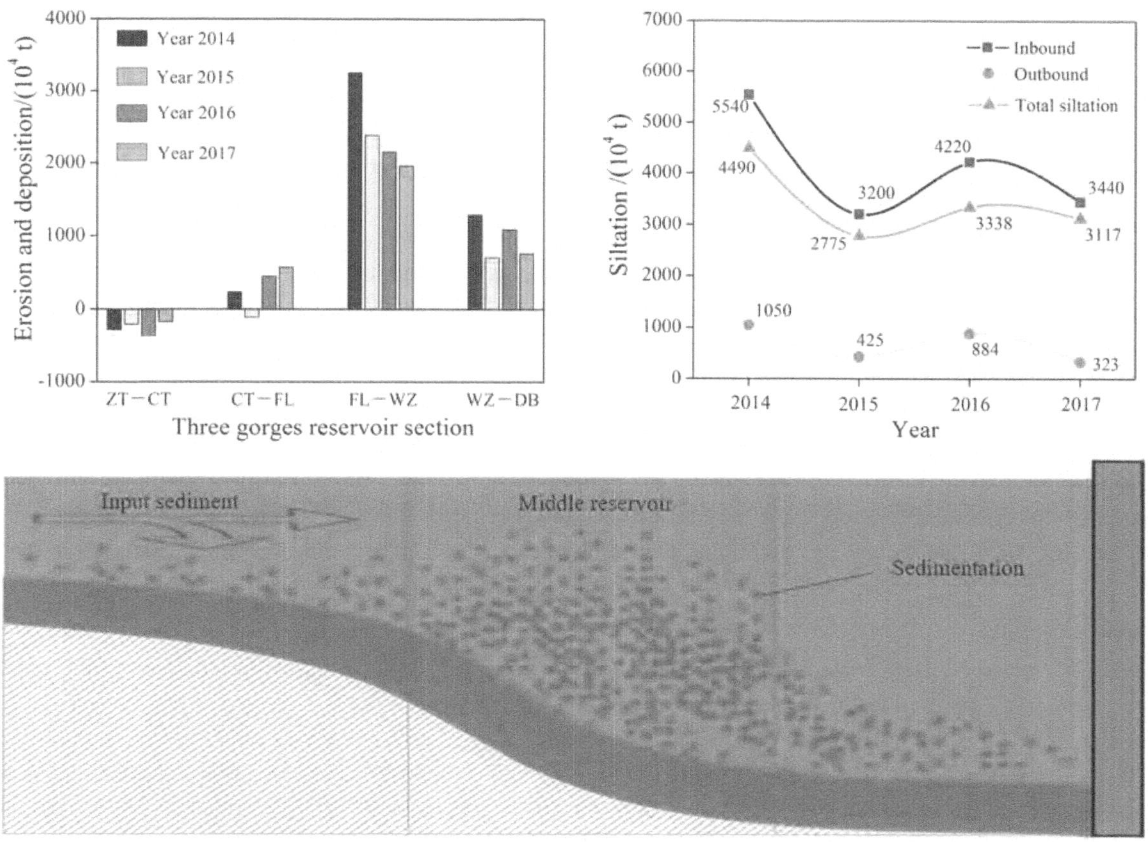

Figure 7 | Schematic diagram of sedimentation, sedimentation and sedimentation in the Three Gorges Reservoir.

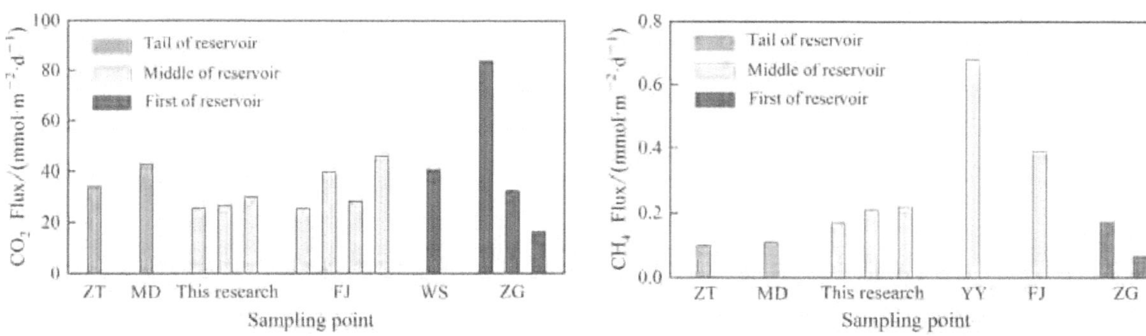

Figure 8 | Comparison of CO_2 and CH_4 fluxes in the first, middle and tail sections of the Three Gorges Reservoir area (Yang *et al.* 2012; Yang 2017; Yao *et al.* 2012; Wei *et al.* 2013; Li *et al.* 2014; Fu *et al.* 2016; Zhao 2016).

CO_2 and CH_4 fluxes at the water–gas interface in the Three Gorges Reservoir are 27.4 ± 10.8 and 0.20 ± 0.09 mmol·m^{-2}·d^{-1}, respectively. Compared to the published literature, as shown in Figure 7, CO_2 and CH_4 fluxes at the water–gas interface in the reservoir area compare with the reservoir head of the Three Gorges Reservoir (Yang *et al.* 2012; Yao *et al.* 2012; Zhao 2016) and the tail of the reservoir (Yang 2017). There was no significant difference in CO_2 flux, but the CH_4 flux in the middle of the reservoir was higher than the flux at the head and tail of the reservoir. The above relationship characteristics are also supported in the correlation analysis of physical and chemical indexes of water quality. As more sediment deposition in the middle section of the reservoir leads to a high CH_4 flux, it is suggested that sediment deposition can be removed by dredging to slow down the CH_4 emission.

3.2.2. Influence of DO

DO has an important effect on the discharge of CH_4 in the reservoir (Schrier-Uijl *et al.* 2011). The content of DO in water determines the process of organic matter degradation in water. Usually, in the process of the carbon cycle, the sediments CH_4 produced in anaerobic conditions, through pore water into the sediment and water interface, and then enter the water upward diffusion, part of CH_4 in the process of diffusion was oxidized to CO_2, and during the period of water diffusion, CH_4 is further oxidation, only a small amount of CH_4 discharge into the atmosphere (Chen *et al.* 2009). Studies have shown that CO_2 flux is significantly negatively correlated with DO in water (Wang *et al.* 2012). Similarly, the concentration of CH_4 is significantly negatively correlated with the concentration of DO in water (Wang *et al.* 2009), indicating that the higher the DO in water, the more CH_4 is oxidized. Biological activities in the water can affect DO concentration in the water, that is, phytoplankton can increase DO in the water through photosynthesis; on the contrary, respiration of aquatic plants and decomposition of organic matter can consume DO in the water. DO produced by photosynthesis also affects GHG production and emissions (Arend *et al.* 2011). In this study, there was a significant negative correlation between DO and CH_4 flux (Furlanetto *et al.* 2012), and CH_4 flux increased with the decrease of DO. In the middle part of the Three Gorges Reservoir, the DO in the low-water-level period is significantly lower than that in the high-water-level period, and the content of DO in the water body decreases, which results in the CH_4 produced by the sediment at the bottom, cannot be fully oxidized during the rising process. Therefore, the content of CH_4 in the low-water-level period is higher than that in the high-water-level period.

3.2.3. Influence of TOC

The decomposition or mineralization of organic matter in lakes or reservoirs is essential for the production of CO_2 and CH_4. The main source of organic carbon in water is related to the system biological residues and exogenous input. Organic carbon is the direct carbon source of methanogenic bacteria decomposition and CO_2 production, and its content has an important influence on the production and emission of CO_2 and CH_4 in the reservoir (Sobek *et al.* 2012). Under anaerobic conditions, the decomposition of most organic matter at the bottom of the reservoir promotes the formation of CH_4 (Roland *et al.* 2017). DOC affects the CO_2 concentration of water to some extent, and this effect varies depending on the concentration, source and geographic location of the DOC (Zhao *et al.* 2008). During the silting process in the middle section of the Three Gorges Reservoir, a large amount of organic carbon is carried, providing a carbon source for methanogens at the bottom. Under the action of the methanogens and the DO in the water column, CO_2 and CH_4 are generated and released into the atmosphere (Jacinthe *et al.* 2012). When the content of organic matter in sediments and water is high, the number of methanogenic bacteria increases, and the increase of CH_4 production leads to the increase of emission flux. In this study, TOC in water was significantly positively correlated with CH_4 flux (Martinez-Cruz *et al.* 2017). As shown in Figure 6, CH_4 flux increased with the increase of TOC (Li *et al.* 2018a). In northern reservoirs with high DOC concentrations, CO_2 concentrations are usually strongly correlated with DOC concentrations, and aquatic metabolism may be the main driver of CO_2 oversaturation (Saidi & Koschorreck 2017).

3.2.4. Influence of TN

TN is often used to indicate the degree of water pollution by nutrients. The increase of water nutrient level is conducive to CH_4 production, mainly due to the increase of nutrient elements and organic carbon content transported from water to sediment, methanogens get more substrates and nutrients, dissolved nitrogen in rivers is a potential contributor to GHG emissions (Qu *et al.* 2017). Previous studies have shown that CH_4 production in lake and reservoir sediments increases with the increase of lake nutrient level (Davidson *et al.* 2015). On the one hand, phytoplankton will absorb nitrogen nutrients in the water under photosynthesis, resulting in a decrease in CO_2 concentration in the water, and atmospheric CO_2 will enter the water (Jiang *et al.* 2012). On the other hand, the increase of nitrogen concentration in water will improve primary productivity, increase water respiration and lead to the increase of CO_2 concentration in water. Thus, the transition leads to changes in the CO_2 'source' and 'sink' processes between photosynthesis and respiration. The increase of nitrogen nutrients in the reservoir ecosystem (Morgane *et al.* 2019) can stimulate primary production and carbon sequestration (Knoll *et al.* 2013). The high concentration of nutrients can provide substrates for aquatic organisms to accelerate respiration, reduce DO in water and form an anaerobic environment, which is conducive to the generation and discharge of CH_4 in water. In this study, TN and CH_4 flux showed a very significant positive correlation, as shown in Figure 6, and CH_4 flux increased with the increase of TN content. The results show that the TN content in the water body is the environmental factor affecting the seasonal variation of CH_4 emission.

4. CONCLUSIONS

1. The average annual CO_2 flux in the middle of the Three Gorges Reservoir is 27.4 ± 10.8 mmol·m^{-2}·d^{-1}. There was no statistical difference in the CO_2 flux between the measured points. The CO_2 flux during the drainage period and the low-water-level period was significantly higher than that during the high-water-level period, and reached the lowest value during the high-water-level period. The average annual CH_4 flux in the middle of the reservoir area is 0.20 ± 0.09 mmol·m^{-2}·d^{-1}. There is no statistical difference in CH_4 flux between the measured points, and CH_4 flux in the drainage stage and the low-water stage is significantly higher than that in the high-water stage.
2. In the middle part of the Three Gorges Reservoir, there is a certain correlation between the spatial distribution and deposition amount of sediment and the emission of carbon flux of GHG. With the increase of sediment deposition, the emission of carbon flux of GHG also tends to increase. The partial pressure and flux of CO_2 and CH_4 in water were significantly positively correlated with TOC and TN and significantly negatively correlated with DO.
3. The high value of CH_4 flux at the water–gas interface in the middle section of the reservoir is related to the spatial distribution and deposition of sediment. In future studies, the GHG emission in the middle section of the reservoir should be further studied in combination with the physical and chemical characteristics of sediment.
4. This research focuses on the macroscopic exploration of the middle section of the Three Gorges Reservoir area (Wanzhou–Fuling section). However, there is still a lack of microscopic observations in this field, such as microbes, phytoplankton and zooplankton related to GHGs, which need to be further studied.
5. The research is based on long-term observation and is heavily influenced by field conditions. Further research can be carried out in the area of laboratory simulation experiment.

ACKNOWLEDGEMENT

This work was supported by the National Natural Science Foundation Project of China (No.51609026) and the Postgraduate Research Innovation Project of Chongqing Jiaotong University (CYS21352).

DATA AVAILABILITY STATEMENT

All relevant data are included in the paper or its Supplementary Information.

REFERENCES

Abril, G., Martinez, J. M., Artigas, L. F., Moreira-Turcq, P., Benedetti, M. F., Vidal, L., Meziane, T., Kim, J.-H., Bernardes, M. C., Savoye, N., Deborde, J., Souza, E. L., Albéric, P., Landim de Souza, M. F. & Roland, F. 2014 Amazon River carbon dioxide outgassing fuelled by wetlands. *Nature* **505** (7483), 395–398.

Alizadeh, M. J., Kavianpour, M. R., Danesh, M., Adolf, J., Shamshirband, S. & Chau, K.-W. 2018 Effect of river flow on the quality of estuarine and coastal waters using machine learning models. *Engineering Applications of Computational Fluid Mechanics* **12** (1), 810–823.

Arend, K. K., Beletsky, D., Depinto, J. V., Ludsin, S. A, Roberts, J. J., Rucinski, D. K., Scavia, D., Schwab, D. J. & Höök, T. O. 2011 Seasonal and interannual effects of hypoxia on fish habitat quality in central Lake Erie. *Freshwater Biology* **56** (2), 366–383.

Barros, N., Cole, J. J., Tranvik, L. J., Prairie, Y. T., Bastviken, D., Huszar, V. L. M., del Giorgio, P. & Roland, F. 2011 Carbon emission from hydroelectric reservoirs linked to reservoir age and latitude. *Nature Geoscience* **4** (9), 593–596.

Chen, X. Y., Dan, H. C., Jiang, D. G., Li, H. & Dai, L. 2009 Research on process of GHG emission from reservoir. *Water Resources and Power* **27** (5), 37–39.

Dash, S. S., Sahoo, B. & Raghuwanshi, N. S. 2019 A SWAT-Copula based approach for monitoring and assessment of drought propagation in an irrigation command. *Ecological Engineering* **127**, 417–430.

Dash, S. S., Sahoo, B. & Raghuwanshi, N. S. 2021 How reliable are the evapotranspiration estimates by Soil and Water Assessment Tool (SWAT) and Variable Infiltration Capacity (VIC) models for catchment-scale drought assessment and irrigation planning? *Journal of Hydrology* **592**, 125838.

Davidson, T. A., Audet, J., Svenning, J. C., Lauridsen, T. L., Søndergaard, M., Landkildehus, F., Larsen, S. E. & Jeppesen, E. 2015 Eutrophication effects on GHG fluxes from shallow-lake mesocosms override those of climate warming. *Global Change Biology* **21** (12), 4449–4463.

Fu, J. N., Cao, M., Deng, B., Ziran, Z., Chen, Y., Dang, C., Wang, Y. & Wang, F. 2016 Partial pressure and diffusion flux of dissolved carbon dioxide in typical mainstream and tributaries of the Three Gorges Reservoir during high water level period. *Earth and Environment* **44** (1), 64–72.

Furlanetto, L. M., Marinho, C., Palma, S. C., Albertoni, E. F., Figueiredo-Barros, M. P. & de Assis Esteves, F. 2012 Methane levels in shallow subtropical lake sediments: dependence on the trophic status of the lake and allochthonous input. *Limnologica* **42** (2), 151–155.

Gong, W. J., Wen, S. L., Wang, H. W. *et al.* 2019 Characteristics of GHG occurrence and emission in summer and autumn of Daheiting Reservoir. *China Environmental Science* **39** (11), 4611–4619.

Hu, J., Yang, S. F., Wang, X. K., Tao, W., Xin, L. & Zhong, J. 2013 Sedimentation in Yangtze River above Three Gorges Project since 2003. *Journal of Sediment Research* **1**, 39–44.

Jacinthe, P. A., Filippelli, G. M., Tedesco, L. P. & Raftis, R. 2012 Carbon storage and GHG emission from a fluvial reservoir in an agricultural landscape. *Catena* **94** (9), 53–63.

Jiang, T., Guo, J. S., Li, Z., Fang, F., Bai, L. & Liu, J. 2012 A preliminary study on the characteristics of greenhouse gas fluxes at the water-air interface of Pengxi River, a tributary of the Three Gorges Reservoir, under different operating conditions. *Environmental Science* **33** (5), 1463–1470.

Kargar, K., Samadianfard, S., Parsa, J., Nabipour, N., Shamshirband, S., Mosavi, A. & Chau, K.-W. 2020 Estimating longitudinal dispersion coefficient in natural streams using empirical models and machine learning algorithms. *Engineering Applications of Computational Fluid Mechanics* **14** (1), 311–322.

Knoll, L. B., Vanni, M. J., Renwick, W. H., Dittman, E. K. & Gephart, J. A. 2013 Temperate reservoirs are large carbon sinks and small CO_2 sources: results from high-resolution carbon budgets. *Global Biogeochemical Cycles* **27** (1), 52–64.

Kolb, B. & Ettre, L. S. 2006 *Static Headspace-Gas Chromatography: Theory and Practice*, 2nd edn. Jone Wiley and Sons, New Jersey, p. 349.

Kumar, A., Yang, T. & Sharma, M. P. 2019 GHG measurement from Chinese freshwater bodies: a review. *Journal of Cleaner Production* **233**, 368–378.

Lemer, J. & Roger, P. 2001 Production, oxidation, emission and consumption of methane by soils: a review. *European Journal of Soil Biology* **37**, 25–50.

Li, Z., Bai, L., Guo, J., Fang, F. & Jiang, T. 2013 Comparative study on water-air CO_2,CH_4 flux in two tributaries in the Three Gorges Reservoir, China. *Environmental Science* **34** (3), 1009–1015.

Li, S., Wang, C. Y., Cao, M., Qian, H., Xu, T., Zhou, Z., Deng, B. & Wang, F. 2014 Partial pressure and diffusion flux of dissolved carbon dioxide in the mainstream and tributary of the central Three Gorges Reservoir in summer. *Environmental Science* **35** (3), 885–891.

Li, W. J., Yang, S. F. & Fu, X. F. 2015a Sedimentation characteristics of the Three Gorges Reservoir during its initial operation. *Advances in Water Science* **26** (5), 676–685.

Li, W. J., Yang, S. F., Hu, J. & Fu, X. 2015b Experimental research on flocculation of fine sediment in Three Gorges Reservoir area. *Journal of Basic Science and Engineering* **23** (5), 851–859.

Li, L., Pu, J. B., Li, J. H., Shi, Y., Qiong, X. & Tao, Z. 2016a Variations of CO_2 exchange fluxes across water-air interface and environmental meaning in a surface stream in subtropical karst area, SW China. *Environment Science* **37** (7), 2487–2495.

Li, W. J., Li, N., Yang, S. F. & Wang, T. 2016b Analysis of the sedimentation in the Three Gorges Reservoir based on the sediment carrying capacity. *Advances in Water Science* **27** (5), 726–734.

Li, S. Y. 2018 CO_2 oversaturation and degassing using chambers and a new gas transfer velocity model from the Three Gorges Reservoir surface. *Science of the Total Environment* **640**, 908–920.

Li, L., Xue, B., Yao, S., Tao, Y. & Yan, R. 2018a Spatial–temporal patterns of methane dynamics in Lake Taihu. *Hydrobiologia* **822** (1), 143–156.

Li, Z., Chen, Y. B., Li, C., Guo, J., Xiao, Y. & Lu, L. 2018b Advances of eco-environmental effects and adaptive management in river cascading development. *Advances in Earth Science* **33** (07), 675–686.

Li, Z., Lu, H. L., Lv, P. Y., Zhang, Z. Y. & Guo, J. S. 2020 Imbalanced stoichiometric reservoir sedimentation regulates methane accumulation in China's Three Gorges Reservoir. *Water Resources Research* **56** (9), 26447.

Liu, Y., Yang, S. F. & Hu, P. 2014 Discussion on the siltation characteristics of continuous bends in the early storage period of the Three Gorges Reservoir area. *Yangtze River* **45** (1), 1–4.

Liu, S., Xi, X. L., Xia, X., Yang, X. & Ran, L. 2017 Hydrological and geomorphological control on CO_2 outgassing from low-gradient large rivers: an example of the Yangtze River system. *Journal of Hydrology* **550**, 26–41.

Liu, W. T., Yao, X. L., Xue, J. Y., Zhao, Z., Zhang, L., Wang, X. & Cai, Y. 2019 Seasonal and spatial variations in GHG (CH_4 and CO_2) emission along the riparian zone of middle and lower reaches of Yangtze River. *Resources and Environment in the Yangtze Basin* **28** (11), 2718–2726.

Luo, J. C., Mao, R. & Li, S. Y. 2018 pCO_2 in the main rivers of the Three Gorges Reservoir and its influencing factors. *Environmental Science* **39** (7), 3134–3141.

Maeck, A., DelSontro, T., McGinnis, D. F., Fischer, H., Flury, S., Schmidt, M., Fietzek, P. & Lorke, A. 2013 Sediment trapping by dams creates methane emission hot spots. *Environmental Science & Technology* **47**, 8130–8137.

Martinez-Cruz, K., Gonzalez-Valencia, R., Sepulveda, J. A., Plascencia-Hernandez, F., Belmonte-Izquierdo, Y. & Thalasso, F. 2017 Methane emission from aquatic ecosystems of Mexico City. *Aquatic Sciences* **79** (1), 159–169.

Mccully, P. 2002 *Flooding the Land, Warming the Earth: GHG Emissions from Dams.* International Rivers Network, Berkeley, CA, USA.

Morales-Pineda, M., Cozar, A., Laiz, I., Úbeda, B. & Gálvez, J. A. 2014 Daily, biweekly, and seasonal temporal scales of pCO_2 variability in two stratified Mediterranean reservoirs. *Journal of Geophysical Research Biogeosciences* **119** (4), 509–520.

Morgane, L. M., Chantal, G.-O., Alain, M., Souchon, Y., Étrillard, C., Levain, A., Moatar, F., Pannard, A., Souchu, P., Lefebvre, A. & Pinay, G. 2019 Eutrophication: a new wine in an old bottle? *Science of the Total Environment* **651**, 1–11.

Nilsson, C. & Berggren, K. 2000 Alterations of riparian ecosystems caused by river regulation. *BioScience* **50** (9), 783–792.

Qin, Y., Zang, Y., Li, Z. & Ma, J. 2017a CH_4 fluxes during the algal bloom in the Pengxi River. *Environmental Science* **35** (11), 1578–1588.

Qin, Y., Yang, B. X., Li, Z., Hao, B. & Du, H. 2017b CO_2 and CH_4 flux across water-air interface in summer in the downstream of Jinsha River, Southwest China. *Journal of Lake Sciences* **29** (04), 991–999.

Qu, B., Aho, K. S., Li, C., Kang, S., Sillanpää, M., Yan, F. & Raymond, P. A. 2017 GHG emissions in rivers of the Tibetan Plateau. *Scientific Reports* **7** (1), 16573.

Rantakari, M., Heiskanen, J., Mammarella, I., Tulonen, T., Linnaluoma, J., Kankaala, P. & Ojala, A. 2015 Different apparent gas exchange coefficients for CO_2 and CH_4: comparing a brown-water and a clear-water lake in the boreal zone during the whole growing season. *Environmental Science & Technology* **49** (19), 11388–11394.

Rik, W. 1992 Relationship between wind speed and gas exchange over the ocean. *Journal of Geophysical Research* **97** (C5), 7373–7382.

Rodhe, H. 1990 A comparison of the contribution of various gases to the greenhouse effect. *Science* **248** (4960), 1217–1219.

Roland, F., Darchambeau, F., Morana, C., Bouillon, S. & Borges, A. V. 2017 Emission and oxidation of methane in a meromictic, eutrophic and temperate lake (Dendre, Belgium). *Chemosphere* **168**, 756–764.

Rosa, L. P., Santos, M. A. D., Matvienko, B., Sikar, E., Lourenço, R. S. M. & Menezes, C. F. 2003 Biogenic gas production from major Amazon reservoirs, Brazil. *Hydrological Processes* **17** (7), 1443–1450.

Saidi, H. & Koschorreck, M. 2017 CO_2 emissions from German drinking water reservoirs. *Science of the Total Environment* **581**, 10–18.

Sawakuchi, H. O., Bastviken, D., Sawakuchi, A. O., Krusche, A. V., Ballester, M. V. R. & Richey, J. E. 2014 Methane emissions from Amazonian Rivers and their contribution to the global methane budget. *Global Change Biology* **20** (9), 2829–2840.

Schrier-Uijl, A. P., Veraart, A. J., Leffelaar, P. A., Berendse, F. & Veenendaal, E. M. 2011 Release of CO_2 and CH_4 from lakes and drainage ditches in temperate wetlands. *Biogeochemistry* **102** (1–3), 265–279.

Shamshirband, S., Nodoushan, E. J., Adolf, J. E., Manaf, A. A., Mosavi, A. & Chau, K.-W. 2019 Ensemble models with uncertainty analysis for multi-day ahead forecasting of chlorophyll a concentration in coastal waters. *Engineering Applications of Computational Fluid Mechanics* **13** (1), 91–101.

Sobek, S., Delsontro, T., Wongfun, N. & Wehrli, B. 2012 Extreme organic carbon burial fuels intense methane bubbling in a temperate reservoir. *Geophysical Research Letters* **39** (1), L01401.

Tremblay, A., Lambert, M. & Gagnon, L. 2004 Do hydroelectric reservoirs emit GHG? *Environmental Management* **33** (S1), S509–S517.

Tremblay, A., Varfalvy, L., Roehm, C., Varfalvy, L. & Garneau, M. 2005 *GHG Emissions: Fluxes and Processes, Hydroelectric Reservoirs and Natural Environments*. Springer, New York, pp. 725–732.

Wang, D. Q., Chen, Z., Sun, W., Hu, B. & Xu, S. 2009 Methane and nitrous oxide concentration and emission flux of Yangtze Delta plain river net. *Science China Chemistry* **52** (5), 126–135.

Wang, C. H., Du Q, Y. & Zhao, D. Z. 2012 Study on the space-time characteristics of CO_2 emission flux in the reservoir and its response to environmental factors. *Journal of Hydroelectric Power* **31** (2), 148–153.

Wei, H. B., Wu, X. Q., Cao, M., Fu, J., Jia, X., Deng, B. & Wang, F. 2013 Partial pressure and diffusion flux of dissolved carbon dioxide in mainstream and tributary (Zhuyi River) of the Three Gorges Reservoir in summer. *Journal of Shanghai University (Natural Science Edition)* **37** (4), 497–504.

Wu, Y. 2012 *Study on GHG Flux from Newly Created Marshes and Carbon Sink in the Three Gorges Reservoir*. Chongqing University, Chongqing.

Wu, M. 2016 *The Characteristics and Influencing Factors of GHG Emissions in Different Water Bodies in Chongzhou*. Sichuan Agricultural University, Ya'an.

Yang, L., Zhang, Y., Hou Pei, Q., Sun, B., Zhang, Q., Wang, X. & Lu, F. 2012 Spatiotemporal variation and mechanism of nitrogen and phosphorus in the middle and lower reaches of the Three Gorges Reservoir. *Resources and Environment in the Yangtze Basin* **21** (6), 732.

Yang, P., Gong, C., He, Q. H. & Huang, J. 2013 Yujiangkou fish and shrimp mixed pond water–gas interface GHG flux and main influencing factors. *Journal of Environmental Science* **33** (5), 1493–1503.

Yang, B. X. & Lv, F. 2016 A summary of research status of headspace balance method. *Chemical Engineering & Equipment* (7), 206–207.

Yang, S., Li, W., Hu, X., Hu, J. & Wang, T. 2016 Study on sediment deposition model and channel passing capacity improvement of three gorges reservoir. *Journal of Chongqing Jiaotong University (Natural Science Edition)* **35** (S1), 41–48.

Yang, B. 2017 *Comparative study on greenhouse gas fluxes at the water-gas interface between the Three Gorges and the lower reaches of the Jinsha River*. ChongQing Jiaotong University.

Yao, C. Z., Zhang, J. & Wu, Y. 2012 Partial pressure of water body dissolved carbon dioxide in mainstream and Xiangxi bay of Three Gorges Reservoir area in autumn. *Chinese Journal of Ecology* **31** (1), 152–157.

Yao X, L. I. Z., Guo, J. S., Lin, C., Chen, Y. & LI, C. 2015 Comparison between closed static chamber method and thin boundary layer method on monitoring air-water CO_2 diffusion flux. *Journal of Lake Sciences* **27** (2), 289–296.

Zhao, X. J., Zhao, T. Q., Zheng, H., Duan, X., Chen, F., Ouyang, Z. & Wang, X. 2008 Emissions of greenhouse gases in reservoirs and their influencing factors. *Environmental Science* **29** (8), 2377–2387.

Zhao, Y., Zeng, Y., Wu, B. F., Wang, Q., Yuan, C. & Xu, Z. 2011 Observation on GHG emissions from Xiangxi River in Three Gorges Region. *Advances in Water Science* **22** (4), 546–553.

Zhao, W. 2016 Temporal-Spatial patterns of methane emission flux and its influence factors in HuBei Part of Three Gorges Reservoir. *Acta Hydrobiologica Sinica* **22** (4), 776–781.

First received 8 March 2021; accepted in revised form 16 June 2021. Available online 6 July 2021

Impact of Climate Change on Hydrology and Water Resources doi: 10.2166/wcc.2021.319

Research on water environmental capacity accounting of the Yongzhou Section of Xiangjiang River Basin based on the SWAT-EFDC coupling model

Li Wu[a,b], Zhe Chen[a], Xuan Ding[b], Hui-ying Liu[c] and Dun-qiu Wang[a,*]

[a] Guangxi Key Laboratory of Environmental Pollution Control Theory and Technology, College of Environmental Science and Engineering, Guilin University of Technology, Guilin 541004, China

[b] Environmental Protection Bureau of Yongzhou Economic and Technological Development Zone, Yongzhou 425000, China

[c] Yongzhou Hydrology and Water Resources Investigation Bureau of Hunan Province, Yongzhou 425000, China

*Corresponding author. E-mail: wangdunqiu@sohu.com

ABSTRACT

In this paper, a coupling model of SWAT (Soil and Water Assessment Tool) and EFDC (Environmental Fluid Dynamics Code) was established, and the relationship between the pollution source and water quality response was identified. Based on the hydrodynamic water quality simulation results and the one-dimensional WEC (water environmental capacity) theoretical formula, the total nitrogen and total phosphorus WEC and the remaining WEC of the Yongzhou Section of Xiangjiang River Basin under the guaranteed rate of 90% and in 2017 were calculated, respectively. It can be seen from the results that the total nitrogen WEC of the Yongzhou Section of Xiangjiang River Basin in 2017 is 27,673.04 t, the total nitrogen WEC under the guaranteed rate of 90% is 19,497.61 t/a and the total phosphorus WEC of the Yongzhou Section of Xiangjiang River Basin in 2017 is 4,877.22 t. The total phosphorus WEC under the guaranteed rate of 90% is 2,936.64 t/a; in 2017, the remaining WECs of total nitrogen and total phosphorus in the entire basin are 14,646.69 and 3,358.67 t, respectively.

Key words: coupling model, EFDC model, SWAT model, water environment capacity, Xiangjiang River Basin

HIGHLIGHTS

- A calculation method of water environment capacity was studied, which can provide a reference for the calculation of water environment capacity in large and medium-sized basins
- The coupling model of SWAT and EFDC fix the lack of watershed sites and measured data, and inaccurate calculation of water environment capacity.
- The coupling model identifies the relationship between pollution sources and water quality response.

1. INTRODUCTION

It is the rapid urbanization that has led to the increasingly serious problem of the river basin water environment, and the environmental protection of watershed water is essential for sustainable economic and social development. The upper reaches of the first-level tributary of the Yangtze River in the Yongzhou Section of Xiangjiang River Basin are also the final receiving water body of non-point source nitrogen and phosphorus pollutants in the basin. The Xiangjiang River Basin would experience water shortages induced by climate change and pollution source emission (Wang *et al.* 2017). It is very important to understand the water environmental capacity (WEC) of nitrogen and phosphorus in the Yongzhou Section of Xiangjiang River Basin for water environmental protection and governance. The concept of WEC was first derived from the TMDL (total maximum daily loads) management plan proposed by the United States in 1972 (USEPA 1999). At present, the commonly used methods for the calculation of WEC include the trial-and-error method, the analytical method, the probabilistic dilution model method, the stochastic theory and the system optimization method (Thomann & Sobel 1964; Burn & Lence 1992; Zhang 2017; Wang *et al.* 2019). With the maturity of various water quality models, the model can identify the relationship between water quality and pollutant response, making it widely used in WEC research and becoming the focus of research (Deng *et al.* 2010; Fang 2015; Feng 2016; Meng 2018). The analytical method is to use a river model verified by actual measurement data to predict the flow and pollutant concentration of each section of the water body and then substitute it into the WEC calculation model for accounting. The method is characterized by strong operability, clear physical concepts and relatively objective calculation results, making it one of the most popular used methods.

Domestic scholars use the WASP model, the EFDC model and other water quality models together with the WEC calculation model (i.e., analytical method) to calculate the WEC. Their studies were often utilized to calculate the temporal and spatial distribution of WEC in small watersheds (Yen *et al.* 2012; Xu 2014, 2018; Tang 2016; Yao *et al.* 2018; Zhao 2018; Wang *et al.* 2019; Zheng *et al.* 2019; Hu *et al.* 2020). Compared with the methods used previously, the analytical method employed in their studies can simulate the transformation and interactions between nitrogen and phosphorus in different forms. The main disadvantages of the analytical method are as follows: (1) A large amount of data has to be used. Unfortunately, the hydrology and water quality stations and monitoring frequency of large-scale river basins in our country can hardly satisfy the needs of water environment capacity calculations. In that case, it is impossible to accurately calculate the true situation of the water environment capacity of the basin by using water quality models according to the measured data of hydrology and water quality at a few sites. (2) Due to the lack of measured water quality data, a large number of parameters of the water environment capacity calculation model have to quote the empirical values, which brings greater uncertainty to the calculation results. (3) The use of river water quality models, such as the EFDC model alone, requires many field surveys on river topography, water depth and other data, which consumes a lot of manpower, material resources and time. It is not realistic to carry out a large-scale survey on a large quantity of river basins from the perspective of time and cost.

At present, the commonly used hydrological models mainly include the HSPF model (Zhang *et al.* 2012), the AGNPS model (Meng & Zhao 2008), the L-THIA model (Shen *et al.* 2007; Hui 2015) and the SWAT model (Li 2012; Jiang 2015; Zhai 2015), which are widely used. The coupling of the SWAT model with the EFDC model, the WASP model and other water quality models in hydrological models shows the good effect, which is widely used in river and lake water bodies (Liu *et al.* 2008; Du *et al.* 2016; He *et al.* 2020). It can be used for the analysis of the input meteorological data, DEM elevation map, land-use-type map, soil distribution map and other data through the simulation of the hydrological process, and the output of the water system of the river basin, the runoff of the sub-basin, the outlet of the sub-basin, the river elevation, the water depth, nitrogen and phosphorus pollution load and other simulation results. The calibrated results can be used as the boundary conditions of water quality models, such as the EFDC model to realize the coupling of hydrology and water quality models, so as to eliminate the disadvantages of insufficient data, large amount of preliminary investigation work and high cost when using water quality models to calculate WEC in large river basins. Based on the hydrodynamic water quality simulation of the SWAT-EFDC coupling model, this paper identifies the relationship between pollution sources and water quality response. Under the premise of the satisfaction of the water quality standards required by water function zoning, the theoretical formula of one-dimensional water environment capacity was used to calculate the WECs of total phosphorus and total nitrogen in 2017, and the WEC and remnant WEC of the total nitrogen and total phosphorus in the Yongzhou Basin, Xiangjiang River under different hydrological conditions, thereby providing references for the calculation of the WEC of large river basins, and establishing a total amount control plan for the Xiangjiang River Yongzhou Basin. It can also provide a basis for the formulation, and data support for the further development and control of the water environment of the river basin.

2. REGIONAL OVERVIEW

2.1. Overview of the research area

The research area is in the upper reaches of the Xiangjiang River which is the largest river in Hunan Province and a major tributary of the Yangtze River. Its west source originates from Haiyang Mountain in Lingchuan County of Guangxi Zhuang Autonomous Region, and its east source originates from Yegouling, Lanshan County, Hunan Province. Xiangjiang River has a total length of 948 km and a drainage area of 94,700 km^2, which has a drainage basin with the largest amount of water resources and the highest utilization rate. Besides, the river belongs to the Dongting Lake water system dominated by a subtropical monsoon climate. The long-term average annual rainfall is between 1,200 and 1,900 mm, and the annual average temperature is between 17.6 and 18.6 °C. Red soil and yellow soil are the main soil types, and the annual average sunshine is between 101.5 and 113 kcal/cm^2 (YCHO 2018).

2.2. Overview of the pollution sources

There are 363 urban point sources in the research area, including 18 urban centralized sewage treatment plants, 143 rural centralized sewage treatment facilities and 202 industrial enterprises (the statistics only include the enterprises that directly discharge wastewater into rivers). At the same time, there are also 2,758 large-scale livestock and poultry farms in 2017.

The total nitrogen emissions are 12,927.12 t/a, and the total phosphorus emissions are 1,501.82 t/a in the area. These emissions are composed of three parts. Most of the emissions are from the planting industry with a total nitrogen emission of 9,464.1 t/a and a total phosphorus emission of 864 t/a. The emissions from the large-scale livestock and poultry farms are in the second place, which have a total nitrogen emission of 2,362.29 t/a and a total phosphorus emission of 575.66 t/a. The least emissions come from the urban point sources with the total nitrogen emission of 1,100.73 t/a and the total phosphorus emissions of 64.32 t/a (YEEB 2020).

3. MODEL ESTABLISHMENT AND RESEARCH METHODS

3.1. Database construction

The database for the SWAT model and the EFDC model is classified into two categories: spatial database and attribute database. See Table 1 for the names, parameters and sources of the main data.

3.2. Model establishment

3.2.1. Establishment of a watershed hydrological model

The SWAT model, also called the distributed watershed hydrological model (Soil and Water Assessment Tool model), consists of three parts, i.e., hydrological process model, soil erosion model and pollution load model. They are used to simulate runoff, sediment, nutrients, pesticides and other substances, respectively.

First, input the spatial data and attribute data required by the database (including weather data, terrain data, soil data and land use/cover data) into the SWAT model, then identify the DEM image and calculate the catchment area of the watershed, and finally extract the river network and divide the sub-basin. By setting the thresholds for land-use type, soil type and slope division to 10, 10 and 15%, respectively, the total area of the watershed is calculated to be 1,8633.6 km^2, and the watershed is divided into 115 sub-basins and 818 HURs.

The research area is a typical area dominated by agroforestry ecosystems, and there are eight types of land. The land-use types that cover the biggest area include forest land, paddy field and dry land in sequence, accounting for 89.3% of the total watershed in the test area, as shown in Figure 1. The division of its sub-basins and the distribution of water systems are shown in Figure 2.

3.2.2. Establishment of a watershed quality model

The EFDC model, also called Environmental Fluid Dynamics Code, integrates multiple mathematical models, including hydrodynamic module, temperature module, salinity module, dye module, water quality module, sediment module and Lagrangian tracer. It can be used in water bodies such as rivers, lakes, estuaries, reservoirs, wetlands and coastal areas, covering the simulation of processes, such as one-dimensional, two-dimensional and three-dimensional flow fields of water systems, material transportation, ecological processes and freshwater inflow.

Firstly, we rely on the Google Earth image map to draw the river base map according to the river channel generated by the SWAT model, and then use the CVLGrid1.1 to generate a grid of all the river channels. After that, we use the elevation and water depth data of each point in the sub-basin file output by the SWAT model and employ the terrain interpolation function of EFDC to set the river topography and water depth of the entire study area. Then, the hydrology and pollution source load data, as well as the measured water temperature and meteorological data required in the .rch file output by the SWAT model, are used as the boundary conditions of hydrodynamics and water quality to realize the coupling of the SWAT simulation and the EFDC model. Subsequently, all the pollution sources in the basin are generalized as point sources and non-point sources, and input into the EFDC model. At last, the generalized pollution source setting interface diagram is obtained, as shown in Figure 3.

3.3. Model operation and calibration

3.3.1. Operation and calibration of the SWAT model

The SWAT model simulation period is from January 2000 to December 2019. To avoid the errors because many variables are zero at the beginning of the model operation, the first 5 years are selected as the model's warm-up period to improve the model accuracy. The runoff, ammonia nitrogen and total phosphorus data of Daoxian Station and Lengshuitan Station were selected in the upstream and downstream, respectively, for parameter calibration and result verification. The spatial calibration was implemented sequentially from upstream to downstream; in terms of calibration object, the calibration of the flow was made first, which was followed by the calibration of the ammonia nitrogen and total phosphorus (Moriasi *et al.* 2015).

Table 1 | Parameters required for model database establishment

Type		Format	Data type	Source	Remarks
Spatial data	Digital elevation model	GRID	Elevation data	Geospatial data sites	STRM with a 12 m × 12 m resolution
	Land use type	GRID	Land-use type	Resource and Environmental Science Center, Chinese Academy of Sciences	Land-use map with a resolution of 1 km × 1 km
	Soil type	GRID	Soil distribution type	Resource and Environmental Science Center, Chinese Academy of Sciences	Soil-type map with a resolution of 30 m × 30 m
Attribute data	Meteorological data	TXT table	Daily rainfall, the maximum and minimum mean temperatures, daily radiation, wind speed, relative humidity and evaporation capacity	China Meteorological Data Network	Term of data for Dao County and Lengshuitan station: 2000–2019
	Hydrology and water quality data	EXCEL table	Monthly runoff, nitrogen and phosphorus concentrations	Yongzhou Hydrographic Bureau	Term of data: 2000–2019
	Pollution source data	EXCEL table	Data of pollution sources from urban point sources, livestock and poultry breeding and planting industry	Yongzhou Ecological Environment Bureau	Data from the second national census of pollution sources
	Data on crops and fertilizers	Word file	Crop species, fertilization data, agricultural management measures	Yongzhou Agricultural and Rural Bureau	Farmland quality survey report
	Soil data		Hydrologic characteristics, density, electrical conductivity, as well as physical and chemical properties of soil	Scientific Data Center for Cold and arid Regions	HWSD soil texture data set
	Management measures		Planting pattern as well as fertilization dosage and time	Field Survey and Relevant Statistical Yearbook	

Daoxian Station uses the period from 2005 to 2017 as the regular rate, while Lengshuitan Station uses the period from 2012 to 2017. And both stations use the period from 2018 to 2019 as the verification period. Because Lengshuitan Station was built in 2012, the observation data started in 2012.

The model selects R^2 (coefficient of determination) and E_{ns} (Nash–Sutcliffe efficiency coefficient) as the indicators for model evaluation. The certainty coefficients of runoff, ammonia nitrogen and total phosphorus of the two stations are greater than 0.68, and the Nash coefficient is greater than 0.61 (refer to Table 2 for details). It is generally believed that if the conditions, i.e., $R^2 > 0.6$ and $E_{ns} > 0.5$, are satisfied at the same time, and the simulation results are acceptable (Zhao *et al.* 2017).

3.3.2. Operation and calibration of the EFDC model

The EFDC model simulation time period is from January 1, 2017 to December 31, 2017, with January 1, 2017 as the 0th day, and December 31, 2017 as the 364th day. The time step of the model is 5 s, and the initial water temperature is set to 6 °C. Using the cold start and based on the EFDC's hydrodynamic module, temperature module and water quality module, the model selects the parameters related to chemical oxygen demand, nitrogen and phosphorus and uses the water quality monitoring data of Daoxian Station from January to May 2017 for adjustment. Besides, it employs the data of water level, water temperature and water quality of Daoxian Station for verification.

In this study, the data of water level, water temperature and water quality of Daoxian Hydrological Station in 2017 were selected for calibration. As shown in Figure 4(a)–4(d), the relative error range of the water level at Daoxian Station in 2017 was between 0.35 and 1.37%, the relative error range of water temperature was between 0.5 and 17.06%, the relative error range of ammonia nitrogen was between 2.67 and 14.8% and the relative error range of total phosphorus was between 1.14 and 23.32%. Because the relative error of all data is within 25%, and the curve fits well with the actual value, the simulation effect of the model is great (Fan 2017).

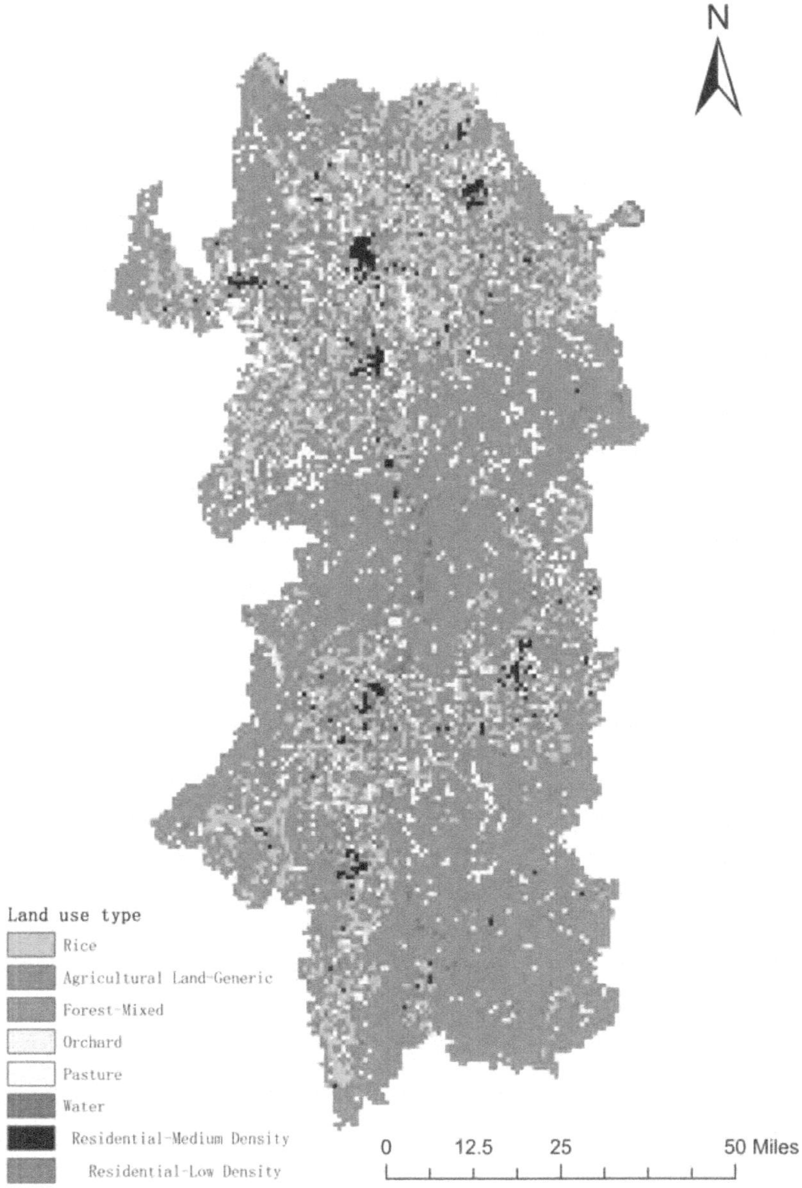

Figure 1 | Land-use types in the Yongzhou Section of Xiangjiang River Basin in 2018.

3.4. Calculation method of water environment capacity

3.4.1. One-dimensional theoretical calculation formula

Assuming that in small and medium-sized rivers with abundant data, the horizontal mixing length is much smaller than the length of the calculated river section when pollutants enter the water body, the pollutants can be mixed uniformly in a short time, and only the vertical concentration changes of the pollutants are considered, a one-dimensional water quality model should be adopted (Li & Zhang 1992). The calculation formula is as follows (Lian 2006):

$$W = 86.4\left[(Q_0 + q)C_s \exp\left[\frac{kx}{86{,}400u}\right] - C_0 Q_0\right] \tag{1}$$

where W refers to the water environment capacity of the river section (kg/d); Q_0 represents the design flow of the river section (m^3/s); C_s denotes the standard concentration of pollutants in the river section (mg/L); C_0 is the concentration of pollutants in the river section (mg/L); k refers to the degradation coefficient of pollutants in the river section (d^{-1}); u represents the average

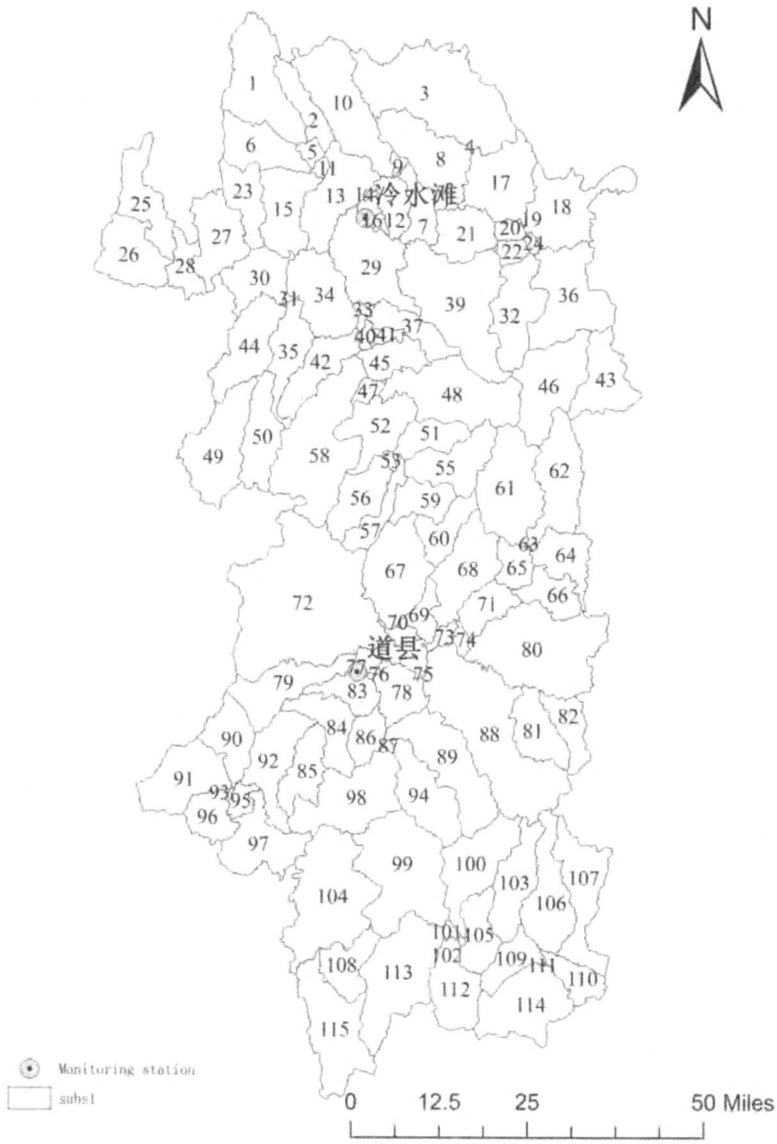

Figure 2 | Division and station distribution of the Yongzhou Section of Xiangjiang River Basin in 2018.

flow velocity of the river section (m^2); x denotes the length of the river section (m) and q is the sewage discharge flow of the river section (m^3/s).

3.4.2. Design hydrological conditions

According to the *Calculation Regulations for Pollution Capacity of Water Area* (GB/T25173-2010) (Li 2019), the design flow rate should be 90% of the driest monthly average flow or the driest monthly average flow for the past 10 years. However, using the average flow rate of the driest month for the past 10 years is too strict for the Xiangjiang River Basin where the water quality is still excellent. Based on the actual situation, 90% of the average flow rate of the driest month is selected as the design flow to calculate the conservative water environment capacity (Ma *et al.* 2021). On the other hand, the pollutant source survey data of this project uses 2017 as the base year. Therefore, the average flow in 2017 is selected as the design flow for the calculation of the WEC and remaining environmental capacity in 2017.

3.4.3. Executive standard

According to the results of the *Yongzhou Water Environment Function Zoning* (YWRB 2014), the Xiangjiang River Basin in Yongzhou is divided into 58 first-level water function areas and 53 second-level water function areas. At present, the current

Figure 3 | Flow boundary and pollution source setting interface diagram after generalization.

water quality of the assessment sections in all functional areas is better than that of Category III. According to the requirements of the Ministry of Ecology and Environment of the People's Republic of China on the cross-sectional assessment of national and provincial water quality control in the Yangtze River Basin and the principle of 'improving only, not deteriorating', the eight source protection areas implement the *Surface Water Environmental Quality Standards (GB3838-2002)* Class II standard, and the rest follows the Class III water quality standard.

Table 2 | Evaluation results of model simulation test area

Site	Index	Calibration period	R^2	E_{ns}	Verification period	R^2	E_{ns}
Daoxian Station	Runoff	2005–2017	0.73	0.7	2018–2019	0.85	0.73
	Ammonia	2005–2017	0.65	0.59	2018–2019	0.68	0.61
	Total phosphorus	2005–2017	0.70	0.68	2018–2019	0.79	0.78
Lengshuitan Station	Runoff	2012–2017	0.82	0.81	2018–2019	0.9	0.61
	Ammonia	2012–2017	0.66	0.56	2018–2019	0.8	0.73
	Total phosphorus	2012–2017	0.79	0.78	2018–2019	0.77	0.68

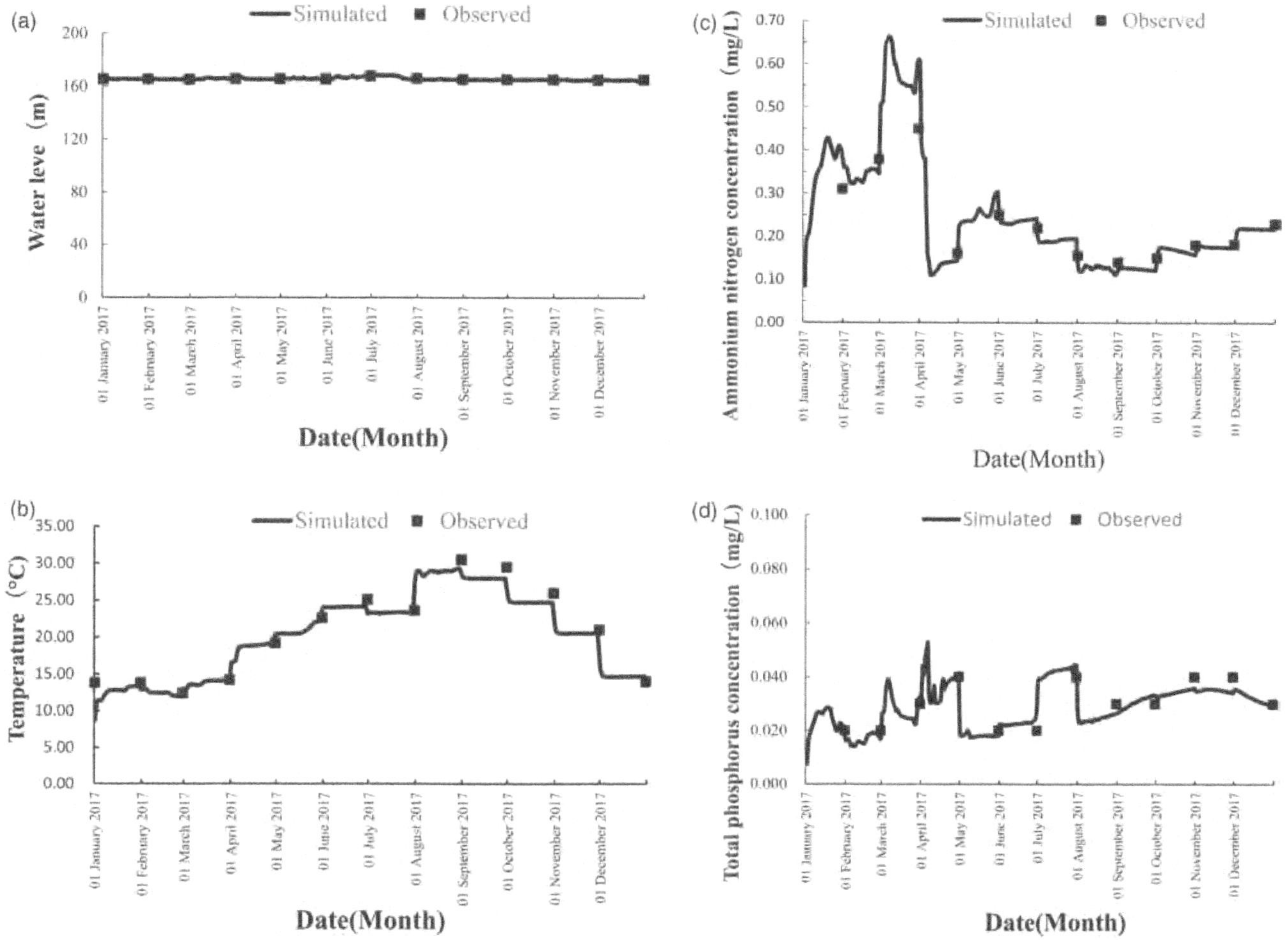

Figure 4 | Data of Daoxian: (a) Daoxian Station Water Level Verification Map; (b) Daoxian Station Water Temperature Verification Map; (c) Daoxian Station Ammonia Nitrogen Verification Map and (d) Daoxian Station Total Phosphorus Verification Map.

3.4.4. Determination of water degradation coefficient

The comprehensive attenuation coefficient (k) mainly reflects the impact of the concentration of pollutants in water with the changes in time, environment, and pollutants themselves. The commonly used calculation methods include the actual measurement method (Li 2019):

$$k = \frac{u}{\Delta x} \ln \frac{C_A}{C_B} \tag{2}$$

where k refers to the comprehensive attenuation coefficient of pollutants (d^{-1}); C_A represents the pollutant concentration of the upper section (mg/L); C_B denotes the pollutant concentration of the lower section (mg/L) and u is the average flow velocity of the lower section of the design flow (m/s).

Note that due to the lack of measured concentration data at the inlet and outlet of each sub-basin, the pollutant concentration value output by the EFDC model was used in this study to determine the degradation coefficient.

4. RESULTS AND DISCUSSION

4.1. Analysis of WECs of total nitrogen and total phosphorus

According to the calculation formula of one-dimensional theoretical water environment capacity, using the total nitrogen and total phosphorus as the accounting factors, the 90% driest monthly average flow and 2017 monthly average flow of each sub-basin were selected as the design hydrological parameters. The concentration results of total nitrogen and total phosphorus at the upstream outlet output by the EFDC model were used as the pollution concentration C_0 of the river section. In the meantime, the WEC of each sub-basin river section was calculated based on the following factors, including the water quality target

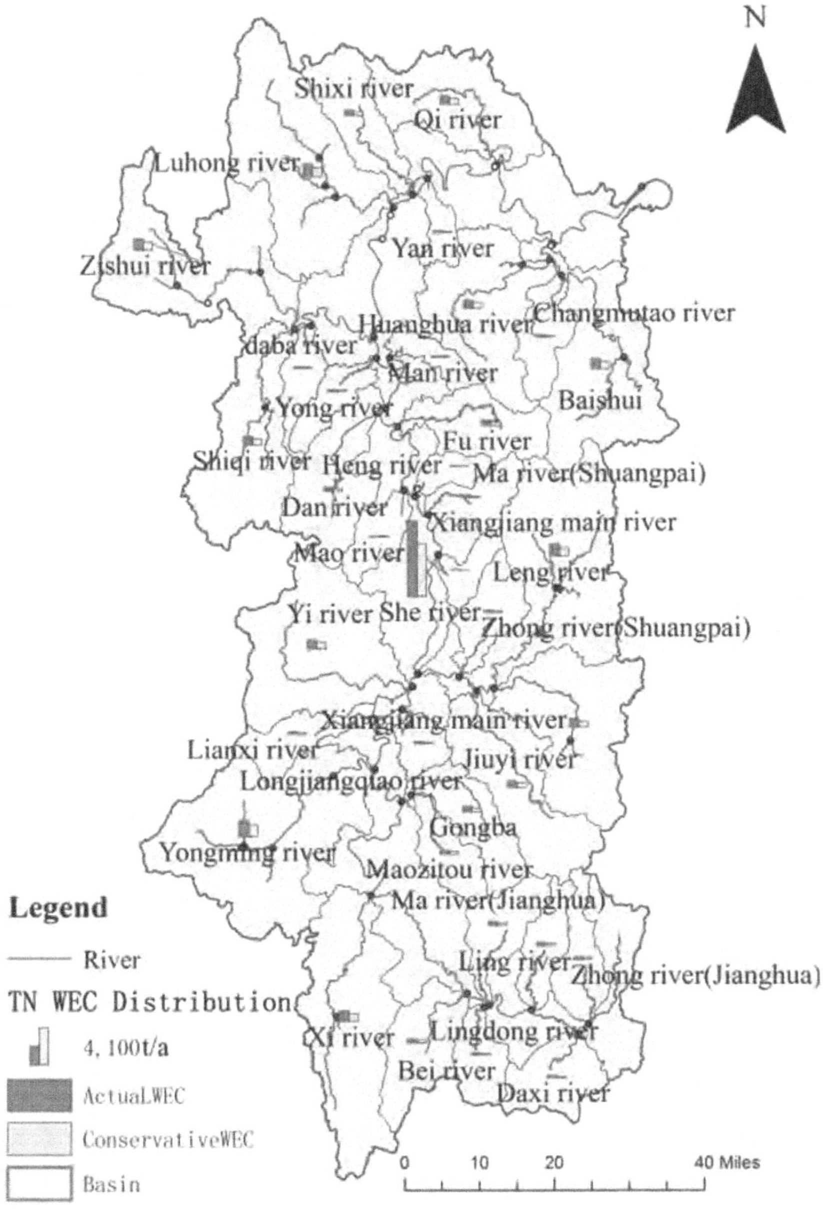

Figure 5 | WEC distribution of TN in the Yongzhou Section of Xiangjiang River Basin.

of the water function zoning as the standard concentration C_s for pollutant control, the sum of the pollutant flows of the sub-basin as the total discharge q of the sub-basin, the flow of the upstream water outlet as the design flow Q_0, the average flow velocity of each grid of the river section as u, and k determined by the empirical attenuation system. According to the classification requirements of water body types, the 115 sub-basins in the Yongzhou Section of Xiangjiang River Basin are divided into 36 secondary basins, and the water environment capacity is shown in Figures 5 and 6. It can be seen from Figures 5 and 6 that the top three water environmental capacities of total nitrogen and total phosphorus in the Yongzhou Section of Xiangjiang River Basin are the main basin of the Xiangjiang River, the Yongming River basin and the Luhong River basin. While the last three are Moujiang River Basin, Hengjiang River Basin and Shejiang River Basin, all of them are located within the territory of Shuangpai, Yongzhou. The main reason is that the three tributary rivers are short with small flow. The WEC of the=total nitrogen in the Yongzhou Section of Xiangjiang River Basin in 2017 was 27,673.04 t, which was 19,497.61 t/a under the 90% guarantee rate, about 0.7 times of the total. The WEC of total phosphorus in the Yongzhou Section of

Figure 6 | WEC distribution of TP in the Yongzhou Section of Xiangjiang River Basin.

Xiangjiang River Basin in 2017 was 4,877.22 t, which was 2,938.64 t/a under the 90% guarantee rate, about 0.6 times of the total. Obviously, the WEC of total phosphorus suffers a stronger impact by hydrological conditions than the total nitrogen does. Furthermore, the result shows no matter what the hydrological conditions were, the river basin with longer river generally had a larger WEC, which was consistent with that of Ma *et al.*'s (2021) paper.

4.2. Analysis of WECs of total nitrogen and total phosphorus in different water periods

According to the long-term hydrological statistics of the Yongzhou Section of Xiangjiang River Basin, the Xiangjiang River has a wet season from May to September, a dry season from December to February, and the rest of the time belongs to the normal season. It can be seen from Figures 7 and 8 that the WECs of total nitrogen and total phosphorus in the secondary

Figure 7 | Different water periods for the WEC distribution of TN in the Yongzhou Section of Xiangjiang River Basin.

watershed of the Yongzhou Basin of Xiangjiang River at different water periods in 2017 are sequenced in the descending order as wet season > normal season > dry season, the WEC had an uncertain range of fluctuations during the year, as noted by Ma *et al.* (2020). The main reason for the largest water environment capacity in the high water period is that the non-point source nitrogen and phosphorus pollution is well controlled. Although the pollution of nitrogen and phosphorus is heavy during the wet season, the WEC increases faster when the river flow increases on a large scale. The WECs of total nitrogen and total phosphorus in the dry season are 6,821.23 and 966.46 t, respectively; those in the normal season are 9,243.36 and 1,622.83 t, respectively and in the wet season, the above values are 11,608.45 and 2,288.93 t, respectively. The WECs of total nitrogen and total phosphorus in the wet season of the Xiangjiang River Yongzhou Basin are about twice that in the dry season, while in the normal season, it is generally about 1.5 times that in the dry season. The WEC

Figure 8 | Different water periods for the WEC distribution of TP in the Yongzhou Section of Xiangjiang River Basin.

of total nitrogen is slightly less than twice and 1.5 times, while the WEC of total phosphorus is slightly larger than twice and 1.5 times, showing that the capacity is large in summer and autumn, and small in spring and winter.

4.3. Analysis of the remaining WECs of total nitrogen and total phosphorus

It can be seen from Figures 9 and 10 that the overall situation of the residual WEC (the residual WEC is equal to the actual WEC minus the current pollutant emissions) of total nitrogen and total phosphorus in the Yongzhou Basin of Xiangjiang River is relatively good. In 2017, the residual WECs of total nitrogen and total phosphorus in the entire basin were 14,646.69 and 3,358.67 t, respectively, and the remaining WECs of all secondary basins were positive. In general, the residual

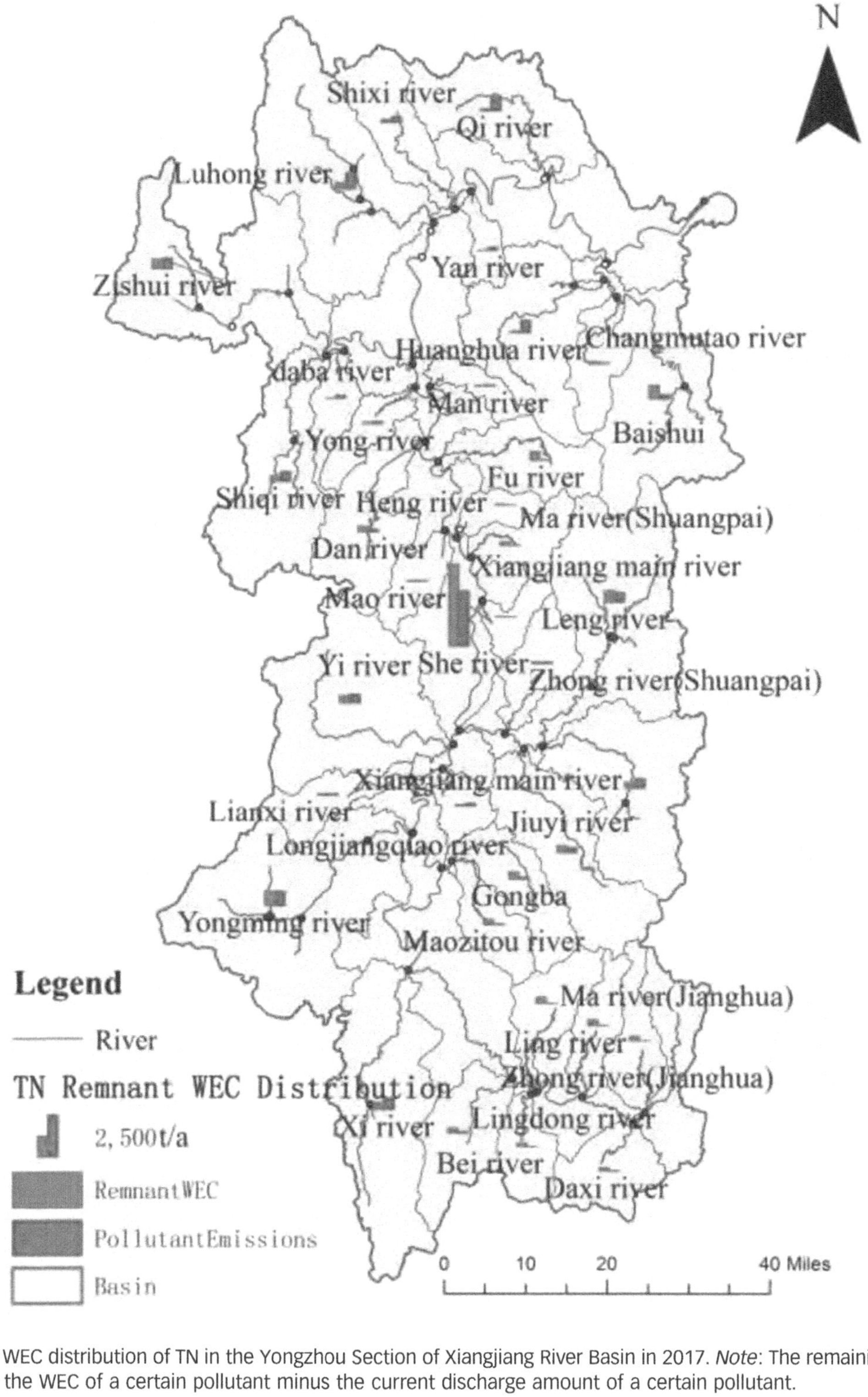

Figure 9 | Remnant WEC distribution of TN in the Yongzhou Section of Xiangjiang River Basin in 2017. *Note*: The remaining WEC of a certain pollutant is equal to the WEC of a certain pollutant minus the current discharge amount of a certain pollutant.

WEC of total phosphorus in the Yongzhou Section of the Xiangjiang River Basin was larger in the southern region and smaller in the northern region. Among the secondary basins in the Yongzhou Section of Xiangjiang River Basin, the remaining WECs of total nitrogen and total phosphorus in the main basin of the Xiangjiang River were the largest, i.e., 4,861.39 and 1,191.11 t/a, respectively. Meanwhile, its current emissions of non-point source nitrogen and phosphorus pollution were

Figure 10 | Remnant WEC distribution of TP in the Yongzhou Section of Xiangjiang River Basin in 2017. *Note:* The remaining WEC of a certain pollutant is equal to the WEC of a certain pollutant minus the current discharge amount of a certain pollutant.

also relatively large. The principle of 'Focusing on governance, improving quality and efficiency' is practiced, and agricultural non-point source pollution prevention and control measures are taken to promote the development of existing agricultural industrialization. The remaining WECs of total nitrogen and total phosphorus in the Yanjiang River Basin were the smallest, i.e., 2.57 and 0.57 t/a, respectively. The development of large-scale agricultural development projects, as well as the livestock and poultry breeding projects in this basin, must be strictly restricted. The residual WECs of total nitrogen and total phosphorus in the Baishui River Basin, Yanjiang River Basin, Yijiang River Basin, Ningyuan River Basin, Jiuyi River Basin, Mengba Basin and Yongming River Basin were relatively large, and its non-point source pollution control was well implemented. When formulating plans for livestock, poultry breeding and agricultural planting in the basin, the government can appropriately expand the scale following the principle of 'Combination of prevention and control, and green development'. The residual WECs of total nitrogen and total phosphorus in the Yanjiang River Basin, Daheba Basin, Maojiang River Basin, Yongjiang River Basin, Hengjiang River Basin, Moujiang Basin, Shejiang River Basin, Zhonghe (Shuangpai) Basin, Longjiangqiao Basin and Lianxi River Basin were relatively small, and their current emission levels of total nitrogen and total phosphorus were relatively high, which were the key areas for non-point source control and water environmental protection in the entire river basin. These river basins are environmentally sensitive areas, and the principle of 'Governing priority, strict control of increments' should be implemented in addition to the utilization of the existing environmental capacity and appropriate development. Moreover, the results of the remaining WEC could be applied to divide the water environment control units in the Yongzhou Section of Xiangjiang River Basin and build a water environment zoning management and control system.

5. CONCLUSION

1. The simulation curve of the SWAT and EFDC coupling model shows good curve fitness with the actual value, and the relative error is within the acceptable range. Therefore, the simulation effect of the model is good.
2. The total nitrogen WEC of the Yongzhou Section of Xiangjiang River Basin in 2017 was 27,673.04 t, which was 19,497.61 t/a under the 90% guarantee rate; while the total phosphorus WEC of the Yongzhou Section of Xiangjiang River Basin in 2017 was 4,877.22 t, which was 2,938.64 t/a under the 90% guarantee rate. Obviously, the WEC of total phosphorus suffers a stronger impact by hydrological conditions than the total nitrogen does.
3. The WEC had an uncertain range of fluctuations in 2017, the WECs of total nitrogen and total phosphorus in the low water period were 6,821.23 and 966.46 t, respectively, those in the normal water period were 9,243.36 and 1,622.83 t, respectively and the values in the high water period were 11,608.45 and 2,288.93 t, respectively.
4. The overall situation of the residual WECs of total nitrogen and total phosphorus in the Yongzhou Section of Xiangjiang River Basin River is relatively good. In 2017, the residual WECs of total nitrogen and total phosphorus in the entire basin were 14,646.69 and 3,358.67 t, respectively. The residual WECs of total nitrogen and total phosphorus in the Yanjiang River Basin, Daheba Basin and Maojiang River Basin were relatively small, and their current emission levels of total nitrogen and total phosphorus were relatively high, which were the key areas for non-point source control and water environmental protection in the entire river basin.

ACKNOWLEDGEMENTS

This work was supported by the National Natural Science Foundation of China (No.51638006); National Natural Science Foundation of China Youth Fund (No.41807136); Guangxi Science and Technology Base and Talent Special Project (Guike AD19110012); Guangxi Key Laboratory of Environmental Pollution Control Theory and Technology Open Fund (Guikeneng 1701K006).

DATA AVAILABILITY STATEMENT

All relevant data are included in the paper or its Supplementary Information.

REFERENCES

Burn, D. H. & Lence, B. J. 1992 Comparison of optimization formulations for waste-load allocations. *Journal of Environmental Engineering* **118** (4), 597–612.

Deng, Y., Zheng, B. H., Guo, F. & Li, Z. 2010 Study on the total water pollutant load allocation in the Changjiang (Yangtze River) estuary and adjacent seawater area. *Estuarine, Coastal and Shelf Science* **86** (3), 331–336.

Du, W., Li Y.-P., Hua, L., Wang. C., Wang, P., Wang, J., Wang. Y., Chen, L., Buoh, R. B., Jepkirui, M., Pan, B., Jiang, Y. & Acharya, K. 2016 Water age responses to weather conditions in a hyper-eutrophic channel reservoir in Southern China. *Water* **8** (9), 372.

Fan, Y. M. 2017 *Based on the EFDC Simulated Sewage Leak Event,the Study on the Effect of Water Quality of Harbin Section of the Songhua River Was Studied.* Harbin Normal University, Harbin.

Fang, Y.-J. 2015 *Water Environment Simulation and Total Control of Ganjiang River Basin Based on SWAT Model*. Nanchang University, Nanchang.

Feng, L.-Z. 2016 *Study on the Dynamic Water Environment Capacity and Risk Assessment of the Hohhot Section of the Yellow River*. Inner Mongolia Agricultural University, Hohhot.

He, J., Wu, X., Zhang, Y., Zheng, B., Meng, D., Zhou, H., Lu, L., Deng, W., Shao, Z. & Qin, Y. 2020 Management of water quality targets based on river-lake water quality response relationships for lake basins – A case study of Dianchi Lake. *Environmental Research* **186**, 1–16.

Hu, K., Wang, Y., Feng, B., Wu, D., Tong, Y. & Zhang, X. 2020 Calculation of WEC of large shallow lakes – a case study of Taihu Lake. *Hydrology Research* **22**, 223–236.

Hui, H. K. 2015 *Spatiotemporal Simulation of Surface Runoff Non-Point Source Pollution Loadbased on L-THIA Model and Its Environmental Effects: Taking Hulan River-Ashi River Basinas an Example*. Harbin Normal University, Heilongjiang.

Jiang, X.-F. 2015 *Analysis of the Distribution Characteristics of Non-Point Source Pollution in the Ash River Basin and Prevention and Control Strategies*. Harbin Institute of Technology, Heilongjiang.

Li, S. 2012 *Simulation of non-Point Source Nitrogen and Phosphorus Pollution in Nansi Lake Basinbased on SWAT Model and Lake Sediment Response*. Shandong Normal University, Shandong.

Li, X.-H. 2019 *Research on Water Environment Carrying Capacity of Dahuofang Basin*. Chong Qing Jiaotong University, Chong Qing.

Li, Y.-W. & Zhang, Y. L. 1992 Research on the calculation method of pollution load allocation. *Environmental Science Research* **02**, 45–48.

Lian, Y.-Y. 2006 *Capacity Measurement Technology*. China Metrology Press, Beijing.

Liu, Z.-J., Hashim, N. B., Kingery, W. L., Huddleston, D. H. & Xia, M. 2008 Hydrodynamic modeling of St. Louis Bay Estuary and watershed using EFDC and HSPF. *Journal of Coastal Research* **10052**, 107–116.

Ma, H. B., Tian, C. C., Huang, D. J. & Wang, L. T. 2020 Study on dynamic water environmental capacity of theriver network in Wenhuang plain based on the hydrodynamic and water quality coupling model. *IOP Conference Series: Earth and Environmental Science* **612**, 012061.

Ma, Q. X., Pang, Y. & Mu, R. H. 2021 Water environmental capacity calculation based on control of contamination zone for water environment functional zones in Jiangsu section of Yangtze river, China. *Water* **13**, 587. https://doi.org/10.3390/w13050587.

Meng, C. 2018 *Research on the Total Amount of Pollutants Control in the Watershed Based on the Water Environment's Pollution Holding Capacity*. North China Electric Power University, Beijing.

Meng, C. H. & Zhao, B. 2008 Application of AGNNPS model to agricultural non-point source pollution in small watersheds. *Journal of Yangtze River Scientific Research Institute* **25** (4), 24–27. 32.

Moriasi, D. N., Gitau, M. W., Pai, N. & Daggu, P. 2015 Hydrologic and water quality models: performance measures and evaluation criteria. *Transactions of the ASABE* **58**, 1763–1785.

Shen, T., Liu, L. Y., Ma, J. F. & Li, C. J. 2007 Spatial distribution characteristics of non-point source pollution in Miyun Reservoir area based on L-THIA model. *Transactions of the Chinese Society of Agricultural Engineering* **23** (5), 62–68.

Tang, G.-P. 2016 *Research on Water Quality Simulation and Water Environment Capacity Basedon WASP Model – Taking the Lower Reaches of Nanhe River as an Example*. University of Science and Technology of Suzhou, Suzhou.

Thomann, R. V. & Sobel, M. J. 1964 Estuarine water quality management and forecasting. *Journal of the Sanitary Engineering Division* **46** (3), 77–82.

USEPA 1999 *Protocol of developing nutrient TMDLs*, EPA 841-B-99-007.

Wang, G. Q., Zhang, J. Y., Xu, Y. P., Bao, Z. X. & Yang, X. Y. 2017 Estimation of future water resources of Xiangjiang River Basin with VIC model under multiple climate scenarios. *Water Science and Engineering* **10** (02), 87–96. https://doi.org/10.1016/j.wse.2017.06.003.

Wang, Q.-R., Liu, R.-M., Men, C., Guo, L. & Miao, Y. 2019 Temporal-spatial analysis of WEC based on the couple of SWAT model and differential evolution algorithm. *Journal of Hydrology* **569**, 155–166.

Xu, A.-N. 2014 *WASP Model and Its Application in Reservoir Water Quality Simulation Research*. North China Electric Power University, Beijing.

Xu, J. 2018 *Research on Water Quality Management Plan Based on WASP Model in Bahe River Basin*. Shaanxi University of Science and Technology, Xian.

Yao, P. C., Pan, W. B., Chen, H. & Chen, Z. 2018 Research on WEC of Changqing River Basin. *Northern Environmental* **030** (001), 134–136.

Yen, C.-H., Chen, K.-F., Shen, Y., Lin, C. C. & Hong, J. J. 2012 Pollution source investigation and water quality management in the Carp Lake Watershed, Taiwan. *CLEAN: Soil, Air, Water* **40** (1), 24–33.

Yongzhou City History Office 2018 *Yongzhou Statistical Year Book*, China Statistics Press, Beijing, China, pp. 468–510.

Yongzhou Ecological Environment Bureau 2020 *General Survey Report of the Second Pollutant in Yongzhou*. China, pp. 108–140.

Yongzhou Water Resources Bureau 2014 *Yongzhou City Water Function Zoning Report*. China, pp. 45– 70.

Zhai, X.-Y. 2015 *Watershed Environmental Hydrological Process and its Numerical Simulation Underchanging Environment*. Wuhan University, Hubei.

Zhang, S.-S. 2017 *Research on Calculation of Water Environment Capacity Based on MIKE11 Water Quality Model*. Hefei University of Technology, Anhui.

Zhang, H., Zeng, F. T., Fang, H. Y., Zhang, H. J., Xiao, J. E. & Lin, S. 2012 Calculation of non-point source load in Danshui riverbasin based on HSPF and regression model. *Acta Scientiae Circumstantiae* **32** (4), 856–864.

Zhao, C.-C. 2018 Research on ammonia nitrogen water quality management plan based on WASP model in bahe river basin. *Journal of Anhui Agricultural University* **45** (2), 302–308.

Zhao, K., Su, B.-L., Shen, M.-M., Guan, Y. T. & Zhou, J. W. 2017 An improved method for SWAT model parameter identification. *South-to-North Water Diversion and Water Conservancy Science and Technology* **15** (04), 49–53.

Zheng, Y., Hu, X.-B., He, Q. & He, J. Y. 2019 Research on Chongqing reservoir area water environment capacity based on EFDC model. *Three Gorges Ecological Environment Monitoring* **4** (02), 9–16.

First received 16 August 2021; accepted in revised form 12 November 2021. Available online 30 November 2021

Impact of Climate Change on Hydrology and Water Resources doi: 10.2166/wcc.2021.055

Potential influence of climate and land-use changes on green water security in a semi-arid catchment

Alireza Abbasi [a,*], Mahdi Amirabadizadeh[a,b], Amirhosein Aghakhani Afshar[c] and Mostafa Yaghoobzadeh[a]
[a] Department of Science and Water Engineering, University of Birjand, Birjand, Iran
[b] Research Group of Drought and Climate Change, University of Birjand, Birjand, Iran
[c] Department of Water Engineering, Faculty of Civil Engineering, University of Tabriz, Tabriz, Iran
*Corresponding author. E-mail: alrz.abbassi@gmail.com

AA, 0000-0002-9966-926X

ABSTRACT

Temporal and spatial changes of green water (GW) security due to climate and land-use/land-cover (LULC) changes can be used to make the best decisions for sustainable GW management. In this study, simultaneous effects of climate and LULC changes on water resources in Kashafrood Basin were evaluated using the Soil and Water Assessment Tool (SWAT). A land change modeler was set up to monitor LULC, assess changes and make predictions. The MIROC-ESM model derived from Coupled Model Intercomparison Project Phase 5 under two representative concentration pathway (RCP) emission scenarios RCP2.6 and RCP8.5 was applied to evaluate the effects of climate change. Two indices of GW-Scarcity and GW-Vulnerability, representing GW-Security, were quantified using the GW-Footprint concept in Kashafrood Basin. The results show that the annual average of blue water was predicted to increase by 142–350%, and GW storage and the annual averages of GW flow were predicted to decrease by 12–65 and 8–20%, respectively, depending on emission scenarios and time. The GW-Security estimates in the entire basin suggest a better condition in the future by indicating 24–45 and 16–52% decreases in GW-Scarcity and GW-Vulnerability, respectively, depending on emission scenarios and time.

Key words: green water security, land change modeler, land-use/land-cover change, MIROC-ESM, Soil and Water Assessment Tool

HIGHLIGHTS

- Land change modeler predicts a decline in agricultural area and a growth in urban and pasture areas.
- Blue water rises by 142–350% under the effect of climate and land-use changes.
- Green water (GW) storage and GW flow decrease by 12–65 and 8–20%, respectively.
- The condition of the basin in terms of GW-Security would improve in the future.
- GW-Scarcity and GW-Vulnerability escalate in southern and some western subbasins.

INTRODUCTION

Climate and land-use/land-cover (LULC) changes play a prominent role in altering the hydrological processes. Climate change influences the regional water balance by affecting precipitation and temperature, and LULC changes are directly linked to changes in evapotranspiration, interception and infiltration (Pan *et al.* 2017; Shao *et al.* 2018). Land degradation, soil erosion, desertification, urbanization, suburbanization, industrialization and poor management of environmental resources induce the LULC changes which decrease available water supplies (Briassoulis 2019). LULC change analysis indicates the qualitative and quantitative characteristics of LULC. This analysis reveals the forces or factors that directly or indirectly induce changes. Moreover, the prediction of LULC changes indicates the spatial and temporal levels of this analysis. A study of LULC changes can provide incentives to make policies for integrated water resources management and sustainable development. Additionally, Global Circulation Models (GCMs) can be used as one of the most reliable tools for climate projections in climate change assessment. Coupled Model Intercomparison Project Phase 5 (CMIP5), with improved capability of simulating climate change, provides numerous GCMs and has a smaller bias than CMIP3 (Jia *et al.* 2019). The Intergovernmental Panel of Climate Change (IPCC), in its Fifth Assessment Report (AR5), released four new greenhouse gas emission scenarios (called representative concentration pathways (RCPs) 2.6, 4.5, 6.0 and 8.5) based on their potential range of radiative forcing values (W/m^2) (Tan *et al.* 2017). Climate change mainly affects the amount of precipitation, while land-use change mainly influences soil water holding capacity and surface evapotranspiration. These changes in hydrological processes threaten the freshwater resources availability for human, irrigation and energy generation (Veettil & Mishra 2016). Water consumption is one of the stressors that expose 80% of world's population to the water security risk (Vörösmarty *et al.* 2010; Rodrigues *et al.* 2014). Approximately 1.8 billion people would encounter water scarcity by 2025 (WWAP 2012; Connor 2015). Thus, it is necessary to have a full perception of water resources components in a sustainable manner. Furthermore, awareness of the effects of climate and LULC changes on water resources can supply a theoretical basis for ecological water resource planning and local water resource management.

The freshwater cycle is classified into 'green' water (GW) and 'blue' water (BW) according to storage types in hydrological processes. BW is described as the total deep aquifer recharge and surface runoff stored in lakes, rivers and shallow aquifers. GW flow (GWF) is the actual evapotranspiration emitted to the atmosphere as a result of evaporation from water bodies and soil and transpiration from vegetation. GW storage (GWS) refers to the portion of precipitation that infiltrates the soil (Falkenmark & Rockström 2006; Rodrigues *et al.* 2014). GW resources play a significant role in sustaining the production and services in land ecosystems (Lyu *et al.* 2019). Considering the water balance in hydrological cycles, GW (consumed by evapotranspiration in grasslands, forests, wetlands and cultivated lands) and BW account for 65 and 35% of the total global precipitation, respectively. Moreover, GW contributes to over 80% of the global grain production (Liu *et al.* 2009). Although the availability of GW is much larger than BW, the allocation of GW is rarely considered. Competition over the consumption of the limited GWF between natural ecosystems and human economy for biomass production leads to GW-Scarcity. These tensions are intensified as the GW demand for biomass grows with the growth of population and increase of GW demands per capita (Schyns *et al.* 2019). Therefore, the scarcity and vulnerability assessments of GW resources can provide a useful basis for the future. The BW and GW consumed by human activities are defined as water footprints (Hoekstra 2003; Hoekstra *et al.* 2011). The BW-Footprint represents the consumption of BW sources for producing goods and services (e.g., industrial, domestic, irrigation and power production), while the GW-Footprint represents the consumption of GW in agricultural lands (e.g., evapotranspiration from cultivated land and pastures) (Hoekstra *et al.* 2011; Veettil & Mishra 2016).

Water security is described as 'capacity to maintain a reliable access to sufficient quantity and good quality of water for livelihood, socio-economic development and human well-being, to ensure protection against water-related disasters and water-borne pollution in order to conserve ecosystems' (UN Water 2013; Marttunen *et al.* 2019). Similar to integrated water resources management, water security suggests a paradigmatic approach toward water systems analysis that covers quantity and quality concerns including water availability and hazards. In the absence of an integrated framing in water security, the greater environmental objectives can be excluded (Cook & Bakker 2012). In water resources discipline-based definition, GW-Security is calculated as a ratio of consumed water to available water (Rodrigues *et al.* 2014; Veettil & Mishra 2016). This ratio is employed to specify the geographic hotspots under water stress (Hoekstra *et al.* 2011; Rodrigues *et al.* 2014). The assessment of GW-Security is a research prerequisite and can contribute to universal analysis of environmental health, food security, poverty and human and economic developments (Hoekstra *et al.* 2011).

Due to poor water resources management, significant levels of water scarcity already exist in Iran. In addition to climate change and LULC changes, some important issues in agricultural policy, which are directly related to water security, are aiming to achieve food self-sufficiency in the country, heavy dependence on irrigation, mismatch between the geographic distribution of water availability and the spatial cropping pattern, low water-use efficiency and a small share of rain-fed agriculture (Karandish & Hoekstra 2017). Kashafrood River, as the longest river in the northeast of Iran, is the main source of surface water supply in Mashhad, which is the major city located within Kashafrood River Basin. Extreme dry events have caused water scarcity and contamination in Kashafrood Basin (Afshar *et al.* 2017). Most of the studies on water resources components in Kashafrood Basin have mainly focused on the projected climate variations, ignoring the future LULC variability (Sayari *et al.* 2013; Afshar *et al.* 2018). A comparison of the simultaneous effects of climate and LULC changes and the effect of climate change can reveal the prominent role of LULC changes in water resources projection.

Given that climate and LULC changes directly affect water balance and water resources components, it is desirable to study their effects on water resources and GW-Security considering the LULC changes. Hence, the objectives of the present study are (a) building a land change modeler (LCM) model, analyzing the changes and predicting the future LULC maps; (b) investigating the spatiotemporal distribution of BW, GWS and GWF in Kashafrood Basin under the proposed scenarios and (c) quantifying the GW-Security and the spatial distribution of GW-Security based on water scarcity and vulnerability. This is known to be the first study that has investigated the effects of climate and LULC changes on water resources components in Kashafrood Basin simultaneously and has assessed water security focusing on GW in the future. The spatiotemporal LULC changes were also analyzed to develop better policies for water resources management. The understanding of variation of freshwater resources, GW-Security and LULC changes in Kashafrood Basin in the future can help policymakers to manage the basin more sustainably.

MATERIALS AND METHODS

Case study and data description

This study was done in Kashafrood Basin (35°40′–36°3′N, 58°2′–60°8′E) located in the northeast of Iran. The altitude of the study area is from 391 m in the southeast to 3,234 m in the northwest (Figure 1). This basin, with a total area of 16,750 km^2, covers major cities such as Mashhad (the second most populous city in Iran), Quchan and Chenaran. It has a semi-arid and cold climate with low annual precipitation and high evapotranspiration rate in summer. The climate data for study area within 1992–2017 are mean precipitation of 340 mm/year, mean annual temperature of 13.6 °C, maximum recorded temperature of 20.6 °C and minimum recorded temperature of 7.1 °C (Afshar *et al.* 2017). The predominant LULC type in the basin in 2017 was pasture, followed by irrigated farmland, rain-fed land, forest, residential areas, barren lands and water bodies (Figure 2). The main crops cultivated in Kashafrood Basin are wheat (spring and winter), barley and sugar beet according to Khorasan Razavi agricultural landscape report.

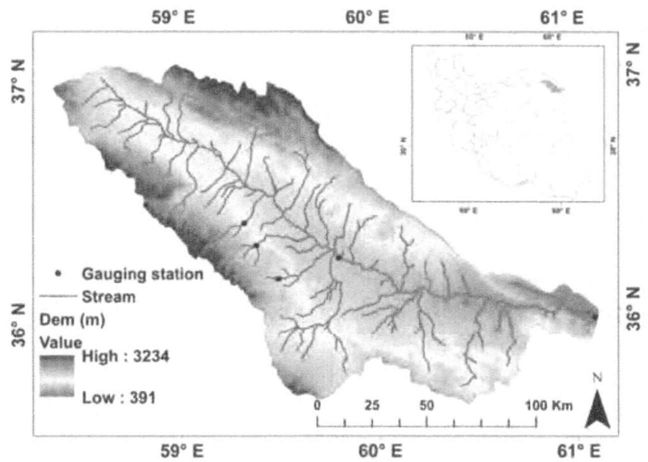

Figure 1 | Location of the study area.

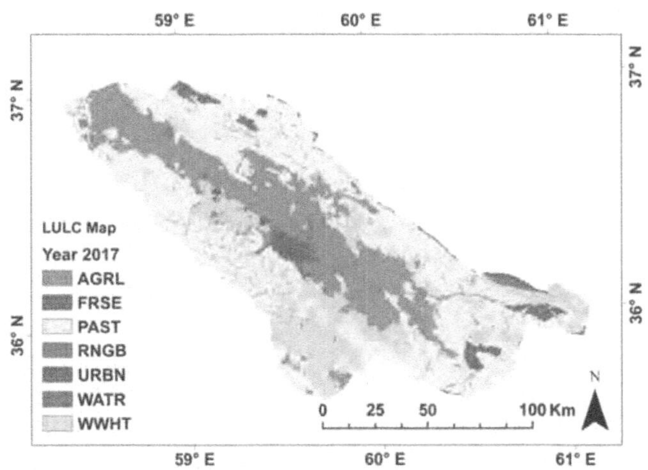

Figure 2 | LULC map of 2017.

The used databases are satellite data, digital elevation model (DEM), geographical data (roads, cities and towns), soil-type data, meteorological data, hydrological data and the future climate data. DEM data, received from the National Cartographic Center of Iran (NCC), were used to generate the digital river network and to delineate the basin into subbasin design. The soil property map was also gained from the Range and Watershed Department (RWD) of Khorasan Razavi. Meteorological data (including daily precipitation and temperature data from 1992 to 2011) were attained from Iran Meteorological Organization (IMO). The monthly runoff data records at five gauging stations (Sar Asiab Shandiz, Zire Band Golestan, Golestan Jaghargh, Hesar Dehbar and Kartian) were collected from Water Resource Management Company (IWRMC). The datasets required to project the future land use were the land-use maps for previous years and the variables affecting land-use changes. Remote sensing was employed to prepare land-use maps. Landsat satellite images from Landsat Thematic Mapper (LTM) in 1987, Enhanced Thematic Mapper (ETM^+) in 2002 and Operational Land Imager (OLI) in 2017 were acquired from the United States Geological Survey (USGS) website (https://earthexplorer.usgs.gov/). Image preprocessing, including radiometric and atmospheric corrections, image mosaicking and image cutting, was performed to obtain accurate classification. Using the maximum-likelihood method, the basin was classified into seven land-use types: pasture (PAST), agricultural land-generic (AGRL), winter wheat (WWHT), forest (FRSE), urban area (URBN), range-brush (RNGB) and water body (WATR). Google Earth and field investigations were applied to examine the accuracy of interpreted images, which helped in preparing the LULC maps for 1987, 2002 and 2017. The geographical data obtained from NCC were applied to prepare the affective variables (distance to road and distance to urban area) by ArcGIS.

Soil and Water Assessment Tool model and simulation setup

The Soil and Water Assessment Tool (SWAT) model was employed to simulate the water resources components. The SWAT model, which is a semi-physically-based, semi-distributed and basin-scale model, is widely applied for water quantity assessment in many countries. For example, it was used in the studies conducted in the United States (Veettil & Mishra 2016), Brazil (Rodrigues *et al.* 2014), India (Gosain *et al.* 2011), China (Zang & Liu 2013; Zhu *et al.* 2018) and Iran (Faramarzi 2010; Afshar *et al.* 2018). In SWAT, a basin can be divided into several subbasins according to the DEM data. Each subbasin can also be divided into different hydrologic response units (HRUs), which contain identical land-use types, slope characteristics and soil types (Arnold *et al.* 2012). Kashafrood Basin was divided into 217 subbasins and 635 HRUs. The surface flow was estimated by the Soil Conservation Service (SCS) curve number method, and the potential evapotranspiration was calculated by the Hargreaves method. Hydrological variables, such as runoff and evapotranspiration, were obtained from each subbasin. The Sequential Uncertainty Fitting program algorithm (SUFI-2) in SWAT–CUP interface was utilized to optimize, validate, calibrate and render uncertainty analysis of the model parameters. In SUFI-2, parameter uncertainty applies to all sources of uncertainty including uncertainty in conceptual model, measured data, driving variables and parameters (Abbaspour 2011). In this study, determination coefficient (R^2) and Nash–Sutcliffe efficiency (NSE) were utilized to assess the model performance. *P*-factor and *r*-factor were also applied in uncertainty analysis. *P*-factor is percentage of the measured data bracketed by 95% predictive uncertainty band, and *r*-factor is the mean width of the band (Abbaspour *et al.* 2004). SWAT performance analysis was done within 19 years from 1992 to 2011. For model evaluation, the dataset of 1992–2011 was divided into calibration period (1992–2000) and validation period (2001–2011). A warm up period (1992–1995) was chosen to achieve a steady state for modeling.

Land change modeler

The LCM for ecological sustainability was designed for the analysis of land-cover changes and the projection of the future land-use maps. The basic principle in this model is to evaluate the trend of changes from one land-use class to another, considering the influence of several variables such as roads, slope and aspect. Generally, LCM predicts a land-use pattern based on the previous change trend. This model was selected because of its efficiency in land-use change prediction (Pickard *et al.* 2017; Anand *et al.* 2018). LCM is a Cellular Automata and Markov-Chain-based built-in module at TerrSet IDRISI and contains the tools for analyzing land-cover change, identifying class transition potential and modeling several transitions at one time to predict the land-use maps. The change analysis tool was applied to recognize the class changes within 1987–2002. This tool assesses the transition among land-use classes in a selected time. Considering the sophisticated land-use changes in Kashafrood Basin, dominant transitions were specified (transitions below 1 km^2 were neglected) and classified into two sub-models: (1) transitions to urban area (AGRL to URBN, PAST to URBN and WWHT to URBN) that directly considers the growth of urbanization and (2) other transitions (AGRL to PAST, FRSE to PAST, PAST to AGRL, PAST to RNGB and WWHT to PAST). Driver variables were tested, and the variables with Cramer's coefficient of higher than 0.1 were selected. Distance to roads and distance to urban area were selected for the first sub-model, while DEM, slope and aspect were selected for the second sub-model. Modelling of transitions at one time in LCM was performed by multi-layer perceptron artificial neural network. A hard modeling process, which relies on the predictions with particular change classifications based on a class designation competitive model identical to a multi-objective decision process, was employed (Armenteras *et al.* 2019). In the present study, the prediction of land-use change was validated by a comparison of the reference map and the prediction map of 2017. The percentage of area differences in land-use classes was less than 8%. Considering this level of agreement, the model was found eligible to predict the changes for 2032, 2062 and 2092 (Saifullah *et al.* 2017).

Description of the scenarios for climate and LULC changes

To survey the simultaneous effects of climate and LULC changes on hydrological components, four time intervals were considered: historical period (1992–2013), near future (2014–2042), intermediate future (2043–2071) and distant future (2072–2100). MIROC-ESM is a Japanese earth system model derived from the CMIP5 to project the climate conditions. Among 14 GCMs, MIROC-ESM has shown a good performance in reproducing the observed climatology in Kashafrood Basin (Afshar *et al.* 2018). The bias-corrected spatial disaggregation (BCSD) method was used to downscale the MIROC-ESM model data. Details of this method have been presented by Afshar *et al.* (2018) and Jafarzadeh *et al.* (2019). After downscaling the climate data and predicting the future land-use maps by LCM, appropriate climatic and land-use data were prepared for each time period. LULC maps of 2002, 2032, 2062 and 2092 were related to historical, near future, intermediate future and distant future periods, respectively. RCPs were labeled based on their radiative forcing level by 2100. Furthermore, the lowest level (RCP2.6) and the highest level (RCP8.5) of radiative forcing were considered to investigate the highest impact and the lowest impact of climate change on water resources. Low representative RCP2.6 scenario is shown to be technically feasible, by participation of all countries in the world in the short run. It can be referred to the broadening participation of the Organization for Economic Cooperation and Development (OECD) countries (Van Vuuren *et al.* 2010). RCP8.5 follows the pathway of a continuously rising radiative forcing and targets 8.5 W/m^2 in 2100 with a further enhanced residual circulation and significant CH_4 increases. In addition, by selecting these two scenarios, the results of this study can be compared with a similar study in Kashafrood Basin conducted by Afshar *et al.* (2018).

In other words, six statuses for the future scenarios and one historical scenario were examined to analyze the long-term values of water resources and GW-Security. Figure 3 shows a schema of the SWAT model used for assessing water resource components and GW-Security in Kashafrood Basin.

Quantification of GW-Security

Vulnerability and scarcity indices were applied to assess GW-Security in the basin. GW-Scarcity indicator was measured as the ratio of GW-Footprint to GW-Availability (Hoekstra *et al.* 2011; Veettil & Mishra 2016). Actual evapotranspiration represents the GW-Footprint. As already depicted, the Hargreaves method was selected for simulating evapotranspiration at the HRU spatial scale. GW-Availability is the amount of available water for sustainable crop growth and can be calculated by subtracting the wilting point from the total soil moisture in the root zone. Wilting point is the minimum soil moisture content at which crop sustainability no longer exists (Rodrigues *et al.* 2014; Veettil & Mishra 2016). Therefore, the initial soil water content obtained from the HRU output files represents the GW-Availability. Closeness of the mean water use to the physical water availability threshold can be represented by the water scarcity indicator (Rodrigues *et al.* 2014). GW-Scarcity can be calculated by the following equation:

$$\mathrm{GW-Scarcity}_{(x,t)} = \frac{\mathrm{GW-Footprint}_{(x,t)}}{\mathrm{GW-Availability}_{(x,t)}} \tag{1}$$

where GW-Availability$_{(x,t)}$ is the amount of initial soil water content in zone 'x' within the period 't', and GW-Footprint$_{(x,t)}$ is the GW consumed in zone 'x' during period 't'.

The GW-Vulnerability indicator can be described as the susceptibility of agricultural water consumption under drought-like conditions or low soil water content (Padowski & Jawitz 2012). Thus, GW-Vulnerability is measured as the

Figure 3 | Schematic flowchart of the used SWAT model.

ratio of GW-Footprint to the historical low GW-Availability (the 30th percentile). GW-Vulnerability is calculated by the following equation:

$$\mathrm{GW-Vulnerability}_{(x,t)} = \frac{\mathrm{GW-Footprint}_{(x,t)}}{\mathrm{GW-Availability}_{(\mathrm{P30})(x,t)}} \quad (2)$$

where GW-Availability$_{(\mathrm{P30})(x,t)}$ is the historical low volume of soil water content at zone '*x*' during the period '*t*', which exceeds 70% of the time, expressed by the 30th percentile of the records.

RESULTS AND DISCUSSION

LULC change analysis and prediction

Figure 4 shows the LULC maps predicted for 2032, 2062 and 2092 by the LCM model. The relative areas in each LULC class for the historical and the future periods are presented in Table 1.

According to Table 1 and Figure 4, the AGRL LULC declines in the specified periods. Population growth, pressure of industry, water shortage, decline of agricultural productivity and change of rural lifestyles are among the factors affecting the

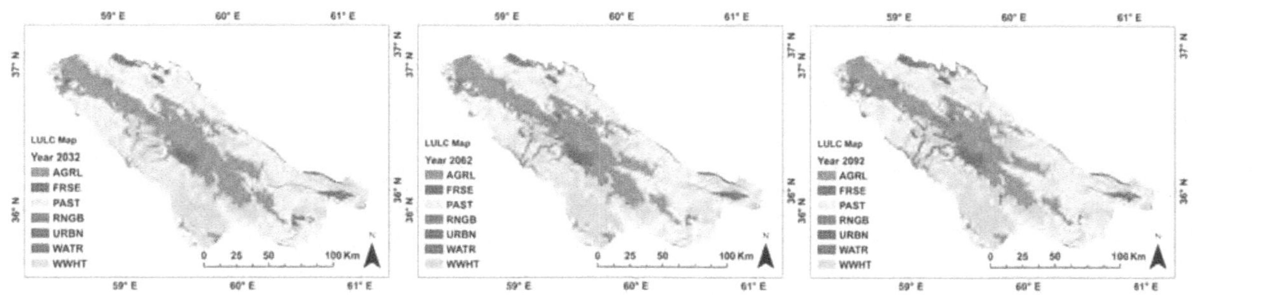

Figure 4 | LULC maps predicted for the future.

Table 1 | LULC areas in the historical and future periods and change percent in the future in Kashafrood Basin

	LULC areas in the historical period (km^2)			LULC areas in the future (km^2)			Change percent (%)		
LULC	**1987**	**2002**	**2017**	**2032**	**2062**	**2092**	**2002–2032**	**2032–2062**	**2062–2092**
AGLR	4,803	4,724	4,626	4,591	4,468	4,354	−0.79	−0.73	−0.68
FRSE	517	491	474	445	404	368	−0.27	−0.24	−0.21
PAST	8,554	8,607	8,644	8,709	8,803	8,886	0.60	0.56	0.49
RNGB	81	89	122	98	108	117	0.05	0.05	0.05
URBN	186	240	326	348	452	553	0.64	0.62	0.60
WATR	1	2	6	2	2	2	0	0	0
WWHT	2,608	2,598	2,547	2,553	2,510	2,468	−0.26	−0.25	−0.25

AGRL class. FRSE LULC also has a declining trend, which can be due to the overexploitation of forests for timber and overgrazing in sub-mountain areas. However, the PAST LULC has an increasing trend. Since the pastures include dense, semi-dense and poor pastures, the conversion of other LULC classes to poor pastures due to the deforestation and reduction of water resources might be responsible for this trend. The RNGB land use covers a small area of the basin and its increase is likely due to conversion from poor pasture and largely due to drought and soil degradation in recent decades. According to the results, the URBN land use has an increasing trend, particularly due to population growth in residential areas. WATR in the study area is confined to the lake of dams and covers a very small area. Due to the water resources scarcity in recent decades, the government has adopted a dam-building policy. The construction of several dams in the area shows an increased water body in the historical LULC maps. After pasture and agricultural lands, the rain-fed area (WWHT) is the most abundant LULC in Kashafrood Basin. This land use has a gentle downward trend. Precipitation reduction, soil degradation and, in some areas, conversion to industrial areas are among the factors involved in the reduction of this LULC.

The 1987 and 2002 LULC layers were employed for the calibration of LCM as the observed data, while the 2017 classified LULC layer was applied for the verification of the simulated map for 2017. By analyzing LULC class changes in two time periods of 1987–2002 and 2002–2017, it could be concluded that the largest area change in the first period was a decline of AGRL class by 0.2% of the basin area. In the second period, the URBN class experienced the highest level of change with a 0.37% increase in the basin area. In both periods, the WATR class had the least change in the basin area due to its very small surface area. Therefore, its change was not taken into account in predicting the LULC maps. A comparison of the changes in the two periods implied that the intensity of changes was greater in the second period than in the first period. The simulation of LULC changes and the prediction of future LULC maps were carried based on the changes in the course of 1987–2002. Moreover, the decreasing or increasing trend of each LULC in future maps was repeated.

Simulation results of SWAT

The validation and calibration of the SWAT model were conducted with the SUFI-2 algorithm at five gauging stations in Kashafrood Basin. The goodness-of-fit between the simulated and observed flows was calculated (Table 2). The average R^2

Table 2 | Goodness-of-fit statistics at five gauging stations (Afshar *et al.* 2018)

	Calibration results				Validation results			
Flow station	R^2	NSE	*P*-factor	*r*-factor	R^2	NSE	*P*-factor	*r*-factor
Sar Asiab Shandiz	0.72	0.71	0.57	1.1	0.65	0.64	0.56	0.96
Zire Band Golestan	0.66	0.65	0.58	1.29	0.84	0.83	0.65	0.92
Golestan Jaghargh	0.66	0.64	0.63	0.95	0.74	0.73	0.64	0.81
Hesar Dehbar	0.63	0.60	0.74	0.92	0.81	0.81	0.78	0.75
Kartian	0.68	0.63	0.66	0.90	0.87	0.87	0.67	0.58

values and NSE in both validation and calibration periods indicated that the SWAT model could satisfactorily simulate streamflow dynamics in Kashafrood Basin with an acceptable accuracy (Moriasi *et al.* 2007).

Spatial distribution of BW, GWF and GWS

For water resources management and planning in any region, it is essential to evaluate the spatial distribution of water resources components (BW and GW) affected by climate and LULC changes and subsequently identify the critical areas. To achieve a full understanding of water resources components, spatiotemporal distributions of mean annual temperature and annual precipitation under RCP2.6 and RCP8.5 were prepared (Figures 5 and 6). As shown in Figures 6 and 7, although the temperature will increase in future periods in the basin, it has a variable spatial distribution at the subsurface level. Furthermore, according to the RCP2.6 and RCP8.5 emission scenarios, precipitation will have a different trend in future periods. In a study conducted by Badou *et al.* (2018) on four subbasins of the Beninese part of the Niger River Basin, the effect of climate change on water resources components has been evaluated using three climatic models (HIRHAM5, RCSM and

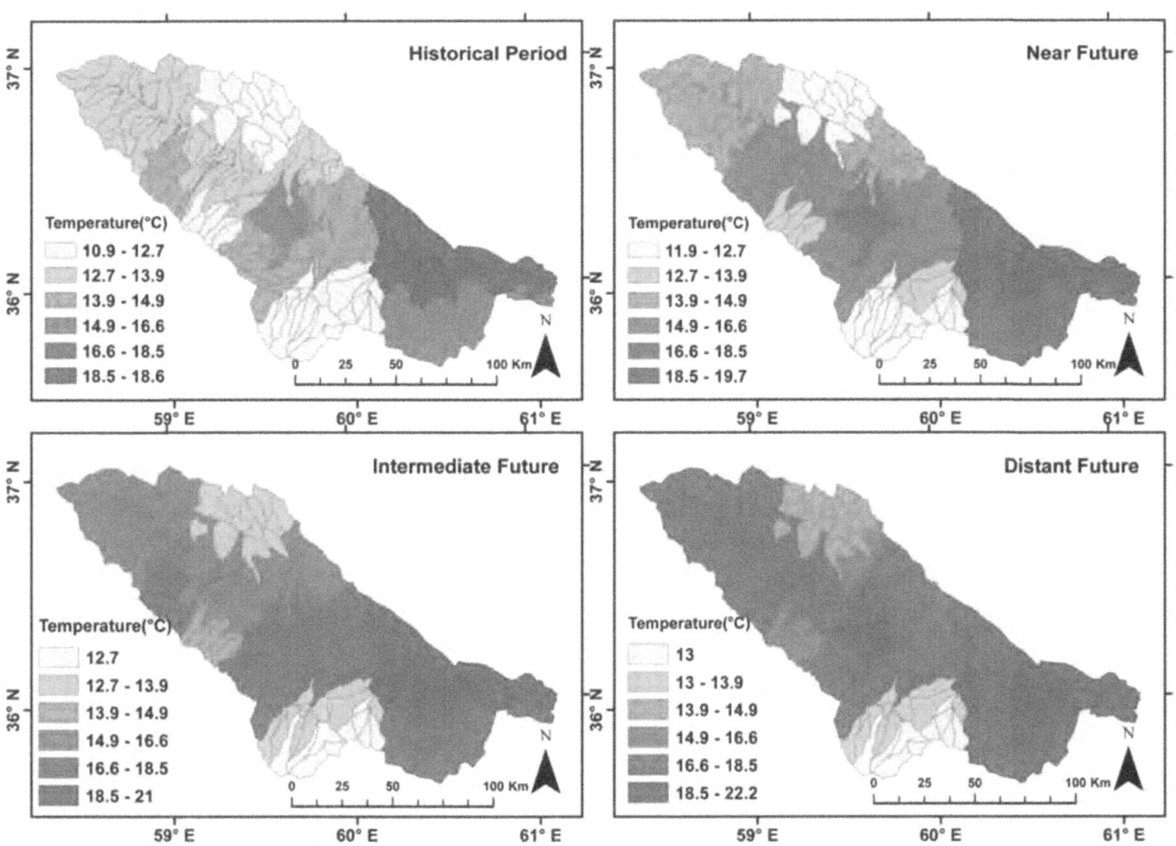

Figure 5 | Spatiotemporal distribution of temperature.

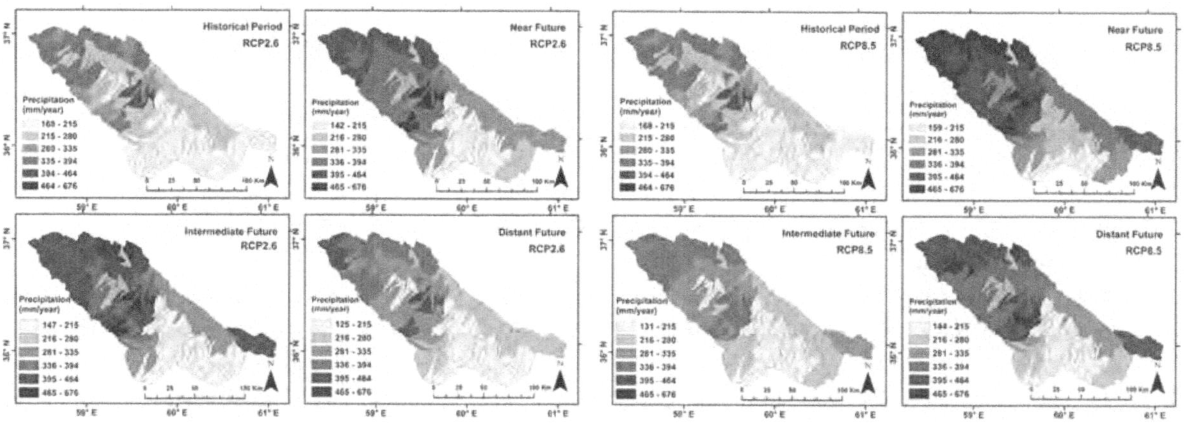

Figure 6 | Spatiotemporal distribution of precipitation under RCP2.6 and RCP8.5.

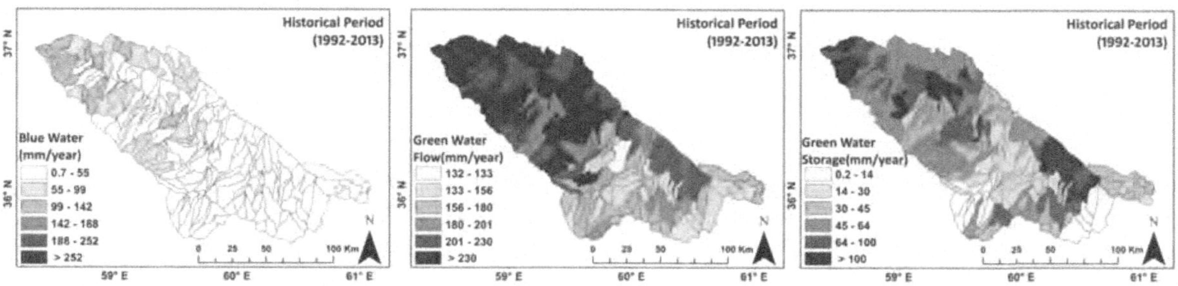

Figure 7 | Spatial distribution of the water resources components in the historical period.

RCA4). A comparison of climate variables between historical and future periods suggests that rainfall will increase for HIRHAM5 and RCSM but shows mixed trends for RCA4. The mean temperature will also increase for HIRHAM5 and RCSM but decrease for RCA4.

As previously mentioned, four time periods were considered to evaluate the water resources components (historical period, near future, intermediate future and distant future), and two RCPs (RCP2.6 and RCP8.5) were applied to illustrate the climate change impacts in the future periods. Figure 7 displays the spatial distribution of the water resources components in the historical period.

According to Figure 7, in the historical period, the BW content was less than 55 mm/year in all the subbasins, except for the northern subbasins and some central and western subbasins. The maximum BW content was observed at the northwestern subbasins, where Kashafrood River enters the basin, while the minimum BW content was recorded at the downstream subbasins in the south. The highest and the lowest levels of BW were 126 and 0.7 mm/year, respectively. Due to the precipitation pattern in the historical period, high precipitation in the northern and northwestern subbasins justifies the high BW content in the subbasins. The dominant forest land cover in some northern basins and the soil texture with higher sand percentages in the northern and northwestern subbasins are the factors that contribute to the increased BW content. GWF content, which represents the actual evapotranspiration, is greater in the north and northwest of the basin than in the southern half of the basin. The maximum and minimum values of GWF in the basin were 323 and 132 mm/year. Since the GWF content directly depends on vegetation, GWF levels were higher in the northern half of the basin, particularly in the areas covered with agricultural lands, pastures and forests, compared to the southern half of the basin. In the southern subbasin, due to less precipitation and lack of water for agriculture, the GWF level is lower. Therefore, the GWF level was less than 180 mm/year in the subbasins with a rain-fed land cover. GWS represents the soil moisture content and depends on the soil type and soil texture. As shown in Figure 7, the GWS content was higher in three parts of the basin, with the values greater than 100 mm/year. The high values of GWS in the relevant subbasins are justified by the soil texture (mainly with higher clay and silt percentages). However, the amount of GWS reached the lowest level in the southwestern part of the basin.

The maximum and minimum amounts of GWS in the subbasins were 164 and 0.3 mm/year. Generally, the mean values of GWF, BW and GWS in the basin were, 204, 39.4 and 54.2 mm/year, respectively. Precipitation rate, as the most important factor in the whole basin, was 254.8 mm/year.

Quantification of future water resource components

The RCP2.6 and RCP8.5 scenarios were employed to estimate the effect of climate change on water resource components. Figures 8 and 9 illustrate the spatial distributions of GWF, BW and GWS in the near, intermediate and distant future under the RCP2.6 and RCP8.5 scenarios.

Referring to Figure 8, under the RCP2.6 scenario, the amount of BW would increase in the majority of the subbasins in the future rather than in the historical period. An increase of BW content would be greater in the northeastern and southeastern subbasins than in other subbasins. The main factor affecting the BW amount is precipitation. Moreover, the LULC conversion of agriculture to pasture (likely poor pasture) at the northeastern part and the conversion of rain-fed land use to poor pasture and barren lands at the southeastern part of the basin reduce infiltration and increase runoff, which in turn increases the BW content. The GWF content would generally decline in most subbasins in all three near future, intermediate future and distant future periods compared to the historical period. Subbasins in the southern part oriented toward the center of Kashafrood Basin show an intense GWF level decrease, which can be due to the agricultural LULC conversion to other LULC classes in the subbasins. The GWS level would be generally decreased in the majority of the subbasins in the three future periods. Such a decrease, however, would not occur in the subbasins with high silt content and high storage capacity. The severity of GWS decline, especially in the northern subbasins, appears to be lower in the intermediate future period than in other future periods. The mean annual levels of water components under the RCP2.6 scenario along with precipitation values in the whole basin discovered during different time periods are illustrated in Table 3.

According to Table 3, under the RCP2.6 scenario, the amount of BW in the whole basin would be higher in the near future rather than in the historical period. The BW amount would also be greater in the intermediate future than in the historical

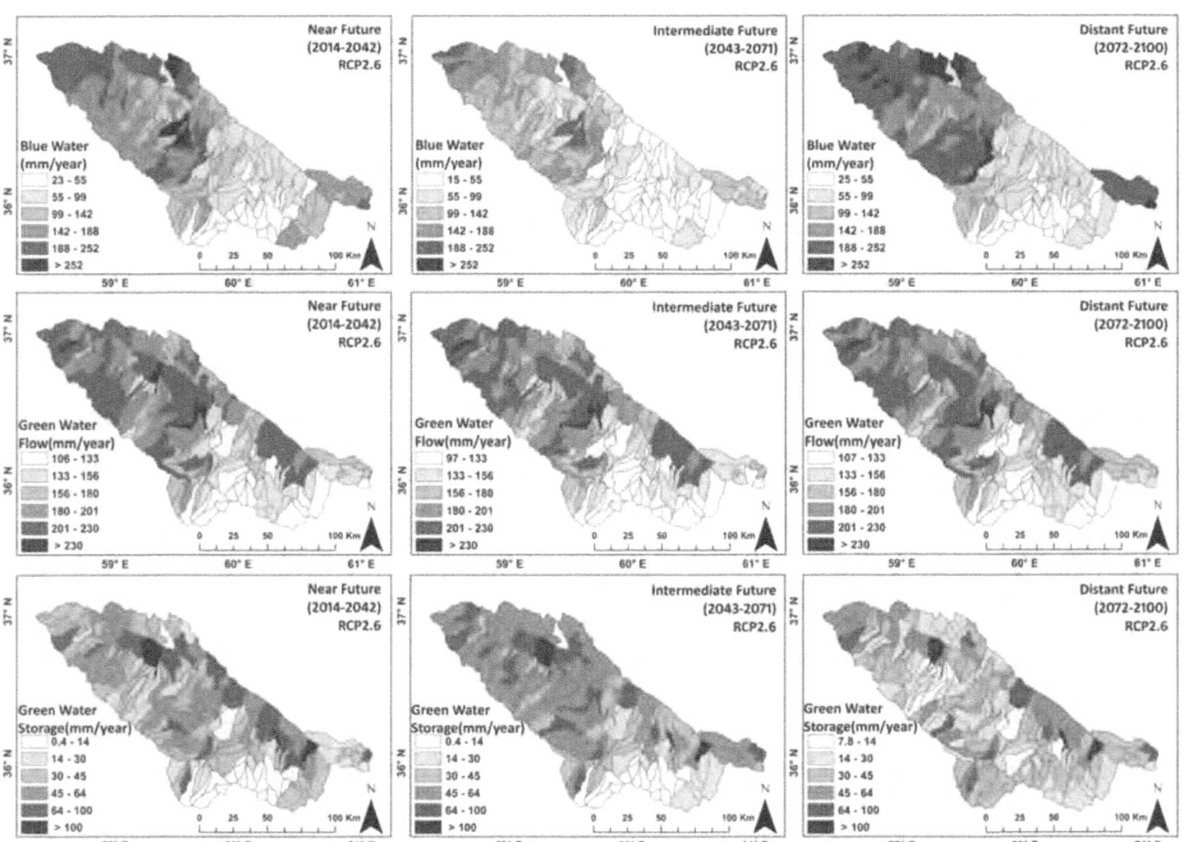

Figure 8 | Spatial distribution of water resources in the future under the RCP2.6 scenario.

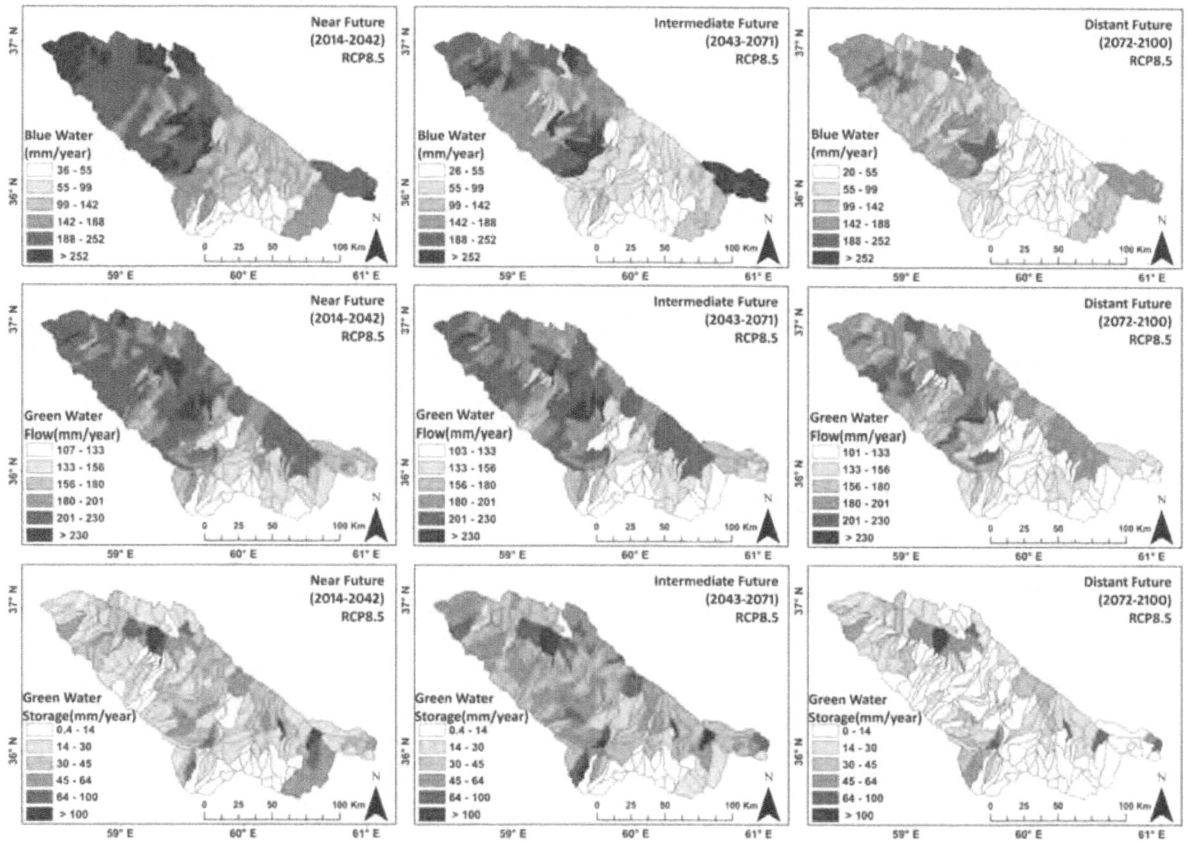

Figure 9 | Spatial distribution of water resources in the future under the RCP8.5 emission scenario.

Table 3 | Levels of water resources components in different time periods under the RCP2.6 emission scenario

Water resource (mm/year)	Historical period (1992–2013)	Near future (2014–2042)	Intermediate future (2043–2071)	Distant future (2072–2100)	CV in time periods
BW	39.4	136.2	95.5	146.3	0.46
GWF	204	176.8	170.7	175.7	0.08
GWS	54.2	41.6	47.4	36.4	0.17
PR	254.8	322	274.1	331.3	0.12

BW, blue water; GWF, green water flow; GWS, green water storage; PR, precipitation; CV, coefficients of variation.

period but lower than the BW amount in the near future. The amount of BW would be far greater in the distant future than in all the preceding periods. The GWF amount would be lower in the near future than in the historical period. The amount of GWF would be lower in the intermediate future than in the historical period and the near future. However, the GWF amount would be lower in the distant future than in the historical and near future periods but higher than the GWF amount in the intermediate future. The GWS amount would be lower in the near future than in the historical period. However, the GWS amount would be lower in the intermediate future than in the historical period and higher in the near future. The amount of GWS would be lower in the distant future than in the other periods.

Figure 9 shows the increase of BW amount in all the future periods and the decrease of BW intensity over time periods under the RCP8.5 scenario. Due to the higher rate of precipitation in the northern half of the basin, the BW amount is higher in the northern subbasins than in the southern subbasins. Also, the conversion of agricultural and rain-fed LULCs to poor pasture LULC contributes to the increase of BW content in the southeastern subbasins. In all the future periods, the GWF content would decline and this decline would be more severe in the southern subbasins due to the decrease of irrigated agricultural lands. The GWS content also declines in all the future periods. This decline is intensified in the southern subbasins and a few northeastern subbasins.

According to Table 4, under the RCP8.5 scenario, the amount of BW in the whole basin would be higher in the near future than in the historical period. The BW amount would also be greater in the intermediate future than in the historical period but lower than the BW amount in the near future. The BW amount would be higher in the distant future than in the historical period but lower than the BW amount in the near and intermediate future periods. The amount of GWF would slightly decrease under both the RCP2.6 and RCP8.5 scenarios over all the future periods. The GWS content would be lower in the near future than in the historical period. The GWS content would also be lower in the intermediate future than in the historical period but higher than the GWS content in the near future. The GWS would be at its lowest level in the distant future, compared to the other periods. As shown in Tables 3 and 4, the coefficients of variation of BW under the RCP2.6 and RCP8.5 scenarios over the four time periods are 46 and 52%, respectively. These coefficients indicate the high intensity of BW changes in both emission scenarios over the time periods. The coefficients of variation of GWF under the RCP2.6 and RCP8.5 scenarios are 8 and 9%, respectively, showing the small change of GWF over the time periods. The coefficients of variation of GWS are 17 and 41%, respectively, indicating that the GWS changes are more severe in the RCP8.5 scenario than in the RCP2.6 scenario. Figure 10 shows the estimation of water resource components under the RCP2.6 and RCP8.5 emission scenarios. According to Figure 10, the BW amount estimated under the RCP8.5 scenario is significantly higher than the BW amount estimated under the RCP2.6 scenario for the near future and intermediate future periods. However, the BW amount estimated under the RCP8.5 scenario is lower than the BW amount estimated under RCP2.6 scenario for the distant future. There is no significant difference between the GWF estimates in the RCP2.6 and RCP8.5 scenarios. However, the RCP8.5 scenario shows a higher estimate in the near and intermediate future periods and a lower estimate in the distant future. Estimates of GWS level are greater in the RCP2.6 scenario than in the RCP8.5 scenario for all the future periods.

In a similar study conducted by Afshar *et al.* (2018) in Kashafrood Basin, they investigated only the effect of climate change on water resources components in the future periods using MIROC-ESM data. However, in this study, the effects of climate and LULC changes have been considered simultaneously. A comparison of the results obtained considering the effect of climate change along with the results obtained considering the effects of both climate and LULC changes has been presented in

Table 4 | Water resources components over time periods under the RCP8.5 emission scenario

Water resources (mm/year)	Historical period (1992–2013)	Near future (2014–2042)	Intermediate future (2043–2071)	Distant future (2072–2100)	CV over time periods
BW	39.4	177.5	144.2	98.7	0.52
GWF	204	186.3	180	161.6	0.09
GWS	54.2	30.4	42.5	19	0.41
PR	254.8	375.1	331.8	265.5	0.18

BW, blue water; GWF, green water flow; GWS, green water storage; PR, precipitation; CV, coefficients of variation.

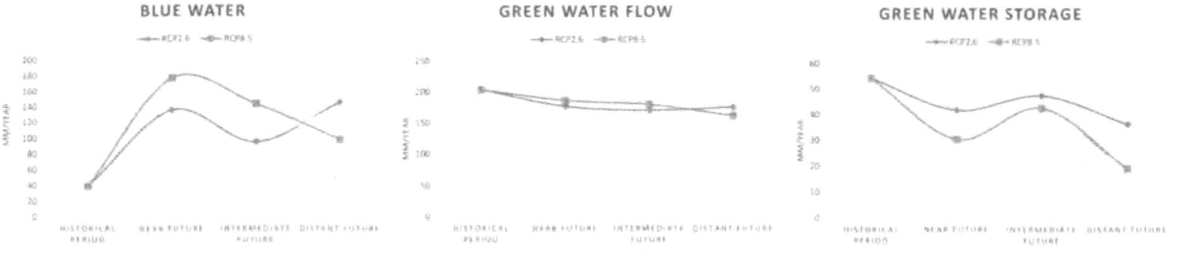

Figure 10 | Comparison of water resources components estimated under the RCP2.6 and RCP8.5 scenarios.

Table 5 | Percentage of difference compared to situation with only climate change effects

Water resource	Difference under RCP2.6 (%)			Difference under RCP8.5 (%)		
	2014–2042	2043–2071	2072–2100	2014–2042	2043–2071	2072–2100
BW	25.8	35.9	18.1	20.3	21.4	30.8
GWF	−9.4	−9.1	−8.8	−9.7	−9.4	−8.0
GWS	−29.6	−25.3	−25.7	−32.6	−33.3	−41.4

Table 5. Obviously, the BW levels are lower and the GWF and GWS values are higher when only the effect of climate change on water resources is considered.

The assessment of the spatiotemporal trend of water resources components in the Ohio River Basin by Du *et al.* (2018) shows that despite the overall volumetric increase of both BW and GW in the entire basin, changes in their annual average values follow a distinctive spatial pattern. Furthermore, climate change in the Ohio River Basin is identified as being influential on BW, whereas land-use change increases GW remarkably, but is counterproductive on BW. In another study on the Weyb River basin in Ethiopia, Serur (2020) investigated the effects of climate change on water resources components. Results revealed a rise of BW and GW in the entire basin and in all the subbasins under different emission scenarios. Also at the subbasin level, the spatial variations of BW and GW-Availability are very high as compared to that of the entire basin analysis under all RCP scenarios. Xu & Wu (2018) estimated the County-Based GW Availability in the United States from effective rain using three different methods: Smith, U.S. Department of Agriculture – SCS (USDA-SCS) and the NHD plus V2 dataset. Their analysis illustrated that the fraction of GW resources availability varies significantly across regions. The water availability index for GW also depends on the precipitation pattern, crop type and spatially differentiated regions. However, the present study shows that the most important factors for BW are precipitation and LULC, while for GW, they are LULC and temperature.

GW-Security

GW-Security was evaluated by calculating GW-Scarcity and GW-Vulnerability. The impact of climate and LULC changes on GW-Security was assessed over four time periods (i.e., one historical and three future periods). Figure 11 displays the spatial distribution of GW-Vulnerability and GW-Scarcity in the historical period. Some central and southern subbasins in the study area faced high levels of GW-Scarcity (Figure 11). Higher temperature and agricultural LULC in the central subbasins induce higher evapotranspiration and consequently cause higher GW-Scarcity. Moreover, the soil texture with higher sand percentage in some of the southwestern subbasins is responsible for lower water availability and thereby higher GW-Scarcity. In the northern subbasins, despite the greater extent of agricultural LULC, GW-Scarcity is lower due to the higher altitude, lower temperatures and higher precipitation. The pattern of GW-Vulnerability is similar to that of GW-Scarcity. However, the vulnerability is more severe in the southeastern basins that have higher temperatures.

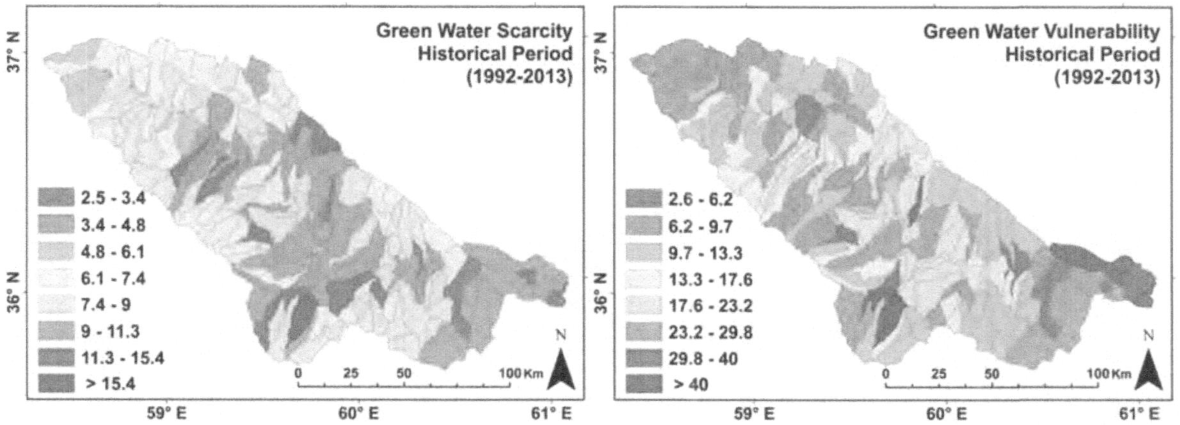

Figure 11 | Spatial distribution of GW-Security in the historical period.

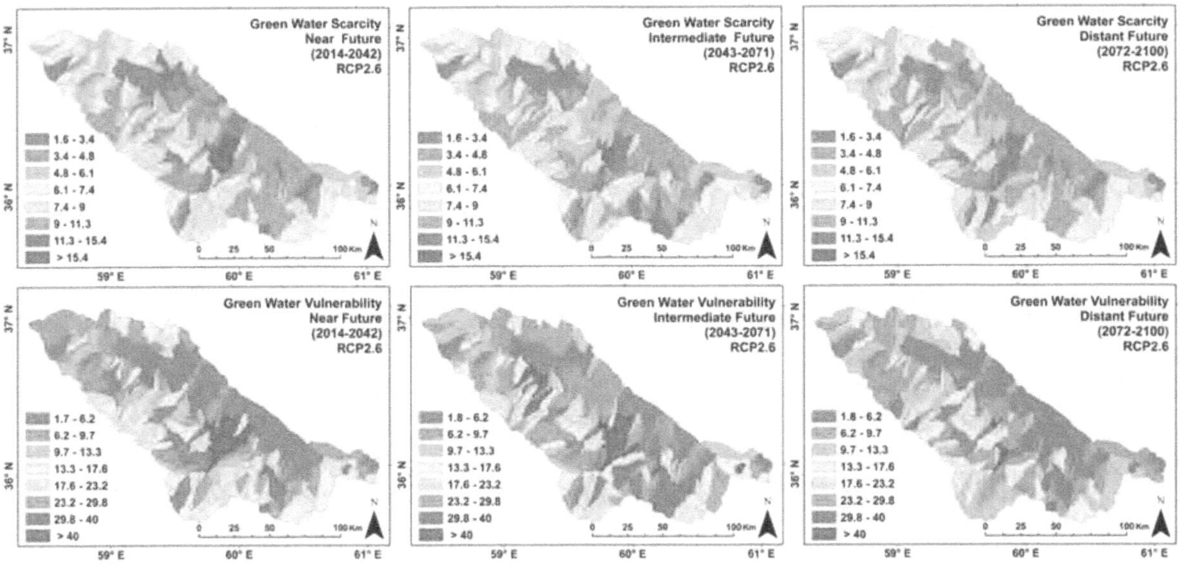

Figure 12 | Spatial distribution of GW-Security in the future under RCP2.6 emission scenario.

Spatial distributions of GW-Vulnerability and GW-Scarcity under RCP2.6 scenario are shown in Figure 12. As can be seen in Figure 12, in most of the subbasins along the northeast to southeast direction and some southwestern subbasins, the GW-Scarcity is lower in the three future periods than in the historical period. An increase of precipitation and LULC conversion to pasture in the future would reduce GWF and increase GW-Availability in the mentioned subbasins. This can consequently lead to less GW-Scarcity. However, in the southern subbasins, GW-Scarcity would be intensified in the future due to lower precipitation and higher temperatures. Another factor is LULC change from forest to pasture, which reduces GWS and GW-Availability. Some central subbasins, with a higher rate of sand in their soil texture, have a higher GW-Scarcity, which is expected to be intensified in the near future. Vulnerability has a similar pattern in the future. Generally, the southern and western subbasins and a few central subbasins would be more vulnerable in the future, with a more severe vulnerability in the intermediate future. Spatial distributions of GW-Scarcity and GW-Vulnerability under the RCP8.5 scenario are shown in Figure 13. As can be seen in Figure 13, the GW-Scarcity in the majority of the subbasins would be lower in the near future

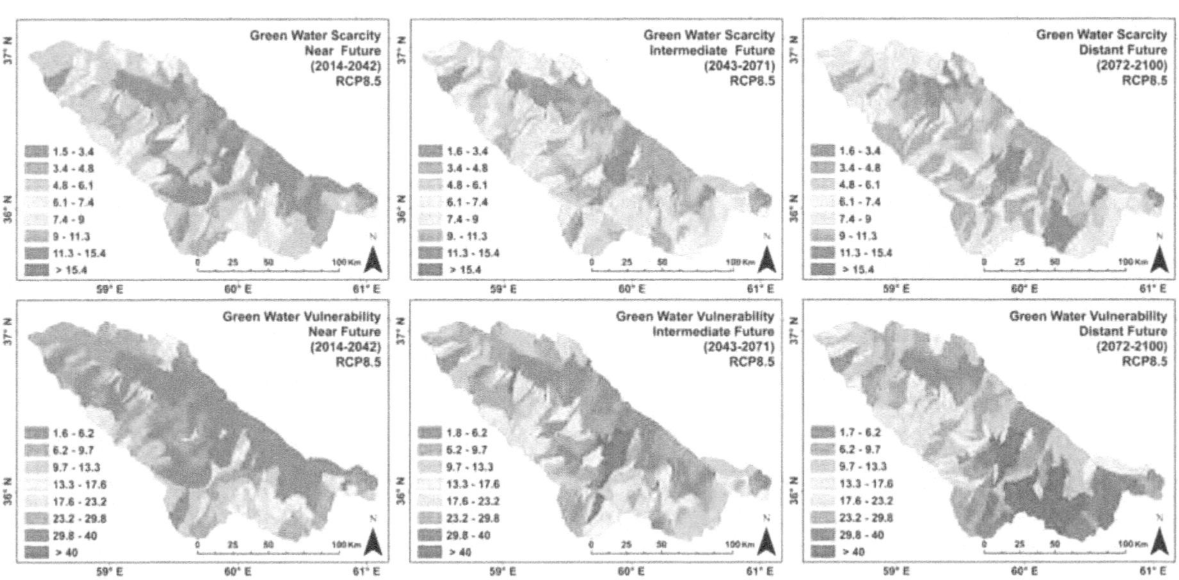

Figure 13 | Spatial distribution of GW-Security in the future under the RCP8.5 emission scenario.

Table 6 | Water security indicators over time periods in Kashafrood Basin

Water security indicator	Historical period 1992–2013	RCP2.6			RCP8.5		
		2014–2042	2043–2071	2072–2100	2014–2042	2043–2071	2072–2100
Scarcity	7.94	5.30	5.57	5.34	4.31	5.24	6.02
Vulnerability	11.23	6.90	7.63	6.77	5.38	7.08	9.43

than in the historical period. While in the intermediate and distant future periods, the GW-Scarcity would increase along the northwest to the southwest direction and in the southern subbasins, with an intensified value in the distant future. GW-Vulnerability in the majority of the subbasins would be lower in the near future than in other periods. However, in the intermediate future, the GW-Vulnerability would be greater in the southern and western subbasins and a few central subbasins compared to other subbasins. Moreover, a greater GW-Vulnerability would be seen in the southern basins in the distant future. As shown in Table 6, in both emission scenarios, water scarcity and vulnerability in the entire basin would be reduced in the future compared to the historical period.

In the Savannah Basin, Veettil & Mishra (2018) concluded that GW was more sensitive to changes in land-use pattern. In addition, the magnitude of various water security indicators is different within each county, suggesting that water scarcity is controlled by various factors within a region. The present study also shows that LULC plays an important role in the values of GW-Security indicators.

CONCLUSION

The SWAT model was employed to calculate the values of water resources components in Kashafrood Basin in the future. The climatic and LULC data for the future were prepared using the MIROC-ESM climatic model and the LCM model, respectively. The MIROC-ESM climatic model data downscaled by the BCSD method was applied to assess the effect of climate change on water resources in the basin under the RCP2.6 and RCP8.5 scenarios. After preparing the historical land-use maps using remote sensing, the LCM model was applied to project the future land-use maps focusing on urban growth. Moreover, the SWAT model was validated and calibrated to evaluate the effects of climate and LULC changes on water resource components. Finally, GW-Security in Kashafrood Basin was assessed using the GW-Vulnerability and GW-Scarcity indicators. According to the results, the main LULC changes in the basin in the future are agricultural, urban and pasture LULCs. Moreover, agricultural lands would be decreased, and urban areas and pasture would be increased in the future. The factors, including population growth, depletion of water resources and deforestation, were found to intensify the changes. Precipitation, as a climate variable, and LULC are the two factors affecting the BW amount in the basin. Compared to the historical period, the BW amount is estimated to be increased in the future, especially in the northern subbasins and in the near future. The GWF would be slightly lower in the future than in the historical period. GWS, which has a smaller share in water resources, will generally decline in the future. The increase of temperature and the reduction of soil permeability due to LULC changes are the most important reasons for the GWS decline. Generally, conditions of the basin would be improved in terms of GW-Scarcity and GW-Vulnerability over the future periods rather than the historical period. Given that vulnerability represents the worst condition of water availability, it was easier to identify the hotspots with higher risks through the spatial distribution of GW-Vulnerability. These hotspots can help the policymakers to adopt the best decisions for improving water conditions in vulnerable subbasins. However, it is suggested to apply a combined transition matrix from two periods in LCM to improve the LULC change trends, and employ other observational data such as evapotranspiration, soil moisture and deep recharge to calibrate the SWAT model in order to obtain accurate GW estimates in further studies. Since soil erosion is inevitable, changes in soil characteristics could be considered in SWAT modeling, and BW security could be evaluated. These constraints should be addressed in future works.

DATA AVAILABILITY STATEMENT

All relevant data are included in the paper or its Supplementary Information.

REFERENCES

Abbaspour, K. C. 2011 *SWAT-CUP4: SWAT Calibration and Uncertainty Programs – A User Manual*. Swiss Federal Institute of Aquatic Science and Technology, Eawag.

Abbaspour, K. C., Johnson, C. & Van Genuchten, M. T. 2004 Estimating uncertain flow and transport parameters using a sequential uncertainty fitting procedure. *Vadose Zone Journal* **3** (4), 1340–1352.

Afshar, A. A., Hasanzadeh, Y., Besalatpour, A. & Pourreza-Bilondi, M. 2017 Climate change forecasting in a mountainous data scarce watershed using CMIP5 models under representative concentration pathways. *Theoretical and Applied Climatology* **129** (1–2), 683–699.

Afshar, A. A., Hassanzadeh, Y., Pourreza-Bilondi, M. & Ahmadi, A. 2018 Analyzing long-term spatial variability of blue and green water footprints in a semi-arid mountainous basin with MIROC-ESM model (case study: Kashafrood River Basin, Iran). *Theoretical and Applied Climatology* **134** (3–4), 885–899.

Anand, J., Gosain, A. K. & Khosa, R. 2018 Prediction of land use changes based on land change modeler and attribution of changes in the water balance of Ganga basin to land use change using the SWAT model. *Science of the Total Environment* **644**, 503–519.

Armenteras, D., Murcia, U., González, T. M., Barón, O. J. & Arias, J. E. 2019 Scenarios of land use and land cover change for NW Amazonia: impact on forest intactness. *Global Ecology and Conservation* **17**, e00567.

Arnold, J. G., Moriasi, D. N., Gassman, P. W., Abbaspour, K. C., White, M. J., Srinivasan, R., Santhi, C., Harmel, R., Van Griensven, A. & Van Liew, M. W. 2012 SWAT: model use, calibration, and validation. *Transactions of the ASABE* **55** (4), 1491–1508.

Badou, D. F., Diekkrüger, B., Kapangaziwiri, E., Mbaye, M. L., Yira, Y., Lawin, E. A., Oyerinde, G. T. & Afouda, A. 2018 Modelling blue and green water availability under climate change in the Beninese Basin of the Niger River Basin, West Africa. *Hydrological Processes* **32** (16), 2526–2542.

Briassoulis, H. 2019 *Analysis of Land Use Change: Theoretical and Modeling Approaches*.

Connor, R. 2015 *The United Nations World Water Development Report 2015: Water for a Sustainable World, 1*. UNESCO Publishing.

Cook, C. & Bakker, K. 2012 Water security: debating an emerging paradigm. *Global Environmental Change* **22** (1), 94–102.

Du, L., Rajib, A. & Merwade, V. 2018 Large scale spatially explicit modeling of blue and green water dynamics in a temperate mid-latitude basin. *Journal of Hydrology* **562**, 84–102.

Falkenmark, M. & Rockström, J. 2006 The new blue and green water paradigm: Breaking new ground for water resources planning and management *Journal of Water Resources Planning and Management* **132**, 129–132.

Faramarzi, M. 2010 *Assessment of Regional Water Endowments, Crop Water Productivity, and Implications for Intra-Country Virtual Water Trade in Iran*. ETH Zurich, Zurich, Switzerland.

Gosain, A., Rao, S. & Arora, A. 2011 Climate change impact assessment of water resources of India. *Current Science* **101** (3), 356–371.

Hoekstra, A. Y. 2003 Virtual water trade: a quantification of virtual water flows between nations in relation to international crop trade. In: *Proceedings of the International Expert Meeting on Virtual Water Trade 12*, 2003, Delft, pp. 25–47.

Hoekstra, A. Y., Chapagain, A. K., Mekonnen, M. M. & Aldaya, M. M. 2011 *The Water Footprint Assessment Manual: Setting the Global Standard*. Routledge, Abingdon, UK.

Jafarzadeh, A., Pourreza-Bilondi, M., Afshar, A. A., Khashei-Siuki, A. & Yaghoobzadeh, M. 2019 Estimating the reliability of a rainwater catchment system using the output data of general circulation models for the future period (case study: Birjand City, Iran). *Theoretical and Applied Climatology* **137** (3–4), 1975–1986.

Jia, K., Ruan, Y., Yang, Y. & Zhang, C. 2019 Assessing the performance of CMIP5 global climate models for simulating future precipitation change in the Tibetan Plateau. *Water* **11** (9), 1771.

Karandish, F. & Hoekstra, A. 2017 Informing national food and water security policy through water footprint assessment: the case of Iran. *Water* **9** (11), 831.

Liu, J., Zehnder, A. J. & Yang, H. 2009 Global consumptive water use for crop production: the importance of green water and virtual water. *Water Resources Research* **45** (5), W05428.

Lyu, L., Wang, X., Sun, C., Ren, T. & Zheng, D. 2019 Quantifying the effect of land use change and climate variability on Green water resources in the Xihe River Basin, Northeast China. *Sustainability* **11** (2), 338.

Marttunen, M., Mustajoki, J., Sojamo, S., Ahopelto, L. & Keskinen, M. 2019 A framework for assessing water security and the water–energy–food nexus – the case of Finland. *Sustainability* **11** (10), 2900.

Moriasi, D. N., Arnold, J. G., Van Liew, M. W., Bingner, R. L., Harmel, R. D. & Veith, T. L. 2007 Model evaluation guidelines for systematic quantification of accuracy in watershed simulations. *Transactions of the ASABE* **50** (3), 885–900.

Padowski, J. C. & Jawitz, J. W. 2012 Water availability and vulnerability of 225 large cities in the United States. *Water Resources Research* **48** (12), W12529.

Pan, S., Liu, D., Wang, Z., Zhao, Q., Zou, H., Hou, Y., Liu, P. & Xiong, L. 2017 Runoff responses to climate and land use/cover changes under future scenarios. *Water* **9** (7), 475.

Pickard, B., Gray, J. & Meentemeyer, R. 2017 Comparing quantity, allocation and configuration accuracy of multiple land change models. *Land* **6** (3), 52.

Rodrigues, D. B., Gupta, H. V. & Mendiondo, E. M. 2014 A blue/green water-based accounting framework for assessment of water security. *Water Resources Research* **50** (9), 7187–7205.

Saifullah, K., Barus, B. & Rustiadi, E. 2017 Spatial modelling of land use/cover change (LUCC) in South Tangerang City, Banten. In: *IOP Conference Series: Earth and Environmental Science*. IOP Publishing, p. 012018.

Sayari, N., Bannayan, M., Alizadeh, A. & Farid, A. 2013 Using drought indices to assess climate change impacts on drought conditions in the northeast of Iran (case study: Kashafrood basin). *Meteorological Applications* **20** (1), 115–127.

Schyns, J. F., Hoekstra, A. Y., Booij, M. J., Hogeboom, R. J. & Mekonnen, M. M. 2019 Limits to the world's green water resources for food, feed, fiber, timber, and bioenergy. *Proceedings of the National Academy of Sciences* **116** (11), 4893–4898.

Serur, A. B. 2020 Modeling blue and green water resources availability at the basin and sub-basin level under changing climate in the Weyb River basin in Ethiopia. *Scientific African* **7**, e00299.

Shao, G., Guan, Y., Zhang, D., Yu, B. & Zhu, J. 2018 The impacts of climate variability and land use change on streamflow in the Hailiutu river basin. *Water* **10** (6), 814.

Tan, M. L., Yusop, Z., Chua, V. P. & Chan, N. W. 2017 Climate change impacts under CMIP5 RCP scenarios on water resources of the Kelantan River Basin, Malaysia. *Atmospheric Research* **189**, 1–10.

Van Vuuren, D. P., Stehfest, E., Den Elzen, M. G., Van Vliet, J. & Isaac, M. 2010 Exploring IMAGE model scenarios that keep greenhouse gas radiative forcing below 3 W/m^2 in 2100. *Energy Economics* **32** (5), 1105–1120.

Veettil, A. V. & Mishra, A. K. 2016 Water security assessment using blue and green water footprint concepts. *Journal of Hydrology* **542**, 589–602.

Veettil, A. V. & Mishra, A. K. 2018 Potential influence of climate and anthropogenic variables on water security using blue and green water scarcity, Falkenmark index, and freshwater provision indicator. *Journal of Environmental Management* **228**, 346–362.

Vörösmarty, C. J., McIntyre, P. B., Gessner, M. O., Dudgeon, D., Prusevich, A., Green, P., Glidden, S., Bunn, S. E., Sullivan, C. A. & Liermann, C. R. 2010 Global threats to human water security and river biodiversity. *Nature* **467** (7315), 555–561.

UN Water 2013 *Water Security and the Global Water Agenda: A UN-Water Analytical Brief*. United Nation University, Ontario, Canada.

WWAP, U. 2012 *World Water Assessment Programme: The United Nations World Water Development Report 4: Managing Water under Uncertainty and Risk*. UNESCO, Paris.

Xu, H. & Wu, M. 2018 A first estimation of county-based green water availability and its implications for agriculture and bioenergy production in the United States. *Water* **10** (2), 148.

Zang, C. & Liu, J. 2013 Trend analysis for the flows of green and blue water in the Heihe River basin, northwestern China. *Journal of Hydrology* **502**, 27–36.

Zhu, K., Xie, Z., Zhao, Y., Lu, F., Song, X., Li, L. & Song, X. 2018 The assessment of green water based on the SWAT model: a case study in the Hai River Basin, China. *Water* **10** (6), 798.

First received 20 February 2021; accepted in revised form 8 May 2021. Available online 10 June 2021

© 2022 The Authors Impact of Climate Change on Hydrology and Water Resources doi: 10.2166/wcc.2021.180

Modeling the effects of land-use and climate change on the performance of stormwater sewer system using SWMM simulation: case study

Salam Naje Hussain[a], Haider M. Zwain[b,*] and Basim K. Nile[a]
[a] College of Engineering, University of Kerbala, Kerbala 56100, Iraq
[b] College of Water Resources Engineering, Al-Qasim Green University, 51013 Al-Qasim Province, Babylon, Iraq
*Correspending author. E-mail: haider.zwain@wrec.uoqasim.edu.iq

HMZ, 0000-0003-4995-2697

ABSTRACT

Flooding of stormwater drainage systems represents a major problem in developing urban areas that could be influenced by land-use and climate change. Flooding problems can be assessed using simulation models such as the stormwater management model (SWMM). In this study, the generation of intensity duration frequency curves (IDF) that integrates climate change effect was conducted for Al-Najaf Governorate in Iraq for the first time. In addition, the effects of land-use and climate change on the stormwater sewer system of Al-Ameer District was simulated using SWMM. The results indicated that by increasing the sub-catchment area from 50 to 100%, an increment in total surface runoff from 20,380 to 37,350 m^3, and total flooding from 10,513 to 26,032 m^3 have occurred, respectively. As a response to climate change, changing the return period from 2 to 5 years has increased the total surface runoff from 14,120 to 27,110 m^3 (representing 48% of raise), and total flooding increased from 5,914 to 17,591 m^3 (accounting 72% of increment). To conclude, flooding locations and magnitude were identified, whilst the system failed to discharge surface runoff at critical conditions, whereas the effect of climate change on the stormwater drainage system was more adverse than the effect of land-use.

Key words: Al-Najaf, climate change, flooding, rainfall intensity, stormwater drainage system, SWMM simulation

HIGHLIGHTS

- An IDF curve was generated for the first time for Al-Najaf Governorate.
- SWMM successfully simulated the stormwater drainage system in the study area.
- The system will not withstand climate change impacts in the next 10 years.
- Change in land-use has increased impervious surfaces, causing floods in downstream.
- Flooding locations and magnitude, and number of flooded manholes were identified.

1. INTRODUCTION

Urban flooding presents a major threat to population, infrastructure, properties, and the environment. In recent years, urbanization is changing land-use, increasing surface runoff generation, and causing urban flooding. Furthermore, extreme rainfall events due to climate change have increased in severity and frequency worldwide (Nile *et al.* 2018). In addition, urban flooding occurs when stormwater drainage systems are insufficient to discharge surface runoff. This runoff moves over surfaces, stores in low-lying areas within the catchment until it infiltrates, drains or evaporates (Hossain Anni *et al.* 2020). However, infiltration rates in the urban areas depends on the land-use, whether it is impervious or previous cover. In the process of urbanization, large non-urban or agricultural lands are transformed into impervious land, leading to change in land-use that totally alters natural hydrological processes (Hu *et al.* 2020).

Moreover, climate change alters rainfall patterns that can increase runoff, resulting in an increase in sewer-related problems such as regular flooding and contamination of receiving waters (Abdellatif *et al.* 2015). Higher short-term precipitation intensities are anticipated as a result of climate change, which could have negative effects such as increased flooding and sewer overflow risks (Olsson & Foster 2014). Climate change is one of the main challenges facing Iraq. It is likely to have a catastrophic impact on the environment and the economy, especially in urban areas. Iraq is located in an arid to semi-arid region, where the prevailing continental climate is usually cold in winter and hot in summer, as well as fluctuating rainfall in winter periods (Hassan & Nile 2020). Iraq suffers from a number of serious problems, especially in Al-Najaf Governorate. This was evidenced by the rapid expansion of the urbanization process in the city, the increase of impervious areas, and frequent and increasingly intense rainstorms. The unpredictable and unprecedented rainfall intensities in the city and lack of networks maintenance are among the main reasons for floods increase (Zwain *et al.* 2021).

Furthermore, sewer systems are often not proportionally designed to accommodate large surface runoff volumes resulting from urbanization and climate change, thus, there is a great need for rainwater management. Therefore, simulation models may rapidly present indicative results to predict flooding information that helps to protect the population, infrastructure, and property (Babaei *et al.* 2018). In order to predict water flooding and water depth in stormwater sewers and manholes, the stormwater management model (SWMM) has great potential to produce relevant relations between many parameters and the efficiency of the stormwater network. The model is widely used for drainage systems in urban areas for planning, research, and design (Rabori & Ghazavi 2018). In several studies, the SWMM program has been used to simulate the flow on catchment surfaces as well as in sewer systems (Jiang *et al.* 2015). Obaid *et al.* (2014) used SWMM software to model the impacts of climate change and increased population during peak tourism season on the sewer system in Karbala city of Iraq. In Al-Anwar suburb of Al-Kut city, Iraq, Hadia *et al.* (2015) used the SWMM program to assess the performance of the sewer system, whereas they reported that flow velocity was sufficient, but pipe diameters were larger than those needed to discharge the sewage flow. To determine the best return time for designing an urban drainage system in Palermo, Italy, Fortunato *et al.* (2014) used SWMM to evaluate the socioeconomic landscape impacts of floods and future solution scenarios to avoid economic losses during future flooding.

However, efficient simulation requires good hydrological analysis that is a prerequisite for any project. It is used to assess the catchment area and determine the design discharge or the amount of runoff that the stream must be designed to transport. However, there is a huge problem in the availability, accuracy and duration of the hydrological data in Al-Najaf Governorate. For example, there is no IDF curve data for any city in the state. This leads to obvious difficulties in preparing a viable and effective design of sewer networks. As a result, the sewerage system suffers from recurring failures. One of the most severe cases can be observed at Al-Ameer District, where flooding is happening repeatedly. Since the stormwater drainage system is already built, it could be possible to reassess its efficiency in order to modify it or develop it with relevant solutions. These improvements should be based on useful and up to date analysis.

Therefore, this study first aims to provide hydrological information that is not available for Al-Najaf Governorate by predicting the precipitation intensity using the intensity-duration-frequency (IDF) curve for the first time. Second, this study aims to assess the current condition of the stormwater drainage system in Al-Ameer District located in Al-Najaf Governorate of Iraq using SWMM simulation. Third, the effects of land-use characteristics and rainfall intensity change as a response to climate change on the performance of stormwater drainage system were also evaluated. After achieving the above objectives, this study is expected to provide a technical solution for experts, engineering and decision makers in order to identify flood problems and evaluate proposed solutions.

2. MATERIALS AND METHODS

2.1. Study area

Al-Najaf Governorate is located 165 km southwest of Baghdad, Iraq (32° 01′ 33.38″ N and 44° 20′ 46.50″ E). The climate is arid to semi-arid with an average temperature of 24 °C, average annual rainfall of 99 mm/year, evaporation of 3,483 mm/year, average wind speed of 10 km/hr, and humidity of 41% (Zwain *et al.* 2021). As shown in Figure 1, the study area of Al-Ameer District is located close to the center of Al-Najaf Governorate, with longitude and latitude of 32° 00′ 28″ N and 44° 21′ 51″ E. The land is on a slope that ranges from 41.78 to 36.008 m above sea level, the total area is about 1.64 km^2 with 34% pervious surface and 66% impervious surface.

2.2. Rainfall intensity analysis

The relationship between precipitation intensity and frequency (IDF) is one of the most widely used tools in many anti-flood engineering projects. The IDF curves express the relationship between precipitation intensity, duration, and return period. In order to construct the IDF curves in Al-Najaf, an historical series of maximum rainfall intensity is required with a higher time accuracy (within 1 minute time interval) (Nhat *et al.* 2006). The steps for an IDF drawing using Easy Fit 5.6 and Microsoft EXCEL 2016 can be illustrated as follows:

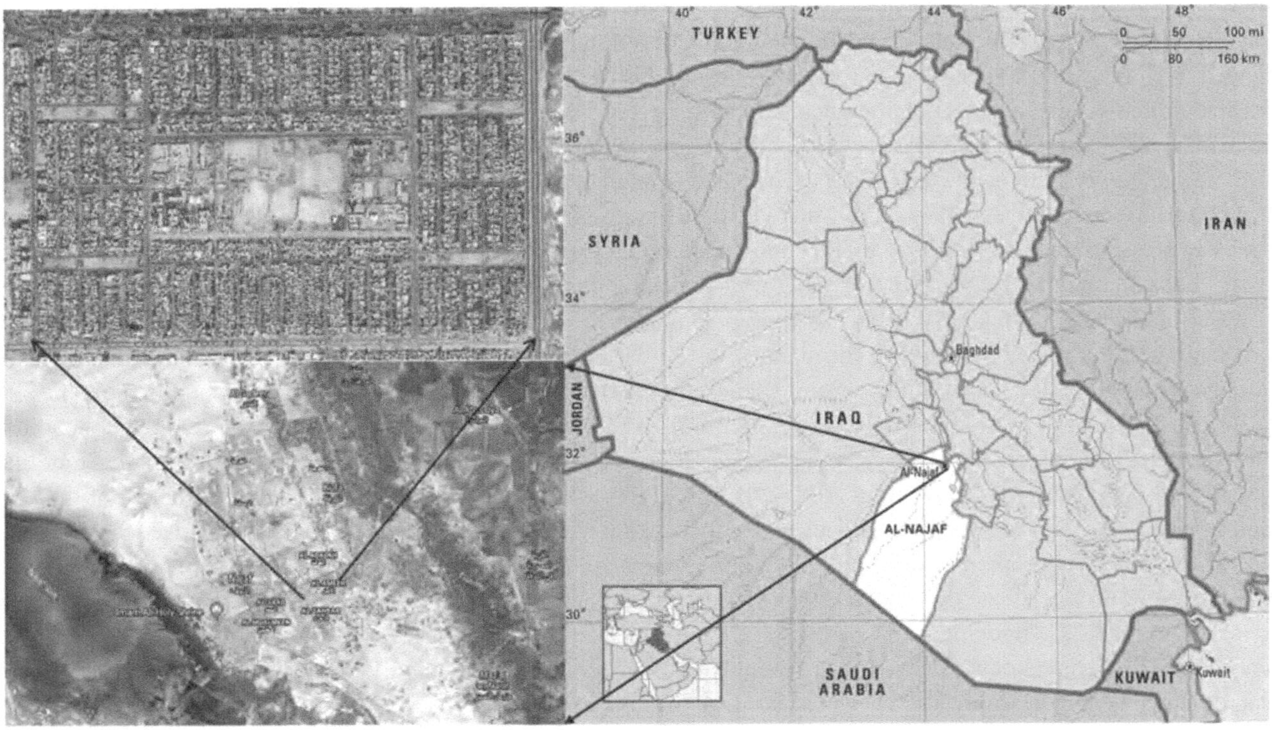

Figure 1 | Case study location of Al-Ameer District in Al-Najaf, Iraq (Google map, 2021).

1. Finding the maximum daily rainfall per year, as shown in Table 1.
2. The maximum precipitation of 5, 10, 20, 30, 60, and 120 minutes are evaluated using the IMD reduction formula (Rathnam *et al.* 2001) as shown below in Equation (1), and calculated data are listed in Table 2.

$$p_t = p_{24} * \left(\frac{t}{24}\right)^{\frac{1}{3}} \tag{1}$$

where:

Pt is required precipitation depth for the specific duration in mm, P_{24} is daily precipitation in mm, and t is the time duration in hours for which precipitation depth is required.

Table 1 | Maximum daily rainfall recorded in Najaf governorate during 1989–2018

Year	Maximum Daily Rainfall (mm)	Year	Maximum Daily Rainfall (mm)	Year	Maximum Daily Rainfall (mm)
1989	5	**1999**	8.2	**2009**	12.6
1990	12	**2000**	10.7	**2010**	10.9
1991	8.8	**2001**	17.3	**2011**	13.8
1992	16.2	**2002**	15.2	**2012**	18.2
1993	34.4	**2003**	19	**2013**	64.5
1994	42	**2004**	8.7	**2014**	22.2
1995	11.5	**2005**	27.7	**2015**	32.9
1996	17.9	**2006**	22.6	**2016**	26.6
1997	21.1	**2007**	12.8	**2017**	8.6
1998	12.4	**2008**	26.8	**2018**	19.3

Table 2 | Maximum daily precipitation depth (mm) for the duration T-minutes

	Precipitation Depth (mm)					
Year	**5 (min)**	**10 (min)**	**20 (min)**	**30 (min)**	**60 (min)**	**120 (min)**
1989	0.76	0.954	1.202	1.376	1.733	2.184
1990	1.82	2.289	2.885	3.302	4.160	5.242
1991	1.33	1.679	2.115	2.421	3.051	3.844
1992	2.45	3.091	3.894	4.458	5.616	7.076
1993	5.21	6.563	8.269	9.466	11.926	15.026
1994	6.36	8.013	10.096	11.557	14.561	18.345
1995	1.74	2.194	2.764	3.164	3.987	5.023
1996	2.71	3.415	4.303	4.925	6.206	7.819
1997	3.2	4.026	5.072	5.806	7.315	9.216
1998	1.88	2.366	2.981	3.412	4.299	5.416
1999	1.24	1.564	1.971	2.256	2.843	3.582
2000	1.62	2.041	2.572	2.944	3.710	4.674
2001	2.62	3.301	4.159	4.760	5.998	7.557
2002	2.30	2.900	3.654	4.182	5.270	6.639
2003	2.88	3.625	4.567	5.228	6.587	8.299
2004	1.32	1.660	2.091	2.394	3.016	3.800
2005	4.2	5.285	6.658	7.622	9.603	12.1
2006	3.42	4.312	5.433	6.219	7.835	9.872
2007	1.938	2.442	3.077	3.522	4.438	5.591
2008	4.058	5.113	6.442	7.374	9.291	11.706
2009	1.908	2.404	3.029	3.467	4.368	5.504
2010	1.651	2.080	2.620	2.999	3.779	4.761
2011	2.090	2.633	3.317	3.797	4.784	6.028
2012	2.756	3.472	4.375	5.008	6.310	7.950
2013	9.767	12.306	15.504	17.748	22.361	28.173
2014	3.362	4.236	5.336	6.109	7.696	9.697
2015	4.982	6.277	7.908	9.053	11.406	14.371
2016	4.028	5.075	6.394	7.319	9.222	11.619
2017	1.302	1.641	2.067	2.366	2.981	3.756
2018	2.923	3.682	4.639	5.311	6.691	8.430

3. Using the software Easy Fit 5.6, data in Table 2, Gumbel Distribution, and selecting the probability of 2, 5, 10 and 25 years, the probability of rain per minute can be obtained, as shown in Table 3.
4. From data in Table 3, an IDF curve for Al-Najaf Governorate is generated.

2.3. SWMM simulation

2.3.1. Model setup

The Storm Water Management Model (SWMM) V.5.1 developed by the US Environmental Protection Agency (USEPA) was operated to assess the performance of stormwater drainage system, quantify flooding, and their locations in Al-Ameer District. SWMM is a dynamic rainfall–runoff simulation model used to simulate runoff quantity and quality from primarily urban watersheds, locate flood areas, test different stormwater management methods, and come up with cost-effective solutions to control stormwater. As illustrated in Figure 2, the whole study area was divided into 43 sub-catchments, a total of 343

Table 3 | The rain intensity is converted to mm/hr

	Rainfall Intensity (mm\hr)					
Return Period (Years)	5 (min)	10 (min)	20 (min)	30 (min)	60 (min)	120 (min)
2	29.55	18.61	11.73	8.95	5.64	3.55
5	47.46	29.90	18.84	14.37	9.06	5.70
10	61.27	38.60	24.32	18.56	11.69	7.36
25	80.79	50.90	32.06	24.47	15.41	9.71

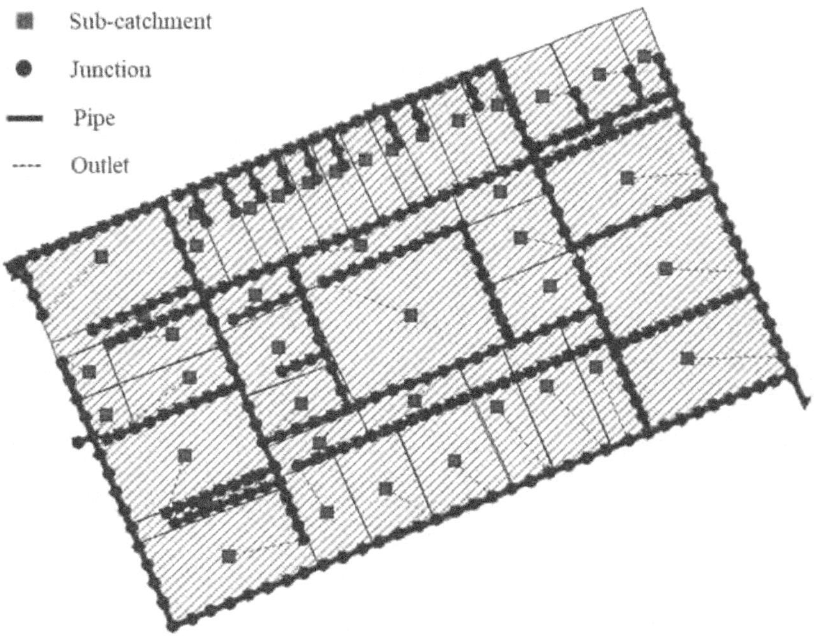

Figure 2 | Distribution of study area sub-catchments, junctions and pipes.

junctions and 343 pipes, and each sub-catchment drains stormwater to the nearest junction. According to SWMM three levels of flow routing sophistication, dynamic flow routing has been used because it has the ability to account for pressurized flow, channel storage, flow reversal, backwater, and entrance/exit losses. Hence, dynamic flow routing considers the most theoretically precise consequences (Rossman 2012), and no surface runoff routing is considered. In SWMM, there are three infiltration methods; the Green-Ampt model was used in this study because it assumes sharp wetting interface in the soil column that is separating the soil where the wetting interface above is completely saturated from the soil below, which is at the initial moisture value.

2.3.2. Determination of model parameters

The total surface area of the catchment is 1.64 km^2, with an estimated 66% of impervious surfaces and 34% of pervious surfaces for all sub-catchments. These ratios were selected based on real condition of sub-catchments surfaces in the study area. The slope of all sub-catchments was calculated in the field, and was roughly found to be 0.5%. Based on the properties of studied area and pipes type used in the system, Manning roughness coefficients were 0.013 and 0.009, respectively. The depth of depression storage on impervious and pervious surfaces, and average of sub-catchments width were determined based on the SWMM user's manual.

Information on pipe diameter and length, pipe inlet and outlet, junction invert elevation, ground elevation, maximum junction depth, minimum pipe cover, and surcharge elevation was supplied by Al-Najaf Sewage Directorate as a geographic information system (GIS) map. The data was converted to shape file using Arc map 10.4.1 software, then processed using storm and sanitary analysis 2015 (SSA) and exported to input data in SWMM to determine the spatial distribution of precipitation. As for rain intensity, SWMM allows the entry of hydrological data in different methods. In this study, time series is one of the most effective methods that depends on the IDF curve, and was applied with return periods of 2, 5, 10, and 25 years. This method offers more accurate simulation of the model because it is based on predicted future rainfall data at different time intervals.

2.3.3. Additional input data

There is additional flow discharged into the stormwater drainage system from neighboring areas. First, this includes surface runoff flowing from the main street of Al-Kufa/Al-Najaf and discharging into three inlet junctions at the north of study area stormwater drainage system, as shown in Figure 3. The street system is collecting the surface runoff generated from rainfall and is considered as a small urban rain system, where rational method is used to calculate the peak discharge values at each junction inlet and were found to be 0.0461 m^3/sec. It is an important factor because the main street of Al-Kufa-Al-Najaf in not served by any stormwater drainage system, and branch roads are acting as open channels discharging stormwater to the study area stormwater drainage system. Second, the stormwater drainage system of the study area is connected to neighboring Districts of Al-Eskan and Al-Ishtiraki, and discharge is estimated to be 0.226 and 0.4 m^3/sec, respectively. All of these additional flows were input data as initial flow in the SWMM simulation.

2.4. Scenarios design

According to the actual land-use of the study area in 2008, 50% of it is designed as a sub-catchment that contributes to the surface runoff. However, this designed percentage is changing over time as the land-use of the area is changing. The area is facing urban expansion; this includes the conversion of public parks into residential areas and commercial centers due to land limitation, and homes yards are transferred into houses due to the high land price and families extension. Therefore, an increase in surface runoff beyond designed quantities is expected to occur over time, but its rate is not exactly specified as the urban expansion is taking place in irregular forms. Hence, to study the effect of land-use and determine expected flooding, the designed sub-catchment area was varied as 50, 75, and 100%. In recent years, Al-Najaf Governorate is witnessing an increase in rainfall intensities within short rainfall duration due to climate change effect. Therefore, the effect of rainfall intensities, as a response to climate change, was simulated using SWMM at different return periods of 2, 5, 10, and 25 years.

2.5. Model calibration and validation

Validation is a process of comparing the model and its behavior to the real system and its behavior. In this study, model validation was conducted by comparing predicted maximum discharge in pipes acquired from the SWMM program with actual pipes discharge acquired from Al-Najaf Sewage Directorate. As recommended by Zwain *et al.* 2020, mean of normal squared error (NMSE) and correlation coefficient (R) were statistical parameters selected for model validation. NMSE (Equation (2)) measures mean relative scatter and reflects both systematic and systematic (random) errors, and R (Equation (3)) reflects the

Figure 3 | The surface runoff from Al-Kufa/Al-Najaf Street to the study area.

linear relationship between two variables and is thus insensitive to either an additive or a multiplicative factor.

$$NMSE = \frac{\overline{(C_{\circ} - C_p)^2}}{\overline{C_{\circ}} - \overline{C_p}} \tag{2}$$

$$R = \frac{\overline{(C_{\circ} - \overline{C_{\circ}})(C_P - \overline{C_p})}}{\sigma C_{\circ}\ \sigma C_P} \tag{3}$$

where:

Cp is predicted data, Co is observed data, $\overline{C}$ is average data over the dataset, and σC is standard deviation over the dataset.

3. RESULTS AND DISCUSSION

3.1. IDF curve for Al-Najaf governorate

Al-Najaf Governorate has a severe lack of hydrological data and various information is unavailable to conduct further research related to environmental engineering and hydrological analysis. In this study, a detailed procedure and data of IDF curve generation are shown in Tables 1–3. An IDF curve has been successfully generated for Al-Najaf Governorate for the first time as shown in Figure 4. From Table 1, it was noticed that there is an increase in the maximum daily rainfall (mm) in the last 10 years. For example, in 2013 and 2015, the maximum daily rainfall reached 64.5 and 33 mm, respectively. This could be due to the effect of climate change in Al-Najaf Governorate. Unfortunately, although extreme rainfall events have increased in several regions worldwide, there are still uncertainty and limited data available. This will affect the infrastructure capability related to human health and safety, particularly sewer and stormwater drainage systems. For this reason, understanding future extreme events are essential for sanitary systems design. Potential non-uniform and climate-induced changes such as heavy rainfall events question the accuracy and adequacy of current infrastructure design concepts, which rely on an assumption of climate data (Cheng & AghaKouchak 2014).

At any return period, the results in Figure 4 indicated that the rainfall intensity at short duration of 5 min was generally three times higher than at 30 min duration. It is similarly noticed at time duration of 30 and 120 min, that the rainfall intensity decreased about three times also. On the other hand, the stormwater drainage system in Al-Ameer District had been designed in 2008 for a maximum rainfall intensity of 20 mm/h, but the study area has been exposed to the climate change impacts, thereby there has been a noticeable increase in rainfall intensities, especially after 2009. According to Figure 5, rainfall intensity of 20 mm/h was exceeded at several occupations that include 2 years return period and below 10 min duration, 5 years return period and below than 20 min duration, 10 years return period and below than 30 min duration, 5 years return period and below than 35 min duration. Hence, it is expected that more areas will be flooded in the coming years, unless some effective solutions are introduced (Hassan *et al.* 2017).

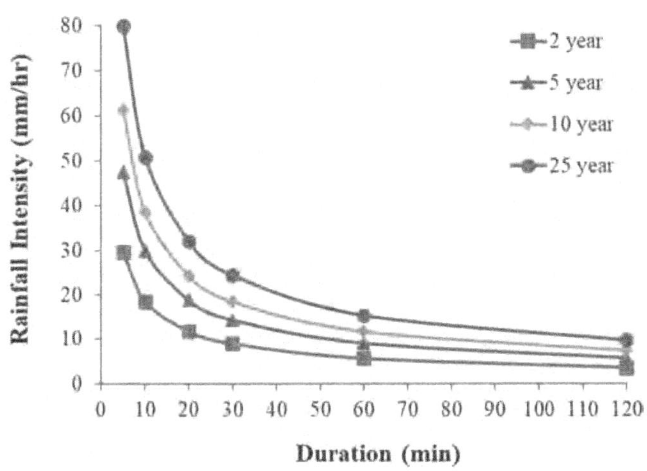

Figure 4 | Intensity Duration Frequency (IDF) Curve for Al-Najaf Governorate.

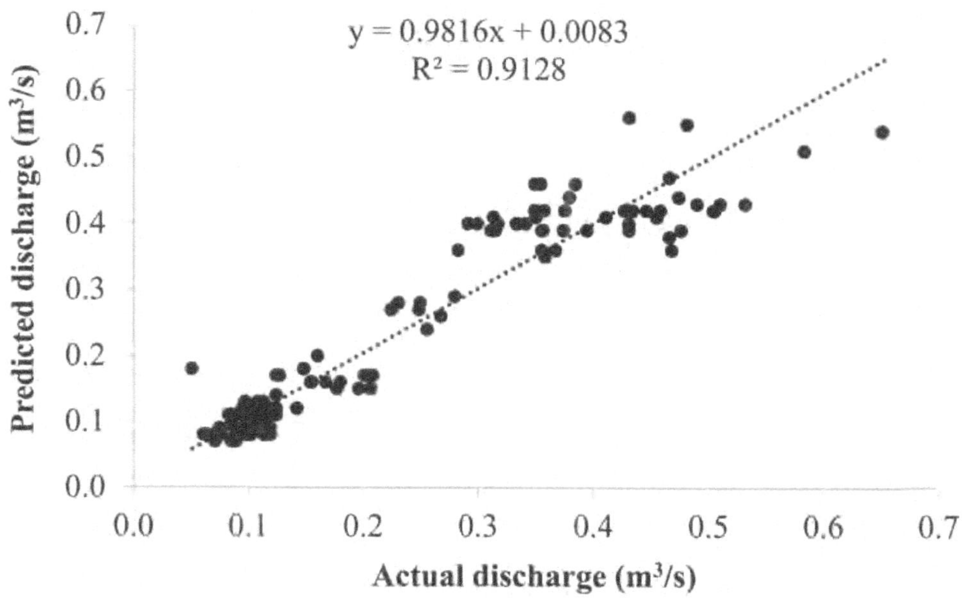

Figure 5 | Linear relationship between predicted data over actual measured data.

3.2. Model validation

The simulation of Al-Ameer district stormwater drainage system was validated using actual discharge data provided by Najaf Sewerage Directorate and predicted data by SWMM. Statistical results of NMSE and R values are shown in Table 4, and predicted data over actual data are shown in Figure 5. The validation results showed that NMSE was 0.002, which is very close to ideal fit of zero. The R value was 0.95, which shows a positive correlation and very close to ideal value of 1. Linear correlation in Figure 4 showed a very well fitted data of predicted data over actual data. Therefore, the results above indicated that the SWMM simulation predicted data very close to actual data, hence, no model calibration is needed at this stage.

3.3. Effect of land-use on stormwater drainage system

In order to study the effect of land-use on the stormwater drainage system in Al-Ameer District, different sub-catchment areas that contribute to the surface runoff were taken into consideration. This is because the actual land-use in Al-Ameer District is not well known due to progressing random development. At a constant rainfall intensity, return period of 10 years and throughout 2 hours of rainfall duration, different sub-catchments areas of 50, 75, and 100% were simulated using SWMM, and the results are shown in Table 5. Increasing the sub-catchment area from 50 to 75% led to an increment of impervious

Table 4 | Linear correlation of predicted data over actual data

Parameter	Value	Ideal fit	NOTE
NMSE	0.002	Zero	It is close to zero, the model fits well.
R	0.95	1	It is close to 1, the model fits well.

Table 5 | Effect of land-use assessment of Al-Ameer District stormwater drainage system throughout 2 hours of rainfall duration

Return Period (Years)	Sub-catchment (%)	Total Surface runoff (m^3)	Maximum flooded manhole volume (m^3)	Total flooding (m^3)	Number of flooded manholes
10	50	20380	1703 (R315)[a]	10513	70
10	75	29150	2074 (R315)[a]	18414	84
10	100	37350	2222 (R315)[a]	26032	95

[a]Name of manhole.

surfaces of 33%, while enlarging the sub-catchment area from 75 to 100% increased the impervious surfaces by 25%. Table 5 details the assessment of Al-Ameer District stormwater drainage system. By increasing the sub-catchment area from 50 to 100%, an increment in total surface runoff from 20,380 to 37,350 m^3, total flooding from 10,513 to 26,032 m^3, and the number of flooded manholes from 70 to 95 have occurred, respectively.

Notably for almost 2 hours, manhole R315 (indicated in Figure 6) was mostly flooded with maximum flooding volume of 1,703, 2,074, and 2,222 m^3 for 50, 75, and 100%, respectively. It's worth reporting that data obtained in Table 5 occurred at any time throughout the 2 hours duration and not after 2 hours. The 2 hours of rainfall duration was selected according to filed survey and sanitary experts in the study area. This period was sufficient to withstand the longest storm event and drain surface runoff. Undrained flows after two hours is considered as a flood and may indicate a problem in the system. It is also important to note that any change in land-use, for example increasing the area of sub-catchment, will lead to extra flooding volumes, in which the stormwater drainage system may not be able to withstand it. Similarly, Kong *et al.* (2017) reported the effect of land-use changes on stormwater management using SWMM simulation. They found that 33.3% reduction in pervious surfaces (increased of impervious surfaces) results in 92.9% increase in surface runoff, 31.7% increase in peak flow, and 35 min earlier of peak runoff time. In addition, Hu *et al.* (2020) observed that urbanization increased impervious surfaces in Beijing's central area, leading to increment in surface runoff and flooding. They correlated impervious surfaces rise with surface runoff increase, leading to flooding risk in the area.

After 2 hours, Figure 6(a)–6(c) illustrates is the locations of flooding in Al-Ameer District, type of flooding, and percentage of flooded manholes, for sub-catchments areas of 50, 75, and 100%, respectively. Knowing flood locations is very important for specifically identifying flood problems. Hence, this study is expected to provide a technical solution for experts, engineering and decision makers in order to identify problems in the system and evaluate proposed solutions. The type of flooding from manholes is divided into five stages according to the SWMM simulation, as recommended by Hassan *et al.* (2017):

- Stage 1 (no flooding) ranges from (0 to 0.001 m^3/s)
- Stage 2 (very light flooding) ranges from (greater than 0.001 to 0.01 m^3/s)
- Stage 3 (medium flooding) ranges from (greater than 0.01 to 0.02 m^3/s).
- Stage 4 (high flooding) ranges from (greater than 0.02 to 0.04 m^3/s).
- Stage 5 (very high flooding) for (greater than 0.04 m^3/s).

Figure 6(a) shows the assessment of stormwater drainage system at 50% of sub-catchment area, 2 hours of rainfall duration, 10 years of return period, and rainfall intensity of 7.36 mm/hour. Figure 6(a) reveals that the number of flooded manholes was 13 manholes out of 343 total manholes, and the ratio of flooded manholes is 4% only and non-flooded is 96%. According to the flooding flowrate, the flooded manholes stages were 21% of medium flooding (Stage 3), 50% of high flooding (Stage 4), and 29% of very high flooding (Stage 5). Among all manholes, R315 at system downstream had the most flooded manholes with flooding stage 5 and remains flooded for 1.7 hour of rainfall duration. The total flooded area in the study area in 2 hours was 25,965 m^2. It is concluded that the system performance at 50% sub-catchment area is fairly average during the flood period of 2 hours.

Figure 6(b) displays the behaviour of stormwater drainage system at 75% of sub-catchment area, 2 hours of rainfall duration, 10 years of return period, and rainfall intensity of 7.36 mm/hour. It is observed that 25 manholes were flooded, constituting 8% of 343 total manholes, in which 16% were slightly flooded with Stage 2, 16% were fairly flooded with Stage 3, 24% were highly flooded with Stage 4, and 44% were the very highly flooded with Stage 5. It's interesting to report that the flooding mostly observed in the downstream of the study area (north-east direction) and total flooded area after 2 hours increased to 33,750 m^2. In this regard, it can be concluded that the system could hardly withstand the surface runoff generated from 75% of sub-catchment, and flood is therefore considered relatively high.

By considering 100% of sub-catchment area, Figure 6(c) reveals a deterioration in the system performance, in which 36 manholes were flooded, representing 11% out of a total 343. Flooding stages were distributed as 14% very light flooding (Stage 2), 11% medium flooding (Stage 3), 32% high flooding (Stage 4), and mostly 43% very high flooding (Stage 5). The flooding condition in this case was the worst due to increased impervious surfaces and reduced green areas. The downstream area (north-east direction) was completely flooded, and total flooding area rose to 54,900 m^2. Flooding water level is very high and might even enter properties, causing damage to property and infrastructure. Accordingly, it's concluded that the system can't withstand a sub-catchment area of 100% and flooding was very high at 2 hours of rainfall. After all, it was observed that the flooding map has expanded as a result of urbanization and change in land-use from pervious to impervious surfaces.

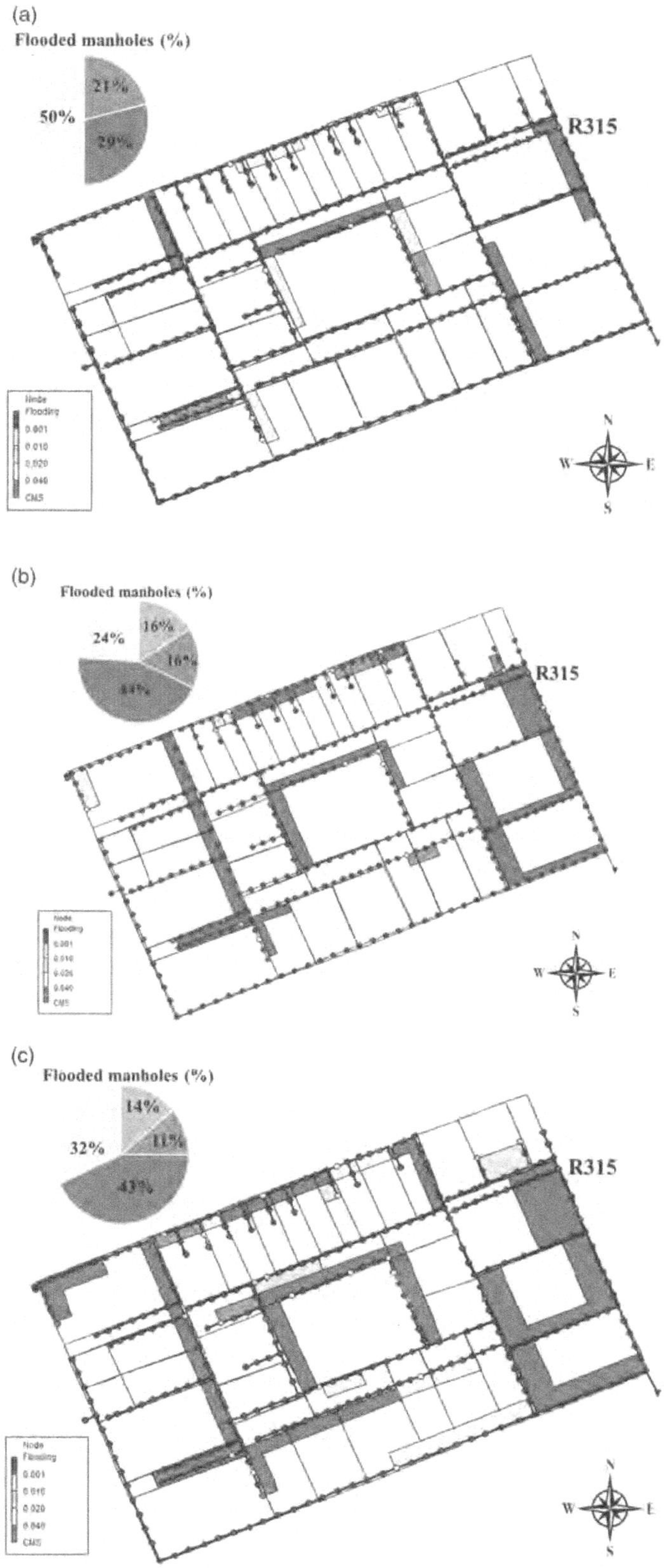

Figure 6 | Effect of land-use on Al-Ameer District stormwater drainage system at 10 years return period and (a) 50% sub-catchment, (b) 75% sub-catchment and (c) 100% sub-catchment.

Likewise, Shanableh *et al.* (2018) revealed a positive correlation between urbanization and floods, in which flooding increased substantially in areas where land-use has changed to residential area. They also reported that flooding increased by 60% in 10 years due to rapid urbanization.

3.4. Effect of climate change on stormwater drainage system

Climate change have a direct influence on increased rainfall intensity that will affect surface runoff generation. Different return periods of 2, 5, 10, and 25 years that represent a series of rainfall intensities were selected as a response to climate change effect. Throughout 2 hours of rainfall duration, the assessment results are tabulated in Table 6, and rainfall intensities can be obtained from the IDF curve in Figure 4. It is observed that increasing the return period has dramatically increased the surface runoff generated in the study area. For example, when the return period shifted from 2 to 5 years, the total surface runoff rose from 14,120 and 27,110 m^3 (representing 48% of raise), and total flooding increased from 5,914 to 17,591 m^3, accounting for 72% of increment. When higher return period is selected, higher flooding volume and number of flooded manholes were observed. For instance at 25 years of return period, the surface runoff and total flooding were 52,006 and 41,230, respectively. This means that 21% only of the surface runoff is discharged by the system and 79% of it is flooded, indicating the failure of the system.

Hence, from the results in Table 6, it is expected that the system may withstand climate change effect only for the next 10 years. In general, the greater the return period, the greater the flooding volume. Similarly, an increase in peak runoff was observed when the total rainfall was increased due to increasing return period (Babaei *et al.* 2018). In support of that, Hossain Anni *et al.* (2020) reported that the number of flooded manholes and volume of flooding increased by 47 and 48% when the return period increased from 10 to 25 years. In like manner, Hassan *et al.* (2017) noticed a positive correlation between climate change and increased rainfall intensity, where they reported an increase in flooding of 47% as an effect of climate change in the last 8 years.

For more realistic evaluation of stormwater drainage system, the system was assessed after 2 hours of rainfall duration. In this duration, it is expected that the stormwater drainage system Al-Ameer District will discharge most of the excess stormwater, but remaining flooding may be considered a risk warning. Figure 7 showed flooded manholes, flooding magnitude, and flooding locations. Flooding may occur whenever that water discharge at manholes or sewers exceeds the maximum designed values. At 2 years of return period, Figure 7(a) displayed that 15 manholes out of a 343 were flooded, consisting of 4% of total manholes. Flooded manholes distribution was 7% of very light flooding, 20% of medium flooding, 40% of high flooding, and 33% very high flooding. Again, R315 manhole was mostly flooded for about 1.51 hr. It is noticed that flooding in small quantities was mainly in the downstream, and therefore surface runoff can be well controlled at this return period. Figure 7(b) assesses the stormwater drainage system at the end of 2 hours rainfall duration and 5 years of return period. The flooded manholes were 26 out of 343 manholes, which is equivalent to 8% of the total manholes. Flooded manholes are classified as 14% very light flooding, 11% medium flooding, 26% high flooding, and 46% very high flooding. Very high flooding locations were scattered mainly in downstream (east direction) and few locations in the center and upstream (west direction). In addition, flooding occurred mostly in main sewer lines, and manhole R315 had the longest duration flood of a very high flood. Increased flooding magnitude is due to increased rainfall intensity, however, the system still can withstand a return period of 5 years with climate change effect.

Figure 7(c) shows the condition of the stormwater drainage system at a return period of 10 years. The number of flooded manholes reached 36 out of 343, representing 11% of total manholes. It is distributed as 13% very light flooding, 11% medium flooding, 32% high flooding, and 44% high flooding. However, it is noticed that the number of flooded manholes

Table 6 | Effect of climate change assessment of Al-Ameer District stormwater drainage system throughout 2 hours of rainfall duration

Sub-catchment (%)	Return Period (Years)	Total Surface runoff (m^3)	Maximum flooded manhole volume (m^3)	Total flooding (m^3)	Number of flooded manholes
100	2	14120	1306 (R315)[a]	5914	26
100	5	27110	1835 (R315)[a]	17591	72
100	10	37350	2222 (R315)[a]	26032	95
100	25	52006	2295 (R319)[a]	41230	106

[a]Name of manhole.

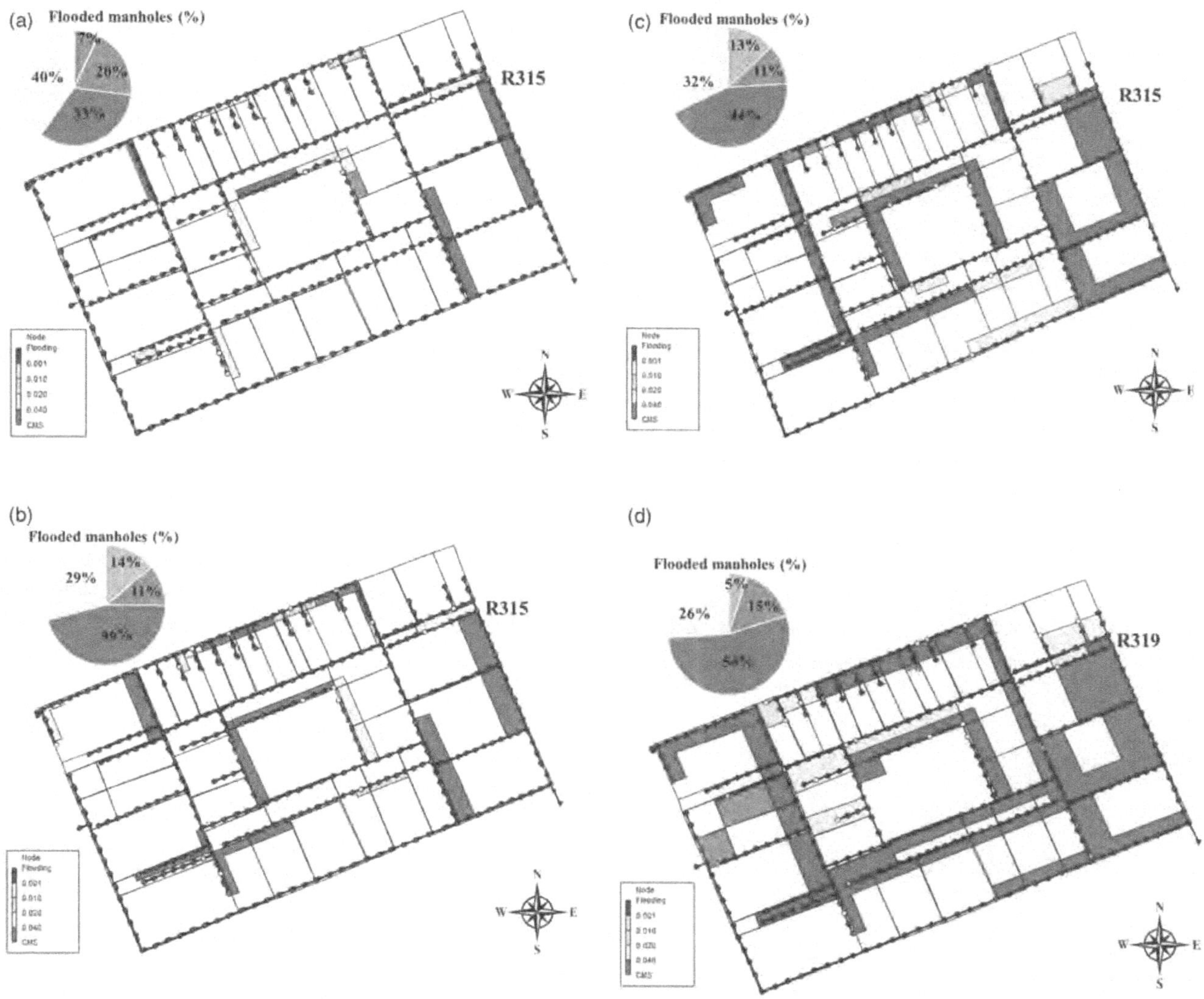

Figure 7 | Effect of climate change on Al-Ameer District stormwater drainage system at 100% sub-catchment and (a) 2 years return period, (b) 5 years return period, (c) 10 years return period, and (d) 25 years return period.

is less important than the magnitude of manholes flooding. For example, the magnitude flooding in manhole 315 (Stage 5) is five times higher than in other manholes (Stage 1). Therefore, flooding area drawing was used to illustrate flooding magnitude, in which maximum flooding occurs in the downstream (north-east and south-east directions) area and in the main sewer that passes through the center of the area. At 10 years return period, the system fails to discharge excess rainfall, resulting in the accumulation of stormwater at downstream. Flooding quantities greatly increased and flooding depth is expected to rise more than 13 cm (curbstone height) and may enter the houses, causing damage to infrastructure and properties.

At a return period of 25 years, Figure 7(d) showed that 48 manholes out of 343 were flooded, consisting of 14% of the total manholes. Again, an increased number of flooding manholes does not affect the area as much as flooding magnitude. Therefore, flooding magnitude were distributed as 5% very light flooding, 15% medium flooding, 26% high flooding, and 54% high flooding. A major system failure was observed, where 26% of Al-Ameer District was flooded. Many locations were flooded, not only in downstream, but also in the center and upstream. Figure 7(d) also showed that stormwater entered the housing, causing damage to infrastructure and properties. It worth reporting that branch sewers were also flooded in addition to main sewers. High quantities of flooding is due to high rainfall intensity of 80 mm/hr at the beginning of a rainstorm. The system fails to discharge surface runoff because it was beyond the design capacity. This indicates that climate change effect was not considered in future prediction of rainfall intensities needed for storm system design. Hence, climate change effect should be integrated in the generation of IDF curves (Noor *et al.* 2018). It is concluded that this prediction of rainfall intensity is one of the most high-risk expected intensities and has severe effects on the system and properties.

Moreover, more areas were flooded when the return period increased further. Likewise, Babaei *et al.* (2018) observed that when the return period increased from 2 to 5 years and then to 10 years, more areas were flooded each time and the drainage system should be expanded by 20% to prevent flooding. On the other hand, most stormwater drainage systems are designed to serve a certain maximum value of rainfall intensity based on previous hydrological data. Meanwhile in recent years, rainfall total amounts are less than before, but rainfall intensities are 4 times higher due to the effect of climate change (Nile *et al.* 2019).

4. CONCLUSION

For the first time in Al-Najaf Governorate of Iraq, an IDF curve that integrated climate change effect was produced with return periods of 2, 5, 10, and 25 years. SWMM successfully simulated the impacts of sub-catchments and climate change on the stormwater drainage system in Al-Ameer District. Without calibration, SWMM model was validated over actual discharge in the system, with NMSE value of 0.002 and R value of 0.95. Random urban development in Al-Ameer District has changed the land-use from pervious to impervious areas; when the sub-catchment increased from 50 to 100%, the total flooding area rose from 25,965 to 54,900 m^2 and ratio of flooded manholes increased from 4 to 11%. It is recommended to preserve green lands to improve infiltration of stormwater and increase pervious surfaces that will lead to reduction in surface runoff, resulting in low flooding volume. On the other hand, the adverse effect of the climate change on the stormwater drainage system was more than effect of land-use. When the return period increased to 25 years, a major system failure was observed, whereas 26% of Al-Ameer District was flooded. Finally, it is expected that the system may withstand climate change effect only for the next 10 years only.

ACKNOWLEDGEMENTS

The authors would like to thank Al-Najaf Sewerage Directorate for providing some information.

CONFLICT OF INTEREST

The authors declare no conflict of interest.

DATA AVAILABILITY STATEMENT

All relevant data are included in the paper or its Supplementary Information.

REFERENCES

Abdellatif, M., Atherton, W., Alkhaddar, R. & Osman, Y. Z. J. H. S. J. 2015 Quantitative assessment of sewer overflow performance with climate change in northwest England. *Hydrological Sciences Journal* **60** (4), 636–650.

Babaei, S., Ghazavi, R. & Erfanian, M. 2018 Urban flood simulation and prioritization of critical urban sub-catchments using SWMM model and PROMETHEE II approach. *Physics and Chemistry of the Earth, Parts A/B/C* **105**, 3–11.

Cheng, L. & AghaKouchak, A. 2014 Nonstationary precipitation intensity-duration-frequency curves for infrastructure design in a changing climate. *Scientific Reports* **4** (1), 7093.

Fortunato, A., Oliveri, E. & Mazzola, M. R. 2014 Selection of the optimal design rainfall return period of urban drainage systems. *Procedia Engineering* **89**, 742–749.

Hadia, M. O., El-Mewafi, M., Metwally, S., Awadd, A. & Suliman, S. 2015 *Evaluation of Wastewater Network of Al-Anwaar in Al-Kut City, Iraq by Using SWMM and GIS Techniques.*

Hassan, W. H. & Nile, B. K. 2020 Climate change and predicting future temperature in Iraq using CanESM2 and HadCM3 modeling. *Modeling Earth Systems and Environment* **7**, 737–748.

Hassan, W. H., Nile, B. K. & Al-Masody, B. A. 2017 Climate change effect on storm drainage networks by storm water management model. *Environmental Engineering Research* **22** (4), 393–400.

Hossain Anni, A., Cohen, S. & Praskievicz, S. 2020 Sensitivity of urban flood simulations to stormwater infrastructure and soil infiltration. *Journal of Hydrology* **588**, 125028.

Hu, S., Fan, Y. & Zhang, T. 2020 Assessing the effect of land use change on surface runoff in a rapidly urbanized city: a case study of the central area of Beijing. *Land* **9** (1), 17.

Jiang, L., Chen, Y. & Wang, H. 2015 Urban flood simulation based on the SWMM model. *Proc. IAHS* **368**, 186–191.

Kong, F., Ban, Y., Yin, H., James, P. & Dronova, I. 2017 Modeling stormwater management at the city district level in response to changes in land use and low impact development. *Environmental Modelling & Software* **95**, 132–142.

Nhat, L. M., Tachikawa, Y. & Takara, K. 2006 Establishment of intensity-duration-frequency curves for precipitation in the monsoon area of Vietnam. *Annuals of Disas. Prev. Res. Inst* **49** (B), 93–103.

Nile, B. K., Hassan, W. H. & Esmaeel, B. A. 2018 An evaluation of flood mitigation using a storm water management model [SWMM] in a residential area in Kerbala, Iraq. In: *IOP Conference Series: Materials Science and Engineering*. Vol. 433, p. 012001.

Nile, B. K., Hassan, W. H. & Alshama, G. A. 2019 Analysis of the effect of climate change on rainfall intensity and expected flooding by using ANN and SWMM programs. *ARPN Journal of Engineering and Applied Sciences* **14** (5), 974–984.

Noor, M., Ismail, T., Chung, E.-S., Shahid, S. & Sung, J. H. 2018 Uncertainty in rainfall intensity duration frequency curves of peninsular Malaysia under changing climate scenarios. *Water* **10** (12), 1750.

Obaid, H., Shamsuddin, S., Basim, K. & Shreeshivadasan, C. 2014 Modeling sewer overflow of a city with a large floating population. *Hydrology: Current Research* **5** (2), 1.

Olsson, J. & Foster, K. J. H. R. 2014 Short-term precipitation extremes in regional climate simulations for Sweden. *Hydrology Research* **45** (3), 479–489.

Rabori, A. M. & Ghazavi, R. 2018 Urban flood estimation and evaluation of the performance of an urban drainage system in a Semi-Arid urban area using SWMM. *Water Environment Research* **90** (12), 2075–2082.

Rathnam, E. V., Jayakumar, K. & Cunnane, C. 2001 *Runoff Computation in A Data Scarce Environment for Urban Stormwater Management-A Case Study*. pp. 446–454.

Rossman, L. A. 2012 *Storm Water Management Model user's manual Version 5.0 U.S.* Environmental Protection Agency, Washington, DC.

Shanableh, A., Al-Ruzouq, R., Yilmaz, A. G., Siddique, M., Merabtene, T. & Imteaz, M. A. 2018 Effects of land cover change on urban floods and rainwater harvesting: a case study in Sharjah, UAE. *Water* **10** (5), 631.

Zwain, H. M., Nile, B. K., Faris, A. M., Vakili, M. & Dahlan, I. 2020 Modelling of hydrogen sulfide fate and emissions in extended aeration sewage treatment plant using TOXCHEM simulations. *Scientific Reports* **10** (1), 22209.

Zwain, H. M., Almurshedi, K. R., Vakili, M., Dahlan, I. & Naje, A. S. 2021 Water quality and radionuclides content assessment of AL-Najaf sea: case study. *Journal of Ecological Engineering* **22** (2), 262–271.

First received 5 May 2021; accepted in revised form 5 August 2021. Available online 23 August 2021

Impact of Climate Change on Hydrology and Water Resources doi: 10.2166/wcc.2021.110

Flood season segmentation and scheme optimization in the Yellow River

Ke Zhou [a,b]
[a] Faculty of Architectural Engineering, Tianjin University, Tianjin 300072, China
[b] Faculty of Arts and Design, North China University of Water Resources and Electric Power, Zhengzhou 450046, China
E-mail: zhouke@ncwu.edu.cn

KZ, 0000-0001-5347-6318

ABSTRACT

River flood season segmentation is a significant measure for flood prevention. This study aims to carry out theoretical analysis on flood season segmentation methods and put forward a framework for proper flood season segmentation through comparison between different segmentation methods. The studied framework consists of a Fisher optimal partition method for determining the optimum numbers of the sub-seasons, an ensemble approach for segmenting a defined flood season, and a nonparametric bootstrap combined with a fuzzy optimum selection method (NPB-FOS) for testing the rationality of the flood season staging schemes. The present research findings show that different methods could result in different staging schemes. It is proved through rational analysis that the staging scheme obtained by probability change point (PCP) is superior to others. The flood season of the downstream reach of the Yellow River can be segmented into three sub-seasons, i.e. early flood season (01 June–20 July), main flood season (21 July–28 September), and late flood season (29 September–08 November). The segmentation results of the flood season should play an active role in flood prevention.

Key words: flood season segmentation, framework, methods optimization, the Yellow River

HIGHLIGHTS

- The conjunctive use of Fisher optimal partition, ensemble approach, nonparametric bootstrap, and fuzzy optimum selection method applied to flood season segmentation.
- The rational methods segment a flood season into several sub-seasons.
- An optimal framework for reasonable flood season segmentation was proposed.
- A case study in the Yellow River was carried out and reasonable results were obtained.

GRAPHICAL ABSTRACT

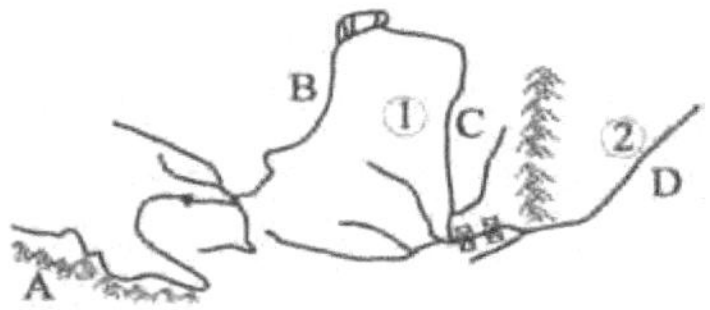

INTRODUCTION

With the increasingly frequent extreme climate change and extreme precipitation events, flood disasters seriously threaten citizens' lives and restrict the sustainable development of the social economy. As estimated by The World Resources Institute, about 21 million people are affected by floods every year, and even worse, the number would reach 54 million in 2030 due to global climate change. However, a flood is not only a kind of disaster but a precious resource. For the important rivers, flood seasonality analyses and appropriate scheme to segment an entire flood season into multiple sub-seasons help reduce flood risk and utilize water resources effectively. In some monsoon-influenced regions, floods often show the characteristics of significant seasonal changes due to the influence of monsoon or climate change. Therefore, how to scientifically study the seasonal characteristics of floods is a key issue in hydrology and water resource development (Cunderlik *et al.* 2004). Macdonald *et al.* (2010) analyzed the spatial and temporal variability of flood seasonality for 30 years in Wales and they stated that noticeable regional variations in timing and length of flood season were evident. Ye *et al.* (2017) believed that understanding the causes of flood seasonality is helpful for better flood management and they examined the seasonality of maximum annual floods and its variation at more than 250 basins across the contiguous USA. Ledingham *et al.* (2019) analyzed the seasonality of storm rainfall and flood runoff based on more than 500 rain gauge stations in Britain and gauged rainfall datasets and found that the seasonal occurrences of annual daily rainfall and flood maxima are significantly different in dry lowland catchments. For a large river channel, the flood limiting water level (FLWL) is an important characteristic water level to control flood risk and should be determined based on the river's design flood criteria. In the traditional regulation of river flood control, the FLWL tends to be fixed throughout the flood season (Jiang *et al.* 2015). However, the fixed FLWL ignores the seasonality of flooding and may result in unreasonable water transfer and management for the river systems. To solve this problem, multi-stage FLWL control has been proposed. And the flood season segmentation is a prerequisite for multi-stage FLWL control which means that the entire flood season must be first divided into sub-seasons.

Generally speaking, there are three main kinds of methods that can be applied to segment the entire flood season, i.e. the qualitative analysis method (Adhikari *et al.* 2016), statistical analysis method (Chen *et al.* 2013, 2015), and clustering analysis method (Mo *et al.* 2018; Al-Jawad *et al.* 2019). The flood season segmentation obtained by the qualitative analysis method relies on the researchers' practical experience and subjective judgment, and only a small amount of quantitative calculation is used to verify the results. Besides, due to its strong subjectivity, segmentation results were often vague and rough and could not be applied to the practical operation (Bender & Simonovic 2000). Statistical analysis is a method that obtains the segmentation by statistically analyzing a large amount of hydrological data in flood season. Although river managers can easily use this kind of method for its explicit mathematical concept and clear calculation steps, there are still some shortcomings, such as the single index, the threshold selection subjectivity, and the sampling error (Esogbue *et al.* 1992). The clustering analysis, which can segment the flood season by calculating the degree of representative indicators, has an advantage due to sufficient mathematical foundation, clear and easy calculation, and more stable results. Therefore, it is a popular method at present. Representative methods of clustering analysis are the fractal analysis method (FAM) (Hajkowicz & Collins 2007), change-point analysis (Liu *et al.* 2010), and set pair analysis (Nicklow *et al.* 2010). In summary, even though many studies have been carried out to obtain the results of flood season staging, there are still some problems that remain to be solved: (1) During the studies for flood season staging, the researchers usually classify a flood season into several sub-flood seasons subjectively, but for a particular river channel, especially large river channels, how many sub-flood seasons should be segmented? This is still waiting to be answered. (2) Different methods or indexes will result in different flood season staging schemes, but which one is the most suitable for a river flood season segmentation? (3) How to test the rationality of flood segmentation is also a key problem.

Therefore, the objectives of this paper are to (1) determine the optimum numbers of segmentation by the Fisher optimal partition method; (2) segment a defined flood season into several sub-seasons by an ensemble approach combined with the FAM, mean change-point (MCP) analysis method, and probability change-point (PCP) analysis method; and (3) test the rationality of the flood season staging scheme and make an optimal selection by a nonparametric bootstrap combined with a fuzzy optimum selection (NPB-FOS) method. This study aims to overcome the existing problems in the previous studies and to realize the reasonable segmentation of river flood season.

METHODS AND OPTIMIZATION PROCEDURE

Fisher optimal partition method

The Fisher optimal partition method is a kind of statistical clustering approach and it aims to determine whether each class has a significant difference in the optimal partition or not. The detailed steps are given as follows (Roach *et al.* 2018).

Step 1. Define the class diameter

Class diameter is the index that can reflect the degree of difference among classes. The greater the diameter, the greater the difference is. Assume that a class is $\{x_s, x_{s+1}, \ldots, x_t\}(t > s)$, and the main equations are as follows:

$$\bar{x}_{ij} = \frac{1}{s-t+1}\sum_{l=s}^{t} x_l \tag{1}$$

$$D(i, j) = \sum_{l=s}^{t} (x_l - \bar{x}_{ij})^T (x_l - \bar{x}_{ij}) \tag{2}$$

where $\bar{x}_{ij}$ is the mean value, $D(i, j)$ is the diameter (the sample of the total sum of squares).

Step 2. Define loss function

Suppose that n samples can be classified into k class:

$$G_1 = \{i_1, i_1 + 1, \ldots, i_2 - 1\}, \ldots, G_k = \{i_k, i_k + 1, \ldots, n\}(1 = i_1 < i_2 < \cdots < i_k < n = i_{k+1} - 1)$$

and the loss function is as follows:

$$L[b(n, k)] = \sum_{l=1}^{k} D(i_l, i_{l+1} - 1) \tag{3}$$

When n and k are known, the optimal segmentation can be described as follows:

$$L^*[b(n, k)] = \min \sum_{l=1}^{k} \mathrm{D}(i_l, i_{l+1} - 1) \tag{4}$$

And the corresponding point is the optimal dividing point.

Step 3. Recursive formula

The optimal segmentation of ordered samples is always based on the optimal k–1 segmentation, and the formulas are as follows:

$$L^*[b(n, 2)] = \min_{2 \le i \le n} \{D(1, i - 1) + D(i, n)\} \tag{5}$$

$$L^*[b(n, k)] = \min_{k \le i \le n} \{L^*[b(i_k - 1, k - 1)] + D(i, n)\} \tag{6}$$

Step 4. Determine the optimum number of the sub-flood seasons

Two ways can be adopted to determine the optimum number of the sub-flood seasons:

(1) Plot the curve variation on the base of the number of the optimal partition k and the minimum loss function $L[b(n, k)]$, and the k at the bend of the curve can be regarded as the optimum number of the sub-flood seasons.

(2) Calculate the nonnegative slope as follows:

$$\delta(k) = \left| \frac{L^*[b(n, k)] - L^*[b(n, k - 1)]}{k - (k - 1)} \right| - \left| \frac{L^*[b(n, k + 1)] - L^*[b(n, k)]}{(k + 1) - 1} \right| \tag{7}$$

If $\delta(k)$ is large, it shows that k classes are better than k–1 classes. If k approaches 0, it is not necessary to continue to classify. Generally, k is the number of the optimal partition when $\delta(k)$ is maximized.

Ensemble methods for flood season segmentation

In this study, an ensemble method combined with several common methods (FAM , MCP analysis method, PCP analysis method) is used to segment the flood season. We mainly introduce the main principles of the above methods.

Fractal analysis method

The FAM was proposed by Mandelbrot (1982). Assume that the number of days in the flood season period is T, and therefore the capacity dimension D_b of this flood season period can be expressed as:

$$N_N(\varepsilon) = \frac{N(\varepsilon)}{N_T} \tag{8}$$

$$D_b = d - b \tag{9}$$

where $N(\varepsilon)$ is the number of samples whose values are larger than a given threshold in the flood season period; N_T is the time scale, $N_T = T/\varepsilon$; ε is the step size; b is the gradient of the straight line in $\ln \varepsilon \sim \ln N_N(\varepsilon)$ diagram; and d is the topological dimension. In this study, $d = 2$ (the samples of runoff series are two-dimensional data).

Calculate the capacity dimension of different flood season periods. If the capacity dimension of one period is similar to another period's capacity dimension, it can be considered that the two periods belong to the same sub-flood season.

MCP analysis method

It is well known that the starting point of the flood season is usually the time when the rainfall changes from less to more, and that the end point of the flood season would be a time when the rainfall varies from more to less, which could be considered as change points. Therefore, the change-point analysis method can be used to determine the corresponding change points. The time series x_i that follows a normal distribution and is mutually independent can be expressed as follows (Lin *et al.* 2020):

$$x_i = \mu_i + e_i \tag{10}$$

where e_i is the random error, and if it is independent and equal variance, $E(e_i) = 0$, assuming that

$$\mu_1 = \mu_2 = \cdots = \mu_{m_{i-1}} = b_1, \mu_{m_q} = \mu_{m_{q+1}} = \cdots, = \mu_n = b_{q+1}(1 < m_1 < m_2 < \cdots < m_q \leq n)$$

Obviously, m_j will be an MCP when $b_{j+1} \neq b_j$.

The least-square method is usually applied to determine the MCP, especially in the estimation of multiple change points. Moreover, an independence test needs to be done before using the MCP method.

PCP analysis method

The PCP analysis method based on peak-over-threshold (POT) sampling is commonly used in flood season analysis due to its advantage of good unbiasedness and effectiveness. The cumulative time method is widely used to obtain the PCP (Efron 1979; Perreault *et al.* 2000), which defines a statistical variable T_k:

$$T_K = K\left(\frac{u_k}{k} - \frac{u_n}{N}\right), \; k = 1, 2, \ldots, N \tag{11}$$

where u_k represents the number of the specified event that occurs during k tests; u_n represents the number of the same event that occurs in all (N times) tests.

Obviously, u_k/k is the frequency of specified event that occurs in k test, and u_N/N is the frequency of the specified event that occurs in all tests.

Based on the definition of the expectation, the $E(T_k)$ can be expressed as follows:

$$E(T_K) = \begin{cases} kN^{-1}(N - m + 1)(p_1 - p_2) & 1 \leq k \leq m - 1 \\ N^{-1}(N - k)(m - 1)(p_1 - p_2) & m \leq k \leq N \end{cases} \tag{12}$$

where p_1 is the probability of the specified event that occurs before the PCP, p_2 is the probability of the specified event that occurs after the PCP, and m is the assumed initial PCP.

NPB-FOS method and optimization procedure

The NPB-FOS method is mainly composed of four theoretical methods: the nonparametric bootstrap, mathematical statistics, relative membership degree (RMD), and two-stage fuzzy optimum selections. The purpose of nonparametric bootstrap is to build a variety of bootstrap samples with the distribution characteristics of original samples. The mathematical statistic is used to compute the relative frequency of annual maximum flood in each sub-flood season and the average peak flood during the main flood season in different staging schemes. The RMD method can calculate the RMD of the relative frequency in each sub-flood season and the average peak flood during the main flood season in different staging schemes. The two-stage fuzzy optimum selection method is applied to integrate the membership degrees to get the two-stage fuzzy membership degree that can test multiple staging schemes' rationality and make an optimum selection. The detailed steps are described as follows.

Step 1. Nonparametric bootstrap

To obtain a variety of approximations that meet the overall distribution characteristics of original samples, Efron (1979) proposed a 're-sampling' approach, called 'Bootstrap'. The nonparametric bootstrap method is a basic method of the bootstrap method, which is randomly sampling with replacement from original samples. Therefore, in this study nonparametric bootstrap is applied to get a great number of new samples without making any hypothesis for original samples.

Step 2. Relative frequency and confidence intervals

Apply annual maximum sampling to obtain N flood events, record the number of annual maximum flood in each sub-flood season, and then calculate the initial relative frequency by the following equation:

$$\mathrm{RF}_i = \frac{b_i}{N} \tag{13}$$

where b_i is the occurrence times of annual maximum flood during the ith sub-flood season, $i = 1, 2, \ldots, M$, and M is the number of sub-flood seasons in a flood season.

Due to the numbers of days in each sub-flood season being different, the initial relative frequency needs to be adjusted and the adjusted relative frequency obtained with the following equation:

$$\mathrm{RF'}_i = \mathrm{RF}_i \frac{a/m}{a_i} = \frac{\mathrm{RF}_i a}{m a_i} \tag{14}$$

where a is the number of days in the whole flood season, and a_i is the number of days in the ith sub-flood season.

To make the total relative frequency in each sub-flood season equal to 1 (assume that the sum is s'), the relative frequency can be expressed as follows:

$$\mathrm{RF''} = \frac{\mathrm{RF'}_i}{s'} \tag{15}$$

The confidence interval estimation formula of the normal distribution can be used to calculate the upper and lower limit of the confidence interval of relative frequency if it follows a normal distribution. The upper limit y_{up} and lower limit y_{down} of the confidence interval are estimated by the following formulas:

$$y_{\mathrm{up}} = \bar{X} + \frac{S}{\sqrt{n}} t_{\alpha/2}(n-1) \tag{16}$$

$$y_{\mathrm{down}} = \bar{X} - \frac{S}{\sqrt{n}} t_{\alpha/2}(n-1) \tag{17}$$

where $\bar{X}$ is the sample mean; $t_{\alpha/2}(n-1)$ is the quantile of $t(n-1)$ distribution, and S is the standard deviation of the samples.

Step 3. Relative membership degree

X is the matrix which is composed of the characteristic index value:

$$X = \begin{bmatrix} x_{11} & x_{12} \ldots & x_{1k} \\ x_{21} & x_{22} \ldots & x_{2k} \\ \ldots & \ldots & \ldots \\ x_{M1} & x_{M2} & x_{Mk} \end{bmatrix} = (x_{ij}) \tag{18}$$

where k is the number of the sub-flood seasons; M is the number of the characteristic index in each sub-flood season.

The RMD (r_{ij}) of each characteristic value in each sub-flood season can be calculated by the following equations:

$$r_{ij} = \frac{x_{ij}}{\max x_{ij}} (x_{ij} \text{ is positive}) \tag{19}$$

$$r_{ij} = \frac{x_{ij} - \min x_{ij}}{\max x_{ij} - \min x_{ij}} (x_{ij} \text{ is negative}) \tag{20}$$

According to the complementary fuzzy set, the relative inferiority degree can be expressed as follows:

$$r'_{ij} = (1 - r_{ij}) \tag{21}$$

Step 4. Two-stage fuzzy membership degree

Usually, we use 'good' or 'bad' to evaluate multiple things; the two-stage fuzzy membership is used to combine the membership degree and inferiority degree to evaluate 'good' and 'bad' of the segmentation, and it can be calculated by the following equation:

$$u_j = \left[1 + \frac{\sum_{i=1}^{m} (w_i r'_{ij})^2}{\sum_{i=1}^{m} (w_i r_{ij})^2} \right]^{-1} \tag{22}$$

where w_i is the weight of i characteristic index. When u_j equals 1, the j segmentation is the optimal one.

Step 5. Optimization procedure

Assume that a flood season is divided into three sub-flood seasons, i.e. early flood season, main flood season, and late flood season. The guidelines for examining the rationality of the multiple flood season staging schemes and making an optimum selection by the proposed methods are as follow:

(1) Use the nonparameter bootstrap method to obtain N bootstrap samples from daily inflow series, then record the number of the annual maximum flood during each sub-flood season in each staging scheme, and calculate the relative frequency RF_{1j}, RF_{2j}, and RF_{3j}.

(2) Calculate the upper and lower limit of the relative frequency during each sub-flood season in each staging scheme and take the mean value of all staging schemes as the upper and lower limit of confidence intervals for each sub-flood season.

(3) Take the upper limit $y_{2\mathrm{up}}$ of main flood season and the lower limit $y_{1\mathrm{down}}$, $y_{3\mathrm{down}}$ of early flood season and late flood season as the benchmark index (confidence level $\alpha = 0.05$). The discrepancy extent between the relative frequency during each sub-flood season in each staging scheme and the benchmark index is represented by d_{ij} (generalized distance), and it can be expressed as follows:

$$d_{ij} = \begin{cases} \sum_{t=1}^{N} (y_{1\mathrm{down}} - \mathrm{RF}''_{1j}) \\ \sum_{t=1}^{N} \mathrm{RF}''_{2j} - y_{2\mathrm{up}} \\ \sum_{t=1}^{N} y_{3\mathrm{down}} - \mathrm{RF}''_{3j} \end{cases} \tag{23}$$

(4) Calculate the RMD (R'_{Fj}) in each staging scheme according to generalized distance, and figure out the RMD(R_{Qj}) of the mean peak flow during the main flood season in different staging schemes.

(5) Put the RMD (R'_{Fj}) of three sub-flood seasons in the same staging schemes into Equation (23), and assign an equal weight w_i as suggested by Chen *et al.* (2015) to each sub-flood season to calculate the two-stage fuzzy membership degree (u_{Fj}) of relative occurrences frequency of annual maximum flood (simplified as the membership degree of relative frequency).

(6) Under the multiple combinations of weight w_F and w_Q, we can obtain frequency-peak flood flow two-stage fuzzy membership degree (u_j) by putting the u_{Fj} and R_{Qj}, which are in the same staging scheme into Equation (23).

(7) Test the rationality and make an optimal selection according to the u_j. As suggested by Liu *et al.* (2010), when u_j is less than 0.8, the deflection between the jth staging schemes and bootstrap samples is large, and therefore, it can be regarded as an unreasonable scheme that should be abandoned. If the staging scheme is greater than 0.8, choose the largest u_j of the staging scheme as the optimal one.

CASE STUDY IN THE DOWNSTREAM REACHES OF THE YELLOW RIVER

In this study, the downstream reaches of the Yellow River were selected as the case study for evaluating the proposed framework.

The downstream reaches of the Yellow River refer to the reaches below the river section of the Huayuankou section, as shown in Figure 1, with a total length of 780 km.

The basin area upper the Huayuankou section is 730,000 km^2, which accounts for 97.1% of the Yellow River basin area. The runoff of the upper Huayuankou section accounts for 96.6% of the Yellow River total runoff. The variation of runoff and sediment in the Huayuankou section plays a leading role in flood control both in the upper and lower reaches of the Yellow River. The annual sediment transport passing through the Huayuankou section amounts to 1.6 billion tons/year, making the flood control situation complex.

There are 110 dangerous works and control projects in the downstream reaches of the Yellow River; the length of the project is 305.2 km, and the length of protection is 261.3 km. There are 2,830 dam stacks that play an important role in controlling the river regime.

In recent years, a lot of research has been carried out on the topic of flood control in the downstream reaches of the Yellow River, but there are few studies on the rational segmentation of the flood season.

In this paper, the maximum flood peak discharge data from 1950 to 2015 at the Huayuankou station were used, as shown in Figure 2.

There are five main control hydrology survey stations in the downstream reaches of the Yellow River, and the design flood on each station section is shown in Table 1 (Cai *et al.* 2021).

The flood prevention in the downstream reaches of the Yellow River is under the Yellow River Conservancy Commission (YRCC) supervision led by the Ministry of Water Resources of China. From the traditional rule, the start and end dates of the flood season are 1 June and 1 October without considering the period features of the entire flood season. The data used in this study are daily rainfall and daily flow. All the data are provided by the YRCC.

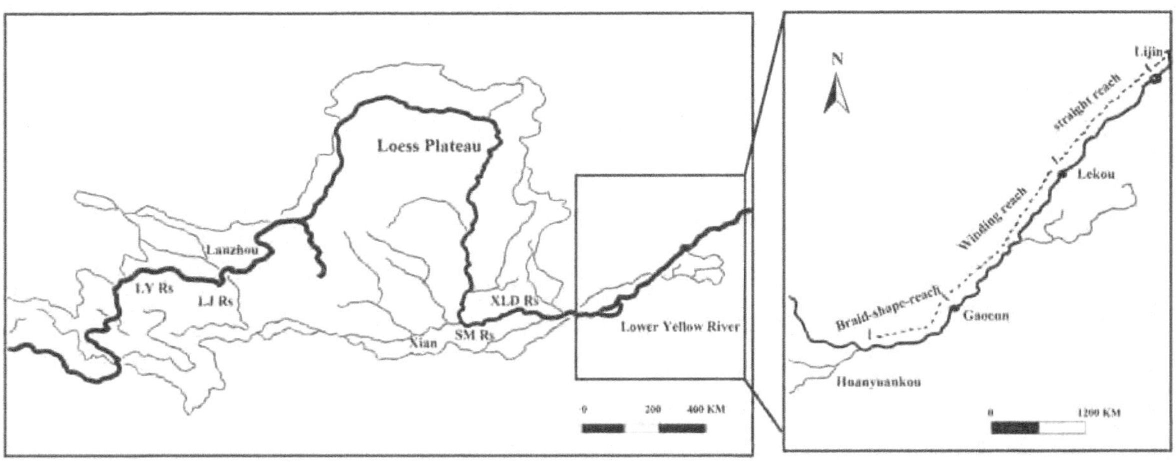

Figure 1 | The general situation and the locations of the different channel pattern reaches of the lower Yellow River.

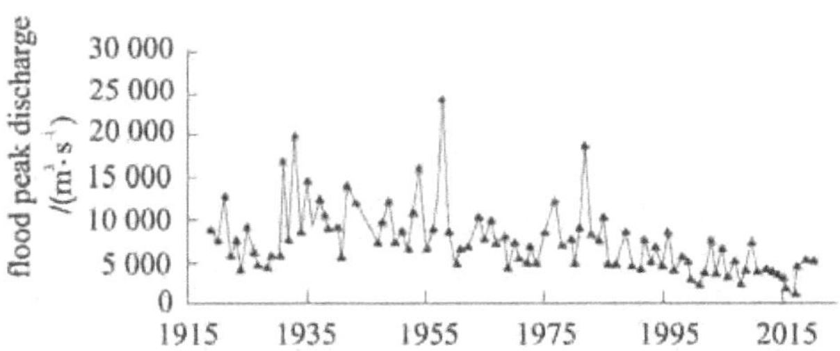

Figure 2 | The maximum annual flood peak discharge at the Huayuankou station.

Table 1 | Design flood discharge on the downstream reaches of the Yellow River (million m^3/s)

Stations	Probabilities (%)				
	0.1	1	2	5	10
Huayuankou	22,600	15,700	12,500	10,000	7,500
Jiahetan	21,000	15,070	12,056	9,645	7,200
Gaocun	20,300	14,400	11,520	9,200	7,000
Sunkou	18,100	13,000	10,400	8,300	6,200
Aishan	10,175	9,775	7,800	6,200	4,500

RESULTS AND DISCUSSION

Optimum numbers of segmentation

As shown in Figure 3(a), the curve was plotted based on the k (the base of the number of the optimal partition) and $L[b(n, k)]$ (the minimum loss function). The inflection of the curve lay at the point where k equals to 3, and where the loss function changed most sharply at the point $k = 3$. Also, as shown in Figure 3(b), when k approached 3, $\delta(k)$ reached its maximum and the value was 3.1806. According to the Fisher optimal partition method principle, the number of the optimal classification is 3. In summary, the flood season would be most appropriately segmented into three stages, i.e. early flood season, main flood season, and late flood season.

Comparison of different flood season segmentation methods

Segmentation by the FAM

Early flood season. The maximum daily rainfall from 1960 to 2019 was used as the fractal index. The average annual precipitation is 652.9 mm. The longest continuous rainfall is 11 days (in 1957). The maximum annual precipitation is 1041.3 mm. The minimum annual rainfall is 375.9 mm. As suggested by Liu *et al.* (2010), the thresholds were 1.1 times the average value of the samples during each stage. The beginning time of the early flood season is 1 June. As shown in Table 2, initially T took the period from 01 June– 30 June, that is to say, $T = 30$ days. Then, the capacity dimension of $T =$ 30–50 days was calculated with 10 days as a step. According to Table 1, when $T = 50$ days (01 June–20 July), the slope of $\ln \varepsilon \sim \ln N_N(\varepsilon)$ has a large deviation, and the maximum deviation was 0.089, which was 5.18% (>5%) of the minimum capacity dimension (1.717), indicating that the capacity dimension mutated in $T = 50$ days. When $T = 30$–45 days, the maximum deviation was 0.015, which was 0.83% of the minimum capacity dimension. The capacity dimensions were approximately close, and therefore, it could be considered that they belong to the same sub-flood season. Thus, the duration of the early flood season was from 1 June to 15 July ($T = 45$ days).

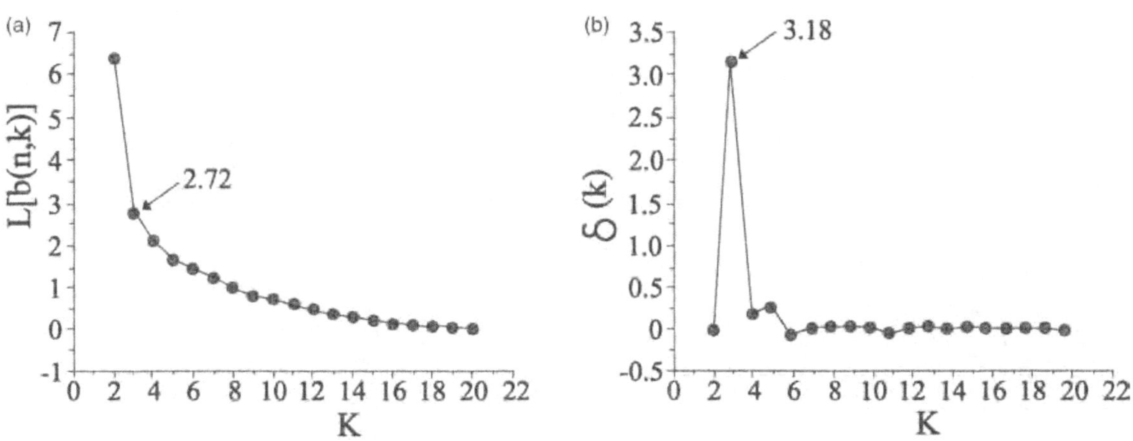

Figure 3 | Analysis results of the optimal partition number.

Table 2 | Determination of early flood season by the FAM

Time period	T (days)	Threshold level	Slope of $\ln \varepsilon \sim \ln N_N(\varepsilon)$	Capacity dimension
01 Jun.–20 Jun.	30	53.6	0.184	1.821
01 Jun.–10 Jul.	40	56.7	0.180	1.836
01 Jun. –15 Jul.	45	57.4	0.195	1.798
01 Jun.–20 Jul.	50	63.5	0.269	1.717

Table 3 | Determination of the main flood season by the FAM

Time period	T (days)	Threshold level	Slope of $\ln \varepsilon \sim \ln N_N(\varepsilon)$	Capacity dimension
16 Jul.–14 Aug.	30	87.9	0.273	1.740
16 July.–24 Aug.	40	85.0	0.267	1.725
16 July.–03 Sept.	50	89.4	0.304	1.687
16 July.–13 Sept.	60	86.3	0.293	1.704
16 July.–23 Sept.	70	80.5	0.283	1.713

Main flood season. Suppose that $T = 30$ days (16 July–14 August), then the capacity dimension of $T = 30$–70 days is calculated with 10 days as a step. As shown in Table 3, in $T = 30$–70 days, the maximum deviation of the slope of $\ln \varepsilon \sim \ln N_N(\varepsilon)$ is 0.037, which is 2.19% of the minimum capacity dimension (1.687), and the capacity dimensions were approximately close. Besides, the capacity dimension of $T = 70$ days was calculated with 5 days as the step size, and the results showed that the slope of $\ln \varepsilon \sim \ln N_N(\varepsilon)$ has a large deviation, and the maximum deviation was 0.109, which is 6.46% (>5%) of the minimum capacity dimension. Therefore, they were considered as the same sub-flood season, and then the duration of the main flood season is from 16 July to 23 September ($T = 70$ days).

Late flood season. In the same way, the late flood season takes 24 September as the beginning date. The initial assumption is $T = 30$ days (24 September–23 October), and then the capacity dimension of $T = 30$–50 days was calculated with 10 days as the step. It could be seen from Table 4 that the slope of each $\ln \varepsilon \sim \ln N_N\ (\varepsilon)$ is close. Moreover, each capacity dimension is approximately close, and hence, they could be considered to be of the same stage. In $T = 30$–50 days, the results showed that the maximum deviation of the slope of each $\ln \varepsilon \sim \ln N_N(\varepsilon)$ is 0.026, which is 1.81% of the minimum capacity dimension (1.436). Therefore, the late flood season was from 24 September to 12 November ($T = 50$ days).

In summary, the FAM was applied to divide the flood season into the early flood season (01 June–15 July, $T = 45$ days), main flood season (16 July–23 September, $T = 70$ days), and late flood season (24 September–12 November, $T = 50$ days).

Segmentation by MCP analysis

SPSS software was applied to generate a series of the maximum daily flow and maximum 7-day flood volume during flood season based on the daily flow and 7-day flood volume at the Huayuankou hydrometric section.

Table 4 | Determination of late flood season by the FAM

Time period	T (days)	Threshold level	Slope of $\ln \varepsilon \sim \ln N_N(\varepsilon)$	Capacity dimension
24 Sept.–23 Oct.	30	50.3	0.552	1.438
24 Sept.–03 Nov.	40	50.5	0.563	1.436
24 Sept.–12 Nov.	50	45.7	0.537	1.455

Table 5 | Segmentation by the MCP method

Index	Early flood season	Main flood season	Late flood season
Maximum daily flow	01 Jun.–15 Jul.	16 Jul.–23 Sept.	'24 Sept.–03 Nov.
Maximum 7-day flood volume	01 Jun.–17 Jul.	18 Jul.–25 Sept.	26 Sept.–05 Nov.

According to the results of the Fisher optimal partition method, the flood season should be divided into three stages, namely, the early flood season, the main flood season, and the late flood season. We take 15 days as the calculation step length and extend the selected duration forward and backward. As a result, the ranges of variation of two points were 18 June–20 July and 18 August–30 September. The MCP analysis is carried out by the least-square method and the stepwise adjustment method to gradually adjust the staging results. The results were shown in Table 5. It can be seen from Table 4 that, with the index of maximum daily flow, the flood season can be segmented as early flood season (01 June–15 July), main flood season (16 July–23 September), and late flood season (24 September–03 November); while with the index of maximum 7-day flood volume, the flood season can be segmented as early flood season (01 June–17 July), main flood season (18 July–25 September), and late flood season (26 September–05 November).

Segmentation by PCP analysis

According to the current situation of river flow and regulation rules, the flows (4,500, 5,500, 6,500, 7,500, and 8,500 m^3) were used as the thresholds for flow POT sampling. Through MCP analysis, the initial two change points were 1 August and 1 October. The results obtained by the PCP analysis method are shown in Table 6. With the threshold of 4,500 m^3, the flood season can be segmented as early flood season (01 June–15 July), main flood season (16 July–23 September), and late flood season (24 September–03 November); with the threshold of 5,500 m^3, the flood season can be segmented as early flood season (01 June–17 July), main flood season (18 July–25 September), and late flood season (26 September–05 November); with the threshold of 6,500 m^3, the flood season can be segmented as early flood season (01 June–20 July), main flood season (21 July–28 September), and late flood season (29 September–08 November); with the threshold of 7,500 m^3, the flood season can be segmented as early flood season (01 June–20 July), main flood season (21 July–23 September), and late flood season (24 September–10 November); with the threshold of 8,500 m^3, the flood season can be segmented as early flood season (01 June–25 July), main flood season (26 July–25 September), and late flood season (26 September–12 November).

Ensemble segmentation approach. The flood season segmentation in the downstream reaches of the Yellow River by an ensemble approach is shown in Table 7.

Rationality examination and optimized selection

To ensure that the bootstrap sampling can correctly reflect the distribution characteristics of the population and reduce the sampling deviation, we extracted 10,000 samples each time with replacement and repeated 10,000 times. According to the NPB-FOS method, u_{Fj} and R_{Qj} were calculated, respectively, and the results were shown in Table 8, with a confidence level of 0.05.

From Table 8, we can see that the best flood season segmentation scheme would be the MCP method with the index of maximum 7-day flood volume if only take the occurrence frequency of annual maximum floods in each sub-flood into

Table 6 | Segmentation by the PCP analysis method

Threshold (m^3/s)	Early flood season	Main flood season	Late flood season
4,500	01 Jun.–15 Jul.	16 Jul.–23 Sept.	24 Sept.–03 Nov
5,500	01 Jun.–17 Jul.	18 Jul.–25 Sept.	26 Sept.–05 Nov.
65,000	01 Jun.–20 Jul.	21 Jul.–28 Sept.	29 Sept.–08 Nov.
7,500	01 Jun.–20 Jul.	21 Jul.–23 Sept.	24 Sept.–10 Nov.
8,500	01 Jun.–25 Jul.	26 Jul.–25 Sept.	26 Sept.–12 Nov.

Table 7 | The flood season segmentation with an ensemble method

Methods	Samples	Early flood season	Main flood season	Late flood season
FAM	Maximum daily rainfall	01 Jun.–15 Jul.	16 Jul.–23 Sept.	24 Sept.– 12 Nov.
MCP	Maximum daily flow	01 Jun.–15 Jul.	16 Jul.–23 Sept.	24 Sept.– 03 Nov.
	Maximum 7-day flood volume	01 Jun.–17 Jul.	18 Jul.–25 Sept.	26 Sept.– 05 Nov.
PCP	4,500	01 Jun.–15 Jul.	16 Jul.–23 Sept.	24 Sept.– 03 Nov.
	5,500	01 Jun.–17 Jul.	18 Jul.–25 Sept.	26 Sept.– 05 Nov.
	65,000	01 Jun.–20 Jul.	21 Jul.–28 Sept.	29 Sept.– 08 Nov.
	7,500	01 Jun.–20 Jul.	21 Jul.–23 Sept.	24 Sept.–10 Nov.
	8,500	01 Jun.–25 Jul.	26 Jul.–25 Sept.	26 Sept.–12 Nov.

Table 8 | RMD of the frequency and average flood peak in main flood season

Methods and samples	u_{Fj}	R_{Qj}
FAM	0.9968	0.9051
MCP (max. daily)	0.9980	0.9408
MCP (max. 7 days)	1.0000	0.9502
PCP (4,500 m^3/s)	0.9996	0.9087
PCP (5,500 m^3/s)	0.9975	0.9454
PCP (6,500 m^3/s)	0.9982	0.9665
PCP (7,500 m^3/s)	0.9990	0.9525
PCP (8,500 m^3/s)	0.6550	1.0000

account. The reason is that this segmentation scheme spans the longest duration during the main flood season, and thus, the annual maximum flood occurs most frequently.

As the mean peak flood flow in the main flood season as concern, the period obtained by the PCP method with a threshold of 4,500 m^3 is the best one, whose duration of the main flood season is the shortest. The shorter duration of the main flood season, the bigger the mean peak flood flow in the main flood season would be. In conclusion, it was difficult to select the optimum segmentation scheme only by the membership degree of relative frequency (u_{Fj}) or the RMD of mean peak flood flow in the main flood season (R_{Qj}). To fully consider the frequency and magnitude of flood occurrence, the NPB-FOS method is applied. In the optimization process, we adjusted their weight in the range from 0.3 to 0.7 with the step size of 0.1 and kept the sum equal to 1 to eliminate the influence of different weight value. The results of the two-stage fuzzy optimum selection are shown in Table 9.

Table 9 | Two-stage fuzzy comprehensive relative membership degree of different segmentations

	u_{Fj}: u_{Qj}				
Main flood season	= 0.3:0.7	= 0.4:0.6	= 0.5:0.5	= 0.6:0.4	= 0.7:0.3
16 Jul.–23 Sept.	0.9911	0.9929	0.9950	0.9971	0.9986
16 Jul.–23 Sept.	0.9967	0.9974	0.9981	0.9989	0.9994
18 Jul.–25 Sept.	0.9977	0.9982	0.9987	0.9992	0.9996
16 Jul.–23 Sept.	0.9919	0.9935	0.9954	0.9973	0.9987
18 Jul.–25 Sept.	0.9970	0.9976	0.9983	0.9990	0.9995
21 Jul.–28 Sept.	0.9983	0.9987	0.9990	0.9994	0.9997
21 Jul.–23 Sept.	0.9979	0.9983	0.9988	0.9993	0.9996
26 Jul.–25 Sept.	0.9800	0.9572	0.9226	0.8795	0.8362

As shown in Table 9, the u_{Fj} of the segmentation for the main flood season, i.e. 21 July–28 September, is the largest regardless of the weight combinations, and therefore, the segmentation obtained by the PCP method is the desirable scheme that takes both the relative frequency of occurrence and the mean peak flood flow into consideration.

CONCLUSIONS

The determination of the optimal number of the sub-flood seasons, the rational test, and the optimal selection of the flood season segmentation schemes have been a major problem for a large river flood prevention and have not been effectively solved. In this study, a scientific and rational flood season segmentation framework was put forward through a comparison of different segmentation methods. A systematic framework that consists of a Fisher optimal partition method, an ensemble approach, and an NPB-FOS was put forward, which can scientifically determine the optimum numbers of the segmented sub-flood seasons, divide a flood season into several sub-seasons, examine the rationality of the flood season segmentation schemes, and select the optimal schemes.

As a case study, through analysis of the Yellow River's downstream reaches, some research findings can be achieved: (1) The entire flood season can be segmented as three sub-flood seasons, i.e early, middle, and late flood seasons. (2) It can be concluded through comparison analysis that segmentation obtained by the PCP method is better than that obtained by other methods. (3) The best segmentation stages for the downstream reach of the Yellow River are early flood season (01 June–20 July), main flood season (21 July–28 September), and late flood season (29 September–08 November).

The results of this study show that September still belongs to the main flood season (the main flood season is from 21 July to 28 September), and the duration (29 September–08 November) belongs to the late flood season which is different from the traditional flood season partition (1 October is the end of flood season). As is well known, the main flood season is prone to heavy rainfall and flood events; if the water level is raised at this stage, it would increase flood risk for the downstream of the river. Therefore, it is suggested that the relevant flood prevention authorities prepare large flood events in this stage so that the downstream safety is ensured under the scientific management.

The proposed Yellow River flood season segmentation framework could be applied to other river flood season segmentation as well as reservoirs. However, there are still some shortcomings waiting to be discussed, such as that the selection of indexes and thresholds inevitably has some subjectivity, which may bring some errors to the result of stages.

FUNDING

The study was supported by the Natural Science Fund of China (No. 50579020).

CONFLICT OF INTEREST

The author declares that there are no known competing financial interests or personal relationships that could have appeared to influence the work reported in this paper.

DATA AVAILABILITY STATEMENT

All relevant data are included in the paper or its Supplementary Information.

REFERENCES

Adhikari, U., Nejadhashemi, A. P., Herman, M. R. & Messina, J. P. 2016 Multiscale assessment of the impacts of climate change on water resources in Tanzania. *J. Hydrol. Eng.* **50**, 05016034.

Al-Jawad, J. Y., Alsaffar, H. M., Bertram, D. & Kalin, R. M. 2019 A comprehensive optimum integrated water resources management approach for multidisciplinary water resources management problems. *J. Environ. Manage.* **239**, 211–224. https://doi.org/10.1016/j.jenvman.2019.03.045.

Bender, M. J. & Simonovic, S. P. 2000 A fuzzy compromise approach to water resources systems planning under uncertainty. *Fuzzy Sets Syst.* **115**, 35–44.

Cai, R., Zhang, H., Bu, H., Shi, Z. & Dai, W. 2021 Flood frequency analysis of the lower Yellow River. *J. Yellow River* **43** (3), 54–57.

Chen, L., Singh, V. P., Guo, S., Fang, B. & Liu, P. 2013 A new method for identification of flood seasons using directional statistics. *Hydrol. Sci. J.* **58** (1), 28–40.

Chen, L., Singh, V. P., Guo, S., Zhou, J., Zhang, J. & Liu, P. 2015 An objective method for partitioning the entire flood season into multiple sub-seasons. *J. Hydrol.* **528**, 621–630.

Cunderlik, J. M., Ouarda, T. B. M. J. & Bobée, B. 2004 Determination of flood seasonality from hydrological records. *Hydrol. Sci. J.* **49** (3), 511–529.

Efron, B. 1979 Bootstrap methods: another look at the jack knife. *Ann. Statist.* **7** (1), 1–26. doi:10.1007/978-1-4612-4380-9_41.

Esogbue, A. O., Theologidu, M. & Guo, K. 1992 The application of fuzzy sets theory to the optimal flood control problem arising in water resources systems. *Fuzzy Sets Syst.* **48** (2), 155–172.

Hajkowicz, S. & Collins, K. 2007 A review of multiple attributes analysis for water resource planning and management. *Water Resour. Manage.* **21** (9), 1553–1566.

Jiang, Z., Sun, P., Ji, C. & Zhou, J. 2015 Credibility theory based dynamic control bound optimization for reservoir food limited water level. *J. Hydrol.* **529**, 928–939.

Ledingham, J., Archer, D., Lewis, E., Fowler, H. & Kilsby, C. 2019 Contrasting seasonality of storm rainfall and flood runoff in the UK and some implications for rainfall-runoff methods of flood estimation. *Hydrol. Res.* **50** (5), 1309–1323.

Lin, P., You, J., Gan, H. & Jia, L. 2020 Rule-based object-oriented water resource system simulation model for water allocation. *Water Resour. Manage.* **34**, 3183–3197.

Liu, P., Guo, S., Xiong, L. & Chen, L. 2010 Flood season segmentation based on the probability change-point analysis technique. *Hydrol. Sci. J.* **55** (4), 540–564.

Macdonald, N., Phillips, I. D. & Mayle, G. 2010 Spatial and temporal variability of flood seasonality in Wales. *Hydrol. Process.* **24** (13), 1806–1820.

Mandelbrot, B. 1982 *The Fractal Geometry of Nature*. Times Books, New York.

Mo, C., Mo, G., Liu, P., Zhong, H., Wang, D., Huang, Y. & Jin, J. 2018 Reservoir operation by staging due to climate variability. *Hydrol. Sci. J.* **63** (6), 926–937. doi:10.1080/02626667.2018.1457220.

Nicklow, J. F., Reed, P., Savic, D., Dessalegne, T., Harrell, L., Chan-Hilton, A., Karamouz, M., Minsker, B., Ostfeld, A., Singh, A. & Zechman, E. 2010 State of the art for genetic algorithms and beyond in water resources planning and management. *J. Water Resour. Plan. Manag.* **136** (4), 412–432.

Perreault, L., Bernier, J., Bobée, B. & Parent, E. 2000 Bayesian change-point analysis in hydrometeorological time series. Part 1. The normal model revisited. *J. Hydrol.* **235** (3), 221–241.

Roach, T., Kapelan, Z. & Ledbetter, R. 2018 Resilience-based performance metrics for water resources management under uncertainty. *Adv. Water Resour.* **116**, 18–28.

Ye, S., Li, H. Y., Leung, L. R., Guo, J., Ran, Q., Demissie, Y. & Sivapalan, M. 2017 Understanding flood seasonality and Its temporal shifts within the contiguous United States. *J. Hydrometeorol.* **18** (7), 1997–2009. doi:10.1175/jhm-d-16-0207.1.

First received 22 March 2021; accepted in revised form 20 May 2021. Available online 14 June 2021

 Impact of Climate Change on Hydrology and Water Resources doi: 10.2166/wcc.2021.133

Flood risk assessment in Ya'an, Sichuan, China based on the emergy theory

Xiaolong Shu[a,b], Yufeng Ren[a,c], Zhe Duan[b], Xing Liu[b,*], Xiaojun Hua[a,c] and Huike Lei[b]

[a] Hubei Key Laboratory for Smart Yangtze River and Hydroelectric Sciences, Yichang, Hubei 443133, China
[b] College of Water Conservancy and Hydropower Engineering, Sichuan Agricultural University, Ya'an, Sichuan 477200, China
[c] China Yangtze Power Co., Ltd, Yichang, Hubei 443133, China
*Corresponding author. E-mail: xingliud178@tom.com

ABSTRACT

The emergy theory provides a new approach for flood risk assessment from an ecological perspective. By employing the emergy method, we used five indicators (rainfall runoff, medical workers and students per 10,000 people, social fixed assets investment, unit land GDP, and land-use types) from three dimensions (natural environment, population, and social economy) and the GIS technique to assess the potential impact and risk of a flood disaster on different regions in Ya'an City. Our findings revealed regional differences in the distribution of flood risks in Ya'an City: Lushan County and Yucheng District face the highest risks, followed by Tianquan County and Mingshan District, and Shimian County has the lowest risk. The index method was employed to analyze the regional differences. By training a back-propagation neural network with data on flood disasters in the study area, we produced a flood risk distribution map. We found that Mingshan District, Lushan County and Yucheng District have higher risks than other regions. The results largely agree with what we obtained using the emergy method. Our study shows that flood risk assessment based on the emergy theory can provide a scientific basis for flood control and disaster relief initiatives.

Key words: assessment indicators, BP neural network, emergy theory, flood risk assessment

HIGHLIGHTS

- Emergy theory generates a new approach for flood risk assessment.
- Research findings were verified by the reality of the study area, ensuring the validity.
- Analysis for BP neural network training was conducted in the area experiencing a flood.
- The flood risks in the study area assessed by using the emergy theory are in agreement with the reality.

1. INTRODUCTION

Floods are one of the most common hazards worldwide, have a wide-ranging impact and present a serious threat to socio-economic development and people's lives and properties (Foudi Osés-Eraso & Tamayo 2015). In July 2020, most parts of Sichuan, Chongqing, Hubei, Anhui and Zhejiang around China experienced multiple rounds of heavy rainfall, resulting in severe flood disasters around many regions, leading to some major losses to the sustainable development of our national economy, jeopardizing the security of people's lives and properties, and affecting social stability. Recently, flood risk assessment has been a heated topic in hydrology and catastrophology. Analysis of the flood hazard is key to disaster risk assessment and control and provides a timely solution to early warning, flood control, and disaster relief.

Currently, most efforts for flood risk assessment in China and abroad have been made with historical flood data, hydrological and hydrodynamic models, and index methods (Tanoue *et al.* 2016; Abdulrazzak *et al.* 2019; Yongzhi *et al.* 2021). Bhuiyan & Baky (2014) employed ArcGIS to map the flood hazard distribution around the low-lying areas of Bangladesh, and used digital elevation model (DEM) data to extract index factors, to work out the flood risk analysis diagram for different scales of floods. Kazakis *et al.* (2015) employed the analytic hierarchy process to determine the weight of total flow, altitude, land-use type and other attributes, and superposed different parameter information following their weighted values, and completed flood risk assessment. Chengguang *et al.* (2015) selected 10 assessment indexes based on disaster-inducing factors, hazard-inducting environment, and hazard-affected bodies, built up a random woodland intelligence algorithm-based flood risk assessment model, and then used the GIS technology to assess the flood risks in the Dongjiang River Basin. Sun *et al.* (2017) used the GIS software to summarize the flood risk analysis mode from the three aspects of historical data, system indexes, and remote sensing data, commented on the features and the short-comings of some common models, and proposed that extended flood risk analysis models and enhanced application of the 3S technology to flood risk analysis would be the direction for us to enhance and extend flood risk analysis in the future. Hanif *et al.* (2020) investigated the impact of short-term and long-term climate changes on the hydrological regime and flood risks, generated a flood map to identify the vulnerability, and formulated some flood adaptive strategies for 43 regions around the globe.

The indexes in these aforementioned research works are usually not interrelated, and for these works have determined the weights by analyzing the importance of the factors to ultimately work out the risk assessment indexes. They have ignored the potential relation between different indexes and would not help to assess the flood risks accurately. The emergy analysis approach, created by American ecologist H. T. Odum, refers to a kind of emergy accounting of the ecological–economic system to assess the status and function of different types of energy in the system, so it is a quantitative research method to assess flood risks (Huang & Odum 1991). This approach converts different types of incomparable energy into the same standard emergy for comparison to quantitatively analyze the natural system and the human socio-economic system, the real value of resources and the environment, and their interrelationship (Luo *et al.* 2019). Wu *et al.* (2018), based on the emergy theory, calculated the rainstorm intensity, susceptibility, and adaptive capacity to assess the vulnerability of each district of Zhengzhou in Henan province to flood disasters, and the results showed that different districts have different demand for rainfall flood control, and the districts of higher vulnerability demand higher. Chang & Huang (2015) pointed out that the emergy approach works to highlight the exposure, susceptibility, and adaptive capacity, and spatialized calculation of the emergy indexes can help assess the risk of floods. Therefore, the infrastructure management and regional planning are critical factors to flood risks. Based on the emergy theory, this paper assesses the flood risks in Ya'an and is expected to provide a new research direction for flood risk assessment in the future.

This paper is organized as follows: First, the introduction introduces some approaches to flood risk assessment in China and abroad and the status quo of research on application of the emergy theory to risk assessment. The Research Methods section elaborates on the principles of the emergy-based risk assessment model, an overview of the survey region, and data sources and processing. The Results and Discussion section uses the emergy method and the back propagation (BP) neural network-based index method to assess the risk of floods and analyze the results. The last section concludes the paper and points out some prospects for future efforts.

2. METHODS

2.1. The basis for application of emergy theory to floor risk assessment modeling

The emergy theory and method proposed by American ecologist H. T. Odum work to convert different types of incomparable energy into standardized comparable values based on the emergy conversion rate. This approach measures the emergy of different types of energies in the ecological system that consists of the human social system and the natural ecological system on the benchmark of solar energy. Based on solar energy, the energy in the ecosystem can be converted into the solar energy of a unified standard through the emergy conversion rate equation (Zhao 2013):

$$E = \tau \times B \tag{1}$$

where E stands for emergy (seJ), τ is the emergy conversion rate (seJ/J or seJ/g), and B is the volume of energy or substance (J/g).

The flood disaster risk energy system is established based on the causal relationship between the systems comprised of the city system and the nature–agriculture system to demonstrate the different energy and material flow between the ecosystem and the economic system (Wangyang *et al.* 2019). When rainfall takes place, the rainwater-stored energy (N1) may endanger the rainfall area. After deducting buildings' and vegetation's retainment, evapotranspiration and infiltration losses, rainfall is then converted into an overland runoff, and the energy stored in the total overland runoff accumulation represents the exposure in this region (N2). The exposure intensity (N3) is the ratio of the emergy stored in the volume of runoff to the emergy stored in the rainfall. When extreme weather takes place, the sum of the energy loss of the city system (N4) and the energy loss of the nature–agriculture system (N5) is called susceptibility (N6). The possibility of flood disaster in the system has positive correlation with the energy it stores. Thanks to extreme weather, the environment also changes accordingly, and the disaster-stricken region is also affected; this impact is denoted by the potential impact index (N7). The

Table 1 | Emergy flow of flood hazard risk

Energy flow (seJ)	Description	Definition
N1	Potential exposure *I*	Rainfall
N2	Exposure *L*	Total volume of runoff
N3	Exposure intensity	N2/N1
N4	Natural–agricultural system susceptibility	Property losses in the natural–agricultural system caused by extreme weather
N5	City system susceptibility	Property losses in the city system caused by extreme weather
N6	Susceptibility *M*	N4 + N5
N7	Potential impact index	N6 × N3
N8	Population indicator	Demographic factor in the survey region
N9	Social and economic indicator	Social and economic factors in the survey region
N10	Adaptive ability *N*	N8 + N9
N11	Risk	N7/N10

potential impact index is related to the exposure and susceptibility and the product of susceptibility and exposure intensity. In addition to the natural environment, the factors affecting a region's disaster risk include the demographic factor (N8) and the socio-economic factor (N9). They together demonstrate the adaptive capacity of a system (N10) and the coping capacity of this region to any natural hazard. Finally, we use the potential impact and adaptive capacity to assess flood disaster risk in the survey region (N11). Table 1 lists the definitions of different energy flows. For the working principles, see 'Risk assessment modeling'.

2.2. Risk assessment modeling

A risk refers to how a system is adversely affected, which can be indicated by exposure, susceptibility, adaptive capacity, and potential impact indexes. By analyzing these constituent factors of risks, and exploring their inter-relations, we can put forward some emergy indicators to represent the risk of floods and thus accurately assess the risk of floods in a region.

(1) Exposure L

Exposure refers to the extent to which a region is involved in a natural disaster; for instance, in a flood, a region that is more susceptible to a flood has higher exposure (Wanzhi *et al.* 2019). When a flood takes place, the exposure of a region to the flood can be represented by the total emergy of the accumulative runoff in this region, i.e., the product of the accumulative runoff and its corresponding emergy conversion rate.

$$L = r \times \tau \tag{2}$$

where L refers to exposure (seJ); r is the runoff depth (mm); τ is the emergy conversion rate (seJ/J or seJ/g).

(2) Susceptibility M

Susceptibility refers to the extent to which the system in a region is affected by extreme weather (Chang & Huang 2015). As a response of a system to extreme weather, susceptibility is not just decided by the scale of a hazard like a flood, but also dependent on relevant features of the disaster-stricken region. Different land-use types stock different energies, so when a flood disaster occurs, the impact on different regions of different land-use types also differs. Thus, we can analyze the susceptibility based on the emergy power intensity of different land types (Jun *et al.* 2015). Therefore, the susceptibility of a region is denoted by the total energy stored by different land types.

$$M = \rho \times A \tag{3}$$

Table 2 | Factors affecting adaptive capacity

Contributing factors	Classification	Definition
Society	Medical workers per 10,000 people	Medical workers/(total population of a region/10,000), indicating local medical capacity
	Enrolled students per 10,000 people	Students enrollment/(total population of a region/10,000), indicating overall education
Economy	Unit land GDP	Region GDP/region area, indicating local economic effectiveness, which has positive correlation to the adaptive capacity
	Total social fixed assets	In positive correlation to the adaptive capacity
Demography	Population density	Year-end resident population/region area, the region of higher-density population distribution seeing lower adaptive capacity
	Proportion of people below 14 and above 65 years old	Number of this group/resident population, this group being highly susceptible, and less adaptive to a natural disaster

where M refers to the susceptibility of the region (seJ), ρ is the emergy power intensity of different land-use types (seJ/m^2), and A is the area of different land-use types (m^2).

(3) Adaptive capacity N

In addition to exposure, the risk of a system is also decided by its adaptive capacity. Adaptive capacity refers to the capacity of a system to make continuous progress and adjustments to adapt itself to the environment when any natural disaster is triggered by extreme weather (Wahab & Ludin 2018). The adaptive capacity of a system constantly changes along with the changing environment, and different regions and groups have different adaptive capacity to the environment. For instance, an underdeveloped rural village has a lower adaptive capacity than a developed region of complete infrastructure, and adults have a much stronger adaptive capacity to the environment than children and the aged. Therefore, the adaptive capacity of a region should be assessed from the three aspects of society, economy, and demographics. The classification and interpretation of each influencing factor are shown in Table 2.

(4) Potential impact index I

When extreme weather occurs, the impact on the system can be denoted by the potential impact index (Yangfan 2018). The risk of a flood disaster is in positive correlation to the potential impact. The product of exposure and susceptibility denotes the potential impact index.

$$I = i \times M \tag{4}$$

$$i = \frac{L}{q} \tag{5}$$

where I refers to potential impact index (seJ), I is exposure intensity, M is the susceptibility of the region (seJ), L is exposure (seJ), and q is potential exposure (seJ).

(5) Risk index V

Risk index refers to the ratio of the potential impact to the adaptive capacity, indicating the damage of the disaster to a system in case of extreme weather (Chao *et al.* 2020). The risk is positively correlated to the system's potential impact index, but in negative correlation to the adaptive capacity.

$$V = \frac{I}{N} \tag{6}$$

where V refers to the index of vulnerability, I is the potential impact index (seJ), and N is the adaptive capacity.

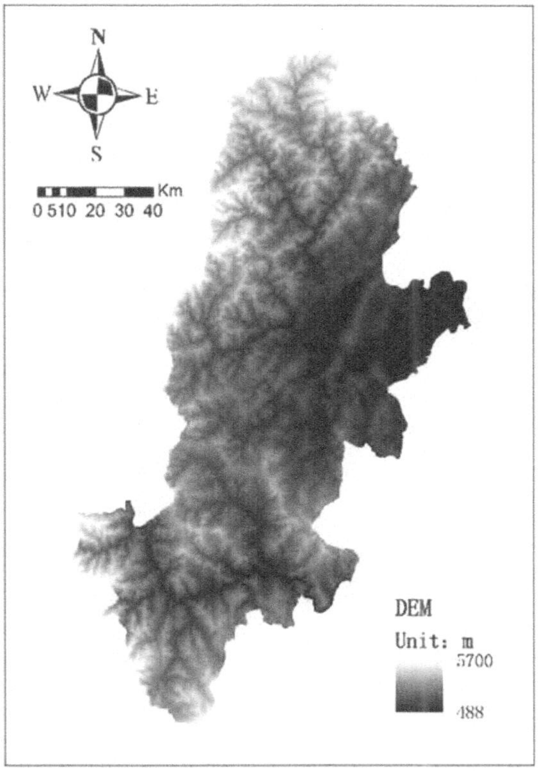

Figure 1 | Elevation map of Ya'an.

2.3. Overview of the study area

Ya'an (101°56′–103°23′E, 28°51′–30°56′N), located on the western margin of the Sichuan Basin, is in the transition zone from the Basin to the Qinggai-Tibet Plateau. It is a long strip stretching about 220 km long from north to south and spreads about 70 km wide from east to west. Ya'an City has jurisdiction over six counties and two districts (Figure 1), covering a total area of 15,046 km^2. The city boasts quite diversified topographic features. Mountains cover the largest area within its borders, with higher terrain in its south, west, and north, and low-lying land in its center and east. Thanks to its unique 'windward slope' and 'horn mouth' terrains to lift up water vapor in a cyclic process for coagulation to form precipitation, the city has 218 annual mean rainfall days and is called the 'Rainy City'. In all four seasons in a year it rains, so all districts and counties in Ya'an make a potentially high-risk region. Statistics showed that during the flood season in 2018, Ya'an experienced 18 regional rainstorms, 465,100 people in 137 townships (towns and sub-districts) in 8 counties (districts) of the city were affected, with a total economic loss up to 2.01 billion yuan. These figures indicate that the rainstorm and flood disasters exert quite a strong impact on the economic development of Ya'an, and are important factors hindering the economic development of Ya'an City.

2.4. Data sources

The data sources for this study and their specifications are shown in Table 3.

2.5. Data processing

The rainfall runoff data for this study were obtained from Ya'an Municipal Communique on Water Resources 2018. The number of medical workers, enrolled students, people below 14 and above 65 years old, social fixed assets investment, and GDP were obtained from *Ya'an Statistics Yearbook* and *Sichuan Statistics Yearbook*. The main land-use data of Ya'an were obtained from Natural Resources and Planning Bureau of Ya'an City, *Ya'an Statistics Yearbook*, and Resource and Environment Science and Data Center, China Academy of Sciences (Figure 2) in combination with the Geographic Information System (GIS). The data for the BP neural network were normalized to remove the dimension before the network

Table 3 | Data presentation

Method	Category	Description	Source	Specification
Emergy theory	Hydrology and water resource	Rainfall runoff	Ya'an Municipal Communique on Water Resources 2018	Various water resources data for Ya'an in 2018
	Society and economy	Medical workers	Ya'an Statistics Yearbook and Sichuan Statistics Yearbook	Number of medical workers in each district or county
		Student enrollment		Number of students enrolled in each district or county
		People below 14 and above 65 years old		Population of this group in each district or county
		Social fixed assets investment		Social fixed assets investment in each district or county
		GDP		Total output value for each district or county
	Land-use type L	Main land-use type data	Natural Resources and Planning Bureau of Ya'an City, Ya'an Statistics Yearbook, and Resource and Environment Science and Data Center, China Academy of Sciences	Various land-use resources data
BP neural network	Disaster-inducing factors	TRMM 3B43 satellite precipitation	NASA, USA https://disc.gsfc.nasa.gov/	Global 0.25° × 0.25° monthly precipitation raster data 1998–2016
		Annual runoff depth isogram	Hydrology and Water Resources Bureau of Sichuan Province http://www.schwr.com/	1956–2000 Synchronization series
		Land cover UMD	University of Maryland, USA https://www.umd.edu/	Global land cover classification at 1 km spatial resolution
		Soil type	Food and Agriculture Organization (FAO) http://www.fao.org/home/en/	Global land cover classification at 1 km spatial resolution
		DEM	Geospatial data cloud http://www.gscloud.cn/	SRTM V4.1 data merged, disposed, and generated as SRTM V4.1 data
	Disaster	Flood-stricken points in Ya'an in history	Flood Control & Drought Relief Headquarters, People's Government of Sichuan Province/ Information collected online	79 villages as of the provincial headquarters and 10 villages collected online

training. Specifically, the data were normalized by Matlab to obtain the weights of various indexes and conduct overlay analysis in the GIS.

Data obtained are arranged as shown in Table 4.

3. RESULTS AND DISCUSSION

3.1. Calculated results

Based on the above data, we calculated the exposure, susceptibility, and adaptive capacity of Ya'an City with the method mentioned in the Methods section.

3.2. Exposure

Exposure is the product of accumulative runoff and its corresponding emergy conversion rate (Jun *et al.* 2015), which can be calculated via Equation (2). The results are shown in Table 5.

3.3. Susceptibility

A region's susceptibility to disaster can be denoted by the total energy stored by different land-use types (Chang & Huang 2015). In this study, the susceptibility of different land-use types can be calculated with Equation (3). The results are shown in Table 6.

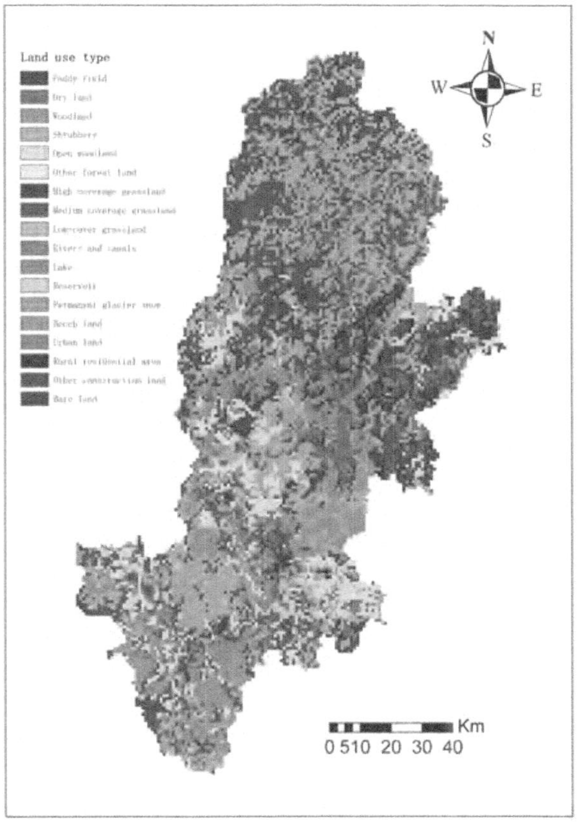

Figure 2 | Types of land use in Ya'an City.

Table 4 | Data of Ya'an City by type

Region	Shimian	Hanyuan	Baoxing	Lushan	Tianquan	Xingyang	Yucheng	Mingshan
Accumulative runoff (100 million m^3)	49.29	27.30	42.33	22.59	60.27	44.98	23.57	9.85
Medical workers (10,000 people)	111.00	65.00	73.00	67.00	102.00	78.00	171.00	56.00
Enrolled students (10,000 people)	1,769.00	1,298.00	1,033.00	1,197.00	1,373.00	1,355.00	1,559.00	1,224.00
Social fixed assets investment (100 million RMB Yuan)	51.04	45.54	25.96	28.59	48.24	47.30	106.49	62.83
Unit land GDP on average (10,000 RMB/km^2)	306.91	373.68	107.86	362.72	268.62	417.05	1,672.62	1,314.24
Farmland (km^2)	87.00	295.00	42.00	84.00	123.00	94.00	192.00	166.00
Woodland (km^2)	798.00	501.00	818.00	256.00	1,971.00	416.00	475.00	147.00
Area of land for mining and industry (km^2)	3.00	3.00	1.00	1.00	2.00	1.00	3.00	8.00
Area of land for residence (km^2)	153.00	17.00	212.00	43.00	50.00	20.00	193.00	43.00
Area of land for transportation purposes (km^2)	0.56	0.61	0.17	0.48	8.08	0.99	3.75	0.55
Water bodies (km^2)	36.00	74.00	4.00	4.00	3.00	3.00	10.00	4.00

Table 5 | Results of exposure calculation

Indicators	Shimian	Hanyuan	Baoxing	Lushan	Tianquan	Xingyang	Yucheng	Mingshan
Accumulative runoff (100 million m^3)	49.29	27.30	42.33	22.59	60.27	44.98	23.57	9.85
Exposure (10^{20} seJ)	27.99	15.51	24.05	12.84	34.24	25.55	13.39	5.60

Table 6 | Sensitivity calculation results

Region	Shimian	Hanyuan	Baoxing	Lushan	Tianquan	Xingyang	Yucheng	Mingshan
Farmland (km^2)	87.00	295.00	42.00	84.00	123.00	94.00	192.00	166.00
Emergy (10^6 seJ)	1.07	3.63	0.52	1.03	1.51	1.16	2.36	2.04
Woodland (km^2)	798.00	501.00	818.00	256.00	1,971.00	416.00	475.00	147.00
Emergy (10^6 seJ)	4.82	3.03	4.94	1.55	11.91	2.51	2.87	0.89
Area of land for residence (km^2)	3.00	5.00	2.00	12.00	5.00	5.00	21.00	6.00
Emergy (10^8 seJ)	3.38	5.64	2.25	13.53	5.64	5.64	23.68	6.76
Area of land for mining and industry (km^2)	3.00	3.00	1.00	1.00	2.00	1.00	3.00	8.00
Emergy (10^7 seJ)	5.44	5.44	1.81	1.81	3.63	1.81	5.44	14.52
Area of land for transportation purposes (km^2)	0.56	0.61	0.17	0.48	8.08	0.99	3.75	0.55
Emergy (10^7 seJ)	1.59	1.73	0.48	1.36	22.90	2.81	10.63	1.56
Water bodies (km^2)	36.00	74.00	4.00	4.00	3.00	3.00	10.00	4.00
Emergy (10^2 seJ)	9.74	20.02	1.08	1.08	0.81	0.81	2.71	1.08
Susceptibility (10^{24} seJ)	4.15	6.43	2.54	13.90	8.44	6.15	25.39	8.42

3.4. Adaptive capacity

When a natural disaster occurs due to extreme weather, the system, to alleviate the resulted losses, can constantly make progress and adjustment to adapt to the changing environment (Wanzhi *et al.* 2019). The results of the adaptive capacity calculation are shown in Table 7.

The above calculated results reveal the following findings:

(1) Table 5 shows that the bigger the accumulation runoff in the system, the more energy it will store, and the higher the exposure will be. Tianquan County has the highest annual precipitation, and generates the largest accumulative runoff, resulting in high emergy storage, so Tianquan County is the region with the highest exposure. Mingshan District sees relatively low annual precipitation, so its accumulative runoff is relatively small, leading to relatively low emergy storage, so the exposure is also relatively low.

(2) Based on the emergy power density of different land-use types, we calculate the emergy of different land-use types for all districts and counties of Ya'an City. It can be seen from the results that areas with high energy values, such as residential areas, industrial and mining areas, and transportation, are highly sensitive, so floods are at greater risk. The areas with a

Table 7 | Results of adaptive capacity calculation

Region	Shimian	Hanyuan	Baoxing	Lushan	Tianquan	Xingyang	Yucheng	Mingshan
Medical workers (10,000 people)	111.00	65.00	73.00	67.00	102.00	78.00	171.00	56.00
Emergy (10^{22} seJ)	1.49	0.87	0.98	0.90	1.37	1.04	2.29	0.75
Enrolled students (10,000 people)	1,769.00	1,298.00	1,033.00	1,197.00	1,373.00	1,355.00	1,559.00	1,224.00
Emergy (10^{22} seJ)	4.14	3.04	2.42	2.80	3.21	3.17	3.65	2.87
Social fixed assets investment (100 million RMB Yuan)	51.04	45.54	25.96	28.59	48.24	47.30	106.49	62.83
Emergy (10^{21} seJ)	3.27	2.92	1.67	1.83	3.09	3.03	6.83	4.03
Unit land GDP on average (10,000 RMB Yuan/ km^2)	306.91	373.68	107.86	362.72	268.62	417.05	1,672.62	1,314.24
Emergy (10^{20} seJ)	1.97	2.40	0.69	2.33	1.73	2.68	10.75	8.45
Adaptive capacity (10^{22} seJ)	5.98	5.34	3.04	3.35	5.65	5.54	12.48	7.36

low energy value, such as water and cultivated land, are less sensitive and less risky. Table 6 shows that compared with other districts and counties, Yucheng District has a relatively high proportion of land for residence, mining, and industrial purposes, and the sensitivity is high, so it is vulnerable to flood threats. However, Hanyuan County and other areas with large areas of cultivated land and woodland are less sensitive. Therefore, these areas are less threatened by flood disasters, thus featuring relatively low susceptibility.

(3) The stronger the system's adaptive capacity is, the stronger capacity it has against a flood disaster. Table 7 shows that different districts and counties in Ya'an City have different adaptive capacity. Yucheng and Mingshan districts boast better economic strength, medical and healthcare conditions, and popular education to see a stronger adaptive capacity. Hanyuan and Tianquan counties see favorable economic growth, but relatively low popular education, so they have a relatively poor adaptive capacity. Baoxing County, which is economically underdeveloped and has poorer medical services and a lower level of popular education, presents the poorest adaptive capacity among districts and counties in Ya'an.

3.5. Risk assessment

Based on the calculated results of exposure, susceptibility, and adaptive capacity, we use the calculation method mentioned in the Methods section to calculate the potential impact index and risk of a flood disaster to various districts and counties of Ya'an City. The calculated results are shown in Table 8.

With the calculated results in Table 8, the GIS software was used to obtain the potential impact indexes and a risk distribution map for various districts and counties of Ya'an City (Figure 3).

The potential impact index is related to the exposure and susceptibility of a region, which indicates the possible losses of a system in case of extreme weather (Yangfan 2018). Figure 3(a) presents the potential impact of a flood disaster on different districts and counties in Ya'an City. Yucheng District, thanks to a high population density, social fixed assets investment, and economic development, boasts a high level of accumulative social assets, so it will see the highest potential impact in case of a flood disaster. Baoxing and Shimian Counties, with a low population density and slow economic growth, see lower potential indices than other regions.

The risk of a system means that in the same case of a flood disaster, the higher risk emergy it has, the higher risk the system will see, and the more serious the impact will be. Figure 3(b) shows that Lushan County sees the highest risk index, followed by Yucheng District, and Shimian County witnesses the lowest risk index. Shimian County sees the lowest risk as this county has more medical workers per 10,000 people than others, and high social fixed assets investment, so effective control measures can be taken in the case of a flood disaster.

3.6. Model effect analysis

We built a BP neural network-based flood risk assessment model to conduct a comprehensive assessment of the emergy theory's performance in flood disaster risk assessment. The artificial neural network uses a lot of simple processing units (neurons) to form a nonlinear dynamic system, which has learning, memorization, calculation, identification, predication, and many intelligent processing functions. It is different from the so-called electronic brain–electronic computer and the AI system deduced based on symbols, being an AI system that imitates our human brain neural system on a different level and to a different extent (Kuang 2017).

The BP neural network is an error back-propagation learning algorithm, and its working principle is to convert an input vector into an output vector via the hidden layer and to map from the input space onto the output space. Its weight works to realize forward mapping using the network input of the current weight and the expected output to meet the mapping requirements for contrastive learning (Jing *et al.* 2020). The BP neural network does not only have an input layer node and an output layer node, but also has one or more hidden layer nodes. After signal input, we should first propagate it forward to the hidden layer nodes; under the activation function, the output signal from the hidden node is propagated to the output node, to

Table 8 | Results of risk assessment for districts and counties of Ya'an City

Region	Shimian	Hanyuan	Baoxing	Lushan	Tianquan	Xingyang	Yucheng	Mingshan
Potential impact index (10^{24} seJ)	3.32	5.14	2.03	11.12	6.75	4.92	20.31	6.74
Vulnerability (10^2)	0.56	0.96	0.67	3.32	1.19	0.89	1.63	0.92

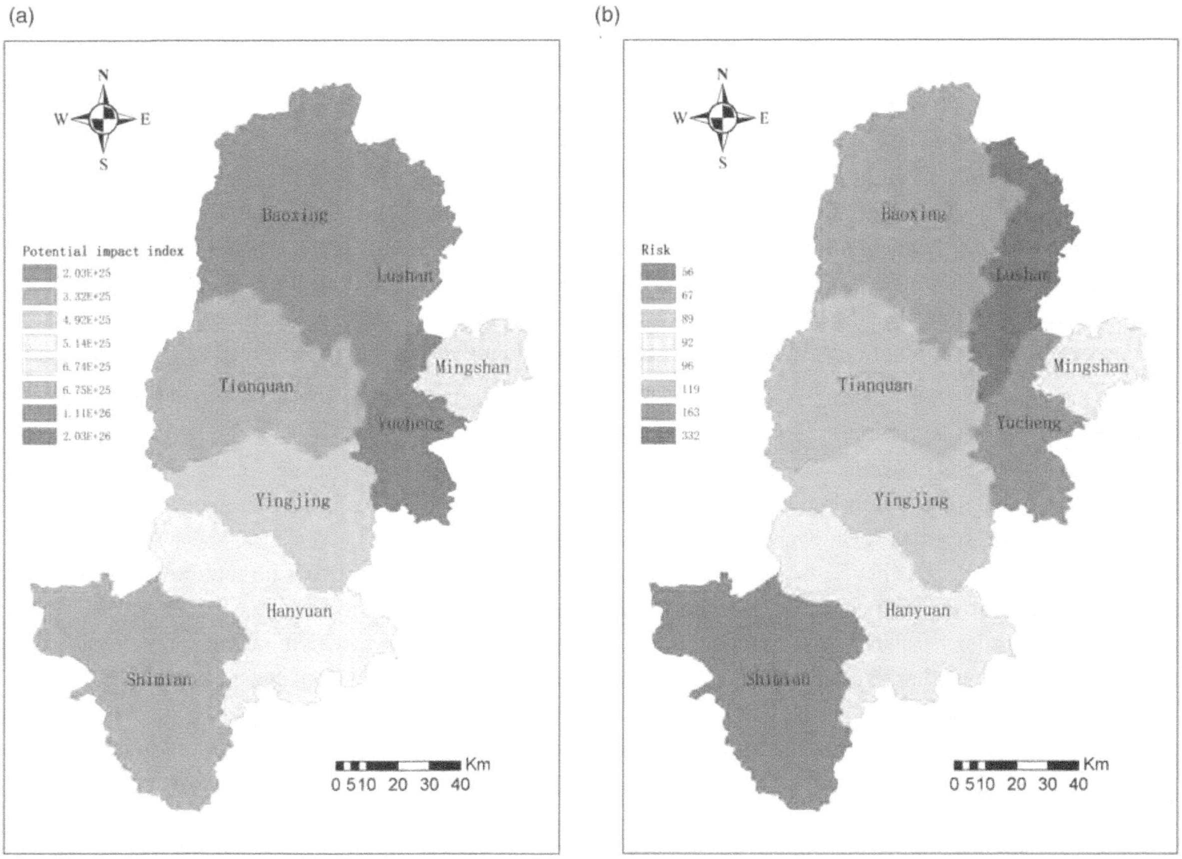

Figure 3 | Potential impact indices and risk profiles.

ultimately output the results. Its main operation procedures are to: (1) provide the neural network with training examples, i.e., learning samples, including input and expected output, (2) determine the permissible error of the network between the actual output and the expected input, and (3) change all connection weight values in the network, to make the generated output closer to the expected output, until it meets the demand for the determined permissible error.

To assess the flood disaster risk in Ya'an City, we first should consider the local natural and geographic environment and the socio-economic conditions to choose the assessment indicator factors from hydro-meteorology and topography geology and society. We select precipitation, runoff depth, drainage basin water storage, slope, elevation, population density, and GDP factors in this paper as assessment indicators to form the assessment indicator system, standardize the data, reduce its dimensions, and divide each indicator into five grades through the natural-break point method (Table 9). After

Table 9 | Range of indicators

Risk ranking	Very low	Low	Medium	High	Very high
Precipitation	0–0.176	0.176–0.282	0.282–0.478	0.478–0.639	0.639–1
Runoff	0–0.231	0.231–0.412	0.412–0.592	0.592–0.792	0.792–1
Drainage basin water storage	0–0.369	0.369–0.522	0.522–0.643	0.643–0.749	0.749–1
Slope	0–0.478	0.478–0.604	0.604–0.718	0.718–0.843	0.843–1
Elevation	0–0.420	0.420–0.576	0.576–0.714	0.714–0.847	0.847–1
Population density	0–0.141	0.141–0.471	0.471–0.510	0.510–0.945	0.945–1
GDP	0–0.146	0.146–0.379	0.379–0.641	0.641–0.829	0.829–1

Table 10 | Weights of indicators

Factor	Rainfall	Runoff depth	Drainage basin water storage	Slope	Elevation	Population density	GDP
Weight	0.271	0.064	0.075	0.084	0.113	0.265	0.128

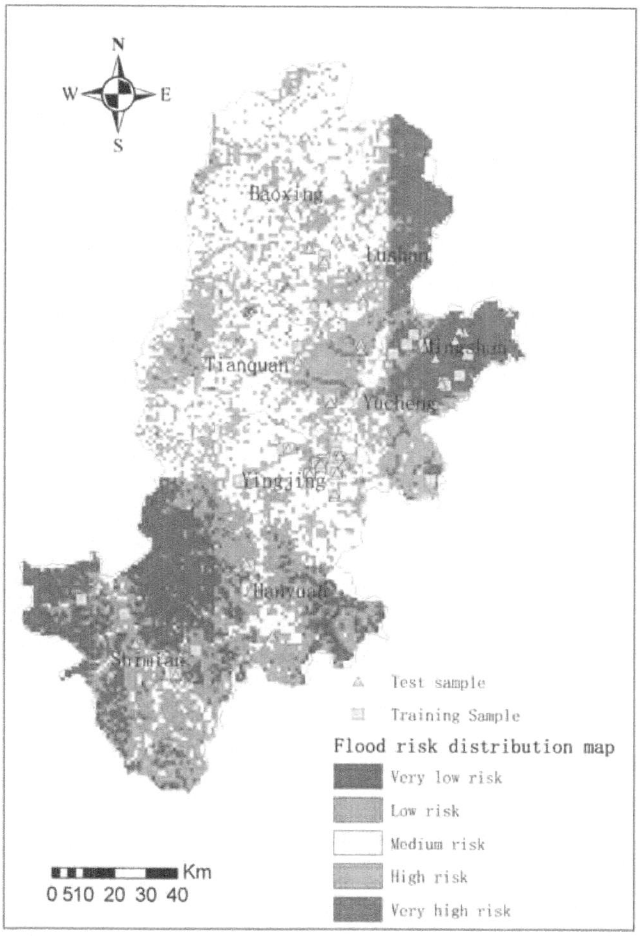

Figure 4 | Flood risk distribution of Ya'an City.

standardization, we use the BP neural network, in combination with the actual flooding position, and conduct learning training to obtain the weights of all assessment indicators (Table 10). Then we use the raster calculator on the GIS platform to overlay all indicator factors according to their respective weights to obtain the hazard distribution map of Ya'an City (Figure 4). The training accuracy based on the BP neural network is 0.768.

The risk distribution map shows that most regions with relatively high flood hazard risk are in the east part of Ya'an City: Lushan County, Yucheng District, and Mingshan District face the highest risks. Yucheng and Mingshan districts boast relatively high-density population distribution and are more likely to be affected by a flood disaster's damage, so corresponding disaster control and relief efforts should be made. Other districts and counties in Ya'an do not feature high risk, but their vegetation should also be protected, to prevent any flood disaster, and provide bases for local governments to carry out flood control and disaster relief. Compared with some previous research efforts, we have found both indicate that a region, more developed in the economy with more accumulative assets, sees higher potential impact and boasts higher adaptive capacity (Wu *et al.* 2018).

In order to verify the objectivity of the research results, we compared the historical flood event in Ya'an City. On 22 August 2019, 47 towns and villages in seven counties (districts) of the city were affected, with 96,000 affected people and 22,500 emergency relocations, resulting in a direct economic loss of 3.078 billion yuan. The areas with the largest losses are Lushan and Yucheng, and the disasters are mainly concentrated in Lushan County. The actual loss is consistent with the experimental results. Overall, the prediction matches the actual flood disaster situation, which represents the actual situation of the region.

4. CONCLUSIONS

Using emergy analysis, we constructed a flood risk assessment indicator system for Ya'an City that involved susceptibility and adaptive capacity of the hazard-affected region. In combination with the BP neural network, we have made a risk assessment and contrastive analysis of the study area. The results are as follows:

(1) Two methods have been applied to various districts and counties of Ya'an City. The results show that more developed regions like Yucheng and Mingshan districts have the highest potential impact index but lower risk than other regions. This is because these two economically developed districts, well-equipped with medical facilities, have a high adaptative capacity and thus can effectively address various emergencies in case of a flood disaster. Lushan County is just the opposite. This study reveals that a system with a strong adaptive capacity can effectively lower its risk of a flood disaster.
(2) This study assesses the flood risks of various districts and counties of Ya'an City. In actual practice, the emergy theory has often been used to investigate the sustainable development capacity of an ecological–economic system. However, studies on its application to flood disaster risk assessment are rare, most of which stay in the stage of theoretical or methodological analysis and show defects in real-world application. This paper selects Ya'an area as the research area, and applies emergy theory to the risk analysis of Ya'an City, making the analysis results more objective. Moreover, it is also of great research value to explore whether the emergy theory could be applied to other fields.

ACKNOWLEDGEMENT

This study is funded by the Yangtze River Collaborative Fund for Hydro-sciences, Natural National Science Foundation of China (SFC; U2040210).

DATA AVAILABILITY STATEMENT

All relevant data are included in the paper or its Supplementary Information.

REFERENCES

Abdulrazzak, M., Elfeki, A., Kamis, A., Kassab, M., Alamri, N., Chaabani, A. & Noor, K. 2019 Flash flood risk assessment in urban arid environment: case study of Taibah and Islamic universities' campuses, Medina, Kingdom of Saudi Arabia. *Geomatics, Natural Hazards and Risk* **10**, 780–796.

Bhuiyan, S. R. & Baky, A. A. 2014 Digital elevation based flood hazard and vulnerability study at various return periods in Sirajganj Sadar Upazila, Bangladesh. *International Journal of Disaster Risk Reduction* **10**, 48–58.

Chang, L. & Huang, S. 2015 Assessing urban flooding vulnerability with an Emergy approach. *Landscape and Urban Planning* **143**, 11–24.

Chao, X., Nan, M., Fei, L., Xiaoming, L. & Zhiyun, O. 2020 Research on Urban Green Infrastructure Management in Macao from the Perspective of Ecosystem Service Demand. Classical Chinese Garden.

Chengguang, L., Chen, X., Shiwei, Z., Wang, Z. & Xushu, W. 2015 Flood risk assessment model based on random forest and its application. *Shuili Xuebao* **46**, 58–66.

Foudi, S., Osés-Eraso, N. & Tamayo, I. 2015 Integrated spatial flood risk assessment: the case of Zaragoza. *Land Use Policy* **42**, 278–292.

Hanif, A., Dhanasekar, A., Keene, A., Li, H. & Carlson, K. 2020 Flood risk assessment methodology for planning under climate change scenarios and the corresponding change in land cover. *Journal of Water and Climate Change* **11** (4), 1370–1382.

Huang, S. & Odum, H. T. 1991 Ecology and economy: emergy synthesis and public policy in Taiwan. *Journal of Environmental Management* **32** (4), 313–333.

Jing, K., Xiaoxi, M., Hailian, W., Lu, W., Kai, S. & Aitao, T. 2020 Research progress of artificial neural networks in material science. *Materials Reports* **34** (21), 21172–21179.

Jun, X., Wei, S., Xinping, L., Si, H., Like, N. & Gippel, C. J. 2015 Revisions on water resources vulnerability and adaption measures under climate change. *Advances in Water Science* **26** (02), 279–286.

Kazakis, N., Kougias, I. & Patsialis, T. 2015 Assessment of flood hazard areas at a regional scale using an index-based approach and analytical hierarchy process: application in Rhodope–Evros region, Greece. *Science of the Total Environment* **538**, 555–563.

Kuang, W. 2017 Overview of the development of artificial neural networks. *Frontiers of Science and Technology* **6** (2), 165–167.

Luo, P., Xiaorong, H., Xiaoyue, W. & Lang, Z. 2019 Research on the sustainability of water eco-economic system in Chengdu based on emergy analysis. *Water Power* **45** (09), 12–16.

Sun, Z., Zhu, X. & Pan, Y. 2017 Flood risk analysis: progress, challenges and prospect. *Journal of Catastrophology* **32**, 125–130.

Tanoue, M., Hirabayashi, Y. & Ikeuchi, H. 2016 Global-scale river flood vulnerability in the last 50 years. *Scientific Reports* **6**, 36021.

Wahab, A. M. & Muhamad Ludin, A. N. 2018 Flood vulnerability assessment using artificial neural networks in Muar Region, Johor Malaysia. *IOP Conference Series, Earth and Environmental Science* **169**, 12056.

Wangyang, Y., Chunbo, J., Jian, L. & Qi, Z. 2019 Hydrologic-hydrodynamic model and its application in floodrisk analysis. *Journal of Hydroelectric Engineering* **38** (08), 87–97.

Wanzhi, L., Di, Y., Xiaoli, F. & Tiaofeng, Z. 2019 Risk assessment of rainstorm and flood disasters based on the hazard grades/indices in Qinghai Province. *Journal of Glaciology and Geocryology* **41** (3), 680–688.

Wu, Z., Shen, Y. & Wang, H. 2018 Assessment of vulnerability to flood disasters based on Emergy theory. *South-North Water Diversion and Water Resources Science and Technology* **16**, 9–14.

Yangfan, X. 2018 *Flood Hazard Assessment Based on Spatial Multi-Criteria Decision Making and Heterogeneous Spatial Data Integration.* Huazhong University of Science & Technology, Wuhan, China.

Yongzhi, L., Wenwen, T., Wenting, Z., Xingnan, Z. & Shuai, N. 2021 Review of flood disaster risk analysis based on disaster chain. *Water Resources Protection* **37** (1), 20–27.

Zhao, L. 2013 *Ecological Economics*. China Economic Publishing House, Beijing.

First received 1 April 2021; accepted in revised form 10 June 2021. Available online 22 June 2021

Impact of Climate Change on Hydrology and Water Resources doi: 10.2166/wcc.2021.195

Assessing the impacts of climate change on drought-prone regions in Bhima sub-basin (India) using the Standard Precipitation Index

Rajashekhar S. Laddimath [a,*], Nagraj S. Patil[b], P. S. Rao[c] and Nagendra[b]

[a] Department of Civil Engineering, KLE Institute of Technology, Hubballi, Karnataka, India
[b] Department of Civil Engineering, Centre for P.G. Studies, Visvesvaraya Technological University, Belagavi, Karnataka, India
[c] Advanced Centre for Integrated Water Resources Management (ACIWRM), Bengaluru, Karnataka, India
*Corresponding author. E-mail: rajashekhar.sl@gmail.com

RSL, 0000-0003-4115-6274

ABSTRACT

In Bhima sub-basin, the water sector is at high demand and in critical stress due to rapid urbanization. The past few decades have witnessed extreme events and seasonal shifts due to anthropogenic activity triggered climate change. Thus, to evaluate the variability of extreme events, assessing the historical and future trends of precipitation in climate change scenarios is vital for developing comprehensive mitigation and adaptation strategies. This paper examines the drought-prone regions by studying spatio-temporal variation of drought scenarios using the Standard Precipitation Index (SPI). The change factor method is used to downscale precipitation data from general circulation model (GCM) outputs under four Representative Concentration Pathway (RCP) scenarios to project future downscaled precipitation, to be input to examine the drought for 12 months. GCM and scenario uncertainty in climate change impact assessments are examined using box-whisker plots. Temporal variation for 12-month SPI shows significant changes over RCP scenarios. For the beginning of the period, 2021 precipitation is scanty for RCP 2.6 and 4.5 scenarios. Mild to moderate and severe drought events for the RCP 2.6 scenario are more predominant. Severe drought events under the RCP 6.0 scenario dominate over others. Lastly, the inconsistent pattern of drought events for RCP 8.5 is reported.

Key words: change factor method, climate change, extreme events, RCP scenarios, spatio-temporal variation of drought, Standard Precipitation Index

HIGHLIGHTS

- Impacts of climate change on drought : regional scenario in Bhima sub-basin.
- Trend analysis using Mann-Kendall test to examine the baseline climate.
- Analysis of drought characteristics by Standardized Precipitation Index (SPI).
- Occurrence of drought events under CMIP5 model/RCP scenarios in different grids of the study area.
- Temporal variation and spatial pattern of the drought characteristics for 12-month SPI.

INTRODUCTION

Climate change in India will significantly impact agricultural production and water will be a major constraint in the future for food production (Aggarwal 2008). Developing countries like India are more vulnerable to the impacts of climate change and its implications on rain-fed agricultural areas have been affecting rural populations owing to the large dependency on agriculture for livelihoods. Agriculture primarily depends on monsoon, and changing precipitation patterns and intensities along with duration will probably influence the balance in water supply and demand mechanisms, thus affecting the hydrological cycle leading to frequent natural disasters (Gleick *et al.* 2010). Global warming is changing our climate, resulting in dire consequences. In the past few decades this has worsened as extreme events and seasonal shifts have bcome a reality. The consequences of global warming may not be uniform everywhere but its adverse effects will certainly affect life-supporting systems. According to Bates *et al.* (2010), the projected increase in greenhouse gas (GHG) emissions will intensify the water cycle and cause extreme events such as droughts and floods.

One of the most important sources of information for impact assessment studies is the output from general circulation models (GCMs) of the climate system (Robock *et al.* 1993). Owing to anthropogenic activities and variability in nature, it has been a challenge for researchers and policymakers to assess natural resources (more importantly water) with minimum uncertainty under perturbed climate conditions. Quite a number of studies in the recent past report the use of GCMs to assess the impacts of climate change at the regional level using statistical downscaling techniques (Woldemeskel *et al.* 2016; Sharma *et al.* 2018; Kang *et al.* 2019; Rehana 2019). GCMs can be used with higher confidence and less uncertainty using systematic evaluation of model performance (Nyunt *et al.* 2016).

The climate change impact assessment on the regional spatial scale (finer resolution) is hampered significantly by a considerable amount of uncertainty (Kundzewicz *et al.* 2018). The inter-modal uncertainty and inter-scenario uncertainty, referred to hereafter as GCM uncertainty and scenario uncertainty, respectively, have influenced the GCM outputs derived using statistical downscaling models. According to an Intergovernmental Panel on Climate Change IPCC report (2007), lack of scientific knowledge on climate processes and initial conditions results in GCM uncertainty. On the other hand, scenario uncertainty considers GHG emission scenarios under IPCC and causes unpredicted forecasting on the socio-economic and human intervention in policymaking. Thus, the uncertainty accumulates at various levels to propagate the large ranges at the regional level (Rowell 2006; Wilby & Harris 2006; Ghosh & Mujumdar 2007; Lindner *et al.* 2014; Fang *et al.* 2018).

The water sector is under high demand as well as under critical stress owing to rapid urbanization and industry needs. Large numbers of dams across the main river and tributaries of Krishna river (Bhima is a sub-basin of Krishna river basin) have changed the natural flow regime to such an extent that, almost zero flow conditions prevailed in the lower catchment area for a longer period of time (Venot *et al.* 2008). Hence, it is critical to further pursue a study on drought trends to examine the severity and its impact on food security and managing the water resource (Mahajan & Dodamani 2015). The mean and maximum temperature in the Krishna basin is rising significantly during all the seasons (Deshpande *et al.* 2016). Evidently, droughts in the Krishna basin have been occurring since 1901 with regular frequency and prolonged periods of ten to 15 years. However, in the 20th century, droughts lasted for a smaller duration of two to five years and were distributed widely and mainly concentrated in the upper Krishna and Bhima river basin (Gaur *et al.* 2007). More recently, studies by Kulkarni *et al.* (2014) and Chanapathi & Thatikonda (2020) carried out work on Krishna basin using hydrological models to assess the impact of climate change on water resources. Approximately 64% of the population in Bhima sub-basin depends on agriculture as a major source of income and has been suffering due to frequent droughts in Maharashtra (upper Bhima sub-basin). The Upper Bhima river basin is facing both periodic and prolonged water shortages due to exhaustive irrigation development (Garg *et al.* 2012). Udmale *et al.* (2014) highlighted several key challenges related to environmental degradation and ecosystem, putting this basin at potential risk regarding climate change. Surinaidu *et al.* (2013) reported a significant decline in groundwater (by −6 m), which for decades has endangered the basaltic aquifers and impacted agricultural activities. Developing countries basically lack an integrated approach to monitoring and effective evolution of agricultural projects. Thus, Biswas (1987) recommends the involvement of local authorities in the monitoring and evaluation process of watershed management to effectively mitigate the frequent droughts in the Upper Krishna basin in Maharashtra (Bhima sub-basin). Hence, robust and resilient strategies for managing the natural resources and climate change mitigation measures for the Bhima sub-basin need to be sensitized to the local administration through policy documents. Watershed management policies set the guidelines and directives for the optimal utilization of resources. For instance, water policy guides the allocation priorities, namely, agriculture, industry, domestic, etc., as well as for awards, incentives and penalties towards conservation and discharge of water resources. Therefore, this paper on the impact assessment of climate change is vital in the framework of watershed management policies.

MATERIALS AND METHODS

Study area

This study is focused on Upper Bhima (K5) and Lower Bhima (K6), as shown in Figure 1, which covers the entire Bhima sub-basin, an important sub-basin of Krishna river basin, the second largest eastward draining river of India (Biggs *et al.* 2007). Bhima sub-basin, with a catchment area of 70,263 km^2, shares the catchment areas of Maharashtra, Karnataka and Telangana states. The Upper Bhima river basin is situated in Maharashtra state surrounded by mountainous chains of the Western Ghats and has a semi-arid tropical climatic condition. This region covers part of the Western Ghats mountain ranges to Deccan Plateau. Owing to the proximity of the Western Ghats, in the past, during 1972 and 2004, this region received the maximum

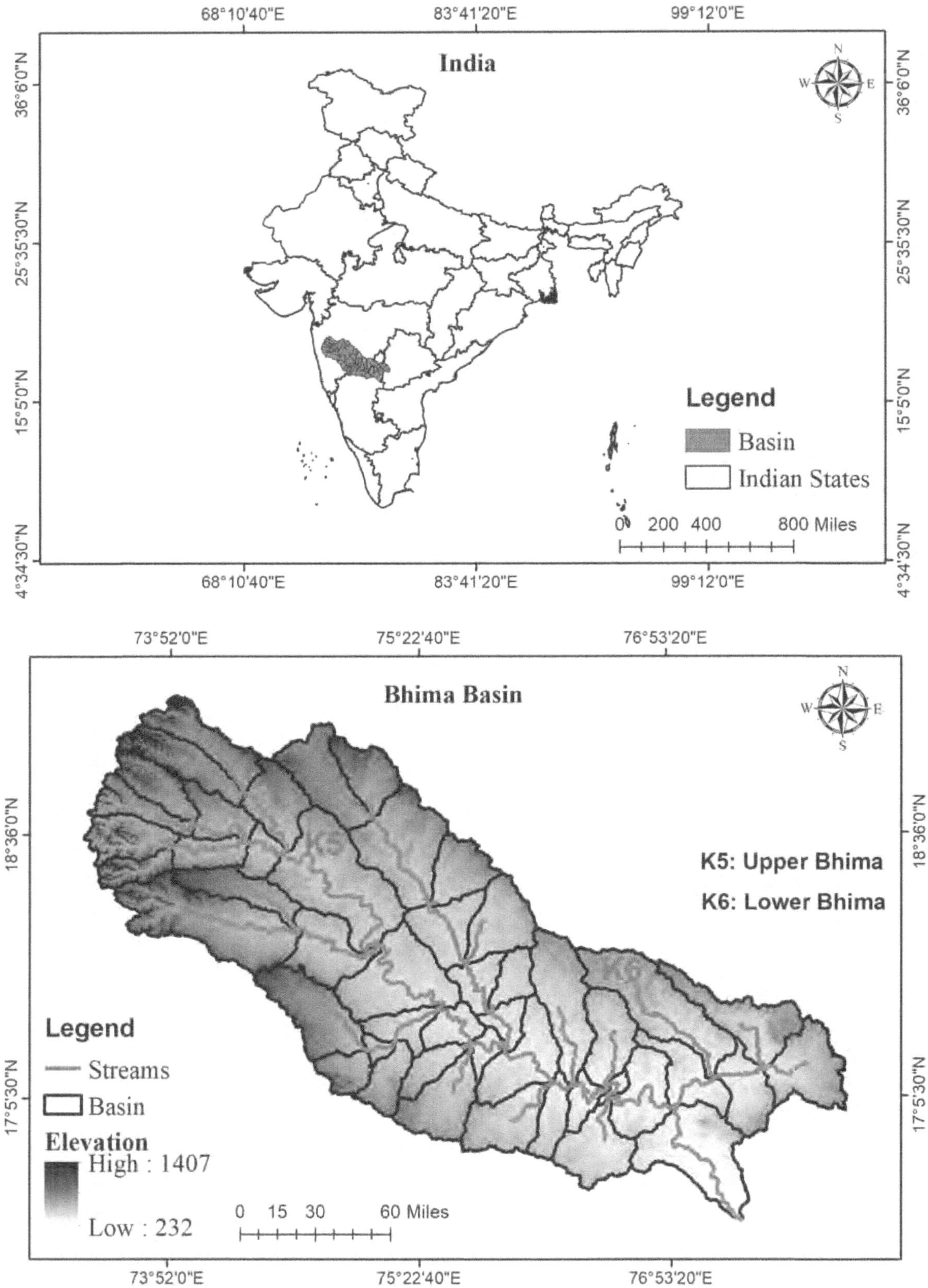

Figure 1 | Location map of the Bhima sub-basin; Upper Bhima (K5) and Lower Bhima (K6) (study area).

rainfall and recorded average annual rainfall of 2,731.11 mm (Biggs *et al.* 2007). Lower Bhima had 678 mm of average annual rainfall meaning there was scanty rainfall in this region (Karnataka and Telangana). The temperature in the upper region (Bhimashanker) most of the times drops to 5–6 °C in the lower reaches of Bhima (Sholapur district of Maharashtra) and sometimes rises up to 46 °C. The Indian Meteorological Department (IMD) Pune reported a variation in annual mean temperature of 25 °C in the upper region to 27.2 °C in the lower region during 1969–2009.

Literature examining the impacts of changing climate on water resources and watershed management for the Bhima sub-basin reported that approximately 64% of the population depends on agriculture as a major source of income and has been suffering due to frequent droughts in Maharashtra (Upper Bhima sub-basin). The Upper Bhima river basin is facing both periodic and prolonged water shortages due to exhaustive irrigation development (Garg *et al.* 2012). Faster growing urbanization and the needs of stakeholders have disturbed the natural regime to hamper the rainfall distribution for the past decades causing an unbalanced supply and demand mechanism. Uneven distribution and uncertainty in rainfall pattern across this river basin has resulted in the stakeholders of this region depending on available groundwater resources, making the basin a water-scarce region. Greater parts of Bhima sub-basin covering Andhra Pradesh, Karnataka, Maharashtra and Telagana are facing massive rainfall deficits, drought scenarios and crop failures. Hence, examining the historical changes in the Bhima sub-basin region is, indeed, an important exercise and so historical trend analysis using the Mann–Kendall test is performed.

Historical trend analysis

Mann (1945) proposed a non-parametric test based upon the application of Kendall's (Kendall 1975) test considering the randomness alongside the time for correlation and it is well-known to the research community as the Mann–Kendall test (MK test).

The mathematical equations for calculating Mann–Kendall statistics *S*, *var* and standardized test statistics Z are as follows:

$$S = \sum_{i=1}^{n-1} \sum_{j=k+1}^{n} \text{sign}(x_j - x_i) \tag{1}$$

where, if $S > 0$ then later observations in the time series tend to be larger than those that appear earlier in the time series, while the reverse is true if $S < 0$.

The variance of *S*, (i.e., *var*) is given by

$$\text{var} = \frac{1}{18}\left[n(n-1)(2n+5) - \sum_t f_t(f_t - 1)(2f_t + 5)\right] \tag{2}$$

where *t* varies over the set of tied ranks and f_t is the number of times (i.e., frequency) that the rank *t* appears.

The MK test uses the following test statistic:

$$z = \begin{cases} (S-1)/se, & S > 0 \\ 0, & S = 0 \\ (S+1)/se, & S < 0 \end{cases} \tag{3}$$

where $se =$ the square root of *var*. If there is no monotonic trend (the null hypothesis), then for time series with more than ten elements, $z \sim N\ (0, 1)$, i.e., *z* has a standard normal distribution.

The present study detects and assesses the significant trends in the spatial and temporal variations of historical data for the Bhima basin. For this purpose, the spatial variation of changes in rainfall IMD 0.25°×0.25° gridded data for the period of 1969 to 2016 is an input for generating the annual mean rainfall over the entire Bhima basin. Figures 2 and 3 illustrate the annual mean rainfall and the maximum daily rainfall observed during 1969–2016, respectively. MK trend test 'tau' value (Figure 4) explains the rainfall trend for the annual rainfall series of the IMD gridded data over the basin. The tau value of the MK test is quite similar and as good as the correlation coefficient with its value ranging from −1 to +1. A positive (+) tau value shows the increasing trend of the series of data (rainfall in this case) and its counterpart negative (−) tau value represents the decreasing trend. Figure 5 illustrates the average annual rainfall over the basin (with tau value). In a similar context, the

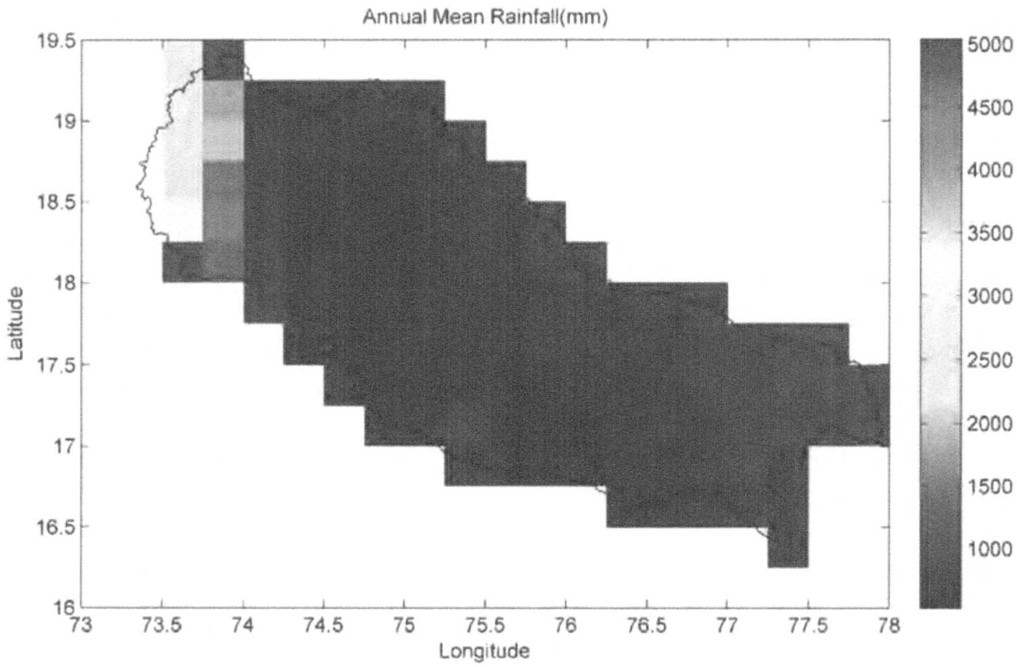

Figure 2 | Annual mean rainfall (mm) in the study area during 1969–2016.

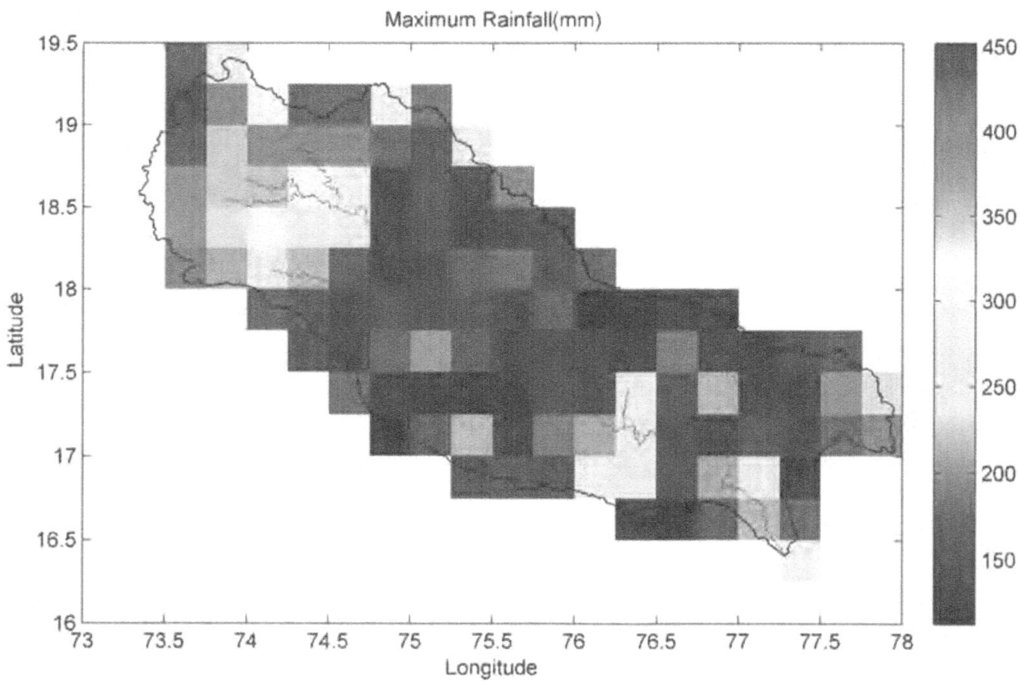

Figure 3 | Maximum daily rainfall (mm) in the study area during 1969–2016.

trend in the annual maximum daily rainfall is analysed and the tau value of the gridded data with the trend is shown in Figures 6 and 7, respectively.

Upon examining the spatial plots and analysing the trends in annual rainfall over the basin, it is clear that there is neither a fixed increase nor a decrease of annual rainfall over the basin. However, the grids towards the Western Ghats region (Maharashtra state) lying in the Upper Bhima (K5) show increasing trends (positive tau values) in rainfall (Figure 6). The annual

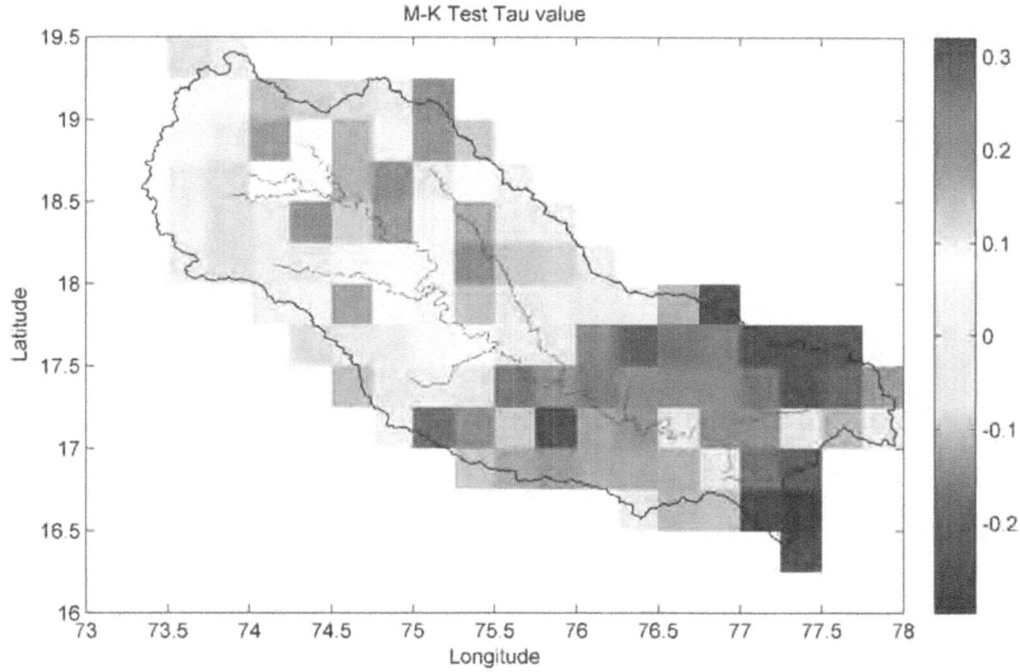

Figure 4 | MK trend test tau values for the annual rainfall series (1969–2016) of IMD grids of the basin.

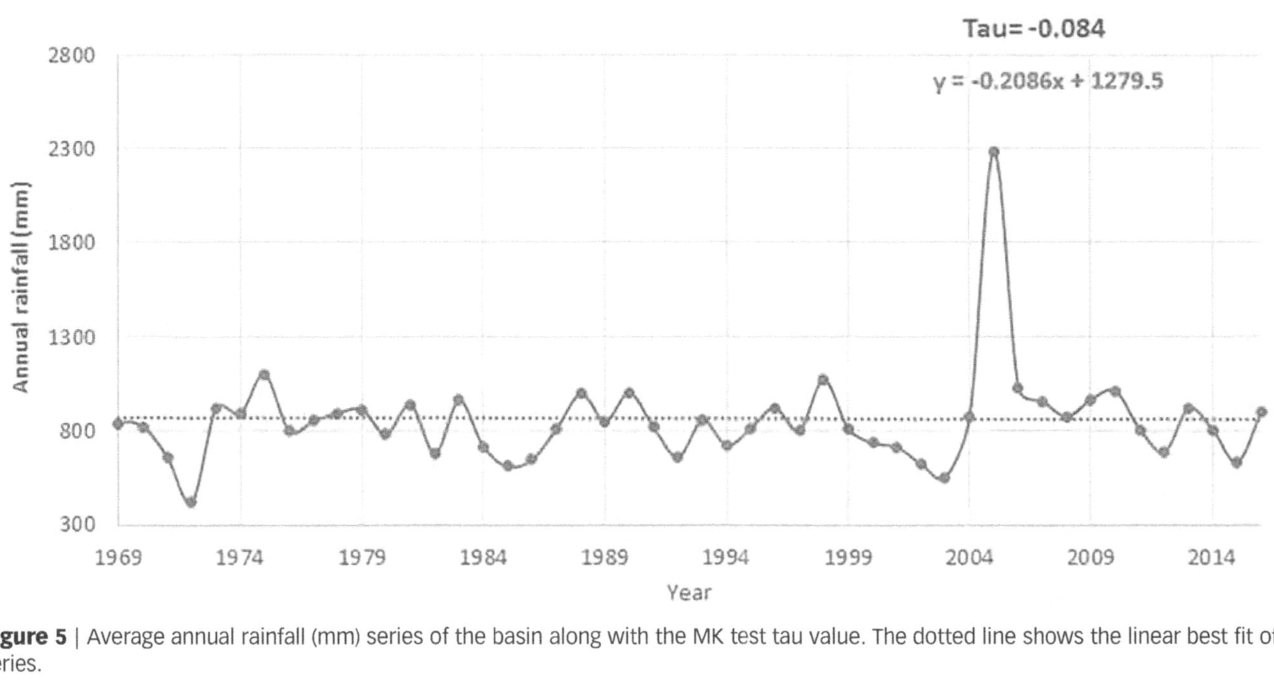

Figure 5 | Average annual rainfall (mm) series of the basin along with the MK test tau value. The dotted line shows the linear best fit of the series.

mean rainfall varies between 2,000 and 2,500 mm (Figure 2), and the maximum rainfall of 450 mm (Figure 3) is also evidence of the occurrence of extreme events. The Lower Bhima (K6) has altogether a decreasing trend of annual rainfall. The spatial grids on the downstream side (Karnataka state) have annual mean rainfall values between 500 and 1,000 mm and the maximum daily rainfall is in the range of 100–150 mm. Further, as we examine the annual rainfall (Figure 5), the best fit line for the range of data has a mean value of 850 mm with a decreasing rainfall trend until 2004. Interestingly, in the year 2005, the

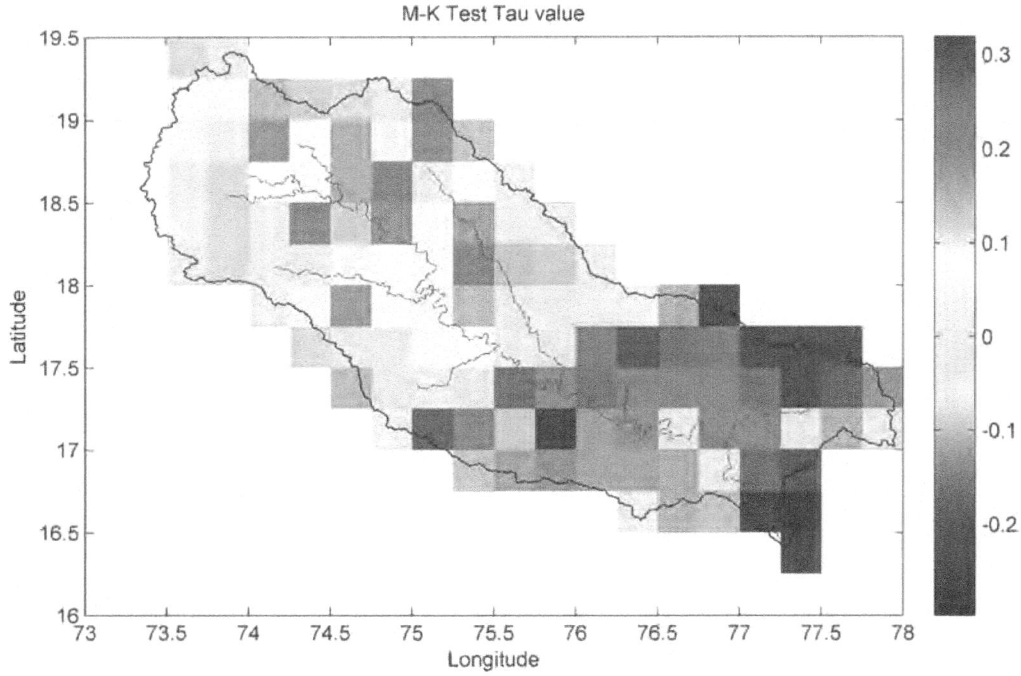

Figure 6 | MK trend test tau values for the annual maximum rainfall series (1969–2016) of IMD grids of the basin.

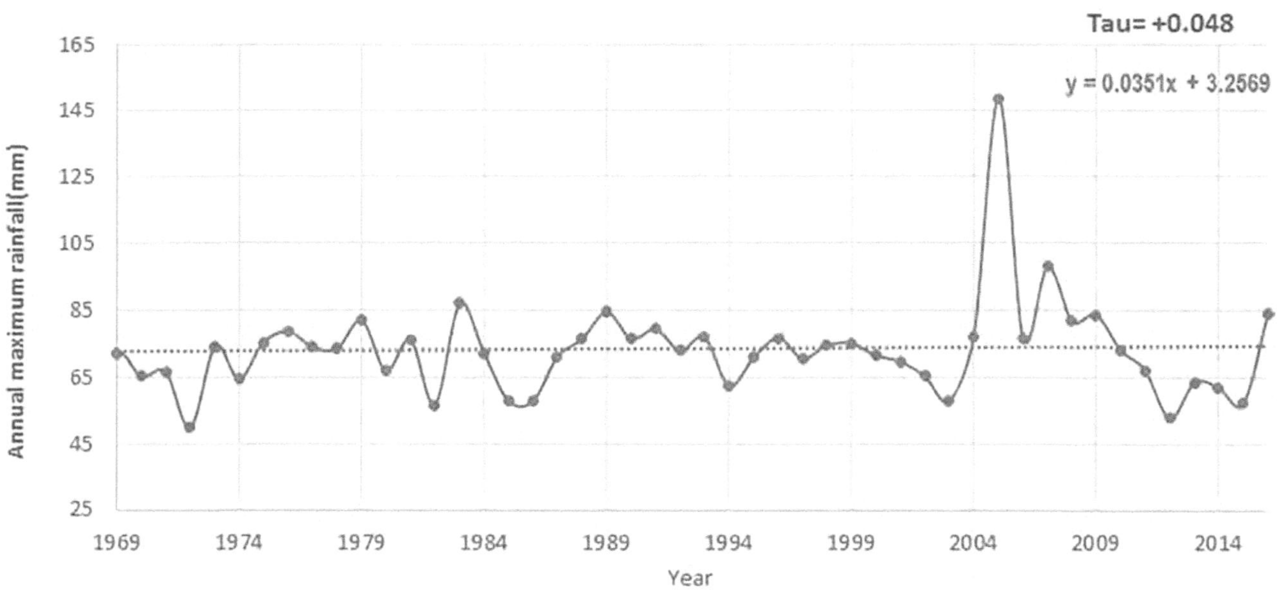

Figure 7 | Average annual maximum rainfall (mm) series of the basin along with the MK test tau value. The dotted line shows the linear best fit of the series.

annual maximum rainfall reaches 2,300 mm and declines subsequently, such that, during 2005, the Bhima sub-basin experienced flash floods on account of extreme events of rainfall causing tangible and intangible effects on the people of this sub-basin. However, the extreme rain events during 2005 are attributed to an increase in the fluctuations of the monsoon westerly winds, due to increased warming in the Arabian Sea (Kulkarni *et al.* 2020). The average slope of yearly time series data as per the linear equation is −0.2086 resulting in a 0.044% decline of annual rainfall for every decade. As the tau values are more predominant in Lower Bhima this may result in drought conditions in the regions of Lower Bhima.

Downscaling of precipitation

Statistical downscaling methods are generally used in climate change impact assessment studies for downscaling climate projections provided by the GCMs. GCMs are robust mathematical tools to estimate the probable changes in climate through the GHG concentrations to respond to future climate systems. Table 1 outlines the GCMs selected for downscaling the precipitation (*rainfall in India is a major form of precipitation and thus, precipitation and rainfall in India are synonymous*). The selection of these five GCMs is upon higher rankings having daily simulation for the future in four different RCPs on the ranking component considered by Raju *et al.* (2016) for the Godavari and Krishna basins of India (Bhima sub-basin is the sub-basin of Krishna). A study by Raju *et al.* (2016) on 45 Coupled Model Intercomparison Project 5 (CMIP5)-based GCMs ranked the GCMs based on several metrics. However, the inclusion of all these GCMs into the current research is quite challenging owing to the non-availability of complete data and challenge that lies in handling the missing data (if any) for considering all or very large numbers of GCMs, such that many of these GCMs are not completely equipped with run data as provided by IPCC. Besides, the exclusion of some GCMs may even narrow the range of uncertainty represented by the remaining models. However, multi-model ensembles have now become a standard procedure to quantify and assess agreement and uncertainties among multiple GCMs in projecting future climate with a fair number of selections in GCMs (Wang *et al.* 2018).

A well-known statistical downscaling technique, the change factor (CF) method, is employed to downscale the precipitation data from GCM outputs under four Representative Concentration Pathways (RCP) scenarios published under the Fifth Assessment Report of the IPCC (van Vuuren *et al.* 2011). Among the wide ranging downscaling techniques, the CF method is a commonly used approach by the research community. CF is easy to implement and assess the impacts of changing climate at regional scale-related applications (Boé *et al.* 2007) and enabling representation of multiple GCMs and emission scenarios with the least minimal computing effort. For impact assessment studies involving complications, the CF technique has been adopted by the research community owing to its simplicity and flexibility of application with direct scaling of local data in line with changes suggested by the GCM simulations most relied upon. More information on downscaling climate variables using the CF method can be found in the papers by Boé *et al.* (2007) and Anandhi *et al.* (2011).

CF is broadly classified as: i) additive change factor (ACF) method and ii) multiplicative change factor (MCF) method. ACF evaluates the numerical difference in GCM variable obtained from present climate simulation and from a future climate scenario considered at the same GCM grid location as shown in Equation (4). This method is typically used for temperature, solar radiation, wind speed and relative humidity climate variables, except precipitation, since precipitation will not have a negative range.

$$CF_{add} = (\overline{GCM_f} - \overline{GCM_b}) \tag{4}$$

The MCF method is the same as ACF except that the ratio between the future and current GCM simulations is calculated, instead of an arithmetical difference, as shown in Equation (5):

$$CF_{mul} = (\overline{GCM_f})/(\overline{GCM_b}) \tag{5}$$

Table 1 | List of GCM models used for the research

Sl. No.	Model name	Model centre
1	MIROCESM_CHEM	Agency for Marine (Japan) - Earth Science and Technology, Atmosphere and Ocean Research Institute for Environmental Studies
2	MIROC5	
3	CSIRO-Mk3.6	Commonwealth Scientific and Industrial Research Organization
4	NorESM1 - M	Bjerknes Climate Research Centre, Norwegian Meteorological Institute Norway
5	GFDL-CM3	Geophysical Fluid Dynamics Laboratory, USA

Source: http://www.ipcc-data.org/sim/gcm_monthly/AR5/Reference-Archive.html.

where CF_{add} and CF_{mul} are additive and multiplicative change factors, respectively. $\overline{GCM_f}$ is a GCM future climate scenario and $\overline{GCM_b}$ is the value from the GCM baseline considered for a temporal domain.

The next step is to obtain the local scaled future values ($LSf_{add,i}$ and $LSf_{mul,i}$) by relating the change factors obtained from Equations (1) and (2).

$$LSf_{add,i} = LOb_i + CF_{add} \tag{6}$$

$$LSf_{mul,i} = LOb_i \times CF_{mul} \tag{7}$$

where LOb_i are the observed values of climate variables considered at the *ith* time step on an individual meteorological station, or the averaged meteorological time series for a catchment for the designated temporal domain. $LSf_{add,i}$ *and* $LSf_{mul,i}$ are the values of future scenarios of climate variables obtained using additive and multiplicative formulation of the CF method. The methodology to estimate future scenarios using CF is best described pictorially in Figure 8.

An output from a single GCM for a particular scenario has predominant uncertainties and should not account for predictions. Rather, to provide superior qualitative and quantitative climate change information it is essential to ascertain the uncertainties in the predicted results. This may be attained by the ensemble methods by minimizing the associated uncertainties in the multi-prediction approach. A multi-model ensemble average will help in addressing the underlying uncertainties by employing a set of climate models and scenarios (Knutti *et al.* 2010; Warner *et al.* 2010) by taking a simple arithmetic average. The IPCC has identified scenarios and, as per the IPCC's Fifth Assessment Report (AR5) these scenarios are called Representative Concentration Pathways (RCPs) with four pathways: RCP 8.5, RCP 6.0, RCP

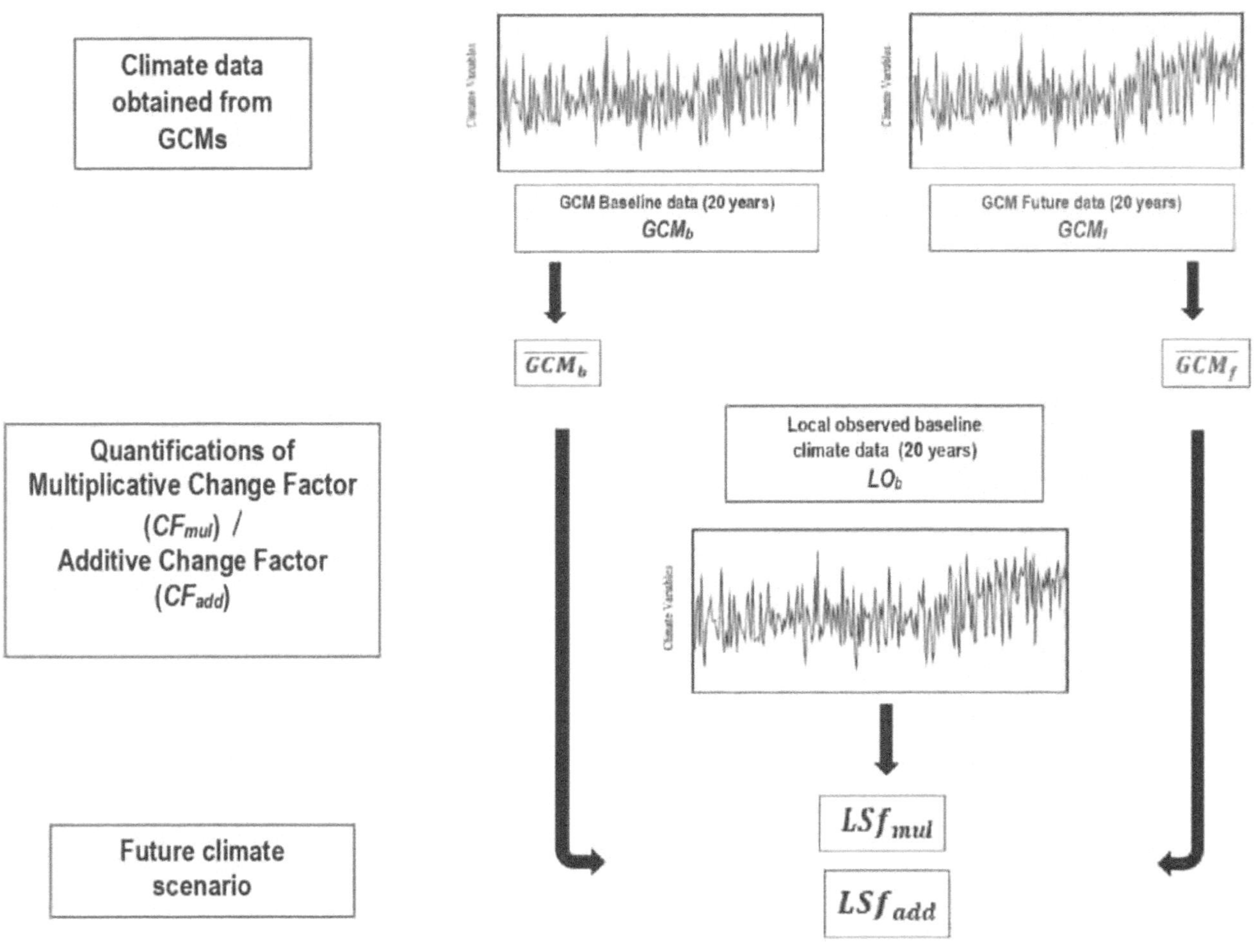

Figure 8 | Methodology to estimate future scenarios using change factor.

4.5 and RCP 2.6. These RCP pathways have been adopted by the climate research community, terrestrial ecosystems and emission inventory experts, who have put forward comprehensive data sets with high spatial resolution until 2100 to designate different climate futures depending upon GHG emissions. The multi-model ensemble of precipitation projections for the four RCPs 2.6, 4.5, 6.0 and 8.5 validated the highest capability for precipitation downscaling (Homsi *et al.* 2019).

The mean annual rainfalls over the basin for the four RCP scenarios are shown in Figures 9–12. It is inferred from these spatial maps and histograms that the percentage (%) change in mean annual rainfall is positive and is increasing for most of the grids lying both in Upper Bhima and Lower Bhima region for all the four RCP scenarios. The percentage change in mean annual rainfall varies from 2% to 30% for the grids in the basin under RCP scenarios RCP 2.6, 4.5 and 6.0. Further, for RCP 8.5 scenario, this variation is slightly higher with the range extending to 35% and, moreover, during 2061–2100 almost all the grids in the basin under RCP 8.5 scenario depict extremity changes (30–35% variation) in the record of mean annual rainfall. However, this may be due to rising radiative forcing pathway RCP 8.5 scenarios leading to very high GHG emissions (IPCC 2013). A histogram, shown below each of the spatial maps, is used to summarize climate data. Histogram plot illustrates a graphic interpretation of numerical data by presenting the number of data points that fall within a quantified range of values. Red line in the histogram indicates mean percentage change.

Standard precipitation index (SPI)

For monitoring drought and guiding early warning, Steinemann (2003) defined drought indicator as a variable to identify and assess drought conditions. The levels of severity in climate can be categorized into 'mild', 'moderate', 'severe', or 'extreme' drought. Generally used drought indicators include Standard Precipitation Index (SPI), Palmer Drought Severity Index (PDSI), Crop Moisture Index (CMI), Surface Water Supply Index (SWSI) and Reclamation Drought Index (RDI) (WMO & GWP 2016).

Most of the aforementioned indicators of drought need numerous input data such as rainfall, temperature, soil moisture, snowpack, reservoir storage, etc. However, SPI proposed by McKee *et al.* (1993) is used as a drought index using a single indicator 'precipitation' as an input variable. SPI has been a popular method owing to its 'ease of use'. SPI in comparison with other indices is an easy method for calculation and it is effective in giving early drought warning and drought damage control (Wang *et al.* 2015).

To assess the impact of drought in the past, the World Meteorological Organization (WMO) had recommended SPI to monitor and account for drought conditions (Hayes *et al.* 1999). This method makes use of the downscaled GCM data for forecasting future scenarios. It is the most popular among the other drought indices (Karabulut 2015) and also recognized widely for illustrating meteorological droughts (Hayes *et al.* 1999).

A recent study on monitoring of drought by Zhao *et al.* (2018) using Gravity Recovery and Climate Experiment (GRACE) data is compared and contrasted with SPI. Machine learning tools adopt the basics of SPI mechanism to predict standardized streamflow index for hydrological drought analysis (Shamshirband *et al.* 2020).

To determine the SPI values, the long-term rainfall data are fitted with a probability distribution. The current study intended to use the Gamma distribution to fit the long-term rainfall record similar to the methods adopted by Tsakiris *et al.* (2007) and Sönmez *et al.* (2005).

The standard procedure for SPI is as follows:

(1) Gamma distribution is fitted to the time series of non-zero rainfall (X) for all the period '*t*' of interest (e.g., 3 months, 6 months, 9 months, 12 months, etc.). Avoid the overlapping between two time periods.

The Gamma probability distribution function (PDF) [$g(X)$] is given as:

$$g(X) = \frac{1}{\beta^{\alpha}\Gamma(\alpha)} X^{(\alpha-1)} e^{-\frac{X}{\beta}} \text{ for } X > 0 \tag{8}$$

where $\alpha > 0$ and $\beta > 0$ are shape and scale parameters, respectively. Fitting a Gamma distribution involves computation of parameters α and β.

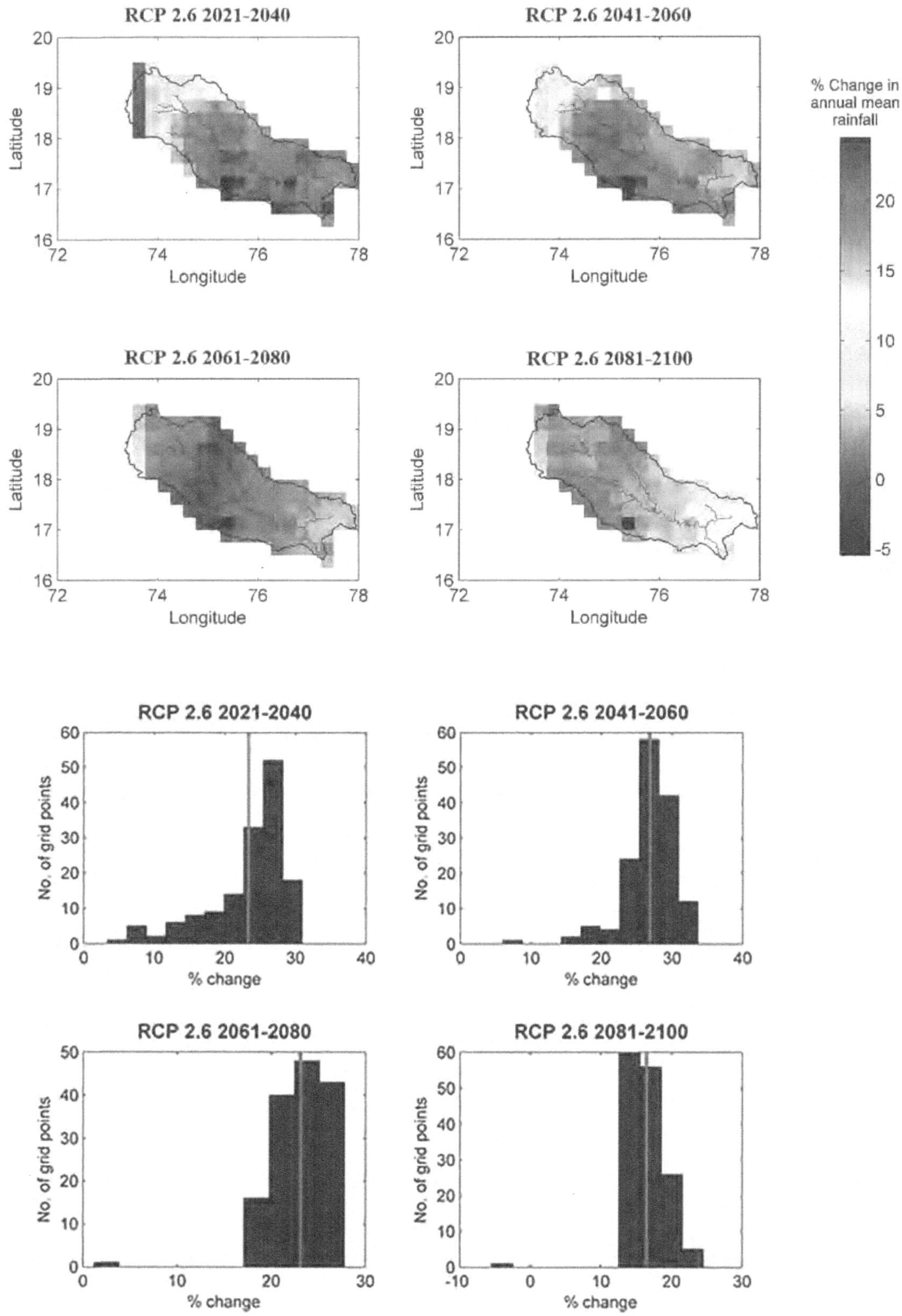

Figure 9 | Spatial maps (top four panels) and histograms (bottom four panels) of percentage change in mean annual rainfall projected from ensemble average in the RCP 2.6 scenario. Red line in the histogram indicates mean percentage change.

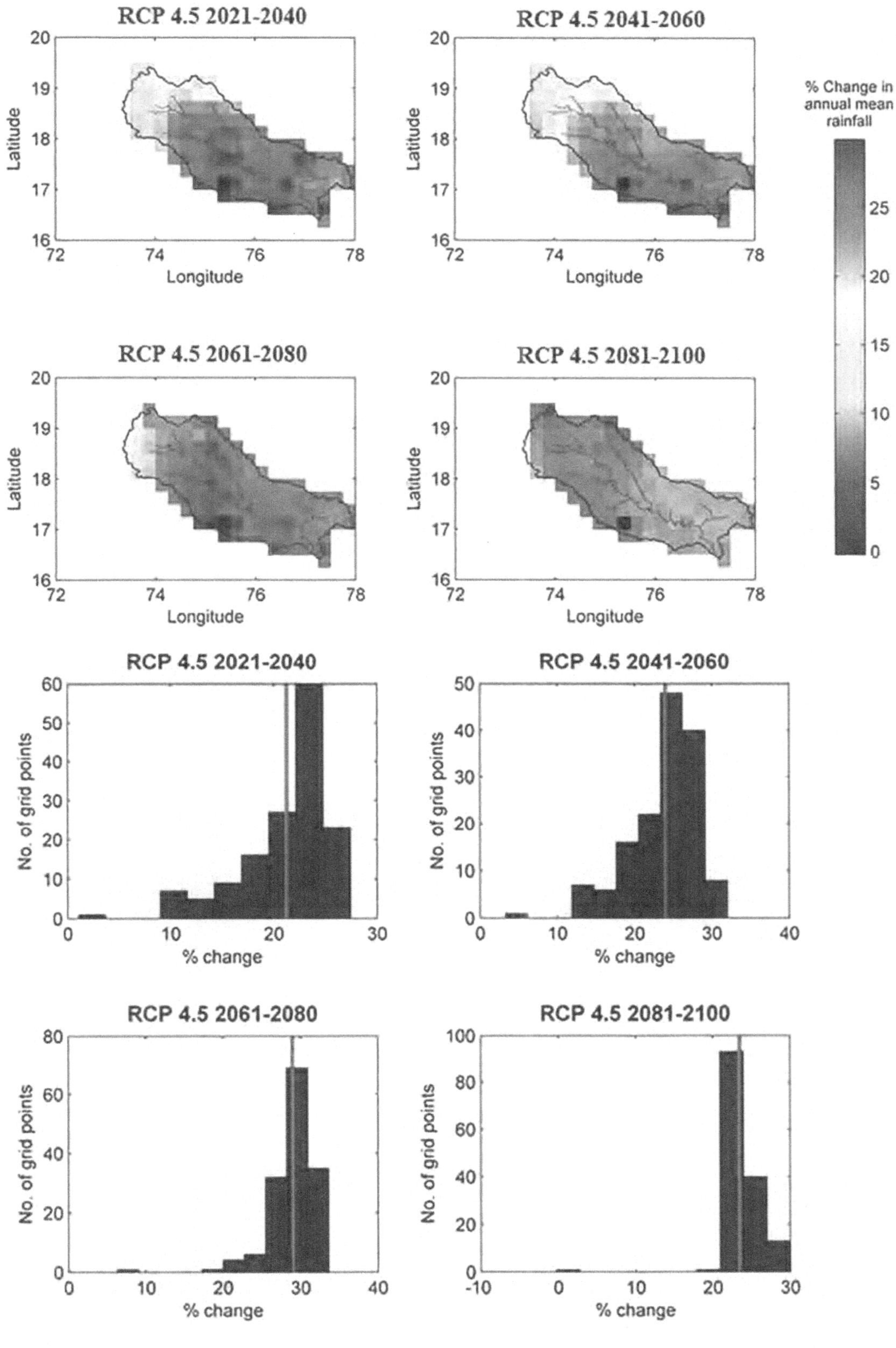

Figure 10 | Spatial maps (top four panels) and histograms (bottom four panels) of percentage change in mean annual rainfall projected from ensemble average in the RCP 4.5 scenario. Red line in the histogram indicates mean percentage change.

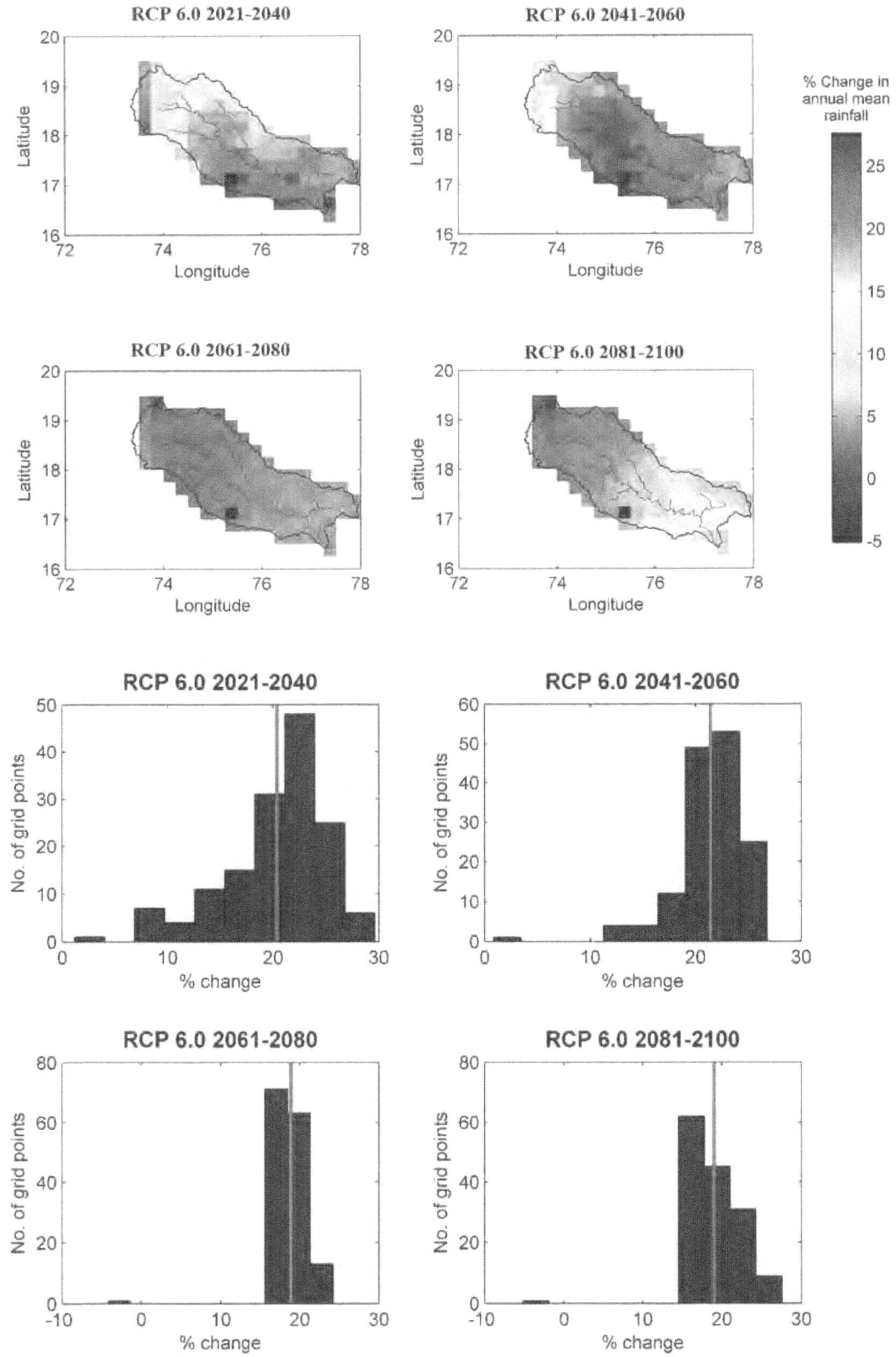

Figure 11 | Spatial maps (top four panels) and histograms (bottom four panels) of percentage change in mean annual rainfall projected from ensemble average in the RCP 6.0 scenario. Red line in the histogram indicates mean percentage change.

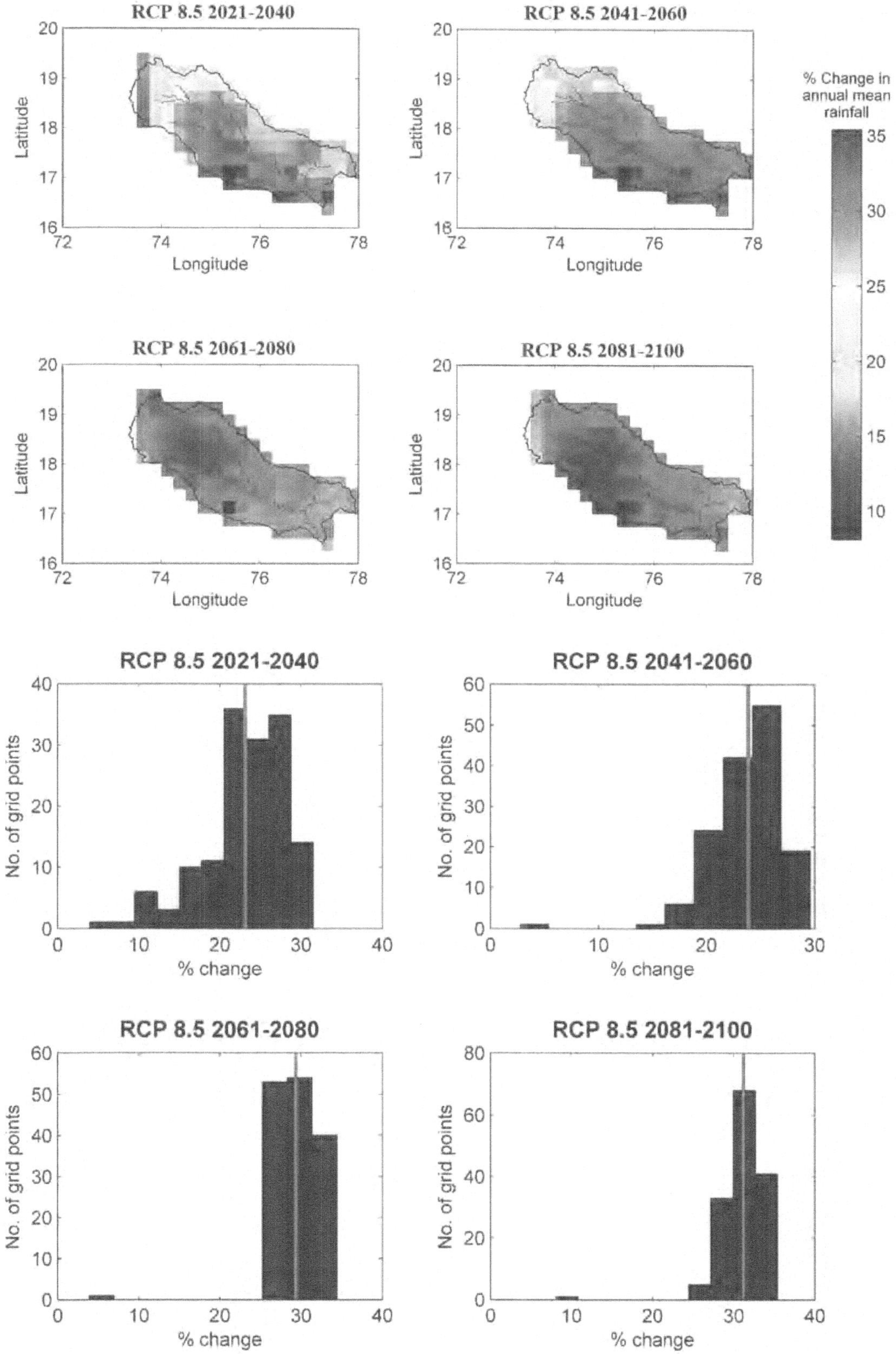

Figure 12 | Spatial maps (top four panels) and histograms (bottom four panels) of percentage change in mean annual rainfall projected from ensemble average in the RCP 8.5 scenario. Red line in the histogram indicates mean percentage change.

Further, shape and scale parameters α and β estimates are given by:

$$\alpha = \frac{1}{4A}\left(1 + \sqrt{1 + \frac{4A}{3}}\right) \tag{9}$$

$$\beta = \frac{\bar{X}}{\alpha} \tag{10}$$

where,

$$A = \ln(\bar{X}) - \frac{\sum \ln(\bar{X})}{n} \tag{11}$$

and n is the number of periods of non-zero rainfall in the time series.

(2) Calculate the value of cumulative distribution function CDF [G(X)] conforming to each value of non-zero rainfall (X):

$$G(X) = \int_0^X g(X)\, dX = \frac{1}{\beta^{\alpha}\Gamma(\alpha)} \int_0^X X^{(\alpha-1)} e^{-X/\beta}\, dX \tag{12}$$

G(X) can be computed using incomplete Gamma function.

(3) Estimate the zero rainfall (or, no-rainfall) probability (q) from the historical time series.

The value of CDF [$H(X)$] for an explicit rainfall (X) is then given by:

$$H(X) = q + (1 - q) \times G(X) \tag{13}$$

(4) The value of SPI for the rainfall is computed as the value of a standard normal deviate corresponding to the value of CDF [H(X)].

McKee *et al.* (1993) used the categories as shown in Table 2 to define drought intensities resulting from the SPI.

RESULTS AND DISCUSSION

Quantification of uncertainty

In order to address both GCM and scenario uncertainties an attempt is made in this section to interpret the outputs obtained from GCMs with four RCP scenarios in assessing the severity of climate change. It would be difficult to present the uncertainty study for each IMD grid in the basin and, thus, basin average value is considered for this research. Likely changes (based on the five GCMs ensemble average) in the aforementioned climate variables for future projected time series ranging from 2021 to 2100 and compared with that of their mean values over the historical period 1986–2005 are presented. Thus, for

Table 2 | Drought categorization based on SPI (McKee *et al.* 1993)

SPI value	Category
SPI>2.00	Extremely wet
1.50 to 1.99	Very wet
1.00 to 1.49	Moderately wet
−0.99 to 0.99	Near normal
0 to −0.99	Mild dry
−1.00 to −1.49	Moderately dry
−1.50 to −1.99	Severely dry
SPI<−2.00	Extremely dry

this work, box plots illustrate the decadal changes in downscaled climate variables. GCM uncertainty in climate change impact on the climate variables is examined with box plots for 20-year time series data.

Figure 13 shows box plots of the GCM and scenario uncertainty in climate change impact on projections of precipitation obtained over the evaluation period. Projections obtained from five GCMs and four RCP scenarios (ensemble average) are indicated as '+' (increase) and '−' (decrease) sign with respect to average value of historical period and tabulated as maximum, mean and minimum percentage change of respective climate variables (Table 3). The historical value of the climate data is denoted by the horizontal line. The variability of the projection represented by the box in the box plot and median of the box plots indicate the deviation of the data.

Inferring from Table 3 and Figure 13, the mean percentage change for precipitation with the values 7.64%, 9.11%, 12.41% and 16.33% shows an increasing tendency along with the time series. The box plot for 2021–2040 and 2041–2060 shows the symmetric distribution of data, while the projections for 2061–2080 and 2081–2100 has skewness with the later one being more unsymmetrical and showing relatively more uncertainty than earlier ones.

Meteorological drought

Precipitation, being the major indicator for most of the themes on applications to link the availability of water, is the key input data and indicator of drought showing the different characteristics of a drought. The future projections of precipitation for the

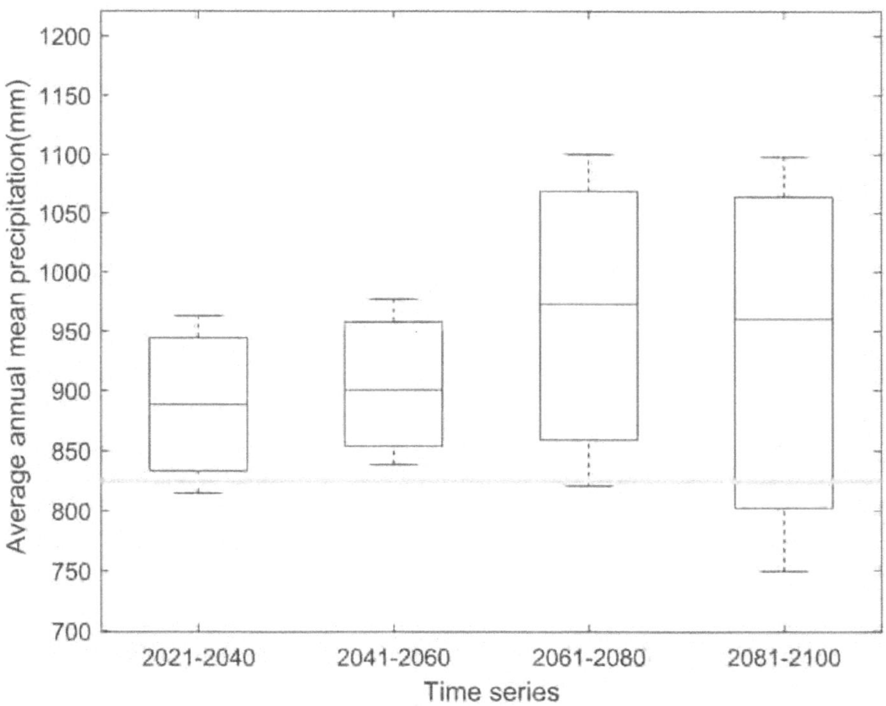

Figure 13 | GCM+scenario uncertainty in climate change impact on average annual mean precipitation over the Bhima sub-basin. The historical value of the climate data is denoted by the horizontal line.

Table 3 | Changes in the average annual mean precipitation (mm) over the Bhima sub-basin (GCM+scenario uncertainty)

Period	Max.	Mean	Min.	Max. % change	Mean % change	Min. % change
2021–2040	962.72	888.19	815	16.67	7.64	−1.23
2041–2060	976.6	900.33	838.4	18.35	9.11	1.6
2061–2080	1,102	927.58	821	33.55	12.41	−0.51
2081–2100	1,098.2	959.9	749.5	33.09	16.33	−9.17

Note: The average annual mean precipitation over the basin during 1986–2005 is 825.180 mm.

period 2021–2100 involved CMIP5 scenarios along with RCPs as accepted in the AR5 released in 2014 by the IPCC. Time series of long-term records (2021–2100) of downscaled monthly precipitation data in the Bhima sub-basin is used as input to the SPI program. The salient features of the places under examination for analysis of 'dry' period are listed in Table 4 along with the grid location of places selected in Bhima sub-basin and with district/administrative boundaries' details in Figure 14.

Table 4 | Details of places of grid location in Bhima sub-basin with district/administrative boundaries

Sl. No	Place	District/Administrative boundaries	Lat.	Long.	Code
1	Ahmednagar	Ahmednagar, Maharashtra	19.1	74.75	G1
2	Pune	Pune, Maharashtra	18.5	73.86	G2
3	Barshi	Osmanabad, Maharashtra	18.2	75.7	G3
4	Baramati	Pune, Maharashtra	18.2	74.58	G4
5	Pandharpur	Sholapur, Maharashtra	17.7	75.33	G5
6	Sholapur	Sholapur, Maharashtra	17.7	75.91	G6
7	Aland	Gulbarga, Karnataka	17.6	76.57	G7
8	Chincholi	Rangareddi, Hyderabad Karnataka	17.5	77.42	G8
9	Kalburgi	Gulbarga, Karnataka	17.3	76.84	G9
10	Yadgir	Gulbarga, Karnataka	16.8	77.14	G10

Figure 14 | Grid location of places selected in Bhima sub-basin along with district/administrative boundaries' details.

Severity of future drought using SPI estimated from the downscaled GCM output

The impact studies on meteorological changes, more importantly ascertaining drought, require quantitative assessment. Thus, SPI as drought index is used to project the future meteorological condition under different RCP scenarios to assess the impact on Bhima basin. To assess its ability to predict the drought conditions, monthly precipitation time series data for Bhima basin (average basin conditions) covering the period of 2021–2100 are considered. The SPI is capable of computing drought intensity over any specified interval, yet the most influential feature of this method is its inherent ability to simultaneously evaluate drought over a set of timescales, e.g., one, three, six, nine, 12, etc., popularly used by the research community. This research work considers SPI 12-month time scale using the algorithm presented in Gimbel *et al.* (2015) with modification using the 'precintcon' package to analyse the precipitation intensity, concentration and anomaly.

SPI based on 12 months compares the precipitation data for the consecutive 12 months' recorded data within the same 12 consecutive months for all the available historical data. Since these particular timescales' data are the cumulative result of smaller periods, they may be above or below normal, making the longer SPI fall towards zero except for a typical dry or wet trend. Long-term precipitation patterns are reflected by these SPIs. These timescales are linked with the applications on stream flows, groundwater phenomena and the reservoir levels considered at the longer duration.

Temporal variation for 12-month SPI presented in Figure 15 shows significant changes over RCP scenarios. For the beginning of the period, 2021–2040 precipitation appears to be scanty for RCP 2.6 and RCP 4.5 scenarios. Mild to moderate and severe drought events for RCP 2.6 scenarios are more predominant as is evident from the summary of the report presented in Table 5. Extreme drought condition with 17.5% occurrence for the basin during 2021–2040 has more impact in the mid-region of the basin under G5, G6 and G7 grid point locations (Figure 16(a)). The upper region of Bhima (K5) under G1 and G2 grid locations has near-normal conditions during this period. When we shift to the next time slot for 2041–2060, the occurrence of extreme drought events drops marginally. For the 2061–2080 time period, there is an increase in the extreme event of drought with 21.5% of the basin falling under the extreme drought category, and the grid locations G1 and G2 which were under near-normal conditions have been falling under this category as is evident from spatial plots. This event is continuous for 2081–2100 with 28% of the basin falling under extreme drought events.

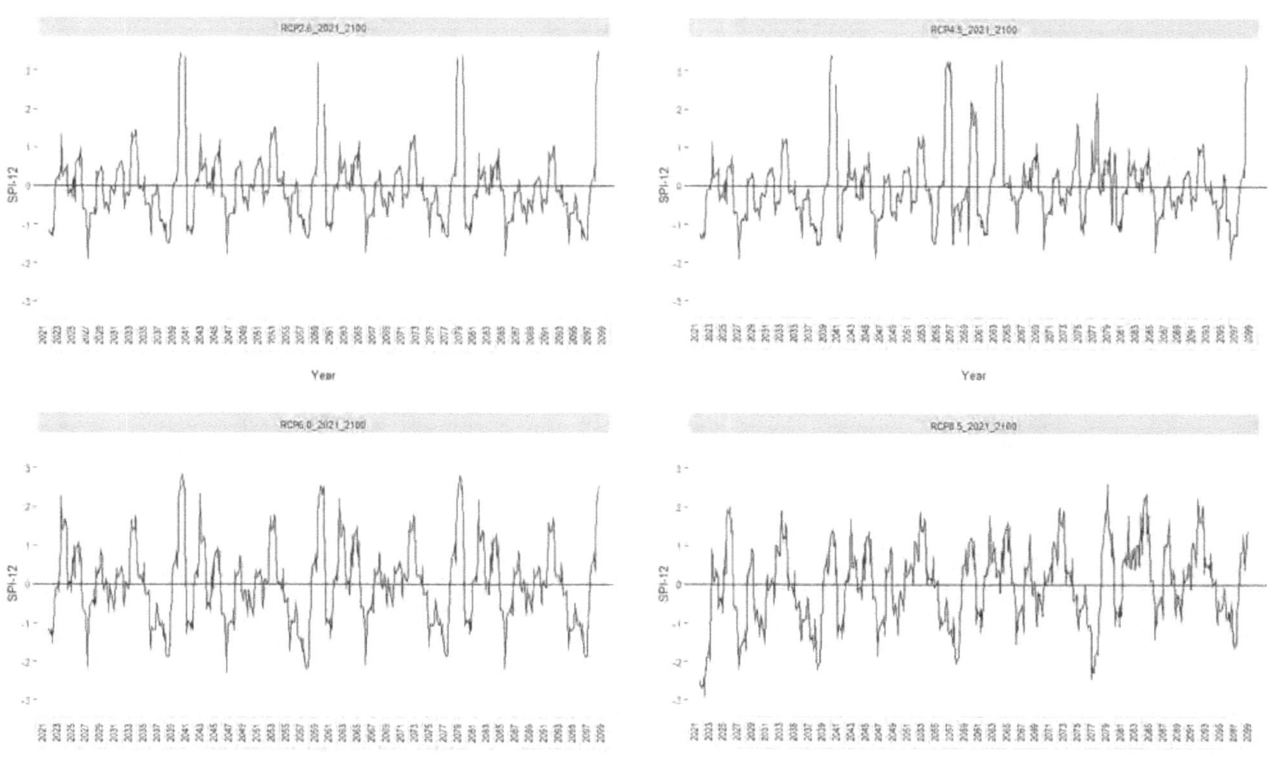

Figure 15 | Temporal variation of drought of 12-month SPI at Bhima basin for the period 2021–2100 under scenarios RCP 2.6, RCP 4.5, RCP 6.0 and RCP 8.5.

Table 5 | Summary of percentage (%) of occurrence of drought events for SPI-12 under scenarios RCP 2.6, RCP 4.5, RCP 6.0 and RCP 8.5

	SPI range	Description	RCP2.6	RCP4.5	RCP6.0	RCP8.5
2012–2040	0 to −0.99	Near normal conditions	7.5	30.5	12.5	8.5
	−1.00 to −1.49	Mild to moderate drought	35.5	16.5	26.5	12
	−1.50 to −1.99	Severe drought	32.5	21.5	24.5	9.5
	−2.00 or less	Extreme drought	17.5	7.5	20.5	14
	>0	Wet conditions	7	24	16	56
2041–2060	0 to −0.99	Near normal conditions	9.5	26	10.5	10.5
	−1.00 to −1.49	Mild to moderate drought	37.5	16.5	28.5	10
	−1.50 to −1.99	Severe drought	21.5	15	29.5	7.5
	−2.00 or less	Extreme drought	15.5	3.5	12.5	18
	>0	Wet conditions	16	39	19	54
2061–2080	0 to −0.99	Near normal conditions	21.5	35	13.5	9.5
	−1.00 to −1.49	Mild to moderate drought	29.5	12.5	33	14.5
	−1.50 to −1.99	Severe drought	17.5	8.5	26.5	8
	−2.00 or less	Extreme drought	21.5	0	8.5	12
	>0	Wet conditions	10	44	18.5	56
2081–2100	0 to −0.99	Near normal conditions	17.5	29.5	20.5	11.5
	−1.00 to −1.49	Mild to moderate drought	29.5	12.5	24.5	20
	−1.50 to −1.99	Severe drought	17.5	26.5	22.5	10.5
	−2.00 or less	Extreme drought	28	2.5	14.5	8.5
	>0	Wet conditions	7.5	29	18	49.5

Near normal conditions are most likely to prevail during 2021–2100 for the RCP 4.5 scenario with the magnitude ranging from 26 to 35% occurrence for the Bhima sub-basin. The variations for different time intervals are shown in the spatial maps (Figure 16(b)). The extreme drought events for RCP 4.5 tend to decline from the 2021 time period. However, during 2061–2080 there are zero occurrence extreme drought events (Table 5). Moderate and severe drought events have predominant impact during 2021–2040 in the lower region at grid locations G8, G9 and G10 and upper region at G1 and G2 grids during 2081–2100. Mild to moderate and severe drought conditions outweigh the other conditions under RCP 6.0. Most parts of upper Bhima are affected with these conditions (Figure 16(c)), with extreme drought events during 2027, 2047, 2058 and 2066 (as seen in the temporal variation plot, Figure 15). Extreme drought events occur with 20.5% intensity in the basin during 2021–2040 and mostly affect the grid locations G5, G6, G7 and G9. Spatial plots of RCP 8.5 scenarios shown in Figure 16(d) have varied occurrence of dry periods across the different time periods. The inconsistent pattern of occurrence of drought events over the time period shows 18% of the basin with extreme drought during 2041–2060 followed by 14%, 12% and 8.5% of its occurrence during 2021–2040, 2061–2080 and 2081–2100, respectively. It is probable that the emission scenarios of GHG considered for RCP 8.5 may likely resemble an inconsistent pattern of drought events.

CONCLUSION AND RECOMMENDATIONS

The observed annual rainfall series in the Western Ghats region (Maharashtra state) lying in the Upper Bhima (K5) shows an increasing trend. The observed annual rainfall series in Lower Bhima (K6) has a decreasing trend. The evaluation results of CF downscaling method in Bhima basin indicate a satisfactory performance. Further, it is observed that the CF downscaling satisfactorily captured the cross-correlation between rainfall grids of Upper Bhima (K5) and Lower Bhima (K6) regions. The comparison of historical and future downscaled precipitation shows that the mean annual precipitation (for the historical period 1986–2005) is likely to increase under climate change scenarios. The maximum, mean and minimum percentage change in the average annual mean precipitation (mm) with respect to the historical data (1986–2005) during 2021–2040 is +16.67%, +7.64% and −1.23%, respectively. During 2041–2060, percentage change in the average annual mean precipitation is +18.35%, +9.11% and +1.6%,

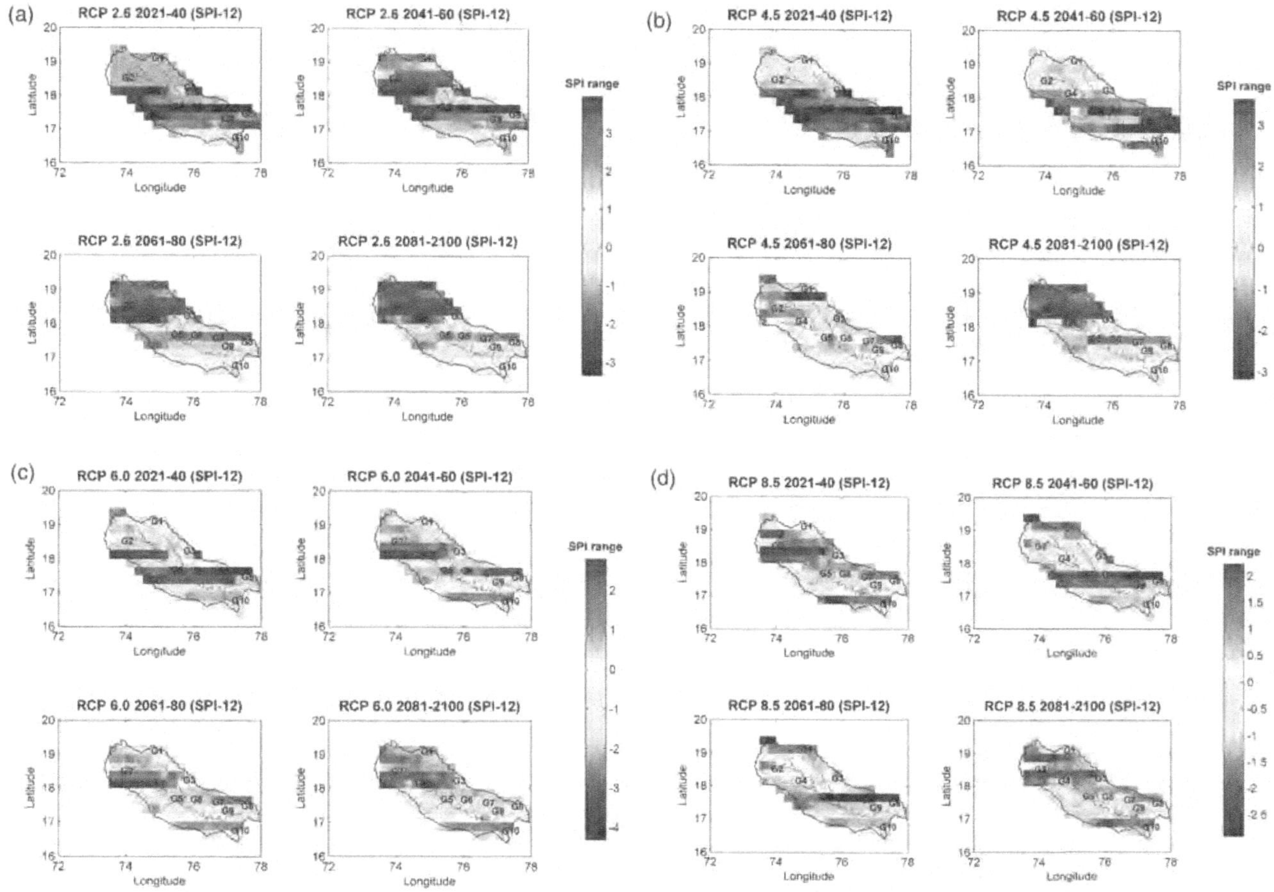

Figure 16 | Spatial pattern maps of drought of 12-month SPI for Bhima basin during 2021–2100 under scenarios RCP 2.6, RCP 4.5, RCP 6.0 and RCP 8.5.

respectively. While, during 2061–2100, the change with respect to the historical data (1986–2005) is +33.55%, +12.41% and −0.51%, respectively. Lastly, for the period 2081–2100, the change is +33.09%, +16.33% and −9.17%, respectively.

The GCM and scenarios' uncertainty encountered are accounted for by quantifying them individually. To make better representation of the climate variable for the uncertainty study for each IMD grid in the basin, the basin average value was considered for this research. Likely changes (based on the five GCMs' ensemble average) in all the climate variables for future projected time series ranging from 2021 to 2100 in the periods of 2021–2040, 2041–2060, 2061–2080 and 2081–2100 were compared with those of their mean values over the historical period 1986–2005. Precipitation being the dominant input variable to the hydrological modeling means uncertainty in this variable plays a major role in climate change impact studies as it is always burdened with such high uncertainties. For policy-framing purposes, the projections provided by ensemble average across GCMs and scenarios may be used as a guideline. Uncertainty increases with time and higher confidence may be attained for nearer future (2021–2040) projections than the far future ones (2081–2100).

Assessing the severity of future meteorological drought using the SPI estimated from the downscaled GCM output is yet another important objective of the research. To assess its ability to predict drought conditions, monthly precipitation time series data for Bhima basin (average basin conditions) covering the period of 2021–2100 are considered. Twelve-month SPI (i.e., SPI-12), the occurrence of mild to moderate drought events under RCP 2.6 and 6.0 scenarios is predominant for the basin, while severe drought events recorded for RCP 6.0 scenario dominates over other RCPs scenarios. However, the extreme drought events occur commonly and extensively under RCP 2.6 and 6.0 scenarios. The results under RCP 8.5 scenario show abnormal and extreme conditions in which GHG mitigation policies are not implemented and GHGs are emitted according to current trends. The current work on meteorological drought assessment will provide valuable information for managers and policymakers to investigate the impacts of climate change on drought conditions. The proposed work in this paper for developing the strategies and mitigation plan for regional assessment will serve as a decision-making tool for effective watershed management plans and policies. Future scope of the research may also include studying the effect of climate change on river water quality management using the projections from GCM outputs with the statistical downscaling methodology.

DATA AVAILABILITY STATEMENT

All relevant data are included in the paper or its Supplementary Information.

REFERENCES

Aggarwal, P. K. 2008 Global climate change and Indian agriculture: impacts, adaptation and mitigation. *Indian Journal of Agricultural Sciences* **78** (11), 911.

Anandhi, A., Frei, A., Pierson, D. C., Schneiderman, E. M., Zion, M. S., Lounsbury, D. & Matonse, A. H. 2011 Examination of change factor methodologies for climate change impact assessment. *Water Resources Research* **47** (3).

Bates, B., Kundzewicz, Z. & Wu, S. 2010 *Climate Change and Water*. Intergovernmental Panel on Climate Change Secretariat. Available from: http://www.nwc.gov.au (accessed 21 May 2021).

Biggs, T., Gaur, A., Scott, C., Thenkabail, P., Gangadhara, R., Parthasaradhi, G., Murali, K., Acharya, S. & Turral, H. 2007 *Closing of the Krishna Basin: Irrigation, Streamflow Depletion and Macroscale Hydrology. IWMI Research Report 111*. International Water Management Institute (IWMI), Colombo, Sri Lanka, p. 38.

Biswas, A. K. 1987 Monitoring and evaluation of irrigated agriculture. *Food Policy* **12** (1), 47–61. https://doi.org/10.1016/0306-9192(87)90047-9.

Boé, J., Terray, L., Habets, F. & Martin, E. 2007 Statistical and dynamical downscaling of the Seine basin climate for hydro-meteorological studies. *International Journal of Climatology* **27** (12), 1643–1655.

Chanapathi, T. & Thatikonda, S. 2020 Investigating the impact of climate and land-use land cover changes on hydrological predictions over the Krishna river basin under present and future scenarios. *Science of the Total Environment* **721**, 137736.

Deshpande, N., Kothawale, D. & Kulkarni, A. 2016 Changes in climate extremes over major river basins of India. *International Journal of Climatology* **36** (14), 4548–4559.

Fang, G., Yang, J., Chen, Y., Li, Z. & De Maeyer, P. 2018 Impact of GCM structure uncertainty on hydrological processes in an arid area of China. *Hydrology Research* **49** (3), 893–907.

Garg, K. K., Bharati, L., Gaur, A., George, B., Acharya, S. & Jella, K. 2012 Spatial mapping of agricultural water productivity using the SWAT model in Upper Bhima Catchment, India. *Irrigation and Drainage* **61** (1), 60–79.

Gaur, A., McCornick, P. G., Turral, H. & Acharya, S. 2007 Implications of drought and water regulation in the Krishna Basin, India. *International Journal of Water Resources Development* **23** (4), 583–594.

Ghosh, S. & Mujumdar, P. P. 2007 Nonparametric methods for modeling GCM and scenario uncertainty in drought assessment. *Water Resources Research* **43** (7).

Gimbel, K. F., Felsmann, K., Baudis, M., Puhlmann, H., Gessler, A., Bruelheide, H., Kayler, Z., Ellerbrock, R. H., Ulrich, A., Welk, E. & Weiler, M. 2015 Drought in forest understory ecosystems-a novel rainfall reduction experiment. *Biogeosciences* **12** (4), 961–975.

Gleick, P. H., Adams, R. M., Amasino, R. M., Anders, E., Anderson, D. J., Anderson, W. W. & Mabogunje, A. L. 2010 Climate change and the integrity of science. *Science* **328** (5979), 689–690.

Hayes, M. J., Svoboda, M. D. & Wilhite, D. A. 1999 Monitoring the 1996 drought using the standardized precipitation index. *Bulletin of the American Meteorological Society* **80** (3), 429–438.

Homsi, R., Shiru, M. S., Shahid, S., Ismail, T., Harun, S. B., Al-Ansari, N., Chau, K.-W. & Yaseen, Z. M. 2019 Precipitation projection using a CMIP5 GCM ensemble model: a regional investigation of Syria. *Engineering Applications of Computational Fluid Mechanics* **14** (1), 90–106.

IPCC 2007 Climate change 2007: the physical science basis. In: *Contribution of Working Group I to the Fourth Assessment Report of the Intergovernmental Panel on Climate Change* (Solomon, S. D., Qin, M., Manning, Z., Chen, M., Marquis, K. B., Averyt, M., Tignor, M. & Miller, H. L., eds). Cambridge University Press, Cambridge, UK. and New York, NY, USA, p. 996.

IPCC 2013 *Climate Change 2013: The Physical Science Basis*. Cambridge University Press, New York, NY, USA.

Kang, S., Zhang, Q., Qian, Y., Ji, Z., Li, C., Cong, Z., Zhang, Y., Guo, J., Du, W., Huang, J., You, Q., Panday, A. K., Rupakheti, M., Chen, D., Gustafsson, O., Thiemans, M. H. & Qin, D. 2019 Linking atmospheric pollution to cryospheric change in the Third Pole region: current progress and future prospects. *National Science Review* **6** (4), 796–809.

Karabulut, M. 2015 Drought analysis in Antakya-Kahramanmaraş Graben, Turkey. *Journal of Arid Land* **7** (6), 741–754.

Kendall, M. G. 1975 *Rank Correlation Methods*. Charles Griffin, London, UK.

Knutti, R., Furrer, R., Tebaldi, C., Cermak, J. & Meehl, G. A. 2010 Challenges in combining projections from multiple climate models. *Journal of Climate* **23** (10), 2739–2758.

Kulkarni, B. D., Deshpande, N. R., Patwardan, S. K. & Bansod, S. D. 2014 Assessing hydrological response to changing climate in the Krishna Basin of India. *Journal of Earth Science & Climatic Change* **5**, 211.

Kulkarni, A., Sabin, T. P., Chowdary, J. S., Rao, K. K., Priya, P., Gandhi, N., Bhaskar, P., Buri, V. K., Sabade, S. S., Pai, D. S., Ashok, K., Mitra, A. K., Niyogi, D. & Rajeevan, M. 2020 Precipitation changes in India. In: *Assessment of Climate Change Over the Indian Region* (Krishnan, R., Sanjay, J., Gnanaseelan, C., Mujumdar, M., Kulkarni, A. & Chakraborty, S., eds). Springer, Singapore, pp. 47–72. https://doi.org/10.1007/978-981-15-4327-2_3

Kundzewicz, Z. W., Krysanova, V., Benestad, R. E., Hov, Ø., Piniewski, M. & Otto, I. M. 2018 Uncertainty in climate change impacts on water resources. *Environmental Science & Policy* **79**, 1–8.

Lindner, M., Fitzgerald, J. B., Zimmermann, N. E., Reyer, C., Delzon, S., van der Maaten, E., Schelhaas, M.-J., Lasch, P., Eggers, J., van der Maaten-Theunissen, M., Suckow, F., Psomas, A., Poulter, B. & Hanewinkel, M. 2014 Climate change and European forests: what do we know, what are the uncertainties, and what are the implications for forest management? *Journal of Environmental Management* **146**, 69–83. https://doi.org/10.1016/j.jenvman.2014.07.030.

Mahajan, D. R. & Dodamani, B. M. 2015 Trend analysis of drought events over Upper Krishna Basin in Maharashtra. *Aquatic Procedia* **4**, 1250–1257.

Mann, H. B. 1945 Nonparametric tests against trend. *Econometrica* **13**, 245–259.

McKee, T. B., Doesken, N. J. & Kleist, J. 1993 The relationship of drought frequency and duration of time scales. In: *Eighth Conference on Applied Climatology, American Meteorological Society*, January 17–23, 1993, Anaheim, CA, USA, pp. 179–186.

Nyunt, C. T., Koike, T. & Yamamoto, A. 2016 Statistical bias correction for climate change impact on the basin scale precipitation in Sri Lanka, Philippines, Japan and Tunisia. *Hydrology and Earth System Sciences Discussions*, 1–32.

Raju, K. S., Kumar, D. N. & Babu, I. N. 2016 Ranking of global climate models for Godavari and Krishna River Basins, India, using compromise programming. *Sustainable Water Resources Planning and Management Under Climate Change*, 87–100. doi:10.1007/978-981-10-2051-3_4.

Rehana, S. 2019 River water temperature modelling under climate change using support vector regression. In: *Hydrology in A Changing World* (Singh, S. K. & Dhanya, C. T., eds). Springer, Cham, Switzerland, pp. 171–183.

Robock, A., Turco, R. P. & Harwell, M. A. 1993 Use of general circulation model output in the creation of climate change scenarios for impact analysis. *Climatic Change* **23**, 293–335.

Rowell, D. P. 2006 A demonstration of the uncertainty in projections of UK climate change resulting from regional model formulation. *Climatic Change* **79** (3), 243–257.

Shamshirband, S., Hashemi, S., Salimi, H., Samadianfard, S., Asadi, E., Shadkani, S., Kargar, K., Mosavi, A., Nabipour, N. & Chau, K.-W. 2020 Predicting standardized streamflow index for hydrological drought using machine learning models. *Engineering Applications of Computational Fluid Mechanics* **14** (1), 339–350.

Sharma, T., Vittal, H., Chhabra, S., Salvi, K., Ghosh, S. & Karmakar, S. 2018 Understanding the cascade of GCM and downscaling uncertainties in hydro-climatic projections over India. *International Journal of Climatology* **38**, e178–e190.

Sönmez, F. K., Koemuescue, A. U., Erkan, A. & Turgu, E. 2005 An analysis of spatial and temporal dimension of drought vulnerability in Turkey using the standardized precipitation index. *Natural Hazards* **35** (2), 243–264.

Steinemann, A. 2003 Drought indicators and triggers: a stochastic approach to evaluation. *Journal of the American Water Resources Association* **39** (5), 1217–1233.

Surinaidu, L., Bacon, C. G. D. & Pavelic, P. 2013 Agricultural groundwater management in the Upper Bhima Basin, India: current status and future scenarios. *Hydrology and Earth System Sciences* **17** (2), 507–517. https://doi.org/10.5194/hess-17-507-2013.

Tsakiris, G., Loukas, A., Pangalou, D., Vangelis, H., Tigkas, D., Rossi, G. & Cancelliere, A. 2007 Drought characterization. *Drought Management Guidelines Technical Annex*, CIHEAM/EC MEDA Water, pp. 85–102.

Udmale, P., Ichikawa, Y., Kiem, A. & Panda, S. 2014 Drought impacts and adaptation strategies for agriculture and rural livelihood in the Maharashtra State of India. *The Open Agriculture Journal* **8**, 41–47.

van Vuuren, D. P., Edmonds, J., Kainuma, M., Riahi, K., Thomson, A., Hibbard, K., Hurtt, G. C., Kram, T., Krey, V., Lamarque, J.-F., Masui, T., Meinshausen, M., Nakicenovic, N., Smith, S. J. & Rose, S. K. 2011 The representative concentration pathways: an overview. *Climatic Change* **109** (1–2), 5–31. https://doi.org/10.1007/s10584-011-0148-z.

Venot, J.-P., Sharma, B. R. & Rao, K. V. G. K. 2008 *The Lower Krishna Basin Trajectory: Relationships Between Basin Development and Downstream Environmental Degradation. IWMI Research Report 125.* International Water Management Institute (IWMI), Colombo, Sri Lanka, p. 30.

Wang, W., Zhu, Y., Xu, R. & Liu, J. 2015 Drought severity change in China during 1961–2012 indicated by SPI and SPEI. *Natural Hazards* **75** (3), 2437–2451.

Wang, B., Zheng, L., Liu, D. L., Ji, F., Clark, A. & Yu, Q. 2018 Using multi-model ensembles of CMIP5 global climate models to reproduce observed monthly rainfall and temperature with machine learning methods in Australia. *International Journal of Climatology* **38** (13), 4891–4902. https://doi.org/10.1002/joc.5705.

Warner, K., Hamza, M., Oliver-Smith, A., Renaud, F. & Julca, A. 2010 Climate change, environmental degradation and migration. *Natural Hazards* **55** (3), 689–715.

Wilby, R. L. & Harris, I. 2006 A framework for assessing uncertainties in climate change impacts: low-flow scenarios for the River Thames, UK. *Water Resources Research* **42** (2).

Woldemeskel, F. M., Sharma, A., Sivakumar, B. & Mehrotra, R. 2016 Quantification of precipitation and temperature uncertainties simulated by CMIP3 and CMIP5 models. *Journal of Geophysical Research: Atmospheres* **121** (1), 3–17.

World Meteorological Organization (WMO) and Global Water Partnership (GWP) 2016 *Handbook of Drought Indicators and Indices* Integrated Drought Management Programme (IDMP), Integrated Drought Management Tools and Guidelines Series 2, Geneva, Switzerland. https://www.droughtmanagement.info/literature/GWP_Handbook_of_Drought_Indicators_and_Indices_2016.pdf

Zhao, C., Huang, Y., Li, Z. & Chen, M. 2018 Drought monitoring of Southwestern China using insufficient GRACE data for the long-term mean reference frame under global change. *Journal of Climate* **31** (17), 6897–6911.

First received 17 May 2021; accepted in revised form 24 August 2021. Available online 24 September 2021

Impact of Climate Change on Hydrology and Water Resources doi: 10.2166/wcc.2021.175

Ecological drought and its state assessment: a case study in the Yellow River estuary

Jinliang Zhang[a], Bojun Liu[a,b,*], Libin Yang[a], Liyuan He[a], Xinyi Cao[b] and Guangwen Shao[c]

[a] Yellow River Engineering Consulting Co., Ltd., Zhengzhou 450003, China
[b] College of Hydrology and Water Resources, Hohai University, Nanjing 210098, China
[c] School of Water Conservancy and Environment, University of Jinan, Jinan 250022, China
*Corresponding author. E-mail: bojun_l689@126.com

ABSTRACT

Water cycle has been intensified by global warming, leading to frequent extreme climate events. Drought is an extreme climate phenomenon. Runoff decrease and human water demand increase aggravate the water shortage of regional ecosystems, affecting regional water and land ecosystems and causing ecological drought, river cutoff and water pollution. Finally, the reverse succession and the imbalance of regional ecological structures take place. The clarification of the concept of ecological drought for effective evaluation of regional ecological drought degree has become an urgent important scientific issue to be resolved. Therefore, in this paper, the typical region of the Yellow River estuary was studied for the analysis of characteristics of regional ecological changes and the definition of the concept and connotation of ecological drought. Based on the representative monitoring and early warning indices to ecological drought, the evaluation method and the classification standard of regional ecological drought were proposed. The regional ecological drought includes four levels: I (Severe), II (General), III (Weak) and IV (None). The indicator thresholds of river runoff, biodiversity and vegetation coverage on different ecological drought levels were quantified. The research results can be technically beneficial for the improvement of global ecological drought emergent support capacity and reducing loss due to drought.

Key words: concept redefinition, ecological drought, leading indicators, state assessment, Yellow River estuary

HIGHLIGHTS

- The concept of ecological drought in areas with important ecological functions is discussed.
- A model with certain applicability for ecological drought assessment is proposed.
- The regional ecological drought can be divided into four levels.
- The thresholds of leading indicators to different ecological drought degrees are quantified.

1. INTRODUCTION

Ecological environment is the basic natural condition related to human survival and development (Wang *et al.* 2020). The maintenance of the benign cycle of ecosystems relies on the rational development and protection of water resources, and full considerations of ecological environment water demand and sustainable utilization of water resources (Kuemmerlen *et al.* 2019; Yu *et al.* 2019). Water is the most important condition for the benign cycle of ecological environment (Rinaldo *et al.* 2018). With the impact of climate change and human activities, regional drought may become more serious, and the drought will further aggravate the shortage of water for ecological environment systems (Ostad-Ali-Askari *et al.* 2019). Drought is divided into meteorological, agricultural, hydrological and socioeconomic drought in different research fields (Bachmair *et al.* 2016; Liang *et al.* 2019). The fifth assessment report from the Intergovernmental Panel on Climate Change (IPCC) states that the frequency, intensity and duration of global drought events have increased significantly in the last 50 years in the 21st century, and this trend will be further developed in the future (Porter *et al.* 2018). In China, drought is one of the natural disasters with high frequency, long duration and wide influence range and causes important loss to livings and production, especially agricultural production (Novoa *et al.* 2019; Yao *et al.* 2019). In addition, the once-in-a-thousand-year drought in Australia from 2002 to 2010 caused ecological loss of 800 million AUD in the Murray-Darling region (Millar & Stephenson 2015). The extreme drought event in the Amazon region in 2010 resulted in a net carbon loss of 2.2 Pg (1 Pg = 1,015 g), basically the annual carbon uptake of global forest ecosystem, in the regional tropical rainforest ecosystem (Goulden & Bales 2019). The identification of drought and the measures to deal with and alleviate drought have become an urgent scientific problem to be solved (Jiang *et al.* 2019).

According to the conventional concepts of agricultural drought and drought resistance, some studies name wetland ecological issues as 'ecological drought', in which the object only refers to wetland, though this explains that drought will cause ecological loss of wetland. With the proposal of the concept of wetland ecological drought, the ecological drought evaluation function of annual average water level was proposed to analyze the early warning index and the level of Baiyangdian Wetland ecological drought in China (Zhang *et al.* 2010; Huang *et al.* 2019). Until 2016, the Science for Nature and People Partnership (SNAPP) of the United States proposed a concept of ecological drought, which is defined as a complex comprehensive process caused by the shortage of natural water supply (changes in hydrological processes caused by natural or artificial management), leading to changes in the growth state of surface vegetation and soil water conditions (Slette *et al.* 2019). It is a xerophytic ecological environment constructed by vegetation affected by the interaction between water stress and living environment. Ecological drought will negatively affect the ecosystem in many ways. This concept of ecological drought is developed to involve the impact of drought on terrestrial plants and soil water cycle. It is still understood as an impact, but whether it will be developed into a disaster and whether this disaster should be taken as ecological drought are still not clear. With more and more attention paid to the impact of drought on ecosystems (Brown & Alastair 2017; Chadd *et al.* 2017), Crausbay *et al.* (2017) defined ecological drought as a 'sudden shortage of available water resources leading to the vulnerability of ecosystem exceeding its threshold and feedback in natural and human systems'. Munson *et al.* (2020) found that ecological drought should involve the sensitivity of ecosystems to the change in drought degree and the response of organisms at different levels. Therefore, ecological drought is defined as a 'shortage of available water leading to the performances of organisms or ecosystems deviating from the upper limits'.

Based on the understanding of this concept of ecological drought, MODIS was adopted to normalize vegetation index and land surface temperature data to establish the temperature vegetation dryness index for the quantitative study on the characteristics of ecological drought in different ecological functional areas and the analysis of the driving factors. In order to evaluate the ecological impact of drought, five watershed-scale drought plans in southwestern Montana were analyzed using the conceptual framework of ecological drought (McEvoy *et al.* 2018). The nonparametric density estimation and the super probability method were adopted to quantitatively assess the water quality risk caused by ecological drought (Kim *et al.* 2019). In certain papers, river ecological and minimum runoff values were taken as the indicators of ecological drought to establish an ecological drought monitoring framework (Park *et al.* 2020). Especially, the minor change in low runoff during drought will cause the violent response of regional aquatic ecosystems. The experimental results in the rivers of New Zealand showed that the impact on macroinvertebrates was relatively small when the runoff was reduced by 10% (James & Suren 2010). In contrast, the experimental results in the rivers of America showed that when the runoff was reduced to 96% of the annual average, the density and growth rate of salmon were reduced to 60–67% of the normal values (Mccargo & Peterson 2010). Therefore, ecological drought can be considered as a gradual disaster. The aquatic ecosystem is relatively fragile and has poor drought resistance. The longer the drought lasts, the more serious the loss will be, and even the aquatic ecosystem will be irreversibly affected.

Currently, the research mostly focuses on wetlands, and terrestrial ecology is also targeted later. And, the global research on regional ecological drought evaluation indices and the quantification of the indices is still undeveloped. In general, the concept and connotation of ecological drought are still not very clear, universal drought assessment indices are still absent and the specific division of ecological drought degree is rarely reported. These issues restrict and interfere with the development of drought resistance. In order to integrate eco-environmental impact into disaster prevention and mitigation planning, the planners should make two topics clear: a more comprehensive concept of ecological drought and how to categorize and evaluate the degree of ecological drought. Therefore, aiming at the dilemmas of ambiguous concepts of ecological drought and difficult effective evaluation, in this paper, the Yellow River estuary was taken as the research object to explore the regional climate, hydrology and ecological characteristics for the analysis of the interaction between regional drought and ecological factors, so a more accurate and clear concept of regional ecological drought would be proposed to resolve the problem of ambiguous cognition of regional ecological drought. An evaluation system was constructed to realize the assessment of ecological drought degree and to provide a scientific and technological scaffold for improving the ecological drought resisting emergency supportability in river basins and regions.

2. STUDY AREA AND MATERIALS

2.1. Study area

The Yellow River estuary refers to the entire region of Dongying city, with a land area of 7,923 km^2 (seen in Figure 1), located in the Yellow River Estuary Delta (Xiao *et al.* 2020). Diatom, dinoflagellate and green algae are distributed in the Yellow River estuary, with an average phytoplankton content of 0.475 mg/L. The regional zooplankton content, 0.295 mg/L, is much higher than those in the upper and middle reaches of the Yellow River. This region possesses low terrain, agreeable temperatures and good water quality with an average benthic biomass of 100 g/m^2. According to the survey, the Yellow River estuary is one of the spawning grounds of fish in the Yellow River Basin, and this region is rich in migratory fish and produces 10 native fish worth protecting: Silurus Lanzhouensis, Cyprinus Carpio, Silurus Huangzhouensis, Copperhead Norvegicus, Trout, Crucian Carp, Hemiculter Leucisclus, Loach, Wheatfish and Kaifeng catfish.

2.2. Data sources

Runoff data of Lijin station in 1956–2016 were obtained from the Yellow River Conservancy Commission. The Landsat TM images of the Yellow River estuary (version of 1986, 1996, 2006, 2015, 2016, 2017, 2018 and 2019, processed by ArcGIS and ENVI software) were collected from the Institute of Water Resources Protection in the Yellow River Conservancy Commission. Assessment index data from 2000 to 2016 were obtained from the Yellow River Shandong Bureau.

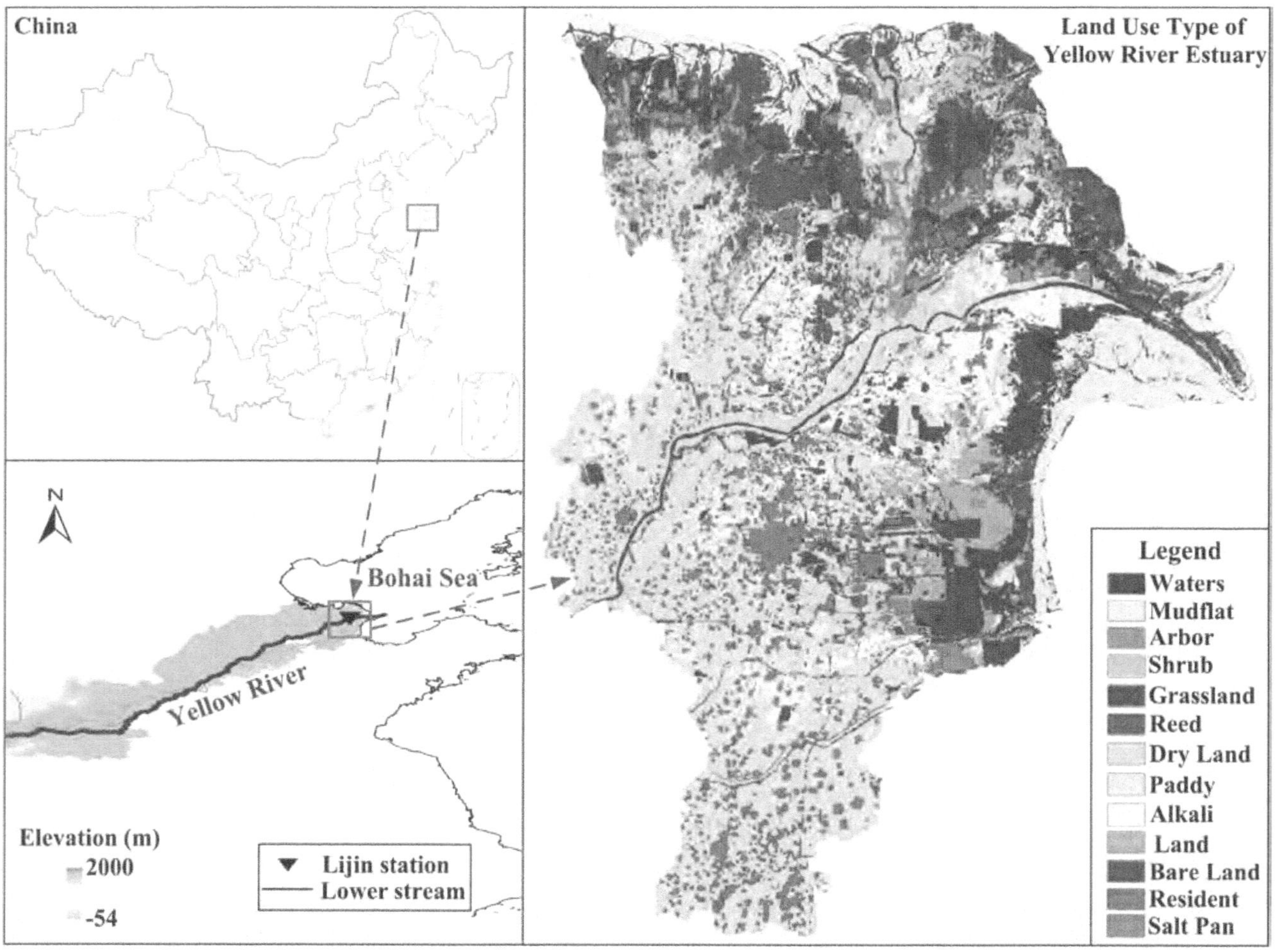

Figure 1 | Geographic location of the Yellow River estuary.

3. DEFINITION OF ECOLOGICAL DROUGHT CONCEPT

3.1. Runoff variation

Based on the measured runoff data of Lijin station from 1956 to 2016, the hydrological evolution characteristics of the estuary were analyzed. Figure 2 shows that the annual average runoff was 27.506 billion m^3, showing an obvious downward trend. In detail, the 1950s and 1960s were wet periods. From the 1970s to 1990s, affected by climate changes and human activities, the runoff decreased continuously. Since 2002, the runoff increased gradually. The annual runoff reduction rate of Lijin station is 19.46 $m^3/(s{\cdot}a)$, with an M–K test value of −5.38, passing the 0.01 significance test, indicating that the runoff reduction trend of Lijin station is very significant. The turning point of runoff took place in 1985. The runoff of Lijin station was mainly concentrated between July and October, and this runoff accounted for 60.24% of the annual runoff. The largest runoff, accounting for 17.34% of the annual runoff, took place in August. The smallest runoff, accounting for only 2.99% of the annual runoff, took place in February. The proportion of average monthly runoff in the dry seasons increased and that in wet seasons decreased. Before 1985, in general, the average monthly runoff of Lijin station increased in winter (December–January of the next year) and summer (June–August), and decreased in spring (March–May) and autumn (September–November). After the turning point, the runoff in 12 months decreased, which was mainly affected by the operation and storage of Xiaolangdi Reservoir and downstream water intake. These results imply that the contradiction between supply and demand of water resources in the Yellow River estuary will be more serious due to the homogenized precipitation and reduced runoff.

3.2. Variation of wetland area

The wetland area data of the Yellow River estuary in 1986, 1996, 2006, 2015, 2016, 2017, 2018 and 2019 were analyzed. Figure 3 shows that the wetland area in the Yellow River estuary presents a volcanic trend: it increased slowly from 1986 to 1996, decreased rapidly after 1996 and was kept constant after 2015. The maximum wetland area is 1,674.84 km^2 in 1996, accounting for 65.2% of the total area of the estuary Delta; the minimum wetland area is 1,314.77 km^2 in 2019, accounting for 53.3% of the total area of the estuary delta. Figure 3 shows that before 2000, the natural wetland area in the Yellow River estuary area accounted for more than 90% of the total wetland area. During this period, the wetland in the Yellow River Delta was principally in a natural state, insignificantly affected by human development. The area of natural wetland decreased slowly and even increased to a small extent from 1992 to 1996. However, from 2015 to 2019, this region was greatly affected by human activities, and some wetlands were transformed into breeding land.

3.3. Ecological water demand analysis

The ecological water demand of the Yellow River estuary mainly includes the water demand of migratory fish in the estuary, offshore water demand (freshwater amount required to maintain the habitat protection of offshore fish near the estuary) and water demand of sediment transport (water amount required to prevent seawater intrusion). According to the report from the *Yellow River Engineering Consulting Co., Ltd*, the suitable discharge amount of the Yellow River from May to September was

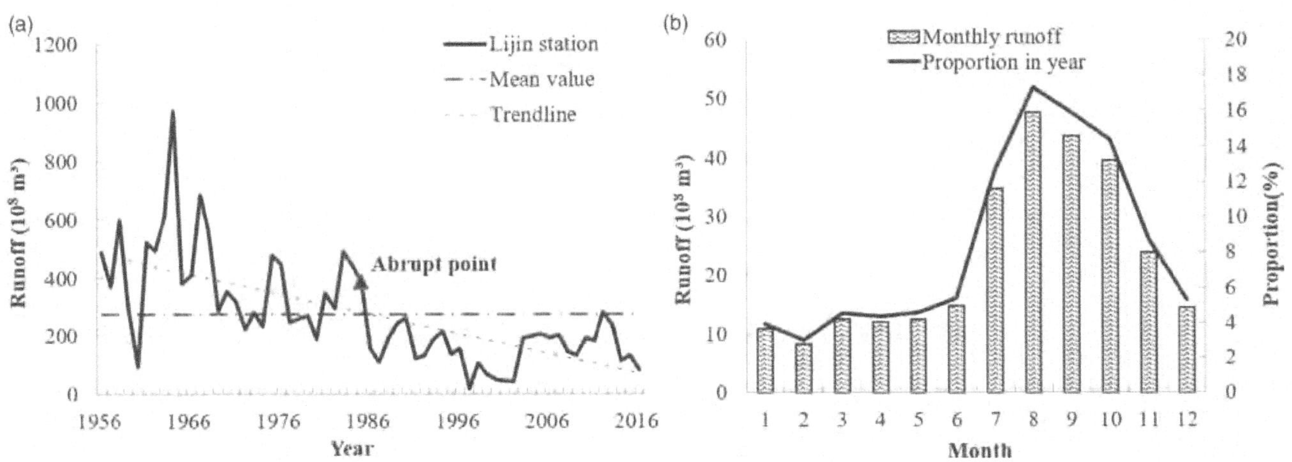

Figure 2 | Changes in annual (a) and inter-annual (b) runoff of the Yellow River estuary.

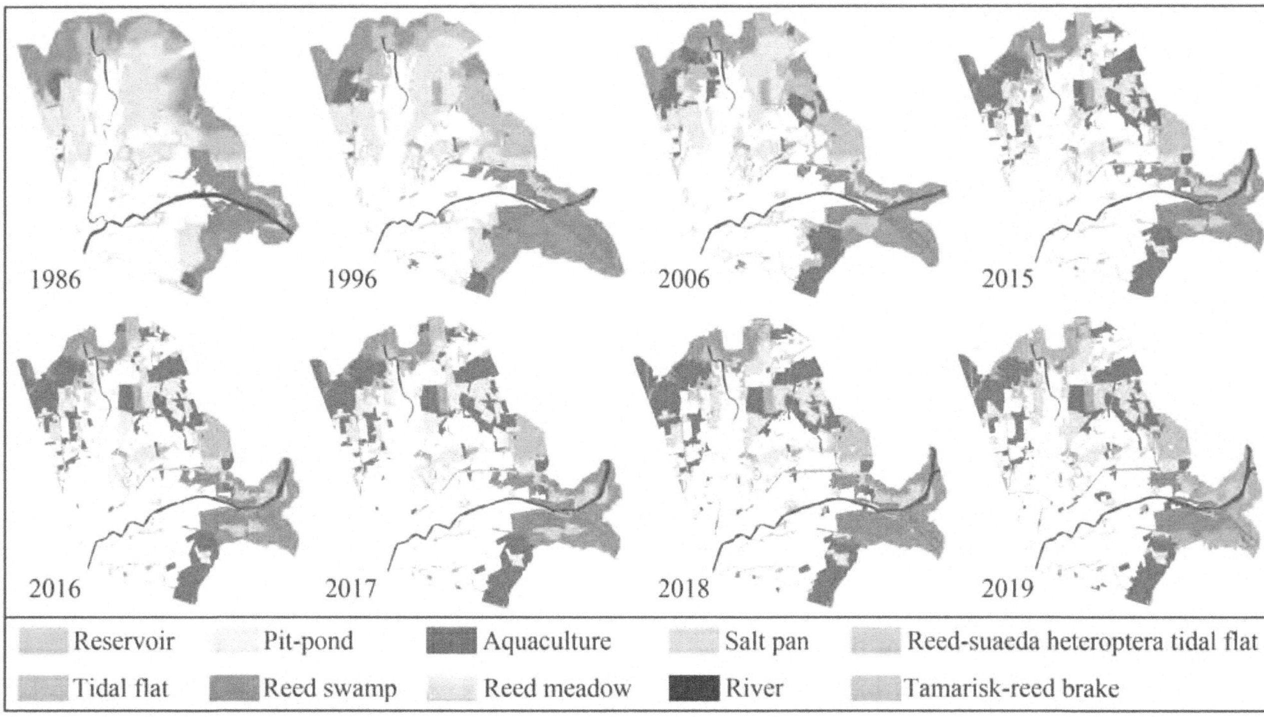

Figure 3 | Spatial distribution change of wetland area in Yellow River estuary delta.

12 billion m^3. 2.2 billion m^3 was selected as the suitable discharge amount for crucial fish spawning from May to June and offshore water demand. The annual average sediment transport water demand of the downstream Yellow River should not be less than 20 billion m^3, and the sediment transport water demand should not be less than 15 billion m^3 in the flood seasons. Based on these assumptions, through the coupling of ecological water demand items, the total ecological water demand of the Yellow River estuary was evaluated (see Table 1).

Based on the measured runoff of Lijin section in non-flood seasons from 2012 to 2016, the ecological water demand and supply of the Yellow River estuary were analyzed. The runoff in 2012 and 2013 met the water demand indices, but the runoff did not meet the water demand indices from 2014 to 2016. The average runoff in these 5 years did not meet the water demand indices, and the degree of satisfaction was only 78.5%.

3.4. Proposed ecological drought concept

Drought, as the driving force, causes water deficit and changes the regional hydrological cycle conditions. For instance, the regional water volume, flow velocity, water surface area, river connectivity at different scales, biomass and species diversity are reduced or decreased (Raheem *et al.* 2019). When drought occurs, the material and energy in the food chain, area and depth of habitat, connections between habitats are reduced, competition between aquatic organisms for food is intensified, water environment is deteriorated and the spatial distribution characteristics of aquatic organisms are changed, leading to the decrease of diversity, reproductive capacity and population size of aquatic organisms (Kovach *et al.* 2019). Moreover, the runoff, flow velocity and sediment-carrying capacity of the flow are decreased or weakened, cutting off the exchange of material, energy and information flow in the horizontal and vertical directions of the river.

Table 1 | Ecological water demand in the Yellow River estuary (Lijin station)

Month	Jan	Feb	Mar	Apr	May	Jun	Jul	Aug	Sep	Oct	Nov	Dec
Minimal value (m^3/s)	75	75	75	75	150	150	Sediments transport water				75	75
Suitable value (m^3/s)	120	120	120	120	250	250					120	120

The continuous reduction of runoff indicates that the Yellow River estuary is threatened by hydrological and socioeconomic drought. The Yellow River estuary possesses the largest delta wetland in the basin, and this wetland accounts for one-third of the Yellow River wetland area, showing that the Yellow River estuary possesses multiple ecological service functions, such as runoff regulation, flood control, water resources operation and storage, groundwater recharge, water quality purification (water pollution uptake), climate regulation and cultural services (such as tourism, entertainment, scientific research and education). Therefore, the estuary is important from an ecological point of view. However, the existing water amount cannot meet the ecological water requirement, which will inevitably have a certain impact on the regional ecological environment. Considering that water is the fundamental reason for the formation of the Yellow River estuary wetland and the main control factor of wetland ecological evolution, the vulnerability and variability of the wetland ecosystem depend on water. Therefore, regional ecological drought refers to a phenomenon of water shortage caused by the imbalance of water acquirement and loss, or water supply and demand, and is a major form threatening regional ecological safety.

To sum up, in this paper, ecological drought in the regions with important ecological functions is defined as follows: when a regional water shortage/pollution accident occurs and the regional water resources cannot meet the water requirement of ecological environment, adverse influence and even a disaster effect are subjected to the regional ecological environment. This situation is called 'ecological drought'.

4. ECOLOGICAL DROUGHT ASSESSMENT MODELING

The assessment of ecological drought is a complex systematic problem containing multiple indices. Human beings and nature are the driving factors of ecosystem changes, and the impact of ecological drought will be transferred to human society through ecosystem services (Xu *et al.* 2017). Considering that ecological drought is affected by natural driving force, ecological structure, ecological service functions and socioeconomic pressure, the Drive-Pressure-State-Impact-Response model (DPSIR) was adopted to build the index system of ecological drought assessment (Kagalou *et al.* 2012). The DPSIR model is an evaluation model constructed by the European Environment Agency by combing the merits of PSR and DSR models. The shuffled frog leaping algorithm and the projection pursuit method are integrated into the DPSIR model. In detail, the projection index functions were optimized with the shuffled frog leaping algorithm, and the projection values of the ecological drought complex system were determined to evaluate the regional ecological drought. The ecological drought evaluation model was established as follows:

(1) Normalization of the indices: Because the assessment of ecological drought is complex and involves multiple indices, in order to eliminate the influence of dimensions of indices, those indices positively correlated to ecological drought degree were normalized as follows:

$$x_{ij} = \frac{(x^*_{ij} - x_{\min j})}{(x_{\max j} - x_{\min j})} \tag{1}$$

Those indices negatively correlated to ecological drought degree were normalized as follows:

$$x_{ij} = \frac{(x_{\min j} - x^*_{ij})}{(x_{\max j} - x_{\min j})} \tag{2}$$

where $x_{\max j}$ and $x_{\min j}$ stand for the maximum and minimum values of index j in the index system, respectively. The x_{ij} values calculated by Equations (1) and (2) are normalized to be an evaluation index positively correlated to ecological drought degree, and the index is in the range of [0,1].

(2) Construction of the projection pursuit index function: The essence of projection pursuit is to find the optimal projection direction (OPD) which can reflect the data characteristics and extract data information to the maximum extent.

$$z_i = \sum_{j=1}^{p} a_j x_{ij} \tag{3}$$

where a_j represents the unit vector. $a_j \in [-1, 1]$ and $\sum_{j=1}^{p} a_j^2 = 1$.

If a ($a = (a_1, a_2, \ldots, a_p)$) is the OPD, the projection value of ecological drought degree can be calculated by using the above equation, and then the ecological drought degree can be quantitatively analyzed. During the combination of projection values, it is expected to extract variation information as much as possible, so a projection index function $Q(a)$ was constructed as the basis for the determination of the OPD. $Q(a)$ can be calculated by the following equation:

$$Q(a) = S_z * D_z \tag{4}$$

$$S_z = \sqrt{\frac{1}{n-1}\sum_{i=1}^{n}(z_i - \bar{z})^2} \tag{5}$$

$$D_z = \sum_{i=1}^{n}\sum_{k=1}^{n}(R - r_{ik})f(R - r_{ik}) \tag{6}$$

where S_z denotes the standard deviation of the projection value z_i, D_z stands for the local density of the projection value z_i, and $\bar{z}$ represents the average value of projection values z_1, z_2 and z_m. R represents the window radius of local density and can be determined by experiments. Generally, R is set to be $0.1S_z$. $f(R - r_{ik})$ represents the unit step function. Under the condition of $R > r_{ik}$, $f(R - r_{ik}) = 1$. If not, $f(R - r_{ik}) = 0$. The larger the intraclass density D_z is, the more significant the classification is. When the index function $Q(a)$ reaches the limit, the OPD is found.

(3) Optimization of the projection index function by the shuffled frog leaping algorithm: When the sampling scheme set is fixed, the index function $Q(a)$ only changes along with the projection direction a. when $Q(a)$ reaches the maximum, the corresponding direction a is the OPD vector. Therefore, the optimization objective function is expressed as follows:

$$\begin{cases} \max Q(a) \\ \|a\| = 1 \end{cases} \tag{7}$$

The objective function is a complex nonlinear optimization problem with $\{a_j | j = 1, 2, \ldots, p\}$ as the optimization variables and can be optimized and solved with the shuffled frog leaping algorithm.

(4) Determination of ecological drought classification standard and evaluation of the adaptability of ecological drought on the basis of the calculated projection values of samples in the study area and projection values of the standard: First, according to the index values corresponding to the evaluation levels and OPD a^* calculated on the basis of the normalized values, the ranges of projection values corresponding to each level will be obtained by using the equations above. In this way, the ecological drought degree evaluation and classification standard $\{x_{ij} | i = 1, 2, \ldots, n; j = 1, 2, \ldots, p\}$ will be determined. After calculations following the equations above, the ecological drought degree projection value y of the estuary will be obtained. According to the ranges of projection values corresponding to each level, the range to which ecological drought belongs will be determined.

A large number of indexes for evaluation can thoroughly reflect the 'driving force', 'pressure', 'state', 'impact' and 'response' characteristics of ecological drought. However, too many indexes will cause the lack of representativeness and independence of the evaluation model and even deviations of the calculation results. Thereby, the rough set method was adopted to thoroughly screen the set of indexes to improve the representativeness and independence of the evaluation indexes and to ensure the diversity of the evaluation indexes (Vafaeyan & Thibault 2009).

5. RESULTS AND DISCUSSION

5.1. Results

After the thorough screening with the rough set method, the optimized assessment indexes were obtained and are listed in Table 2.

The data of indices of the Yellow River estuary from 2014 to 2016 were taken as the input for modeling and calculations. The regional ecological drought was divided into four levels: I (Severe), II (General), III (Weak) and IV (None). The OPD of each index was obtained (see Table 3).

Accordingly, the evaluation levels of ecological drought in the Yellow River estuary are shown in Table 4.

Table 2 | Assessment indexes of ecological drought

Framework	Index	ID	Index direction
Drive	SPI index	D_1	−
	Natural growth rate of population	D_2	−
	Proportion of primary industry	D_3	−
	River runoff	D_4	+
	Urbanization rate	D_5	−
Pressure	Water resources per capita	P_1	+
	Average water resources per mu	P_2	+
	Arable land area per capita	P_3	−
	Green space area per capita in cities and towns	P_4	+
	Water consumption per capita	P_5	+
	Domestic water consumption of urban residents	P_6	+
State	Industrial added value water consumption	S_1	−
	Groundwater development and utilization rate	S_2	−
	Proportion of agricultural water	S_3	−
	Leakage rate of pipe network	S_4	−
	Average irrigation water consumption per mu	S_5	−
	Water consumption per unit GDP	S_6	−
Impact	Vegetation coverage	I_1	+
	Quality compliance rate of drinking water source	I_2	+
	Urban green space area per capita	I_3	+
	Proportion of eco-environmental water consumption	I_4	+
	Biodiversity	I_5	+
Response	Reuse rate of reclaimed water	R_1	+
	Ratio of water-saving irrigation area	R_2	+
	Popularization rate of urban water-saving appliances	R_3	+
	Popularization rate of urban tap water	R_4	+
	Utilization coefficient of irrigation water	R_5	+
	Reuse rate of industrial water	R_6	+

The data of indices of the Yellow River estuary from 2000 to 2016 were taken as the input for the evaluation system model, and the results are shown in Table 5. From 2000 to 2002, the degree of ecological drought in the Yellow River estuary was serious, and the ecological environment situation was urgent. In the years of 2008, 2009, 2014 and 2015, the degree of ecological drought was weak. In 2016, the degree of ecological drought reached the level of General. The reason is that with the social and economic development of the Yellow River estuary, the supportability of water intake, water use and water resources allocation was increased. Even if the river runoff decreased, the regional ecological drought resisting capacity could deal with the drought. Nevertheless, the problem of water resources supply and demand in the Yellow River estuary is still prominent and has had a certain impact on the regional ecological environment. The main reasons for the regional ecological drought are the increase of water resources demand and the decrease of wetland area caused by the rapid economic and social development and continuous increase of population. It is urgent to alleviate the ecological drought in the Yellow River estuary by optimizing the allocation of water resources, transferring water from other basins, saving and intensively using water resources, aquatic ecological environment remediation and other measures.

5.2. Discussion

The entropy method was adopted to calculate the weight of evaluation index (Yazdi 2018). Based on the theory and method of information entropy, the entropy method determines the weight of each index according to the different degree of index values data. In general, the greater the different degree of the index values of an index is, the better the order is, the smaller the entropy value is and the greater the final weight is. The original index data of the Yellow River estuary are expressed as $S = \{X_{ij}\}_{17 \times 28}$, in which X_{ij} represents the original data of the jth index in the ith year. The value i is in the range of 17 years from 2000 to 2016. The value j is in the range of 28 different types of indices in the ecological drought evaluation model. The calculation steps are presented as follows:

Table 3 | Assessed rating for selected indexes

ID	Dimension	Assessed rating and its value range: I	II	III	IV	OPD
D_1	/	≤ -2.0	(−2.0, − 1.5]	(−1.5,0.0]	≥ 0.0	0.182
D_2	‰	>20	15–20	2–15	<2	0.153
D_3	%	>30	15–30	3–15	<3	0.132
D_4	m^3	<50	50–100	100–150	>150	0.187
D_5	%	<20	20–50	50–80	>80	0.181
P_1	m^3	<500	500–1,000	1,000–1,700	>1,700	0.210
P_2	m^3/hm^2	<400	400–700	700–1,500	>1,500	0.161
P_3	hm^2	<0.1	0.1–0.6	0.6–1	>1	0.186
P_4	m^2	<20	20–100	100–180	>180	0.171
P_5	m^3/a	>1,000	800–1,000	500–800	<500	0.196
P_6	L/d	>200	150–200	100–150	<100	0.248
S_1	$m^3/10^4$ currency	>90	60–90	30–60	<30	0.160
S_2	%	>100	60–100	30–60	<30	0.141
S_3	%	>75	55–75	40–55	<40	0.164
S_4	%	>30	15–30	5–15	<5	0.178
S_5	m^3/hm^2	>400	300–400	200–300	<200	0.193
S_6	$m^3/10^4$ currency	>100	80–100	50–80	<50	0.185
I_1	%	<20	20–40	40–60	>60	0.205
I_2	%	<90	90–98	98–100	100	0.166
I_3	m^2	<7	7–8	8–10	>10	0.198
I_4	%	<2	2–5	5–10	>10	0.185
I_5	Category	<10	10–50	50–100	>100	0.167
R_1	%	<10	10–20	20–30	>30	0.162
R_2	%	<30	30–50	50–75	>75	0.151
R_3	%	<50	50–70	70–90	>90	0.195
R_4	%	<70	70–80	80–90	>90	0.206
R_5	/	<0.45	0.45–0.5	0.5–0.6	>0.6	0.171
R_6	%	<40	40–60	60–80	>80	0.181

Table 4 | Ecological drought assessment

Z-value range	Rating	Ecological drought degree	Statement
$Z < 1.312$	I	Severe	The functions of ecosystem are devastated by frequent aquatic ecological crises and disaster events.
$1.312 \leq Z < 2.267$	II	General	The functions of ecosystem are partially destroyed by increased aquatic ecological crises and disaster events.
$2.267 \leq Z < 3.126$	III	Weak	The functions of ecosystem are almost normal with fewer aquatic ecological crises and disaster events.
$Z \geq 3.126$	IV	None	The functions of ecosystem are good with rare aquatic ecological crises and disaster events.

Table 5 | Assessment results of the Yellow River estuary

Year	Z-value	Rating	Ecological drought degree
2000	0.796	I	Severe
2001	0.848	I	Severe
2002	0.670	I	Severe
2003	4.653	IV	None
2004	4.069	IV	None
2005	4.896	IV	None
2006	3.453	IV	None
2007	4.786	IV	None
2008	2.319	III	Weak
2009	2.647	III	Weak
2010	3.555	IV	None
2011	4.350	IV	None
2012	5.316	IV	None
2013	5.643	IV	None
2014	2.847	III	Weak
2015	2.541	III	Weak
2016	1.953	II	General

(1) Nondimensionalization:

$$\begin{cases} A_{ij(+)} = \dfrac{X_{ij} - X_{\min}}{X_{\max} - X_{\min}} \\ A_{ij(-)} = \dfrac{X_{\max} - X_{ij}}{X_{\max} - X_{\min}} \end{cases} \tag{8}$$

where $A_{ij(+)}$ represents dimensionless equations of positive indices; $A_{ij(-)}$ represents dimensionless equations of negative indices. In the case of the occurrence of ln0 during the calculation of information entropy, the value of 0 was replaced with 0.000001.

(2) Calculate the proportion P_{ij} of the jth index value in the sum of the jth index values of all the objects evaluated.

$$P_{ij} = \frac{A_{ij(+/-)}}{\sum_{i=1}^{17} A_{ij(+/-)}} \tag{9}$$

(3) Calculate the information entropy e_j of the jth index.

$$e_j = -\frac{1}{\ln i}\sum_{i=1}^{17} P_{ij} \times \ln P_{ij} \tag{10}$$

(4) Calculate the redundancy g_j and weight w_j of the jth index. The larger the entropy value is, the smaller the redundancy and the smaller the weight.

$$\begin{cases} g_j = 1 - e_j \\ w_j = \dfrac{g_j}{\sum_{j=1}^{28} g_j} \end{cases} \tag{11}$$

According to the analysis results, among the 28 evaluation indexes (weight characteristics shown in Figure 4), the three indexes with the largest weight values are river runoff (0.1273), biodiversity (0.1231) and vegetation coverage (0.1031),

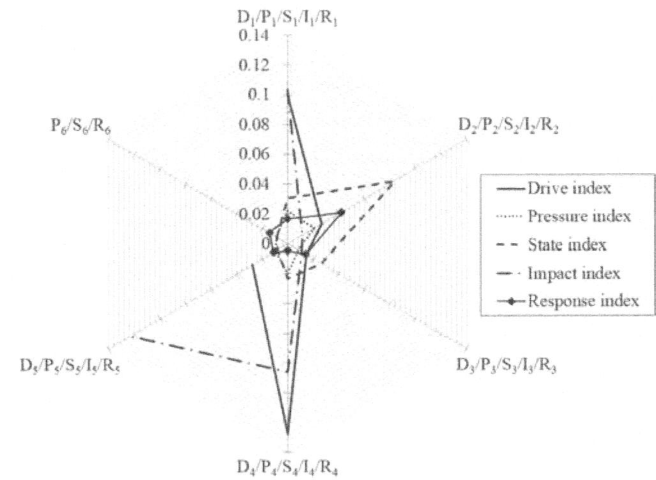

Figure 4 | Distribution characteristics of index weight.

indicating that these three evaluation indexes have great influence on regional ecological drought. Namely, sufficient river runoff, abundant biodiversity and high vegetation coverage are important factors to improve the ecological drought resisting capacity in the Yellow River estuary. In other words, the ecological drought can be estimated preliminarily based on the threshold changes of river runoff, regional biodiversity and vegetation coverage.

River runoff (R), biodiversity (B) and vegetation coverage (VC) are selected as the leading indicators to express the different ecological drought degree, and thus, the dynamic thresholds of leading indicators under different ecological drought degrees can be seen in Table 6. It is implied that natural factors have potential influence on the evolution of ecological drought, and most of the natural environmental factors are stable and even contribute to the ecological self-regulation mechanism. Human activity factors have great potential to damage the ecological environment, and unscientific production and living styles can accumulate a large number of damage factors in a short time, to directly or indirectly damage the natural environment and to interfere with the regulation ability of ecosystems. As a consequence, the degree of ecological drought is worsened. Therefore, it is difficult for a single remedial measure to effectively alleviate the regional ecological drought. We should focus on integrated, comprehensive, practical and macroscopic strategies such as water source protection, ecological restoration, ecological water supply and improvement of water resources allocation capacity.

For the areas with ecological drought similar to the Yellow River estuary, it is suggested to formulate a systematic and comprehensive macro strategy, especially paying attention to the work of multi-water source joint commissioning and supply and biodiversity habitat protection. From the systematic point of view, for the regions with important ecological functions, ecological drought shows a strong feedback effect under the dual influence of human beings and nature. Reasonable construction of regional evaluation index models and scientific selection of indices are the effective guarantee for ecological drought evaluation. On the basis of existing studies, the ecological drought evaluation model constructed in this paper, through case study, systematizes and simplifies ecological drought, which is abstract and difficult to define. Moreover, the qualitative ecological drought evaluation indexes are quantified and digitized, and individual problems are integrated. These characteristics play a great role in disaster prevention and mitigation. The effectiveness of the evaluation model and its practicability and operability in the actual regional ecological drought evaluation process were also verified by real cases. Thereby, this model can serve as an effective evaluation tool for the policymakers.

Table 6 | Thresholds of leading indicators to different ecological drought degrees

Ecological drought degree	Thresholds of leading indicators
I (Severe)	①$R < -10.2\%$; ②$B < -10\%$; ③$VC < -9\%$
II (General)	①$R \in [-10.2\%, -7.6\%)$; ②$B \in [-10\%, -7\%)$; ③$VC \in [-9\%, -6\%)$
III (Weak)	①$R \in [-7.6\%, -3.4\%)$; ②$B \in [-7\%, -5\%)$; ③$VC \in [-6\%, -3\%)$
IV (None)	①$R \in [-3.4\%, +3.4\%]$; ②$B \in [-5\%, +5\%]$; ③$VC \in [-3\%, +3\%]$

Nonetheless, because the data used in this paper were on a time scale, the degree of ecological drought on the scale of spatial distribution in the Yellow River estuary cannot be reflected. Meanwhile, the regional ecosystem constantly changes under the effect of natural environment and human disturbance. The evaluation of ecological drought can only represent the basic state at a specific moment, and the evolution mechanism of ecological drought is not explained. Therefore, the dynamic assessment of ecological drought on the regional grid scale should be highlighted in the future. It is believed that with the deep study on ecological drought, the evaluation indices will be clearer, and the model constructed in this paper will be more meaningful.

6. CONCLUSIONS

Based on the current concept of ecological drought, the potential characteristics of ecological drought were analyzed in this paper, and the concept of ecological drought in areas with important ecological functions was discussed. Based on the clarified concept of ecological drought, an evaluation model of regional ecological drought degree was constructed. Taking the Yellow River estuary of China as the case study, the usability of the model was verified. This paper draws the following conclusions:

1. The concept of ecological drought in the region with important ecological functions was defined.
2. The regional ecological drought was divided into four levels: I (Severe), II (General), III (Weak) and IV (None). From 2000 to 2016, three severe ecological droughts happened due to water shortage; thus, water resources is one of the key elements for the mitigation of ecological drought.
3. The indicator thresholds of river runoff, biodiversity and vegetation coverage on different ecological drought levels were quantified to predict an ecological drought in potential. The threshold values of representative indexes can play a role of quick judgment for ecological drought.
4. To a certain extent, the research results eliminate the blind area and disunity for the concept of ecological drought and realize the evaluation of ecological drought by modeling. This work can provide theoretical scaffolds and technical strategies for planners to formulate disaster prevention and mitigation policies.

ACKNOWLEDGEMENTS

The research would not have been possible without the interest and cooperation of Hohai University. We also extend our thanks to Mrs Ya Li for giving a great many constructive suggestions for paper figures.

FUNDING

The paper is jointly supported by the National Key R&D Program of China (No. 2018YFC1508706), Project funded by the China Postdoctoral Science Foundation (No. 2019M652551) and the Science and Technology Program of the University of Jinan (No. XBS2013).

DECLARATION OF COMPETING INTEREST

The authors declare that they have no known competing financial interests or personal relationships that could have appeared to influence the work reported in this paper.

DATA AVAILABILITY STATEMENT

All relevant data are included in the paper or its Supplementary Information.

REFERENCES

Bachmair, S., Svensson, C., Hannaford, J., Barker, L. J. & Stahl, K. 2016 A quantitative analysis to objectively appraise drought indicators and model drought impacts. *Hydrology and Earth System Sciences* **20** (7), 2589–2609.

Brown, A. 2017 Ecological resilience: drought sensitivity. *Nature Climate Change* **7** (2), 96–96.

Chadd, R. P., England, J. A., Constable, D., Dunbar, M. J., Extence, C. A., Leeming, D. J., Murray-Bligh, J. A. & Wood, P. J. 2017 An index to track the ecological effects of drought development and recovery on riverine invertebrate communities. *Ecological Indicators* **82**, 344–356.

Crausbay, S. D., Ramirez, A. R., Carter, S. L., Cross, M. S. & Sanford, T. 2017 Defining ecological drought for the twenty-first century. *Bulletin of the American Meteorological Society* **98** (12), 2543–2550.

Goulden, M. L. & Bales, R. C. 2019 California forest die-off linked to multi-year deep soil drying in 2012–2015 drought. *Nature Geoscience* **12** (8), 632–637.

Huang, B. B., Yan, D. H. & Wang, H. 2019 Research and prospect on effects of drought on river water ecosystem. *Journal of Water Resources and Water Engineering* **30** (2), 12–18 (in Chinese).

James, A. & Suren, A. M. 2010 The response of invertebrates to a gradient of flow reduction – an instream channel study in a New Zealand lowland river. *Freshwater Biology* **54** (11), 2225–2242.

Jiang, S., Wang, M., Ren, L., Xu, C. Y., Yuan, F., Liu, Y., Lu, Y. & Shen, H. 2019 A framework for quantifying the impacts of climate change and human activities on hydrological drought in a semiarid basin of Northern China. *Hydrological Processes* **33** (7), 1075–1088.

Kagalou, I., Leonardos, I., Anastasiadou, C. & Neofytou, C. 2012 The DPSIR approach for an integrated river management framework. a preliminary application on a Mediterranean site (Kalamas River – NW Greece). *Water Resources Management* **26** (6), 1677–1692.

Kim, J. S., Jain, S., Lee, J. H., Chen, H. & Park, S. Y. 2019 Quantitative vulnerability assessment of water quality to extreme drought in a changing climate. *Ecological Indicators* **103**, 688–697.

Kovach, R. P., Dunham, J. B., Robert, A. C., Snyder, C. D., Letcher, B. H., Young, J. A., Beever, E. A., Pederson, G. T., Lynch, A. J., Hitt, N. P., Konrad, C. P., Jaeger, K. L., Rea, A. H., Sepulveda, A. J., Lambert, P. M., Stoker, J., Giersch, J. J. & Muhlfeld, C. C. 2019 An integrated framework for ecological drought across riverscapes of North America. *BioScience* **69** (6), 418–431.

Kuemmerlen, M., Reichert, P., Siber, R. & Schuwirth, N. 2019 Ecological assessment of river networks: from reach to catchment scale. *Science of the Total Environment* **650**, 1613–1627.

Liang, C., Li, D., Yuan, Z., Liao, Y., Nie, X., Huang, B., Wu, X. & Xie, Z. 2019 Assessing urban flood and drought risks under climate change, China. *Hydrological Processes* **33** (9), 1349–1361.

Mccargo, J. W. & Peterson, J. T. 2010 An evaluation of the influence of seasonal base flow and geomorphic stream characteristics on coastal plain stream fish assemblages. *Transactions of the American Fisheries Society* **139** (1), 29–48.

McEvoy, J., Bathke, D. J., Burkardt, N., Cravens, A., Haigh, T., Hall, K. R., Hayes, M. J., Jedd, T., Podebradska, M. & Wickham, E. 2018 Ecological drought: Accounting for the non-human impacts of water shortage in the upper Missouri headwaters basin, Montana, USA. *Resources* **7** (1), 1–14.

Millar, C. I. & Stephenson, N. L. 2015 Temperate forest health in an era of emerging megadisturbance. *Science* **349** (6250), 823–826.

Munson, S. M., Bradford, J. B. & Hultine, K. R. 2020 An integrative ecological drought framework to span plant stress to ecosystem transformation. *Ecosystems* **9** (19), 1–16.

Novoa, V., Ahumada-Rudolph, R., Rojas, O., Saez, K., Francisco, D. & Arumi, J. L. 2019 Understanding agricultural water footprint variability to improve water management in Chile. *Science of the Total Environment* **670**, 188–199.

Ostad-Ali-Askari, K., Krause, S., Zalewski, M. & Boogaard, F. C. 2019 The impact of future climate change and human activities on hydro-climatological drought, analysis and projections: using $CMIP_5$ climate model simulations. *Water Environment Research* **4**, 71–88.

Park, S. Y., Sur, C., Lee, J. H. & Kim, J. S. 2020 Ecological drought monitoring through fish habitat-based flow assessment in the gam river basin of Korea. *Ecological Indicators* **109**, 105830.1–105830.10.

Porter, A. J., Kuhn, T. R. & Nerlich, B. 2018 Organizing authority in the climate change debate: IPCC controversies and the management of dialectical tensions. *Organization Studies* **39** (7), 873–898.

Raheem, N., Cravens, A. E., Cross, M. S., Crausbay, S. & Suberu, P. 2019 Planning for ecological drought: integrating ecosystem services and vulnerability assessment. *Wiley Interdisciplinary Reviews: Water* **3**, e1352.

Rinaldo, A., Gatto, M. & Rodriguez-Iturbe, I. 2018 River networks as ecological corridors: a coherent ecohydrological perspective. *Advances in Water Resources* **112** (3), 27–58.

Slette, I. J., Post, A. K., Awad, M., Even, T., Punzalan, A., Williams, S., Smith, M. D. & Knapp, A. K. 2019 How ecologists define drought, and why we should do better. *Global Change Biology* **25** (10), 3193–3200.

Vafaeyan, S. & Thibault, J. 2009 Selection of pareto-optimal solutions for process optimization using rough set method: a new approach. *Computers & Chemical Engineering* **33** (11), 1814–1825.

Wang, J., He, G., Fang, H. & Han, Y. 2020 Climate change impacts on the topography and ecological environment of the wetlands in the middle reaches of the Yarlung Zangbo-Brahmaputra River. *Journal of Hydrology* **590** (7), 125419.

Xiao, R., Wu, X., Du, J., Deng, B. & Xing, L. 2020 Impacts of anthropogenic forcing on source variability of sedimentary organic matter in the Yellow River estuary over the past 60 years. *Marine Pollution Bulletin* **151**, 110818.1–110818.6.

Xu, H., Wang, X. & Yang, T. 2017 Trend shifts in satellite-derived vegetation growth in central Eurasia, 1982–2013. *Science of the Total Environment* **579**, 1658–1674.

Yao, N., Li, L., Feng, P., Feng, H. & Li, Y. 2019 Projections of drought characteristics in China based on a standardized precipitation and evapotranspiration index and multiple GCMs. *Science of the Total Environment* **704**, 135245.

Yazdi, J. 2018 Water quality monitoring network design for urban drainage systems, an entropy method. *Urban Water Journal* **15**, 227–233.

Yu, H., Kong, B., He, Z. W., Wang, G. & Wang, Q. 2019 The potential of integrating landscape, geochemical and economical indices to analyze watershed ecological environment. *Journal of Hydrology* **583**, 124298.

Zhang, L. L., Yin, J. X. & Hou, Z. C. 2010 Ecological drought evaluation function of Baiyangdian based on fuzzy membership degree. *Journal of Hohai University (Natural Sciences)* **38** (3), 252–257 (in Chinese).

First received 29 April 2021; accepted in revised form 30 August 2021. Available online 17 September 2021

© 2022 The Authors Impact of Climate Change on Hydrology and Water Resources doi: 10.2166/wcc.2021.244

Impact of meteorological drought on agriculture production at different scales in Punjab, Pakistan

Muhammad Waseem[a], Tahira Khurshid[b], Adnan Abbas[c,*], Ijaz Ahmad[a] and Zeeshan Javed[d]

[a] Center of Excellence in Water Resources, University of Engineering and Technology, Lahore 54890, Pakistan
[b] College of Hydrology and Water Resources, Hohai University, 1 Xikang Road, Nanjing 210098, China
[c] Land Science Research Center, Nanjing University of Information Science & Technology, Nanjing 210044, China
[d] Jiangsu Key Laboratory of Atmospheric Environment Monitoring and Pollution Control, School of Environmental Science and Engineering, Nanjing University of Information Science and Technology, Nanjing 210044, China
*Corresponding author. E-mail: adnanabbas@nuist.edu.cn

ABSTRACT

The present study aimed to quantify the impacts of the meteorological drought on the production of maize crops, using district-level observed precipitation and yield data of 21 districts across Punjab, Pakistan from 2001 to 2020. The overall analysis showed that the Standardized Precipitation Index (SPI) effectively reflects the variations in drought characteristics in Punjab on spatiotemporal scales. It also resulted that in south Punjab, the maize yield was negatively affected by the meteorological drought, and yield was sensitive to short-term (1 and 3 months) drought during the critical growth period of the crop. The overall analysis depicted that the meteorological drought was associated with about 27% of overall yield variations. Moreover, all of the southern districts and few districts from Central Punjab were becoming increasingly sensitive to meteorological drought where significant spatial variations in drought effects and sensitivity exist over time. Conclusively, this study showed a spatiotemporal pattern of drought and its impact on maize yield, indicating that the districts where variability in maize production was significantly associated with drought and recommend adoption of the management strategies and mitigation measures.

Key words: agriculture, climate change, drought, SPI, yield

HIGHLIGHTS

- Extreme climate events like drought can cause the greatest impacts especially in agronomic fields.
- Agricultural yield is sensitive to short-term drought during the critical growth period of the crop and is also significantly associated with overall yield variations.
- Management policies that barrier against short-term drought may be most effective at sustaining long-term crop productivity.

INTRODUCTION

The agriculture system is very responsive and sensitive to the fluctuation in rainfall patterns, and due to its dependency on water, it is the extremely affected sector by extreme events like droughts (UNISDR 2009). Droughts are a global phenomenon; however, their effects are more vulnerable in developing countries because of lesser potential to mitigate the impacts of such extreme events. Approximately, two billion people are affected by drought, and seven to eight billion dollars have been spent as a drought-related cost since 1990 (Zipper *et al.* 2016). Considering the hydrological cycle, drought is classified into different four classes, i.e., meteorological drought, hydrological drought, and agricultural and socio-economic drought (Waseem *et al.* 2020). Among these different types of droughts, meteorological drought is the most important form of drought as surface water resources, and most importantly, the agriculture of a country is heavily dependent on meteorological information. It is mainly as meteorological drought can cause a significant reduction in agriculture yield especially in the arid region of a country.

Drought influencing factors, e.g., global warming, have caused an increase in the frequency and intensity of meteorological drought in different areas of the world and has produced vulnerable effects on crop production, which ultimately threaten food security both at a global and regional scale (Godfray *et al.* 2010). For this reason, the important factor affecting the growth and development of crops is changes in rainfall, both spatial and temporal patterns, and its impacts on the crop water cycle (Tao *et al.* 2003; Hanif & Ali 2014). Similarly, a short burst of rising temperature will affect the water cycle and cause soil moisture deficiency, infertility and damage crop yields (McKeown *et al.* 2005). Recent studies (Janjua *et al.* 2010; Mahmood *et al.* 2012; Mendelsohn 2014; Kirby *et al.* 2016) have also indicated the impacts of climate variations on food production in both Pakistan and globally. These studies recommended detailed accounts of agricultural production and yield variation under extreme events for making sure the sufficient and sustainable food supply (Van Ittersum *et al.* 2013) to the rapidly growing global population under climate change.

Most of the developing countries are agriculture-based economies and have direct exposure to climate change and extreme events (Mendelsohn 2014). It is also expected that extreme events (e.g., heatwaves, hot extremes, and droughts) could become relatively frequent and intense in the near future (Zaman *et al.* 2020). Like other countries, Pakistan has also experienced several extreme events especially the droughts, a dramatic shift in the rainfall, and a steady rise in temperature for the past 30 years (Zahid & Rasul 2011; Abbas 2013; Xie *et al.* 2013; Ali *et al.* 2015; Khan *et al.* 2019, 2020; Nawaz *et al.* 2019). For instance, the 1998–2002 drought, which was one of the worst droughts during the last 50 years, affected 1.2 million people, caused the death of 2 million animals, and major agricultural loss. The southwest region of Pakistan was mainly affected by this 1998–2002 drought and other drought events (e.g., 2004–2005 and 2009), as the annual rainfall of this region mostly fails to meet the annual evaporation. This imbalance is tremendously increasing and causing perpetual drought conditions. In this context, several studies (i.e., Sheikh 2001; Zahid & Rasul 2011; Xie *et al.* 2013; Ahmed *et al.* 2018; Muhammad *et al.* 2018; Khan *et al.* 2020; Waseem *et al.* 2020) have been conducted in Pakistan to assess the intensity, duration, and frequency of the low precipitation or drought.

Droughts in Pakistan have seriously affected the economy, agricultural system, cropping pattern, and annual productivity in most parts of the country, and this situation could be worse in the future (Akmal *et al.* 2014; Khan *et al.* 2020). Hence, the agricultural sector of Pakistan is under serious threat under this kind of climate variation (Janjua *et al.* 2010). These facts highlight the dire need to understand the resilience of agricultural systems to drought so that effective plans and policies can be made to mitigate the impacts of drought in the near future (Zipper *et al.* 2016).

Punjab, the province of Pakistan, is considered as the food basket of Pakistan but climate-induced changes have placed a greater risk to its agricultural systems. So, the present study focuses to assess the impacts of drought on the yield of one of the major food crops (i.e., maize) grown in Punjab province. Maize is one of the food and cash crops of the Punjab agricultural sector and serves as feed and silage as well. It is a high-yield cereal crop globally and, in Pakistan, it is the third main cereal crop after wheat and rice. It is sown in two seasons, i.e., spring (locally called Rabi) and autumn (locally called Kharif). In spring, it is sown from February till March and the crop completes its tasseling and silking stages in April–early May. Till the end of May, it reaches its grain filling and maturity stage and is harvested in June. While the autumn maize crop is sown from July to August and harvested between September and December. The life cycle of the maize crop depends upon the availability of water, and drought stress at the reproductive or maturity stage can damage the grain yield of the maize crop. Water deficiency during pre-flowering, flowering, and post-flowering stages can cause yield reduction by 25, 50, and 21%, respectively (Sah *et al.* 2020). According to an estimation, climate change has caused a decline in global maize production by 3.8% between 1980 and 2010 (Lobell *et al.* 2011).

In view of this, it would be essential to understand the spatiotemporal variation of drought impacts on maize crop yield so that the negative consequences of future drought can be minimized. Particularly, the timing (i.e., time of the year drought occurs) and timescale (i.e., drought duration) at which the strongest impacts of drought on crop yield need to be quantified, and this information will help the government organizations and stakeholders for designing the drought mitigation strategies. Furthermore, understanding the spatial pattern of drought sensitivity for crops and variations in their patterns over time can help stakeholders plan for regional shifts in actual and realized crop production under the stress of future climate change (Zipper *et al.* 2016). Hence, in the current study, to find the impact of meteorological drought at the local scale, the following research questions are framed: (1) to what extent does the meteorological drought exert influence on maize yield, and how does the spatial pattern of this relationship vary and (2) what timescale and timing of the drought put the strongest influence on the maize yield?

METHODOLOGY

Study area

Punjab province (Figure 1) is the food basket of Pakistan with fairly flat alluvial land and fertile soil for agricultural activities. It is irrigated by five rivers (i.e., Jehlum, Ravi, Sutlej, Chenab, and Indus) and is considered as the lifeline of Pakistan. The total

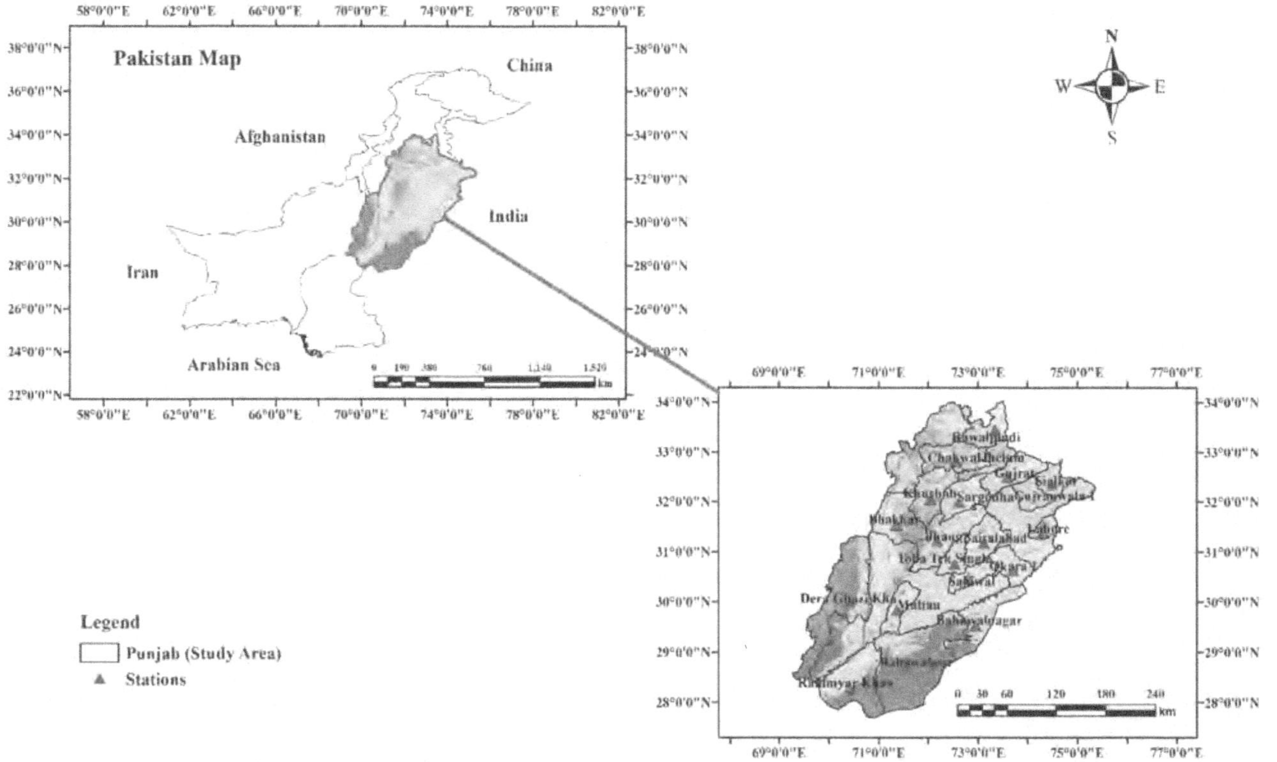

Figure 1 | Study area map.

area of Punjab is 205,344 km^2 (79,284 sq mi). Climatically, there are four seasons in Punjab, i.e., hot weather (April to June), cold weather/mild weather (October to March), and rainy season (July to September). Average rainfall varies from 460 to 960 mm and temperature ranges from −2 to 45 °C, but in some regions of Punjab, temperature exceeds 50 °C. The topography of Punjab also includes several mountainous ranges, such as the Suleiman range in Southwest Punjab, Margalla Hills in North and Salt range (Pothohar Plateau), and deserts including Rajasthan in Southern Punjab and parts of Thal and Cholistan also lie in the south of Punjab. Punjab's agricultural sector comprises major crops including maize, rice, cotton, wheat, and sugarcane and minor crops such as onion, potato, chilies, mash, mung, and masoor with two cropping seasons, i.e., Autumn (Kharif) and Spring (Rabi) seasons are observed (Amjad *et al.* 2008). In the current study, 36 districts of Punjab were divided into the southwestern region (Dera Ghazi Khan), southeastern region (Bahawalpur, Bahawalnagar), south Punjab (Multan), and north region of Punjab/Pothohar region (Jehlum, Chakwal, Rawalpindi, and Khushab). Moreover, this study is based on secondary data covering the period from 2001 to 2020. Monthly precipitation data were only available from 21 weather stations, established in 21 districts out of 36 total districts, so in the current study, only these gauging stations were considered for further analysis. The meteorological data were collected from the Pakistan Meteorological Department (https://www.pmd.gov.pk/en/). District-wise yearly maize yield and area data were collected from the Crop Reporting Service Government of Punjab (http://www.crs.agripunjab.gov.pk/) for the same districts and the years for which precipitation data were available.

Calculation of drought severity

The Standardized Precipitation Index (SPI) was developed by McKee (1995) for meteorological drought monitoring and to assess the changes in climatic conditions at different timescales (Liu *et al.* 2021). It is a widely used drought index (Tigkas *et al.* 2015) and has also been applied in studies related to the quantification of drought effects (Yildirak & Selcuk-Kestel 2015; Bautista-Capetillo *et al.* 2016). In the current study, SPI was used at 1-, 3-, 6-month, and annual timescales. SPI is used for the identification of meteorological drought. In general, for estimation of SPI, long-term rainfall data are first transformed into a most fitted probability distribution (e.g., Gamma distribution) and then further converted into a normal distribution, such that the mean SPI is zero (Edwards 1997).

$$\mathrm{SPI} = \frac{X_i - \bar{X}}{\sigma} \tag{1}$$

where X is the precipitation observation, $\bar{X}$ is the long-term mean, and σ is the standard deviation

The SPI value ranges from -3 to 3, where the positive value of SPI indicates wet condition while the negative value represents the drought condition. A region is said to be in a drought condition if the index value of SPI is constantly negative. That drought event finishes when the value of SPI converts to positive. Therefore, each drought occurrence has a period determined by its start and end, and the severity for each month.

The SPI can be estimated at different timescales, e.g., 1-, 3-, 6- and 12-month reference duration. Moreover, the cumulative sums of precipitation for 3-, 6-, 9-, and 12-month timescales were used to calculate SPI-3, SPI-6, SPI-9, and SPI-12, respectively. For example, the cumulative sum for October while calculating SPI-3 was obtained by adding previous 2-month (i.e., August and September) data to the October data. Similarly, for November, September, and October data were added. For SPI-6, the cumulative sum was obtained by adding 6 months from May to October. Moreover, the time series were fitted to the Gamma probability distribution, and the cumulative probability distribution of the series data was transformed into the normal distribution.

Agricultural yield data processing

District-wise maize yield data were obtained from the Cropping Reporting Service Government of Punjab from 2001 to 2020. The length of the crop yield record was the same for all the selected districts. In addition, the districts were categorized by keeping in view the crop growing seasons (i.e., Kharif and Rabi crop). It was observed that only five districts have only Kharif maize crop, whereas the rest of the districts have both season's maize crop. This division helped to calculate the impacts of drought stress and shift in drought sensitivity for a particular month of the crop cycle. Before further analysis, the crop yield data were also auto-scaled using linear regression for the elimination of the gradual trend due to non-climatic factors, e.g., agronomic and genetic improvements (Potop *et al.* 2010). The auto-scaled analysis was carried out due to the reason that in some districts an increase in yield area was observed, whereas in some others yield area was decreased over time and a few districts have shown exceptional behavior. Moreover, crop yield variability with respect to time is measured by a coefficient of variation (CV). The CV is a statistical measure of the relative spread of data points around the mean (μ) in a data series.

$$\mathrm{CV} = \frac{\sigma}{\mu} \tag{2}$$

where σ represents the standard deviation.

Statistical modeling of drought and agricultural yield

The statistical modeling between meteorological drought and agriculture yield has been investigated using regression analysis. Regression analysis is a statistical tool used to estimate the level of effect of any independent time series (hereafter meteorological drought index) on any dependent time series (hereafter agriculture yield). Regression analysis is generally performed to understand the relationships between independent time series and dependent time series. The mathematical expression of regression between any independent time series and any dependent time series is as follows:

$$Y = \beta_0 + \alpha X + \varepsilon \tag{3}$$

where Y is the dependent time series, α is the regression coefficient (slope), X is the independent time series, β_0 is the intercept, and ε is an error term.

Drought sensitivity could be defined as the slope of linear regression (α) between detrended crop yield anomalies and SPI (Vicente-Serrano *et al.* 2013). A positive value of slope will indicate the low yield in case of severe drought and vice versa, as a more negative value of SPI shows more severe drought. Furthermore, to assess the temporal shifts in drought sensitivity, a linear regression was employed using monthly, 3-, 6-month, and annual timescales for SPI to the first 10 years (i.e., 2001–2010) and the last 10 years (2011–2020) of data, and significant differences of slope between these two spans were observed.

Districts were labeled as no change where slopes did not show a significant difference ($p < 0.05$) in either period. In this way, the behavior of other slopes was observed to find if they shifted toward the steeper slopes, i.e., higher value (negative to positive) or shallow slopes, i.e., the lower value (positive to negative). A positive slope indicates lower yield during meteorological drought and a negative slope indicates higher yield during meteorological drought. Moreover, for each district, a separate linear relationship was developed between maize yield and SPI at different timescales, i.e., monthly, 3- , 6-month, and annual timescales. For each relationship, R^2, slope, and p-value were calculated using the two-tailed t-test, and R^2 was used to indicate the best relationship between crop yield and SPI (Vicente-Serrano *et al.* 2013; Zipper *et al.* 2016). The R^2 formula is calculated by dividing the sum of the first errors by the sum of the second errors and subtracting the derivation from (1). Here is what the R^2 equation looks like:

$$R^2 = 1 - \frac{R_{SS}}{T_{SS}} \tag{4}$$

where R^2 represents the coefficient of determination, T_{SS} is the total sum of squares, and R_{SS} is the sum of squares of residuals.

RESULTS

Spatiotemporal variation of maize yield

During the past two decades (2001–2020), Punjab's annual maize yield has gradually increased over time with the rising trend (mean trends of 0.8140 tonnes acre^{-1} year^{-1}), accompanied by an increasing trend in the range of spatial variability (Figure 2(a)). Furthermore, crop yield variability with respect to time is measured by a CV. The analysis showed that CV has a lower value in the districts with higher productivity and increasing yielding trends, and a higher value of CV in the districts having low yield with mild increasing yield trends. The overall analyses resulted in the strongest increasing yield trend and the lowest yield variability in Sahiwal, Okara, Jhang, Toba Tak Singh, and Faisalabad (1.841, 1.476, 1.460, 1.188, and 1.037 tonnes acre^{-1} year^{-1}, respectively), whereas the weakest increasing trend (0.020 tonnes acre^{-1} year^{-1}) and higher yield variability were observed in districts of South Punjab. These variations are mainly due to climatic factors, fertility, and other soil properties. Figure 2(b) and 2(c) shows the spatial pattern of maize production and long-term trend in Punjab.

Spatial variability of drought impact

The variability analysis of drought indicated significant spatial variations in the relationship between meteorological drought and maize yield. The strength of the relationship between SPI-1, SPI-3, SPI-12, and maize yield helped to figure out the districts that were more sensitive to drought. The R^2 value for the fitted relationship ranged from 0 to 37% for Rabi maize and 0 to 41% for Kharif maize (Figure 3). Southwestern region (Dera Ghazi Khan), southeastern region (Bahawalpur and Bahawalnagar), south Punjab (Multan), and north region of Punjab (Jehlum, Chakwal, Rawalpindi, and Khushab) were identified as the drought-sensitive region with the moderate to strong correlation between SPI-1, SPI-3, and maize yield (i.e., severe drought was associated with a large negative yield, indicating the increase in drought sensitivity), whereas the weak or even negative correlation with negative or very low slope values was observed in Northeastern and Central Punjab (Gujranwala, Gujrat, Sialkot, Lahore, Mandi Bahaudin, Toba Tak Singh, Jhang, Sahiwal, and Okara), indicating reduced drought sensitivity or high yield during the drought period as shown in Figure 3, except Faisalabad, where the moderate correlation was found during the growing season of Kharif and Rabi crop.

Slopes of the relationship between detrended Maize yield and SPI-1, SPI-3, and SPI-12 are shown in Figure 4. It will be noteworthy to mention that a positive slope reveals yield losses during drought and a negative slope shows yield gain in response to drought conditions. Moreover, 15 districts out of 21 districts of this study are fully or partially irrigated areas except for Pothohar Plateau. Hence, results reveal that irrigation reduces the negative effects of weather unevenness on maize yield, particularly in the Northeastern and Central regions of Punjab, and therefore eliminates the yield sensitivity to meteorological drought. Maize yield data reveal that about 88–97% of potential agriculture is reported from irrigated zones, and only 3–12% of maize production comes from the non-irrigated zones of these districts. In other words, irrigation decouples the yield from inter-annual drought. But, despite having an irrigation system, some districts in south Punjab (e.g., Dera Ghazi Khan and Bahawalpur) and north Punjab (Jehlum) had shown a significant positive correlation between drought and maize yield during the growing season of Kharif and Rabi maize crops.

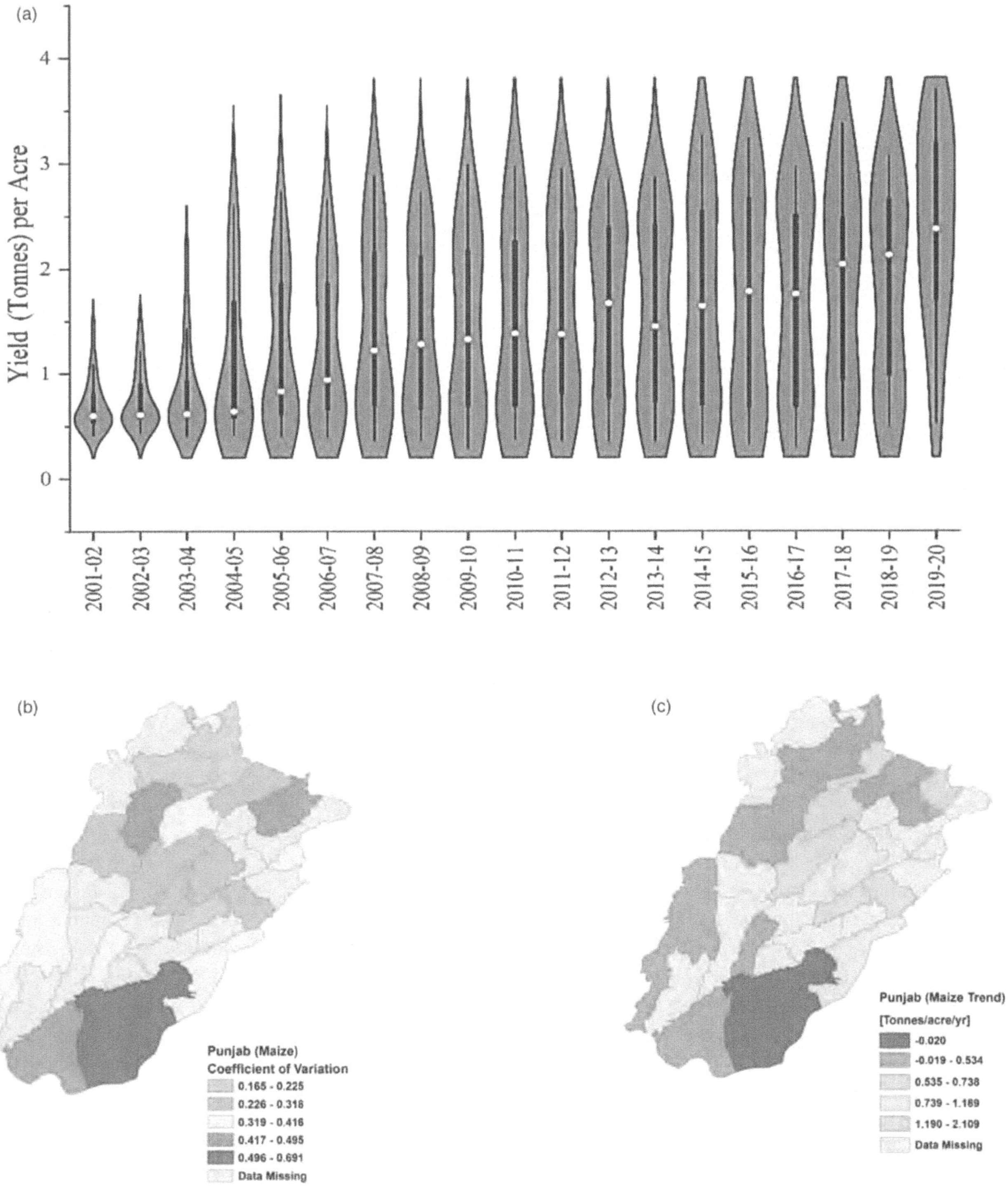

Figure 2 | (a) Annual distribution of yield in Punjab Pakistan; (b) CV of maize yield over the study period of 2001–2020; and (c) long-term trend in maize yield (Tonnes/acre).

Timescale and timings of drought impact on maize yield

As maize is sown in two seasons, namely, spring and autumn, all districts bearing diverse topography and water resources availability and, hence, give separate responses to the meteorological drought timescale during different stages of Kharif and Rabi maize crop cycles (Figure 5). The analysis showed that maize yield is sensitive to the drought occurring at the

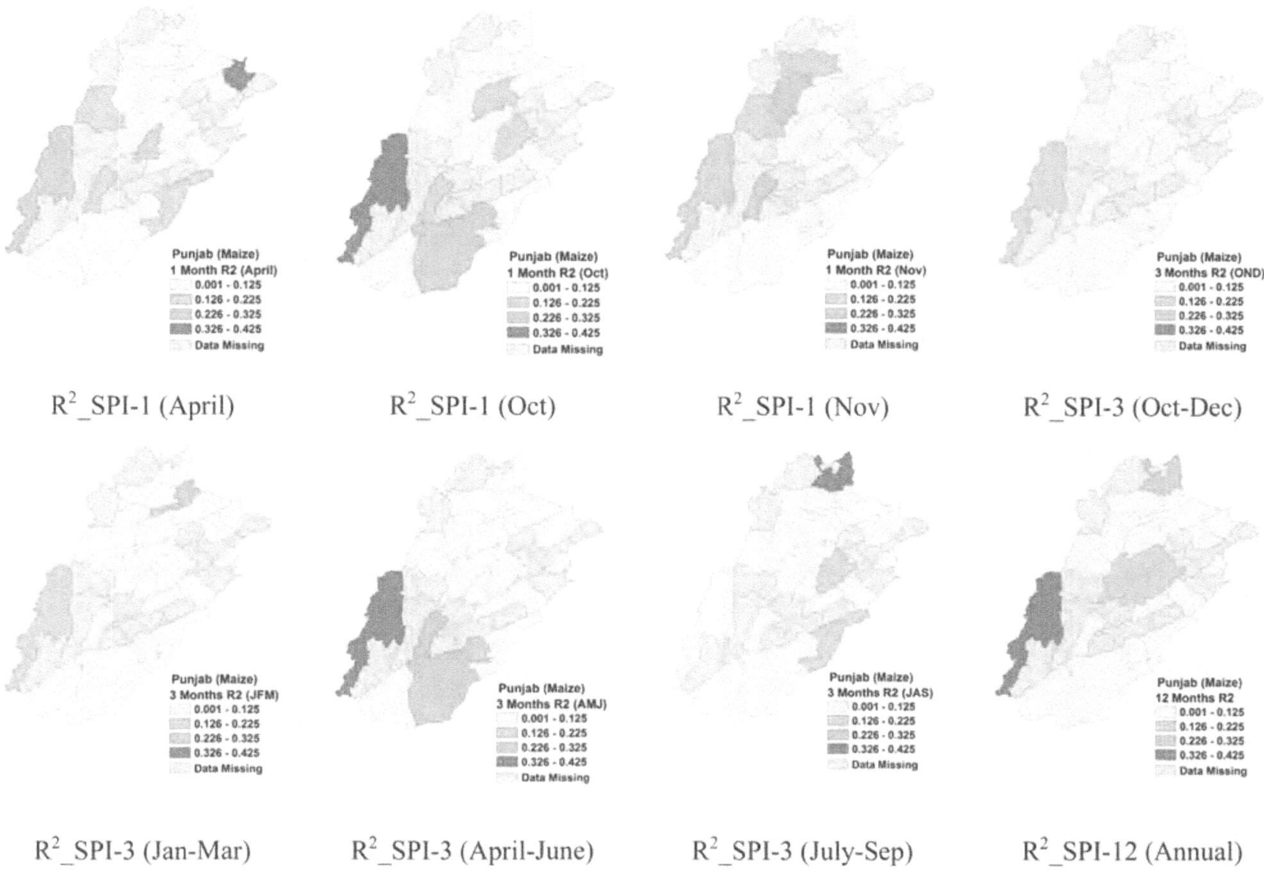

Figure 3 | Spatial variations in R^2 of the relationship between yield and SPI at the different timescale.

growing stage. The spring maize is sown in February–March and completes its tasseling and silking stages in April–early May. Toward the end of May, the crop reaches its grain filling and maturity stage, and after passing through the grain formation stage, it is harvested in early June. Spring maize is generally grown in Jhelum (North Punjab) and Central Punjab. Maize yield in Jehlum (North Punjab) was more sensitive to a 1-month drought occurring in February, which corresponds to the sowing stage of the crop, whereas yield from the districts of South Punjab (Bahawalpur, Bahawalnagar, Multan, and Rahim Yar Khan), and Central Punjab (Faisalabad, Toba Tak Singh, and Sialkot) are more sensitive to a 2-month drought (i.e., April and May), which are the growing season of spring maize crop.

Moreover, autumn maize is grown in all districts of Punjab and is more susceptible to meteorological drought as compared to spring maize. Autumn maize is sown in July–August and harvested between September till December. September, October, and November are probably the growing stage of the autumn maize depending upon the early or late sowing of the crop. The districts lying in the North and South Punjab exhibit more sensitivity for a growing stage of maize crop at a 2–3-month timescale (i.e., September to November) of drought, with the moderate to strong correlation between SPI and maize yield. Dera Ghazi Khan, Bahawalpur, and Multan (Southern Punjab) are identified as more drought-stressed districts for autumn maize and are strongly correlated with the 1-month SPI of October and November.

Drought sensitivity shift quantification over time

The results indicated a substantial spatial pattern in drought sensitivity across Punjab for maize yield between period 1 (2001–2010) and period 2 (2011–2020) as shown in Figure 6. It was observed that some districts of Central and Northeastern Punjab (such as Gujranwala, Sialkot, Sargodha, Faisalabad, and Okara) exhibit the shifts toward increased drought sensitivity (which can be defined as a higher slope of the relationship between SPI and maize yield) for the 1-month SPI during the growing season of the Kharif maize crop. During the 3-month SPI (October–December and January–March), reduced drought sensitivity has been prevalent in South Punjab including districts of Dera Ghazi Khan, Bahawalpur, and Rahim Yar Khan, whereas

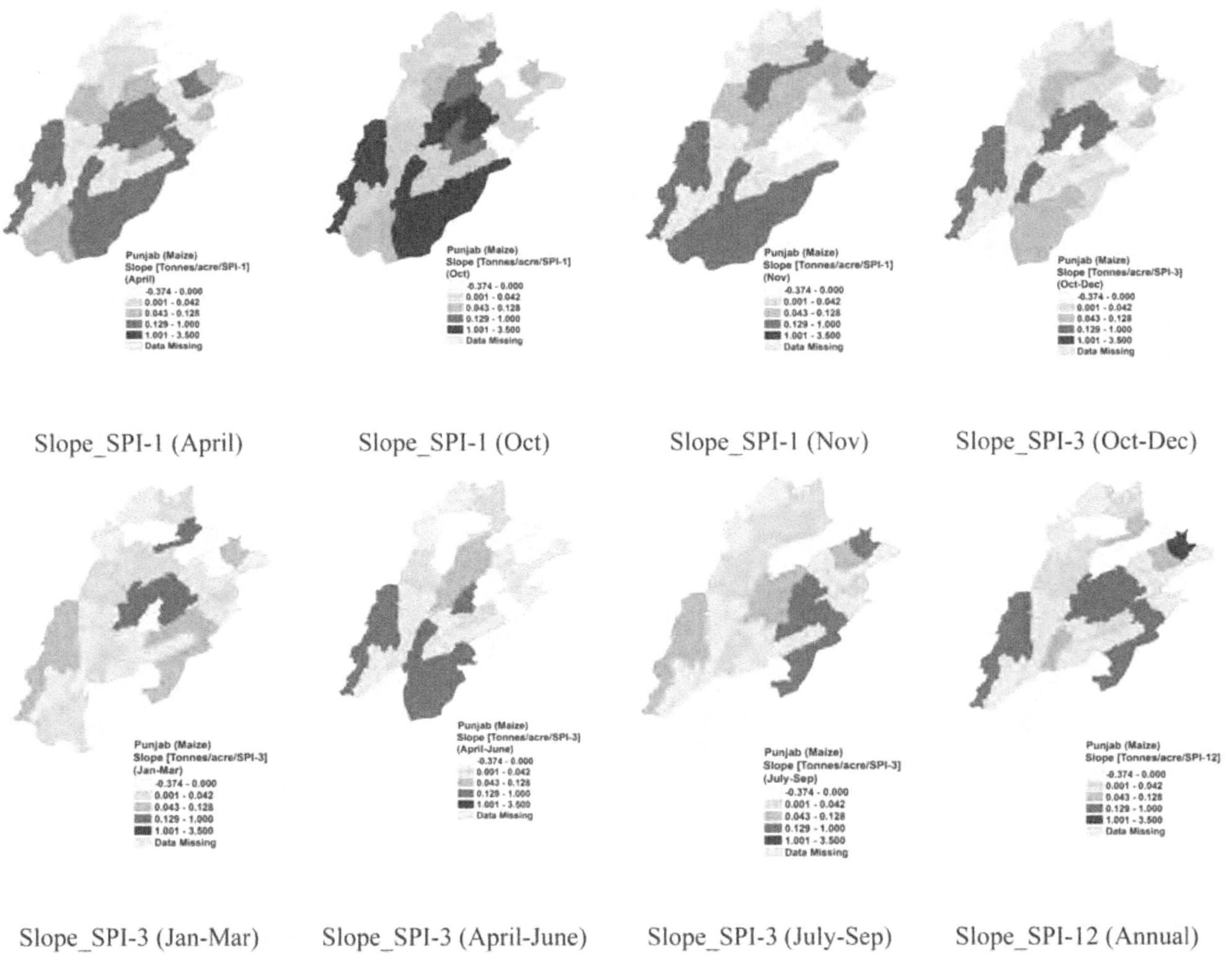

Figure 4 | Variations in the slope of the relationship between yield and SPI at the different timescale.

increased sensitivity was common in some districts of Central Punjab (Gujranwala, Gujrat, Sialkot, Sargodha, and Okara). During the 3-month SPI (April–June), which is the growing season for the Rabi maize crop, a scatter pattern of a shift in drought sensitivity is observed across Punjab. On the other hand, the 12-month SPI shows increased drought sensitivity for Gujranwala, Sargodha (Central Punjab), and Bakhar (Southwestern Punjab). In Figure 6, the legend 'No change' indicates no significant changes among the two periods, shallower represents the slope of the relationship was lower in the second period as compared to the first period, and 'Pos. To Neg.' shows the shift of slope from a positive value to a negative value. 'Steeper' signposts the increase in slope. 'Neg. To Pos.' shows the shift of slope from negative to positive, and the 'missing data' indicates the districts with missing data.

DISCUSSIONS

Considering the food, energy, and water nexus, one of the major challenges is to ensure food security at the global level for a rapidly increasing population while conserving water and land resources. For achieving this goal, it is essentially required to improve the food productivity, manage existing agricultural land, and other strategies, e.g., minimize the agricultural inputs and food waste (McLaughlin & Kinzelbach 2015). To make the agricultural system better adaptive to climate variation, it is important to recognize the impacts of climate extremes events (e.g., drought) on the agricultural system (Chen *et al.* 2014). Hence, this study attempted to quantify the spatiotemporal variations of drought sensitivity of maize yield for the identification of both when and where the meteorological drought is strongly associated with the variations in maize yield.

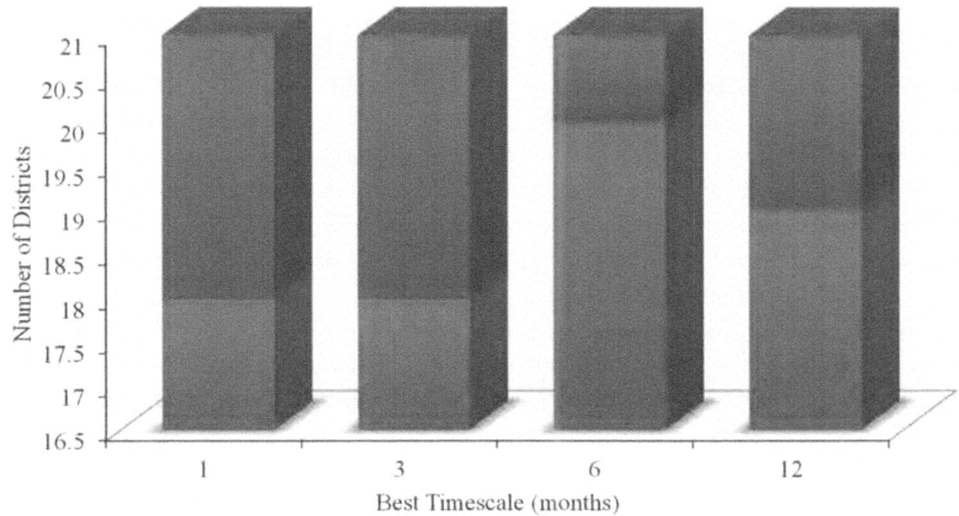

Figure 5 | Graph showing the number of best-fit relationships at the different timescales of drought (red color indicating the significant relations with $p > 0.05$). Please refer to the online version of this paper to see this figure in colour: http://dx.doi.10.2166/wcc.2021.244.

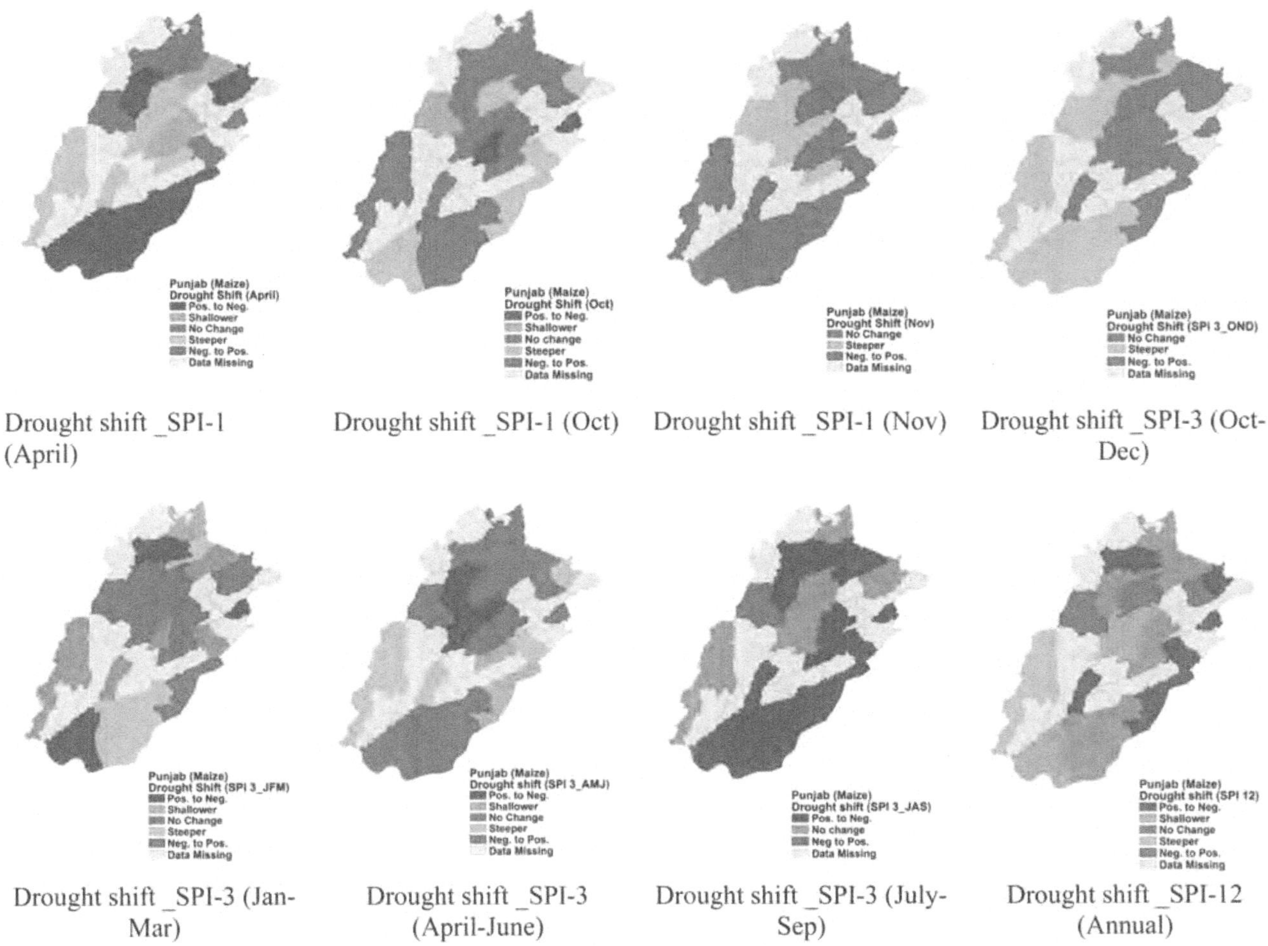

Figure 6 | Shift in maize yield in response to drought between (2001–2010) and (2011–2020).

The results showed large spatial variations in drought sensitivity and its impacts on maize yield. It was hypothesized that district-level variability in drought sensitivity and impacts could be controlled by local physical characteristics such as the availability of groundwater or soil water retention ability, as the SPI does not consider the soil moisture or groundwater storage (Ma *et al.* 2014). For instance, the districts of southern Punjab have sandy soil, which may be the cause of an increase in drought sensitivity for this region due to the poor water retention capacity. The results also reveal the negative drought sensitivity in the districts of Central Punjab, where supplemental water sources in the form of shallow groundwater or irrigation are available for the crops and could help in the mitigation of drought, however, ignored in the SPI calculation. In other words, poorly drained soils and the shallow water table provide a buffer from drought. Previous studies also showed that crops could receive groundwater as a subsidy (Zipper *et al.* 2015), mainly in a drought period; hence, an increase in yield could possibly be due to the availability of groundwater. Another extensive supportive source of water for crops is irrigation (Hain *et al.* 2015). The results of R^2 in the case of the districts in Central Punjab such as Gujranwala, Gujrat, Sahiwal, Okara, Mandi Bahaudin, Lahore, and Jhang indicate that irrigation might have decoupled the drought effects on crop yield. These findings agree with the preceding research studies (e.g., Troy *et al.* 2015; Araujo *et al.* 2016), which describe that irrigation acts as a mitigation tool in reducing crop yield sensitivity to dry spells. Although irrigation might be an effective source for decreasing drought sensitivity, at the same time, groundwater abstraction already goes beyond the sustainable yield in various aquifers in numerous areas of the world, including Pakistan (Iqbal *et al.* 2016). Therefore, a careful approach and practices in addition to the irrigation as a mitigation tool for reducing the drought impacts are required, such as genetics for more drought-resistant crops, and advances in plant breeding may also reduce the drought sensitivity in the future (Manavalan *et al.* 2009).

The analysis also indicated that short-term drought (i.e., 1–3 months) events during the growing season (i.e., April–May for Rabi maize crop and September–October for Kharif maize crop) tend to be a dominant factor for maize yield variability in the south region of Punjab. Short-term drought is more important for the study of annual crop yield responses because crop growth and yield are more responsive to the fluctuation in short-term weather events that could extensively and rapidly change the soil moisture conditions (Wu & Wilhite 2004).

Moreover, spatial shifts in drought sensitivity were observed across Punjab and resulted in more drought sensitivity in Central Punjab especially during the flowering and post-flowering maize season of the Kharif crop. The present study also indicates that the regions, such as South Punjab and some districts of Potohar Plateau (North Punjab), might also be at high risk from droughts, particularly when drought comes about at the flowering stage of maize development. Moreover, it will be noteworthy here to mention that as SPI does not take into account the soil moisture conditions and, hence, may not explain the actual soil moisture regime experienced by crops, hence, some soil moisture based methods for drought quantification (Carrão *et al.* 2016) could enhance impact assessment analysis, which was lacking in the current study due to limited availability of observed soil moisture data.

Overall, this study highlights the challenge of sustainable food supply amidst the meteorological drought. These results contribute to understanding the climate-induced yield variability by quantifying the variability in drought sensitivity both at spatial and temporal scales across Punjab for one of the major cereal crops, i.e., maize. Such spatiotemporal understanding helps us to find both, i.e., *when* and *where* meteorological droughts have strong impacts on maize yield, and how yield sensitivity to drought has changed over time, which can guide management, stakeholders, and farmer's drought responses and mitigation at district and regional levels.

CONCLUSION

Conclusively, this study resulted that meteorological drought has been associated with Punjab maize yield variability with an average of 27% over the past two decades (2001–2020), and short-term droughts (i.e., 1 and 3 months) during the critical growth stage of the crop cycle were more correlated with yield variations having a higher value of R^2. Moreover, substantial spatial variations in drought effects with respect to the magnitude might be due to the local physical characteristics (e.g., topography, soil texture, and water table) and agricultural management such as surface or groundwater irrigation. The results also illustrated that maize yield from South Punjab was more sensitive (having a higher slope) to droughts as compared to Central Punjab, where there is the availability of an extensive irrigation network. In addition to the above, drought shift analysis revealed that Central Punjab might also be severely affected by drought shift, particularly during the growing season of the

Kharif maize crop. These results emphasize the need to examine the resilience and sustainability of the current irrigation system in order to mitigate the drought impacts in the future.

Furthermore, the adoption of multi-pronged approaches, e.g., the growing of more drought-resistant crops, and the adoption of agronomic mitigation practices (irrigation) are required to cope with future drought challenges. This study presents a detailed spatiotemporal analysis of drought impacts on the maize yield of the Punjab province and describes the need for research, management, and policy changes to buffer future maize production from anticipated variations in the future drought intensity, frequency, and duration. As Punjab maize yield contributes significantly to national maize production, and therefore, this study has considerable implications for national agricultural output and, hence, food security. At the same time, efforts would be needed to introduce measures (structural and nonstructural including bio-engineering measures) to protect natural resources from climate-related hazards that are expected to increase due to climate change and may well outpace the positive impacts of these climate changes.

DATA AVAILABILITY STATEMENT

All relevant data are included in the paper or its Supplementary Information.

REFERENCES

Abbas, F. 2013 Analysis of a historical (1981–2010) temperature record of the Punjab province of Pakistan. *Earth Interactions* **17** (15), 1–23.

Ahmed, K., Shahid, S. & Nawaz, N. 2018 Impacts of climate variability and change on seasonal drought characteristics of Pakistan. *Atmospheric Research* **214**, 364–374.

Akmal, M., Ahmad, N., Khan, A., Bibi, F. & Ali, J. 2014 *Climate Change and Adaptation–Farmer's Experiences from Rainfed Areas of Pakistan*. Climate Change Centre. The Agricultural University, Peshawar, p. 40.

Ali, N., Ahmad, I., Chaudhry, A. G. & Raza, M. A. 2015 Trend analysis of precipitation data in Pakistan. *Science International* **27**, 803–808.

Amjad, R., Arif, G. M. & Mustafa, U. 2008 *Does the Labor Market Structure Explain Differences in Poverty in Rural Punjab?* MPRA Paper 62369, University Library of Munich, Germany.

Araujo, J. A., Abiodun, B. J. & Crespo, O. 2016 Impacts of drought on grape yields in Western Cape, South Africa. *Theoretical and Applied Climatology* **123** (1), 117–130.

Bautista-Capetillo, C., Carrillo, B., Picazo, G. & Júnez-Ferreira, H. 2016 Drought assessment in Zacatecas, Mexico. *Water* **8** (10), 416.

Carrão, H., Russo, S., Sepulcre-Canto, G. & Barbosa, P. 2016 An empirical standardized soil moisture index for agricultural drought assessment from remotely sensed data. *International Journal of Applied Earth Observation and Geoinformation* **48**, 74–84.

Chen, H., Wang, J. & Huang, J. 2014 Policy support, social capital, and farmers' adaptation to drought in China. *Global Environmental Change* **24**, 193–202.

Edwards, D. C. 1997 *Characteristics of 20th Century Drought in the United States at Multiple Time Scales*. Air Force Institute of Technology, Wright-Patterson AFB, OH.

Godfray, H. C. J., Beddington, J. R., Crute, I. R., Haddad, L., Lawrence, D., Muir, J. F., Pretty, J., Robinson, S., Thomas, S. M. & Toulmin, C. 2010 Food security: the challenge of feeding 9 billion people. *Science* **327** (5967), 812–818.

Hain, C. R., Crow, W. T., Anderson, M. C. & Yilmaz, M. T. 2015 Diagnosing neglected soil moisture source–sink processes via a thermal infrared–based two-source energy balance model. *Journal of Hydrometeorology* **16** (3), 1070–1086.

Hanif, M. & Ali, J. 2014 *Climate Scenarios 201-2014*. Inter-Cooperation Pakistan, Haripur, Sawabi, Attock, Chakwal-Pakistan, p. 27.

Iqbal, N., Hossain, F., Lee, H. & Akhter, G. 2016 Satellite gravimetric estimation of groundwater storage variations over Indus Basin in Pakistan. *IEEE Journal of Selected Topics in Applied Earth Observations and Remote Sensing* **9** (8), 3524–3534.

Janjua, P. Z., Samad, G., Khan, N. U. & Nasir, M. 2010 Impact of climate change on wheat production: a case study of Pakistan [with comments]. *The Pakistan Development Review* **49**, 799–822.

Khan, N., Shahid, S., Ismail, T., Ahmed, K. & Nawaz, N. 2019 Trends in heat wave related indices in Pakistan. *Stochastic Environmental Research and Risk Assessment* **33** (1), 287–302.

Khan, M. I., Zhu, X., Arshad, M., Zaman, M., Niaz, Y., Ullah, I., Anjum, M. N. & Uzair, M. 2020 Assessment of spatiotemporal characteristics of agro-meteorological drought events based on comparing Standardized Soil Moisture Index, Standardized Precipitation Index and Multivariate Standardized Drought Index. *Journal of Water and Climate Change* **11** (S1), 1–17.

Kirby, J., Mainuddin, M., Mpelasoka, F., Ahmad, M., Palash, W., Quadir, M., Shah-Newaz, S. & Hossain, M. 2016 The impact of climate change on regional water balances in Bangladesh. *Climatic Change* **135** (3–4), 481–491.

Liu, C., Yang, C., Yang, Q. & Wang, J. 2021 Spatiotemporal drought analysis by the standardized precipitation index (SPI) and standardized precipitation evapotranspiration index (SPEI) in Sichuan Province, China. *Scientific Reports* **11** (1), 1–14.

Lobell, D. B., Schlenker, W. & Costa-Roberts, J. 2011 Climate trends and global crop production since 1980. *Science* **333** (6042), 616–620.

Ma, M., Ren, L., Yuan, F., Jiang, S., Liu, Y., Kong, H. & Gong, L. 2014 A new standardized Palmer drought index for hydro-meteorological use. *Hydrological Processes* **28** (23), 5645–5661.

Mahmood, N., Ahmad, B., Hassan, S. & Bakhsh, K. 2012 Impact of temperature ADN precipitation on rice productivity in rice-wheat cropping system of Punjab province. *Journal of Animal and Plant Sciences* **22**, 993–997.

Manavalan, L. P., Guttikonda, S. K., Phan Tran, L.-S. & Nguyen, H. T. 2009 Physiological and molecular approaches to improve drought resistance in soybean. *Plant and Cell Physiology* **50** (7), 1260–1276.

McKee, T. B. 1995 Drought monitoring with multiple time scales. In: *Proceedings of 9th Conference on Applied Climatology*, 1995, Boston.

McKeown, A., Warland, J. & McDonald, M. R. 2005 Long-term marketable yields of horticultural crops in southern Ontario in relation to seasonal climate. *Canadian Journal of Plant Science* **85** (2), 431–438.

McLaughlin, D. & Kinzelbach, W. 2015 Food security and sustainable resource management. *Water Resources Research* **51** (7), 4966–4985.

Mendelsohn, R. 2014 The impact of climate change on agriculture in Asia. *Journal of Integrative Agriculture* **13** (4), 660–665.

Muhammad, W., Yang, H., Lei, H., Muhammad, A. & Yang, D. 2018 Improving the regional applicability of satellite precipitation products by ensemble algorithm. *Remote Sensing* **10** (4), 577.

Nawaz, Z., Li, X., Chen, Y., Guo, Y., Wang, X. & Nawaz, N. 2019 Temporal and spatial characteristics of precipitation and temperature in Punjab, Pakistan. *Water* **11** (9), 1916.

Potop, V., Türkott, L., Kožnarová, V. & Možný, M. 2010 Drought episodes in the Czech Republic and their potential effects in agriculture. *Theoretical and Applied Climatology* **99** (3), 373–388.

Sah, R. P., Chakraborty, M., Prasad, K., Pandit, M., Tudu, V. K., Chakravarty, M. K., Narayan, S. C., Rana, M. & Moharana, D. 2020 Impact of water deficit stress in maize: phenology and yield components. *Scientific Reports* **10** (1), 2944.

Sheikh, M. M. 2001 Drought management and prevention in Pakistan. In: *COMSATS 1st Meeting on Water Resources in the South: Present Scenario and Future Prospects*, Islamabad, pp. 1–2.

Tao, F., Yokozawa, M., Hayashi, Y. & Lin, E. 2003 Future climate change, the agricultural water cycle, & agricultural production in China. *Agriculture, Ecosystems & Environment* **95** (1), 203–215.

Tigkas, D., Vangelis, H. & Tsakiris, G. 2015 DrinC: a software for drought analysis based on drought indices. *Earth Science Informatics* **8** (3), 697–709.

Troy, T. J., Kipgen, C. & Pal, I. 2015 The impact of climate extremes and irrigation on US crop yields. *Environmental Research Letters* **10** (5), 054013.

UNISDR G 2009 *Global Assessment Report on Disaster Risk Reduction*. United Nations International Strategy for Disaster Reduction Secretariat, Geneva, Switzerland.

Van Ittersum, M. K., Cassman, K. G., Grassini, P., Wolf, J., Tittonell, P. & Hochman, Z. 2013 Yield gap analysis with local to global relevance – a review. *Field Crops Research* **143**, 4–17.

Vicente-Serrano, S. M., Gouveia, C., Camarero, J. J., Beguería, S., Trigo, R., López-Moreno, J. I., Azorín-Molina, C., Pasho, E., Lorenzo-Lacruz, J. & Revuelto, J. 2013 Response of vegetation to drought time-scales across global land biomes. *Proceedings of the National Academy of Sciences* **110** (1), 52–57.

Waseem, M., Ahmad, I., Mujtaba, A., Tayyab, M., Si, C., Lü, H. & Dong, X. 2020 Spatiotemporal dynamics of precipitation in southwest arid-agriculture zones of Pakistan. *Sustainability* **12** (6), 2305.

Wu, H. & Wilhite, D. A. 2004 An operational agricultural drought risk assessment model for Nebraska, USA. *Natural Hazards* **33** (1), 1–21.

Xie, H., Ringler, C., Zhu, T. & Waqas, A. 2013 Droughts in Pakistan: a spatiotemporal variability analysis using the Standardized Precipitation Index. *Water International* **38** (5), 620–631.

Yildirak, K. & Selcuk-Kestel, A. S. 2015 Adjusting SPI for crop specific agricultural drought. *Environmental and Ecological Statistics* **22** (4), 681–691.

Zahid, M. & Rasul, G. 2011 Frequency of extreme temperature and precipitation events in Pakistan 1965–2009. *Science International* **23** (4), 313–319.

Zaman, M., Ahmad, I., Usman, M., Saifullah, M., Anjum, M. N., Khan, M. I. & Uzair Qamar, M. 2020 Event-based time distribution patterns, return levels, and their trends of extreme precipitation across Indus Basin. *Water* **12** (12), 3373.

Zipper, S. C., Soylu, M. E., Booth, E. G. & Loheide, S. P. 2015 Untangling the effects of shallow groundwater and soil texture as drivers of subfield-scale yield variability. *Water Resources Research* **51** (8), 6338–6358.

Zipper, S. C., Qiu, J. & Kucharik, C. J. 2016 Drought effects on US maize and soybean production: spatiotemporal patterns and historical changes. *Environmental Research Letters* **11** (9), 094021.

First received 20 June 2021; accepted in revised form 12 August 2021. Available online 30 August 2021

© 2022 The Authors Impact of Climate Change on Hydrology and Water Resources doi: 10.2166/wcc.2021.241

Disaggregation of future GCMs to generate IDF curves for the assessment of urban floods

H. Tayşi [a,*] and M. Özger [b]

[a] Faculty of Engineering, Department of Civil Engineering, Gebze Technical University, Gebze, Kocaeli 41400, Turkey
[b] Faculty of Civil Engineering, Department of Hydraulics and Water Resources Engineering, Istanbul Technical University, Sariyer, Istanbul 34467, Turkey
*Corresponding author. E-mail: h.taysi@gtu.edu.tr

HT, 0000-0002-4302-7364; MÖ, 0000-0001-9812-9918

ABSTRACT

Urbanization and industrialization cause an increase in greenhouse gas emissions, which in turn causes changes in the atmosphere. Climate change is causing extreme rainfalls and these rainfalls are getting stronger day after day. Floods are threatening urban areas, and short-duration rainfall and outdated drainages are responsible for urban floods. Intensity–Duration–Frequency (IDF) curves are crucial for both drainage system design and assessment of flood risk. Once IDF curves are determined from historical data, they are assumed to be stationary. However, IDF curves must be non-stationary and time varying based on preparation for extreme events. This study generates future IDF curves with short-duration rainfalls under climate change. To represent future rainfall, an ensemble of four Global Climate Models generated under Representative Concentration Pathways (RCP) 4.5 and 8.5 were used in this study. A new approach to the HYETOS disaggregation model was applied to disaggregate daily future rainfall into sub-hourly using disaggregation parameters of hourly measured rainfalls. Hence, sub-hourly future rainfalls will be obtained capturing historical rainfall patterns instead of random rainfall characteristics. Finally, historical and future IDF curves were compared. The study concludes that increases in short-duration rainfalls will be highly intensified in both the near and distant futures with a high probability.

Key words: climate change, disaggregation, GCM, IDF curves, RCP, urban floods

HIGHLIGHTS

- Climate change impacts on rainfall intensities are assessed by comparing historical IDF curves with future IDF curves considering different GCMs and RCPs.
- Short-duration rainfalls under climate change impacts were simulated for the assessment of urban floods.
- Future daily rainfall data can be disaggregated into sub-hourly rainfall with a new approach to HYETOS using disaggregation parameters of hourly measured rainfalls.

GRAPHICAL ABSTRACT

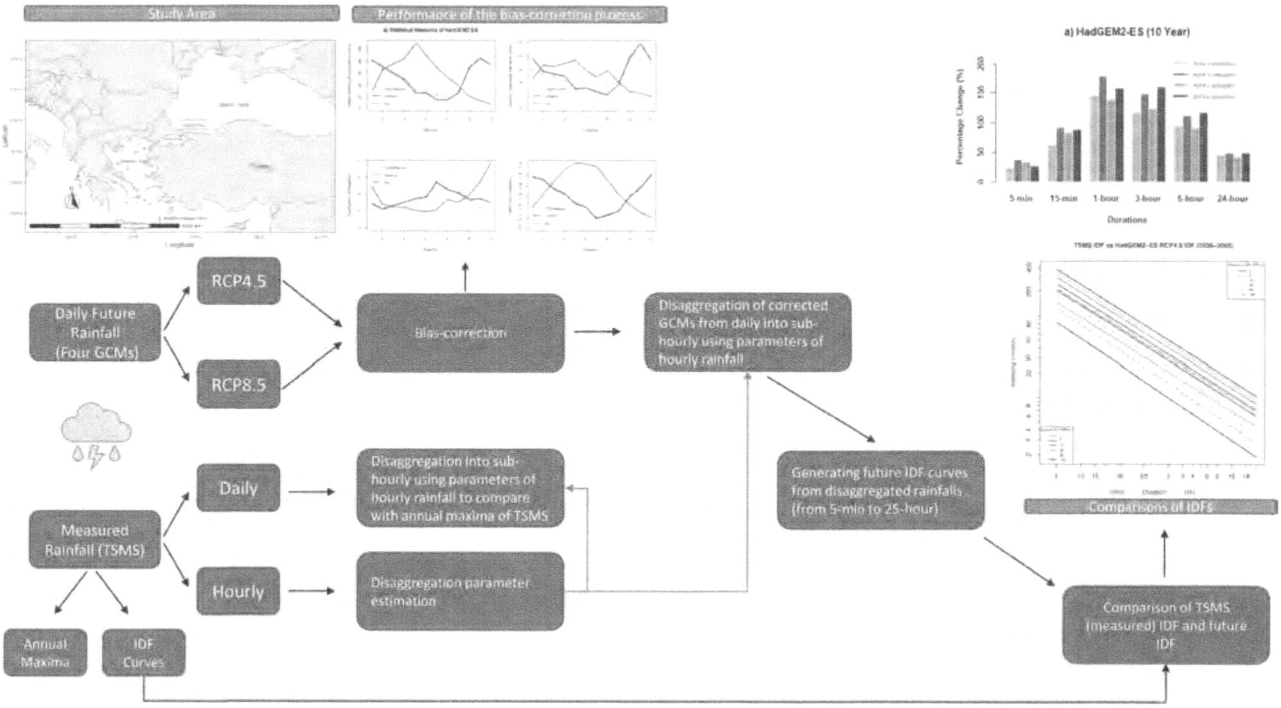

1. INTRODUCTION

Dealing with heavy rainfall events that can cause flooding, loss of life, crops, and properties, is challenging for urban areas. High intensity rainfall is a key factor in the occurrence of flooding events. Urban floods in particular are increasing day after day in many countries. Flash floods are one of the most common flood types that occur in urban areas. Short-duration and high intensity rainfalls are usually responsible for flash floods. These floods have destructive effects on people, cities, drainage, bridges, and the economy. Other factors also cause an increase in the destructiveness of flash floods. The number of impermeable surfaces increases continuously due to urbanization. Hence, drainage systems are becoming ineffective against fast runoffs caused by flash floods, and their response time is decreasing (Hosseinzadehtalaei *et al.* 2020). Water is then flowing through cities due to overflow from drainage systems. Since the mentioned factors are predominantly occurring in urban areas, the effective management and awareness of both urban drainage systems and short-duration rainfalls are significant. As mentioned before, serious management of drainage and short-duration rainfalls is necessary. Rainfall Intensity–Duration–-Frequency (IDF) curves play a crucial role in assessing flood risk and quantifying heavy rainfalls. IDF curves are also effective tools in designing and operating urban drainage systems. Hence, IDF curves are required in designing hydraulic structures such as sewers, drainages, gutters, and culverts to cope with floods, because understanding extreme rainfalls and their frequency will help drainage systems against floods. IDF curves give a possible rainfall intensity in selected durations for return periods. Using IDF curves in design water facilities allows engineers to be ready for extreme events, hence possible damage can be decreased (Sarhadi & Soulis 2017). IDF curves are widely applied in many water-related projects, flood forecasting and management, and water management (Cheng & AghaKouchak 2014). In addition, IDF curves help not only drainage systems but also other structures vulnerable to water such as railways and telecommunications networks (Hosseinzadehtalaei *et al.* 2020).

Irrepressible growth of industrial activities, urbanization and population enhances greenhouse gases (carbon dioxide, methane, aerosols, etc.) emissions. This enhancement causes major variations in climate and leads to the necessity to deal with a serious challenge in the future: climate change (Mirhosseini *et al.* 2013). Climate change causes global warming by increasing global temperatures, and this causes enhancement of evapotranspiration and water vapour in the atmosphere, hence more extreme events. Extreme rainfall is one of the most serious consequences of these changes. Extreme rainfall events cause flooding and damage to sewers, drainage systems, bridges, and infrastructure (Singh *et al.* 2016; Sarhadi &

Soulis 2017). However, current IDF curves assume that extreme rainfall events will not change in the future climate conditions (Cheng & AghaKouchak 2014). Hence, developing advanced IDF curves, which are successful at the representation of both historical and future climate conditions, is a huge necessity. With these types of IDF curves, it will be possible to deal with extreme rainfall events under non-stationary climate conditions. To generate IDF curves, theoretical probability distribution functions such as Gumbel, Generalized Extreme Value (GEV), Log-Normal, and Log Pearson Type III are applied (Nwaogazie *et al.* 2019). Although IDF curves based on historical rainfall events are used frequently, they are still not fully sufficient against a rapidly changing environment. Historical rainfall-based IDF curves are stationary; therefore, they are ineffective in catching climate change conditions (Singh *et al.* 2016). Many studies have been performed to update IDF curves considering future conditions. Mirhosseini *et al.* (2013) evaluated 3-hourly precipitation data generated by six Global Climate Models (GCMs) and Regional Climate Models (RCMs) to examine changes in future IDF curves compared with current ones in Alabama. They performed bias-correction using the quantile-based mapping method. They concluded that rainfalls are expected to be less for short durations (less than 4 h) in the future. Therefore, short-duration rainfalls are serviceable for designing municipal infrastructures. Hajani (2020) generated future stationary and non-stationary IDF curves to compare them with standard regional ARR (Australian Rainfall-Runoff) IDF curves. She also exhibited the impact of climate change on these curves. GEV distribution was applied to arrange the stationarity of the curves by keeping GEV's parameters constant or dependent on time. The study concluded that standard regional IDF curves better match the new stationary IDF curves when compared with non-stationary IDF curves. In the study of Zhu *et al.* (2012), they investigated the generation of IDF curves that were affected by climate change to evaluate rainfall extremes by considering historical and future climate data. Future IDF curves were developed for six locations that represent various climatic conditions in the United States, and the annual maximum series were calculated for 3, 6, 9, 12, 18, 24, 48, and 96 h. An approach called Areal Reduction Factor (ARF) was proposed to relate maximum rainfall intensity measured at a station with the maximum areal average gridded rainfall intensity. They concluded that climate change impacts are more effective at the regional level; hence, evaluations need to be performed locally. Short-duration and high intensity rainfalls are greatly affected by climate change, according to their study. De Souza Costa *et al.* (2020) performed a study on IDF curves using HadGEM2-ES, CanESM2, and MIROC5 models generated under RCP4.5 and RCP8.5 scenarios for the Amazon region, Brazil. They used future daily maximums, measured sub-daily maximums and historical daily maximums. This study concluded that annual rainfalls are decreasing on CanESM2 and HadGEM2-ES models while they are increasing for the MIROC5 model. MIROC5 has the largest difference and CanESM2 has the smallest difference when compared with existing IDF values.

Flash floods are more challenging to deal with compared with slower floods occurring from long-duration rainfalls. Therefore, knowledge of short-duration rainfall (i.e., rainfall durations of less than 4 h) is crucial for the design of urban drainage systems and for urban hydrology studies. Hence, IDF curves must be generated considering short-duration rainfalls for a healthier environment in cities. However, measuring short-duration rainfall is not always possible due to its cost, limitations of a station's capacity, climatic, and geographic conditions. On the other hand, daily rainfalls exist due to its convenience for measurement. It is vital to use some methods to simulate these kinds of rainfalls from longer durations (i.e., daily). In this study, a method called 'disaggregation' was applied to obtain sub-hourly rainfalls.

The main purpose of this study is to generate future rainfall IDF curves using short-duration rainfalls to assess urban floods using an ensemble of four GCMs and then compare them with IDF curves generated by measured rainfalls. Florya meteorological station operated by the Turkish State Meteorological Service (TSMS) in Istanbul was selected as the study area since TSMS provides IDF curves and a wide range of daily measured rainfalls (1980–2005) for the station. In addition, TSMS provides hourly rainfall records for the 14-year period (2005–2018). To assess future rainfall events for the period 2036–2065 and 2066–2095 considering climate change impacts, GCMs generated under Representative Concentration Pathway (RCP) scenarios RCP4.5 and RCP8.5 were provided. Criteria for selecting these periods were as follows: (i) We aimed to evaluate future rainfall intensities for the near future, which is 2036–2065 and for the distant future, which is 2066–2095. Analyses for a wide range are not always a reliable method since there will be significant variations and uncertainty. Therefore, the analyses could mislead the results. Besides, separating years into near and distant futures will help us to take precautions for cities considering the closeness of disasters. (ii) Since four GCMs were used in the study, it was necessary to select common periods for all GCMs. HadGEM2-ES developed by the Met Office Hadley Centre Institute (MOHC), ICHEC-EC-EARTH developed by the Irish Centre for High-End Computing (ICHEC), IPSL-IPSL-CM5A-MR developed by Institut Pierre-Simon Laplace (IPSL), and NorESM1-M developed by Norwegian Climate Center (NCC) were selected as GCMs to generate future IDF curves. When the future conditions are considered, there is a huge uncertainity about variations. Using only one climate model can be insufficient to capture changes in rainfall. Hence, more reliable and efficient IDF analyses were intended using an ensemble of GCMs in the study.

In the study, future rainfall intensities (mm/h) from each GCM were computed for selected durations (5, 10, 15, and 30 min; 1, 2, 3, 4, 5, 6, 8, 12, 18, and 24 h) and for each return period of 2, 5, 10, 25, 50, and 100 years. The Gumbel function was selected as the probability distribution function (PDF) for annual maximum rainfall events to generate IDF curves since the past studies reported that the Gumbel distribution is a very effective method to fit annual extreme rainfalls (Oyebande 1982). However, GCMs are not suitable to use directly due to biases between measured and historical raw rainfall data. The distribution mapping method was applied to correct these biases.

As mentioned before, providing short-duration rainfall is challenging due to various factors. Provided future rainfall events were in daily form; hence, a Barlett–Lewis model-based disaggregation model called 'HYETOS' developed by Koutsoyiannis & Onof (2001) was applied for the disaggregation of daily rainfall into sub-hourly to generate IDF curves with short and long durations, which are in the range of 5 min and 24 h. Bartlett–Lewis disaggregation model parameters are required for the disaggregation process and obtained using some statistics of hourly rainfall data. Since future rainfall data is daily, it is not possible to compute HYETOS parameters for future rainfalls. Hence, parameters computed by Florya station's hourly measured rainfalls were used as the disaggregation parameters of future data. Since the disaggregation process is based entirely on rainfall statistics, the most accurate way to conduct the disaggregation process is to use the parameter sets obtained from measured rainfall patterns. In this case, the future rainfalls of Florya station were disaggregated using separately obtained disaggregation parameters for each month from the measured rainfalls in order to perform the most accurate disaggregation. Thus, even if the rainfalls change in the future, the disaggregation can be performed by imitating the past rainfall. All computations, analyses, and plottings were made on the R Studio platform in this study.

This study has six main objectives: (i) Generating more reliable and effective future IDF curves under various climate change scenarios for urban area drainage systems by evaluating short-duration future rainfall data. To date, only a few studies have generated IDF curves for short-duration rainfalls considering climate change. (ii) Disaggregation of future daily rainfalls into sub-hourly rainfalls with a new approach to the HYETOS disaggregation model. Providing sub-hourly rainfall data is not always possible even for simulated rainfalls. Besides, measuring this kind of short-duration data is highly challenging for meteorological stations. HYETOS model usually uses random parameters for the disaggregation process. In this study, disaggregation parameters were obtained from 14-year measured hourly rainfall and parameter sets were determined for each month using some statistical measures. Afterwards, the determined parameters were applied for the disaggregation of measured (applied to test the method) and future daily rainfalls. This process gives a chance for future data to capture historical patterns of a selected region's rainfall instead of random rainfall characteristics during the disaggregation process. Briefly, with this method, it will be possible to disaggregate any daily rainfalls of a large region or city using only short records of hourly rainfall. (iii) To verify the accuracy of the selected disaggregation model by comparing maximum values of disaggregated rainfalls with measured ones. (iv) Applying an effective bias-correction process to GCMs to obtain more reliable future rainfalls. (v) Evaluation of the climate change impact on rainfall intensities by comparing historical and future IDF curves. The IDF curves obtained from the ensemble of four GCMs under RCP4.5 and RCP8.5 scenarios were also compared with each other. (vi) Assessment of climate change impacts on short-duration rainfall using IDF curves. Finally, the study will provide a useful methodology for designing urban drainage systems against floods.

2. WORKFLOW OF THE STUDY

The first step in the process was to estimate disaggregation parameters from hourly measured rainfalls provided by TSMS. As mentioned before, 1-h rainfall is necessary to obtain HYETOS disaggregation parameters. After obtaining parameters, daily measured rainfalls of TSMS were disaggregated into sub-hourly rainfalls (5, 10, 15, and 30 min) using the parameters of hourly rainfalls. Afterwards, the sub-hourly rainfalls were aggregated until 24 h. Maximum rainfalls for each selected duration (5, 10, 15, and 30 min; 1, 2, 3, 4, 5, 6, 8, 12, 18, and 24 h) and for the 1980–2005 periods were supplied by TSMS. Maximum values of rainfalls disaggregated by daily measured rainfall were computed for each selected duration. To verify the accuracy of the disaggregation method, maximum values of both measured and disaggregated rainfalls were then compared. In the next phase of the study, biases of all four GCMs were corrected with the help of TSMS (measured) daily rainfall. Again, the corrected daily GCMs were disaggregated into sub-hourly rainfalls using the previously estimated disaggregation parameters of TSMS hourly measured rainfall. IDF curves for each scenario and each GCM were generated considering all durations (from

5 min to 24 h) and all return periods (2, 5, 10, 25, 50, and 100 years) using the Gumbel method. Finally, future IDF curves were compared with the IDF curves of measured rainfalls (TSMS IDF) provided by TSMS. Future IDF curves under different GCMs and scenarios were compared with each other to deduce the behaviour of models and scenarios under climate change. Figure 1 summarizes the work done for the study.

3. STUDY AREA

The study area of Istanbul is located in north-western Turkey. The city is located in the Marmara region with a total area of 5,343 km^2 and a population of 15,519,267. The geographical location of the city is 41°00′49″N 28°57′18″E. One of the most important characteristics of this city is that it separates Europe and Asia. Thus, the city has lands in both Europe and Asia. The Black Sea and the Marmara Sea are connected in Bosphorus. Istanbul has the highest population in Turkey and Europe. The hottest months are July and August, with an average temperature of 23.8 °C, and January is the coldest month with an average temperature of 6.0 °C. In the summer months, the average rainfall amount is 106.2 mm, the average rainfall for the other 9 months is 87.7 mm. Istanbul is rainy on average 130 days a year. The average total rainfall per year is 817.4 mm. The wettest month is December with 122.6 mm rainfall and the driest month is July with 31.6 mm. January has the highest number of wet days with 17.5 days, and July has the lowest number of wet days with 4.2 days. The maximum rainfall measured in a day is 227 mm (BBC Weather 2020; TSMS 2020). Rainfall data were provided for Florya meteorological station operated by TSMS in Istanbul. IDF curves and rainfalls in daily and hourly resolution were required for the study. Florya station was selected since the required data are existing. Future climate data obtained from four GCMs were generated for the selected station (Figure 2).

Table 1 includes the station properties and sensor types that W is wind, T is temperature, M is moisture, R is rainfall, ST is soil temperature and P is pressure. Data are collected from Automatic Meteorological Observation Station (AMOS) and daily climate (DC) stations.

4. DATA

4.1. Measured data

TSMS is a meteorological service that provides all kinds of meteorological data. In this study, (i) measured daily rainfall data, (ii) measured hourly rainfall data, (iii) maximum values for the selected durations, and (iv) IDF curves generated under measured rainfalls were provided by TSMS.

The HYETOS parameters are required to have at least 1-h resolution rainfall data to determine disaggregation parameters because they are calculated using the statistics of 1, 6, 12, and 24-h rainfalls. Hourly rainfall data are available for Florya station for the period 2005–2018. These rainfall data were used to obtain disaggregation parameters to be used for the disaggregation of future rainfalls (2036–2095) and daily measured rainfalls (1980–2005). In addition, the maximum values of

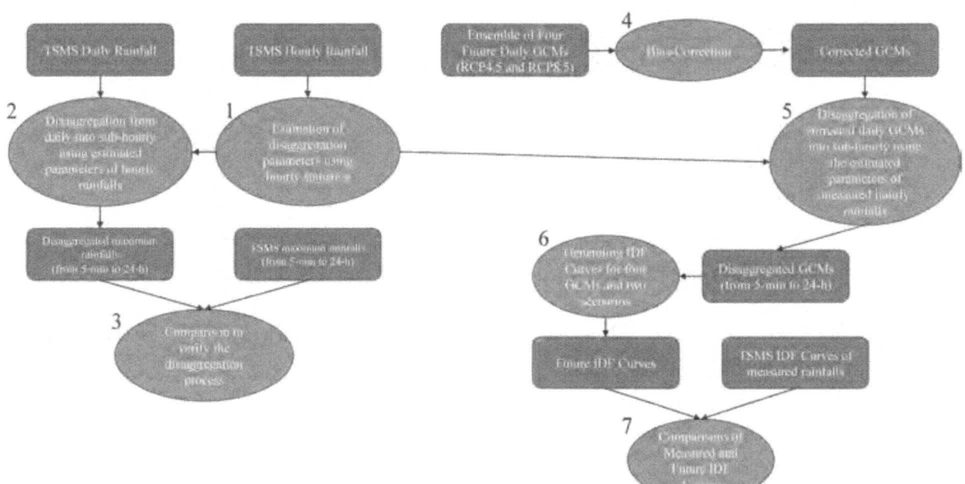

Figure 1 | Workflow of the analyses.

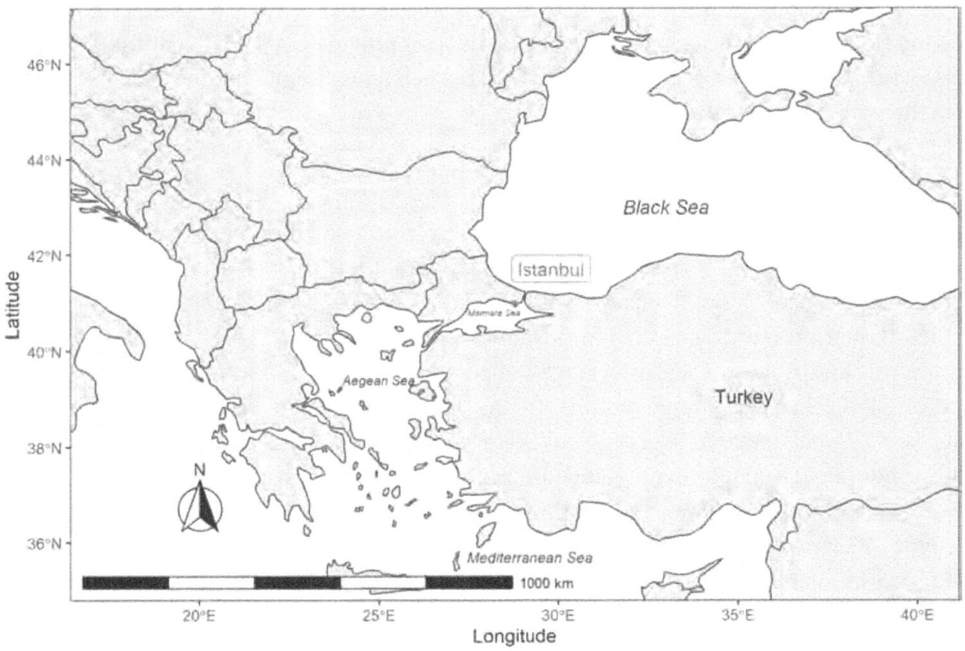

Figure 2 | The location of the selected station in Istanbul.

Table 1 | Station properties

Station	Station number	Coordinates	Sensors	Observation type	Elevation (m)
Florya	17636	40°58′32.9″N 28°47′11.4″E	W, T, M, R, ST, P	AMOS-DC	37

measured rainfalls for each selected duration for the period of 1980–2005 were provided by TSMS. The accuracy of the selected disaggregation method was evaluated by comparing these TSMS maximum rainfalls with maximum rainfalls disaggregated from daily measured rainfalls for each duration. Finally, the IDF curves generated by TSMS using 78 years of rainfall record (1938–2015) were used for the comparisons with future IDF curves for each scenario and GCM.

Monthly rainfall characteristics of Florya station are given in Table 2.

Table 2 | Rainfall characteristics of monthly rainfalls of Florya station

Statistics	January	February	March	April	May	June
Mean (mm)	77.16	73.14	55.6	40.91	28.04	32.24
Min (mm)	0	0	0.2	0	0.4	0.2
Max (mm)	158.4	164.4	117.2	115.2	77	127.6
Standard Deviation (mm)	50.49	52.54	32.75	33.35	21.87	33.69
	July	**August**	**September**	**October**	**November**	**December**
Mean (mm)	19.69	14.31	35.68	63.78	53.51	60.37
Min (mm)	0	0	0	0.2	0	0
Max (mm)	65.4	72	135.6	238	122.8	136.4
Standard Deviation (mm)	23.19	21.08	44.5	62.8	38.07	52.97

As mentioned in previous sections, GCMs are not available to use directly due to biases between measured and historical raw data. Historical raw data, measured data, and future data were used together for the bias-correction method. As measured data to be used in the bias-correction process, daily rainfalls for the period of 1980–2005 were supplied by the TSMS.

4.2. Global climate model (GCM)

Climate models represent the climate system under various carbon emission scenarios to understand future changes. GCM can be used to simulate both historical and future rainfall events. As mentioned before, dealing with changes in the future using a single GCM is not effective, because each model has advantages and disadvantages, and these disadvantages can create some trouble about capturing rainfalls. Therefore, this study employs an ensemble of four GCMs to provide a more reliable IDF generation. The historical raw and future data were obtained from the Earth System Grid Federation ESGF – Lawrence Livermore National Laboratory (LLNL) website. Daily historical raw rainfalls for the period 1980–2005 were used in the study. Daily future rainfall data are also available for 2036–2065 and 2066–2095 with RCP4.5 and RCP8.5 scenarios for selected GCMs. As mentioned before, separating future into the near and distant future will give more reliable results, because the selection of a wide range period for analyses will lead to significant variations. In addition, the selected periods were common for all GCMs. GCMs were provided from the Coordinated Regional Climate Downscaling Experiment (CORDEX) Europe program. GCMs HadGEM2-ES, ICHEC-EC-EARTH, IPSL-IPSL-CM5A-MR, and NorESM1-M with a 12.5 km (0.11°×0.11°) resolution were preferred in this study. The reason for selecting these GCMs was their availability for the desired factors (the presence of both RCP4.5 and RCP8.5 scenarios, Europe domain, and 12.5 grid resolution). Outputs from the GCMs were downscaled to Florya station in this study. To correct the biases between measured and historical raw data, the bias-correction method distribution mapping was employed.

The Intergovernmental Panel on Climate Change (IPCC) published the Fifth Assessment Report (IPCC 2014) to assess climate change in the future using RCP scenarios. RCPs are used to define emissions of air pollutants, greenhouse gases, and atmospheric concentrations. Radiative forcing is 4.5 W/m^2 for RCP4.5 and 8.5 W/m^2 for RCP8.5. These values are valid on a global scale, and they give approximate total radiative forcing for each RCP scenario in 2100 (Edenhofer *et al.* 2015). Watts per square meter (W/m^2) is the unit of solar radiation falling per unit area. It also represents the energy imbalance in the atmosphere. Table 3 shows each GCM and RCM, centre of models, resolutions, and scenarios, respectively. Figure 3 presents all the data types used in the study with their periods.

5. METHODS

5.1. GCM data bias-correction

GCM outputs are not suitable for direct use in hydrological studies due to the need for bias-correction. These biases originate from inconsistencies between measured and historical raw rainfall (Rathjens *et al.* 2016). Measured high rainfalls and the number of dry days cannot be well-represented if biases exist. Seasonal alterations and extreme temperatures are predicted poorly due to biases. GCMs simulate low rainfall days instead of dry days (Teutschbein & Seibert 2010). The Climate Model Data for Hydrologic Modeling (CMhyd) software developed by Texas A&M University (TAMU) which is available online was chosen as a bias-correction process. The general framework of the bias-correction process was described by Rathjens *et al.* (2016) as shown in Figure 4. First, biases between measured rainfall and historical raw climate data are

Table 3 | Description of selected climate models

Driving GCM	RCM	GCM Model centre	RCM Model centre	Resolution	Scenario
HadGEM2-ES	CCLM4-8-17	Met Office Hadley Centre	The Climate Limited-area Modeling Community	0.11°× 0.11°	Historical, RCP4.5 and RCP8.5
ICHEC-EC-EARTH	KNMI-RACMO22E	Irish Centre for High-End Computing	Royal Netherlands Meteorological Institute	0.11°× 0.11°	Historical, RCP4.5 and RCP8.5
IPSL-IPSL-CM5A-MR	IPSL-WRF381P	Institut Pierre-Simon Laplace	Institut Pierre-Simon Laplace	0.11°× 0.11°	Historical, RCP4.5 and RCP8.5
NorESM1-M	GERICS-REMO2015	Norwegian Climate Center	Climate Service Center Germany	0.11°× 0.11°	Historical, RCP4.5 and RCP8.5

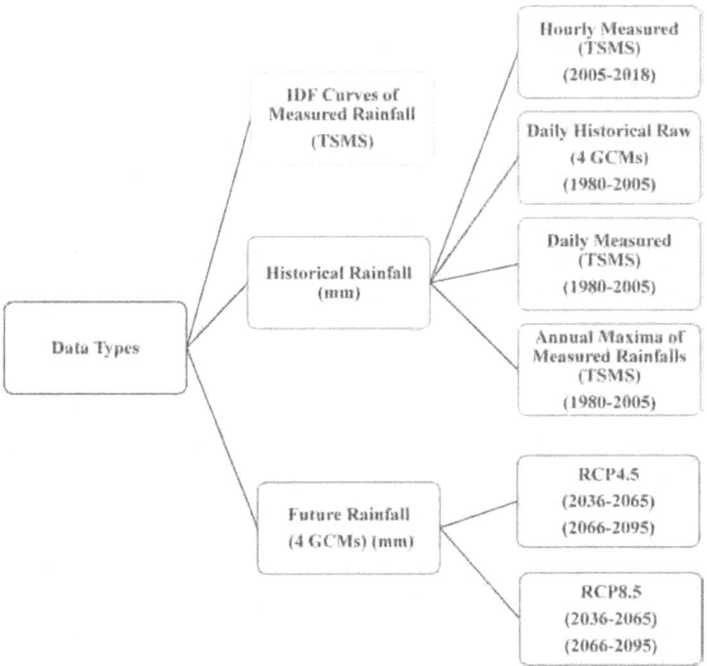

Figure 3 | Used data types.

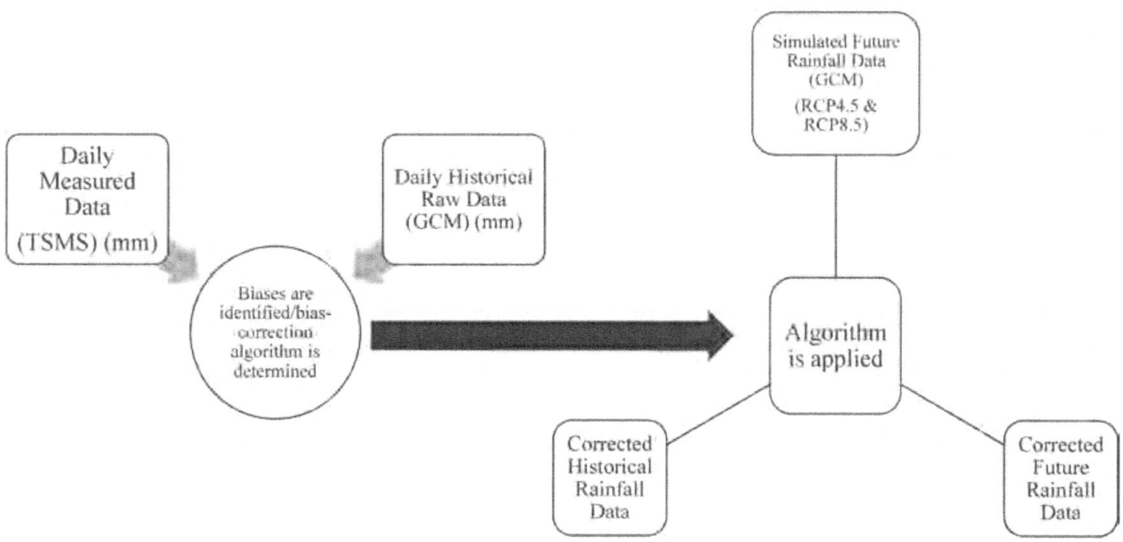

Figure 4 | Framework of bias-correction.

identified, and the parameters of bias-correction algorithm are then determined. This algorithm is then applied to future climate data to correct biases. This way, one can finally obtain corrected historical and future climate data. Bias-correction helps users to use GCMs in hydrological studies by bringing simulated data closer to real data. Several bias-correction methods, including distribution mapping, were explained in the study of Teutschbein & Seibert (2010). In this study, the distribution mapping method was employed as the bias-correction method and biases between overlapping periods of measured and historical raw rainfalls were corrected.

Measured TSMS data and GCM historical raw data between 1980 and 2005 were used as the calibration period for the bias-correction process.

5.1.1. Distribution mapping

Teutschbein & Seibert (2010) applied this method in their studies. 'Probability mapping', 'quantile matching', 'statistical downscaling', and 'histogram equalization' terms can be used for distribution mapping in the literature. Using distribution mapping, the distribution function of simulated climate data is corrected to coordinate the measured data distribution function. To perform this analysis, a transfer function is used to shift the distribution of simulated data. It is assumed that the biases are stationary under climate change for this method (Teutschbein & Seibert 2010). Distribution mapping employs the Gamma distribution to remove biases. Thom (1958) expressed the Gamma distribution with shape parameter k, and scale parameter β. The Gamma distribution is applicable to the distribution of rainfall data (Teutscbein & Seibert 2010).

$$f_y = \frac{1}{\beta^k \Gamma(k)} x^{k-1} e^{-x/\beta}; \; x \geq 0; \; \beta, k > 0 \tag{1}$$

where β is the scale parameter, k is the shape parameter, Γ is the gamma function, and x is the normalized daily rainfall. Each grid and month has its own scale and shape parameter. With this method, mean, variance, skew, and frequency of rainfall events are corrected. The distribution profile is managed by shape parameter k. Three conditions are considered for the value of k. When $k < 1$, it defines the exponentially shaped Gamma distribution, $k = 1$ describes the exponential distribution, and $k > 1$ indicates a skewed uni-modal distribution. The scale parameter β dictates dispersion of the Gamma distribution. Commonly $k > 1$ situation is used for measured daily rainfall. If the scale parameter β is small, it eventuates to a more compressed distribution, and this ends up with lower probabilities of extreme rainfall. If the β is large, this causes a stretched distribution, and this is the reason for higher probabilities of extreme events (Teutscbein & Seibert 2010). The study by Teutschbein & Seibert (2010) showed that gamma distribution parameters fitted to simulated climate data show similar patterns for the selected catchments in the study area. They reported that the level of commitment of the distribution parameters (k/β) defines the skill for the GCM to reproduce rainfall. As mentioned before, Teutschbein & Seibert (2010) compared several bias-correction methods including linear scaling, local intensity scaling, power transformation, variance scaling, and distribution mapping considering the skills of methods to arrange the statistics of the respective measured climate data. The study concluded that distribution mapping is the best method for rainfall with the minimum MAE (minimum absolute error). They also concluded that the method is applicable to both current and future climate data.

5.2. Disaggregation of daily rainfalls into sub-hourly rainfalls

Hydrological studies such as designing drainage systems require high-resolution rainfall data for the generation of IDF curves. This need arises from the fact that maximum values of finer scales of rainfall are responsible for urban floods. However, providing sub-hourly rainfall data is not always possible even for simulated rainfalls. Besides, measuring this kind of short-duration data is highly challenging for meteorological stations due to the limitations of a station's capability, costs, and geographic conditions. To cope with this shortcoming of sub-hourly data, disaggregation methods, which are able to calculate finer-scale data (i.e., hourly and sub-hourly) from coarser-scales (i.e., daily data), are used. In this study, we intend to generate future IDF curves for short and long durations ranging from 5 min to 24 h. Hence, the disaggregation process was used to disaggregate GCMs into sub-hourly rainfalls.

Koutsoyiannis & Onof (2001) developed a computer programme called HYETOS based on the Bartlett–Lewis model, and they implemented a disaggregation scheme in an R package called 'HYETOSMinute'. The Bartlett–Lewis model was constructed by Rodriguez-Iturbe *et al.* (1987) to overcome the inefficiency of simple Poisson models. In this study, the HYETOS disaggregation model was applied.

The original Bartlett–Lewis model has five parameters (β, γ, μ_x, η, and λ) for the disaggregation process. Storm origins are developed by λ, cell origins are developed by β, cell arrivals end after a specific time, and the time is exponentially distributed with γ. Each cell has a duration exponentially distributed with η. Uniform intensity for each cell is distributed exponentially with μ_x. Hanaish *et al.* (2011) explained the original Bartlett–Lewis rectangular pulses model in their study.

Rodriguez-Iturbe *et al.* (1988) adjusted the original model to boost the flexibility of the model to generate larger diversity of rainfall. This modified model is called the Modified Bartlett–Lewis Rectangular Pulse Model (MBLRPM). With the Gamma distribution, η varies for each storm.

In this model, β and γ also altered, therefore ratios $k{=}\beta/n$ and $\varphi{=}\gamma/\eta$ stay constant, so that the MBLRPM model has six parameters (α, φ, μ_x, k, λ, v). An enhanced version of the Evolutionary Annealing-Simplex Method (EAS) is applied to estimate Bartlett–Lewis model parameters. Four historical statistical values (mean, variance, auto-covariance lag-1, and the proportion of dry days) for 1-, 6-, 12-, and 24-h time scales of rainfall data are used to make the estimation. MBLRPM parameters are used as inputs to the disaggregation method. Each month also has its own parameters for the disaggregation process. Bartlett–Lewis parameters cannot be calculated for future data due to the absence of hourly future rainfall data. In this study, six different disaggregation parameters were obtained from rainfall statistics for each month of 14 years (2005–2018). Afterwards, mean values of parameters were computed, and histograms were plotted for parameter values of 12 months, separately. Twelve different parameter sets were determined for each month with the help of histogram plots and mean values. For instance, 12 parameter sets containing six parameters (α, φ, μ_x, k, λ, v) were determined for each month. Each parameter reflects the dominant rainfall characteristics of each month. Thus, each GCM has parameters for the disaggregation of future rainfall data. Afterwards, the determined parameters were applied for the disaggregation of respective months of measured (applied to test the method) and future daily rainfalls.

For the assessment of accuracy of the selected disaggregation method HYETOS, daily measured rainfalls provided by TSMS were disaggregated using disaggregation parameters of hourly rainfalls. Since IDF curves are generated with annual maximums of each duration, maximum values of disaggregated rainfalls were computed for each selected duration. Therefore, evaluating the relationship of maximum values was sufficient to deduce the performance of the disaggregation method. A comparison was made between the maximum rainfalls of TSMS and the maximum of disaggregated rainfalls. Results showed that maximum values of disaggregated and measured rainfalls are in good agreement with each other. This shows that the MBLRPM can be used as a disaggregation method.

5.3. Generating future IDF curves

IDF curves represent a relationship between rainfall intensity and storm duration for given return periods (T). Each IDF curve value shows the probability of occurring on average $1/T$ for each year. For example, if an IDF curve value is calculated for a 100-year return period and a 6-h duration, this extreme rainfall value can occur in any given year with 1/100 (% 1) chance. Periods of 2, 5, 10, 25, 50, and 100 years were selected as a return period, and 5, 10, 15, 30 min, 1, 2, 3, 4, 5, 6, 8, 12, 18, and 24 h were used as standard durations.

The GCMs generated under RCP4.5 and RCP8.5 climate change scenarios were used to generate future IDF. In determining future IDF curves, the annual maximum rainfall values for each standard duration for both the near future and distant future (2036–2065 and 2066–2095) are required. In this study, future IDF curves of GCMs HadGEM2-ES, ICHEC-EC-EARTH, IPSL-IPSL-CM5A-MR and NorESM1-M under two climate change scenarios RCP4.5 and RCP8.5 for two selected periods 2036–2065 and 2066–2095 were generated. Finally, 16 different IDF tables were generated to be compared with each other to present (i) climate change impact on rainfall intensities, (ii) differences between GCMs and RCP scenarios, and (iii) behaviour of short- and long-duration rainfalls under climate change.

PDFs are used to generate IDF curves. Generally, the Gumbel probability distribution function is used for the determination of IDF curves. Oyebande (1982) reported that the Gumbel distribution is a very effective method to fit annual extreme rainfalls. Other extreme value distributions such as Frechet and Weibull distributions require three parameters: location, scale, and shape. When the shape parameter is set to zero, they transform to Gumbel. Therefore, Gumbel has only two parameters: location and scale. The function of Gumbel PDF is defined as:

$$F(x) = \frac{1}{\beta} e^{\frac{x-\alpha}{\beta}} e^{-e^{\frac{x-\alpha}{\beta}}} \tag{2}$$

where α is the location, and β is the scale parameter. In this study, Method of Moments (MoM) was applied for the estimation of distribution parameters. Calculating rainfall intensities requires a Gumbel frequency factor for each return period. The mean and standard deviation of annual maximum values for each duration are then calculated. The Gumbel frequency factor K_T is calculated using the following equation (Nwaogazie *et al.* 2019):

$$K_T = \frac{\sqrt{6}}{\pi} \left[0.5772 + \ln \left[\ln \left[\frac{T}{T-1} \right] \right] \right] \tag{3}$$

where T is the return period.

The value of random variable R, which is rainfall (mm) for this study, was found with the equation given by Chow (1951):

$$R = M + K_T S \tag{4}$$

where R is rainfall (mm), M and S are the mean and standard deviation of measured maximum rainfall for the current duration, respectively, and K_T is the Gumbel frequency factor for each return period. Hence, rainfall values are calculated for the current duration at different return periods. Rainfall intensity I (mm/h) can be calculated by dividing rainfall R by selected duration d (in hours).

$$I = \frac{R}{d} \tag{5}$$

Then, the process is performed for each duration, and maximum rainfall intensities are obtained for each duration and return period.

Briefly, the steps to generate IDF curves can be given as follows:

1. Annual maximum values of rainfall data for each standard duration (5, 10, 15, and 30 min, and 1, 2, 3, 4, 5, 6, 8, 12, 18, and 24 h) and periods (2036–2065 and 2066–2095) are calculated.
2. MoM is applied to obtain Gumbel parameters.
3. Gumbel frequency factors are determined for each return period.
4. Mean and standard deviation of measured maximum rainfall values are calculated for each duration.
5. Rainfall values are obtained following Equation (4) and rainfall intensity is calculated by dividing rainfall amounts by durations.
6. The process is repeated for each duration.
7. IDF curves are plotted with calculated rainfall intensities for each duration and return period.

6. RESULTS AND DISCUSSION

6.1. Evaluation of the bias-correction process

The distribution mapping method was applied to correct biases between measured and historical raw data. For the process, measured and historical raw data for periods 1980–2005 were imported for calibration. To evaluate the performance of the bias-correction method, different statistics and plots were used. Root mean square error (RMSE), Nash–Sutcliffe efficiency (NSE) (Nash & Sutcliffe 1970) and the percent bias (PBIAS) were used to test variation between measured rainfalls, raw, and corrected GCM. RMSE measures the estimation error. NSE is an efficient measure to reveal the matching level between measured and simulated. The optimal value for the NSE is 1, which indicates a perfect fit. The average trend of simulated data to measured data can be found by PBIAS. Obtaining negative values from PBIAS shows an overestimation; otherwise, positive values indicate an underestimation. 0 is the optimal value for PBIAS (Berhanu *et al.* 2016). Monthly statistical values, including maximum, mean, median, standard deviation (SD), coefficient of variation (CV), 90th (Q) percentile, and wet day probability (WDP), were calculated as summarized in Table 4 to conduct comparisons.

Afterwards, cumulative distribution functions (CDFs) of all concerned data for May are plotted in Figure 5.

Mean monthly rainfalls, coefficient of variations, standard deviations, and wet day probabilities for each month of measured, raw, and corrected were plotted separately for each GCM (Figure 6). Briefly, using several techniques, measured, historical raw and historical corrected rainfall data were compared to prove the convenience of the selected bias-correction method.

Table 4 shows several statistical measures and tests between measured, raw, and corrected data. When monthly mean rainfalls were compared, there were wide differences between raw and measured data for all GCMs, while corrected data presented closer relationships. All raw data of GCMs underestimate mean monthly rainfalls except IPSL-IPSL-CM5A-MR. As can be seen from the NSE column in Table 4, all corrected GCM data have a highly strong relationship with measured data in the range of 0.87 and 0.98. Raw data of GCMs show NSE values far than 1; however, raw data of NorESM1-M have the strongest relationship with NSE=0.8 when compared with other raw data. Similarly, in RMSE comparisons, there are significant differences for all raw data of GCMs compared with corrected data. Negative PBIAS values indicate

Table 4 | Mean monthly statistical values and tests of measured, historical raw and corrected data

GCMs and Scenarios	Maximum (mm)	Median (mm)	Mean (mm)	SD (mm)	CV	90th *Q* (mm)	WDP	RMSE (mm)	NSE	PBIAS (%)
Measured	147.00	46.61	54.75	5.07	3.23	5.32	0.31			
HadGEM2-ES Raw	119.32	40.85	45.62	4.91	3.54	5.36	0.38	54.05	−3.0	0.4
HadGEM2-ES Corrected	187.60	29.18	49.69	5.23	3.52	3.46	0.30	9.48	0.87	−9.3
ICHEC-EC-EARTH Raw	93.06	37.40	40.25	3.84	3.10	4.82	0.65	14.39	0.70	−13.9
ICHEC-EC-EARTH Corrected	149.30	51.43	42.44	4.82	3.27	4.81	0.31	3.72	0.98	−5.90
IPSL-CM5A-MR Raw	117.07	57.38	60.77	4.47	2.61	6.98	0.50	25.89	0.03	30.20
IPSL-CM5A-MR Corrected	149.71	45.78	50.39	4.78	3.24	4.44	0.31	4.48	0.97	−7.90
NorESM1-M Raw	110.36	34.60	39.00	4.37	3.83	4.49	0.56	11.39	0.81	−16.4
NorESM1-M Corrected	134.41	50.85	52.55	4.84	3.27	5.02	0.31	4.32	0.97	−3.90

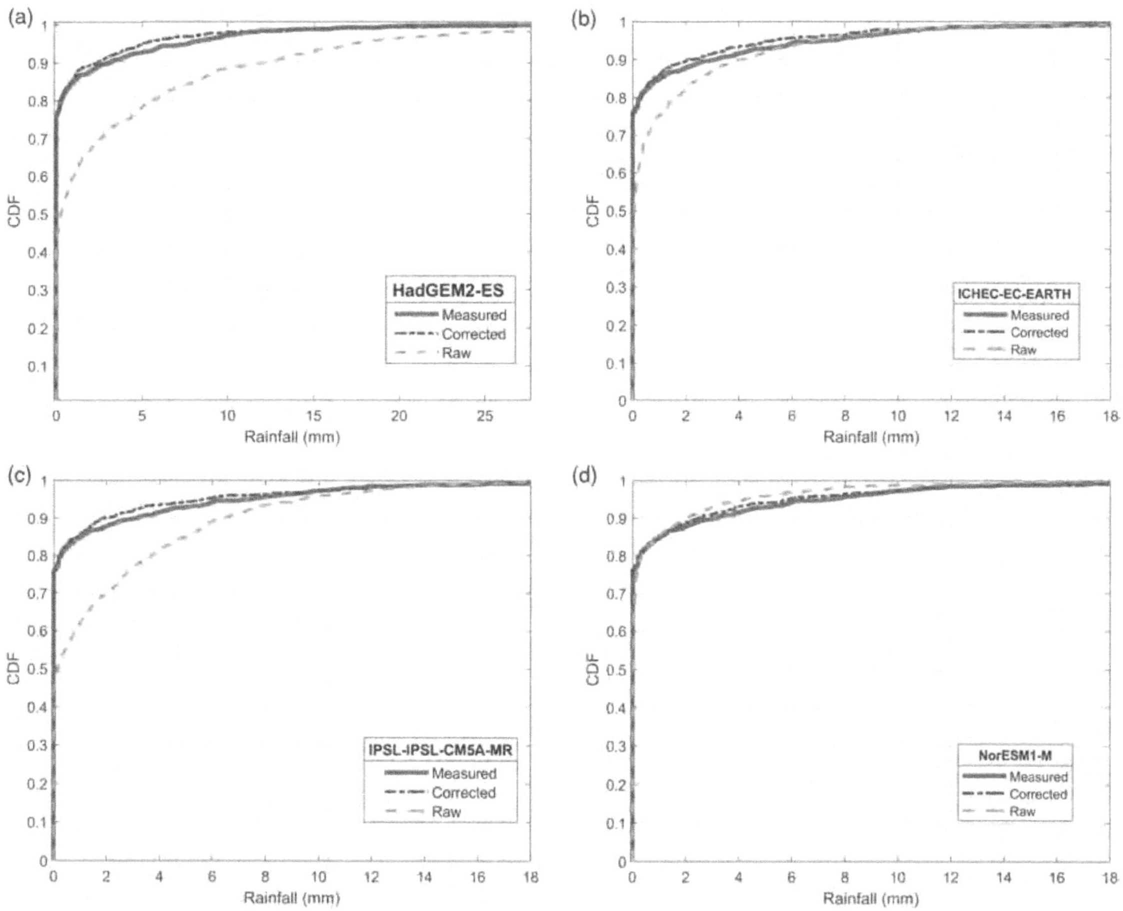

Figure 5 | CDF plots for measured, historical raw, and corrected data for May of (a) HadGEM2-ES, (b) ICHEC-EC-EARTH, (c) IPSL-IPSL-CM5A-MR, and (d) NorESM1-M.

an overestimation of measured data. What stands out in Table 4 is the corrected data of GCMs overestimate measured data, however, all corrected data estimate better than raw data except HadGEM2-ES. The most surprising aspect of the PBIAS values is that raw data from HadGEM2-ES have a closer value to zero. This can be considered surprising since all test values and statistical measures of HadGEM2-ES corrected data are in highly strong relationships with measured data except PBIAS. Figure 5 shows the CDFs of each GCM for May. According to CDF plots, all corrected data are in close relationship with measured data. Figure 6 reveals the variations of raw and corrected data compared with measured data

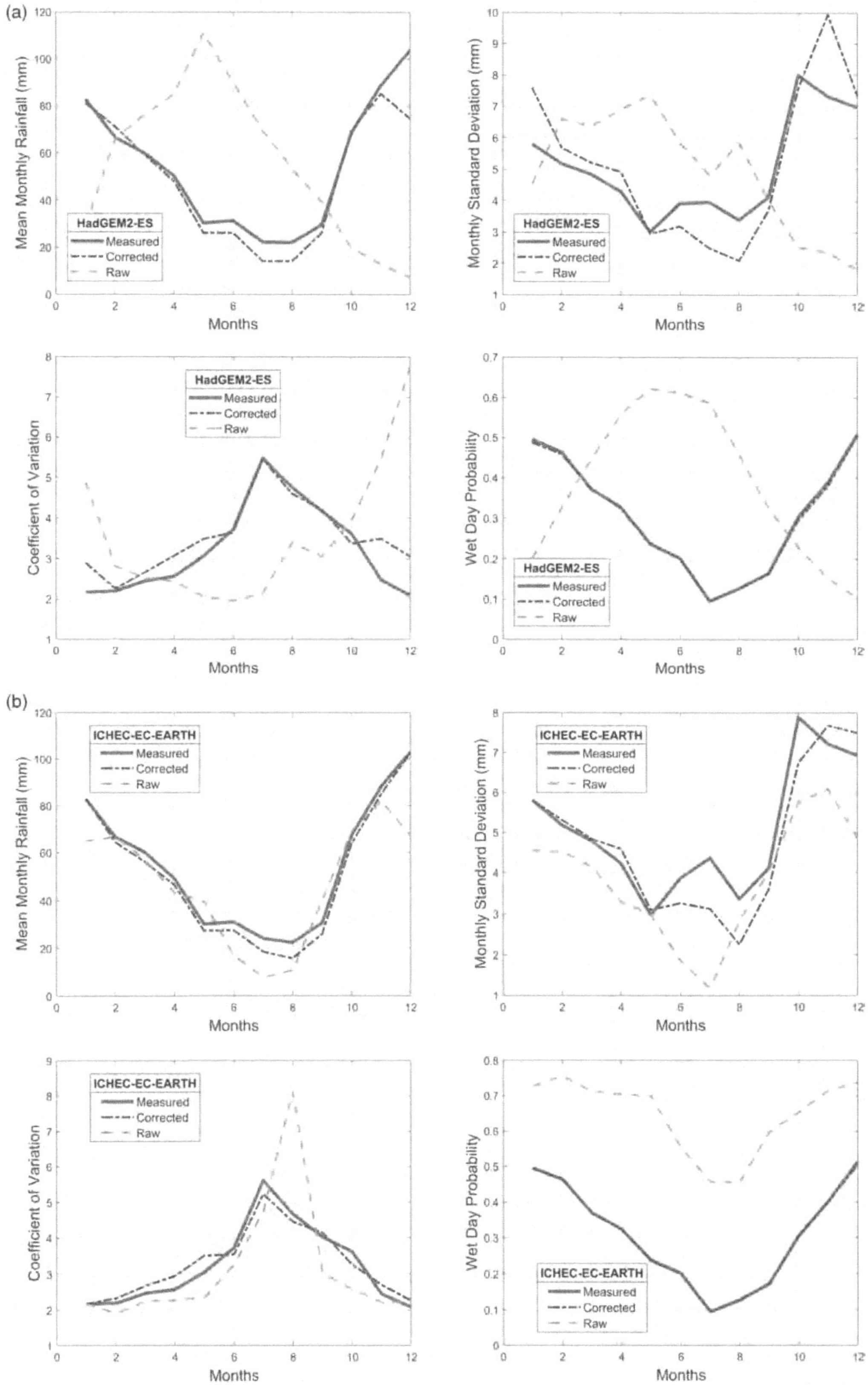

Figure 6 | Plots of monthly statistical measures of each GCM (a) HadGEM2-ES, (b) ICHEC-EC-EARTH, (c) IPSL-IPSL-CM5A-MR, and (d) NorESM1-M (*Continued*).

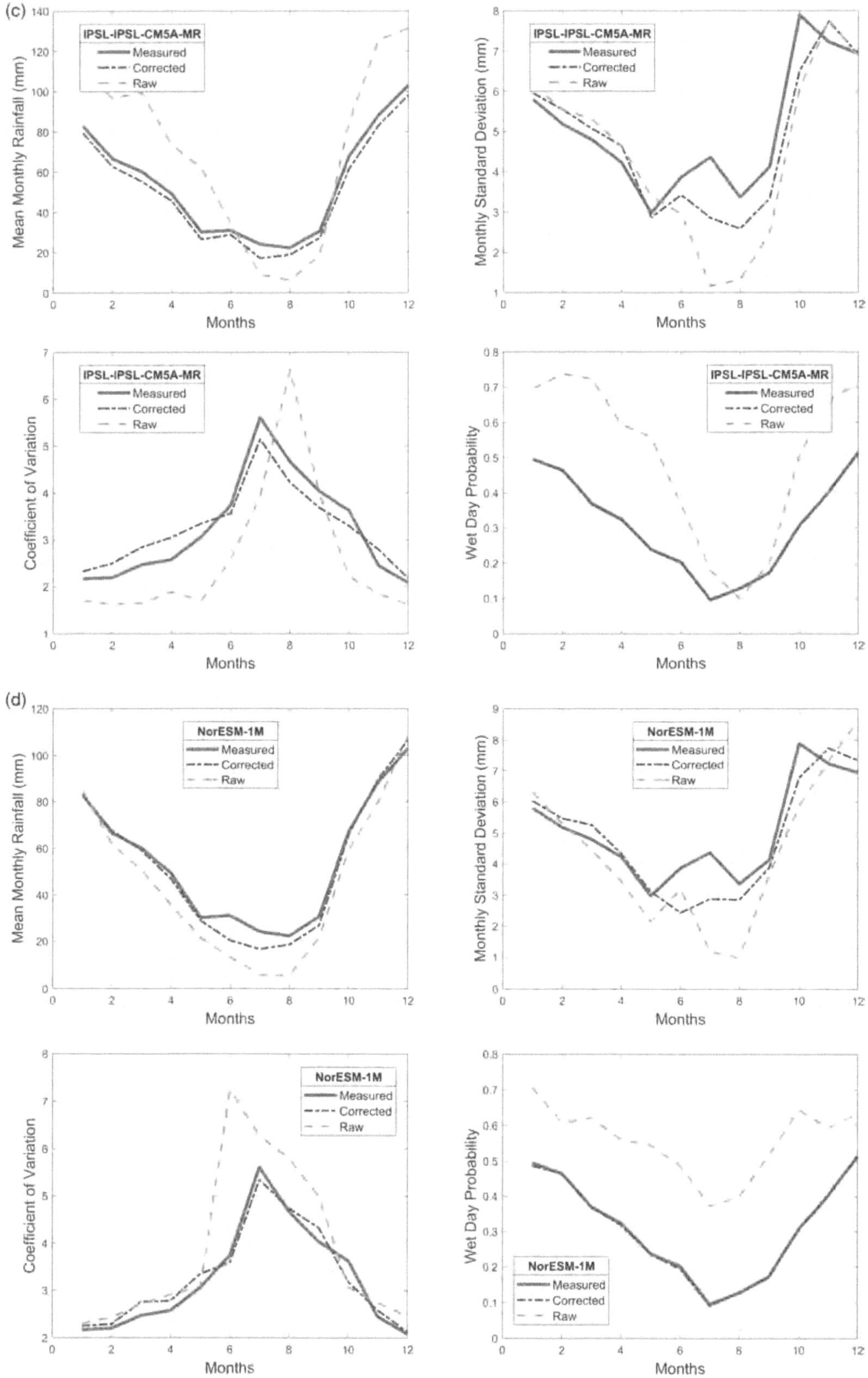

Figure 6 | (Continued.).

by plotting the statistical measures. As mentioned before, as we can see from the first chart of each GCM, mean monthly rainfalls of raw data show a very weak relationship to measured data. In a similar manner, SDs and CVs for corrected data show a better representation of measured data. As is very clear from WDP charts, corrected and measured have almost the same tendency, while raw data often overestimates wet days. Results conclude that the corrected data of HadGEM2-ES give the worst results compared with others. Even though the corrected data of the other three GCMs show a strong relationship with NSE ranging between 0.97 and 0.98, it is difficult to select the best one between them since they exhibit close values.

The findings of bias-correction analyses have shown that all GCMs show a tremendous representation of measured data when considered substantial measures and tests such as NSE, RMSE, PBIAS, mean monthly rainfalls and other statistics. Briefly, mean, standard deviation, and coefficient of variations were adjusted for almost all months thanks to distribution mapping. It can be said that GCM raw data are improved, and biases are corrected between measured and historical raw data. These results prove that distribution mapping is an effective method for bias-correction. Therefore, future GCMs can be used without any doubt to generate future IDF curves.

6.2. Performance of the disaggregation model

Before applying the disaggregation process for future GCMs, it is necessary to test the performance of the disaggregation model. Since IDF curves are generated using annual maxima for each duration, capturing great correlation between maximum values of measured and disaggregated rainfalls is sufficient for verification. Therefore, in this section, comparisons between maximum values of disaggregated and measured values were performed. Annual maximum rainfalls for each duration for the period of 1980–2005 were provided by TSMS. TSMS daily measured rainfalls (1980–2005) were disaggregated into sub-hourly rainfalls (5, 10, 15, and 30 min) using disaggregation parameters obtained from statistics of hourly rainfalls. Afterwards, these sub-hourly rainfalls were aggregated to longer durations (1, 2, 3, 4, 5, 6, 8, 12, 18, and 24 h). After that, scatter diagrams containing all maximum values of measured and disaggregated rainfalls are plotted in Figure 7. Correlation coefficients between the maxima of disaggregated rainfall and measured rainfall were found at 0.82. The range of correlation coefficients is between −1 and +1. A correlation coefficient higher than 0.8 indicates a fairly strong relationship. These comparisons revealed that there is a high correlation between measured and disaggregated rainfall intensities. Since the selected disaggregation model shows a good performance to obtain sub-hourly data from daily data, the process was applied for the disaggregation of daily future rainfall data, as well. The trend of IDF curves generated by TSMS using measured rainfall, and the trend of IDF curves generated by disaggregated rainfalls are plotted together in Figure 8.

6.3. Changes in rainfall intensities under future climate conditions

IDF curves were generated using the Gumbel distribution for selected return periods and durations for an ensemble of four GCMs using the methodology explained in 'Generating Future IDF Curves'. HadGEM2-ES, ICHEC-EC-EARTH, IPSL-IPSL-

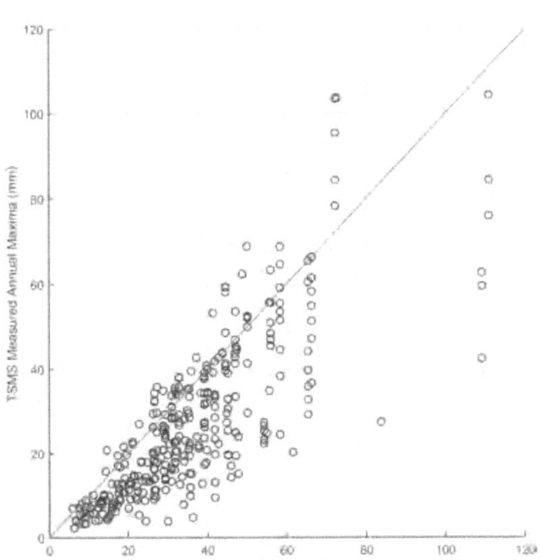

Figure 7 | Scatter plot of annual maxima of disaggregated and measured rainfalls for years 1980-2005.

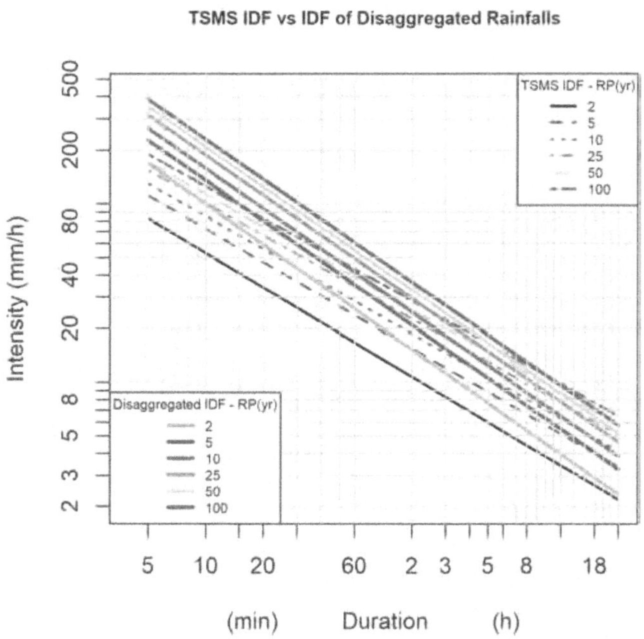

Figure 8 | IDF curve trends of TSMS and disaggregated rainfalls.

CM5A-MR, and NorESM1-M models were preferred. This section deals with the variations of future IDF curves of the near future (2036–2065) and distant future (2066–2095) with respect to TSMS IDF curves (generated by TSMS using measured rainfalls of 1938–2015). Besides, differences between different GCMs and RCP scenarios were exhibited. RCP4.5 and RCP8.5 scenarios were employed to generate GCM. IDF values of measured rainfalls (TSMS) for Florya station are given in Table 5.

In total, 16 different future IDF tables were generated for Florya station (four GCMs, two RCPs, and two future periods). IDF curves were generated for all return periods mentioned before. However, to increase the simplicity of understanding of

Table 5 | TSMS IDF values of Florya Station

Rainfall intensities (mm/h)

	Return periods					
Durations	**2-year**	**5-year**	**10-year**	**25-year**	**50-year**	**100-year**
5 min	73	107	127	151	168	183
10 min	52	77	95	116	131	146
15 min	42	63	76	93	105	117
30 min	28	42	51	64	73	83
1 h	17	25	31	38	44	49
2 h	10	15	19	24	27	31
3 h	8	11	14	18	21	25
4 h	6	9	12	15	18	20
5 h	5	8	10	13	16	18
6 h	5	7	9	12	14	16
8 h	4	6	7	9	11	13
12 h	3	4	5	7	8	10
18 h	2	3	4	5	6	8
24 h	2	3	3	4	5	6

analyses, tables and figures were shown for a small return period (10 years) and a longer return period (100 years). The first findings of the analyses are summarized in Table 6 for both periods of 2036–2065 and 2066–2095. Table 6 includes average percentage changes in the future IDF values of each GCM and RCP compared with TSMS IDF curves for 10-year and 100-year return periods.

As seen in Table 6, future rainfall intensities tend to increase for all return periods except the 100-year return period of NorESM1-M under RCP8.5 for 2036–2065, and 100-year return period for ICHEC-EC-EARTH under RCP4.5 for 2066–2095. The relative change of the 10-year return period ranges between 17 and 88.6%; changes range between −1.7 and 66.7% for the 100-year return period for 2036–2065. For 2066–2095, 10-year changes range between 7.3 and 106.2%, while 100-year ranges between −11.6 and 61.4%. HadGEM2-ES shows the highest increase for both future periods and for both return periods. Under RCP8.5 scenarios of HadGEM2-ES, rainfall intensities show slight increases compared with RCP4.5 except 10-year for 2066–2095. The 100-year return period of the ICHEC-EC-EARTH RCP4.5 scenario shows the highest decline. For small periods (10-year) return periods, the increase in future rainfall intensities is higher than longer periods (100-year). 10-year intensities tend to increase more for the distant future (2066–2095) than the near future (2036–2065). Most of the RCP8.5 scenarios show a higher increase compared with RCP4.5. This also supports the previous findings in the literature (Xin *et al.* 2013; Nasim *et al.* 2018). However, for some GCMs and return periods, it is possible to see higher intensities under RCP4.5 compared with RCP8.5 as in the study of Singh *et al.* (2016). Similarly, most of the future rainfalls are increasing compared with measured ones; however, there are reductions for several periods in rainfall intensities in the future. A possible explanation for this might be that some GCMs can underestimate rainfalls due to various factors such as diversity of climatic conditions of different regions in the world and uncorrected biases as mentioned in the study of Mirhosseini *et al.* (2013).

Tables 7 and 8 show the average percentage changes in rainfall intensities for both 10-year and 100-year return periods for all GCMs and RCPs compared with measured values in terms of short- and long-duration rainfall intensities for the first future period (2036–2065) and the second future period (2066–2095), respectively. Short-duration rainfall indicates rainfall durations of less than 4 h, and long durations indicate rainfall durations 4 h and longer than 4 h as is suggested in the study by Mirhosseini *et al.* (2013). It is apparent from Tables 6 and 7 that short-duration rainfalls have a rising tendency for all return periods and GCMs in both the near and distant futures. As mentioned before, HadGEM2-ES scenarios have the highest increase compared with other GCMs for both short and long durations for the two return periods. The highest increase in the near future is in HadGEM2-ES under the RCP8.5 scenario with 96.06% and is HadGEM2-ES under RCP4.5 with 116.76% for the distant future. For 2036–2065, IPSL-IPSL-CM5A-MR RCP4.5 has the lowest increase with 29.12% for 10-year short-duration rainfalls, while it is ICHEC-EC-EARTH under the RCP4.5 scenario with 17.31% for 2066–2095. For the 100-year return period in 2036–2065, IPSL-IPSL-CM5A-MR RCP4.5 has the lowest increase with 16.27%. For the 100-year return period, ICHEC-EC-EARTH RCP4.5 is increasing less than other GCMs and RCPs with 5.11% for

Table 6 | Average percentage changes of future IDF values (mm/h) compared with TSMS IDF (mm/h) values for 10-year and 100-year return periods for future periods (2036–2065) and (2066–2095)

Percentage change in rainfall intensities (%)

	(2036–2065) (%)		(2066–2095) (%)	
	Return periods		Return periods	
GCMs and Scenarios	10-year	100-year	10-year	100-year
HadGEM2-ES RCP4.5	88.3	66.3	106.2	61.4
HadGEM2-ES RCP8.5	88.6	66.7	105.8	65
ICHEC-EC-EARTH RCP4.5	25.4	9.9	7.3	−11.6
ICHEC-EC-EARTH RCP8.5	24.1	7.6	27.8	10.7
IPSL-IPSL-CM5A-MR RCP4.5	19.9	1.3	30.9	14.1
IPSL-IPSL-CM5A-MR RCP8.5	28.9	12.4	31.4	12.7
NorESM-1M RCP4.5	25.1	10.3	26.5	7.5
NorESM-1M RCP8.5	17	−1.7	31	9.7

Table 7 | Average percentage changes of future IDF values (mm/h) compared with TSMS IDF values (mm/h) in terms of short-duration rainfall (SDR) and long-duration rainfall (LDR) for the first period (2036–2065)

	Change in SDR (%)		Change in LDR (%)	
	Return periods		Return periods	
GCMs and Scenarios (2036–2065)	10-year	100-year	10-year	100-year
HadGEM2-ES RCP4.5	91.16	80.33	85.48	52.21
HadGEM2-ES RCP8.5	96.06	87.07	81.16	46.33
ICHEC-EC-EARTH RCP4.5	44.85	36.75	5.94	−16.94
ICHEC-EC-EARTH RCP8.5	35.48	26.92	12.69	−11.77
IPSL-IPSL-CM5A-MR RCP4.5	29.12	16.27	10.77	−13.58
IPSL-IPSL-CM5A-MR RCP8.5	42.97	33.55	14.88	−8.80
NorESM-1M RCP4.5	43.58	36.95	6.62	−16.29
NorESM-1M RCP8.5	32.30	19.53	1.79	−22.84

Table 8 | Average percentage changes of future IDF values (mm/h) compared with TSMS IDF values (mm/h) in terms of short-duration rainfall (SDR) and long-duration rainfall (LDR) for the second period (2066–2095)

	Change in SDR (%)		Change in LDR (%)	
	Return periods		Return periods	
GCMs and Scenarios (2066–2095)	10-year	100-year	10-year	100-year
HadGEM2-ES RCP4.5	116.76	82.66	95.71	40.14
HadGEM2-ES RCP8.5	113.60	85.81	97.99	44.16
ICHEC-EC-EARTH RCP4.5	17.31	5.11	−2.74	−28.29
ICHEC-EC-EARTH RCP8.5	30.96	22.31	24.72	−0.97
IPSL-IPSL-CM5A-MR RCP4.5	41.99	33.18	19.81	−5.02
IPSL-IPSL-CM5A-MR RCP8.5	41.38	30.69	21.39	−5.31
NorESM-1M RCP4.5	39.48	27.79	21.39	−5.31
NorESM-1M RCP8.5	44.73	29.98	17.26	−10.58

2066–2095. RCP4.5 rainfall intensities could decrease below and increase above RCP8.5. However, RCP8.5 scenarios generally have higher intensities than RCP4.5. Similar to Table 6, rainfall intensities for smaller periods will increase more than longer periods for short-duration rainfalls. When both the near and distant futures are compared, 10-year rainfall intensities in the distant future tend to increase more than the near future for all RCPs of HadGEM2-ES and ICHEC-EC-EARTH. For other GCMs, there is no constant tendency for the increases since they are changing depending on RCPs and GCMs.

For long durations, we can see that HadGEM2-ES models have the highest increases compared with other GCMs. The highest increase in the near future for 10-year and 100-year belongs to HadGEM2-ES RCP4.5 with 85.48 and 52.21%, respectively. For the distant future, HadGEM2-ES RCP8.5 scenarios show the highest change for 10-year and 100-year return periods with 97.99 and 44.16%, respectively. Contrary to short durations, all GCMs except HadGEM2-ES tend to see a reduction in rainfall intensities for the 100-year return period for both future periods. For the 10-year return period, only ICHEC-EC-EARTH under RCP4.5 is showing a decline for 10-year in the distant future. Therefore, we see higher increases for smaller return periods compared with the 100-year return period for longer durations. This finding also supports the findings of Table 6. NorESM1-M has the highest decrease in the near future, while ICHEC-EC-EARTH under RCP4.5 is decreasing more than others for the distant future. The range of changes in 10-year short-duration rainfall intensities are between 29.12 and 96.06%, and 30.96 and 116.76% for the near future and distant future, respectively. In the short duration of the 100-year return period, the range of changes in the near future is between 16.27 and 87.07%, and for the distant future, the range is between 5.11 and 85.81%.

Average percentage changes for two return periods (10-year and 100-year) and six durations (5 min, 15 min, 1 h, 3 h, 6 h and 24 h) are shown in Figure 9. It shows the percentages for selected durations and return periods for each GCM. For comparison, 5, 15 min and 1, 3-h rainfalls were selected to represent short-duration, 6 and 24-h rainfalls were selected for long-

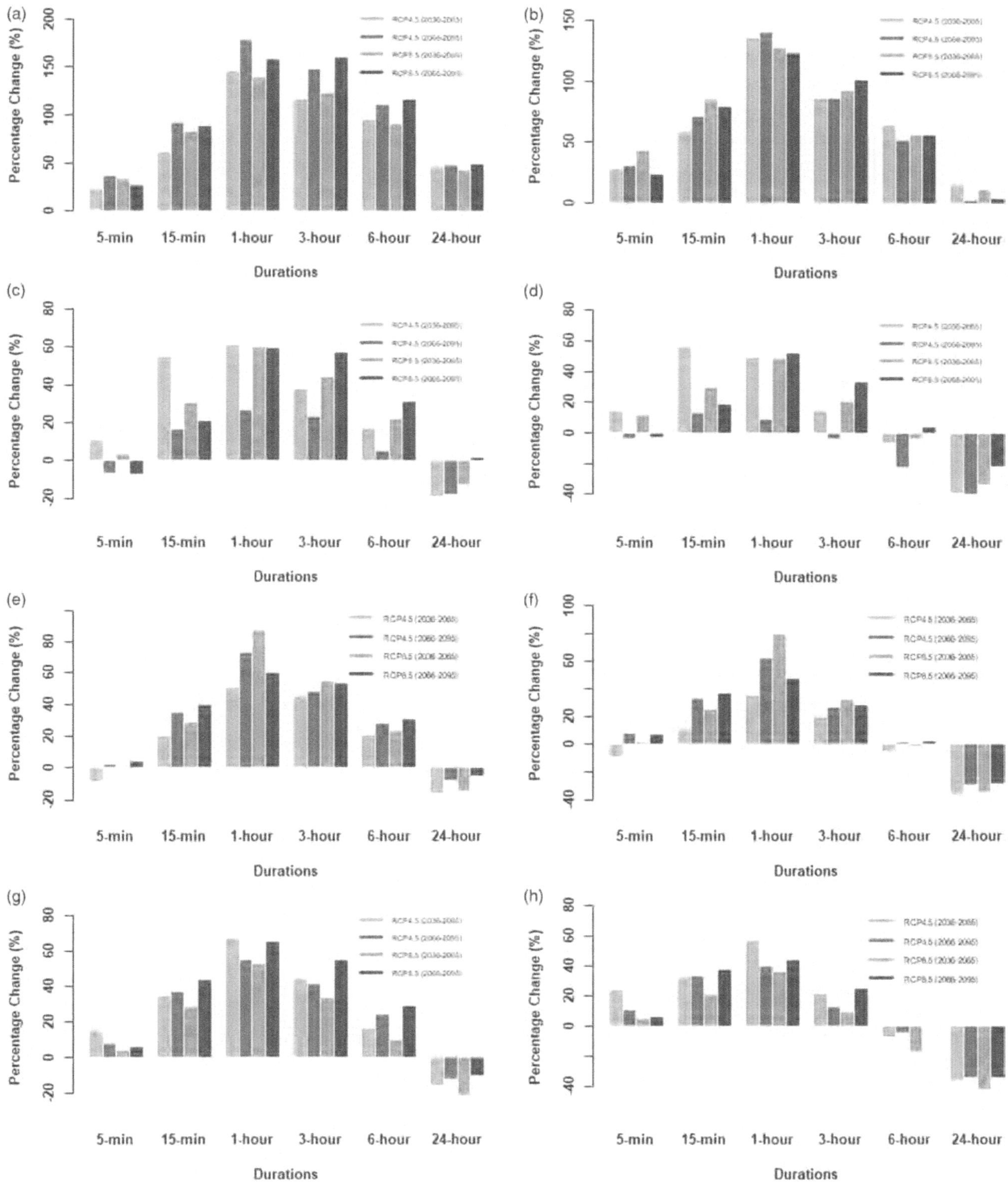

Figure 9 | Average percentage changes for different durations, periods and RCPs for (a) HadGEM2-ES 10-year return period, (b) HadGEM2-ES 100-year return period, (c) ICHEC-EC-EARTH 10-year return period, (d) ICHEC-EC-EARTH 100-year return period, (e) IPSL-IPSL-CM5A-MR 10-year return period, (f) IPSL-IPSL-CM5A-MR 100-year return period, (g) NorESM1-M 10-year return period and (h) NorESM1-M 100-year return period.

duration rainfalls. As mentioned in the findings of Tables 6–8, a great majority of short-duration rainfall intensities except several GCMs tend to increase in the future. It is also apparent that long-duration rainfall intensities, especially in 24-h, are encountering a great decline for all GCMs. However, 6-h values are increasing for most GCMs. As seen from Figure 9, HadGEM2-ES is the only one with a continuous increase. Another significant finding in Figure 9 is that 1-h values show increases more than other selected durations for all GCMs.

These analyses conclude that we will see highly intensified rainfalls for short-duration (less than 4 h) both in the near and distant futures, while the same cannot be said for long durations. Besides, increases in 10-year return periods are higher than the 100-year return period, which means that rainfall intensities that we are more likely to see will be greater than the rare rainfall intensities for both short and long durations, and for both future periods. The chance of rainfall intensities decreasing in the future is low. The most interesting aspect of the findings is we can see much higher short-duration rainfalls with a high possibility in both the near and distant futures.

Since there will be 16 IDF curve plots, we decided to reduce plot numbers to be shown by eliminating half of them. Rainfall intensity trends for TSMS IDF and future IDF curves are shown in Figure 10. The figure shows the various combinations of IDF curves generated with different RCPs and future periods (2036–2065) and (2066–2095) for all GCMs.

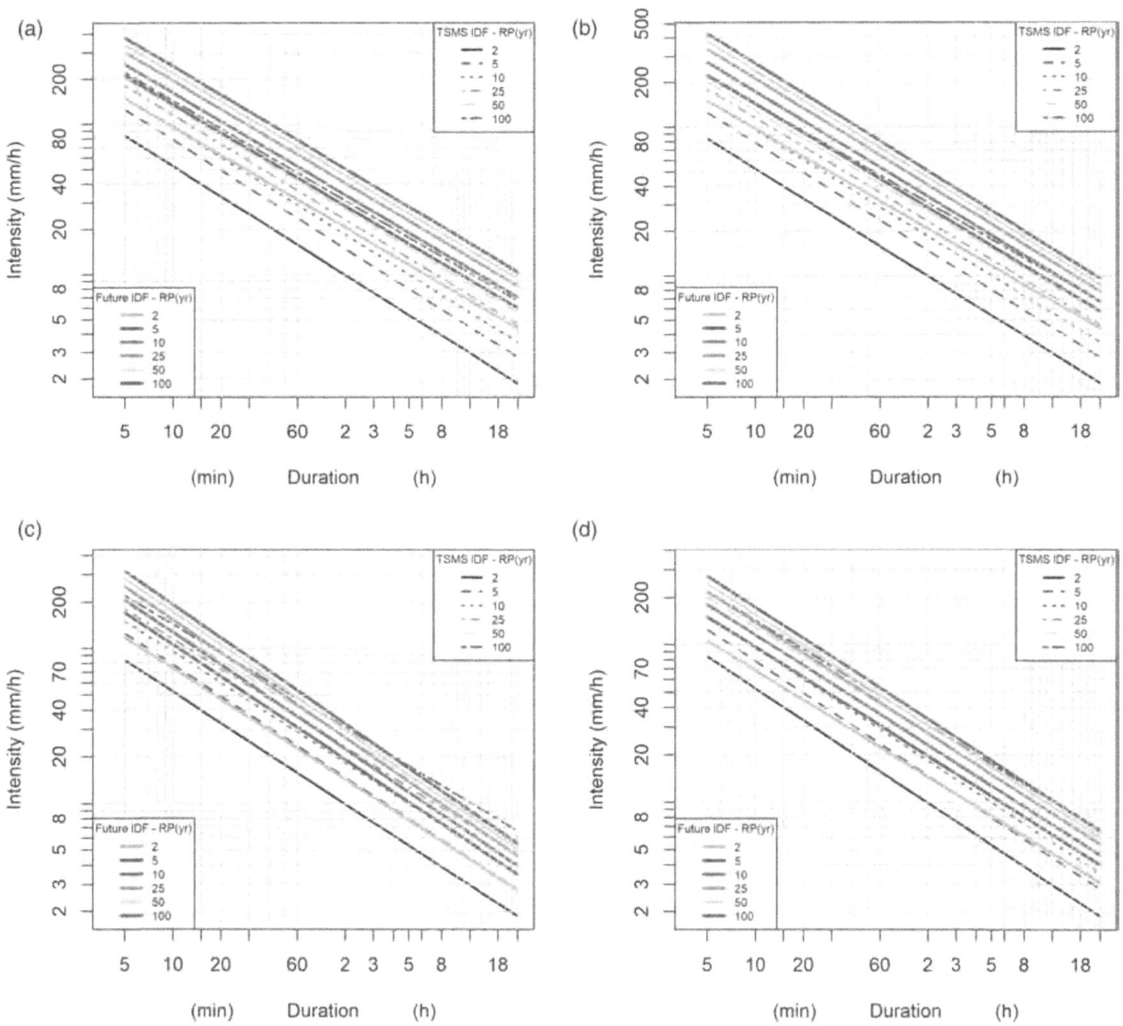

Figure 10 | IDF curve trends for selected future GCMs, periods, and RCPs compared to TSMS IDF curves: (a) TSMS vs HadGEM2-ES RCP4.5 (2036-2065), (b) TSMS vs HadGEM2-ES RCP8.5 (2036-2065), (c) TSMS vs ICHEC-EC-EARTH RCP8.5 (2036-2065), (d) TSMS vs ICHEC-EC-EARTH RCP8.5 (2066-2095), (e) TSMS vs IPSL-IPSL-CM5A-MR RCP4.5 (2066-2095), (f) TSMS vs IPSL-IPSL-CM5A-MR RCP8.5 (2066-2095), (g) TSMS vs NorESM1-M RCP4.5 (2036-2065) and (h) TSMS vs NorESM1-M RCP4.5 (2066-2095) (*Continued*).

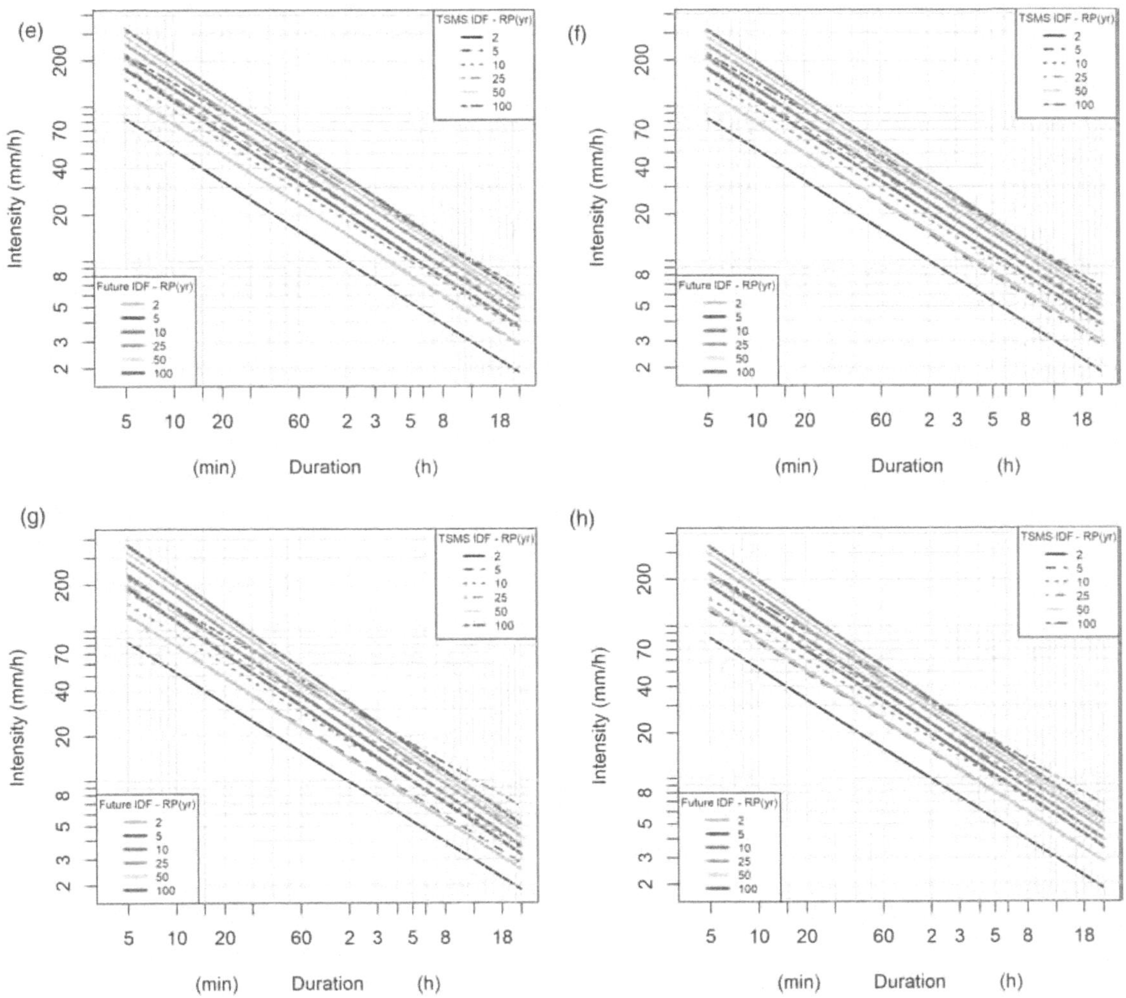

Figure 10 | *(Continued.)*.

7. CONCLUSION

The aim of the present study was to generate reliable and updated rainfall IDF curves to be used as a serviceable tool in designing urban drainage systems against urban floods. Rainfall IDF curves play a crucial role in assessing flood risk and quantifying heavy rainfalls. IDF curves are also effective tools in designing and operating urban drainage systems. Since flash floods occur due to short-duration and high intensity rainfalls, the assessment of short-duration rainfalls was a major necessity. Hence, the first objective of the study was to generate IDF curves with both short-duration (less than 4 h) and long-duration (4 h and longer than 4 h) rainfalls. However, generating stationary IDF curves is not reliable in the assessment of floods since the climate of our world is changing continuously. Rainfalls are getting heavier due to climate change and floods are getting more hazardous for human life and cities. Besides, urbanization is also increasing impermeable surfaces and enabling floods to be more destructive. Therefore, IDF curves were generated especially for short-duration rainfalls to cope with all negative effects of climate change, urbanization, and insufficient drainage systems.

In the study, an ensemble of four GCMs, HadGEM2-ES, ICHEC-EC-EARTH, IPSL-IPSL-CM5A-MR and NorESM1-M, were provided for two different future periods 2036–2065 and 2066–2095. RCP4.5 and RCP8.5 scenarios were selected to obtain GCMs to represent future daily rainfall data. Since GCMs are not suitable to use directly due to biases, a bias-correction process was done. The distribution mapping method performed well (NSE ranges between 0.87 and 0.98) to adjust biases between measured and historical raw data.

In this study, disaggregation parameters were obtained from 14-year measured hourly rainfall, and parameter sets were determined for each month using some statistical measures. Afterwards, the determined parameters were applied for the disaggregation of measured (applied to test the method) and future daily rainfalls. Hence, this process gives a chance for future data to capture historical patterns of selected region's rainfall instead of random rainfall characteristics. The results concluded that there is a close relationship between maximum values of TSMS measured rainfalls and maximum values of disaggregated rainfalls. Briefly, it is possible to disaggregate any daily rainfalls of a large region or city using only short records of hourly rainfall.

The results have shown that we will see much higher increases in short-duration rainfall intensities compared with longer durations for both the near and distant futures. Long-duration rainfalls tend to increase for 10-year return periods, while there are declines for the 100-year period for all GCMs except for HadGEM2-ES. These increases in short-duration rainfall intensities are valid for smaller and longer return periods. However, increases in smaller return periods are higher than longer return periods. This means that rainfall intensities that we are more likely to see will be greater than the rare rainfall intensities. Increases in short-duration rainfall intensity rise to 96.06% for 2036–2065 and 116.76% for 2066–2095, respectively. The main conclusion of the study is that short-duration rainfalls will be intensified in the near and distant future with a high possibility.

The importance of short-duration rainfalls on urban floods was explained in the previous sections. Long-duration rainfalls usually cause slower and less hazardous floods compared with short-duration floods. On the contrary, flash floods are increasing and getting more hazardous day after day. Hence, based on the results that the short-duration rainfalls will be more intensified, it is strongly recommended to use updated IDF curves in the assessment of flood risk, designing, and operating of urban drainage systems. In addition, flood management standards in cities must be rearranged considering IDF curves.

This study once again proved the need to use an updated IDF curve, which is generated under future climate conditions, for hydrology, hydraulic, and other water-related applications. The results also prove the necessity of selecting several GCMs for the analyses since every GCM shows major differences to others. Likewise, different disaggregation methods can simulate sub-hourly rainfall data in different ways. Therefore, future studies can be performed for more stations to enrich the awareness of climate change by evaluating more GCMs, disaggregation methods, and distribution functions.

DATA AVAILABILITY STATEMENT

Data cannot be made publicly available; readers should contact the corresponding author for details.

REFERENCES

BBC Weather 2020 *Weather – Istanbul*. Available from: https://www.bbc.com/weather/745044.

Berhanu, B., Seleshi, Y., Demisse, S. S. & Melesse, A. M. 2016 Bias correction and characterization of climate forecast system re-analysis daily precipitation in Ethiopia using fuzzy overlay. *Meteorological Applications* **23** (2), 230–243.

Cheng, L. & AghaKouchak, A. 2014 Nonstationary precipitation intensity-duration-frequency curves for infrastructure design in a changing climate. *Scientific Reports* **4** (1), 1–6.

Chow, V. T. 1951 A general formula for hydrologic frequency analysis. *Eos, Transactions American Geophysical Union* **32** (2), 231–237.

de Souza Costa, C. E. A., Blanco, C. J. C. & de Oliveira-Júnior, J. F. 2020 IDF curves for future climate scenarios in a locality of the Tapajós Basin, Amazon, Brazil. *Journal of Water and Climate Change* **11** (3), 760–770. https://doi.org/10.2166/wcc.2019.202.

Edenhofer, O., Pichs-Madruga, R., Sokona, Y., Minx, J. C., Farahani, E., Kadner, S. & Seyboth, K. 2015 Climate Change 2014 Mitigation of Climate Change Summary for Policymakers and Technical Summary, Working Group III Contribution to the Fifth Assessment Report of the Intergovernmental Panel on Climate Change (IPCC). In: *Annual Review of Environment and Resources*, Vol. 40, Issue 1. Available from: http://www.annualreviews.org/doi/10.1146/annurev-environ-021113-095626.

Hajani, E. 2020 Climate change and its influence on design rainfall at-site in New South Wales State, Australia. *Journal of Water and Climate Change* **11** (S1), 251–269.

Hanaish, I. S., Ibrahim, K. & Jemain, A. A. 2011 Daily rainfall disaggregation using HYETOS model for Peninsular Malaysia. *Matrix* **2**, 1.

Hosseinzadehtalaei, P., Tabari, H. & Willems, P. 2020 Climate change impact on short-duration extreme precipitation and intensity–duration–frequency curves over Europe. *Journal of Hydrology* **590**, 125249.

IPCC. 2014 *Climate Change 2014: Synthesis Report. Contribution of Working Groups I, II and III to the Fifth Assessment Report of the Intergovernmental Panel on Climate Change* (R. K. Pachauri & L.A. Meyer eds.). IPCC, Geneva, Switzerland.

Koutsoyiannis, D. & Onof, C. 2001 Rainfall disaggregation using adjusting procedures on a Poisson cluster model. *Journal of Hydrology* **246** (1–4), 109–122. https://doi.org/10.1016/S0022-1694(01)00363-8.

Mirhosseini, G., Srivastava, P. & Stefanova, L. 2013 The impact of climate change on rainfall Intensity–Duration–Frequency (IDF) curves in Alabama. *Regional Environmental Change* **13** (S1), 25–33. https://doi.org/10.1007/s10113-012-0375-5.

Nash, J. E. & Sutcliffe, J. V. 1970 River flow forecasting through conceptual models. Part I—a discussion of principles. *Journal of Hydrology* **10** (3), 282–290.

Nasim, W., Amin, A., Fahad, S., Awais, M., Khan, N., Mubeen, M., Wahid, A., Rehman, M. H., Ihsan, M. Z. & Ahmad, S. 2018 Future risk assessment by estimating historical heat wave trends with projected heat accumulation using SimCLIM climate model in Pakistan. *Atmospheric Research* **205**, 118–133.

Nwaogazie, I. L., Sam, M. G., Enciso, R. Z. & Gonsalves, E. 2019 Probability and non-probability rainfall intensity-duration-frequency modeling for Port-Harcourt metropolis, Nigeria. *International Journal of Hydrology* **3** (1). https://doi.org/10.15406/ijh.2019.03.00164.

Oyebande, L. 1982 Deriving rainfall intensity-duration-frequency relationships and estimates for regions with inadequate data. *Hydrological Sciences Journal* **27** (3), 353–367.

Rathjens, H., Bieger, K., Srinivasan, R. & Arnold, J. G. 2016 *CMhyd User Manual Documentation for Preparing Simulated Climate Change Data for Hydrologic Impact Studies.*

Rodriguez-Iturbe, I., Cox, D. R. & Isham, V. 1987 Some models for rainfall based on stochastic point processes. *Proceedings of the Royal Society of London A: Mathematical and Physical Sciences* **410** (1839), 269–288.

Rodriguez-Iturbe, I., Cox, D. R. & Isham, V. 1988 A point process model for rainfall: further developments. *Proceedings of the Royal Society of London A: Mathematical and Physical Sciences* **417** (1853), 283–298.

Sarhadi, A. & Soulis, E. D. 2017 Time-varying extreme rainfall intensity-duration-frequency curves in a changing climate. *Geophysical Research Letters* **44** (5), 2454–2463.

Singh, R., Arya, D. S., Taxak, A. K. & Vojinovic, Z. 2016 Potential impact of climate change on rainfall intensity-duration-frequency curves in Roorkee, India. *Water Resources Management* **30** (13), 4603–4616. https://doi.org/10.1007/s11269-016-1441-4.

Teutschbein, C. & Seibert, J. 2010 Regional climate models for hydrological impact studies at the catchment scale: a review of recent modeling strategies. *Geography Compass* **4** (7), 834–860. https://doi.org/10.1111/j.1749-8198.2010.00357.x.

Thom, H. C. S. 1958 A note on the gamma distribution. *Monthly Weather Review* **86** (4), 117–122.

Turkish State Meteorological Service 2020 *Resmi İstatistikler (İl ve İlçelerimize Ait İstatistiki Veriler).* Available from: https://web.archive.org/web/20180422234324/https:/mgm.gov.tr/veridegerlendirme/il-ve-ilceler-istatistik.aspx?k=A&m=ISTANBUL.

Xin, X., Zhang, L., Zhang, J., Wu, T. & Fang, Y. 2013 Climate change projections over East Asia with BBC_CSM1.1 climate model under RCP scenarios. *Journal of the Meteorological Society of Japan* **91** (4), 413–429. https://doi.org/10.2151/jmsj.2013-401.

Zhu, J., Stone, M. C. & Forsee, W. 2012 Analysis of potential impacts of climate change on intensity–duration–frequency (IDF) relationships for six regions in the United States. *Journal of Water and Climate Change* **3** (3), 185–196. https://doi.org/10.2166/wcc.2012.045.

First received 15 June 2021; accepted in revised form 21 October 2021. Available online 29 October 2021

Impact of Climate Change on Hydrology and Water Resources doi: 10.2166/wcc.2021.348

Temporal change of extreme precipitation intensity–duration–frequency relationships in Thailand

N. Yamoat[a,*], R. Hanchoowong[b], S. Sriboonlue [c] and A. Kangrang [d]

[a] Department of Civil and Environmental Engineering Technology, King Mongkut's University of Technology North Bangkok, 1518 Pracharat 1 Road, Wongsawang, Bangsue, Bangkok 10800, Thailand
[b] Department of Civil Engineering, Mahanakorn University of Technology, 140 Cheum-Sampan Road, Nong Chok, Bangkok 10530, Thailand
[c] Arcadis U.S., Suite 200, Irvine, CA 92602, USA
[d] Faculty of Engineering, Mahasarakham University, Khamriang, Kantarawichai, Maha Sarakham 44150, Thailand
*Corresponding author. E-mail: nirat.y@cit.kmutnb.ac.th

SS, 0000-0002-4927-5326; AK, 0000-0003-2575-264X

ABSTRACT

Due to climate change, many research studies have derived the updated extreme precipitation intensity–duration–frequency relationship (IDF curve) from forecasted sub-hourly rainfall intensity time series, which is one of the most important tools for the planning and designing of hydraulic infrastructures. In this study, the IDF curves (1990–2016) of the six regions and procedures are used in accordance with those of the Royal Irrigation Department (RID)'s study (1950–1988). Each set of IDF relationships consists of 81 intensity values which are the combination of nine durations and nine return periods. The intensity ratios of this study and RID are compared. A greater-than-1 ratio indicates extreme intensity increment from the past to the present. Considering 81 ratios for each region, the number of greater-than-1 ratios for the North, Northeast, Central, East, West, and South regions are 8, 2, 31, 34, 6, and 7, respectively. These ratio numbers are far below 81 which means that the majority of extreme rainfall intensities do not increase from the past to the present. The study found that using accurate historical sub-hourly rainfall time series to create a set of IDF curves would be more reliable than using forecasted rainfall modeling.

Key words: extreme precipitation intensity, IDF relationship, temporal change, Thailand

HIGHLIGHTS

- Using accurate historical sub-hourly rainfall time series to create a set of IDF curves would be more reliable than using forecasted rainfall modeling.
- The future period intensities from eight global climate models are closer to the base period intensities than the ones from a global climate model.
- The majority of recent and past IDF curves did not increase as expected.

INTRODUCTION

Expected rainfall intensity is required and used in the planning and design of engineering and environmental infrastructures in that it is termed the design storm (Balbastre-Soldevila *et al.* 2019; Cook *et al.* 2020). The appropriate design storm must be neither too large nor too small. When too large, the concerned infrastructure will be unnecessarily overdesigned and expensive, while when too small it will risk the damage caused by flood or heavy rain. A suitable design storm can be estimated from the relationship of extreme rainfall intensity, its duration, and frequency of occurrence. Generally, the shorter duration of the storm, the heavier the rainfall. Also, the heavier the storm, the less frequently it occurs. The imperative tool for planners and engineers to choose the appropriate design storm is, therefore, an accurate maximum rainfall intensity–duration–frequency (IDF) relationship (Mirhosseini *et al.* 2012; Sun *et al.* 2019). It is actually the relationship of the maximum rainfall intensity (or depth) at a specific duration and a specific return period. It can be in the form of the table of maximum intensity values at several durations and return periods or a set of curves of maximum intensity as a function of duration at several return periods, a so-called IDF curve. A return period is an average length of time that a specified, or higher, rainfall intensity occurs at a particular place with a specified duration.

An IDF curve was initiated by Sherman (1931) and Bernard (1932) and since then it has been extensively used in the USA (Hershfield 1962) and is spreading worldwide (Bezak *et al.* 2018). In Thailand, Mustonen (1969) originally proposed 14 curves, i.e., five curves for the Northeast, two each for the Northern, Central, Eastern, and Southern regions, and just one curve for the West. The Royal Irrigation Department (RID), the main responsible institute for irrigation and drainage in Thailand, issued a detailed study of IDF curves and specified a total of 59 curves for the whole country. They are 20 for the North, 10 for the Northeast, 5, 8, 4, and 12 for the Central, East, West, and South, respectively (Bumpenkit 1999). The average representative IDF curves for each of the five regions were also presented.

An IDF relationship is usually derived from historical data of sub-hourly rainfall intensity time series, usually 15 min in Thailand, for many years of records (Bumpenkit 1999; Shrestha *et al.* 2017). Annual maximum intensity values at each of several durations, e.g., 0.25, 0.5, 0.75, 1, 2, 3, 6, 12, and 24 h are extracted from the time series and then being used for constructing IDF curves (Koutsoyiannis *et al.* 1998). After the last few decades, Thailand as well as other countries experienced flooding more frequently in both urban and rural areas and climate change was thought to be one of the causes (Shrestha *et al.* 2017; Choi *et al.* 2019). It has been suspected that the official IDF curves for each country might be underestimated for the designed storms that cause more frequent flooding due to the global warming. Climate change with the increase of greenhouse gas is however expected to result in higher maximum rainfall intensity in the future, since a warmer atmosphere raises its water holding capacity by 7% for every 1% of rising temperature (Trenberth 2011; Cheng & AghaKouchak 2014; Ganguli & Coulibaly 2017). The earth's mean temperature has been rising for 1.2–1.5 °C since a the pre-industrialized period (Hu & Ayub 2019; Jacobson *et al.* 2019). This impelled a lot of researchers to create new sets of IDF curves in parallel with future rainfall data which are not real and not yet existing (Noor *et al.* 2018). Thus, we should create the most accurate and precise future rainfall data from the global climate modeling (GCM).

Future rainfall data can be created via GCM or regional climate modeling (RCM) (Agilan & Umamahesh 2016; Kohnova *et al.* 2019). Both types of models are similar in all aspects except that the global models cover or nearly cover the whole world, while the regional models involve only portions of the earth. An RCM is always derived from a GCM to be able to apply for an area of interest. Several institutes produced future climate data either from GCMs or RCMs depending on their interest (Moriondo & Bindi 2006; Chokkavarapu & Mandla 2019). They simulate the interaction of atmosphere, ocean, land, and sea ice, which are changing all the time. Only transient GCMs (not the steady state ones) can forecast the future climate data of which only precipitation is of interest to this study (White *et al.* 2009). GCMs are make of numerical methods to solve the equations of conservations of mass, energy, and momentum with three-dimensional grids and a regular time step. A historical global gridded precipitation data set is needed for appropriate initial and boundary conditions and bias correction.

A daily time step is usually the highest temporal resolution obtained from GCM simulations; however, to produce IDF relationships, a higher resolution of 15 min or less is needed. Therefore, a procedure to disaggregate the daily to 15 min time step is need (Sun *et al.* 2019). The highest spatial resolution is normally about 1° × 1° lat/lon, about 100 km, which is too large and needs to be downscaled (Wilby *et al.* 2002; Cook *et al.* 2020). Having obtained 15-min rainfall intensity time series, the annual maximum series of rainfall intensity for several durations up to 24 or 48 h can be derived. For this derivation, we always assume that the annual maximum series data are stationary, meaning that the statistics of the data are temporally constant. Naturally, it is not so, therefore, nonstationary annual maximum series of rainfall intensity in the future are to be assumed. The future rainfall data of annual maximum series must be created in order to derive an accurate IDF curve (AghaKouchak *et al.* 2018; Ganguli & Coulibaly 2019). Furthermore, we must estimate the future atmospheric composition scenarios from very low greenhouse gas emission or concentration to very high.

The lengthy procedure for the future IDF curve modeling makes their accuracy and precision skeptical (Bellprat *et al.* 2019; Thompson & Smith 2019; Buchanan 2020; Cook *et al.* 2020). Each step of calculation produces errors and uncertainties which can add up tremendously. Can we be confident that the derived sets of maximum rainfall are accurate and precise? The accuracy of a current IDF relationship would not be known beforehand let alone that of the future IDF. This is because the predicted values of maximum rainfall intensity are not the actual values which have not yet existed at the time of the infrastructure being designed. For example, the value of a rainfall intensity at a 25-year return period means that the actual value will be known at least in 25 years to come or even 100 years or more. A majority of previous research found that the return levels for future IDF curves were always higher than the present ones (Liew *et al.* 2013). Shrestha *et al.* (2017), for example, created the IDF curves of periods 20112030 and 20462065 for Bangkok, the capital city of Thailand. They found that most of the future return levels are larger than that of the present ones. As aforementioned, the lengthy procedure of future IDF curve productions involves numerous assumptions, approximations, and uncertainties that they cannot be trusted. Several researchers recently have tried to use the reliable past data to predict future extreme events (Shepherd *et al.* 2018; Woo & Johnson 2018; Sillmann *et al.* 2021; and references therein). However, they are still not accurate and precise enough. This study attempts to derive new sets of IDF curves for all regions of Thailand (data periods: 1990–2016) and compares them to those of the original Royal Irrigation Department (1950–1998) using exactly the same methodology. The number of stations and their record lengths may not be the same for each pair of corresponding regions but average values will be used for comparison.

The aim of the study is, therefore, to compare the IDF relationships of the current study to those of the national standard in Bumpenkit (1999) to quantify the change of extreme rainfall intensities of the two periods to determine the acceptability of the official IDF curves.

STUDY AREA

Thailand is a part of mainland Southeast Asia situated between 97.1–106.0 E and 5.4–20.6 N and surrounded by Myanmar, Laos, Cambodia, and Malaysia. It is divided into six regions, namely Northern, Northeastern, Central, Western, Eastern, and Southern (Figure 1(a)). A series of mountain ranges run in the north-south direction extending from the Northern region passing the Western to the Southern one, these mountain ranges are on the west and border Thailand with Myanmar (Figure 1(b)). The trans-boundary Mekong River that runs along the eastern boundary divides Thailand from Lao and Cambodia. The Northeast region is a plateau with the Mekong on to the east. The Central region is a lowland receiving flood flow

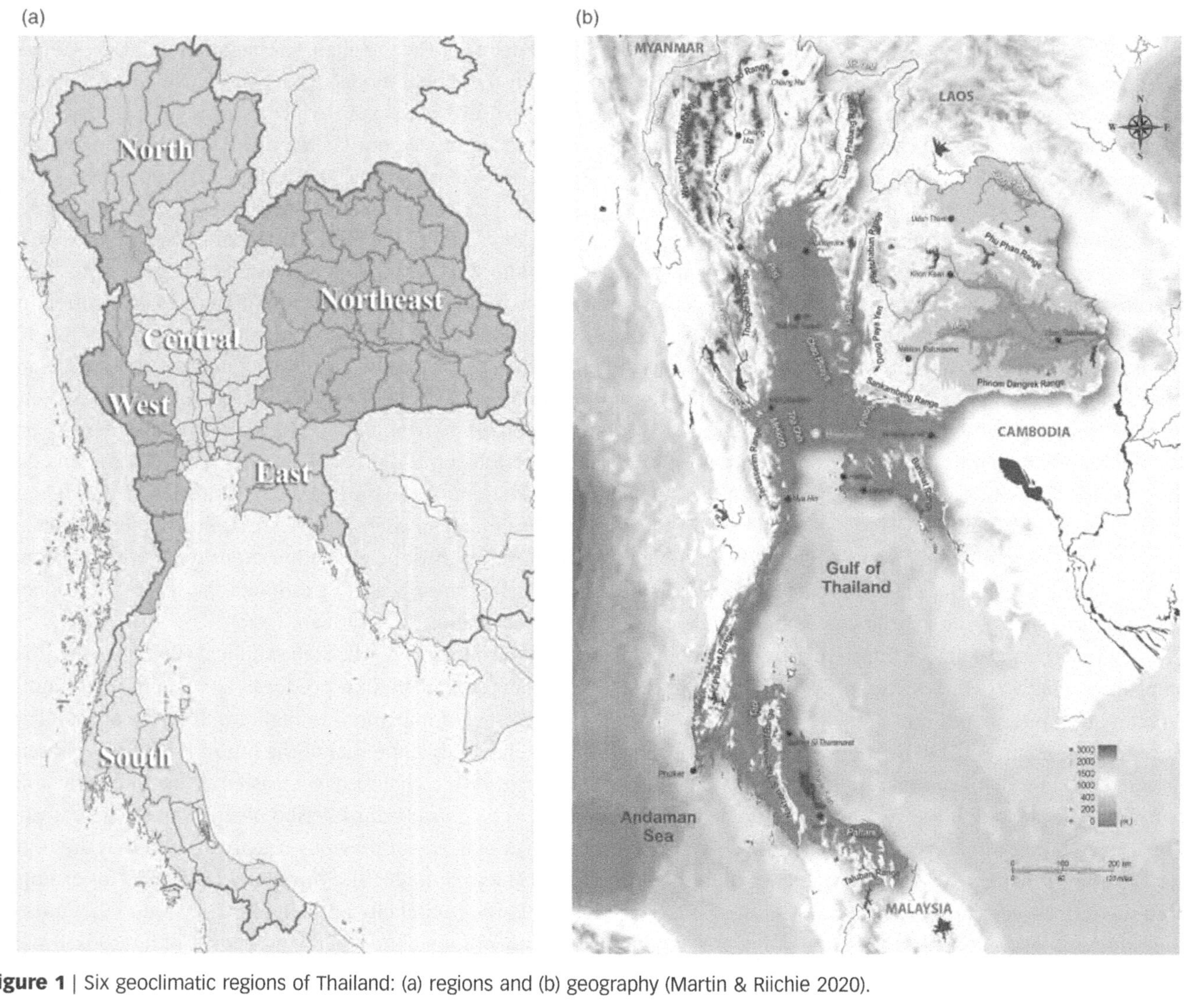

Figure 1 | Six geoclimatic regions of Thailand: (a) regions and (b) geography (Martin & Riichie 2020).

from the North and draining to the Gulf of Thailand. The Central normally includes the western and eastern regions; however, since their precipitation characteristics are different, we separate them for the purposes of this study. The Southern region of Thailand is a peninsula extending southward from the Central region to Malaysia with the Gulf of Thailand (of Pacific Ocean) on the east and Andaman Sea (of Indian Ocean) on the west. (Figure 1(b)).

The climate characteristics of Thailand are governed by the dynamics of the Intertropical Convergence Zone (ITCZ), a low atmospheric pressure band occurring from the combination of the solar radiation intensity and the inclination of the earth axis. During a few months at the beginning of the year, the ITCZ is located far below Malaysia; therefore, Thailand is without rainfall. The ITCZ moves rapidly northward to the western, central, and eastern parts in May and slowly moving to the north passing the Central, Northeastern, and Northern departing the northern boundary in mid-July. During the period of the ITCZ passing, it induces humidity from the Indian Ocean and the Gulf of Thailand, as southwest monsoon, to the upper part of the country. After a few weeks, the ITCZ comes back from China to the northern part of Thailand again, leaving the country's upper part to the south in mid-October. From July to October, the humidity from the southwest monsoon from the Indian Ocean plus the tropical cyclone from the South China Sea and the Pacific Ocean moves to Thailand's upper part creating heavy rainfall. After departing the upper part, the ITCZ moves across the peninsula from October to the New Year causing heavy rain in the Southern region. Accordingly, precipitation patterns of Thailand can be separated into two main regions: the upper and lower parts. The upper includes the Northern, Western, Northeastern, Central, and Eastern regions, whereas the lower is the Southern region. The lower part consists of two seasons, that is, the dry season from January to April and the rainy season from May to December. The upper part has three seasons: summer from February to April, rainy season from May to October, and cool season from November to January. The rainy season can also be separated into two periods by a drought spell: the former rainy season from May to July and the latter rainy season with a higher peak from August to October.

METHODS

The maximum rainfall IDF relationship (IDF curve) of a homogeneous region is always presented as a set of curves of maximum average intensity as a function of duration in that each curve has a fixed return period. To create the relationship, the data of annual maximum rainfall intensities at several corresponding durations of rain from several automatic rain gauges are required. For example, for 20 years of data from a rainfall station, there are 20 maximum intensity values for each of a number of durations, i.e., 15, 30, 45 min, 1 h, 2, 3, 6, 12, and 24 h in this study. An automatic rain gauge usually gives a time series of rainfall depths of a given temporal resolution, e.g. 15 min, which is common in Thailand, for each year. These rainfall depth time series are divided by the time resolution to obtain the time series of average rainfall intensities. The maximum average intensity of each year for each of the nine durations has to be evaluated. Let $\xi_{15}(t)$ be a time series of average rainfall intensities for the 15-min resolution from the beginning to the end of a year; therefore, the maximum average intensity, $i(15)$, of the year is as follows:

$$i(15) = \max\{\xi_{15}(t)\}, \quad t = 15, 30, 45\,\text{min}, \ldots \tag{1}$$

where t is the resolution time step. The time series of average rainfall intensities for the durations, d, other than 15 min, which are integer multiples, N, of the resolution, can be calculated from the following equation (Koutsoyiannis *et al.* 1998):

$$\xi_d = \frac{d \sum \xi_\delta(t - i\delta)}{d} \tag{2}$$

where δ is the resolution duration which is 15 min in this study, the summation is from $i = 0$ to $N - 1$, and N is an integer multiple of the resolution. Then the maximum intensity for durations other than 15 min can be evaluated from the time series $\xi_d(t)$ of a year as:

$$i(d) = \max\{\xi_d(t)\} \tag{3}$$

The procedures described in Equations (1)–(3) are applied to the time series of average rainfall intensities of 15 min resolutions for all annual data sets. Then for each of the rainfall stations, there are nine sets of annual maximum rainfall intensity series, denoted by i_{15}, i_{30}, i_{45}, …, i_{24} for the durations of 15, 30, 45 min, …., 24 h, respectively.

Each set of annual maximum rainfall intensities was assumed to follow the distribution of the extreme value type I (EV I) or Gumbel distribution which is the popular one for extrapolating to higher return levels (Koutsoyiannis *et al.* 1998; da Silva *et al.* 2019). To model the EV I distribution, we need two parameters, namely location parameter, u, and scale parameter, β, which can be evaluated using the method of moment as follows (Chow *et al.* 1988):

$$u = \mu - 0.5772\beta \tag{4}$$

$$\beta = \sqrt{6}\frac{\sigma}{\pi} \tag{5}$$

where μ and σ are the mean and standard deviations of the annual maximum intensity series. From the cumulative distribution function of the EV I model, for each fixed rain duration, the return level or corresponding intensity, i, can be evaluated from the following equation (Chow *et al.* 1988),

$$i = u - \beta \ln\left[\ln\left(\frac{T}{(T-1)}\right)\right] \tag{6}$$

where T is the return period. Equation (6) shows that only intensity and return period are the random variables excluding the duration.

The first step in evaluating maximum intensity $i(d,T)$ involved collecting annual intensity maxima at nine rain durations of 0.25, 0.5, 0.75, 1, 2, 3, 6, 12, and 24 h from each of automatic rain gauge stations. There are several automatic stations scattered all over Thailand and we separated them into six geoclimatic regions, namely the North (N), Northeast (Ne), Central (C), West (W), East (E), and South (S). Each region has a number of maximum intensity data of varied length as shown in Table 1 which shows that the basic data of the two studies (RID and this study) were not identical. The average and standard deviation of each data set were calculated, and then they were spatially averaged for each region (see also Figure 2).

The average values of the mean and standard deviation of each region were used to estimate the spatially average location and scale parameters from Equations (4) and (5), respectively. At each of nine rain durations, we used Equation (6) to estimate the corresponding maximum intensities of the specified return periods of 2, 5, 10, 25, 50, 100, 200, 500, and 1,000 years for each region.

RESULTS AND DISCUSSION

The derivation results of IDF relationships of this study are presented and then they were shown against the study of RID. Lastly, the ratios of the corresponding maximum intensities of this study to those of the RID were presented and discussed.

Table 1 | Basic information from the RID study (Bumpenkit 1999) and this study

Publication	Region	Number	Latitude	Longitude	Period
RID	North	20	16.07–18.84	98.75–101.16	1953–1995
	Northeast	10	14.97–17.48	101.73–104.86	1954–1998
	Central	5	13.79–15.70	99.54–100.66	1950–1998
	East	8	12.86–13.69	100.81–102.51	1960–1998
	West	4	11.81–14.34	99.49–99.95	1959–1998
	South	12	6.42–10.49	98.31–101.83	1954–1998
This study	North	8	17.62–19.96	97.98–100.78	1990–2015
	Northeast	11	14.88–17.87	102.03–105.02	1990–2015
	Central	7	13.73–16.79	99.53–101.15	1990–2015
	East	3	12.92–13.37	100.80–100.98	1990–2008
	West	6	13.00–14.24	98.55–100.06	1990–2015
	South	14	6.42–12.59	98.25–101.82	1990–2016

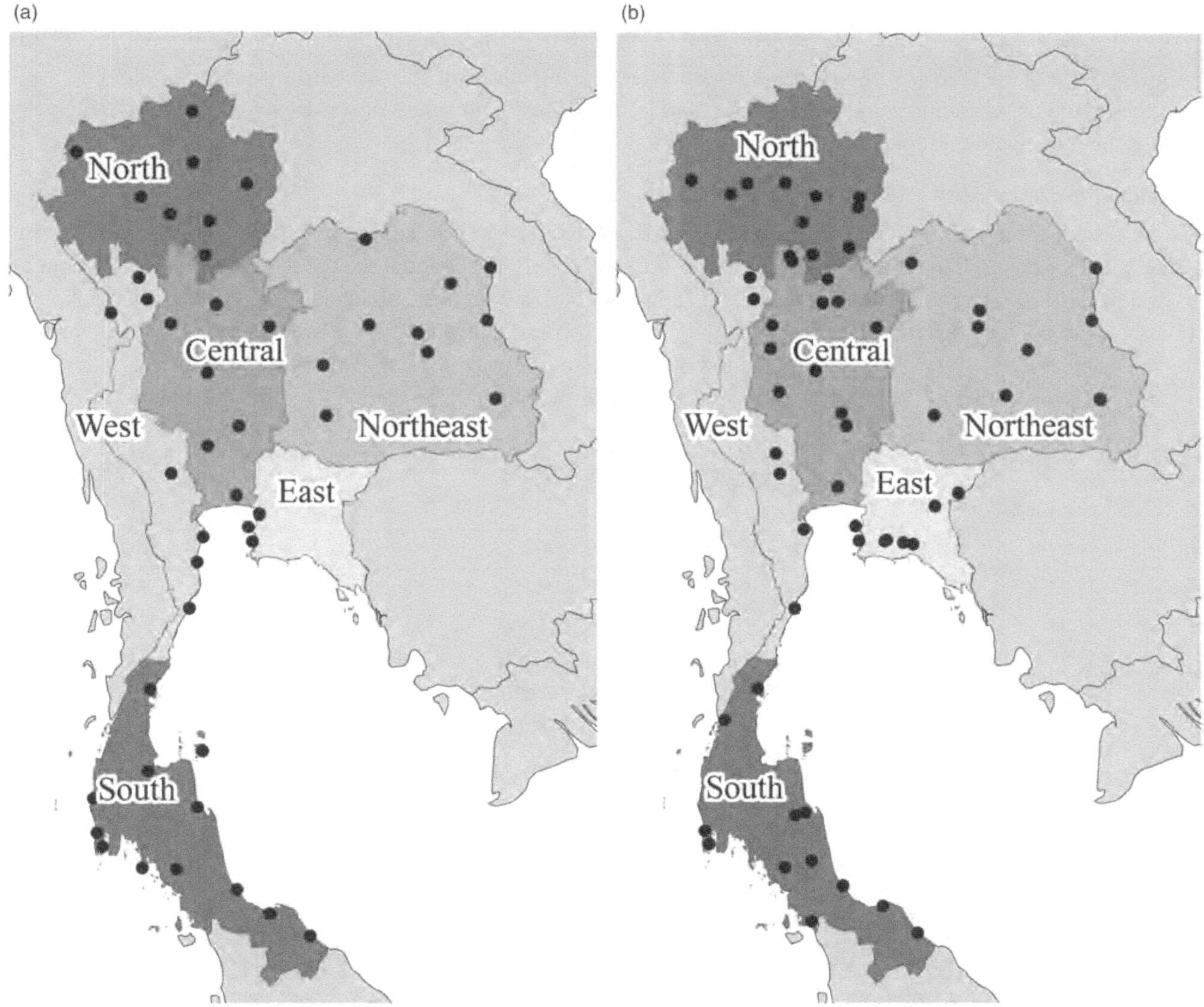

Figure 2 | Location of the stations that are used in this study and the RID study (Bumpenkit 1999): (a) this study and (b) RID's study.

Gumbel distribution parameters

Two parameters of Gumbel distribution, namely location parameter, u, and scale parameter, β, were evaluated from the means and the standard deviations of the annual maximum intensity series of each of the durations and regions. The mean values of the maximum rainfall intensity for each of the regions and each of durations are illustrated in Table 2 in

Table 2 | Spatial average values of the mean intensities at several durations for every region

	Duration (h)								
Region	**0.25**	**0.50**	**0.75**	**1.0**	**2.0**	**3.0**	**6.0**	**12.0**	**24.0**
N	91.99	63.83	50.43	42.05	23.85	16.98	9.20	5.06	3.13
Ne	61.58	43.62	35.65	30.22	18.13	13.14	7.30	4.02	2.80
C	76.84	55.33	44.23	37.00	21.17	15.08	8.32	4.44	3.07
E	124.29	80.81	60.81	48.59	27.45	19.22	10.35	5.38	3.43
W	87.83	59.93	46.53	38.58	23.20	17.48	9.78	5.35	3.30
S	106.20	77.71	62.24	52.04	31.53	23.00	13.30	7.82	5.42

which the maxima in durations 15 and 30 min belong to the East region and those in higher durations are in the South. For the standard deviations, the minimum spreading values belong to the North for all durations except the smallest duration which goes to the West (Table 3). Since the β parameters vary directly with the standard deviation, its minima are the same as the minimum spreading values of σ (Tables 3 and 4). The maximum values of u parameter follow the same pattern as those of the mean values (Tables 2 and 5).

IDF relationship comparison

In this subsection, we present IDF relationships from our study for all regions in Table 6 for the benefit of those who would like to use them for determining rough design storms or further research. We also add a set of IDF relationships from the RID study (Bumpenkit 1999) for the Central region only in Table 7. The reason for presenting only a single set of IDF relationships of RID is because one can determine the rest from the ratios in the next subsection. For a brief overview, the study presents only a single set of IDF curves of our study for the Central region in Figure 3. However, Figures 4–6 compare the Central region curves of this study to those of RID's study for the return periods of 2, 5, 10, 25, 50, 100, 200, 500 and 1,000 years.

Table 3 | Spatial average values of the standard deviations of the maximum intensities at several durations

	Duration (h)								
Region	**0.25**	**0.50**	**0.75**	**1.0**	**2.0**	**3.0**	**6.0**	**12.0**	**24.0**
N	41.48	22.49	18.45	16.40	9.14	6.66	3.59	1.94	1.18
Ne	44.58	30.43	25.27	22.16	14.04	10.47	6.01	3.33	2.02
C	52.97	36.51	29.30	24.61	13.96	10.01	5.55	3.06	1.73
E	48.22	28.25	20.86	17.65	11.10	8.66	5.15	2.75	1.33
W	36.98	25.92	20.03	17.65	11.53	9.63	6.1	3.63	2.35
S	38.46	26.01	20.83	17.81	11.84	8.67	5.87	3.64	2.26

Table 4 | Spatial average values of scale parameters, β

	Duration (h)								
Region	**0.25**	**0.50**	**0.75**	**1.0**	**2.0**	**3.0**	**6.0**	**12.0**	**24.0**
N	32.342	17.535	14.385	12.787	7.126	5.193	2.799	1.513	0.920
Ne	34.759	23.726	19.703	17.278	10.947	8.163	4.686	2.596	1.575
C	41.301	28.467	22.845	19.188	10.885	7.805	4.327	2.386	1.349
E	37.597	22.026	16.264	13.762	8.655	6.752	4.015	2.144	1.037
W	28.833	20.210	15.617	13.762	8.990	7.508	4.795	2.830	1.832
S	30.011	20.280	16.241	13.886	9.232	6.760	4.577	2.838	1.762

Table 5 | Spatial average values of location parameters, u

	Duration (h)								
Region	**0.25**	**0.50**	**0.75**	**1.0**	**2.0**	**3.0**	**6.0**	**12.0**	**24.0**
N	73.322	53.709	42.127	34.669	19.737	13.983	7.634	4.187	2.599
Ne	41.517	29.925	24.277	20.247	11.811	8.428	4.595	2.521	1.891
C	53.001	38.899	31.044	25.924	14.887	10.575	5.822	3.063	2.291
E	102.589	68.096	51.422	40.647	22.455	15.323	8.032	4.142	2.831
W	71.187	48.265	37.516	30.637	18.011	13.146	7.012	3.716	2.242
S	88.878	66.004	52.866	44.025	26.202	19.098	10.658	6.182	4.403

Table 6 | IDF relationship of this study (extreme intensity in mm/h)

		Return period (years)								
D (h)	**Region**	**2**	**5**	**10**	**25**	**50**	**100**	**200**	**500**	**1,000**
0.25	North	85.2	121.8	146.1	176.8	199.5	222.1	244.6	274.3	296.7
	Northeast	54.3	93.7	119.7	152.7	177.1	201.4	225.6	257.5	281.6
	Central	68.1	114.9	145.9	185.1	214.2	243.0	271.7	309.6	338.3
	East	116.4	159.0	187.2	222.8	249.3	275.5	301.7	336.2	362.3
	West	81.8	114.4	136.1	163.4	183.7	203.8	223.9	250.3	270.3
	South	99.9	133.9	156.4	184.9	206.0	226.9	247.8	275.4	296.2
0.5	North	60.1	80.0	93.2	109.8	122.1	134.4	146.6	162.7	174.8
	Northeast	38.6	65.5	83.3	105.8	122.5	139.1	155.6	177.4	193.8
	Central	49.3	81.6	103.0	130.0	150.0	169.9	189.7	215.8	235.5
	East	76.2	101.1	117.7	138.5	154.0	169.4	184.7	205.0	220.2
	West	55.7	78.6	93.7	112.9	127.1	141.2	155.3	173.8	187.9
	South	73.4	96.4	111.6	130.9	145.1	159.3	173.4	192.0	206.1
0.75	North	47.4	63.7	74.5	88.1	98.3	108.3	118.3	131.5	141.5
	Northeast	31.5	53.8	68.6	87.3	101.2	114.9	128.6	146.7	160.4
	Central	39.4	65.3	82.5	104.1	120.2	136.1	152.0	173.0	188.8
	East	57.4	75.8	88.0	103.4	114.9	126.2	137.6	152.5	163.8
	West	43.2	60.9	72.7	87.5	98.5	109.4	120.2	134.6	145.4
	South	58.8	77.2	89.4	104.8	116.2	127.6	138.9	153.8	165.0
1	North	39.4	53.8	63.4	75.6	84.6	93.5	102.4	114.1	123.0
	Northeast	26.6	46.2	59.1	75.5	87.7	99.7	111.7	127.6	139.6
	Central	33.0	54.7	69.1	87.3	100.8	114.2	127.5	145.2	158.5
	East	45.7	61.3	71.6	84.7	94.3	104.0	113.5	126.2	135.7
	West	35.7	51.3	61.6	74.7	84.3	93.9	103.5	116.1	125.7
	South	49.1	64.9	75.3	88.4	98.2	107.9	117.6	130.3	139.9
2	North	22.3	30.4	35.8	42.5	47.5	52.5	57.5	64.0	69.0
	Northeast	15.8	28.2	36.4	46.8	54.5	62.2	69.8	79.8	87.4
	Central	18.9	31.2	39.4	49.7	57.4	65.0	72.5	82.5	90.1
	East	25.6	35.4	41.9	50.1	56.2	62.3	68.3	76.2	82.2
	West	21.3	31.5	38.2	46.8	53.1	59.4	65.6	73.9	80.1
	South	29.6	40.0	47.0	55.7	62.2	68.7	75.1	83.6	90.0
3	North	15.9	21.8	25.7	30.6	34.2	37.9	41.5	46.2	49.9
	Northeast	11.4	20.7	26.8	34.5	40.3	46.0	51.7	59.2	64.8
	Central	13.4	22.3	28.1	35.5	41.0	46.5	51.9	59.1	64.5
	East	17.8	25.5	30.5	36.9	41.7	46.4	51.1	57.3	62.0
	West	15.9	24.4	30.0	37.2	42.4	47.7	52.9	59.8	65.0
	South	21.6	29.2	34.3	40.7	45.5	50.2	54.9	61.1	65.8
6	North	8.7	11.8	13.9	16.6	18.6	20.5	22.5	25.0	27.0
	Northeast	6.3	11.6	15.1	19.6	22.9	26.2	29.4	33.7	37.0
	Central	7.4	12.3	15.6	19.7	22.7	25.7	28.7	32.7	35.7
	East	9.5	14.1	17.1	20.9	23.7	26.5	29.3	33.0	35.8
	West	8.8	14.2	17.8	22.3	25.7	29.1	32.4	36.8	40.1
	South	12.3	17.5	21.0	25.3	28.5	31.7	34.9	39.1	42.3
12	North	4.7	6.5	7.6	9.0	10.1	11.1	12.2	13.6	14.6
	Northeast	3.5	6.4	8.4	10.8	12.7	14.5	16.3	18.7	20.5
	Central	3.9	6.6	8.4	10.7	12.4	14.0	15.7	17.9	19.5
	East	4.9	7.4	9.0	11.0	12.5	14.0	15.5	17.5	19.0
	West	4.8	8.0	10.1	12.8	14.8	16.7	18.7	21.3	23.3
	South	7.2	10.4	12.6	15.3	17.3	19.2	21.2	23.8	25.8
24	North	2.9	4.0	4.7	5.5	6.2	6.8	7.5	8.3	9.0
	Northeast	2.5	4.3	5.4	6.9	8.0	9.1	10.2	11.7	12.8
	Central	2.8	4.3	5.3	6.6	7.6	8.5	9.4	10.7	11.6
	East	3.2	4.4	5.2	6.1	6.9	7.6	8.3	9.3	10.0
	West	2.9	5.0	6.4	8.1	9.4	10.7	11.9	13.6	14.9
	South	5.0	7.0	8.4	10.0	11.3	12.5	13.7	15.4	16.6

D, duration.

Table 7 | IDF relationship of the Central from RID's study (extreme intensity in mm/h)

	Return period (years)								
Duration (h)	**2**	**5**	**10**	**25**	**50**	**100**	**200**	**500**	**1,000**
0.25	96.3	127.4	148	173.9	193.3	212.6	231.4	256.6	275.5
0.5	72.6	92.6	105.9	122.6	135	147.4	159.7	175.8	188
0.75	56.8	72.2	82.5	95.4	105	114.5	124	136.5	145.9
1	48.4	62.4	71.7	83.5	92.2	100.8	109.4	120.8	129.3
2	28.4	37.8	44	51.8	57.6	63.4	69.1	76.7	82.4
3	20.6	27.3	31.9	37.7	42	46.3	50.6	56.2	60.4
6	11.4	15.6	18.4	22	24.6	27.3	29.9	33.3	35.9
12	6	8.5	10.2	12.3	13.9	15.4	17	19	20.6
24	3.4	5	6	7.3	8.3	9.3	10.3	11.5	12.5

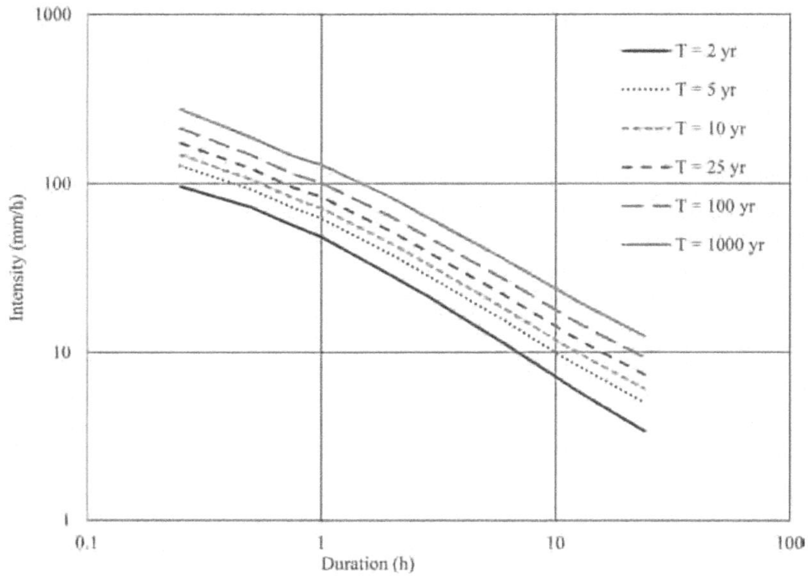

Figure 3 | IDF curve for the Central from this study.

They describe that the maximum intensity values from our study are all lower than those from RID for a 2-year return period; nonetheless, only a few for a 100-year return period at a higher duration are as shown in Figure 7.

Intensity ratios between this study and RID's values

It is always presumed that, due to climate change, the maximum rainfall intensity at a specific duration and the return period should progressively increase in accordance with time. Here we compare the rainfall intensities from two data sets: the older one from RID (1950–1998) to the more recent one in this study (1990–2016) (Table 1). These two data sets, even with some overlap, exhibit a range of time difference about 15–25 years. Therefore, the rainfall intensity values from this study should likely be higher than those from the RID study due to climate change. The rainfall intensity ratios of this study compared to the RID study are used to determine their change. If the ratio is higher than 1, it means the intensity is increasing as expected and vice versa. The intensity ratios of the North, Northeast, Central, East, West, and South are illustrated in Table 8, and the statistical summary of the intensity ratio is shown in Table 9.

Each region comprises 81 ratios, as the combination of nine durations and nine return periods is taken into account. The ratios that are higher than 1.0 show the increase in rainfall intensity from the past to the present. The statistical summary in Table 9 shows that the highest number of the higher-than-1 ratio is in the East region for 34 ratios and the lowest in the

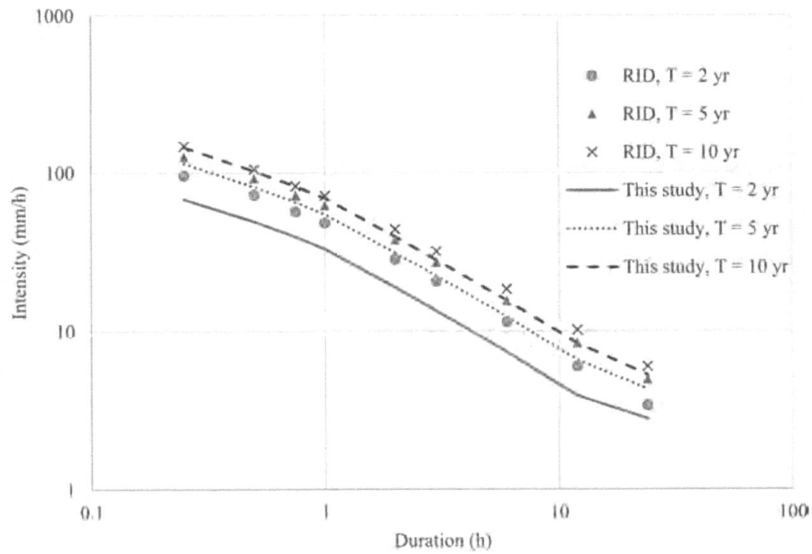

Figure 4 | IDF curves comparison between those from the RID study and this study for the Central region at return periods of 2, 5 and 10 years.

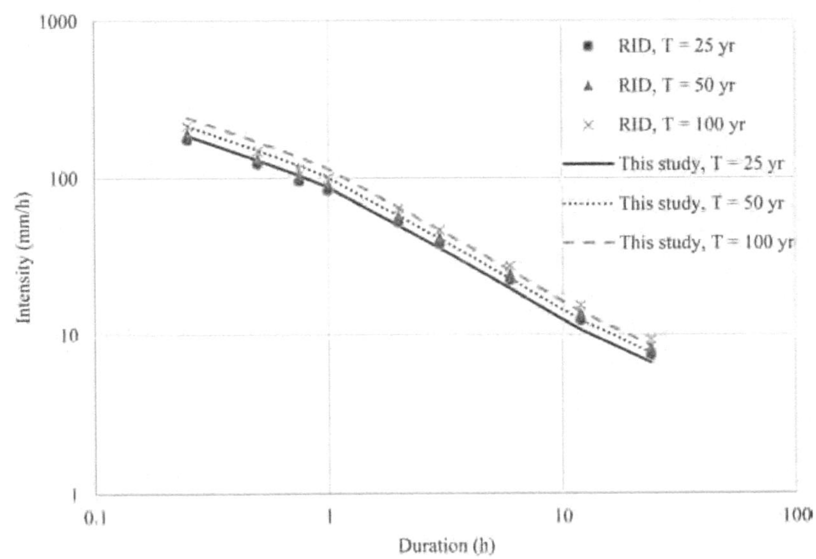

Figure 5 | IDF curves comparison between those from the RID study and this study for the Central region at return periods of 25, 50 and 100 years.

Northeast at 2 ratios. Considering the mean values of ratio and numbers of higher-than-1 ratio (Table 9), we can classify all regions into three groups, (1) Northeast, (Ne), (2) North, West, and South (NWS), and (3) Central and East (CE). The Northeast group (Ne) consists of a single region called the Khorat plateau. Its landscape is undulating terrain with a range of Phu Phan dividing the region into two basins, i.e., the Sakon Nakhon Basin in the North and the Khorat Basin in the South. A majority of the region drains its into the Mekong River. The second group consisting of the NWS is mountainous with mountain ranges lying along the north-south direction extending from southernmost to northernmost. The rainfall of the area derives mainly from the Southwest monsoon and it always causes flash floods and landslides. The last group, CE, is the combination of the Central lowland and the low hilly eastern part. Most of the drainage from this region goes to the Gulf of Thailand.

The plateau, the Northeast group, has two greater-than-1 ratios at 0.25 h duration for 500- and 1,000-year return periods with the mean intensity value of 0.81 (Tables 8 and 9). The mountainous regions, the NWS, have the greater-than-1 ratios of 8, 6, and 7, with the mean intensity values of 0.89, 0.90, and 0.90 for the North, the West, and the South, respectively. The ratios for the

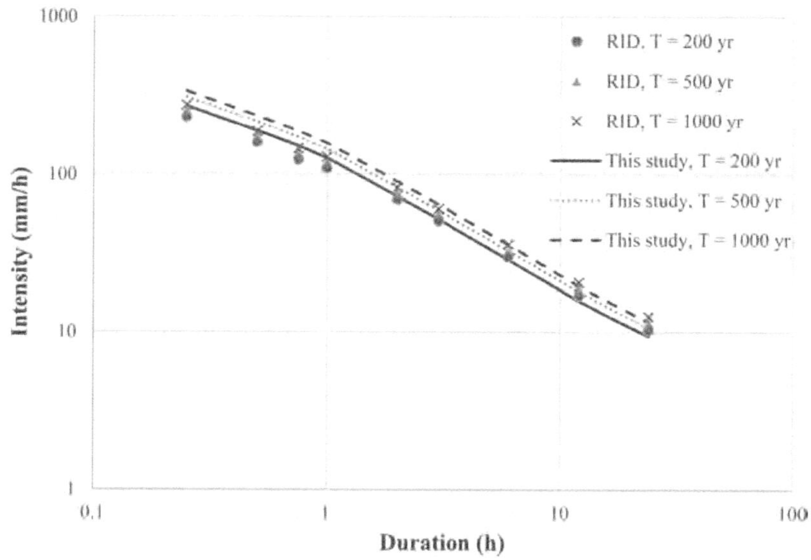

Figure 6 | IDF curves comparison between those from the RID study and this study for the Central region at return periods of 200, 500 and 1,000 years.

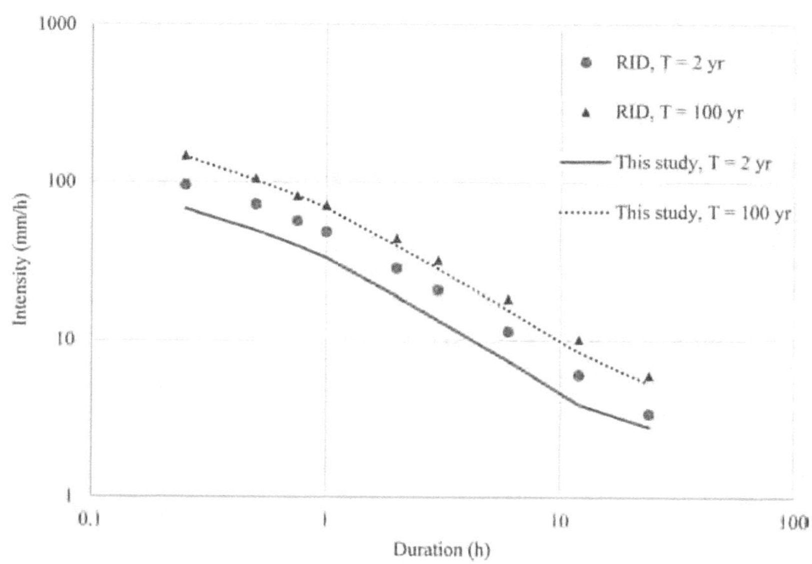

Figure 7 | IDF curves comparison between those from the RID study and this study for the Central region at return periods of 2 and 100 years.

North and the South are only from the smallest duration, 0.25 h. The North has eight ratios at 5-, 10-, 25-, 50-, 100-, 500-, and 1,000-year return periods, while the South has seven ratios from 10- to 1,000-year return periods. The West region, which is located between the North and the South, has six ratios with five from the smallest duration of 0.25 h and the return periods between 50 and 1,000 years as well as one ratio from the next smallest duration of 0.5 h with return periods of 1,000 years.

The last groups, which are the Central and the East, have 31 and 34 ratios, respectively (Tables 8 and 9). For the Central region, the number of greater-than-1 ratios is four each from the 25- and 50-year return periods with the durations of 0.25, 0.5, 0.75, and 1 h; five ratios from the 100-year return period with the durations of 0.25–2 h; and six ratios each for the return periods of 200, 500, and 1,000 years with the durations of 0.25–3 h (Table 8).

The East region has the largest number of 34 ratios with irregular distribution. There are nine ratios each of 0.25 and 0.5 h durations for all return periods, that is, one ratio of 2-year return period for 0.75 durations, 4 each of 3 and 12 h durations for 100- to 1,000-year return periods, and seven ratios of 6 h duration for 10- to 1,000-year return periods (Table 8). The analysis of the Eastern region in this study, however, was the least reliable because only three data sets were used.

Table 8 | Intensity ratios of this study to RID's values (bold values show greater-than-1 ratios)

Return Period (years)	Region	Duration (h) 0.25	0.5	0.75	1	2	3	6	12	24
2	N	0.95	0.90	0.89	0.89	0.88	0.87	0.86	0.83	0.89
	Ne	0.53	0.52	0.52	0.53	0.52	0.52	0.52	0.50	0.60
	C	0.71	0.68	0.69	0.68	0.66	0.65	0.65	0.66	0.82
	E	**1.18**	**1.07**	**1.01**	0.98	0.95	0.95	0.93	0.91	1.00
	W	0.82	0.76	0.74	0.74	0.77	0.79	0.76	0.74	0.75
	S	0.95	0.91	0.91	0.90	0.89	0.88	0.85	0.83	0.87
5	N	**1.05**	0.91	0.90	0.92	0.87	0.84	0.83	0.80	0.85
	Ne	0.71	0.68	0.68	0.70	0.68	0.68	0.66	0.65	0.73
	C	0.90	0.88	0.90	0.88	0.83	0.82	0.79	0.78	0.86
	E	**1.17**	**1.07**	1.00	0.98	0.96	0.98	0.99	0.96	0.97
	W	0.91	0.85	0.81	0.82	0.83	0.87	0.84	0.82	0.85
	S	0.99	0.93	0.92	0.91	0.90	0.89	0.85	0.82	0.83
10	N	**1.09**	0.90	0.91	0.93	0.87	0.83	0.82	0.77	0.82
	Ne	0.80	0.75	0.75	0.77	0.74	0.74	0.73	0.70	0.79
	C	0.99	0.97	1.00	0.96	0.90	0.88	0.85	0.83	0.89
	E	**1.16**	**1.07**	1.00	0.97	0.96	0.99	**1.01**	0.97	0.92
	W	0.95	0.89	0.84	0.85	0.86	0.90	0.87	0.85	0.87
	S	**1.01**	0.94	0.92	0.92	0.91	0.89	0.85	0.82	0.81
25	N	**1.15**	0.92	0.90	0.94	0.86	0.83	0.81	0.76	0.80
	Ne	0.88	0.82	0.82	0.84	0.80	0.79	0.78	0.75	0.84
	C	**1.06**	**1.06**	**1.09**	**1.05**	0.96	0.94	0.89	0.87	0.90
	E	**1.16**	**1.07**	0.99	0.97	0.95	1.00	**1.03**	0.99	0.93
	W	0.99	0.93	0.87	0.89	0.80	0.92	0.90	0.87	0.90
	S	**1.03**	0.98	0.93	0.93	0.91	0.89	0.85	0.81	0.80
50	N	**1.16**	0.92	0.91	0.94	0.86	0.82	0.80	0.75	0.79
	Ne	0.92	0.86	0.86	0.88	0.83	0.83	0.81	0.78	0.86
	C	**1.11**	**1.11**	**1.14**	**1.09**	1.00	0.98	0.92	0.89	0.91
	E	**1.15**	**1.07**	0.99	0.97	0.96	1.00	**1.04**	1.00	0.92
	W	**1.01**	0.95	0.88	0.90	0.89	0.94	0.92	0.86	0.91
	S	**1.04**	0.95	0.93	0.94	0.91	0.90	0.85	0.81	0.79
100	N	**1.18**	0.93	0.92	0.95	0.86	0.82	0.80	0.74	0.79
	Ne	0.96	0.89	0.90	0.92	0.86	0.85	0.84	0.80	0.89
	C	**1.14**	**1.15**	**1.19**	**1.13**	**1.02**	1.00	0.94	0.91	0.91
	E	**1.15**	**1.07**	0.99	0.97	0.96	**1.01**	**1.05**	**1.01**	0.92
	W	**1.03**	0.97	0.90	0.92	0.90	0.95	0.93	0.90	0.92
	S	**1.05**	0.95	0.93	0.94	0.93	0.90	0.85	0.81	0.78
200	N	**1.19**	0.95	0.92	0.95	0.86	0.81	0.80	0.74	0.78
	Ne	0.99	0.92	0.92	0.94	0.88	0.87	0.85	0.82	0.91
	C	**1.17**	**1.19**	**1.23**	**1.17**	**1.05**	**1.03**	0.96	0.92	0.92
	E	**1.15**	**1.07**	0.99	0.96	0.96	**1.01**	**1.05**	**1.02**	0.90
	W	**1.04**	0.98	0.91	0.93	0.91	0.96	0.93	0.91	0.93
	S	**1.06**	0.96	0.93	0.94	0.92	0.90	0.85	0.81	0.78
500	N	**1.21**	0.93	0.92	0.95	0.86	0.81	0.79	0.73	0.77
	Ne	**1.03**	0.95	0.95	0.97	0.90	0.89	0.88	0.84	0.93
	C	**1.21**	**1.23**	**1.27**	**1.20**	**1.08**	**1.05**	0.98	0.94	0.93
	E	**1.15**	**1.06**	0.99	0.96	0.96	**1.02**	**1.06**	**1.02**	0.90
	W	**1.06**	1.00	0.92	0.95	0.92	0.98	0.94	0.91	0.94
	S	**1.07**	0.96	0.93	0.95	0.92	0.90	0.85	0.81	0.78
1,000	N	**1.22**	0.93	0.93	0.96	0.86	0.81	0.79	0.73	0.77
	Ne	**1.05**	0.97	0.97	0.99	0.92	0.90	0.88	0.85	0.94
	C	**1.23**	**1.25**	**1.29**	**1.23**	**1.09**	**1.07**	0.99	0.95	0.93
	E	**1.15**	**1.07**	0.99	0.96	0.96	**1.02**	**1.07**	**1.02**	0.90
	W	**1.07**	**1.01**	0.93	0.95	0.93	0.98	0.95	0.92	0.95
	S	**1.07**	0.96	0.93	0.95	0.92	0.90	0.85	0.81	0.77

Table 9 | Statistical summary of intensity ratios

Region	No. of ratio > 1	Mean	SD	Min.	Max.
N	8	0.89	0.1099	0.73	1.22
Ne	2	0.81	0.1329	0.50	1.05
C	31	0.98	0.1611	0.65	1.29
E	34	1.01	0.0683	0.90	1.18
W	6	0.90	0.0734	0.74	1.07
S	7	0.90	0.0704	0.77	1.07

Table 10 | Comparison of rainfall intensities of IDF studies from Weesakul *et al.* (2017) and Shrestha *et al.* (2017)

Publication		Weesakul *et al.* (2017)			Shrestha *et al.* (2017)		
T (years)	D (h)	BP 1990–2014	FP 2015–2039	R	BP 1981–2010	FP 2011–2030	R
2	1	51.5	92.0	1.8	37.6	41.4	1.1
	3	21.5	30.0	1.4	31.2	33.0	1.1
	6	10.0	16.5	1.5	14.8	18.3	1.2
	12	5.5	8.1	1.5	8.3	8.7	1.0
	24	3.2	4.7	1.5	4.8	5.0	1.0
5	1	60.8	93.0	1.5	47.2	48.8	1.0
	3	28.8	39.3	1.4	38.6	43.6	1.1
	6	15.5	21.1	1.4	20.4	25.4	1.2
	12	8.0	10.2	1.3	10.6	13.3	1.3
	24	4.5	6.2	1.4	6.0	6.9	1.2

T, return periods (years); D, durations (h); BP, base period; FP, future period; R, ratio of FP to BP.

For the practical planning and design of hydraulic infrastructures, we have to choose appropriate rainfall intensity from IDF curves by first considering rainfall duration and then return period. Rainfall duration is chosen from the size and shape of the watershed or from its time of concentration, while a small watershed normally gives a shorter duration (Chow *et al.* 1988). The choice of return period depends on the size of watershed and the risk of failure (Chow *et al.* 1988).

Extreme outflow discharge from a small watershed does not always cause serious damage compared to that from a large watershed. Table 8 shows that the greater-than-1 ratios mostly exist at short durations especially at higher return periods. At return periods of 500 and 1,000 years, the greater-than-1 ratios exist at 15 min for all groups. For lower return periods at 50, 100, and 200 years, the ratios exist at 15 min for all except group 1 (Ne). The planning and design of hydraulic structures of larger watersheds with longer rainfall durations and involving higher risk require longer return periods. Therefore, the findings presented in Table 8 imply the acceptability of the national standard IDF curves from RID for all of Thailand. However, for the last group, namely the East and the Central regions, at the return periods from 50 years onward, the planning and design of hydraulic structures should be conducted with care.

Table 9 summarizes the number of greater-than-1 ratios for all regions which are classified into three groups, i.e., Ne, NWS, and CE. The first group, Ne, comprises only two greater-than-1 ratios from 81 ratios. The NWS group shows average greater-than-1 ratios of 7.3. Regarding the CE group, its average number of greater-than-1 ratios is 34.5 from 81 ratios. These results suggest that Thailand's IDF curves that were derived from the accurate historical rainfall data can render reliable design storms. To confirm this finding, two independent studies of IDF curves of the same rainfall station in Bangkok, Thailand, were investigated.

Weesakul *et al.* (2017) and Shrestha *et al.* (2017) studied future IDF curves from the same data set but with different methodology. The former assumed the annual maximum intensity series following Gumbel (EV 1) distribution, while the latter assumed the general extreme value (EV3) distribution. The former acquired the future precipitation time series during 2015–2039 from a global climate model using the GEM 2-ES model with the greenhouse gas scenario of RCP4.5, whereas the latter derived the future ones during 2011–2030 from eight models with the scenario of SRA1B. The two scenarios are similar in the greenhouse gas of low to medium concentration. For bias correction, the former used the current data sets of 1990–2014, while the latter used those of 1981–2010.

The values of extreme rainfall intensities in Table 10 were digitized from Weesakul *et al.* (2017) and Shrestha *et al.* (2017). The two studies used the same rainfall data set from the Sukhumvit area in Bangkok. The base periods (BP) and the forecast periods (FP) of the two studies are somewhat similar (Table 10). The significant difference between both studies is that the latter used eight global climatic models, while the former used only one to generate the future daily rainfall data sets. The results of the latter, therefore, should be more reliable. When considering the ratios of FP intensities to BP intensities, the former values are between 1.3 and 1.8 with the mean of 1.47 and the latter values are in the range of 1.0–1.3 averaging at 1.12. This implies that the more reliable method renders the closer of the future values to the base values, even though larger-than-1 ratios present in all of the results. All climatic models assume an increase in greenhouse gas which may trigger higher rainfall intensity in the future. We cannot predict the long-term rate of the greenhouse gas emission. The COVID-19 pandemic is one recent example that can temporarily reduce the emission rate. Some studies and attempts have been implemented to reduce and stabilize the emission rate such as the green new deal energy plan and the like (Jacobson *et al.* 2019; Nielsen *et al.* 2020). This study, therefore, suggests that using an accurate historical data set could be favorable to that using lengthy synthesized future data set for generating the IDF relationships.

CONCLUSIONS

An accurate IDF curve is crucial for hydraulic infrastructure design. The accuracy of an IDF curve is not known beforehand because the historical set of precipitation data were used to predict the future rainfall information. Due to climate change, new IDF curves are expected to change as a result of increasing rainfall intensities. We have compared a set of past IDF curves (RID's study) with that of the recent ones (this study) and found that the majority of them did not increase as expected. Another shortcoming of this finding is the relatively small time-lapse of about 15–20 years between the two data sets from the RID study and this study.

ACKNOWLEDGEMENT

This research was funded by the College of Industrial Technology, King Mongkut's University of Technology North Bangkok (Grant No. Res-CIT0270/2021).

DATA AVAILABILITY STATEMENT

All relevant data are included in the paper or its Supplementary Information.

REFERENCES

AghaKouchak, A., Ragno, E., Love, C. & Moftakhari, H. 2018 *Projected Changes in California's Precipitation Intensity-Duration-Frequency Curves*. Report California Energy Commission, California, USA.

Agilan, V. & Umamahesh, N. V. 2016 Is the covariate based non-stationary rainfall IDF curve capable of encompassing future rainfall changes? *Journal of Hydrology* **541** (B), 1441–1455.

Balbastre-Soldevila, R., Garcia-Bartual, R. & Andrea-Domenech, I. 2019 A comparison of design storms for urban drainage system applications. *Water* **11**, 757.

Bellprat, O., Guemas, V., Doblas-Reyes, F. & Donat, M. G. 2019 Towards reliable extreme weather and climate event attribution. *Nature Communications* **10**, 1732.

Bernard, M. M. 1932 Formulas for rainfall intensities of long durations. *Transactions of the American Society of Civil Engineers* **96**, 592–624.

Bezak, N., Sraj, M., Rusjan, S. & Mikos, M. 2018 Impact of the rainfall duration and temporal rainfall distribution defined using the Huff curves on the hydraulic flood modelling results. *Geosciences* **8**, 69.

Buchanan, M. 2020 The limit of a model. *Nature Physics* **16**, 605.

Bumpenkit, P. 1999 *The Relationship of Rainfall Intensity Duration Frequency of Difference Regions in Thailand*. Report Royal Irrigation Department (RID), Thailand.

Cheng, L. & AghaKouchak, A. 2014 Nonstationary precipitation intensity-duration-frequency curves for infrastructure design in a changing climate. *Scientific Reports* **4**, 7093.

Choi, J., Lee, O., Jang, J., Jang, S. & Kim, S. 2019 Future intensity-depth-frequency curves estimation in Korea under representative concentration pathway scenarios of Fifth assessment report using scale-invariance method. *International Journal of Climatology* **39**, 887–900.

Chokkavarapu, N. & Mandla, V. R. 2019 Comparative study of GCMs, RCMs, downscaling and hydrological models: a review toward future climate change impact estimation. *SN Applied Sciences* **1**, 1698.

Chow, V. T., Maidment, D. R. & Mays, L. W. 1988 *Applied Hydrology*. McGraw-Hill, New York, USA.

Cook, L. M., McGinnis, S. & Samaras, C. 2020 The effect of modelling choices on updating intensity-duration-frequency curves and stormwater infrastructure designs for climate change. *Climatic Change* **159**, 289–308.

da Silva, L. V., Casaroli, D., Evangelista, A. W. P., Junior, J. A. & Battisti, R. 2019 Rainfall intensity-duration-frequency relationships for risk analysis in the region of Matopiba, Brazil. *Revista Brasileira de Meteorologia* **34** (2), 247–254.

Ganguli, P. & Coulibaly, P. 2017 Does nonstationary in rainfall require nonstationary intensity-duration-frequency curves? *Hydrology and Earth System Sciences* **21**, 6461–6483.

Ganguli, P. & Coulibaly, P. 2019 Assessment of future changes in intensity-duration-frequency curves for Southern Ontario using North American (NA)-CORDEX models with nonstationary methods. *Journal of Hydrology: Regional Studies* **22**, 100587.

Hershfield, D. 1962 Extreme rainfall relationships. *ASCE Journal of the Hydraulics Division* 88, 73–92.

Hu, H. & Ayub, B. M. 2019 Machine learning for projecting extreme precipitation intensity for short durations in a changing climate. *Geosciences* **9**, 209.

Jacobson, M. Z., Delucchi, M. A., Cameron, M. A., Coughlin, S. J., Hay, C. A., Manogaran, I. P., Shu, Y. & von Krauland, A.-K. 2019 Impacts of green new deal energy plans on grid stability, cost, jops, health, and climate in 143 countries. *One Earth* **1**, 449–463.

Kohnova, S., Foldes, G., Labat, M. M. & Hlavcova, K. 2019 Changes in the IDF curves of short-term rainfall in the mountainous and lowland areas of Slovakia. *Earth and Environmental Science* **362**, 012083.

Koutsoyiannis, D., Kozonis, D. & Manetas, A. 1998 A mathematical framework for studying rainfall intensity-duration-frequency relationships. *Journal of Hydrology* **206**, 118–135.

Liew, S. C., Raghavan, S. V. & Liong, S.-Y. 2013 How to construct future IDF curves, under changing climate, for sites with scare rainfall records? *Hydrological Processes* **28** (8), 3276–3287.

Martin, S. A. & Riichie, R. J. 2020 Sourcing Thai geography literature for ASEAN and international education. *Singapore Journal of Tropical Geography* **41** (1), 61–85.

Mirhosseini, G., Srivastava, P. & Stefanova, L. 2012 The impact of climate change on rainfall intensity-duration-frequency (IDF) curves in Alabama. *Regional Environment Change*. doi:10.1007s10113-012-0375-5.

Moriondo, M. & Bindi, M. 2006 Comparison of temperature simulated by GCMs, RCMs and statistical downscaling: potential application in studies of future crop development. *Climate Research* **30** (2), 149–160.

Mustonen, S. E. 1969 *Rainfall-Intensity-Frequency Curves for Some Stations in Thailand*. Thai Meteorological Department, Bangkok, Thailand.

Nielsen, K. S., Stern, P. C., Dietz, T., Gilligan, J. M., van Vuuren, D. P., Figueroa, M. J., Folke, C., Gwozdz, W., Ivanova, D., Reisch, L. A., Vandenbergh, M. P., Wolske, K. S. & Wood, R. 2020 Improving climate change mitigation analysis: a framework for examining feasibility. *One Earth* **3**, 325–336.

Noor, M., Ismail, T., Chung, E.-S., Shahid, S. & Sung, J. H. 2018 Uncertainty in rainfall intensity duration frequency curves of Peninsular Malaysia under changing climate scenarios. *Water* **10**, 1750.

Shepherd, T. G., Boyd, E., Calel, R. A., Chapman, S. C., Dessai, S., Dima-West, I. M., Fowler, H. J., James, R., Maraun, D., Martius, O., Senior, C. A., Sobel, A. H., Stainforth, D. A., Tett, S. F. B., Trenberth, K. E., van den Hurk, B. J. J. M., Watkins, N. W., Wilby, R. L. & Zenghelis, D. A. 2018 Storylines: an alternative approach to representing uncertainty in physical aspects of climate change. *Climatic Change* **151**, 555–571.

Sherman, C. W. 1931 Frequency and intensity of excessive rainfall at Boston. *Transactions of the American Society of Civil Engineering* **95**, 951–960.

Shrestha, A., Babel, M. S., Weesakul, S. & Vojinovic, Z. 2017 Developing intensity-duration-frequency (IDF) curve under climate change uncertainty: the case of Bangkok, Thailand. *Water* **9**, 145.

Sillmann, J., Shepherd, T. G., van den Hurk, B., Hazeleger, W., Martius, O., Slingo, J. & Zscheischler, J. 2021 Event-based storylines to address climate risk. *Earth's Future* **9**, e2020EF001783.

Sun, Y., Wendi, D., Kim, D. E. & Liong, S.-Y. 2019 Deriving intensity-duration-frequency (IDF) curves using downscaled in situ rainfall assimilated with remote sensing data. *Geoscience Letters* **6**, 17.

Thompson, E. L. & Smith, L. A. 2019 Escape from model-land. *Economics* **13**, 40.

Trenberth, K. E. 2011 Changes in precipitation with climate change. *Climate Research* **47** (1), 123–138.

Weesakul, U., Chaowiwat, W., Rehan, M. M. & Weesakul, S. 2017 Modification of a design storm pattern for urban drainage systems considering the impact of climate change. *Engineering and Applied Science Research* **44** (3), 161–169.

White, P., Hilario, F. D., Guzman, R. G. & Cinco, T. A. 2009 *A Review of Climate Change Model Predictions and Scenario Selection for Impacts on Asian Aquaculture*. PAGASA, Philippine.

Wilby, R. L., Dawson, C. W. & Barrow, E. M. 2002 SDMS – a decision support tool for the assessment of regional climate change impacts. *Environmental Modeling and Software* **17** (2), 17–159.

Woo, G. & Johnson, N. F. 2018 Stochastic modelling of possible pasts to illuminate future risk. In: *Oxford Handbook of Complex Disaster Risks* (Shultz, J., Rechkemmer, A. & Johnson, N. F., eds). Oxford University Press, UK.

First received 23 June 2021; accepted in revised form 8 September 2021. Available online 24 September 2021

© 2022 The Authors Impact of Climate Change on Hydrology and Water Resources doi: 10.2166/wcc.2021.247

Remote sensing inversion characteristic and driving factor analysis of wetland evapotranspiration in the Sanmenxia Reservoir area, China

Shiyan Wang*, Chang Liu, Yiqian Tan, Jie Wang, Fei Du, Zhen Han, Zhi Jiang and Liang Wang
State Key Laboratory of Simulation and Regulation of Water Cycle in River Basin, and Department of Water Ecology and Environment, China Institute of Water Resources and Hydropower Research, Beijing 100038, China
*Corresponding author. E-mail: shiyan_wang188@outlook.com

ABSTRACT

Evapotranspiration in the hydrologic cycle realizes the energy and water transport in the atmosphere. Evapotranspiration differences concerning land-use types provide data for studying the evapotranspiration of river basins. To investigate the evapotranspiration in the reservoir under artificial regulation, we selected the river basin in the Sanmenxia Reservoir as the study area. Data sources are two-period Landsat8 OIL_TIRS remote sensing images during the growing season of wetland plants. Based on meteorological data in this river basin, we investigated the evapotranspiration differences of different land-use types using the Surface Energy Balance Algorithm for Land model. The FAO Penman–Monteith formula verified the remote sensing inversion results. Analysis shows that significant differences were manifested between wetland and non-wetland landscapes in evapotranspiration among different land-use types. Non-wetland landscapes accounted for about 97.23% of the river basin's total area, but their daily average evapotranspiration was only 7.26 mm/d. Those of wetland landscapes were 2.77% and 12.17 mm/d. In this river basin, the differences between the wetland and non-wetland landscapes in evapotranspiration are mainly associated with plant diversity, vegetation coverage and surface temperature (beyond other driving factors like meteorological and hydrological solar radiation factors).

Key words: evapotranspiration, land utilization, river basin in Sanmenxia Reservoir area, SEBAL model, wetland

HIGHLIGHTS

- Evapotranspiration data of different land types that support study on the hydrological cycle process in the watershed.
- SEBAL model based on satellite remote sensing data and geographic information system technology.
- The model simulation results have been verified by the Penman–Monteith formula.

1. INTRODUCTION

The global hydrologic cycle is greatly affected by climate change. The hydrologic cycle links surface processes and atmospheric processes. It realizes the dynamic circulation of solid, liquid and gaseous water through various physical forms (e.g., precipitation, runoff and evapotranspiration) and regulates global water resources' spatial and temporal patterns. Moreover, the hydrologic cycle plays a vital role in forming, developing and succession of wetlands. Evapotranspiration, a key process in the hydrologic cycle, enables the transfer of energy and water on the land surface and in the atmosphere (Wu 2016) and accelerates the succession of wetland ecosystems while facilitating the hydrologic cycle. The wetland ecosystem, one of three major global ecosystems (forest and ocean), provides the carrier of material conversion for the evapotranspiration process.

Moreover, wetlands play irreplaceable roles in climate regulation, pollutant degradation and floodwater storage by constituting a unique ecosystem formed under the water and land interaction. Therefore, the research on the evolution patterns of wetland evapotranspiration is a significant part of the research on the global hydrologic cycle due to its particularity, importance and complexity. The accurate estimation of evaporation capacity would be useful for drought monitoring and evaluation, wetland vegetation growth, crop water demand and production management, soil moisture status monitoring and groundwater level monitoring.

With the development and perfection of satellite remote sensing and geographic information system technology, the limitations of the traditional point measurement method have been resolved by introducing remote sensing data into the evapotranspiration model and combining it with meteorological data, thus realizing the 'point-to-plane' conversion of evapotranspiration estimation. Evapotranspiration research is developing in the direction of combining experimental science with theoretical science. In the 1960s, handheld infrared thermometer was used for the first time to measure crop temperature and evaluate crop growth status. The measured temperature value could be affected by crop structure (vegetation coverage, leaf area index, crop height and surface roughness), light condition (azimuth angle and elevating angle) and observation angle. For a long time, evaporation research has been mainly at the site scale measured by evaporation instruments. In the 1970s, the thermal infrared image data generated by satellite-borne thermal infrared scanner was also applied to calculate the difference between brightness temperature and air temperature (Allen *et al.* 1998). Bastiaanssen developed the Surface Energy Balance Algorithm for Land (SEBAL) to calculate latent heat flux and estimate the actual evapotranspiration, evaporation ratio and surface impedance in the Mediterranean region. He reported that the surface impedance was closely correlated with the surface soil moisture content, but the correlation varied with the spatial and temporal changes. After a correction through field experiment, the correlation between the surface impedance and surface soil moisture content could be used to estimate the soil moisture content in other regions (Bastiaanssen 1995). Based on Landsat and observed meteorological data, Liu *et al.* (2020) utilized the SEBAL model to estimate the spatial and temporal patterns of wetland evapotranspiration in the Liaohe River Delta in the growing season. Zheng *et al.* (2020) explored the evapotranspiration in the Jinghe River basin based on the SEBS model. The methods proposed and optimized as aforementioned have provided verification data for studying the evapotranspiration at the traditional point scale. They enriched the theoretical research foundation for the river basin evapotranspiration and gradually realized the estimation of global evapotranspiration.

Studies have shown that about 65% of global precipitation returns to the atmosphere in evaporation and continues to participate in the water cycle. Therefore, the estimation of evaporation can determine regional water exchange and is the key to the closure of the regional water cycle. The global environmental change affects regional landscape patterns. Human activities further intensify the change of landscape patterns in the global environment by changing local land-use and landscape patterns. The landscape pattern change is not a simple one-to-one relationship with the driving factors but the result of the interaction of various factors. The complex mechanism of action between landscape patterns and the hydrological process is essential geoscience research and closely relates to the hydrological process. Human activities have significantly changed the most apparent type of area, leading to changes in evapotranspiration, such as vegetation coverage, surface roughness, leaf area index and surface albedo. They are affected by complex underlying surface conditions and climate changes, being challenging to estimate accurately. Ekwueme & Agunwamba (2020) found through statistical significance study that evaporation, sunshine hours, solar radiation and rainfall have little influence on annual runoff because they are not statistically significant. It can be inferred from the results that identifying the factors that control runoff change can enhance the understanding of long-term processes of climatic runoff. Therefore, it is of great significance for promoting the rational distribution of regional water resources to find out the spatio-temporal variation law of evaporation and to define the evaporation under different wetland landscape patterns and vegetation cover conditions. Evapotranspiration, as an index to characterize wetland ecological characteristics, can reflect the self-development and evolution ability of the wetland ecosystem and serve as the basis for ecological water supply and restoration and reconstruction of degraded wetlands in the reservoir area. It also provides a relevant reference basis for ecological restoration and reconstruction of degraded wetlands in the reservoir area to promote ecological conservation and biodiversity protection of wetlands in the Sanmenxia Reservoir area. Moreover, this study analyzes remote sensing inversion characteristics and driving factors of wetland evapotranspiration in the reservoir area on a large scale. Through literature research, it is found that there are few relevant studies considering human disturbance, which makes up for the deficiency of wetland research in the reservoir area to a certain extent.

The main body of this paper includes five parts: the introduction of the research area, the introduction of the research method, the results and analysis, the discussion and the conclusion.

2. PROFILE OF STUDY AREA

Located within the territory of Henan Province and Shanxi Province, the river basin in the Sanmenxia Reservoir area starts from Sanmenxia hydro-junction in the east, extends to the Weinan City of Shaanxi Province in the west, adjoins Pinglu and Ruicheng of Shanxi Province and the Tielian Mountain in Dali of Shaanxi Province in the north and leans against Yixiao Mountain and Xiaoqinling Mountain Range in the south (Figure 1). The Sanmenxia Dam at 20 km to the east of Sanmenxia

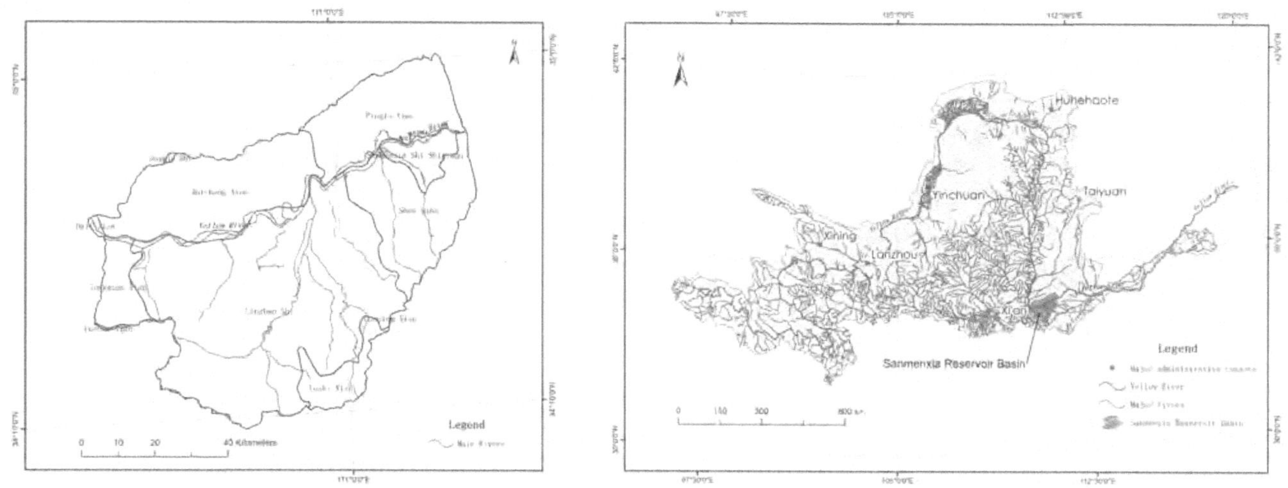

Figure 1 | Location and administrative division of river basin in the Sanmenxia Reservoir area.

City was used as the control node. The basin area is 688,400 km^2, accounting for 91.5% of the total area of the Yellow River basin. This river basin is in the intermountain (Zhongtiao Mountain and Qinling Mountains) basin. The Yellow River valley from the west of Sanmenxia (Henan) to Fenglingdu (Shanxi) is relatively wide, and its two banks are tableland formed by loess and plain on lower reaches of the Weihe River. The part from Jiamakou in Linyi County to Fenglingdu in Ruicheng County in Shanxi is a wandering-type silted river channel. The whole river basin is characterized by a semiarid continental monsoon climate, with uneven spatial and temporal distribution of precipitation. Specifically, precipitation is less in winter and spring, and a rainstorm is concentrated in summer and autumn. In this reservoir area, the annual average precipitation is 587 mm, and precipitation is mainly concentrated during July–September, accounting for 50% of the annual total precipitation. The annual average evaporation capacity is 2,221 mm, up to a maximum value of 2,665 mm.

3. METHODS

3.1. Data source and data preprocessing

The remote sensing data selected in this study were Landsat8 OIL_TIR from the United States Geological Survey (USGS) (http://earthexplorer.usgs.gov/). Two-period remote sensing images in the growing season of wetland plants were selected for a comparative analysis of changes. The images were acquired on July 7, 2014 and July 5, 2019, respectively, during which the cloud cover was small and surface plants expanded further.

After the remote sensing data were preprocessed through radiometric calibration, atmospheric correction and image fusion. We then chose eCognition, which is based on the object-oriented remote sensing image processing software, to segment the image with the multiresolution segmentation algorithm and classified the image's land use with the feature indexes of the image classification.

The selected meteorological data of Sanmenxia Meteorological Station within the river basin in 2014 and July 2019 were derived from the Meteorological Data Center of China Meteorological Administration (http://data.cma.cn/). The daily potential evapotranspiration capacities of Sanmenxia Meteorological Station throughout July were estimated by combining meteorological factors (i.e., wind velocity, air temperature, precipitation, temperature and humidity) with the latest corrected Penman–Monteith formula proposed by the Food and Agriculture Organization of the United Nations (FAO).

3.2. Basic principles of the SEBAL model

Tsanov *et al.* (2020) combine web and MATLAB. It is mentioned in the article that most methods to alleviate water pressure take into account the temporal and spatial characteristics of water resources and simulate practical problems by applying scientific knowledge rather than considering the problems as a whole. The SEBAL model is a monolayer model developed by Bastiaanssen (1995) based on the energy balance on the land surface to calculate latent heat flux. This model regards soil and vegetation as a whole and neglects the energy and material transport between vegetation and soil. It has been extensively applied to the remote sensing-based estimation of actual evapotranspiration under clear physical significance and high calculation accuracy. Zeng *et al.* (2008) described the SEBAL model-based remote sensing inversion of evapotranspiration in detail using Landsat TM/ETM+ data, which guided the application of the SEBAL model in the evapotranspiration inversion. By optimizing the selection of cold and hot spots in the SEBAL model, Xia (Cheng & Brutsaert 2005) improved its inversion accuracy. With high vegetation coverage and relatively homogeneous overall terrain, the river basin in the Sanmenxia Reservoir area is suitable for evapotranspiration inversion *via* the SEBAL model.

In this study, the SEBAL model was applied for inverting some basic parameters, such as ground reflectivity, NDVI and surface emissivity within seven wavebands of Landsat8 OIL_TIRS multispectral data. Next, the inversion of surface temperature was conducted using the atmospheric correction method to operate the thermal infrared bands. Finally, the surface energy-balance equation was established by combining the meteorological data in the river basin:

$$R_n + G + H + \lambda \mathrm{ET} = 0 \tag{1}$$

where R_n is the net surface radiation flux ($\mathrm{W \cdot m^{-2}}$), G is the soil heat flux ($\mathrm{W \cdot m^{-2}}$), H is the latent heat flux ($\mathrm{W \cdot m^{-2}}$), λ is the latent heat of water vaporization ($\mathrm{W \cdot m^{-2} \cdot mm^{-1}}$) and ET is the evapotranspiration capacity (mm).

After calculating the major model parameters (R_n, G, H, λ and ET) in Equation (1), the constant evaporation ratio method was introduced to extend the instantaneous evapotranspiration on time scale to obtain the evapotranspiration capacity in 24 h. The technical process of SEBAL model-based evapotranspiration inversion is displayed in Figure 2.

3.3. Model and determination of important parameters

3.3.1. Net surface radiation flux: R_n

Net surface radiation flux, which serves as the primary energy source in the transport and exchange process of energy, momentum and moisture on the earth surface, reflects the relationship of solar shortwave radiation absorbed and reflected within the unit surface area with atmospheric long-wave radiation in the following formula:

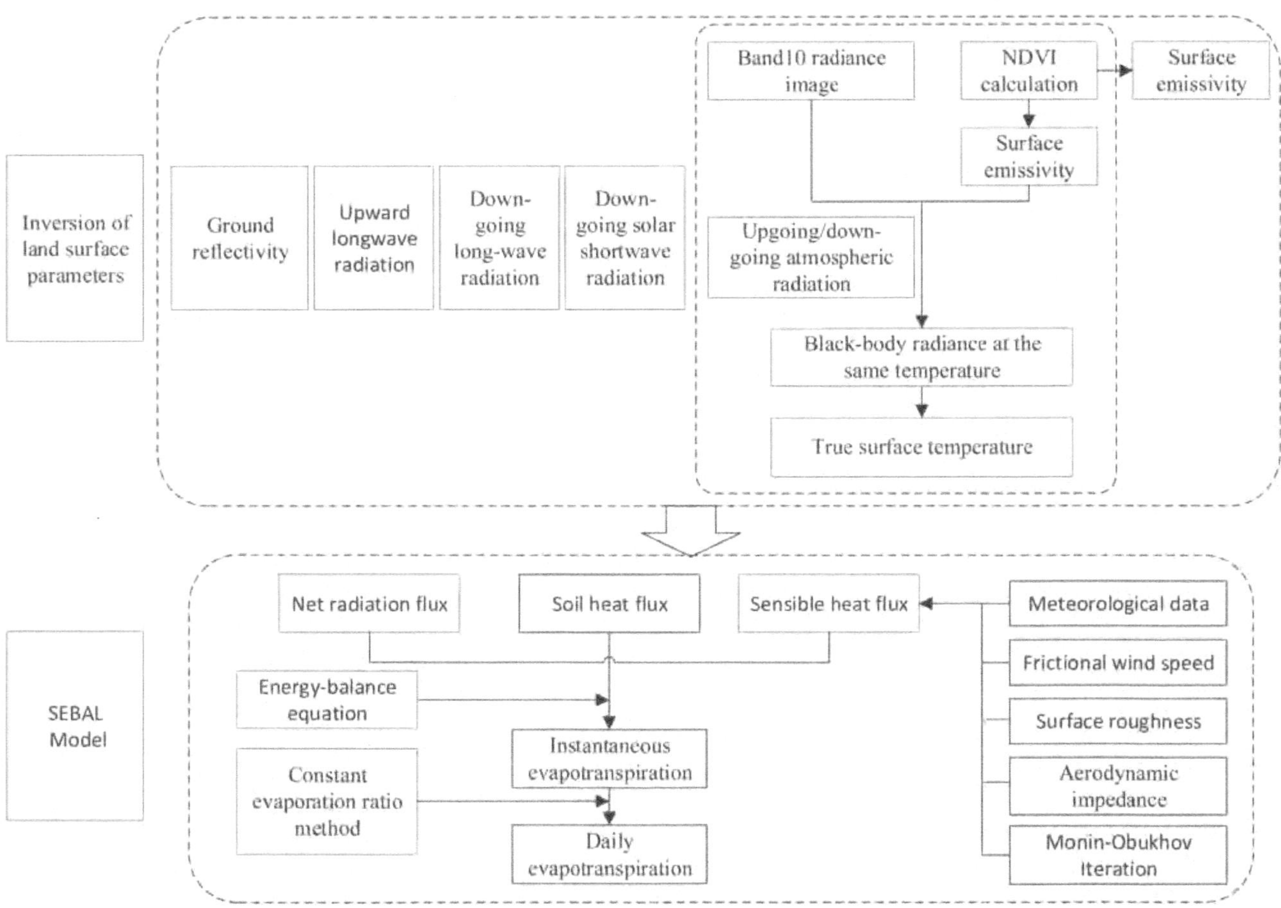

Figure 2 | Flowchart of the SEBAL model.

$$R_n = (1 - \alpha)R_{s\downarrow} + \mathrm{R}_{L\downarrow} - R_{L\uparrow} - (1 - \varepsilon)R_{L\downarrow} \tag{2}$$

where α is the ground reflectivity, $R_{s\downarrow}$ is the down-going solar shortwave radiation, $R_{L\downarrow}$ is the down-going long-wave radiation, $R_{L\uparrow}$ is the upgoing long-wave radiation and ε is the surface emissivity.

3.3.2. Soil heat flux: *G*

Soil heat flux is the energy transferred by topsoil to deep soil in the form of heat conduction. Despite the small soil heat flux in comparison with R_n, H and λET, it plays a significant role in maintaining the energy balance (Bastiaanssen *et al.* 1998).

$$G = \frac{T_s - 273.16}{\alpha} \times \left| 0.0032 \times \frac{\alpha}{c_{11}} + 0.0062 \times \left| \frac{\alpha}{c_{11}} \right|^2 \right| \times (1 - 0.978\mathrm{NDVI}^4) \times R_n \tag{3}$$

where c_{11} denotes the influence of satellite transit time on G, and it is taken as 1.0 if the satellite transits through the territory at a local time of 12:00–14:00. NDVI is the normalized differential vegetation index, and T_s is the surface temperature (K), which is obtained through inversion using the atmospheric correction method.

3.3.3. Sensible heat flux: *H*

Sensible heat flux is the energy dissipated in the atmosphere during the near-surface thermal momentum transfer process. By combining parameters like wind velocity, surface roughness, air density and aerodynamic resistance, the Monin–Obukhov similarity theory (Cheng & Brutsaert 2005) was introduced to correct the aerodynamic resistance through multiple iterative operations until a steady H value was obtained.

$$H = \frac{\rho_{\mathrm{air}} C_p dT}{r_{\mathrm{ah}}} \tag{4}$$

where ρ_{air} is the air density ($\mathrm{kg \cdot m^{-3}}$), C_p is the heat constant in air (taken as $1{,}004\,\mathrm{J \cdot kg^{-1} \cdot K^{-1}}$; $dT = T_{0.01} - T_2$), $T_{0.01}$ represents the temperature at the height of 0.01 m, T_2 is the temperature at the height of 2 m and r_{ah} is the aerodynamic resistance ($\mathrm{s \cdot m^{-1}}$).

In the SEBAL model, dT presents the following linear relation with surface temperature:

$$dT = aT_s + b \tag{5}$$

In order to obtain stable a and b values in Equation (5), the energy and mass transfer process between the ground and air was simulated *via* the Monin–Obukhov similarity theory. In the SEBAL model, we took the image elements with sufficient moisture supply, complete vegetation coverage and low temperature as 'cold spots', and the high-temperature image elements without surface vegetation coverage as 'hot spots'. As the energy dissipated by cold-spot image elements from the surface to the atmosphere was nearly 0, $H = 0$, and as the evapotranspiration capacity of hot-spot image elements was nearly 0, $\lambda\mathrm{ET} = 0$. The intense subjectivity in selecting 'cold spots' and 'hot spots' significantly impacted the inversion accuracy. Based on the above assumption, stable values of a, b, r_{ah} and H were obtained through the loop iterative calculation of Monin–Obukho length, namely, $L = -\rho_{\mathrm{air}} C_p u^{*3} T_s / KgH$ (u^* is the friction velocity under neutral stability).

3.3.4. Calculation of evaporation ratio and evapotranspiration capacity

According to the constant evaporation ratio method principle, the ratio (namely instantaneous evaporation ratio Λ) of latent heat flux to the difference value between net surface radiation flux and sensible heat flux within a day is stable (Harrison 1963). The evaporation ratio can be calculated using the following formula:

$$\Lambda = \frac{\lambda\mathrm{ET}}{R_n - G} = \frac{R_n - G - H}{R_n - G} \tag{6}$$

The latent heat flux in 24 h are given by the following equation:

$$\lambda \mathrm{ET}_{24} = \Lambda\ (R_{n24} - G_{24}) \tag{7}$$

The evapotranspiration capacity in 24 h can be estimated using the following equation:

$$\mathrm{ET}_{24} = \frac{86,400\Lambda\ (R_{n24} - G_{24})}{\lambda} \tag{8}$$

where G_{24} can be ignored in the estimation of the evapotranspiration capacity in 24 h.

λ is the latent heat of water vaporization (J/kg) (Harrison 1963), and its calculation formula is as follows:

$$\lambda = (2.501 - 0.002361(T_s - 273.15)) \times 10^6 \tag{9}$$

R_{n24} is the daily net radiation flux (W/m^2), and it can be calculated using the following formula:

$$R_{n24} = (1 - \alpha)R_{a24} - 110\tau_{\mathrm{sw}} \tag{10}$$

$$R_{a24} = \frac{G_{\mathrm{sc}}}{\pi \times d_r}[w_s \sin\emptyset \sin\delta + \cos\emptyset \cos\delta \sin w_s] \tag{11}$$

$$w_s = \arccos[-\tan\emptyset \tan\delta] \tag{12}$$

$$\delta = 0.409\sin\left(\frac{2\pi}{365}J - 1.39\right) \tag{13}$$

where G_{sc} is the solar constant (1,367 W/m^2), d_r is the relative sun-to-earth distance, $\emptyset$ is the geographic latitude of the image element (rad) and δ is the solar declination.

3.4. Model verification *via* the Penman–Monteith formula

To verify the accuracy of the SEBAL model, the FAO Penman–Monteith formula was used to verify the potential evapotranspiration at Sanmenxia Meteorological Station. That formula is as follows:

$$\mathrm{ET}_0 = \frac{0.408\Delta(R_n - \mathrm{G}) + \gamma\dfrac{900}{T + 273}u_2(e_s - e_a)}{\Delta + \gamma(1 + 0.34u_2)} \tag{14}$$

where ET_0 is the potential evapotranspiration capacity (mm/d), R_n is the net surface radiation (MJ/(m$^2 \cdot$ d)), γ is the psychrometer constant (KPa/°C), u_2 is the wind velocity at the height of 2 m, e_s is the saturation vapor pressure (KPa), e_a is the actual vapor pressure (KPa) and Δ is the slope of saturation vapor pressure–temperature relation curve (KPa/°C).

Based on the verification data in 2014 and Equation (14), the potential evapotranspiration capacity at Sanmenxia Meteorological Station in the same period was calculated as 5.032 mm. By contrast, the result obtained through the SEBAL model inversion was 4.66 mm at the meteorological station. The error was 7.39% and not large. Hence, obtaining the evapotranspiration capacity in this area through the SEBAL model inversion is quite reliable.

4. RESULTS AND DISCUSSION

4.1. Results

4.1.1. Spatial distribution characteristics of evapotranspiration capacities in different land-use types

The land-use types in the Sanmenxia river basin include artificial wetland, urban land, farmland, beach land, forest land, grassland, lake, marsh wetland and river (Yi 2013). The total area of Sanmenxia river basin is 5,999.48 km^2, and the total wetland (including artificial wetland, beach land, lake, marsh wetland and river) areas in 2014 and 2019 were 150.37 and 166.42 km^2, respectively, accounting for 2.5 and 2.77% of the total basin area, respectively. The total non-wetland (urban land, farmland, forest land and grassland) areas were, respectively, 5,849.54 and 5,838.30 km^2, accounting for 97.5 and 97.23% of the total basin area as shown in Figure 3(a).

Affected by soil moisture status, meteorological factors, vegetation growth, the daily evapotranspiration capacities in different land-use types within the whole study area in 2014 and 2019 were 0–10.61 and 0–13.392 mm/d, respectively. In 2014, the daily average evapotranspiration capacities in descending order were lake (10.61 mm/d), river (9.33 mm/d), forest land (8.91 mm/d), artificial wetland (8.11 mm/d), marsh (8.07 mm/d), beach land (6.43 mm/d), grassland (5.30 mm/d), farmland (3.40 mm/d) and urban land (2.71 mm/d). In 2019, the daily average evapotranspiration capacities in descending order were river (13.49 mm/d), lake (13.25 mm/d), forest land (9.91 mm/d), marsh (9.60 mm/d), artificial wetland (9.44 mm/d), beach land (9.09 mm/d), grassland (6.01 mm/d), farmland (5.36 mm/d) and urban land (4.82 mm/d). The daily evapotranspiration distributions in the entire study area are shown in Table 1 and Figure 4(a) and 4(b).

4.1.1.1. Urban land. Wang *et al.* (2017) found that urbanization changed the landscape pattern of wetlands, resulting in a continuous decrease in wetland area, landscape fragmentation, increased heterogeneity and decreased connectivity of wetlands. Urbanization also changes the hydrological conditions of wetlands, weakens the natural regulation and storage capacity, strengthens the runoff coefficient, increases the flood peak discharge and aggravates soil erosion. This study area is located within the territory of Sanmenxia City, and many cities and towns are distributed along the Yellow River. The urban land area in 2019 was increased by 7.36 km^2 in comparison with that in 2014. The daily average evapotranspiration capacity was 2.11 mm/d higher than that in 2014, driven by the high-temperature weather in July 2014 (with daily average temperature of up to 37.5 °C) relative to that in 2019.

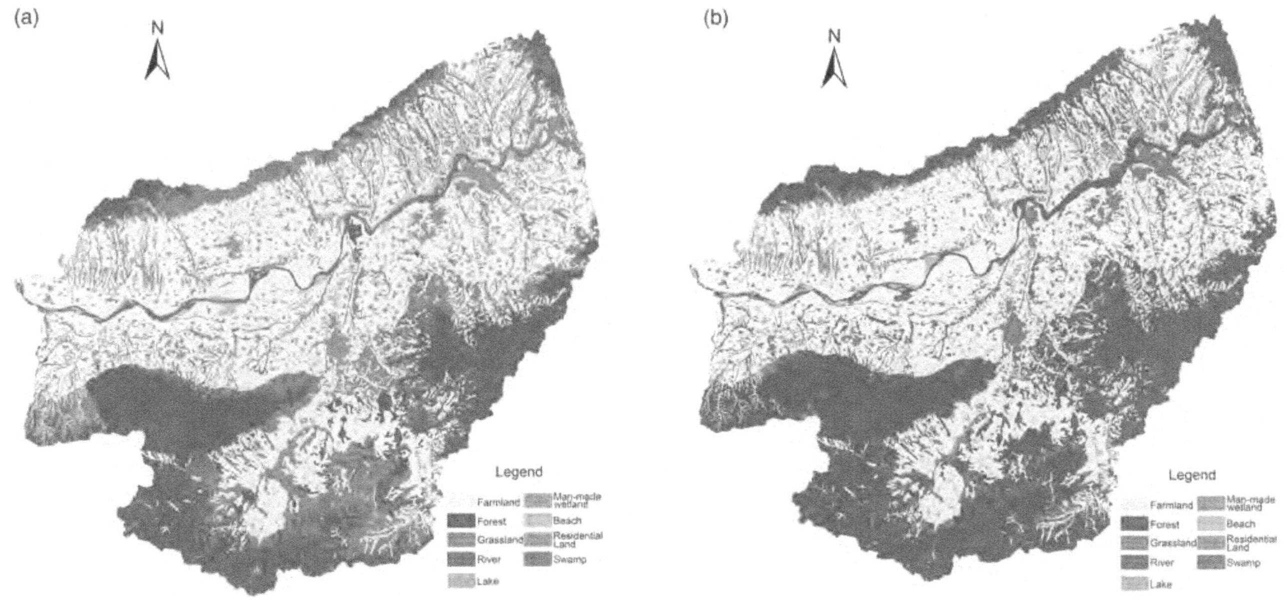

Figure 3 | Land-use types in the Sanmenxia river basin in July 2014 (a) and land-use types in the Sanmenxia river basin in July 2019 (b).

Table 1 | Daily average evapotranspiration capacities of wetlands and non-wetlands (mm/d)

		Wetland					Non-wetland			
Year	**Item**	**Lake**	**River**	**Artificial wetland**	**Beach land**	**Marsh**	**Forest land**	**Farmland**	**Urban land**	**Grassland**
2014	Area (km^2)	9.16	83.00	1.17	46.39	10.64	2,090.28	2,657.46	233.71	868.08
	Daily evapotranspiration	10.61	10.33	8.11	6.43	8.07	8.31	3.40	2.71	5.30
2019	Area (km^2)	9.32	87.19	1.26	29.18	12.50	2,402.93	2,658.18	241.065	557.84
	Daily evapotranspiration	13.25	13.49	9.44	9.09	9.60	9.91	5.36	4.82	6.01

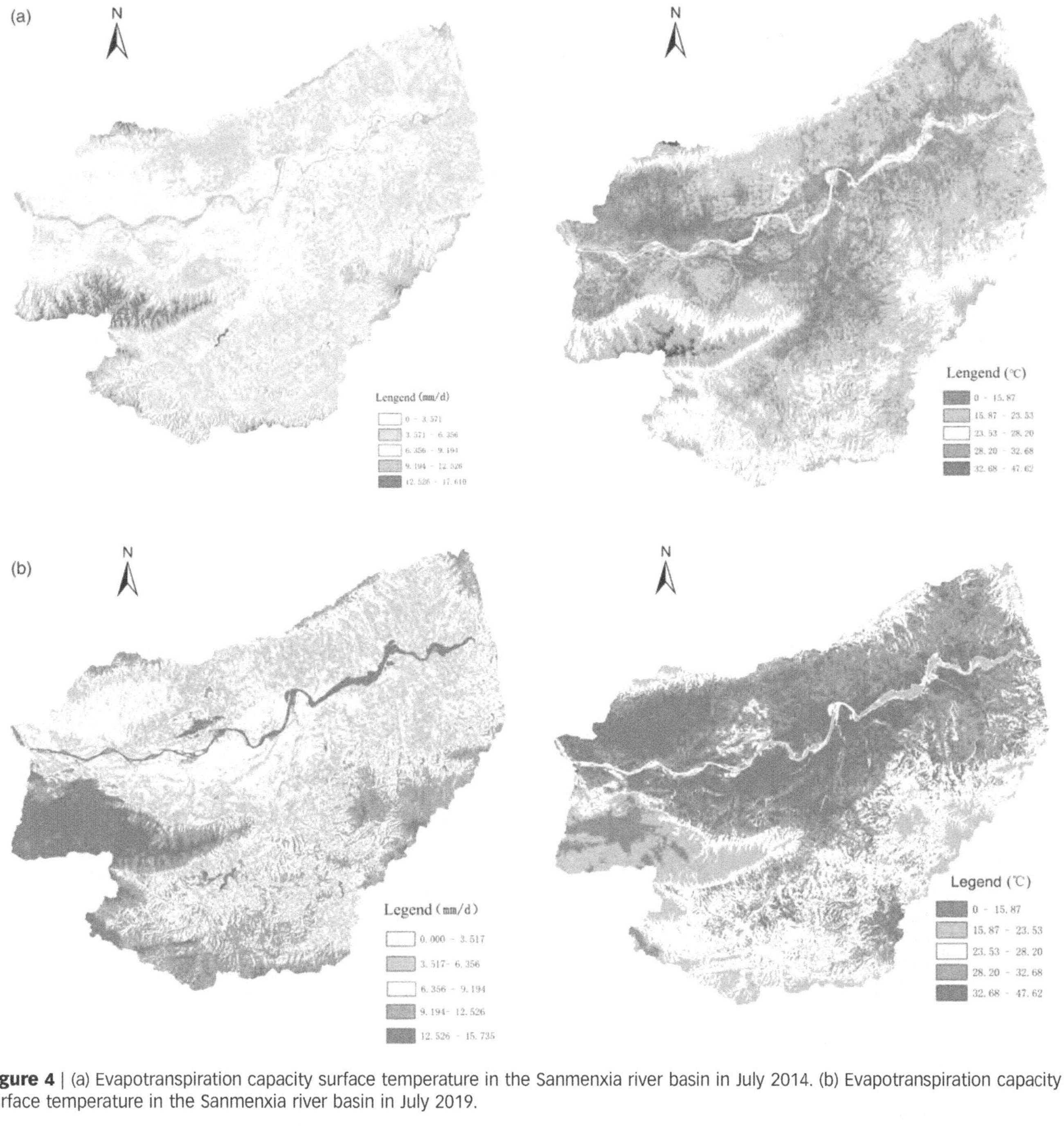

Figure 4 | (a) Evapotranspiration capacity surface temperature in the Sanmenxia river basin in July 2014. (b) Evapotranspiration capacity surface temperature in the Sanmenxia river basin in July 2019.

4.1.1.2. Forest land, farmland and grassland. Cao *et al.* (2017) found that the spatial distribution characteristics of forest evapotranspiration contribution in four seasons were significant. High temperature, abundant precipitation and sufficient light in summer were the periods of vigorous growth of forest land and the highest evapotranspiration contribution of forest land in a year occurred in summer. The maximum contribution value of farmland to annual evapotranspiration is also in summer when the crop coefficient reaches the maximum, and the opposite is true in winter.

Forest land is mainly located in the northern and southern mountainous areas in the study area. The daily average evapotranspiration capacities in 2014 and 2019 were 8.31 and 9.91 mm/d, respectively. The agricultural production area is mainly distributed in flat zones on the north and south of the Yellow River, covering approximately 7.86% of the total basin area. The daily average evapotranspiration capacities of the agricultural production area in 2014 and 2019 were 3.40 and 5.36 mm/d, respectively. Grasslands presented dispersed distribution in the study area; the northern grassland patch area was large, while the southern grassland patch area was small. The daily average evapotranspiration capacities in 2014 and 2019 were 5.30 and 6.01 mm/d, respectively.

4.1.1.3. River and lake. Cui *et al.* (2021) found that the annual average actual evapotranspiration, vegetation transpiration and vegetation evaporation after water storage under artificial regulation were increased compared with those before water storage. On the other hand, ground evaporation decreased after water storage compared with before water storage. In addition, Zhou *et al.* (2014) found that among the changes of many climate factors, precipitation is the key factor to maintain the water quantity of wetland, and the decrease of precipitation will obviously intensify the shrinkage of wetland area, and even change the wetland type. The main river within this river basin is the Tongguan-Sanmenxia section of the Yellow River. Affected by the water-level regulation and precipitation of Sanmenxia hydro-junction, the river area (about 1.45% of the total basin area) changed little in 2014 and 2019. The daily average evapotranspiration capacities in 2014 and 2019 were 10.33 and 13.49 mm/d, respectively.

Lake is an important part of the wetland ecosystem. The lakes in the study area mainly include Dinghu Bay and Taijisheng Lake, accounting for about 0.16% of the total basin area. The main vegetation types are *Phragmites australis* (Cav.) *Trin. ex Steud* and *Typha angustifolia*. The daily average evapotranspiration was only second to that of the river wetland landscape, being 10.61 and 13.25 mm/d in 2014 and 2019, respectively.

4.1.1.4. Beach land, marsh and artificial wetland. Lv *et al.* (2018) found that the wetland area in the Beijing-Tianjin-Hebei region showed a decreasing trend in the past 30 years, and the wetland area change showed a trend from a slight increase to a rapid decrease. Among the five types of natural wetland, the area of lakes and channels increased, while the area of tidal flats, tidal lands and marshland decreased. The main reasons for the impact are the rapid growth of population and economy and other factors, and the regions with serious wetland damage also show differences on time scale. Beach land is located at the two banks of the river and the river island, including the hydro-fluctuation belt generated due to the water-level change in the reservoir area. Influenced by the water-level regulation in Sanmenxia hydro-junction, large areas of beach lands were formed along the Yellow River, and their unique geographical location and hydrologic process have contributed to unique wetland soil and vegetation landscapes. The inversion results show that the reservoir regulation significantly affected the beach land area, but the less disturbed by human activities. The beach land accounted for about 0.49% of the total basin area, and the daily average evapotranspiration capacities in 2014 and 2019 were 6.43 and 9.09 mm/d, respectively.

As important wetland types, marsh and beach land are interchangeable with the water-level variation in the reservoir area. The marshland within the river basin is mainly distributed along the river, with the main vegetation community of *P. australis (Cav.) Trin. ex Steud*, and its area is 0.2% or so of the total basin area. The marsh area is affected by the changes of the river's main current to a great extent. Yi (2013) believed that the rechanneling and diversion of river, in particular, have changed the physiochemical properties and hydrological conditions of soil, thus affecting the vegetation succession of marsh wetland. Located in a densely populated area, the marsh wetland in the Sanmenxia river basin is greatly disturbed by human activities. Under the dual impacts of water surface evaporation and plant transpiration, the daily average evapotranspiration capacities in 2014 and 2019 were 8.07 and 9.60 mm/d, respectively.

In the river basin, the artificial wetlands include reservoirs, pit ponds and canals, most of which concentrate on aquaculture and embody the human disturbance to nonnatural wetlands, and the artificial wetlands account for about 0.021% of the total basin area. The daily average evapotranspiration capacities in 2014 and 2019 were 8.11 and 9.44 mm/d, respectively.

4.1.2. Study of factors driving the evapotranspiration process

The evapotranspiration process in the basin is affected by many factors. For instance, the changes in meteorological conditions, solar radiation and hydrologic processes can significantly alter the potential evaporation of the underlying surface in an area and influence the water deprivation due to the evapotranspiration in the basin. This can modify the production and domestic water-use pattern of the whole basin (Yuan *et al.* 2012). The analysis indicates that:

1. According to some previous evapotranspiration studies (Zeng *et al.* 2008; Liu *et al.* 2009; Zhang 2011, 2013; Li *et al.* 2012), the daily evapotranspiration capacity in the water area (lake and river) was the highest. However, in the evapotranspiration results in July 2014 and July 2019, the daily evapotranspiration capacity of the water area often exceeded that of

artificial wetland. This was because the artificial wetland was dried up, and the wetland vegetations experienced poor growth status and even withered due to the persistent drought. Therefore, by comparing the differences of water area and other land-use types (beach land, marsh and artificial wetland) in evapotranspiration, we deduce that the water surface area is one of the driving factors influencing evapotranspiration.

2. The main reasons why the daily evapotranspiration capacity of urban land and grassland was higher than that of farmland were as follows. First, there were a small quantity of green belts and woods in the urban area in the land-use classification. Thus, the evapotranspiration capacity obtained through the inversion was high. Second, the main crop planted in the agricultural region of the Sanmenxia basin is winter wheat, and following the last 10 days of May when winter wheat enters the heading stage is the most critical period for its growth with high water demand. Nevertheless, the precipitation was relatively small after July 2019, so the overall crop growth in the whole area was poor, and the difference between grassland and farmland in daily average evapotranspiration was quite small due to drought. However, the vegetation type of farmland is simple, mainly being wheat, and the primary vegetation type in grassland is bush, with a relatively high vegetation diversity. The difference between the two land-use types in the evapotranspiration manifests that the plant diversity and their growth status in the area constitute one of the driving factors influencing the evapotranspiration.
3. The study shows that surface temperature and vegetation coverage in the river basin are key factors influencing the evapotranspiration process. The daily average evapotranspiration capacity of forest land was close to wetlands (e.g., marsh, beach land and artificial wetland). Meanwhile, its value was about twice those of farmland and grassland. By analyzing the surface temperatures (Figure 3(c)) obtained through the remote sensing inversion in 2019, it is shown that farmland and grassland were areas with high surface temperatures (both exceeded 23.5 °C), while the areas with low surface temperatures (below 23.5 °C) mainly appeared in the southeast mountainous area, with forest land being the primary land-use type. By comparing the surface temperature results in July 2014, the average surface temperatures of both farmland and grassland were over 37.7 °C, and that of forest land was 32.1 °C. This indicates that surface temperature is the primary factor influencing the evapotranspiration process, and the evapotranspiration capacity is relatively small in the area with relatively high surface temperature. In addition, the vegetation coverage of forest land within the mountainous area is higher than that of farmland and grassland. The dense forest and favorable vegetation growth can facilitate the evapotranspiration of forest land, which is consistent with the conclusion of Gu *et al.* (2021).

4.2. Discussion

4.2.1. Comparative analysis of wetland landscapes and non-wetland landscapes in evapotranspiration capacity

The remote sensing images in both periods reflect that the evapotranspiration capacity of wetland landscapes is higher than that of non-wetland landscapes. According to the evapotranspiration results obtained from remote sensing in this area in 2019, the daily average evapotranspiration capacity corresponding to different landscape types could be divided into three gradient ranges: 4.82–6.01 mm/d (farmland, urban land and grassland), 9.09–9.91 mm/d (artificial wetland, beach land, marsh and forest land) and 13.25–13.49 mm/d (lake and river). Except for forest land, the daily average evapotranspiration capacity of all wetland types was greater than 9.09 mm/d, and that of all non-wetland types was smaller than 6.01 mm/d. In the whole river basin, wetland's daily average evapotranspiration capacity was 12.17 mm/d, while that of non-wetland was 7.26 mm/d (accounting for only 59.7% of the former), indicating a great difference between the two land-use types (Figure 5). A similar feature was manifested in the remote sensing inversion results in the same period in 2014, the daily average evapotranspiration capacity of wetland was 8.71 mm/d and that of non-wetland was 4.93 mm/d, accounting for merely 56.6% of the former (Figure 5).

By comparing the daily evapotranspiration results in July 2014 and July 2019, wetland landscapes' overall daily average evapotranspiration capacity in 2014 was 3.46 mm/d which was lower than that in 2019. On the other hand, the daily evapotranspiration capacity of non-wetland landscapes in July 2014 was 2.33 mm/d which was smaller than that in July 2019. Overall, the daily evapotranspiration of wetland landscapes was higher than that of non-wetland landscapes, and the latter was approximately 60% of the former.

The evapotranspiration capacity of wetland landscapes is higher than that of non-wetland landscapes because of sufficient water supply to wetland vegetations, i.e., the higher the temperature (within a certain range of threshold) is, the faster the evapotranspiration process will be. The area of lakes and rivers in the Sanmenxia basin accounts for 67.5% of the total wetland area, and the broad water surface area has accelerated the evapotranspiration process and facilitated the hydrologic cycle of the whole wetland ecosystem. This resulted in a relatively high evapotranspiration capacity of wetland landscapes.

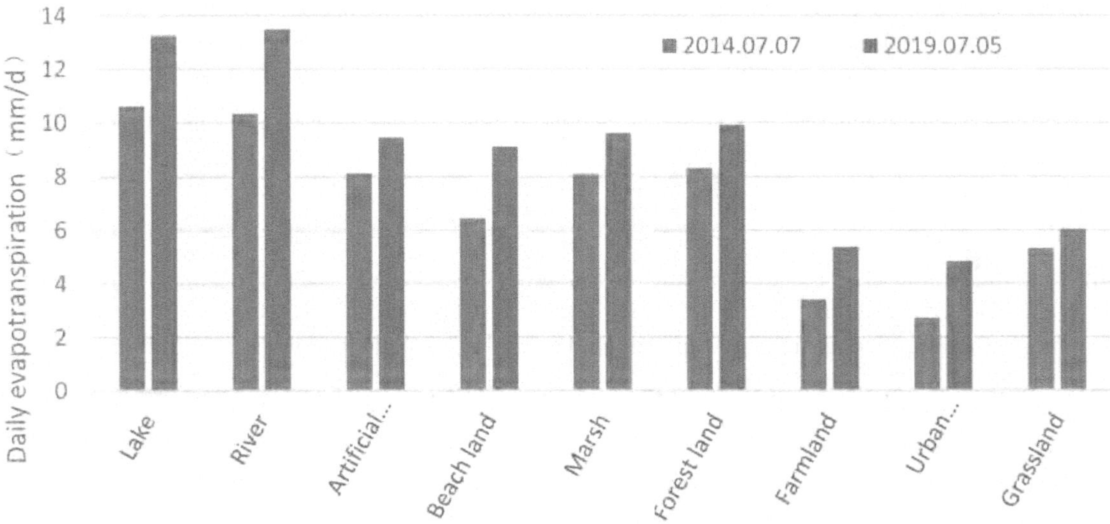

Figure 5 | Comparison of daily evapotranspiration capacity for wetland landscapes and non-wetland landscapes.

4.2.2. Evapotranspiration difference analysis under landscape patterns in different periods

Based on a comparison of area changes of landscape types in 2014 and 2019, we found that the area changes of farmland, urban land, artificial wetland and marsh were little, but the areas of forest land and grassland changed to the greatest extent. In addition, they experienced mutual transformation. In the southwest and northern mountainous areas, the forest land area in 2019 was 312.65 km^2 more than that in 2014, while the grassland area was 310.24 km^2 less than that in 2014. Influenced by the water-level regulation of the Sanmenxia hydro-junction project (Mao *et al.* 2006; Wang *et al.* 2011), the areas of beach land and river wetland along the river on the lower reaches of the reservoir were mutually transformed. Namely, partial beach land was gradually converted into grassland and marsh.

By analyzing the evapotranspiration inversion results in the same month of the two periods, it can be found that the evapotranspiration of forest land was higher than that of grassland, while that of river was higher than that of beach land. This indicates that the evapotranspiration capacity of different surface features is unrelated to their areas but correlated with land cover and soil texture. The evapotranspiration capacities of all surface features in July 2014 were lower than those in July 2019. Specifically, the evapotranspiration changes of river, lake, beach land and urban land were the most obvious. However, the relative difference values among marsh, forest land and grassland were small. They reflect that other than soil moisture status and meteorological factors (e.g., daily average air temperature), the evapotranspiration capacity is also greatly affected by the interannual variation of vegetation cover characteristics, which is consistent with the conclusion of Chen & Zeng (2013).

5. CONCLUSIONS

The daily evapotranspiration capacities of wetlands and non-wetlands in the Sanmenxia river basin were quantitatively estimated *via* the SEBAL model, and the remote sensing inversion results were verified using the FAO Penman–Monteith formula. The verification results demonstrate that the SEBAL model is relatively accurate and feasible in studying evapotranspiration in the Sanmenxia river basin.

Significant differences exist between wetland landscapes and non-wetland landscapes in different land-use types in the aspect of evapotranspiration capacity. The daily average evapotranspiration capacities of river and lake are the highest, followed by those of forest land, marsh, beach land and artificial wetland, and then those of grassland, urban land and farmland. The non-wetland landscapes (forest land, grassland, farmland and urban land) account for 97.23% of the total basin area, with the daily average evapotranspiration capacity of only 7.26 mm/d. On the other hand, the wetland landscapes (river, lake, marsh, beach land and artificial wetland) account for 2.77% of the total basin area, with the daily average evapotranspiration capacity of 12.17 mm/d. In a sense, the daily evapotranspiration capacity of the non-wetland area is about 59.7% of the wetland area.

Beyond meteorological, hydrological and solar radiation factors, the interannual variations of water surface area, vegetation diversity, vegetation growth status and vegetation coverage, along with surface temperature, are also important driving factors of wetland evapotranspiration.

It is worth mentioning that using remote sensing to estimate evapotranspiration of wetlands has important scientific significance:

1. Evapotranspiration (ET) includes the evaporation of water from the surface and plant surfaces and the transpiration of water through the surface and inside plants, which is an essential part of the hydrological cycle. At the same time, evaporation and evapotranspiration are also the main heat balance items because changing water into gas requires heat absorption. Surface heat and water balance largely determine the composition and evolution of geographical environment. A clear evapotranspiration process is helpful to understand the energy balance and water cycle. At the same time, it also helps to correctly assess the impact of climate and human activities on farmland ecosystems.
2. Wetland is an essential part of the ecosystem. Wetland hydrological process controls the biological and chemical processes of wetland which is the most important factor affecting the function and evolution of wetland. Evaporation is an essential part of the wetland water cycle. Water loss is the leading way to correspond with evaporation. The latent heat flux is the primary energy expenditure wetland evaporation, so to speak to a great extent, affecting the wetland water heat balance for the wetland system. We investigated its relationship between rainfall and evaporation, quantitative estimation of evaporation amount. It has health significance to the whole wetland ecosystem.
3. Rapid acquisition of evapotranspiration from remote sensing is of great significance for ecosystem protection, which also helps decision-makers to make rapid judgments.

ACKNOWLEDGEMENTS

This work was supported by the National Natural Science Foundation of China (NSFC) Program (Nos 51679256 and 51179207), the Major National Science and Technology Program of Water Pollution Control and Governance (No. 2018ZX0711103), the Science and Technology Program of China Institute of Water Resources and Hydropower Research (Nos WE0145B012021 and WE0163A052018) and the Open Research Fund of Guangxi Key Laboratory of Water Engineering Materials and Structures, Guangxi Institute of Water Resources Research (No. GXHRI-WEMS-2020-13).

CONFLICT OF INTEREST

No conflict of interest exits in the submission of this manuscript.

DATA AVAILABILITY STATEMENT

All relevant data are included in the paper or its Supplementary Information.

REFERENCES

Allen, R. G., Pereira, L. S. & Raes, D. 1998 *Crop Evapotranspiration – Guidelines for Computing Crop Water Requirements*. FAO Irrigation and Drainage Paper 56. FAO, p. 56.

Bastiaanssen, W. 1995 *Regionalization of Surface Flux Densities and Moisture Indicators in Composite Terrain: A Remote Sensing Approach Under Clear Skies in Mediterranean Climates*. Wageningen Agriculture University, Netherland, p. 273.

Bastiaanssen, W. G. M., Menenti, M., Feddes, R. A. & Holtslag, A. A. M. 1998 A remote sensing surface energy balance algorithm for Land (SEBAL) 1. Formulation. *Journal of Hydrology* **212–213**, 198–212.

Cao, Y., Gao, L. & Guo, M. 2017 Characteristics of evapotranspiration for different types of land use in Liaoning Province. *Advances in Science and Technology of Water Resources* **37** (02), 14–19.

Chen, H. & Zeng, X. 2013 Impact of vegetation interannual variability on evapotranspiration. *Acta Ecologica Sinica* **33** (14), 4343–4353.

Cheng, Y. M. B. & Brutsaert, W. 2005 Pathology of Monin-Obukhov similarity in the stable boundary layer. *Journal of Geophysical Research* **110** (D6), 25.

Cui, H., Wang, L. C., Wang, H. J., Xiao, W. H., Hou, B. D. & Gao, B. 2021 Temporal and spatial changes of actual evapotranspiration and its relationship with meteorological factors in the Three Gorges Reservoir area. *Research of Soil and Water Conservation* **193** (04), 202.

Ekwueme, B. N. & Agunwamba, J. C. 2020 Modeling the influence of meteorological variables on runoff in a tropical watershed. *Civil Engineering Journal* **6** (12), 2344–2351.

Gu, J., Xue, H., Dong, G., Zhou, L., Li, J., Dang, S. & Li, S. 2021 Effects of NDVI/land-use on spatiotemporal changes of evapotranspiration in the Yellow River Basin. *Arid Land Geography* **44** (01), 158–167.

Harrison, L. P. 1963 Fundamental concepts and definitions relating to humidity. In: *Humidity and Moisture*, Vol. 3 (Wexler, A., ed.). Reinhold Publishing Company, New York.

Li, H., Wang, H., Kong, Y. & Li, L. 2012 Estimation of evapotranspiration in Yellow River delta wetland based on two-source energy balance (TSEB) model. *Remote Sensing Technology and Application* **01**, 58–67.

Liu, C., Gao, W. & Gao, Z. 2009 Estimation of regional surface evapotranspiration using MODIS products. *Advances in in Water Science* **06**, 782–788.

Liu, M., Hu, D., Yu, C. & Wang, S. 2020 Temporal and spatial change characteristics of growing season evapotranspiration its cause analysis in Liaohe River delta wetland, China. *Acta Ecologica Sinica* **40** (02), 701–710.

Lv, J., Jiang, W., Wang, W., Chen, K., Deng, Y., Chen, Z. & Li, Z. 2018 Wetland landscape pattern change and its driving forces in Beijing-Tianjin-Hebei region in recent 30 years. *Acta Ecologica Sinica* **12**, 4492–4503.

Mao, Z., Peng, W., Wang, S. & Zhou, H. 2006 Study on effects of water level variation of the Sanmenxia reservoir on wetland hydrological processes. *Journal of China Institute of Water Resources and Hydropower Research* **01**, 36–41.

Tsanov, E., Ribarova, I., Dimova, G., Ninov, P. & Makropoulos, C. 2020 Water stress mitigation in the Vit River basin based on WEAP and Matlab simulation. *Civil Engineering Journal* **6** (11), 2058–2071.

Wang, X., Wang, J., Ge, Y. & Zhang, P. 2011 Environmental impact analysis on shore groundwater in the case of low water level of Sanmenxia Reservoir. *Yellow River* **10**, 39–41.

Wang, J.-Y., Han, Y.-P., Liu, L.-L., Wang, L., Bing, L.-F., Yin, Y. & Xi, F.-M. 2017 Progress in studies of the impacts of urbanization on wetland. *Science & Management* **13** (03), 61–65.

Wu, W. 2016 Estimation and discussion of regional evapotranspiration in Anhui Province based on remote sensing model. In: *The 33rd Annual Meeting of the Chinese Meteorological Society S21 Development and Application of new Generation Meteorological Satellite Technology*. Chinese Meteorological Society, p. 4.

Yi, W. 2013 *A Study of Changes on Landscape Pattern and Driving Factors of Wetland in Sanmenxia Area Based on Remote Sensing*. China University of Geosciences, Beijing (S2), 88.

Yuan, Y., Yan, D., Jia, Y. & Hu, D. 2012 The impact of scale change of land use on evapotranspiration in Nenjiang River basin. *Journal of Hydraulic Engineering* **43** (12), 1440–1446.

Zeng, L., Song, K., Zhang, B. & Du, J. 2008 Retrieval of evapotranspiration in Zhalong wetland based on SEBAL model. *Agrometeorology of China* **04**, 420–426.

Zhang, D. 2011 SEBAL Model Improved and Study on the Relationship between ET and luce. Beijing Forestry University, Beijing.

Zhang, X. 2013 Estimation of evapotranspiration in wetland and surrounding area of Poyang Lake based on remote sensing technology. In: *Proceedings of the Annual Conference – S2 Lake Management, Development and Protection*. Chinese Hydraulic Engineering Society, p. 8.

Zheng, Q.-q., Dai, P.-c., Zhang, J.-y. & Wu, Z.-p. 2020 Evapotranspiration in the Jinghe River Basin based on the surface energy balance system. *Arid Zone Research* **37** (06), 1378–1387.

Zhou, B., Yin, J., Jin, B. & Zhu, L. 2014 Degradation of Wuchsng Lake wetland and its causes during 1980–2010. *Acta Geographica Sinica* **11**, 1697–1706.

First received 23 June 2021; accepted in revised form 17 September 2021. Available online 7 October 2021

© 2022 The Authors Impact of Climate Change on Hydrology and Water Resources doi: 10.2166/wcc.2021.432

An estimation of the evapotranspiration of typical steppe areas using Landsat images and the METRIC model

Jun Wang[a,b,*], Heping Li[a,b] and Haiyuan Lu[a,b]

[a] Yinshanbeilu Grassland Eco-hydrology National Observation and Research Station, China Institute of Water Resources and Hydropower Research, Beijing 100038, China

[b] China Institute of Water Resources and Hydropower Research, Institute of Water Resources for Pastoral Area, Hohhot 010020, China

*Corresponding author. E-mail: slwj1988@163.com

ABSTRACT

Remote sensing excels in estimating regional evapotranspiration (ET). However, most remote sensing energy balance models require researchers to subjectively extract the characteristic parameters of the dry and wet limits of the underlying surfaces. The regional ET accuracy is affected by wrong determined ideal pixels. This study used Landsat images and the METRIC model to evaluate the effects of different dry and wet pixel combinations on the ET in the typical steppe areas. The ET spatiotemporal changes of the different land cover types were discussed. The results show that the surface temperature and leaf area index could determine the dry and wet limits recognition schemes in grassland areas. The water vapor flux data of an eddy covariance system verified that the relative error between the $ET_{d,METRIC}$ and $ET_{d,GES}$ of eight DOY_s (day of the year) was 18.8% on average. The ET_{METRIC} values of the crop growth season and the ET_{IMS} of eight silage maize irrigation monitoring stations were found to have a relative error of 11.1% on average. The spatial distribution of the ET of the different land cover types in the study area was as follows: $ET_{water} > ET_{arable\ land} > ET_{forest\ land} > ET_{unutilized\ land} > ET_{grassland} > ET_{urban\ land}$.

Key words: dry and wet limits, evapotranspiration, METRIC model, remote sensing, typical steppe areas

HIGHLIGHTS

- Landsat images and the METRIC model estimating regional ET had obtained reliable and accurate results in typical steppe areas.
- The water vapor flux data of an eddy covariance system conducted the verification.
- The surface temperature and leaf area index could determine the dry and wet limits recognition schemes in grassland areas.

1. INTRODUCTION

Steppe ecoregions are important natural resources in northern China, and are mainly distributed in the semi-arid regions of Eurasia where the vast majority of the rivers are seasonal inland rivers (Batnasan 2003; Tong *et al.* 2004). Due to this hydrogeographical feature, it can be expected that the precipitated water will result in only minimal discharges to the rivers or ground water. For example, the water exchange between the grassland areas and the atmosphere through precipitation and evapotranspiration (ET) will be the main branch of the hydrological cycle in steppe environments (Yamanaka *et al.* 2007). In recent years, large-scale human activities have changed the gall atmosphere, surfaces, soil, and underground processes of the natural water cycle, which had led to the hydrologic cycles presenting obvious 'natural-artificial' dual properties in the modern environmental settings (Wang *et al.* 2006). ET is an important component of the prairie hydrologic cycles (Allen *et al.* 1998). When the ET is disturbed, the results will cause changes in the regional hydrological balances, which may have major impacts on the regional water storage processes (Bradley 1999). Therefore, the accurate estimations of the ET in grassland areas are of great significance to hydrological process change simulations and water resource management decision-making, as well as grassland drought monitoring and evaluation processes (Liou & Kar 2014; Pedro-Monzonís *et al.* 2015; Numata *et al.* 2017; Talsma *et al.* 2018).

Many ET monitoring and estimation methods have emerged with related theoretical research. These include the water balance method, aerodynamics methods, and scintilla meter measurements (Bowen 1926; Thornthwaite 1948; Howell *et al.* 1991; Rana & Katerji 2000). These methods have displayed the ability to describe the characteristic rules of the ET processes at micro and farmland scales in detail. In addition, they have also provided important references and basic information for water vapor exchanges on underlying surfaces, decision-making concerning farmland irrigation, comprehensive water resource management, and so forth (Long & Singh 2013; Liou & Kar 2014; French *et al.* 2015; Xu *et al.* 2015). However, in terms of regional ET estimation, owing to fluctuations in weather conditions and the dynamic nature of water-heat transfer processes (Gao *et al.* 2008), it has been determined that these traditional point-scale monitoring methods and approaches cannot fully reflect the actual characteristics of spatial changes in ET (Su *et al.* 2003; Cai *et al.* 2009; Monteith & Unsworth 2013; Taherparvar & Pirmoradian 2018). It has been determined that remote sensing technology can offer a new approach for regional ET estimation. Because of the spatial continuity and large span characteristics of remote sensing, the water that is consumed during ET can be derived directly, without the need to quantify other complex hydrological processes (Trezza *et al.* 2013).

ET models using remote sensing technology have been developed in recent years, including empirical methods and mechanism methods. Many achievements in theory innovation and method techniques have been presented (Khand *et al.* 2017a). Among these proposed methods, two-source energy balance (TSEB), surface energy balance index (SEBI), simplified surface energy balance index (S-SEBI), surface energy balance algorithm for land (SEBAL), mapping evapotranspiration at high resolution with internalised calibration (METRIC) model have been widely used to calculate regional ET based on the theory of energy balance (Menenti 1993; Norman *et al.* 1995; Bastiaanssen *et al.* 1998a; Roerink *et al.* 2000; Su 2002; Allen *et al.* 2007b; Yang *et al.* 2010). As a typical representative of estimating regional ET by using remote sensing, the METRIC model is a variant of the SEBAL model that implements the residual method to calculate regional ET (Spiliotopoulos *et al.* 2017; Zamani Losgedaragh & Rahimzadegan 2018). The METRIC model has been successfully used in many areas, as a physical model for high-precision estimation of regional ET (Allen *et al.* 2007a; Allen *et al.* 2013; Morton *et al.* 2013; Paço *et al.* 2014; Reyes-González *et al.* 2017; Spiliotopoulos *et al.* 2017).

The METRIC model is very similar to the SEBAL model (Zamani Losgedaragh & Rahimzadegan 2018), in which the computing core uses two limit pixels (dry and wet pixels) for solving the aerodynamic resistance and geothermic step above zero plane displacement. This method has been found to greatly increase the calculation efficiency of the sensible heat flux. However, the limit pixels which are subjectively chosen by the model users must be the upper and lower critical limits of the sensible heat flux/latent heat flux in the study area (Foolad *et al.* 2018), to ensure the calculation accuracy of the regional ET (Allen *et al.* 2007a, 2007b). During the previous practical application processes, different research studies tended to diversify the understanding of the underlying surface characteristics in study areas using inconsistent subjective selection criteria of the dry and wet limits. This resulted in the model's calculation accuracies being negatively influenced by the lack of ideal dry and wet pixel determinations (Allen *et al.* 2011a, 2011b, 2013), which then triggered space ambiguities in the estimations of the ET using remote sensing technology. For example, different research studies have used the same basic data and methods but obtained different ET estimation results. To address this problem, Tasumi and Khand suggested that the dry and wet limits should be selected according to the surface temperature (T_s) and leaf area index (LAI), or the normalized difference vegetation index (NDVI) related trend charts (Tasumi 2003; Khand *et al.* 2017b). Scientist established the dry and wet limit conditions for given pixels and periods by assuming the relatively stable energy differences between the dry and wet limits within certain periods (for example, daytime or eight day periods) (Scientist *et al.* 2013). Allen proposed that the dry and wet limits should be selected by CIMEC processes and based on setting the relative location conditions of the weather stations and farmlands in a study area (Allen *et al.* 2013). Related scholars have found that surface temperature has an important impact on evapotranspiration (Jiang *et al.* 2020; Hu *et al.* 2021). Similar research results also showed that LAI was one of the important influencing factors of evapotranspiration (Xue *et al.* 2020; Volik *et al.* 2021; Wang *et al.* 2021). For the grassland region with few meteorological stations, arid climate, and sparse vegetation, there are still few studies by relevant scholars and experts (Yu *et al.* 2020). In the process of understanding the METRIC model in similar situations, we have not found more scholars to explore the relevance. Therefore, we try to use a new dry-wet boundary method to verify and determine the corresponding dry-wet boundary.

In general, it is easy for researchers to make subjective errors when selecting sites independently. To some extent, the errors caused by personal factors are avoided based on actual T_s and LAI, and the selection is made from an objective perspective as far as possible. The objectives of this study were to identify the dry and wet limits in grassland areas by analyzing the spatial distribution characteristic relationship of T_s and LAI; the water vapor flux data obtained by an eddy covariance system and the crop water consumption data of an irrigation monitoring station calculated by the water balance method were used for evaluating and validating the ET estimation results using Landsat images and the METRIC model. On this basis, the spatiotemporal distribution of the regional ET during the crop growth season of 2011 in different land cover types was studied using eight effective sunshine images of Landsat-5 and Landsat-7.

2. MATERIALS AND METHODS

2.1. Study area

The Xilin River Basin is a typical inland river basin in the Inner Mongolia typical steppe areas of China (Chen 2002). The basin, with an area of approximately 11,172 km^2 (115.53° – 117.25°E, 43.41° – 44.63°N), has a typical continental temperate semi-arid climate. In accordance with the statistical data of the Xilinhot Station detailed on the National Meteorological Data website, the multi-year (period 1971–2011) average precipitation, ET from the water surface, air temperature, and wind speed are 263.5 mm, 1,904 mm, 3 °C, and 3.4 m s^{-1}, respectively. The average precipitation gradually decreases from the southeast to northwest (Sheng 2017). The National Scientific Research Station of the Xilin Gol Grassland Ecosystem in Inner Mongolia (GES) and the irrigation monitoring station of the Institute of Water Resources for Pastoral Areas (IMS) are both located within the study area (Figure 1).

It was observed that the land cover in the study area included six types of arable land, forest land, grassland, water, urban land and unutilized land (Figure 1). The arable land mainly referred to the irrigated artificial grassland used for planting silage maize and a small amount of alfalfa. The crops were found to be influenced by climatic conditions, which generally resulted in a short growth season from May to September. The forest land was mainly distributed at the southern foothills of the Greater Khingan Mountains in the southeastern section of the study area. The grassland areas were the main land cover type in the study area, and accounted for more than 90% of the entire basin, which mainly included Stipa grandis and Chinensis (Bai *et al.* 2004; Li *et al.* 2016). The water areas mainly comprised the Xilin River, which was observed to transverse the entire basin. The unutilized land, such as bare and wetland areas, was mainly distributed on both sides of the river channels. Xilinhot City, which was located in the middle sections of the river basin, was the main urban land area.

2.2 Data

(1) Landsat data

In this study, Landsat-5 and Landsat-7 images were chosen to estimate ET. In the data bands of these images, visible light and near-infrared bands have a high spatial resolution of 30 m, which satisfies the validation requirements for the spatial accuracy of ET estimation results in this study. In combination with the cloud cover of the study area, the data were divided into a calibration period and a verification period. The effective 8-day sunshine data during the crop growth season from 2006 to 2009 were used to study the selection scheme of the dry and wet pixels in the METRIC model, and the effective 8-day

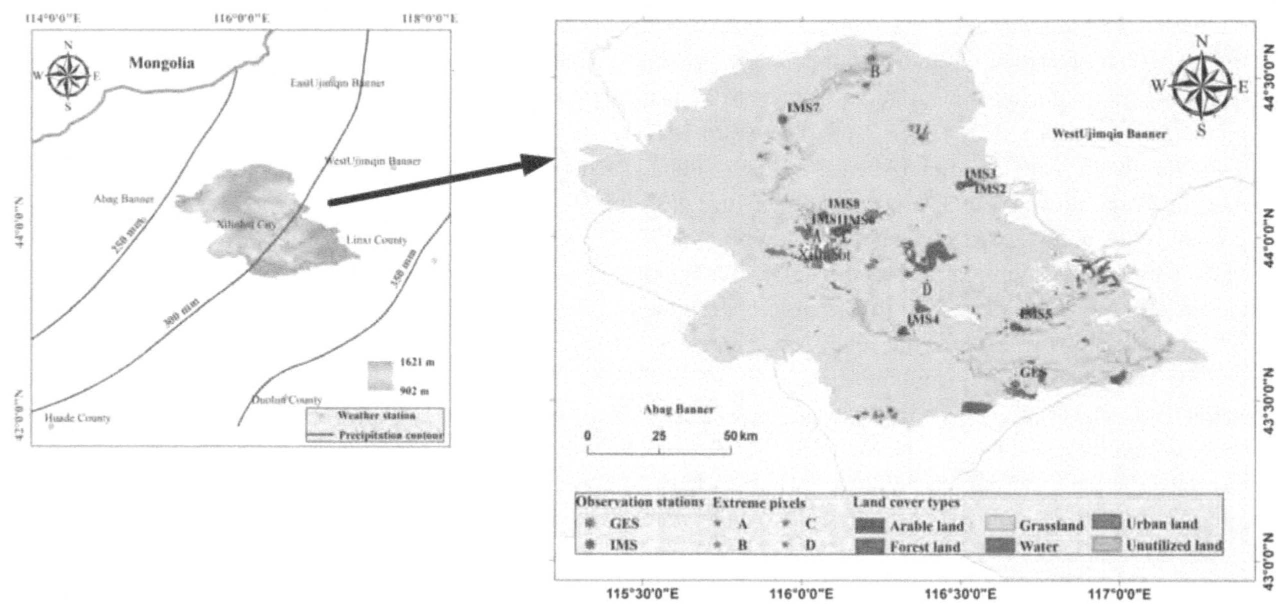

Figure 1 | Study area location.

sunshine data during the crop growth season (from May to September) of 2011 (day of the year from 121 to 273) were used to calculate the long-time scale of ET (Table 1). All the data were obtained from the United States Geological Survey (http://glovis.usgs.gov/).

(2) Meteorological data

Meteorological data included air temperature, wind velocity, water vapor pressure, solar radiation, humidity, and precipitation in the seven weather stations of East Ujimqin Banner, Abag Banner, Huade County, West Ujimqin Banner, Xilinhot City, Linxi County, and Duolun County (Figure 1). Among these, the air temperature data were used to calculate the surface temperature and net radiation by the METRIC model; the wind velocity data to calculate the aerodynamic resistance by the METRIC model; and the water vapor pressure data to calculate atmospheric transmittance and surface albedo by the METRIC model (Bastiaanssen *et al.* 1998a; Allen *et al.* 2007a). Other parameters such as solar radiation and humidity were used to calculate reference crop evapotranspiration (ET_0) by FAO Penman-Monteith equation (Allen *et al.* 1998). The meteorological data were obtained from the National Meteorological Data Information Center (http://data.cma.cn/).

(3) Evapotranspiration measurement data

The ET measurement data were obtained from an eddy covariance system in GES, and the eight irrigation monitoring stations (IMS) (Figure 1). (a). An eddy covariance system was an open system composed of a CSAT-3 three-dimensional ultrasonic anemometer and a LI7500 infrared analyzer. (b). The crop cultivated in the monitoring station was silage maize. The average volumetric weight of soil was 1.54–1.82 g cm-3 for a thickness varying from 0 to 100 cm. The field water holding capacity (θ_f) was 14.3%–19.0% (accounting for the dry soil weight), and the buried depth of the groundwater level was over 3 m. Based on the water balance method, the water consumption data of the silage maize growth period were calculated to evaluate the results of the METRIC model.

2.3. Theoretical basis

2.3.1. METRIC model

The SEBAL model is the theoretical basis of the METRIC model (Bastiaanssen *et al.* 1998a, 1998b; Allen *et al.* 2007b). In the model, by considering the interference of the underlying surface elevation, slope gradient, slope direction, and other factors, the surface information obtained through remote sensing are converted to surface albedo, surface emissivity, vegetation index, roughness length, surface temperature, and other characteristic parameters. This model is combined with the meteorological observational data of the underlying surfaces to calculate the regional ET (Allen *et al.* 2007b; González-Dugo *et al.* 2012). The theoretical equation is given as follows:

$$LE = R_n - G - H \tag{1}$$

where LE is the latent energy consumed by ET; R_n is the net radiation; G is the soil heat flux; and H is the sensible heat flux. Each energy component is generally expressed in W m^{-2}.

Table 1 | Landsat-5/7 TM data information (PATH 124/ROW 29 and 30)

Calibration period				Verification period			
date	**day of the year (DOY_s)**	**cloud coverage**	**sensor**	**date**	**day of the year (DOY)**	**cloud coverage**	**sensor**
2006/8/4	216	0%	Landsat-5	2011/4/12	102	1%	Landsat-5
2006/9/21	264	0%	Landsat-5	2011/5/14	134	8%	Landsat-5
2007/4/17	107	14%	Landsat-5	2011/5/22	142	23%	Landsat-7
2007/7/6	187	0%	Landsat-5	2011/8/2	214	0%	Landsat-5
2008/8/25	238	0%	Landsat-5	2011/8/10	222	0%	Landsat-7
2008/9/26	270	0%	Landsat-5	2011/9/11	254	0%	Landsat-7
2009/6/25	176	6%	Landsat-5	2011/9/19	262	2%	Landsat-5
2009/8/12	224	0%	Landsat-5	2011/10/5	278	6%	Landsat-5

(1) Net radiation

R_n denotes the difference in the whole-band radiation quantity between the downward projection from the sky (including the sun and the atmosphere) and the upward projection from the Earth's surface (including soil, plants, and water). This parameter represents the main power source for the energy and water transmission of the underlying surfaces (Allen *et al.* 2007b):

$$R_n = (1 - \alpha)K_{in} + (L_{in} - L_{out}) - (1 - \varepsilon)L_{in} \tag{2}$$

where α is the surface albedo (dimensionless); K_{in} is the incoming short-wave radiation (W m^{-2}); L_{in} is the incoming long-wave radiation (W m^{-2}); L_{out} is the outgoing long-wave radiation (W m^{-2}); and ε is the broad-band surface emissivity (dimensionless).

(2) Soil heat flux

G refers to the part of the energy stored in vegetation and soil due to conduction, and its changes are mainly affected by net radiation, surface temperature, and the vegetation cover on underlying surfaces. For the comprehensive analysis in this study, the achievements of Wright and Bastiaanssen are referred as follows (Wright 1982; Bastiaanssen *et al.* 1998a):

$$G = \frac{T_s - 273.15}{\alpha}(0.0038\alpha + 0.0074\alpha^2) \times (1 - 0.978NDVI^4)R_n \tag{3}$$

where T_s is the surface temperature (K); and NDVI is the normalized difference vegetation index (dimensionless).

(3) Sensible heat flux

H is a parameter describing the exchange of energy between land surfaces and atmosphere, in which energy is transferred from the Earth's surface to the atmosphere through convection or conduction. The calculation of H in the METRIC model is based on the following assumptions: dry and wet pixels exist in the study area. There is a linear relationship between the temperature difference dT ($dT = T_1 - T_2$) at z_1 and z_2 above the height of the zero-plane displacement and T_s. The space interpolation of the meteorological data on the underlying surfaces can then be avoided. Furthermore, the errors due to T_s are corrected to calibrate the calculation results of each pixel in remote sensing imaging through the elevation, slope gradient, and slope direction:

$$H = \rho_{air} c_p \frac{dT}{r_a} \tag{4}$$

where ρ_{air} is the air density (kg m^{-3}); c_p is the air specific heat capacity at a constant pressure (1,004 J kg^{-1} K^{-1}); and r_a is the aerodynamic resistance between two near-surface heights, which has been calculated by the Monin–Obukhov theory (Allen *et al.* 2007b). The calculation equations are given as follows:

$$dT = aT_{s,adtum} + b \tag{5}$$

$$T_{s,adtum} = T_s + 0.0065\Delta Z \tag{6}$$

where a and b are empirical constants; $T_{s,adtum}$ is the surface temperature adjusted to a common elevation data for each image pixel by using a digital elevation model and customized lapse rate (K); for non-planar areas, it is necessary to use the numerical elevation difference ΔZ for the corrections; and ΔZ is the difference between the elevation of each pixel and the reference elevation (m).

Dry and wet pixels are the two extreme pixels required for calculation using the METRIC model. Dry pixels refer to the dry idle wasteland or bare land areas without vegetation cover; these pixels are characterized by high temperature and almost zero ET (Bastiaanssen *et al.* 1998a; Allen *et al.* 2007b). In this study, dry pixels approximately satisfied LE $\approx$ 0. The dry end of the dT function is estimated by rearranging (Equation (4)) for a selected dry pixel in the image:

$$dT_{dry} = \frac{(R_n - G)_{dry} r_{adry}}{\rho_{airdry} c_p} \tag{7}$$

where $r_{a\ dry}$ is r_a computed for the roughness and stability conditions of the dry pixel; and $\rho_{air\ dry}$ is ρ_{air} calculated at the dry pixel.

Wet pixels refer to areas with adequate water supply, dense vegetation growth, low temperatures and potential ET levels in the remote sensing images (Bastiaanssen *et al.* 1998a; Allen *et al.* 2007b). These could include completely covered areas or open water masses with good plant growth. For the wet pixels, $H \approx 0$ is usually set in the SEBAL model, and such H values are considered in the METRIC model (Allen *et al.* 2007b).

$$dT_{wet} = \frac{(R_n - G - H)_{wet} r_{awet}}{\rho_{airwet} c_p} \tag{8}$$

where $r_{a\ wet}$ is ra computed for the roughness and stability conditions of the wet pixel; $\rho_{air\ wet}$ is ρ_{air} calculated at the wet pixel.

The distribution of dT is obtained by extracting the information regarding the dry and wet pixels (Daniel *et al.* 2017). The values for a and b in (Equation (5)) are estimated from these two pairs of dT and associated $T_{s,adtum}$.

$$a = \frac{dT_{dry} - dT_{wet}}{T_{s,adtum,dry} - T_{s,adtum,wet}} \tag{9}$$

$$b = dT_{dry} - aT_{s,adtum,dry} \tag{10}$$

(4) Surface parameters

Surface parameters contain α, ε, NDVI, T_s and momentum roughness length (Z_{om}) which are crucial in the atmospheric-surface interaction and consequently affect the ET estimation. Among them, α is computed by integrating band reflectance within the short-wave spectrum using a weighting function (Allen *et al.* 2007b); ε is computed using an empirical equation (Tasumi 2003); NDVI is the ratio of the differences in reflectivity for the near-infrared band and red band to their sum (Allen *et al.* 2007b); T_s is the single window algorithm proposed by Qin (Qin *et al.* 2010); Z_{om} for each image is specified by the corresponding NDVI (Allen *et al.* 2007b).

(5) Long-time scale of the daily evapotranspiration

In addition, it is assumed that the instantaneous ET ratio (EF_{inst}, $EF_{inst} = \mathrm{LE_{inst}}/(R_n\text{-}G)$inst) is equal to the daily ET ratio (EF_d, $EF_d = \mathrm{LE_d}/(R_n\text{-}G)_\mathrm{d}$), when the instantaneous ET is used to calculate the daily ET ($\mathrm{ET_d}$) (Bastiaanssen *et al.* 1998a, 1998b). In this study, to estimate long-time scale of the daily ET, the daily reference ET ratio ($\mathrm{ET_rF}$ is the ratio between $\mathrm{ET_d}$ and daily $\mathrm{ET_0}$, $\mathrm{ET_rF} = \mathrm{ET_d}/\mathrm{ET_0}$) of the eight $\mathrm{DOY_s}$ were calculated, and the crop growth season was divided into seven intervals (Table 1). Each interval consisted of two $\mathrm{DOY_s}$ of effective sunshine satellite images. Thereafter, by assuming that the $\mathrm{ET_rF}$ change between two adjacent effective sunshine days is linearly correlated with the DOY, interpolation was used to calculate the daily ET during the period:

$$ET_{d,DOY} = (c \times DOY + d) \times ET_{0,DOY} \tag{11}$$

$$c = \frac{ET_rF_n - ET_rF_m}{DOY_n - DOY_m} \tag{12}$$

$$d = ET_rF_n - c \times DOY_n \tag{13}$$

where c and d are constants; $\mathrm{ET_{d,DOY}}$ and $\mathrm{ET_{0,DOY}}$ are the daily ET and $\mathrm{ET_0}$ between two adjacent effective sunshine days respectively; and $\mathrm{ET_rF_n}$ and $\mathrm{ET_rF_m}$ are the $\mathrm{ET_rF}$ when the DOY is n and DOY is m, respectively (DOY_n: 278,262,254,222,214,142,134; DOY_m: 262,254,222,214,142,134,102).

A flow chart of the METRIC model is shown in Figure 2 (Chang *et al.* 2017).

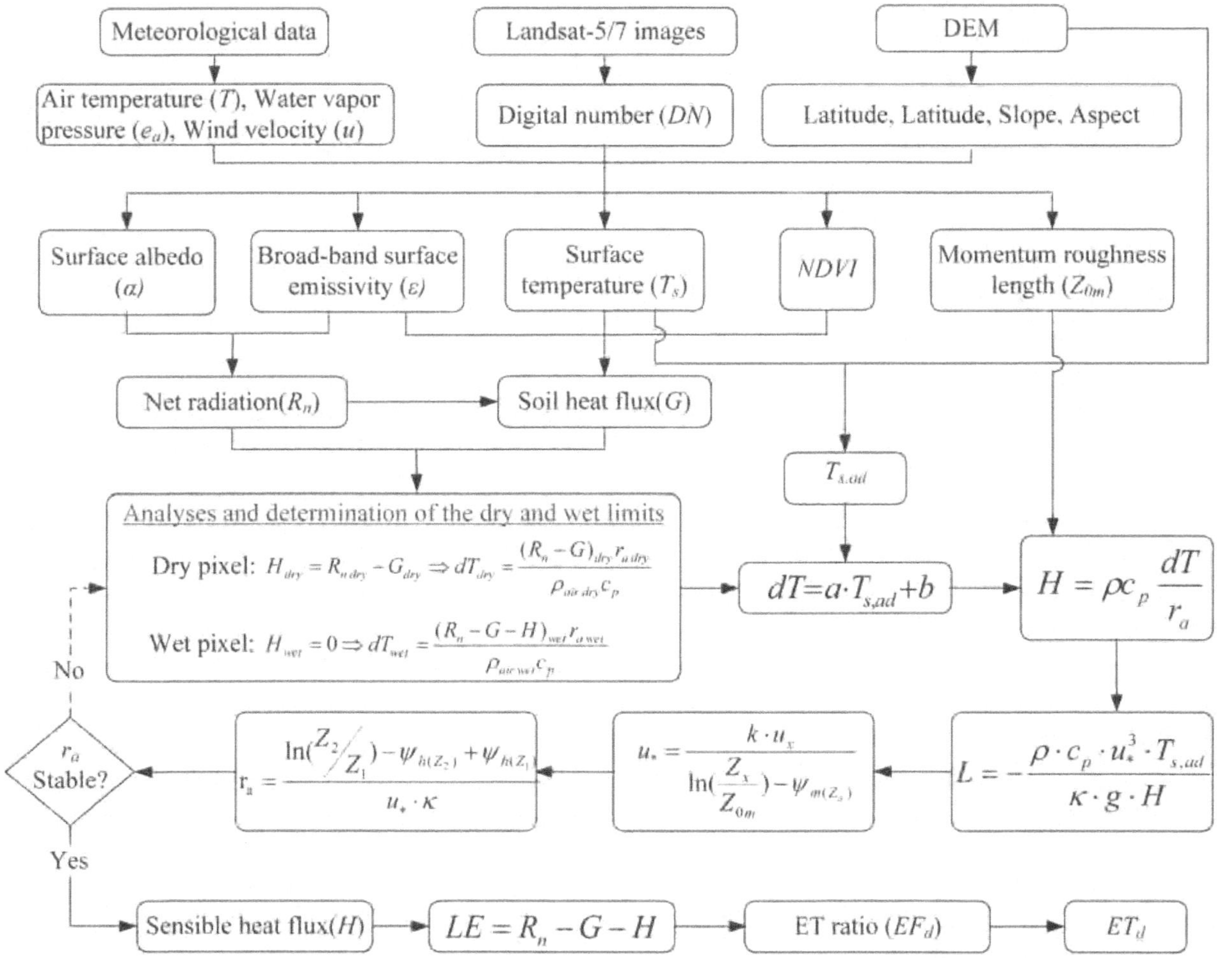

Figure 2 | Flowchart of the METRIC model. L is the Monin-Obukhov length, and ψ_m and ψ_h are the stability correction factors for momentum and sensible heat, respectively. Z is the elevation of each pixel (m). u_* is the friction velocity (m s^{-1}). κ is von Karman's constant (0.41).

2.3.2. Evapotranspiration of silage maize at irrigation monitoring stations

The evapotranspiration of silage maize consumed by the crops was calculated using the water balance method as follows (Kang 2007):

$$ET_{IMS} = P_a + I - L + K - \Delta W \tag{14}$$

where ET_{IMS} is the ET value of IMS; P_a is the effective precipitation; I denotes irrigation; L is the amount of deep leakage; and K is the groundwater recharge amount. Because the buried depth of the groundwater was greater than 3 m, as per the observations of a buried negative pressure meter, groundwater recharge was observed to be minimal; therefore, $K \approx 0$. ΔW represents the variation in planned soil moisture layer during the growth cycle of silage maize. All the above parameters are in units of mm.

The water balance equation can be used to calculate the water consumption of silage maize during the crop growth of the IMS. The result of this calculation is then used to verify the regional ET estimated by the METRIC model.

2.3.3. Comparison of ET_{METRIC} and ET_{IMS}

ET_{METRIC} was accurately estimated by computing the mean absolute error (MAE) and mean relative error (MRE).The values of MAE and MRE were computed as follows (Mayer & Butler 1993):

$$MAE = \frac{\sum_{i=1}^{t} |ET_{METRIC} - ET_{IMS}|}{t} \tag{15}$$

$$MRE = \frac{1}{t}\sum_{i=1}^{t}\left|\frac{ET_{METRIC} - ET_{IMS}}{ET_{IMS}}\right| \times 100\% \tag{16}$$

where ET_{METRIC} is the estimated ET by the METRIC model; and t is the number of observations.

3. RESULTS AND DISCUSSION

3.1. Analyses and determination of the dry and wet limits

The dry and wet limits in the METRIC model refer to the dry and wet pixels. Among these, the dry pixel refers to the dry idle wasteland or bare land areas without vegetation cover, and the pixels are characterized by high temperatures and almost zero ET (Bastiaanssen *et al.* 1998a; Allen *et al.* 2007b). In this study, the dry pixel approximately satisfied $H \approx R_n\text{-}G$, $LE \approx 0$. The wet pixels refer to the pixels with adequate water supply, dense vegetation growth, low temperatures and potential ET levels in the remote sensing images. These could include completely covered areas or open water masses with good plant growth. For the wet pixels, $H \approx R_n\text{-}G\text{-}LE$, of which $H \approx 0$ is usually set in the METRIC model. The distribution of the dT was obtained by extracting the information regarding the dry and wet pixels (Daniel *et al.* 2017). In the process of selecting dry and wet points, the image brought by underlying surface factors should be fully considered. The selection of dry and wet points should conform to the similar conditions of the surrounding environment to avoid the occurrence of other ground classes in the vicinity, so as to bring errors to the calculation results. Then, in order to reduce the ET estimation errors caused by spatial ambiguity in the extraction processes, for the typical steppe areas in the Xilin River Basin with the scarce meteorological station and sparse vegetation on the underlying surface, the identification scheme of this study was based on a spatial distribution feature relation graph of the T_s and LAI (Figure 3(a)). Then, by identifying the A, B, C and D limit pixels in the spatial distributions of the T_s and LAI, the different dry and wet pixel combination plans were set. The water vapor flux data measured by this study's eddy covariance system were used to evaluate the accuracy of the regional ET estimations of the different dry and wet pixel combinations, in order to further determine the dry and wet limit recognition scheme in the grasslands of the study area.

The recognition of the regional dry and wet pixels required the calculation of the T_s and LAI. Among these, a single-window algorithm was used to calculate the T_s that ranged between 280.31 and 312.09 K, with an average value of 302.67 K in the study area (Figure 4(a)). The soil regulatory vegetation index SAVI was applied to calculate LAI that ranged between 0.00 and 6.00, with an average value of 0.60 in the study area (Figure 4(b)) (Allen *et al.* 2007b). The T_s and LAI of 150 pixels were randomly extracted in this study. The A, B, C, and D points of the study area were determined in combination with the land coverage types (Figure 1). Among these, Point A was a bare land section in the study area, where the pixel type was mainly exposed and there was minimal vegetation. Point B was a wetland section in the study area. Point C was a farm-growing area, and the pixels in this area were artificially irrigated grassland with sufficient water. Point D was a dry

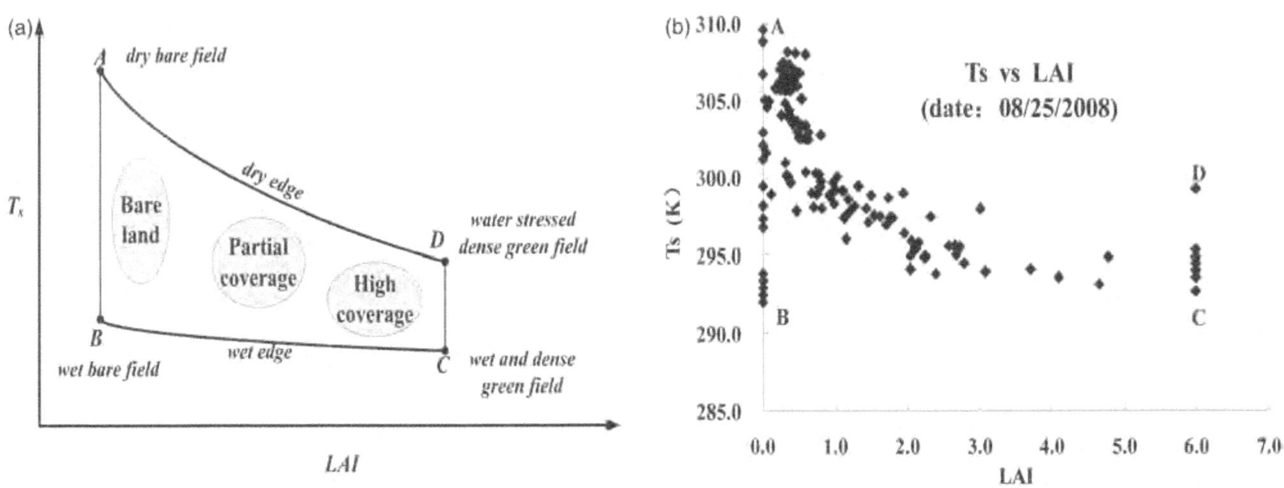

Figure 3 | General trend of surface temperature (T_S) and leaf area index (LAI).

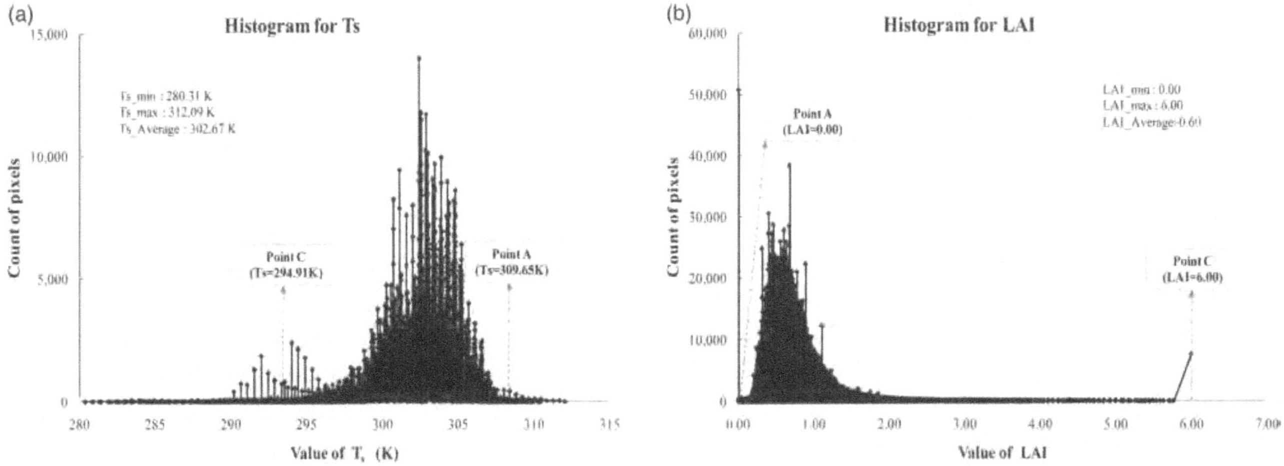

Figure 4 | Frequency histogram for surface temperature (T_S) and leaf area index (LAI).

agricultural field in the study area, where the dry agricultural land had high vegetation coverage due to the planting of crops. To test the influences of the different dry and wet limits on the estimation accuracy of the H and LE, and in accordance with the T_s-LAI related trend chart of the study area (Figure 3(b)), four dry and wet limit combinations $M_{dry,wet}$ were set ($M_{A,B}$;$M_{A,C}$; $M_{D,B}$; and $M_{D,C}$). Then, H, LE and basin ET on the satellite crossing day were calculated (Table 2).

Table 2 details the basin ET results which were calculated by the METRIC model on August 25, 2008. When compared with the measured data of the water vapor flux in the eddy covariance system of the GES, and under the premise of constant characteristic parameters and other energy components, the METRIC model's estimates of the regional ET results of four dry and wet limit combinations displayed major differences. Among the four combinations, the ET estimation results of $M_{A,C}$ were determined to be the most ideal. The daily ET estimation results ($ET_{d,METRIC}$) had a relative error of 9.2% with the measured data of water vapor flux by the eddy covariance system ($ET_{d,GES}$), followed by the $M_{A,B}$ combination with a relative error of 27.7%. The results of the $M_{D,B}$ and $M_{D,C}$ combinations were found to have the poorest estimation accuracies, with relative errors of approximately 100%. It was found that when the constraint of the ET was considered to be greater than 0, $ET_{d,METRIC} \approx 0$ was estimated by the two combinations. The reasons for these results were that in the two combinations of $M_{A,C}$ and $M_{A,C}$ with Point D as the dry pixel, and the mean value of the stable H was greater than (R_n-G) through the Monin-Obukhov similarity theory, which led to an unreasonable phenomenon of LE. In conclusion, the $M_{A,C}$ combination formed by Points A and C was determined to be a more suitable selection as the dry and wet pixels in the study area.

The calculation results revealed that the T_s of the dry pixels (Point A) were distributed in the front 5% of the frequency histogram, and those of the wet pixels (Point C) were distributed in the rear 5% of the frequency histogram (Figure 4(a)).

Table 2 | Effects of the different dry and wet limit combinations on the ET estimation

$M_{dry,wet}$	$T_{s,adtum,dry}$ $T_{s,adtum,wet}$ K	LAI_{dry} LAI_{wet} /	$(R_n\text{-}G)_{dry}$ $(R_n\text{-}G)_{wet}$ W m^{-2}	*a* /	*b* /	H^* W m^{-2}	$(R_n\text{-}G)^*$ W m^{-2}	$ET_{d,METRIC}$ mm d^{-1}	$ET_{d,GES}$ mm d^{-1}	Relative error %
$M_{A,B}$	309.65 291.09	0 0	355.72 411.23	0.56	− 159.8	261.69	396.17	3.13	4.32	28%
$M_{A,C}$	309.65 294.91	0 6	355.72 480.21	0.74	− 215.61	222.36	396.17	3.93	4.32	9%
$M_{D,B}$	299.28 291.09	6 0	449.53 411.23	1.12	− 323.14	939.76	396.17	≈ 0	4.32	100%
$M_{D,C}$	299.28 294.91	6 6	449.53 480.21	2.17	− 635.87	759.58	396.17	≈ 0	4.32	100%

Note: ① *a* and *b* are empirically determined constants from the dT calculation for a given satellite image. ② * is mean data of all of the image pixels when using dry and wet limits combinations.

According to this study's calculation results, the LAI of the dry pixels (Point A) were distributed in the rear 3% of the frequency histogram, and those of the wet pixels (Point C) were distributed in the front 3% of the frequency histogram (Figure 4(b)). Based on the comprehensive consideration of the land cover status on the underlying surfaces of the study area, the high-temperature areas which accounted for the front 5% of the T_s frequency histogram, and the bare land which accounted for the rear 1–3% of the LAI frequency histogram, were selected as the dry pixels in this study. The low-temperature areas which accounted for the rear 5% of the T_s frequency histogram, and the farmland which accounted for the front 1–3% of the LAI frequency histogram, were selected as the wet pixels. The specific operational steps were as follows: With the land cover as the base map, GIS was used to overlap the T_s and LAI land areas. Then, in accordance with the set constraints of the above-mentioned scheme, GIS or ENVI software were used to automatically identify the dry and wet pixels (the percentage was the data with invalid values eliminated according to the season and cloud cover).

3.2. Analysis of the ET in the calibration period

The results of the regional ET estimated by the dry and wet limits identification scheme need to be verified and analyzed in order to determine if the characteristics of the ET and water consumption on the underlying surfaces had been accurately reflected. In previous related studies, it had been determined that the eddy covariance system of GES was one of the most reliable means to monitor ET (Siedlecki *et al.* 2016; Consoli *et al.* 2018). Therefore, it was utilized to measure water vapor flux data in order to evaluate the accuracy of the estimated calibration period of this study's METRIC model (Table 3).

$ET_{d,METRIC}$ and $ET_{d,GES}$ of the natural grassland areas with flat terrain and uniform growth in the study area during eight different DOY_s are presented in Table 3. The $ET_{d,METRIC}$ was combined with field survey results, land cover status and the NDVI feature analysis, in order to extract the average ET of nine pixels (0.81 ha) within a 3×3 window centered on the location of the eddy covariance system. The $ET_{d,\ GES}$ was the actual measurement of the water consumption data by the water vapor flux of the GES. It was found that, when compared with the measured $ET_{d,GES}$, the absolute error of the eight DOY_s of the $ET_{d,METRIC}$ ranged between −0.02 and 0.60 mm/d, with an average value of 0.17 mm/d. The relative error was determined to be between 4.0 and 37.9%, with $MRE = 18.8\%$. Meanwhile, it was found that among the eight DOY_s, the 6-day $ET_{d,METRIC}$ results were higher than the measured data of the $ET_{d,GES}$. Therefore, when combined with the conclusion that the eddy covariance method often underestimates the ET (due to the unclosed energy balance equation obtained by Massman (Massman & Lee 2002)), it was considered in this study that the daily ET results of the natural grassland areas simulated by the METRIC model were acceptable.

In addition, linear regression was conducted on the ET simulated using this study's METRIC model, as well as the ET measured by the eddy covariance system (Figure 5). As viewed from the results, the simulated and measured values displayed relatively high goodness of fit. The determinant R^2 of the regression equation was determined to be 0.96, and the RMSE was 0.32 mm/d, which was considered to be a high fitting degree. In addition, the slope of the linear equation was 0.94, which was close to the standard value of 1.0. The ET values estimated by the METRIC model were observed to be consistent with the actual measured data of the eddy covariance system in the overall change trend.

Table 3 | Comparisons of the ET values observed by eddy covariance with those estimated using METRIC for different periods on nature grassland in GES

date	day of the year (DOY)	$ET_{d,GES}$ (mm d^{-1})	$ET_{d,\ METRIC}$ (mm d^{-1})	Absolute error (mm)	Relative error (%)
2006/8/4	216	1.45	1.72	0.27	18.7
2006/9/21	264	1.18	1.41	0.22	18.9
2007/4/17	107	0.66	0.90	0.25	37.9
2007/7/6	187	1.07	1.36	0.29	27.0
2008/8/25	238	4.32	3.93	−0.40	9.2
2008/9/26	270	0.53	0.51	−0.02	4.0
2009/6/25	176	3.22	3.82	0.60	18.7
2009/8/12	224	0.82	0.95	0.13	16.2

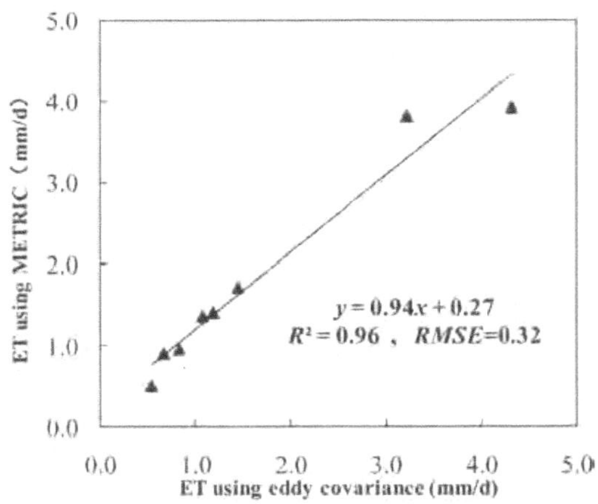

Figure 5 | The fitted line of the ET estimated by the METRIC model and the ET measured by the eddy covariance system.

3.3. Estimated accuracy of evapotranspiration by METRIC model

All the crops grown at the eight monitoring stations were silage maize, with an average growth period of 97 days. The irrigation methods included central pivot sprinkler irrigation (CPSI), drip irrigation (DI), low-pressure pipeline irrigation (LPPI), and semi-fixed irrigation (SFI). The average actual water consumption of silage maize during the growth period was determined to be 443.20 mm using (Equation (14)). The average value of ET_{METRIC} of silage maize in the eight stations during the growth period was determined to be 428.73 mm, in which MAE = 49.15 mm and MRE = 11.1%. By summarizing the previous research results, Kalma reviewed the various estimation methods of different remote sensing ET, and observed that when compared with the ground-based ET measurement value, the majority of the relative errors ranged from 15.0% to 30.0% (Kalma *et al.* 2008). Therefore, in this study it was believed that the accuracy of long-term scale ET estimation was reasonable and consistent with the previous research related conclusions (Allen *et al.* 2011b; Lian & Huang 2015) (Table 2).

The estimation results summarized in Table 4 show that the MRE of CPSI and DI is lower than that of SFI and LPPI by approximately 21.5% and the cause of this phenomenon was analyzed in this study. The ET_{METRIC} is the regional comprehensive water consumption of the underlying surface. Within the corresponding time period, the researchers randomly sampled the field water consumption data, and CPSI and DI was employed to reflect the actual situation of field water consumption more objectively than SFI and LPPI, because historical measured data are applied to confirm that the uniformity of CPSI and DI is better than that of SFI and LPPI. Therefore, ET_{IMS} by SFI and LPPI may not accurately reflect the regional comprehensive water consumption as CPSI and DI do.

3.4. Spatial characteristic of the ET

The spatial distribution of the ET in the Xilin River Basin during the crop growth season ranged from 62.31 to 778.86 mm, with an average value of 461.51 mm, and an average daily ET intensity of 3.02 mm/d (Figure 6). In terms of the spatial

Table 4 | Comparison of ET_{METRIC} and ET_{IMS} during the growth period of silage maize

No.	Irrigation type	Growth cycle (DOY_m-DOY_n)	P_a (mm)	I (mm)	L (mm)	ΔW (mm)	ET_{IMS} (mm)	ET_{METRIC} (mm)	Absolute error (mm)	Relative error (%)
IMS1	CPSI	140–243	133.40	450.00	/	36.38	547.03	511.26	35.77	6.5
IMS2	CPSI	145–243	127.30	337.50	/	5.28	459.52	470.32	10.80	2.3
IMS3	CPSI	145–243	127.30	375.00	/	35.48	466.83	490.83	24.00	5.1
IMS4	CPSI	152–243	127.30	405.00	/	33.62	498.69	469.55	29.14	5.8
IMS5	CPSI	152–243	127.30	375.00	/	41.00	461.31	479.02	17.72	3.8
IMS6	DI	156–250	127.30	142.00	1.73	– 10.54	278.11	308.53	30.42	10.9
IMS7	LPPI	145–243	127.3	120	/	– 8.46	255.76	311.47	55.71	21.8
IMS8	SFI	145–243	127.3	450	/	– 1.26	578.56	388.86	189.7	32.8

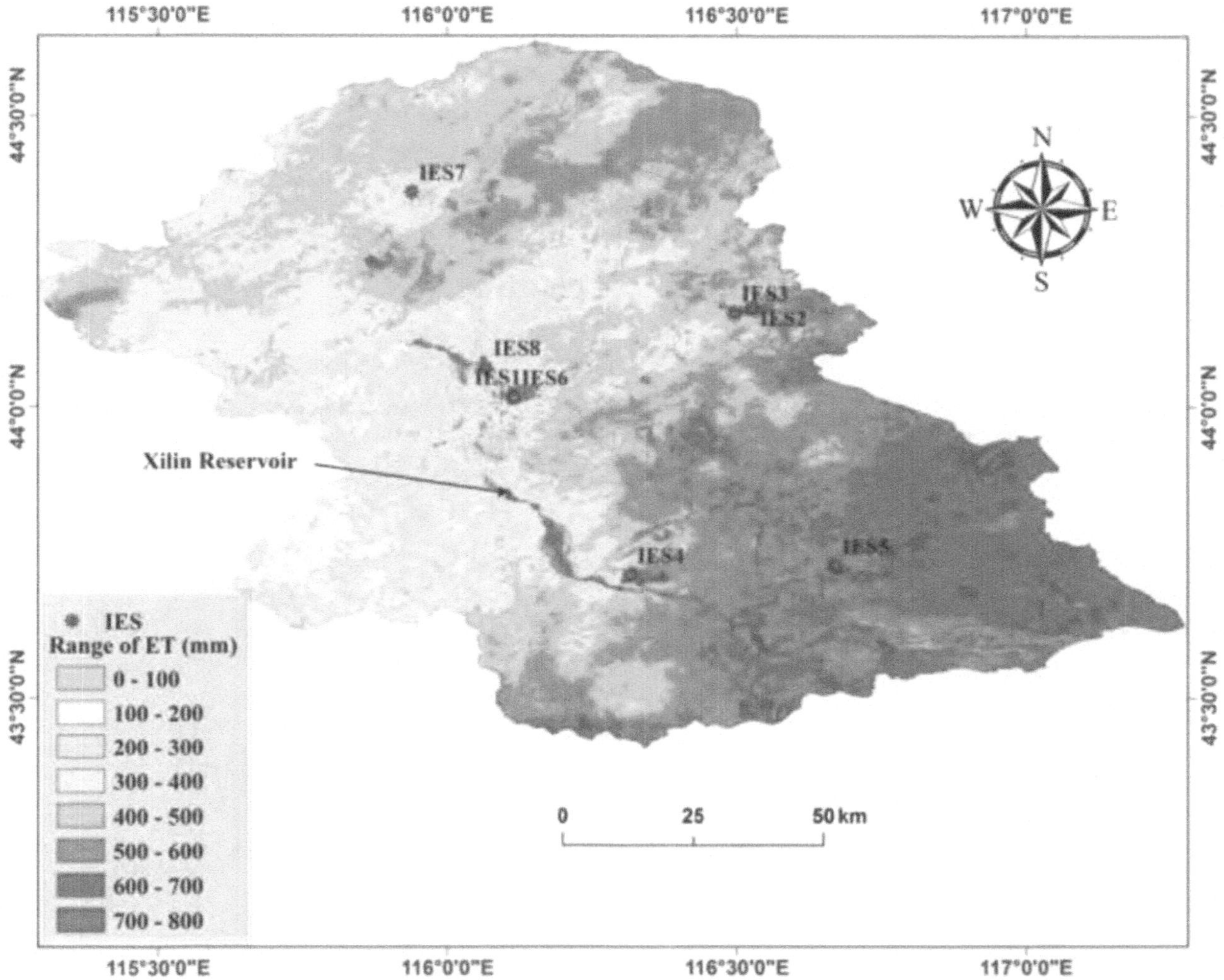

Figure 6 | Estimating the ET value of crop growth season by the METRIC model.

distribution, the ET displayed obvious spatial distribution traits from the forest-grass land intersections in the southeastern section of the Xilin River Basin, to the low-covered grassland areas in the northwestern section. These traits were found to be due to the comprehensive influences of climatic conditions, soil moisture supply status, vegetation cover, and other factors. In this study, based on the analysis of the ET results detailed in Figure 6 and land cover types detailed in Figure 1, it was found that except for the water masses with high ET characteristics (such as the Xilin Reservoir), the ET during the crop growth season displayed an overall declining change trend from the southeast to the northwest. The southeastern part of the basin was a shrubby forest area at the southern foot of the Great Khingan Mountains. This area was characterized by relatively abundant rainfall, favorable growth of forest-grass vegetation, and good ET conditions. Moreover, in recent years, the small amount of natural grassland in the basin has developed to become irrigated artificial grassland for silage maize cultivation. As a result, the ET of the forest-grass land intersections, and the irrigated artificial grassland planting areas in the southeast, was observed to be somewhat higher than that in the western and northern natural grass areas of the basin.

The ET of crop growth season was extracted according to the vegetation cover classification. For the different land cover types, the sequence was as follows: $ET_{water} > ET_{arable\ land} > ET_{forest\ land} > ET_{unutilized\ land} > ET_{grassland} > ET_{urban\ land}$ (Figure 7). The sufficient water levels had created favorable ET conditions for the water mass areas. During the crop growth season, the ET_{water} exceeded 740 mm (Figure 7(d)), and the average daily ET intensity reached 4.92 mm d^{-1}, which was the maximum among all the land types. In addition, a standard deviation of 7.28 mm indicated that the water masses were evenly distributed

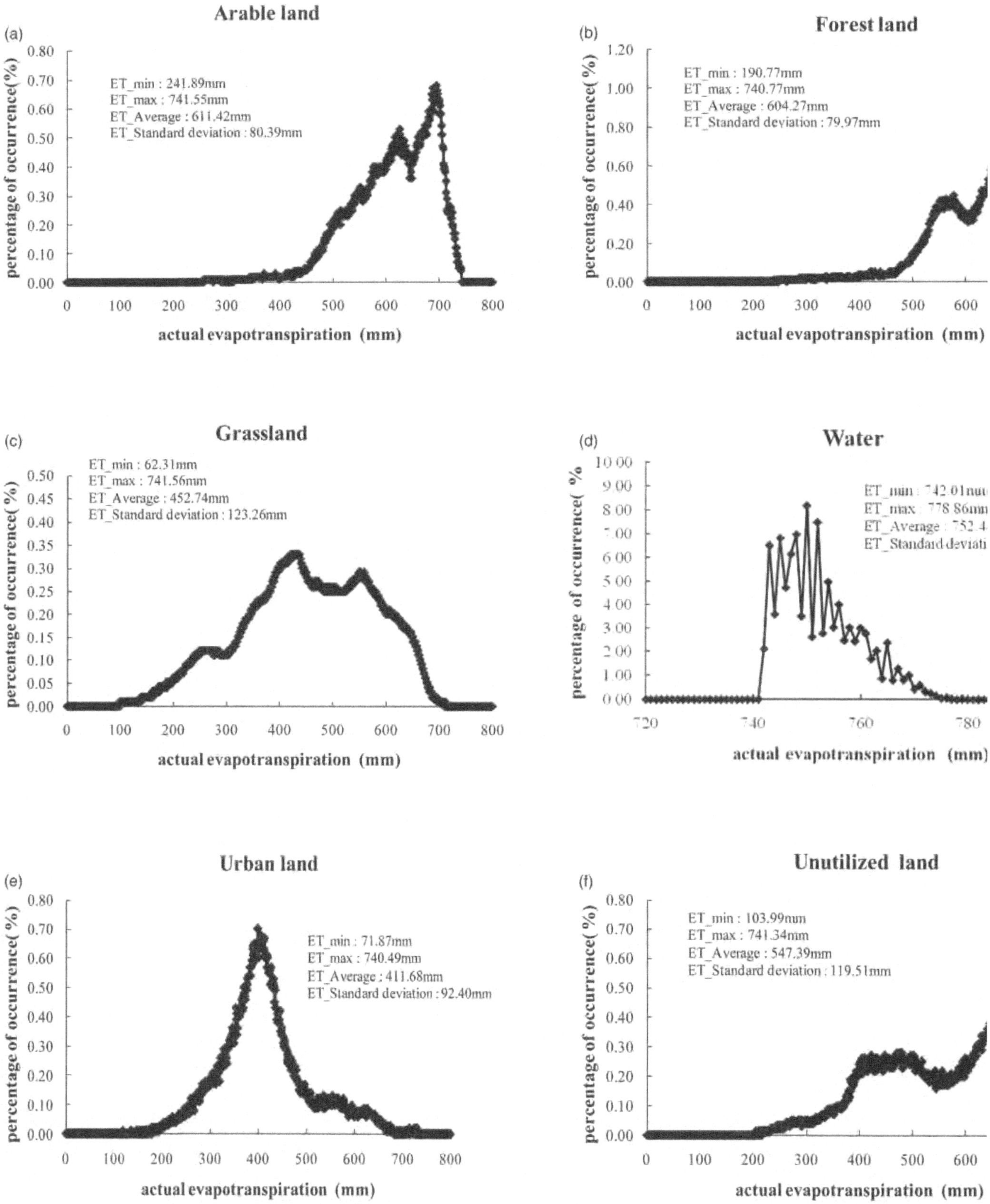

Figure 7 | Frequency distribution of the ET_{METRIC} values for different land cover types.

in space. The average ET value of the arable land in the basin was determined to be 611.42 mm (Figure 7(a)), and the average ET intensity was 4.00 mm d^{-1}. The artificially irrigated grasslands cultivated from the natural grasslands were supported by modern irrigation technology to ensure that the water demands for crop growth were met. The forest land was mostly distributed in the southeastern mountainous areas and characterized by abundant precipitation. The relatively sufficient precipitation and moist underlying surface environments guaranteed adequate water for plant growth, and the ET level was similar to that of hearable land (Figure 7(b)). The distribution of the unutilized land was mainly in the low-lying areas of the Xilin River Valley (Figure 1), as well as the exposed highlands along the coast. The average ET of the crop growth season was 547.39 mm. However, the mean variance calculation result of 119.51 mm indicated that the ET of the unutilized land in the regions was quite different (Figure 7(f)). According to the field investigations and previous research results, due to the precipitation reductions caused by climate change in the entire basin, especially in the western region (Wang *et al.* 2017), coupled with grassland degradation induced by overgrazing in recent years (Tong *et al.* 2004), some of the typical steppe areas have degenerated into desertification grassland or sandy land, with low vegetation coverage, drought, and little water. These factors jointly restrict the activities of the ET in grassland areas. The results showed that the ET intensity of the grassland (2.96 mm d^{-1}) was significantly lower than that of the majority of the other land types, and only higher than that of the urban land areas (2.69 mm d^{-1}) (Figure 7(c) and 7(e)).

4. CONCLUSIONS

The goal of this research study was to estimate the spatial and temporal distributions of the ET in typical steppe areas using a remote sensing energy balance method. In order to solve the spatial ambiguity problems of identifying the dry and wet limits in remote sensing ET estimations, the effects of the different dry and wet limit combinations on the ET estimation in typical steppe areas of the Xilin River Basin were estimated and compared based on Landsat images and the METRIC model. According to the test results of the water vapor flux data measured by an eddy covariance system, the error percentage of the ET in the eight DOY_s regions estimated by the dry and wet limit pixels and comprehensively selected from the T_s and LAI, was 18.8% on average. The simulated values were observed to have high goodness of fit with the measured values. On this basis, the spatiotemporal distribution of the regional ET during the crop growth season of 2011was studied using eight effective sunshine images of Landsat-5 and Landsat-7. The relative error was found to be 11.1%, which was verified by the measurement data in eight silage maize irrigation monitoring stations using the water balance method. Then, combined with the verification results of this study's eddy covariance system and water balance method, as well as the previous research conclusions of the regional ET estimated by remote sensing, a certain reliability was proven in the estimations of the ET of the typical steppe areas in the Xilin River Basin when Landsat images and the METRIC model were applied. Meanwhile, to analyze the spatial distributions of the ET, the ET results of the crop growth season were extracted according to the vegetation cover classification. It was concluded in this study that the spatial distribution of the ET of the different land cover types in the study area were as follows: $ET_{water} > ET_{arable\ land} > ET_{forest\ land} > ET_{unutilized\ land} > ET_{grassland} > ET_{urban\ land}$.

ACKNOWLEDGEMENTS

This work was supported by the special fund of China Institute of Water Resources and Hydropower Research (Grant NO. MK2020J03), the Natural Science Foundation of Inner Mongolia (Grant NO. 2020MS05057). The authors are grateful to the National scientific research station of Xilin Gol Grassland Ecosystem in Inner Mongolia for providing the eddy covariance data and meteorological data. We greatly appreciate the help of Chen shiping in providing ground-based measurements of evapotranspiration.

CONFLICT OF INTEREST STATEMENT

The authors declare no conflict of interest.

DATA AVAILABILITY STATEMENT

All relevant data are included in the paper or its Supplementary Information.

REFERENCES

Allen, R. G., Pereira, L. S., Raes, D. & Smith, M. 1998 Crop Evapotranspiration: Guidelines for Computing Crop Water Requirements. FAO Irrigation and Drainage Paper No. 56. FAO, 56.

Allen, R. G., Tasumi, M., Morse, A., Trezza, R., Wright James, L., Bastiaanssen, W., Kramber, W., Lorite, I. & Robison Clarence, W. 2007a Satellite-based energy balance for mapping evapotranspiration with internalized calibration (METRIC)–applications. *Journal of Irrigation and Drainage Engineering* **133** (4), 395–406.

Allen, R. G., Tasumi, M. & Trezza, R. 2007b Satellite-Based energy balance for mapping evapotranspiration with internalized calibration (METRIC)–Model. *Journal of Irrigation and Drainage Engineering* **133** (4), 380–394.

Allen, R. G., Irmak, A., Trezza, R., Hendrickx, J. M. H., Bastiaanssen, W. & Kjaersgaard, J. 2011a Satellite-based ET estimation in agriculture using SEBAL and METRIC. *Hydrological Processes* **25** (26), 4011–4027.

Allen, R. G., Pereira, L. S., Howell, T. A. & Jensen, M. E. 2011b Evapotranspiration information reporting: I. Factors governing measurement accuracy. *Agricultural Water Management* **98** (6), 899–920.

Allen R. G., Burnett, B., Kramber, W., Huntington, J., Kjaersgaard, J., Kilic, A., Kelly, C. & Trezza, R. 2013 Automated calibration of the METRIC-Landsat evapotranspiration process. *JAWRA Journal of the American Water Resources Association* **49** (3), 563–576.

Bai, Y., Han, X., Wu, J., Chen, Z. & Li, L. 2004 Ecosystem stability and compensatory effects in the Inner Mongolia grassland. *Nature* **431** (7005), 181–184.

Bastiaanssen, W. G. M., Menenti, M., Feddes, R. A. & Holtslag, A. A. M. 1998a The surface energy balance algorithm for land (SEBAL): part 1 formulation. *Journal of Hydrology* **212** (98), 801–811.

Bastiaanssen, W. G. M., Pelgrum, H., Wang, J., Ma, Y., Moreno, J. F., Roerink, G. J. & Wal, T. V. D. 1998b A remote sensing surface energy balance algorithm for land (SEBAL): part 2: validation. *Journal of Hydrology* **212** (1–4), 213–229.

Batnasan, N. 2003 Freshwater issues in Mongolia. In *Proceeding of the National Seminar on IRBM in Mongolia*, Ulaanbaatar.

Bowen, I. S. 1926 The ratio of heat losses by conduction and by evaporation from any water surface. *Physical Review* **27** (6), 779–787.

Bradley, R. S. 1999 Paleoclimatology: reconstructing climates of the quaternary. *Arctic Antarctic & Alpine Research* **3** (31), 329–332.

Cai, X., Xu, Z., Su, B. & Yu, W. 2009 Distributed simulation for regional evapotranspiration and verification by using remote sensing (in Chinese with English abstract). *Transactions of the Chinese Society of Agricultural Engineering* **25** (10), 154–160.

Chang, Y., Ding, Y., Zhao, Q. & Zhang, S. 2017 Remote estimation of terrestrial evapotranspiration by Landsat 5 TM and the SEBAL model in cold and high-altitude regions: a case study of the upper reach of the Shule River Basin, China. *Hydrological Processes* **31** (3), 514–524.

Chen, S. 2002 *Study the Land use/Land Cover Change and Cycle of Xilin River Basin by Remote Sensing and GIS (in Chinese)*. University of Chinese Academy of Sciences (the Institute of Remote Sensing Applications), Beijing.

Consoli, S., Milani, M., Cirelli, G., Barbagallo, S., Marzo, A., Vanella, D. & Toscano, A. 2018 Energy and water balance of a treatment wetland under Mediterranean climatic conditions. *Ecological Engineering* **116**, 52–60.

Daniel, D. L. F., Ortegafarías, S., Fonseca, D., Ortegasalazar, S., Kilic, A. & Allen, R. 2017 Calibration of METRIC model to estimate energy balance over a drip-irrigated apple orchard. *Remote Sensing* **9** (7), 670.

Foolad, F. , Blankenau, P., Kilic, A., Allen, R. G., Huntington, J. L., Erickson, T. A., Ozturk, D., Morton, C. G., Ortega, S., Ratcliffe, I., Franz, T. E., Thau, D., Moore, R., Gorelick, N., Kamble, B., Revelle, P., Trezza, R., Zhao, W. & Robison, C. W. 2018 Comparison of the Automatically Calibrated Google Evapotranspiration Application–EEFlux and the Manually Calibrated METRIC Application. Preprints, 2018070040.

French, A. N., Hunsaker, D. J. & Thorp, K. R. 2015 Remote sensing of evapotranspiration over cotton using the TSEB and METRIC energy balance models. *Remote Sensing of Environment* **158**, 281–294.

Gao, Y. C., Long, D. & Li, Z. L. 2008 Estimation of daily actual evapotranspiration from remotely sensed data under complex terrain over the upper Chao river basin in North China. *International Journal of Remote Sensing* **29** (11), 3295–3315.

González-Dugo, M., Gonzalez-Piqueras, J., Campos, I., Balbontín, C. & Calera, A. 2012 Estimation of surface energy fluxes in vineyard using field measurements of canopy and soil temperature. *IAHS-AISH Publication* **352**, 59–62.

Howell, T. A., Schneider, A. D. & Jensen, M. E. 1991 History of Lysimeter Design and Use for Evapotranspiration Measurements, Lysimeters for Evapotranspiration and Environmental Measurements, pp. 1–9.

Hu, X., Shi, L., Lin, G. & Lin, L. 2021 Comparison of physical-based, data-driven and hybrid modeling approaches for evapotranspiration estimation. *Journal of Hydrology* **601**, 126592.

Jiang, Y., Tang, R. & Jiang, X. 2020 Assessing the directional effects of remotely sensed land surface temperature on evapotranspiration estimation. In: *IGARSS 2020–2020 IEEE International Geoscience and Remote Sensing Symposium*. pp. 4558–4561.

Kalma, J. D., Mcvicar, T. R. & Mccabe, M. F. 2008 Estimating land surface evaporation: a review of methods using remotely sensed surface temperature data. *Surveys in Geophysics* **29** (4–5), 421–469.

Kang, S. Z. 2007 *Conspectus of Agricultural Soil and Water Engineering* (in Chinese). Chinese Agricultural Press, Beijing, pp. 86–89.

Khand, K., Kjaersgaard, J., Hay, C. & Jia, X. 2017a Estimating impacts of agricultural subsurface drainage on evapotranspiration using the Landsat imagery-based METRIC model. *Hydrology* **4** (4), 1–16.

Khand, K., Numata, I., Kjaersgaard, J. & Vourlitis, G. L. 2017b Dry season evapotranspiration dynamics over human-impacted landscapes in the southern Amazon using the Landsat-based METRIC model. *Remote Sensing* **9** (7), 706.

Li, F., Zheng, J., Wang, H., Luo, J., Zhao, Y. & Zhao, R. 2016 Mapping grazing intensity using remote sensing in the Xilingol steppe region, Inner Mongolia, China. *Remote Sensing Letters* **7** (4), 328–337.

Lian, J. & Huang, M. 2015 Evapotranspiration estimation for an Oasis area in the Heihe River Basin using Landsat-8 images and the METRIC model. *Water Resources Management* **29** (14), 5157–5170.

Liou, Y.-A. & Kar, K. S. 2014 Evapotranspiration estimation with remote sensing and various surface energy balance algorithms – a review. *Energies* **7** (5), 2821.

Long, D. & Singh, V. P. 2013 Assessing the impact of end-member selection on the accuracy of satellite-based spatial variability models for actual evapotranspiration estimation. *Water Resources Research* **49** (5), 2601–2618.

Massman, W. J. & Lee, X. 2002 Eddy covariance flux corrections and uncertainties in long-term studies of carbon and energy exchanges. *Agricultural & Forest Meteorology* **113** (1–4), 121–144.

Mayer, D. G. & Butler, D. G. 1993 Statistical validation. *Ecological Modelling* **68** (1–2), 21–32.

Menenti, M. 1993 Parameteraization of land surface evaporation by means of location dependent potential evaporation and surface temperature range. Bolle, Feddes and Kalma, Exchange processes at the land surface for a range of space and time scales.

Monteith, J. L. & Unsworth, M. H. 2013 Chapter 8 - microclimatology of radiation: (iii) Interception by Plant Canopies and Animal Coats. In: *Principles of Environmental Physics (Fourth Edition)* (Monteith, J. L. & Unsworth, M. H., eds.). Academic Press, Boston, pp. 111–133.

Morton, C. G., Huntington, J. L., Pohll, G. M., Allen, R. G., McGwire, K. C. & Bassett, S. D. 2013 Assessing calibration uncertainty and automation for estimating evapotranspiration from agricultural areas using METRIC. *JAWRA Journal of the American Water Resources Association* **49** (3), 549–562.

Norman, J. M., Kustas, W. P. & Humes, K. S. 1995 Source approach for estimating soil and vegetation energy fluxes in observations of directional radiometric surface temperature. *Agricultural and Forest Meteorology* **77** (3–4), 263–293.

Numata, I., Khand, K., Kjaersgaard, J., Cochrane, M. & Silva, S. 2017 Evaluation of Landsat-based METRIC modeling to provide high-spatial resolution evapotranspiration estimates for Amazonian forests. *Remote Sensing* **9** (1), 46.

Paço, T. A., Pôças, I., Cunha, M., Silvestre, J. C., Santos, F. L., Paredes, P. & Pereira, L. S. 2014 Evapotranspiration and crop coefficients for a super intensive olive orchard. An application of SIMDualKc and METRIC models using ground and satellite observations. *Journal of Hydrology* **519**, 2067–2080.

Pedro-Monzonís, M., Solera, A., Ferrer, J., Estrela, T. & Paredes-Arquiola, J. 2015 A review of water scarcity and drought indexes in water resources planning and management. *Journal of Hydrology* **527**, 482–493.

Qin, Z., Karnieli, A. & Berliner, P. 2010 A mono-window algorithm for retrieving land surface temperature from Landsat TM data and its application to the Israel-Egypt border region. *International Journal of Remote Sensing* **22** (18), 3719–3746.

Rana, G. & Katerji, N. 2000 Measurement and estimation of actual evapotranspiration in the field under Mediterranean climate: a review. *European Journal of Agronomy* **13** (2–3), 125–153.

Reyes-González, A., Kjaersgaard, J., Trooien, T., Hay, C. & Ahiablame, L. 2017 Comparative analysis of METRIC model and atmometer methods for estimating actual evapotranspiration. *International Journal of Agronomy* **2017**, 1–16.

Roerink, G. J., Su, Z. & Menenti, M. 2000 S-SEBI: A simple remote sensing algorithm to estimate the surface energy balance. *Physics and Chemistry of the Earth, Part B: Hydrology, Oceans and Atmosphere* **25** (2), 147–157.

Scientist, G. B. S. P., Analyst, S. B. G., Scientist, R. K. S., Scientist, N. M. V. & Student, H. A. 2013 Operational evapotranspiration mapping using remote sensing and weather datasets: a new parameterization for the SSEB approach †. *Jawra Journal of the American Water Resources Association* **49** (3), 577–591.

Sheng, X. 2017 *Handbook of Characteristic Values of Rivers and Lakes in Inner Mongolia Autonomous Region (in Chinese). Xilinguole League.* Inner Mongolia University Press, Inner Mongolia, China, 248 pp.

Siedlecki, M., Pawlak, W., Fortuniak, K. & Zieliński, M. 2016 Wetland evapotranspiration: eddy covariance measurement in the Biebrza Valley, Poland. *Wetlands* **36** (6), 1055–1067.

Spiliotopoulos, M., Holden, N. M. & Loukas, A. 2017 Mapping evapotranspiration coefficients in a temperate maritime climate using the METRIC model and Landsat TM. *Water* **9** (1), 23.

Su, Z. 2002 The surface energy balance system (SEBS) for estimation of turbulent heat fluxes. *Hydrology & Earth System Sciences* **6** (1), 85–99.

Su, Z. , Yacob, A., Wen, J., Roerink, G., He, Y., Gao, B., Boogaard, H. & van Diepen, C. 2003 Assessing relative soil moisture with remote sensing data: theory, experimental validation, and application to drought monitoring over the North China Plain. *Physics and Chemistry of the Earth, Parts A/B/C* **28** (1–3), 89–101.

Taherparvar, M. & Pirmoradian, N. 2018 Estimation of rice evapotranspiration using reflective images of Landsat satellite in sefidrood irrigation and drainage network. *Rice Science* **25** (2), 111–116.

Talsma, C. J., Good, S. P., Jimenez, C., Martens, B., Fisher, J. B., Miralles, D. G., McCabe, M. F. & Purdy, A. J. 2018 Partitioning of evapotranspiration in remote sensing-based models. *Agricultural and Forest Meteorology* **260–261**, 131–143.

Tasumi, M. 2003 *Progress in Operational Estimation of Regional Evapotranspiration Using Satellite Imagery.*

Thornthwaite, C. W. 1948 An approach toward a rational classification of climate. *Geographical Review* **38** (1), 55–94.

Tong, C., Wu, J., Yong, S., Yang, J. & Yong, W. 2004 A landscape-scale assessment of steppe degradation in the Xilin River Basin, Inner Mongolia, China. *Journal of Arid Environments* **59** (1), 133–149.

Trezza, R., Allen, R. G. & Tasumi, M. 2013 Estimation of actual evapotranspiration along the Middle Rio Grande of New Mexico using MODIS and Landsat imagery with the METRIC model. *Remote Sensing* **5** (10), 5397–5423.

Volik, O., Kessel, E., Green, A., Petrone, R. & Price, J. 2021 Growing season evapotranspiration in boreal fens in the Athabasca oil sands region: variability and environmental controls. *Hydrological Processes* **35** (2), e14020.

Wang, H., Wang, J. H., Qin, D. Y. & Jia, Y. W. 2006 Theory and methodology of water resources assessment based on dualistic water cycle model (in Chinese with English abstract). *Journal of Hydraulic Engineering* **12**, 1496–1502.

Wang, J., Li, H. P., Lu, H. Y. & Li, J. 2017 Variation of precipitation and runoff in typical grassland area (in Chinese). *Journal of China Hydrology* **37** (04), 86–90.

Wang, L., Liu, Z., Guo, J., Wang, Y., Ma, J., Yu, S., Yu, P. & Xu, L. 2021 Estimate canopy transpiration in larch plantations via the interactions among reference evapotranspiration, leaf area index, and soil moisture. *Forest Ecology and Management* **481**, 118749.

Wright, J. L. 1982 New evapotranspiration crop coefficients. *Journal of the Irrigation & Drainage Division* **108** (IR2), 57–74.

Xu, D., Liu, Y., Yang, D. & Zhang, B. Z. 2015 *Evapotranspiration Scale Effect and Temporal-Spatial Scale Expansion (in Chinese)*. Science China Press, Beijing.

Xue, J., Bali, K. M., Light, S., Hessels, T. & Kisekka, I. 2020 Evaluation of remote sensing-based evapotranspiration models against surface renewal in almonds, tomatoes and maize. *Agricultural Water Management* **238**, 106228.

Yamanaka, T., Kaihotsu, I., Oyunbaatar, D. & Ganbold, T. 2007 Summertime soil hydrological cycle and surface energy balance on the Mongolian steppe. *Journal of Arid Environments* **69** (1), 65–79.

Yang, D., Chen, H. & Lei, H. 2010 Estimation of evapotranspiration using a remote sensing model over agricultural land in the North China Plain. *International Journal of Remote Sensing* **31** (14), 3783–3798.

Yu, D., Li, X., Cao, Q., Hao, R. & Qiao, J. 2020 Impacts of climate variability and landscape pattern change on evapotranspiration in a grassland landscape mosaic. *Hydrological Processes* **34** (4), 1035–1051.

Zamani Losgedaragh, S. & Rahimzadegan, M. 2018 Evaluation of SEBS, SEBAL, and METRIC models in estimation of the evaporation from the freshwater lakes (Case study: Amirkabir dam, Iran). *Journal of Hydrology* **561**, 523–531.

First received 26 August 2021; accepted in revised form 9 December 2021. Available online 30 December 2021

© 2022 The Authors Impact of Climate Change on Hydrology and Water Resources doi: 10.2166/wcc.2021.078

Temporal and spatial variations of extreme precipitation in the Guangdong-Hong Kong-Macao Greater Bay area from 1961 to 2018

Yueying Zhou, Zhijian Luo, Shenlin Li, Zufa Liu, Yanpeng Shen and Wenshan Zhuo

ABSTRACT

Disasters caused by extreme precipitation under global warming are anticipated to have a strong negative impact on urban construction and social security. In this study, daily grid precipitation datasets of the Guangdong-Hong Kong-Macao Greater Bay Area (GBA) for the period 1961–2018 were extracted to explore the temporal and spatial characteristics of extreme precipitation by using regression analysis, moving average and kriging interpolation. The frequency and intensity indices showed an increasing trend, whereas a decreasing trend was found for the persistence indices, which indicates that GBA tends to slowly become wetter. The mean values of extreme precipitation indices (EPIs) in GBA generally increased from west to east and from north to south. Except for the indices of consecutive wet days and consecutive dry days, other EPIs showed an upward trend in most regions, especially in coastal cities where floods are more likely to occur. Principal component analysis and regression analysis showed that the correlations between the EPIs mostly passed the 0.05 significance test, which suggests that they had a good indicator of extreme precipitation in GBA. This study provides a theoretical basis for extreme precipitation disaster prevention and control within the urban agglomerations of the GBA.

Key words | extreme precipitation indices (EPIs), extreme precipitation, Guangdong-Hong Kong-Macao Greater Bay Area (GBA), spatial and temporal variation

Yueying Zhou
Zhijian Luo
Department of Resources and the Urban Planning,
Xinhua College of Sun Yat-sen University,
Guangzhou 510520,
PR China

Shenlin Li
Guangdong Research Institute of Water Resources and Hydropower,
Guangzhou 510630,
PR China

Zufa Liu
School of Civil Engineering,
Sun Yat-sen University,
Guangzhou 510275,
PR China

Yanpeng Shen
Qingdao Huayi Environmental Protection Science & Technology Co. Ltd,
Qingdao 266001,
PR China

Wenshan Zhuo (corresponding author)
Instrumental Analysis and Research Center,
Sun Yat-sen University,
Guangzhou 510275,
PR China
E-mail: *zadeozws@mail.sysu.edu.cn*

HIGHLIGHTS

- Nine indices were applied to spatio-temporally analyze extreme precipitation trends.
- Time-series analysis and spatial interpolation were used to examine the characteristics of extreme precipitation in the GBA.
- Except for CDD and CWD, the other seven EPIs had an increasing tendency.
- Coastal regions generally showed a stronger upward trend of extreme precipitation.

doi: 10.2166/wcc.2021.078

INTRODUCTION

Extreme precipitation usually refers to precipitation of high intensity but with a low occurrence probability. Disasters caused by extreme precipitation can have a serious impact on the natural environment and human life. According to the fifth Climate Change Assessment Report by the Intergovernmental Panel on Climate Change (IPCC), global mean surface temperature (GMST) increased by 0.85 °C between 1880 to 2012. Global warming has accelerated the hydrological cycle and increased atmospheric water vapor fluxes, leading to substantial changes in the intensity and frequency of extreme hydrological events (Papalexiou & Montanari 2019). Research on extreme precipitation is of great importance to social development and human health.

The past decade has seen many studies devoted to the temporal and spatial characteristics of extreme precipitation. Extreme precipitation events have occurred increasingly frequently around the globe during the 20th century. Sun *et al.* (2021) examined changes in extreme precipitation (measured by such indicators as the annual maxima of 1-day (RX1day) and 5-day (RX5day)) based on high-quality station data from 1900 to 2018, and found that extreme precipitation had increased at about two-thirds of the stations. It has been found that extreme precipitation has a statistically significant connection to the GMST, and extreme precipitation is anticipated to become dramatically stronger with further global warming (Myhre *et al.* 2019). There have been regional studies that yielded different patterns of extreme precipitation (Soltani *et al.* 2016; Li *et al.* 2020; Tian *et al.* 2020). In Alaska, the magnitude of the projected increase of RX1day was the largest in the southern area, although the percentage increase was largest in the north of Alaska (Bennett & Walsh 2015). However, the great majority of extreme precipitation indices (EPIs) in the Chichaoua-Mejjate region in Morocco have shown a non-significant downward trend, although temperature-based indices of all stations showed increasing trends (Hadri *et al.* 2020). In the West Rapti River basin of Nepal, EPIs are anticipated to increase significantly in future, especially RX1day and RX5day (Talchabhadel *et al.* 2020).

With the rising temperature in China, the characteristics of extreme precipitation of the country have also changed. Cao *et al.* (2020) analyzed the daily data of 554 meteorological stations throughout China from 1960 to 2017 and found that the magnitude and intensity indices of extreme precipitation were increasing, while the indices of sustainability were decreasing, and these indices are included in the nine indices in this paper. To date, there are few studies focusing on the temporal and spatial changes of extreme precipitation in the Guangdong-Hong Kong-Macao Greater Bay Area (GBA), and most of the related studies focused on areas either larger or smaller than it. Zeng *et al.* (2017) found that extreme precipitation in most parts of western and southeastern China showed an upward trend during 1961–2010, such as the northwest river basin, Huai River basin, eastern Yangtze River basin, southeast river basin, southwest river basin and eastern Pearl River basin. Yu *et al.* (2018) selected the summer daily precipitation data from 429 stations with no missing data over eastern China between 1961 and 2012, and found that the precipitation thresholds roughly increased from northwest to southeast. In addition, using the generalized Pareto distribution fitting, they found higher probabilities of extreme precipitation in South China. Wang *et al.* (2016) examined extreme precipitation in Guangdong Province, and discovered that in most regions the frequency and intensity indices continued to increase from 1960–2013. Liu *et al.* (2018) found that the intense short-term rainfall (above the rain rate of 20 mm/h) in Foshan stayed around a certain level in the April-May-June rainy season but showed an increasing trend in July-August-September. There have been several studies on extreme precipitation in relation to global warming and urbanization in recent years (Papalexiou & Montanari 2019; Jiang *et al.* 2020), focusing on the impact of urbanization on extreme precipitation.

Previous studies have implied that extreme precipitation is likely to increase in GBA, which is located in southern China and home to large cities, but the spatio-temporal characteristics of extreme precipitation in this region are yet to be explored. This study attempts to identify the spatio-temporal characteristics of extreme precipitation in GBA, with a vision to provide a theoretical basis for the water resources management in response to climate changes, as well as technical support for improving water resources regulation and flood risk management capabilities in the GBA.

DATA AND METHOD

Study area

The GBA (111.35°E–115.43°E, 21.56°N–24.40°N) is located in the central-southern part of Guangdong Province and the estuary area of the Pearl River Basin in China. This region is humid and the water cycle is extremely active, with dense river networks and abundant precipitation. The temporal and spatial distribution of rainfall is extremely uneven in the GBA. As one of the four largest bays in the world, GBA is home to nine cities (Shenzhen, Huizhou, Dongguan, Guangzhou, Jiangmen, Zhongshan, Zhuhai, Foshan and Zhaoqing) in the Pearl River Delta, Hong Kong and Macau. It covers an area of approximately 56,000 km^2 and has a population of approximately 70 million. Figure 1 shows the location of GBA and the distribution of meteorological stations in this region. The GBA boasts high economic vitality and inclusiveness, and plays a very important role in China's social development. In 2019, *The Outline of the Development Plan for the Guangdong-Hong Kong-Macao Greater Bay Area* issued by China's State Council specified the goal of building this area into a world-class bay area and creating a high-quality development model. However, the rapid economic development of GBA requires greater urban flood and drainage control.

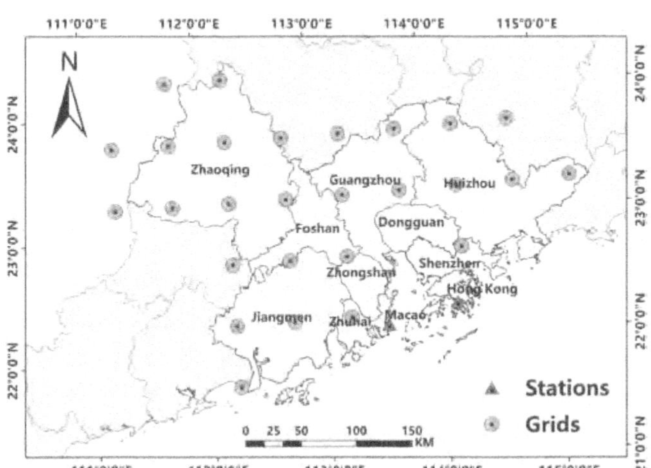

Figure 1 | Distribution of the meteorological stations and precipitation grid in the GBA.

Precipitation data

The precipitation data were taken from the $0.5° \times 0.5°$ daily grid precipitation datasets (DGPD) in China for the period 1961–2018, established by the National Meteorological Information Center, Meteorological Data Room. The locations of the grid points are shown in Figure 1. The datasets can be downloaded from The China Meteorological Data Service Center (https://data.cma.cn/). We extracted the 27 grids of the nine cities in the Pearl River Delta by using R language. Because the meteorological stations used to build the DGPD do not include Hong Kong and Macau, the Hong Kong precipitation data were taken from the Hong Kong Observatory (HKO) (https://www.hko.gov.hk) and the Macao data were from the Macao Meteorological and Geophysical Bureau (http://www.smg.gov.mo).

According to the 2,472 national meteorological stations across China, the National Meteorological Information Center established the DGPD by using the thin plate spline method and a digital elevation model for interpolation. The root mean square error (RMSE) between the grid data and the actual precipitation is 0.49 mm, and the relation coefficient amounts to 0.93. Especially in southeast China, the relative bias error is mostly between −10% and 10% (Zhao *et al.* 2014).

Research methods

Nine EPIs were selected in this paper and the annual average of each index was calculated. Regression analysis and the moving average method were used to analyze the temporal trend of each index, and kriging spatial interpolation was used to analyze the average value of each index.

EPIs

Due to the differences in climatic conditions and topographical conditions among regions, there is generally no uniform standard for extreme precipitation. Referring to most scholars, in this study, we selected nine EPIs (Table 1) from the extreme climate index system defined

Table 1 | The EPIs applied within this paper

Type	Indicator	Symbol	Definition	Unit
Persistence	Consecutive dry days	CDD	Maximum number of consecutive days with PRCP <1 mm	d
	Consecutive wet days	CWD	Maximum number of consecutive days PRCP ≥1 mm	d
	Annual total wet-day precipitation	PRCPTOT	Annual total precipitation in wet days (defined as PRCP ≥1 mm)	mm
Frequency	Number of slightly heavy precipitation days	R10	Annual count of days when PRCP ≥10 mm	d
	Number of heavy precipitation days	R25	Annual count of days when PRCP ≥25 mm	d
Intensity	Simple daily intensity index	SDII	Annual total precipitation divided by the number of wet days in the year	mm/d
	Very wet days	R95p	Annual total precipitation when PRCP ≥95th percentile	mm
	Max 1-day precipitation amount	RX1day	Annual maxima of 1-day precipitation	mm
	Max 5-day precipitation amount	RX5day	Annual maxima of 5-day precipitation	mm

PRCP, daily precipitation.

by the Expert Team on Climate Change Detection and Indices (ETCCDI), and divided them into persistence, frequency and intensity indices (Soltani *et al.* 2016; Talchabhadel *et al.* 2020).

Regression analysis method

In the regression analysis method, a mathematical model was established to describe the statistical relationship between the EPIs x and the year T as follows:

$$x_i = a + b \times T_i \quad i = 1, 2, 3, \ldots, n \tag{1}$$

where a is the constant value and b is the regression coefficient.

The sign of the b indicates the tendency of x. When $b > 0$, it means that with the increase of time T, x shows an upward trend, whereas when $b < 0$, it means that with the decrease of time T, x shows a downward trend. $b \times 10$ is called the climate tendency rate, and its unit is mm/10a. The regression coefficient b and constant a can be estimated by the least square method.

Moving average method

The principle of the moving average method is to calculate the moving average by sequentially increasing and decreasing the new and old data, so as to eliminate accidental changes, find the development trends and make predictions. Given a random sequence of n samples, the moving average sequence can be expressed as:

$$\hat{x}_j = \frac{1}{k}\sum_{i=1}^{k} x_{i+j-1} \tag{2}$$

where k is the moving length, $j = 1, 2, \cdots, n-k+1$.

Kriging interpolation

In a random process under certain conditions, the optimal linear unbiased estimator can be obtained by kriging interpolation in geostatistics (Le & Zidek 2006). The spatial distribution changes of extreme precipitation were presented using the Kriging spatial interpolation method in the ArcGIS data processing platform.

RESULTS AND DISCUSSION

Temporal changes

Temporal changes of mean annual values of different EPIs and its mean 5-year moving values (M5YMV) in GBA for 1961–2018 are shown in Figure 2. The trends of all indices were tested for significance with a confidence level of 0.05.

In the persistence indices, CDD showed a downward trend (−0.2 d/10a), with the highest value in 2005 (61 d) and the lowest value in 2017 (21 d). There was no clear

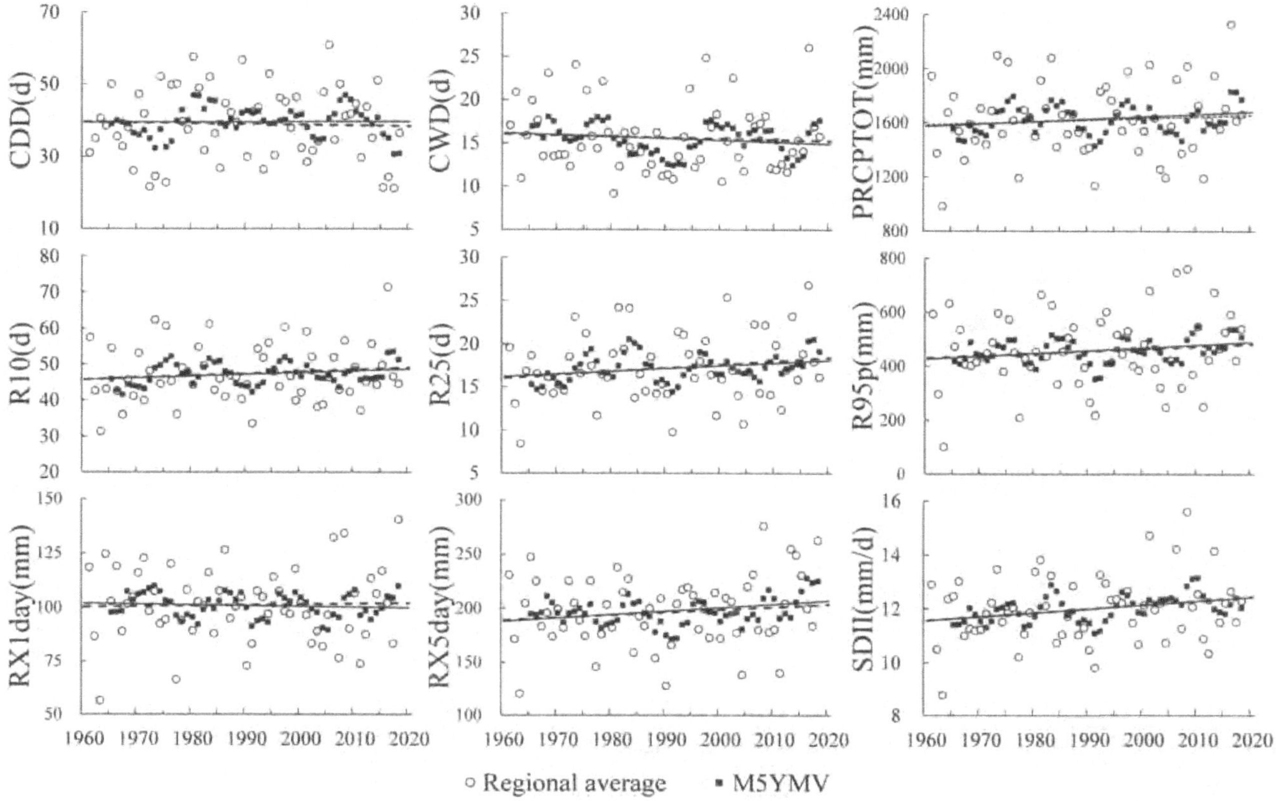

Figure 2 | Temporal variation of each EPI in GBA for 1961–2018.

trend in the M5YMV. The CWD showed a downward trend (−0.1 d/10a), with the lowest value in 1980 (9 d) and the highest value in 2016 (26 d). The trend of the M5YMV was decreasing overall for CWD (−0.2 d/10a). CDD is slightly larger than that of CWD, indicating that the drought of the GBA has eased a little. PRCPTOT showed an upward trend (17.9 mm/10a), with the lowest value in 1963 (983 mm), and the highest value in 2016 (2,330 mm). The M5YMV PRCPTOT trend was increasing (12.4 mm/10a), indicating that the total precipitation in GBA has increased.

In the frequency indices, R10 and R25 both showed an upward trend (0.6 d/10a and 0.4 d/10a), as well as the M5YMV (0.5 d/10a and 0.3 d/10a). R10 had its lowest value in 1963 (31 d) and its highest value in 2016 (72 d), while R25 had its lowest value in 1963 (8 d) and its highest value in 2016 (27 d).

In the intensity indices, SDII, R95p, RX1day and RX5day all presented an upward trend, with trends of 0.1 mm/(d·10a), 10.5 mm/10a, 0.3 mm/10a and 3.2 mm/10a, respectively. The trends of the M5YMV were 0.1 mm/(d·10a), 8.6 mm/10a, −0.3 mm/10a and 2.4 mm/10a, respectively. RX5day had a larger amplitude than RX1day, and the frequency of extreme precipitation events has increased. SDII had the minimum value (8.8 mm/d) in 1963, and the maximum (15.6 mm/d) in 2008. R95p had the minimum value (102 mm) in 1963, and the maximum (762 mm) in 2008. The minimum RX1day (56 mm) occurred in 1963 and the maximum (141 mm) occurred in 2018. The minimum RX5day (120 mm) occurred in 1963 and the maximum (276 mm) occurred in 2008. Except for RX1day, the relative increase of intensity indices is larger than that of PRCPTOT, indicating that precipitation in the GBA is becoming more concentrated, which is not conducive to water resources management.

According to the temporal mean annual value of EPIs, the persistence, frequency and intensity of extreme precipitation in GBA generally increased. Except for CDD and CWD, the other seven EPIs showed a non-significant upward trend, indicating that the GBA slowly became wetter from 1961 to 2018 and its extreme precipitation was gradually increasing. The overall drought in the GBA had slowed down, which was consistent with the national trend, but slighter (Cao *et al.* 2020). The highest value of the high-intensity precipitation indices occurred after 2008, usually in 2008 and 2016, when climate anomalies happened, and Wang *et al.* (2016) also found that the highest value of extreme precipitation event intensity appeared in 2008. The increase of the long-lasting index RX5day was greater than that of the short-lasting index RX1day, indicating that the risk of extreme precipitation in GBA has increased in recent years, because multiple days of precipitation can easily cause floods. The ratio of the mean annual R95p to the mean annual PRCPTOT was 28.2%, indicating that nearly one third of the total annual precipitation was contributed by extreme precipitation, which was less than the corresponding value in the entire Guangdong Province.

Spatial changes

Spatial changes of mean values of the indices

The spatial distribution of mean annual values of different EPIs is shown in Figure 3. In most areas, CDD was within the range 36.5–45.2 d, with the highest values in the southeast, whereas in the north-central part of Zhaoqing, the value was below 36.4 d (Figure 3(a)). In general, CWD was within the range 13.9–18.2 d, and the values were the highest in the north and northeast GBA, whereas the values were less than 13.8 days in southeast Hong Kong

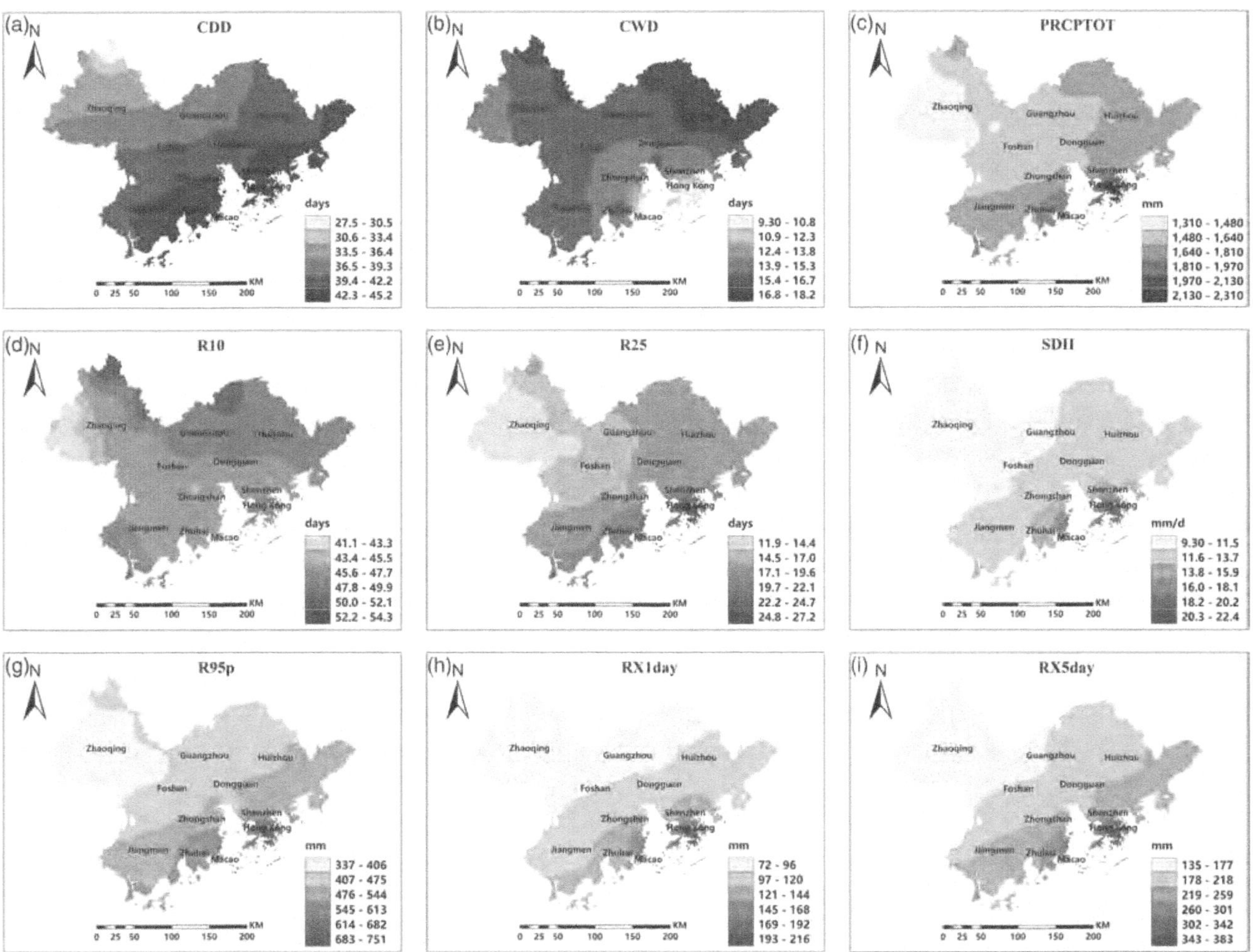

Figure 3 | Spatial variation of the mean values of each EPI in the GBA for 1961–2018.

and Macau (Figure 3(b)). In the majority of the GBA, PRCPTOT was around 1,310–1,810 mm, but in the southern coastal area, including Zhuhai, Hong Kong and Macau, it could exceed 1,970 mm (Figure 3(c)).

The R10 was irregularly distributed (Figure 3(d)). In general, the values in the northern and southeastern GBA were higher than 47.8 d, whereas the values in some parts of Zhaoqing were below 45.5 d. R25 (Figure 3(e)) was higher than 19.7 d in Zhuhai, Macau and Hong Kong, located in the southern coastal GBA, while the values in some parts of the west GBA were less than 14.4 d.

SDII in Hong Kong and Macau was above 18.2 mm/d, and in Zhaoqing was generally below 11.5 mm/d (Figure 3(f)). R95p was mostly within the range 407–544 mm, and in the south of Zhuhai, Hong Kong and Macau, it could exceed 614 mm, whereas it was less than 406 mm in the west of Zhaoqing (Figure 3(g)). RX1day (Figure 3(h)) was mostly within the range 72–144 mm, but it could exceed 169 mm in Hong Kong and Macau. RX5day (Figure 3(i)) was within the range 135–259 mm, but it could exceed 343 mm in Hong Kong.

Overall, the extreme precipitation in GBA decreased from the southeast coastal areas to the northwest inland areas, and CDD also showed similar changes, indicating that the precipitation concentration in the southeast coastal cities was relatively high, which could lead to more floods in the wet season and higher risk of drought in the dry season, especially in Hong Kong, Macau and Zhuhai.

Spatial changes in the EPIs trends

The spatial distribution of trends and tendencies of different EPIs is shown in Figure 4. The trends of all indices of each

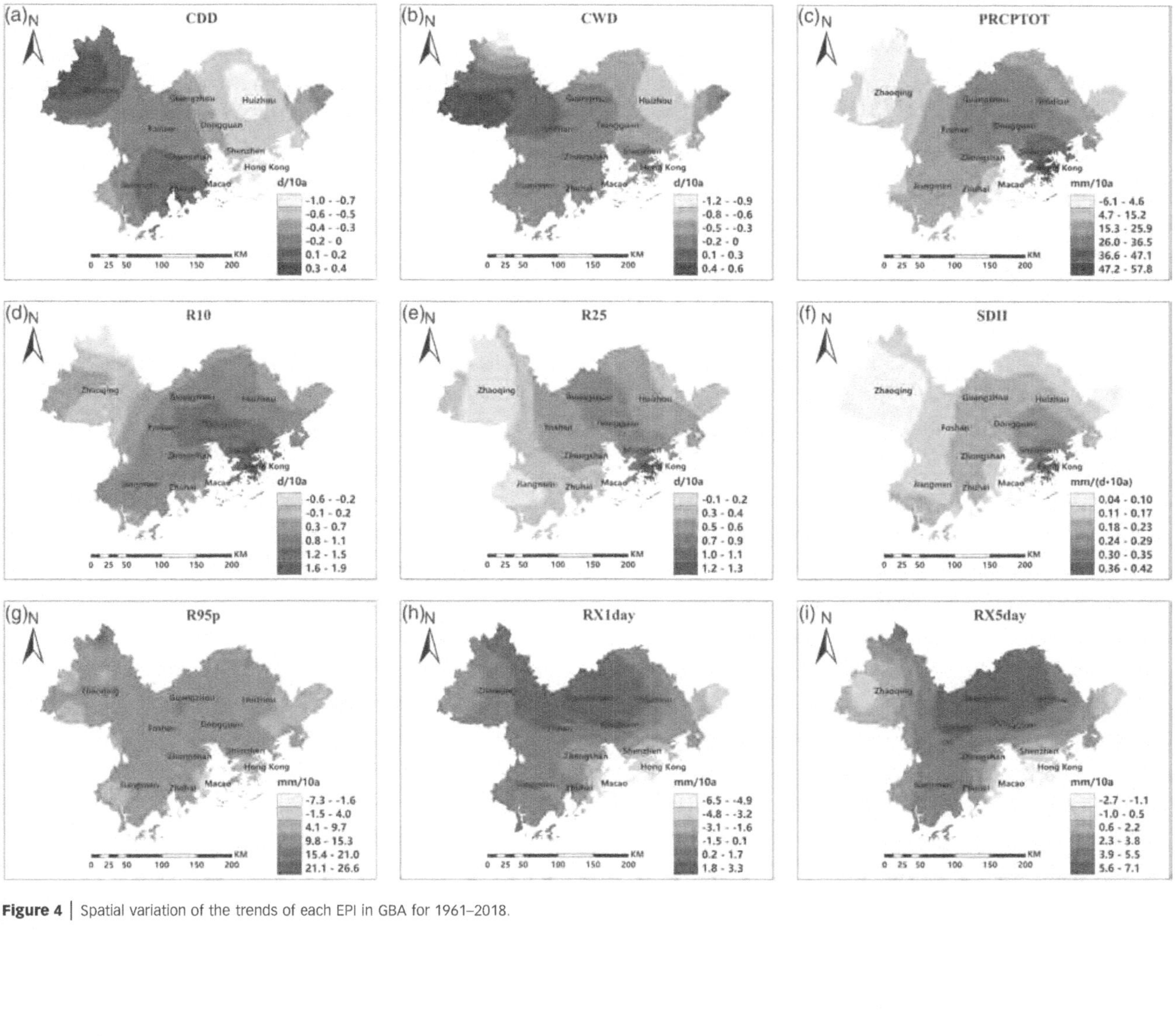

Figure 4 | Spatial variation of the trends of each EPI in GBA for 1961–2018.

grid were tested for significance with a confidence level of 0.05. The variation range of CDD (Figure 4(a)) was mostly around −1.0 to 0.4 d/10a, and no changes passed the significance tests. Overall, 65.5% of the grids showed a downward trend, whereas the northwestern part of Zhaoqing, Zhuhai, and parts of Macau showed an upward trend. CWD (Figure 4(b)) varied from −1.2 to 0.6 d/10a, and 62.1% of the grids showed a downward trend. Among them, CWD in the southwestern part of Zhaoqing increased significantly, whereas it decreased significantly in the northern part. PRCPTOT varied from −6.1 to 57.8 mm/10a (Figure 4(c)), and 96.6% of the grids showed a non-significant upward trend. The areas with an obvious upward trend were mainly in Shenzhen and Hong Kong in the southeast GBA.

The variation in R10 was from −0.6 to 1.9 d/10a (Figure 4(d)), and 86.2% of the grids showed an upward trend, with significant increases in Shenzhen and Hong Kong. The variation in R25 was within −0.1 and 1.3 d/10a (Figure 4(e)), and 89.7% of the grids showed an upward trend, with significant increases in the central part of GBA, including Guangzhou and Dongguan, and in Shenzhen and Hong Kong.

The changes in SDII (Figure 4(f)) varied from 0.04 to 0.42 mm/(d·10a), and the entire study area showed an upward trend. The northern Zhaoqing, Guangzhou and Shenzhen increased significantly, and the upward trend in Hong Kong was also obvious. The changes in R95p ranged from −7.3 to 26.6 mm/10a (Figure 4(g)) and 96.6% of the grids showed a non-significant upward trend. The areas with an obvious upward trend were located in the northern part of Zhaoqing. RX1day (Figure 4(h)) was between −6.5 and 3.3 mm/10a with 62.1% of the grids showing an upward trend. The areas with an obvious upward trend were mainly in Guangzhou and other areas in the northern part of the GBA; the northwestern part of Guangzhou increased significantly, and the eastern part of Zhaoqing decreased significantly. The variation in RX5day was between −2.7 and 7.1 mm/10a (Figure 4(i)), and 79.3% of the grids showed an upward trend, with significant increases in Guangzhou and other areas of the northern part of the GBA.

The strengthening of the subtropical high pressure causes the East Asian summer monsoon to continue to enhance, and more water vapor is transported into China. Most of the EPIs in the GBA showed an upward trend, with a significant increase in some areas. The annual precipitation in northern and southeastern coastal areas showed an increase, as well as the precipitation concentration and the frequency of extreme precipitation events. According to the temporal changes, it could be seen that the increase in extreme precipitation in the GBA was mainly manifested in the southeast coastal area, where the task of urban disaster prevention and mitigation will be more difficult than in other areas. However, although the overall drought in the GBA declined, the CDD in the northwestern part of Zhaoqing showed an upward trend, while other indices mainly showed a downward trend, indicating that the drought increased in this area. In addition, except for RX1day and RX5day, the increase rate of the indices in coastal cities was greater than that of inland areas. The conclusions above are consistent with those achieved by Wang *et al.* (2016), Zeng *et al.* (2017), and Cao *et al.* (2020).

As urbanization continues, greenhouse gas emissions have increased. Moreover, the hardening of the underlying surface has reduced infiltration and increased evaporation. Many studies have shown that urbanization changes urban extreme precipitation (Chang *et al.* 2018; Paul *et al.* 2018; Papalexiou & Montanari 2019; Jiang *et al.* 2020), but the magnitude of this effect varies among cities. As a world-class urban agglomeration, the GBA is one of the most economically vigorous regions in China. As in Figure 5, from the statistical yearbook data of each city, the urbanization rate of the GBA was above 47.7% at the end of 2018, and

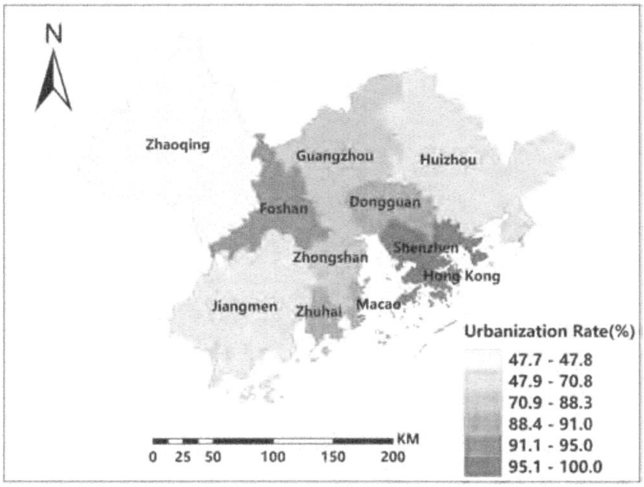

Figure 5 | The spatial distribution of urbanization rate in the GBA.

some cities were even close to 100%. The high urbanization and intensive human activities will inevitably influence the temporal and spatial distribution of extreme precipitation within the region. According to this paper, the extreme precipitation in most parts of the GBA was slowly increasing, and the coastal areas such as Hong Kong, Macau, Zhuhai and the southern part of Shenzhen were increasing more rapidly. These areas, where urbanization rates are more than 90%, play an important role in the development of the GBA, therefore flood prevention should not be underestimated. For example, monitoring and early warning of extreme weather, analyzing the causes and mechanisms of extreme precipitation events, and taking effective countermeasures to reduce the losses caused by disasters. In addition, although the urbanization rate of Zhaoqing was the lowest, there was large room for development. According to the results of this paper, Zhaoqing may have a trend of increasing drought, so attention should be paid to the reasonable allocation of water resources, including promoting water saving measures and optimizing water utilization structure.

This paper lacks the further discussion of factors that affect extreme precipitation. In addition to climate warming and urbanization, there are other factors in connection with extreme precipitation, such as El Niño, and the relationship between them needs to be further studied, using data such as temperature, underlying surface, time of occurrence of El Niño, to obtain more substantial results by using cycle analysis, mutation analysis and global climate mode (Chang *et al.* 2018; Paul *et al.* 2018; Yu *et al.* 2018; Sun *et al.* 2021; Talchabhadel *et al.* 2020). Further discussions on Chinese anomalies in 2008 and 2016 are also necessary. Besides, when there is a small range of heavy precipitation, the precipitation intensity of grid data is likely to weaken (Zhao *et al.* 2014). Therefore, it is necessary to conduct comprehensive inter-comparison and analysis of observation data from the meteorological stations and grid datasets.

Correlation of EPIs

The principal component analysis method was used to analyze the EPIs. Two consistent components with an eigenvalue variance greater than 1 were selected as the principal component. The trend of PRCPTOT and each EPI was shown to have good consistency (Table 2).

Correlation analysis among the EPIs showed a good correlation between each EPI and PRCPTOT (Table 3), except

Table 2 | Principal component load and explanatory variance of EPIs in GBA for 1961–2018

Component	PRCPTOT	CDD	CWD	R10	R25	R95p	SDII	RX1day	RX5day	Variance (%)
1	0.956	−0.185	0.432	0.875	0.930	0.944	0.877	0.675	0.781	61.016
2	−0.207	0.595	−0.610	−0.353	−0.123	0.180	0.220	0.484	0.390	15.290

Table 3 | Correlation analysis for EPIs in the GBA between 1961 and 2018

	PRCPTOT	CDD	CWD	R10	R25	R95p	SDII	RX1day	RX5day
PRCPTOT	1.000								
CDD	−0.239*	1.000							
CWD	0.497**	−0.175	1.000						
R10	0.960**	−0.203	0.573**	1.000					
R25	0.940**	−0.185	0.345**	0.892**	1.000				
R95p	0.864**	−0.179	0.213	0.717**	0.865**	1.000			
SDII	0.754**	−0.057	0.168	0.668**	0.836**	0.887**	1.000		
RX1day	0.527**	−0.017	0.108	0.351**	0.427**	0.736**	0.584**	1.000	
RX5day	0.627**	0.068	0.240*	0.514**	0.607**	0.746**	0.695**	0.724**	1.000

Note: * and ** represent a significant correlation at the 0.05 level and the 0.01 level, respectively.

for CDD which was significantly correlated at the 0.05 level, the others were significantly correlated at the 0.01 level. The correlation coefficients between the two persistence indices (CDD and CWD) and the four intensity indices were relatively small, indicating that there was a weak influence among them. Except for CDD and CWD, the correlations between the EPIs was relatively high, all passing the 0.01 significance test. Therefore, the nine EPIs selected in this paper have a good indicator effect for temporal and spatial variations of extreme precipitation in the GBA.

CONCLUSIONS

Based on the daily grid precipitation datasets of GBA for 1961–2018 and the precipitation data for Hong Kong and Macau, we analyzed the temporal and spatial changes of extreme precipitation in GBA with the nine EPIs recommended by ETCCDI.

The temporal changes of EPIs in GBA were generally gentle, with a generally wetting tendency overall. Among the persistence indices, CDD and CWD showed a downward trend, whereas PRCPTOT showed an upward trend. The frequency indices (R10 and R25) and the intensity index (R95p, SDII, RX1day and RX5day) both showed an upward trend.

The mean multi-year spatial distribution of the EPIs in the GBA increased from west to east and north to south. Except for CWD, the EPIs showed an upward trend in many areas, and coastal regions increased significantly. There were clear upward trends of PRCPTOT, R10, R25, and SDII, which usually occurred in Guangzhou, Dongguan, Shenzhen and Hong Kong. SDII had an increasing tendency throughout the GBA. The increasing areas of R95p, RX1day and RX5day were mainly in the northern part of the GBA. The increasing trend of extreme precipitation was greater in regions with more developed economies and denser populations.

Principal component analysis and regression analysis showed that PRCPTOT had a good consistency with each EPI. The nine EPIs selected in this paper therefore were good indicators of precipitation in the GBA.

While vigorously developing the economy, attracting talent and investment, the GBA should devote a certain budget to early warning and forecasting of extreme precipitation as well as to responding to disasters that may be caused by extreme precipitation, such as heavy rains, floods and continuous droughts, so as to reduce the economic losses. Subsequently, we can truly create a high-quality life circle that is suitable for living, business and traveling. The findings of this study provide a theoretical basis for GBA to formulate a water resource management strategy to address climate change, and provide technical support for improving the regional water resources regulation and control capabilities as well as the ability to reduce the impacts of droughts and floods.

ACKNOWLEDGEMENTS

This study was supported by the Scientific Research Project approved by the Department of Education of Guangdong Province (No. 2020KQNCX125) and the Scientific Research Foundation of Xinhua College of Sun Yat-sen University (No. 2019KYYB08).

DATA AVAILABILITY STATEMENT

All relevant data are included in the paper or its Supplementary Information.

REFERENCES

Bennett, K. E. & Walsh, J. E. 2015 Spatial and temporal changes in indices of extreme precipitation and temperature for Alaska. *International Journal of Climatology* **35** (7), 1434–1452. doi:10.1002/joc.4067.

Cao, Q., Hao, Z., Fu, X., Hao, J. & Lu, C. 2020 Analysis of spatial-temporal changes of extreme climate elements in China from 1960 to 2017. *Yellow River* **42** (2), 11–17. doi:10.3969/j.issn.1000-1379.2020.02.003.

Chang, X., Xu, Z., Zhao, G., Cheng, T. & Song, S. 2018 Spatial and temporal variations of precipitation during 1979–2015 in Jinan City, China. *Journal of Water and Climate Change* **9** (3), 540–554. doi:10.2166/wcc.2017.029.

Hadri, A., Saidi, M. E. M., Saouabe, T. & El Alaoui El Fels, A. 2020 Temporal trends in extreme temperature and precipitation events in an arid area: case of Chichaoua Mejjate region (Morocco). *Journal of Water and Climate Change* 1–21. doi:10.2166/wcc.2020.234.

Jiang, X., Luo, Y., Zhang, D. L. & Wu, M. 2020 Urbanization enhanced summertime extreme hourly precipitation over the Yangtze River Delta. *Journal of Climate* **33** (13), 5809–5826. doi:10.1175/jcli-d-19-0084.1.

Le, N. D. & Zidek, J. V. 2006 *Statistical Analysis of Environmental Space-Time Processes[M]*. Springer Science & Business Media, Berlin, pp. 101–134.

Li, M., Han, T. & Hao, X. 2020 Reginonal characteristics of extreme precipitation events in Eurasia. *Transactions of Atmospheric Sciences* **43** (4), 687–698. doi:10.13878/j.cnki.dqkkxxb.20191202001.

Liu, S., Lei, Y., Li, Y. & Zhang, Y. 2018 Analysis of the climatological characteristics of short-term intense rainfall of Foshan City and Its causal factors. *Guangdong Meteorology* **40** (4), 24–27. doi:10.3969/j.issn.1007-6190.2018.04.006.

Myhre, G., Alterskjær, K., Stjern, C. W., Hodnebrog, Ø., Marelle, L., Samset, B. H., Sillmann, J., Schaller, N., Fischer, E., Schulz, M. & Stohl, A. 2019 Frequency of extreme precipitation increases extensively with event rareness under global warming. *Scientific Reports* **9** (1), 345–357. doi:10.1038/s41598-019-52277-4.

Papalexiou, S. M. & Montanari, A. 2019 Global and regional increase of precipitation extremes under global warming. *Water Resources Research* **55** (6), 4901–4914. doi:10.1029/2018WR024067.

Paul, S., Ghosh, S., Mathew, M., Devanand, A., Karmakar, S. & Niyogi, D. 2018 Increased spatial variability and intensification of extreme monsoon rainfall due to urbanization. *Scientific Reports* **8** (1), 1–10. doi:10.1038/S41598-018-22322-9.

Soltani, M., Laux, P., Kunstmann, H., Stan, K., Sohrabi, M. M., Molanejad, M., Sabziparvar, A. A., Ranjbar SaadatAbadi, A., Ranjbar, F., Rousta, I., Zawar-Reza, P., Khoshakhlagh, F., Soltanzadeh, I., Babu, C. A., Azizi, G. H. & Martin, M. V. 2016 Assessment of climate variations in temperature and precipitation extreme events over Iran. *Theoretical and Applied Climatology* **126** (3/4), 775–795. doi:10.1007/s00704-015-1609-5.

Sun, Q. H., Zhang, X. B., Zwiers, F., Westra, S. & Alexander, L. V. 2021 A global, continental, and regional analysis of changes in extreme precipitation. *Journal of Climate* **34** (1), 243–258. doi:10.1175/JCLI-D-19-0892.1.

Talchabhadel, R., Aryal, A., Kawaike, K., Yamanoi, K. & Nakagawa, H. 2020 A comprehensive analysis of projected changes of extreme precipitation indices in West Rapti River basin, Nepal under changing climate. *International Journal of Climatology* 1–19. doi:10.1002/joc.6866.

Tian, Q. Y., Li, Z. L. & Sun, X. L. 2020 Frequency analysis of precipitation extremes under a changing climate: a case study in Heihe River basin, China. *Journal of Water and Climate Change* 1–15. doi:10.2166/wcc.2020.170.

Wang, W., Yan, J. & Liu, Y. 2016 Variation characteristics of extreme precipitation events in Guangdong province. *Bulletin of Soil and Water Conservation* **36** (2), 293–299. doi:10.13961/j.cnki.stbctb.

Yu, J., Qi, M., Sun, Q. & Tao, L. 2018 Statistical characteristics of summer extreme rainfall over eastern China and its relation with EL Niño. *Trans Atmos Sci* **41** (1), 77–84. doi:10.13878/j.cnki.dqkxxh.20151027001.

Zeng, Y., Tan, Y., Qiu, X. & Liu, C. 2017 Analysis on the tendency of different precipitation in Ten major watersheds in China. *Jiangsu Agricultural Sciences* **45** (14), 189–195. doi:10.15889/j.issn.1002-1302.2017.14.052.

Zhao, Y., Zhu, J. & Xu, Y. 2014 Establishment and assessment of the grid precipitation datasets in China for the past 50 years. *Journal of the Meteorological Sciences* **34** (04), 414–420. doi:10.3969/2013jms.0008.

First received 3 March 2021; accepted in revised form 8 May 2021. Available online 21 May 2021

 Impact of Climate Change on Hydrology and Water Resources doi: 10.2166/wcc.2022.359

Deterministic and probabilistic projections and their credibility in analyzing future precipitation variations in the Yellow River Basin, China

Zhouliang Sun[a,b], Yanli Liu[b,c,d], Jianyun Zhang[a,b,c,d,*], Hua Chen[a], Zhangkang Shu[b], Tiesheng Guan[b,c,d], Guoqing Wang[b,c,d], Junliang Jin[b,c,d], Zhenxin Bao[b,c,d] and Cuishan Liu[b,c,d]

[a] State Key Laboratory of Water Resources and Hydropower Engineering Science, Wuhan University, Wuhan 430072, China
[b] State Key Laboratory of Hydrology-Water Resources and Hydraulic Engineering, Nanjing Hydraulic Research Institute, Nanjing 210029, China
[c] Yangtze Institute for Conservation and Development, Nanjing 210098, China
[d] Research Center for Climate Change, Ministry of Water Resources, Nanjing 210029, China
*Corresponding author. E-mail: jyzhang@nhri.cn

ZS, 0000-0002-0388-9621

ABSTRACT

It remains a key challenge to obtain reliable future precipitation estimates and their reliability under different climate scenarios. In this study, the deterministic projection of future precipitation in the Yellow River Basin (YRB) was obtained within the Bayesian model averaging (BMA) framework. A probability estimation method based on the BMA weighting scheme was proposed to obtain the probabilistic projection of precipitation. We also analyzed the credibility of these two projections. The results showed that four indexes projected by the BMA method showed an increasing trend with a higher probability. The probabilities of increasing with varying degrees were more than those for decreasing for all the precipitation indexes. The credibility of the precipitation estimation under specific climate scenarios was testified by the lower E_D (the mean of long-term annual relative simulation deviation) and V_D (the variance of long-term annual relative simulation deviation). The estimation based on the BMA model is more trustworthy than any other model. For the four precipitation indicators, the accuracy between the calculated VR (Variation range, to describe the interval of variation of the indicators) with the greatest likelihood and the actual VR was 38.31–53.74%. In 81.93–94.70% of grids, the deviations were smaller than one level. Both the deterministic and probabilistic projections have high geographic distribution and variation trend consistency.

Key words: Bayesian model averaging, climate change, credibility assessment, precipitation indexes, probability estimation, Yellow River Basin

HIGHLIGHTS

- A probability estimation method was proposed for the probabilistic projection of future climate.
- The deterministic projection and probabilistic projection of future precipitation were consistent.
- The projection accuracy was positively correlated with historical occurrences of precipitation.
- The probability of increase is greater than that of decrease for the future precipitation of YRB.

GRAPHICAL ABSTRACT

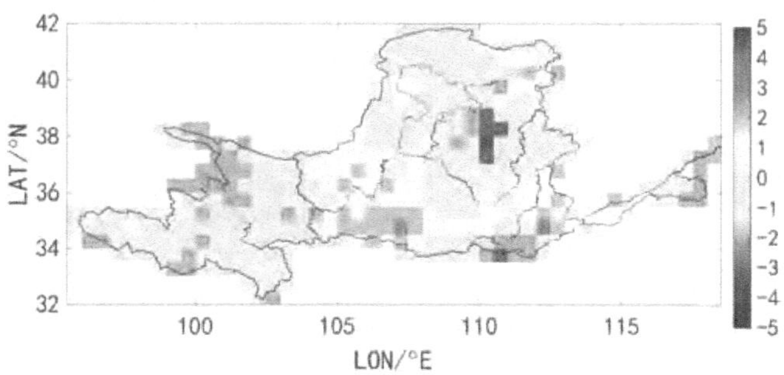

1. INTRODUCTION

The precipitation over the Yellow River Basin (YRB) has undergone continuous and drastic variation in the past decades (Liu *et al.* 2020), especially in terms of extremes and potential redistribution of water resources in time and space, threatening the lives of residents in the whole basin. Another concern lies in the future precipitation variation in the forthcoming decades and centuries under global environmental change. Studies have confirmed the incredible impact of climate change on people and the environment (Meshram *et al.* 2018, 2020; Almazroui *et al.* 2021a, 2021b), but it is still a significant challenge to reliably estimate future climate and its credibility under different climate scenarios and climate models.

The vulnerable natural system in the YRB is of interest to researchers worldwide because it reflects the evolution of water resources under climate change and human activity in the future (Wu *et al.* 2016; Wang *et al.* 2019b). Moreover, the precipitation in the YRB has changed in time and space for decades (Lu *et al.* 2014; Zhang *et al.* 2014, 2019; Iqbal *et al.* 2018; Zhao *et al.* 2019), arousing many concerns and prompting research on future precipitation change as well as its impact on hydrology and water resources. The GCMs provide scenario determination and forcing data, forming the basis of climate change projection. They are an original estimation of long-term precipitation without specific precipitation indexes, but most of the projected precipitation indexes derive from the GCMs data. Typical indexes include the mean over time and extreme value. The key to obtaining a reliable future precipitation projection is to solve uncertainty in multiple models and get valid information. Multi-model ensemble, including equal-weighted or unequal weighted averaging, is the most common and effective solution (Radermacher & Tomassini 2012; Chen *et al.* 2017; Guo *et al.* 2017). For the YRB previous studies mainly focused on the impact of future precipitation change, while there was little research on future precipitation change itself. Of the research that has been done in this area, Wang *et al.* (2017a) and Chen *et al.* (2019) investigated the impacts of climate change on water resources (indicated by runoff) by using the projected precipitation and temperature, applying the Variable Infiltration Capacity (VIC) model, in the whole YRB and the source region of the YRB. The impacting mechanisms usually come from comparing runoff in the baseline period (the model output) and runoff change caused by the precipitation change (the model input). As for the impacts of future precipitation change on hydrology and water resources, Yin *et al.* (2016) analyzed the future extreme precipitation with different return periods in the Huang-Huai-Hai region using five GCMs multi-scenarios. The result indicated that the climate models had errors in simulating extreme historical precipitation. Liu *et al.* (2017) studied the projected climatic year types (CYTs) throughout China between 2015 and 2099 from spatial patterns, temporal changes, and frequency, demonstrating better performance and superiority of multi-model ensembles than individual models in projecting changes in CYTs by offsetting systematic errors. Nevertheless, all future water resources projections depend on the future climate scenarios, and the reliability of conclusions is subject to the credibility of the GCMs. Thus, the main challenge is evaluating and generating more reliable precipitation from GCMs.

Multi-model ensembles are a practical solution for hydrological, meteorological, and climatological predictions or forecasts. The ensemble projection could be understood as a weighted model average from the member models. Raftery *et al.* (2005) applied Bayesian model averaging (BMA) to forecast the surface temperature and sea level pressure in the Pacific Northwest for better and sharper forecasting, confirming the excellent performance of BMA in the multi-model ensemble when the ensemble members are overdispersed. Duan *et al.* (2007) also used BMA to acquire more accurate hydrological predictions from some original competing predictions in terms of more satisfying daily root mean square error and daily absolute mean error, as well as a better description of total predictive uncertainty than the members. These results make it a typical case of ensemble hydrological forecasting using BMA. From research on climate change, the ensemble members mainly come from climate models, including Regional Climate Models (RCMs) and GCMs. Dars *et al.* (2017) proved that BMA performed better than individual models in multi-model ensemble prediction with a weighted average of four CMIP5 models to assess the impacts of climate change on future precipitation. The result will significantly help related managers, experts, and policymakers. A similar study on temperature (Min & Hense 2006; Miao *et al.* 2014) inferred that BMA enjoys more reliability and less uncertainty than other model averaging methods or individual models. The application of BMA in climate change has yielded encouraging results and has been widely recognized (Duan & Phillips 2010; Notaro *et al.* 2016). Irrespective of whether the ensemble members come from hydrological models or climate models, the consistent process in the ensemble is to capture the practical information from those member models for a more reliable future projection.

Researchers have focused on the uncertainty and credibility of climate change projections due to enormous uncertainty in climate models and downscaling data processing. The output of climate scenarios is also uncertain for the complicated processes in the atmospheric system. It has been challenging to acquire desired results and assess the credibility of climate projection evaluations, especially without a practical evaluation system. The study (Fricker *et al.* 2013) pointed out validation of performance measures for the credibility of future climate projections. These issues about credibility aroused our keen interest in the credibility evaluation of future precipitation projection over water-scarce areas in the YRB. We aim to improve the projection level of climate elements by using appropriate indicators to measure the precipitation projection ability of GCM models.

In this paper, we proposed a probability precipitation estimation method based on the BMA method and the measurement of its credibility. For the future precipitation projection of the YRB, there would be two significant questions to resolve: (1) How to ensure good performance for BMA in terms of the precipitation indexes evaluation compared to the original climate models over the YRB and (2) How to ensure the credibility of the precipitation estimation based on BMA. To find the answer, we used the daily observed precipitation data during 1986–2005 and future precipitation projections selected from five GCMs during 2006–2099. Four precipitation indexes were adopted to describe the precipitation situation. Two indicators were used to assess the performance and credibility of precipitation ensemble projection conducted by the BMA method.

2. STUDY AREA AND DATA

2.1. The Yellow River Basin

With a length of 5,464 km and a drainage area of 795,044 km^2, the YRB (Figure 1) is the second-largest river in China. It originates in the Tibetan Plateau and runs through four geomorphic units – the Qinghai-Tibetan Plateau, the Inner Mongolia Plateau, the Loess Plateau, and the North China Plain. Supplying water resources to 100 million people, the YRB runs through an arid region, a semi-arid region, and a semi-humid region, before reaching the Bohai Sea (Wu *et al.* 2017; Wang *et al.* 2019a). These regions range in elevation between 1,000 and 4,000 m, have annual precipitation ranging from 123 to 1,021 mm, have annual evaporation varying from 700 to 1,800 mm, and have dramatic temperature fluctuations in time and space (Wang *et al.* 2019a). The long-time average annual value of precipitation, runoff (engaging a drastic decrease), and annual evapotranspiration were estimated to be about 460, 450, and 1,100 mm on average, respectively. Since the YRB is located in a temperate continental climate zone and is regulated by the East Asian monsoon, precipitation is the primary controlling source and restrictive factor of the runoff and water resources. This directly affects the water supply for 12% of China's population, 15% of China's arable land, and 8% of China's GDP approximately, even though the scarce runoff accounts for only 2% of China's total (Guan *et al.* 2016; Yin *et al.* 2017; Wang *et al.* 2019a). Therefore, the variation of precipitation plays a vital role in sustainable development and even in the daily life of the people of the YRB. Both in terms of time and space, the precipitation in the YRB is remarkably uneven, with June to October accounting for about 70% of annual total precipitation and the precipitation in the lower and middle reaches two times that of the upper reaches (Mu *et al.* 2013). Over the past 50 or 60 years, with the increase of air temperature, the precipitation of the YRB has fallen, and so too has runoff, exacerbating a profound contradiction between water demand and water supply and worsening the drought situation (Wu *et al.* 2017; Wang *et al.* 2018, 2019a).

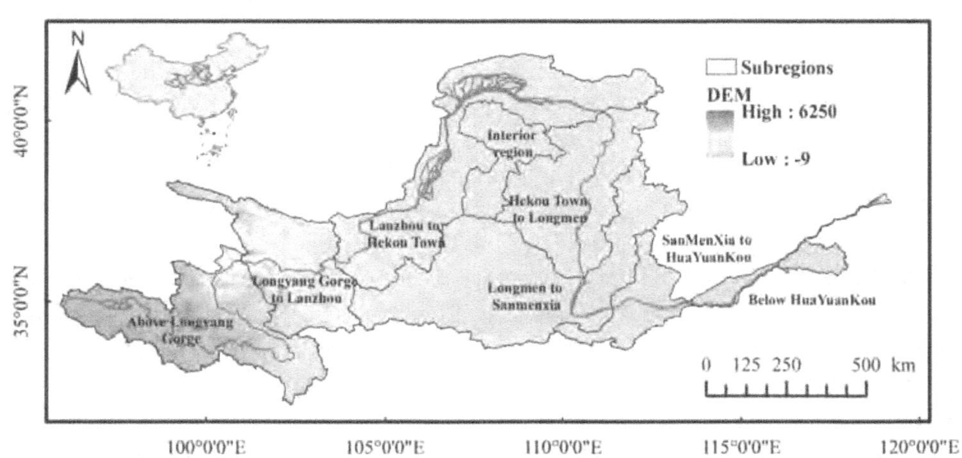

Figure 1 | Study area (the Yellow River Basin of China).

2.2. Data

The daily observed precipitation data during 1986–2015 came from the Meteorological Data Center of the China Meteorological Administration (CMA) (http://data.cma.cn/). It was gridded into the spatial resolution of $0.5° \times 0.5°$, the data of the YRB were extracted and put into use.

After a comprehensive multi-criteria assessment, Zhou & Han (2018) selected 5 models (Table 1) from 18 CMIP5 GCMs. Therefore, the total rank of these selected models is more satisfactory. In the model evaluation of Zhou & Han (2018), several statistics on precipitation and temperature were calculated. The statistics consisted of climatological mean, interannual variation, seasonal cycle, inter-annual modes, and probability distribution function. They were assessed by Mean Absolute Error, Correlation Coefficient, Root Mean Square Error, and S-value (for the degree of overlap between the simulated probability distribution and the observed probability distribution). The eighteen models were ranked by Rank Score (RS) performance, with these five models ranking better than other models. These models were corrected by the Quantile-Mapping method (Tong *et al.* 2017; Han *et al.* 2018).

Each model consists of daily simulated climate data during 1986–2005 and daily projected climate data during 2006–2099, including climatic elements of precipitation and temperature data under three emissions scenarios of RCP 2.6, RCP 4.5, and RCP 8.5. The detailed information of the selected models is shown in Table 1. The gridded climate data in the YRB were extracted as the observed precipitation data. Notably, for comparing the precipitation variation in a different time in the future, we chose two representative periods from the future period, the period of 2021–2050 (the near future scenario) and 2061–2090 (the distant future scenario). Given the homogeneity of different climatic elements and emissions scenarios, the climatic element of precipitation under emissions scenarios from RCP 4.5 was only analyzed in this study.

Considering the non-correspondence of the spatial resolution between the observed precipitation and the climate data, we unified the grids between observed data and climate data by matching the longitude and latitude of each grid, confirming that there are 415 coincident grids in both observation data and climate data.

3. METHODOLOGY

In this section, we state our chosen precipitation indexes. The research methods mainly included BMA and the probability estimation method. BMA was used to calculate the optimal weight of each model for the ensemble and to project each precipitation index deterministically (in Section 4.3). The probability estimation method calculated the probabilistic projections of future variations of each precipitation index based on the BMA weights (in Section 4.4).

3.1. Precipitation indexes

There are four precipitation indexes adopted in this study (Table 2). The PY indicates the total quantity of precipitation in the whole year over the YRB, while the PS indicates summer precipitation, which affects the water resources in the YRB. The R_{10} indicates the frequency of heavy rain that occurs over the YRB. It reflects the extreme attributes of precipitation, in other words, the frequency of rainstorm events. The PI indicates the concentration of precipitation. It shows the concentration of precipitation events in time in a year. The PY and the PS are calculated from the annual precipitation, while the R_{10} and the PI are daily. It is noted that the annual precipitation is the mean value of annual precipitation in all grids, and so is the daily precipitation.

Table 1 | The selected GCMs

No.	Model name	Temporal resolution	Spatial resolution
1	CNRM-CM5 (CN)	Daily	$0.25° \times 0.25°$
2	CSIRO-Mk3-6-0 (CS)	Daily	$0.25° \times 0.25°$
3	EC-EARTH (EC)	Daily	$0.25° \times 0.25°$
4	MIROC-ESM-CHEM (MI)	Daily	$0.25° \times 0.25°$
5	NorESM1-M (No)	Daily	$0.25° \times 0.25°$

Table 2 | The definition of precipitation indexes

Index name	Description	Reference	Unit
Annual mean precipitation (*PY*)	The mean value of annual precipitation	Libanda & Nkolola (2019)	mm
Summer precipitation (*PS*)	The mean value of summer precipitation	Qing *et al.* (2017)	mm
Heavy rain index (R_{10})	The mean number of days with daily precipitation is more than 10 mm.	Hanittinan *et al.* (2020)	day
Annual mean precipitation intensity (*PI*)	The mean ratio of annual precipitation and the number of days with daily precipitation greater than 1.0 mm.	Wu *et al.* (2019)	mm/day

3.2. Bayesian model averaging

BMA is a widely used probabilistic method for an ensemble of referable members (Hoeting *et al.* 1999), synthesizing the valuable information into a comprehensive and superior output from several comparable alternatives. BMA is typically applied to multi-model ensemble prediction of hydrometeorology, with temperature (Duan *et al.* 2007), streamflow (Yan *et al.* 2020), and precipitation (Liu & Xie 2014) among the typical cases. BMA can also integrate future projections from different climate models (Duan & Phillips 2010). In analyzing the changing mechanism of a variable under multiple factors, BMA can facilitate the desired results (Duan & Phillips 2010; Liu & Xie 2014). Wang *et al.* (2017b) applied the BMA ensemble method to obtain the multi-model ensembles projection of Kaifeng in the YRB from four GCMs under four emissions scenarios (RCP2.6, RCP4.5, RCP6.0, and RCP8.5). BMA was compared with the simple averaging method and individual models, and the merged projection could better capture the changing behaviors of climatic factors due to its lowest bias. These pieces of evidence show that BMA has advantages in dealing with multi-model issues. According to their simulation or reproduction competency to our object variables, the effect of BMA is usually regarded as assigning weights to each member for its extraordinary ability to reduce the uncertainty of different outputs. In this study, BMA is used for future precipitation projection. All the variables that appear below are directly related to the background of this study. In other words, the definition and description of variables are from the perspective of this study of precipitation projection based on multiple climate models.

If Pr is the predicted precipitation, $D = [Pr_1^o, Pr_2^o, \ldots, Pr_T^o]$ is the observed precipitation data set used for BMA training with data length T, and $M = [M_1, M_2, \ldots, M_K]$ is the ensemble of precipitation projections from K models, then, with the law of total probability, the BMA probabilistic projection of Pr can be expressed as:

$$p(Pr|D) = \sum_{k=1}^{K} p(M_K|D) \cdot p_k\ (Pr|M_K, D) \tag{1}$$

where $p(Pr|D)$ is the predicted probability of Pr given observed precipitation D, $p(M_k|D)$ is the posterior probability of a model projection M_k given observed precipitation D, and $p_k(Pr|M_k, D)$ is the posterior distribution of Pr given model projection M_k and observed precipitation D there is always $\sum_{k=1}^{K} p(M_k|D) = 1$.

Then, the posterior mean precipitation of the BMA projection is:

$$E(Pr|D) = \sum_{k=1}^{K} p(M_K|D) \cdot E[p_k(Pr|M_K, D)] = \sum_{k=1}^{K} w_k M_k \tag{2}$$

where $E(Pr|D)$ is the expected precipitation of BMA projection, w_k refers to $p(M_k|D)$, and we can obtain $\sum_{k=1}^{K} w_k = 1$, and w_k is the weight of the kth model as well as the right objective of the BMA method.

Considering the requirement of Gaussian distribution for $p_k(Pr|M_k, D)$, we transform both the observed precipitation data and model projection data using the Box-Cox transformation first, and all the data would be transformed with the same parameter.

Define $\theta = [\{w_k, \sigma_k, k = 1, 2, \ldots, K\}]$ as the parameters to be solved, then the log-likelihood function is:

$$L(\theta) = log\left(\sum_{k=1}^{K} w_k \cdot p_k(Pr|M_K, D)\right) = \log\left(\sum_{k=1}^{K} w_k \cdot g(Pr|M_K, \sigma_k^2)\right) \tag{3}$$

where σ_k^2 is the variance of model projection M_k concerning observation D.

Generally, w_k is optimized by the Expectation-Maximization (EM) algorithm with the same distribution requirement as BMA. Since the transformed precipitation data approximately obeys the Gaussian distribution, the algorithm can be implemented legitimately. In brief, the EM algorithm consists of the E step and M step. The best solution of w_k is generated through the alternation between these two steps. The following steps describe the procedure of the EM algorithm in detail.

Step 1: Initialization

$$\text{Iter} = 0,\ w_k^{\text{Iter}} = \frac{1}{K},\ \sigma_k^{2(\text{Iter})} = \frac{1}{K}\sum_{t=1}^{T}\frac{\sum_{k=1}^{K}(Pr_t - M_{k,t})^2}{T} \tag{4}$$

where T is the total number of data points in the training period.

Step 2: Computing the initial likelihood

$$L(\theta)^{(0)} = \sum_{t=1}^{T}\log\left(\left(\sum_{k=1}^{K} w_k^{(0)} \cdot g(Pr_t^{\text{obs}}|M_k^t, \sigma_k^{2(0)})\right)\right) \tag{5}$$

where g() denotes the Gaussian distribution.

Step 3: Computing a latent variable known as the expectation step

$$\text{Iter} = \text{Iter} + 1,\ \hat{z}_{k,t}^{\text{Iter}} = \frac{g(Pr_t|M_{k,t}, \sigma_k^{(\text{Iter}-1)})}{\sum_{k=1}^{K} g(Pr_t^{\text{obs}}|M_{k,t}, \sigma_k^{(\text{Iter}-1)})} \tag{6}$$

where $z_{k,t}$ is a latent variable.

Step 4: Computing the weights known as the maximization step

$$w_k^{\text{Iter}} = \frac{1}{T}\sum_{t=1}^{T} z_{k,t}^{\text{Iter}} \tag{7}$$

Step 5: Computing the projection variance

$$\sigma_K^{2(\text{Iter})} = \frac{\sum_{t=1}^{T}\hat{z}_{k,t}^{\text{Iter}} \cdot (Pr_t^{\text{obs}} - M_{k,t})^2}{\sum_{t=1}^{T}\hat{z}_{k,t}^{\text{Iter}}} \tag{8}$$

Step 6: Updating the likelihood using Equation (7)

Step 7: Checking convergence

If $L(\theta^{\text{Iter}}) - L(\theta^{\text{Iter}-1})$ is no more than a prespecified tolerance level, stop; otherwise, go back to Step 3 for a new optimization circle.

3.3. Probabilities in different variation ranges

Based on the BMA weights, the probability of variation refers to the probability that the variation of a precipitation index locates in a particular range. According to the future changes of each index, six variation ranges are selected for each

Table 3 | Variation ranges (VRs) of precipitation indexes (Inf. means infinity, −Inf. means infinitesimal)

Indexes name	VR1	VR2	VR3	VR4	VR5	VR6
PY (mm)	−Inf. to − 50	− 50 to − 25	− 25 to 0	0–25	25–50	50 to Inf.
PS (mm)	−Inf. to − 50	− 50 to − 50	− 25 to 0	0–25	25–50	50 to Inf.
R_{10} (day)	−Inf. to − 4	− 4 to − 2	− 2 to 0	0–2	2–4	4 to Inf.
PI (mm/day)	−Inf. to − 1	− 1 to − 0.5	− 0.5 to 0	0–0.5	0.5–1	1 to Inf.

index. The variation ranges are shown in Table 3. In this paper, the variation ranges (VRs) are defined as events. For example, if the variation of *PY* is between −50 and −25 mm, a VR2 event is considered to have occurred in PY (the event level is 2).

The BMA method is based on the total probability formula. The principle is to solve the possibility of various causes (if they are mutually exclusive) through the result that has already occurred or find out the most likely cause for this result. Theoretically, the result must be caused by one of those causes. The cause of the current result must be found among all those causes that have been known in advance. Therefore, there is a prerequisite: every possible cause has an objective explanation for the result. For the precipitation projection, the observed precipitation is the result, and the precipitation in each climate model is the cause. There is no evidence to suggest that precipitation in these climate models can perfectly (or partially) simulate observed precipitation. However, we proposed that climate models can explain the observed precipitation process under the current understanding. There were two reasons for this. First, these climate models were designed to simulate precipitation as closely as possible, and researchers had considered the physical mechanisms of precipitation. Second, the climate model used in this study had been strictly evaluated, and the models with the best performance of precipitation simulation had been screened from many climate models. We believed that this hypothesis was reasonable and necessary for these two reasons.

Given the principle of the BMA method (to solve the probabilities of each cause based on the result) and the reasonable assumption, we can use the obtained probabilities (the BMA weights) of each cause (the models) to predict the future variations in precipitation and their probabilities. In this study, five climate models were applied. The probability of variation means the proportion of the projected results within a specific range. This can be defined as:

$$P^{\mathrm{LVR} \leq \Delta_i < \mathrm{UVR}} = \sum_{\mathrm{LVR} \leq \Delta_{i,j} < \mathrm{UVR}} w_j \tag{9}$$

where P is the probabilities of variation of i (i means one of the precipitation indexes), Δ_i is the variation of the ith index, $\Delta_{i,j}$ is the variation of the ith index in the jth model, w_j is the BMA weight of the jth model, LVR and UVR is the lower bound and upper bound in the current variation ranges for index i (Table 3), respectively. For example, if the projected future variations of *PY* ($i = 1$) from the five models ($\Delta_{i,j}$ or $\Delta_{PY,j}$) are 37, −6, 18, 20, and −9 mm, and the BMA weights of the five models (w_j) are 0.14, 0.23, 0.38, 0.09, and 0.16. Then, the probabilities of different variations of *PY* in the future are:

$$\begin{aligned}
P^{\Delta_{PY}<-50} &= \sum_{\Delta_{PY,j}<-50} w_j = 0 \\
P^{-50\leq\Delta_{PY}<-25} &= \sum_{-50\leq\Delta_{PY,j}<-25} w_j = 0 \\
P^{-25\leq\Delta_{PY}<0} &= \sum_{-25\leq\Delta_{PY,j}<0} w_j = \sum_{j=2,5} w_j = 0.39 \\
P^{0\leq\Delta_{PY}<25} &= \sum_{0\leq\Delta_{PY,j}<25} w_j = \sum_{j=3,4} w_j = 0.47 \\
P^{25\leq\Delta_{PY}<50} &= \sum_{25\leq\Delta_{PY,j}<50} w_j = \sum_{j=1} w_j = 0.14 \\
P^{50\Delta_{PY}} &= \sum_{50\Delta_{PY}} w_j = 0
\end{aligned} \tag{10}$$

3.4. Performance assessment of climate models

To better reflect the simulation and projection performance of all models and their ensembles precipitation in the whole YRB, the mean value of all grids in the basin was taken as the precipitation assessment index for the basin. The BMA weight of each model was calculated by taking the projected annual precipitation of each model as the input and the observed annual precipitation as the output. The superiority of the ensemble was evaluated by Root Mean Squared Error (*RMSE*) and total deviation (*TD*). The smaller the *RMSE* is, the better the model's performance in process simulation. In contrast, the smaller the *TD* is, the better the model's performance in quantitative simulation. The *RMSE* and *TD* are defined as:

$$RMSE = \sqrt{\frac{1}{m}\sum_{i=1}^{m}(M_i - H_i)^2} \tag{11}$$

$$TD = \frac{1}{m}\sum_{i=1}^{m}M_i - \frac{1}{m}\sum_{i=1}^{m}H_i \tag{12}$$

where M_i is the *i*th historical simulation of each model, including the BMA, H_i is the *i*th historical observation of each model, including the BMA, m is the total number of years.

Considering the primary purpose of this study is to explore the variation of precipitation indexes in the future based on the BMA model, it is necessary to investigate the simulation performance of the BMA model for precipitation indexes. Therefore, to accurately describe the simulation deviation of a model, the mean of the absolute value of simulation deviation (*MASD*) is applied, the *MASD* here is defined as:

$$MASD = \frac{1}{m}\sum_{i=1}^{m}|P_{si} - P_{oi}| \tag{13}$$

where the P_{si} is the simulated precipitation index of the *i*th year and the P_{oi} is the observed precipitation index of the *i*th year.

The historical simulation performance of the climate model determines whether a model is satisfactory. In other words, the credibility of precipitation in the future predicted by a climate model rests with its historical performance, and simulation deviation is the right indicator to explain the historical performance of a climate model. The simulation deviation is always fluctuant, so the mean indicates its long-term average level of deviation from observation, while the variance indicates the stability of the simulation deviation. The following formula is used to calculate the mean and variance of annual relative simulation deviation:

$$E_D = \frac{1}{m}\sum_{i=1}^{m}D_i \tag{14}$$

$$V_D = \frac{1}{m-1}\sum_{i=1}^{m}(D_i - E_D)^2 \tag{15}$$

$$D_i = \frac{|M_i - H_i|}{H_i} \tag{16}$$

where E_D is the mean long-term annual relative simulation deviation, V_D is the variance of long-term annual relative simulation deviation, and M_i is the *i*th historical simulated value of each model, H_i is the *i*th historical observed value of each model, D_i is the absolute value of relative simulation deviations, and i is the number of years.

4. RESULTS

4.1. BMA weight of each model and the superiority of ensemble

The BMA weight of each model was calculated, and the historical simulating performance was assessed by *RMSE* and *TD* simultaneously (Table 4). The CS model has the maximum weight while the CN model has the minimum. The weight of the other models stands between the two. From a process perspective, the CS model has the minimum *RMSE*, and the

Table 4 | BMA weights and historical precipitation simulating assessment of each model

Model name	CN	CS	EC	MI	No	BMA
RMSE (mm)	65.9	63.9	69.5	85.2	83.2	51.7
TD (mm)	4.2	11.1	12.1	8.6	0.4	7.8
Weight	0.144	0.258	0.149	0.233	0.216	/

CN model and the EC model have a similar level to the CS model, while the MI model has the maximum *RMSE* and the No model is close to it. From a quantitative perspective, the No model has the minimum *TD*, the EC model has the maximum, and the CS model is close. The BMA model has the minimum *RMSE* among all the models (since the calculation of the BMA weights is based on process performance). It has better performance than most models (better than 3/5 of the models) from the quantitative perspective, though it is not the best.

The BMA model simulates the average state of historical precipitation indexes more precisely with the smaller *MASD* than any single model (Figure 2). Each model has different simulation performances in different precipitation indexes, but the BMA model consistently has the best performance in all four precipitation indexes with the minimum MASD. At the same time, every model among the five has its best or worst performance for different indexes. For instance, the CN model performs nicely for the *PY* and R_{10}, but not so for the *PS* and *PI*, and even worse for the *PS*. Thus, it is necessary to do a model ensemble to simulate and project the precipitation over the YRB.

Climate models are inherently uncertain, so is the BMA model derived from climate models. Therefore, credibility is a crucial reference to precipitation index estimation. The credibility of precipitation indexes estimation is quantitatively described by the E_D (mean long-term annual relative simulation deviation, Equation (14)) from the perspective of competence and the V_D (variance of long-term annual relative simulation deviation, Equation (15)) from the robustness perspective, respectively. The E_D and V_D on each grid in Figures 3 and 4 are based on the relative simulation deviation instead of the absolute simulation deviation. It is worth noting that, for the BMA model in Figures 3 and 4, the E_D and V_D are calculated from the annual time series of precipitation indexes ensembled by BMA weights. The blank grids in R_{10} are since the observed results of these grids are 0, which leads to abnormal calculation results.

There is a significant improvement in *PY*, *PS*, and *PI* estimation. Furthermore, from an E_D perspective (Figure 3), the simulation deviation of the BMA model is smaller than that of any other model. Finally, from a V_D point of view (Figure 4), the BMA model has a more stable simulation deviation than any other model.

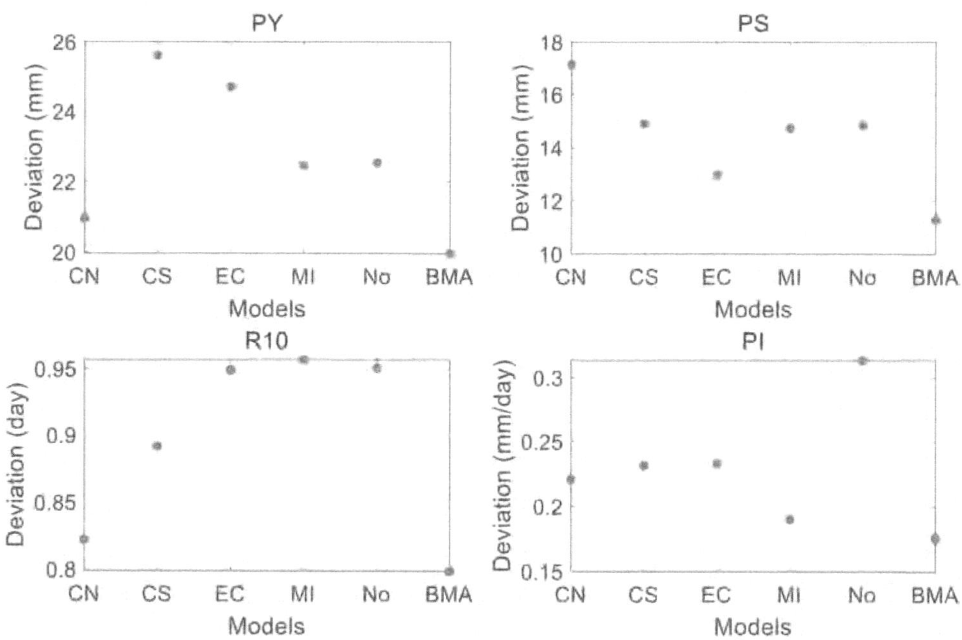

Figure 2 | The *MASD* of four precipitation indexes in the YRB by different models.

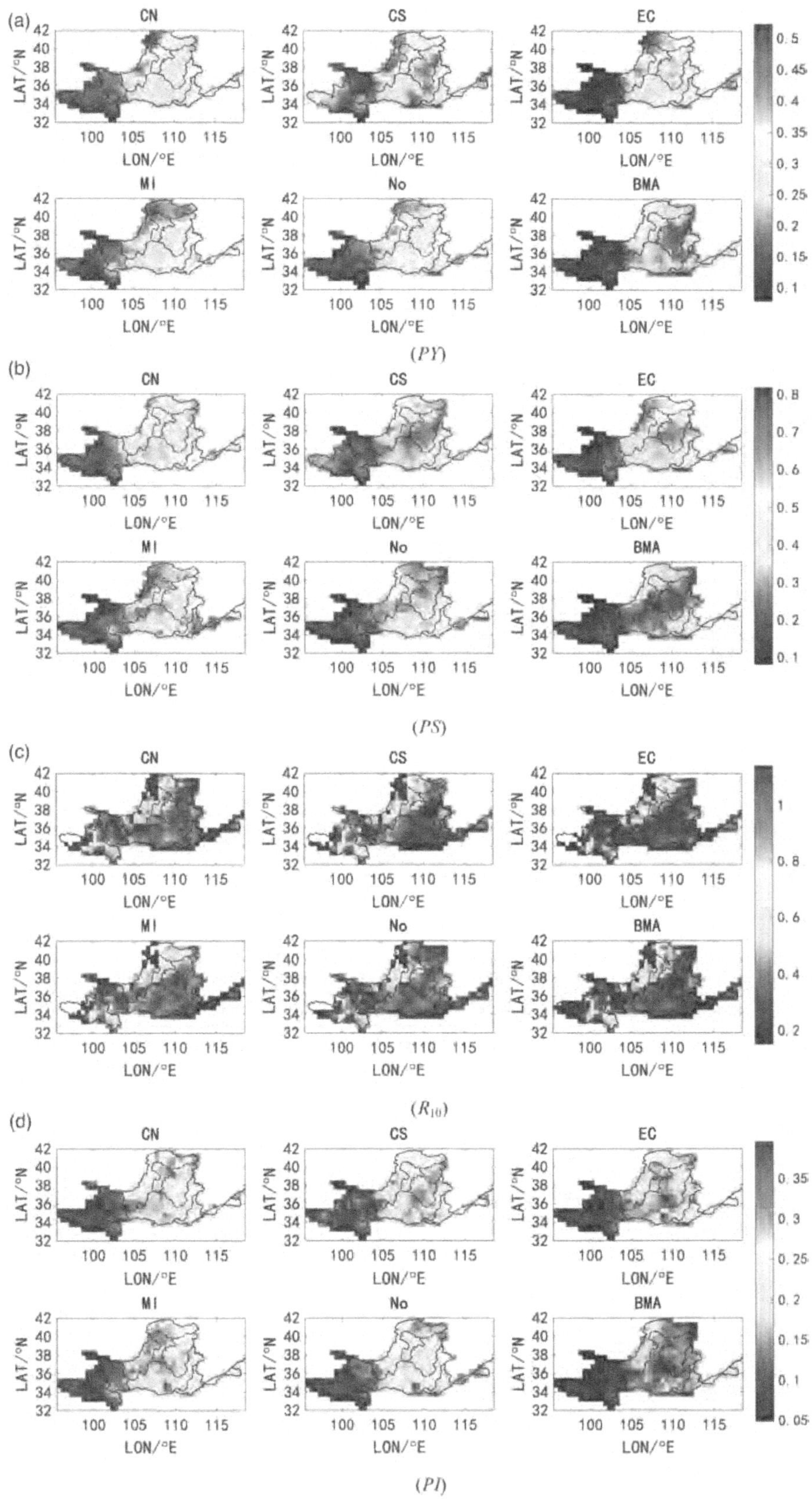

Figure 3 | Comparison of E_D (mean of long-term annual relative simulation deviation) between the BMA model and other models. The smaller the E_D in the BMA model, the smaller the deviation of the BMA model.

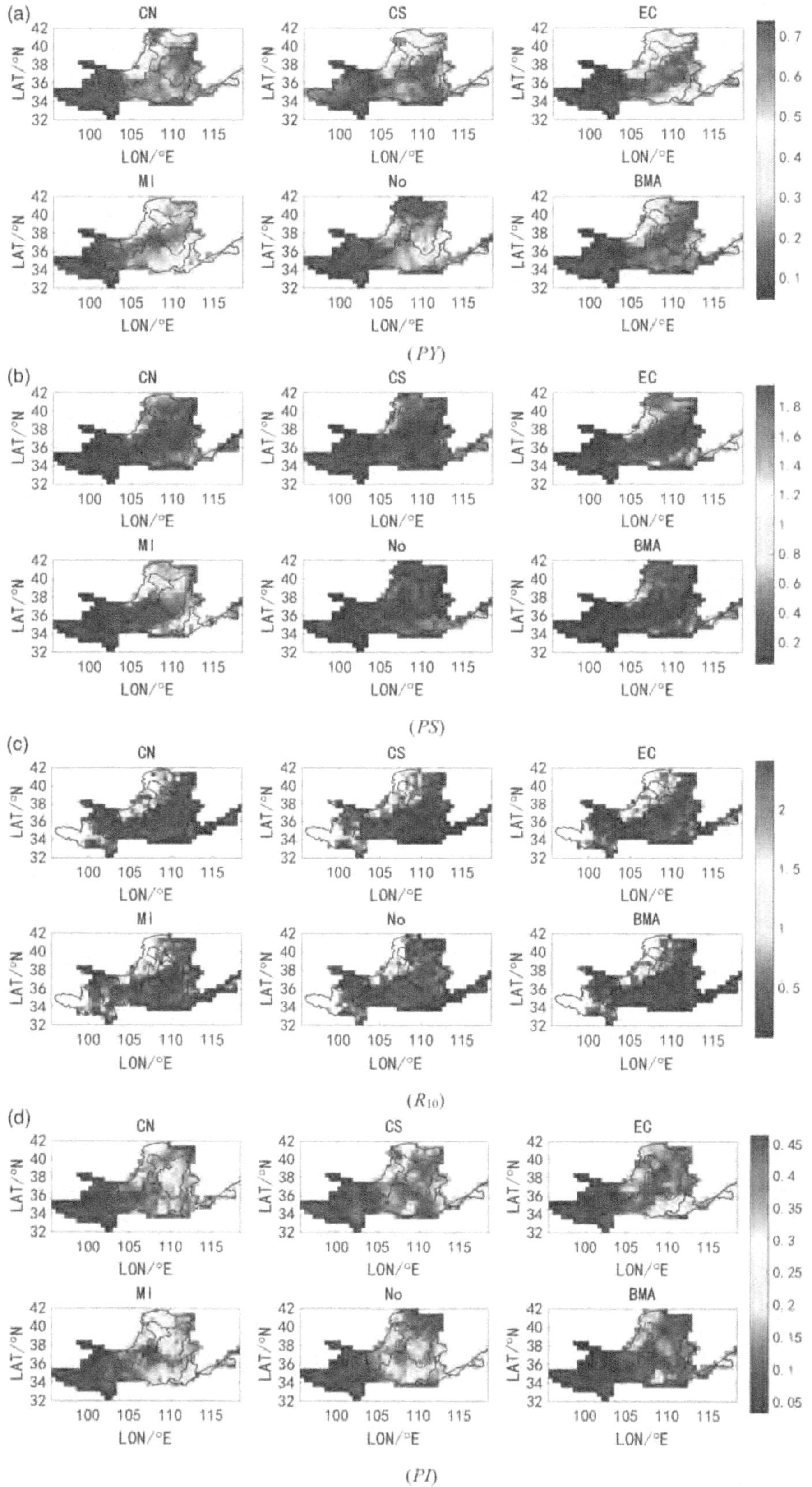

Figure 4 | Comparison of V_D (variance of long-term annual relative simulation deviation) between the BMA model and other models. The smaller the V_D in the BMA model, the more stable the deviation of the BMA model.

Concerning the *PY* (Figure 3(a)), the relatively large E_D in the north part of the CN, CS, EC, and MI models are as high as 0.5. However, that of the BMA model is no larger than 0.35. Thus, though the No model has less E_D than the other models, it is still higher than the BMA model. The same conclusion could be seen with the *PS* (Figure 3(b)) and *PI* (Figure 3(d)), as well as in the east of the YRB. In all models, the E_D in the west is smaller than in the east, coming in below 0.2 generally in the west and more significant than 0.3 in the east.

The R_{10} (Figure 3(c)) has a different result, and it seems that the E_D from the BMA model is not necessarily less than the other individual models. However, as can be seen from R_{10}, the result of the BMA model is not worse overall.

As with the V_D, the BMA model is not worse or better than any other model. Taking *PY* (Figure 4(a)) as a case, in the north YRB from the CN, CS, and EC model, and so with the north and southeast of the YRB from the MI model, the V_D of these regions is more significant than 0.5 and even up to 0.8. However, it is correspondingly less than 0.4 with the BMA model. In short, the performance of the BMA model is superior to that of other individual models on R_{10} (Figure 4(c)) and *PI* (Figure 4(d)). The *PS* (Figure 4(b)) is more acceptable than the EC and MI models while performing equivalent with the CN, CS, and No models.

In the projection of future precipitation, it is crucial to understand the credibility of future projection. The credibility here refers to whether the BMA model is more credible than any single model in the deterministic projection. In this study, the assessment indicators are set as E_D, the mean long-term annual relative simulation deviation, and V_D, the variance of long-term annual relative simulation deviation. As mentioned above, the simulation deviation is expressed as the absolute value of relative simulation deviation since it is interested in the deviation of both directions (i.e., positive and negative). The lower E_D means the smaller range of deviation, and the lower V_D, the more stable the deviation. As shown in Figures 2–4, compared with each model, the performance of reconstructing the historical precipitation situation of the BMA model is more satisfying in terms of lower *MASD*, E_D, and V_D. Thus, the BMA model has more significant advantages in predicting future climate under specific scenarios. Consequently, we get more rational ensemble weights and probability estimation of different changes in precipitation indexes. Based on the assumption of weight as a probability in the BMA model, the application of BMA improves the flexibility and the credibility of the projection.

Therefore, it is possible to conclude that the BMA model synthetically performs better than any single climate model in simulating the historical precipitation in the YRB due to valid information from each pattern. Thus, the precipitation in the future projected by the BMA model will be more rational than any single model.

4.2. Evaluation of the probability estimation method

We took 2006–2015 as the validation period to verify the probability estimation method used in this study. This period was the historical observation period and the projection period in the models, making it suitable to be the validation period. We compared the estimated variation with the observed variation and quantified the deviation of the method and its spatial distribution for each precipitation index. The detailed validation steps are as follows:

1. As above, we defined the variation ranges of precipitation index (VR) as the difference of annual mean value between the validation period (2006–2015) and the historical period (1986–2005).
2. According to the observed precipitation data, the variation of each precipitation index was calculated, and they were converted into VRs (observed VR, Figure 5) according to Table 3.
3. Similarly, for the projected precipitation data (GCMs), the estimated variations of each model (2006–2015) were calculated, respectively. Then, the probability of different variations of each precipitation index was calculated according to the probability estimation method. The VR with the highest probability (estimated VR) in each grid was selected as the estimated variation of each precipitation index.
4. The difference between the estimated VR and the observed VR was taken as the deviation of probability estimation. In each grid, the estimation deviation of each precipitation index was calculated separately. The proportion of deviations of different sizes in the whole basin was classified (the proportion of grids with various deviations in the total number of YBR grids). The statistical results of deviations are shown in Table 5.

We needed to learn about the observed precipitation variation during the validation period. There was no apparent spatial characteristic in the observed variation of each precipitation index shown in Figure 5. The VRs in the four precipitation indexes differed significantly in terms of area. The VR6 occurred in 35.1% of the grids in *PY*, while the VR1 (0%) and the

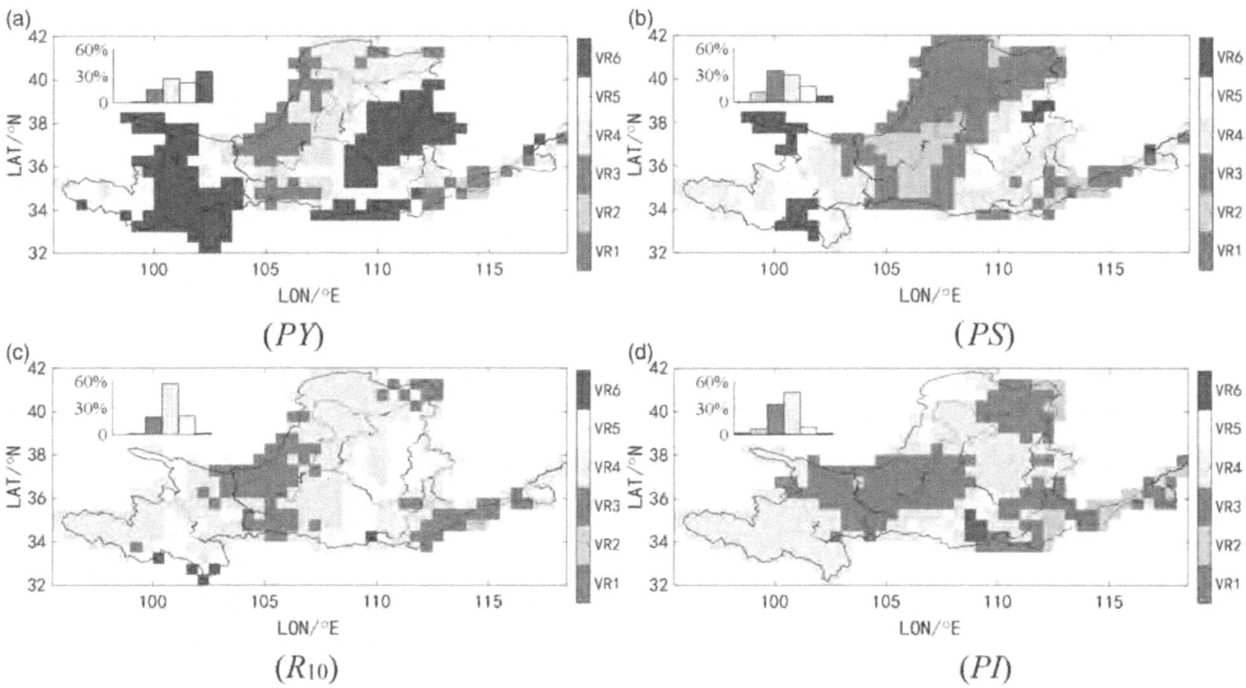

Figure 5 | Observed variation of four precipitation indexes in 2006–2015. This classified representation is based on every precipitation index's variation range (VR) at each grid. For example, if the variation of *PY* in a grid is VR5, the average variation of *PY* is between 25 and 50 mm (see Table 3 for VRs). The proportions in the graph refer to the percentages of the number of grids occurring in each VR.

Table 5 | Proportions (%) of different deviations between observed VR and estimated VR

Deviation	−5	−4	−3	−2	−1	0	1	2	3	4	5
PY	0	0.24	1.21	10.60	25.31	38.31	18.31	5.54	0.48	0	0
PS	0	0	0	5.78	16.87	30.84	34.70	11.57	0.24	0	0
R_{10}	0	0	0	3.37	29.88	47.47	17.35	1.93	0	0	0
PI	0	0	0	3.37	13.74	53.74	18.31	6.51	2.89	1.20	0.24

The difference in the number of VR, if a grid was estimated as VR2 and observed as VR4, the deviation is −2. Negative deviations mean underestimating and positive deviations mean overestimating. A deviation of 0 indicates an accurate estimation.

VR2 (0.72%, only three grids) had no or almost none. In *PS*, the VR3 occurred in 34.9% of the grids, while the VR1 (0%) and the VR2 (6.5%) did not occur or very little. The VR4 occurred in 57.4% of grids in R_{10}, while the VR1 (0%), the VR2 (0.7%), and the VR6 (1.2%) did not or very little. The VR4 occurred in 48.0% of the grids in *PI*, while the VR1 (0%), the VR2 (6.5%), the VR5 (8.2%), and the VR6 (1.2%) had no or very few occurrences.

Similarly, we counted the estimated variation of the precipitation during the validation period, as shown in Figure 6. Compared with Figure 5, there was no apparent spatial relationship (in grids) between the observed results and the estimated results. Each VR (whether observed or estimated) had no apparent spatial distribution characteristics. In each precipitation index, the distribution of the estimated proportions was relatively close to the observed results for each VR, among which the *PY*, R_{10}, and *PI* were in good agreement. However, the estimated and observed results of *PS* were a bit different. The VR2 and the VR3 were underestimated by 8.9 and 20%, respectively (absolute value, not relative to the VR2 or the VR3, the same as below), and the VR4 was overestimated by 30.9%.

The deviation of the estimated result is a point of concern. Table 5 shows the proportion of grids with different estimation deviations in all grids of the whole basin. Among the four precipitation indexes of *PY*, *PS*, R_{10}, and *PI*, the grids with a deviation of 0 accounted for 38.31, 30.84, 47.47, and 53.74% of the whole basin, respectively, meaning that the VR with the highest estimated probability in these grids was consistent with observed VR. In other words, if the VR with the highest

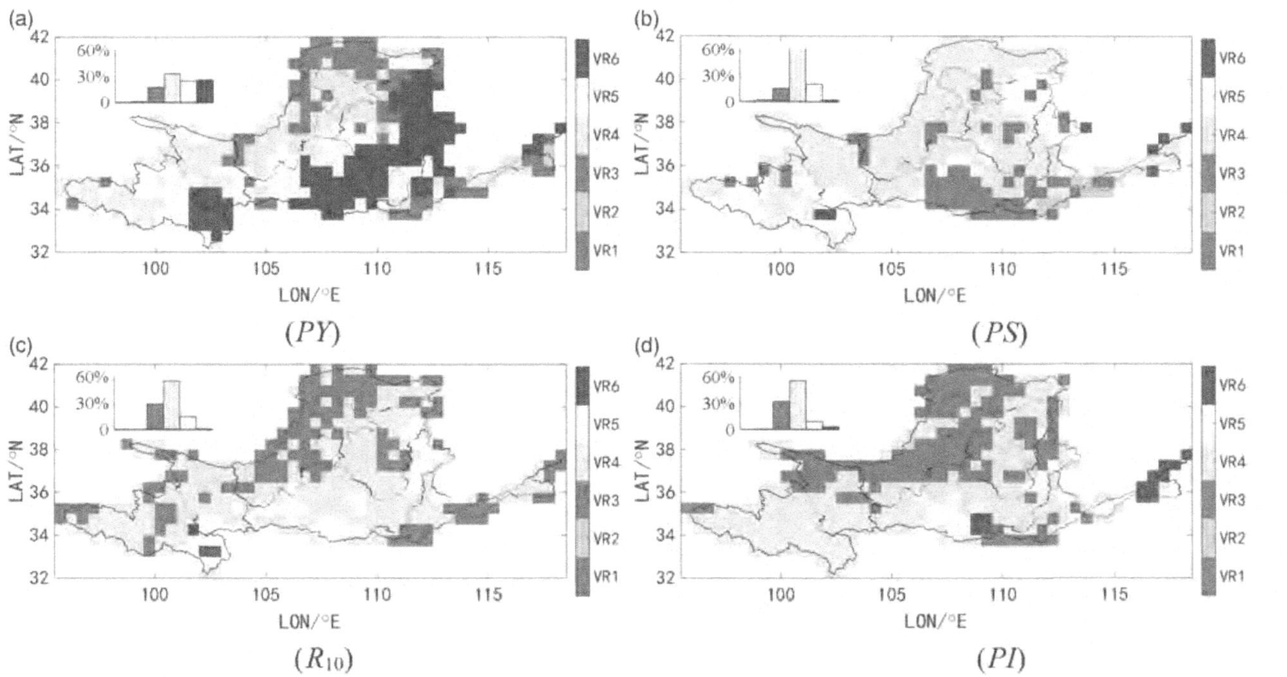

Figure 6 | Estimated average variation of four precipitation indexes in 2006–2015. As in Figure 5, the average variation is expressed as the classified VRs, but the VR at each grid has the highest estimated probability among six VRs.

estimated probability was used as the estimated result, the estimation accuracies of these four indexes were between 30.84 and 43.74%. The estimation deviations of 81.93, 82.41, 94.70, and 85.79% of the four precipitation indexes were between −1 and 1. According to the current classification of precipitation indexes (Table 3), if the deviation of 1 level was taken as the allowable deviation, the probability estimation method could capture the actual occurrence of precipitation indexes in a significant probability (81.93–94.70%). The grids with an estimated deviation of −1 accounted for 13.74–29.88% of the whole basin, which meant that the VR with the highest estimated probability in these grids was underestimated by 1 level than the observed VR. The grids with an estimated deviation of 1 in the whole basin are between 17.35 and 34.70%. Therefore, the VR with the highest estimated probability in these grids was overestimated by 1 level than the observed VR. The proportions of grids with a deviation of 2 (overestimated by two levels) and −2 (underestimated by two levels) were between 1.93 and 11.57%. The total proportion of grids with overestimated or underestimated deviation in the four precipitation indexes of more than three levels was less than 4.33%.

Figure 7 shows the spatial distribution of the estimation deviation of each precipitation index and the main distribution areas of overestimation and underestimation. The overestimation of *PY* mainly appeared in the southern part of Subregion 3 and the western part of Subregion 5 and Subregion 7. The underestimation was mainly in the middle of Subregion 1, Subregion 2, and Subregion 6. The overestimation of *PS* was mainly in Subregion 3, Subregion 4, the west of Subregion 5, and the northwest of Subregion 6, while the underestimation was mainly in the southeast of Subregion 1, the northwest of Subregion 2, the east of Subregion 5, and the southeast of Subregion 6. The overestimation of R_{10} was mainly in the west of Subregion 5 and the east of Subregion 7. The underestimation was mainly in the middle of Subregion 1, the northwest of Subregion 2, the north of Subregion 3 and Subregion 6. The overestimation of *PI* was mainly in Subregion 7 and Subregion 8, the central and western parts of Subregion 5 and Subregion 3, and the northeast of Subregion 6, and the underestimation was mainly in Subregion 4 and the north of Subregion 3. Therefore, the projection of future precipitation variation in these regions should be focused on and treated with caution.

Analyzing the estimated composition of each observed VR (that is, which VR is estimated by the probability estimation method for each observed VR) helps us determine the deviations of the overestimation or underestimation in Figure 7. Figure 8 shows the composition of the estimation of each observed VR in different precipitation indexes. In the probability estimation method, the observed VR in a grid could be estimated as any VR1–VR6 (taking the VR with the highest estimated probability). For example, in *PY*, all grids where VR6 had occurred accounted for 35.1% of the whole basin (Figure 5(a)). In

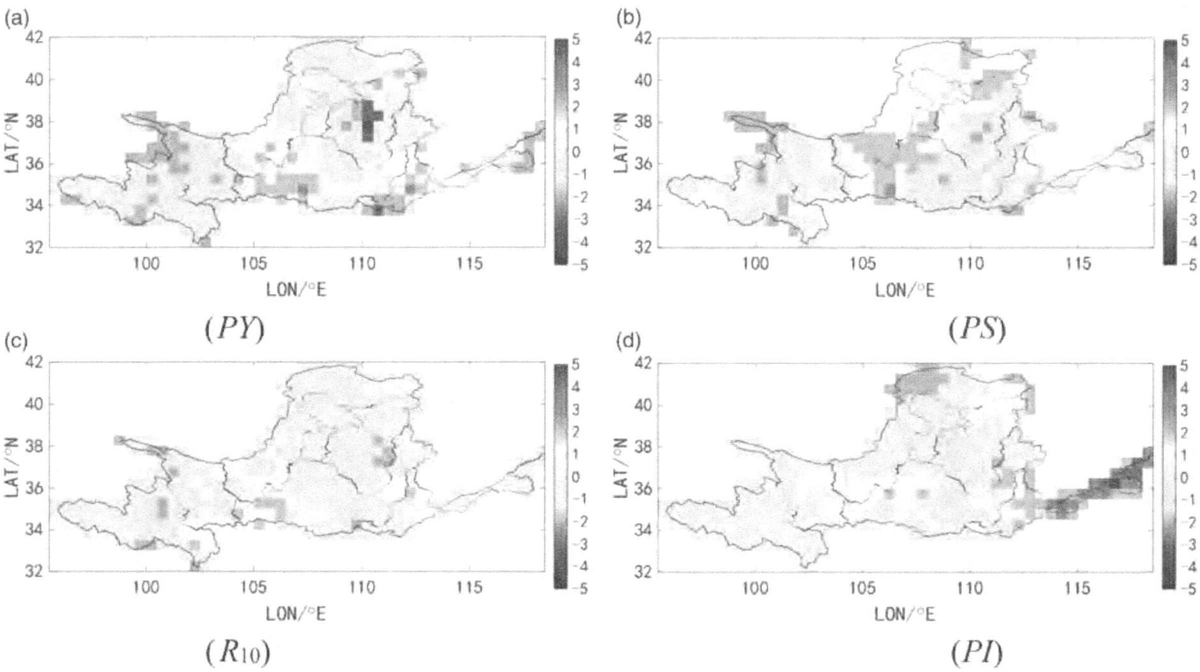

Figure 7 | Deviation between the observed VR and estimated the VR. The estimated VR has the highest probability among all VRs at each grid. The deviation is represented by the difference between the estimated VR number and the observed VR number. Where green means the deviation is 0 (accurate estimation), yellow means the deviation is 1 (overestimate by 1 level), and cyan means the deviation is 1 (underestimate by 1 level). Please refer to the online version of this paper to see this figure in colour: http://dx.doi.org/10.2166/wcc.2022.359.

the estimation results of these grids, the proportions of VR1–VR6 are 0, 0.7, 3.4, 21.2, 29.5, and 45.2%, respectively. The accuracy of the probability estimation method for VR6 in *PY* is 45.2%.

Figure 9 shows the accuracy of each VR in all precipitation indexes, corresponding to the occurrence proportion of each VR. PY, PS, and R_{10} showed an apparent positive correlation between the estimation accuracy and the observed occurrence proportion. We also noticed that there were some differences in *PI*. This was due to the less observed occurrence proportion of VR1 (1.9%), VR2 (6.5%), and VR6 (1.2%), which led to large estimation deviations. There was a positive correlation between the estimation accuracy and the observed occurrence proportion of other VRs in *PI*.

We can explain why some areas in Figure 7 were overestimated or underestimated. For all the precipitation indexes in Table 5, the area of overestimation by 1 level accounted for the most significant proportion among the proportions of different degrees of overestimation, and the same is true in the underestimation. Since the proportion of overestimated or underestimated of each VR was based on the occurrence proportion of this VR, we would not consider those VRs with very few occurrences. For example, the occurrence proportion of VR6 in *PI* was 1.2% (five grids), although there were 20% underestimated by 1 level, and only one grid (5*20%) was underestimated. After statistical analysis, we decided not to consider the VRs that occurred in less than ten grids. In Table 6, both the underestimated and overestimated by 1 level came from the grids where the VR with a high proportion of occurrences. However, among VRs that occurred more often, higher-level VRs tended to be underestimated, while lower-level VRs tended to be overestimated. Therefore, in Figure 7, the proportions underestimated by 1 level and overestimated by 1 level mainly came from those VRs with higher occurrences.

In the probabilistic projection, credibility is mainly reflected in two aspects. The first is the deviation, the difference between the projection with the highest probability and the observation. Precipitation varies from place to place, leading to the difference in the ability of climate models to simulate precipitation in different places, which is the reason for the spatial heterogeneity of deviations. In Figure 7, the spatial distribution of the simulated deviations is a semi-quantitative measure of credibility. The larger the classified deviations, the lower the credibility is. For example, in the future projection of PI in Figure 7(d), the deviation of the source region (Subregion 1) is 0, it can be considered that the credibility of this region is relatively higher, and the deviation of the downstream region (Subregion 8) is between 2 and 5 (overestimated), it indicates that the credibility of this region is relatively lower.

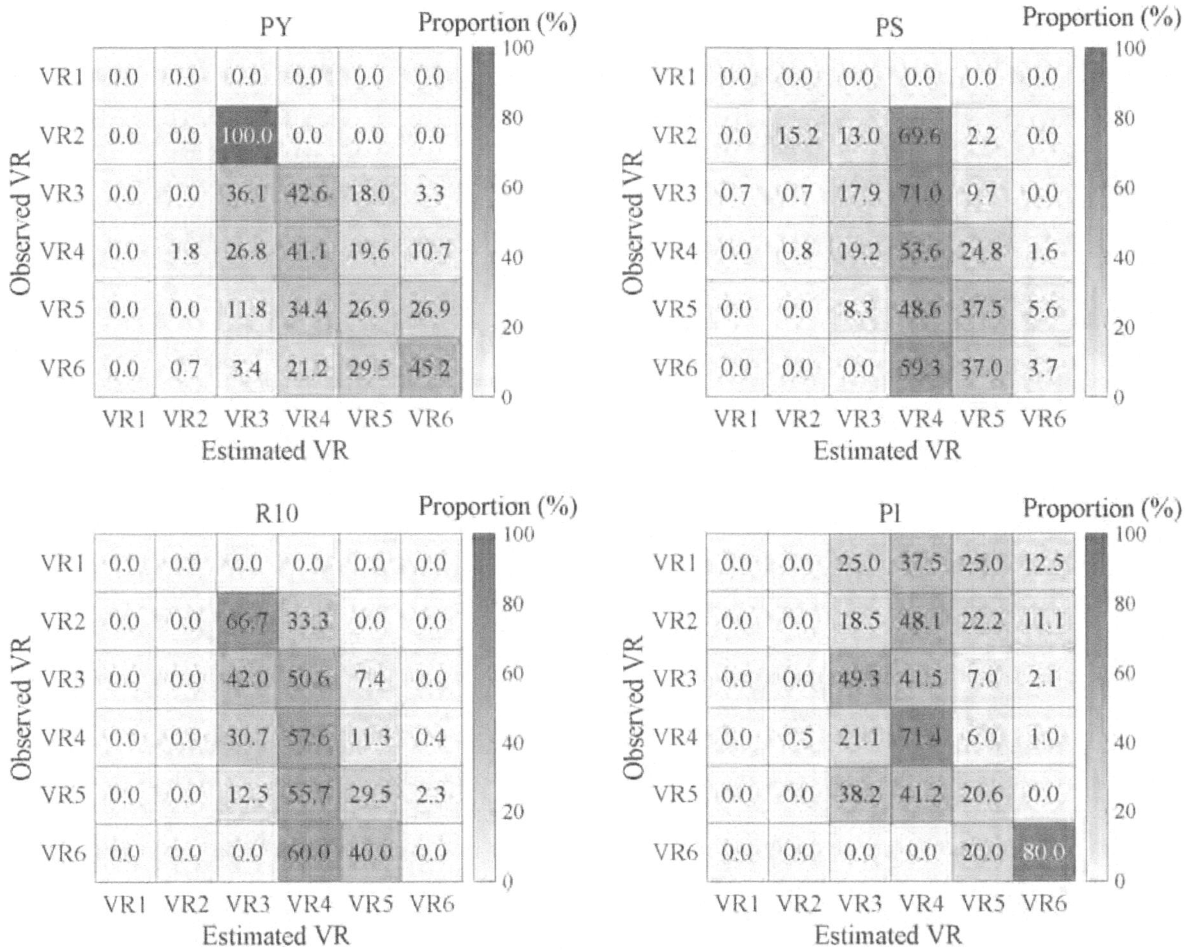

Figure 8 | Estimated composition of each observed VR in each precipitation index. The numbers represent the proportion of the number of grids estimated as different VRs in the observed VR. The numbers on the diagonal represent the estimation accuracy of the VR. In general, the sum of the numbers in a row of a VR is 100% in the subgraph of each precipitation index. If the sum is 0, the VR did not occur in reality.

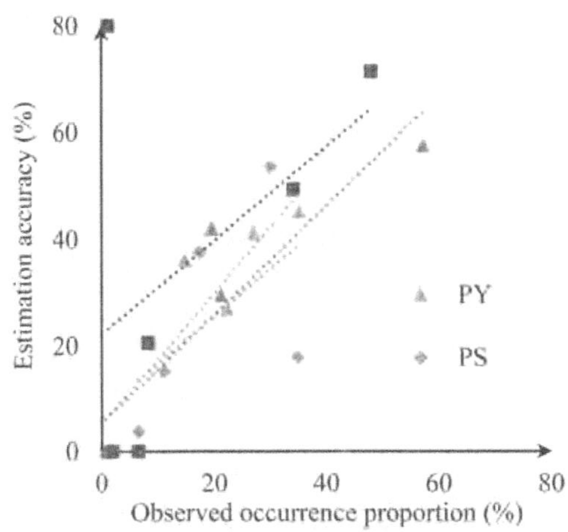

Figure 9 | Relationship between estimation accuracy and observed occurrence proportion for four precipitation indexes (in one precipitation index, the points come from all the VRs occurred).

Table 6 | VRs in main areas that were underestimated by 1 level and overestimated by 1 level in each precipitation index

Precipitation indexes	Underestimated by 1 level (the number of grids)	Overestimated by 1 level (the number of grids)
PY	VR4(30), VR5(32), VR6(43)	VR3(26), VR4(22), VR5(25)
PS	VR4(24), VR5(35), VR6(10)	VR3(103), VR4(31)
R_{10}	VR4(73), VR5(49)	VR3(41), VR4(27)
PI	VR4(42), VR5(14)	VR3(59), VR4(12)

We also analyzed the accuracy of the probability projection method for different precipitation events (VRs). For each estimation, we can find the accuracy of the result in Figure 8, and we can also know the probability that the estimation might be for other VRs. The probability here is obtained by normalizing the numbers in Figure 8 by column, slightly different from the estimated composition discussed above. It is worth noting that the results in Figure 8 were mainly for the VRs that occurred more frequently in the past (as discussed above) because the more occurrences of VRs, the higher the accuracy of the projection would be (Figure 9).

4.3. Deterministic projection: the precipitation evolution in the future

The ensembled projected precipitation is calculated based on the BMA weights on all models. Figure 10 illustrates the estimation of the average variation for every precipitation index of the YRB in two 30-year periods (taking 2021–2050 as the near future scenario and 2061–2090 as the distant future scenario).

The *PY* in the YRB shows an increasing trend in both 2021–2050 (Figure 10(a-1)) and 2061–2090 (Figure 10(a-2)), and the annual precipitation variation in the northwest and the east (the downstream) of the YRB will be obviously and consistently more minor than the other region. The increase in annual precipitation will be greatest in the southwest of the YRB and up to more than 80 and 120 mm locally under the two future scenarios. There is a spatial pattern of low-high-low (L-H-L) for the distribution of *PY* variation from the northwest to the southeast of the YRB.

Compared with the *PY*, the *PS* (Figure 10(b-1) and 10(b-2)) in the YRB will vary. In terms of the overall trend, there will be a bit of a decreasing trend locally under both the future scenarios. The decrease in *PS* will reach 10 and 20 mm, respectively. Unlike the *PY*, the *PS* in the southwest and northeast tends to increase, but the southeast will be opposite, and both increases and decreases in 2061–2090 are more significant than that in 2021–2050. In particular, there is a spatial pattern of high-low-high (H-L-H) for the distribution of *PS* variation from the southwest to the northeast of the YRB.

The R_{10} will have a trend of increasing in most areas of the YRB with an increment of 1–2 in 2021–2050 (Figure 10(c-1)) and 2–4 in 2061–2090 (Figure 10(c-2)). Dramatically, there will be a more significant increase of R_{10} in the southwest of the YRB, where the R_{10} will be up to 4 and 7 in 2021–2050 and 2061–2090, respectively. Thus, spatially, the distribution pattern is somewhat like the *PY*.

The *PI* of the YRB (Figure 10(d-1) and 10(d-2)) will have a slightly increasing trend in both periods. There is no recognizable spatial pattern of *PI* variation. Still, *PI* variation in the northwest of the YRB will be more stable wholly based on a consistent increase or decrease of less than 0.5.

With the future estimation of precipitation indexes in Figure 10 and the definition of these indexes in Table 2, the precipitation evolution in the future could be projected sequentially. The projection can be implemented from the four indexes. The annual precipitation is more concerning due to its dominant effect. In all areas of the YRB, the annual precipitation will increase. The increase from 2061 to 2090 is more significant than 2021–2050, while the annual increase in the southwestern region will be more significant than that in other regions. The summer precipitation is another focus for the impact on the flood problem. In the southwest and east regions, the increment will be dramatically higher than in both scenarios. Comparing the precipitation variation of *PY* and *PS*, the increment of annual precipitation mainly occurs in summer, which means the risk of flooding will increase in the future. The R_{10}, one of the most direct indicators of rainstorm disaster, will also increase on average about four to seven times per year in the southwest and two to four times in the midland. With the *PY*, *PS*, and R_{10}, the southwest of the YRB, namely the upstream, needs more attention with the most increment of these indexes. The situation will be more difficult in the distant future (2061–2090). By contrast, the *PI* will increase slightly, mainly in the south of the YRB, still increasing the whole YRB.

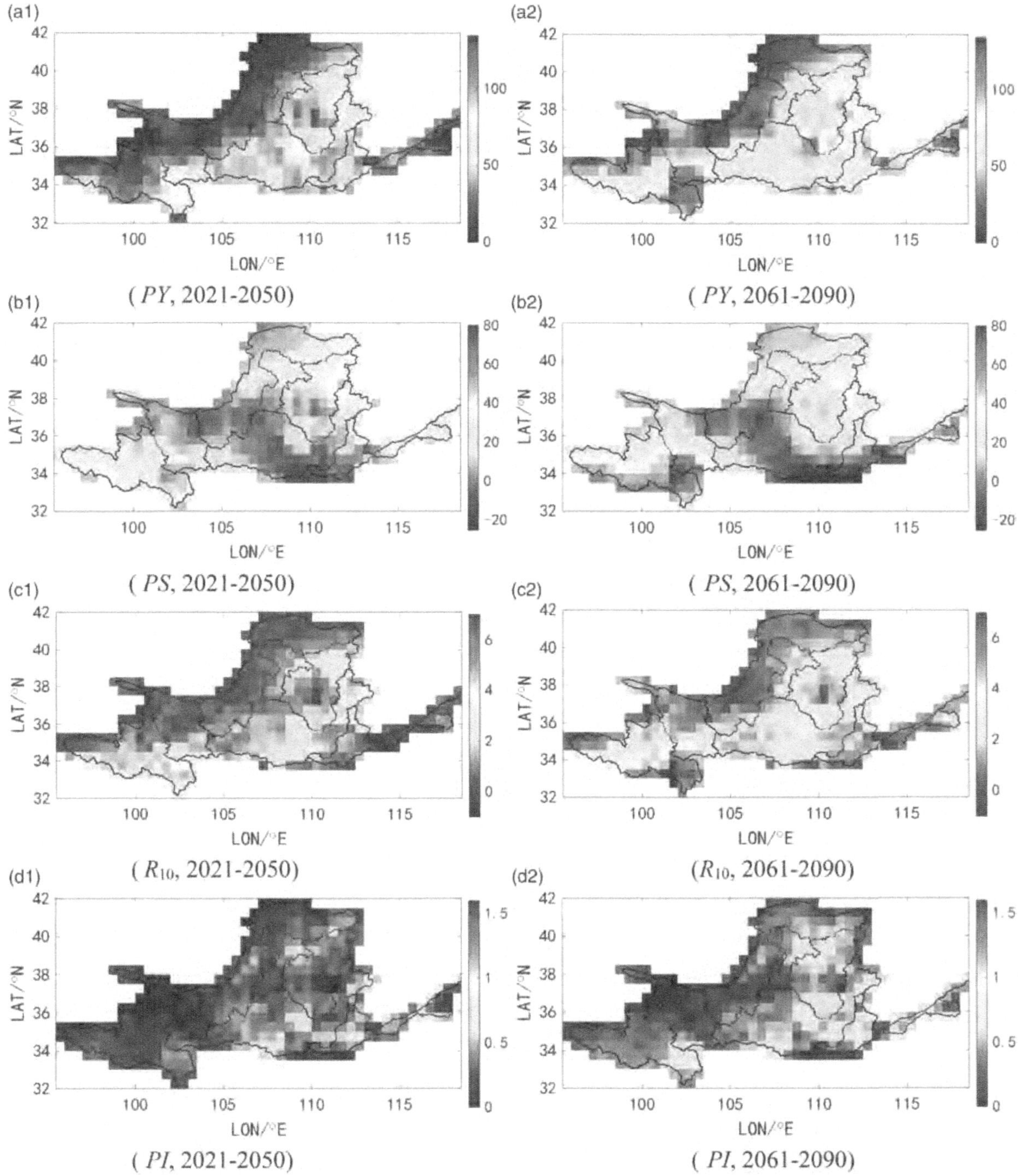

Figure 10 | Estimated average variation for four precipitation indexes of the YRB in the coming 30 years.

4.4. Probabilistic projection: probability estimation of future precipitation variation

As shown in Figure 11, the VR in every subgraph means the corresponding index has changed in a specific range. The meaning of VR corresponding to each indicator is shown in Table 3. For one grid, the sum of the probability values of all six VRs is 1, where the sum of the probabilities of VR1–VR3 is the probability of decreasing, and the sum of the probabilities of VR4–VR6 is the probability of increasing. Obviously, among all the indexes under both scenarios in Figure 11, the probability of increasing is greater than that of decreasing in most regions of the YRB.

In most regions of the YRB, the probability of *PY* increasing by more than 50 mm in both scenarios is much higher, and the probability in 2021–2050 (Figure 11(a-1)) is greater than that in 2061–2090 (Figure 11(a-2)), especially in the south of the

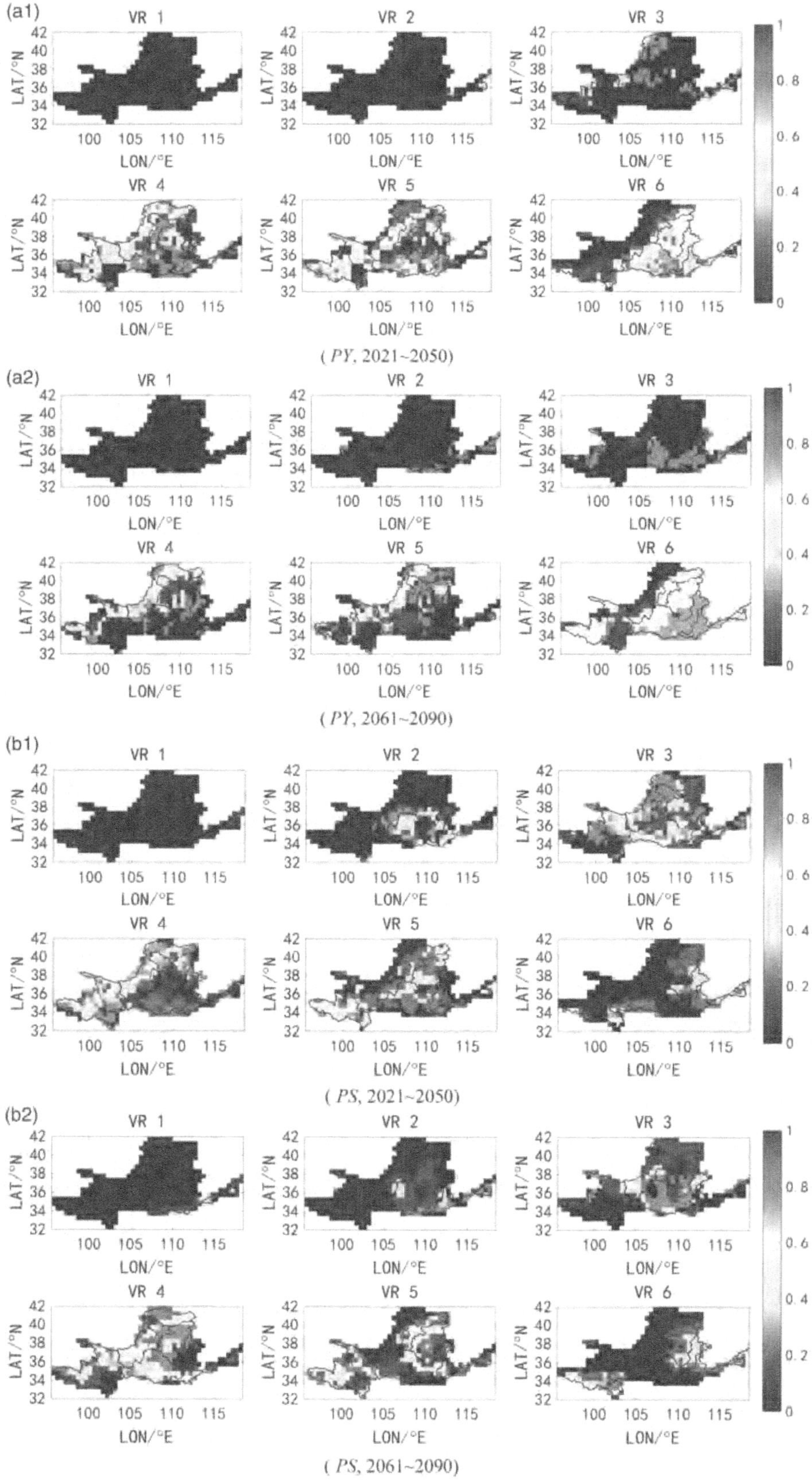

Figure 11 | Probability estimation of future variation in different ranges for precipitation indexes. Each VR represents a specific change in the precipitation index, VR1–VR3 represent the decreased change, and VR4–VR6 represent the increased change. The specific representations are shown in Table 3 (*continued.*).

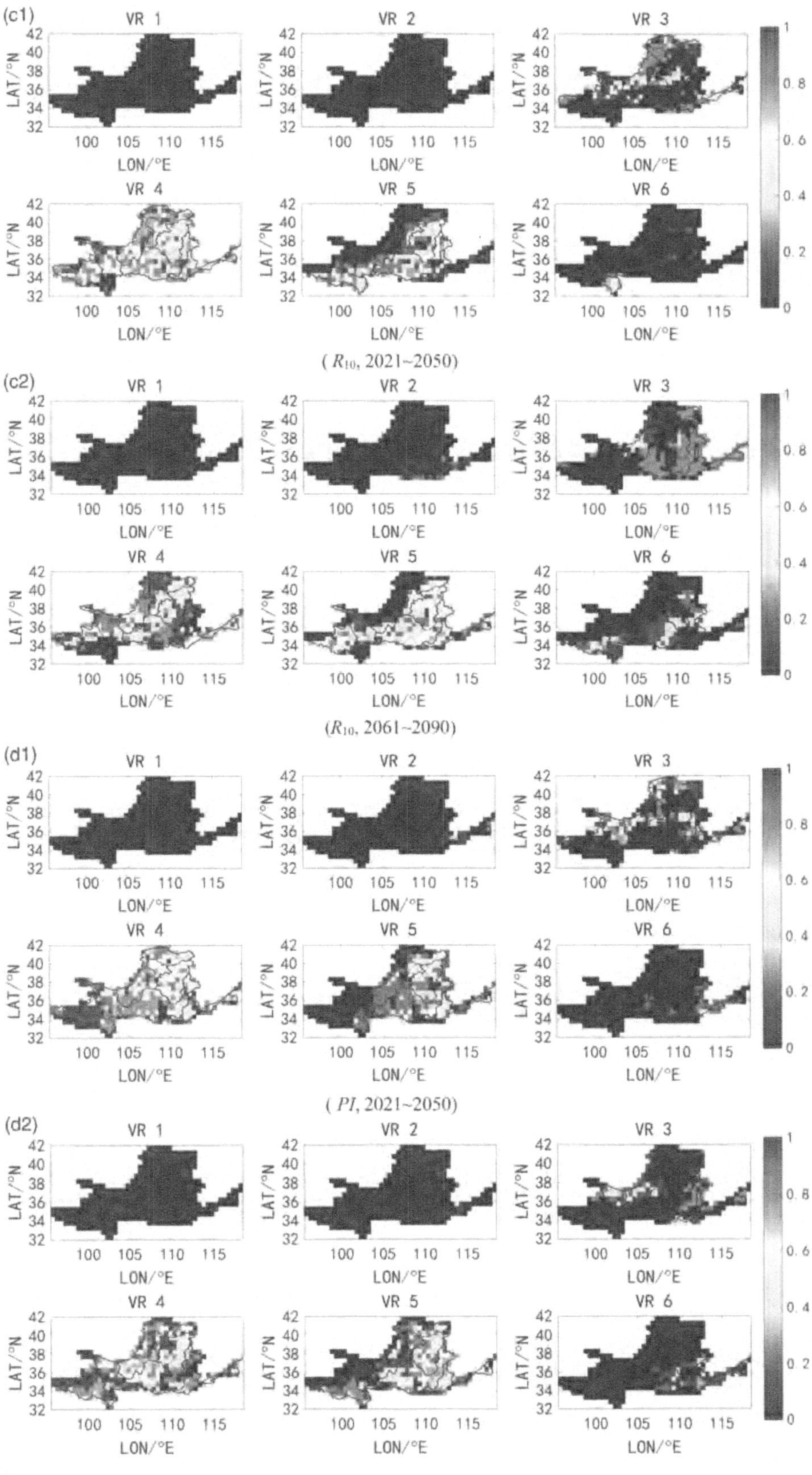

Figure 11 | Continued.

YRB. In the northwest of the YRB, it may increase 0–50 mm with a probability of 40–60%, whatever the scenario. The probability of *PY* decreasing by more than 25 mm in almost all regions (VR1 and VR2 in Figure 11(a-1) and 11(a-2)) is close to 0 under both scenarios.

PS has a more decentralized probability distribution than *PY* and is the largest among all indexes. By comparison with *PY*, the probabilities of *PS* increasing by 0–25 mm or decreasing by 0–25 mm will be proportionable under both scenarios. In particular, the difference in the spatial distribution pattern of probability between the two scenarios is slight. The probability of increasing by 0–25 mm in the north of the YRB is closer to 100% than 2021–2050 (Figure 11(b-1)). The same is true for the southwest part.

The probabilities of R_{10} changing concentrate more on a specific variation range than either *PY* or *PS*, overwhelmingly occurring to increase by 0–2 days and by 2–4 days. Compared with in 2021–2050 (Figure 11(c-1)), there are more significant and more recognizable probabilities of R_{10} increasing by 0–2 days in the northwest regions in 2061–2090 (Figure 11(c-2)). Like *PY*, the probability of R10 decreasing by more than two days in most regions of the YRB is 0.

The probability distribution in each VR of *PI* is similar to R_{10}. Most probability occurs in VR4 (*PI* increasing by 0–0.5 mm/day) and VR5 (*PI* increasing by 0.5–1 mm/day). Under both scenarios, the probability of *PI* increasing by 0–0.5 mm/day will be up to more than 0.8 in most regions in the west and north of the YRB, while the probabilities of *PI* increasing by 0.5–1 mm/day are between 20 and 40%. It mainly occurs in the east and south. The probabilities of other VRs are negligible.

5. DISCUSSION: THE CONSISTENCY BETWEEN THE DETERMINISTIC AND PROBABILISTIC PROJECTION

There is an apparent consistency between the deterministic projection in Section 4.3 and the probabilistic projection in Section 4.4, mainly manifested in spatial distribution and variation. In the deterministic projection in Figure 10(a-1), the increase of the *PY* of regions such as the southeast of Subregion 1 and Subregion 2 and Subregions 5–7 was significantly higher than that of other regions. In the probabilistic projection in Figure 11(a-1), the occurrences of VR6 (increased by more than 50 mm) in these regions were relatively high. However, as shown in Figure 10(a-1), the increases in the northwest of Subregions 1–2, Subregion 3, and Subregion 4 are mostly less than 50 mm. The variations of these regions are more likely to be between 0–25 mm or 25–50 mm, as shown in Figure 11(a-1).

We can also see that the future variations of the southeast of Subregions 1–2 in Figure 10(a-1) are between 50 and 100 mm, and the probability of increasing more than 50 mm in the future in these regions in Figure 11(a-1) is close to 1. Similarly, in the northeast of Subregion 5 and Subregion 6, the future variations of these regions are between 50 and 100 mm (Figure 10(a-1)), and the probabilistic projections indicate that these regions have the greatest probability of increasing by more than 50 mm in the future (Figure 11(a-1)). This shows consistency in the future variation. The projections of the two periods in the future have these consistencies in both aspects (Figures 10(a-2) and 11(a-2)).

In this study, we also found that the accuracy of the projection of a precipitation event (VR) is related to its occurrences (Figures 8 and 9). This leads us to think that this may be the best method to define precipitation events for each precipitation index to obtain more comprehensive and specific accuracy information of probability estimates. At present, we have considered four typical precipitation indexes. If we consider more different precipitation indexes, we can understand the ability of the probability estimation method to project other precipitation characteristics. This study is based on the data of five GCMs. If more models are applied, a higher estimation accuracy may be obtained. We will discuss the above issues in the subsequent work and constantly improve the probability estimation method of future precipitation.

6. CONCLUSION

This paper proposed an estimation method of future precipitation variation using the BMA method, constituting the deterministic and probabilistic projection of future precipitation. It was applied in the YRB. First, the spatial distribution of variations of four precipitation indexes (namely annual mean precipitation (*PY*), summer precipitation (*PS*), heavy rain index (R_{10}), and annual mean precipitation intensity (*PI*)) over the YRB was analyzed sequentially. Then, based on the precipitation projection and BMA weights of all selected models, the probability of variation in different ranges for all precipitation indexes at different times in the future was estimated spatially. Then, the credibility of the deterministic projection based on BMA was evaluated with E_D (mean of long-term annual relative simulation deviation) and V_D (variance of long-term annual relative simulation deviation). The deviation and accuracy evaluated the credibility of the probabilistic projection. The main conclusions are:

1. The BMA model performs well in reconstructing the precipitation conditions because of the lowest *MASD* (the mean of the absolute value of simulation deviation) in simulating the observed historical precipitation indexes. In addition, the relatively lower E_D (the mean of long-term annual relative simulation deviation) and V_D (the variance of long-term annual relative simulation deviation) in simulating precipitation situations under a given climate scenario give the BMA model more credibility than the individual models.
2. The probability estimation method proposed in this study can give sufficient probability information of different future precipitation variations. Among the four precipitation indexes, the accuracy between the estimated VR with the highest probability and the observed VR is 38.31–53.74%. The deviations in 81.93–94.70% of grids were no more than 1 level. Thus, the estimation accuracy of different precipitation events (VRs) varies greatly, and there was a positive correlation between the accuracy and the occurrence of events.
3. The precipitation situations indicated by four indexes of *PY*, *PS*, $R_{10,}$ and *PI* have an increasing tendency in the whole YRB for a distinct higher probability of increasing with varying degrees rather than of decreasing, which suggests an increased risk of summer flooding in the future, compared to the near future. The increment of all the precipitation indexes will be even more significant in the distant future, as will the probability for a range of variation.
4. There is a good consistency between the deterministic projection and the probabilistic projection of the future precipitation in the YRB, which is reflected in the remarkable coincidence of spatial distribution and variation trend. With a better method to define precipitation events (classification method) and applying more and more reliable climate models, it is expected that more reliable precipitation projection can be obtained in the future.

ACKNOWLEDGEMENTS

This work was supported by grants from the National Key Research and Development Program of China (2018YFC1508104), the National Natural Science Foundation of China (52079079, 51779146, 51809174), the Natural Science Foundation of Jiangsu Province (BK20191129), and the Natural Science Foundation of Zhejiang Province (2021C03017). The authors are also grateful to the National Climate Center and China Meteorological Administration for providing data and the helpful comments by the editor and the reviewers. Han *et al.*'s early model evaluation work offered outstanding support for our research.

DECLARATION OF INTEREST STATEMENT

The authors declare that they have no known competing financial interests or personal relationships that could have appeared to influence the work reported in this paper.

DATA AVAILABILITY STATEMENT

All relevant data are available from daily surface precipitation: http://data.cma.cn/. CMIP5: https://esgf-node.llnl.gov/search/cmip5.

REFERENCES

Almazroui, M., Ashfaq, M., Islam, M. N., Rashid, I. U., Kamil, S., Abid, M. A., O'Brien, E., Ismail, M., Reboita, M. S., Sörensson, A. A., Arias, P. A., Alves, L. M., Tippett, M. K., Saeed, S., Haarsma, R., Doblas_Reyes, F. J., Saeed, F., Kucharski, F., Nadeem, I., Silva_Vidal, Y., Rivera, J. A., Ehsan, M. A., Martínez_Castro, D., Muñoz, Á. G., Ali, M. A., Coppola, E. & Sylla, M. B. 2021a Assessment of CMIP6 performance and projected temperature and precipitation changes over South America. *Earth Systems and Environment* **5**, 155–183. https://doi.org/10.1007/s41748-021-00233-6.

Almazroui, M., Islam, M. N., Saeed, F., Saeed, S., Ismail, M., Ehsan, M. A., Diallo, I., O'Brien, E., Ashfaq, M., Martínez-Castro, D., Cavazos, T., Cerezo-Mota, R., Tippett, M. K., Gutowski, W. J. J., Alfaro, E. J., Hidalgo, H. G., Vichot-Llano, A., Campbell, J. D., Kamil, S., Rashid, I. U., Sylla, M. B., Stephenson, T., Taylor, M. & Barlow, M. 2021b Projected changes in temperature and precipitation over the United States, Central America, and the Caribbean in CMIP6 GCMs. *Earth Systems and Environment* **5**, 1–24. https://doi.org/10.1007/s41748-021-00199-5.

Chen, H. P., Sun, J. Q. & Li, H. X. 2017 Future changes in precipitation extremes over China using the NEX-GDDP high-resolution daily downscaled data-set. *Atmospheric and Oceanic Science Letters* **10** (6), 403–410. https://doi.org/10.1080/16742834.2017.1367625.

Chen, L., Chang, J. X., Wang, Y. M. & Zhu, Y. L. 2019 Assessing runoff sensitivities to precipitation and temperature changes under global climate-change scenarios. *Hydrology Research* **50** (1), 24–42. https://doi.org/10.2166/nh.2018.192.

Dars, G. H., Najafi, M. R. & Qureshi, A. L. 2017 Assessing the impacts of climate change on future precipitation trends based on downscaled CMIP5 simulations data. *Mehran University Research Journal of Engineering and Technology* **36** (2), 385–394. https://doi.org/10.22581/muet1982.1702.16.

Duan, Q. Y. & Phillips, T. J. 2010 Bayesian estimation of local signal and noise in multi-model simulations of climate change. *Journal of Geophysical Research: Atmospheres* **115** (18), 1–15. https://doi.org/10.1029/2009JD013654.

Duan, Q. Y., Ajami, K. N., Gao, X. G. & Sorooshianb, S. 2007 Multi-model ensemble hydrologic prediction using Bayesian model averaging. *Advances in Water Resources* **30** (5), 1371–1386. https://doi.org/10.1016/j.advwatres.2006.11.014.

Fricker, T. E., Ferro, C. A. T. & Stephenson, D. B. 2013 Three recommendations for evaluating climate predictions. *Meteorological Applications* **20** (2), 246–255. https://doi.org/10.1002/met.1409.

Guan, X. J., Liang, S. X. & Meng, Y. 2016 Evaluation of water resources comprehensive utilization efficiency in the Yellow River Basin. *Water Science and Technology: Water Supply* **16** (6), 1561–1570. https://doi.org/10.2166/ws.2016.057.

Guo, J. H., Huang, G. H., Wang, X. Q., Li, Y. P. & Lin, Q. G. 2017 Investigating future precipitation changes over China through a high-resolution regional climate model ensemble. *Earth's Future* **5** (3), 1–19. https://doi.org/10.1002/eft2.187.

Han, Z. Y., Tong, Y., Gao, X. J. & Xu, Y. 2018 Correction based on quantile mapping for temperature simulated by the RegCM4. *Climate Change Research* **14** (4), 331–340. (in Chinese) https://doi.org/10.12006/j.issn.1673-1719.2017.156.

Hanittinan, P., Tachikawa, Y. & Ram-Indra, T. 2020 Projection of hydroclimate extreme indices over the Indochina region under climate change using a large single-model ensemble. *International Journal of Climatology* **40** (6), 2924–2952. https://doi.org/10.1002/joc.6374.

Hoeting, J. A., Madigan, D., Raftery, A. E. & Volinsky, C. T. 1999 Bayesian averaging models. *Statistical Science* **14** (4), 382–417.

Iqbal, M., Wen, J., Wang, S. P., Tian, H. & Adan, M. 2018 Variations of precipitation characteristics during the period 1960–2014 in the source region of the Yellow River, China. *Journal of Arid Land* **10** (3), 388–401. https://doi.org/10.1007/s40333-018-0008-z.

Libanda, B. & Nkolola, N. B. 2019 Skill of CMIP5 models in simulating rainfall over Malawi. *Modeling Earth Systems and Environment* **5** (4), 1615–1626. https://doi.org/10.1007/s40808-019-00611-0.

Liu, J. & Xie, Z. 2014 BMA probabilistic quantitative precipitation forecasting over the Huaihe Basin using TIGGE multi-model ensemble forecasts. *Monthly Weather Review* **142** (4), 1542–1555. https://doi.org/10.1175/MWR-D-13-00031.1.

Liu, J., Du, H. B., Wu, Z. F., He, H. S., Wang, L. & Zong, S. W. 2017 Recent and future changes in the combination of annual temperature and precipitation throughout China. *International Journal of Climatology* **37** (2), 821–833. https://doi.org/10.1002/joc.4742.

Liu, J., Jin, L. J., Huang, C. X. & Zhang, L. N. 2020 Variations of seasonal precipitation in the Yellow River Basin and its relationship to general circulation and SST. *Proceedings of the International Association of Hydrological Sciences* **383**, 5–12. https://doi.org/10.5194/piahs-383-5-2020.

Lu, H. L., Jing, W. L., Zhao, J. C., Liu, X. J. & Huang, Z. 2014 Characteristics of the temporal variation in temperature and precipitation in China's Lower Yellow River Region. *Advances in Meteorology* **2014**. https://doi.org/10.1155/2014/186823.

Meshram, S. G., Singh, S. K., Meshram, C., Ravinesh, C., Deo, R. C. & Ambade, B. 2018 Statistical evaluation of rainfall time series in concurrence with agriculture and water resources of Ken River basin, Central India (1901–2010). *Theoretical and Applied Climatology* **134**, 1231–1243. https://doi.org/10.1007/s00704-017-2335-y.

Meshram, S. G., Kahya, E., Meshram, C., Ghorbani, M. A., Ambade, B. & Mirabbasi, R. 2020 Long-term temperature trend analysis associated with agriculture crops. *Theoretical and Applied Climatology* **140**, 1139–1159. https://doi.org/10.1007/s00704-020-03137-z.

Miao, C. Y., Duan, Q. Y., Sun, Q. H., Huang, H., Kong, D. X. & Yang, T. T. 2014 Assessment of CMIP5 climate models and projected temperature changes over Northern Eurasia. *Environmental Research Letters* **9** (5), 055007. https://doi.org/10.1088/1748-9326/9/5/055007.

Min, S. K. & Hense, A. 2006 A Bayesian approach to climate model evaluation and multi-model averaging with an application to global mean surface temperatures from IPCC AR4 coupled climate models. *Geophysical Research Letters* **33** (8), 1–5. https://doi.org/10.1029/2006GL025779.

Mu, J. X., Liu, Q. C., Gabriel, H. F., Xu, D., Xu, J. D., Wu, C. L. & Ren, H. J. 2013 The impacts of climate change on water stress situations in the yellow river basin, China. *Irrigation and Drainage* **62** (5), 545–558. https://doi.org/10.1002/ird.1784.

Notaro, V., Liuzzo, L. & Freni, G. 2016 A BMA analysis to assess the urbanization and climate change impact on urban watershed runoff. *Procedia Engineering* **154**, 868–876. https://doi.org/10.1016/j.proeng.2016.07.461.

Qing, Y., Ma, Z. & Xu, B. 2017 Modulation of monthly precipitation patterns over East China by the Pacific Decadal Oscillation. *Climatic Change* **144**, 405–417. https://doi.org/10.1007/s10584-016-1662-9.

Radermacher, C. & Tomassini, L. 2012 Thermodynamic causes for future trends in heavy precipitation over Europe based on an ensemble of regional climate model simulations. *Journal of Climate* **25** (21), 7669–7689. https://doi.org/10.1175/JCLI-D-11-00304.1.

Raftery, A. E., Gneiting, T., Balabdaoui, F. & Polakowski, M. 2005 Using Bayesian model averaging to calibrate forecast ensembles. *Monthly Weather Review* **133** (5), 1155–1174. https://doi.org/10.1175/MWR2906.1.

Tong, Y., Gao, X. J., Han, Z. Y. & Xu, Y. 2017 Bias correction of daily precipitation simulated by RegCM4 model over China. *Chinese Journal of Atmospheric Sciences* **41** (6), 1156–1166 (in Chinese). https://doi.org/10.3878/j.issn.1006-9895.1704.16275.

Wang, G. Q., Zhang, J. Y., Jin, J. L., Weinberg, J., Bao, Z. X., Liu, C. S., Liu, Y. L., Yan, X. L., Song, X. M. & Zhai, R. 2017a Impacts of climate change on water resources in the Yellow River basin and identification of global adaptation strategies. *Mitigation and Adaptation Strategies for Global Change* **22** (1), 67–83. https://doi.org/10.1007/s11027-015-9664-x.

Wang, W. G., Ding, Y. M., Shao, Q. X., Xu, J. Z., Jiao, X. Y., Luo, Y. F. & Yu, Z. B. 2017b Bayesian multi-model projection of irrigation requirement and water use efficiency in three typical rice plantation region of China based on CMIP5. *Agricultural and Forest Meteorology* **232**, 89–105. https://doi.org/10.1016/j.agrformet.2016.08.008.

Wang, G. J., Pan, J., Shen, C. C., Li, S. J., Lu, J., Lou, D. & Hagan, D. F. T. 2018 Evaluation of evapotranspiration estimates in the Yellow River Basin against the water balance method. *Water* **10** (12), 1–18. https://doi.org/10.3390/w10121884.

Wang, F., Wang, Z. M., Yang, H. B., Zhao, Y., Zhang, Z. Z., Li, Z. H. & Hussain, Z. 2019a Copula-based drought analysis using Standardized Precipitation Evapotranspiration Index: a case study in the Yellow river basin, China. *Water* **11** (6), 1–20. https://doi.org/10.3390/w11061298.

Wang, Y. P., Zhao, W. W., Wang, S., Feng, X. M. & Liu, Y. X. 2019b Yellow River water rebalanced by human regulation. *Scientific Reports* **9** (1), 1–10. https://doi.org/10.1038/s41598-019-46063-5.

Wu, X. S., Wang, Z. L., Zhou, X. W., Lai, C. G., Lina, W. X. & Chen, X. H. 2016 Observed changes in precipitation extremes across 11 basins in China during 1961–2013. *International Journal of Climatology* **36** (8), 2866–2885. https://doi.org/10.1002/joc.4524.

Wu, H., Li, X. & Qian, H. 2017 Detection of anomalies and changes of rainfall in the Yellow River Basin, China, through two graphical methods. *Water* **10** (1). https://doi.org/10.3390/w10010015.

Wu, S. Y., Wu, Y. J. & Wen, J. H. 2019 Future changes in precipitation characteristics in China. *International Journal of Climatology* **39** (8), 3558–3573. https://doi.org/10.1002/joc.6038.

Yan, Z. Q., Zhou, Z. H., Liu, J. J., Han, Z. Y., Gao, G. & Jiang, X. T. 2020 Ensemble projection of runoff in a large-scale basin: modeling with a global BMA approach. *Water Resources Research* **56** (7), e2019WR026134. https://doi.org/10.1029/2019WR026134.

Yin, J., Yan, D. H., Yang, Z. Y., Yuan, Z., Yuan, Y. & Zhang, C. 2016 Projection of extreme precipitation in the context of climate change in Huang-Huai-Hai region, China. *Journal of Earth System Science* **125** (2), 417–429. https://doi.org/10.1007/s12040-016-0664-3.

Yin, Y. Y., Tang, Q. H., Liu, X. C. & Zhang, X. J. 2017 Water scarcity under various socio-economic pathways and its potential effects on food production in the Yellow River basin. *Hydrology and Earth System Sciences* **21** (2), 791–804. https://doi.org/10.5194/hess-21-791-2017.

Zhang, Q., Peng, J. T., Singh, V. P., Li, J. F. & Chen, Y. D. 2014 Spatio-temporal variations of precipitation in arid and semi-arid regions of China: The Yellow River basin as a case study. *Global and Planetary Change* **114**, 38–49. https://doi.org/10.1016/j.gloplacha.2014.01.005.

Zhang, Y., Xia, J. & She, D. X. 2019 Spatiotemporal variation and statistical characteristic of extreme precipitation in the middle reaches of the Yellow River Basin during 1960–2013. *Theoretical and Applied Climatology* **135** (1–2), 391–408. https://doi.org/10.1007/s00704-018-2371-2.

Zhao, Y., Wang, M. Z., Li, J., Yang, X. J., Zhang, N. & Chen, H. 2019 Diurnal variations in summer precipitation over the Yellow River Basin. *Advances in Meteorology* **2019**, 1–10. https://doi.org/10.1155/2019/2482656.

Zhou, W. C. & Han, Z. Y. 2018 Assessing CMIP5 climate simulations and objective selection of models over the Yellow River basin. *Journal of Meteorology and Environment* **34** (06), 42–55 (in Chinese). https://doi.org/10.3969/j.issn.1673-503X.2018.06.005.

First received 20 September 2021; accepted in revised form 21 February 2022. Available online 8 March 2022

Impact of Climate Change on Hydrology and Water Resources doi: 10.2166/wcc.2021.221

Prediction of S12-MKII rainfall simulator experimental runoff data sets using hybrid PSR-SVM-FFA approaches

Sandeep Samantaray and Dillip Kumar Ghose *

Department of Civil Engineering, National Institute of Technology, Silchar, Assam, India
*Corresponding author. E-mail: dillip@civil.nits.ac.in

SS, 0000-0003-3379-6087; DKG, 0000-0001-7427-474X

ABSTRACT

Effective prediction of runoff is a substantial feature for the successful management of hydrological phenomena in arid regions. The present research findings reveal that a rainfall simulator (RS) can be a valuable instrument to estimate runoff as the intensity of rainfall is modifiable in the course of an experimental process, which turns out to be of great advantage. The rainfall-runoff process is a complex physical phenomenon caused by the effect of various parameters. In this research, a new hybrid technique integrating PSR (phase space reconstruction) with FFA (firefly algorithm) and SVM (support vector machine) has gained recognition in various modelling investigations in contrast to the principle of empirical risk minimization through ANN practices. Outcomes of SVM are contrasted against SVM-FFA and PSR-SVM-FFA models. The improvements in NSE (Nash–Sutcliffe efficiency), RMSE (Root Mean Square Error), and WI (Willmott's Index) by PSR-SVM-FFA over SVM models specify that the prediction accuracy of the hybrid model is better. The established PSR-SVM-FFA model generates preeminent WI values that range from 0.97 to 0.98, while the SVM and SVM-FFA models encompass 0.93–0.95 and 0.96–0.97, respectively. The proposed PSR-SVM-FFA model gives more accurate results and error limiting up to 2–3%.

Key words: PSR-SVM-FFA, rainfall simulator, runoff, SVM, SVM-FFA

HIGHLIGHTS

- The models developed in this study using three algorithms are SVM, SVM-FFA, and hybrid PSR-SVM-FFA models.
- SVM, SVM-FFA, and PSR-SVM-FFA were used, while the earlier studies used simple machine learning approaches for predicting runoff.
- Interactions of all techniques along with five different scenarios are presented.
- Watershed has different slopes; hence, six different types of slope conditions have been considered.

1. INTRODUCTION

The principal purpose of a rainfall simulator (RS) is to simulate natural precipitation precisely and specifically. Simultaneously, RSs regulate the intensity and duration of rainfall. They are beneficial as rainfall can be created swiftly on demand wherever essential, without waiting for natural rainfall at a desired intensity and interval, thus eradicating the unpredictable and erratic unevenness of natural rain (Wildhaber *et al.* 2012; Iserloh *et al.* 2013). The necessity for short- and long-term runoff simulation is significant for managing watersheds which comprises surplus runoff control, managing and using runoff for detailed tenacities. The complexity of the rainfall–runoff process and its unpredictability depending on watershed features and rainfall patterns makes it difficult to predict and estimate with desired accuracy (Abudi *et al.* 2012; Samantaray *et al.* 2019, 2021). Yet, hydrologists have recently developed several approaches and models varying from empirical- to physical-based relationships. Physical models have been ascertained to be improved for simulating runoff; however, they require a huge quantity of data. Hence, the necessity to develop alternate models for runoff and sediment yield simulation utilizing accessible data has taken primacy (Samantaray & Sahoo 2020a; Xu *et al.* 2020; Jimmy *et al.* 2021; Mohanta *et al.* 2021). The runoff modelling process is a key constituent in climatological engineering. Therefore, an improved and consistent smart model has to be explored for runoff forecasting that is vital for water resources engineering.

Pérez-Latorre *et al.* (2010) developed two diverse RS designs for obtaining rainfall intensities having drop size and energy like that of natural rainfall. Comparing the efficiency of two single simulators, RS-2 gave a more accurate drop-size distribution in the middle part and also a lesser budget than that of RS1. Akbarzadeh *et al.* (2009) investigated the decline of overland flow and splash detachment of clayey soil after casing it with various synthetic geotextiles utilizing RS as well as neuro-fuzzy and ANN. It was observed that the neuro-fuzzy model gives better estimates followed by ANN and multivariate regression for runoff and splash prediction, respectively. Elhakeem & Papanicolaou (2009) provided comprehensive operational phases for estimating *in situ* runoff curve number (CN) for certain agronomic fields in Iowa State using RS. Observed results revealed that RSs are valuable devices to estimate *in situ* runoff CN because the intensity of rainfall was modifiable at the time of an experimental run. Wildhaber *et al.* (2012) assessed the appropriateness of a new field hybrid RS which combined spray nozzle and dropped former features of RS for soil erosion study and for quantifying the effect of vegetation and soil arrangement permanency on runoff and erosion in sub-alpine parkland region. Outcomes demonstrated that vegetation cover plays a vital part in the soil structure stability of alpine soils. Abudi *et al.* (2012) designed and tested a movable RS to simulate precipitations that persuade runoff and soil erosion. They observed that the field RS acts as a potent device for field penetration, crusting, and erosion of soil. Iserloh *et al.* (2013) characterized non-natural rain produced by 13 RSs, constructed in different European research organizations, utilizing Laser Precipitation Monitor and rainfall accumulators in every simulation for ensuring comparability of outcomes. Aksoy *et al.* (2012) developed a laboratory-gauge non-movable RS to spray rain on a flume which can be provided with both lateral and longitudinal slopes. Performance of RS is assessed based on the intensity of rainfall and its consistency, raindrop size dissemination, and kinetic energy of produced rainfall. Cea *et al.* (2010) analysed the accuracy and efficiency of 2D vibrant wave models for computing rainfall–runoff directly from rainfall data sets within distinct urban geometries. They also determined different model constraints that are appropriate while using an experimental set-up in a laboratory RS. Fister *et al.* (2012) evaluated wind and precipitation features of Movable Wind and RS. The adequacy of simulation eminence and condition of suitability for relative measurement of soil erosion in field was assessed. Jin *et al.* (2009) assessed the impact of residue cover and intensity of rainfall on the runoff loss of soil organic matter directly, by quantifying sediment loss over time and variations in sediments' particle size distribution, using a field RS in the laboratory.

Sivapragasam *et al.* (2001) suggested a simple and effective technique based on Singular Spectrum Analysis integrated with SVM for predicting runoff data of the Tryggevaelde catchment, Denmark. Results obtained were contrasted with that of the non-linear prediction (NLP) model and observed that the projected method gives considerably greater accuracy in predicting runoff compared with NLP. Bray & Han (2004) utilized the SVM model, for forecasting flood events, focusing on identifying a suitable model arrangement and its important constraints to model rainfall–runoff process. Behzad *et al.* (2009) proposed a new machine learning tool called SVM, for runoff modelling of the Bakhtiyari River watershed, Iran. Outcomes of the proposed model were compared with ANN and that of ANN combined with GA and it was observed that SVM performs the best as compared to other models. Okkan & Serbes (2012) examined the applicability of least squares SVM for predicting runoff of the Gordes and Tahtali catchments, located in Turkey, and compared its outcomes with ANN and other regression models. Findings from their study showed that the hybrid model helps in making run time substantially quicker with greater accuracy in runoff prediction. Misra *et al.* (2009) and Sharma *et al.* (2015) applied SVM and ANN models for predicting both runoff and sediment concentration of different watersheds. An outcome of SVM was assessed against ANN and simple regression, and it was observed that the SVM model was more competent for predicting runoff and sediment concentration under equivalent prediction accuracy. Jajarmizadeh *et al.* (2015) compared outcomes of the Soil and Water Assessment Tool (SWAT) with SVM for predicting the monthly streamflow of the Roodan catchment situated in Southern Iran. The SVM model gave better values for average flow as compared with SWAT, while SWAT performed better for predicting entire runoff volume having lesser error during validation. Granata *et al.* (2016) comparatively studied the modelling of the rainfall–runoff process amid SVM technique and Storm Water Management Model (SWMM) at two gauging stations situated in the northern part of Italy. SVR showed abundant application prospectives for applications in the urban hydrologic field as compared to SWMM. Ouyang *et al.* (2016) applied SVR for forecasting monthly precipitation at a climate station in Changchun, China, and implemented the phase space reconstruction (PSR) method for designing input vectors. Outcomes indicated that the projected amalgam model is reasonable for forecasting monthly precipitation and performs better. Sedighi *et al.* (2016) studied the performance of ANN and SVM models for simulating the rainfall–runoff process prejudiced by the height of snow water equivalent (SWE) in Roodak catchment, Tehran region, Iran. Contemplation of SWE improves the performance and accuracy of SVM. Wang *et al.* (2017) projected a novel amalgam model based on variation mode decomposition, PSR, and wavelet NN augmented by genetic algorithm (GA) to forecast multistep ahead wind speed. Experimental outcomes showed that the projected model outclasses all other models used in the study and performs the best. Sun & Wang (2018) proposed a pioneering amalgam model to forecast wind speed, comprising fast collaborative empirical mode disintegration, model entropy, PSR, and back-propagation NN for enhancing the accuracy of wind speed prediction, and observed that the projected model performed better to forecast short-term wind speed. Lekscha & Donner (2018) compared time interval entrenching and differential entrenching techniques aimed at PSR by examining two fine-premeditated paleo climate sets of data from Ecuador and Mexico. Liu *et al.* (1998) developed a PSR map from observed streamflow indicators based on the dynamic system concept, for modelling and predicting streamflow on a daily basis. Results revealed that the suggested PSR model could be utilized for making a timid peculiarity amid a noisy and deterministic frenzied indicator. Tao *et al.* (2018a) projected the hybridization of the SVM model with FFA coupled with PSR for a monthly precipitation forecast in Chhattisgarh, India. The projected model significantly improved the forecasting accuracy of the model and it was found to be a very vigorous smart model, which can be used for the Indian provincial precinct. Mehr *et al.* (2019) developed a new approach based on the integration of SVR and FFA to forecast monthly precipitation of Tabriz and Urmia substations, Iran. SVR-FFA gives promising accuracy in forecasting precipitation in the semi-arid province as compared to other models. Soleymani *et al.* (2016) developed an amalgam method comprising Radial Basis Function and FFA for predicting the water level of Selangor River, Malaysia. The performance of RBF-FFA was examined using simulated and real-time water level data and it was found that RBF-FFA is a proficient technique for accurately predicting river water level. Samantaray & Ghose (2020) applied RBFN and BPNN techniques for runoff prediction at the Ong river basin, India. Xu *et al.* (2020) used FFA with deep learning for improving the prediction efficacy of hydrologic processes, optimizing parameters of SVR spontaneously, and establishing a prediction model for Huangfuchuan in the Fugu region of Shanxi State. Experimental outcomes revealed that the hybrid model accomplished enhanced prediction values. Bozorg-Haddad *et al.* (2018) implemented GA, SVM, ANN, and dynamic programming for calculating optimal release of hydropower reservoirs of Karoon dam, Iran, under forecasting and non-forecasting scenarios. The SVM model output was compared with other applied models and it was found that SVM and ANN performed best to produce optimal reservoir operation. Tao *et al.* (2018b) investigated the ability of a hybrid model combining fuzzy model with FA to predict daily reference evapotranspiration over Burkina Faso province. Outcomes showed the effect of using FFA for significantly improving the performance of the conventional ANFIS model.

The objective of this research is to develop a novel hybrid PSR-SVM-FFA model for runoff prediction considering runoff data generated through experimental set-up (RS). The prediction performance of the proposed robust model is assessed against the hybrid SVM-FFA and benchmark SVM model based on different rainfall intensities with respect to basin slope. The application of PSR-SVM-FFA for runoff prediction is a novel approach of the present study.

2. METHODOLOGY

2.1. Support vector machine

SVM is a set of connected supervised learning techniques utilized for classifications and regressions (Vapnik 1998). SVM can be characterized in a double layer network (where given weights are non-linear in principal and linear in the succeeding layer). SVMs choose constraints for the first layer as training input vectors, as this minimizes Vapnik Chervonenkis (VC) dimension. The network structure of SVM is specified in Figure 1.

In mathematical terms, an elementary function for statistical learning procedure is:

$$y = f(x) = \sum_{i=1}^{M} \alpha_i \varphi_i(x) = w\varphi(x) \tag{1}$$

where output is a linear sum of M. Non-linear alteration in Equation (1) is done by $\varphi(x)$. This comprises a big range of models. The decision function of SVM, which is one among the model range, is signified in Equation (2)

$$y = f(x) = \left\{ \sum_{i=1}^{N} \alpha_i K(x_i, x) \right\} - b \tag{2}$$

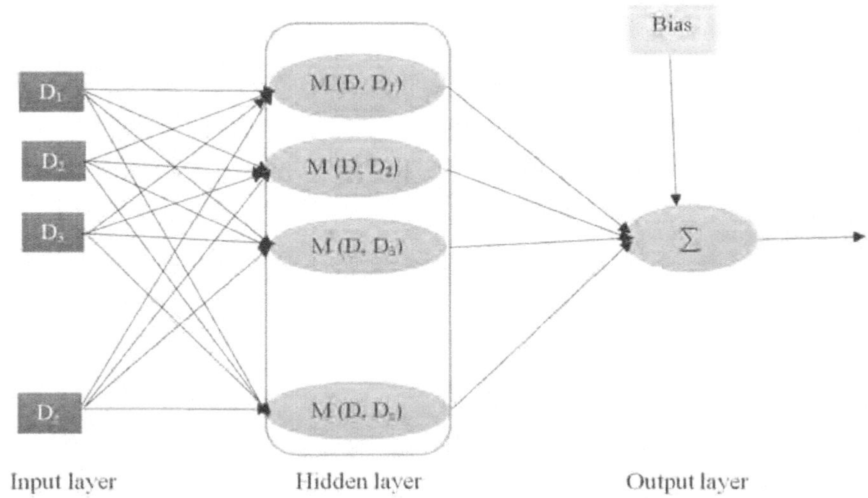

Figure 1 | Architecture of SVM.

where K is the kernel function, α_i and b are constraints obtained by maximizing its objective function, N is the number of training data, x_i are vectors utilized during training, and x is the independent vector.

It utilizes training data to calibrate the model for estimating model constraints, but also retains most vital parts of the input vector in its model (Samantaray & Sahoo 2020b). This vector is known as a support vector. Distinctive structures of kernel functions utilized to transform non-linear input vectors allow SVM to be free from many training vectors, so as to obtain a very small resultant model.

2.2. Firefly algorithm

Yang (2010) introduced a firefly algorithm (FFA) motivated by the societal conduct of fireflies, based on their blinking characteristics. Three important norms in FFA architecture depend on the features of real-world fireflies irrespective of their sex; (i) they are presumed to be unisex, therefore, every firefly can attract another; (ii) luminous intensity regulates the attraction of a single firefly to another; and (iii) flashing intensity is proportionate to the magnitude of light produced by a firefly. On the basis of these norms, the FFA model's objective function is signified by brightness and light intensity produced from a firefly. Intensity and desirability are given in Equations (3) and (4), respectively. Every firefly has its own desirability β with regard to its strength and how it entices other swarm members (Yang 2010).

$$I = I_0 e^{-\gamma r^2} \tag{3}$$

$$w(r) = w_0 e^{-\gamma r^2} \tag{4}$$

where $w(r)$ and I signify desirability and light intensity at distance r from firefly, in respective order. w_0 and I_0 are desirability and light intensity at distance $r = 0$ from firefly and γ is the coefficient of light immersion. Distance r_{ij} amid any two fireflies i and j is articulated in Equation (5) (Yang 2010):

$$r_{ij} = \|x_i + x_j\| = \sqrt{\sum_{k=1}^{d} (x_{i,k} - x_{j,k})} \tag{5}$$

where x_i and x_j are positions of fireflies i and j. FFA is modest, supple, and adaptable, which is most effective to solve a varied array of miscellaneous real-world glitches. FFA has the capability to control its modality and control the scale constraint for adapting to a problematic background (Sahoo *et al.* 2021). For any meta-heuristic algorithm, a moral equilibrium amid utilization and investigation throughout pursuit processes must be sustained for achieving better outcomes. FFA acts as a global method in optimization problems, and hence is designed for exploring search space and probably to give an optimum/near-optimum result, if utilized solitarily (Fister *et al.* 2013). The proposed hybrid SVM-FFA model flowchart is presented in Figure 2.

2.3. Phase space reconstruction

The theory of the PSR approach helps in the possible recognition of the chaotic, deterministic, and stochastic nature of dynamic classifications (Sivakumar *et al.* 2001). PSR is primarily utilized for reconstructing single-variable time series to avoid any univariable in time series. The system's state is signified at each point, while its evolution time is signified at each flight route, corresponding to changing preliminary constraints. Each point or group of points in a system consists of distinctive patterns to attract flight routes toward themselves. This arrangement of pattern designs is identified as attractors. For estimating the intricacy of attractors in the system, the reconstruction of the phase system is very much needed; hence, attractor behaviour could be defined and enumerated as being multifaceted or not. With assistance of the time delay entrenching concept, a PSR could be fabricated based on univariable dimension time series. PSR of any specified time series can be articulated as follows:

$$X_t = (X_t, X_{t-1}, X_{t-2}, \ldots, X_{t-(m-1)\tau}), \quad t = 1, 2, \ldots, M; \; M = n - (m-1)\tau \tag{6}$$

where t is the time delay, m represents embedding dimensions, while M signifies the number of points in PSR. PSR amid X_t and $X_{t-(m-1)\tau}$, when rebuilt, can graphically depict the existence of an attractor aimed at deterministic chaos in a time series. A phase space sequence of m-dimensions can be abridged in the subsequent manner (Delafrouz *et al.* 2017):

$$PhS = \begin{bmatrix} x_1 & x_{1-\tau} & x_{1-2\tau} & x_{1-3\tau} & \cdots & x_{1-(m-1)\tau} \\ \cdots & \cdots & \cdots & \cdots & \cdots & \cdots \\ x_{n_m} & x_{n_{m-\tau}} & x_{n_{m-2\tau}} & x_{n_{m-3\tau}} & \cdots & x_{n_m-(m-1)\tau} \end{bmatrix} \tag{7}$$

Assortment of an optimum delay time t along with dimensions of embedding m is a very significant stage during the construction of a PSR. Mutual information function (MIF) is one of the common methods to determine optimal delay time in context to the scheming of an attractor. An illustration of the hybrid extravagance model is demonstrated in Figure 3.

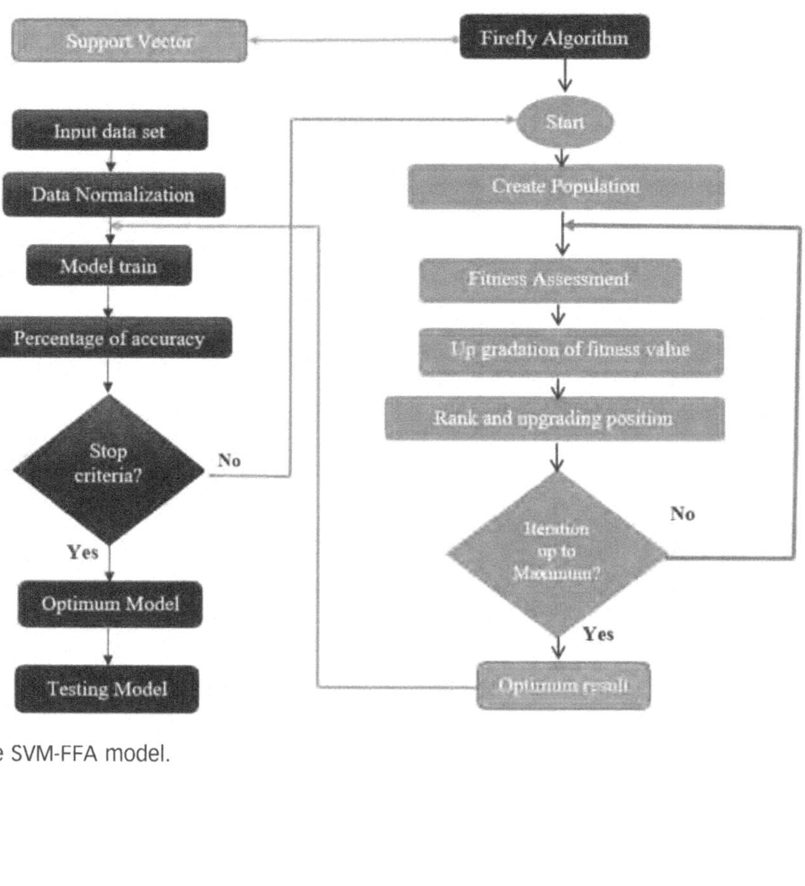

Figure 2 | Flowchart of the SVM-FFA model.

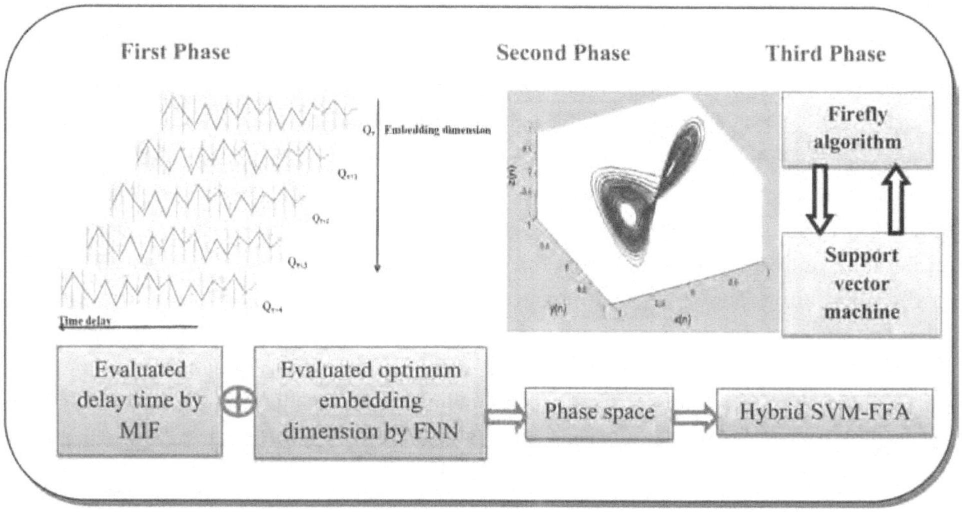

Figure 3 | Graphical representation of the hybrid PSR-SVM-FFA.

2.4. Experimental set-up (description of RS)

RS consists of a sand tank that is straddling on a sustenance frame with essential services, features, and arrangements for facilitating studies of groundwater flow, groundwater abstraction, flood hydrographs, and fluvial mechanics. The sump tank and centrifugal pump straddle a casing underneath the sand tank that delivers water for numerous experimentations. Water leaving the sand tank from various outlets proceeds to the sump tank under gravitational force for reprocessing (Pérez-Latorre *et al.* 2010). A centrifugal pump draws water from the sump tank via drumming at the tank bottom. A pressure regulator in every provision ensures that the flow is unaffected by changes in the other feed provided that the regulator is adjusted to suit. The outlet from the centrifugal pump incorporates a pressure relief valve to limit the system pressure to a maximum of approximately 3.0 barg and prevent the pump from overheating, if the flow from the two outlets is restricted or stopped. Water discharging from the relief valve is returned to the sump tank via a connection in the sidewalls of the tank (Elhakeem & Papanicolaou 2009). When demand from the two outlets is high, the relief valve will remain closed to maintain the pressure in the system. When demand is reduced and the system pressure rises above 3.0 barg, the valve will open to relieve the excess flow.

The shallow sand tank is contrived from stainless steel to resist corrosion and should be filled with sand or other granular material as appropriate to the studies. Collection of tapping sockets in the sand tank is linked with multiple tube manometer which permits the determination of water table surface. The level in each tube can be read by sliding the common scale along the track at the top of the manometer. Before using the manometer to measure water levels, it is important to expel air from flexible tubes connecting manometer tubes to tapping points. Each tapping in the sand tank incorporates a filter mesh to retain the sand while allowing the water to flow. Valves and pipework below the sand tank allow water draining from each well to flow back to the sump tank. Supple outlet tubes permit water to be extracted to an accumulating container with the motivation of gauging rates of volumetric flow. The two wells are purposely designed to be short in length so that they can be left in position without affecting the surface flow experiments. The plug of sand directly above each well can be removed if required for abstraction experiments, but the effect on the results will be negligible if the sand is left in place.

Water and sediment exiting the sand tank via the weir chute/diffuser is dropped to the outlet accumulating container that measures water flow and assembles any sediment eroded from the sand chamber. This tank is fabricated from clear acrylic. Rain is delivered to the watershed region by two rows of four sprig spouts overhead the chamber, straddling a sustenance structure. The height of the sprig spouts overhead the sand chamber can be speckled for optimizing experimentation by regulating sustenance structure height. Solitary persons at both ends of the equipment must grip sustenance structure during modification performance. A separating valve upstream of every nozzle permits essential design alteration. Since the flow rate through each nozzle is dependent on the pressure, if the appropriate pressure regulator is adjusted to give the required flow rate, then the flow through each nozzle will remain constant when other nozzles are turned on or off. To achieve this, the feed flow control valve should be opened fully and the pressure regulator adjusted to give the required flow through the nozzles. The flexible tube from the arrangement of spray nozzles is connected to one of the water feeds, when required, using the self-sealing quick-release connector. The height of nozzles should be adjusted at a required flow rate to give adequate coverage over the surface of sand without excessive spray over the sides of the sand tank.

River inlet tank straddling on the right side of sand tank permits water stream flowing to the surface of the sand, simulating stream from a watercourse upstream. The river inlet tank is fabricated from stainless steel and is bolted to the end wall of the sand tank adjacent to the shallow cut-out. Water comes in at the tank bottom, courses uphill through a glass marble bed for minimizing any turbulence, and then streams sideward to the sand surface through a rectangular section. An anti-erosion mat is supplied to reduce any local scour where the water enters the sand tank. This mat is buried just beneath the sand surface adjacent to the river inlet tank outlet. The flexible tube from the base of the sand tank is connected to one of the water feeds when required, using the self-sealing quick-release connector.

It has been found through the experimentation that well-graded sand in the range 16/30 mesh (1,000 to 500 μm) will allow all of the experiments described in the teaching manual to be carried out without the need to change the sand between runs. To minimize the cost of filling the tank, using a 16/30 mesh swimming pool silica filter grit is suggested. To fill the tank, 550 kg of sand will be required. Figure 4(b) shows the pictographic view of the RS during the experiment. Statistical parameters of the experimental data set (Burgan & Aksoy 2018) for both training and testing phases are given in Table 1.

2.5. Processing and preparation of data

Runoffs are collected from RS with respect to various rainfall intensities of the pump. To enrich the maximum amount of runoff, first switch on the power supply of the pump, and then switch it off. Data are collected in 5 s intervals up to a constant

(a)

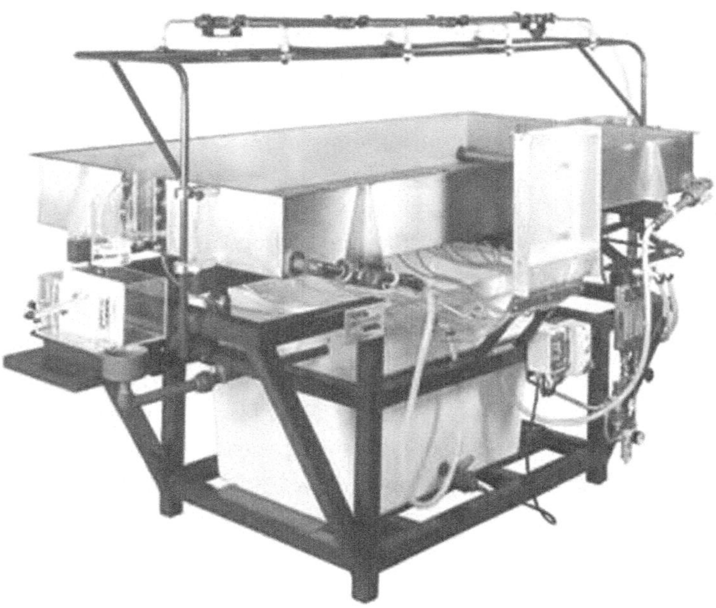

(b)

Figure 4 | Comprehensible delineation of the PSR-SVM-FFA model. (a) Graphical view of S12-MKII rainfall simulator and (b) pictographic view of watershed during the experiment.

Table 1 | Statistical parameters of experimental data set

Statistical parameters	Training set (480)	Testing set (240)	Total data set (720)	Training set (480)	Testing set (240)	Total data set (720)
Slope 0^0				Slope 1^0		
Min	0.04	0.04	0.04	0.04	0.04	0.04
Max	0.79	0.81	0.81	0.94	0.92	0.94
Mean	0.41325	0.432199	0.419361	0.420958	0.396792	0.412903
Kurt	−0.96964	−1.04858	−1.01077	−0.98548	−0.967	−0.9918
Skew	−0.02542	−0.21722	−0.08664	−0.01955	0.150961	0.037441
SD	0.211402	0.215723	0.212565	0.217727	0.213765	0.216546
Slope 2^0				Slope 3^0		
Min	0.04	0.04	0.04	0.04	0.04	0.04
Max	0.82	0.79	0.82	0.79	0.79	0.79
Mean	0.416792	0.4445	0.426028	0.412417	0.438452	0.420639
Kurt	−1.11488	−1.1812	−1.16005	−1.14219	−1.30635	−1.19653
Skew	0.024677	−0.23992	−0.06299	−6.4E-05	−0.04127	−0.00169
SD	0.219708	0.225516	0.221911	0.217343	0.235875	0.223684
Slope 4^0				Slope 5^0		
Min	0.04	0.04	0.04	0.04	0.04	0.04
Max	0.88	0.79	0.88	0.79	0.79	0.79
Mean	0.400417	0.454268	0.417889	0.400333	0.462218	0.420472
Kurt	−0.97373	−1.26826	−1.13257	−1.00105	−1.32073	−1.16801
Skew	0.146643	−0.264	0.014725	0.110759	−0.25404	0.001526
SD	0.213393	0.233795	0.220812	0.211888	0.240142	0.222677

limit of runoff generation. For a period of 1 h, runoff data are taken for model formulation. Data composed from 40 min (480 data) are employed for training and the remaining 20 min (240 data) for the testing model. The following arrangements are applied as input:

Model	Input	SVM	SVM*FFA
M*1	$P_{t:1}$	SVM1	SVMI*FFA
M*2	$P_{t:1}$, $P_{t:1.5}$	SVM2	SVMII*FFA
M*3	$P_{t:1}$, $P_{t:1.5}$, $P_{t:2}$	SVM3	SVMIII*FFA
M*4	$P_{t:1}$, $P_{t:1.5}$, $P_{t:2}$, $P_{t:2.5}$	SVM4	SVMIV*FFA
M*5	$P_{t:1}$, $P_{t:1.5}$, $P_{t:2}$, $P_{t:2.5}$, $P_{t:3}$	SVM5	SVMV*FFA

where $P_{t:1}$ is rainfall when pump input is 1 L/min; $P_{t:1.5}$ is rainfall when pump input is 1.5 L/min; $P_{t:2}$ is rainfall when pump input is 2 L/min; $P_{t:2.5}$ is rainfall when pump input is 2.5 L/min; and $P_{t:3}$ is rainfall when pump input is 3 L/min.

Input and output data before training are put between a quantified range (0–1). This process is identified as normalization. The equation utilized for putting data in the range is

$$Y_i = \frac{Y - Y_{\min}}{Y_{\max} - Y_{\min}} \tag{8}$$

where Y_i is the converted data set, Y is the actual data set, $Y_{\min}$ is the minimum of actual data set, and $Y_{\max}$ is the maximum of the actual data set.

2.6. Model performance evaluation

Indicators are NSE, RMSE, and WI used to assess the performance of model efficiency. The formulae can be articulated as

$$\mathrm{NSE} = 1 - \left[\frac{\sum_{i=1}^{N}(A_i - B_i)^2}{\sum_{i=1}^{N}(A_i - \overline{A_i})^2}\right] \quad -\infty < \mathrm{NS} \leq 1 \tag{9}$$

$$\mathrm{RMSE} = \sqrt{\frac{1}{N}\sum_{i=1}^{N}(B_i - A_i)^2} \tag{10}$$

$$\mathrm{WI} = 1 - \left[\frac{\sum_{i=1}^{N}(A_i - B_i)^2}{\sum_{i=1}^{N}(|B_i - \overline{A_i}| + |A_i - \overline{A_i}|)^2}\right] \tag{11}$$

where A_i and B_i are measured and predicted ith runoffs and $\overline{A_i}$ is the mean of measured runoff.

3. RESULTS AND DISCUSSION

3.1. Experimental result

Runoff is generated by using RS for six various slopes (Slope $= 0^0$, Slope $= 1^0$, Slope $= 2^0$, Slope $= 3^0$, Slope $= 4^0$, Slope $= 5^0$) of watersheds in context to different input pump efficiencies (1, 1.5, 2, 2.5, and 3 L/min). As rainfall intensity has an impact on runoff, here a detailed description of runoff for axiom rainfall intensity condition (3 L/min) is given.

When rainfall occurs at maximum intensity in a sloppy horizontal bed modelled with sand, the peak hyetograph is established and achieves the highest flow rate. After achieving the peak steady flow rate with the highest epoch, the rising limb fall and rainfall intensity decrease and reach the minimum flow rate with a step-down decrement of rainfall intensity. Here, the hyetograph achieves the highest flow rate of 0.8 L/min for the 5^0 slope of the designed bed. Similarly, peak hyetograph is established with the highest flow rate of 0.88 m^3/s for 4^0 slope, 0.78 m^3/s for 3^0, 0.82 m^3/s for 2^0, 0.94 m^3/s for 1^0 slope, and 0.81 m^3/s 0^0 slope.

3.2. Model performance analysis

The performance assessment of various input scenarios utilizing the SVM algorithm is presented in Table 2. Statistical measures like RMSE, WI, and NSE are used for the performance evaluation of models (training and testing phase). Amid five models, SVM5 provides the best performance value for all sloping conditions. While considering the testing period, the best WI values are 0.9412, 0.9344, 0.9378, 0.9516, 0.9518, and 0.9408 for SL-0^0, SL-1^0, SL-2^0, SL-3^0, SL-4^0, and SL-5^0, respectively. Similarly, preeminent values of WI are 0.9392, 0.9266, 0.9297, 0.9408, 0.9415, and 0.9353 for various slopes in training phases.

Among six slopes, the SVM5-FFA model indicates the best result having a WI value of 0.9543 during training and 0.9716 during the testing phase of SL-4^0. For rest slopes, the SVM5-FFA model provides the best WI values during both periods among all simulations as presented in Table 3. Model outcomes of SVM-FFA based on NSE, RMSE, and WI for both periods are provided in Table 3.

Results of the PSR-SVM-FFA technique during training and testing periods are given in Table 4. When **SL-4^0** is considered, the obtained WI value shows the best performing results. In **SL-4^0**, the best values of WI for training and testing periods are 0.9623 and 0.9812. The WI values for other sloping conditions (SL-0^0, SL-1^0, SL-2^0, SL-3^0, and SL-5^0) are presented in Table 4. Based on obtained results, a possible inference regarding the dominance of variables on runoff dynamics can be that dynamics reveal a low-dimension chaotic behaviour, and hence, appropriateness of a low-dimension chaotic method. Manifestation of an ideal embedding dimension, i.e. prediction accuracy increases with an increase in embedding dimension to reach its best, i.e. $m_{\mathrm{opt}} = 14$, and thereafter, reduces with further increase in embedding dimension and best prediction outcomes attained with a small number of neighbours support this argument.

Table 2 | SVM model results

Positions	Model	Training NSE	Training RMSE	Training WI	Testing NSE	Testing RMSE	Testing WI
Slope 0^0	SVM1	0.521	92.51	0.8276	0.493	90.38	0.8462
	SVM2	0.555	88.23	0.8491	0.517	85.34	0.8697
	SVM3	0.591	84.72	0.8717	0.557	81.25	0.8936
	SVM4	0.629	79.91	0.8913	0.596	77.77	0.9233
	SVM5	**0.691**	**76.55**	**0.9392**	**0.661**	**73.16**	**0.9412**
Slope 1^0	SVM1	0.505	93.64	0.8049	0.459	91.81	0.8326
	SVM2	0.535	89.28	0.8382	0.504	87.98	0.8618
	SVM3	0.569	85.43	0.8651	0.545	83.71	0.8891
	SVM4	0.613	81.48	0.8857	0.562	78.56	0.8995
	SVM5	**0.662**	**77.59**	**0.9266**	**0.625**	**73.86**	**0.9344**
Slope 2^0	SVM1	0.513	92.71	0.8145	0.486	90.54	0.8401
	SVM2	0.539	88.77	0.8398	0.509	86.48	0.8654
	SVM3	0.582	85.04	0.8667	0.549	82.45	0.8909
	SVM4	0.615	81.34	0.8871	0.584	78.24	0.9144
	SVM5	**0.671**	**76.84**	**0.9297**	**0.628**	**73.79**	**0.9378**
Slope 3^0	SVM1	0.526	92.47	0.8313	0.494	89.89	0.8541
	SVM2	0.556	88.08	0.8518	0.521	84.94	0.8719
	SVM3	0.605	84.65	0.8784	0.561	80.45	0.8947
	SVM4	0.651	78.93	0.8959	0.609	75.79	0.9256
	SVM5	**0.701**	**75.36**	**0.9408**	**0.667**	**69.01**	**0.9516**
Slope 4^0	SVM1	0.529	91.79	0.8358	0.498	88.78	0.8581
	SVM2	0.559	87.03	0.8622	0.525	84.79	0.8739
	SVM3	0.606	83.81	0.8834	0.561	80.38	0.8966
	SVM4	0.656	78.89	0.8991	0.621	75.77	0.9272
	SVM5	**0.704**	**73.58**	**0.9415**	**0.669**	**69.59**	**0.9518**
Slope 5^0	SVM1	0.518	92.61	0.8179	0.492	90.41	0.8415
	SVM2	0.544	88.37	0.8443	0.515	85.87	0.8668
	SVM3	0.591	84.79	0.8717	0.549	81.65	0.8915
	SVM4	0.619	81.32	0.8902	0.592	78.12	0.9162
	SVM5	**0.689**	**76.69**	**0.9353**	**0.656**	**73.27**	**0.9408**

3.3. Assessment of results for applied models

For all slopes, the assessment of the PSR-SVM-FFA model during training and testing periods is represented in Figures 5 and 6. Among all slopes, the best value of WI is obtained in Slope 4^0. Here, WI values for SVM5, SVM5-FFA, and PSR-SVM-FFA-SL-4^0 models are 0.9412, 0.9668, and 0.9812, respectively. Hence, it can be observed that prediction models proposed in the present research achieve enhanced prediction performances in comparison to models discussed in the literature. As in the case of any model, runoff predictions provided by a newly applied PSR-SVM-FFA model are related to a certain degree of uncertainty that needs to be enumerated. Uncertainty related to the model structure is assessed for three data-driven models (SVM, SVM-FFA, and PSR-SVM-FFA) considered for best input combination model M#5.

A linear scale plot of actual versus predicted runoff for the proposed model of projected slopes is shown in Figure 7. Results illustrate that estimated peak runoffs are 0.7624, 0.7832, and 0.7903 m^3/s for SVM, SVM-FFA, and PSR-SVM-FFA against an actual peak 0.81 m^3/s for the slope 0^0. The approximated peak runoff is 0.8783, 0.903, and 0.9147 m^3/s for SVM, SVM-FFA, and PSR-SVM-FFA adjacent to actual peak 0.94 m^3/s for slope 1^0 condition. Similarly, for slope 2^0, tangible runoff is 0.82 m^3/s aligned with predicted runoffs 0.7689, 0.7881, and 0.7988 m^3/s for SVM, SVM-FFA, and PSR-SVM-FFA, respectively. Correspondingly, apprised peak runoffs are 0.7517, 0.765, and 0.7715 m^3/s for SVM, SVM-FFA, and PSR-SVM-FFA against the actual peak 0.79 m^3/s for the slope 3^0. Similarly, approached peak runoffs are 0.8375, 0.855, and 0.8634 m^3/s for SVM, SVM-FFA, and PSR-SVM-FFA adjacent to actual peak 0.88 m^3/s for slope 4^0. Consistently, for slope 5^0, the evaluated peak runoff is 0.79 m^3/s associated with predicted runoffs 0.7432, 0.7609, and 0.7703 m^3/s for SVM, SVM-FFA, and PSR-SVM-FFA, respectively. Simulations exhibit that the proposed technique can efficiently improve

Table 3 | SVM-FFA model results

		Training period			Testing period		
Positions	**Model**	**NSE**	**RMSE**	**WI**	**NSE**	**RMSE**	**WI**
Slope 0^0	SVM*FFA	0.696	90.24	0.8442	0.664	88.67	0.8521
	SVMI*FFA	0.719	86.01	0.8698	0.694	83.82	0.8777
	SVMII*FFA	0.754	82.45	0.9084	0.729	78.91	0.9193
	SVMIII*FFA	0.789	77.77	0.9378	0.755	75.52	0.9482
	SVMIV*FFA	**0.821**	**73.81**	**0.9487**	**0.791**	**69.75**	**0.9668**
Slope 1^0	SVM*FFA	0.687	91.69	0.8273	0.656	90.41	0.8349
	SVMI*FFA	0.702	87.82	0.8574	0.677	86.14	0.8703
	SVMII*FFA	0.745	83.72	0.8946	0.713	80.77	0.9085
	SVMIII*FFA	0.777	79.33	0.9329	0.741	76.53	0.9444
	SVMIV*FFA	**0.802**	**74.22**	**0.9473**	**0.766**	**70.58**	**0.9606**
Slope 2^0	SVM*FFA	0.689	91.55	0.8313	0.657	89.14	0.8453
	SVMI*FFA	0.705	87.64	0.8672	0.681	84.82	0.8736
	SVMII*FFA	0.751	83.54	0.8975	0.718	80.66	0.9111
	SVMIII*FFA	0.781	78.03	0.9338	0.744	75.79	0.9468
	SVMIV*FFA	**0.808**	**74.14**	**0.9474**	**0.768**	**69.97**	**0.9611**
Slope 3^0	SVM*FFA	0.699	89.88	0.8537	0.671	88.51	0.8652
	SVMI*FFA	0.722	85.89	0.8714	0.697	83.67	0.8793
	SVMII*FFA	0.761	80.89	0.9162	0.731	78.64	0.9227
	SVMIII*FFA	0.791	76.68	0.9418	0.759	73.54	0.9493
	SVMIV*FFA	**0.823**	**70.41**	**0.9491**	**0.799**	**68.35**	**0.9684**
Slope 4^0	SVM*FFA	0.701	89.57	0.8553	0.674	86.34	0.8681
	SVMI*FFA	0.725	85.55	0.8786	0.706	82.57	0.8947
	SVMII*FFA	0.767	80.41	0.9231	0.734	78.49	0.9382
	SVMIII*FFA	0.798	76.54	0.9489	0.762	73.35	0.9554
	SVMIV*FFA	**0.831**	**70.23**	**0.9543**	**0.802**	**67.74**	**0.9716**
Slope 5^0	SVM*FFA	0.693	90.42	0.8343	0.662	88.87	0.8482
	SVMI*FFA	0.716	86.67	0.8682	0.688	84.21	0.8757
	SVMII*FFA	0.753	83.39	0.9016	0.724	79.02	0.9123
	SVMIII*FFA	0.783	77.78	0.9352	0.751	75.62	0.9478
	SVMIV*FFA	**0.815**	**73.95**	**0.9477**	**0.787**	**69.93**	**0.9632**

Table 4 | Values of τ, m with results of PSR-SVM-FFA model

		Training period			Testing period		
Model	**Optimal parameters of τ, m for PSR**	**NSE**	**RMSE**	**WI**	**NSE**	**RMSE**	**WI**
PSR-SVM-FFA-SL-0^0	($m = 14$, $\tau = 5$)	0.926	52.89	0.9565	0.878	60.38	0.9757
PSR-SVM-FFA-SL-1^0	($m = 14$, $\tau = 5$)	0.883	59.34	0.9546	0.836	63.21	0.9731
PSR-SVM-FFA-SL-2^0	($m = 14$, $\tau = 5$)	0.899	57.09	0.9555	0.854	62.49	0.9742
PSR-SVM-FFA-SL-3^0	($m = 14$, $\tau = 5$)	0.943	49.77	0.9584	0.889	59.23	0.9767
PSR-SVM-FFA-SL-4^0	($m = 14$, $\tau = 5$)	**0.958**	**45.01**	**0.9623**	**0.904**	**55.56**	**0.9812**
PSR-SVM-FFA-SL-5^0	($m = 14$, $\tau = 5$)	0.901	55.13	0.9556	0.865	61.65	0.9751

the forecasting accurateness of many conventional techniques in terms of all applied evaluation indices. Therefore, a robust data-driven model with powerful performances is proposed for forecasting experimental runoff.

The box plot of actual and estimated runoff values was drawn as presented in Figure 8. It can be observed that data ranges, quantiles, and medians for predicted runoff values were analogous to those of actual runoff values. Interquartile ranges of all models show similarity with actual runoff. In the present study, outliers were not mentioned in Figure 8. They signify an extreme runoff value (maximum and minimum peak) – that is, runoff in the course of high flow and low flow periods.

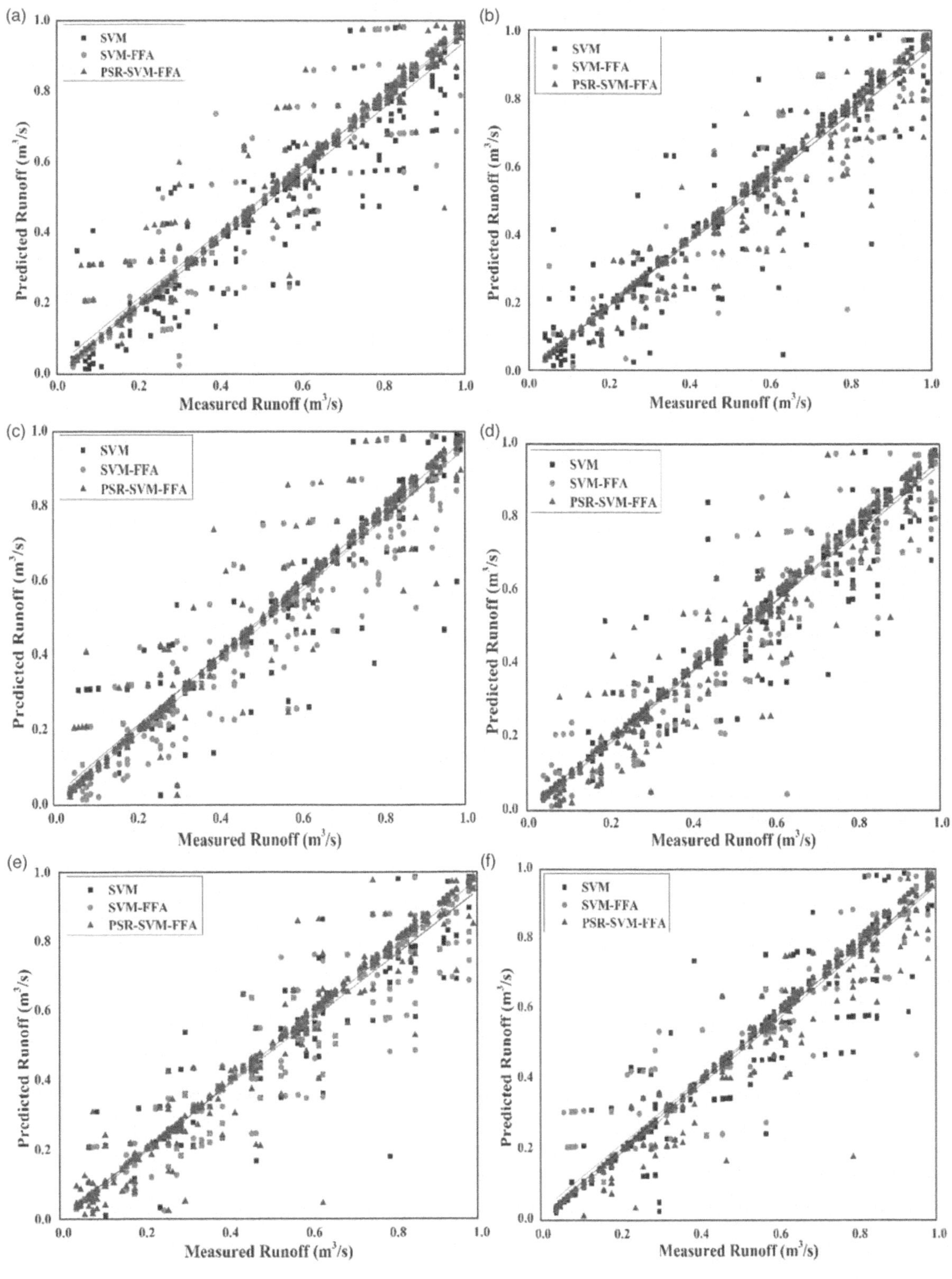

Figure 5 | Actual versus predicted runoff for (a) Slope 0^0, (b) Slope 1^0, (c) Slope 2^0, (d) Slope 3^0, (e) Slope 4^0, and (f) Slope 5^0 during training phase.

PSR-SVM-FFA performed better in apprehending extreme runoff, while SVM and SVM-FFA could not capture either top, bottom, or both extremes. The reason is that SVM and SVM-FFA techniques applied in the present study had a comparatively less complex structure and fewer operational parameters than the PSR-SVM-FFA model.

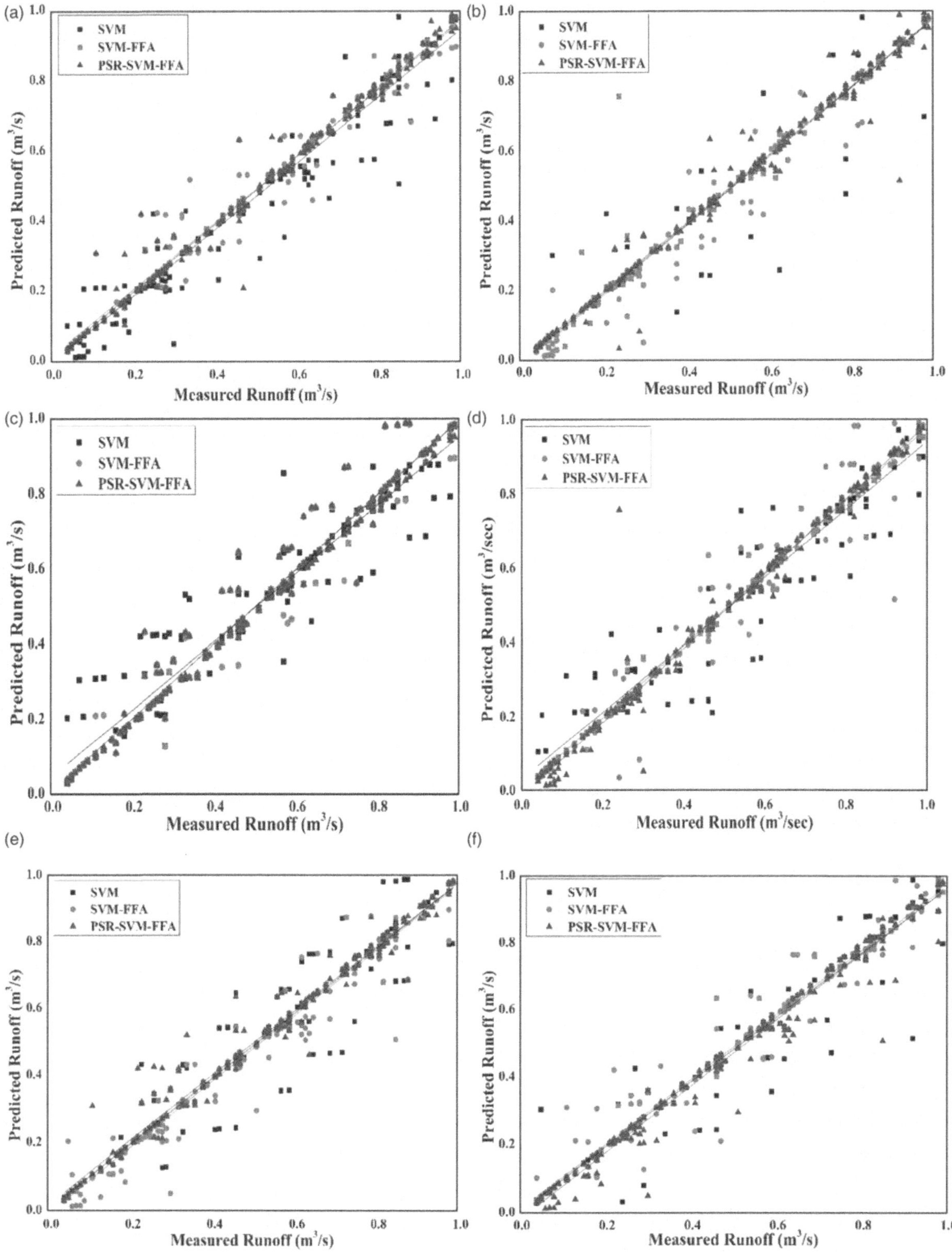

Figure 6 | Actual versus predicted runoff for (a) Slope 0^0, (b) Slope 1^0, (c) Slope 2^0, (d) Slope 3^0, (e) Slope 4^0, and (f) Slope 5^0 during testing phase.

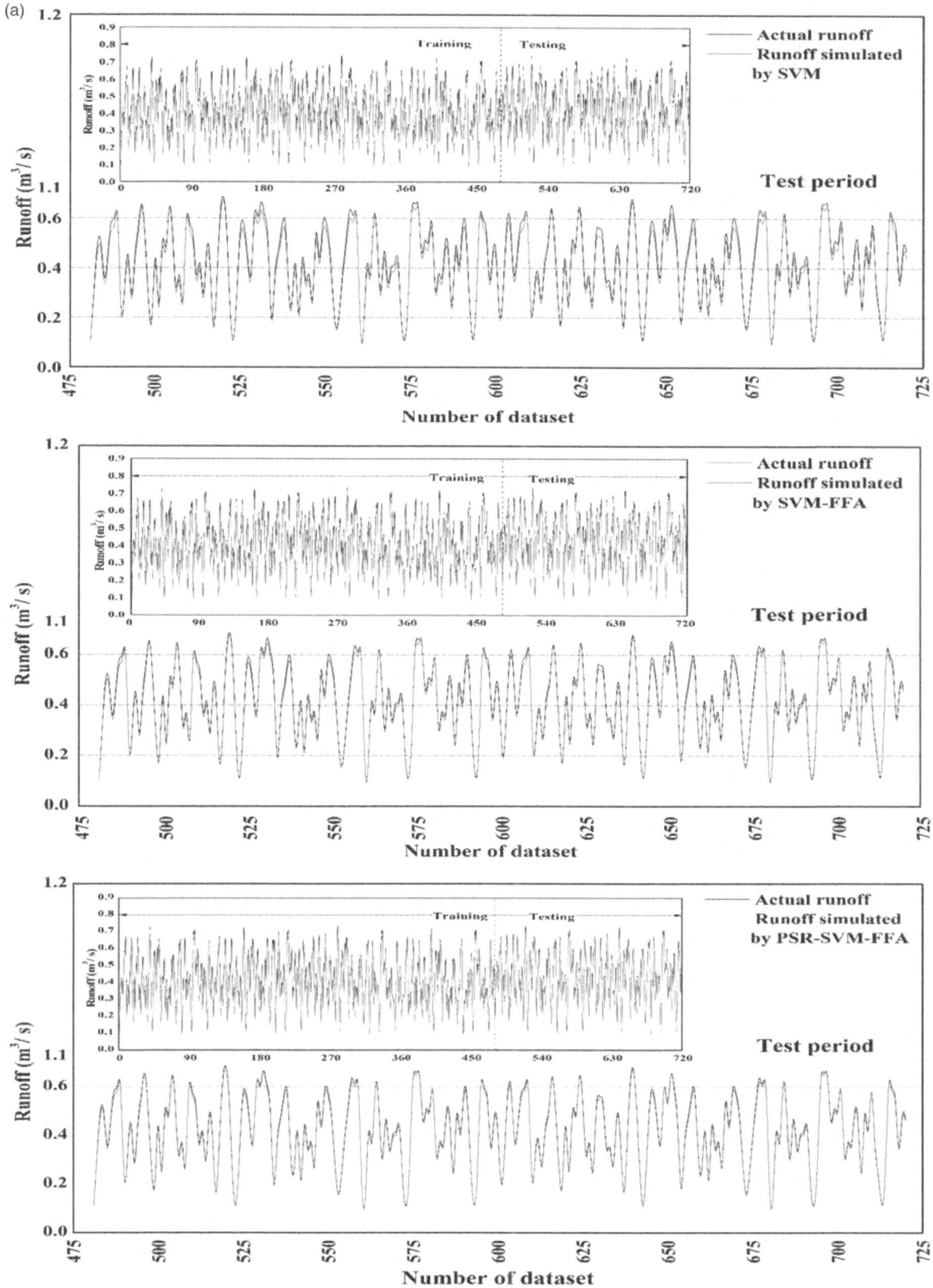

Figure 7 | Deviation of actual and predicted runoff at (a) Slope 0^0, (b) Slope 1^0, (c) Slope 2^0, (d) Slope 3^0, (e) Slope 4^0, and (f) Slope 5^0 condition during testing phase. *(continued.).*

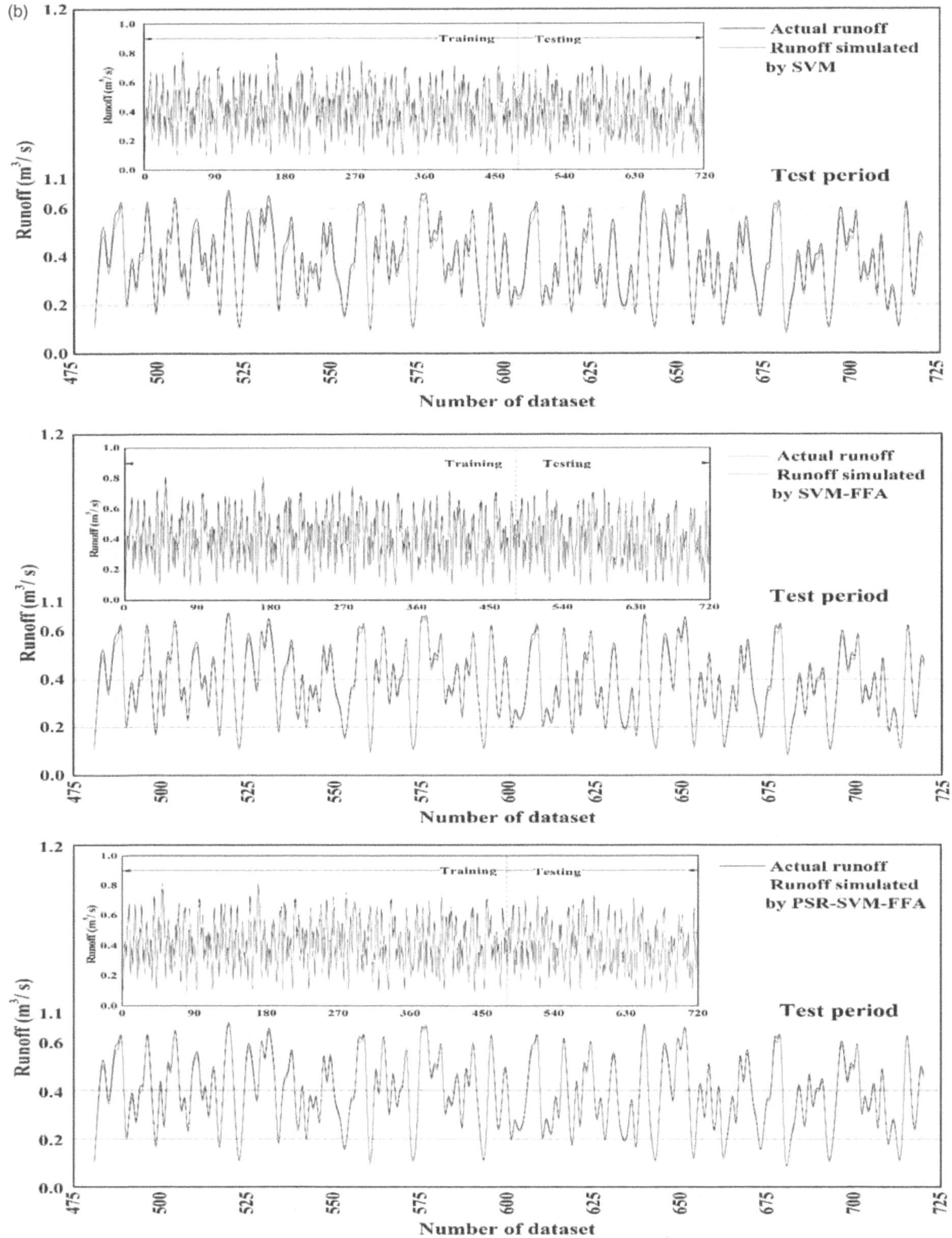

Figure 7 | Continued.

To have better knowledge with regard to the prediction accuracy of three challenging techniques, SVM, SVM-FFA, and PSR-SVM-FFA, a comparison is done between runoff values of varied observations, ranges, and predictions. Histogram plots of observed and estimated runoff values are demonstrated in Figure 9. Plots in Figure 9 signify the occurrence

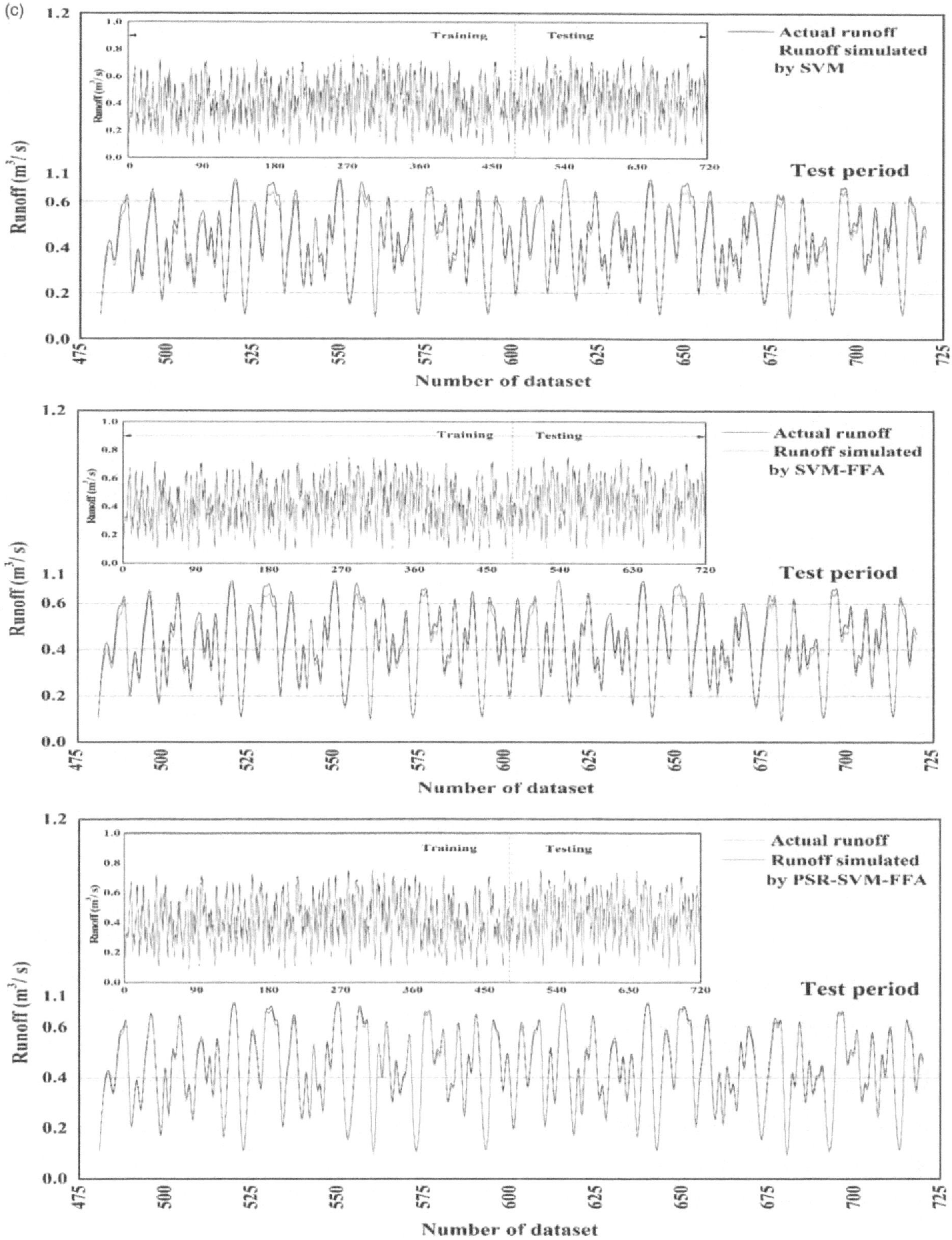

Figure 7 | Continued.

(d)

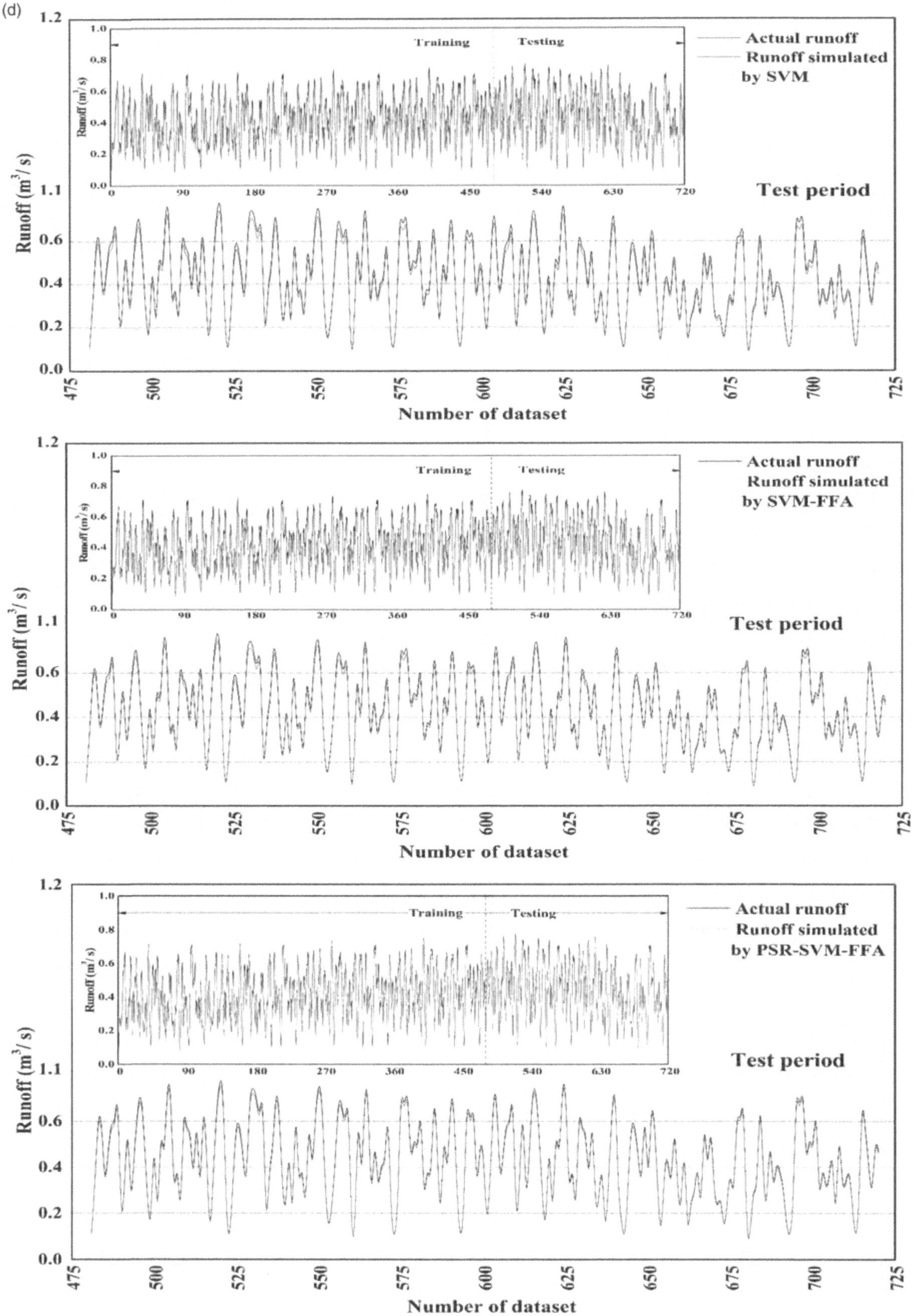

Figure 7 | Continued.

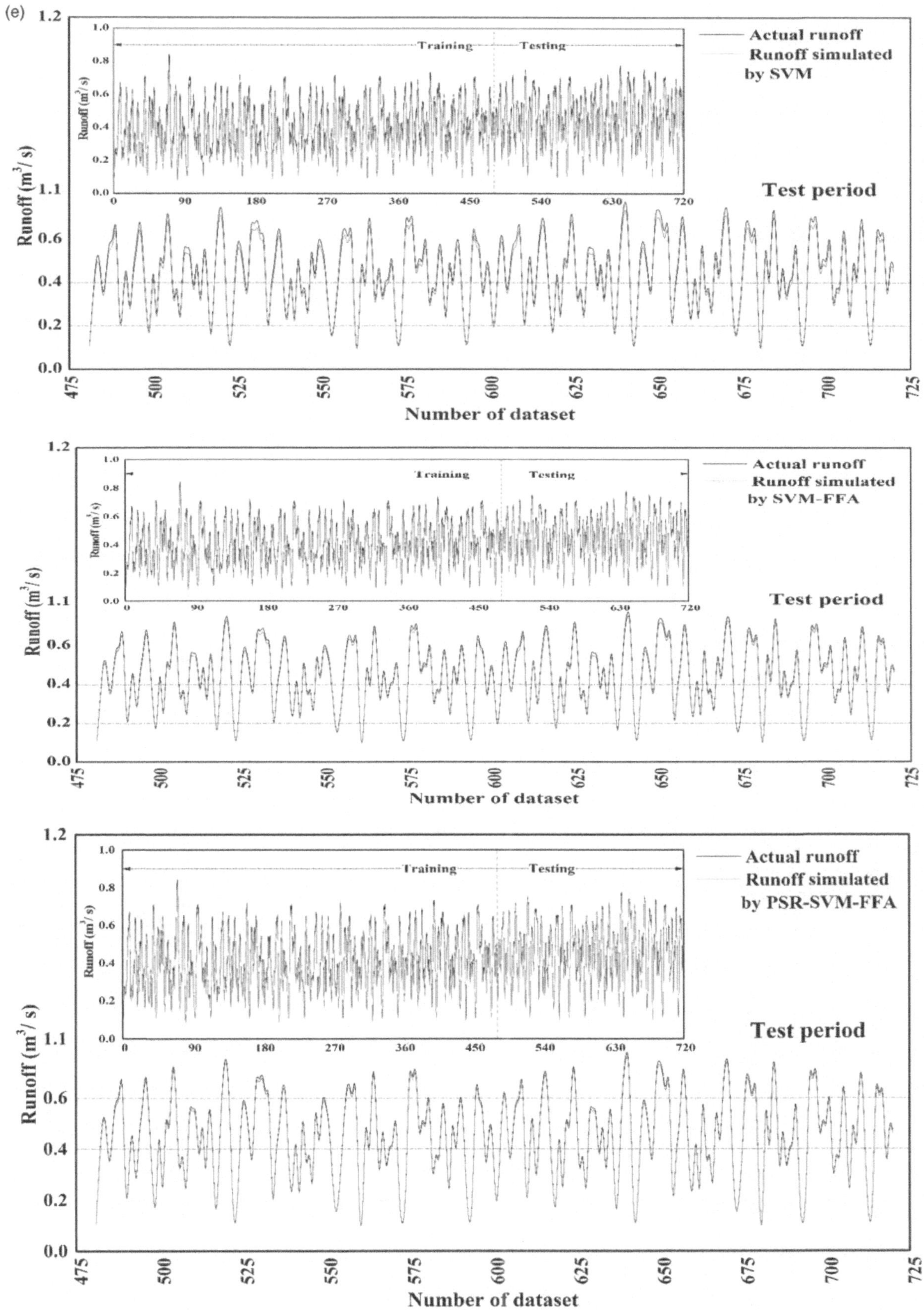

Figure 7 | Continued.

(f)

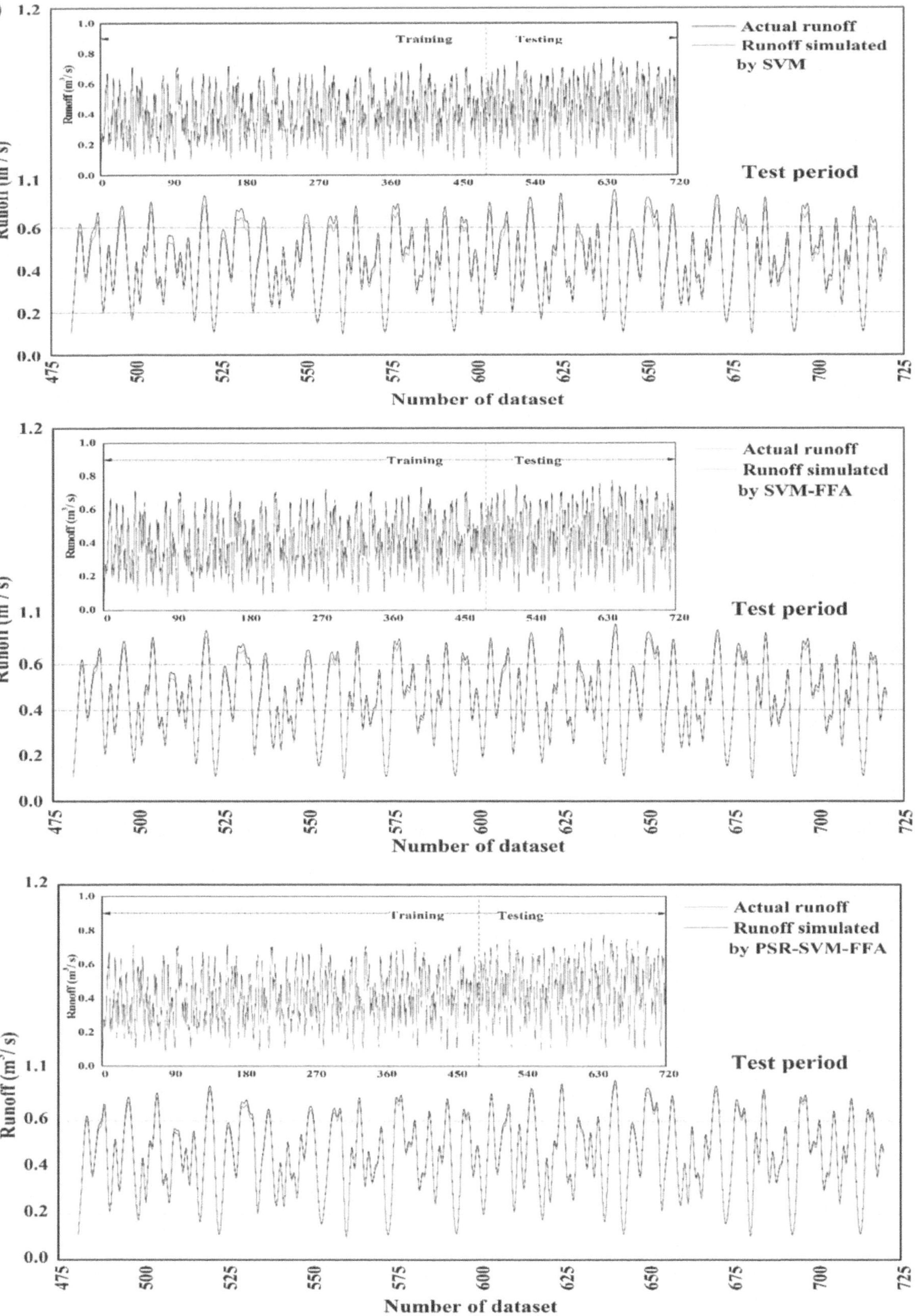

Figure 7 | Continued.

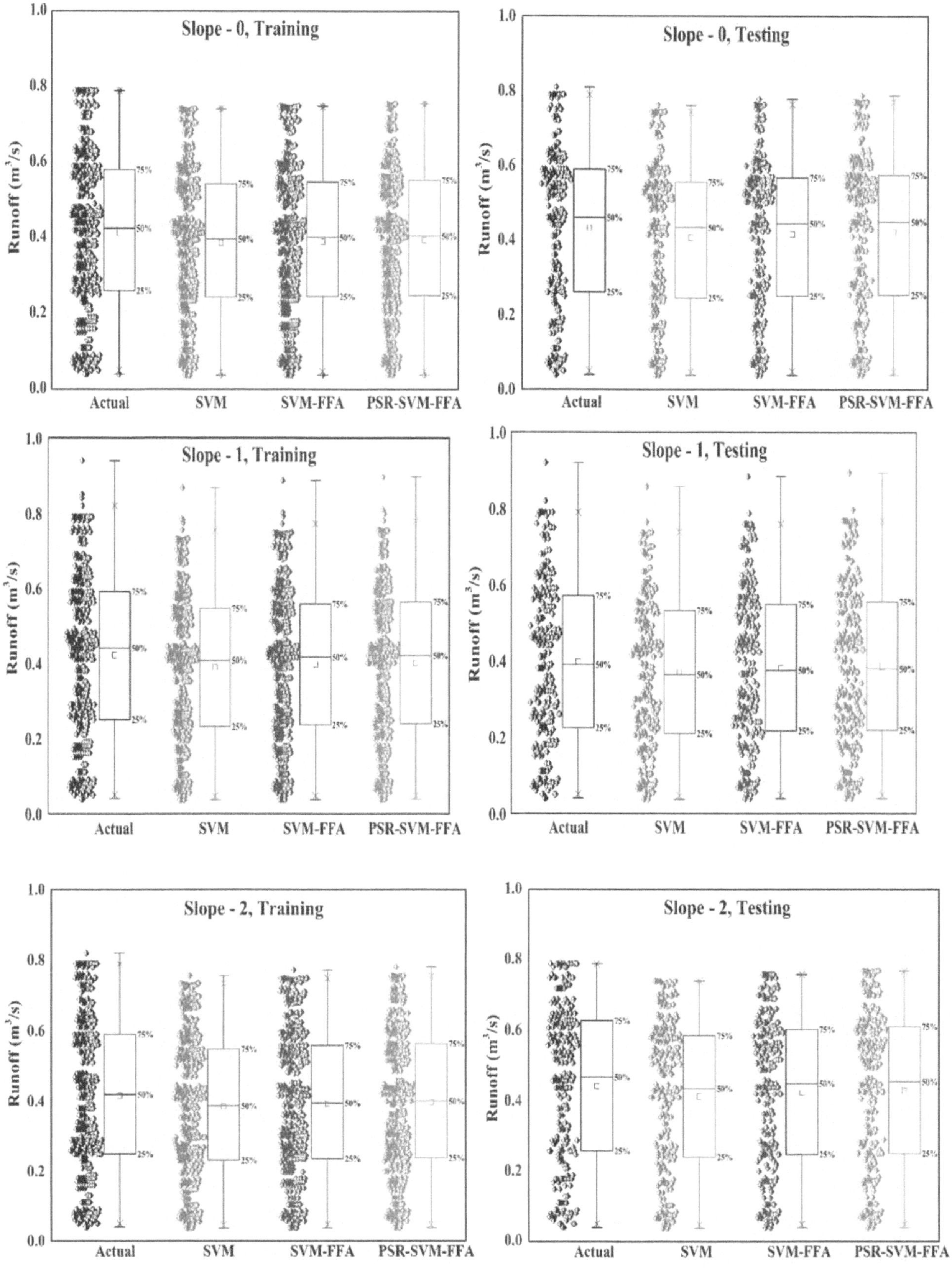

Figure 8 | Box plots of runoff estimates of various ML models and observed runoff. *(continued.).*

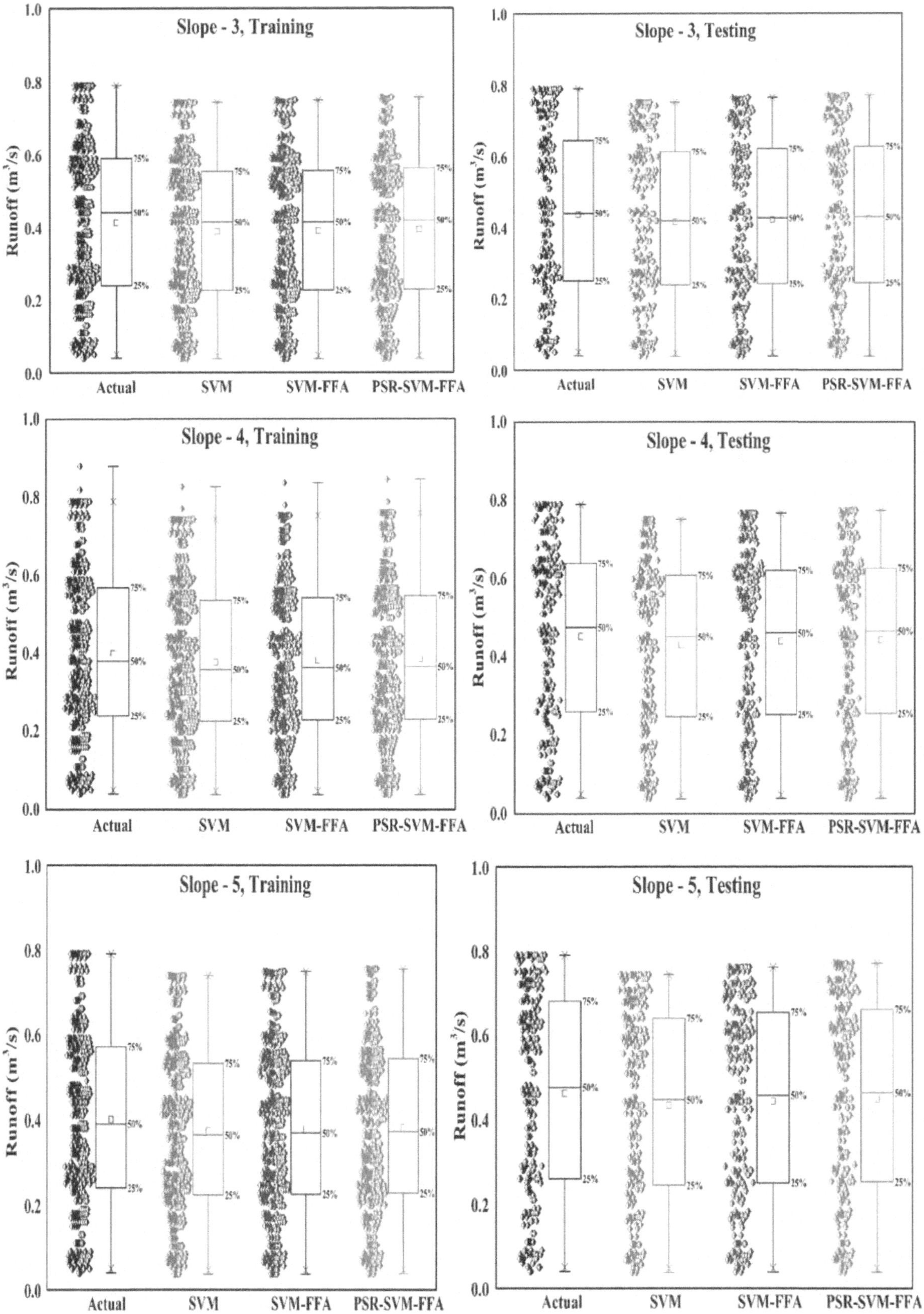

Figure 8 | Continued.

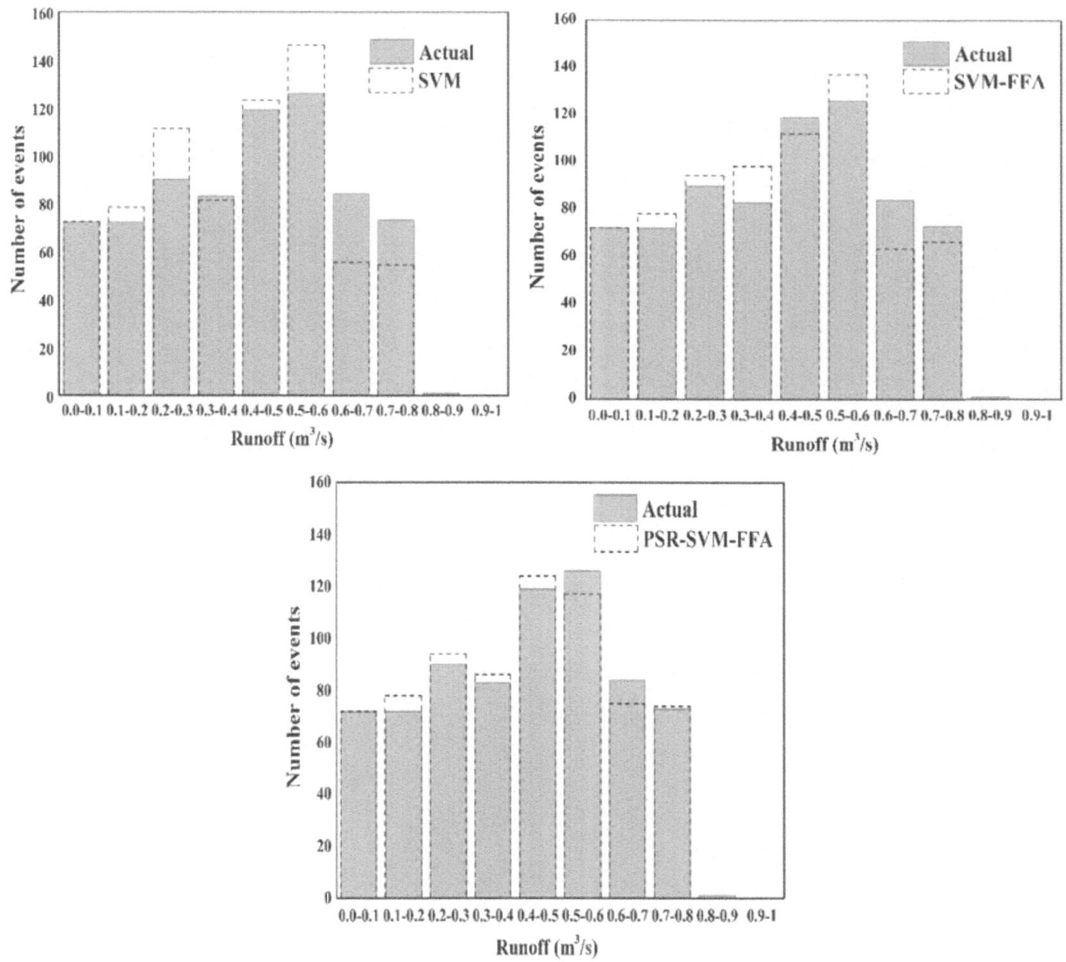

Figure 9 | Histogram plot of actual versus SVM, SVM-FFA, and PSR-SVM-FFA for (a) Slope 0^0, (b) Slope 1^0, (c) Slope 2^0, (d) Slope 3^0, (e) Slope 4^0, and (f) Slope 5^0 condition. *(continued.).*

probability of runoff between a precise range. The estimation of runoff for all slope conditions shows that for the lowest (0.0–0.1 m^3/s) and the highest (0.7–0.8 m^3/s) ranges of runoff. The frequency (number of events) of precise estimations by the PSR-SVM-FFA model illustrates improved agreement with the frequency of observed values than that of the frequency of precise estimations by the SVM-FFA and stand-alone SVM model.

3.4. Assessment of various models

Statistical indicators like NSE, RMSE, and WI are used to evaluate the performance of SVM, SVM-FFA, and PSR-SVM-FFA models for various slope conditions. The assessment of performance measures is indicated in Table 5, which shows the effectiveness of each model. Runoff evaluation is significant, and hence, methods applied here are important to demonstrate runoff information. It is evident that PSR-SVM-FFA performs the best when compared with SVM-FFA and SVM. The assessment of each model is demonstrated in Figure 10 in terms of linear bar scheme. These experimental results revealed that the PSR-SVM-FFA model obtained more reliable prediction runoff values and had a clear advantage over other proposed models in runoff prediction. Hence, the proposed PSR-SVM-FFA is fit for the prediction intervals of a certain period in the runoff series and provides supporting information for decision-makers in water resources management.

In the past, many investigations on runoff processes were not appropriate and accessible for an experimental set-up. It is understood that the findings of runoff prediction from the experimental study will provide a new horizon of understanding the rainfall–runoff process precisely. The PSR-SVM-FFA model is established to be a suitable model for runoff forecasting in an arid watershed. The research was carried out over various slopes of watershed, which will help non-gauging arid watershed

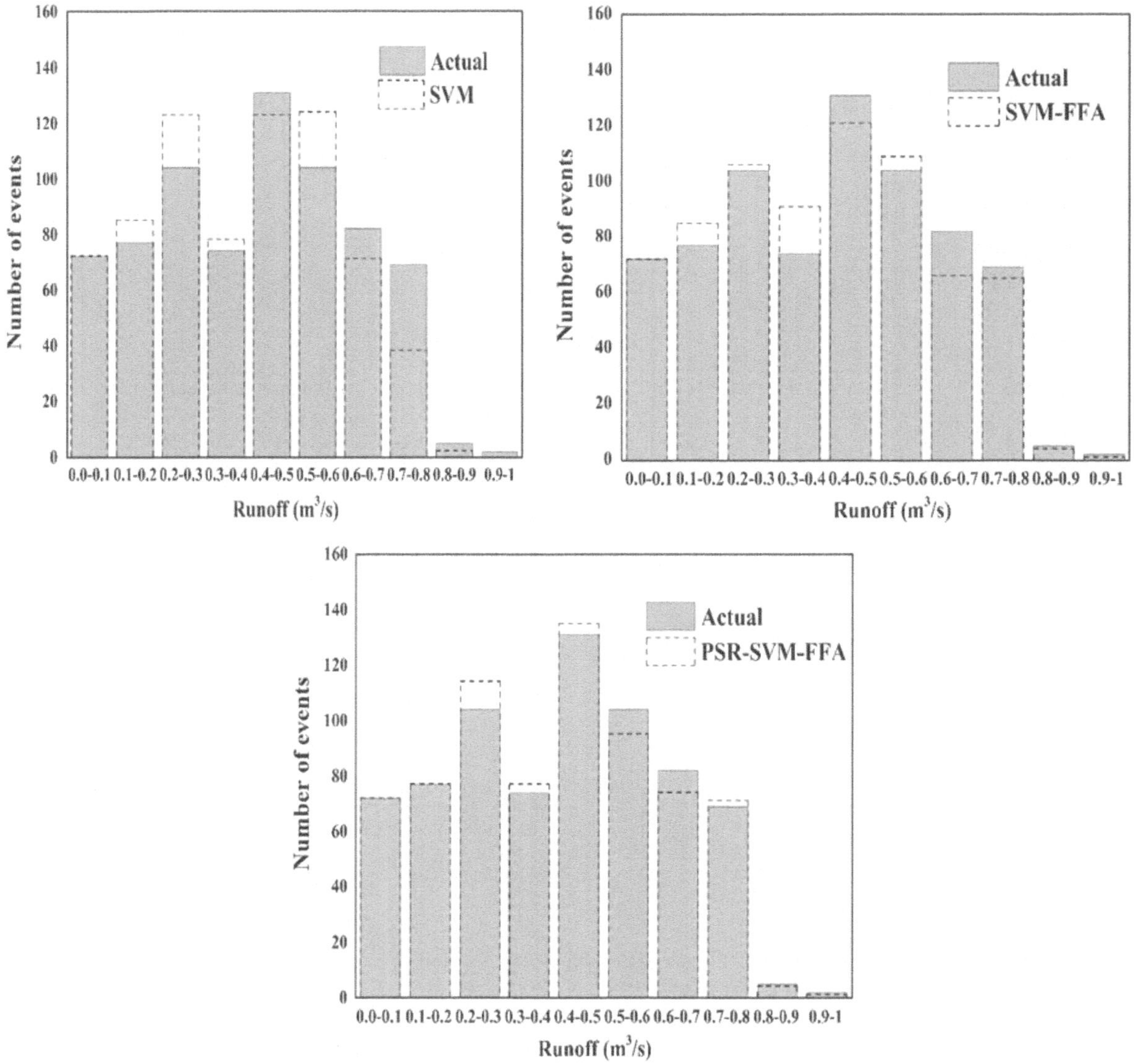

Figure 9 | Continued.

for wide-spread good performance to predict runoff in similar conditions. Also, the present research findings can be used for irrigation and water resources engineering to design a hydraulics structure for agriculture. Basically, this study will be helpful where runoff measurement is impossible or simply for un-gauged catchments. In that situation, if a researcher/scientist knows the watershed/region slope, rainfall amount, and type of soil available there, then, with the help of RS, runoff data are generated, which will be helpful for runoff prediction. In reference to the objective of the present study (which is to present a newly developed method of runoff prediction), it is indicated that at least one case study is needed to verify the applicability of the presented model, and hence, others might apply for other scenarios. Similarly, the proposed approach can be applied to two or more different case studies for observing the validity and performance of models under different circumstances.

The advantages of proposed data-driven models are that all models attained suitable simulation outcomes; yet, their modelling complexities and costs vary. Conventional models necessitate a huge volume of input data; hence, data pre-processing, distribution, and validation processes are restricted by unsolved data assortment. In addition, conventional models utilize formulae for imitating certainty; however, imitation can never precisely match certainty. In general, simplification of real-world presents ambiguity and drops the accuracy of results. Artificial intelligence models have flexible model architectures and controlling input data parameters, which accelerate modelling procedures by reducing modelling uncertainty. Hybrid data-driven models also attained better performances, with much simpler procedures and at a lower cost. In brief, the PSR-SVM-FFA

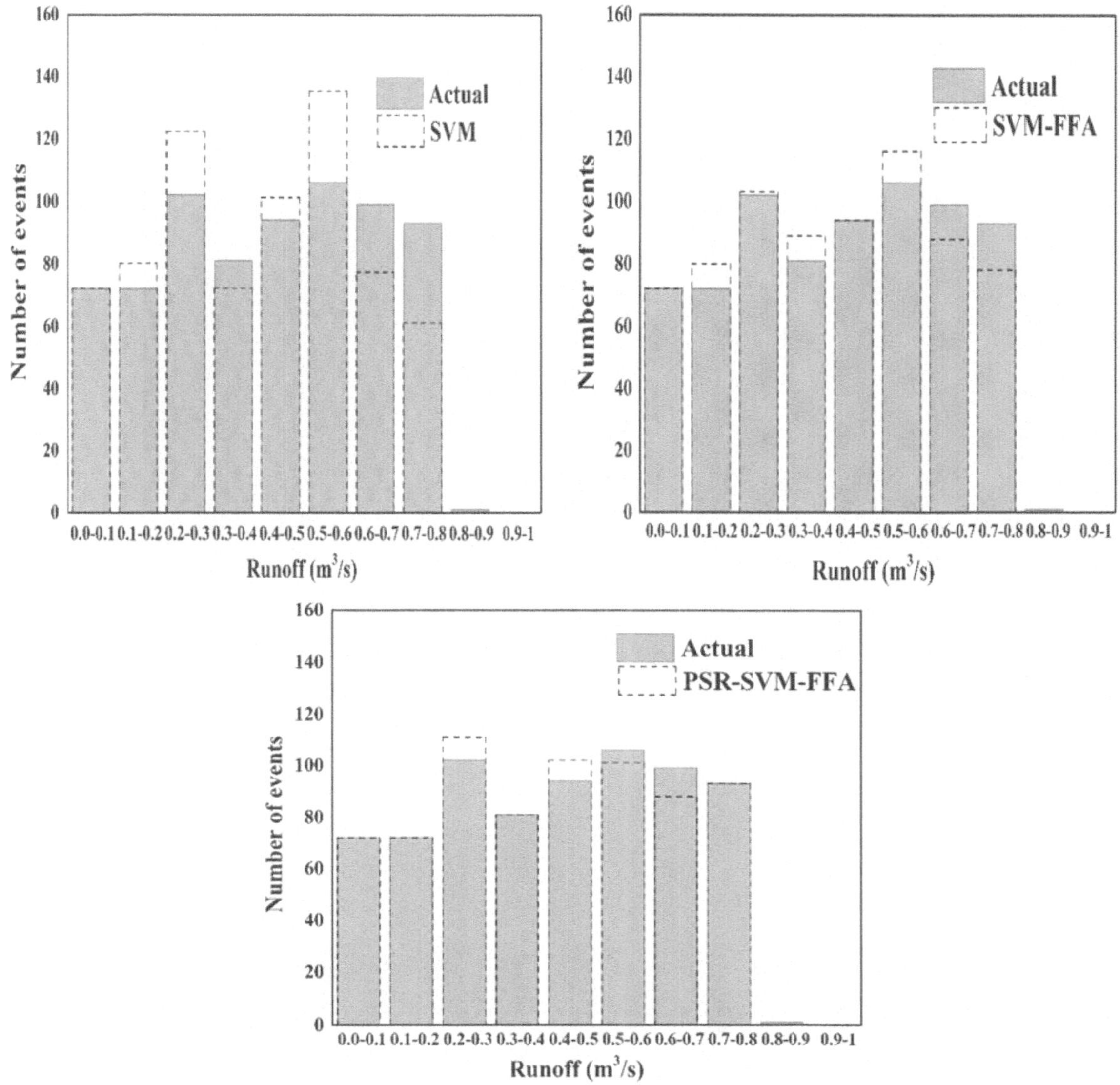

Figure 9 | Continued.

Table 5 | Performance index of various approaches for runoff prediction

	SVM			SVM-FFA			PSR-SVM-FFA		
Watershed bed slope	**NSE**	**RMSE**	**WI**	**NSE**	**RMSE**	**WI**	**NSE**	**RMSE**	**WI**
Slope 0^0	0.661	73.16	0.9412	0.791	69.75	0.9668	0.878	60.38	0.9757
Slope 1^0	0.625	73.86	0.9344	0.766	70.58	0.9606	0.836	63.21	0.9731
Slope 2^0	0.628	73.79	0.9378	0.768	69.97	0.9611	0.854	62.49	0.9742
Slope 3^0	0.667	69.01	0.9516	0.799	68.35	0.9684	0.889	59.23	0.9767
Slope 4^0	**0.669**	**69.59**	**0.9518**	**0.802**	**67.74**	**0.9716**	**0.904**	**55.56**	**0.9812**
Slope 5^0	0.656	73.27	0.9408	0.787	69.93	0.9632	0.865	61.65	0.9751

model is simpler and more appropriate. Previous studies have all emphasized the high efficiency and accuracy of the various hybrid models (Van *et al.* 2020; Han & Morrison 2021; Ouma *et al.* 2021; Reddy *et al.* 2021) in runoff prediction. In the current study, it was shown that the proposed PSR-SVM-FFA approach found more efficacy and superiority.

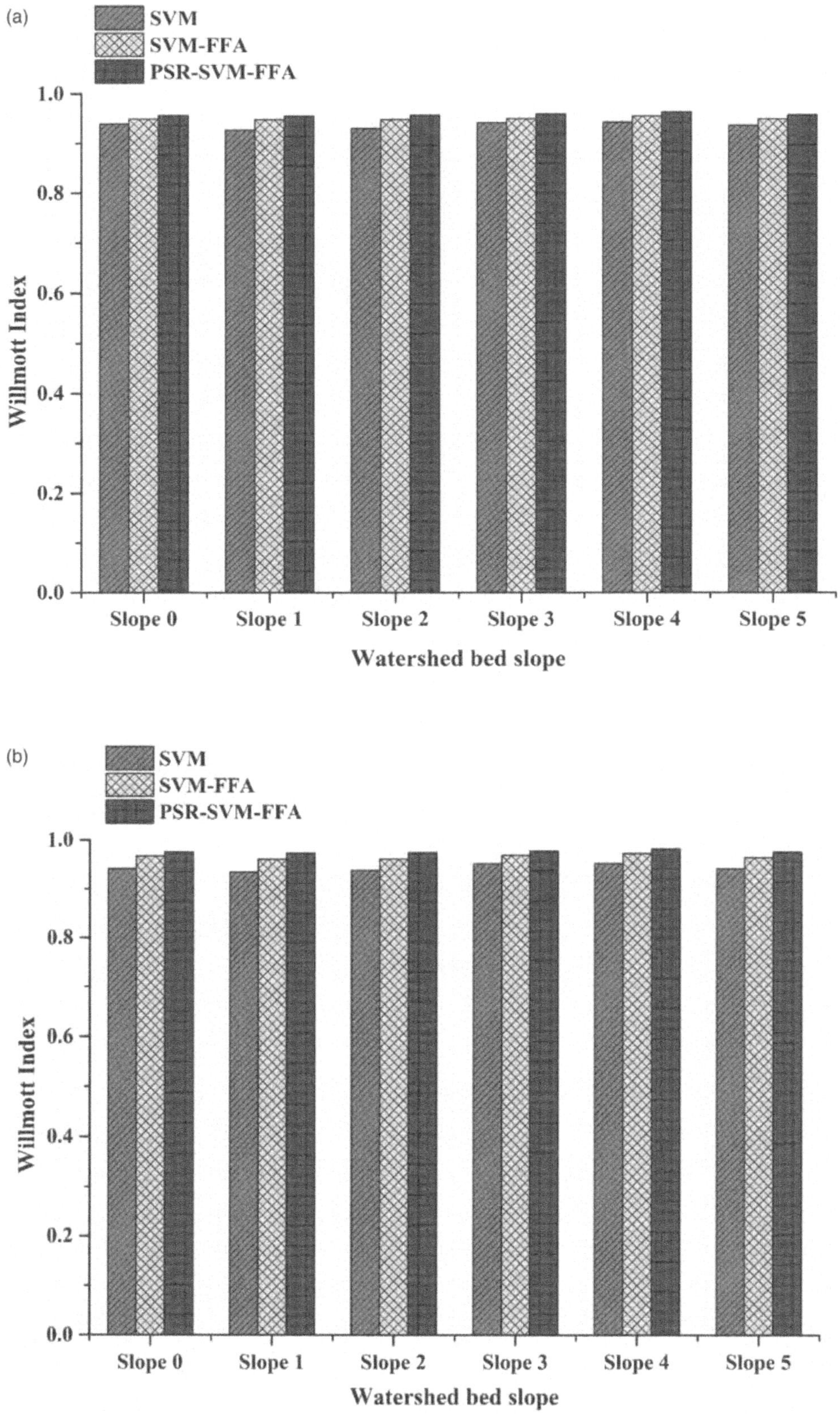

Figure 10 | Comparison of the performance index for the proposed watershed (a) training and (b) testing phases.

One limitation of this study is the consideration of experimental set-up within the laboratory room (i.e. at constant room temperature). Future research will include testing of the proposed approach on field study (River gauge station). Also, emphasis should be given to evaluating the usage of additional input variables for hydrological predictions produced from weather data (collected climatological data) utilizing sediment load, groundwater level, and flood modelling, as an alternative prediction performance. Additionally, satellite data can be assimilated as an external instructive characteristic to the prediction matrix where more precise prediction accuracy can be attained.

4. CONCLUSION

The major benefit of an RS is its compressed structure which fits well in a laboratory room for constructing the economic model. In the present study, the potential of the PSR-SVM-FFA model is compared with the SVM-FFA and standard SVM model for runoff prediction, considering distinguished experimental runoff data emerged from RS. Results obtained from this study specify that the PSR-SVM-FFA method is helpful for modelling rainfall–runoff series that consist of six different slopes of watershed, which gives good prediction performance compared to other proposed models. Outcomes of various performance indices demonstrate that PSR-SVM-FFA has a better ability for non-linear complex mapping. The results indicate a minimal change in the amount of predicted runoff volume and peaks, at different rainfall intensities. This study recommends that the proposed hybrid method can be adopted for modelling other hydrological parameters. Also, this study can be a basis for future studies in different parametric conditions like infiltration and evapotranspiration losses. Application of other robust data-driven models seeking runoff prediction can also be considered for further analysis.

FUNDING

The research has not been supported through any funds.

CONFLICTS OF INTEREST

The authors declare that they have no conflict of interest.

CONSENT TO PARTICIPATE

The authors have made a significant contribution to this manuscript, have seen, and approved the final manuscript.

AUTHORS' CONTRIBUTIONS

S. S.: Conceptualization, Roles/Writing – original draft, Methodology, Formal analysis, Experiment, Data curation, Formal analysis, Writing – review and editing. D. K. G.: Visualization, Project administration.

DATA AVAILABILITY STATEMENT

Data cannot be made publicly available; readers should contact the corresponding author for details.

REFERENCES

Abudi, I., Carmi, G. & Berliner, P. 2012 Rainfall simulator for field runoff studies. *Journal of Hydrology* **454–455**, 76–81.

Akbarzadeh, A., Mehrjardi, R. T., Rouhipour, H., Gorji, M. & Rahimi, H. G. 2009 Estimating of soil erosion covered with rolled erosion control systems using rainfall simulator (neuro-fuzzy and artificial neural network approaches). *Journal of Applied Sciences Research* **5** (5), 505–514.

Aksoy, H., Unal, N. E., Cokgor, S., Gedikli, A., Yoon, J., Koca, K., Inci, S. B. & Eris, E. 2012 A rainfall simulator for laboratory-scale assessment of rainfall-runoff-sediment transport processes over a two-dimensional flume. *Catena* **98**, 63–72.

Behzad, M., Asghari, K., Eazi, M. & Palhang, M. 2009 Generalization performance of support vector machines and neural networks in runoff modeling. *Expert Systems with Applications* **36** (4), 7624–7629.

Bozorg-Haddad, O., Aboutalebi, M., Ashofteh, P. S. & Loáiciga, H. A. 2018 Real-time reservoir operation using data mining techniques. *Environmental Monitoring and Assessment* **190** (10), 1–22.

Bray, M. & Han, D. 2004 Identification of support vector machines for runoff modelling. *Journal of Hydroinformatics* **4**, 265–280.

Burgan, H. I. & Aksoy, H. 2018 Annual flow duration curve model for ungauged basins. *Hydrology Research* **49** (5), 1684–1695.

Cea, L., Garrido, M. & Puertas, J. 2010 Experimental validation of two-dimensional depth-averaged models for forecasting rainfall-runoff from precipitation data in urban areas. *Journal of Hydrology* **382**, 88–102.

Delafrouz, H., Ghaheri, A. & Ghorbani, M. A. 2017 A novel hybrid neural network based on phase space reconstruction technique for daily river flow prediction. *Soft Computing* **22** (7), 2205–2215.

Elhakeem, M. & Papanicolaou, A. N. 2009 Estimation of the runoff curve number via direct rainfall simulator measurements in the state of Iowa, USA. *Water Resources Management* **23**, 2455–2473.

Fister, W., Iserloh, T., Ries, J. B. & Schmidt, R. 2012 A portable wind and rainfall simulator for *in situ* soil erosion measurements. *Catena* **91**, 72–84.

Fister, I., Fister, I., Yang, X. S. & Brest, J. 2013 A comprehensive review of firefly algorithms. *Swarm and Evolutionary Computation* **13**, 34–46.

Granata, F., Gargano, R. & de Marinis, G. 2016 Support vector regression for rainfall-runoff modeling in urban drainage: a comparison with the EPA's storm water management model. *Water (Switzerland)* **8**, 1–13.

Han, H. & Morrison, R. R. 2021 Data-driven approaches for runoff prediction using distributed data. *Stochastic Environmental Research and Risk Assessment* **2021**, 1–19.

Iserloh, T., Ries, J. B., Arnáez, J., Boix-Fayos, C., Butzen, V., Cerdà, A., Echeverría, M. T., Fernández-Gálvez, J., Fister, W., Geißler, C. & Gómez, J. A. 2013 European small portable rainfall simulators: a comparison of rainfall characteristics. *Catena* **110**, 100–112.

Jajarmizadeh, M., Lafdani, E. K., Harun, S. & Ahmadi, A. 2015 Application of SVM and SWAT models for monthly streamflow prediction, a case study in South of Iran. *KSCE Journal of Civil Engineering* **19**, 345–357.

Jimmy, S. R., Sahoo, A., Samantaray, S. & Ghose, D. K. 2021 Prophecy of Runoff in a River Basin Using Various Neural Networks. In: *Communication Software and Networks* (S. C. Satapathy, V. Bhateja, M. R. Murty, N. G. Nhu & J. Kotti eds.), Vol. 134. Springer, Singapore, pp. 709–718.

Jin, K., Cornelis, W. M., Gabriels, D., Baert, M., Wu, H. J., Schiettecatte, W., Cai, D. X., De Neve, S., Jin, J. Y., Hartmann, R. & Hofman, G. 2009 Residue cover and rainfall intensity effects on runoff soil organic carbon losses. *Catena* **78** (1), 81–86.

Lekscha, J. & Donner, R. V. 2018 Phase space reconstruction for non-uniformly sampled noisy time series. *Chaos: An Interdisciplinary Journal of Nonlinear Science* **28** (8), 085702.

Liu, Q., Islam, S., Rodriguez-Iturbe, I. & Le, Y. 1998 Phase-space analysis of daily streamflow: characterization and prediction. *Advances in Water Resources* **21**, 463–475.

Mehr, A. D., Nourani, V., Khosrowshahi, V. K. & Ghorbani, M. A. 2019 A hybrid support vector regression–firefly model for monthly rainfall forecasting. *International Journal of Environmental Science & Technology* **16**, 335–346.

Misra, D., Oommen, T., Agarwal, A., Mishra, S. K. & Thompson, A. M. 2009 Application and analysis of support vector machine based simulation for runoff and sediment yield. *Biosystems Engineering* **103** (4), 527–535.

Mohanta, N. R., Biswal, P., Kumari, S. S., Samantaray, S. & Sahoo, A. 2021 Estimation of Sediment Load Using Adaptive Neuro-Fuzzy Inference System at Indus River Basin, India. In: *Intelligent Data Engineering and Analytics*, Vol. 1177 (S. C. Satapathy, Y-D, Zhang, V. Bhateja & R. Majhi eds.). Springer, Singapore, pp. 427–434.

Okkan, U. & Serbes, Z. A. 2012 Rainfall-runoff modeling using least squares support vector machines. *Environmetrics* **23**, 549–564.

Ouma, Y. O., Cheruyot, R. & Wachera, A. N. 2021 Rainfall and runoff time-series trend analysis using LSTM recurrent neural network and wavelet neural network with satellite-based meteorological data: case study of Nzoia hydrologic basin. *Complex & Intelligent Systems* **2021**, 1–24.

Ouyang, Q., Lu, W., Xin, X. & Zhang, Y. 2016 Monthly rainfall forecasting using EEMD-SVR based on phase-space reconstruction. *Water Resources Management* **30** (7), 2311–2325.

Pérez-Latorre, F. J., de Castro, L. & Delgado, A. 2010 A comparison of two variable intensity rainfall simulators for runoff studies. *Soil and Tillage Research* **107**, 11–16.

Reddy, B. S. N., Pramada, S. K. & Roshni, T. 2021 Monthly surface runoff prediction using artificial intelligence: a study from a tropical climate river basin. *Journal of Earth System Science* **130** (1), 1–15.

Sahoo, A., Samantaray, S. & Ghose, D. K. 2021 Prediction of flood in Barak River using hybrid machine learning approaches: a case study. *Journal of the Geological Society of India* **97** (2), 186–198.

Samantaray, S. & Ghose, D. K. 2020 Modelling runoff in a river basin, India: an integration for developing un-gauged catchment. *International Journal of Hydrology Science and Technology* **10** (3), 248–266.

Samantaray, S. & Sahoo, A. 2020a Prediction of runoff using BPNN, FFBPNN, CFBPNN algorithm in arid watershed: a case study. *International Journal of Knowledge-Based and Intelligent Engineering Systems* **24** (3), 243–251.

Samantaray, S. & Sahoo, A. 2020b Estimation of runoff through BPNN and SVM in Agalpur Watershed. In: *Frontiers in Intelligent Computing: Theory and Applications*, Vol. 1014 (S. C. Satapathy, V. Bhateja, B. L. Nguyen, N. G. Nguyen & D-N. Le eds.). Springer, Singapore, pp. 268–275.

Samantaray, S., Sahoo, A. & Ghose, D. K. 2019 Assessment of runoff via precipitation using neural networks: watershed modelling for developing environment in arid region. *Pertanika Journal of Science and Technology* **27** (4), 2245–2263.

Samantaray, S., Sahoo, A., Mohanta, N. R., Biswal, P. & Das, U. K. 2021 Runoff Prediction Using Hybrid Neural Networks in Semi-Arid Watershed, India: A Case Study. In: *Communication Software and Networks*, Vol. 134 (S. C. Satapathy, V. Bhateja, M. R. Murty, N. G. Nhu & J. Kotti eds.). Springer, Singapore, pp. 729–736.

Sedighi, F., Vafakhah, M. & Javadi, M. R. 2016 Rainfall–runoff modeling using support vector machine in snow-affected watershed. *Arabian Journal for Science and Engineering* **41**, 4065–4076.

Sharma, N., Zakaullah, M., Tiwari, H. & Kumar, D. 2015 Runoff and sediment yield modeling using ANN and support vector machines: a case study from Nepal watershed. *Modeling Earth Systems and Environment* **1** (3), 1–8.

Sivakumar, B., Berndtsson, R. & Persson, M. 2001 Monthly runoff prediction using phase space reconstruction. *Hydrological Sciences Journal* **46**, 377–387.

Sivapragasam, C., Liong, S. Y. & Pasha, M. F. K. 2001 Rainfall and runoff forecasting with SSA-SVM approach. *Journal of Hydroinformatics* **3**, 141–152.

Soleymani, S. A., Goudarzi, S., Anisi, M. H., Hassan, W. H., Idris, M. Y. I., Shamshirband, S., Noor, N. M. & Ahmedy, I. 2016 A novel method to water level prediction using RBF and FFA. *Water Resources Management* **30** (9), 3265–3283.

Sun, W. & Wang, Y. 2018 Short-term wind speed forecasting based on fast ensemble empirical mode decomposition, phase space reconstruction, sample entropy and improved back-propagation neural network. *Energy Conversion and Management* **157**, 1–12.

Tao, H., Sulaiman, S. O., Yaseen, Z. M., Asadi, H., Meshram, S. G. & Ghorbani, M. A. 2018a What is the potential of integrating phase space reconstruction with SVM-FFA data-intelligence model? Application of rainfall forecasting over regional scale. *Water Resources Management* **32** (12), 3935–3959.

Tao, H., Diop, L., Bodian, A., Djaman, K., Ndiaye, P. M. & Yaseen, Z. M. 2018b Reference evapotranspiration prediction using hybridized fuzzy model with firefly algorithm: regional case study in Burkina Faso. *Agricultural Water Management* **208**, 140–151.

Van, S. P., Le, H. M., Thanh, D. V., Dang, T. D., Loc, H. H. & Anh, D. T. 2020 Deep learning convolutional neural network in rainfall–runoff modelling. *Journal of Hydroinformatics* **22** (3), 541–561.

Vapnik, V. 1998 *Statistical Learning Theory*. John Wiley & Sons, Toronto.

Wang, D., Luo, H., Grunder, O. & Lin, Y. 2017 Multi-step ahead wind speed forecasting using an improved wavelet neural network combining variational mode decomposition and phase space reconstruction. *Renewable Energy* **113**, 1345–1358.

Wildhaber, Y. S., Bänninger, D., Burri, K. & Alewell, C. 2012 Catena Evaluation and application of a portable rainfall simulator on subalpine grassland. *Catena* **91**, 56–62.

Xu, L., Zhao, J., Li, C., Li, C., Wang, X. & Xie, Z. 2020 Simulation and prediction of hydrological processes based on firefly algorithm with deep learning and support vector for regression. *International Journal of Parallel, Emergent and Distributed Systems* **35** (3), 288–296.

Yang, X.-S. 2010 Firefly algorithm, stochastic test functions and design optimisation. *International Journal of Bio-Inspired Computation* **2**, 78–84.

First received 4 June 2021; accepted in revised form 11 October 2021. Available online 28 October 2021

Impact of Climate Change on Hydrology and Water Resources doi: 10.2166/wcc.2021.246

Runoff modeling in Kolar river basin using hybrid approach of wavelet with artificial neural network

Deepak Kumar Tiwari *, Hari Lal Tiwari and Raman Nateriya

Department of Civil Engineering, Maulana Azad National Institute of Technology, Bhopal 462003, Madhya Pradesh, India
*Corresponding author. E-mail: deepaktiwaricivil7sept@gmail.com

DKT, 0000-0002-9253-1153; HLT, 0000-0002-6817-6999

ABSTRACT

In this paper, the Kolar River watershed, Madhya Pradesh is taken as the study area. This study area is located in Narmada River in Central India. The data set consists of monthly rainfall of three meteorological stations, Ichhawar, Brijesh Nagar, and Birpur rainfall stations from 2000 to 2018, runoff data at Birpur and temperature data of Sehore district. In this paper, radial basis function neural network models have been studied for generation of rainfall–runoff modeling along with wavelet input and without wavelet input to the RBF neural network. A total of 15 models were developed in this experiment based on various combinations of inputs and spread constant of RBF model. The evaluation criteria for the best models selected are based on R^2, AARE, and MSE. The best predicting model among the networks is model 8, which has input of R(t-1), R(t-2), R(t-3), R(t-4), and Q(t-1). For the RBFNN model, the maximum value of R^2 is 0.9567 and the lowest values of AARE and MSE are observed. Similarly, for the WRBFNN model, the maximum value of R^2 is 0.9889 and the lowest values of AARE and MSE are observed. WRBF performs better than RBF with any data processing techniques which shows the proposed model possesses better predictive capability.

Key words: ANN, rainfall–runoff modeling, RBF, watershed management, wavelets

HIGHLIGHTS

- Fifteen ANN model used for analysis.
- New data processing technique of wavelets used.
- Analysis of Kolar river basin, main tributary of Narmada in Central India.
- Radial basis function used for modeling.
- Suitable method for data scarce region and semiarid environment.

INTRODUCTION

Rainfall–runoff modeling is a versatile tool for water resource planning and management, city planning, flood, land use, etc. (Ghumman *et al.* 2011). It also helps in minimizing the effect of drought-related issues on water resources. Owing to change in climatic conditions in recent decades because of global warming, the hydrological cycle in India has been adversely affected (Sonali & Nagesh Kumar 2016a) and due to anthropogenic activities there is an increase in global surface temperature which is clear from various evidence (Sonali & Nagesh Kumar 2016b). Many hydrological models have been developed since 1850, ranging from mathematical relations between them, empirical models, conceptual ones in which the physical processes are based on physical laws existing in nature and physical models which are small-scale prototypes of the models (Todini 2007). Conceptual and physical models account for all the physical processes involved in the catchment but they are very data-intensive and time-consuming (Sušanj *et al.* 2016). These types of models are not suitable for areas which suffer from data scarcity and poorly managed data sites. General time series models such as auto regressive integrated moving average (ARIMA) are popularly used for hydrological time series forecasting (Nourani *et al.* 2009), but Mujumdar & Kumar (1990) suggested that ARIMA should be avoided due to an increase in variance continuously on differencing the series. Also, these models are linear in nature and do not account for the non-stationarities and non-linearities in hydrologic time series data. ARIMA is commonly used for the hydrological time series data which have periodicity over time (Zhang *et al.* 2011). Moreover, seasonal variability is also one of the climatic factors for runoff variation (Bekele *et al.* 2021).

Nowadays, ANN models are popularly used to develop the rainfall–runoff (RR) relationship and they are black-box models which are data-driven and give the relation between rainfall and runoff without considering the physical explanation of processes involved (Todini 2007). These are mathematical models with repetitive iterations which help in the development of some non-linear relationships between two hydrological phenomena, rainfall and runoff (Poonia & Tiwari 2020). They also do not require prior knowledge of physical processes and morphometry of the basin for prediction. ANN consists of three layers: input layer, hidden layer, and output layer (Figure 1). The input layer consists of the number of nodes equal to input parameters. The hidden layer consists of the mathematical formula for processing of inputs to find which matches best to parameters which are to be predicted. Many times, the input data non-stationarity or dynamic space time variability for data makes it difficult to predict (Sharghi *et al.* 2018). Also, predicting the capability of hybrid models, data processing techniques combined with ANN gives better results (Barman & Bhattacharjya 2020). Among the publications regarding ANN used for runoff prediction, most of them have focused on back propagation neural networks (Asadi *et al.* 2013). Multilayer perceptron (MLP) and radial basis function (RBF) are the two most widely used neural networks in rainfall runoff analysis (Phukoetphim *et al.* 2016; Tayyab *et al.* 2016; Shoaib *et al.* 2018; Poonia & Tiwari 2020), but RBFNN techniques provide better solutions as compared to MLP methods (Kumar & Yadav 2011). Moreover, data quality is key to better predicting capability of the model. In recent years, wavelet transforms have emerged as an excellent tool for predicting (Krishna *et al.* 2011; Badrzadeh *et al.* 2015; Alizadeh *et al.* 2017a). Wavelet transform reduces the data noising in natural data and soothes data range by driving data to multiresolution and on different scales.

Radial basis function models have been used for flow forecasting during the last two decades using Gaussian function for network processing (Dawson *et al.* 2002). Lin & Chen (2004) applied the RBF network for runoff analysis for hourly prediction and successfully applied it for determining the complex relationship between rainfall and runoff. Later, a study on the internal functioning of RBF and its hydrological interpretation was conducted (Achela *et al.* 2009). The results of the study show that a single hidden layered RBFNN is an effective tool to forecast daily flows and that the activations of the hidden layer nodes are far from arbitrary, but appear to represent flow components of the predicted hydrograph. Miaoli *et al.* (2020) used the Levenberg–Marquardt (LM) algorithm for training RBF neural networks which was really an innovative approach to RBF networks. It was found that the LM algorithm makes a network increase in efficiency by carrying out more error value, improves convergence speed, reduces the storage space, and can be successfully applied to problems in various fields. Poonia & Tiwari (2020) also applied the RBF network for runoff simulation. However, with wavelet processed input data, RBFNN models are one of the innovative approaches attempted in this study. Researchers have identified wavelets as excellent data processing techniques to increase the efficiency of any network (Nourani *et al.* 2011).

In this study, RBF, a branch of ANN, has been employed to predict the RR relationship for Kolar River, Madhya Pradesh, India. Also, to counter the non-stationarity, wavelet preprocessing is performed to input data and compared with simple RBF models. A total of 15 models based on input combinations of runoff, rainfall, and temperature have been developed for RBF models and wavelet RBF (WRBF) and compared based on three evaluation criteria, namely, coefficient of determination (R^2), mean square error (MSE), and root mean square error (RMSE).

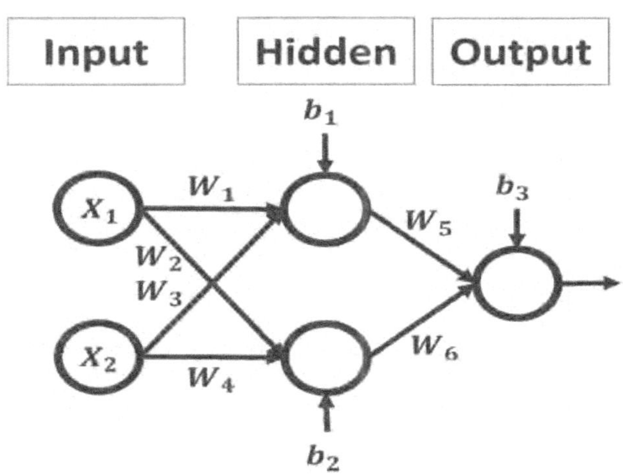

Figure 1 | A general schematic view of the three-layered artificial neural network.

STUDY AREA AND DATA SET

The study area is Kolar river catchment, which rises in the Vindhya range of Sehore district and flows in a south-westerly direction to meet the Narmada near Nasrullahganj in the Raisen district of Madhya Pradesh. The Kolar river basin has an upper elevation of 600 m, elevation of 432 m at the dam and downwards until Narmada, and 350 m at the downstream. The main purpose of the Kolar dam is to supply water for Bhopal city, irrigation for Jholiapur area of Raisen district, and fishing activities. It is also a tourist attraction for people nearby. Kolar catchment is situated in the Sehore district in the state of Madhya Pradesh, India. The salient features for the catchment are given in Table 1, and Figure 2 presents the Kolar catchment geographical map.

Data collection

The present study uses the daily data of rainfall for 30 years from 1988 to 2018 for three sites, namely, Birpur, Brijesh Nagar, and Ichhawar, in millimeters. Also, average daily data of temperature is used for the same period in degrees Celsius. Data of average daily discharge in cubic meters per second were available from October 1999 to October 2018 (19 years and 3 months) and provided by Kolar dam authority, a state government body associated with the management of the dam. Figure 3 shows the daily rainfall measurements of three rainfall stations, Birpur, Brijesh Nagar, and Ichhawar, in millimeters. Here, places with high spatiotemporal rainfall variability, such as mountain regions as in Kolar region, have a variety of slopes from very steep to mild, thus, input data can be a large source of uncertainty in hydrological modeling. Modeling of mountain regions is highly uncertain as it contains data of spatially distributed rainfall and also depends on the density of rain gauge stations. For the temperature data set, the nearest daily data were available at Birpur site. These temperature data were validated with the nearest available data at Bhopal airport, which is about 30 km from the site.

METHODOLOGY

In this study, the RBFNN model is used for RR modeling of Kolar river basin, located around 20 km from Bhopal, the capital city of Madhya Pradesh state, India. First, daily data for rainfall, runoff, and temperature were converted into monthly scale and then normalized in the range of 0.1 to 0.9 to overcome any abnormalities due to data range (Seo *et al.* 2015). After normalization of input, the RBFNN model is applied with 15 different combinations of inputs. Again, the input data are post-processed with wavelet transformation application and the RBFNN model is used for RR modeling. This network is termed the WRBFNN model. These models are less complex, with a lower number of input nodes, single hidden layers of ten neurons, and output layer with a single neuron. Data required for the model are divided into 70, 15, and 15% for training, testing, and validation stages as standard percentages given in MATLAB. Other researchers have used different percentages of 60, 20, and 20% also (Nourani & Komasi 2013).

Normalization

Normalization of input and output data safeguards enhanced computation efficiency, data integrity, and redundancy elimination (Nawi *et al.* 2013). Before applying rainfall, discharge, and temperature data to the models, data were normalized between 0.1 and 0.9 to remove any error caused due to null value of normalized data as given by Equation (1):

$$X_n = \frac{X_i - X_{min}}{X_{max} - X_{min}} \tag{1}$$

where X_n is the normalized value of the data set, X_i is the actual value of the data set, and X_{min} and X_{max} are the minimum and maximum values of the data set, respectively.

Table 1 | Characteristics of the three weather stations

Station	Latitude	Longitude	Elevation (m)	Avg Tmax (c)	Avg Tmin (c)	Rainfall (mm)	Climate
Birpur	22°58′	77°20′	441	31.88	18.76	940	Semi-arid
Brijesh Nagar	22°57′	77°08′	505	30.66	16.45	1,370	Humid
Ichhawar	23°01′	77°01′	515	32.24	17.8	1,032	Dry – Sub humid

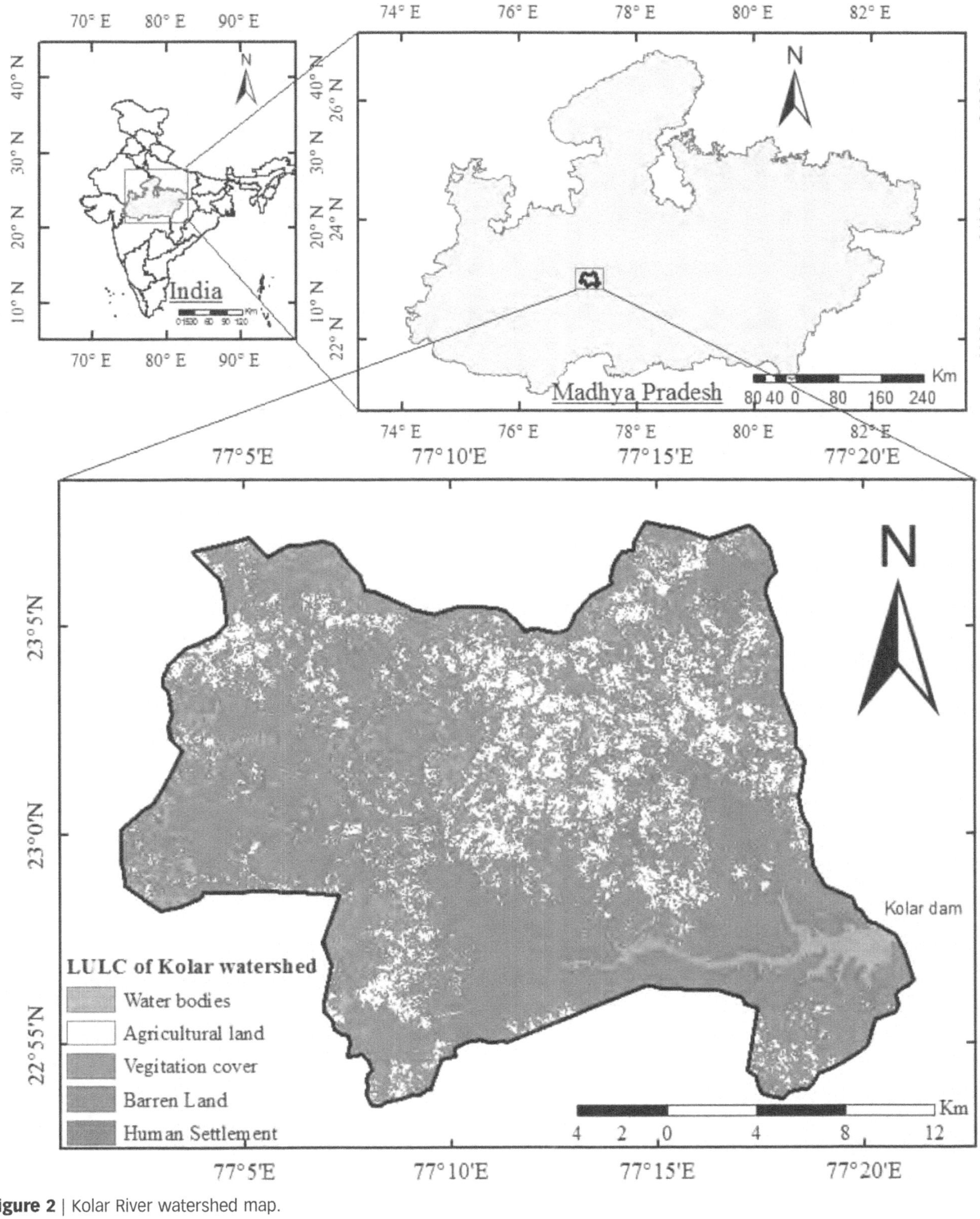

Figure 2 | Kolar River watershed map.

Radial basis function NN (RBFNN)

The radial basis function network (RBFN) has been used for non-linear pattern recognition because of its simple structural arrangement and its ability to predict how learning proceeds in an explicit manner (Shamseldin 2009). RBFN was initially introduced to solve the real multi-variable interpolation problem (Powell 1987). Broomhead & Lowe (1988) proposed the

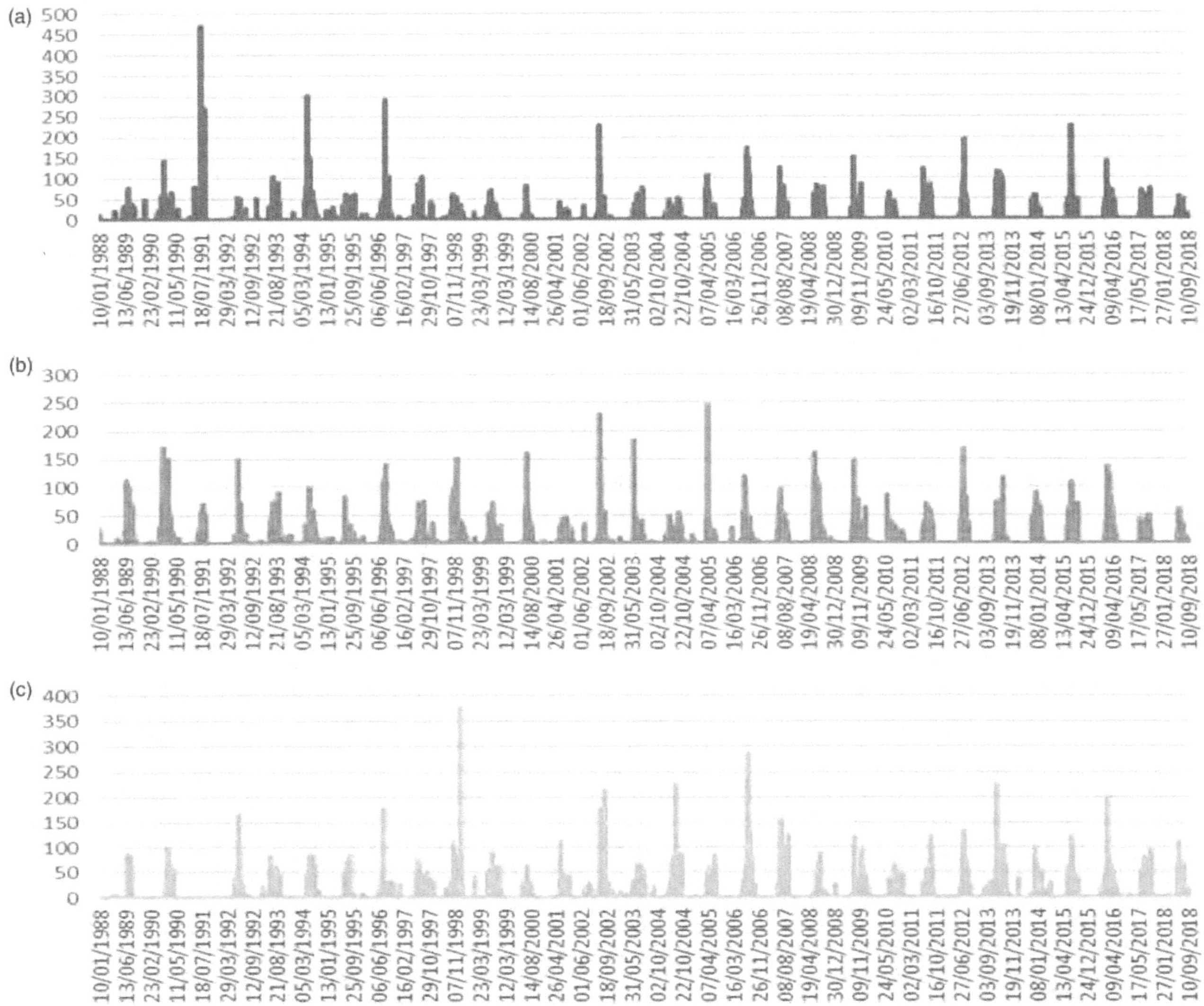

Figure 3 | Daily rainfall data of (a) Birpur, (b) Brijsh Nagar, and (c) Ichhawar rain gauge stations.

RBF network which consists of three layers: an input layer, a hidden layer, and an output layer. A general RBF model is shown in Figure 4. In this network, the quantity of input and output nodes depends upon the number of input and output variables. X is the input variable and Y is the output variable. φ is radial basis function and c is the center of each hidden node, corresponding to observed input data. Mostly, sigmoidal and Gaussian kernel functions are used in RBF systems. This is represented by Equation (2):

$$Y = \sum_{i=1}^{n} W\emptyset(IIX - CII) \tag{2}$$

where W is weights associated with each input parameter, φ is radial basis function, c is the center of each hidden node, and n is number range of input variable. This RBF Gaussian function is given as Equation (3):

$$f(y) = e^{-y^2} \tag{3}$$

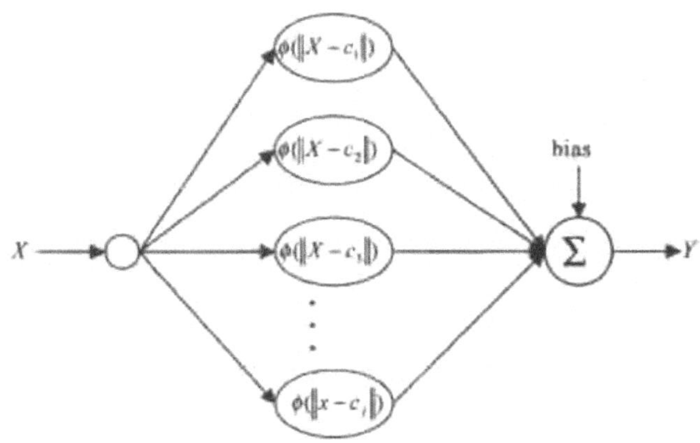

Figure 4 | General structure of RBFNN (Lin & Chen 2004).

Wavelet transforms

Wavelet decomposition is an advanced tool in data processing techniques. Its use has increased nowadays in data handling, signal processing, image processing, and optical engineering (Seo *et al.* 2015). Wavelet transform is used to discover, denoise, and smooth time series. Wavelet transform breaks the original time series into various time series which have sinusoidal wave patterns output and these outputs are used as input for the neural network. The end result will be a collection of time series frequencies which represent the original signal with different scales and resolution. The main advantage of decomposing the time series is that the hidden collective frequency in the time series will be released and various parameters such as variations in mode and its temporal change can be easily identified (Kim *et al.* 2017). It will help tremendously in data synthetization and analysis. Also, here, the time period of data is on a monthly basis and discrete in nature. The wavelet transform of a time series f(t) is defined in Equation (4) as:

$$f(x, y) = \frac{1}{\sqrt{x}} \int_{-\infty}^{\infty} f(t)t\Psi\left(\frac{t-y}{x}\right) dt \tag{4}$$

where $\Psi(t)$ is the basic wavelet having time series length t. The variables are x and y, where x is the scale or dilation factor that determines the characteristic frequency so that its variation gives rise to a 'spectrum'; and y is the translation in time so that its variation represents the 'sliding' of the wavelet over f(t) (Krishna *et al.* 2011). The original signal X(n) passes through two complementary filters (low pass and high pass filters) and emerges as two signals as approximations (A) and details (D). The approximations are the high-scale, low frequency components of the signal. The details are the low-scale, high frequency components. Normally, the low frequency content of the signal (approximation, A) is the most important part. It demonstrates the signal identity (Figure 5). To find out the best decomposition level for optimum performance of the model, levels at 2, 3, and 4 decomposition level were tested and the best performance was found for decomposition level of 4. This is in accordance with Equation (5) (Nourani *et al.* 2014):

$$L = \text{int}\{\log(n)\} \tag{5}$$

where L and n denote decomposition level and data number (time series length), respectively.

Input data preparation

The daily data of rainfall, runoff, and temperature are available from 2000 to 2018. Rainfall, runoff, and temperature data are utilized for developing the RBFNN network. First, 15 ANN models were developed based on input combinations and results were obtained. After that, input variables were normalized and wavelet decomposed at level 3 and these wavelet transforms were used to feed as input signals in the WRBFNN network. This normalization is done due to large variation in values of rainfall and runoff. Moreover, logistic functions (Equation (5)) vary between 0 and 1, due to which this normalization is justified. Data of runoff and rainfall during rainy seasons are only considered for network architecture for better results and used

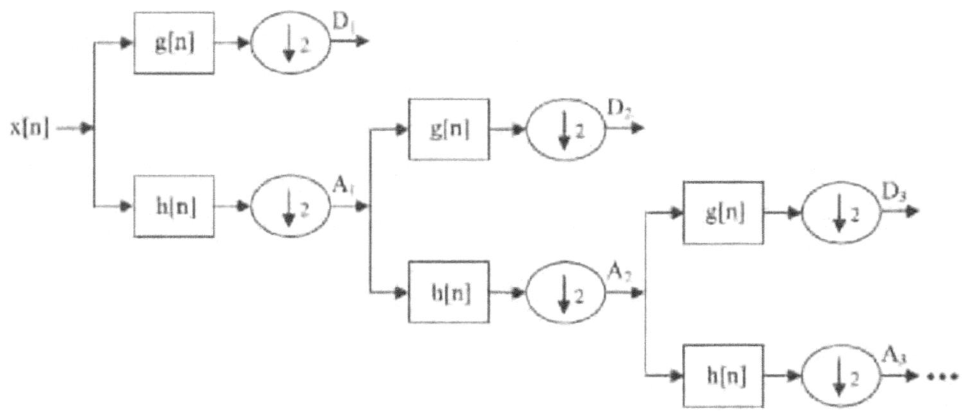

Figure 5 | Diagram of wavelet decomposition of high and low filter pass.

for training, testing, and validation of the neural network, that is, from 15 June to 15 October, which are monsoon months in India.

ANN model architecture

In the present study, 15 models are developed using several input combinations of daily antecedent precipitation of the last 4 days, i.e., R, R(t-1), R(t-2), R(t-3), and R(t-4), where these are precipitation on the same day, previous day, 2 days before, 3 days before, and 4 days before the present day and Q(t-1), Q(t-2), and Q(t-3) are precipitation on the previous day, 2 days before, and 3 days before the present day. Also mean monthly temperature, T(t-1), which is the previous day's temperature, is also used for the network. In the case of RBFNN network models, input combinations are used just after normalization, but in the case of WRBFNN, after normalization, discrete wavelet decompositions of input data up to level 2 are corelated with output and used for feeding as input signals. Table 2 shows the various input combinations for both the RBFNN and WRBFNN networks.

In Table 2, based on various combinations of input variables, a total of 15 models are generated. These models are used for analysis for both the networks. These network variations are based on the trial-and-error method. Other combinations can also be used, but due to time constraints only 15 models are selected.

Table 2 | Model variants for radial basis function neural network (RBFNN) and wavelet radial basis function neural network (WRBFNN)

Model No.	Input combinations	Spread constant	Output layer
M1	R(t-1)	1	Q(t)
M2	R(t-1), R(t-2)	1	Q(t)
M3	R(t-1), R(t-2), R(t-3)	1	Q(t)
M4	R(t-1), R(t-2), R(t-3), R(t-4)	1	Q(t)
M5	R(t-1), Q(t-1)	1	Q(t)
M6	R(t-1), R(t-2), Q(t-1)	1	Q(t)
M7	R(t-1), R(t-2), R(t-3), Q(t-1)	1	Q(t)
M8	R(t-1), R(t-2), R(t-3), R(t-4), Q(t-1)	1	Q(t)
M9	R(t-1), Q(t-1), Q(t-2)	1	Q(t)
M10	R(t-1), R(t-2), Q(t-1), Q(t-2)	1	Q(t)
M11	R(t-1), R(t-2), R(t-3), Q(t-1), Q(t-2)	1	Q(t)
M12	R(t-1), R(t-2), R(t-3), R(t-4), Q(t-1), Q(t-2)	1	Q(t)
M13	R(t-1), R(t-2), R(t-3), R(t-4), Q(t-1), T(t-1)	1	Q(t)
M14	R(t-1), R(t-2), R(t-3), R(t-4), Q(t-1), Q(t-2), T(t-1)	1	Q(t)
M15	R(t-1), R(t-2), R(t-3), R(t-4), T(t-1)	1	Q(t)

Evaluation criteria

Performance of the models is evaluated based on three criteria, namely, coefficient of determination (R^2), average absolute residual error (AARE), and mean square error (MSE), which are given by Equations (6), (7), and (8), respectively:

$$R^2 = \frac{n(\sum Q_{obs}*Q_{sim}) - (\sum Q_{obs})*(\sum Q_{sim})}{\sqrt[\frac{1}{2}]{\left[n(\sum Q_{obs}^2) - (\sum Q_{obs})^2\right] X \left[n(\sum Q_{sim}^2) - (\sum Q_{sim})^2\right]}} \tag{6}$$

$$AARE = \sum_{i=1}^{n} \frac{IQ_{sim} - Q_{obs}I}{n} \tag{7}$$

$$MSE = \frac{1}{n}\sum_{1}^{n}(Q_{sim} - Q_{obs})^2 \tag{8}$$

where Qobs is observed value of discharge and Qsim is simulated value of discharge; n is number of observations. The range of R^2 lies between 0 and 1, which represents no correlation and perfect correlation between observed and simulated value. A lower value of AARE represents greater predictive capability of the model. MSE is calculated at the output layer, and if MSE is within acceptable limits the training of the NN model is then assumed to be completed, otherwise the feed backward pass is continued for updating of synaptic weights between hidden and output layers by using the learning algorithm. The synaptic weights between input and hidden layers can also be updated in an analogous method.

RESULTS AND DISCUSSION

The selection of variables used for input inspires the performance of the rainfall–runoff model. These variables applied for selection in the rainfall–runoff model or network depend on either number of time lagged input variables (Tokar & Markus 2000; Riad & Mania 2004; Wang *et al.* 2017) or input variables based on most correlated variables in lag time series data (Sudheer *et al.* 2002; Ali *et al.* 2010). The authors have applied the first approach to the study due to there being a large number of modeling networks. The daily observed rainfall data of Kolar catchment is transformed by the DWT and applied at level 4. With $N = 11{,}202$ testing/validation rainfall data points in the current study, the input rainfall data can be decomposed into approximations and details. The simulated values of the runoff components, averaged over the length of the data, are shown in Table 3. In the paper, two models, RBFNN and WRBFNN, were compared based on

Table 3 | RBFNN and WRBFNN models' simulation results for given period

	RBFNN			WRBFNN		
Model No.	**R^2**	**AARE**	**MSE**	**R^2**	**AARE**	**MSE**
M1	0.0312	0.0843	0.01501	0.0856	0.0813	0.01463
M2	0.0414	0.0798	0.01311	0.1567	0.0768	0.01273
M3	0.353	0.0753	0.01245	0.4516	0.0723	0.01207
M4	0.4315	0.0716	0.01059	0.5149	0.0686	0.01021
M5	0.0563	0.0811	0.01487	0.2156	0.0781	0.01449
M6	0.1127	0.0804	0.01456	0.3895	0.0774	0.01418
M7	0.6784	0.0489	0.00741	0.7916	0.0459	0.00703
M8	0.9567	0.0356	0.00088	0.9889	0.0097	0.00051
M9	0.5423	0.0512	0.00959	0.6846	0.0482	0.00922
M10	0.6458	0.0498	0.00815	0.7829	0.0468	0.00777
M11	0.7856	0.0417	0.00453	0.8357	0.0387	0.00415
M12	0.9483	0.0388	0.00186	0.9786	0.0114	0.00148
M13	0.8856	0.0401	0.00316	0.9046	0.0371	0.00278
M14	0.8623	0.0395	0.00245	0.8938	0.0365	0.00207
M15	0.3827	0.0738	0.01122	0.5109	0.0708	0.01084

15 model combinations. They are tested at the spread constant of 1 to observe the correlation between output and input parameters. These values are also better as antecedent rainfall. The trend increases in R^2 value until model 4 for both the cases of RBFNN and WRBFNN. For model 5, the value of R^2 decreases for the previous day's antecedent rainfall and previous day's discharge. Again, it started increasing for model 6 onwards until model 8, which has highest R^2 for both the models. However, the value of R^2 in the case of the WRBFNN model is greater than the RBFNN model which shows that use of wavelet transforms of the input in neural networks makes the network more efficient.

In the last three models of the NN network, when temperature is used as input for the best performing model in the first 12 models, the value of coefficient of determination decreased significantly. Therefore, it may be concluded that for this Kolar river watershed, temperature is not a significant factor for prediction of runoff.

This best performing model for both the networks is used for prediction for one-year ahead runoff. Figures 6 and 7 show the regression for predicted value and observed value of both the models.

Figure 6 shows the regression plot of the RBFNN network of model 8. Coefficient of determination between observed and predicted value for this model shows better predicting capability of the model. However, during regression analysis of WRBFNN model 8, better predicting capability has been seen, as is evident from training, testing, and validation of the model. Based on results of both the models, it can be concluded that RBF is an efficient neural network for the rainfall–runoff modeling and the efficiency can be further increased with use of data processing technique of wavelet transforms.

These results were compared with almost similar research work in the recent past, which shows that the present work is also novel and efficient in the field of rainfall–runoff modeling. Alizadeh *et al.* (2017) used the ANN network alone and along

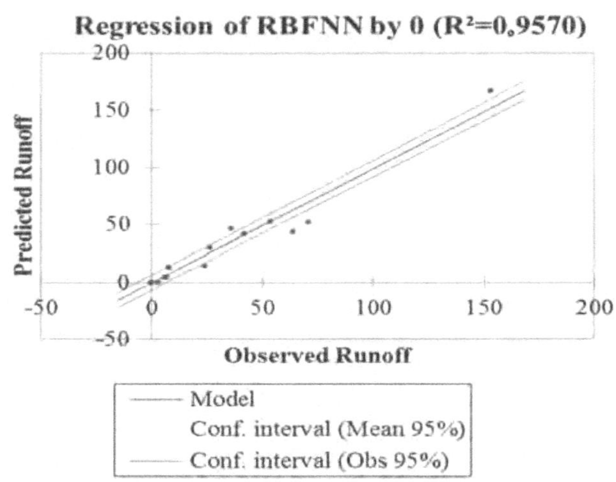

Figure 6 | The runoff prediction for 1 day ahead for the best performing network model of RBFNN scatter plot (runoff in m^3/sec).

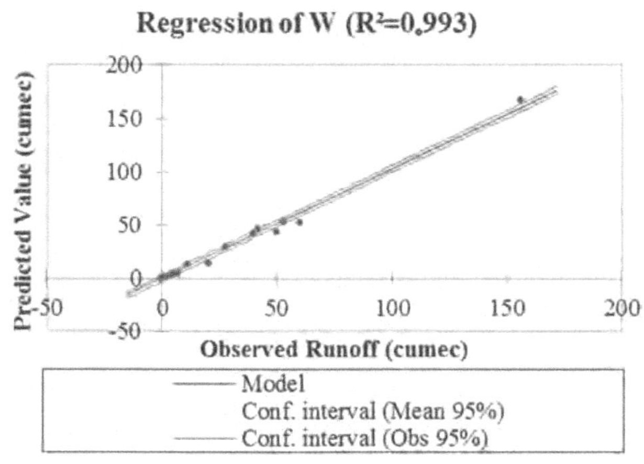

Figure 7 | The runoff prediction for 1 day ahead for the best performing network model of WRBFNN scatter plot (runoff in m^3/sec).

with wavelets (WANN) for rainfall–runoff prediction for Tolt River, Washington, USA. The best performing model in this work has an R^2 value of 0.71 for ANN and 0.97 for WANN. These models were applied on a monthly scale and showed better predicting capability in the multi-model approach. The number of models in the present study is also greater as compared to them. The present models performed better than this model. Poonia & Tiwari (2020) also used feed forward back propagation (FFBP) network and RBF simultaneously and found that RBF performs better, which also supports the results of the findings of the present study.

CONCLUSION

The rainfall–runoff correlation rests on climatic besides physical factors comprising daily variations in precipitation, catchment slope, elevation, land cover, soil humidity, underground water storage, etc. Owing to being dependent on multivariables, the RR relationship shows a complicated non-linear relationship. Many models are available for simulation of relationships but are highly catchment dependent. Owing to its ability to model complicated non-linear relations without any need for a high number of parameters, the RBF model is proposed for Kolar catchment and, to remove bias in data signals, wavelets are proposed for data processing of inputs. In this paper, a new prediction model for rainfall–runoff modeling is proposed which has the advantages of a self-learning dynamic neural network. The main aim of this modeling is to predict the runoff generated in the Kolar river basin. Two methods were applied for the simulation of the network. The first method, RBFNN, is applied with radial basis function neural network for 15 models differing in input combination for the rainfall–runoff modeling. In the second model, WRBFNN, data processing techniques of wavelet which remove the bias in data signals and give better predicting capability were applied for the prediction of runoff. For the first time, RBFNN is applied with wavelets for the central region of India for Narmada basin and for 15 models in one study. The best predicting model in these two networks is model 8, which has inputs of R(t-1), R(t-2), R(t-3), R(t-4), and Q(t-1). For the RBFNN model, the maximum value of R^2 is 0.9567 and the least value of AARE and MSE is observed. Similarly, for the WRBFNN model, the maximum value of R^2 is 0.9889 and least values of AARE and MSE are observed. It shows the importance of discharge and time lagged rainfall data for predicting the runoff generated during the hydrological process. These models were applied for one-year ahead data for the next year and gave promising predicting values which shows the model to be useful for prediction for the next year. Also, the use of temperature data in prediction shows no significant contribution and, on the contrary, leads to a decrease in coefficient of determination when introduced with the best performing models. This model has the potential to be improved further with a greater number of input combinations and data from different river basins. This result cannot be made generic to all the basins but further research is required. Most of the models have performed better than the same models with wavelet transformed inputs. This method can be applied to data-scarce regions where data are available for shorter duration and manually recorded which invites human errors. For the network to perform better, data with high accuracy are required to train the network better. Other data, like evapotranspiration, ground water recharge, morphometry of basin, solar radiation, wind velocity, etc. are also required to be taken into account for the runoff analysis. Models like SWAT, PCSWMM, etc. can also be used for better prediction.

DATA AVAILABILITY STATEMENT

All relevant data are included in the paper or its Supplementary Information.

REFERENCES

Achela, D., Fernando, K. & Shamseldin, A. Y. 2009 Investigation of internal functioning of the radial-basis-function neural network river flow forecasting models. *Journal of Hydrologic Engineering* **14** (3), 286–292.

Ali, S., Ghosh, N. C. & Singh, R. 2010 Rainfall–runoff simulation using a normalized antecedent precipitation index. *Hydrological Sciences Journal* **55** (2), 266–274.

Alizadeh, M. J., Kavianpour, M. R., Kisi, O. & Nourani, V. 2017 A new approach for simulating and forecasting the rainfall-runoff process within the next two months. *Journal of Hydrology* **548**, 588–597.

Asadi, S., Shahrabi, J., Abbaszadeh, P. & Tabanmehr, S. 2013 A new hybrid artificial neural networks for rainfall-runoff process modeling. *Neurocomputing* **121**, 470–480.

Badrzadeh, H., Sarukkalige, R. & Jayawardena, A. W. 2015 Hourly runoff forecasting for flood risk management: application of various computational intelligence models. *Journal of Hydrology* **529**, 1633–1643.

Barman, S. & Bhattacharjya, R. K. 2020 ANN-SCS-based hybrid model in conjunction with GCM to evaluate the impact of climate change on the flow scenario of the River Subansiri Swapnali Barman and Rajib Kumar Bhattacharjya. *Journal of Water and Climate Change* **11** (4), 1150–1164.

Bekele, W. T., Haile, A. T. & Rientjes, T. 2021 Impact of climate change on the streamflow of the Ardjo-Didessa catchment under RCP scenarios. *Journal of Water and Climate Change* **12**, 2325–2337.

Broomhead, D. S. & Lowe, D. 1988 Multivariable functional interpolation and adaptive networks. *Complex Systems* **2**, 321–355.

Dawson, C. W., Harpham, C., Wilby, R. L. & Chen, Y. 2002 Evaluation of artificial neural network techniques for flow forecasting in the River Yangtze, China. *Hydrology and Earth System Sciences* **6** (4), 619–626.

Ghumman, A. R., Ghazaw, Y. M., Sohail, A. R. & Watanabe, K. 2011 Runoff forecasting by artificial neural network and conventional model. *Alexandria Engineering Journal* **50** (4), 345–350.

Kim, B., Jeong, H., Kim, H. & Han, B. 2017 Exploring wavelet applications in civil engineering. *KSCE Journal of Civil Engineering* **21** (4), 1076–1086.

Krishna, B., Rao, Y. R. S. & Nayak, P. C. 2011 Time series modeling of river flow using wavelet neural networks. *Journal of Water Resource and Protection* **3** (1), 50–59.

Kumar, M. & Yadav, N. 2011 Multilayer perceptrons and radial basis function neural network methods for the solution of differential equations: a survey. *Computers and Mathematics with Applications* **62** (10), 3796–3811.

Lin, G. F. & Chen, L. H. 2004 A non-linear rainfall-runoff model using radial basis function network. *Journal of Hydrology* **289** (1–4), 1–8.

Miaoli, M., Xiaolong, W. & Honggui, H. 2020 Accelerated Levenberg-Marquardt algorithm for radial basis function neural network. In: *Proceedings - 2020 Chinese Automation Congress, CAC 2020.* Institute of Electrical and Electronics Engineers Inc., pp. 6804–6809.

Mujumdar, P. P. & Kumar, D. N. 1990 Stochastic models of streamflow: some case studies. *Hydrological Sciences Journal* **35**, 395–410.

Nawi, N. M., Atomi, W. H. & Rehman, M. Z. 2013 The effect of data pre-processing on optimized training of artificial neural networks. *Procedia Technology* **11**, 32–39.

Nourani, V. & Komasi, M. 2013 A geomorphology-based ANFIS model for multi-station modeling of rainfall-runoff process. *Journal of Hydrology* **490**, 41–55.

Nourani, V., Komasi, M. & Mano, A. 2009 A multivariate ANN-wavelet approach for rainfall-runoff modeling. *Water Resources Management* **23** (14), 2877–2894.

Nourani, V., Kisi, Ö. & Komasi, M. 2011 Two hybrid Artificial Intelligence approaches for modeling rainfall-runoff process. *Journal of Hydrology* **402** (1–2), 41–59.

Nourani, V., Baghanam, A. H., Adamowski, J. & Kisi, O. 2014 Applications of hybrid Wavelet-Artificial Intelligence models in hydrology. A review. *Journal of Hydrology* **514**, 358–377.

Phukoetphim, P., Shamseldin, A. Y. & Adams, K. 2016 Multimodel approach using neural networks and symbolic regression to combine the estimated discharges of rainfall-runoff models. *Journal of Hydrologic Engineering* **21** (8), 04016022.

Poonia, V. & Tiwari, H. L. 2020 Rainfall-runoff modeling for the Hoshangabad Basin of Narmada River using artificial neural network. *Arabian Journal of Geosciences* **13** (944), 1–10.

Powell, W. W. 1987 Hybrid organizational arrangements: new form or transitional development? *California Management Review* **30**, 67–87.

Riad, S. & Mania, J. 2004 Rainfall-runoff model using an artificial neural network approach. *Mathematical and Computer Modelling* **40**, 839–846.

Seo, Y., Kim, S. & Singh, V. P. 2015 Multistep-ahead flood forecasting using wavelet and data-driven methods. *KSCE Journal of Civil Engineering* **19** (2), 401–417.

Shamseldin, A. Y. 2009 Investigation of internal functioning of the radial-basis-function neural network river flow. *Journal of Hydrologic Engineering* **14** (3), 286–292.

Sharghi, E., Nourani, V., Najafi, H. & Molajou, A. 2018 Emotional ANN (EANN) and wavelet-ANN (WANN) approaches for Markovian and seasonal based modeling of rainfall-runoff process. *Water Resources Management* **32** (10), 3441–3456.

Shoaib, M., Shamseldin, A. Y., Khan, S., Khan, M. M., Khan, Z. M., Sultan, T. & Melville, B. W. 2018 A comparative study of various hybrid wavelet feedforward neural network models for runoff forecasting. *Water Resources Management* **32** (1), 83–103.

Sonali, P. & Nagesh Kumar, D. 2016a Spatio-temporal variability of temperature and potential evapotranspiration over India. *Journal of Water and Climate Change* **7** (4), 810–822.

Sonali, P. & Nagesh Kumar, D. 2016b Detection and attribution of seasonal temperature changes in India with climate models in the CMIP5 archive. *Journal of Water and Climate Change* **7** (1), 83–102.

Sudheer, K. P., Gosain, A. K. & Ramasastri, K. S. 2002 A data-driven algorithm for constructing artificial neural network rainfall-runoff models. *Hydrological Processes* **16** (6), 1325–1330.

Sušanj, I., Ožanić, N. & Marović, I. 2016 Methodology for developing hydrological models based on an artificial neural network to establish an early warning system in small catchments. *Advances in Meteorology* **2016**, 9125219.

Tayyab, M., Zhou, J., Zeng, X. & Adnan, R. 2016 Discharge forecasting by applying artificial neural networks at the Jinsha River Basin, China. *European Scientific Journal* **12** (9), 108.

Todini, E. 2007 Hydrological catchment modelling: past, present and future. *Hydrology and Earth System Sciences* **11** (1), 468–482.

Tokar, A. S. & Markus, M. 2000 Precipitation-runoff modeling using artificial neural networks and conceptual models. *Journal of Hydrologic Engineering* **5** (2), 156–161.

Wang, W., Chau, K., Xu, D., Qiu, L. & Liu, C. 2017 The annual maximum flood peak discharge forecasting using hermite projection pursuit regression with SSO and LS method. *Water Resources Management* **31** (1), 461–477.

Zhang, Q., Bin, B. W., Yong, H. & Ren, M. 2011 Singular spectrum analysis and ARIMA hybrid model for annual runoff forecasting. *Water Resources Management* **25**, 2683–2703.

First received 16 June 2021; accepted in revised form 26 November 2021. Available online 30 December 2021

Impact of Climate Change on Hydrology and Water Resources doi: 10.2166/wcc.2021.189

Multi-scale investigation on streamflow temporal variability and its connection to global climate indices for unregulated rivers in India

Pavan Kumar Veditha[a], Tarun Pant[b], Maheswaran Rathinasamy[a] and Ankit Agarwal [b,*]

[a] Department of Civil Engineering, MVGR College of Engineering, Vizianagaram 535005, India
[b] Department of Hydrology, Indian Institute of Technology Roorkee, Roorkee, Uttarakhand 247667, India
*Corresponding author. E-mail: ankit.agarwal@hy.iitr.ac.in

AA, 0000-0001-8572-7046

ABSTRACT

With the increasing stress on water resources for a developing country like India, it is pertinent to understand the dominant streamflow patterns for effective planning and management activities. This study investigates the spatiotemporal characterization of streamflow of six unregulated catchments in India. Firstly, Mann Kendall (MK) and Changepoint analysis were carried out to detect the presence of trends and any abrupt changes in hydroclimatic variables in the chosen streamflows. To unravel the relationships between the temporal variability of streamflow and its association with precipitation and global climate indices, namely, Niño 3.4, IOD, PDO, and NAO, continuous wavelet transform is used. Cross-wavelet transform and wavelet coherence analysis were also used to capture the coherent and phase relationships between streamflow and climate indices. The continuous wavelet transforms of streamflow data revealed that intra-annual (0.5 years), annual (1 year), and inter-annual (2–4 year) oscillations are statistically significant. Furthermore, a better understanding of the in-phase relationship between the streamflow and precipitation at intra-annual and annual time scales were well-captured using wavelet coherence analysis compared to cross wavelet transform. Furthermore, our analysis also revealed that streamflow observed an in-phase relationship with IOD and NAO, whereas there was a lag correlation with Niño 3.4 and PDO indices at intra-annual, annual and interannual time scales.

Key words: continuous wavelet transform, cross wavelet spectrum, streamflow, time-frequency characterization, wavelet, wavelet coherence

HIGHLIGHTS

- Unravelled the spatiotemporal characterization of streamflow of six unregulated catchments.
- Cross-wavelet transform and wavelet coherence analysis used to disentangle the coherent and phase relationships between streamflow and climate indices.
- Streamflow observed an in-phase relationship with IOD and NAO, whereas there was a lag correlation with Niño 3.4 and PDO indices at intra-annual, annual and interannual time scale.

1. INTRODUCTION

Human-induced global warming is changing the frequency and magnitude of hydrological variables altering the water cycle (Fanta *et al.* 2001; Adeloye & Montaseri 2002; Burn & Hag Elnur 2002; Cunderlik & Burn 2002; Kahya & Kalaycı 2004; Cherinet *et al.* 2019; Sridhar *et al.* 2019; Zhong *et al.* 2019). Due to abrupt changes in the water cycle, water resources' occurrence and availability are deeply affected. These variations in hydroclimatic processes trigger extreme events, endangering the availability of water resources and affecting socioeconomic conditions (Brutsaert & Parlange 1998; Kiely *et al.* 1998; Lins & Slack 1999; Khatiwada *et al.* 2016). Variability of these hydroclimatic variables is pertinent to map the hydrological cycle dynamics (Zhao *et al.* 2014). Streamflow is an important hydroclimatic parameter that aids in understanding the water cycle dynamics (Tamaddun *et al.* 2016). Many studies have investigated the variations in streamflow (trends, seasonality, and change-point detection) around the globe, confirming that streamflow is affected by the wide range of local (precipitation) and global (El Niño Southern Oscillation, Indian Ocean Dipole) climatic drivers (Hu *et al.* 2011; Masih *et al.* 2011; Panda *et al.* 2013; Abeysingha *et al.* 2016; Fathian *et al.* 2016). For instance, Burn & Hag Elnur 2002 identified that Pacific

Decadal Oscillation (PDO), North Atlantic Oscillation (NAO) North Pacific index (NP) affect the timing of runoff in the Mackenzie river basin in northern Canada. Sahu *et al.* 2012 evaluated IOD and ENSO's influence on extreme streamflow in the Citarum River in Indonesia. The results indicated that IOD could capture extreme events and a high correlation between seasonal IOD and seasonal streamflow. Nalley *et al.* (2016) used NAO, PDO, and ENSO to understand the variability of streamflow in Quebec and Ontario. Furthermore, significant water cycle changes lead to extreme drought and floods (Kustu *et al.* 2010; Miguez-Macho & Fan 2012; Himayoun & Roshni 2019; Sahu *et al.* 2020).

For a country like India with an extensive river system and streamflow being the primary source of water supply for irrigation, power generation, industrial use, and others, it is crucial to understand the streamflow dynamics. Quantifying the influences of local and global climate drivers on streamflow at basin scale is of utmost importance to develop probabilistic flood and drought forecasting models in assessing hydrologic risk. It would further help in decision-making for disaster management across the basin in India. Maity & Nagesh Kumar 2008 used climate indices to develop forecasting models for the Mahanadi river in India. Their results indicated that climatic indices help improve the predictability of models at the basin scale.

Similarly, Panda *et al.* (2013) have analyzed using the Mann-Kendall test and climate models to understand climate variability on streamflow in the Mahanadi river basin. The results showed the correlation of ENSO on precipitation and streamflow in the basin. Finally, Sharma *et al.* (2019) evaluated ENSO and IOD's influence on streamflow variability for the Tapi river basin, India. They found that the IOD influences streamflow compared to ENSO using trend analysis.

Significant progress has been made in quantifying the impact of global drivers on streamflow. However, there are only a few studies dedicated to the quantification of multi-scale variability in streamflow processes. Further, in conventional approaches to studying hydrological processes, this work focuses on one particular scale. The effects of finer scales are understood through the constitutive relation, and the consequences of coarser scales are neglected by assuming that the system is homogeneous at larger scales (Agarwal *et al.* 2019). In recent years, wavelet analysis has emerged as an essential tool in spectral analysis owing to its ability to deal with non-stationary data. Lau & Weng (1995) highlighted that wavelet is an effective hydrological time series analysis method. Adamowski *et al.* (2013) used wavelet analysis for quantifying the spatial-temporal variability of annual streamflow and meteorological changes in eastern Ontario and southwestern Quebec. Taye & Willems (2013) used wavelet analysis for identifying sources of temporal variability in hydrological extremes of the upper Blue Nile basin. Canchala *et al.* (2020) investigated the relationship of the streamflow variability with ENSO of Colombian Pacific Basins at interannual and decadal timescales and revealed significant coherencies at interannual (2–7 years) and decadal-scale (8–14 years). It is important to note that all the mentioned studies were based on catchments which are regulated. In the regulated catchment, hydropower plant, dams, or any other hydraulic structure might alter the natural flow regime and affect the streamflow variability by increasing annual base flow and decreasing peak flow magnitude (Swain & Patra 2017; Milner *et al.* 2019). However, for better understanding of the natural streamflow variability, and its relation with other hydrologic variables, unregulated catchments must be considered. Therefore, the main aim of this study is to investigate the streamflow variability of six different unregulated streamflow stations at different time scales and quantifies the influence of precipitation and global climate indices on them. For this purpose, we have considered six virgin catchments from the Indian subcontinent and analyse the streamflow variability using different statistical methods such as wavelets analysis, wavelet coherence analysis and Mann Kendall Test. The rest of the paper is arranged in the following manner. Section 2 provides information about the study area and the data used. The methods used in the study are described in Section 3. Results and discussion are provided in Section 4 and 5 respectively, whereas Section 6 gives the concluding remarks from the study.

2. STUDY AREA AND DATA USED

2.1. Study area description

India consists of 58 major river basins with varying flows and climatic regimes. Six unregulated streamflow stations from six different major river basins viz Brahmani, Cauvery, Mahanadi, Narmada, Godavari, and Subarnarekha basin are selected (Figure 1). Around 74 gauging stations are operational in the selected six river basins. In addition, one unregulated gauging station is determined based on the availability of long-term continuous data without missing values. All the selected river basins directly influence the water resources in their respective regions in irrigation, power generation and flood risk hazards in their flow regions. Figure 1 represents the geographical location of selected gauging stations in the river basins chosen for analysis, and Table 1 shows the details of chosen river basins.

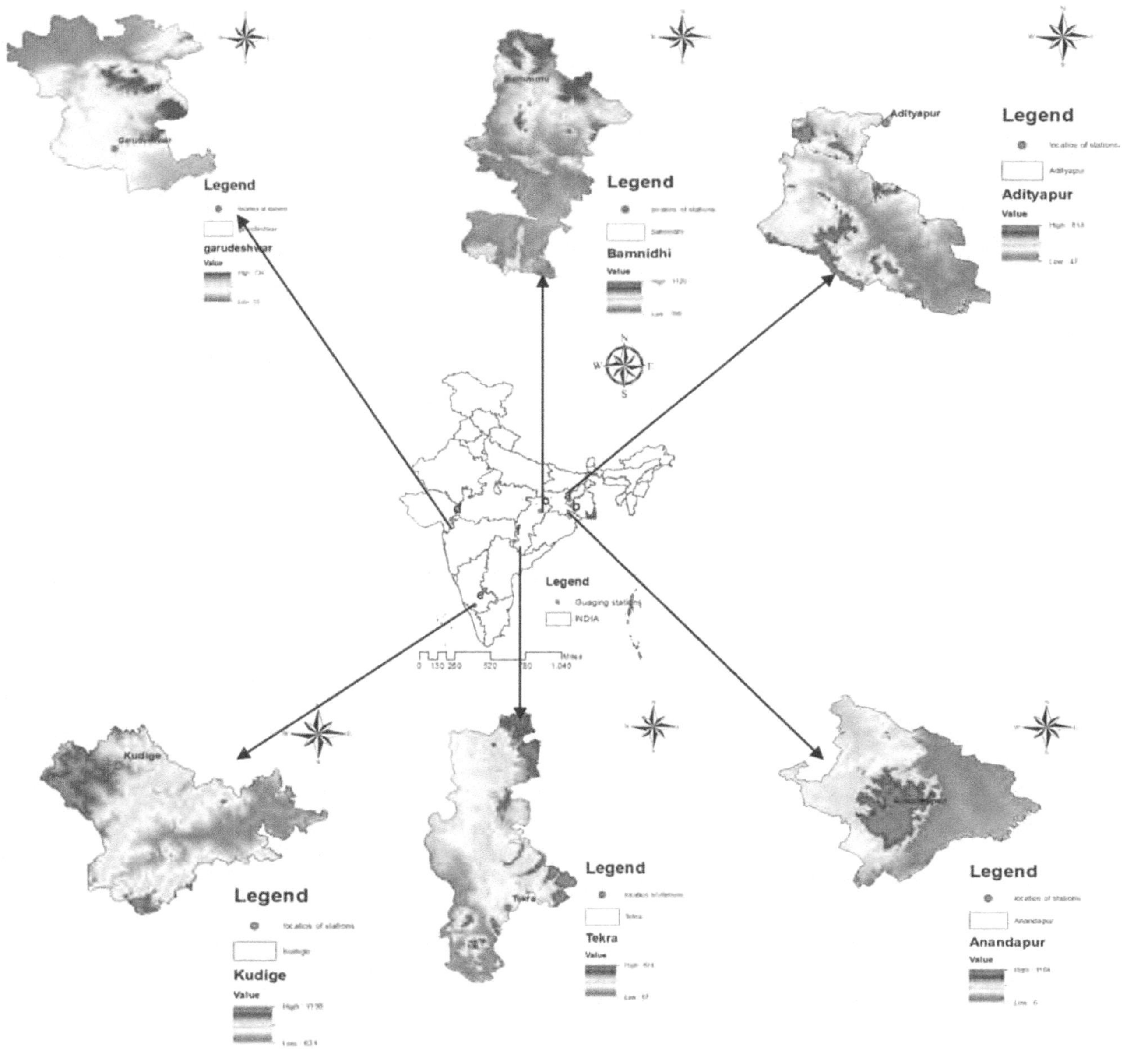

Figure 1 | Locations of gauging stations and catchments of the chosen river basins.

Table 1 | Details of the six catchments analyzed in this study

Sno	River	Station name	Indicated as	Latitude	Longitude	Catchment Area (Km^2)	River length (Km)	Annual average discharge (m^3/s)
1	Brahmani	Anandapur	BA	21.78	86.3	8,570	799	678
2	Cauvery	Kudige	CK	12.45	76.24	1,934	805	677
3	Mahanadi	Bamnidhi	MB	22.45	83.12	9,730	1,465	3061.8
4	Narmada	Garudeshwar	NG	21.54	73.66	87,892	900	2,119
5	Godavari	Tekra	GT	19.28	80.38	108,780	1,312	1,447
6	Subarnarekha	Adityapur	SA	22.95	86.52	6,309	395	392

2.2. Streamflow data

Streamflow data sets for the six selected stations were obtained from the WRIS India website http://indiawris.gov.in/wris/ for a 41-year period from 1975 to 2015. The daily streamflow data are converted to monthly using the downsample toolbox in MATLAB 2019a.

2.3. Precipitation data

The precipitation data set used a high-resolution of 0.25°x 0.25°daily gridded data available from 1901 to 2019. This data set is developed by Indian Metrological Department (IMD)for the spatial domain of 66.5° to 100°E and 6.5° to 38.5°N covering entire India (Pai *et al.* 2014). The data set used in this study is from 1975 to 2015, 41 years of daily precipitation and was downloaded from http://www.imdpune.gov.in/Clim_Pred_LRF_New/Grided_Data_Download.html

2.4. Global climate indices

Global climate indices influence various hydrological variables, including precipitation and streamflow around the world. Based on previous studies, Indian Ocean Dipole (IOD), North Atlantic Oscillation (NAO), El Niño Southern Oscillation (Niño3.4), and Pacific Decadal Oscillation (PDO) are major indices found to influence the Indian subcontinent. The brief details on these indices are as follows:

(1) IOD index represents sea surface temperature between the western equatorial Indian ocean (50° E- 70°E and 10° S- 10° N) and the southeastern Indian ocean (90°E - 110° E and 10° S- 10°N). Several studies showed that IOD plays a vital role in the climate of the Indian subcontinent. Hence, it is also referred to as Indian Niño. The IOD data was obtained from the Bureau of Metrology, Australia, from the website https://www.esrl.noaa.gov/psd/gcos_wgsp/Timeseries/Data/dmi.long.data. The data is available at monthly scale from the period of 1948 to 2018.
(2) NAO refers to the North Atlantic Ocean fluctuations, calculated based on the subpolar low and subtropical high. The NOAA climate prediction centre (CPC) prepares data for NAO, and it is available at the monthly level from 1948 to 2018.
(3) Niño 3.4 index represents the anomalies in sea surface temperature at East Central Tropical Pacific in the region bounded between 5° N to 5°S and 170° W to 120° W. Data is available from the period 1870 to 2019 at monthly scale.
(4) PDO is a climate index that occurs at a larger scale with its patterns primarily observed in the North Pacific, and its phases are classified into warm and cool based on temperature changes. The length of the data available is from 1948 to 2018.

Data for NAO, NIÑO 3.4, and PDO were obtained from the NOAA website https://www.esrl.noaa.gov/psd/data/climateindices/list/.

3. METHODS

This section describes trend analysis using the Mann-Kendall test and temporal analysis using continuous wavelet transform. It also describe the cross wavelet analysis and wavelet coherence used for investigating the connection between streamflow, precipitation, and climate indices. The detailed frame of the adopted methodology in this study is represented in Figure 2.

3.1. Mann-Kendall test (MK test)

The Mann-Kendall (MK) test determines the trend for each gauging station considered in this study. The MK test capabilities were robust in detecting trends based on Kendall 1975 and Mann 1945. Some of the studies that used this trend detection test to determine trends in hydroclimatic variables include works by Burn & Hag Elnur 2002; Sayemuzzaman & Jha 2014; and Guntu *et al.* 2020. In this test, based on the *p*-value, the presence or absence of a trend is determined. Equations 1 to 3 govern the detection of the trend in a series.

$$S = \sum_{i=1}^{n-1} \sum_{k=i+1}^{n} \operatorname{sgn}(X_k - X_i) \tag{1}$$

$$Var(S) = \frac{1}{18}\left[n(n-1)(2n+5) - \sum_{k=1}^{q} t_k(t_k-1)(2t_k+5)\right] \tag{2}$$

$$\text{If } S > 0,\ Z = \left\{\frac{S-1}{\text{std}(S)}\right\};\ S = 0,\ Z = 0;\ S < 0,\ Z = \left\{\frac{S+1}{\text{std}(S)}\right\}$$

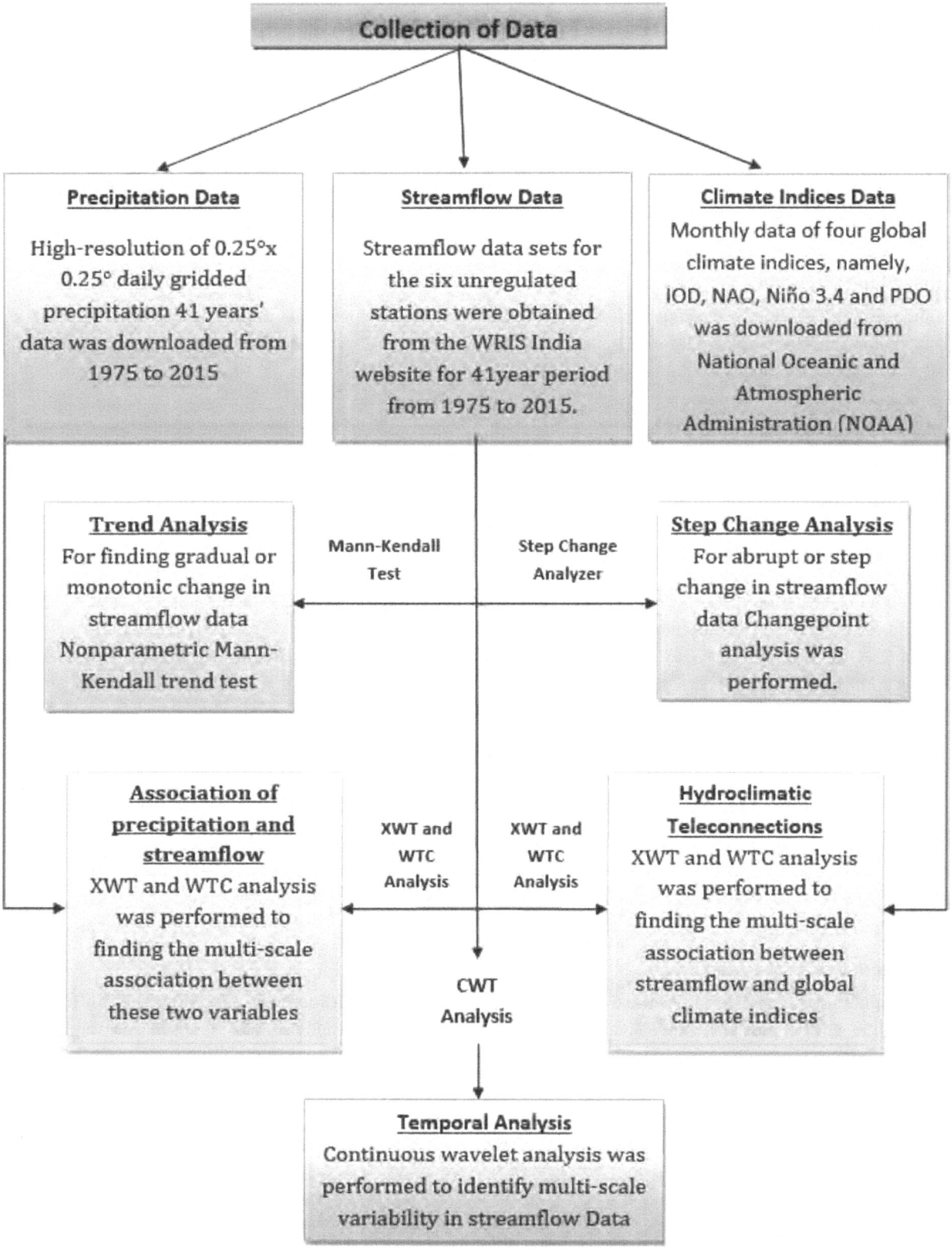

Figure 2 | Framework of the adopted methodology.

where S represents the value of the test statistic, n represents the length of the time series, X_k, X_i represents the sequential data points, $sgn(p)$ is a sign function for which if $p > 0, p = 0$ and $p < 0$, then $sgn(p)$ takes the values of $-1, 0, 1$. Z represents the standardized statistic value. t_k is the number of ties for k^{th} value and q is the number of tie values.

Based on the value of standardized statistic (Z), the p-value (Y) is calculated, which can be mathematically represented by Equation (3)

$$Y = 2(1 - nrmcdf(Z)) \tag{3}$$

where Y is the p-value, *nrmcdf* denotes the normalized cumulative distribution function of standardized statistics. The level of significance (confidence interval) is denoted by α and if the value of $Z > Z_{1-\alpha/2}(=1.96)$ it determines that the null hypothesis is rejected along with an indication of the presence of a trend. If not, it indicated acceptance of the null hypothesis, in the absence of a trend. In this work, the confidence interval value α is assumed to be 0.05 and, i.e., the probability of the null hypothesis to be wrong is 5% and using this value and the value of test statistic S, limits of the Z statistic are determined as -1.96 to 1.96.

In this work the Mann-Kendall test is carried using software MATLAB2019a. Suppose the obtained value for Z statistic lies between $-1.96 < Z < 1.96$, it denotes that there is no significant presence of a trend. If the p-value obtained from the MK test is less than alpha (0.05, i.e., 95% probability of null hypothesis being correct), it can be stated that there is a significant presence of a trend. If the value of Z statistic and P-value are negative, it represents a negative trend; if the positive side represents a positive trend.

3.2. Change point analysis

Change points are abrupt variations in time series data. Such abrupt changes may represent transitions that occur between states. Detection of change points is helpful in modeling and predicting time series. Time series data are sequences of measurements over time, describing the behavior of systems. These behaviors can change over time due to external events and systematic internal distribution changes (Kawahara & Sugiyama 2012). Change point detection (CPD) is the problem of finding abrupt changes in data when a property of the time series changes (Reeves *et al.* 2007).

In this study, change point analysis is carried out in Matlab 2019a; based on the streamflow values in each river, step changes were observed. Suppose there is a change in trend at a particular period considered in the analysis and observed to be a decreasing trend. In that case, a $-$ve value is assigned to the change point, whereas if it is observed to be a positive trend, $+$ve value is assigned.

3.3. Wavelet analysis

Mathematical tools used to analyze a time series in both time and frequency domain due to the localization capabilities are referred to as wavelets (Daubechies 1990). Due to wavelets' multi-resolution and localization capabilities, they are widely applied for time series analysis and prediction (Maheswaran & Khosa 2012; Agarwal *et al.* 2017). Decomposition of wavelets helps reveal the hidden low frequency and high-frequency components in the observed time series, which helps identify the features helpful for good analysis and predictions. In recent years wavelets have been widely used in domains of hydrology and water resources due to their ability to study non-stationarity in a time series (Grossmann & Morlet 1984; Kim & Valdés 2003; Zhou *et al.* 2008; Maheswaran & Khosa 2012). Wavelets are broadly classified into continuous wavelet transforms (CWT), and discrete wavelet transforms (DWT).

Continuous wavelet transforms work on all the scales to analyse a time series ((Mallat 1999). Continuous Wavelet Transform (CWT) is defined as the sum overall time of the signal multiplied by scaled, shifted versions of the wavelet function ψ. The coefficients of the wavelet transform of continuous signal $f(t)$ are defined by a linear integral operator (Maheswaran & Khosa 2012).

$$W(a, \tau) = \int_{\infty}^{-\infty} f(t)\psi_{a,\tau}(t)dt \tag{4}$$

where

$$\psi_{a,\tau}(t) = \frac{1}{\sqrt{a}}\psi\left(\frac{t-\tau}{a}\right)$$

The function $\psi(t)$, which can be real or complex, plays a convolution kernel called a wavelet. The parameter a can be interpreted as a dilation (a > 1) or contraction ($a < 1$) factor of the wavelet function $\psi(t)$ corresponding to different scales of observation. The parameter 'τ' can be interpreted as a temporal translation or shift of the function $\psi(t)$ which allows the study of the signal $f(t)$ locally around the time 'τ'.

The wavelet function $\psi(t)$ has the following properties.

(1) The function integrates to zero:

$$\int_{-\infty}^{\infty} \psi(t)dt = 0 \tag{5}$$

(2) The function is square integral or, equivalently, has finite energy

$$\int_{-\infty}^{\infty} |\psi(t)|^2 dt < \infty \tag{6}$$

The wavelet transform $W(a, \tau)$ are also called *wavelet coefficients* in which the dilation parameter is equivalent to the window size in the windowed Fourier transform. By changing the dilation parameter, the information contained in different data frequencies can be analyzed independently. A small dilation value may be used to analyze high-frequency information, while a larger value may be used to analyze a process with low-frequency components. The wavelet transform, as defined by Equation (4), is called the continuous wavelet transform (abbreviated CWT) because the scale and time parameters, a and τ, assume continuous values. It provides a redundant representation of a signal as the CWT of a function at scale 'a' and location 'τ' can be obtained from the continuous wavelet transform of the same function at other scales and locations. Since the CWT behaves like orthonormal basis decomposition, it can be shown that it is also isometric as it preserves the overall energy content of the signal and, thereby, allows recovery of the function $f(t)$ from its transform by using the following reconstruction formula:

$$f(t) = \frac{1}{C_\psi} \int_{-\infty}^{\infty} \int_{0}^{\infty} a^{-2} W(a, \tau) \psi_{a,\tau}(t) da d\tau \tag{7}$$

where C_ψ is a constant and depends on the choice of the wavelet. Equation (7) suggests that the function $f(t)$ may be seen as a superposition of signals at different scales. It can be obtained by varying the scale parameter 'a'. Continuous wavelet transforms are more efficient at analyzing hydro-climatic variables than non-normal distributions.

3.4. Cross-wavelet transform

Cross-wavelet transform (XWT) explains the coherent relationships between two different time series and the phase relationships. In XWT, the dominant frequency or the lowest frequency may affect the two different time series' coherency. Suppose two discrete time series X ($n = 1 \ldots N$) and Y ($n = 1 \ldots N$), the XWT can be calculated as

$$W^{XY}(a) = W^X(a) \times W^{Y^*}(a) \tag{8}$$

Here $W^x(a)$ is the CWT of the time series X considered. The complex conjugate of $W^X(a)$ the CWT of time series Y. Cross wavelet constructs the relationship between two CWTs with high common power (covariance) of the two-time series (Grinsted *et al.* 2004). The significant covariance of these timeseries can be observed as red noise in the wavelet scalograms.

3.5. Wavelet coherence

Cross-wavelet transform sometimes may not give precise results in terms of coherency. Wavelet coherence is used to overcome this problem, which provides accurate results about coherent relationships. Wavelet coherence exhibits the correlation of two-time series, which is based on time-frequency analysis. Wavelet coherence aids in understanding the relationship between low power frequencies, which cannot be correctly analyzed using a cross wavelet (Agarwal *et al.*, 2016). Consider two-time series y and x; then the wavelet coherence is given by

$$R(y, x) = \frac{\varsigma[W(y, x)]}{\sqrt{\varsigma[W(y)]\varsigma[W(x)]}.} \tag{9}$$

$$R^2(y, x) = R(y, x).R(y, x)^*; \tag{10}$$

In the above equation; R(y,x) – wavelet coherence between y and x; R^2(y,x)-squared wavelet coherence between y and x; W(y,x) denotes the corresponding cross-wavelet transforms; ς denotes smoothing factor.

Wavelet coherence ranges from 0 to 1. Thus, if the correlation coefficient value is closer to 1, there is more correlation between the two-time series and vice versa.

4. RESULTS

In this section, the long-term trend and step-change analysis of streamflow of six unregulated catchments are presented. Furthermore, the precipitation and global climatic indices are decomposed to investigate the association with streamflow at different timescales.

4.1. Long-term trend and change point detection

Firstly, estimation of the general pattern of a streamflow timeseries over a period between 1975 and 2015 for six ungauged catchments. Table 2 shows the Z value, *p*-value, and step-change year obtained through the MK test and step-change detection technique. Our results showed a significant long-term negative trend for station CK and NG at a 5% significance level. However, other stations viz BA, GT, MB, and SA had a significant trend in their respective streamflow data. Figure 3 illustrates the trend lines and change point for each considered streamflow station. In the case of change point analysis, the result indicated a significant abrupt change for all the stations. For instance, significant negative abrupt changes were observed for the station BA, CK, MB, and NG in 2015, 1985, 2007, and 2006, respectively. Whereas stations like GT and SA observed significant positive abrupt changes in their respective streamflow data in 1992 and 1985.

4.2. Wavelet analysis of streamflow

Wavelet transformation (WT) is used to determine the scale-specific dominant parameters that might cause the trend and step change in streamflow. Daily streamflow data of six considered stations are decomposed using continuous wavelet transforms (CWT), and scalograms are plotted (Figure 4(a)–4(f)) for each station. In general, all considered stations (BA, CK, GT, NG, MB, and SA) showed high variability at intra-annual, annual and inter-annual scales (0.5, 1, 2, and 4 years). Interestingly, significant intra-annual oscillations of 0.5 years occurred mostly up to the 1990s for stations BA (Figure 4(a)), MB (Figure 4(d)), NG (Figure 4(e)), and SA (Figure 4(f)). In contrast, such oscillations were absent at station CK (Figure 4(b)) and GT (Figure 4(c)) at this particular scale. Annual oscillations were active for all the stations throughout the study period, except station CK that showed vanishing features from 1985 to 2000. At the inter-annual scale (2 years and 4 years), BA, MB, NG, and SA observed significant oscillations mainly in the 1990s. Scalograms showed the streamflow variability at

Table 2 | Results of trend analysis and step change of streamflow for the chosen gauging stations in India's six river basins

River	Station	River/Station	Z value	*P*-value	Significance	Step change
Brahmani	Anandapur	BA	−0.21	0.83	*	2015 –
Cauvery	**Kudige**	**CK**	**− 3.95**	**7.69e-05**	**	**1985 –**
Godavari	Tekra	GT	−1.67	0.09	*	1992 +
Mahanadi	Bamnidhi	MB	0.35	0.72	*	2007 –
Narmada	**Garudeshwar**	**NG**	**− 2.91**	**0.01**	**	**2006 –**
Subarnarekha	Adityapur	SA	0.86	0.38	*	1985 +

Note: '* 'represents no significant trend and '**' represents a significant negative trend. Stepchange denotes the year in which change has occurred and '-'indicates a negative change trend, and ' + ' indicates a positive change trend.

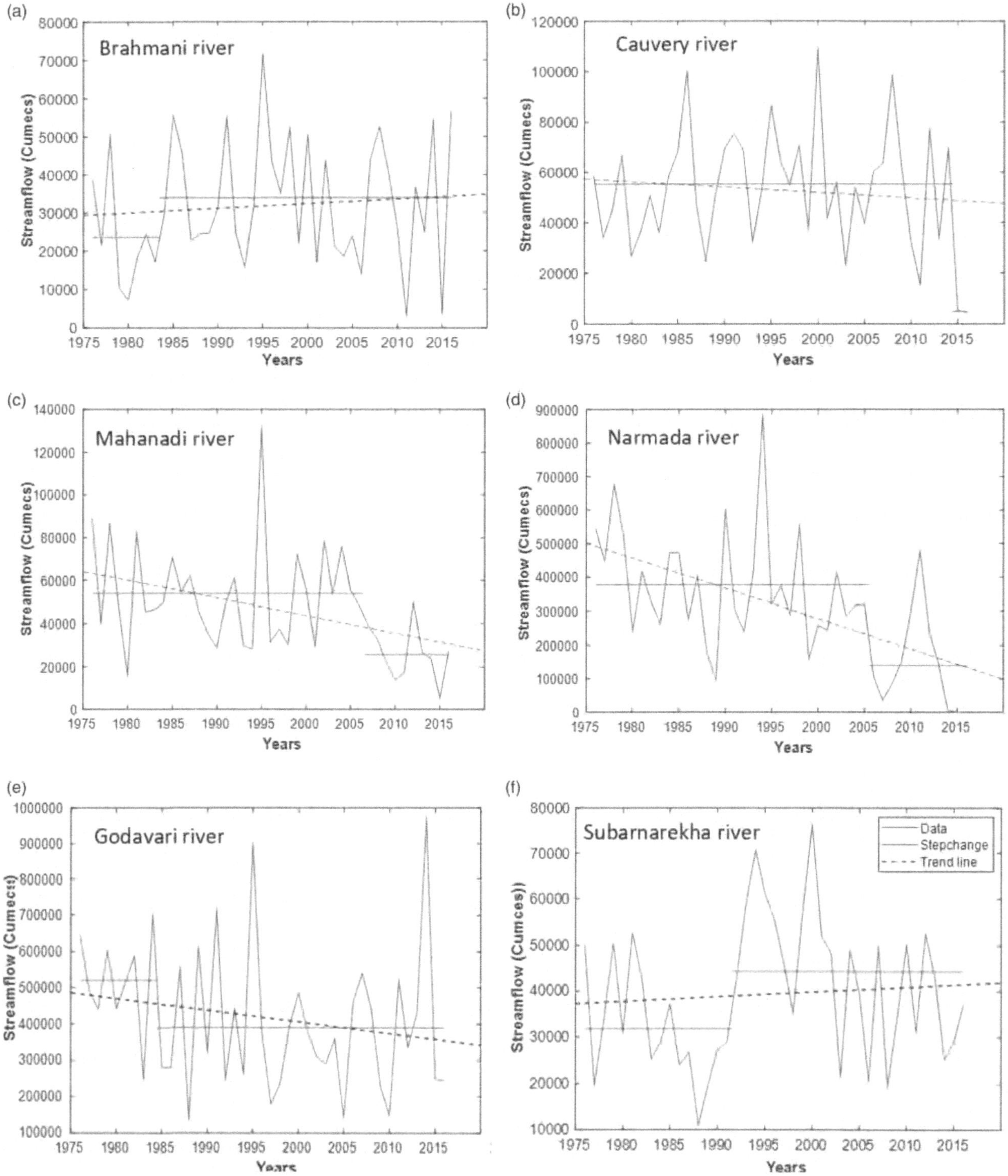

Figure 3 | Annual streamflow trend analysis for gauging stations in chosen river basins.

different time scales; however, the causes of variability in its patterns cannot explain using only CWT. In the next section, quantification of correlation between streamflow with precipitation and also global climatic parameters is carried out using wavelet coherence analysis (WCA) and cross-wavelet analysis (XWT).

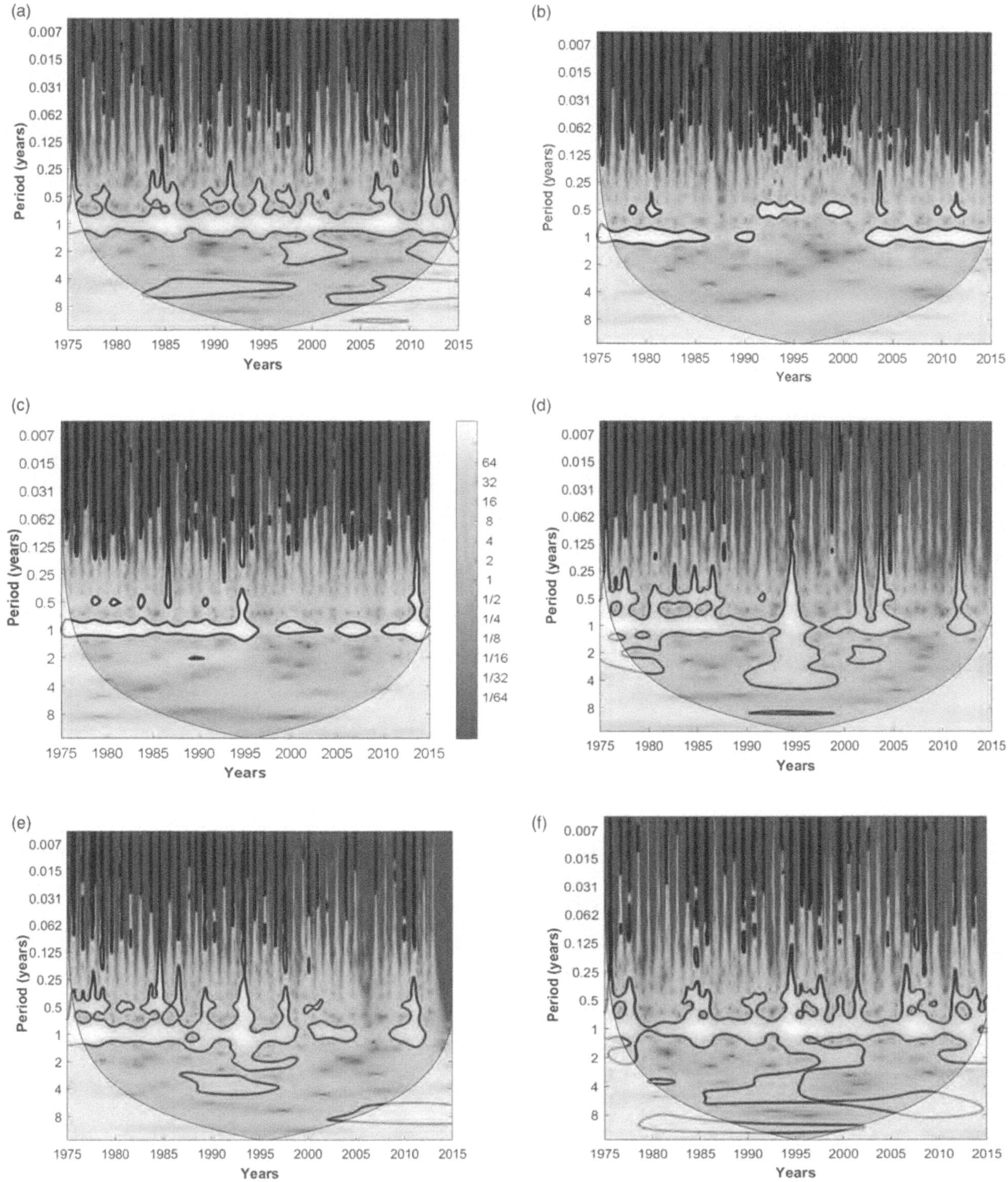

Figure 4 | Continuous wavelet analysis of streamflow at all the stations, (a) BA, (b) CK, (c) GT, (d) MB, (e) NG, and (f) SA. The cone of influence indicates the area affected by the boundary assumption.

4.3. Association of precipitation and streamflow at multiple timescales

Cross-wavelet analysis (XWT, Figure 5(a)–5(f)) and wavelet coherence analysis (WCA, Figure 6(a)–6(f)) are used to quantify the association between rainfall and streamflow. In general, XWT and WCA plot of streamflow of all the stations (BA, CK,

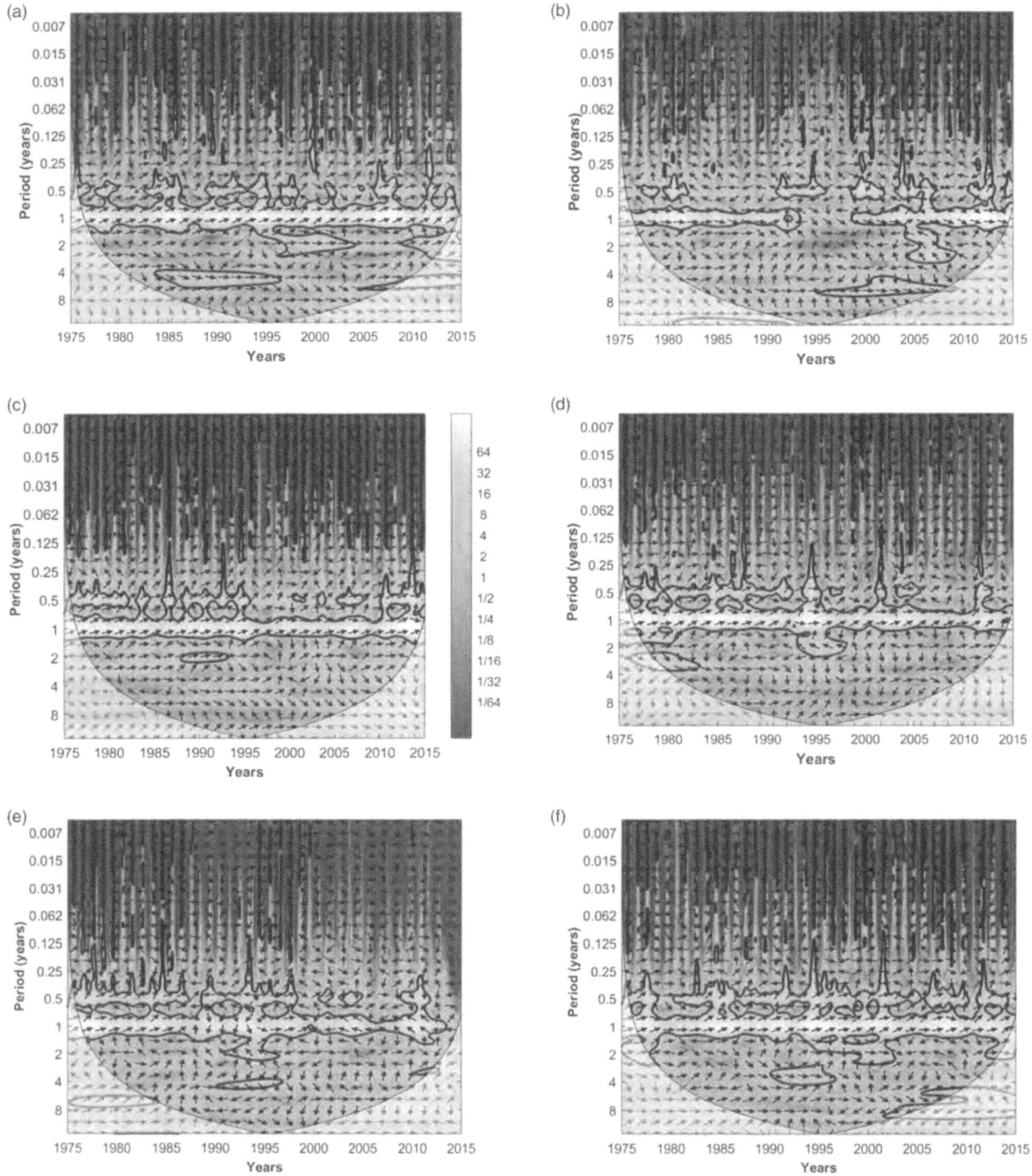

Figure 5 | Cross-wavelet analysis of precipitation & streamflow at all the stations. (a) BA, (b) CK, (c) GT, (d) MB, (e) NG and (f) SA. The sections marked by black color contours are the significant sections indicating a high correlation between precipitation and streamflow. The cone of influence indicates the area affected by the boundary assumption.

GT, MB, NG, and SA) showed an in-phase relationship with precipitation; however, stations GT, MB, and NG also observed lag correlation at 0.25, 0.5, and 1 year of scale respectively. Similarly, all stations except the station CK showed similar associations at intra-annual, annual, and inter-annual time scales in both XWT and WCA. For CK station WCA showed a strong

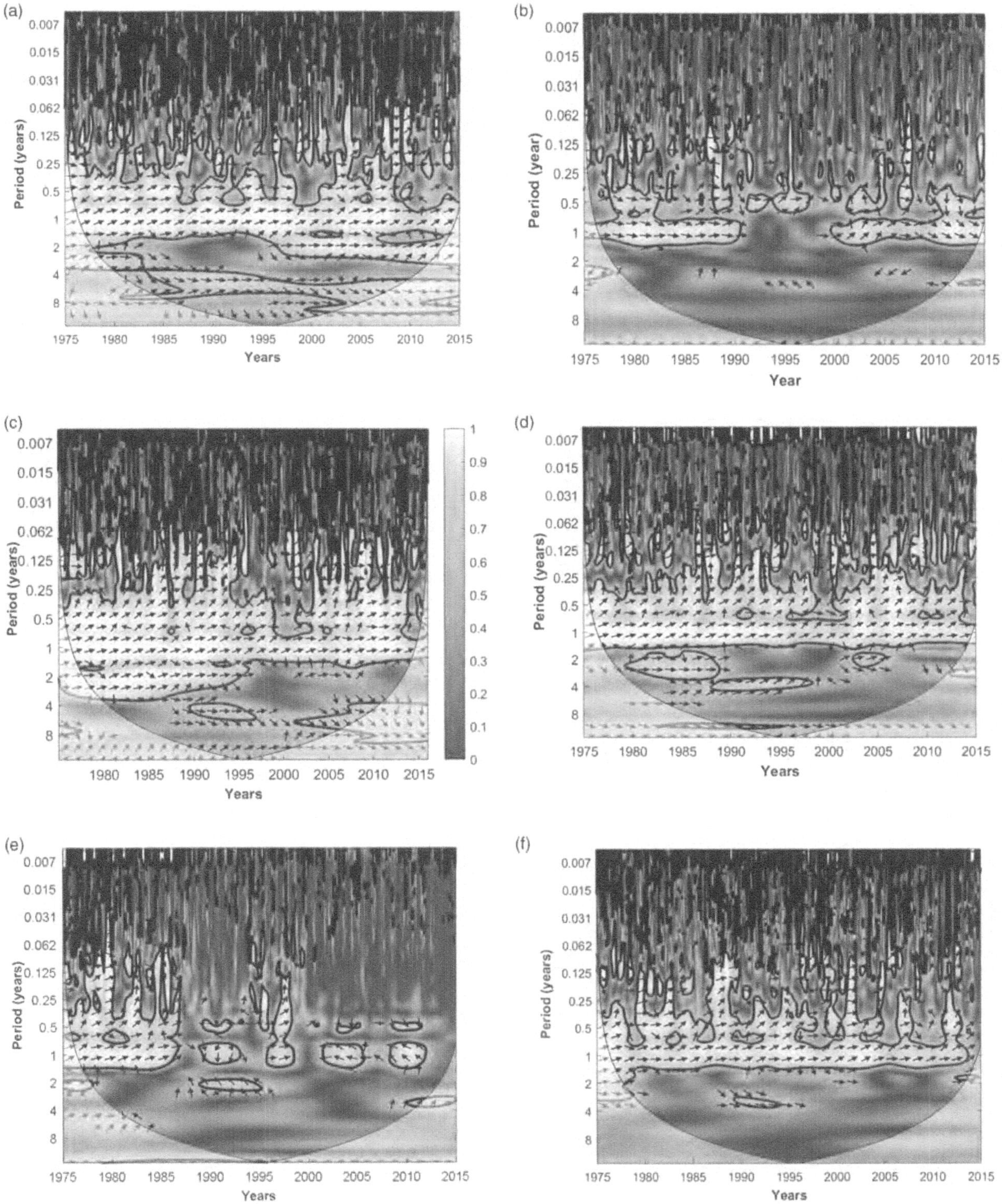

Figure 6 | Wavelet coherence analysis of precipitation & streamflow at all the stations (a) BA, (b) CK, (c) GT, (d) MB, (e) NG, and (f) SA. The sections marked by black color contours are the significant sections indicating high coherence between precipitation and streamflow. The cone of influence indicates the area affected by the boundary assumptions.

feature at 4 years of scale; however, the same is absent in XWT. A significant intra-annual coherence between these two variables can be observed at 0.125, 0.25, and 0.5 years of scale; however, these coherences are short duration and transient in nature. On an annual scale, except for station CK, all other stations observed significant coherence throughout the study period. In the 1980s and 1990s, some stations like BA, GT, and MB also observed significant inter-annual coherence at 2–4 years of scale. The association between rainfall-runoff at a small time scale may be due to the catchment characteristics and local climate. In contrast, the association at 2–4 and 4–8 years may be deriving the relationship from the global connections characterized by large-scale climate oscillations.

4.4. Hydroclimatic teleconnections

Figures 7–14 show the cross-wavelet and wavelet coherence plot for six selected river basins and four chosen climate indices: IOD, NAO, NIÑO 3.4, and PDO.

XWT plot (Figure 7(a)–7(f)) showed a significant association between IOD and streamflow. All stations viz BA, CK, GT, MB, NG, and SA observed a significant correlation at an annual scale from 1980 to 2010. Interestingly, the annual correlation seems to be vanishing from 2000 to 2005 and nonexistent in recent years. Stations MB and NG observed significant inter-annual correlation at 2–4 years of scale from 1990 to 2000. The analysis based on WTC (Figure 8(a)–8(f)) supports our

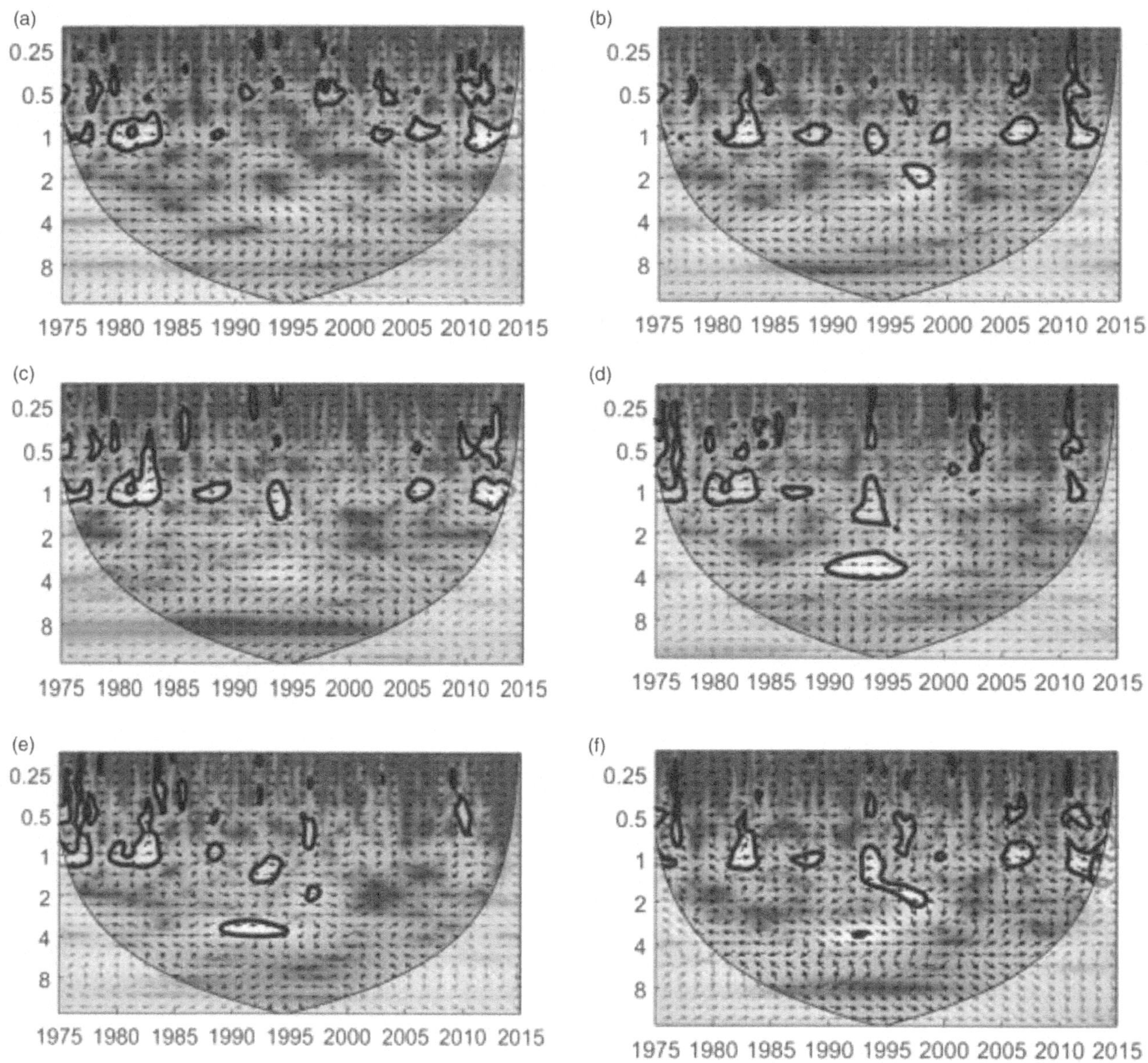

Figure 7 | Cross-wavelet analysis of IOD & streamflow at all the stations. (a) BA, (b) CK, (c) GT, (d) MB, (e) NG and (f) SA. The sections marked by black color contours are the significant sections indicating a high correlation between precipitation and streamflow. The cone of influence indicates the area affected by the boundary assumption.

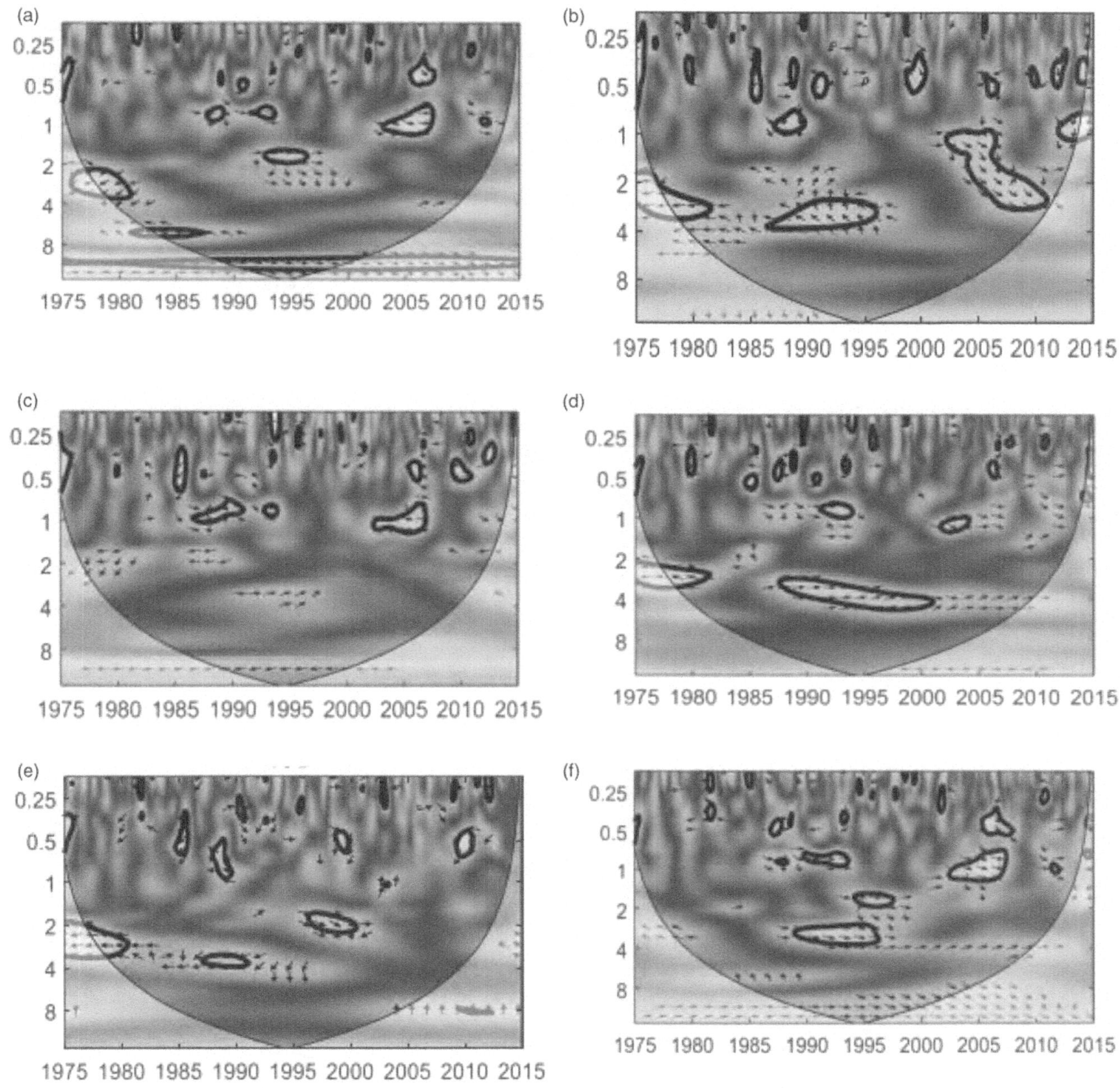

Figure 8 | Wavelet coherence analysis of IOD & streamflow at all the stations (a) BA, (b) CK, (c) GT, (d) MB, (e) NG, and (f) SA. The sections marked by black color contours are the significant sections indicating high coherence between precipitation and streamflow. The cone of influence indicates the area affected by the boundary assumptions.

findings and observed coherence at 2–4 years of scale. Indeed, Indian Ocean Dipole plays an essential role as a modulator of Indian monsoon rainfall (Ashok *et al.* 2001); however, at the regional scale, the influence of IOD on streamflow variability is different (Gouda *et al.* 2017). Our analysis also unraveled a mixed relationship of IOD with different basins and different scales and periods. There was an in-phase relationship for station BA, GT, MB, and SA. In contrast, station NG observed an anti-phase relationship, and station CK showed an anti-phase, lag, and lead relationship at starting (1977–1982), middle (1985–1995), and end year (2000–2010) of the study period. In-phase association at intra-annual (0.5 years) and annual (1 year) scale were observed between streamflow and North Atlantic Oscillation (NAO) in XWT analysis (Figure 9(a)–9(f)) from 1980 to 2010 with breaks in 1985, 1995, 2000 and 2005. The WTC analysis (Figure 10(a)–10(f)) support obtained similar patterns; however, stations like GT, MB, and NG also observed significant correlation at 2 and 4 years of scale. Compared with other indices, NIÑO 3.4 demonstrates maximum significance association with streamflow at an

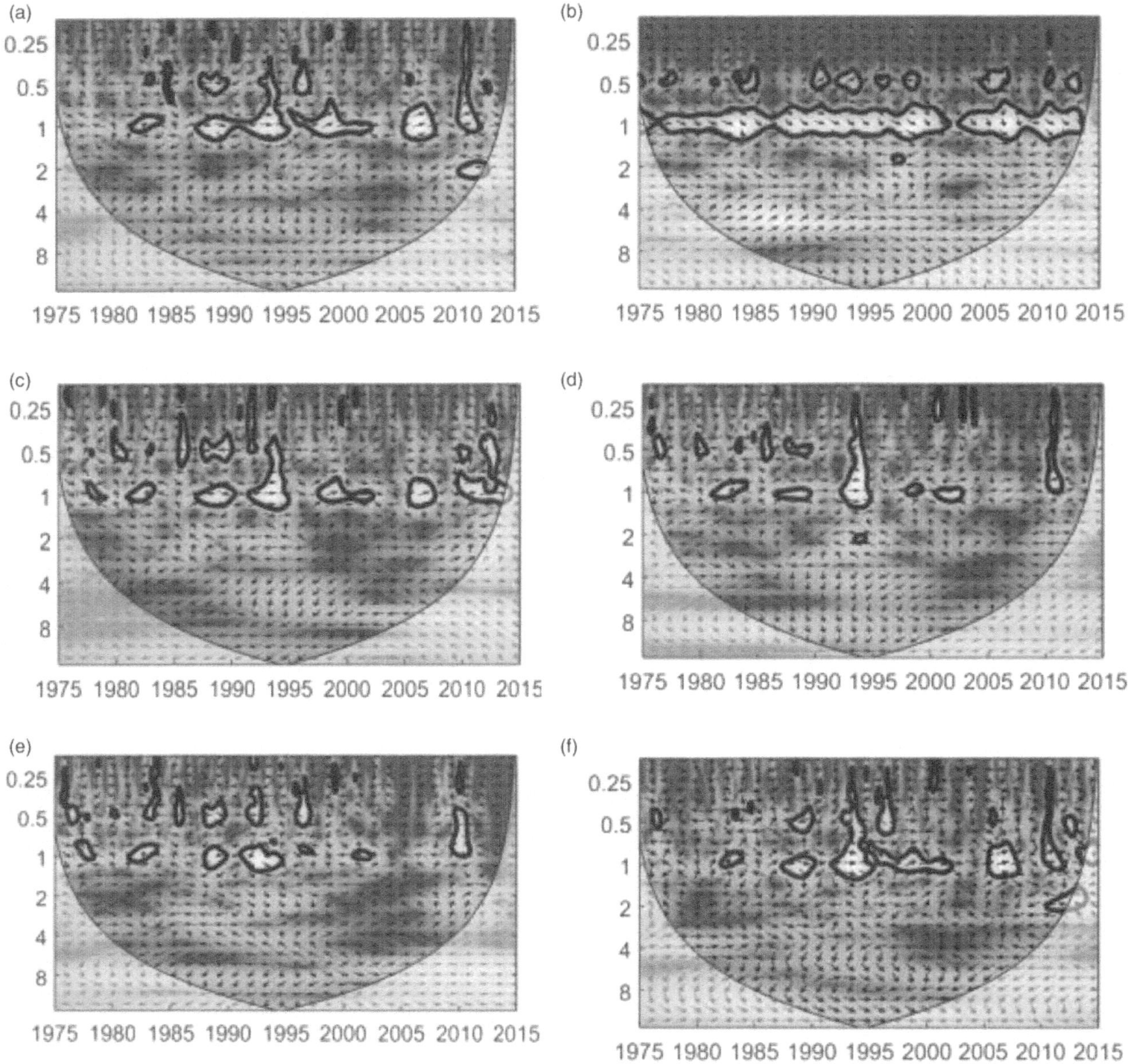

Figure 9 | Cross-wavelet analysis of NAO & streamflow at all the stations. (a) BA, (b) CK, (c) GT, (d) MB, (e) NG and (f) SA. The sections marked by black color contours are the significant sections indicating a high correlation between precipitation and streamflow. The cone of influence indicates the area affected by the boundary assumption.

annual scale. In unity with the statement that summer monsoon rainfall and streamflow over the Indian basin is best correlated with temperature anomaly from NIÑO 3.4 (Smith & Reynolds 2003). Analysis based on the XWT plot (Figure 11(a)–11(f)) showed that stations like BA, GT, MB, and SA showed correlation throughout the period (1975–2010), whereas station GT and NG observed some vanishing features 1995–2005. The WTC plot (Figure 12(a)–12(f)) observed periodicities at intra-annual, inter-annual, and intra-decadal scale along with the annual scale. Interestingly, 4–8 years of cycle were observed in the last decade. The arrows in both scalograms are not pointing in one direction, but most of them showed a leading relationship between streamflow and NIÑO 3.4 at an annual scale. Like NIÑO 3.4, PDO indices also showed a leading association with streamflow at 1 year of scale in the XWT plot (Figure 13(a)–13(f)). In the case of the WTC plot (Figure 14(a)–14(f)), along with 1 year of scale, stations GT, MB, and SA also observed these periodicities at 2, 4 and 8 years of scale. These associations can be shown mostly from 1985 to 2000.

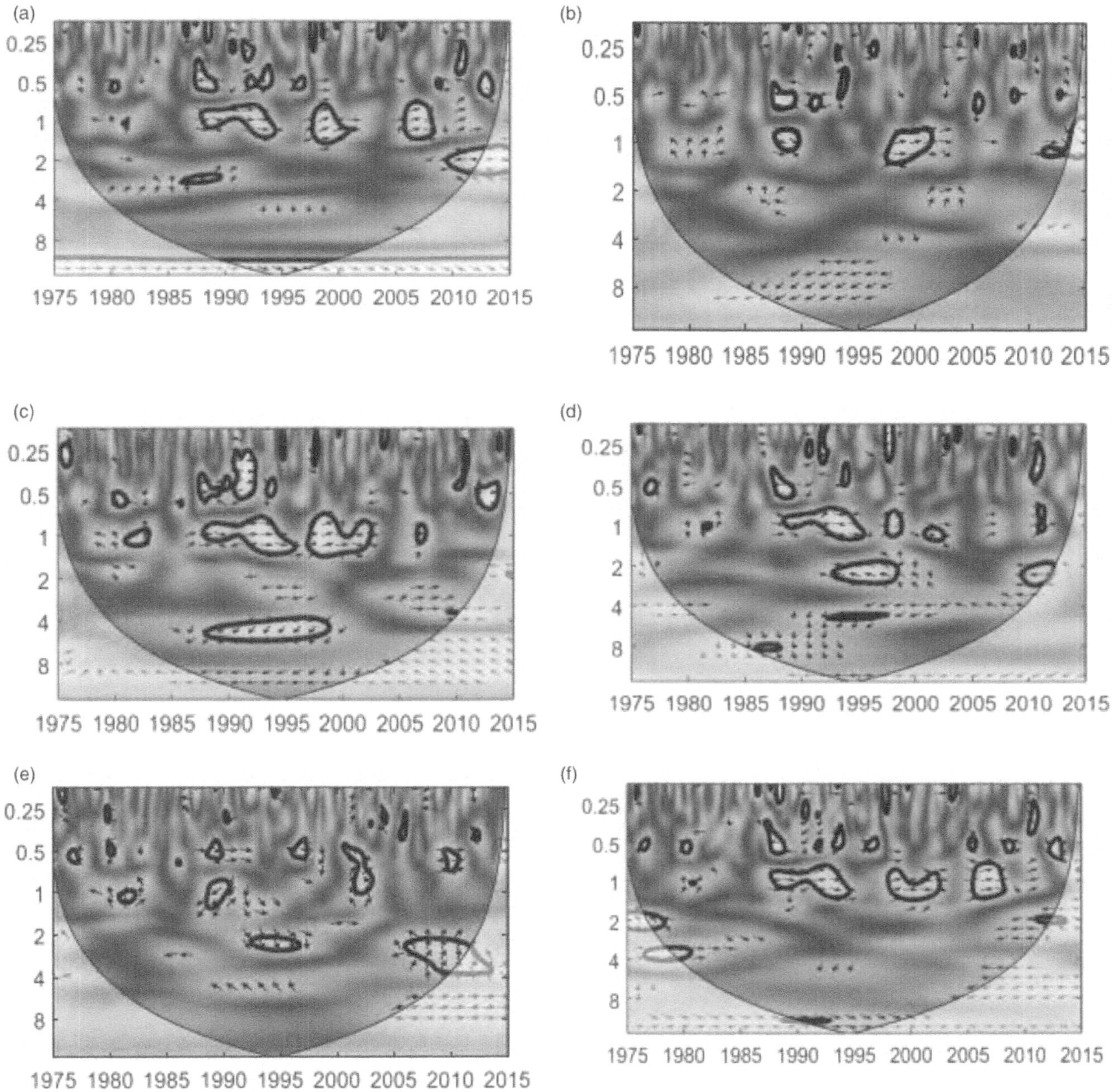

Figure 10 | Wavelet coherence analysis of NAO & streamflow at all the stations (a) BA, (b) CK, (c) GT, (d) MB, (e) NG, and (f) SA. The sections marked by black color contours are the significant sections indicating high coherence between precipitation and streamflow. The cone of influence indicates the area affected by the boundary assumptions.

5. DISCUSSION

The temporal characteristics of six unregulated stations are disentangled by analyzing monotonic and abrupt trends of streamflow and their teleconnections with precipitation and global climate indices. Based on the results of the Mann-Kendall test, only Cauvery and Narmada showed significant flow variability changes. Interestingly, both the rivers indicated a long-term negative trend congruent with findings by Jain *et al.* (2017). On the contrary, Das (2019), Kuriqi *et al.* (2020), Murthy *et al.* (2017), Panda *et al.* (2013), and Rao (1995) reported significant trends for Mahanadi, Godavari, and Subarnarekha basin; however, these results report that the trends are insignificant. Intriguingly, unlike the gradual change, an abrupt shift in streamflow for all the stations was also observed. The change point detection observed major step changes from 1985 to 2015. Some stations like Anandapur (Brahmini) and Garudeshwar (Narmada) observed a significant drop in

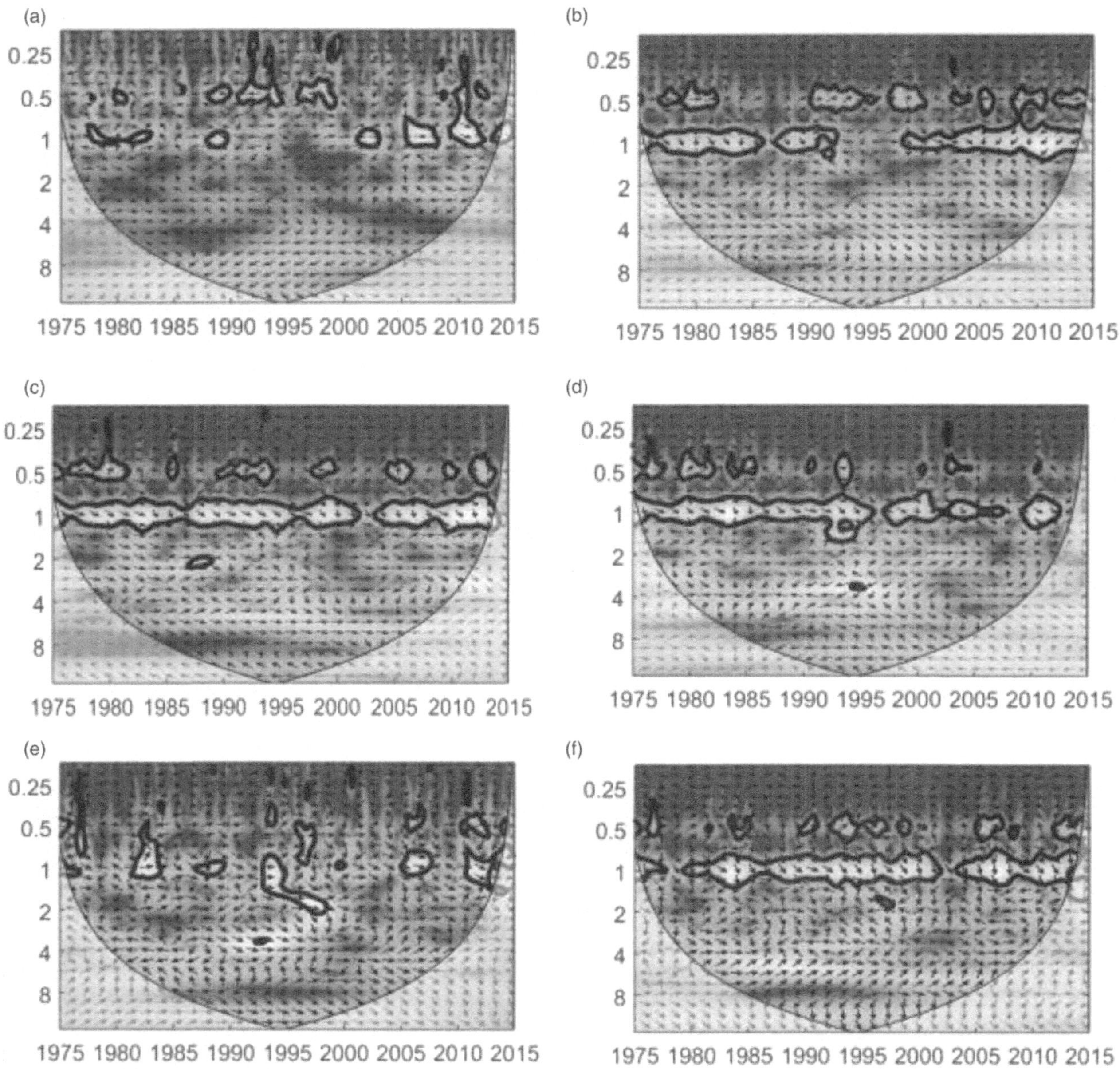

Figure 11 | Cross-wavelet analysis of NIÑO 3.4 & streamflow at all the stations. (a) BA, (b) CK, (c) GT, (d) MB, (e) NG and (f) SA. The sections marked by black color contours are the significant sections indicating a high correlation between precipitation and streamflow. The cone of influence indicates the area affected by the boundary assumption.

streamflow of about 55,000 m^3/sec and 200,000 m^3/sec in 2015 and 2006. Despite unregulated catchments that are not supposed to be influenced by hydropower plants, dams, or other reservoirs, such drastic change in streamflow presents an idea that there may be some other predominating factor for variations streamflow for these basins. Das *et al.* (2018) reported that urbanization and change in land use land cover could have impacted the streamflow alteration. Besides, they also discussed increasing population density over these basins, causing the transformation of cropland into builds up-lands which reduces the infiltration and increase the runoff. Many other researchers (Islam *et al.* 2012; Chawla *et al.* 2018; Vandana *et al.* 2019) found changes in precipitation and climatic patterns responsible for this streamflow alteration.

Decomposition of streamflow data using Continuous Wavelet Transform (CWT) is done to get further insight into streamflow variability. All the stations observed robust features at intra-annual (0.5 years) and annual (1 year) time scale. In addition, most of the stations showed a vital feature throughout the study period. Cross-wavelet and Wavelet Coherence Analysis plot between streamflow and precipitation showed an in-phase relationship for almost all the stations. Overall, our analysis

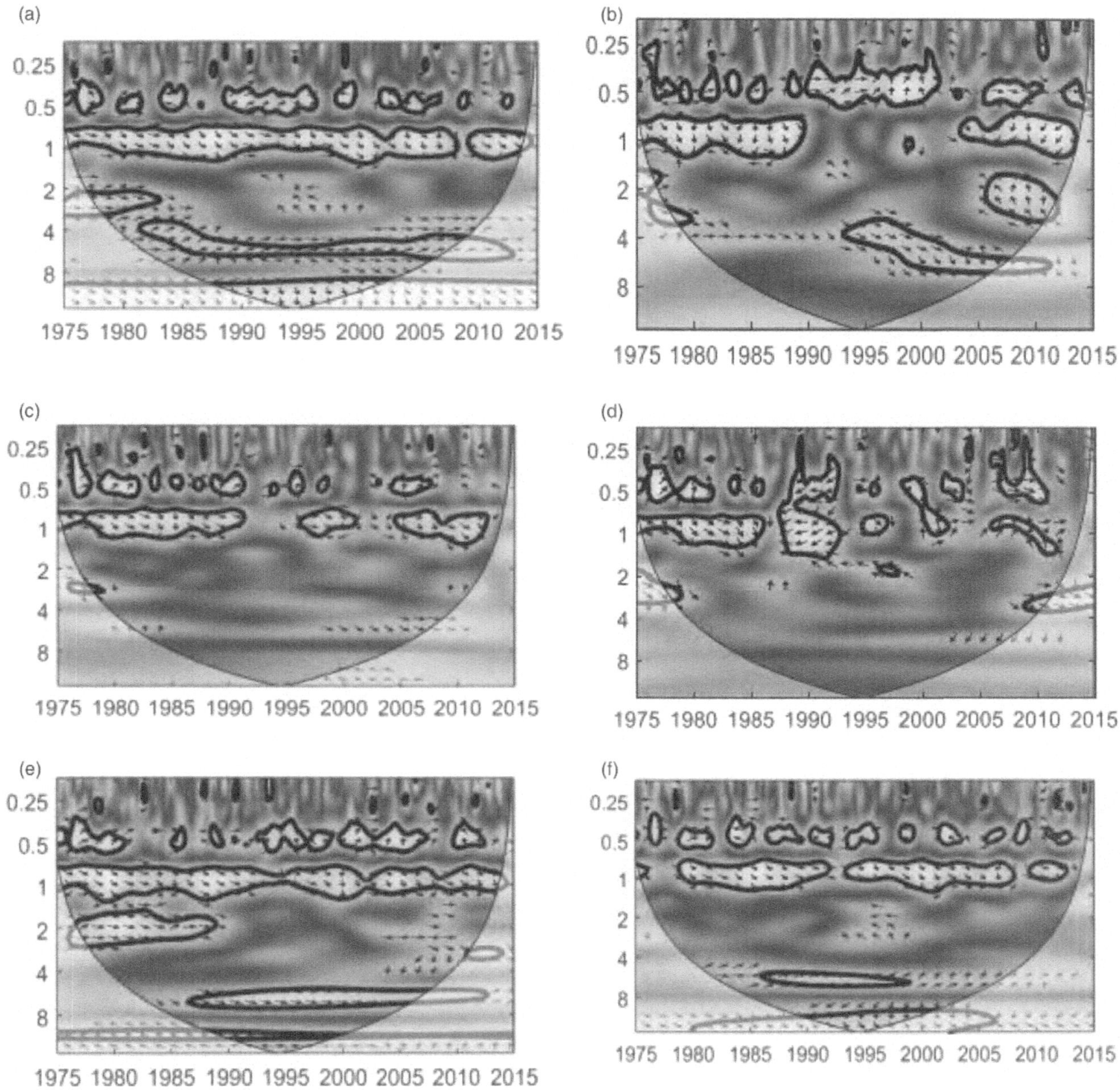

Figure 12 | Wavelet coherence analysis of NIÑO 3.4 & streamflow at all the stations (a) BA, (b) CK, (c) GT, (d) MB, (e) NG, and (f) SA.The sections marked by black color contours are the significant sections indicating high coherence between precipitation and streamflow. The cone of influence indicates the area affected by the boundary assumptions.

observed an in-phase relationship between streamflow and precipitation that is also reported by Jena *et al.* (2014), Maity *et al.* (2007), and Panda *et al.* (2013).

Hydroclimatic teleconnections between streamflow and four climate indices unravel maximum correlation with NIÑO 3.4 followed by PDO index. PDO influences to six basins were similar to NIÑO 3.4, highlighting the possibility of proxy connections as highlighted by Rathinasamy *et al.* 2019. These reported results are also in line with Agarwal *et al.* (2020). However, the association of PDO indices to the selected basin is not much explored yet except the research by Krishnamurthy & Krishnamurthy (2014), who reported that the warm and cold phases of PDO influence the condition of ENSO and thus affect the streamflow and precipitation over Indian basins. This may be due to the warm phase of PDO, which increases the sea surface temperature of the Pacific Ocean, affecting the trade winds and strengthening the Walker Circulation and thus amplifying the NIÑO condition (Agarwal 2019). Ashok *et al.* (2001) also described in their study how the NIÑO

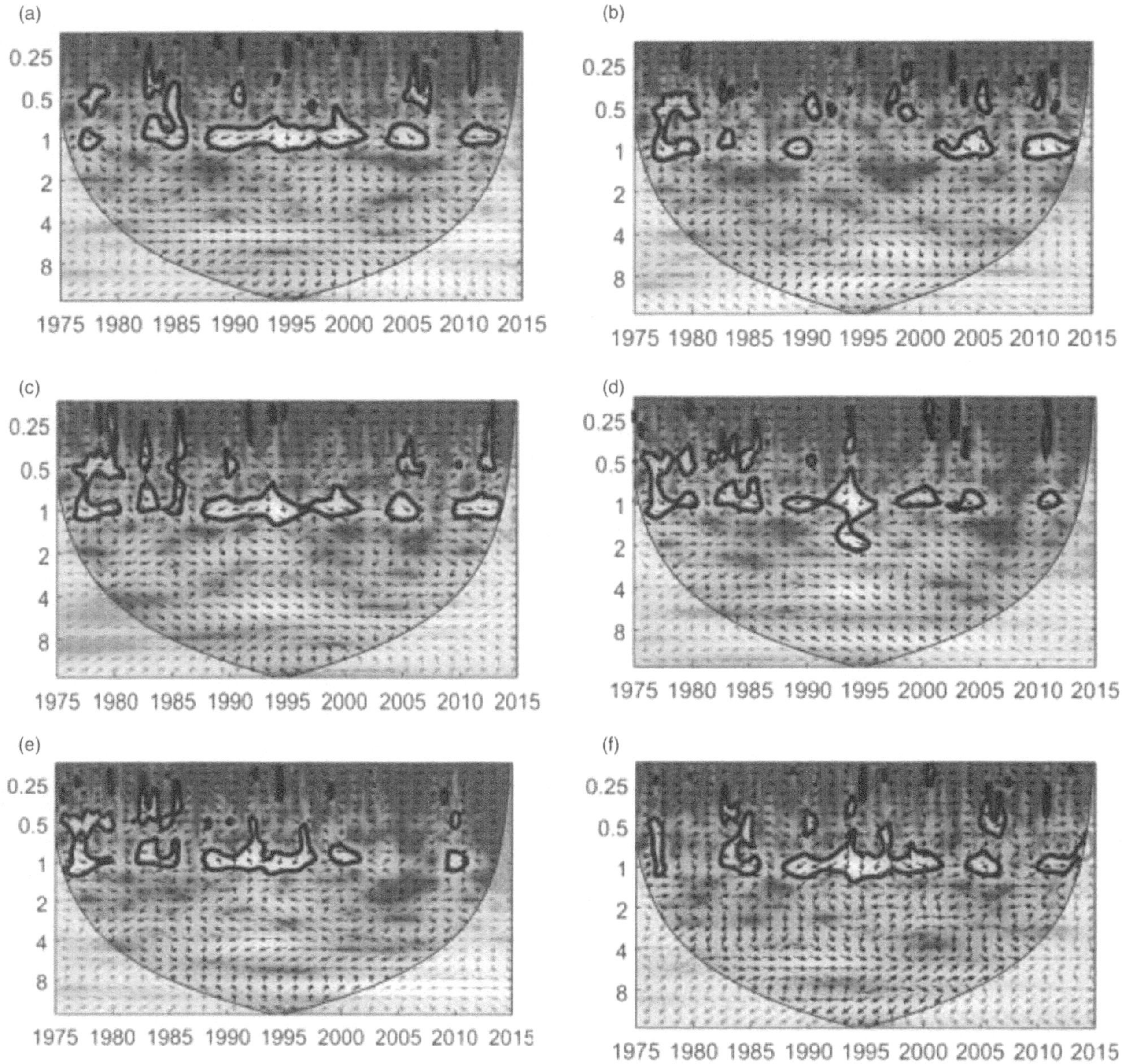

Figure 13 | Cross-wavelet analysis of PDO & streamflow at all the stations. (a) BA, (b) CK, (c) GT, (d) MB, (e) NG and (f) SA. The sections marked by black color contours are the significant sections indicating a high correlation between precipitation and streamflow. The cone of influence indicates the area affected by the boundary assumption.

condition affects the Indian summer monsoon rainfall. Due to pressure and the temperature difference between oceanic surface and landmass of Indian subcontinent moisture driven current move towards land during the normal condition. However, during NIÑO condition (warm phase of ENSO), sea surface temperature increases. The temperature difference between landmass and sea gets minimized, further hindering the adequate moisture movement towards a landmass. Our analysis, in congruence with other studies (Maity *et al.* 2007; Panda *et al.* 2013; Jena *et al.* 2014), observed a positive association between streamflow and precipitation, concluding PDO (positive phase) and NIÑO 3.4 both negatively impact the summer monsoon rainfall and thus reduced the streamflow in the study area.

The Indian Ocean Dipole plays an essential role in deriving hydrologic variables like streamflow and precipitation and acts as a modulator of the Indian monsoon (Ashok *et al.* 2001). However, observed results showed little influence in streamflow and precipitation on considered basins, highlighting Indian streamflow teleconnections spatial diversity. This may be due to the fact that IOD alone does not influence the hydrologic variables of the Indian basin as compared with ENSO, which is also

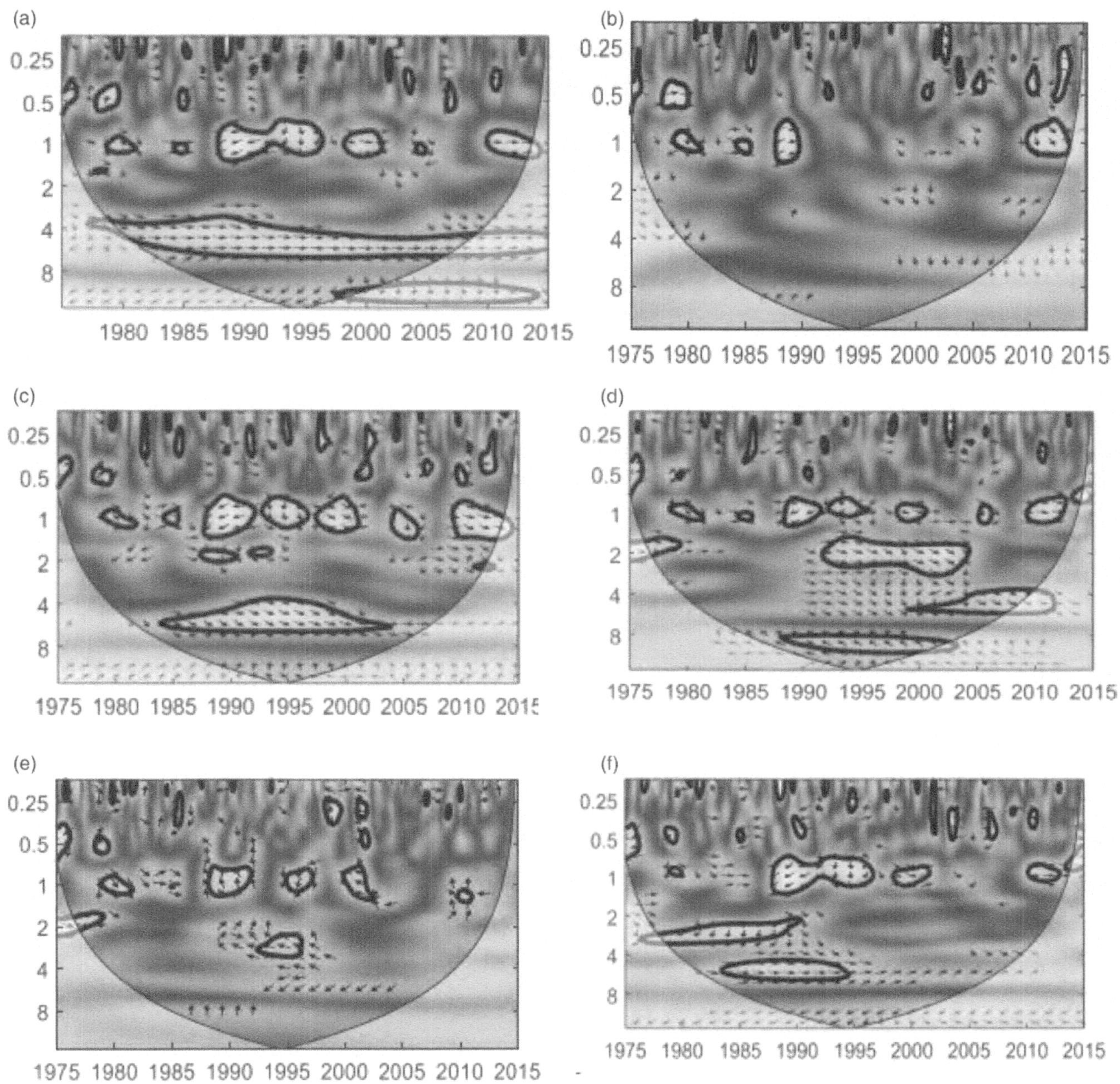

Figure 14 | Wavelet coherence analysis of PDO & streamflow at all the stations (a) BA, (b) CK, (c) GT, (d) MB, (e) NG and (f) SA. The sections marked by black color contours are the significant sections indicating high coherence between precipitation and streamflow. The cone of influence indicates the area affected by the boundary assumptions.

accepted by many researchers like Ashok *et al.* (2003), Marchant *et al.* (2007), Saji & Yamagata (2003), Ashok *et al.* (2003), Saji & Yamagata (2003), and Marchant *et al.* (2007). Recently Crétat *et al.* (2017) and Rathinasamy *et al.* (2019) studied the impact of IOD on the Indian Summer Monsoon by removing the influence of ENSO and found that it did not force the monsoon circulation in the absence of ENSO. Pokhrel *et al.* 2012 also reported the Indian summer monsoon's better prediction by ENSO and IOD's combined influence. This indicated the need to examine further the influence of these two indices' combined effect on the streamflow variability on the Indian peninsular basin.

Like IOD, it is observed that the significant areas of the strong feature in XWT and WTC are less with NAO than other indices, indicating lower NAO dominance in these basins. This is well supported by a previous study by Yadav *et al.* (2009). During the positive phase of NAO, the pressure at the polar region decreases and mid-latitude (Indian subcontinent) increases. This positive phase of NAO intensifies the westerly jet stream over the Middle East, strengthening the western disturbance ((Nageswararao *et al.* 2016). These disturbances mainly affect the northwest region of the Indian subcontinent by heavy rainfall in the winter season. Our study area is around the peninsular region that observed significantly less effect of western disturbance and hence was not affected by NAO. Kurths *et al.* (2019) also confirmed a weaker linkage between NAO and hydrometeorologic variables of the southern region of India.

6. CONCLUSIONS

In this study, India's six different virgin catchments were analyzed in terms of runoff variability and its association with precipitation and climate indices. The Mann-Kendall test and step-change analysis were carried out to detect the gradual long-term changes in the streamflow. Only Narmada and Cauvery showed significant negative gradual changes, whereas a significant step change was observed after 1985 in all basins. Wavelet analysis was used to understand the multi-scale association of basin streamflow with the precipitation and climatic indices. Results revealed the effect of precipitation and climate indices on the streamflow data of the selected river basin. Precipitation showed an in-phase relationship, whereas climate indices NIÑO 3.4 and PDO observed lead, IOD, and NAO showed an in-phase relationship with most stations. Indeed, our preliminary results dispense a better understanding of the interrelationship between the streamflow, precipitation and climate indices. However, the analysis must be further extended on more stations in the basin to strengthen the findings, especially hydroclimatic teleconnections. Similarly, along with precipitation and climate indices, other variables may trigger the river basin changes. For instance, land use and land cover changes or any other anthropogenic changes in the respective regions should be further explored.

ACKNOWLEDGEMENTS

Rathinasamy Maheswaran gratefully acknowledges the financial assistance from SERB under the scheme Early Career Research Award (ECR/16/1721). Furthermore, AA acknowledges the funding support provided by the Indian Institute of Technology Roorkee through Faculty Initiation Grant number IITR/SRIC/1808/F.I.G and COPREPARE project funded by UGC and DAAD under the IGP 2020-2024.

COMPETING INTERESTS

The authors declare that they have no conflict of interest.

DATA AVAILABILITY STATEMENT

All relevant data are available from http://www.imdpune.gov.in/Clim_Pred_LRF_New/Grided_Data_Download.html.

REFERENCES

Abeysingha, N. S., Singh, M., Sehgal, V. K., Khanna, M. & Pathak, H. 2016 Analysis of trends in streamflow and its linkages with rainfall and anthropogenic factors in Gomti River basin of North India. *Theoretical and Applied Climatology* **123**, 785–799.

Adamowski, J., Adamowski, K. & Prokoph, A. 2013 Quantifying the spatial temporal variability of annual streamflow and meteorological changes in eastern Ontario and southwestern Quebec using wavelet analysis and GIS. *Journal of Hydrology* **499**, 27–40.

Adeloye, A. J. & Montaseri, M. 2002 Preliminary streamflow data analyses prior to water resources planning study/Analyses préliminaires des données de débit en vue d'une étude de planification des ressources en eau. *Hydrological Sciences Journal* **47**, 679–692.

Agarwal, A. 2019 *Unraveling Spatio-Temporal Climatic Patterns Via Multi-Scale Complex Networks*. Universität Potsdam, Potsdam.

Agarwal, A., Maheswaran, R., Sehgal, V., Khosa, R., Sivakumar, B. & Bernhofer, C. 2016 Hydrologic regionalization using wavelet-based multiscale entropy method. *Journal of Hydrology* **538**, 22–32.

Agarwal, A., Marwan, N., Rathinasamy, M., Merz, B. & Kurths, J. 2017 Multi-scale event synchronization analysis for unravelling climate processes: a wavelet-based approach. *Nonlinear Processes in Geophysics* **24**, 599–611.

Agarwal, A., Caesar, L., Marwan, N., Maheswaran, R., Merz, B. & Kurths, J. 2019 Network-based identification and characterization of teleconnections on different time scales. *Scientific Reports* **9**, 8808. https://www.nature.com/articles/s41598-019-45423-5.

Agarwal, A., Marwan, N., Maheswaran, R., Ozturk, U., Kurths, J. & Merz, B. 2020 Optimal design of hydrometric station networks based on complex network analysis. *Hydrology and Earth System Sciences* **24**, 2235–2251.

Ashok, K., Guan, Z. & Yamagata, T. 2001 Impact of the Indian Ocean dipole on the relationship between the Indian monsoon rainfall and ENSO. *Geophysical Research Letters* **28**, 4499–4502.

Ashok, K., Guan, Z. & Yamagata, T. 2003 Influence of the Indian Ocean Dipole on the Australian winter rainfall: INFLUENCE OF THE INDIAN OCEAN DIPOLE. *Geophysical Research Letters* **30**.

Brutsaert, W. & Parlange, M. B. 1998 Hydrologic cycle explains the evaporation paradox. *Nature* **396**, 30–30.

Burn, D. H. & Hag Elnur, M. A. 2002 Detection of hydrologic trends and variability. *Journal of Hydrology* **255**, 107–122.

Canchala, T., Loaiza Cerón, W., Francés, F., Carvajal-Escobar, Y., Andreoli, R., Kayano, M., Alfonso-Morales, W., Caicedo-Bravo, E. & Ferreira de Souza, R. 2020 Streamflow variability in Colombian pacific basins and their teleconnections with climate indices. *Water* **12**, 526.

Chawla, I., Osuri, K. K., Mujumdar, P. P. & Niyogi, D. 2018 Assessment of the weather research and forecasting (WRF) model for simulation of extreme rainfall events in the upper Ganga Basin. *Hydrology and Earth System Sciences* **22**, 1095–1117.

Cherinet, A. A., Yan, D., Wang, H., Song, X., Qin, T., Kassa, M. T., Girma, A., Dorjsuren, B., Gedefaw, M., Wang, H. & Yadamjav, O. 2019 Impacts of recent climate trends and human activity on the land cover change of the Abbay River Basin in Ethiopia. *Advances in Meteorology* **2019**, 1–14.

Crétat, J., Terray, P., Masson, S., Sooraj, K. P. & Roxy, M. K. 2017 Indian Ocean and Indian summer monsoon: relationships without ENSO in ocean–atmosphere coupled simulations. *Climate Dynamics* **49**, 1429–1448.

Cunderlik, J. M. & Burn, D. H. 2002 Analysis of the linkage between rain and flood regime and its application to regional flood frequency estimation. *Journal of Hydrology* **261**, 115–131.

Das, S. 2019 Four decades of water and sediment discharge records in Subarnarekha and Burhabalang basins: an approach towards trend analysis and abrupt change detection. *Sustainable Water Resources Management* **5**, 1665–1676.

Das, P., Behera, M. D., Patidar, N., Sahoo, B., Tripathi, P., Behera, P. R., Srivastava, S. K., Roy, P. S., Thakur, P., Agrawal, S. P. & Krishnamurthy, Y. V. N. 2018 Impact of LULC change on the runoff, base flow and evapotranspiration dynamics in eastern Indian river basins during 1985–2005 using variable infiltration capacity approach. *Journal of Earth System Science* **127**, 19.

Daubechies, I. 1990 The wavelet transform, time-frequency localization and signal analysis. *IEEE Transactions on Information Theory* **36**, 961–1005.

Fanta, B., Zaake, B. T. & Kachroo, R. K. 2001 A study of variability of annual river flow of the southern African region. *Hydrological Sciences Journal* **46**, 513–524.

Fathian, F., Dehghan, Z., Bazrkar, M. H. & Eslamian, S. 2016 Trends in hydrological and climatic variables affected by four variations of the Mann-Kendall approach in Urmia Lake basin, Iran. *Hydrological Sciences Journal* **61**, 1–13.

Gouda, K., Sahoo, S., Samantray, P. & Shivappa, H. 2017 Comparative study of monsoon rainfall variability over India and the Odisha State. *Climate* **5**, 79.

Grinsted, A., Moore, J. C. & Jevrejeva, S. 2004 Application of the cross wavelet transform and wavelet coherence to geophysical time series. *Nonlinear Processes in Geophysics* **11**, 561–566.

Grossmann, A. & Morlet, J. 1984 Decomposition of hardy functions into square integrable wavelets of constant shape. *SIAM Journal on Mathematical Analysis* **15**, 723–736.

Guntu, R. K., Rathinasamy, M., Agarwal, A. & Sivakumar, B. 2020 Spatiotemporal variability of Indian rainfall using multiscale entropy. *Journal of Hydrology* **587**, 124916.

Himayoun, D. & Roshni, T. 2019 Spatio-temporal variation of drought characteristics, water resource availability and the relation of drought with large scale climate indices: a case study of Jhelum basin, India. *Quaternary International* **525**, 140–150.

Hu, Y., Maskey, S., Uhlenbrook, S. & Zhao, H. 2011 Streamflow trends and climate linkages in the source region of the Yellow River, China. *Hydrological Processes* **25**, 3399–3411.

Islam, A., Sikka, A. K., Saha, B. & Singh, A. 2012 Streamflow response to climate change in the Brahmani River Basin, India. *Water Resources Management* **26**, 1409–1424.

Jain, S. K., Nayak, P. C., Singh, Y. & Chandniha, S. K. 2017 Trends in rainfall and peak flows for some River Basins in India. *Current Science* **112**, 1712.

Jena, P. P., Chatterjee, C., Pradhan, G. & Mishra, A. 2014 Are recent frequent high floods in Mahanadi basin in eastern India due to increase in extreme rainfalls? *Journal of Hydrology* **517**, 847–862.

Kahya, E. & Kalaycı, S. 2004 Trend analysis of streamflow in Turkey. *Journal of Hydrology* **289**, 128–144.

Kawahara, Y. & Sugiyama, M. 2012 Sequential change-point detection based on direct density-ratio estimation. *Statistical Analysis and Data Mining* **5**, 114–127.

Khatiwada, K., Panthi, J., Shrestha, M. & Nepal, S. 2016 Hydro-climatic variability in the Karnali River Basin of Nepal Himalaya. *Climate* **4**, 17.

Kiely, G., Albertson, J. D. & Parlange, M. B. 1998 Recent trends in diurnal variation of precipitation at Valentia on the west coast of Ireland. *Journal of Hydrology* **207**, 270–279.

Kim, T.-W. & Valdés, J. B. 2003 Nonlinear model for drought forecasting based on a conjunction of wavelet transforms and neural networks. *Journal of Hydrologic Engineering* **8**, 319–328.

Krishnamurthy, L. & Krishnamurthy, V. 2014 Influence of PDO on South Asian summer monsoon and monsoon–ENSO relation, 14.

Kuriqi, A., Ali, R., Pham, Q. B., Montenegro Gambini, J., Gupta, V., Malik, A., Linh, N. T. T., Joshi, Y., Anh, D. T., Nam, V. T. & Dong, X. 2020 Seasonality shift and streamflow flow variability trends in central India. *Acta Geophysica* **68**, 1461–1475.

Kurths, J., Agarwal, A., Shukla, R., Marwan, N., Rathinasamy, M., Caesar, L., Krishnan, R. & Merz, B. 2019 Unravelling the spatial diversity of Indian precipitation teleconnections via a non-linear multi-scale approach. *Nonlinear Processes in Geophysics* **26**, 251–266.

Kustu, M. D., Fan, Y. & Robock, A. 2010 Large-scale water cycle perturbation due to irrigation pumping in the US high plains: a synthesis of observed streamflow changes. *Journal of Hydrology* **390**, 222–244.

Lau & Weng, 1995 lau1995.pdf.

Lins, H. F. & Slack, J. R. 1999 Streamflow trends in the United States. *Geophysical Research Letters* **26**, 227–230.

Maheswaran, R. & Khosa, R. 2012 Comparative study of different wavelets for hydrologic forecasting. *Computers & Geosciences* **46**, 284–295.

Maity, R. & Nagesh Kumar, D. 2008 Basin-scale stream-flow forecasting using the information of large-scale atmospheric circulation phenomena. *Hydrological Processes* **22**, 643–650.

Maity, R., Kumar, D. N. & Nanjundiah, R. S. 2007 Review of hydroclimatic teleconnection between hydrologic variables and large-scale atmospheric circulation patterns with Indian perspective. *ISH Journal of Hydraulic Engineering* **13**, 77–92.

Mallat, S. G. 1999 *A Wavelet Tour of Signal Processing*, 2nd edn. Academic Press, San Diego.

Marchant, R., Mumbi, C., Behera, S. & Yamagata, T. 2007 The Indian Ocean dipole? the unsung driver of climatic variability in East Africa. *African Journal of Ecology* **45**, 4–16.

Masih, I., Uhlenbrook, S., Maskey, S. & Smakhtin, V. 2011 Streamflow trends and climate linkages in the Zagros Mountains, Iran. *Climatic Change* **104**, 317–338.

Miguez-Macho, G. & Fan, Y. 2012 The role of groundwater in the Amazon water cycle: 2. Influence on seasonal soil moisture and evapotranspiration: AMAZON GROUNDWATER AND SEASONAL ET. *Journal of Geophysical Research: Atmospheres* **117**, n/a–n/a.

Milner, V. S., Yarnell, S. M. & Peek, R. A. 2019 The ecological importance of unregulated tributaries to macroinvertebrate diversity and community composition in a regulated river. *Hydrobiologia* **829**, 291–305.

Murthy, D. S., Jyothy, S. A. & Mallikarjuna, P. 2017 Trend analysis of annual maximum daily streamflows at the gauging sites of Godavari River Basin. *Hydrology Journal* **40 & 41** (1–4), 52–57.

Nageswararao, M. M., Mohanty, U. C., Ramakrishna, S. S. V. S., Nair, A. & Prasad, S. K. 2016 Characteristics of winter precipitation over Northwest India using high-resolution gridded dataset (1901–2013). *Global and Planetary Change* **147**, 67–85.

Nalley, D., Adamowski, J., Khalil, B. & Biswas, A. 2016 Inter-annual to inter-decadal streamflow variability in Quebec and Ontario in relation to dominant large-scale climate indices. *Journal of Hydrology* **536**, 426–446.

Pai, D. S., Sridhar, L., Rajeevan, M., Sreejith, O. P., Satbhai, N. S. & Mukhopadhyay, B. 2014 Development of a new high spatial resolution (0.25° × 0.25°) long period (1901–2010) daily gridded rainfall data set over India and its comparison with existing data sets over the region, 18.

Panda, D. K., Kumar, A., Ghosh, S. & Mohanty, R. K. 2013 Streamflow trends in the Mahanadi River basin (India): linkages to tropical climate variability. *Journal of Hydrology* **495**, 135–149.

Pokhrel, S., Chaudhari, H. S., Saha, S. K., Dhakate, A., Yadav, R. K., Salunke, K., Mahapatra, S. & Rao, S. A. 2012 ENSO, IOD and Indian Summer Monsoon in NCEP climate forecast system. *Climate Dynamics* **39**, 2143–2165.

Rao, P. G. 1995 Effect of climate change on streamflows in the Mahanadi River Basin, India. *Water International* **20**, 205–212.

Rathinasamy, M., Agarwal, A., Sivakumar, B., Marwan, N. & Kurths, J. 2019 Wavelet analysis of precipitation extremes over India and teleconnections to climate indices. *Stochastic Environmental Research and Risk Assessment* **33**, 2053–2069.

Reeves, J., Chen, J., Wang, X. L., Lund, R. & Lu, Q. Q. 2007 A review and comparison of changepoint detection techniques for climate data. *Journal of Applied Meteorology and Climatology* **46**, 900–915.

Sahu, N., Behera, S. K., Yamashiki, Y., Takara, K. & Yamagata, T. 2012 IOD and ENSO impacts on the extreme stream-flows of Citarum river in Indonesia. *Climate Dynamics* **39**, 1673–1680.

Sahu, N., Panda, A., Nayak, S., Saini, A., Mishra, M., Sayama, T., Sahu, L., Duan, W., Avtar, R. & Behera, S. 2020 Impact of indo-pacific climate variability on high streamflow events in Mahanadi River Basin, India. *Water* **12**, 1952.

Saji, N. & Yamagata, T. 2003 Possible impacts of Indian ocean dipole mode events on global climate. *Climate Research* **25**, 151–169.

Sayemuzzaman, M. & Jha, M. K. 2014 Seasonal and annual precipitation time series trend analysis in North Carolina, United States. *Atmospheric Research* **137**, 183–194.

Sharma, P. J., Patel, P. L. & Jothiprakash, V. 2019 Impact of rainfall variability and anthropogenic activities on streamflow changes and water stress conditions across Tapi Basin in India. *Science of The Total Environment* **687**, 885–897.

Smith, T. M. & Reynolds, R. W. 2003 Extended reconstruction of global Sea surface temperatures based on COADS data (1854–1997). *Journal of Climate* **16**, 16.

Sridhar, V., Kang, H. & Ali, S. A. 2019 Human-induced alterations to land use and climate and their responses for hydrology and water management in the Mekong River Basin. *Water* **11**, 1307.

Swain, J. B. & Patra, K. C. 2017 Streamflow estimation in ungauged catchments using regional flow duration curve: comparative study. *Journal of Hydrologic Engineering* **22**, 04017010.

Tamaddun, K., Kalra, A. & Ahmad, S. 2016 Identification of streamflow changes across the continental United States using variable record lengths. *Hydrology* **3**, 24.

Taye, M. T. & Willems, P. 2013 Identifying sources of temporal variability in hydrological extremes of the upper Blue Nile basin. *Journal of Hydrology* **499**, 61–70.

Vandana, K., Islam, A., Sarthi, P. P., Sikka, A. K. & Kapil, H. 2019 Assessment of potential impact of climate change on streamflow: a case study of the Brahmani River basin, India. *Journal of Water and Climate Change* **10**, 624–641.

Yadav, R. K., Rupa Kumar, K. & Rajeevan, M. 2009 Increasing influence of ENSO and decreasing influence of AO/NAO in the recent decades over northwest India winter precipitation. *Journal of Geophysical Research* **114**, D12112.

Zhao, G., Tian, P., Mu, X., Jiao, J., Wang, F. & Gao, P. 2014 Quantifying the impact of climate variability and human activities on streamflow in the middle reaches of the Yellow River basin, China. *Journal of Hydrology* **519**, 387–398.

Zhong, F., Cheng, Q. & Ge, Y. 2019 Relationships between spatial and temporal variations in precipitation, climatic indices, and the normalized differential vegetation index in the upper and middle reaches of the Heihe River Basin, Northwest China. *Water* **11**, 1394.

Zhou, H., Peng, Y. & Liang, G. 2008 The research of monthly discharge predictor-corrector model based on wavelet decomposition. *Water Resources Management* **22**, 217–227.

First received 11 May 2021; accepted in revised form 9 October 2021. Available online 26 October 2021

© 2022 The Authors

Impact of Climate Change on Hydrology and Water Resources doi: 10.2166/wcc.2021.331

Variability and trends of climate extremes indices from the observed and downscaled GCMs data over 1950–2020 period in Chattogram City, Bangladesh

Lia Pervin * and Md. Sabbir Mostafa Khan

Department of Water Resources Engineering, Bangladesh University of Engineering and Technology (BUET), Dhaka 1000, Bangladesh
*Corresponding author. E-mail: liapervin@gmail.com

LP, 0000-0002-9799-7585

ABSTRACT

This study was intended to evaluate the variability and trends of climate extremes by incorporating daily data from Chattogram station and from the high-resolution Coordinated Regional Climate Downscaling Experiment (CORDEX) for two different time series. Here, we also focused on evaluating the performance of the selected RCMs (CanESM2, CSIRO, and GFDL from CORDEX) using Taylor diagrams and heat map analysis. Twenty-two extreme climate indices from ETCCDI were computed for the 1950–1989 and 1990–2020 periods. Mann–Kendall and Sen's slope test were performed to estimate the trends from the indices from both station and RCMs data. Highly significant increasing trend for the warm days and warm nights' frequencies were found, whereas the frequency of cold days and cold nights indicated a significantly decreasing trend. On the other hand, a mild increasing trend in 1-day and 5-day maximum rainfall was detected. Also, the average annual precipitation has increased by 6% from the 1950–1989 to 1990–2020 period. During the last three decades, the region has experienced heavier rainfall in the monsoon but increased water stress in the dry season. The two-fold effects of climate change on the local hydrology revealed by this study need to be addressed properly for the sustainable development of this region.

Key words: climate change, climate extremes, heat map, trends, variability

HIGHLIGHTS

- Performance of the selected RCMs was evaluated using Taylor diagrams and Heatmaps analysis for the Chittagong region.
- Variability of climate extremes from 1950–1989 to 1990–2020 was assessed.
- A total of 22 extreme climate indices were computed and their changes over time were investigated.
- Trend analysis was performed for the extreme rainfall and temperature indices from both station and RCMs data.

GRAPHICAL ABSTRACT

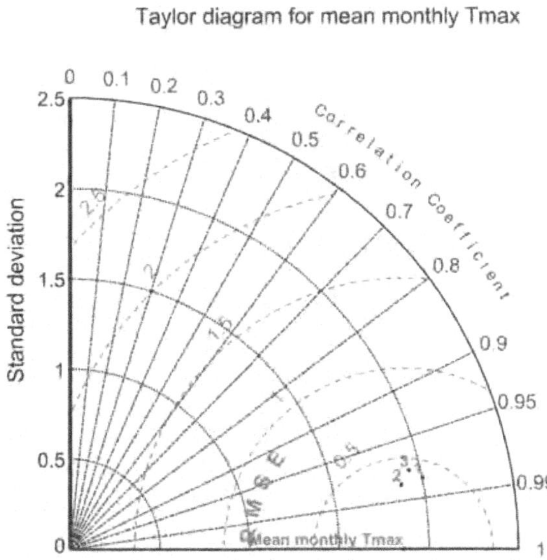

Taylor diagram for Monthly mean Tmax over 1990-2020

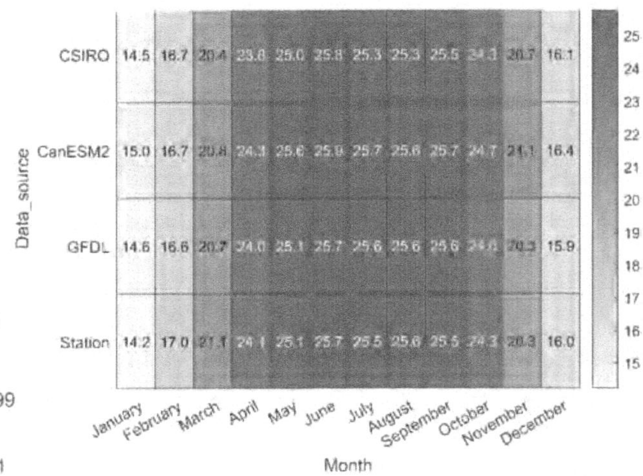

Heat map for mean monthly Tmin over 1990-2020 period

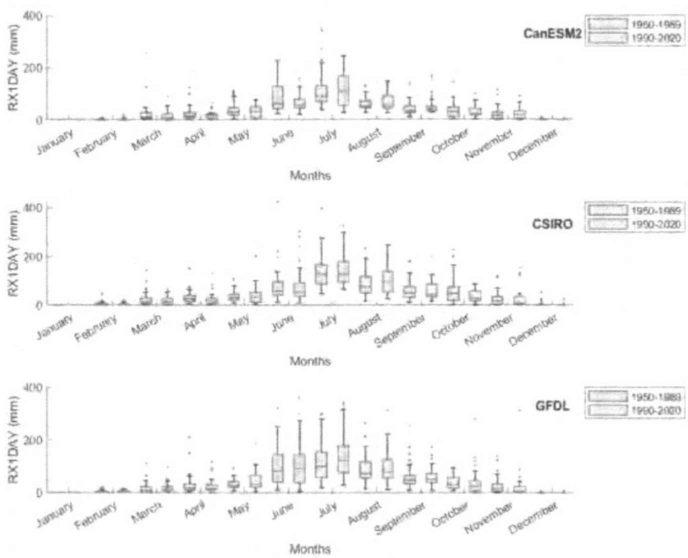

Box-whisker plots for RX1DAY over 1990-2020 period from CanESM2, CSIRO and GFDL

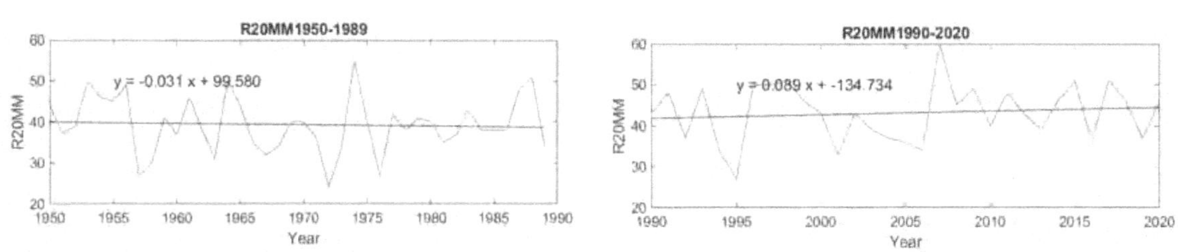

Plot of extreme rainfall indices R20MM over 1950-1989 and 1990-2020 period

INTRODUCTION

Chattogram, the second-largest city and the main port city of Bangladesh, has a history of suffering from floods and waterlogging in every monsoon. Frequent and prolonged waterlogging due to intense rainfall is becoming more common and is causing losses to the economy, as well as to the environment (Tang *et al.* 2018). Various factors, including unplanned urbanization, deterioration of natural outlets, and the poor drainage conditions in Chattogram affect this waterlogging condition. The impact of climate change in recent years has changed the magnitude and variability of the occurrences of extreme events (Raihan *et al.* 2015), which has elevated the waterlogging problems in Chattogram. This claim became stronger when Akter *et al.* (2017) found that more locations in Chattogram City have been experiencing frequent and prolonged waterlogging in recent years. Islam & Raja (2021) prepared a waterlogging inventory map to show the spatial variations of the risk by means of hazard intensity, exposure, and vulnerability to waterlogging in Chattogram, which indicated about 3.03% of the city area is highly vulnerable to waterlogging. The city has a pre-existing drainage condition, which has been getting worse in recent decades. It is suspected that climate change has altered the characteristics of extreme temperature and rainfall events over time.

Bowden *et al.* (2013) reported that globally extreme weather events are changing in terms of their extensiveness, severity, and frequency of occurrence. Changes in mean temperature, accompanied by a rise in minimum temperature, and increase in the frequency of extreme temperatures associated with heatwaves were reported in many regions, including Asia (Klein Tank *et al.* 2006). In Bangladesh, changes in temperature and rainfall patterns have also been reported (Shahid 2010; Shahid *et al.* 2012; Hossain *et al.* 2014; Nowreen *et al.* 2015; Chowdhury *et al.* 2019). An increase in annual and seasonal temperatures as well as climate-related extreme events were detected by Shahid *et al.* (2016) over the 1958–2012 period, whereas Basher *et al.* (2018) identified a decreasing trend in the seasonal total rainfall and consecutive wet days over northeast Bangladesh for the 1984–2016 period. Mullick *et al.* (2019) stated that the climate of Bangladesh is getting cooler and drier for winter and, on the other hand, warmer and wetter for the rest of the year, after analyzing temperature and rainfall data from 1966 to 2015. Rahman & Lateh (2017) found upward trends in minimum temperature in the northern, northwestern, northeastern, central, and central southern parts while the greatest warming in the maximum temperature was in the southern, southeastern, and northeastern parts during 1971–2010. An upward trend of annual rainfall and downward pre-monsoon and post-monsoon rainfall trends was detected during this period.

As reflected in the literature, in recent years, climate change has created new challenges. However, not many studies were reported focusing on the extreme indices and variabilities of local climate in Chattogram. Besides, how this trend and variabilities could have been affected by the ongoing climate change also remain undetermined. Comprehensive scientific knowledge-based studies are required to understand the effect of climate change on the local hydroclimatic system. As noted, there are certain limitations and inconsistencies in the previous studies. Most of the studies were conducted based on various station data to identify the trends of temperature and/or rainfall over a certain period of time, however, their findings were not aligned. It seems more appropriate to assess the temperature and rainfall characteristics from two different epochs (past and present) to gain more comprehensive knowledge on the rate and pattern of changes. Dash & Maity (2019) evaluated precipitation-based climate change indices over three 35-year epochs and analyzed the spatio-temporal trend to extract the characteristics of changes for India. However, our extensive literature review found no studies for our region to investigate both temperature and rainfall indices and trends from two different time series and characterize their changes over time.

In recent years, powerful general circulation models (GCMs), commonly known as global climate models, have been developed, which can model the historical climate of Earth reasonably well. Therefore, the GCMs are considered as the base to project the future climate; however, downscaled GCMs or regional climate model (RCM) data are more competent for local climate modeling than the raw GCMs data (Pervin & Gan 2021). Still, not many studies have utilized the combination of RCMs and station data to investigate the long-term changes in the climate extremes for this region. On the other hand, selecting the appropriate RCMs for any region is the key to assessing the future climate change over that region. However, lack of information about the applicability of particular RCMs for climate change analysis was noticed for this region.

The focus of this study was to investigate the extreme climate indices and to evaluate the trends and variabilities of temperature and rainfall over two different periods, incorporating the most recent available data sets (1950–2020), from the station as well as from the high-resolution (25 km × 25 km) Coordinated Regional Climate Downscaling Experiment (CORDEX). Another objective was to evaluate the performance of the selected RCMs, and thus to offer some options for choosing the regional climate models in this region. This study will help to minimize the study gap on applicability of different RCMs for climate change analysis in this region. Evaluation of indices from two different time periods will help to compare the pace of climate changes and to recognize their effects on the extreme climate events. The outcome from this study will contribute to the knowledge base for understanding and quantifying the potential impact of climate change and thus will identify the vulnerabilities associated with this change. Moreover, the policymaker may find the results useful to adopt revised water management policies.

STUDY AREA AND DATA

Bangladesh is located in South Asia and the geographic coordinates are between 20°34′ to 26°38′ N and 88°01′ to 92°41′ E. The country experiences high temperatures and humidity; it also receives a great deal of annual rainfall (average 2,900 mm/year in the last 30 years), where most of the rainfall (almost 80%) occurs during the monsoon season (June to October). Chattogram (previously called Chittagong), being the second largest city, located in the southeastern region, is also the main port city of Bangladesh. It is also considered the major economic zone of Bangladesh. The geographical location of Chattogram is 22.13° to 22.28° N latitude and 91.45° to 91.54° E longitude. The floodplain and piedmont plains (landform created at the foot of a hill) of Chattogram are poorly drained and subject to waterlogging and flash floods during monsoon season, and each year, the climate extremes are linked with disasters, costing lives and properties in Chattogram (Brammer 2012). Figure 1(a) and 1(b) illustrate the study area location map and the rain gauge location.

In the entire Chattogram district, there are only two meteorological stations: one at Chattogram City and the other at Cox's Bazar. The Chattogram station rainfall records have 2.3% missing data, whereas the Cox's Bazar station has fewer (1.3%) missing data over the 1958–2007 period (Shahid 2011). However, after 2007, the missing data problems do not persist in the Chattogram station, since regular maintenance and continuous data measurement have been carried out. Shah Amanat International Airport meteorological station is the main meteorological station in the Chattogram City Corporation

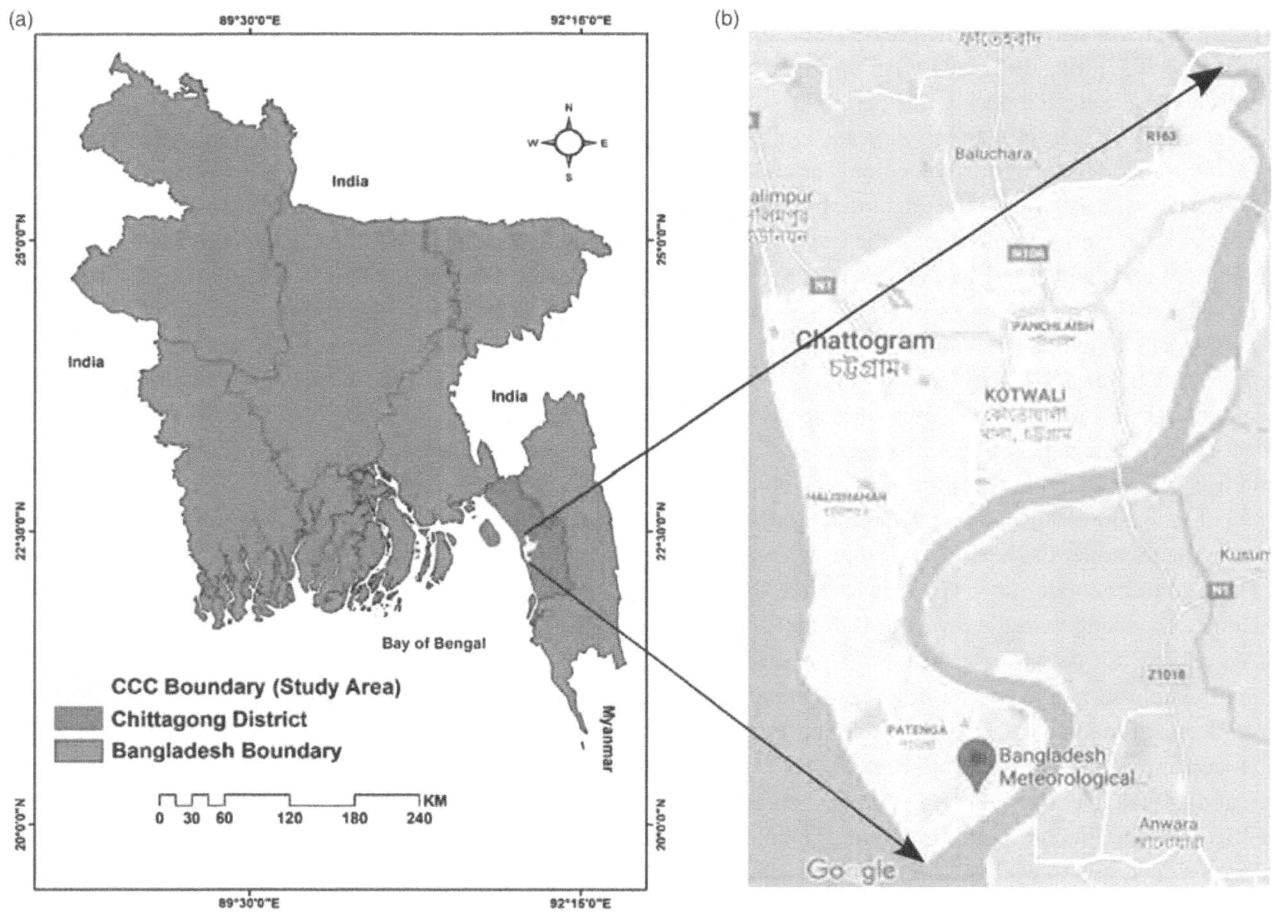

Figure 1 | The study area location map: (a) Chattogram (Chittagong) City Corporation boundary and (b) Bangladesh meteorological station at Shah Amanat International Airport.

area and is shown in Figure 1(b). We have collected the daily rainfall data and the daily maximum and minimum temperature data for the 1950–2020 period from the Shah Amanat International Airport meteorological station. However, meteorological data at this station are found missing for several days in 1971. A commonly used simple method for estimation of missing data suggested by Sattari *et al.* (2016) was applied to fill in missing meteorological data. The arithmetic average of the data corresponding to the nearest weather stations was taken to compute the missing value. Data quality control was carried out before the estimation of indices. A number of checks were carried out for quality controls of data such as precipitation values below 0 mm, winter rainfall higher than 100 mm, and more than 6 consecutive dry days during monsoon. Histograms of the data reveal no existing problems for the Chattogram station rainfall distribution. The Student's t-test was applied to the rainfall time-series data to check the homogeneity, and no significant variation was found at the 95% level of confidence.

Regional climate model data

The GCMs simulate global conditions with spatial scales of more than hundreds of kilometers; however, much finer resolution data are required for analyzing the local hydrology. CORDEX is an international effort to downscale global climate model data into a regional scale which is much finer in resolution. In recent years, studies have used the CORDEX data to assess the climate indices in various regions including India (Adeyeri *et al.* 2019; Nengzouzam *et al.* 2020). The downscaled climate data from CORDEX are publicly accessible at http://cccr.tropmet.res.in/home/nex_gddp_india.jsp. The dataset includes downscaled projections for Representative Concentration Pathway (RCP) 4.5 and RCP 8.5 from the 21 models and scenarios, for which daily scenarios were produced and distributed under CMIP5. Each of the climate projections include daily maximum temperature, minimum temperature, and precipitation for the periods from 1950 to 2100. The spatial resolution of the dataset is 0.25 degrees (~25 km×25 km).

Several studies have been carried out in recent years to review the criterion for the selection of GCMs or multi-model ensembles over the Indian sub-continent (Pandey *et al.* 2019; Sreelatha & Anand Raj 2019). However, we have learned about some weaknesses of the multi-model ensembles. Bannister *et al.* (2017) cautioned that the multi-model ensembles are not capable of considering the relative strengths and weaknesses of each model as an ensemble invariably hides the substantial variations between the individual models. Alexandru & Sushama (2015) found the CanESM2-driven simulations of precipitation were more consistent than the other selected models (MPI-ESM-LR)-driven simulations in India. Panjwani *et al.* (2019) evaluated 12 CMIP5-GCMs over India for precipitation, minimum temperature, and maximum temperature. They the fuzzy analytic hierarchy process (FAHP) and reliability index to assess their ability. Indicators employed were agreement index (AI), RMSE, and CC. FAHP was found suitable to rank GCMs: NorESM1-M for maximum temperature; MIROC5, GFDLCM3, FIO-ESM, and IPSL-CM5A-LR for minimum temperature; IPSL-CM5A-LR, GFDL-ESM2M, HadGEM2, MIROC5, and CSIRO for precipitation were found suitable. However, no studies were found to suggest the most suitable GCMs for Bangladesh, since the application of GCMs and use of climate change scenarios from high-resolution RCMs are very limited in Bangladesh. As Bangladesh is a part of the Indian sub-continent, based on Indian literature we consider CanESM2, GFDL, and CSIRO for the current study.

The CanESM2 was developed by the Canadian Centre for Climate Modelling and Analysis (CCCma) of Environment and Climate Change Canada and is one of the high performing GCMs used by many researchers around the world (Gao *et al.* 2021; Javaherian *et al.* 2021; Pervin *et al.* 2021). The CSIRO model was first developed in Queensland Climate Change Centre, Australia and has been used by many researchers around the globe (Silva *et al.* 2017). The GFDL model results were extensively evaluated by Held *et al.* (2019) against observations from various locations and found to have high-performing results. We preferred these three individual RCMs based on their performance to represent the long-term climate over the Indian region. The historical data from CanESM2, CSIRO, and GFDL are available for the 1950–2005 period and for the 2006–2020 period the data are considered as projected. In this study, we have utilized the climate change scenario RCP 4.5 for the 2006–2020 period, as RCP 4.5 signifies moderate warming level. It was reported that the International Energy Agency (IEA) scenarios agree much more closely with scenarios like RCP 4.5 than the high-end no-policy RCP 8.5 scenario (https://www.pnas.org/content/117/45/27791, last accessed 21/10/2021). Moreover, Hausfather & Peters (2020) found that both near-term and long-term fossil CO_2 emissions are overestimated in emission scenarios associated with 8.5 W/m^2 forcing pathway.

Analysis techniques

Climate change indices

'Confidence has increased that some extremes will become more frequent, more widespread and/or more intense during the 21st century' (IPCC 2014). As a result, the need for scientific research-based information on weather and climate extremes is rising. To gain a uniform perspective on observed changes in weather and climate extremes, the joint Expert Team on Climate Change Detection and Indices (ETCCDI) has defined a core set of descriptive indices of extremes. The indices describe particular characteristics of extremes, including frequency, amplitude, and persistence. The core set includes 27 extremes indices for temperature and precipitation. The ETCCDI resource and information website (http://etccdi.pacificclimate.org/) is maintained by the University of Victoria, Canada. We have carefully chosen 22 ETCCDI indices for this study; their short descriptions are given in Tables 1 and 2.

Trend analysis

In most reported studies on hydrological time series analysis, the regression and/or Mann–Kendall (MK) tests are applied for trend detection (Machiwal & Jha 2012). The MK test is the classic non-parametric test used in the literature to assess monotonic trends in a time series (Mann 1945). The non-linear trend, as well as the turning point, can be derived from Kendall test statistics (Kendall 1975). The MK test has been found to be an excellent tool for trend detection and to find whether the existing trend is statistically significant or not (Ilori & Ajayi 2020; Praveen *et al.* 2020). Details of regression test for linear trend and the MK non-parametric trend test for hydrological time series data analysis can be found in Maity (2018). If a linear trend is present in a time series, then the true slope of the trend is estimated by using a simple non-parametric procedure proposed by Sen (1968) and Theil (1950).

RESULTS AND DISCUSSION

Evaluation of the selected RCMs

We have extracted the daily maximum temperature, daily minimum temperature, and daily rainfall data from the downscaled CanESM2, CSIRO, and GFDL models for the 1950–2020 period over the Chattogram station location. The data are analyzed and compared with the measured daily Tmax, daily Tmin, and daily rainfall data from the Chattogram station for the same period. Taylor diagrams are plotted using the subsequent statistics from each data set. Figure 2(a) illustrates the Taylor diagram for the mean monthly maximum temperature (Tmax) using the station data as the reference value for 1990–2020

Table 1 | Definitions of 11 temperature indices used in this study

Indices notation	Indices name	Units	Description
TXX	Monthly maximum value of daily maximum temperature	°C	Let TX_{kj} be the daily maximum temperatures in month *k*, period *j*. Then, TXX = max (TX_{kj})
TNX	Monthly maximum value of daily minimum temperature	°C	Let TN_{kj} be the daily minimum temperatures in month *k*, period *j*. Then, TNX = max (TN_{kj})
TXN	Monthly minimum value of daily maximum temperature	°C	Let TX_{kj} be the daily maximum temperatures in month *k*, period *j*. Then, TXN = min (TX_{kj})
TNN	Monthly minimum value of daily minimum temperature	°C	Let TN_{kj} be the daily minimum temperatures in month *k*, period *j*. Then, TNN = min (TN_{kj})
TN10p	Cold nights	% days	Percentage of days when TN <10th percentile of the threshold
TN90p	Warm nights	% days	Percentage of days when TN >90th percentile of the threshold
TX10p	Cold days	% days	Percentage of days when TX <10th percentile of the threshold
TX90p	Warm days	% days	Percentage of days when TX >90th percentile of the threshold
WSDI	Warm spell duration index	days	Annual count of days with at least 6 consecutive days when TX >90th percentile of the threshold
CSDI	Cold spell duration indicator	days	Annual count of days with at least 6 consecutive days when TN <10th percentile of the threshold
DTR	Diurnal temperature range	°C	Monthly mean difference between daily maximum and minimum temperatures

Table 2 | Definitions of 11 rainfall indices used in this study

Indices notation	Indices name	Units	Description
PRCPTOT	Annual total wet-day precipitation	mm	Annual total PRCP in wet days (RR ≥1 mm)
R95p	Very wet days	mm	Annual total PRCP when RR >95th percentile of the threshold
R99p	Extremely wet days	mm	Annual total PRCP when RR >99th percentile of the threshold
RX1day	Max 1-day precipitation amount	mm	Let RR_{ij} be the daily precipitation amount on day i in period j. The maximum 1-day value for period j is $R \times 1day = max (RR_{ij})$
RX5day	Maximum 5-day precipitation	mm	Let RR_{kj} be the precipitation amount for the 5-day ending k, period j. The maximum 5-day values for period j are: $R \times 5day = max (RR_{kj})$
R10 mm	Number of heavy precipitation days	days	Annual count of days when precipitation ≥10 mm
R20 mm	Number of very heavy precipitation days	days	Annual count of days when precipitation ≥20 mm
R100 mm	Number of extreme precipitation days	days	Annual count of days when precipitation ≥100 mm
CWD	Maximum length of wet spell	days	Let RR_{ij} be the daily precipitation amount on day i in period j. Count the largest number of consecutive days where: $RR_{ij} \geq 1$ mm
CDD	Maximum length of dry spell	days	Maximum number of consecutive days with precipitation <1 mm
SDII	Precipitation intensity index	mm	Let RR_{wj} be the total daily precipitation amount on wet days w in period j. If W represents number of wet days in j. Then, $SDII = (\sum RR_{wj})/W$

period. The station data for the 1990–2020 period have fewer missing values than that of the 1950–1989 period, and hence the recent data set was considered for evaluation purposes. Similarly, Figure 2(b) and 2(c) represent the Taylor diagrams for the mean monthly minimum temperature (Tmin) and mean monthly rainfall, respectively.

In Figure 2(a), the CanESM2, CSIRO, and GFDL data statistics are denoted by points 1, 2, and 3, respectively. The correlation coefficient for CanESM2 mean monthly maximum temperature (Tmax in °C) is 0.98, standard deviation 2.0, and the root mean square error is around 0.4. Whereas, the correlation coefficient for CSIRO mean monthly maximum temperature (Tmax in °C) is around 0.98, standard deviation 1.85, and RMSE is around 0.39, and the same statistics for GFDL are 0.97, 1.9, and 0.45, respectively. Figure 2(b) illustrates the correlation for mean monthly minimum temperature (Tmin in °C) using the CanESM2, CSIRO, and GFDL data sets which are denoted by numbers 1, 2, and 3, respectively. It is observed from Figure 2(b) that there exist good correlations between the observed data and the model data. For all the three data sets (CanESM2, CSIRO, and GFDL) the correlation coefficients are around 0.99, whereas the standard deviations are around 4 and the RMSEs are around 0.3. In the case of mean monthly rainfall (in mm), the correlation coefficients are around 0.97, and the standard deviation for CanESM2 is 250, for CSIRO it is 255, and for GFDL it is 280. The RMSE is 50, 55, and 60 for CanESM2, CSIRO, and GFDL, respectively. The evaluation of RCMs using the standard matrix, as mentioned above, signifies reasonable agreement with the measured data.

Heat maps are created using the mean of monthly maximum temperature, minimum temperature, and the rainfall for the station data, CanESM2 data, CSIRO data, and GFDL data over the 1990–2020 period. Figure 3(a)–3(c) illustrate the heat map for each month, showing the mean value of monthly maximum temperature (°C), monthly minimum temperature (°C), and rainfall (mm), respectively. From Figure 3(a)–3(c), it is observed that the mean values from the station data and from the RCMs are quite comparable for mean Tmax and mean Tmin. However, mean monthly rainfall from CanESM2, CSIRO, and GFDL reveal some variations from the observed data sets, mostly in heavy rainfall season. Figure 3(c) illustrates that the RCMs slightly overestimate the rainfall for July and August, and underestimate for May and June. As observed here, the performance of daily rainfall from RCMs is not as high as temperatures; similar observations were made by Wu *et al.* (2020).

Owing to a lack of studies on performance evaluation of various RCMs in Bangladesh, we have considered the studies from India. Using three standard metrics, the correlation coefficient (COR), standard deviation, and the root mean square error

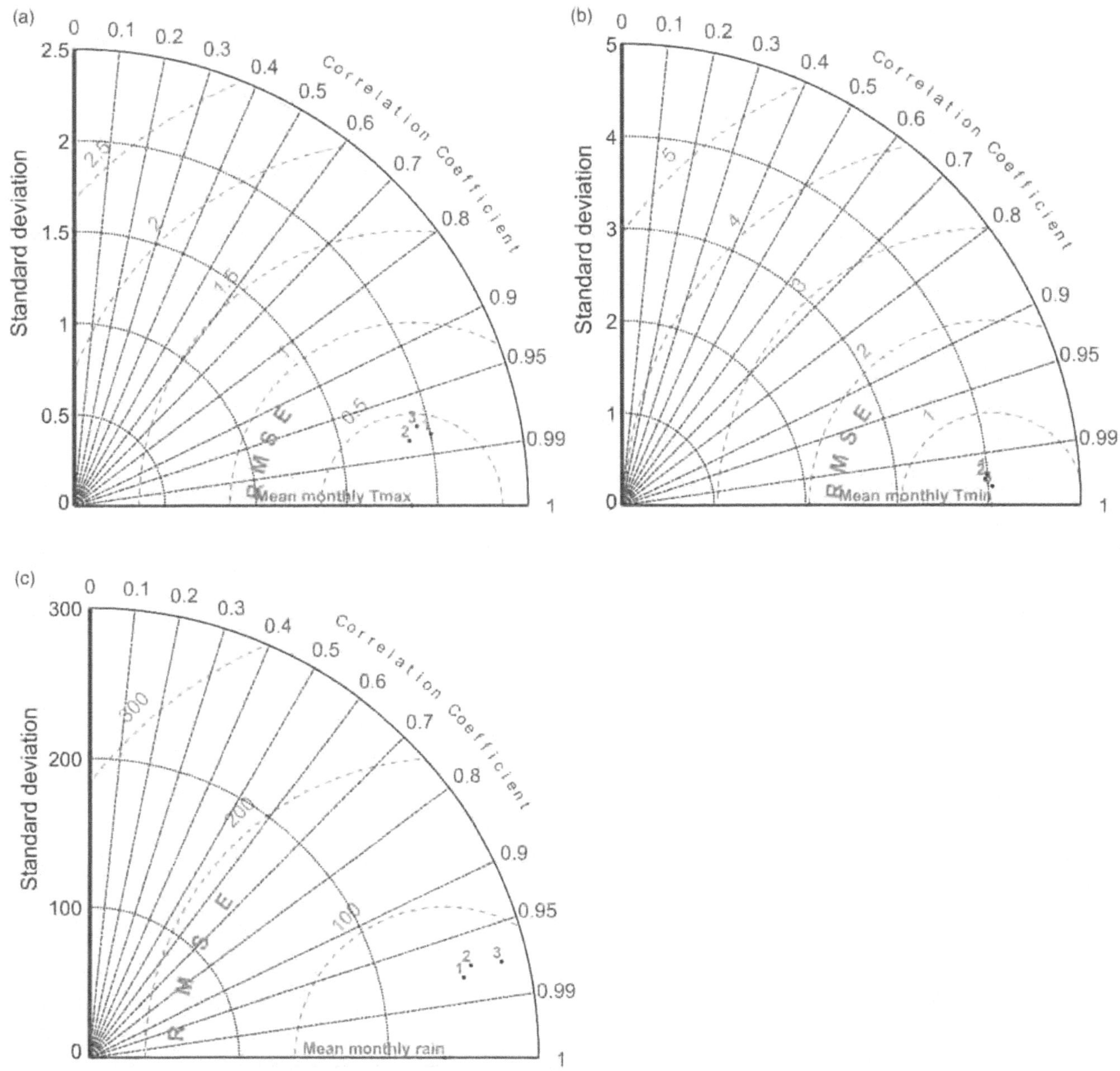

Figure 2 | Taylor diagram for (a) mean monthly Tmax over 1990–2020 period (b) mean monthly Tmin over 1990–2020 period and (c) mean monthly rainfall over 1990–2020 period.

(RMSE), the performance of RCMs (CanESM2, CSIRO, and GFDL) was found to be generally consistent with the findings from previous studies over the Indian region (Alexandru & Sushama 2015; Panjwani *et al.* 2019).

Observed climate variability in Chattogram

The observed daily maximum temperature (Tmax in °C), daily minimum temperature (Tmin in °C), and daily rainfall (mm) data from Chattogram station were analyzed over the 1950–1989 period and the 1990–2020 period. Some climate change studies separated the 1990–2020 period as the recent period (Hausfather & Peters 2020). In this study, the 1990–2020 period is taken as the recent period, hence, the remaining available period 1950–1989 is treated as the historical period. To investigate the variability of Tmax, Tmin, and rainfall from the historical period to the recent period, box plots were generated for each data set. Box plot, a graphic display of distributions of data, is capable of characterizing the scatterings of data efficiently (Graedel 1977). Figure 4(a) illustrates the box plots using the measured daily Tmin for two different time periods,

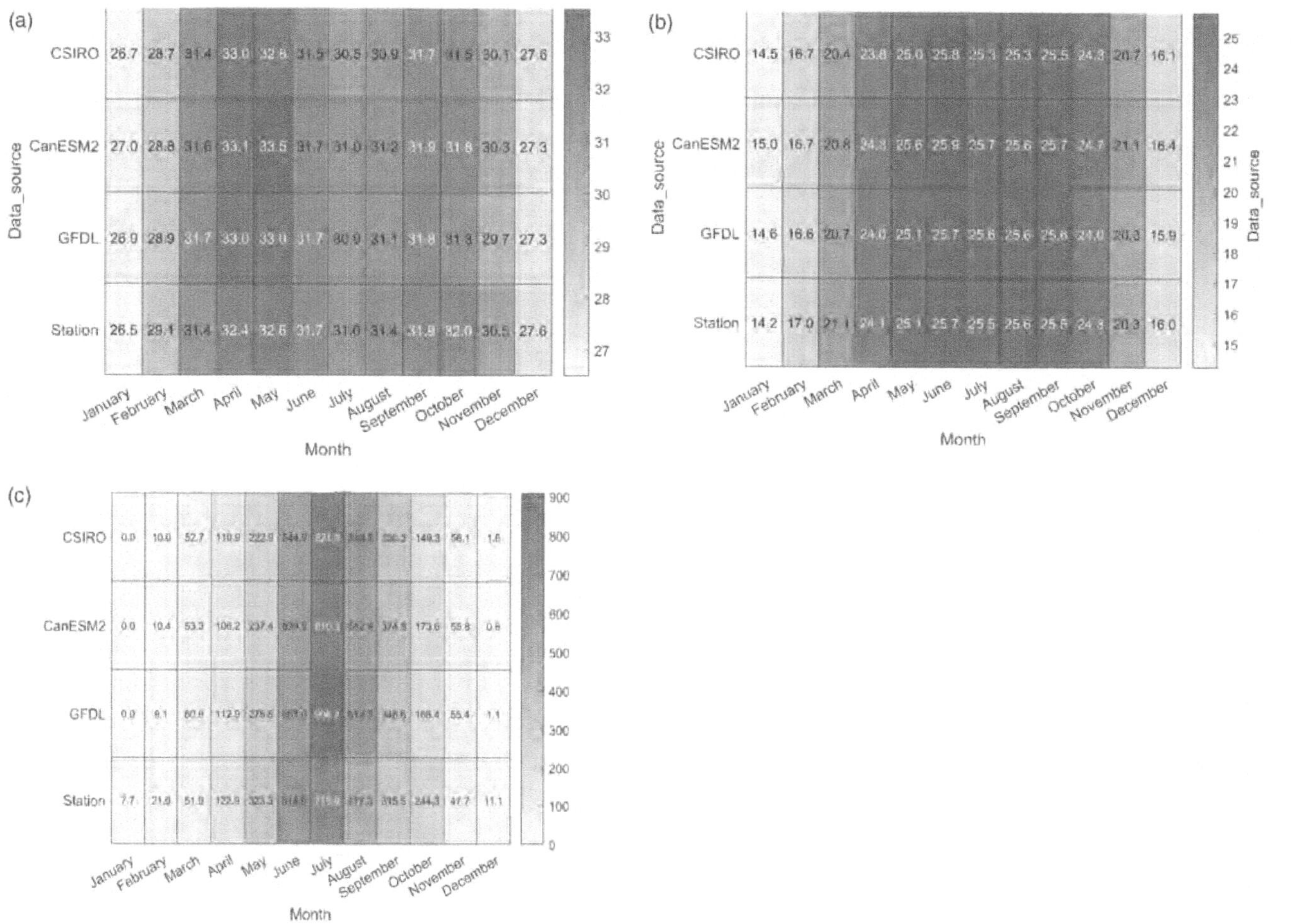

Figure 3 | Heat map for (a) mean monthly Tmax over 1990–2020 period (b) mean monthly Tmin over 1990–2020 period (c) mean monthly rainfall over 1990–2020 period.

1950–1989 and 1990–2020. Similarly, Figure 4(b) and 4(c) depict the box plots for daily Tmax and daily rainfall over the 1950–1989 and 1990–2020 period, respectively. It is noticeable from Figure 4(a) that, in the recent period, the measured daily minimum temperature has increased and the median of Tmin for each month has scaled up; the increment ranges from 0.2 °C to 1.1 °C over the months for the 1990–2020 period. Figure 4(b) illustrates the variability of mean daily maximum temperature, where the median has increased through all the months, ranging from 0.32 °C to 1.32 °C in recent years from the historic period. In the case of mean daily rainfall, it is seen from Figure 4(c) that some months got more rainfall in recent years than before, while some months have lower rainfall than the historic period. April, August, and November are the months having less rainfall in recent years. The mean rainfall anomaly ranges from −2.0 mm/day to 2.87 mm/day for different months; however, the average change of mean daily rainfall is positive over the 1990–2020 period. The average annual rainfall has increased about 6% from 1950–1989 to 1990–2020.

A brief summary of the salient features for the measured temperature and rainfall changes in Chattogram station is provided in Tables 3–5, where Table 3 gives the maximum value, minimum value, mean and the standard deviations of daily Tmin for 1950–1989 and 1990–2020 periods. Tables 4 and 5 provide the same information but for daily Tmax and daily rainfall, respectively, over the 1950–1989 and 1990–2020 periods. Table 3 reveals that the mean of the minimum daily temperature has increased in each month, while the month of February has the highest increment (1.09 °C) from the historic period to the recent period. Again, the maximum and minimum of the daily Tmin are higher in the recent period than the historic period for each month. It is observed from Table 4 that the mean of maximum daily temperature has increased over all the months in the 1990–2020 period from the 1950 to 1989 period. Rise of temperature for both summer and winter seasons was observed and the highest increment of daily Tmax was found in November (1.32 °C) and then in February

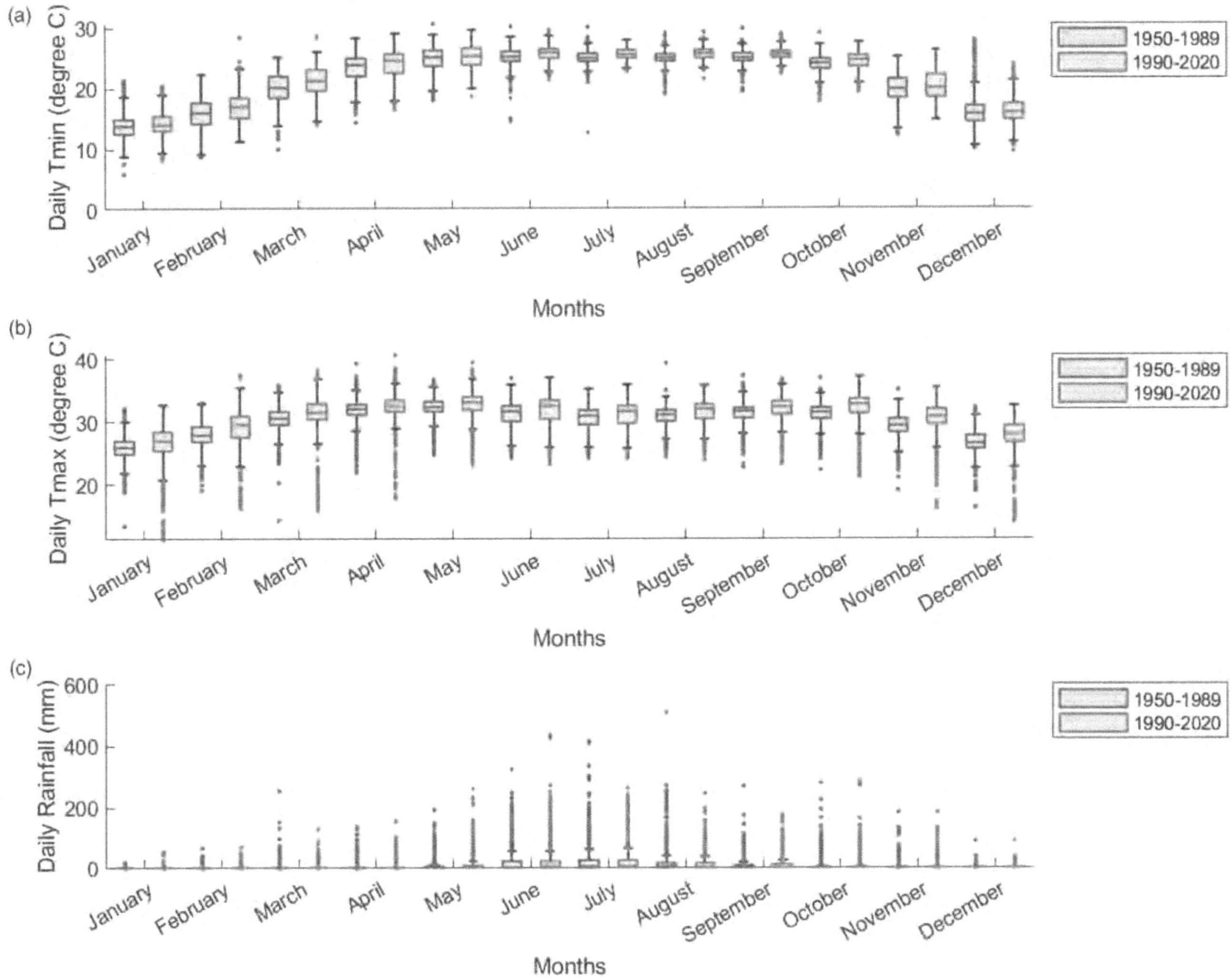

Figure 4 | Box plots using the measured daily data for two different time periods, 1950–1989 and 1990–2020: (a) for Tmin, (b) for Tmax, and (c) for rainfall.

Table 3 | Summary statistics for daily Tmin from Chattogram station over 1950–1989 and 1990–2020 period

Stat	Period	Jan	Feb	Mar	Apr	May	Jun	July	Aug	Sep	Oct	Nov	Dec
Max	1950–1989	21.4	22.3	25.1	28.3	30.6	30.2	30	29	29.8	29	25.2	27.8
	1990–2020	20.5	28.4	28.5	29	29.5	29.5	28	29.2	29	27.5	26.2	23.8
Min	1950–1989	5.95	8.8	10	14.3	18	14.5	12.6	19	19.4	17.8	12.2	10
	1990–2020	8.3	11.2	14	16.5	18.6	21.4	22.7	21.5	22.3	19.2	14.8	9.5
Mean	1950–1989	13.9	15.9	20.1	23.4	24.8	25.1	24.9	24.9	24.96	23.88	19.74	15.81
	1990–2020	14.22	17.01	21.09	24.06	25.15	25.69	25.55	25.59	25.47	24.27	20.26	16.00
Std	1950–1989	1.90	2.38	2.35	2.11	1.99	1.41	0.98	0.91	0.95	1.33	2.27	2.50
	1990–2020	1.93	2.44	2.30	2.17	1.92	1.30	0.88	0.80	0.86	1.43	2.41	2.04

All the temperature data are in °C.

(1.12 °C). Hence, it is clear from the analysis that the winter is warming up more pronouncedly than summer in recent years and the temperature for both Tmax and Tmin has increased significantly. On the other hand, daily rainfall has changed both positively and negatively, although most of the months, especially the monsoon season, have experienced increased mean daily rainfall, while the dry season has faced reduced daily rainfall in recent years.

Table 4 | Summary statistics for daily Tmax from Chattogram station over 1950–1989 and 1990–2020 periods

Stat	Period	Jan	Feb	Mar	Apr	May	Jun	July	Aug	Sep	Oct	Nov	Dec
Max	1950–1989	32.2	33.1	35.8	39.3	36.7	37	35.2	39.3	37.3	37	35.1	32.2
	1990–2020	32.80	37.50	38.40	40.70	39.50	37.00	36.00	35.80	36.80	37.20	35.50	32.60
Min	1950–1989	13.55	19.10	14.45	22.00	24.70	24.30	24.20	24.40	22.80	22.30	19.10	16.40
	1990–2020	11.40	16.30	16.00	17.80	23.00	23.20	24.10	24.00	23.20	21.20	16.30	14.20
Mean	1950–1989	25.86	27.97	30.59	31.82	32.27	31.12	30.58	30.75	31.29	31.03	29.17	26.52
	1990–2020	26.50	29.09	31.37	32.35	32.59	31.70	31.02	31.41	31.90	32.01	30.49	27.58
Std	1950–1989	1.70	2.04	1.80	1.64	1.63	2.13	1.88	1.74	1.67	1.88	1.78	1.71
	1990–2020	3.16	3.12	3.05	2.66	2.40	2.52	2.28	2.03	2.20	2.73	2.62	2.89

All the temperature data are in °C.

Table 5 | Summary statistics for daily rainfall from Chattogram station over 1950–1989 and 1990–2020 periods

Stat	Period	Jan	Feb	Mar	Apr	May	Jun	July	Aug	Sep	Oct	Nov	Dec
Max	1950–1989	20.00	68.00	254.0	138.0	194.00	325.00	417.00	511.00	269.00	281.0	185.0	91.00
	1990–2020	55.00	72.00	130.0	156.0	262.00	438.00	264.00	247.00	177.00	284.0	113.0	37.00
Min	1950–1989	0.00	0.00	0.00	0.00	0.00	0.00	0.00	0.00	0.00	0.00	0.00	0.00
	1990–2020	0.00	0.00	0.00	0.00	0.00	0.00	0.00	0.00	0.00	0.00	0.00	0.00
Mean	1950–1989	0.16	0.43	1.68	3.54	7.56	20.15	22.80	17.40	8.91	6.67	1.81	0.23
	1990–2020	0.25	0.78	1.67	4.10	10.43	20.50	23.09	15.40	10.52	7.88	1.59	0.36
Std	1950–1989	1.27	3.45	11.51	12.21	18.53	38.81	42.85	35.81	20.39	20.25	10.84	3.07
	1990–2020	2.68	5.02	8.34	13.63	25.24	41.24	39.30	28.50	21.86	22.52	9.01	2.78

All the rainfall data are in mm.

Analysis of climate variability from the RCMs

The selected RCMs from CORDEX were analyzed to investigate the variability of the temperature and rainfall between the 1950 and 2020 periods. The means of the minimum temperature per month (TMINMEAN) are plotted in Figure 5(a)–5(c) for the 1950–1989 and 1990–2020 periods. From Figure 5 it is observed that the average minimum temperature has increased significantly for each month in the 1990–2020 period; for the CanESM2 data the average increment is about 1.5 °C. Similar results were obtained utilizing the CSIRO and GFDL data, where the average minimum temperature has increased about 1.2 °C and 1.0 °C, respectively, in recent decades. From Figure 6(a)–6(c) a predominant increase in the median value along with highest values of the mean of the maximum temperature per month (TMAXMEAN) was found over the 1990–2020 period. Consequently, TMAXMEAN from the CanESM2 data exhibits, on average, around 2.0 °C warming in the 1990–2020 period from the 1950 to 1989 period. Similarity has been detected for CSIRO and GFDL temperature data; in both cases the highest and lowest values have risen along with the median value. From the box and whiskers plot (Figure 6), it can be stated that the median value of TMAXMEAN for CSIRO and GFDL has increased about 2.0 °C and 1.8 °C, respectively, from the historical period.

From this analysis, it is notable that the mean Tmax and Tmin for each month in the 1990–2020 period have higher values than that of the historical period. However, the monthly mean of the 1-day maximum rainfall (RX1DAY) has not developed any particular pattern; most of the months got a higher amount of rainfall while a few other months exhibited a lower amount of rainfall in the recent period. From Figure 7(a), the box and whisker plots from the CanESM2 data sets reveal that May and July have become wetter, while June has become drier, and in some other months there are no significant changes. A similar trend was also observed from CSIRO and GFDL data sets (Figure 7(b) and 7(c)); there are some months that have experienced heavier mean daily rainfall while a few other months have experienced lighter mean daily rainfall in recent decades. Therefore, from this analysis, it can be stated that July and August have more rainfall than before, and the overall maximum increment is about 20–30 mm in a month over the 1990–2020 period. The results obtained from the CanESM2, CSIRO, and GFDL are comparable with the results from the station data over the same period.

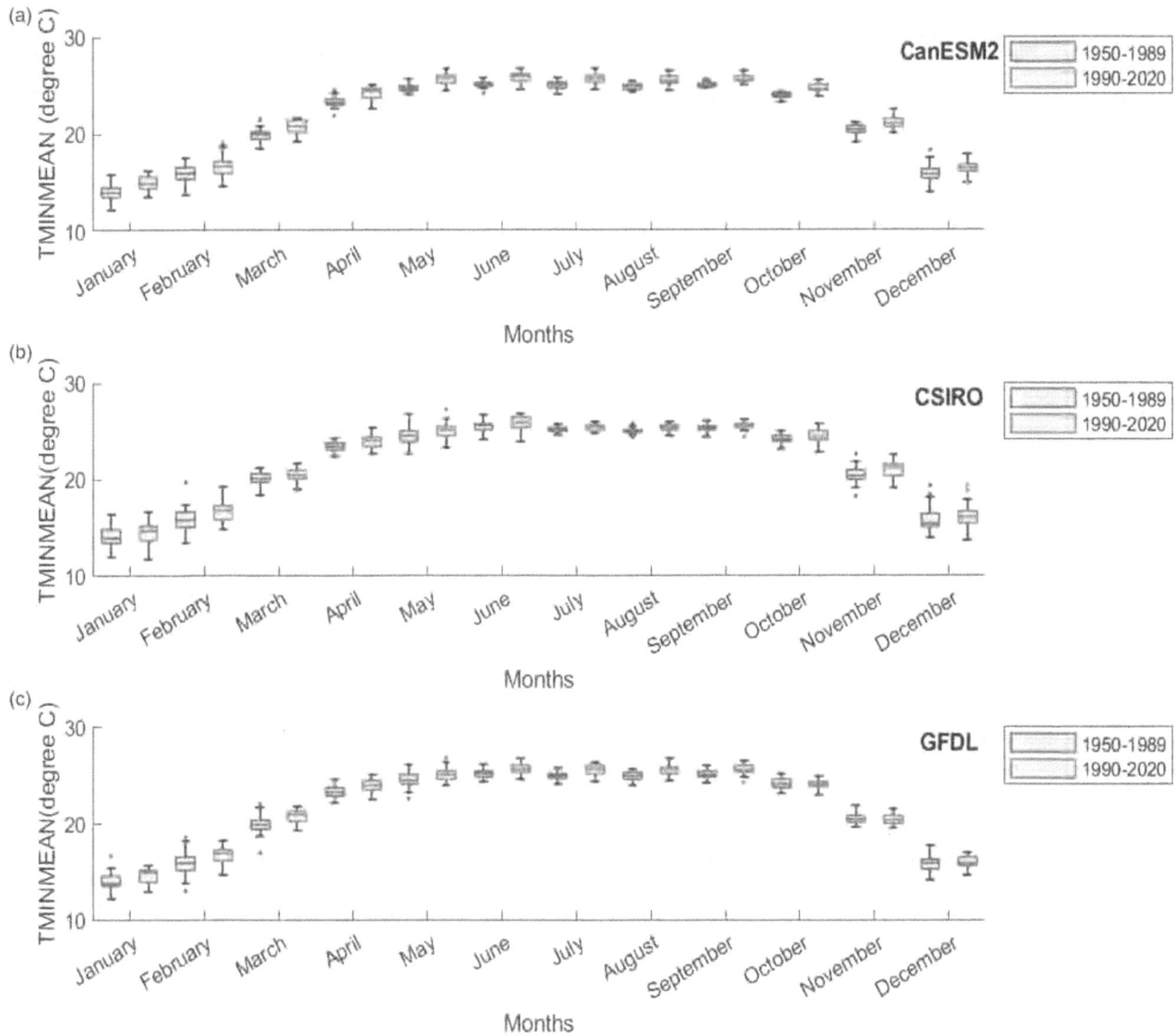

Figure 5 | Box and whisker plots for TMINMEAN over 1950–1989 and 1990–2020 periods from (a) CanESM2, (b) CSIRO, and (c) GFDL.

Extreme indices evaluation from the observed data

Temperature indices

Extreme temperature indices, mentioned in Table 1, were calculated annually using the daily maximum and minimum temperature data from the Chattogram station for 1950–1989 and 1990–2020 periods. The RClimDex package in R programming language was utilized to estimate all 22 climate indices. The present period, 1990–2020, was considered for threshold. Figure 8(a)–8(j) illustrate the variation of different indices with time, where a linear regression line was also added to visualize its tendency. Figure 8(a) and 8(b) demonstrate that the monthly maximum value of daily maximum temperature (TXX) and monthly minimum value of daily maximum temperature (TXN), which were flat or downward in the historic period, have gained a positive slope in recent years, indicating both maximum and minimum temperature have increased in the 1990–2020 period.

The percentage of warm days (TX90p) has grown and shown increasing trends while the percentage of cold days (TX10p) has reduced, indicating decreasing trends in Figure 8(c) and 8(d), respectively. It can be stated that the increasing warm days, and declining cold days, are further evidence of temperature rise and effects of climate change in recent years. The indices TNX and TNN in Figure 8(e) and 8(f) represent the behavior of monthly maximum value of daily minimum temperature and

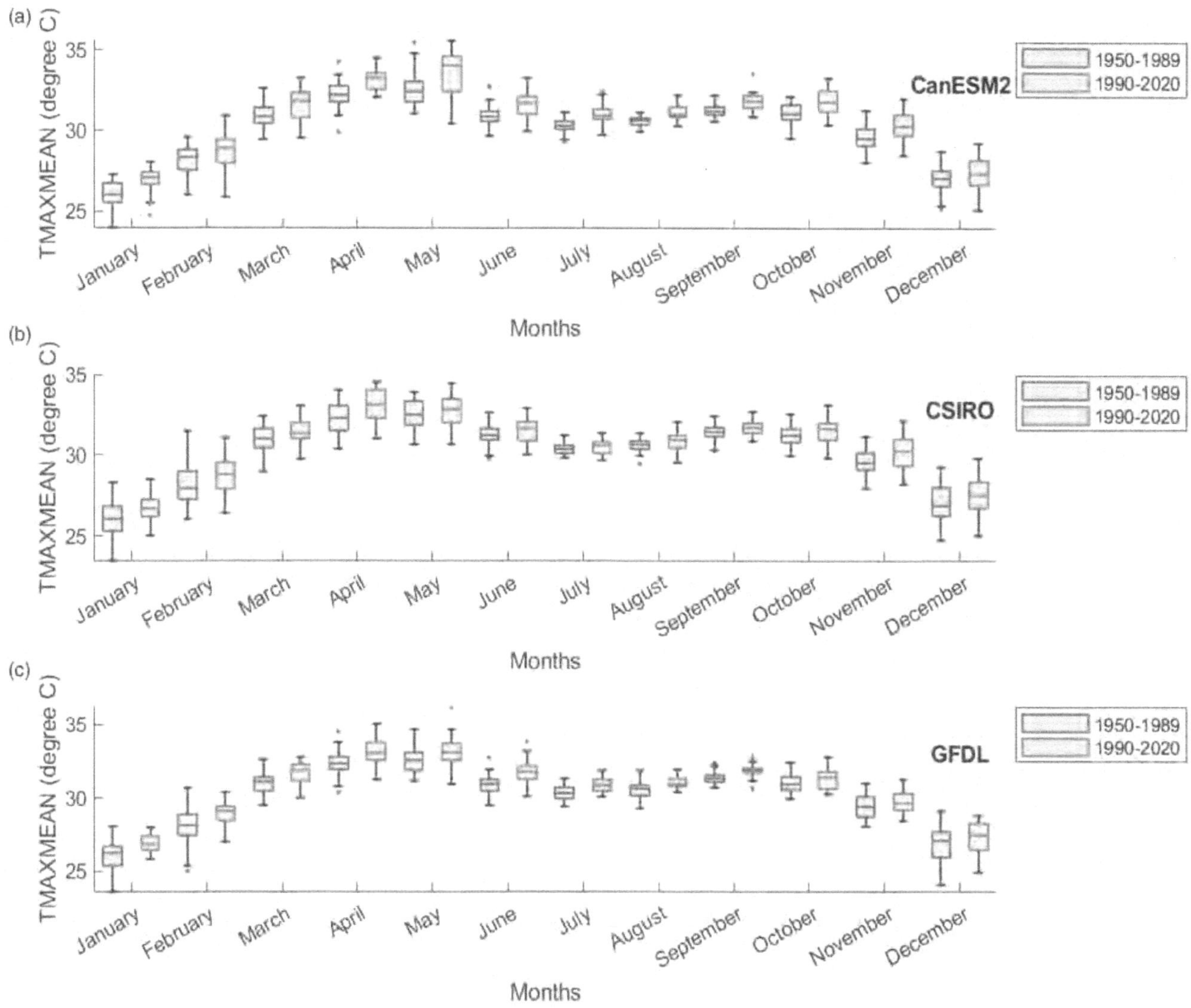

Figure 6 | Box and whisker plots for TMAXMEAN over 1950–1989 and 1990–2020 periods from (a) CanESM2, (b) CSIRO, and (c) GFDL.

monthly minimum value of daily minimum temperature, respectively. From the analysis, it is notable that in both cases the average values have increased from the historic period, whereas TNX has a mild decreasing slope and TNN has a mild increasing slope. On the other hand, the percentage of warm nights (TN90p) has increased and exhibits increasing trends in Figure 8(g). However, Figure 8(h) shows that the cold nights (TN10p) have decreased with a sharp downward slope.

Observed maximum and minimum temperatures recorded in recent decades have their values elevated compared to the previous decades. This is confirmed by verifying the annual diurnal temperature range (DTR) in Figure 8(i); it has increased significantly and exhibits positive trends in the percentage of days with the lowest (below 10th percentile) and highest (above 90th percentile) maximum and minimum temperature. Figure 8(j) and 8(k) reveal the behavior of indices related to the duration of cold (CSDI) and warm periods (WSDI). The negative trend of consecutive days with Tmin below the 10th percentile, associated with the positive trends of the number of consecutive days in which Tmax exceeded the 90th percentile, suggests that over the last decades prolonged and strong heat waves dominated over this region. The analysis of the above-mentioned indices indicates more accelerating trends of daily maximum temperature when compared to daily minimum temperature in this region.

Rainfall indices

Both increasing and decreasing patterns were observed for the extreme rainfall indices (calculated annually). From Figure 9(a), it is found that the observed data from the Chattogram station correspond to an increased extreme rainfall in

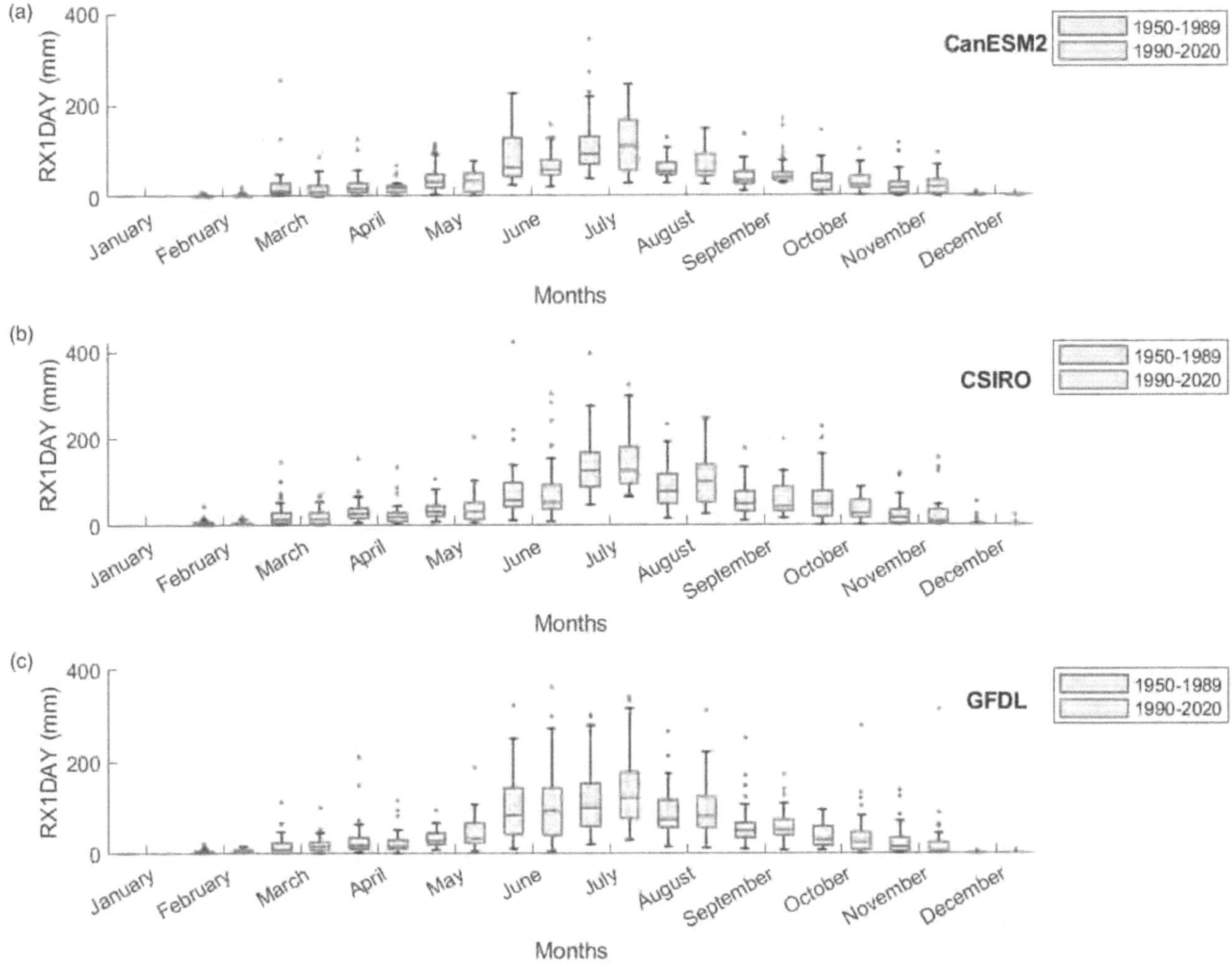

Figure 7 | Box and whisker plots for RX1DAY over 1950–1989 and 1990–2020 periods from (a) CanESM2 (b) CSIRO and (c) GFDL.

a single day (RX1DAY), while the index RX5DAY represents a positive trend in 5-day accumulated rainfall per month over the last three decades in Figure 9(b). These indices offer evaluation of rainfall extremes at a particular station, hence, it can be stated that in recent decades the extreme rainfall events are intensifying. The index for yearly total rainfall (PRCPTOT) is the one that best represents the areas with increasing or decreasing trends of rainfall. In Figure 9(c), the PRCPTOT index reveals a mild positive slope, indicating an increasing tendency of total rainfall per year. The patterns among R100 mm (Figure 9(d)), R20 mm (Figure 9(e)), and R10 mm (Figure 9(f)) were found to be very similar, with positive slope dominating the region in recent decades; however, they exhibited some negative slope in historic periods. The contribution to the sum of the annual rainfall of the days in which the daily rainfall exceeds the 99 percentiles (R99P) indicates an upward trend of extremely wet days in the recent period (Figure 9(g)). Therefore, it is an important understanding that, in the last three decades, the heavy rainfall events have risen significantly, contributing to the increment of total annual rainfall. Also, in the recent period, the consecutive wet days CWD are computed with mild increasing slope, although it was a slightly negative slope in the historic period (Figure 9(h)). Figure 9(i) reveals that the consecutive dry days, CDD, have risen in recent decades, which was the opposite in the past. From this study, it is observed that the total yearly rainfall has increased, including heavy and extremely heavy rainfall events. The number of extremely wet days in recent decades also exhibits an increasing tendency, ultimately causing the extreme rainfall indices to rise in the recent period.

As identified from the above illustrations, the patterns for most of the indices have changed with time. In the 1990–2020 period, the changes of extreme indices are sharper than the historical period. It is evident that climate change is impacting the extreme climate disproportionately by adding frequent and intense heavy rainfall events and causing higher vulnerabilities to floods and waterlogging in this region.

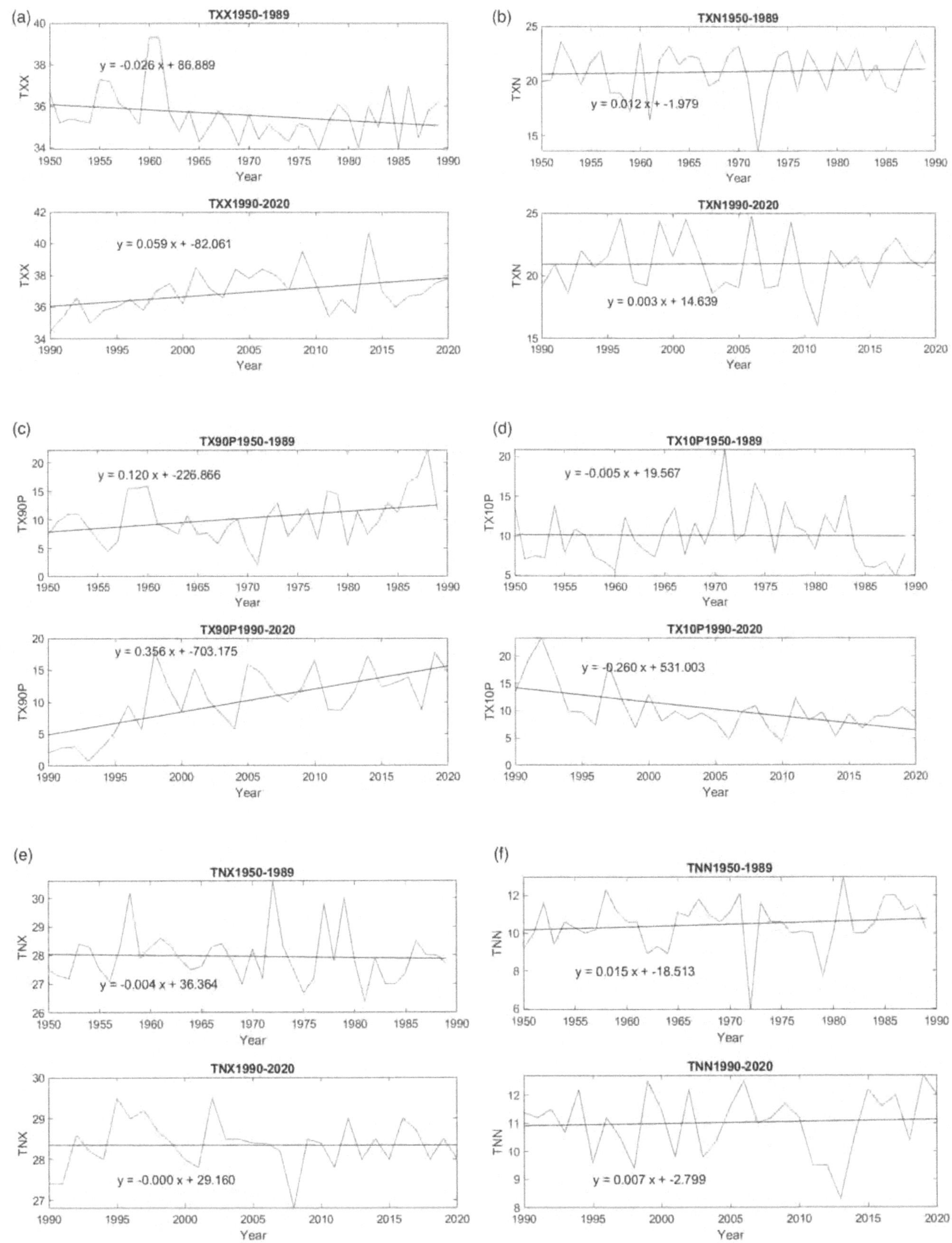

Figure 8 | (a)–(k) Plot of temperature indices over 1950–1989 and 1990–2020 periods. (*continued*).

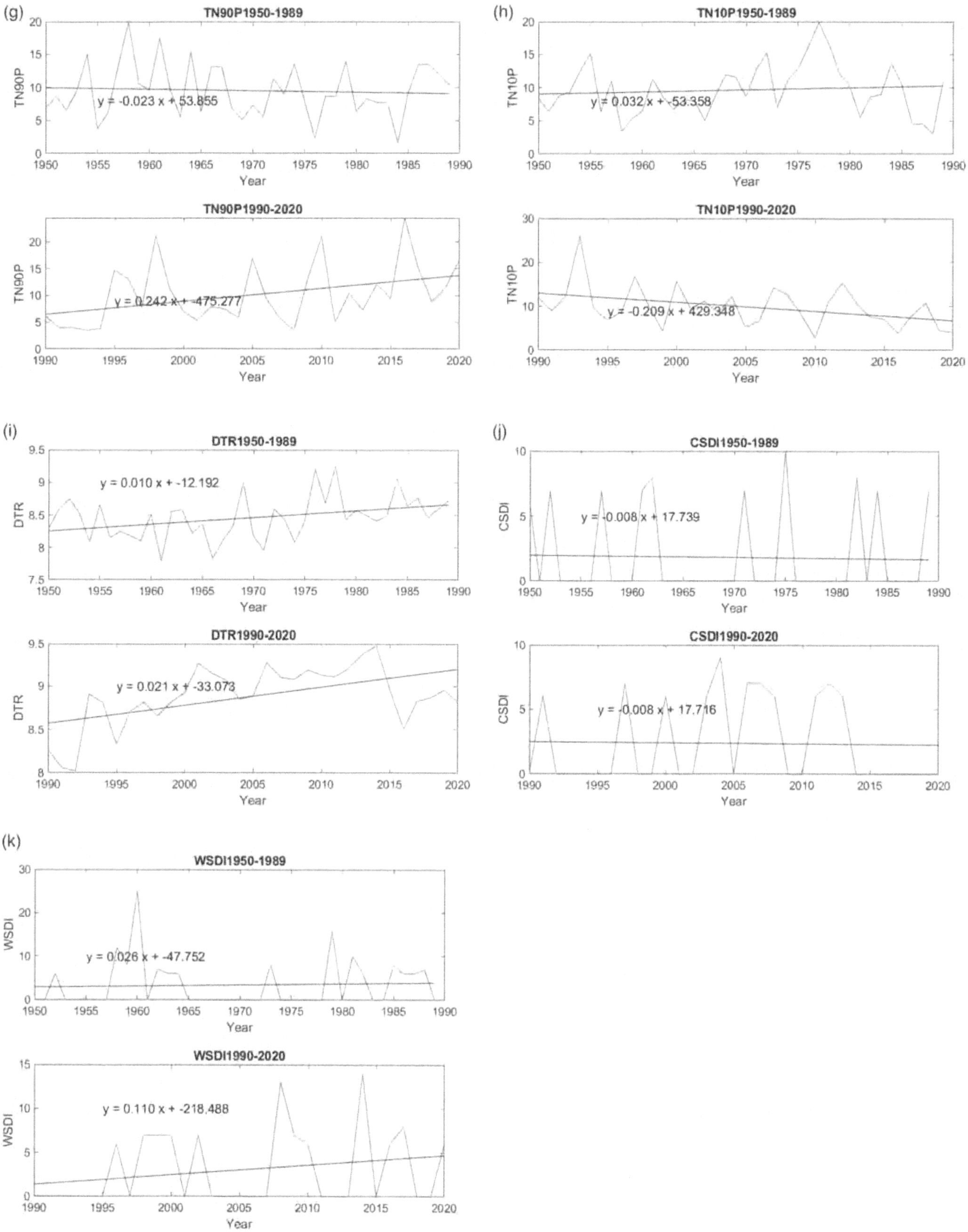

Figure 8 | continued.

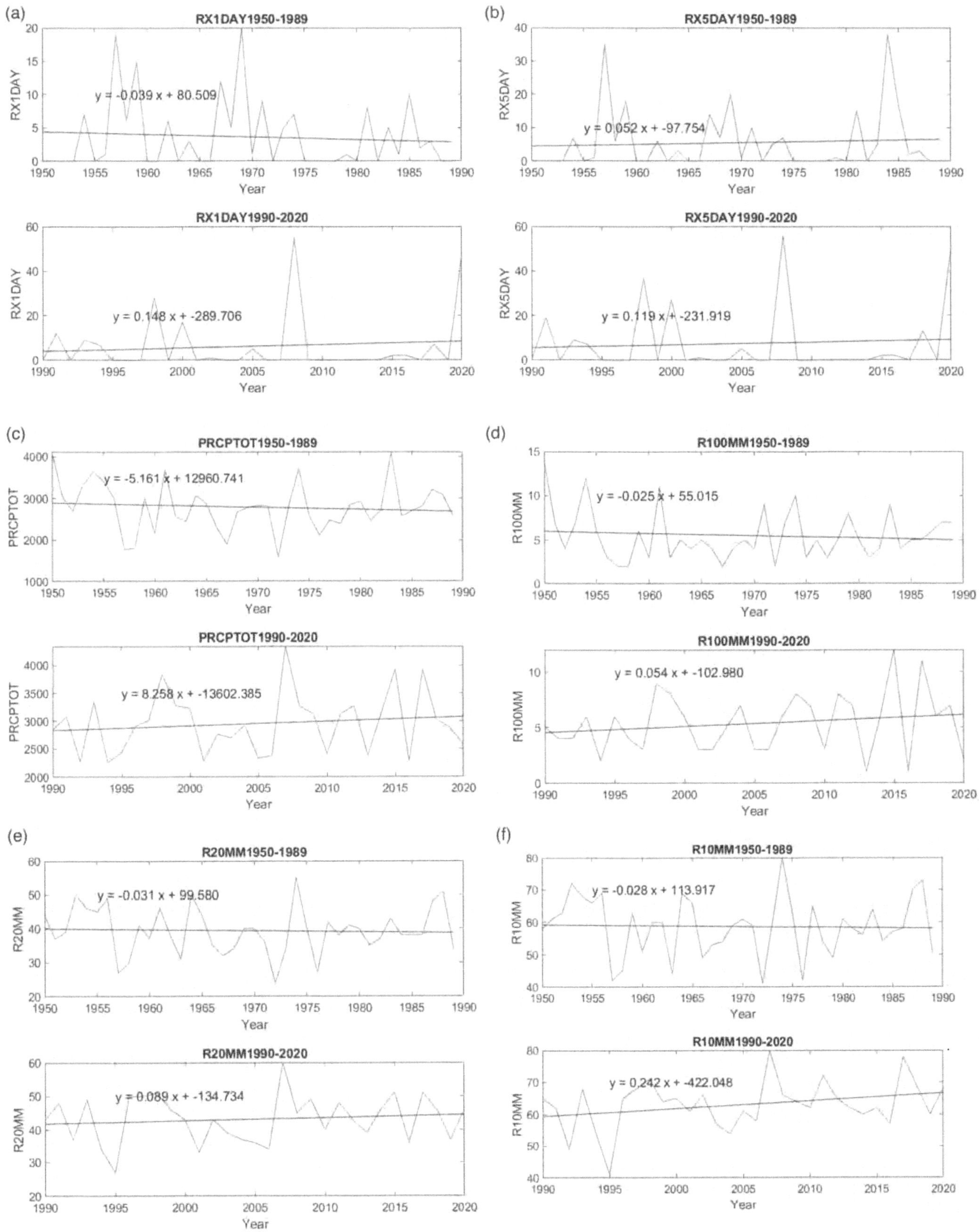

Figure 9 | (a)–(i) Plot of rainfall indices over 1950–1989 and 1990–2020 periods. (*continued*).

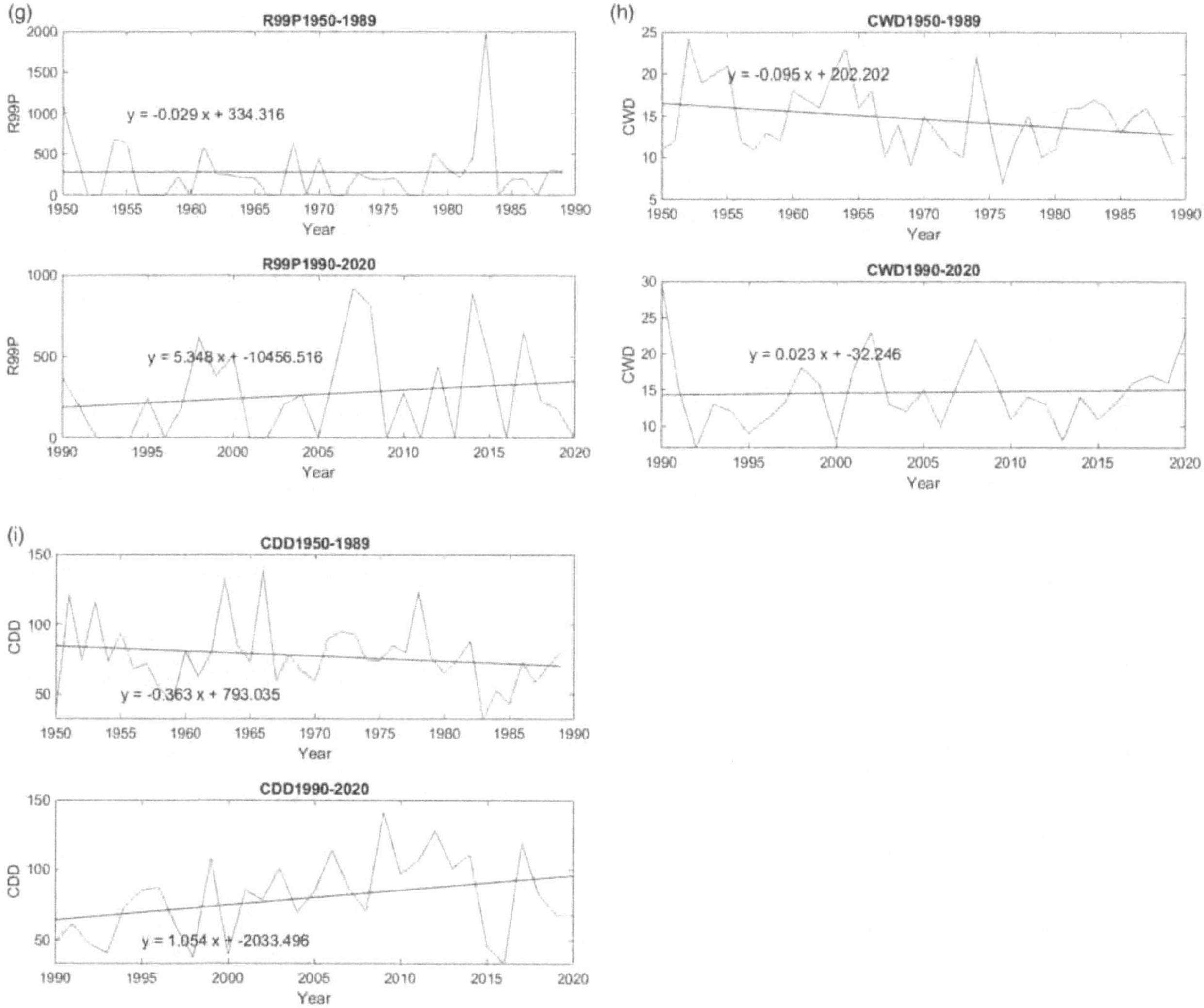

Figure 9 | (*continued*).

Trend analysis of extreme indices from the observed data

Trend analysis was performed utilizing the annually calculated temperature and rainfall indices from the station data. The MK non-parametric trend tests were conducted based on the values of extreme indices to detect the temporal trends and significant levels for the recent (1990–2020) period. In this study 5% two-sided significance level was used to classify the trends as 'significant increasing,' 'significant decreasing,' and 'no significant' trend. Another non-parametric test known as Sen's slope estimator was used to determine the magnitude of change while the MK test reveals the direction of change or trend. Sen's method is used to find the true slope of an existing trend, if there is any. This analysis will eventually help to detect the impact of climate change on the trends of extreme climate indices.

Table 6 depicts the results of the MK trend test and the Sen's slope at 5% significance level for the temperature and rainfall indices. MK test results across the 1990–2020 period reveal that the percentage of warm days (TX90p) has highly significant increasing trend ($p < 0.0005$), and the Sen's slope is estimated as (0.3817%day/year). It is found that the diurnal temperature range (DTR) also exhibits a rising trend which is statistically significant ($p < 0.005$), the value of the Sen's slope is 0.0209 °C/year. The trend of monthly maximum value of daily maximum temperature (TXX) implies statistically significant increment with Sen's slope 0.0666 °C/year. However, the percentage of cold days TX10p possesses a significant decreasing trend ($p < 0.01$), as expected, and the Sen's slope is calculated as −0.175%day/year. Of other extreme temperature indices, the percentage of warm nights, TN90p, also demonstrates a significant increasing trend ($p < 0.05$), with Sen's slope 0.2154%day/year, whereas the percentage of cold nights, TN10p, possesses significant decreasing trend ($p < 0.05$) with Sen's slope −0.1804%

Table 6 | Summary of MK test and the Sen's slope estimator at 5% significant level for extreme temperature and rainfall indices from station data over 1990–2020 period

Indices	z	p	Sen's slope	Remarks
TXX	2.4005	0.0163	0.0666	**Significant increasing trend**
TNX	− 0.1543	0.8773	0	Not significant decreasing trend
TXN	0.2216	0.8246	0	Not significant increasing trend
TNN	0.49435	0.6211	0.01	Not significant increasing trend
TN10p	− 2.1755	0.0295	− 0.1804	**Significant decreasing trend**
TN90p	2.2609	0.0237	0.2154	**Significant increasing trend**
TX10p	− 2.6174	0.0088	− 0.175	**Significant decreasing trend**
TX90p	3.7228	0.0001	0.3817	**Significant increasing trend**
WSDI	1.1528	0.249	0	Not significant increasing trend
CSDI	− 0.1407	0.8881	0	Not significant decreasing trend
DTR	2.8573	0.0042	0.0209	**Significant increasing trend**
PRCPTOT	0.6118	0.5406	5.2	Not significant increasing trend
R95p	0.5098	0.6101	5.9375	Not significant increasing trend
R99p	0.7368	0.4612	6.84	Not significant increasing trend
RX1day	0.4589	0.6463	0.8888	Not significant increasing trend
RX5day	0.5098	0.6101	1.5454	Not significant increasing trend
R10 mm	0.7337	0.4631	0.1111	Not significant increasing trend
R20 mm	0.4097	0.682	0.0476	Not significant increasing trend
R100 mm	0.7046	0.4811	0	Not significant increasing trend
CWD	0.9226	0.3562	0.0833	Not significant increasing trend
CDD	1.7174	0.0859	1.3461	Not significant increasing trend
SDII	0.30611	0.7595	0.036363	Not significant increasing trend

day/year. The other temperature indices TNN, TXN, WSDI demonstrate an upward trend; however, they are not statistically significant. On the other hand, TNX and CSDI exhibit a decreasing trend, although they cannot be claimed as statistically significant.

Annual rainfall amount, PRCPTOT, in Chattogram generally exhibits an upward trend, which is not statistically significant. Increasing trends are observed for annually calculated 1-day maximum rainfall (RX1day) and the 5-day maximum rainfall (RX5day). However, none of the indices have significant trend at 95% confidence level. The largest rate of increment of 1-day maximum rainfall, RX1day (0.888 mm/year) and 5-day maximum rainfall, RX5day (1.5454 mm/year) were estimated from Sen's slope test. From the analysis it was found that all the extreme precipitation indexes, R10MM (the number of 10 mm or more precipitation days in a year), R20MM (number of 20 mm or more precipitation days in a year), R100MM (number of 100 mm or more precipitation days in a year), R95P (very wet days), and R99P (extremely wet days) exhibit increasing trends over the 1990–2020 period; however, they are not statistically significant at 95% confidence interval. On the other hand, consecutive wet days (CWD), consecutive dry days (CDD), and daily intensity index, SDII (these three indexes were obtained for annual daily precipitation) reveal increasing tendency during the 1990–2020 period. Although the extreme rainfall indices examined here do not have statistically significant trends, they clearly indicate the rising tendency of the extreme events in the recent period. Similar results were identified by Akter *et al.* (2017). Thus, the observed changes in characteristics of extreme climate indices could be described as the potential impact of climate change and the increased heavy rainfall events could be linked to the frequent floods and prolonged waterlogging problems in Chattogram.

Trend analysis of extreme indices from the RCMs data

Trends of all the 22 indices were examined for the three selected RCMs. Bias correction methods were not considered here as the focus was on investigating the competency of the selected high-resolution raw RCMs to assess the climate change. Like station data, all the indices from RCMs are estimated annually. Significant rising trends are obtained from Table 7 using CanESM2 data for TXX and TNX with Sen's slope value 0.0466 °C/year and 0.0492 °C/year, respectively. TN90P and TX90P exhibit significantly increasing trends, which are similar to those from the station data. In contrast to significant decreasing trend for CSDI, considerable increasing trend was observed for WSDI. In the case of TN10P and TX10P, significant decreasing trend was established; however, TXN and DTR have no significant increasing trend here. The observed pattern of rainfall extremes using the CanESM2 data also reveals rising trends over the 1990–2020 period. The other rainfall indices, R95P, R99P, R100MM, RX5DAY, and SDII have positive trends, although they are not statistically significant. However, R10MM, R20MM, CWD, and CDD show non-significant decreasing trend over the present period.

Analysis of trends from the temperature and rainfall indices using CSIRO and GFDL data sets were also performed. Table A1 shows the MK and Sen's slope test results for CISRO and Table A2 represents the MK and Sen's slope test results for GFDL. Significant positive trend for TNX and TN90 was found from both CSIRO and GFDL indices; however, TXN and TNN do not exhibit any significant rising trend for both cases. TN10p and TX10p revealed a substantial decreasing trend from both CSIRO and GFDL data sets. However, variations of trends from these two data sets were observed. DTR have a mild increasing trend from Table A1, whereas a mild decreasing trend was observed from Table A2. In the case of rainfall indices from CSIRO, increasing but not significant trend was observed for R95P, R99P, RX1DAY, RX5DAY, CWD, and R100MM and mild negative trends for PRCPTOT, R10MM, R20MM, and SDII. On the other hand, GFDL shows only increasing trends for all the rainfall indices over the 1990–2020 period.

Comparison of trends from the observed and the RCMs

Extreme temperature and rainfall indices calculated from the station data, CanESM2, CSIRO, and GFDL data were analyzed to find the existing trends over the 1990–2020 period, and the results are compared in Table A3. Table A3 is self-explanatory

Table 7 | Summary of MK test and the Sen's slope estimator at 5% significance level for extreme temperature and rainfall indices from CanESM2 data for 1990–2020 period

Indices	z	p	Sen's slope	Remarks
TXX	2.0736	0.03812	0.04659	**Significant increasing trend**
TNX	4.691	2.7×10^{-6}	0.04927	**Significant increasing trend**
TXN	1.9376	0.05267	0.06235	Not significant increasing trend
TNN	2.0736	0.03812	0.03953	**Significant increasing trend**
TN10p	−4.2151	2.5×10^{-5}	−0.56	**Significant decreasing trend**
TN90p	4.8465	1.26×10^{-6}	0.86778	**Significant increasing trend**
TX10p	−3.4848	4.9×10^{-4}	−0.547	**Significant decreasing trend**
TX90p	4.691	2.7×10^{-6}	0.72154	**Significant increasing trend**
WSDI	3.5423	3.9×10^{-4}	0.8125	**Significant increasing trend**
CSDI	−3.7924	1.49×10^{-4}	−0.5294	**Significant decreasing trend**
DTR	0.6295	0.529	0.0025	Not significant increasing trend
PRCPTOT	0.16996	0.865	3.84	Not significant increasing trend
R95p	1.3937	0.1634	11.225	Not significant increasing trend
R99p	0.99916	0.3177	3.14	Not significant increasing trend
RX1day	0.44191	0.6586	0.4254	Not significant increasing trend
RX5day	1.3257	0.1849	3.19908	Not significant increasing trend
R10 mm	−0.79972	0.4239	−0.26667	Not significant decreasing trend
R20 mm	−0.28934	0.7723	−0.0714	Not significant decreasing trend
R100 mm	0.85428	0.393	0	Not significant increasing trend
CWD	−0.98712	0.3236	−0.4782	Not significant decreasing trend
CDD	−0.34045	0.7335	−0.1666	Not significant decreasing trend
SDII	0.8163	0.4143	0.04	Not significant increasing trend

and compares the examined trends of all 22 extreme climate indices from the four data sets. It is recognized from all the data sets that significant increasing trends are dominant among the extreme temperature indices. It implies accelerated heatwaves and more warm days were experienced in the last three decades. Extreme rainfall indices investigated here also exhibit positive trends from both station and RCMs data sets. However, some discrepancies in trends are observed from the temperature and rainfall indices using the observed and the RCMs data. The two RCMs, CanESM2 and GFDL, produced similar regional precipitation-related climate extremes, however, CSIRO showed some disagreement with the observed. Therefore, CanESM2 and GFDL can be considered as high-performing RCMs among those selected. This evaluation of extreme climate indices from the selected RCMs revealed the strength of representing the local climate extremes by the high-resolution regional climate models. Hence, the promising results from the selected RCMs demonstrate their competency to assess the impact of climate change in this region.

DISCUSSION

With the available observed data set at Chattogram station, the three RCMs from CORDEX were implemented to investigate the effects of climate change on the extreme temperature and rainfall over two different time periods (1950–1989 and 1990–2020). It was a priority to evaluate the selected RCMs CanESM2, CSIRO, and GFDL before considering them for assessing the climate change indices and finding the trends. As noted earlier, there is a lack of information about the applicability of different RCMs in Bangladesh, and hence this study intended to offer some guidance for selecting the RCMs in this region to perform climate change analysis. The evaluation of model performance makes use of three standard metrics, the correlation coefficient and the root mean square error and the standard deviation, where all the three RCMs demonstrated promising performance. Using these three RCMs along with the station data, the variabilities of daily temperatures and rainfall over 1950–2020 were investigated. Analysis confirmed that the mean temperature has increased sharply, and the winter warming is more pronounced in recent decades (1990–2020) than the historical period (1950–1989). The variations in rainfall patterns are quite substantial as well, mostly the monsoon season received more rainfall, while the dry season became drier in the recent decades. Similar observations were made by Mullick *et al.* (2019), who found the wet season is becoming wetter and the dry season is becoming drier with the increment of temperature by analyzing the observed temperature and rainfall data from 1966 to 2015 over Bangladesh.

By comparing different indices between the two periods, their changing characteristics were revealed. In recent decades most of the extreme indices show sharper changes than the historical period, which signifies the growing frequency and extensiveness of extreme climate in this region. Increased climate-related extreme events over Bangladesh were detected by Shahid *et al.* (2016) during the 1958–2012 period using station data. However, no studies were reported to investigate the pattern of changes for extreme indices from two epochs: historical to present. Moreover, we investigated the trends of climate change indices from both station data and the RCMs data. The selected RCMs were validated, and they provided reliable results representing the local climate. By comparing the outcomes of trend analysis from the 22 climate change indices, we found that CanESM2 and GFDL have more promising results in resembling the observed trends than CSIRO. The overall performance of the selected RCMs is consistent, and they can be recommended for further climate change analysis in this region.

This study offers a better understanding of the effects of climate change on extreme indices, and how their patterns have changed over time. The increasing trends of extreme temperature and rainfall indices are steeper in recent decades than the historical period, justifying the causes of frequent and prolonged water logging situations in the monsoon, and scarcity of water during the dry season in Chattogram. The observed changes in different attributes from the computed daily temperature and rainfall indices exposed a brand-new challenge in managing the regional hydroclimatic risks in drainage and agricultural sectors. The findings here suggest the need for better regional water management strategies at the national level.

SUMMARY AND CONCLUSION

Chattogram has a long history of waterlogging in the monsoon season due to intense rainfall. Evidence shows that climate change is adding more challenges and this region is likely to undergo immense pressure from climate change on its rainfall and temperature pattern. To improve the Chattogram City waterlogging condition, better planning and management of the city drainage system is crucial, which requires a scientific knowledge-based understanding of its local climate pattern and the recent trends and variabilities.

Therefore, in this study, an attempt has been made to evaluate various aspects of temperature and rainfall extremes including the variability and trends. This study also intended to incorporate the most recent available data sets (1950–2020) from the station as well as from the RCM's high-resolution data sets from CORDEX, to evaluate the performance of the selected RCMs as well as to analyze the effect of climate change on the extreme rainfall and temperature pattern. By analyzing different characteristics of temperature and rainfall from the 1950 to 1989 and 1990–2020 periods, the observations are as follows:

(i) The overall performance of CanESM2, CSIRO, and GFDL are found compatible to represent the local climate of the Chattogram region.
(ii) The consistent incremental trends of monsoon rainfall were observed, while declining trends of dry season rainfall were uncovered.
(iii) Increment of heavy rainfall events in the monsoon acted as one of the key driving forces for floods and the prolonged waterlogging situation of Chattogram City.
(iv) Significant increment of mean daily temperature was observed, where the rise of the mean minimum temperature is higher than the mean maximum temperature over the last three decades.
(v) Both TMINMEAN and TMAXMEAN from CanESM2, CSIRO, and GFDL have scaled up in each month, ranging from 1.0 °C to 1.5 °C and 1.8 °C to 2.0 °C, respectively, from the 1950–1989 to 1990–2020 period.
(vi) Different sets of indices plotting have indicated that the extreme temperature indices have increased in the recent period from the historical period, demonstrating accelerating tendency for both minimum and maximum temperature. However, the extreme precipitation indices indicated mild increasing tendency over the 1990–2020 period.
(vii) The MK non-parametric trend test and the Sen's estimator for 1990–2020 period revealed that warm days (TX90p) and warm nights' frequencies (TN90p) have significantly increased (0.3817%day/year) and (0.2154%day/year), respectively. Again, the diurnal temperature range (DTR) also exhibits a significant increasing trend (0.0209 °C/year). As expected, the frequency of cold days and cold nights exhibited a significantly decreasing trend (−0.175%day/year) and (−0.1804%day/year), respectively. Other extreme indices TNN, TXN, TR20, CDD, and CWD demonstrate the upward trend; however, they are not statistically significant. On the other hand, mild increasing trend in 1-day and 5-day maximum rainfall (0.888 mm/year) and (1.5454 mm/year) was detected and total annual precipitation in Chattogram was also found to be rising, although the trend was not statistically significant.
(viii) MK and Sen's analysis from the RCMs also indicated increasing trends for most of the extreme temperature and rainfall indices, where CanESM2 and GFDL show more competency than CSIRO to represent the realistic trends.

This study revealed that the climate variability and trends from the selected RCMs resembled the trends and variability from the observed data. Therefore, CanESM2, GFDL, and CSIRO from CORDEX can be considered for climate change analysis in this region. From the detected trends and variability of temperature and rainfall, it can be stated that impacts of climate change on extreme events are recognizable and should be considered thoroughly. The knowledge earned from this study will help to understand vulnerabilities associated with the change of extreme temperature and rainfall patterns in Chattogram City. Thus, the city planners may find this study useful for considering the climate change impact in planning and management strategy.

CONFLICTS OF INTEREST

The authors declare no potential conflict of interests.

DATA AVAILABILITY STATEMENT

Data cannot be made publicly available; readers should contact the corresponding author for details.

REFERENCES

Adeyeri, O. E., Lawin, A. E., Laux, P., Ishola, K. A. & Ige, S. O. 2019 Analysis of climate extreme indices over the Komadugu-Yobe basin, Lake Chad region: past and future occurrences. *Weather and Climate Extremes* **23**, 100194. https://doi.org/10.1016/j.wace.2019.100194.

Akter, A., Mohit, S. A. & Chowdhury, M. A. H. 2017 Predicting urban storm water-logging for Chittagong city in Bangladesh. *International Journal of Sustainable Built Environment* **6** (1), 238–249. https://doi.org/10.1016/j.ijsbe.2017.01.005.

Alexandru, A. & Sushama, L. 2015 Current climate and climate change over India as simulated by the Canadian Regional Climate Model. *Climate Dynamics* **45** (3), 1059–1084. doi: 10.1007/s00382-014-2350-y.

Bannister, D., Herzog, M., Graf, H.-F., Hosking, J. S. & Short, C. A. 2017 An assessment of recent and future temperature change over the Sichuan Basin, China, using CMIP5 climate models. *Journal of Climate* **30** (17), 6701–6722. https://doi.org/10.1175/JCLI-D-16-0536.1.

Basher, M. A., Stiller-Reeve, M. A., Saiful Islam, A. K. M. & Bremer, S. 2018 Assessing climatic trends of extreme rainfall indices over northeast Bangladesh. *Theoretical and Applied Climatology* **134** (1–2), 441–452. https://doi.org/10.1007/s00704-017-2285-4.

Bowden, J. H., Nolte, C. G. & Otte, T. L. 2013 Simulating the impact of the large-scale circulation on the 2-m temperature and precipitation climatology. *Climate Dynamics* **40** (7–8), 1903–1920.

Brammer, H. 2012 *The Physical Geography of Bangladesh*, 1st edn. The University Press Limited, Dhaka, Bangladesh.

Chowdhury, A. F. M. K., Kar, K. K., Shahid, S., Chowdhury, R. & Rashid, M. M. 2019 Evaluation of spatio-temporal rainfall variability and performance of a stochastic rainfall model in Bangladesh. *International Journal of Climatology* **39** (11), 4256–4273. https://doi.org/10.1002/joc.6071.

Dash, S. & Maity, R. 2019 Temporal evolution of precipitation-based climate change indices across India: contrast between pre- and post-1975 features. *Theoretical and Applied Climatology* **138** (3), 1667–1678. doi: 10.1007/s00704-019-02923-8.

Gao, D., Chen, T., Yang, K., Zhou, J. & Ao, T. 2021 Projecting the impacts of climate change on streamflow in the upper reaches of the Yangtze River basin. *Journal of Water and Climate Change* **12**, 1724–1743. https://doi.org/10.2166/wcc.2020.082.

Graedel, T. E. 1977 The wind boxplot: an improved wind rose. *Journal of Applied Meteorology and Climatology* **16** (4), 448–450. https://doi.org/10.1175/1520-0450(1977)016 < 0448:TWBAIW > 2.0.CO;2.

Hausfather, Z. & Peters, G. P. 2020 Emissions – the 'business as usual' story is misleading. *Nature* **577**, 618–620. https://www.nature.com/articles/d41586-020-00177-3.

Held, I. M., Guo, H., Adcroft, A., Dunne, J. P., Horowitz, L. W., Krasting, J., Shevliakova, E., Winton, M., Zhao, M., Bushuk, M., Wittenberg, A. T., Wyman, B., Xiang, B., Zhang, R., Anderson, W., Balaji, V., Donner, L., Dunne, K., Durachta, J. & Zadeh, N. 2019 Structure and performance of GFDL's CM4.0 climate model. *Journal of Advances in Modeling Earth Systems* **11** (11), 3691–3727. https://doi.org/10.1029/2019MS001829.

Hossain, M. S., Roy, K. & Datta, D. K. 2014 Spatial and temporal variability of rainfall over the south-west coast of Bangladesh. *Climate* **2** (2), 28–46. https://doi.org/10.3390/cli2020028.

Ilori, O. W. & Ajayi, V. O. 2020 Change detection and trend analysis of future temperature and rainfall over West Africa. *Earth Systems and Environment* **4** (3), 493–512. https://doi.org/10.1007/s41748-020-00174-6.

Intergovernmental Panel on Climate Change (IPCC) 2014 Summary for policymakers. In: *Climate Change 2013 – The Physical Science Basis: Working Group I Contribution to the Fifth Assessment Report of the Intergovernmental Panel on Climate Change*. Cambridge University Press, Cambridge, pp. 1–30. doi:10.1017/CBO9781107415324.004.

Islam, M. R. & Raja, D. R. 2021 Waterlogging risk assessment: an undervalued disaster risk in coastal urban community of Chattogram, Bangladesh. *Earth* **2**, 151–173. https://doi.org/10.3390/earth2010010.

Javaherian, M., Ebrahimi, H. & Aminnejad, B. 2021 Prediction of changes in climatic parameters using CanESM2 model based on RCP scenarios (case study): Lar dam basin. *Ain Shams Engineering Journal* **12** (1), 445–454. https://doi.org/10.1016/j.asej.2020.04.012.

Kendall, M. G. 1975 *Rank Correlation Methods*, 4th edn. Charles Griffin, London, UK.

Klein Tank, A. M. G., Peterson, T. C., Quadir, D. A., Dorji, S., Zou, X., Tang, H., Santhosh, K., Joshi, U. R., Jaswal, A. K., Kolli, R. K., Sikder, A. B., Deshpande, N. R., Revadekar, J. V., Yeleuova, K., Vandasheva, S., Faleyeva, M., Gomboluudev, P., Budhathoki, K. P., Hussain, A. & Spektorman, T. 2006 Changes in daily temperature and precipitation extremes in central and south Asia. *Journal of Geophysical Research: Atmospheres* **111** (D16). https://doi.org/10.1029/2005JD006316.

Machiwal, D. & Jha, M. K. 2012 *Methods for Time Series Analysis. Hydrologic Time Series Analysis: Theory and Practice*. Springer, Dordrecht, The Netherlands. https://doi.org/10.1007/978-94-007-1861-6_4.

Maity, R. 2018 *Statistical Methods in Hydrology and Hydroclimatology*. Springer, Singapore. https://doi.org/10.1007/978-981-10-8779-0.

Mann, H. B. 1945 Nonparametric tests against trend. *Econometrica* **13**, 245–259. http://dx.doi.org/10.2307/1907187.

Mullick, M. R. A., Nur, R. M., Alam, M. J. & Islam, K. M. A. 2019 Observed trends in temperature and rainfall in Bangladesh using pre-whitening approach. *Global and Planetary Change* **172**, 104–113. https://doi.org/10.1016/j.gloplacha.2018.10.001.

Nengzouzam, G., Hodam, S., Bandyopadhyay, A. & Bhadra, A. 2020 Spatial and temporal trends in high resolution gridded rainfall data over India. *Journal of Earth System Science* **129**, 232. https://doi.org/10.1007/s12040-020-01494-x.

Nowreen, S., Murshed, S. B., Islam, A. K. M. S., Bhaskaran, B. & Hasan, M. A. 2015 Changes of rainfall extremes around the Haor basin areas of Bangladesh using multi-member ensemble RCM. *Theoretical and Applied Climatology* **119** (1–2), 363–377. https://doi.org/10.1007/s00704-014-1101-7.

Pandey, B. K., Khare, D., Kawasaki, A. & Mishra, P. K. 2019 Climate change impact assessment on blue and Green water by coupling of representative CMIP5 climate models with physical based hydrological model. *Water Resources Management* **33** (1), 141–158. https://doi.org/10.1007/s11269-018-2093-3.

Panjwani, S., Naresh, K. S., Ahuja, L. & Islam, A. 2019 Prioritization of global climate models using fuzzy analytic hierarchy process and reliability index. *Theoretical and Applied Climatology* **137** (3), 2381–2392. https://doi.org/10.1007/s00704-018-2707-y.

Pervin, L. & Gan, T. Y. 2021 Sensitivity of physical parameterization schemes in WRF model for dynamic downscaling of climatic variables over the MRB. *Journal of Water and Climate Change* **12**, 1043–1058. https://doi.org/10.2166/wcc.2020.036.

Pervin, L., Gan, T. Y., Scheepers, H. & Islam, M. S. 2021 Application of the HBV model for the future projections of water levels using dynamically downscaled global climate model data. *Journal of Water and Climate Change* **12**, 2364–2377. https://doi.org/10.2166/wcc.2021.302.

Praveen, B., Talukdar, S., Shahfahad Mahato, S., Mondal, J., Sharma, P., Islam, A. R. M. T. & Rahman, A. 2020 Analyzing trend and forecasting of rainfall changes in India using non-parametrical and machine learning approaches. *Scientific Reports* **10** (1), 1–21. https://doi.org/10.1038/s41598-020-67228-7.

Rahman, M. R. & Lateh, H. 2017 Climate change in Bangladesh: a spatio-temporal analysis and simulation of recent temperature and rainfall data using GIS and time series analysis model. *Theoretical and Applied Climatology* **128** (1), 27–41. https://doi.org/10.1007/s00704-015-1688-3.

Raihan, F., Li, G. & Harrison, S. P. 2015 Detection of recent changes in climate using meteorological data from south-eastern Bangladesh. *Journal of Climatology and Weather Forecasting* **3** (1), 127. doi: 10.4172/2332-2594.1000127.

Sattari, M. T., Rezazadeh, J. A. & Kusiak, A. 2016 Assessment of different methods for estimation of missing data in precipitation studies. *Hydrology Research* **48** (4), 1032–1044. https://doi.org/10.2166/nh.2016.364.

Sen, P. K. 1968 Estimates of the regression coefficient based on Kendall's tau. *Journal of the American Statistical Association* **63**, 1379–1389. http://dx.doi.org/10.1080/01621459.1968.10480934.

Shahid, S. 2010 Rainfall variability and the trends of wet and dry periods in Bangladesh. *International Journal of Climatology* **30** (15), 2299–2313. https://doi.org/10.1002/joc.2053.

Shahid, S. 2011 Trends in extreme rainfall events of Bangladesh. *Theoretical and Applied Climatology* **104** (3), 489–499. https://doi.org/10.1007/s00704-010-0363-y.

Shahid, S., Harun, S. B. & Katimon, A. 2012 Changes in diurnal temperature range in Bangladesh during the time period 1961–2008. *Atmospheric Research* **118**, 260–270. https://doi.org/10.1016/j.atmosres.2012.07.008.

Shahid, S., Wang, X. J., Harun, S. B., Shamsudin, S. B., Ismail, T. & Minhans, A. 2016 Climate variability and changes in the major cities of Bangladesh: observations, possible impacts and adaptation. *Regional Environmental Change* **16** (2), 459–471. https://doi.org/10.1007/s10113-015-0757-6.

Silva, R. S., Kumar, L., Shabani, F. & Picanco, M. C. 2017 Assessing the impact of global warming on worldwide open field tomato cultivation through CSIRO-Mk3·0 global climate model. *The Journal of Agricultural Science* **155** (3), 407–420. https://doi.org/doi: 10.1017/S0021859616000654.

Sreelatha, K. & Anand Raj, P. 2019 Ranking of CMIP5-based global climate models using standard performance metrics for Telangana region in the southern part of India. *ISH Journal of Hydraulic Engineering* **27**, 556–565. https://doi.org/10.1080/09715010.2019.1634648.

Tang, X., Shu, Y., Lian, Y., Zhao, Y. & Fu, Y. A. 2018 Spatial assessment of urban waterlogging risk based on a Weighted Naïve Bayes classifier. *Science of the Total Environment* **630**, 264–274. doi: 10.1016/j.scitotenv.2018.02.172.

Theil, H. 1950 A rank invariant method of linear and polynomial regression analysis, i, ii, iii. *Proceedings of the Koninklijke Nederlandse Akademie Wetenschappen, Series A Mathematical Sciences* **53**, 386–392. 521–525, 1397–1412.

Wu, J., Han, Z., Xu, Y., Zhou, B. & Gao, X. 2020 Changes in extreme climate events in China under 1.5 °C–4 °C global warming targets: projections using an ensemble of Regional Climate Model simulations. *Journal of Geophysical Research: Atmospheres* **125** (2). https://doi.org/10.1029/2019JD031057

First received 31 August 2021; accepted in revised form 18 November 2021. Available online 28 December 2021

© 2022 The Author

Impact of Climate Change on Hydrology and Water Resources doi: 10.2166/wcc.2021.181

Performance assessment of six bias correction methods using observed and RCM data at upper Awash basin, Oromia, Ethiopia

Bekan Chelkeba Tumsa

Faculty of Civil and Environmental Engineering, Jimma Institute of Technology, Jimma University, Oromia, Ethiopia
E-mail: bekanchelkeba@gmail.com

ABSTRACT

Selecting a suitable bias correction method is important to provide reliable inputs for evaluation of climate change impact. Their influence was studied by comparing three discharge outputs from the SWAT model. The result after calibration with original RCM indicates that the raw RCM are heavily biased, and lead to streamflow simulation with large biases (NSE = 0.1, $R^2 = 0.53$, MAE = 5.91 mm/°C, and PBIAS = 0.51). Power transformation and linear scaling methods performed best in correcting the frequency-based indices, while the LS method performed best in terms of the time series-based indices (NSE = 0.87, $R^2 = 0.78$, MAE = 3.14 mm/°C, PBIAS = 0.24) during calibration. Meanwhile, daily translation was underestimating simulated streamflow compared with observed and was considered as the least performing method. The precipitation correction method has higher visual influence than temperature, and its performance in streamflow simulations was consistent and considerable. Power transformation and variance scaling showed highly qualified performance compared to others with indicated time series values (NSE = 0.92, $R^2 = 0.88$, MAE = 1.58 mm/°C and PBIAS = 0.12) during calibration and validation of streamflow. Hence, PT and VARI were the dominant methods to remove bias from RCM models at Akaki River basin.

Key words: Akaki River, bias correction methods, climate change, performance assessment

HIGHLIGHTS

- This paper clearly identifies that bias correction methods' performances are watershed-dependent.
- Bias correction methods should be separately applied to each watershed.
- Also SWAT Model clearly identifies that bias correction methods have not been adjusted for all uncertainty related to the models.
- Bias of RCM should be removed with caution.
- Power Transformation and Variance scaling methods performed best.

GRAPHICAL ABSTRACT

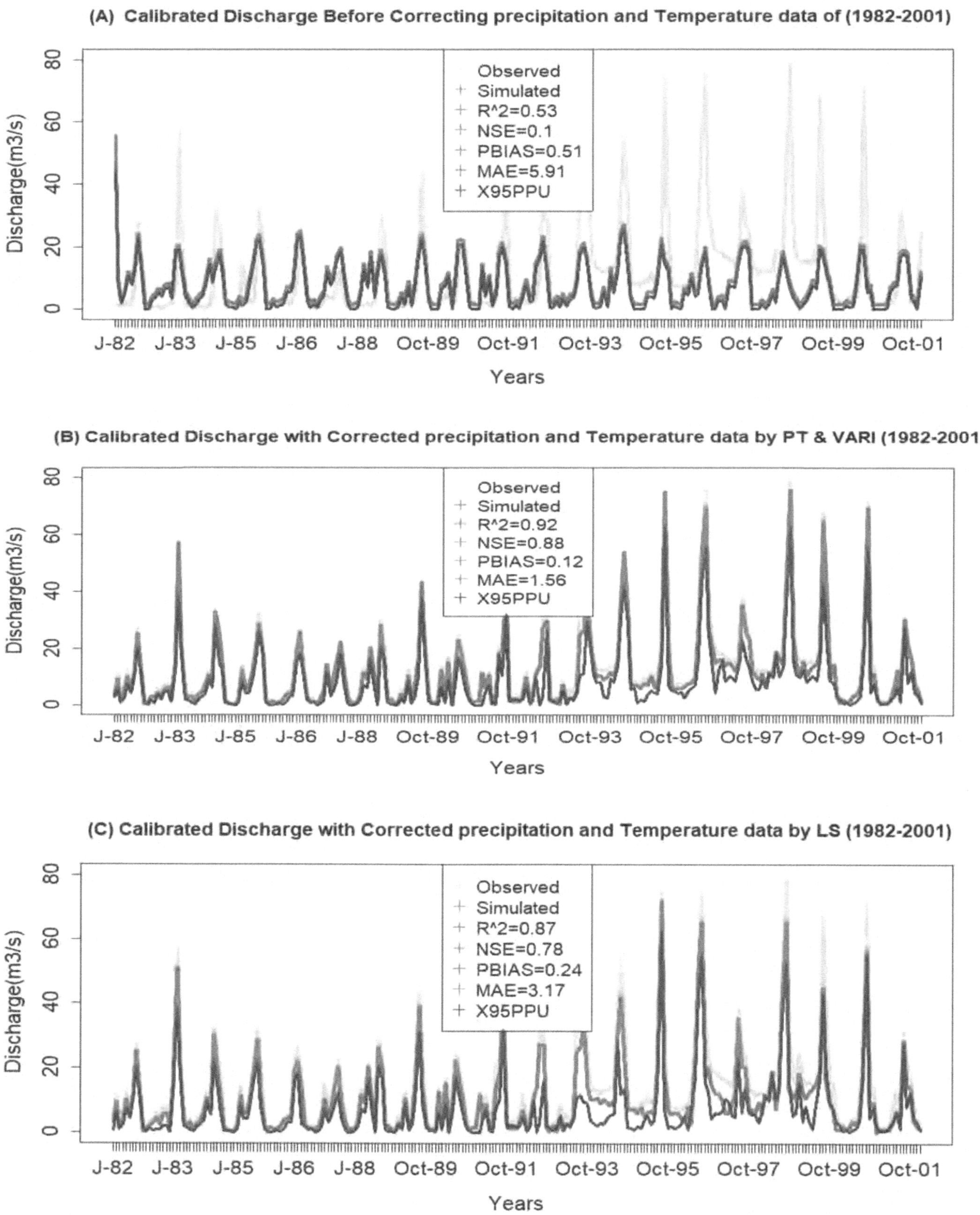

1. INTRODUCTION

Water is the most crucial and significant resource from which life is sustained over the Earth's surface, and saving and protecting this resource is not something that has been easily considered and recognized. In the recent era, particularly the 21st century, in which the world is growing in terms of technology and industrial civilizations, emissions from industry have brought climate change and the resulting variability has exposed water resource depletion. Currently, climate change is recognized in most parts of the globe and modeling it will lead to reducing the crises relating to food security, drought, and loss of energy sectors in the future by developing mitigation policies. Awash basin is the most irrigable area in Ethiopia and many farmers rely on this basin as a guarantee of their food security. However, climate change has become one of the causes for depletion of agricultural products in this region and impacts many farmers and their economic productivity and self-sufficiency (Taye 2018). For climate changes and variability assessments, general circulation models (GCMs) are the most appropriate and advanced tools, evaluating the impacts of climate change by removing its biasness (Haile 2017).

GCM and regional climate models (RCMs) are crucial models to predict, simulate, and project the future climate changes of the Earth globally and regionally (IPCC 2014). Since the 1950s, GCMs have been gradually developing and there have been many changes and improvements to model Earth systems' phenomena through climate change impact modeling (Smith *et al.* 2014). In the Coupled Model Intercomparison Project – Phase 5 (CMIP5), enormous GCMs have been developed into Earth systems' models including biogeochemical and hydrological cycles with predetermined verifications (Dale *et al.* 2017). In many studies related to climate change assessments, changes in precipitation and temperature, which have a direct involvement in global and regional hydrological processes, are the main factors that influence climate change outcomes (Endris *et al.* 2013). On the other hand, the development of new RCMs leads to another opportunity for scientists to analyze the effects of climate change as they inclusively consider higher spatial resolution and further reliable outcomes on a regional scale compared to GCMs (Kang *et al.* 2020). As a result, the GCM predicts the global scale mean surface temperature and periodical circulation of sea-ice extent of the oceans (Fang *et al.* 2015a). The Intergovernmental Panel on Climate Change (IPCC) investigations have been revealed and suggest that climate variables, specifically precipitation and temperature, are more closely related to global warming and are mostly impacting regional- or catchment-scale hydrological processes (Rakhimova *et al.* 2020).

Therefore, pre-processing of RCM outputs such as removing uncertainty (bias) should be the first step of most climate change impact studies to further estimate and predict the climate change consequences and suggest anticipation mechanisms throughout the global community (Chen *et al.* 2013). Generally, GCM or RCM modeled data are not directly used because of their bias as natural phenomena cannot be predicted accurately. Furthermore, this uncertainty and bias were developed during the advancement of circulation models by scientists due to a lack of an absolute and concrete idea about the nature of atmospheric circulation (Tan *et al.* 2020). All these facts suggest that GCM models may not estimate climate variables in the necessary way to reduce the differences, and there is always a deviation between observed and simulated climate variables even if the difference is insignificant. Therefore, it is very important to count on bias correction methods to remove bias from GCM/RCM outputs for predicting climate change impacts over the world's climate regions (Luo *et al.* 2018; Tan *et al.* 2020). The application of appropriate and suitable bias correction methods to the climate model enables researchers to be more confident of the hydrological models' outcome as large errors are expected to be removed. Several studies in this regard, such as that by Ezéchiel *et al.* (2016), have been developed and successfully applied in different climate regions in the world. Moreover, several methods, ranked from simple to more sophisticated approaches, were developed by different researchers depending on the characteristics of catchments up to basin scale for global climate change analysis of regions (Teutschbein 2013). In fact, the selection and application of appropriate bias correction methods to semi-arid climate regions is the most challenging issue due to the behavior of rainfall in such regions being very scarce, erratic, and infrequent in nature according to Mendez *et al.* (2020).

Furthermore, on average, several both simulated and corrected multi-model packages have much better performance in comparison to separate GCM/RCM and provide a reliable approach to the development of climate change scenarios (Olsson *et al.* 2015). In climate change impact analysis, it is more advantageous to select appropriate and suitable bias correction methods to realize the input data are free of bias to some extent. Regarding this concern, Teutschbein & Seibert (2013) have applied six bias corrections to correct precipitation and temperature data relying on statistical metrics with 11 RCM outputs in five typical catchments in Sweden and selected crucial methods that remove bias from RCM models. The main objective of selecting bias correction climate models prior to using them as input for hydrological related models is to suit the simulation of discharge or groundwater recharge by taking into account the accuracy of models' simulations with observed hydrological variables. This is aimed at reducing misleading results from simulation of hydrological processes in unrealistic magnitude, and even though bias correction methods are a relevant mechanism of reducing bias from climate models they are also susceptible to developing another source of uncertainty in the model chain (van Griensven *et al.* 2006).

Most of the time hydrological modeling is based on meteorological and hydrological data that are very sensitive to bias, and it is essential to find out its uncertainty which discourages a realistic result when conducting research (Xiang Soo *et al.* 2020). Since the performance of bias correction methods is watershed, catchment and sub-basin scale dependent, selecting appropriate bias correction methods is necessary to provide reliable inputs for global climate impact analysis (Fang *et al.* 2015b). In reality, there are different bias correction methods that are adapted to different regions and climate conditions and this issue is fundamentally the target of this research. Accordingly, the aim of this study is to clearly evaluate and investigate that the performance of bias correction methods is watershed dependent; and also to clearly identify that the power transformation (PT), linear scaling (LS), and variance scaling methods perform well in removing bias from RCM models.

2. MATERIALS AND METHODS

2.1. Description of the study area

The study was conducted on upper Awash catchment which is located in central Ethiopia along the western margin of the Main Ethiopian Rift (Figure 1). Awash basin is a large river basin which covers a huge area of farmland in the country. This basin is crucial for the farmers whose agricultural productivity serves them as the source of their daily life. It covers a total land area of 110,000 km^2 and is home to 10.5 million inhabitants (Getahun & Gebre 2015). The watershed is geographically bounded between 8°42′ to 9°10′N latitude and 38°24′ to 39°06′E longitude, covering an area of about 9,526 km^2. The entire Akaki sub-basin is surrounded by the valley in the north direction and by the Addis Ababa to Ambo road in the west.

2.2. Climate condition

Great Awash River watershed has a subtropical highland climate. Geographically, the study area is found near to the equator and due to this the temperature is remarkably constant from month to month. It is challenging to find climate data specific to the site, but Addis Ababa is found nearby the catchment. Based on monthly averages for 30 years (1981–2010), the mean monthly minimum and maximum temperatures vary from 7 °C to 11 °C and 21 °C to 28 °C, respectively. The lowest temperature of the study area was 7 °C, which was registered in November and December, and the maximum temperature was 28 °C, registered in March and May. The main rainy season for Akaki watershed is late June to early September. It is characterized by a dry winter, which is the dry season of the area. Generally, the project area has an average annual precipitation of 1,965 mm/year.

2.3. Meteorological data

Meteorological data gained from four stations in the catchment and simulated data from models were applied for the future to evaluate the bias correction method performance in removing bias. Depending on these fundamental aims, climate and hydrological models need meteorological data which are used as raw input for both bias correction and hydrological modeling processes to simulate climate phenomena. These meteorological and hydrological data were gathered from the Ethiopian National Meteorological Agency and Ministry of Water, Energy and Electricity office found in Addis Ababa. Four meteorological stations with relatively long periods of records within the catchment were used for this study and all stations contain every variable for each of them (Table 1).

2.4. Hydrological data

Akaki River flow was required for calibrating and validating the SWAT model. The main gauging stations were along the river basin inside the Akaki catchment and had continuous records at the outlet for 25 years (1982–2007) which were collected from the Ministry of Water, Energy and Electricity of Ethiopia. The purpose of the selected reference climate data set was to provide a reliable database for the study of all aspects of climate. These aspects range from global- to local-scale studies of fluctuations, trends, changes in variability and changes in the numbers of extreme events. Understanding and informing past changes and variations in climate are vitally important to improve the ability to predict, project, and prepare for future events to happen through climate change impacts.

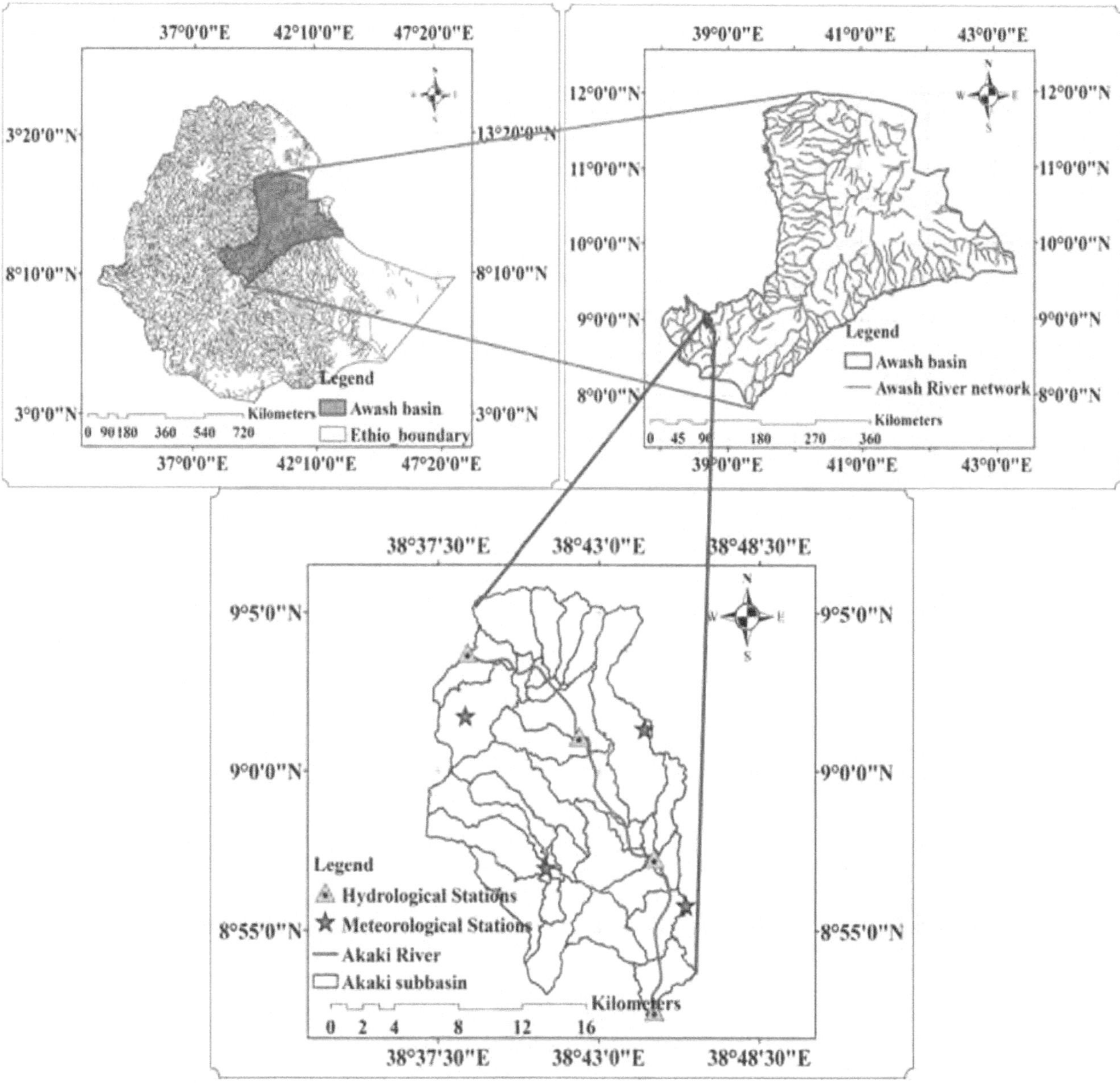

Figure 1 | Map of the study area and Akaki River basin's gauging stations.

Table 1 | The used input data, sources of data, and lengths of records based on the selected stations

Data types	Stations	Length of records	Source of data
Meteorological	Addis Ababa	1981–2010	National Meteorological Agency
and	Boneya		
Simulated data	Bulbula		CORDEX-(CanSEM2) and (RCA4)
	Sebeta		
Streamflow data		1982–2007	Ministry of Water, Energy and Electricity
LULC and soil data		2013	Ministry of Water, Energy and Electricity

2.5. Digital elevation model

Nowadays, the traditional method has been replaced by automatic extraction from a digital elevation model (DEM) due to the release of different types of high-resolution satellites in space at different times and the production of high-quality data. DEM data are used to describe topographic characteristics such as contour, slope, elevation difference, aspect, hill shade, and others. For this study, DEM is the main data set used for development of the sub-basin model components and geometrical data in the SWAT models as well as development of the physical model. The DEM data (Figure 2) used in this study were extracted from http://vertex.daac.asf.alaska.edu, with 12.5×12.5 m resolution.

2.6. Land use land cover and soil classification

Soil texture and land use land cover information are the primary input data for SWAT models in hydrological simulation (Figure 3). These parameters have been considered as the factors affecting the characteristics of watershed including hydrological impacts (the amount of runoff and peak discharge rates, and base flow are altered), physical impacts (streamflow morphology and temperature are changed), water quality impacts (nutrient and pollutant loads increase), and biological impacts (stream biodiversity decreases) (Du *et al.* 2020). The general land use/cover pattern of Akaki river basin was broadly classified into: annual and perennial crop, bare soil, grassland, shrubland, dense and sparse forest, residential, salt pan, water body, wetland, and woodland. On the other hand, the soil development in the study area was mostly due to the physical disintegration and chemical decomposition of volcanic rocks composed of different soil classes such as Calcic Xerosols, Chromic Luvisols, Chromic Vertisols, Eutric Nitisols, Orthic Solonchaks, Pellic Vertisols, and Vertic Cambisols.

2.7. CORDEX climate model

Different climate models have been developed containing different sets of components which, as a result, describe climate systems differently, and are expected to produce different results from their simulation packages. This development of climate models has considered the global climate phenomena throughout the world, but may not free models of uncertainty since nature by itself is unpredictable (Worku *et al.* 2020). On the other hand, while selecting hydrological models for evaluating hydrological simulation, the climate models selected should take into account the simplicity, availability, and familiarity of climate models with the objectives to be considered (Matiu *et al.* 2020). The Canadian Center for Climate Modeling and Analysis (GCM-CCCma) with a high resolution of $3.75° \times 3.75°$ and SMHI as the deriving model, was used for this study as it was preferred for the high resolution which can incorporate other GCM models (Mekonnen *et al.* 2011).

The Second Generation Canadian Earth System (CanSEM2) model and Rossby Centre Regional atmospheric (RCA4) model and with 0.44°/0.11° resolution were used as the raw RCM in this study for its preference to clearly consider an atmosphere-ocean general circulation model, a land-vegetation model, and terrestrial as well as oceanic interactive carbon cycle

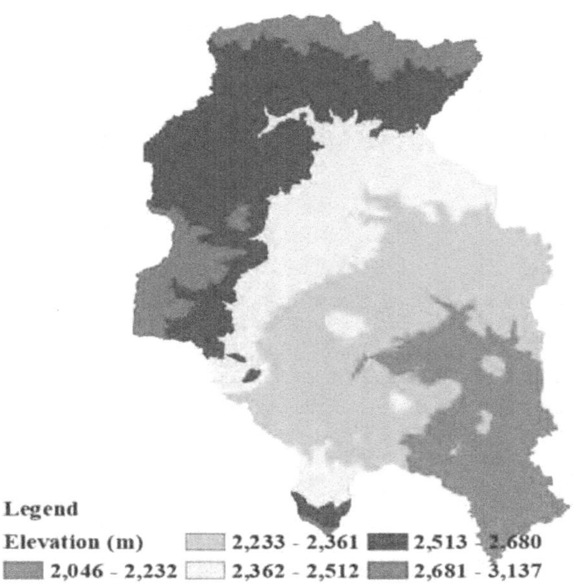

Figure 2 | Digital elevation model of the study area.

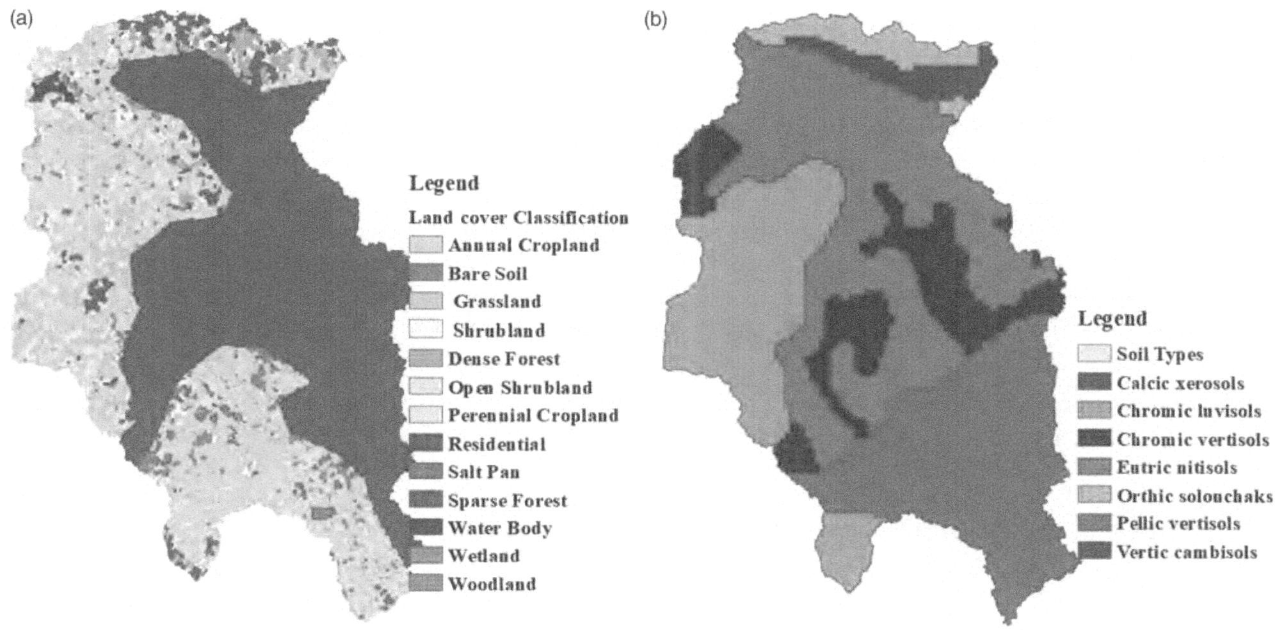

Figure 3 | (a) Land use land cover map and (b) soil map of Akaki River basin.

which basically cause climate variability (Chylek *et al.* 2011). Related to the RCA4 model, the data set is unique, not only as it is large but also as it takes lateral boundary conditions from a very large number of different GCMs with different resolutions. The spatial resolution of observed data was 0.28° × 02.8°, which was already included in the selected GCM models with stated resolution. Therefore, the types of hydrological models used (SWAT in this case), bulkiness of the data, and six bias correction methods being applied with this study, and high resolution with inclusiveness of boundary condition from many other GCMs by this model in consideration was used for the selection of those GCM and RCM models.

2.8. SWAT model and uncertainty

The impacts of corrected precipitation and temperature by bias correction methods in hydrological processes were evaluated by the SWAT hydrological model. This model is a semi-distributed, physically based, and time-continuous hydrological model that uses spatial information such as land use, soil texture, meteorological and topographic data to develop hydrological processes, particularly on climate change assessments. It divides a basin into sub-basins that include topographic information. The sub-basin information was further divided into a minimum of 78 hydrologic response units (HRUs) that uniquely combine land use, slope, and soil type (Zhang *et al.* 2016). This study uses 25 years (1982–2007) of daily streamflow data of Akaki River which was used for model calibration (1982–2001) and validation (2002–2007). This model is encrypted with ARGIS as its extension and performs different hydrological process functions in prediction of Earth system phenomena which is highly necessary to address. However, there are some uncertainties involved with SWAT model projection such as parameter uncertainty, model prediction uncertainty, and simulation effects which are fundamentally overcome by uncertainty analysis methods like 95% prediction uncertainty, SUFI-2, and GLUE during simulation, particularly in calibrating streamflow (Zhang 2019). In general, uncertainty of hydrologic models (SWAT) may emerge from discharge simulation, GCM emission scenario models, and hydrologic model parameterization as well as the hydrologic process in projecting future streamflow (Feng & Beighley 2020).

2.9. Description of bias correction methods

Based on their performance in removing bias from RCMs, there are a number of bias correction methods which are ranked according to their tendency to adjust mean, variance, coefficient of variation, and standard deviation climate variables. For instance, linear scaling is best at adjusting mean of climate variables, whereas power transformation corrects both mean and frequency of climate variables (Smitha *et al.* 2018). These statistical tools play an important role in estimating the differences between recorded and simulated data during climate change impact analysis. The main objective of this study was to verify and identify the best bias correction methods, from six selected methods, upon precipitation and temperature in removing and adjusting bias from RCM models for upper Awash River basin. These bias correction methods consist of linear scaling for both precipitation and temperature, power transformation for precipitation and variance scaling method for temperature, and daily translation (DT) for precipitation and temperature (Table 2). They are classified as major types of existing bias correction methods which enable researchers to remove unnecessary uncertainty from RCMs. All of them were conducted on a daily basis for SWAT models and monthly basis of meteorological data to analyze the result for each calendar month during the period 1981–2010. All bias correction methods were applied to daily values for all methods except daily translation methods which rely on a monthly basis, as described by Ezéchiel *et al.* (2016). Therefore, strong bias correction methods are to be identified before using bias-corrected methods to be applied on climate models' output for future climate change impact assessment. Thus, it is critical to develop strong climate change information that can be used for hydrological modeling in climate change impact assessment and to build optimal adaptation decisions in the water resource sector of the sub-basin. This study was planned to evaluate the performance of six different bias correction methods in adjusting the RCM rainfall and temperature with observed precipitation and temperatures at four gauging stations in Akaki River basin (Figure 4).

2.10. Performance evaluation by time series and frequency-based metrics

During the assessment and evaluation of the performance of each bias correction method, it is critical to consider the capacity of each method in generating discharge, precipitation, and temperature by identifying the daily average, maximum and minimum temperature based on frequency- and time-based indices (Worku *et al.* 2020). These time series metrics like R^2, NSE, PBIAS, and MAE were used in the explanations about the goodness of fit for bias correction methods' performance in correcting the bias from RCM models in comparison with observed meteorological and hydrological series. Meanwhile, frequency-based metrics were used to analyze the mean, coefficient of variation and standard deviation for both precipitation and temperature to estimate the difference between observed and simulated series to indicate further correction if necessary (Luo *et al.* 2018).

$$\mathrm{R}^2 = \left\{\sum_{i=1}^{N}(P_{si} - \bar{P}_s)(P_{oi} - \bar{P}_O)\right\} \Big/ \surd\left\{\sum_{i=1}^{N}(P_{Si} - \bar{P}_S)^2 \sum_{i=1}^{N}(P_{Oi} - \bar{P}_O)^2\right\}, \; 0 \le R^2 \le 1 \tag{1}$$

$$\mathrm{NSE} = 1 - \frac{\sum_{i=1}^{N}(P_{Oi} - P_{Si})^2}{\sum_{i=1}^{N}(P_{Oi} - \bar{P}_O)^2}, \quad -\infty \le NSE \le 1 \tag{2}$$

$$\mathrm{PBIA} = \frac{\sum_{i=1}^{N}(P_{Oi} - P_{Si})}{\sum_{i=1}^{N}P_{Oi}} \tag{3}$$

$$\mathrm{MAE} = \frac{\sum_{i=1}^{N}|P_{Oi} - P_{Si}|}{n} \tag{4}$$

where P_{si} and P_{oi} represent the RCM models' precipitation and observed series at time step i, and P_s and P_o are the corresponding average values for RCM and observed variables, respectively.

Table 2 | Selected bias correction methods to be evaluated for precipitation and temperature series

Bias correction methods for precipitation	Bias correction methods for temperature
Power transformation function	Variance scaling
Linear scaling	Linear scaling
Daily translation	Daily translation

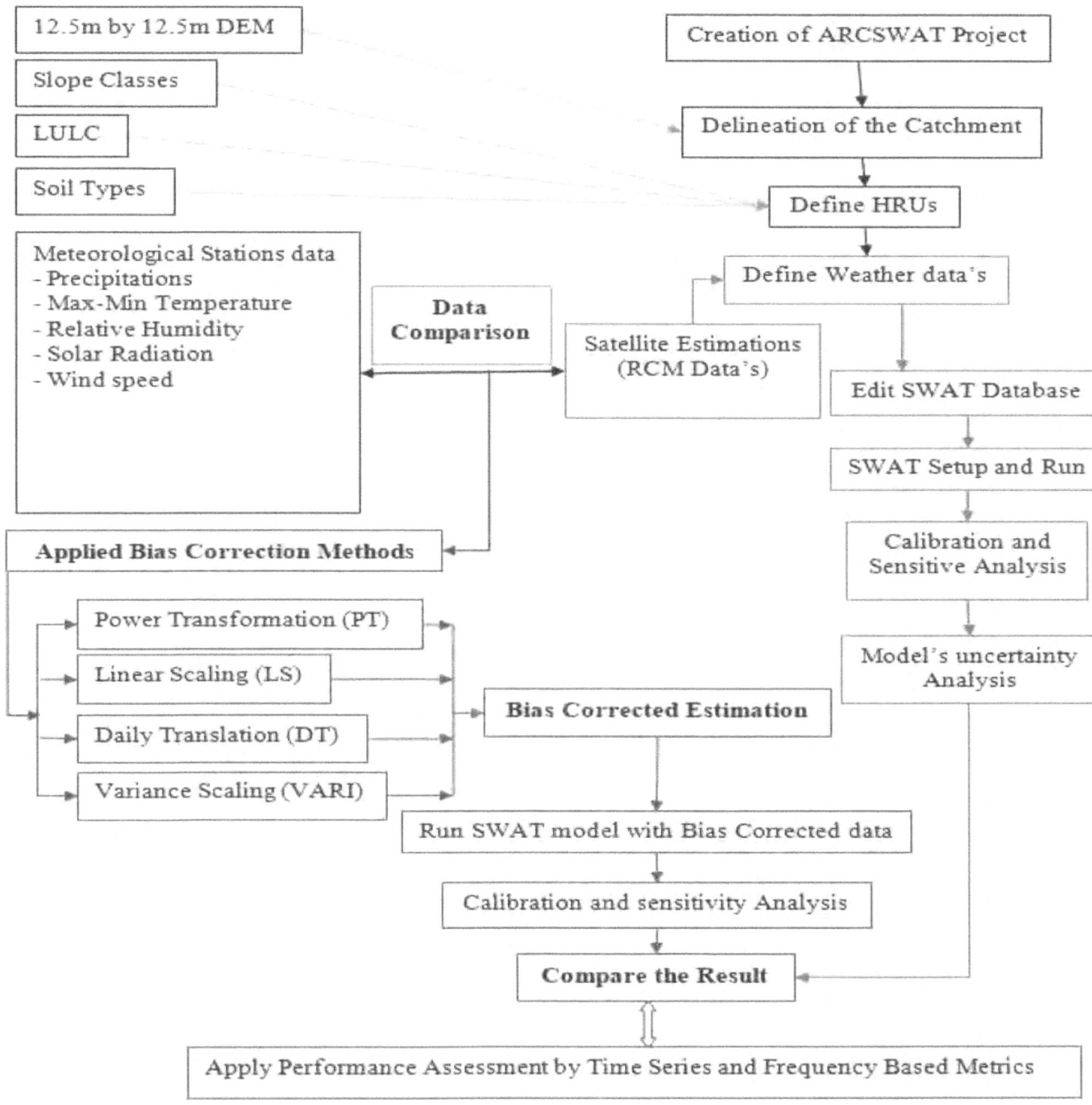

Figure 4 | Flow chart of the methodology.

3. RESULTS AND DISCUSSION

3.1. Performance of RCM models' outputs in reproducing discharges

Most of the time, climate variables modeled from different RCMs do not match with the recorded one because of significant bias during simulation of precipitation or temperature. In this case, evaluation of original RCM data in simulating discharge from precipitation and temperature data series was directly applied to the SWAT model and clearly shows great deviation from June to December when compared to observed discharge, as indicated in Figure 5. In a different study with a similar concept, it was stated that if biased RCM output in modeling discharge is considered small enough to be adjusted through calibration, it is not necessary to apply bias correction (Velasquez *et al.* 2020). However, since the deviation between RCM output and observed discharge had significant bias, the application of bias correction was necessary but intolerable as it aimed to mislead the result. Therefore, the next step to improve the bias of RCM models was to consider the reference,

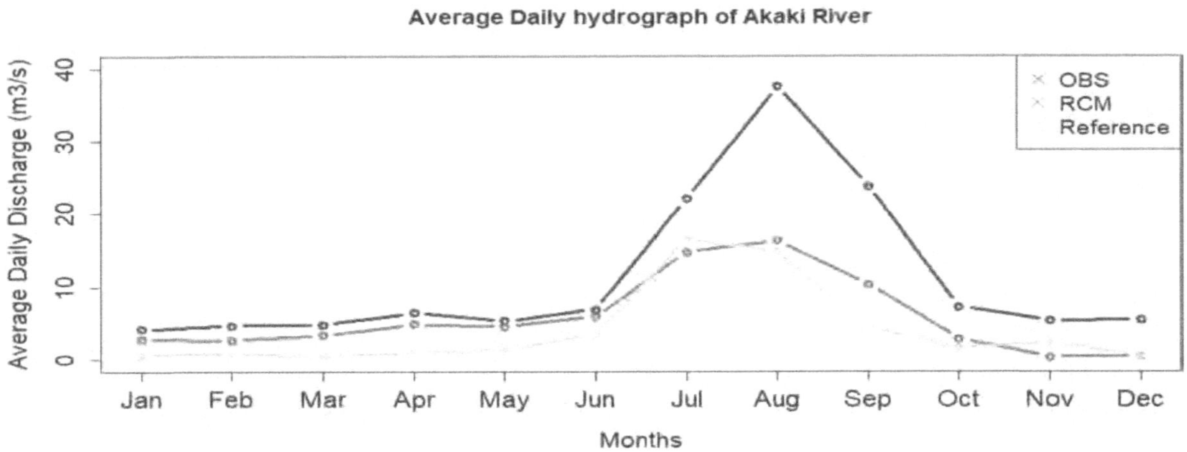

Figure 5 | Daily hydrograph of Akaki River comparing observed, simulated and refence discharge.

simulated and observed discharge with necessary and appropriate bias correction methods depending on their capacity for removing bias. As seen from the developed daily hydrograph in Figure 5, the simulated discharge along with the observed precipitation and temperature was highly biased and underestimated by RCM models' output.

3.2. Validation of original precipitation and temperature of RCM

Assessment mainly focused on the performance of original precipitation and temperature in simulating average discharge from simulated model output which proved that the original RCM output is not advisable to use directly by hydrological models. The study conducted related to precipitation bias correction in Britain also concluded that using original simulated RCM model output directly in modeling discharge was not advisable as it misled the final result from hydrological models' output (Lafon *et al.* 2013). This study also reveals that different bias correction methods correct originally simulated RCM output at different levels of certainty in different climate regions. Accordingly, as seen from Figure 6(a) on the basis of exceedance probability, original RCM output underestimated precipitation compared to others. At the same time, the corrected average precipitation by daily translation method was overestimated with a maximum probability of 0.3 in January to May and underestimated with a probability range between 0.32 and 0.9 during the summer season (June, July, and August). In addition, power transformation function and linear scaling methods overestimated the average precipitation in the summer season (June, July, and August) and almost corrected exactly in the other months in comparison to the observed precipitation. Hence, linear scaling and power transformation function were best performed in bias correction of precipitation at Addis Ababa station.

The simulated precipitations by RCM at Sebeta meteorological station indicate that the model simulates precipitation below the observed and bias corrected in all aspects. This simply indicates that the simulated precipitation by the RCM model used in this study was highly biased and is not advisable to be directly used by hydrological models. Meanwhile, power transformation, daily translation, and linear scaling methods overestimated corrected precipitation with maximum probability of 0.35 from April to May and underestimated it at the beginning of January but discontinued and began new underestimation from May to June. This variation of underestimation or overestimation according to season fundamentally happened from the model capacity during simulation and was corrected by bias correction methods to the required level of removing bias. Hence, power transformation and linear scaling methods were recognized as the best performing methods. The RCM – model outputs for temperature obtained at the Boneya and Bulbula stations were more accurate than other stations as RCM – was overestimated at the two stations compared to others in almost all months except July and November. The probability distribution of maximum temperature of RCM models at Addis Ababa and Sebeta stations was overestimated as seen from Figure 6(e) when compared to observed temperature.

However, linear scaling and daily translation methods corrected the bias of simulated temperature very well, and variance scaling methods also adjusted the bias even though underestimation and overestimation was seen along the observed data. The RCM model overestimated the maximum temperature with probabilities above 0.92. At the same time, underestimation with average exceedance probability of 0.2 by variance scaling method was also verified. However, the exceedance

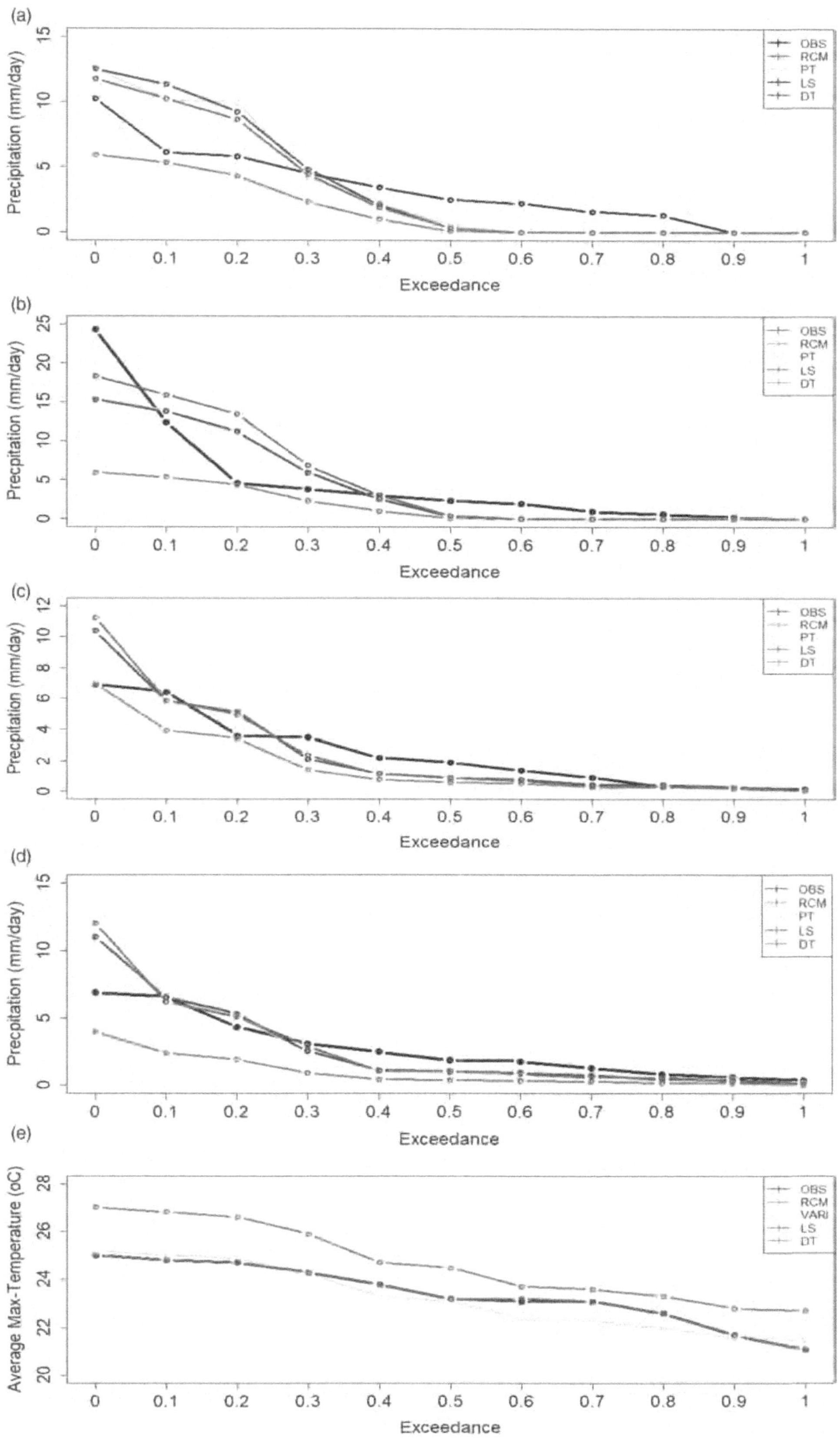

Figure 6 | Exceedance probabilities for (a) Addis Ababa, (b) Boneya, (c) Bulbula, (d) Sebeta stations with the observed, RCM, and bias-corrected precipitation and (e) for temperature at Akaki River basin.

probability of maximum temperature in overestimation or underestimation by linear scaling and daily translation methods was very rare and corrected to the maximum level of its requirement. Hence, daily translation and linear scaling performed very well in correcting bias of the RCM models.

3.3. Validation of corrected precipitation with frequency and time series-based metrics

The clear importance of bias correction methods should be to compare and evaluate to achieve the reliability of input data for hydrological models based on time series and frequency-based metrics. These two statistical methods that were used to verify the capability of bias correction methods used the final SWAT model output to analyze their performance in removing bias from the RCM models. Accordingly, all bias correction methods used in this study performed highly in adjusting mean values and coefficient of variation of climate variables on the daily basis. However, linear scaling methods cannot analyze extreme events due to their unique scaling character in heavy rainfall seasons as a result of their underestimations compared to the observed data. A similar result was arrived at by Piani *et al.* 2010. As seen from Table 3, monthly mean and standard deviation were corrected relative to observed climate variables by all methods, while coefficient of variation was adjusted relative to the RCM (simulated) climate variables. However, power transformation and linear scaling methods were adjusted exactly on the monthly average of precipitation at all stations. This clearly indicates that power transformation and linear scaling methods performed very well in terms of frequency-based metrics (mean) but only slightly in adjusting standard deviation as they overestimated at Addis Ababa station. While evaluating these bias correction methods for precipitation in terms of coefficient of variation, all methods were adjusted to RCM. Hence, when comparing the corrected precipitation adjusted by those methods based on frequency-based metrics, all bias correction methods performed best by adjusting daily precipitation than monthly. These facts demonstrate that bias correction methods were watershed- and season-dependent as they were directly involved with regional watershed characteristics and hydrological processes. Therefore, power transformation function performed best according to the result obtained from the analysis using frequency-based metrics. Generally, this result shows that daily translation and linear scaling methods performed less well in adjusting STD and CV of monthly values than daily at all stations. However, the power transformation method consistently and uniquely adjusted mean and standard deviation correctly in comparison with observed precipitation data at all stations. Hence, this method was very suitable for removing bias from GCM/RCM at this sub-basin.

On the other hand, time series-based metrics are the most imperative statistical method for the evaluation of bias correction methods. Depending on these statistical metrics, different bias correction corrects or adjusts the bias of RCM along with the observed one to different levels of capability. As stated in Table 4, the performance of RCM in simulating precipitation was very poor and needed to be adjusted to use as raw input data for hydrological models' simulation. According to these time series-based metrics, MAE of 2.82 mm, NSE of -1.12, PBIAS of 26.4%, and R^2 of 0.51 at Addis Ababa station shows the simulated precipitation by RCM model is very biased. The daily translation method still underestimates the corrected precipitation

Table 3 | Frequency-based metrics of evaluation of the performance of bias correction methods for original and corrected precipitation

Parameters	Catchment	Frequency indices	OBS	RCM	PT	LS	DT
PRECIP	Addis Ababa	Std	312.13	162.3	336.7	343.6	332.4
		Mean	185.14	87.46	185.1	185.2	183.07
		CV	1.69	1.86	1.82	1.86	1.82
	Boneya	Std	232.26	162.3	243	242.3	245.7
		Mean	130.55	87.47	130.6	130.6	130.5
		CV	1.78	1.86	1.86	1.86	1.88
	Bulbula	Std	243.99	99.06	275.3	275.3	259.3
		Mean	141.93	91.07	141.9	141.9	139.3
		CV	1.72	1.94	1.94	1.94	1.86
	Sebeta	Std	477.63	162.3	420.1	418.8	501.7
		Mean	225.68	87.47	225.7	225.7	223.9
		CV	2.12	1.86	1.86	1.86	2.24

Table 4 | Time series-based evaluation of the performance of bias correction methods for original and corrected precipitation at four stations

Parameters	Stations	Methods	MAE (mm)	PBIAS (%)	NSE (–)	R^2 (–)
Precipitation	Addis Ababa	RCM	2.82	26.4	−1.12	0.51
		PT	1.74	0.53	0.76	0.93
		LS	1.04	0.58	1.7	0.65
		DT	1.74	0.53	−2.2	0.58
	Boneya	RCM	1.37	15.02	−0.67	0.58
		PT	0.34	0.45	0.72	0.92
		LS	0.45	0.62	1.69	0.79
		DT	0.12	0.32	−1.3	0.52
	Bulbula	RCM	1.79	8.03	−1.36	0.49
		PT	0.42	0.51	0.67	0.90
		LS	2.53	0.42	0.64	0.84
		DT	0.31	1.02	−6.2	0.45
	Sebeta	RCM	1.41	10.2	−2.31	0.40
		PT	0.52	0.68	0.72	0.90
		LS	0.44	0.07	0.83	0.86
		DT	0.26	2.3	−1.32	0.55

when considering the average of all stations, with PBIAS, R^2, and MAE, respectively, rising to 23.16%, 0.58, and 0.60 mm. In general, the optimum values of times series-based metrics recommended for precipitation correction with MAE, PBIAS, NSE, and R^2 ranged between 0.12 mm/2.53 mm, 1.02%/0.68%, 2.31/6.2, and 0.45/0.93, respectively. Depending on these time series-based metrics, the performance of power transformation and linear scaling methods in correcting precipitation is much better than frequency-based metrics. While it is important to prioritize these methods based on both frequency-based and time series-based metrics, power transformation and linear scaling methods perform very well in both time series- and frequency-based metrics. Furthermore, in this study, which mainly concentrated on Akaki sub-basin, power transformation function is the best performing method in removing the RCM/GCM bias compared to the other methods.

3.4. Validation of corrected temperatures by frequency- and time series-based metrics

In the case of temperature bias correction, the linear scaling and variance scaling (VARI) methods demonstrate good performance. This clearly means that both methods were capable of correcting bias of RCM models for temperature when compared with observed data in terms of frequency-based and time series-based metrics. On the other hand, the daily translation method mostly underestimates and slightly overestimates temperature correction at almost all stations relatively. This is because, most probably, the time structure problem does not exist in temperature series. A study by Gumindoga *et al.* (2019) arrived at the same conclusion. When these two-pillar meteorological data were used in simulation of discharge by a hydrological model like SWAT, in this case the result was very good compared to when it was forced by RCM outputs except the daily translation method which performed poorly in removing bias from RCM models. As the analysis was applied to each method in terms of exceedance probability, frequency-based and time series-based metrics to predict the bias correction method's capability in simulating discharges, the result obtained by the daily translation method differed greatly between the simulated and observed discharge which showed the poor performance of this method.

On the same subject matter, Chen *et al.* (2013) arrived at a similar conclusion in which the daily translation method even underestimated compared with the discharge driven by RCM output. Hence, the daily translation method failed to show the expected performance in discharge modeling from corrected RCM precipitation and temperature. According to this analysis, corrected temperature by linear scaling and, especially by VARI methods, was acceptable for hydrological modeling because of their high capability in removing large biases from RCM models. Indeed, temperature exhibits a slightly worse performance than precipitation at the Addis Ababa station, which seriously overestimated when compared to the observed one, with an MAE of 2.57 °C and PBIAS of 42.3% (Table 5). According to the result obtained during calibration, all temperature bias

Table 5 | Prioritization of bias correction methods based on time series for original and corrected temperature on daily average scale at four stations

Parameters	Stations	Methods	MAE (mm)	PBIAS (%)	NSE (–)	R^2 (–)
Temperature	Addis Ababa	RCM	2.57	42.3	5.32	0.43
		VARI	1.3	−0.05	−0.74	0.79
		LS	1.01	−0.1	0.3	0.68
		DT	1.27	−0.05	−0.14	0.54
	Boneya	RCM	3.03	37.21	7.13	0.4
		VARI	2.4	−0.02	−1	0.43
		LS	1.08	0	0.04	0.82
		DT	0.85	0.49	−0.1	0.50
	Bulbula	RCM	1.23	−3.12	−4.62	0.42
		VARI	0.67	0.01	−0.28	0.75
		LS	0.42	−0.13	−0.02	0.69
		DT	0.26	−0.34	−0.18	0.52
	Sebeta	RCM	3.62	29.16	6.39	0.38
		VARI	1.02	0.08	1.75	0.78
		LS	0.38	−0.06	−0.25	0.71
		DT	1.02	−0.29	0.42	0.51

correction methods improve the performance of RCM-simulated temperature as PBIAS shows low underestimation and low overestimation to an acceptable value. The PBIAS was close to zero and the performance of these methods was good. Generally, the performance of the variance scaling method was much better at adjusting both frequency- and time series-based metrics in removing bias of temperature variable from RCM models.

In the same way, variance scaling and linear scaling methods were evaluated as the best temperature correction methods by frequency-based metrics. Especially when mean and coefficient of variation among those correction methods was considered, these two methods were ranked as the best performing. However, these methods overestimated standard deviation, not a great deal but to some extent at all stations, except for being exactly adjusted at Addis Ababa station by linear scaling. On the other hand, the daily translation method overestimates all frequency-based metrics (mean, STD, and CV) when compared to that of the observed one. This result clearly shows that daily translation performed less in adjusting mean, STD, and CV at all stations and was not suitable to remove bias for this river basin. Hence, variance scaling and linear scaling methods were ranked as the best performing temperature bias correction methods by frequency indices.

3.5. The performance of bias correction methods in simulating streamflow

In generating streamflow discharge through simulation, precipitation and maximum temperature large bias correction methods have been used in much research to remove bias from climate models. These methods were graded according to the capability of reducing bias from RCM in world climate regions with different characteristics and climate conditions. Depending on the regional climate scenario, bias correction methods most of the time assessed, prioritized, and evaluated through their performance to predict and generate discharge, precipitation and temperature variables by eventually minimizing bias under frequency- and times series-based metrics.

Based on the output of statistical methods, the discharges were simulated from the combined precipitation and temperature data forcibly applied to a SWAT model to evaluate the variation of results before and after the bias correction had been made and to suggest the consequences for water resource policymakers. These corrected precipitation and temperature data with linear scaling and daily translation methods for the simulation of discharge were significantly different to the reference. However, they are much better than the original RCM model that was too biased and underestimated the discharge. The simulated discharge driven by corrected discharge with power transformation for precipitation and variance scaling for temperature performed best. Furthermore, both climate variables corrected by linear scaling method apparently simulated the discharge much better than daily translation and the original RCM. Generally, power transformation and variance scaling methods for precipitation and temperature, respectively, were the best performing methods in removing bias from RCM/GCM of Akaki River basin meteorological data stations. Using the SWAT model output, the calibration and validation was done, which is to overcome any uncertainty remaining in bias correction on monthly time steps using the average measured streamflow, simulated before bias and after the application of bias correction.

Accordingly, these time series performed much better with monthly time series compared to the daily times series. As concluded from the result of calibrated discharge (Figure 7(a)) with uncorrected precipitation and temperature it possesses poor performance with RCM/RCA4 data of $R^2 = 0.53$, NSE $= 0.1$, PBIAS $= 0.51$, and MAE $= 5.91$ which was less than the minimum requirement. On the other hand, precipitation and temperature data corrected by daily translation method to simulate discharge was slightly better than RCM/RCA4, but not an acceptable result as $R^2 = 052 < 0.6$, NSE $= -3.62$, PBIAS $= 1.05$, MAE $= 4.08$ did not fit with the observed data. However, power transformation with the variance and linear scaling methods has the best performance in simulating discharge by removing an expected uncertainty from input data to the models. The time series metrics values of $R^2 = 0.87$, NSE $= 0.78$, PBIAS $= 0.24$, and MAE $= 3.17$ (Figure 7(c)) indicate that the linear scaling method is very good at removing bias from RCM models to reduce the misleading result. Furthermore, power

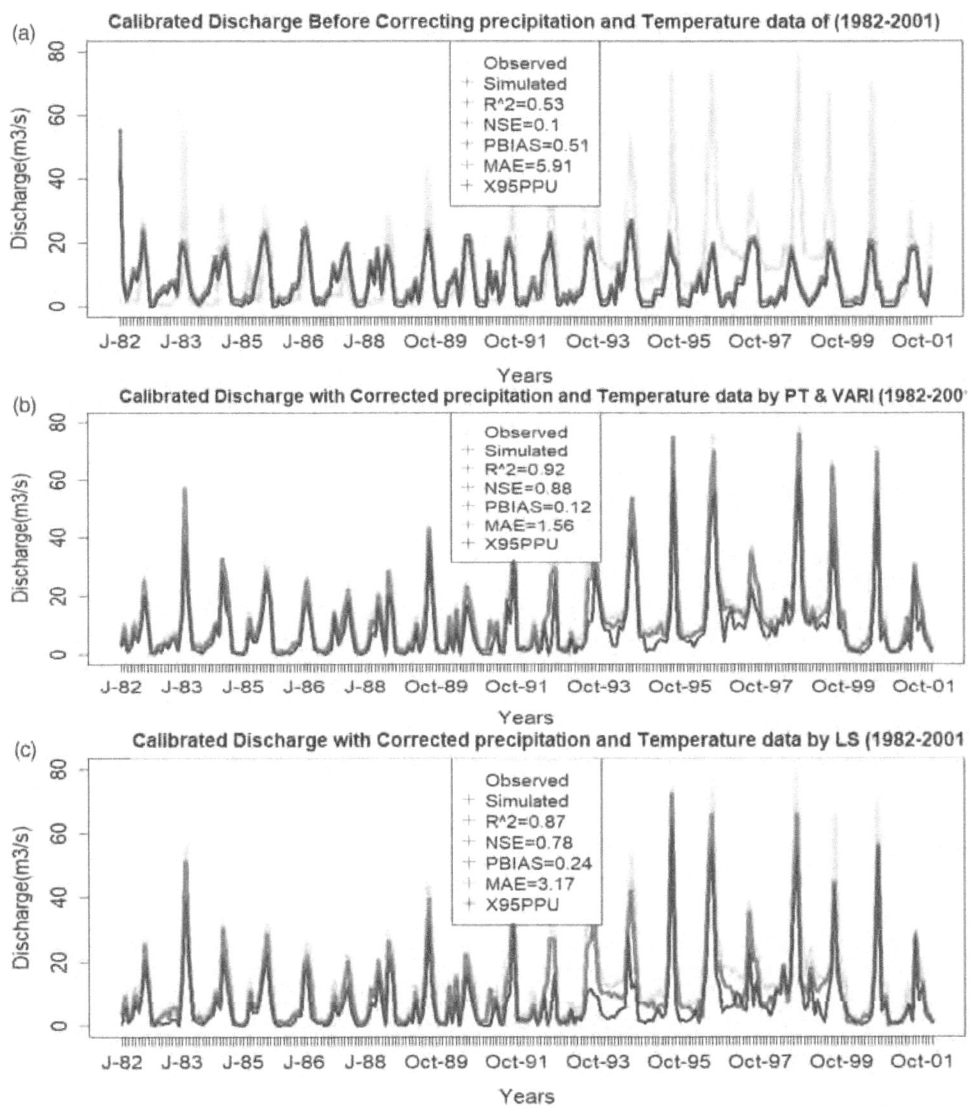

Figure 7 | The calibrated discharge with (a) original RCM, (b) after bias being removed by power transformation and variance scaling, (c) by linear scaling method.

transformation with variance scaling method was the best performing method in removing bias from RCM/RCA4 according to the result obtained during calibration and validation for streamflow.

Moreover, the time series-based metrics value provided here provided evidence for the performance of this method during calibration (Figure 7(b)) with $R^2 = 0.92$, NSE = 0.88, PBIAS = 0.12, and MAE = 1.56 and validation (Figure 8) $R^2 = 0.95$, NSE = 0.9, PBIAS = 0.09, and MAE = 0.9. In general, bias correction methods should be applied consciously with great care in order not to mislead the expected result when hydrological models are used. Even though it should be recognized that a good temporal structure of the original RCM simulations provides considerable assurance in model output, great care is needed in order not to be exposed to a misleading result for policymakers in mitigating climate change.

3.6. Performance of bias correction methods through developing spatial map

Spatial map is the most important way to identify deviation between the observed and simulated precipitation and temperature through developing a spatiotemporal map based on annual or monthly average rainfall and maximum–minimum temperature of the watershed. It is one of the most deterministic models in spatial interpolation which is relatively fast and easy to compute, and straightforward to interpret. Hence, the time series-based values indicated show that RCM performance in representing the sampled location from the neighboring station through simulation was poor. Other bias correction methods show relatively good performance in simulating and removing bias from RCM/GCM models (Table 6).

By applying six bias correction methods on rainfall and temperature spatial map analysis, power transformation with variance and linear scaling methods perform relatively better in removing uncertainty from models. This clearly shows that the simulated area with a limited range of precipitation and temperature values on the map corrected and represented by linear scaling was slightly underestimated. But the corrected value from power transformation and variance scaling function was better as the evidence of time series-based metrics ($R^2 = 0.68$, NSE = 1, PBIAS = 0, MAE = 0) and temperature ($R^2 = 0.72$, NSE = 0.8, PBIAS = −0.02, MAE = 0.08) gives good results.

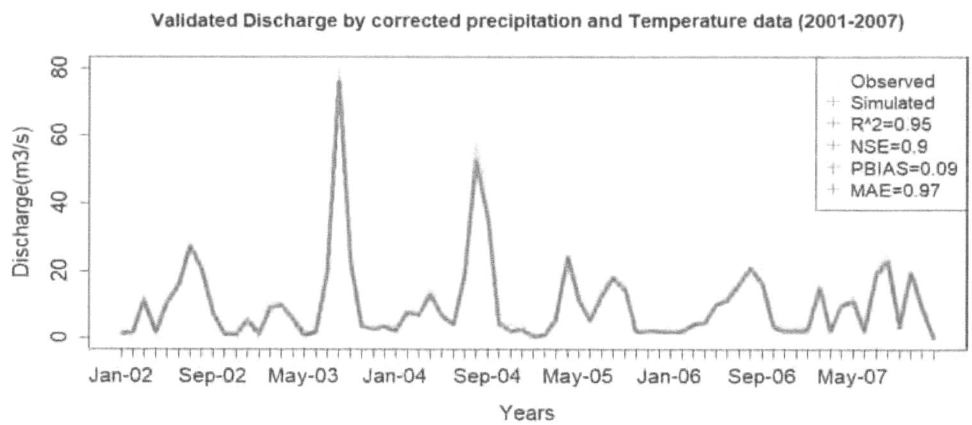

Figure 8 | Validated discharge after bias correction made for temporal precipitation and temperature under power transformation and variance scaling methods.

Table 6 | Performance of bias correction methods by developing spatial map from corrected and original RCM based on annual average precipitation and temperature

Statistical parameter	For precipitation (mm)				For temperature (°C)			
Time series metrics	Bias correction methods				Bias correction methods			
Series	RC	PT	LS	DT	RCM	VARI	LS	DT
R^2	0	0.68	0.57	0.40	0	0.72	0.64	0.52
NSE	−5.4	1	−1.62	0.748	0.36	0.8	0.37	−1.17
PBIAS	0.5	0	−0.11	0.096	−0.01	−0.02	0.01	0.02
MAE	80.5	0	35.69	14.243	0.21	0.08	0.20	0.06

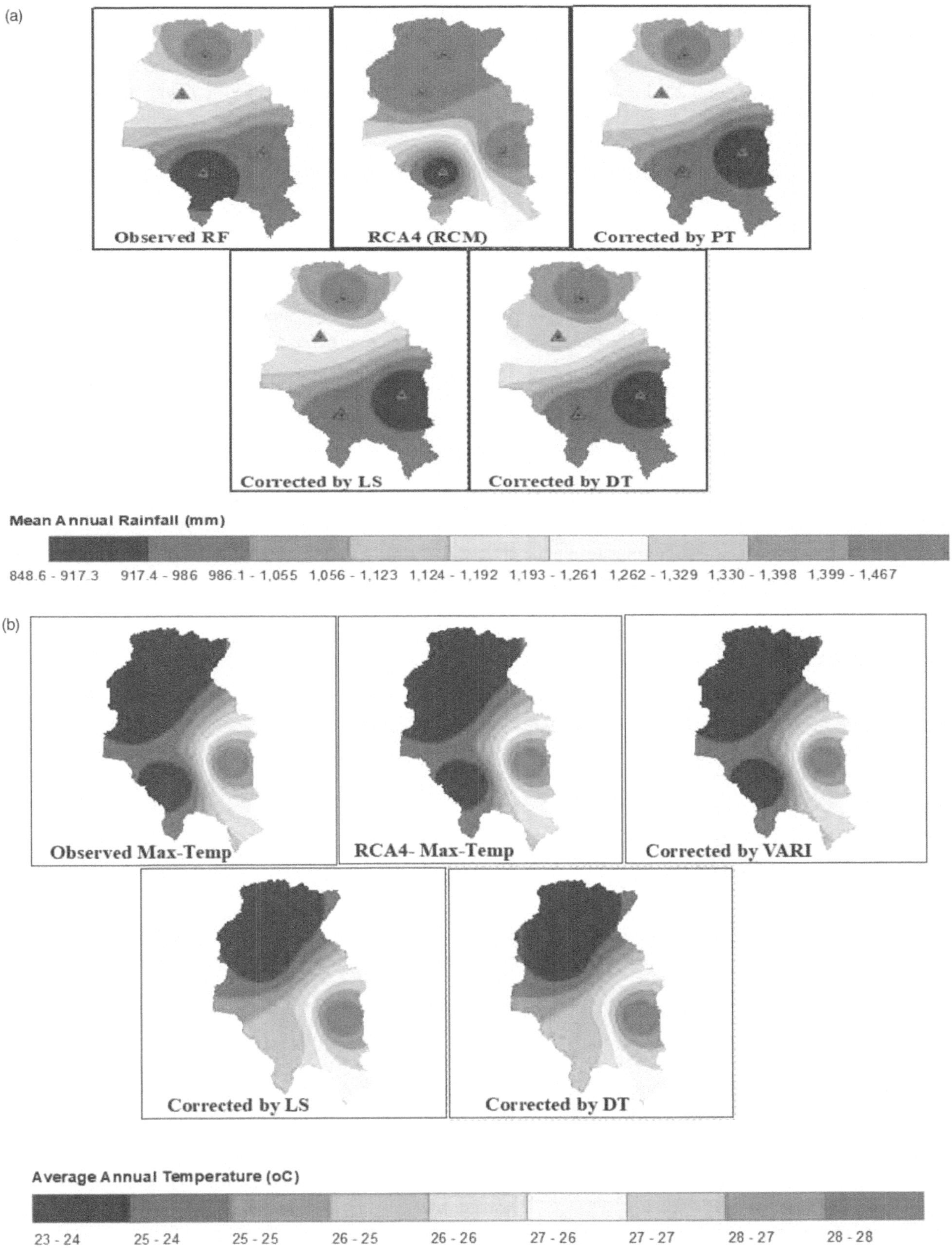

Figure 9 | Spatial map developed from (a) average precipitation and (b) temperature for Addis Ababa, Bulbula, Boneya, and Sebeta meteorological stations.

However, when the values of these time series calculated from the spatial map are compared to others like calibrated and validated values, they are slightly underestimated due to the spatial map being developed from annual average precipitation and temperatures entities (Figure 9). Furthermore, the variation clearly shows that the performance of bias correction methods mainly depends on seasonal values of meteorological parameters rather than corresponding average values. This means bias correction methods performed slightly better when monthly metrological data were used as opposed to daily, when times series metrics are used to evaluate their performance. Meanwhile, when daily metrological data were used, the bias correction methods performed very well when frequency-based metrics were used to evaluate them.

4. CONCLUSION

In this paper, six bias correction methods, three precipitations and three temperatures, were compared and evaluated by frequency-based and time series-based metrics considering their capability of removing bias from climate models for hydrological processes and simulations. In generating streamflow discharge through simulation, precipitation and maximum temperature, large bias correction methods were used in much research to remove bias from climate models. These methods were ranked according to the capability of reducing bias from RCM in world climate regions of different characteristics and climate conditions.

The SWAT model was used to identify the performances of bias correction methods by using original and corrected RCM data as raw inputs. The output of this hydrological method was analyzed by statistical methods along with recorded streamflow and realized the existence of significant bias. This bias is fundamentally merged from regional climate models during simulating precipitation and temperature over the river basin. Referring to this evidence, bias correction methods are very important for fixing the deviation between simulated and observed climate variables in order not to mislead the hydrological model outputs for policymakers to better mitigate climate change consequences. Therefore, this study concludes power transformation and variance scaling method are the most suitable bias correction methods for precipitation and temperature at upper Awash basin, respectively.

ACKNOWLEDGEMENTS

I would like to thank Jimma university Institute of Technology, Research and Publication director office for their great willingness in supporting me with financial aid, Ethiopian Meteorological Agency, Water, Irrigation and Electricity Institute for their great willingness in giving me any necessary data for the accomplishment of this research. Bekan Chelkeba (Msc) and supportive team together collected all necessary data and wrote the paper.

DECLARATION OF COMPETING INTEREST

The author declares no conflicts of interest at all.

DATA AVAILABILITY STATEMENT

All relevant data are included in the paper or its Supplementary Information. All necessary data were incorporated and included in the paper.

REFERENCES

Chen, J., Brissette, F. P., Chaumont, D. & Braun, M. 2013 Finding appropriate bias correction methods in downscaling precipitation for hydrologic impact studies over North America. *Water Resources Research* **49** (7), 4187–4205. doi:10.1002/wrcr.20331.

Chylek, P., Li, J., Dubey, M. K., Wang, M. & Lesins, G. 2011 Observed and model simulated 20th century Arctic temperature variability: Canadian earth system model CanESM2. *Atmospheric Chemistry and Physics* **11**, 22893–22907. doi:10.5194/acpd-11-22893-2011.

Dale, A., Fant, C., Strzrepek, K., Lickley, M. & Solomon, S. 2017 Climate model uncertainty in impact assessments for agriculture: a multi-ensemble case study on maize in sub-Saharan Africa. *Earth's Future* **5** (3), 337–353. doi:10.1002/2017EF000539.

Du, X., Goss, G. & Faramarzi, M. 2020 Impacts of hydrological processes on stream temperature in a cold region watershed based on the SWAT equilibrium temperature model. *Water (Switzerland)* **12** (4). doi:10.3390/W12041112.

Endris, H. S., Omondi, P., Jain, S., Lennard, C., Hewitson, B., Changa, L., Awange, J. L., Dosio, A., Ketiem, P., Nikulin, G., Panitz, H.-J., Buchner, M., Stordal, F. & Tazalika, L. 2013 Assessment of the performance of CORDEX regional climate models in simulating East African rainfall. *Journal of Climate* **26** (21), 8453–8475. doi:10.1175/JCLI-D-12-00708.1.

Ezéchiel, O., Eric, A. A., Josué, J. C., Eliézer, B. I. & Amédée, C. 2016 Comparative study of seven bias correction methods applied to three Regional Climate Models in Mekrou catchment (Benin, West Africa). *International Journal of Current Engineering and Technology* **6** (5), 1831–1840.

Fang, G., Yang, J., Chen, Y., Zhang, S., Deng, H., Liu, H. & De Maeyer, P. 2015a Climate change impact on the hydrology of a typical watershed in the Tianshan Mountains. *Advances in Meteorology* **2015**. doi:10.1155/2015/960471.

Fang, G. H., Yang, J., Chen, Y. N. & Zammit, C. 2015b Comparing bias correction methods in downscaling meteorological variables for a hydrologic impact study in an arid area in China. *Hydrology and Earth System Sciences* **19** (6), 2547–2559. doi:10.5194/hess-19-2547-2015.

Feng, D. & Beighley, E. 2020 Identifying uncertainties in hydrologic fluxes and seasonality from hydrologic model components for climate change impact assessments. *Hydrology and Earth System Sciences* **24** (5), 2253–2267. doi:10.5194/hess-24-2253-2020.

Getahun, Y. S. & Gebre, S. L. 2015 Flood hazard assessment and mapping of flood inundation area of the Awash River basin in Ethiopia using GIS and HEC-GeoRAS/HEC-RAS model. *Journal of Civil & Environmental Engineering* **5** (4). doi:10.4172/2165-784x.1000179.

Gumindoga, W., Rientjes, T. H. M., Tamiru Haile, A., Makurira, H. & Reggiani, P. 2019 Performance of bias-correction schemes for CMORPH rainfall estimates in the Zambezi River basin. *Hydrology and Earth System Sciences* **23** (7), 2915–2938. doi:10.5194/hess-23-2915-2019.

Haile, A. T. 2017 Assessment of climate change impact on flood frequency distributions in Baro Basin, Ethiopia. *Atmospheric Research* **161–162**, 1305–1321. Available from: http://dx.doi.org/10.1016/j.atmosres.2015.03.013%0A; https://doi.org/10.1080/02626667.2017.1365149.

IPCC 2014 *Climate Change 2014: Mitigation of Climate Change. Summary for Policymakers and Technical Summary*. Part of the Working Group III Contribution to the Fifth Assessment Report of the Intergovernmental Panel on Climate Change. doi:10.1017/CBO9781107415416.005.

Kang, Y., Gao, J., Shao, H. & Zhang, Y. 2020 Quantitative analysis of hydrological responses to climate variability and land-use change in the hilly-gully region of the loess plateau, China. *Water (Switzerland)* **12** (1). doi:10.3390/w12010082.

Lafon, T., Dadson, S., Buys, G. & Prudhomme, C. 2013 Bias correction of daily precipitation simulated by a regional climate model: a comparison of methods. *International Journal of Climatology* **33** (6), 1367–1381. doi:10.1002/joc.3518.

Luo, M., Liu, T., Meng, F., Duan, Y., Frankl, A., Bao, A. & De Maeyer, P. 2018 Comparing bias correction methods used in downscaling precipitation and temperature from regional climate models: a case study from the Kaidu River Basin in Western China. *Water (Switzerland)* **10** (8). doi:10.3390/w10081046.

Matiu, M., Pettina, M., Notarnicola, C. & Zebisch, M. 2020 Evaluating snow in EURO-CORDEX regional climate models with observations for the European Alps: biases and their relationship to orography, temperature, and precipitation mismatches. *Atmosphere* **11** (1), 46.

Mekonnen, D. G., Moges, M. A., Mulat, A. G. & Shumitter, P. 2011 The impact of climate change on mean and extreme state of hydrological variables in Megech watershed, Upper Blue Nile basin, Ethiopia. In: *Extreme Hydrology and Climate Variability*. Elsevier Inc., Amsterdam, the Netherlands, pp. 123–135. doi:10.1016/B978-0-12-815998-9.00011-7.

Mendez, M., Maathuis, B., Hein-Griggs, D. & Alvarado-Gamboa, L. F. 2020 Performance evaluation of bias correction methods for climate change monthly precipitation projections over Costa Rica. *Water (Switzerland)* **12** (2). doi:10.3390/w12020482.

Olsson, T., Jakkila, J., Veijalainen, N., Backman, L., Kaurola, J. & Vehviläinen, B. 2015 Impacts of climate change on temperature, precipitation and hydrology in Finland – studies using bias corrected regional climate model data. *Hydrology and Earth System Sciences* **19** (7), 3217–3238. doi:10.5194/hess-19-3217-2015.

Piani, C., Haerter, J. O. & Coppola, E. 2010 Statistical bias correction for daily precipitation in regional climate models over Europe. *Theoretical and Applied Climatology* **99** (1–2), 187–192. doi:10.1007/s00704-009-0134-9.

Rakhimova, M., Liu, T., Bissenbayeva, S., Mukanov, Y., Gafforov, K. S., Bekpergenova, Z. & Gulakhmadov, A. 2020 Assessment of the impacts of climate change and human activities on runoff using climate elasticity method and general circulation model (GCM) in the Buqtyrma River Basin, Kazakhstan. *Sustainability (Switzerland)* **12** (12). doi:10.3390/su12124968.

Smith, M. J., Palmer, P. I., Purves, D. W., Vanderwel, M. C., Luytsarev, V., Calderhead, B., Joppa, L. N., Bishop, C. M. & Ermmott, S. 2014 Changing how earth system modeling is done to provide more useful information for decision making, science, and society. *Bulletin of the American Meteorological Society* **95** (9), 1453–1464. doi:10.1175/BAMS-D-13-00080.1.

Smitha, P. S., Narasimhan, B., Sudheer, K. P. & Annamalai, H. 2018 An improved bias correction method of daily rainfall data using a sliding window technique for climate change impact assessment. *Journal of Hydrology* **556**, 100–118. doi:10.1016/j.jhydrol.2017.11.010.

Tan, Y., Guzman, S. M., Dong, Z. & Tan, L. 2020 Selection of effective GCM bias correction methods and evaluation of hydrological response under future climate scenarios. *Climate* **8** (10), 1–21. doi:10.3390/cli8100108.

Taye, M. T. 2018 Climate change impact on water resources in the Awash Basin, Ethiopia. *Water* **10** (11), 1560. doi:10.3390/w10111560.

Teutschbein, C. 2013 *Hydrological Modeling for Climate Change Impact Assessment: Transferring Large-Scale Information From Global Climate Models to the Catchment Scale*. PhD dissertation, Department of Physical Geography and Quaternary Geology, Stockholm University, Stockholm, Sweden.

Teutschbein, C. & Seibert, J. 2013 Is bias correction of regional climate model (RCM) simulations possible for non-stationary conditions. *Hydrology and Earth System Sciences* **17** (12), 5061–5077. doi:10.5194/hess-17-5061-2013.

van Griensven, A., Meixner, T., Grunwald, S., Bishop, T., Diluzio, M. & Srinivasan, R. 2006 A global sensitivity analysis tool for the parameters of multi-variable catchment models. *Journal of Hydrology* **324** (1–4), 10–23. doi:10.1016/j.jhydrol.2005.09.008.

Velasquez, P., Messmer, M. & Raible, C. C. 2020 A new bias-correction method for precipitation over complex terrain suitable for different climate states: a case study using WRF (version 3.8.1). *Geoscientific Model Development* **13** (10), 5007–5027. doi:10.5194/gmd-13-5007-2020.

Worku, G., Teferi, E., Bantider, A. & Dile, Y. T. 2020 Statistical bias correction of regional climate model simulations for climate change projection in the Jemma sub-basin, upper Blue Nile Basin of Ethiopia. *Theoretical and Applied Climatology* **139** (3–4), 1569–1588. doi:10.1007/s00704-019-03053-x.

Xiang Soo, E. Z., Wan Jaffar, W. Z., Lai, S. H., Othman, F., Elshafie, A., Islam, T., Srivastava, P. & Othman Hadi, H. S. 2020 Evaluation of bias-adjusted satellite precipitation estimations for extreme flood events in Langat river basin, Malaysia. *Hydrology Research* **51** (1), 105–126. doi:10.2166/nh.2019.071.

Zhang, L. 2019 Model uncertainty analysis methods for semi-arid watersheds with different characteristics: a comparative SWAT case study. *Water* **11** (6), 1177.

Zhang, L., Nan, Z., Xu, Y. & Li, S. 2016 Hydrological impacts of land use change and climate variability in the headwater region of the Heihe River Basin, northwest China. *PLoS ONE* **11** (6), 1–25. doi:10.1371/journal.pone.0158394.

First received 5 May 2021; accepted in revised form 11 October 2021. Available online 15 November 2021

Impact of Climate Change on Hydrology and Water Resources doi: 10.2166/wcc.2022.242

Spatiotemporal variations in the elasticity of runoff to climate change and catchment characteristics with multi-timescales across the contiguous United States

Xinglong Gong, Aiqi Xu, Shuping Du* and Yuehan Zhou

School of Water Conservancy & Civil Engineering, Northeast Agricultural University, Harbin 150030, Heilongjiang Province, China
*Corresponding author. E-mail: 1437968102@qq.com

XG, 0000-0003-0686-7994

ABSTRACT

There is a lack of understanding of the impact of timescales on runoff elasticity. In this study, gauged river flow and meteorological data time-series from 1916 to 2015 and 1941 to 2015 across the contiguous United States (CONUS) for 188 catchments were collected. We studied spatiotemporal variations of climate and catchment elasticity at multi-timescales (11, 20, 40, and 60 a) across the CONUS, and the relationships between the elasticity of runoff and climatic factors were also analyzed. Median descriptive statistics and Budyko-coupled climate elasticity methods were used to calculate runoff elasticity in each timescale. The original Mann–Kendall trend test was used to test the temporal trend significance in four timescales, respectively. The results show that (1) the trend and values of runoff elasticity may not be affected by timescales. (2) Catchment characteristics in some western regions may be the second-most sensitive factor for runoff following the maximum temperature. The roles of potential evaporation, maximum temperature, and catchment characteristics on runoff gradually become important from east to west. (3) Runoff elasticity is independent of precipitation, and wet regions tend to have lower elasticity of runoff. There are strong relationships between elasticity and most factors, except for precipitation, which can illustrate well its spatial pattern.

Key words: catchment characteristics, climate change, runoff elasticity, spatiotemporal variation, trend analysis

HIGHLIGHTS

- Runoff elasticity including trend and value is stable at space and timescale.
- Apart from precipitation, extreme temperature and catchment characteristics are also important factors for runoff alteration.
- Runoff elasticity can be explained by the aridity index, runoff coefficient, runoff, and watershed properties.
- The distinct spatial pattern occurs in the contiguous United States.

GRAPHICAL ABSTRACT

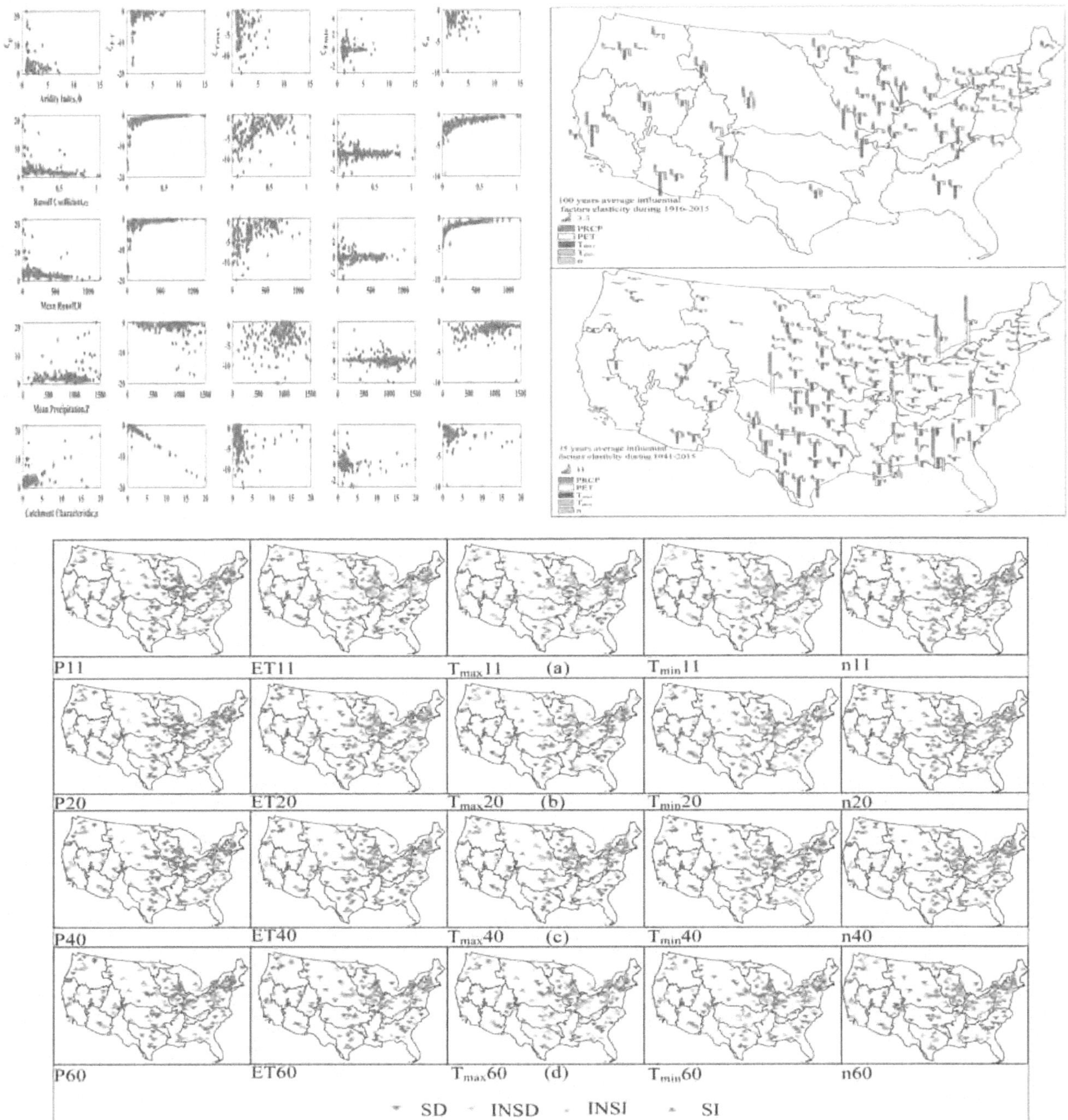

INTRODUCTION

Runoff, the most important component of the hydrological cycle, undergoes an alteration over space and time due to climate change and human activities (Dey & Mishra 2017). Understanding quantitively climate change- and human activities-induced runoff changes in long-term periods at a regional and/or basin scale is of importance for regional water planning and management. Runoff elasticity, which was defined by Schaake (1990) as the ratio of proportional change in streamflow to proportional change in climatic variables, can well reflect the sensitivity of runoff to multiple influential factors including climate change and anthropogenic activities, and which can further obtain runoff changes. Therefore, investigating spatiotemporal variations in runoff elasticity is vital for understanding spatiotemporal variation characteristics of runoff influenced by climate change and human activities.

Under a changing environment, total changes in runoff can be obtained by the difference between natural and changed periods in the measured long-term runoff (Sun *et al.* 2014; Wang *et al.* 2019). However, individual factor-induced changes (climatic factors and human activities-induced changes) may not be directly calculated by comparing the natural runoff with the changed runoff. An essential step for obtaining an individual factor-induced change in runoff is identifying clearly the impact of climate change and human activities on runoff; in other words, we need to distinguish the climatic or anthropogenic component from total runoff changes. Presently, there are a large number of methods for separating the individual impacts of climate variability and human activities on streamflow (Dey & Mishra 2017). These methods can be categorized as hydrological modeling (Wang *et al.* 2010; Chang *et al.* 2015), conceptual approaches (Tomer & Schilling 2009; Jiang *et al.* 2015), analytical approaches (Sun *et al.* 2014; Yang *et al.* 2014), and experimental approaches (Brown *et al.* 2005; Zhao *et al.* 2010). The conceptual approaches include applications of the Budyko hypothesis (Budyko 1974) and the Tomer–Schilling framework (Tomer & Schilling 2009). Budyko (1974) assumed the ratio of the mean annual actual evapotranspiration to the mean annual precipitation as a function of the ratio of the mean annual potential evapotranspiration to the mean annual precipitation and other watershed properties. The analytical approach includes climate elasticity and hydrological sensitivity methods. Schaake (1990) defined elasticity as the ratio of proportional change in streamflow to proportional change in climatic variables. However, these two approaches have their own shortcomings in applications, for example, a constant value of parameter represented catchment properties throughout the analysis was used based on the Budyko hypothesis for the former method (Jiang *et al.* 2015), and the major shortcoming of the latter is that the change in streamflow due to human activities cannot be expressed explicitly in terms of elasticity coefficients (Wang *et al.* 2016; Dey & Mishra 2017). Therefore, considering the temporal variation of catchment characteristic parameters in the Budyko hypothesis is vital for analyzing accurately the runoff alteration.

The Budyko hypothesis-coupled climate elasticity method was commonly used to detect influential factors-induced changes (Zheng *et al.* 2009; Wang & Hejazi 2011; Patterson *et al.* 2013). For example, Yang *et al.* (2014) estimated the climate elasticity and evaluated the contribution of climate change to runoff for 210 catchments to understand the spatial variation of hydrologic response to climate change, and further calculated the parameter n in the Budyko hypothesis. They have shown that the value of n is different between catchments, and the contribution of climate change to runoff has a distinct spatial pattern. Wang *et al.* (2019) used climate elasticity and hydrological sensitivity analysis to estimate the contributions of climate change and human activities in the Beichuan River Basin in the northeastern Tibetan Plateau, China, and they found the dominant factor is human activities. However, the elasticity of runoff is usually calculated by long-term average hydrometeorological data, and the minimum data length is not less than 11 years because soil water storage can be neglected (Shao *et al.* 2012). So far, the impact of data length or timescales on climate elasticity is not clear. Although some scholars have shown that climate elasticity is catchment-based, for example, Sankarasubramanian *et al.* (2001) found that precipitation elasticity is independent of precipitation but sensitive to model structure and calibration error. Climate elasticity acts as a function of climate and catchment conditions, and it may also be explained by other factors such as the aridity index and catchment characteristics. Therefore, apart from analyzing the impact of data length on climate elasticity, relationships of climate elasticity to other factors are also necessary.

The selection of influential factors is a primary question for attribution analysis of runoff alteration. Whether the selection of influential factors is reasonable is directly related to whether the cause of runoff variation can be accurately found. Recently, the factors causing runoff alteration have mainly been climatic factors and anthropogenic activity and have been gradually developed from single to comprehensive. The study area expanded from a small basin to a large basin or a small region to a large region when evaluating the impact of these two types of factors on runoff (Cheng *et al.* 2019; Li *et al.* 2020). For instance, while analyzing the impact of climatic factors on runoff in the 1990s, only the effect of precipitation was considered. With the research development, climatic factors such as potential evaporation, air temperatures (mean or extreme temperatures), ocean–atmosphere interactions that lead to interannual and decadal quasiperiodic behavior (e.g., Pacific Decadal Oscillation (PDO) and Atlantic Multidecadal Oscillation (AMO)) were found to have a non-negligible effect on runoff (Prokoph *et al.* 2012; Nalley *et al.* 2019). Yang & Yang (2011) further combined the first-order differential of the Penman equation and the elasticity of runoff to precipitation, net radiation, air temperature, wind speed, and relative humidity to derive and separate the contributions of different climatic variables. Wang *et al.* (2016) further explored the impact of maximum and minimum temperatures on runoff based on Yang & Yang's (2011) research and found that extreme temperatures may have a greater influence on runoff than precipitation and potential evaporation. Therefore, maximum and minimum temperatures and catchment characteristic elasticity should be considered.

Over the contiguous United States, there are many interesting studies in quantifying streamflow variations using different methods. For example, paired catchment approaches and the Distributed Hydrology Soil 38 Vegetation Model, embedded with a multi-purpose reservoir module (DHSVM-Res), were used to simulate streamflow and reservoir storage in south-central Texas, and they found that urbanization and climate change are the major drivers of variations in the monthly maximum streamflow. Ravindranath & Devineni (2020) applied an extended sequent peak algorithm to streamflow reconstructions in the Upper Missouri River Basin. A monthly water balance model was validated and applied by Li *et al.* (2020) to quantify the hydrologic responses of 81,900 12-digit Hydrologic Unit Code (HUC) watersheds to historical and projected land use/land change (LULC) in 2000, 2010, 2050, and 2100 in the contiguous United States (CONUS). Stepwise and geographically weighted regression models were used to identify key factors controlling the spatially varied hydrologic impacts across the CONUS. Davenport *et al.* (2020) quantified the size of streamflow peaks in response to both seasonal and event-specific rain-fraction using stream gage observations from watersheds across the western United States. Also, some scholars quantified runoff changes caused by precipitation, potential evaporation, and human activities over the contiguous United States (Wang & Hejazi 2011; Patterson *et al.* 2013; Rice *et al.* 2015; Sang *et al.* 2020). In our study, we investigated the spatiotemporal pattern of elasticity of runoff to multiple factors, as an index reflecting the sensitivity of runoff to factors, especially for extreme temperature elasticity, because the temperature has a profound effect on the hydrological process (little analysis for previous studies). Our special analytical perspective can illustrate spatiotemporal pattern characteristics of runoff changes in some way, and provide a reference to quantify runoff changes using the climate elasticity method.

The primary objectives of this study are (1) to analyze the temporal trend and spatial pattern of elasticity of runoff to multiple factors (precipitation, potential evaporation, maximum and minimum temperatures, and catchment characteristics) at different timescales, and (2) to explore the relationships of elasticity of runoff with other factors (mean rainfall, mean runoff, aridity index, runoff coefficient, and catchment characteristic). We select four slipping windows (11, 20, 40, and 60 a) according to our length of data, and then the elasticity of runoff series in each timescale can be obtained by calculating the elasticity of runoff in each timescale in 188 catchments of the CONUS. The original Mann–Kendall method is used to test the trend significance of the series. This work is helpful to determine in detail the variation process of sensitivity of runoff to influential factors under climate change and human interference.

MATERIALS AND METHODOLOGY

Streamflow and meteorological data

Daily streamflow data were obtained from the United States Geological Survey (USGS). The USGS has collected water resource data at approximately 1.5 million sites in all 50 states. Those sites were selected from the Geospatial Attributes of Gages for Evaluating Streamflow, version II (GAGES-II) database. The GAGES-II database contains many basin-specific characteristics for 9,322 basins throughout the United States, including climatic, hydrologic, topographic, land cover and land use, and geologic attributes.

Daily streamflow data from the climatic year 1916 through the climatic year 2015 were selected, and we also considered daily streamflow from the climatic year 1941 through the climatic year 2015. We selected the water year (1 April to 31 March of the next year), not the calendar year, as the yearly scale. These daily streamflow data were downloaded from the USGS National Water Information System (NWIS). We downloaded data according to the standard in Dudley *et al.* (2019). The selection criteria resulted in 2,482 total study gauges: 1,408 for the 1941–2015 period and 203 for the 1916–2015 period. We extracted a large number of catchments according to the stream network and digital elevation of the United States and then filtered some excess sites that were not distributed in the selected catchments. The above process resulted in 128 runoff gauges, and those sites were generally located at the outlet of the catchment from 1941 to 2015 and 60 from 1916 to 2015.

The Global Historical Climatology Network-Daily (GHCN-D) was produced and archived by the National Oceanic and Atmospheric Administration (NOAA) National Climatic Data Center. It comprises over 96,000 sites worldwide that capture all or a subset of daily maximum and minimum temperatures, precipitation, snowfall, and snow depth. The record period varies by the station from less than 1 year to 177 years, with the average precipitation record spanning 33.1 years. We also required that each year's data records be at least 80% complete in two periods (1916–2015 and 1941–2015). In this study, we need daily precipitation and maximum and minimum temperature records. Based on these selection criteria, we finally obtained 993 meteorological stations for 1916–2015 and 1,116 meteorological stations for 1941–2015. Maximum and minimum temperatures were averaged to yield mean daily temperature, and daily meteorological element values (mean daily precipitation and daily temperature) were multiplied by the total days of each year to obtain annual total amounts. The process that filters extra rainfall gauges is similar to the method of selecting runoff gauges, and there are a number of precipitation sites in selected catchments. The above process resulted in 186 rainfall sites from 1941 to 2015 and 103 from 1916 to 2015.

Study area

The continental United States was divided and subdivided into successively smaller hydrologic units classified into four levels: regions, subregions, accounting units, and cataloging units. The hydrologic units are arranged or nested within each other, from the largest geographic area (regions) to the smallest geographic area (cataloging units). The watershed boundary dataset (WBD) consists of the newest and the most recent HUC delineation and further divides the HUC levels into fifth and sixth levels. Hydrologic unit boundaries in the WBD are determined based on topographic, hydrologic, and other relevant landscape characteristics without regard for administrative, political, or jurisdictional boundaries. We obtained hydrologic units, which are shapefile or geodatabase data types, across the contiguous United States from the National Hydrography Dataset (NHD) in the USGS and downloaded national 30 m digital elevation data from the Soil and Water Hub (SWH, https://soilandwaterhub.brc.tamus.edu/Home/GISData). We extracted the streamflow network and then selected the catchment according to the streamflow gauges and meteorological stations mentioned in the 'Streamflow and meteorological data' section based on having obtained hydrological units with varied levels. A catchment defines the areal extent of drainage to an outlet point (streamflow gauges) on a dendritic stream network or multiple outlet points where the stream network is not dendritic and one or more meteorological stations are distributed throughout the catchment of interest by using the hydrological analysis tool in ArcGIS 10.5. The above procedure resulted in 60 study catchments for 1916–2015 and 128 for 1941–2015 (Figure 1).

Methodology

The Food and Agricultural Organization (FAO) of the United Nations improved and upgraded the methodologies for reference evapotranspiration (ET_0) estimation by introducing the reference crop (grass) concept, described by the FAO Penman–Monteith ($PM - ET_0$) equation (Allen *et al.* 1998), and this approach was tested well under different climates and time step calculations and is currently adopted worldwide (Almorox *et al.* 2015). However, the high cost and maintenance of agrometeorological stations and the high number of sensors required to estimate ET_0 make it non-plausible, especially in rural areas (Moratiel *et al.* 2020). For this reason, the estimation of ET_0 using air temperature in places where wind speed, solar radiation,

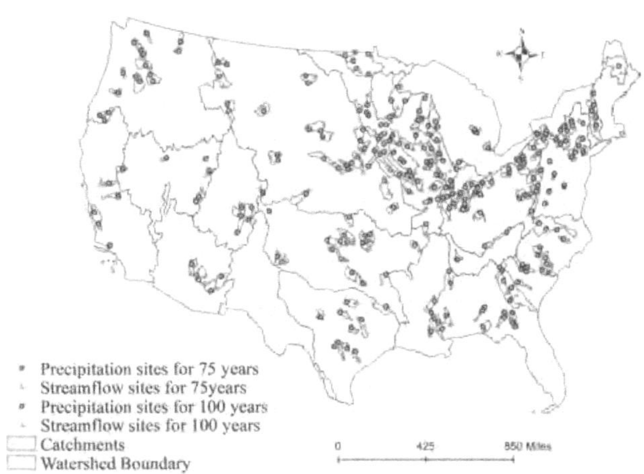

Figure 1 | Study area. The solid black line represents hydrological regions (18 regions). The red line represents selected catchments (total of 188 catchments: 60 for 100 years and 128 for 75 years). Green dots in the catchment represent precipitation stations from 1941 to 2015 (186 rainfall sites). The green triangle represents the streamflow station from 1941 to 2015 (128 runoff sites). Blue dots in the catchment represent precipitation stations from 1916 to 2015 (103 rainfall sites). The blue triangle in the catchment represents the streamflow station from 1916 to 2015 (60 runoff sites). Please refer to the online version of this paper to see this figure in color: http://dx.doi.org/10.2166/wcc.2022.242.

and air humidity data are not readily available is particularly attractive (Moratiel *et al.* 2020). Methods of estimating ET_0 based on only the air temperature have been recommended by many researchers. Allen *et al.* (1998), in the guidelines for $PM - ET_0$, recommended two approaches: (a) using the equation of Hargreaves–Samani (Hargreaves *et al.* 1985) and (b) using the Penman–Monteith temperature (PMT) method that requires temperature data to estimate net radiation and vapor pressure deficit to obtain ET_0. Our study only obtained temperature data (daily maximum and minimum temperatures). Therefore, the PMT method is applied in this study to estimate ET_0, and its calculation process for more details can be seen in the Supplementary material.

The Mann–Kendall test (Mann 1945) ($p \leq 0.05$) is applied to calculate the statistical significance of trends for time-series data because this method is distribution-free, robust against outliers, and has higher power for non-normally distributed data. The significance of trends over time is easily affected by assumptions of whether the time-series data are independent, have short-term persistence (STP), or have long-term persistence (LTP) (Sagarika *et al.* 2014). The Mann–Kendall test assumes the input data to be serially independent, but the presence of positive serial correlation in the time-series data can overestimate the significance of trends (Yue *et al.* 2002). To overcome the effect of serial correlation, the Mann–Kendall test after trend-free pre-whitening, as suggested by Yue *et al.* (2002), and the modified Mann–Kendall test, as suggested by Hamed (2008), can be adopted. Apart from STP, the presence of LTP behavior in the time-series data led to an underestimation of serial correlation in the data structure and an overestimation of the significance of the Mann–Kendall test. Hamed (2008) proposed a technique to incorporate the LTP behavior in the Mann–Kendall test, but the method is suitable for serial data with lengths of at least 100 years. In this study, our data series, data record lengths less than 100, are elasticities of climate and catchment characteristics of streamflow during each sliding window; therefore, we adopted the original Mann–Kendall method with assumption-independent time-series data.

In this study, the Budyko-coupled climate elasticity method was used to estimate the elasticity of runoff to precipitation, potential evaporation, and catchment characteristics. Budyko (1974) assumed the ratio of the mean annual actual evapotranspiration to the mean annual precipitation (E/P) as a function of the ratio of the mean annual potential evapotranspiration to the mean annual precipitation (E_T/P) and other watershed properties:

$$\frac{E}{P} = f\left(\frac{E_T}{P}, n\right) \tag{1}$$

where E is the actual evapotranspiration (mm); P is the annual rainfall (mm); E_T is the potential evapotranspiration (mm); and n represents watershed properties. Since the theory was published by Budyko in 1974, its mathematical function forms have been provided by many researchers (Fu 1981; Yang *et al.* 2008; Wang & Tang 2014). In our study, Yang *et al.*'s (2008) analytical derivation of the Budyko hypothesis was applied, and the function forms as follows:

$$E = \frac{E_T \times P}{\left(E_T^n + P^n\right)^{1/n}} \tag{2}$$

We defined Equation (2) as $f = (E_T, P, n)$, and we can express the total differential as follows:

$$dE = \frac{\partial f}{\partial P} dP + \frac{\partial f}{\partial ET} dE_T + \frac{\partial f}{\partial n} dn \tag{3}$$

The large area (>1,000 km^2) and the long period (>11 years) allow the water balance equation to be made such that the watershed storage is approximately zero (Patterson *et al.* 2013), so the water balance equation can be simplified as $P = E + R$ for a watershed. Thus, substituting Equation (3) into the water balance equation and leading to the differential form:

$$dQ = \left(1 - \frac{\partial f}{\partial p}\right) dP - \frac{\partial f}{\partial E_T} dE_T - \frac{\partial f}{\partial n} dn \tag{4}$$

Dividing Equation (4) by $Q = P - E$ yields the following expression to show the relative change in Q:

$$\frac{dQ}{Q} = \left(1 - \frac{\partial f}{\partial p}\right)\frac{P}{P - E}\frac{dP}{P} - \frac{\partial f}{\partial E_T}\frac{E_T}{P - E}\frac{dE_T}{E_T} - \frac{\partial f}{\partial n}\frac{n}{P - E}\frac{dn}{n} \tag{5}$$

which can be denoted as follows:

$$\frac{dQ}{Q} = \varepsilon_P\frac{dP}{P} + \varepsilon_{E_T}\frac{dE_T}{E_T} + \varepsilon_n\frac{dn}{n} \tag{6}$$

where ε_P, ε_{E_T}, and ε_n are the precipitation elasticity, potential evaporation elasticity, and catchment characteristic elasticity of runoff, respectively. Thus, we can calculate climate elasticity (precipitation and potential evaporation) and watershed characteristic elasticity of runoff by substituting annual climatic variable values into Budyko-type curves. Unlike the calculation process of precipitation and potential evapotranspiration elasticity, maximum and minimum temperature elasticities in this study were estimated by median descriptive statistics due to insufficient meteorological data (Sankarasubramanian *et al.* 2001):

$$\varepsilon = \text{median}\left[\frac{(R_i - \bar{R})/\bar{R}}{(X_i - \bar{X})/\bar{X}}\right] \tag{7}$$

where ε is the climatic factor (maximum and minimum temperatures) elasticity of runoff; R_i is the annual mean runoff (mm); $\bar{R}$ is the multiyear average runoff (mm); X denotes the climatic variables; and $\bar{X}$ denotes the mean of any climatic variable. $[(R_i - \bar{R})/\bar{R}/(X_i - \bar{X})/\bar{X}]$ is calculated for each pair of R_i and X_i, and the median of these values is the non-parametric estimator of the climate elasticity of runoff. However, there is a mathematical error when $X_i = \bar{X}$. Thus, Zheng *et al.* (2009) suggested the following least squares estimator:

$$\varepsilon = \frac{\bar{X}}{\bar{R}}\frac{\sum(X_i - \bar{X})(R_i - \bar{R})}{\sum(X_i - \bar{X})^2} \tag{8}$$

Thus, maximum and minimum temperature elasticities can be obtained as follows:

$$\varepsilon_{T_{\max}} = \frac{\overline{T_{\max}}}{\bar{R}}\frac{\sum(T_{\max_i} - \overline{T_{\max}})(R_i - \bar{R})}{\sum(T_{\max_i} - \overline{T_{\max}})^2} \tag{9}$$

$$\varepsilon_{T_{\min}} = \frac{\overline{T_{\min}}}{\bar{R}}\frac{\sum(T_{\min_i} - \overline{T_{\min}})(R_i - \bar{R})}{\sum(T_{\min_i} - \overline{T_{\min}})^2} \tag{10}$$

where $\varepsilon_{T_{\max}}$ and $\varepsilon_{T_{\min}}$ are the maximum and minimum temperature elasticities of runoff, respectively.

Therefore, by substituting climatic factors, runoff series, and their mean values in each slipping window into Equations (6), (9), and (10), we can obtain the climate, watershed characteristic elasticity of runoff in each slipping window.

RESULTS

Temporal trend of climate and catchment characteristic elasticity of runoff

First, we analyzed the volatility of climate and catchment characteristic elasticity of runoff over time in each timescale. The standard deviation (STD) was used to reflect the degree to which these elasticity time-series deviated from the mean value, which was calculated within each timescale, and a box plot was depicted to show the STD distribution of multiple elasticities of runoff in all catchments (Figure 2). The results show that precipitation, potential evapotranspiration, and watershed characteristic elasticity have a similar and tiny STD in each slipping window and far less than the STD of maximum and minimum temperature elasticities during both 1941–2015 and 1916–2015. The STD value of watershed characteristic elasticity is the lowest among them. These conclusions illustrate that precipitation, potential evapotranspiration, and watershed

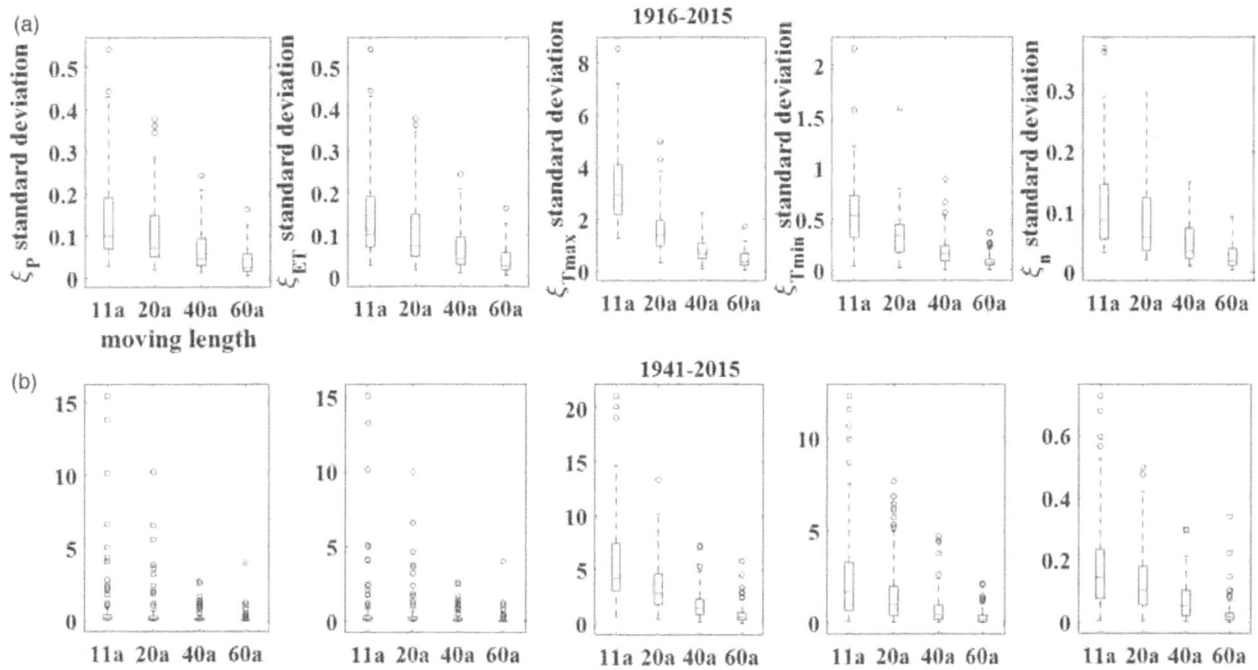

Figure 2 | Standard deviation box plot of multiple elasticities in each slipping window (11, 20, 40, and 60 a) during periods (a) 1916–2015 and (b) 1941–2015. ξ_P, ξ_{E_T}, $\xi_{T_{max}}$, $\xi_{T_{min}}$, and ξ_n represent the precipitation, potential evapotranspiration, maximum temperature, minimum temperature, watershed characteristic elasticity of runoff, respectively. The *Y*-axis in a column represents the same meaning. For the box plots, the horizontal lines of the box represent the lower quartile, median, and upper quartile. Lines extending from the box represent 1.5 times the interquartile range. Data points outside 1.5 times the interquartile range are outliers and represented by blue circles. Please refer to the online version of this paper to see this figure in color: http://dx.doi.org/10.2166/wcc.2022.242.

characteristic elasticity are relatively steady over time compared to maximum and minimum temperature elasticities; the large STD value for extreme temperature may be caused by disturbance from climate change and human activities. Watershed characteristic elasticity is the steadiest among all elasticities of runoff. In addition, there are higher outliers of the STD value and elasticity value in 1941–2015 compared with 1916–2015 in Figures 2 and 3, and the reason why this difference may be runoff is (1) more vulnerable to interference from human activity and (2) shorter lengths of data during 1941–2015. Wang & Hejazi (2011) showed that human activities caused the mean annual streamflow to experience upward changes and those anthropogenic activities have played a vital role in runoff change since the 1970s in the contiguous United States. Timescale means the time period used for calculating elasticity, also known as data length. We must recognize that the STD values of elasticity within different timescales, i.e., we suggest the 20-year data length, are used to quantify runoff changes, because the short data length causes large STD, whereas the long data length is not easily obtained. We can also now demonstrate that the longer time period, the steadier elasticity, and further steadier runoff changes. Therefore, long-term average changes in runoff are reasonable in previous studies.

Due to its low STD for the elasticity of runoff, we compared the mean value of impact factor elasticity of runoff under each slipping window in all catchments with that of the long-term average value (100 and 75 years) during the two study periods, and a box plot was depicted (Figure 3). The results show that the average values for the same factor elasticity of runoff in the four slipping windows are approximately close to each other and that the long-term average value, in other words, the selection of the calculation period, has no impact on the influential factor elasticity of runoff value in the catchment. However, some small differences in the distribution of the long-term (75 and 100 a) average value are more dispersed than the average value in the four slipping windows for the maximum and minimum temperature elasticities of runoff during 1916–2015, which may be caused by the large fluctuation in temperature, especially extreme temperature. Sankarasubramanian *et al.* (2001) demonstrated that precipitation elasticity is fixed in basins. In our study, we found that not only precipitation elasticity but also other climate elasticities (potential evapotranspiration and maximum and minimum daily temperature elasticities) and watershed elasticity are stable over time. Comparing Figures 2 and 3, we should recognize that there is a relatively large STD compared to a long time period even though there is a close average value of elasticity within different time

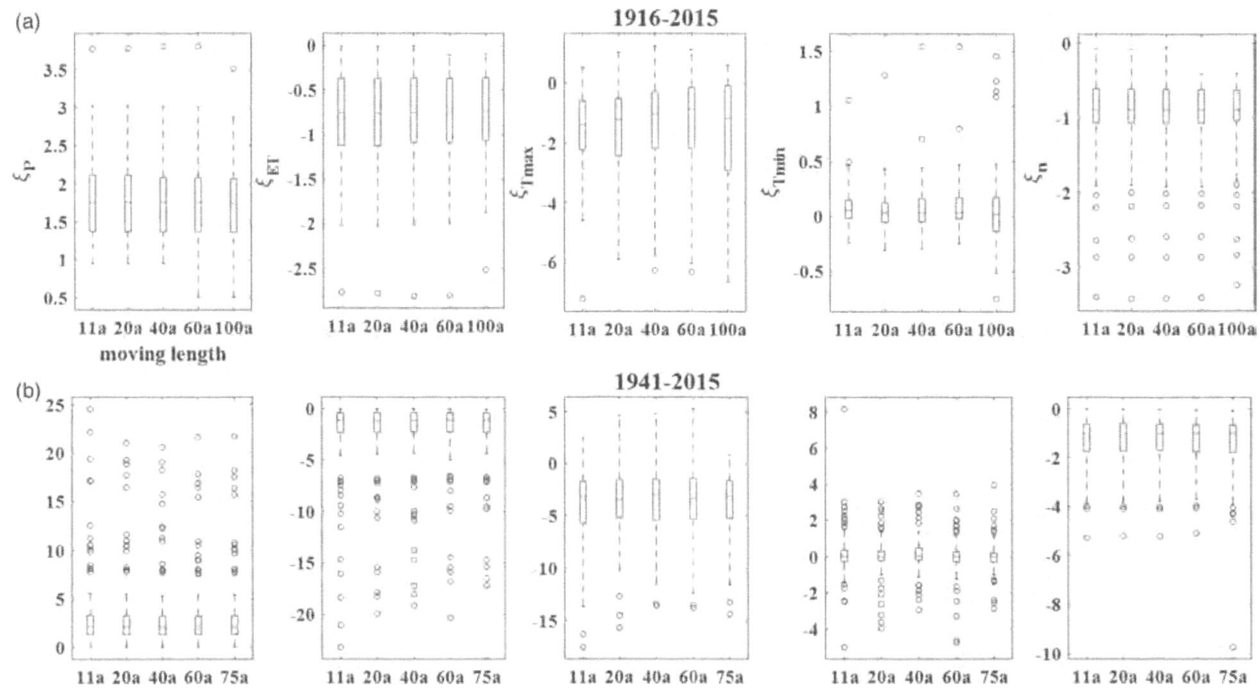

Figure 3 | Box plot of the average value of impact factor elasticity under each slipping window (11, 20, 40, and 60 a) and the entire study period (100 and 75 a). ξ_P, ξ_{E_T}, $\xi_{T_{\max}}$, $\xi_{T_{\min}}$, and ξ_n represent the precipitation, potential evapotranspiration, maximum daily temperature, minimum daily temperature, and watershed characteristic elasticity, respectively. The *Y*-axis in a column represents the same meaning.

periods, which indicate that the long-term average elasticity of runoff in previous studies may hide volatility, further affecting the spatial pattern, which can be further studied in the future.

Second, to further identify the trend significance of multiple influential factor elasticity of runoff over time in four slipping windows, the original Mann–Kendall method was used to test the trend significance ($p \leq 0.05$) of multiple influential factor elasticity of runoff over time in each slipping window during two periods (1916–2015 and 1941–2015) (Figure 4). We find that the elasticity of runoff with different factors has different trends in the United States. Precipitation elasticity of runoff mainly presents a significant declining trend in all slipping windows in the middle north regions of the United States, especially in the Upper Mississippi Region. The significant increasing trends are heterogeneously distributed in other regions of the United States. However, we can see roughly that the number of catchments with a significant increasing trend in the eastern region is more than that in the western region, southern parts are more than northern parts (Figure 4 and Table 1), and a negative trend but similar spatial distribution pattern for evapotranspiration elasticity. The distribution of catchments with trend significance for the maximum temperature elasticity is similar to the minimum temperature elasticity. The number of catchments with a significant increasing trend is a little higher than that with a significant declining trend (Table 1), catchments with a significant decreasing trend are mainly located in eastern regions of the United States for maximum and minimum temperature elasticities, especially the Mid-Atlantic Region and the South Atlantic-Gulf Region, and a significant increasing trend is mainly distributed in northern regions of the United States, especially the Upper Mississippi Region. This conclusion is opposite to the trend of precipitation elasticity (Figure 4). Watershed characteristic elasticity mainly presents a significant increasing trend, those catchments are mainly distributed in the middle and eastern regions (Figure 4), and few catchments present a significant declining trend in the western parts (Table 1).

All in all, we can demonstrate that the elasticity of runoff is not only stable over time but also its trend is almost not affected by calculation timescales.

Spatial pattern of multiple factor elasticity of runoff

To investigate the spatial distribution of the impact factor elasticity of runoff, bar charts of influential factor elasticity (precipitation, potential evapotranspiration, maximum and minimum daily temperatures, and watershed characteristics) of runoff in

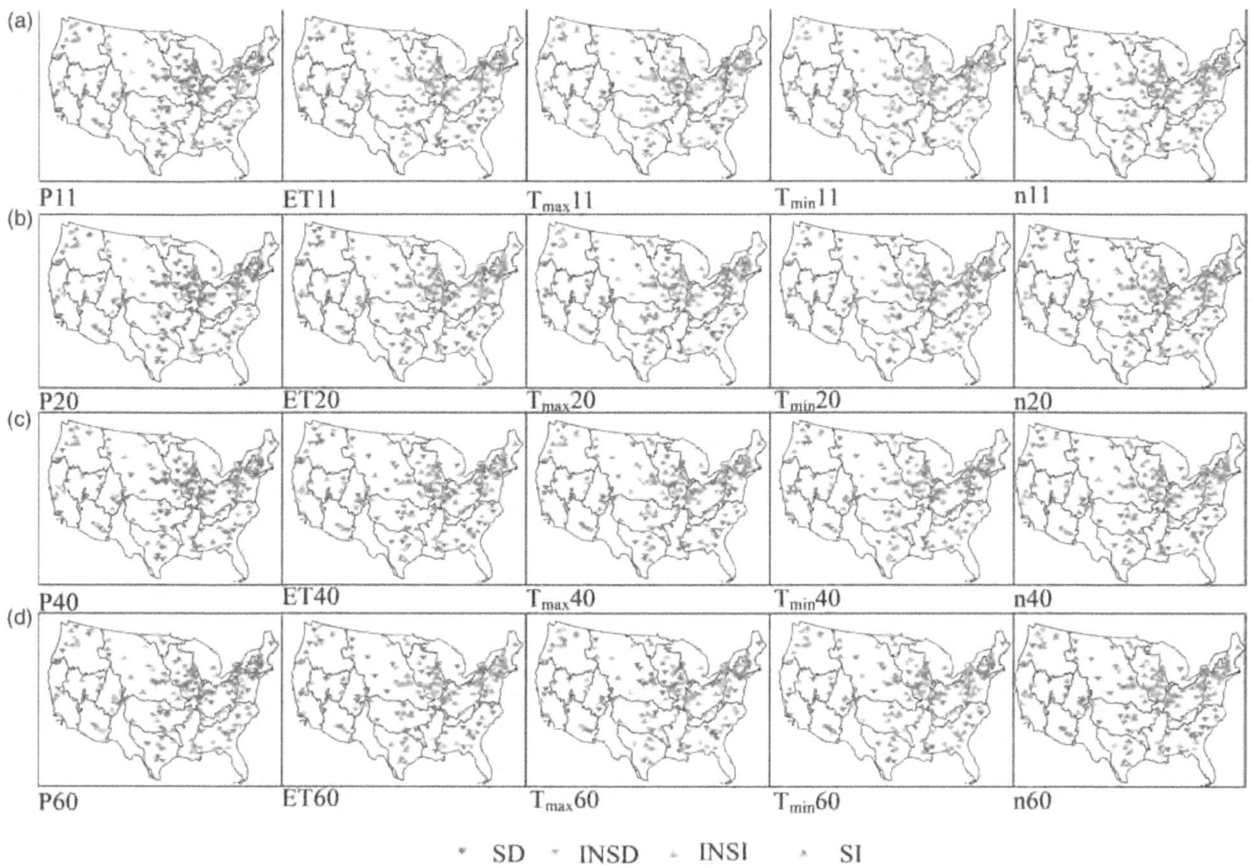

Figure 4 | Trend significance ($p = 0.05$) of multiple elasticities of runoff over time in each slipping window during both periods 1916–2015 and 1941–2015 and (a), (b), (c), and (d) rows represent 11-, 20-, 40-, and 60-year moving averages, respectively. $P11$, ET11, $T_{max}11$, $T_{min}11$, and $n11$ represent the time-series of precipitation, potential evapotranspiration, maximum daily temperature, minimum daily temperature, and watershed characteristic elasticity over time for an 11-year moving average, respectively, and other symbols represent similar meanings but for different slipping windows. 'SD,' 'INSD,' 'INSI,' and 'SI' in the legend represent the significant decline, insignificant decline, insignificant increase, and significant increase trends, respectively.

Table 1 | Number of catchments with a significant trend for elasticity

	11 a		20 a		40 a		60 a	
	SD	SI	SD	SI	SD	SI	SD	SI
ε_P	85	58	95	56	98	61	87	56
ε_{E_T}	54	82	55	96	57	97	57	85
$\varepsilon_{T_{max}}$	39	49	57	62	60	79	64	51
$\varepsilon_{T_{min}}$	36	52	50	61	60	74	45	77
ε_n	25	126	26	138	25	143	40	109

ε_P, ε_{E_T}, $\varepsilon_{T_{max}}$, $\varepsilon_{T_{min}}$, and ξ_n represent the precipitation, potential evapotranspiration, maximum temperature, minimum temperature, and watershed characteristic elasticity, respectively. 'SD' and 'SI' represent the significant decreasing and significant increasing trends, respectively. 11, 20, 40, and 60 a represent 11-, 20-, 40-, and 60-year slipping windows, respectively.

188 catchments of the CONUS were drawn (Figure 5) during 1916–2015 and 1941–2015. The 'Spatiotemporal change of catchment characteristic n' section has shown that influential factor elasticity of runoff has no difference in various time windows; thus, we analyzed the long-term average impact factor elasticity of runoff in the whole study period (75 and 100 a). The results show that most of the factor elasticities of runoff are negative apart from precipitation in the CONUS. The sensitivity of runoff to precipitation is the greatest in the northeastern CONUS. However, the most sensitive factor converts to the

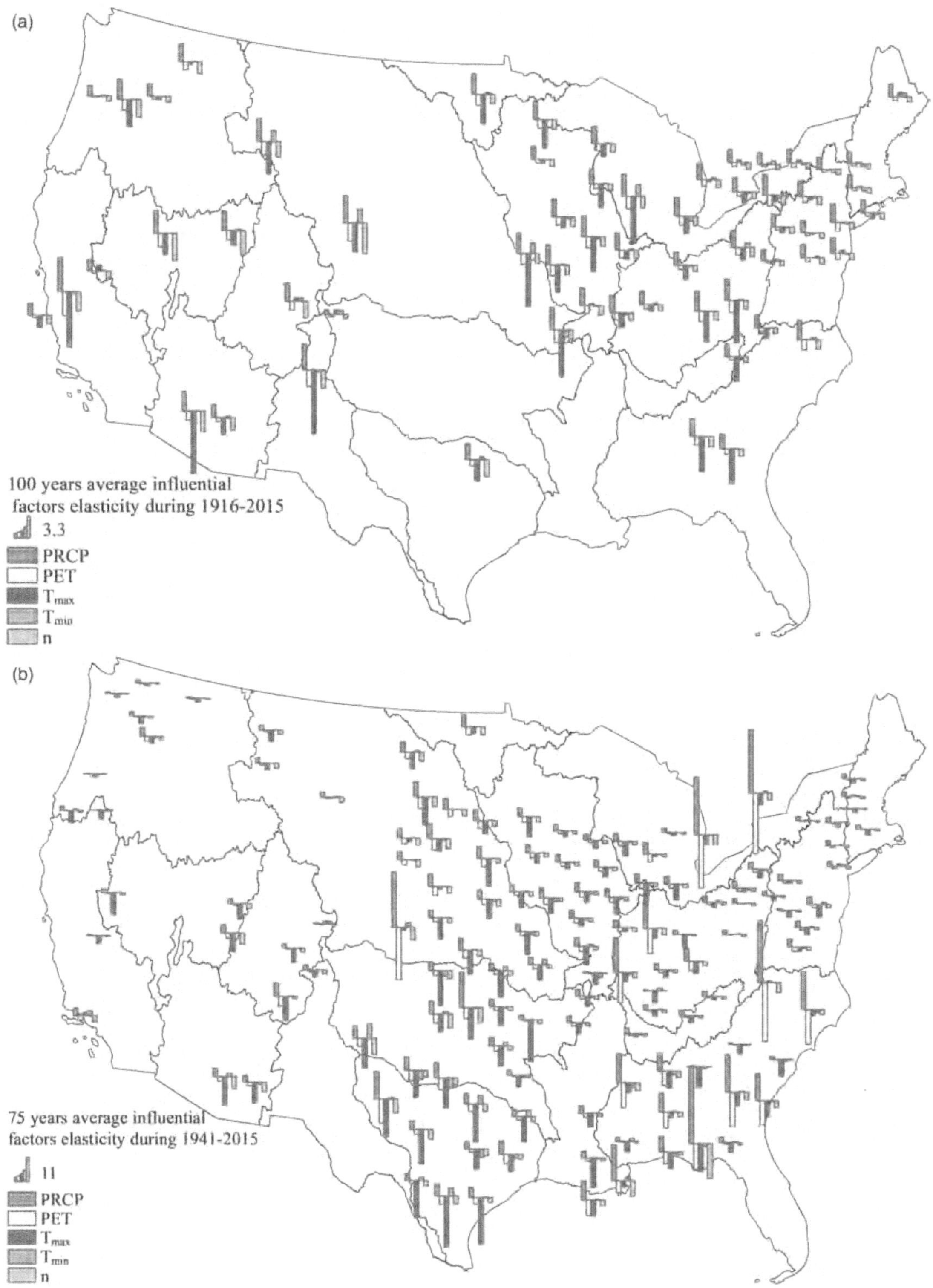

Figure 5 | Spatial patterns of multiple factor elasticity of runoff in 188 catchments (60 for 1916–2015 and 128 for 1941–2015) for two periods: (a) 1916–2015 and (b) 1941–2015. Upward and downward columns represent positive and negative elasticities of annual runoff to influential factors, respectively. The columns are scaled according to the respective elasticity value of multiple influential factors. 'PRCP', 'PET', 'T_{max}', 'T_{min}', and 'n' represent the precipitation, potential evapotranspiration, maximum daily temperature, minimum daily temperature, and catchment characteristics, respectively.

maximum temperature in the central and western regions. The impact of potential evapotranspiration, maximum temperature, and catchment characteristics on runoff gradually becomes important from east to west. In some western regions, catchment characteristics may be the second-most sensitive runoff factor, following the most sensitive maximum temperature factor. The sensitivity of runoff to the minimum temperature is the smallest among all influential factors. This indicates that temperature is also an important factor that affects runoff.

Relationships between multiple elasticities and other factors

We concluded that the elasticity of runoff presents a distinct spatial distribution. To further investigate whether the elasticity of runoff can be explained by other factors, the relationships between the elasticity of runoff and P (precipitation), R (runoff), aridity index (φ), runoff coefficient (α), and catchment characteristics (n) were studied, and a graph in the form of a correlation for their relationships was depicted in our study (Figure 6). Figure 6 shows that arid regions have higher precipitation, potential evaporation, maximum and minimum temperatures, and catchment characteristic elasticity. Maximum temperature and catchment characteristic elasticity scatter against the aridity index compared to the elasticity of the other three factors. Wang & Hejazi (2011) plotted direct human-induced changes against the climate dryness index and the evaporation ratio to further investigate whether the estimated climate- and direct human-induced changes can be explained by other factors, and they found that arid regions (i.e., water-limited) are more vulnerable to climate change and direct human interferences than wet regions (i.e., energy limited). The relationship between the elasticity of runoff and the runoff coefficient for multiple factors presents an exponential trend and a high elasticity value in regions with a low runoff index. Meanwhile, we can also find that the relationship between the multiple factor elasticity and the mean runoff is consistent with the relationship between the elasticity of runoff and the runoff index. However, this relationship tends to be a horizontal line trend for mean rainfall. This result indicates that multiple climate factors (precipitation, potential evaporation, and maximum and minimum temperatures) and catchment characteristic elasticity are independent of precipitation and that wet regions tend to have lower elasticity of runoff values. This conclusion can illustrate the spatial distribution of the multiple factor elasticity of runoff in the United States in Figure 5 and is the same as the result of Sankarasubramanian *et al.* (2001), who used Kendall's τ test to document that estimates of precipitation elasticity are independent of P at each site and

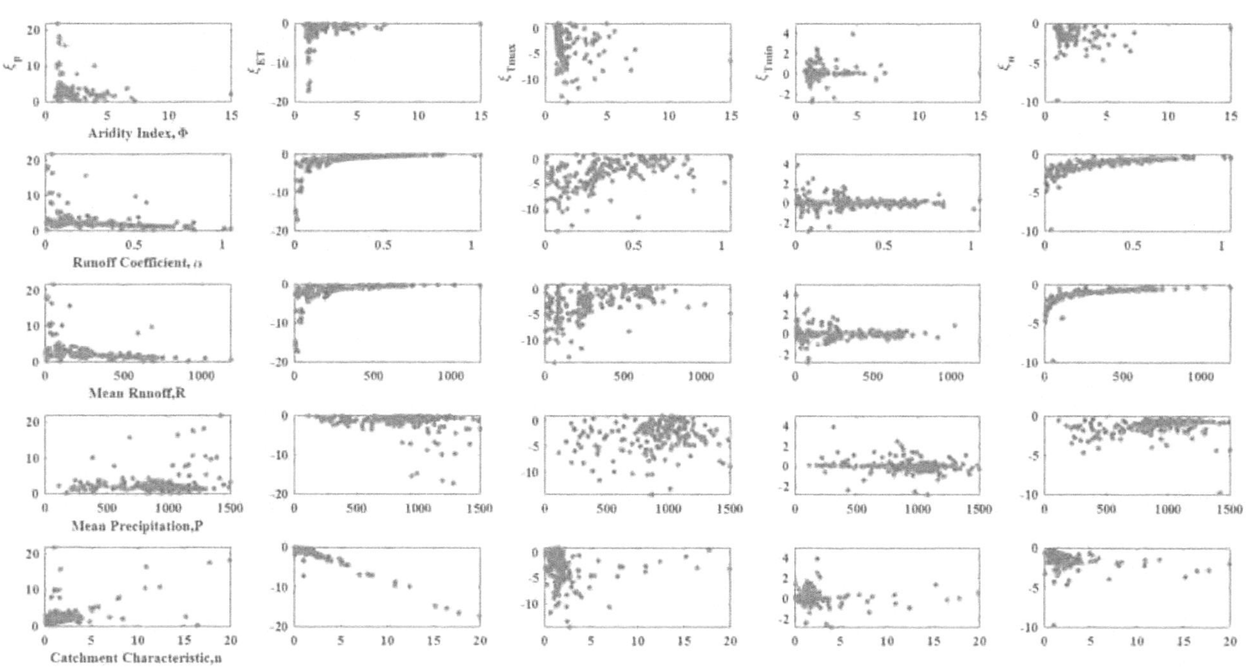

Figure 6 | Multiyear average factor (aridity index, runoff coefficient, mean runoff (mm), mean precipitation (mm), and catchment characteristic) to the elasticity of runoff (precipitation, potential evaporation, maximum and minimum temperatures, and catchment characteristic elasticity) relationships for 188 catchments during two periods 1916–2015 and 1941–2015. ξ_P, ξ_{E_T}, $\xi_{T_{max}}$, $\xi_{T_{min}}$, and ξ_n represent the precipitation, potential evaporation, maximum temperature, minimum temperature, and catchment characteristic elasticity, respectively. The Y-axis in a column represents the same meaning, and the X-axis in a row has the same meaning.

showed that humid basins tend to have significantly lower values of precipitation elasticity and tend to be much more homogeneous in terms of precipitation elasticity than arid regions. The relationship between catchment characteristics and the catchment characteristic elasticity of runoff tends to be a line trend and the values of multiple factor elasticity of runoff increase with the ascent of n. Catchment characteristic n reveals inter-related dynamics between vegetation type, soil properties, and topography (Dey & Mishra 2017). Its value impacts the evaporation of the catchment. Higher values of n indicate that the watershed characteristics favor evapotranspiration (for densely vegetated basins). In this study, we can demonstrate that the evaporation ratio and the multiple factor elasticity are affected by catchment characteristic n.

DISCUSSION

Spatiotemporal change of catchment characteristic *n*

Catchment characteristics were first mentioned by Budyko (1974) who assumed the ratio of the mean annual actual evapotranspiration to the mean annual precipitation evapotranspiration as a function of the ratio of the mean annual potential evapotranspiration to the mean annual precipitation and other watershed properties (represented by n or other parameters in functions), which reflected LULC, vegetation coverage changes, and topography in the watershed. No matter n or other parameters in different Budyko type curves, they all represent catchment characteristics and reveal inter-related dynamics between vegetation type, soil properties, and topography. The increase in the catchment characteristic parameter is thus regarded to result from increases in vegetation, which accordingly caused an increase in evaporation and a decline in runoff (Xu *et al.* 2014). Porporato *et al.* (2004) related the catchment characteristic to the relative soil water holding capacity, average rooting depth, and rain depth. Yang *et al.* (2007) related the relative infiltration capacity, the catchment average slope, and the relative soil storage capacity. In other words, catchment characteristics are synthesized parameters that represent catchment properties (soil, topography, vegetation, etc.), also indirect human activities which changed catchment physical properties, further causing runoff alteration in the runoff generation process. Vegetation coverage, LULC, and soil types change with time; therefore, the value of n changes with the seasonal scale or the annual scale. In general, the value of n is obtained by using the long-term average precipitation, runoff, and actual evaporation in a single watershed, which reflected a long-term average condition in the watershed. Thus, we assumed that the value of n is fixed, and studies on the impact of different values of n on hydrological variables are lacking. For instance, Donohue *et al.* (2011) assessed the influence of n on runoff and demonstrated that the uncertainty in the value of n for the study catchment produces only a small uncertainty in the modeled runoff. Jiang *et al.* (2015) built the relationship of the parameter w (also representing catchment characteristic) in each Budyko-type equation with climatic and human factors. They revealed the time-variation process of w by using an 11-year moving window and found that the parameter was a function of variables of both climate conditions and human activities improving the performance of simulating the mean annual runoff based on all the Budyko-type equations. Zhang *et al.* (2001) defined w as the plant-available water coefficient and it represents the relative difference in the way plants use soil water for transpiration. Li *et al.* (2013) found that catchment characteristic has regional pattern variability. Additionally, Wang *et al.* (2016) revealed that from low to high, and in turn to low, a broadly north-middle and middle-south gradient could be found in the spatial distribution of n. Thus, we can demonstrate that watershed characteristic possesses spatiotemporal variation and that its influence on mean runoff also has the same variation feature.

To investigate the spatiotemporal change characteristics of catchment characteristic n, we analyzed the temporal trend of catchment characteristic n during different slipping windows and the spatial distribution of its mean value across 188 catchments in the United States (Figures 7 and 8). Considering that the average value of n under the four slipping windows is similar, here, we analyzed the spatial distribution of the long-term average n during the two study periods (75 and 100 a) (Figure 8). Figures 7 and 8 show that a significant decreasing trend for n occurs mainly in the northeastern regions of the United States, catchments with the same trend are seldom concentrated in the south, and a significant increasing trend for n mainly occurs in the western regions. The value of n in the south is larger than that in the north, the n value in the east is larger than that in the west, and the largest value n occurs mainly in the eastern and southeastern regions.

Method of estimating climate elasticity

In this study, precipitation and potential evaporation elasticity of runoff were obtained by conceptual approaches, i.e., Budyko hypothesis. Maximum and minimum temperature elasticities were calculated by a non-parametric estimation method due to insufficient meteorological data. The major advantage of the former method is that it can estimate the sensitivity of runoff to different climatic factors, and the latter method has low biasness and is non-parametric. There are many

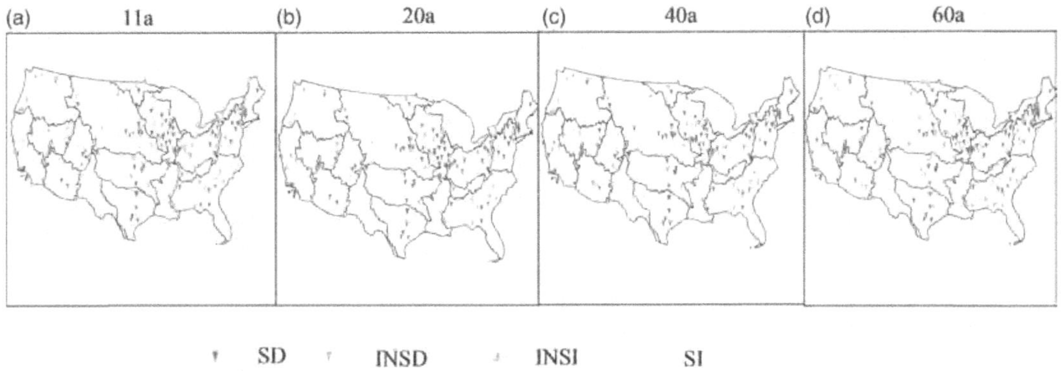

Figure 7 | Trend significance of catchment characteristic *n* during four slipping windows: (a) 11, (b) 20, (c) 40, and (d) 60 years. 'SD,' 'INSD,' 'INSI,' and 'SI' represent the significant declines, insignificant declines, insignificant increases, and significant increasing trends, respectively.

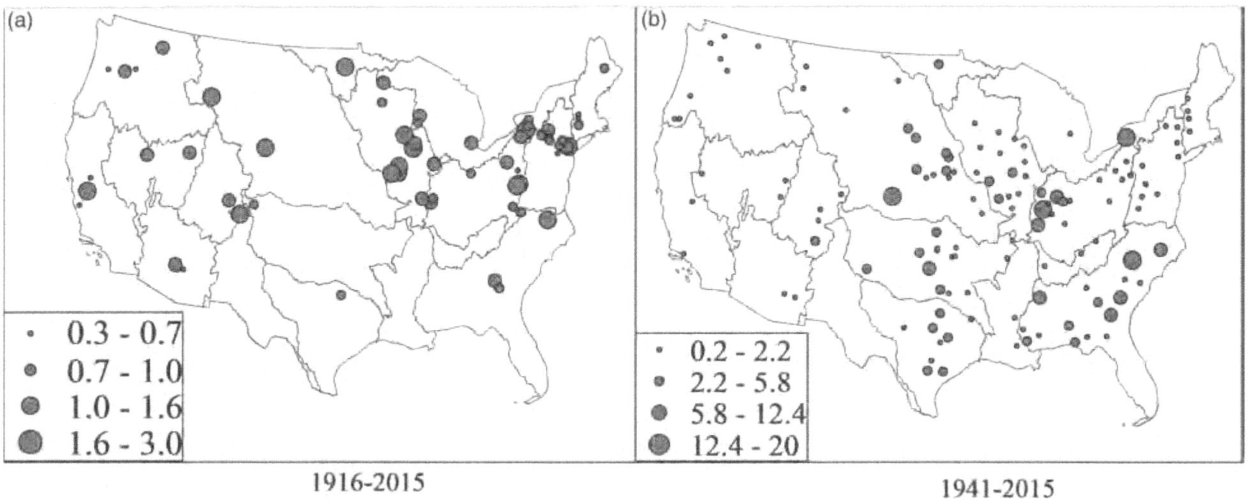

Figure 8 | Spatial distribution of long-term average *n* during (a) 1916–2015 and (b) 1941–2015.

researchers who adopted these two methods in the contribution analysis of runoff. For example, Zheng *et al.* (2009) used a non-parametric method to assess the impacts of climate and land surface change on the streamflow. Results show that for the period 1960–2000, the elasticities of streamflow in relation to precipitation and potential evapotranspiration are 2.10 and −1.04, respectively. Li *et al.* (2012) adopted the sensitivity-based approach (including a non-parametric model and six Budyko framework-based models) and the hydrological modeling approach (using Xinanjiang and SIMHYD models) for three medium-sized catchments in Australia to investigate the impacts of increase or decrease in plantations and climate variability on streamflow. They found the sensitivity-based approaches are only applicable where long-term datasets are available and they only provide results at a mean annual timescale, and small variability in the reduction or increase in streamflow is estimated by these methods. Yang & Yang (2011) compared these two methods in the Futuo River catchment in the Hai River basin and found that the total changes in runoff were essentially different (−11% changes for the partial differential Budyko hypothesis-based method and −18% for the non-parametric estimator approach). The authors thought the reason for this difference was the short length of the data (1961–2000). We also found that precipitation and potential evaporation elasticity are similar to each other using the two methods, but largely biased for extreme temperature elasticity (for some catchments, values for the non-parametric method are higher than those for the conceptual method, here, results are not shown). At present, these two methods are commonly used to analyze runoff changes caused by climate change and human activities, and no one pointed out the inapplicability of the methods. A detailed reason for this difference has not been displayed, and efforts may be made by related scholars.

CONCLUSIONS

In this study, spatial and temporal variations in the elasticity of runoff to multiple climatic factors and catchment characteristics during 1916–2015 and 1941–2015 in the 188 catchments of the CONUS were first investigated. Then, we explored the relationships of the long-term average elasticity of runoff with other factors (aridity index, runoff coefficient, mean rainfall, mean runoff, and catchment characteristic). We found that (1) not only the elasticity of runoff to climatic factors and catchment characteristics but also its trend over time may not be affected by timescales. (2) Precipitation elasticity mainly presents a decreasing trend in the mid-north region and an increasing trend in the same location for the potential evaporation elasticity, an increasing trend for the minimum temperature and the catchment characteristic elasticity, and the heterogeneous trend over the United States for the maximum temperature elasticity. (3) The elasticity of runoff to multiple influential factors has a distinct spatial pattern over the CONUS. The most sensitive factor is precipitation for runoff in northeastern regions, and the roles of potential evaporation, maximum temperature, and catchment characteristics in runoff gradually become important from east to west. In some western regions, catchment characteristics may be the second-most sensitive runoff factor, following the maximum temperature factor. The sensitivity of runoff to the minimum temperature is the smallest among all influential factors in the United States. (4) Multiple climate factors (precipitation, potential evaporation, and maximum and minimum temperatures) and the catchment characteristic elasticity are independent of precipitation, and wet regions tend to have lower elasticity of runoff. The relationship of multiple factor elasticity of runoff with runoff coefficient presents an exponential trend, which is consistent with the relationship of multiple factor elasticity of runoff with mean runoff, but this relationship tends to be a horizontal line trend for mean rainfall. The relationship between the catchment characteristics and the elasticity of runoff presented a positive linear trend.

ACKNOWLEDGEMENT

This study is supported by the Natural Science Foundation of Heilongjiang Province of China (No. E2017009).

AUTHOR CONTRIBUTIONS

X.G. conducted the supervision, writing, and review & editing of resources. S.D. conducted the conceptualization, methodology, validation, formal analysis, and writing – original draft. Y.Z. conducted the writing – review & editing. A.X. conducted the language and flow improvement.

DATA AVAILABILITY STATEMENT

All relevant data are available from streamflow data from http://esapubs.org/Archive/ecol/E091/045/default.htm and https://waterdata.usgs.gov/nwis/; watershed boundary data from https://www.nrcs.usda.gov/wps/portal/nrcs/main/national/water/watersheds/; climate data from https://www.ncei.noaa.gov/products/land-based-station/global-historical-climatology-network-daily/.

REFERENCES

Allen, R. G., Pereira, L. S., Raes, D. & Smith, M. 1998 *Crop Evapotranspiration: Guidelines for Computing Crop Water Requirements*. Irrigations and Drainage Paper No. 56.

Almorox, J., Quej, V. H. & Martí, P. 2015 Global performance ranking of temperature-based approaches for evapotranspiration estimation considering Köppen climate classes. *Journal of Hydrology* **528**, 514–522. doi:10.1016/j.jhydrol.2015.06.057.

Brown, A. E., Zhang, L., McMahon, T. A., Western, A. W. & Vertessy, R. A. 2005 A review of paired catchment studies for determining changes in water yield resulting from alterations in vegetation. *Journal of Hydrology* **310** (1–4), 28–61. doi:10.1016/j.jhydrol.2004.12.010.

Budyko, M. I. 1974 *Climate and Life*. Academic, New York, p. 508.

Chang, J., Wang, Y., Istanbulluoglu, E., Bai, T., Huang, Q., Yang, D. & Huang, S. 2015 Impact of climate change and human activities on runoff in the Weihe River Basin, China. *Quaternary International* **380–381**, 169–179. doi:10.1016/j.quaint.2014.03.048.

Cheng, Q., Zuo, X., Zhong, F., Gao, L. & Xiao, S. 2019 Runoff variation characteristics, association with large-scale circulation and dominant causes in the Heihe River Basin, Northwest China. *Science of the Total Environment* **688**, 361–379. doi:10.1016/j.scitotenv.2019.05.397.

Davenport, F. V., Herrera-Estrada, J. E., Burke, M. & Diffenbaugh, N. S. 2020 Flood size increases nonlinearly across the Western United States in response to lower snow-precipitation ratios. *Water Resources Research* **56** (1). https://doi.org/10.1029/2019wr025571.

Dey, P. & Mishra, A. 2017 Separating the impacts of climate change and human activities on streamflow: a review of methodologies and critical assumptions. *Journal of Hydrology* **548**, 278–290. doi:10.1016/j.jhydrol.2017.03.014.

Donohue, R. J., Roderick, M. L. & McVicar, T. R. 2011 Assessing the differences in sensitivities of runoff to changes in climatic conditions across a large basin. *Journal of Hydrology* **406** (3–4), 234–244. doi:10.1016/j.jhydrol.2011.07.003.

Dudley, R. W., Hirsch, R. M., Archfield, S. A., Blum, A. G. & Renard, B. 2019 Low streamflow trends at human-impacted and reference basins in the United States. *Journal of Hydrology* **580**. doi:10.1016/j.jhydrol.2019.124254.

Fu, B. 1981 On the calculation of the evaporation from land surface. *Scientia Atmospherica Sinica* **5** (1), 23–31. (in Chinese).

Hamed, K. H. 2008 Trend detection in hydrologic data: the Mann–Kendall trend test under the scaling hypothesis. *Journal of Hydrology* **349** (3–4), 350–363. doi:10.1016/j.jhydrol.2007.11.009.

Hargreaves, G. L., Hargreaves, G. H. & Riley, J. P. 1985 Irrigation water requirements for Senegal River Basin. *Journal of Irrigation and Drainage Engineering* **111**, 265–275.

Jiang, C., Xiong, L., Wang, D., Liu, P., Guo, S. & Xu, C.-Y. 2015 Separating the impacts of climate change and human activities on runoff using the Budyko-type equations with time-varying parameters. *Journal of Hydrology* **522**, 326–338. doi:10.1016/j.jhydrol.2014.12.060.

Li, H., Zhang, Y., Vaze, J. & Wang, B. 2012 Separating effects of vegetation change and climate variability using hydrological modelling and sensitivity-based approaches. *Journal of Hydrology* **420–421**, 403–418. doi:10.1016/j.jhydrol.2011.12.033.

Li, D., Pan, M., Cong, Z., Zhang, L. & Wood, E. 2013 Vegetation control on water and energy balance within the Budyko framework. *Water Resources Research* **49** (2), 969–976. doi:10.1002/wrcr.20107.

Li, C., Sun, G., Caldwell, P. V., Cohen, E., Fang, Y., Zhang, Y., Oudin, L., Sanchez, G. M. & Meentemeyer, R. K. 2020 Impacts of urbanization on watershed water balances across the conterminous United States. *Water Resources Research*. doi:10.1029/2019wr026574.

Mann, H. B. 1945 Nonparametric tests against trend [J]. *The Econometric Society* **13** (3), 245–59.

Moratiel, R., Bravo, R., Saa, A., Tarquis, A. M. & Almorox, J. 2020 Estimation of evapotranspiration by the Food and Agricultural Organization of the United Nations (FAO) Penman–Monteith temperature (PMT) and Hargreaves–Samani (HS) models under temporal and spatial criteria – a case study in Duero basin (Spain). *Natural Hazards and Earth System Sciences* **20** (3), 859–875. doi:10.5194/nhess-20-859-2020.

Nalley, D., Adamowski, J., Biswas, A., Gharabaghi, B. & Hu, W. 2019 A multiscale and multivariate analysis of precipitation and streamflow variability in relation to ENSO, NAO and PDO. *Journal of Hydrology* **574**, 288–307. doi:10.1016/j.jhydrol.2019.04.024.

Patterson, L. A., Lutz, B. & Doyle, M. W. 2013 Climate and direct human contributions to changes in mean annual streamflow in the South Atlantic, USA. *Water Resources Research* **49** (11), 7278–7291. doi:10.1002/2013wr014618.

Porporato, A., Daly, E. & Rodriguez-Iturbe, I. 2004 Soil water balance and ecosystem response to climate change. *The American Naturalist* **164**, 625–632.

Prokoph, A., Adamowski, J. & Adamowski, K. 2012 Influence of the 11 year solar cycle on annual streamflow maxima in Southern Canada. *Journal of Hydrology* **442–443**, 55–62. doi:10.1016/j.jhydrol.2012.03.038.

Ravindranath, A. & Devineni, N. 2020 Quantifying streamflow regime behavior and its sensitivity to demand. *Journal of Hydrology* **582**. https://doi.org/10.1016/j.jhydrol.2019.124423.

Rice, J. S., Emanuel, R. E., Vose, J. M. & Nelson, S. A. C. 2015 Continental U.S. streamflow trends from 1940 to 2009 and their relationships with watershed spatial characteristics. *Water Resources Research* **51** (8), 6262–6275. https://doi.org/10.1002/2014wr016367.

Sagarika, S., Kalra, A. & Ahmad, S. 2014 Evaluating the effect of persistence on long-term trends and analyzing step changes in streamflows of the continental United States. *Journal of Hydrology* **517**, 36–53. doi:10.1016/j.jhydrol.2014.05.002.

Sang, Y.-F., Sivakumar, B. & Zhang, Y. 2020 Is there an underestimation of long-term variability of streamflow across the continental United States? *Journal of Hydrology* **581**. https://doi.org/10.1016/j.jhydrol.2019.124365.

Sankarasubramanian, A., Vogel, R. M. & Limbrunner, J. F. 2001 Climate elasticity of streamflow in the United States. *Water Resources Research* **37** (6), 1771–1781. doi:10.1029/2000wr900330.

Schaake, J. C. 1990 From climate to flow. In: *Climate Change and U.S. Water Resources* (Waggoner, P. E., ed.). John Wiley, New York, pp. 177–206.

Shao, Q., Traylen, A. & Zhang, L. 2012 Nonparametric method for estimating the effects of climatic and catchment characteristics on mean annual evapotranspiration. *Water Resources Research* **48** (3). doi:10.1029/2010wr009610.

Sun, Y., Tian, F., Yang, L. & Hu, H. 2014 Exploring the spatial variability of contributions from climate variation and change in catchment properties to streamflow decrease in a mesoscale basin by three different methods. *Journal of Hydrology* **508**, 170–180. doi:10.1016/j.jhydrol.2013.11.004.

Tomer, M. D. & Schilling, K. E. 2009 A simple approach to distinguish land-use and climate-change effects on watershed hydrology. *Journal of Hydrology* **376** (1–2), 24–33. doi:10.1016/j.jhydrol.2009.07.029.

Wang, D. & Hejazi, M. 2011 Quantifying the relative contribution of the climate and direct human impacts on mean annual streamflow in the contiguous United States. *Water Resources Research* **47** (10). doi:10.1029/2010wr010283.

Wang, D. & Tang, Y. 2014 A one-parameter Budyko model for water balance captures emergent behavior in Darwinian hydrologic models. *Geophysical Research Letters* **41** (13), 4569–4577. doi:10.1002/2014gl060509.

Wang, J., Hong, Y., Gourley, J., Adhikari, P., Li, L. & Su, F. 2010 Quantitative assessment of climate change and human impacts on long-term hydrologic response: a case study in a sub-basin of the Yellow River, China. *International Journal of Climatology* **30** (14), 2130–2137. doi:10.1002/joc.2023.

Wang, W., Zou, S., Shao, Q., Xing, W., Chen, X., Jiao, X., Luo, Y., Yong, B. & Yu, Z. 2016 The analytical derivation of multiple elasticities of runoff to climate change and catchment characteristics alteration. *Journal of Hydrology* **541**, 1042–1056. doi:10.1016/j.jhydrol.2016.08.014.

Wang, X., He, K. & Dong, Z. 2019 Effects of climate change and human activities on runoff in the Beichuan River Basin in the northeastern Tibetan Plateau, China. *Catena* **176**, 81–93. doi:10.1016/j.catena.2019.01.001.

Xu, X., Yang, D., Yang, H. & Lei, H. 2014 Attribution analysis based on the Budyko hypothesis for detecting the dominant cause of runoff decline in Haihe basin. *Journal of Hydrology* **510**, 530–540. doi:10.1016/j.jhydrol.2013.12.052.

Yang, H. & Yang, D. 2011 Derivation of climate elasticity of runoff to assess the effects of climate change on annual runoff. *Water Resources Research* **47** (7). doi:10.1029/2010wr009287.

Yang, D., Sun, F., Liu, Z., Cong, Z., Ni, G. & Lei, Z. 2007 Analyzing spatial and temporal variability of annual water-energy balance in nonhumid regions of China using the Budyko hypothesis. *Water Resources Research* **43** (4). doi:10.1029/2006wr005224.

Yang, H., Yang, D., Lei, Z. & Sun, F. 2008 New analytical derivation of the mean annual water-energy balance equation. *Water Resources Research* **44** (3). doi:10.1029/2007wr006135.

Yang, H., Qi, J., Xu, X., Yang, D. & Lv, H. 2014 The regional variation in climate elasticity and climate contribution to runoff across China. *Journal of Hydrology* **517**, 607–616. doi:10.1016/j.jhydrol.2014.05.062.

Yue, S., Pilon, P., Phinney, B. & Cavadias, G. 2002 The influence of autocorrelation on the ability to detect trend in hydrological series. *Hydrological Processes* **16** (9), 1807–1829. doi:10.1002/hyp.1095.

Zhang, L., Dawes, W. R. & Walker, G. R. 2001 Response of mean annual evapotranspiration to vegetation changes at catchment scale. *Water Resources Research* **37** (3), 701–708. doi:10.1029/2000wr900325.

Zhao, F., Zhang, L., Xu, Z. & Scott, D. F. 2010 Evaluation of methods for estimating the effects of vegetation change and climate variability on streamflow. *Water Resources Research* **46** (3). doi:10.1029/2009wr007702.

Zheng, H., Zhang, L., Zhu, R., Liu, C., Sato, Y. & Fukushima, Y. 2009 Responses of streamflow to climate and land surface change in the headwaters of the Yellow River Basin. *Water Resources Research* **45** (7). doi:10.1029/2007wr006665.

First received 14 June 2021; accepted in revised form 20 January 2022. Available online 8 February 2022

Impact of Climate Change on Hydrology and Water Resources doi: 10.2166/wcc.2021.330

Urban water dissipation calculation based on the improved water balance models

Ke Zhou

ABSTRACT

The urban water dissipation models are developed based on observation and different urban construction patterns. The regional water dissipation is selected in Zhengzhou, the capital city of Henan Province, China. According to model calculation results, the annual water dissipation was obtained (820.0 mm), which is significantly higher than that compared with the traditional methods (494. 4 mm). The results show that the water dissipation contribution rate of green space, inside building, hardened ground and pavement, and water surface are 40.58, 38.70, 18.32, and 2.40%, respectively. The water dissipation of hardened ground and pavement, green space, and water surface is greatly influenced by hydrological and meteorological factors. The water dissipation inside a building is proportional to the number of building layers and residents. The results show that water dissipation inside a building cannot be ignored and it is the main component of artificial water dissipation. Under the same rainfall conditions, the water dissipation of the hardened ground and pavement is inversely proportion to the runoff coefficient.

Key words | binary water cycle, mathematical models, mechanism, urban hydrology, water dissipation

Ke Zhou
Faculty of Architectural Engineering,
Tianjin University,
Tianjin 300072,
China
and
Faculty of Arts and Design,
North China University of Water Resources and Electric Power,
Zhengzhou 450046,
China
E-mail: *zhouke@ncwu.edu.cn*

HIGHLIGHTS

- Urban water dissipation was classified according to different urban facilities.
- Water dissipation models were proposed based on different urban water dissipation sectors.
- The water dissipation contribution rate of different urban facilities was calculated.
- The results will be an important technical support for urban water planning and utilization.

GRAPHICAL ABSTRACT

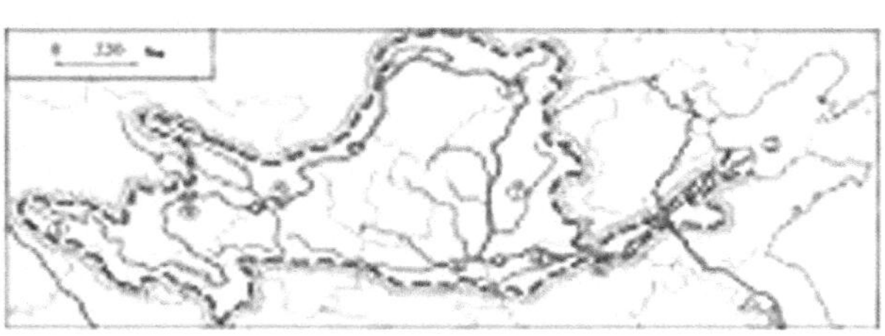

doi: 10.2166/wcc.2021.330

INTRODUCTION

Under the influence of global climate change and rapid urbanization, urban water-logging disasters occur frequently. Urban water environment, water resources, and water ecological problems are becoming more and more serious. As a result, urban water cycle mechanisms and water flux variation have become hot issues in urban hydrological science research (Zhang *et al.* 2014). Climate change and human activities have become two major driving factors affecting the hydrological cycle (Song *et al.* 2013). At the 7th International Congress of Hydrology Science, scholars pointed out that the impact of urbanization on the hydrological process is the focus of urban hydrology and water resource research (Xia & Zuo 2006). Urban hydrology is facing many important challenges (Fletcher *et al.* 2013). For instance, the proportion of urban population in China was 10.64% in 1949. The proportion of urban population exceeded 50% in 2011. In 2019, the proportion of urban population reached 60.77. The urban water issues are becoming more prominent owing to the increase of urban population, urban area, and the total amount of water consumption. In the north regions of China, variation of underlying land surface due to human activities reduces the runoff in many rivers (Zhang *et al.* 2013), increases the frequency of short duration heavy rainfall, intensifies rainfall events, and heightens the cumulative rainfall in the areas affected by urbanization (Reng *et al.* 2016). In Zhengzhou, the capital city of Henan Province, the average precipitation in 2018 was 640.0 mm, and the total water supply was 20.23×10^8 m^3. The urbanization area reached 1,100 km^2, but the water dissipation mechanism is vague and different dissipation components are unclear (Zhang *et al.* 2012).

Urban water dissipation is an important component in the hydrological process, and the magnitude of water dissipation directly affects the size of water flux in the circulation process. The urban hydrological process includes both the 'natural' cycle that occurs on the complex underlying ground surface and the 'social' cycle that concentrates on the urban water supply process (Liu *et al.* 2014).

In the aspect of urban 'natural' water dissipation, the soil–plant–atmosphere system model (SPAC) (Zhang *et al.* 2011) and remote sensing are widely used in the calculation of ground surface evapotranspiration. Experimental and analytical methods are used in plant water evapotranspiration research. There are relatively few studies on water dissipation in the urban 'society' sector (Di *et al.* 2011; Livesley *et al.* 2016). The current research is mainly focused on urban water demand prediction and water requirement management (Li *et al.* 2012; Ni 2013; Giacomoni & Berglund 2015; Loubet *et al.* 2015; Muhammad *et al.* 2015). Where the existing research findings are concerned (Minhua *et al.* 2021), the urban water consumption study is mainly focused on the 'natural' side (Ouyang *et al.* 2021), such as transpiration and dissipation characteristics of urban trees, lawns, and water surfaces. There are few studies on evaporation and dissipation of urban buildings, hardened ground, and pavement.

The water consumption research on the 'social' side is mainly focused on the sector of water resource allocation, and there are few studies on the mechanism of water dissipation, especially inside buildings, which is closely related to the residents' daily life and production. This is the reason why the current theories and models cannot solve the problems of urban high-intensity water consumption.

Therefore, the main purposes of this paper include studying the comprehensive calculation methods by conjunctive use of experiment, investigation and statistical data; carrying out water dissipation mechanism analysis on each water component; constructing water dissipation models through analysis of different urban construction components, and selecting a typical example to carry out urban water dissipation calculation.

MATERIALS AND METHODS

Framework of the urban water dissipation model

In this paper, the urban water dissipation pattern is classified based on urban water requirement, land surface features, and urban hydrological circulation routes. In order to explore the urban water dissipation mechanism, the corresponding mathematical calculation model and water dissipation system model are established. In the models, the urban space is classified as four types, i.e., buildings, hardened ground and pavement, green space, and water surface. Figure 1 is a framework of urban water dissipation. The water dissipation in the model is the mean annual water dissipation per unit area, in units of mm.

Dissipation models for hardened ground, pavement, and roof

Hardened ground and pavement could be classified as impervious (such as concrete, cement, and asphalt road) and permeable pavement. The building roof, as an impervious ground, cannot be penetrated, and the source of water dissipation is only from natural precipitation. It could be considered that the retained precipitation on building roofs could be totally dissipated after a short period of rainfall. For permeable pavement, the intercepted precipitation could be simultaneously dissipated both in evaporation and infiltration. The infiltrated water could be evaporated again. For the annual scale, usually, it is considered that the rain water intercepted by permeable ground surface and building roofs could be evaporated completely (Daba & You 2020). The dissipation module consists of effective rainfall, as follows:

$$W_Y = P_2 + (1 - \psi)(P - P_2) \tag{1}$$

in which, W_Y is dissipation from hardened ground, pavement surface, and building roof, mm; P_2 is the accumulated precipitation volume (daily rainfall less than 2, mm); ψ is the annual runoff coefficient, ≤ 1; and P is the annual precipitation, mm.

For urban sprinkling roads, the sources of water dissipation are from both artificial sprinkle and natural precipitation. The water dissipation from natural precipitation could be calculated with Equation (1). This model assumes that all artificial sprinkles on roads are evaporated and dissipated without surface runoff and infiltration. Therefore, for water dissipation on the sprinkle road, artificial sprinkle water volume W_A should be added.

Dissipation model of urban river and lake surface

Considering the meteorological factors affecting rivers and lakes (Doulgeris *et al.* 2020), we can develop the dissipation model for urban river and lake surfaces as follows:

$$W_W = \frac{\Delta}{\Delta + \gamma}(R_n + A_h) + \frac{\gamma}{\Delta + \gamma}\frac{6.43(1 + 0.536U_2)D}{\lambda} \tag{2}$$

in which W_W is the dissipation capacity of urban river and lake surface; A_h is the energy transported to the water body in the form of advection, MJ/(m^2·d); R_n is the net radiation, MJ/(m^2·d); U_2 is the wind speed at the height of 2 m, m/s; D is the saturated vapor pressure difference, kP_a; Δ is the slope of temperature and the saturated vapor pressure curve, $kP_a/^\circ C$; γ is a constant in wet and dry meters, kPa/$^\circ$C; λ is the latent heat of dissipation, MJ/kg, $\lambda = 2.50$.

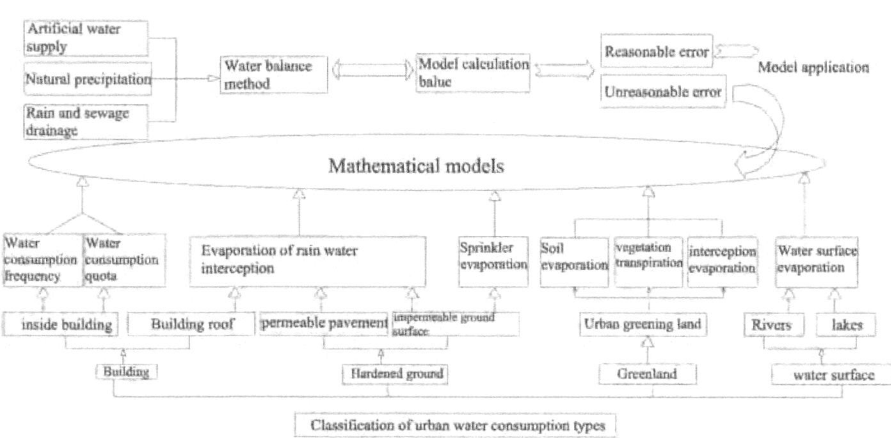

Figure 1 | Framework of the urban water dissipation model.

Dissipation model of urban green space

The water dissipation sources of urban green space are from artificial water supply and natural precipitation. Natural precipitation could be intercepted, evaporated, and infiltrated by plants and soil in the urban green space. A dissipation model for urban green space is as follows:

$$W_G = W_{EI} + W_{ET} + W_{EO} \tag{3}$$

in which, W_G is dissipation from green space; W_{EI} is vegetation rainfall interception dissipation; W_{ET} is vegetation transpiration; W_{EO} is soil dissipation in planting area, mm.

The model is built according to the existing mathematical model. The vegetation rainfall interception dissipation W_{EI} could be calculated with the Noilhan–Planton equation.

$$W_{EI} = V_{EG}\delta W_W \tag{4}$$

where V_{EG} is the crop coverage area, m^2 and δ is the area ratio of wet leaf surface.

$$W_{ET} = K_{cb}E_p \tag{5}$$

$$W_{EO} = K_e E_p \tag{6}$$

where K_{cb} is basic crop coefficient; K_e is the soil water evaporation coefficient, which could be selected according to FAO Irrigation and Drainage Paper No. 56 (Allan *et al.* 1998). E_p is the potential evapotranspiration capacity of vegetation (calculated by the Penman–Monteith method).

$$E_p = \frac{0.408\Delta(R_n - G) + \gamma\left(\frac{900}{(T_{mean} + 273)}\right)U_2(e_s - e_a)}{\Delta + \gamma(1 + 0.34U_2)} \tag{7}$$

where G the is net land surface radiation, MJ/(m^2·d); T_{mean} is the daily mean temperature, °C; e_s is the saturated vapor pressure, kP_a; e_a is the actual vapor pressure, kP_a.

Water dissipation model inside buildings

The dissipation inside buildings is an important component in urban water dissipation. In this study, the principle of bionics was adopted to simulate the different buildings as 'the concrete forest' with 'different height and physiological pattern'. The living buildings are simulated as 'life trees', the office buildings are simulated as 'office trees', and the production buildings are simulated as 'production trees'. Different kinds of trees are composed of their own cells. Each cell has its own water dissipation cycle and behavior. For example, the 'cell' of 'life trees' is a 'family', corresponding to a house; its water dissipation behavior includes laundry, cooking, floor cleaning, window cleaning, wiping tables, bathing, and so on.

According to different functions and mechanisms, urban buildings could be classified as residential buildings, office buildings, commercial buildings, hotels, restaurants, hospitals, teaching buildings, public service buildings (libraries, theatres, and museums), and production workshops. The relevant water dissipation is simulated as a real tree. According to the formula of vegetation transpiration, a general equation for calculating water dissipation inside buildings is established as follows:

$$W_i = C_i N_i A_i W_{DI} \delta_I \tag{8}$$

in which W_i is the total water dissipation inside the ith building, mm; C_i is the dissipation ratio of the ith building to the total buildings, %; N_i is the total number of the ith building layers; A_i is the land area occupied by the ith building, m^2; W_{DI} is the potential water dissipation inside the ith building, mm; and δ_i is the area ratio of wet floor inside the ith building, %.

The parameters in the model are determined according to the building purposes and framework. The area ratio of wet floor is the ratio of wet area to the total area of each building. The area occupied by the building refers to the horizontal area. Potential water dissipation refers to the daily water dissipation capacity per unit area of the building, which depends on the building purpose and water consumption types inside the building. Water dissipation frequency is determined through the questionnaire survey. The water dissipation quota is determined through the quantitative experiment.

Water balance model (Nunes *et al.* 2020)

The urban water use comes from artificial water supply (W_S) and natural precipitation (P). Urban water dissipation can be classified as two parts: outdoor water dissipation (W_{OUT}) and dissipation inside buildings (W_{IN}).

The outdoor water dissipation source comes from two parts: artificial water supply and natural precipitation, while artificial water supply is the only source of water dissipation inside buildings. The natural dissipation from precipitation could be classified as vegetation rainfall interception evaporation, vegetation transpiration, soil evaporation, water surface evaporation, hardened ground, and pavement rain water evaporation and building roof rainfall interception evaporation. Artificial outdoor water supply is mainly used for garden and green space irrigation, road sprinkling, and guardrail cleaning. It is believed that except for a small amount of seepage on the green space irrigation, the majority of outdoor water supply is totally dissipated without ground surface runoff and infiltration. Water consumption in an urban building mainly includes drinking water, face washing, bathing, cooking, washing clothes, flushing toilets, floor wetting, and humidifier. Different building types have different water consumption components and different water consumption quotas.

Based on the circulation process of urban water systems, the water balance model is established as follows:

$$W_S + P = W_{IN} + W_{OUT} + D + W_{SUR} + \Delta W_{GRO} \tag{9}$$

in which W_S is the total amount of annual artificial water supply for a city.

$$W_S = W_{SI} + W_{SO} \tag{10}$$

where W_{SI} is the annual total water supply for urban buildings and is equal to the sum of water supply for each building. W_{SO} is the annual outdoor water supply, mainly including municipal water use (road sprinkling, guardrail cleaning, green space water use, and water recharge to urban rivers and lakes); P is annual precipitation; W_{IN} is the total amount of annual water consumption inside buildings.

$$W_{IN} = \sum_{i=1}^{n} W_i \tag{11}$$

W_{OUT} is the total annual outdoor water consumption.

$$W_{OUT} = W_Y + W_A + W_G + W_W \tag{12}$$

D is the water drainage; W_{SUR} is the annual precipitation surface runoff; and ΔW_{GRO} is the annual underground water storage variation. The unit in the water balance model is m^3.

CASE STUDY

The basic quota of urban water dissipation is obtained by means of case observation, including 'natural' (such as meteorological data and hydrological data) and 'society' sides (such as per quota of water consumption). The frequency of water consumption per person per day is analyzed by a questionnaire survey. The urban underlying ground surface could be classified according to water consumption features. According to the features of different water consumption, the corresponding dissipation model is established. The calculation is made for different water dissipation components using different models. The integrated dissipation is achieved by using the area weighted method. Finally, verification is made with the water balance method.

We selected a region of Zhengzhou city in Henan Province, China, to carry out a water dissipation case study. The region selected covers 352 hm^2 with 60,000 residents and low buildings. Each building generally has 13 layers. The green rate in the region is 54.8%. The annual mean precipitation is 640.5 mm. All data come from the authority department, Henan Province Hydrology and Water Resources Bureau.

Water dissipation patterns

There are many water dissipation components in the urban area. In order to make reasonable water dissipation calculation and according to the model requirement and the water balance model proposed above, water dissipation could be classified as four different components, i.e., water dissipation inside buildings, on the pavement, in green space, and on the water surface (Table 1).

Table 1 | Area ratio of different water dissipation in the study region

Surface type	Area (m^2)	Ratio (%)
Building body	1,137,031	32.31
Pavement	1,796,180	51.04
Green space	517,268	14.70
Water surface	68,421	1.95
Total	3,518,900	100.0

According to the real situation, the hardened pavement includes impervious pavement (such as asphalt road) and permeable pavement (such as pedestrian).

There is no artificial sprinkler on roads in the study area. Hence, the only source of water dissipation comes from rainwater interception.

Investigation on the frequency of living water consumption

Based on the water dissipation model inside buildings (Equation (8)), the frequency of living water consumption was investigated by the questionnaire. Three hundred questionnaires were distributed, and 272 questionnaires were effectively collected, covering the residents and staff of the main buildings in the area. The proportion of male-to-female respondents to the questionnaire is nearly 1:2, ranging in age from 18 to over 60, covering different age groups.

Seven questions were asked about the different frequencies of water consumption in daily life. Statistics were made in m^3 per day per capita. Table 2 shows the content of the questionnaire, and Table 3 shows the daily average water consumption.

Experimental measurement of water dissipation quota

Temperature, humidity, wind speed, precipitation, radiation, and other meteorological data were collected from the meteorological stations. The period of meteorological data is from January 1 to December 31. The 'natural' water consumption was calculated based on the collected meteorological data, including vegetation transpiration, vegetation intercepted water evaporation, soil water evaporation, water surface evaporation, hardened ground, and pavement precipitation interception evaporation.

Table 2 | Contents of the questionnaire

No.	Content	Item selection
1	Gender	A; male; B female
2	Age	A under 18; B 18–25; C 26–60; D over 60.
3	Daily face washing times	A 1; B 2; C 3; D 4; F -
4	Toilet frequency daily	A 1(4); B 2(5); C 1(6); D 1(7); F -
5	Number of washing clothes daily	A 1; B 2; C 3; D 4; E 5; F -
6	Daily bathing times	A 1; B 2; C 1/2; D 1/3; E 1/4; F –
7	Daily cooking times	A 1; B 2; C 3; D 4; F -

Table 3 | Frequency and average living water dissipation

Items	Daily frequency (average)	Mean (L/times)
Flower irrigation	1	0.40
Washing (face, hands)	2	1.00
Humidifiers	1	2.00
Toilet	5	0.25
Washing clothes	2	0.85
Bath	1/2	4.00
Cooking	3	2.20

According to the water balance model (Equations (9)–(11)), five typical families were selected to carry out an experimental investigation. The weighing method was adopted for water dissipation measurement. For instance, through comparing towels weight before (dry) and after (wet) washing face, hands, and bath, the water dissipation of washing face was measured. The water amount consumed in washing clothes is equal to the weight of wet clothes minus the weight of dry clothes. Cooking water consumption refers to the water amount used in cooking, including dissipation during the cooking process and water converted into cooked food. The scale accuracy in the experiment is 1 g. For the wetting floor, humidifier and other water consumption, the calculation is based on the unit area. Table 3 shows the average water dissipation per household.

DISCUSSIONS

Calculation results for different kinds of water dissipation

Using the relevant methods and the equations, water dissipation patterns were classified first in the region. Then, different kinds of water dissipation were calculated.

Hardened ground, pavement, and roof water dissipation

Due to flat building roofs in the experimental area, the roof precipitation–runoff coefficient ψb is 0.90. According to the Urban Drainage Engineering Planning Code (The Ministry Housing and Construction China 2017) and the Outdoor Drainage Design Code (The Ministry of Housing and Construction China 2016), the runoff coefficient ψp of impermeable ground or pavement is 0.90. The runoff coefficient of permeable ground or pavement *ψs* is 0.55. Based on Equation (1), the daily precipitation greater than 2.0 mm was considered as the effective precipitation. According to calculated results using Equation (1), the runoff coefficient of roof water dissipation and impermeable ground and pavement interception dissipation is 189.9 mm. The runoff coefficient of water dissipation of permeable ground or pavement is 320.70 mm.

Water surface evaporation (dissipation)

The lake surface dominates the water surface in the study area. The calculation of water surface evaporation assumes that the water surface is still. Assuming that the study area's water body is enough for evaporation, according to calculation results using Equation (2), the annual water dissipation per unit water surface area is 1,010.7 mm.

Green space water dissipation

Grassland, shrubs, and trees are the main vegetation in the study area. The water dissipation includes three parts: soil evaporation, vegetation transpiration, and plant rainwater interception evaporation. Soil evaporation and vegetation transpiration can be calculated based on plant potential evaporation, and crop potential evaporation can be calculated according to Equation (7). Plant interception rain water evaporation can be calculated according to water surface evaporation. The total area of evapotranspiration equals the total area of the region multiplied by the vegetation coverage rate (i.e., $V_{eg} = 0.55$). The area of soil evaporation equals the green land area. The final calculated soil evaporation of green space is 258.80 mm. Vegetation transpiration dissipation is 488.80 mm, and the evapotranspiration of vegetation rain water interception is 48. 90 mm.

Water dissipation inside a building

Water dissipation inside a building is the largest component in urban water dissipation, closely related to residential life and work. The water source inside the building comes from an artificial water supply through pipelines. Omitting the lost water transfer process, the water dissipation calculation can be classified into two parts, i.e., daily life and floor wetting. The calculation unit of water dissipation in life and work is in per capita. The dissipation calculation of the wetting floor is in the unit area (m^2). According to calculation results, water dissipation inside the building per unit area is 260.6 mm, in which water dissipation of daily life by residents is 151.7 mm, and water dissipation by floor wetting is 108.9 mm.

Water dissipation mechanism analysis

Table 4 shows the different water dissipation items in the study area, in which the water dissipation inside a building

Table 4 | Different water dissipation intensities

Items	Volume (mm)
Inside building	260.6
On roof	189.9
On impermeable pavement	189.9
On permeable pavement	320.7
River, lake surface	1,010.7
Vegetation transpiration	488.8
Interception evaporation	48.9
Soil water evaporation	258.8

is representative of the 'social' sector. The amount of water dissipation is determined by the residents' living and working habits, and the water dissipation varies from person to person and is less affected by meteorological factors. The water dissipation outside the building is greatly affected by meteorological factors.

The building roof water dissipation is the same as impermeable ground or pavement because of the same runoff coefficient. Due to different runoff coefficient, water dissipation on roof is different from that on permeable pavement. The runoff coefficient determines rain water interception evaporation intensity on hardened ground and pavement.

The calculation of green space water dissipation is divided into three parts, in which vegetation interception evaporation depends on the shape of vegetation leaves and the characteristics of precipitation. Vegetation transpiration and soil evaporation represent water dissipation in the urban 'natural' sector, which depends mainly on factors such as temperature and precipitation, sunshine and solar radiation, and other meteorological factors. The amount of water surface evaporation is mainly determined by meteorological factors such as temperature, relative humidity, and solar radiation.

Contribution rate of different water dissipation

The contribution rate of each water dissipation is equal to the water dissipation ratio to the total water dissipation in the study area. Table 5 indicates the different water dissipation rates.

It can be seen from Table 5 that the green space water dissipation rate (40.58%) ranks first. The reason is that water dissipation of green space happens at three layers in vertical direction, i.e., soil evaporation, vegetation transpiration, and intercepted rain water evaporation on plant leaves, and the latter two is much larger than the area of green space.

Table 5 | Different water dissipation contribution rates

Surface types	Inside building	Pavement	Green space	Water surface	Total
Rate (%)	38.70	18.32	40.58	2.40	100.00

This is due to the plantation of trees and shrubs on both sides of the road, the calculated area of which depends on the degree of vegetation coverage. The ratio of calculation area equals to the green space rate, which is 14.70%.

The area ratio occupied by buildings in the region is 32.31%. However, because the water dissipation of the building is distributed at multi-layers in the vertical direction, which is proportional to the number of the building layers, and considering the intercepted rainwater evaporation on roofs, it is concluded that the contribution rate of the total water dissipation of the building is 38.70%, which can be divided into two parts, i.e., water dissipation inside buildings and rain water interception evaporation on building roofs, the former is 31.22%, and the latter is 7.48%.

For the hardened ground or pavement, the area ratio is 51.04%, in which the permeable surface accounts for 40.73%, and the impermeable ground accounts for 10.31%. However, water dissipation occurs only in the surface layer, which is closely related to rainfall characteristics, and the final calculated contribution rate is 18.32%.

The water surface's dissipation contribution rate is 2.40% because of the small proportion of water surface area in the study region.

Water balance verification

Figure 2 shows the total water dissipation and water supply in the region. Assuming that there is no change of surface water volume in the experimental area, the drainage capacity is calculated according to the drainage coefficient, and the groundwater recharge is equal to the leakage of the pipeline, according to the calculation with Equation (9), the relative error accounts for 4.25% in the total water supply. The reason is that there is an error in measuring the variation of groundwater volume, and only the leakage loss of the pipeline is considered in the study.

Comparison analysis on the studied results

The calculation of urban water dissipation is mostly limited to the natural water consumption, and the binary attribute of natural and social sectors is ignored. The urban water dissipation process has obvious dual attributes (natural–social).

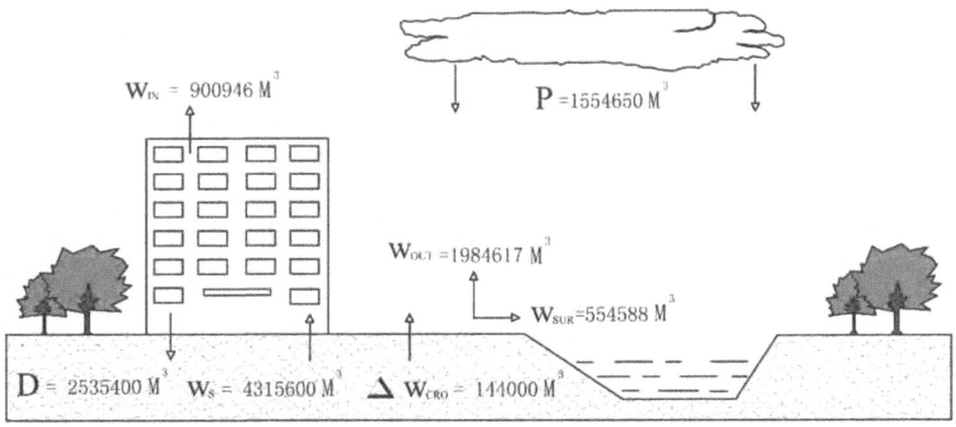

Figure 2 | Schematic of water components in the water balance system.

To accurately calculate urban water dissipation and water circulation, this paper classifies urban water dissipation into five types based on different land use, i.e., green space, water surface, soil surface, buildings, and hardened ground. The analysis framework of urban natural–social dual water dissipation is put forward, and the urban dual water dissipation calculation model is constructed.

Based on the proposed methods, the annual water dissipation in Zhengzhou is obtained (820.0 mm), which is significantly higher than that compared with the traditional methods (494.4 mm). The results show that the water dissipation contribution rate of green space, inside the building, hardened ground and pavement, and water surface are 40.58, 38.70, 18.32 and 2.40%, respectively. It is shown that water dissipation inside the building cannot be ignored, and it is the main component of artificial water dissipation.

The study results show that with the increasing urbanization level, the comprehensive water dissipation intensity increases in urban areas, and the water dissipation proportion of the social sector increases. Population density and building density are the main factors affecting the water dissipation intensity of the urban social side. The increase of urbanization level would reduce the seasonal difference of urban water dissipation.

CONCLUSIONS

In this paper, the water dissipation models were set up based on urban construction patterns and the data from authority departments, which are used to quantitatively calculate water dissipation for different types of urban facilities. The calculated rain water dissipation on roof and impermeable ground is 189.9 mm. Rain water dissipation on permeable ground or pavement is 320.70 mm. The annual water evaporation of water surface is 1,010.7 mm. The vegetation transpiration dissipation is 488.80 mm, and the transpiration from rain water intercepted by vegetation is 48.90 mm. Water dissipation inside buildings per unit area is 260.6 mm, in which water dissipation of daily life by residents is 151.7 mm, and water dissipation by floor wetting is 108.9 mm.

Compared with traditional hydrological models that can only consider the evaporation caused by precipitation and neglect artificial dissipation, more detailed information on water circulation could be obtained using the models proposed in this paper and will play an essential role in urban construction.

The results show that the water dissipation inside buildings cannot be ignored in urban water planning and allocation. This kind of water dissipation is proportional to the number of building layers, and its potential water dissipation is mainly determined by human activities, and is less influenced by hydrological and meteorological factors.

Under the same rainfall conditions, the water dissipation of the hardened ground and pavement is inversely proportional to the runoff coefficient. The water dissipation of hardened ground, pavement, green space, and water surface is greatly influenced by hydrological and meteorological factors.

According to the calculated contribution rate of different water dissipation components to the total water dissipation, the green space water dissipation rate is 40.58%, ranking the first. The building water dissipation contribution rate is 38.70%, which ranks second. For the hardened ground or pavement, the final calculated water dissipation contribution rate is 18.32%. The water dissipation contribution rate of different construction patterns to the total water dissipation is determined by the area ratio of different construction patterns to the total area. The larger the water dissipation, the larger the area, the greater the water dissipation contribution rate would be.

It is sure that the present findings should be helpful to scientifically measure the dissipative flux of urban hydrological cycle, provide theoretical and technical support for urban water planning, and lay a scientific basis for urban water demand management and water saving.

ACKNOWLEDGEMENTS

This project was supported by grants from the National Natural Science Foundation of China (No. 50579020).

DECLARATION OF COMPETING INTEREST

The author declares that there should be no known competing financial interests or personal relationships that could have appeared to influence the work reported in this paper.

DATA AVAILABILITY STATEMENT

All relevant data are included in the paper or its Supplementary Information.

REFERENCES

Allan, R. G., Pereira, L. S. & Raes, D. 1998 *Crop Evapotranspiration Guidelines for Computing Crop Water Requirements*. FAO Irrigation and Drainage Paper. FAO, Rome, pp. 53–56.

Daba, M. H. & You, S. 2020 Assessment of climate change impacts on river flow regimes in the upstream of Awash Basin, Ethiopia: based on IPCC fifth assessment report (AR5) climate change scenarios. *Hydrology* **7** (4), 98–98.

Di, S. C., Wu, W. Y., Liu, H. L. & Pan, X. 2011 Water consumption estimation and evapotranspiration inversion based on remote sensing technology. *Transactions of the Chinese Society of Agricultural Engineering* **28** (10), 98–104.

Doulgeris, C., Panagiota, K. & Pantazis, G. 2020 Assessment of minimum water level in lakes and reservoirs based on their morphological and hydrological features. *Hydrology* **7** (4), 83–83.

Fletcher, T., And Rieu, H. & Hamel, P. 2013 Understanding, management and modelling of urban hydrology and its consequences for receiving waters: a state of the art. *Journal of Advances in Water Resources* **51**, 261–279.

Giacomoni, M. H. & Berglund, E. Z. 2015 Complex adaptive modeling framework for evaluating adaptive demand management for urban water resources sustainability. *Journal of Water Resources Planning and Management* **141** (11), 1–12.

Li, C. M., Wang, J. H., Wang, H. & Zhang, S. G. 2012 *Urban Water Consumption Long Term Prediction Model Based on the Water Price Elasticity in China*. World Environmental and Water Resources Congress, Albuquerque: ASCE, Vol. 2, 2491–2500.

Liu, J. H., Wang, H., Gao, X. R. & Chen, S. L. 2014 Review on urban hydrology. *Journal of Chinese Science Bulletin* **59** (36), 3581–3590.

Livesley, S. J., Mcpherson, G. M. & Calfapietr, A. C. 2016 The urban forest and ecosystem services: impacts on urban water, heat, and pollution cycles at the tree, street, and city scale. *Journal of Environmental Quality* **45** (1), 119–124.

Loubet, P., Roux, P., Guerin-Schneider, L. & Bellon-Maurela, V. 2015 Life cycle assessment of forecasting scenarios for urban water management: a first implementation of the Water model on Paris suburban area. *Journal of Water Research* **90**, 128–140.

Minhua, L., Han, H., Wei, X. & Lv, C. 2021 Temporal and spatial distributions of precipitation on the Huang-Huai-Hai Plain during 1960–2019, China. *Journal of Water and Climate Change*. https://doi.org/10.2166/wcc.2021.313.

Muhammad, A., Zah Rani, A. L. & Amin, A. M. 2015 Urban residential water demand prediction based on artificial neural networks and time series models. *Journal of Water Resource Management* **29** (10), 3651–3662.

Ni, G. 2013 Statistical interpretation of the daily variation of urban water consumption in Beijing, China. *Journal of Hydrological Sciences* **59** (1), 181–192.

Nunes, G., Minoti, R. T. & Koide, S. 2020 Mathematical modeling of watersheds as a subsidy for reservoir water balance determination: the case of Paranoá Lake, Federal District, Brazil. *Hydrology* **7** (4), 85–85.

Ouyang, Y., Wan, Y., Jin, W., Leininger, T. D., Feng, G. & Han, Y. 2021 Impact of climate change on groundwater resource in a region with a fast depletion rate: the Mississippi embayment. *Journal of Water and Climate Change* **2021** (1), https://doi.org/10.2166/wcc.2021.326.

Reng, Y., Liu, Y. J., Sun, X. B., Che, Y. & Sun, X. 2016 Spatial and temporal patterns of precipitation variability over mainland China: III: causes for recent trends. *Journal of Advances in Water Science* **27** (3), 1–25.

Song, X. M., Zhang, J. Y., Zhan, C. S. & Liu, C. Z. 2013 Review for impacts of climate change and human activities on water cycle. *Journal of Hydraulic Engineering* **44** (7), 779–790.

The Ministry of Housing and Construction, China 2016 *Water Drainage Design Criterion Outdoor*. GB 50014-2006(2016). China Architecture Industry Publishing Press, Beijing.

The Ministry of Housing and Construction, China 2017 *Urban Water Drainage Project Planning Criterion*. China Architecture Industry Publishing Press, Beijing. GB50318-2017

Xia, J. & Zuo, Q. T. 2006 Advances in international hydrological science research. *Journal of Advances in Earth Science* **21** (3), 256–261.

Zhang, Y., Zheng, X. L., Wu, C. C. & Zhang, S. H. 2011 Experimental study of evapotranspiration in Liaohe Estuary. *Journal of Advances in Water Science* **22** (3), 351–358.

Zhang, S. F., Meng, X. J. & Liao, Q. 2012 Research on water resources and water balance in Beijing. *Journal of Geographical Research* **31** (11), 1991–1997.

Zhang, J. Y., Her, M., Qi, J., Liu, C., Wang, G. Q. & Jin, J. L. 2013 A new perspective on water issues in North China. *Journal of Advances in Water Science* **24** (3), 303–310.

Zhang, J. Y., Song, X. M., Wang, G. Q., He, R. M. & Wang, X. J. 2014 Development and challenges of urban hydrology in a changing environment: I: hydrological response to urbanization. *Journal of Advances in Water Science* **25** (4), 594–605.

First received 27 November 2020; accepted in revised form 28 March 2021. Available online 13 April 2021

© 2022 The Authors Impact of Climate Change on Hydrology and Water Resources doi: 10.2166/wcc.2021.144

Holocene environmental evolution history based on sporopollenin and micropaleontological reconstruction of KY-01 in the Yellow River Delta

Wang Kui-Feng[a,b,c,*], Yu Xue-Feng[a], Xia Jiang-Bao[c], Xu Shu-Jian[b], Zhang Tai-Ping[a], Xu Yan[a] and Zhang Hong-Jun[a]

[a] Shandong Key Laboratory of Geological Process and Resource Utilization in Metallic Minerals, Key Laboratory of Gold Mineralization Process and Resource Utilization Subordinate to the Ministry of Land and Resource, Shandong Institute of Geological Sciences, Jinan 250013, China
[b] Shandong Provincial Key Laboratory of Water and Soil Conservation and Environmental Protection, Linyi University, Linyi 276000, China
[c] Shandong Key Laboratory of Eco-Environmental Science for Yellow River Delta, Binzhou University, Binzhou 256600, China
*Corresponding author. E-mail: maplewkf@126.com

ABSTRACT

The Yellow River Delta adjacent to the Bohai Sea is located in the monsoon region of northern China, which is sensitive to global climate change. The core data of KY-01 borehole in the Yellow River Delta and the published records were used to reconstruct the climate environment. Based on the analysis of carbonate content, magnetic susceptibility, Ostracoda, foraminifera, sporopollenin and AMS ^{14}C dating on the KY-01 borehole sediment core, the evolution of both the climate and sedimentary environment has been discussed. The results show that: 7656–4145 cal.aBP, seawater moved toward the land surface and the climate was generally warm and humid, when there were small-scale extreme climate fluctuations; 4145–2544 cal.aBP, transgression and recession occurred, the climate changed from warm-wet to dry-cold and there were small-scale extreme climatic events; and 2544 cal.aBP–1855 AD, the sea level was relatively low, transgression and retreat alternately occurred, the climate was drier and colder than the previous stage and the warming and cooling alternated. During the middle Holocene, cold-dry events occurred between 5737–5422 and 4451–4081 cal.aBP, which is consistent with both Chinese and worldwide climate records.

Key words: Holocene, Ostracoda and foraminifera, sporopollenin, Yellow River Delta

HIGHLIGHTS

- Reconstruction of Holocene environment in the Yellow River Delta using multi-proxy indices.
- Two extreme climate events in the study area since Holocene were analyzed.
- Compared with the Holocene climate and environmental changes in Shandong Province.

GRAPHICAL ABSTRACT

Map showing the location of the Yellow River Delta

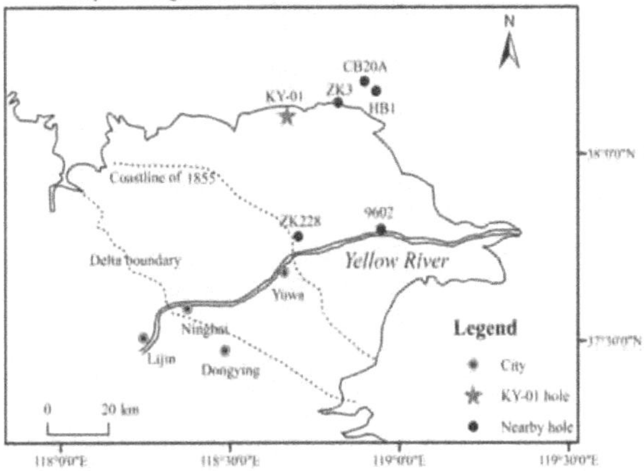

Sediment core from KY-01 in the Yellow River Delta

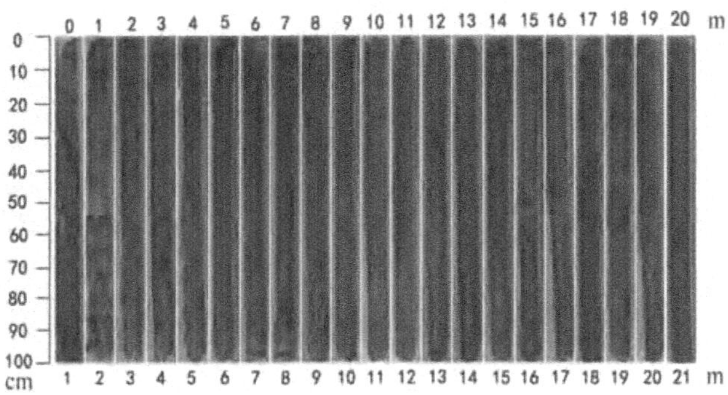

Climatic stages in the Yellow River Delta since the Holocene.

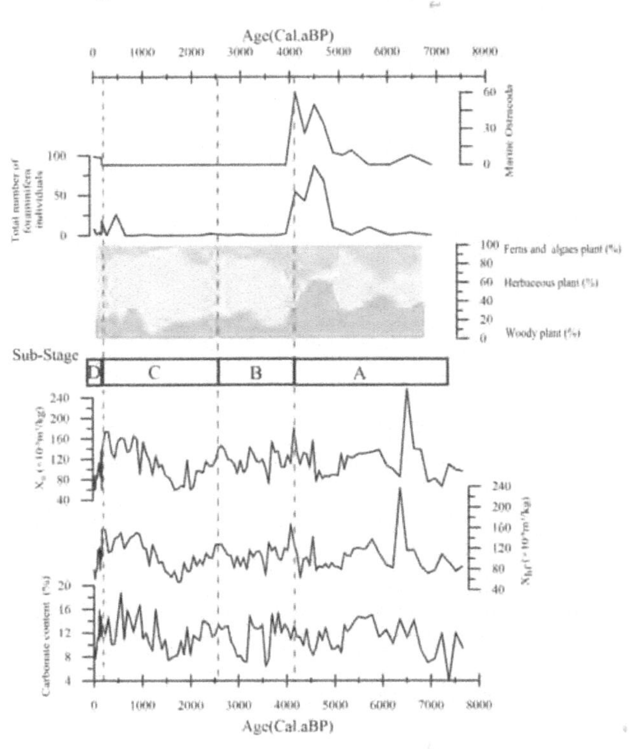

INTRODUCTION

Holocene is an important geological period in geology, also known as the post-glacial period. It is also a geological period most affected by human activities and closely related to human evolution and development. The climate change during Holocene reflects the future trend of climate change and has become the focus of PAGES (Burciaga *et al.* 2019; Chen *et al.* 2020; Han *et al.* 2020; Javadinejad *et al.* 2020). Due to the comprehensive influence of solar radiation, atmospheric circulation, physical and geographical environments in different regions, the climate change in different regions has its own characteristics since Holocene, and there are some differences among regions, so the regional research of Holocene climate change is worthy of attention. The living environment of humans is facing the increasingly dangerous threat of climate change. To predict the future tendency of climate change, it is necessary to study temporal and spatial changes in climate, the driving mechanism of past climate change and the impact of climate change on the terrestrial environment (Gabriela *et al.* 2019). With the deepening of research theories and content, the research methods and technologies of paleoclimate reconstruction tend to be diversified. In terms of research methods of environmental proxy indicators, Zhu used a variety of proxy indicators to reveal the important role of climate change and natural disasters such as floods and transgressions in cultural evolution (interruption, extinction) since the mid-Holocen (Zhu *et al.* 2006). Vött combined the comparative and integrated studies of geomorphology, sedimentology, geochemical elements and micropaleontology, and confirmed the cycle of multiple tsunami activities in the Lefkada Strait and adjacent areas of Greece in the middle ages (Vött *et al.* 2008). In addition, many new dating techniques are also widely used in the reconstruction of the Holocene natural environment, such as AMS ^{14}C dating, ^{137}Cs and ^{210}Pb, and cosmic genesis ^{10}Be, ^{26}Al and ^{36}Cl dating (Guo *et al.* 2018). In the application of new technical means, the emergence of analysis methods, such as stalagmite record (Huang *et al.* 2019), Phytolith record (Zhang *et al.* 2019), bone carbon/hydrogen/oxygen isotope (Ma *et al.* 2019) and biological mitochondrial DNA (Railsback 2018), provided more perfect ideas for the study of climate and environmental changes.

During the Holocene, sea-level changes and different rivers interacted to form different types of deltas. For a river delta located on a plain coast, the division sequence not only has significance for sedimentology but also for studying sea-level changes, river evolution, temperature changes, and other environmental changes (Nian *et al.* 2018). The Yangtze River Delta was characterized by a pattern of incised valleys that form many delta lobes from west to east, as well as different stratigraphic structures that recorded paleoenvironmental information (Mao 2012). The Red River Delta (Zhang *et al.* 2016) and the Mekong River Delta (Hoang *et al.* 2016) were formed during the period of sea-level rise in the middle of the Holocene when the seawater transgressed along the Pearl River. With a rapid rise in the sea level and delta accumulation, the coastline, river type and geomorphic type all change correspondingly. Studies had shown that the Yellow River Delta, located in the transitional zone between land and sea, is controlled by land/sea interactions and is extremely sensitive to changes in the environment and climate (Wang *et al.* 2020; Zhao *et al.* 2020). Therefore, studying the evolution of paleoclimate and paleoenvironment in the Yellow River Delta has become an important means to study the local, regional and global climate and environmental changes in the Yellow River Delta. According to the existing research, most of the research areas are located in Northwest, Central, Northeast and Southeast China, while there is less research in East China, especially in the coastal land–sea interaction zone of East China. Therefore, the Holocene environment of the Yellow River Delta was reconstructed by using a multi-agent index, and the evolution law of paleoclimate and paleoenvironment in the sedimentary core area of the Yellow River Delta was proposed, the process of environmental change in the study area since 7656 cal.aBP was reconstructed, and then two extreme climate events since Holocene were explored. Combined with the existing paleoclimate data of Shandong Province, a comprehensive comparative analysis has also been carried out in order to provide a reference for exploring the Holocene environmental evolution of Shandong Province.

In this paper, the study of the Yellow River Delta mainly includes sedimentological, chronological, carbonate and magnetic susceptibility analyses in sediments, ostracods and foraminifera analysis in sediments and pollen analysis in sediments. The Holocene environmental evolution, paleoclimate reconstruction and abrupt climate change events in the Yellow River Delta were discussed, At the same time, the Holocene climate and environmental changes and the regional sea level in Shandong Province were compared and analyzed.

MATERIALS AND METHODS

Study area

The Yellow River Delta is located at the mouth of the Yellow River on the west bank of the Bohai Sea (117°31′00″–119°18′00″ E, 36°55′00″–38°16′00″ N) (Figure 1(a)–1(c)). It is an alluvial plain formed by the deposition of a large amount of sediment in the estuary carried by the Yellow River, as it passes through the Loess Plateau. Due to the high soil salinity in the Yellow River Delta, the distribution of forest trees is limited by soil type and topography.

Coring and sediment core processing

21 m sediment core was drilled at a latitude of 38° 06′ 08″ N, – a longitude of 118° 40′ 08″ E and an altitude of 3.026 m, with the number of KY-01. The core was extracted using an XY-100 drilling machine from a hole of 21 m in depth (Figure 1(d)–1(g)). The 100 m drilling machine was selected to ensure the core rate. The sediment core segments were sealed and preserved in PVC core tubes, stored in shock-absorbing boxes and transported back to the laboratory (Figure 2).

Radiocarbon dating

In this study, the sedimentary age was mainly determined by the Radiocarbon Dating Laboratory for large organic sediment samples. Choosing peat with high organic carbon content, peat with high organic layer or lacustrine clay as radiocarbon dating samples could more accurately determine the sedimentary age. AMS ^{14}C dating was performed on bulk organic sediment samples at the Beta Analytic Radiocarbon Dating Laboratory. The AMS ^{14}C dates were converted into calendar ages using the program Calib 7.02 based on the INTCAL 13 calibration.

Carbonate content analysis

The carbonate content of the sediment was measured using a GMY-4A carbonate content analyzer. Samples were taken every 10 cm, and a total of 200 samples were obtained. All of the samples were fully ground and sieved through a 0.25-mm sieve and were uniformly weighed at 200 mg. Samples were fully reacted with a 1:3 (V/V) hydrochloric acid solution using a GMY-4A carbonate analyzer to convert the pressure of the generated CO_2 gas into an electrical signal, which was used by the computer to automatically calculate the carbonate content of the sample.

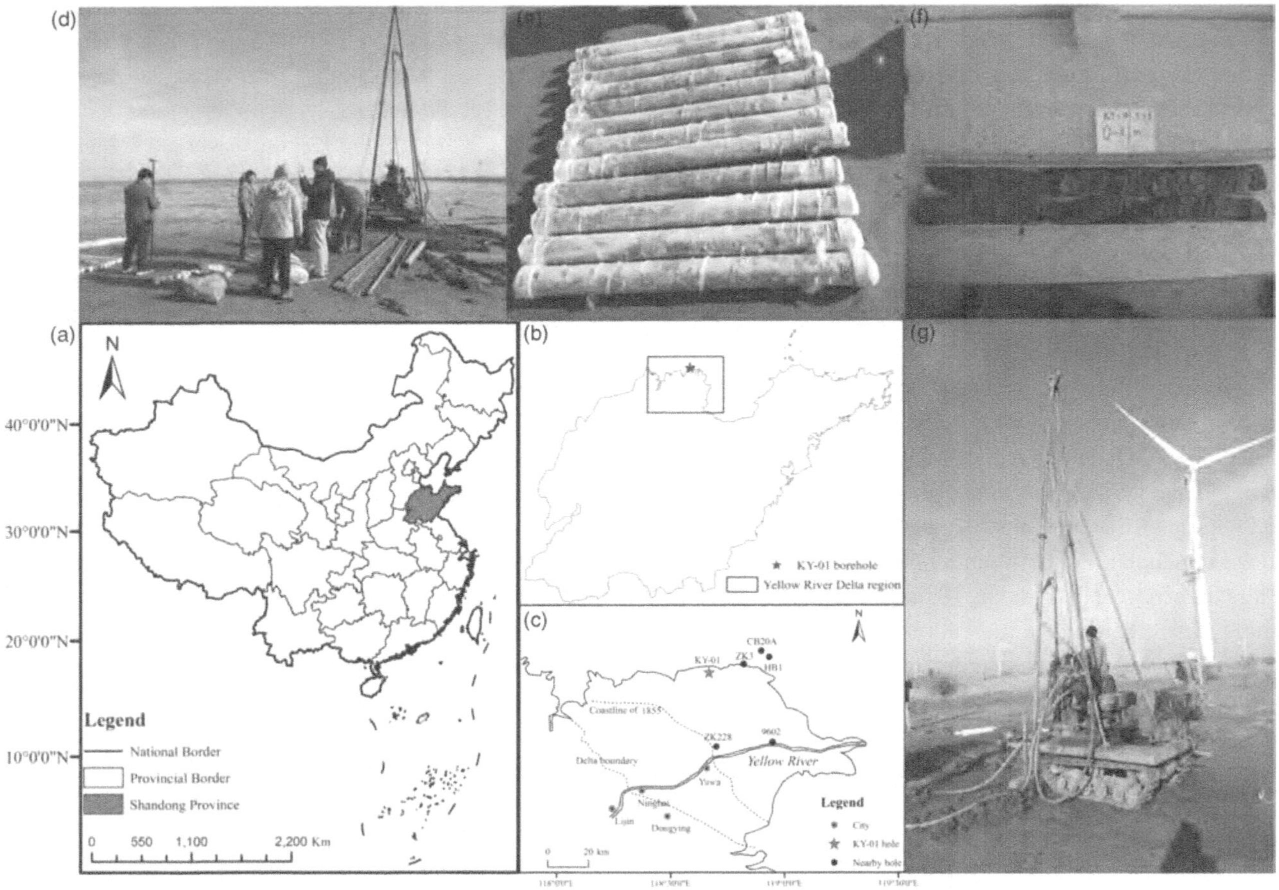

Figure 1 | (a) Map showing the location of the Yellow River Delta, (b,c) location of KY-01 borehole and other boreholes in the Yellow River Delta mentioned in the paper. (d–g) are the photo of cores drilled in the Yellow River Delta.

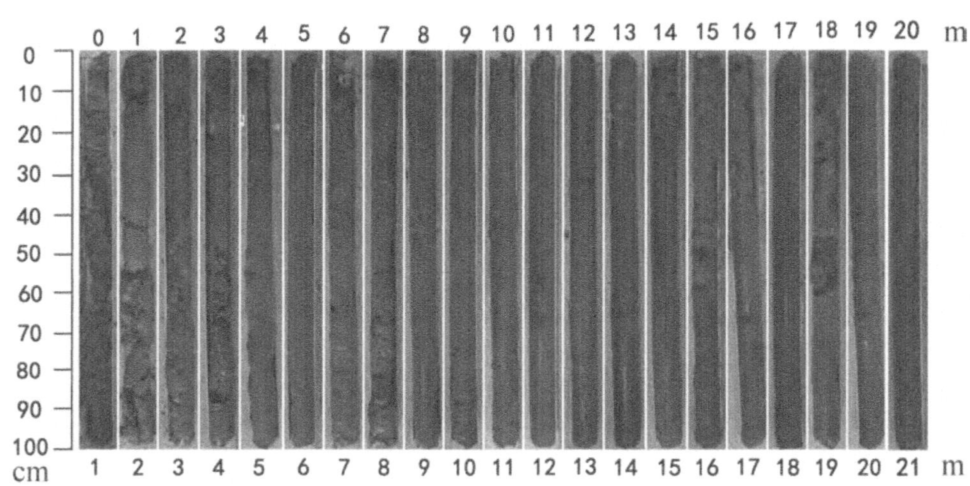

Figure 2 | Photograph of a 21-m continuous sediment core from KY-01 in the Yellow River Delta.

Magnetic susceptibility analysis

The magnetic susceptibility of the sediments was measured indoors using the MS-2B magnetic susceptibility meter produced by Bartington, UK. Samples were taken every 10 cm, and a total of 200 samples were obtained for magnetic susceptibility analysis. After all of the samples were dried and ground thoroughly (without damaging the granularity), the soil samples were compacted and weighed in a 1 cm^3 non-magnetic sample box and then tested by the MS-2B magnetic susceptibility meter.

Pollen analysis

Pollen was measured at the Nanjing Institute of Geology and Palaeonotology, Chinese Academy of Sciences. Samples were taken every 30 cm, and a total of 66 samples were obtained for pollen analysis. In accordance with the standard analytical method of the petroleum and natural gas industry of the People's Republic of China, SY/T5245-91, samples were treated with fluoric acid to remove minerals, then oxidized with concentrated hydrochloric acid and finally sieved to concentrate the spore powder into tubes. Identification under the microscope was based on the oil and gas industry standard of the People's Republic of China, SY/T5915-94.

Foraminifera and Ostracoda analysis

Species assemblages of foraminifera and Ostracoda were analyzed for a total of 55 samples, taken from different intervals based on lithological criteria. Due to their varying preferences of particular habitats, these species provide valuable information for reconstructing sedimentary environments. All of the samples were tested at the Nanjing Institute of Geology and Paleontology, Chinese Academy of Sciences.

Research route

See Figure 3 for the research flow chart.

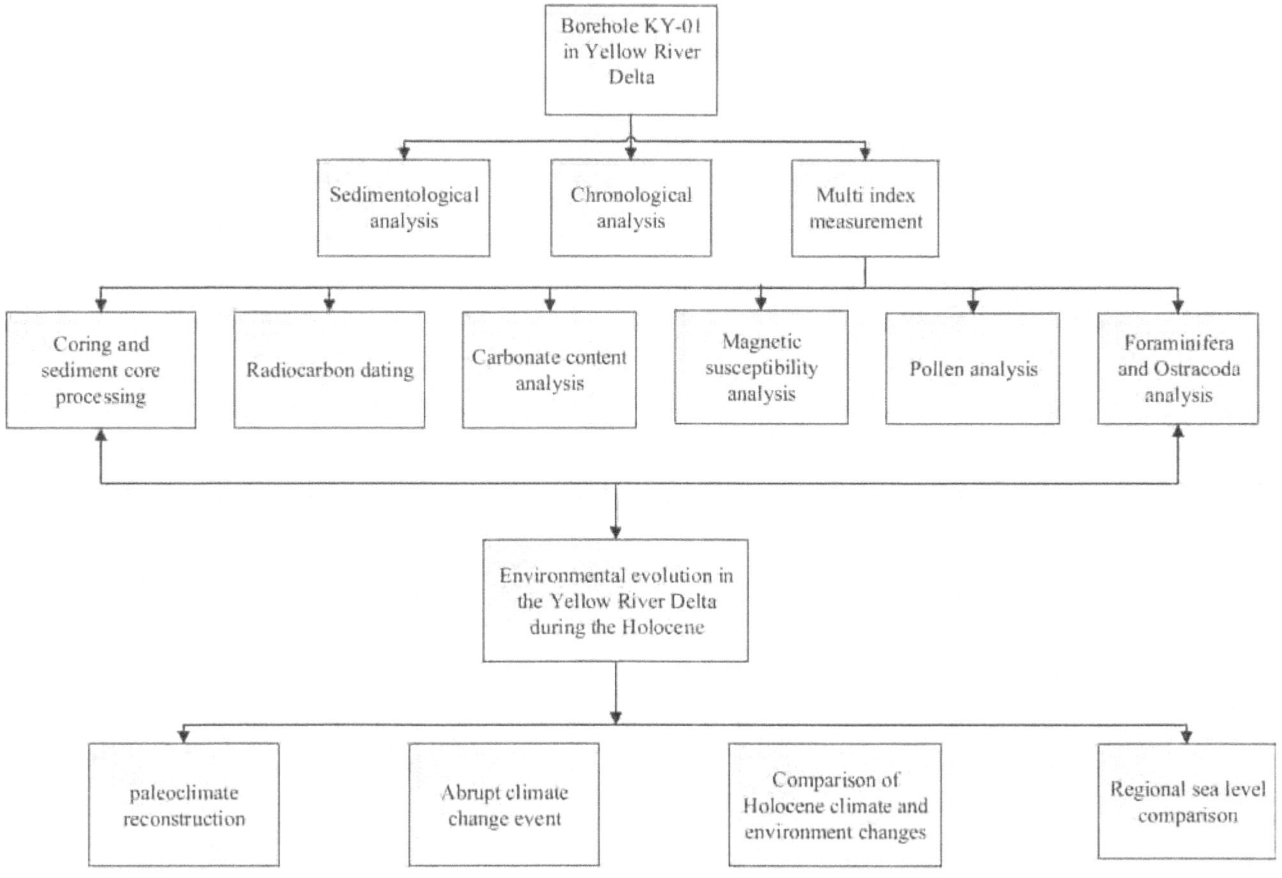

Figure 3 | Research flow chart.

RESULTS

Sedimentological analysis

In this study, the sampling depth is 21 m, but the indexes of carbonate content, magnetic susceptibility, sporopollenin and ostracodes are all analyzed for cores of 0–20 m. According to the lithological characteristics, the KY-01 section was divided into four stratigraphic units from top to bottom (Figure 4). Basal unit 1 (0–1,040 cm) was 1,040 cm thick, the upper layer is silty clay layer and the lower layer is yellow and grayish-yellow silty sand, occasionally with pea roots. Basal unit 2 (1,040–1,410 cm) was 370 cm thick, and it is mainly gray-black and gray coarse silty sand, containing a small amount of shell fragments, without plant roots. Basal unit 3 (1,410–1,640 cm) was 230 cm thick, most of them are interbedded by gray coarse silt and gray-black silty mud, and the shell fragments are evenly distributed as a whole. Basal unit 4 (1,640–2,100 cm) was 460 cm thick, and it is mainly gray silty clay with a thin layer of clayey silty sand, rich in foraminifera and abundant shell debris, such as small oyster debris.

Chronological analysis

AMS ^{14}C dating results

During the Holocene, sediments in the Yellow River Delta were continuously uninterrupted. To establish a time series for the Yellow River Delta, two samples selected from depths of 1,850 and 2,085 cm in the KY-01 sediment core were used as test materials for ^{14}C dating (Table 1).

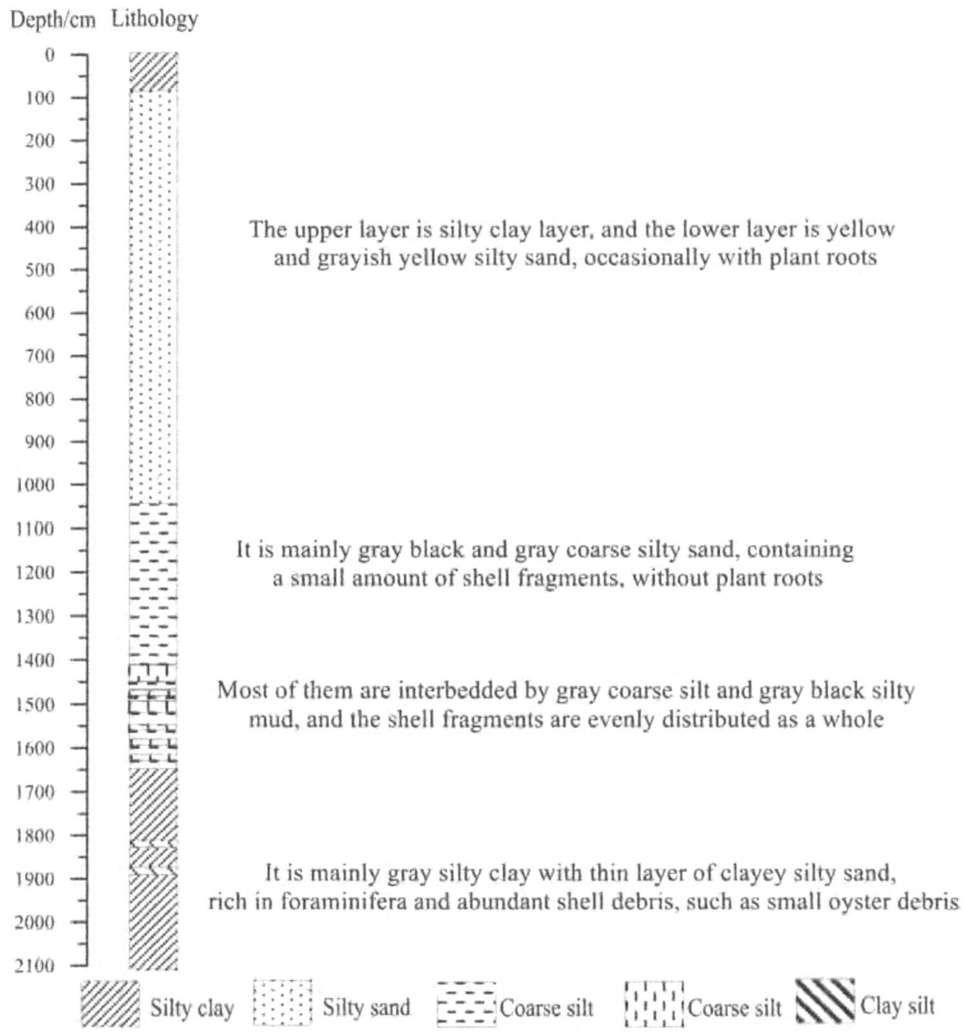

Figure 4 | Histogram of KY-01 borehole section in the Yellow River Delta.

Table 1 | ^{14}C dating results of the KY-01 sediment core in the Yellow River Delta

Sample no.	Depth (cm)	Material	Radiocarbon age (aBP)	Cal age (cal.aBP)
KY-01 184	1,845	Organic sediment	4,630 ± 30	5,418 ± 30
KY-01 208	2,085	Organic sediment	8,000 ± 30	8,884 ± 30

Comparison to other borehole cores in the Yellow River Delta

To analyze the sedimentary environment of the Yellow River Delta and establish an accurate chronological sequence, the core lithology and sedimentary facies of the KY-01 borehole were compared with those of other existing boreholes in the Yellow River Delta. In consulting the relevant literature, it was found that KY-01 pores have been well studied, as well as ZK228 (Zhou *et al.* 2014), 9602 (Hanebuth *et al.* 2011), ZK3 (Zhao *et al.* 2015), HB1 (Liu *et al.* 2009) and CB20A (Xu *et al.* 2006) pores in the northern part of the Yellow River Delta. The comparison of sediment cores from different locations is shown in Figure 5.

By comparison, we found that the lithologic zonation of the KY-01 section has a high degree of consistency with the surrounding layering. Through comparative analysis of lithology, the age at 1,040 cm of the KY-01 section is determined to be 1855 AD.

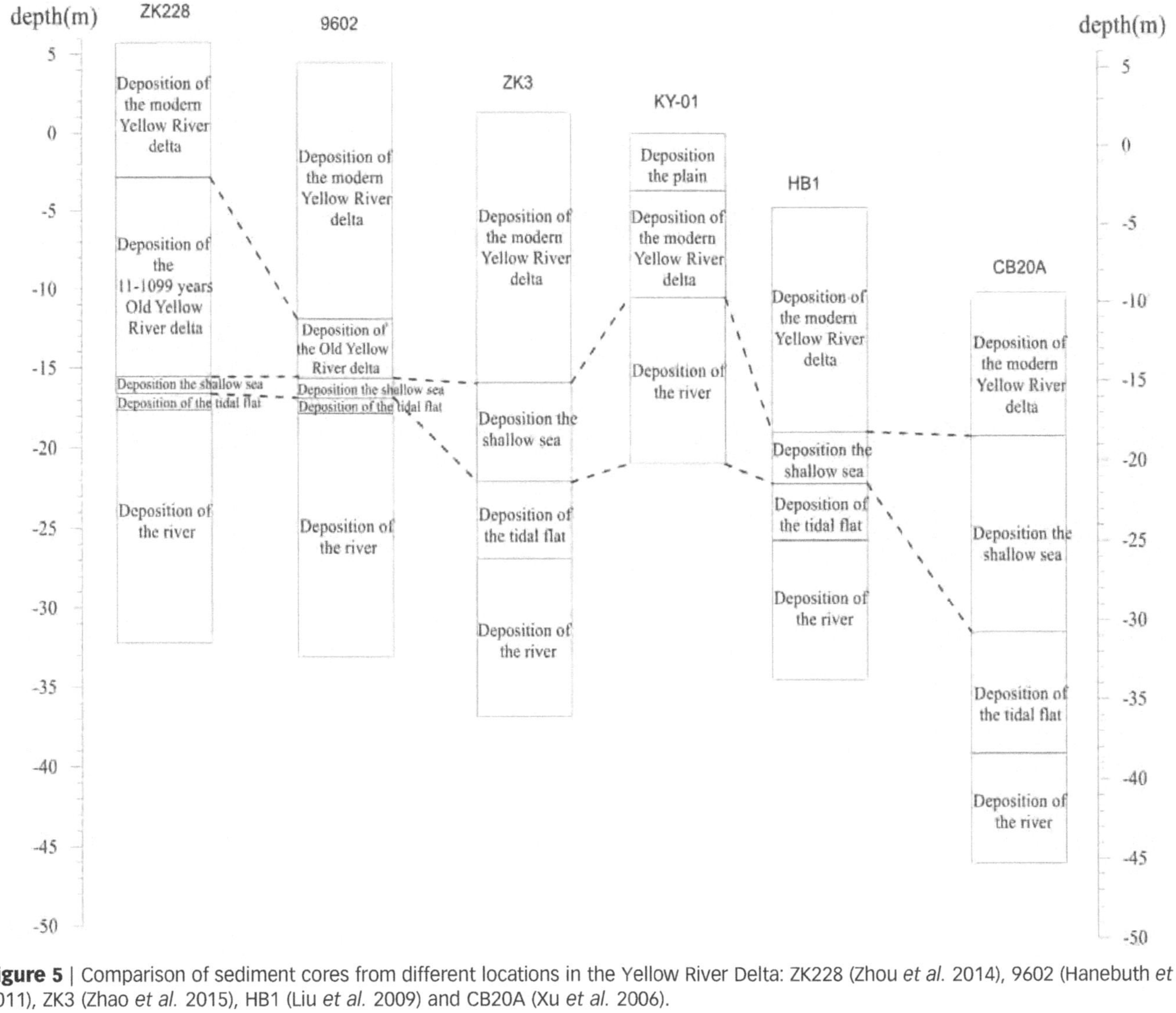

Figure 5 | Comparison of sediment cores from different locations in the Yellow River Delta: ZK228 (Zhou *et al.* 2014), 9602 (Hanebuth *et al.* 2011), ZK3 (Zhao *et al.* 2015), HB1 (Liu *et al.* 2009) and CB20A (Xu *et al.* 2006).

Based on the chronology of the lithology and sedimentary facies and the AMS ^{14}C dating results, time series was established for the KY-01 borehole. The deposition rate of the entire profile is roughly divided into three segments, with the highest deposition rate from 0 to 1,040 cm, with an average of 63.4 mm/a; the deposition rate from 1,040 to 1,845 cm slowed down significantly to 1.5 mm/a; and the deposition rate from 1,845 to 2,085 cm decreased from the previous stage, reaching 0.7 mm/a. The deposition rate curve is shown in Figure 6. Through the deposition rate of different stages, we can determine the starting time of the four stages in this paper (Table 2).

Carbonate analysis in sediments

The carbonate content throughout the KY-01 borehole from the Yellow River Delta varied greatly ranging from 3.7 to 18.8%, with an average of 11.03% (Figure 7), which shows a trend of increasing slowly and then decreasing. The carbonate content ranged between 3.9 and 14.7% from 2,000 to 1,640 cm, with an average value of 11.05%, which shows a trend of first rising, then slightly declining and finally becoming stable; 6.4 and 15.4% from 1,640 to 1,410 cm, with an average value of 11.44%, the carbonate content fluctuates frequently in this stage; 7.4 and 18.8% from 1,410 to 1,040 cm, with an average value of 11.82%, this stage fluctuates violently; and 1,040–0 cm is the sediment deposition of the modern Yellow River Delta, so it will not be studied in this paper.

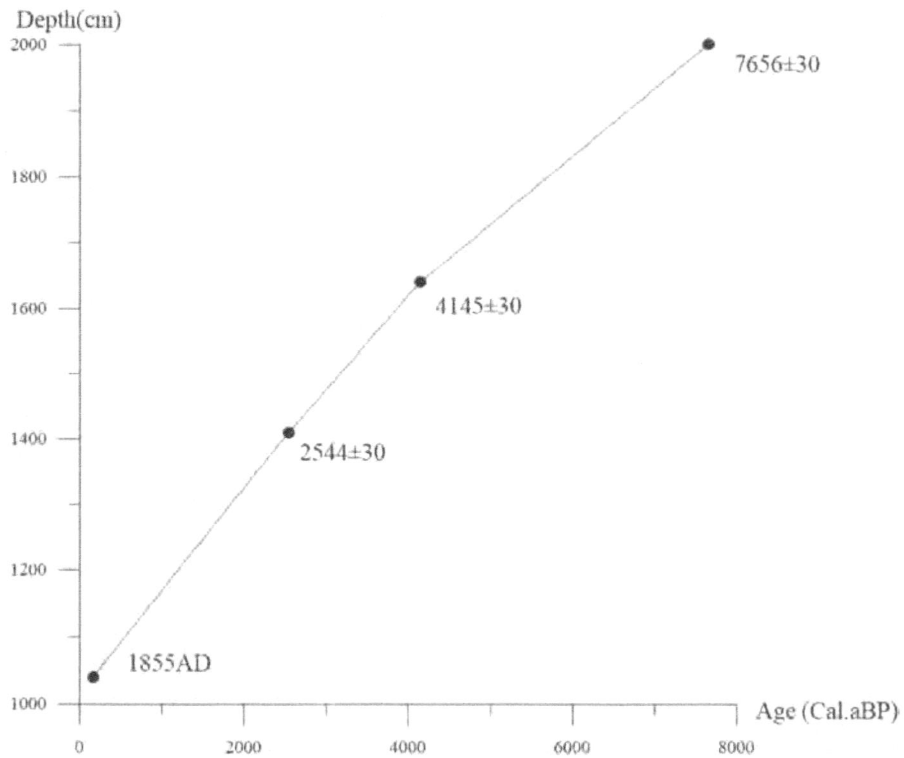

Figure 6 | Sedimentation rate curve of KY-01 in Yellow River Delta.

Table 2 | The starting time of the four stages

Depth (cm)	Age
1,040	1855 AD
1,410	2,544 ± 30 cal.aBP
1,640	4,145 ± 30 cal.aBP
2,000	7,656 ± 30 cal.aBP

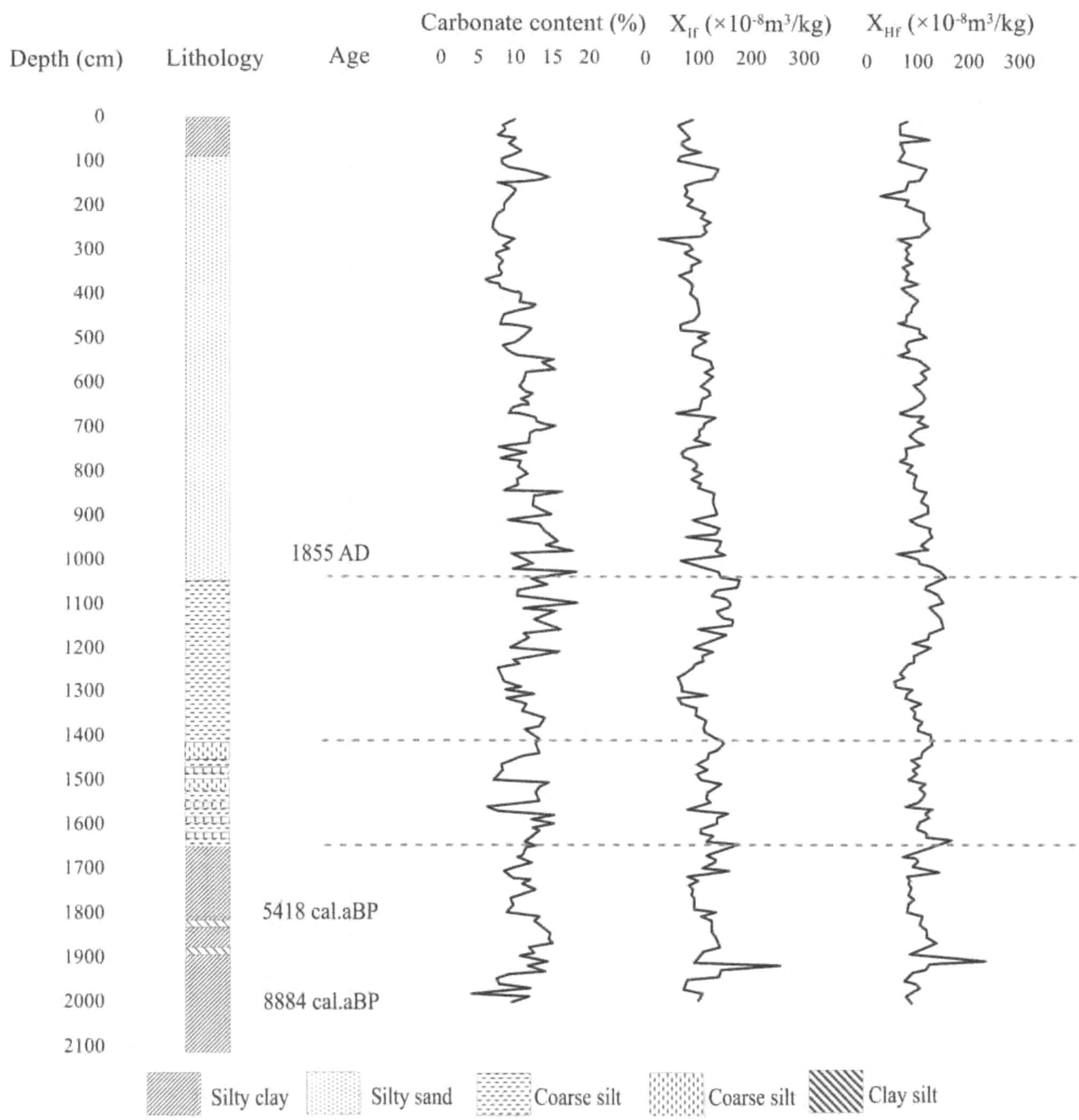

Figure 7 | Variation in carbonate content and magnetic susceptibility in the KY-01 borehole.

Magnetic susceptibility analysis in sediments

The fluctuation in high-frequency susceptibility and low-frequency susceptibility was the most significant in the section from 2,000 to 1,640 cm (Figure 7). In this section, the range in high-frequency susceptibility was 63–237 × 10^{-8} m^3/kg, with an average value of 103.71 × 10^{-8} m^3/kg, and the low-frequency susceptibility ranged between 75 and 257 × 10^{-8} m^3/kg, with an average value of 118.71 × 10^{-8} m^3/kg. For the 1,640–1,410 cm section, the range in high-frequency susceptibility was 68–167 × 10^{-8} m^3/kg, with an average value of 109.67 × 10^{-8} m^3/kg, and the low-frequency susceptibility ranged between 72 and 146 × 10^{-8} m^3/kg, with an average value of 118.71 × 10^{-8} m^3/kg. For the 1,410–1,040 cm section, the range in high-frequency susceptibility was 56–156 × 10^{-8} m^3/kg, with an average value of 107.75 × 10^{-8} m^3/kg, and the low-frequency susceptibility ranged between 60 and 174 × 10^{-8} m^3/kg. The 1,040–0 cm section is the sediment deposition of the modern Yellow River Delta, which will not be studied in this paper.

Ostracoda and foraminifera analysis in sediments

The Ostracoda and foraminifera pore analysis in the KY-01 sediment core from the Yellow River Delta, as well as statistical results on microflora, is presented below. The 0–1,040 cm section is the sediment deposit of the Yellow River since 1855AD, which will not be studied in this paper; 1,040–2,000 cm is the shallow sea sediment (Figure 8).

For the 2,000–1,640 cm section, from 1,655 to 1,895 cm, yellow-brown clay and gray-black clay were interbedded, and from bottom to top, the appearance of the micro-organisms decreased. In the section of 1,685–1,775 cm, the microsomal diversity (species number) was obvious with *Spiroloculina communis* Cushman et Todd, *Quinqueloculina akneriana rotunda*, *Q. argunica* and *Triloculina pentagonalis* observed. Vitreous shell microsomes were mainly *Ammonia beccarii* vars. and *Nonion glabrum*. In the gray–black clay sample obtained at 1745 cm, an unidentified Scaphopoda Dentaliumwas of Laevidentalidae gen. et sp (Figure 9) was observed as well as a shell and large individual foraminifera and Ostracoda (Figure 10). An obvious decrease in the number of microflora was observed in the section of 1,805–1,895 cm, which included *Q. akneriana rotunda*, *N. glabrum* Ho, *A. beccarii* vars. and the wetland species *Pseudogyroidina sinensis*. Ostracods consisted of *Sinocytheridea impressa* and *Cytherepteron miurense*.

As for 1,640–1,410 cm, at 1,415 cm, the presence of peat-containing matter increased; however, only one *Q. argunica* and one *A. beccarii* were found in brown clay containing a small amount of peat. At 1,445 cm and 1,505 cm, one *N. glabrum* and one *Q. akneriana rotunda* were found. After washing, the residues of brown clay from 1,565 cm were all mica fragments, and no micro-organisms were found.

As for 1,410–1,040 cm, at 1,055 cm, fourteen *Q. akneriana rotunda* and four *Q. argunica* are found; at 1,115 cm, twenty-five *Q. akneriana rotunda* are found again; and at 1,205 cm, one *A. beccarii* vars. are found.

Pollen analysis in sediment

Based on the fossil analysis, borehole KY-01 can be divided into the following five stages from bottom to top, 1,040–0 cm is the sediment deposition of modern Yellow River Delta, which will not be studied in this paper (Figure 11).

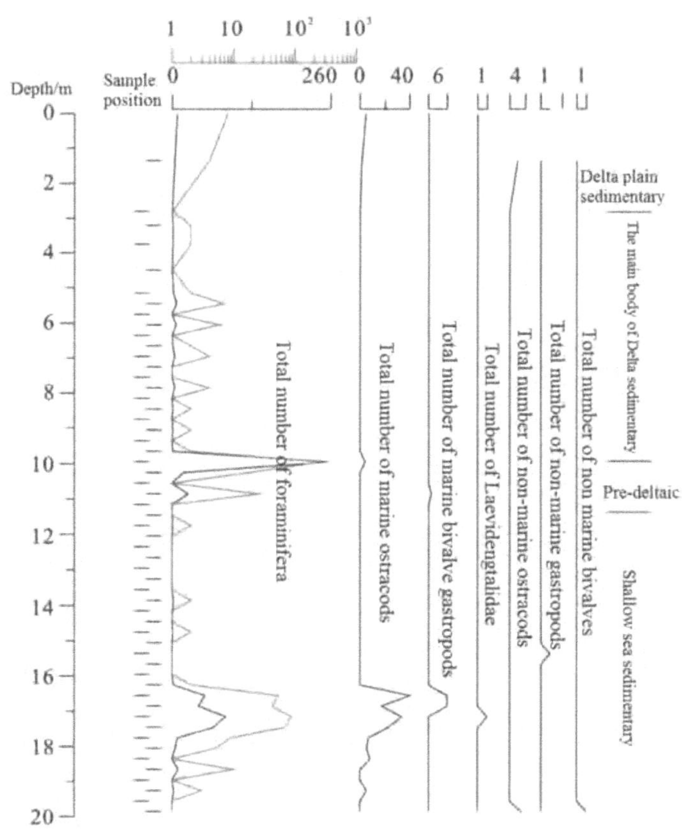

Figure 8 | Analysis of microfossils in the Yellow River Delta section.

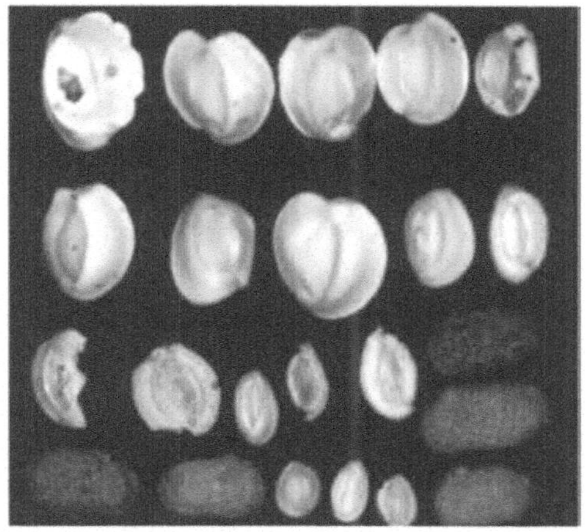

Figure 9 | Microbodies with larger and smaller proportions in the sample (1,745 cm).

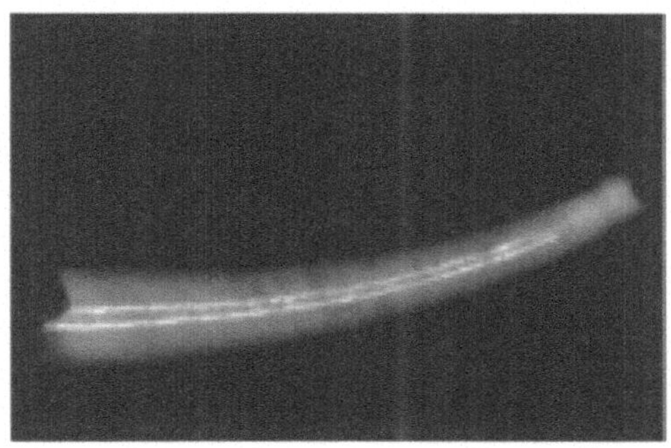

Figure 10 | Laevidentalidae gen. et sp. (1,745 cm).

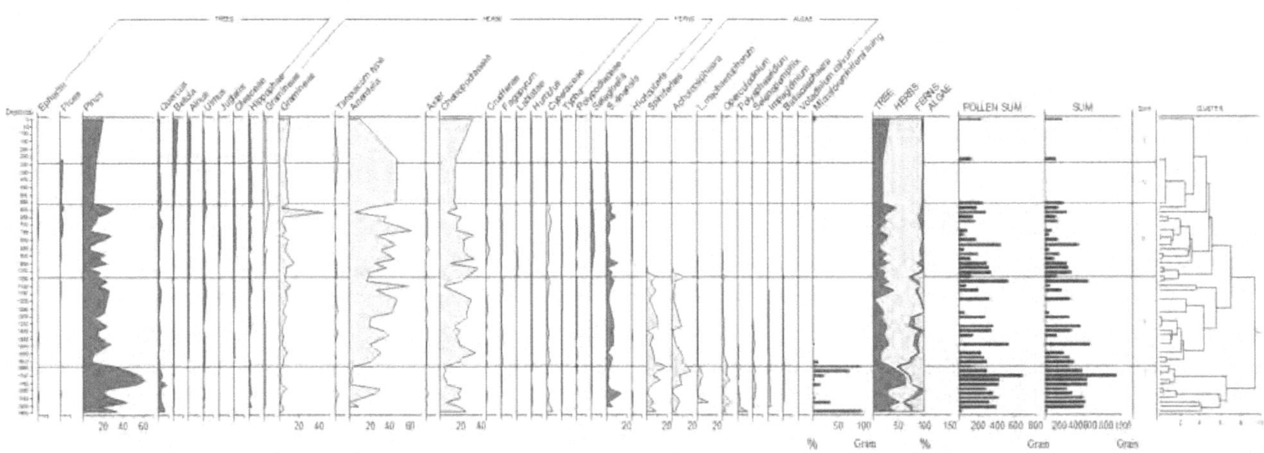

Figure 11 | Sporopollenin fossil percentage chart for borehole KY-01 from the Yellow River Delta, China.

As for 2,000–1,640 cm, the pollen of woody plants was dominant, with an average content of 43.66%, and the pinus pollen of Pinaceae with an average content of 38.28%, the other Pinaceae pollens such as Picea (0.35%), Abies (0.11%) and Tsuga (0.07%) were sporadic. The average pollen content of herbaceous plants was 26.27%, mainly composed of pollen from Artemisia (7.7%) and Chenopodiaceae (11.82%) of the family Chenopodiaceae, and Gramineae (2.4%) and Cyperaceae (1.76%) of the family Poaceae in small amounts. At this stage, many types of algal fossils were observed with an average content of 25.5%, of which *Gonyaulax catenella* was the main alga, followed by Peridinioids. *G. catenella* algae were dominated by

Spiniferites (7.08%) and Achomosphaera (6.61%). The spore content of ferns was low, the average content was 4.55%, mainly *Selaginella sinensis*, the average content was 3.85%, Polypodiaceae, riciopteris, Osmundaceae and Lycopodium of the family sinophyllaceae were less than 1%.

As for 1,640–1,410 cm, a total of eight samples were found in this stage. Herbaceous pollen was dominant, with an average content of 56.3%. The pollen contents of Artemisia and Chenopodiaceae were 28.9 and 16.6%, respectively. Gramineae (5.63%), other Compositae (2.10%) and Cyperaceae (1.06%) were also common. The pollen content of woody plants was slightly lower, with an average content of 18.65% consisting mainly of *Pinus* genus. The spore content of pteridophytes was very low, with an average content of 2.39%. The average content of algae was 12.6%, which was still dominated by *G. catenella*. The contents of Spiniferites and Achomosphaera were relatively high, with average contents of 6.04 and 4.44, respectively. There were few freshwater algae, and the genus *Pediastrum* and the genus *Concentrics* were occasionally found. The inner membrane of microsome foraminifera was rarely seen.

As for 1,410–1,040 cm, a total of 12 samples were found in this stage. Herbaceous pollen was dominant, with an average content of 69.5%. Artemisia and Chenopodiaceae were abundant, with average contents of 33.77 and 22.43%, respectively. The average pollen content of woody plants was 16.4%, consisting mainly of *Pinus* with an average content of 13.6%. The average spore content of pteridophytes was 5.6% dominated by *S. sinensis* with an average content of 4.1%. The average content of algae was generally low (3.68%).

DISCUSSION

Environmental evolution and paleoclimate reconstruction in the Yellow River Delta during the Holocene

Based on the collective results of the carbonate, magnetic susceptibility, Ostracoda, foraminifera and pollen analyses in sediments from the KY-01 borehole in the Yellow River Delta, the environmental evolution of the Yellow River Delta from 7656 cal.aBP to 1855AD can be divided into three stages (Figure 12).

Stage A (2,000–1,640 cm, 7656–4145 cal.aBP)

Overall, carbonate content first increased and then stabilized, with the pollen of woody plants (43.66%) dominating in pollen fossils. The average content of algal fossils was 25.5%. In addition to spores and algae, a large number of microsomal foraminifer inner membranes were also found at this stage. The high abundance of microsomal foraminifer inner membranes reflect a shallow coastal environment suitable for foraminifer development, and the abundance of dinoflagellate and a large number of marine foraminifer inner membranes indicate that the well section was marine sediment. In addition, pollen analysis for the period 4964–4281 cal.aBP showed that during this period woody plants accounted for the largest proportion, the proportion of algal plants gradually increased, the largest number of foraminifera was noted and the number of marine foraminifera increased rapidly. In summary, during this stage, there was marine deposition, the climate overall was warm-humid representing the Holocene warm period, but there were some events of extreme climate fluctuations.

Stage B (1,640–1,410 cm, 4145–2544 cal.aBP)

During this stage, carbonate content fluctuated more frequently, and there were two more obvious content mutations: they were 4.2 kaBP in 4451–4081 cal.abp and 4.2 kaBP in 4451–4081 cal.abp, the magnetic susceptibility fluctuated more intensely than in the previous stage and showed a trend of decreasing fluctuations, the pollen of woody plants and algal plants was significantly reduced and freshwater algae appeared in small amounts. In this stage, the content and diversity of algae were lower than in the previous stage, but Spiniferites and Achomosphaera were still the dominant species, while the number of Polychaetes and micro-foraminifer endomembranes was greatly reduced. Therefore, it is speculated that transgression declined during this stage, with high salinity seawater gradually receding. In addition, in the early period of this stage, the pollen of woody plants and algae decreased significantly, and the number of micro-foraminifer inner membranes decreased significantly, indicating that there were some extreme cooling events during this period. In summary, during this stage, transgression declined, the depositional environment was coastal beach, the climate changed from warm-wet to dry-cold and there were small-scale extreme climate events.

Stage C (1,410–1,040 cm, 2544 cal.aBP–1855 AD)

The fluctuation in carbonate content was more pronounced than in the previous stage, the magnetic susceptibility showed a decreasing trend initially and then increased and the proportion of pollen from woody plants and algae was still low. Pollen

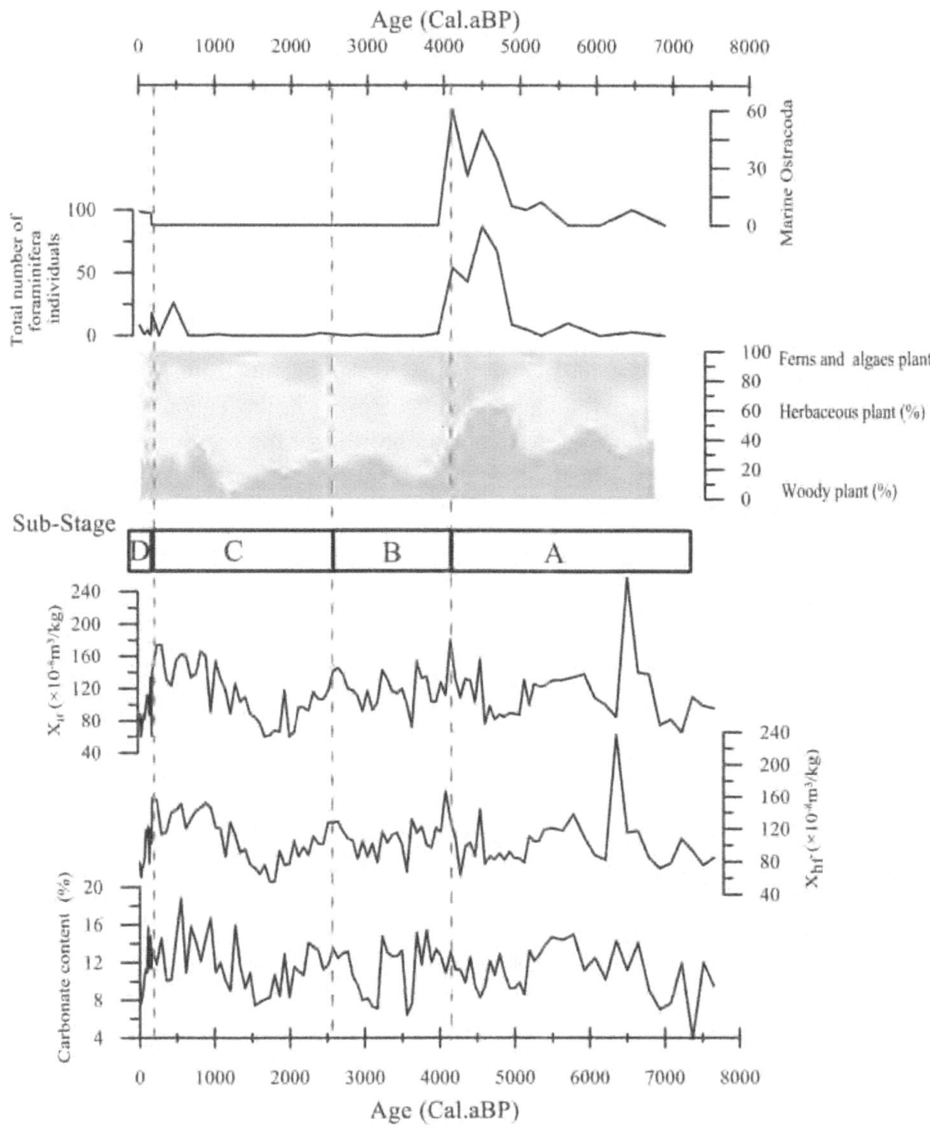

Figure 12 | Climatic stages in the Yellow River Delta since the Holocene.

from herbaceous plants, such as Artemisia and Chenopodiaceae suited to saline-alkali soil, was dominant and more abundant than in the previous stage, which indicates that overall the climate was drier and colder than in the previous stage. Large areas of flat land or saline-alkali land were formed during this period, and the distribution of freshwater in some areas was greater than in the previous stage. It was thus speculated that transgression and regression alternately occurred during this stage, and the sea level was relatively low. In conclusion, the sea level was relatively low, transgression and regression alternately occurred, the climate was drier and colder, the temperature alternately increased and decreased and fluctuations in climate were more intense.

Abrupt climate change event

A comprehensive analysis of the KY-01 borehole magnetic susceptibility, carbonates, spores and microbiological indicators revealed that the climate in the Yellow River Delta began to change around 5737–5422 and 4451–4081 cal.aBP, with significant changes as the temperature decreased and the environment deteriorated.

Cooling events during 4451–4081 cal.aBP

From 4451 to 4081 cal.aBP, the pollen of woody plants and algal plants in the KY-01 sediment core decreased significantly, while that of herbaceous plants increased rapidly to a dominant proportion, consisting mainly of pollen from Artemisia and

Chenopodiaceae, and the number of micro-foraminifera decreased significantly. All lines of evidence indicated that the temperature decreased, the precipitation decreased and the climate became dry and cold during this time. The 4.2 kaBP event is a climate cooling event that occurred during the warm period of the middle Holocene. Records from the middle and low latitudes and subtropical regions around the world, such as Asia, Europe, Africa and North America, indicate a similar start time, duration and strength. This incident has also been commonly recorded in China (Figure 13). In Central China, Zhu studied the Dajiu Lake peat profile in Shennongjia and found that the $\delta^{18}O$ value around 4200 cal.aBP showed a rapid positive deviation of 1‰ over about 200 years (Zhu *et al.* 2006). In the Qinghai-Tibet Plateau (Wu *et al.* 2006), evidence from Cuoe Lake sediments shows that the Rb/Sr ratio of trace elements reflecting the degree of weathering peaked at 4300–4100 cal.aBP, revealing that the carbon isotope of effective precipitation was reduced. In Southwest China (Wang *et al.* 2010), the content of $\delta^{18}O$ in Chongqing Xinyadong during the period 4400–4100 cal.aBP increased by about 1.61‰ within a time scale of 200 years. In North China (Jin & Liu 2002), the $\delta^{13}C$ value of peat increased significantly from 4800 to 4200 cal.aBP in Taishizhuang, Hebei Province. By comparison, we find that the abrupt climate event in the Yellow River Delta during the period 4451–4081 cal.aBP is strongly consistent with the cold events of 4.2 kaBP, and the time differences between regions may be related to the different proxy indexes and dating methods selected for the reconstruction of the paleoclimate, or to differences in landform, geographical environment or climatic conditions.

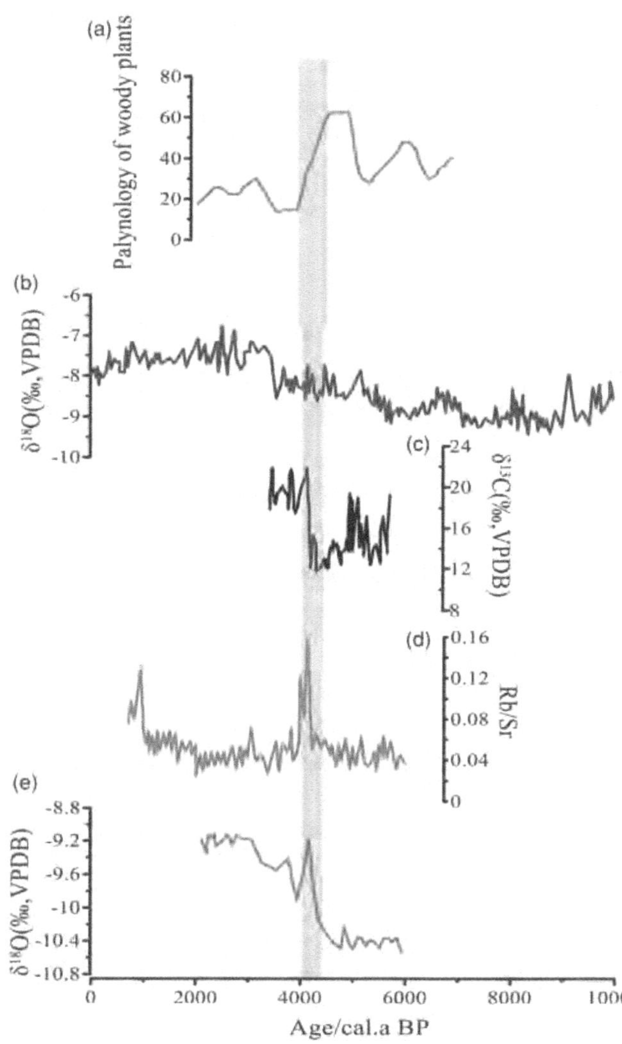

Figure 13 | Typical records of the 4.2 kaBP cold event in the monsoon area of China (gray bars indicate dry-cold period). (a) The pollen content of woody plants in the KY-01 sediment core from the Yellow River Delta; (b) the $\delta^{18}O$ record of Xinyadong stalagmites in Southwest China (Wang *et al.* 2010); (c) the $\delta^{13}C$ record of peat in Taishizhuang, North China (Jin & Liu 2002); (d) the Rb/Sr record of Cuoe Lake sediments in the Tibetan Plateau, Northwest China (Wu *et al.* 2006) and (e) the $\delta^{18}O$ record of Shennongjia stalagmites, Central China (Zhu *et al.* 2006).

Cooling events during 5737–5422 cal.aBP

From 5611 to 5422 cal.aBP, the pollen content of woody plants decreased sharply in the KY-01 sediment core, the pollen content of herbaceous plants increased significantly, and the Ostracod content decreased slightly, which reveals that the temperature decreased greatly and the climate became dry and cold. Many studies in many regions of the world have also recorded climate cooling and drying events in the middle of the Holocene, which is generally referred to as the 5.5 kaBP event. Magny *et al.* (2006) summarized a series of marine and land environmental proxy indicators in the southern and northern hemispheres, which proved that the 5.5 kaBP event had global characteristics. There were also many regions in China that responded to this incident (Figure 14). In South China, the concentration of chlorophyll in the sediments of Lake Mar reached their lowest value between 5,500 and 5000 cal.aBP. In Northeast China, Hanebuth *et al.* (2011) studied Jinchuan peat and found that the δ^{13}C value began to rise around 5000 cal.aBP, revealing drying of the climate. In the Southwest and Northwest of China (Yao & Thompson 2017), the δ^{18}O values in the Dunde ice core and Donggedong stalagmite increased at 5500–6000 cal.aBP, indicating that the temperature suddenly decreased and the climate became dry. Through comparison, it is found that the cooling event noted in the Yellow River Delta between 5611 and 5422 cal.aBP is consistent with those of other regions that occurred during the same period.

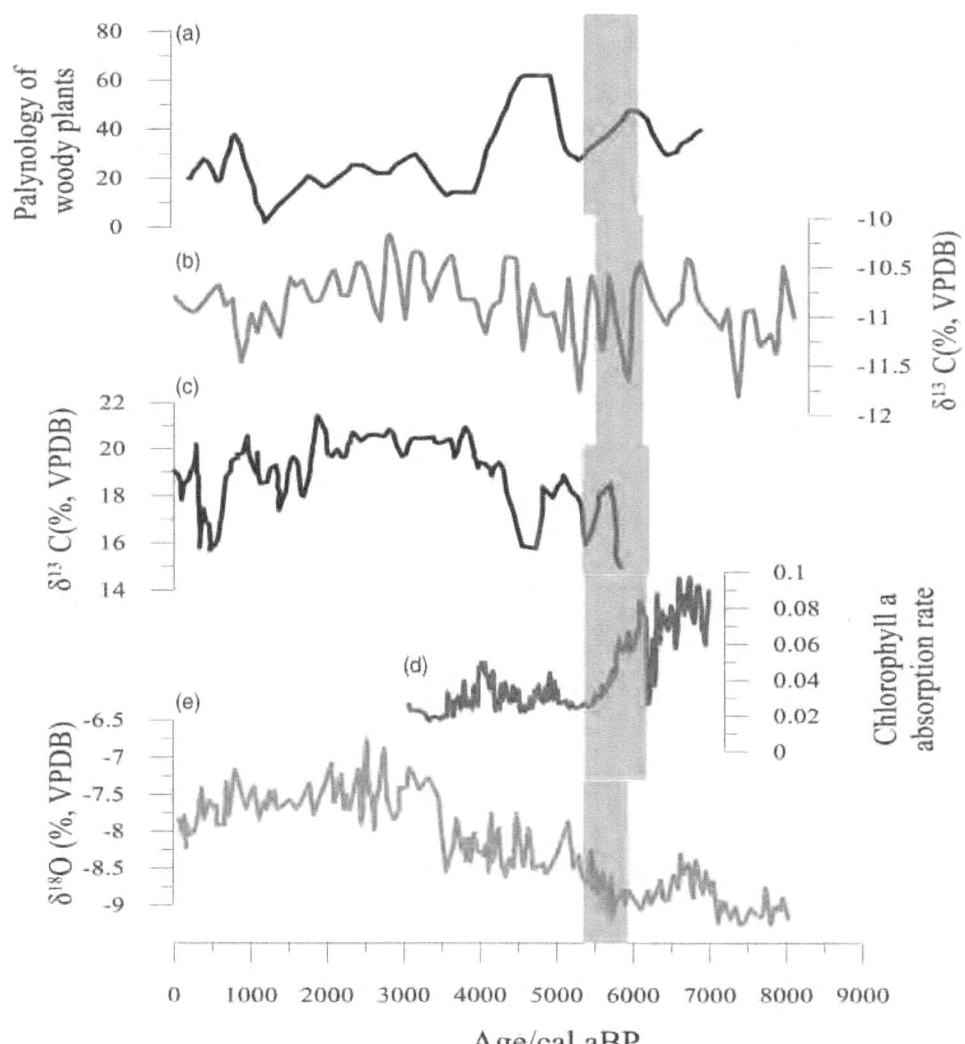

Figure 14 | Typical records of the 5.5 kaBP cold event in the monsoon area of China (gray bars indicate dry-cold period). (a) The pollen content of woody plants in the KY-01 sediment core from the Yellow River Delta; (b) the δ^{18}O record of the Dunde ice core, Northwest China (Yao & Thompson 2017); (c) the δ^{13}C record of Jinchuan peat, Northeast China (Hoang *et al.* 2016) and (d) the absorption rate of chlorophyll *a* in sediments of Huguangyan Maar Lake.

Comparison of Holocene climate and environment changes in Shandong Province

In the study of Holocene environmental evolution in other areas of Shandong Province, we can find the similar research results with this paper. Based on the analysis of sporopollenin data of three boreholes in Shandong Province, Bian (2004) thinks that 7000–3000 cal.aBP, subtropical vegetation types are generally growing in Shandong Province with warm-humid climate, and there are two large-scale cooling events during this period. Ding *et al.* (2011) analyzed the magnetic susceptibility, chromaticity and grain size of the Shanchengcun section in Zhangqiu, Shandong Province, 8500–3100 aBP, the average grain size and coarse silt content were the lowest in the whole section, the magnetic susceptibility value further decreased in the middle and a lower part and the redness and yellowness value increased, which was the highest value in the whole section, indicating that the temperature increased and the precipitation increased in this stage. Zhang *et al.* (2005) analyzed the sporopollenin, foraminifera and grain size in the northern plain of Shandong Province of two boreholes (A1 and A5), 10000–4000 aBP, the aquatic herbaceous plants are dominant, the warm and humid vegetation in the vegetation assemblage increases significantly and reaches peak in the late stage and a large number of marine foraminifera are found, indicating that this stage is a transitional sedimentary environment between land and sea, with warm-humid climate. From the above analysis, it can be seen that the results of carbonate content, magnetic susceptibility, Ostracoda, foraminifera and sporopollenin analyses of hole KY-01 in the Yellow River Delta are consistent with those of Shandong and other regions in China.

Regional sea-level comparison

Since the middle of the 20th century, a large number of scholars have carried out long-term and systematic research on the Holocene sea-level changes in eastern China and obtained a great deal of basic results. Based on the analysis of the sedimentary structure and foraminiferal characteristics of eight boreholes in the South Bank of Laizhou Bay, Tian *et al.* (2016) and others established the Holocene relative sea-level change history of Laizhou Bay; Huang *et al.* (2016) and others have discussed the sedimentary paleoenvironment and paleoclimate history of the Pearl River Delta; Yang *et al.* (2015) and others analyzed the lithology, particle size and burning loss characteristics of Na6 and na9 borehole sections in Qingaowan, Southeast Guangdong; Xian & Jiang (2005) and others studied the evolution characteristics of Holocene sedimentary environment, climate and sea level in the Yellow River Delta by using the grain size, mineral and element geochemical testing information of borehole sediments. From the above analysis, we can see that the results of ostracods, foraminifera and sporopollenin analysis analyses of KY-01 well in the Yellow River Delta reflect the changes in the sea level, which are consistent with the research results of other areas in China.

CONCLUSIONS

From information on the evolution of climate and sea level recorded by environmental proxy indexes in sediment since the Holocene (7656 cal.aBP), it can be seen that the evolution of the regional climate in the Yellow River Delta is affected by global environmental change. However, due to the unique geographical location, regional geographical features and the characteristics of the Yellow River Basin, environmental change in this region has more complexity, and the environmental evolution has regional characteristics. Changes in the climate of the Yellow River Delta from 7656 cal.aBP to 1855AD can be divided into the following three stages.

During 7656–4145 cal.aBP, the climate was warm-humid, and seawater moved onto the land. The total number of marine ostracods was the largest. During 4145–2544 cal.aBP, the pollen of woody and algal plants decreased significantly, while the pollen is suited to saline and alkaline land conditions increased rapidly; the total number of marine ostracods decreased significantly, and the climate changed from warm-humid to dry-cold. During 2544 cal.aBP–1855AD, transgression and regression alternately occurred, the sea level was relatively low, beach land or saline-alkali land formed in large areas and the climate was drier and colder than in the previous stage. In addition, the climatic characteristics of the Yellow River Delta recorded the obvious cold dry events during the middle of the Holocene between 5737–5422 and 4451–4081 cal.aBP, which is consistent with the geological and climatic records of the whole country and even the whole world.

ACKNOWLEDGEMENTS

This work was supported by the National Natural Science Foundation Project (Nos 41602356, 41977262 and 41471160) and Open Research Fund Program of Shandong Key Laboratory of Depositional Mineralization & Sedimentary Minerals (No. DMSM2018024), Study Abroad Program by the Government of Shandong Province, China (No. 201902006), Open Research

Fund Program of Shandong Provincial Key Laboratory of Water and Soil Conservation and Environmental Protection (No. STKF201917), Open Research Fund Program of Shandong Key Laboratory of Eco-Environmental Science for Yellow River Delta (Binzhou University) (No. 2019KFJJ03), Major scientific and technological innovation projects in Shandong Province (No. 2018CXGC0307), Geological Exploration Fund of Shandong Province (Nos 2016(7), 2013(55) and 2018(8)), Determination Technology of Sedimentary Environment Since Holocene in the Yellow River Delta, the Determination Technology of Accurate Age of Sedimentary Strata in the Yellow River Delta and the Key Technology Research of the Division of Sedimentary Facies in the Yellow River Estuary Area. The authors thank the reviewers for their very valuable comments, which greatly improved the quality of the paper.

DATA AVAILABILITY STATEMENT

All relevant data are included in the paper or its Supplementary Information.

REFERENCES

Bian, X. 2004 The Holocene palaeoclimate change sequence in Shandong Province and its correlation with the prehistoric cultural development stage. *Shandong Normal University*. (In Chinese). doi:10.7666/d.Y598966.

Burciaga, U. M., Sáez, P. V. & Ayón, F. J. H. 2019 Strategies to reduce CO_2 emissions in housing building by means of CDW. *Emerging Science Journal* **3** (5), 274–284. doi:10.28991/esj-2019-01190.

Chen, F., Zhang, J., Liu, J., Cao, X., Hou, J., Zhu, L., Xu, X., Liu, X., Wang, M., Wu, D., Huang, L., Zeng, T., Zhang, S., Huang, W., Zhang, X. & Yang, K. 2020 Climate change, vegetation history, and landscape responses on the Tibetan Plateau during the Holocene: a comprehensive review. *Quaternary Science Reviews* **243**, 106444. doi:10.1016/j.quascirev.2020.106444.

Ding, M., Peng, S., Pang, J., Zhang, W., Chen, D. & Yang, H. 2011 Holocene environmental evolution and human cultural development in Central Shandong. *Chinese Journal of Soil Science* **42** (6), 1281–1287. (In Chinese). doi:10.19336/j.cnki.trtb.2011.06.001.

Gabriela, F., Kendrick, J., Vachel, A., Petr, Siim, V. & Angelica, F. 2019 Holocene rapid climate changes and ice-rafting debris events reflected in high-resolution European charcoal records. *Quaternary Science Reviews* **222**, 105877. doi:10.1016/j.quascirev.2019.105877.

Guo, L., Xiong, S., Ding, Z., Jin, G., Wu, J. & Ye, W. 2018 Role of the mid-Holocene environmental transition in the decline of late Neolithic cultures in the deserts of NE China. *Quaternary Science Reviews* **190**, 98–113. doi:10.1016/j.quascirev.2018.04.017.

Han, T., Zin, W. & Kyi, C. 2020 Analysis of streamflow response to changing climate conditions using SWAT model. *Civil Engineering Journal* **6** (2), 194–209. doi:10.28991/cej-2020-03091464.

Hanebuth, T. J. J., Voris, H. K., Yokoyama, Y., Saito, Y. & Okuno, J. 2011 Formation and fate of sedimentary depocentres on Southeast Asia's Sunda Shelf over the past sea-level cycle and biogeographic implications. *Earth Science Reviews* **104** (1–3), 92–110. doi:10.1016/j.earscirev.2010.09.006.

Hoang, T., Van, L., Oanh, T. & Jiro, T. 2016 The influence of delta formation mechanism on geotechnical property sequence of the late Pleistocene-Holocene sediments in the Mekong River Delta. *Heliyon* **2** (11), e00165. doi:10.1016/j.heliyon.2016.e00165.

Huang, K., He, J. & Zong, Y. 2016 Holocene paleoenvironment reconstruction based on pollen data in the Sanshui Basin, Northern Pearl River Delta. *Tropical Geography* **36** (3), 364–373. (In Chinese). doi:10.13284/j.cnki.rddl.002849.

Huang, R., Chen, C., Li, T., Wu, Y., Xiao, S., Zhang, J., Wang, T. & Li, J., 2019 Comparison of climatic events in the late Holocene, based on stalagmite records from monsoon regions near by the east and west Pacific. *Quaternary Sciences* **39** (3), 742–754. (In Chinese). doi:10.11928/j.issn.1001-7410.2019.03.19.

Javadinejad, S., Dara, R. & Jafary, F. 2020 Climate change scenarios and effects on snow-melt runoff. *Civil Engineering Journal* **6** (9), 1715–1725. doi:10.28991/cej-2020-03091577.

Jin, G. & Liu, D. 2002 Mid-Holocene climate change in North China, and the effect on cultural development. *Chinese Science Bulletin* **47** (5), 408–413. doi:10.1515/CHAR.2005.5.1.236.

Liu, J., Saito, Y., Wang, H., Zhou, L. & Yang, Z. 2009 Stratigraphic development during the Late Pleistocene and Holocene offshore of the Yellow River Delta, Bohai Sea. *Journal of Asian Earth Sciences* **36** (4–5), 318–331. doi:10.1016/j.jseaes.2009.06.007.

Ma, W., Liu, Y. & Yang, Z. 2019 Characteristics of ore-forming fluids of Lietinggang-Leqingla Pb-Zn-Fe-Cu-Mo polymetallic deposit in Tibetan: evidence from fluid inclusions and stable isotope compositions. *Earth Science* **44** (6), 1957–1973. (In Chinese).

Magny, M., Leuzinger, U., Sigmar, B. & Jean, N. H. 2006 Tripartite climate reversal in Central Europe 5600–5300 years ago. *Quaternary Research* **65** (1), 3–19. doi:10.1016/j.yqres.2005.06.009.

Mao, L. 2012 Holocene palynological records of biodiversity and climatic events in the Yangtze River delta, China. *Quaternary International* **279–280**, 304. doi:10.1016/j.quaint.2012.08.835.

Nian, X., Zhang, W., Wang, Z., Sun, Q., Chen, J. & Chen, Z. 2018 Optical dating of Holocene sediments from the Yangtze River (Changjiang) Delta, China. *Quaternary International* **467**, 251–263. doi:10.1016/j.quaint.2018.01.011.

Railsback, L. 2018 A comparison of growth rate of late Holocene stalagmites with atmospheric precipitation and temperature, and its implications for paleoclimatology. *Quaternary Science Reviews* **187**, 94–111. doi:10.1016/j.quascirev.2018.03.002.

Tian, L., Tao, Y. & Jiang, X. 2016 Reconstruction of the Holocene relative sea level change for the south coast of Laizhou Bay. *Geological Bulletin of China* **35** (10), 1679–1691.

Vött, A., Brückner, H. & Brockmüller, S. 2008 Traces of Holocene tsunamis across the Sound of Lefkada, NW Greece. *Global and Planetary Change* **66** (1), 112–128. doi:10.1016/j.gloplacha.2008.03.015.

Wang, J., He, X. & Li, Q. 2010 Climate change since 4.5 ka B.P. recorded by trace elements in a stalagmite from Xinya Cave in Chong. *Geographica Sinica* **17** (6), 910–915. (In Chinese). doi:10.13249/j.cnki.sgs.2010.06.007.

Wang, F., Zhang, W., Nian, X., Roberts, A. P., Zhao, X., Shang, Y., Ge, C. & Dong, Y. 2020 Magnetic evidence for Yellow River sediment in the late Holocene deposit of the Yangtze River Delta, China. *Marine Geology* **427**, 106274. doi:10.1016/j.margeo.2020.106274.

Wu, Y., Wang, S. & Hou, X. 2006 Chronology of Holocene lacustrine sediments in Co Ngoin, central Tibetan Plateau. *Science in China Series D Earth Sciences* **49** (9), 991–1001. doi:10.1007/s11430-006-0991-3.

Xian, B. & Jiang, Z. 2005 Environment evolution and eustatic change of Holocene in the Yellow River Delta. *Marine Geology & Quaternary Geology* (3), 1–7. (In Chinese). doi:10.16562/j.cnki.0256-1492.2005.03.001.

Xu, J., Meng, Y., Zhang, X. & Liu, D. 2006 The evolution of the paleogeographic environment of the Yellow River estuary since the late Pleistocene. *Quaternary Science* **26** (3), 327–333. (In Chinese). doi:10.3321/j.issn:1001-7410.2006.03.003.

Yang, B., Yu, F. & Zheng, Z. 2015 Changes in Holocene depositional environment of Qin Ao island inferred from sediment grain-size and loss-on-ignition. *Marine Geology & Quaternary Geology* **35** (6), 41–51. (In Chinese). doi:10.16562/j.cnki.0256-1492.2015.06.005.

Yao, T. & Thompson, L. G. 2017 Trends and features of climatic changes in the past 5000 years recorded by the Dunde ice core. *Annals of Glaciology* **16**, 21–24. doi:10.1017/S0260305500004766.

Zhang, Z., Nie, X., Liu, E. & Xin, L. 2005 The accumulation records of environmental evolution on the salt-water intruded area south of Laizhou Bay since late Pleistocene. *Geographical Research* **24** (1), 105–112. (In Chinese). doi:10.11821/yj2005010012.

Zhang, Z., Wang, G., Jiang, M., Lu, X. & Liu, X. 2016 The impact of Holocene climate changes on Honghe wetland in NE China. *Ecological Engineering* **96**, 72–78. doi:10.1016/j.ecoleng.2016.05.040.

Zhang, X., Du, Y., Ma, C., Ping, S., Feng, C. & Cui, A. 2019 Climatic controls on peat swamp formation and evolution since 1300 year BP as recorded by phytoliths in the Xishan Mountains, Jiangxi Province, China. *Palaeogeography, Palaeoclimatology, Palaeoecology* **522**, 76–88. doi:10.1016/j.palaeo.2019.03.004.

Zhao, G., Ye, S., Li, G., Ding, X. & Yuan, H. 2015 Late quaternary strata and carbon burial records in the Yellow River Delta, China. *Oceanic and Coastal Sea Research* **14** (03), 446–456. doi:10.1007/s11802-015-2773-z.

Zhao, W., Chen, C., Jiang, Q., Ji, M., Zhen, J., Yang, Y., Ning, D., Li, X. & Shen, J. 2020 Holocene hydroclimate in the source region of the Yellow River: a new sediment record from Lake Gyaring, NE Tibetan Plateau. *Journal of Asian Earth Sciences* **205**, 104601. doi:10.1016/j.jseaes.2020.104601.

Zhou, L., Liu, J., Saito, Y., Liu, P. & Li, G. 2014 Fluvial system development and subsequent marine transgression in Yellow River (Huanghe) delta and its adjacent sea regions during last glacial maximum to early Holocene. *Continental Shelf Research* **90** (1), 117–132. doi:10.1016/j.csr.2014.06.012.

Zhu, C., Ma, C., Zhang, W., Zheng, C. & Tang, L. 2006 Pollen record from Dajiuhu basin Shennongjia and environmental changes since 15.753 kaB.P. *Quaternary Sciences* **26** (5), 814–826. doi:10.1016/S1872-2040(06)60041-8.

First received 12 April 2021; accepted in revised form 15 July 2021. Available online 5 August 2021